REFRIGERATION and AIR-CONDITIONING

OTHER BOOKS BY THE SAME AUTHOR

- 9789381068618 Strength of Materials-I
- 9789381068656 Fluid Mechanics
- 9789381068809 Strength of Materials
- 9789382609605 Khanna's Mechanical Objective Type
- 9789382609612 Khanna's Mechanical Conventional Type
- 9789382609285 Fluid Machinery (Hydraulic Machines)
- 9789382609513 Machine Design Data Book
- 9789382609575 Machine Design
- 9789386173973 Khanna's Conventional & Objective Type Questions & Answers in Mechanical Engineering

REFRIGERATION and AIR-CONDITIONING

FOR ALL TYPES OF EXAMINATIONS & INTERVIEWS

[U.P.S.C. Engineering Services Examination, I.A.S. Exams. Graduate Aptitude Test in Engineering (GATE) Conducted by I.I.T's for admission to M. Tech. Courses, Objective Tests for Public Undertaking and Other Organisations, M.Tech Entrance Exams., Interviews and for Quick Review of the subject.]

DR. SADHU SINGH

Professor & Head, Mechanical Engineering Department (Former)

&

Dean, Faculty of Engineering & Technology (Former)

Govind Ballabh Pant University of Agriculture & Technology,

Pantnagar-263145 (Uttarakhand)

Director (Colleges), Punjab Technical University, Jalandhar (Former)

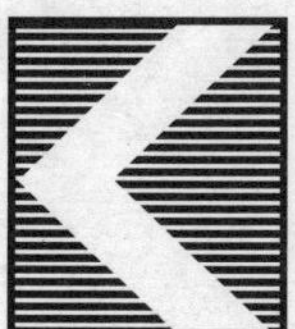

KHANNA BOOK PUBLISHING CO. (P) LTD.

Publisher of Engineering and Computer Books

4C/4344, Ansari Road, Darya Ganj, New Delhi-110002

Phone : 011-23244447-48 **Mobile:** +91-9910909320

E-mail : contact@khannabooks.com

Website : www.khannabooks.com

Dear Readers,

To prevent the piracy, this book is secured with HIGH SECURITY HOLOGRAM on the front title cover. In case you don't find the hologram on the front cover title, please write us to at contact@khannabooks.com or whatsapp us at +91-99109 09320 and avail special gift voucher for yourself.

Specimen of Hologram on front Cover title:

Moreover, there is a SPECIAL DISCOUNT COUPON for you with EVERY HOLOGRAM.

How to avail this SPECIAL DISCOUNT:

Step 1: Scratch the hologram
Step 2: Under the scratch area, your "coupon code" is available
Step 3: Logon to www.khannabooks.com
Step 4: Use your "coupon code" in the shopping cart and get your copy at a special discount
Step 5: Enjoy your reading!

Thanks,
Khanna Publishing House

Price: ₹ 499.00

REFRIGERATION AND AIR CONDITIONING

Dr. Sadhu Singh

ISBN: 978-93-86173-08-9

Edition: 2017
Reprint: 2021

Published by:

KHANNA BOOK PUBLISHING CO. (P) LTD.
4C/4344, Ansari Road, Darya Ganj, New Delhi-110 002
Phone: 011-23244447-48 **Mobile:** +91-9910909320
E-mail: contact@khannabooks.com

Printed in India by: SPS Enterprises, Delhi

PREFACE

Refrigeration and Air – Conditioning is a compulsory subject for the B Tech Mechanical Engineering and other allied branches. It is offered in the final year of the degree program. It is also taught to Diploma students. With the global problem of environment pollution, warming and ozone layer depletion, the importance of this subject has increased manifold in the engineering education The subject is also important from the point of view of job opportunities as it is an essential part of any competitive examination.

Many books on this subject are available in the market but most of these books do not cover the various aspects of this subject. The author has a vast experience of teaching and administration in reputed Institutions and Universities of the country. The author has also long experience of association with the National level examining bodies like UPSC and State Level Public Service Commissions to conduct examinations for National and State Services. This vast experience of the author has been utilized in writing the book.

The book contains seventeen chapters on : Basic Concepts of Thermodynamics, Fundamentals of Refrigeration, Air Refrigeration Cycles, Aircraft Refrigeration Systems, Simple Vapour Compression Systems, Multi - Stage Vapour Compression Systems, Vapour Absorption refrigeration Systems, Refrigerants, Refrigeration Equipment, Cryogenics, Steam Jet Refrigeration System, Psychrometry. Comfort Air-Conditioning, Cooling and Heating Load Estimation, Ducts Design, Fans, and Applications of Refrigeration and Air - Conditioning. A large number of Multiple - Choice Questions from GATE and Civil Services Examinations have been given in the form of Appendices. Useful Tables and Charts have been given at the end of the book. Numerical Problems from Indian Engineering Services Examinations conducted by the UPSC are given chapter wise.

It is hoped that the book in its present form shall be liked by the readers. Further suggestions from the readers for its improvement shall be welcome and duly acknowledged.

I am highly thankful to the Khanna Book Publishing Company for bringing out the bogk in such a short time. My special thanks are due to the Director of the Company, Mr. Mukul Seth for his support and cooperation at all times. The supporting staff of the company has been very helpful for which the author has all appreciations for them.

My son Pritpal Singh, daughter-in-law Narinderpal Kaur, grand children Kanupreet and Amitoej gave me lot of support and encouragement during writing of the book. I express my appreciations for all of them.

15th August, 2016

Dr. Sadhu Singh

BRIEF TABLE OF CONTENTS

Chapter – 1 : Basic Concepts of Thermodynamics	**1–71**
Chapter – 2 : Fundamentals of Refrigeration	**72–89**
Chapter – 3 : Air Refrigeration Cycles	**90–132**
Chapter – 4 : Aircraft Refrigeration Systems	**133–188**
Chapter – 5 : Simple Vapour Compression Systems	**189–311**
Chapter – 6 : Multi–stage Vapour Compression Systems	**312–358**
Chapter – 7 : Vapour Absorption Refrigeration Systems	**359–400**
Chapter – 8 : Refrigerants	**401–418**
Chapter – 9 : Refrigeration Equipment	**419–457**
Chapter – 10 : Cryogenics	**458–477**
Chapter – 11 : Steam Jet Refrigeration	**478–494**
Chapter – 12 : Psychrometry	**495–579**
Chapter – 13 : Comfort Air-conditioning	**580–614**
Chapter – 14 : Cooling and Heating Load Estimation	**615–665**
Chapter – 15 : Ducts Design	**666–700**
Chapter – 16 : Fans	**701–707**
Chapter – 17 : Applications of Refrigeration and Air-conditioning	**708–727**
Appendix – I to IV	**729–798**

DETAILED TABLE OF CONTENTS

CHAPTER – 1 : BASIC CONCEPTS OF THERMODYNAMICS 1–71

1.1 Introduction 1
1.2 Thermodynamic System 1
1.2.1 Types of Systems 2
1.2.2 Control Volume (CV) 2
1.2.3 Concept of Continuum 3
1.3 Some Definitions 3
1.4 Thermodynamic Properties 3
1.4.1 Macroscopic and Microscopic Approach 4
1.4.2 Condition for a Function to be a Property 4
1.5 Phase 5
1.6 State and Path 6
1.7 Process and Cycle 6
1.7.1 Types of Processes 7
1.8 Energy and its Forms 8
1.9 Work 9
1.10 Heat 10
1.11 Thermodynamic Equlibrium 10
1.12 Point Function and Path Function 11
1.13 Concept of an Ideal Gas 11
1.13.1 Characteristics Gas Equation 11
1.14 Zeroth Law of Thermodynamics 12
1.15 First Law of Tahermodynamics 13
1.15.1 Verification of the First Law 13
1.15.2 Corollaries of the First Law 14
1.16 Enthalpy 15
1.17 Specific Heats 15
1.17.1 Specific Heat at Constant Volume, c_p 15
1.17.2 Specific Heat at Constant Pressure, c_P 16
1.17.3 Relation Between the Specific Heats 16
1.18 Perfect Gas Laws 17

1.19 Steady Flow Processes 18
1.19.1 Steady Flow Energy Equation – SFEE 18
1.20 Closed System : Non-flow Porcesses 19
1.20.1 Constant Volume (or Isochoric) Process 19
1.20.2 Constant Pressure (or Isobaric) Process 20
1.20.3 Constant Temperature (or Isothermal) process 20
1.20.4 Reversible Adiabatic (or Isentropic) Process 21
1.20.5 Polytropic Process 24
1.20.6 Constant Internal Energy (or Free Expansion) Process 25
1.21 Throttling Process 26
1.22 Work Done in a Steady Flow Process 26
1.22.1 Constant Volume Process (Isochoric Process) 27
1.22.2 Constant Pressure Process (Isobaric Process) 27
1.22.3 Constant Temperature Process (Isothermal Process) 28
1.22.4 Adiabatic Process 28
1.22.5 Polytropic Process 29
1.23 Limitations of First Law of Thermodynamics 29
1.24 Fundamental Concepts 30
1.25 Statements of Second Law of Thermodynamics 31
1.26 Equivalence of Kelvin and Clausius Statments 32
1.27 Carnot Cycle 33
1.28 Heat Engine 33
1.28.1 Efficiency of Carnot Engine 34
1.29 Heat Pump 35
1.30 Consequences of Carnot Cycle 36
1.31 Concept of Entropy 36
1.31.1 Entropy as a Property 36
1.31.2 Entropy Change in an Irreversible Process 37
1.31.3 Entropy Change for Adiabatic Process 38
1.31.4 Entropy Change for Isolated System 38
1.31.5 Entropy Change of the Universe 38
1.32 Temperature – Entropy Diagram (T-S Diagram) 39
1.33 Carnot Cycle on T-S Diagram 39
1.34 Third Law of Thermodynamics 40
1.35 Pure Substance 41
1.36 Phase 41
1.37 Temperature-Entropy (T-S) Diagram for a Pure Substance 42
1.38 Available and Unavailable Energy 42
1.39 Available Energy Referred to a Cycle 43
1.40 Second Law Efficiency (Effectiveness) 44
1.41 Modes of Heat Transfer 46
1.41.1 Heat Conduction 46
1.41.2 Newton-Rikhman Law for Heat Convection 50

1.41.3 Heat Transfer from Fin 51
1.41.4 Fin Insulated at the Tip 53
1.41.5 Fin Performance 54
1.41.6 Radiation 55
1.42 Modes of Mass Transfer 57
1.43 Mass Transfer by Molecular Diffusion 57
1.43.1 Fick's Law of Diffusion 57
1.43.2 Diffusion Coefficient 58
1.43.3 Convective Mass Transfer 58
1.43.4 Analogy between Momentum and Heat Transfer 58
1.43.5 Analogy between Momentum and Mass Transfer 59
1.43.6 Analogy between Heat and Mass Transfer 60
Multiple – Choice Questions 61
Short Answer Type Qusetions 66
Exercises 68

CHAPTER – 2 : FUNDAMENTALS OF REFRIGERATION **72–89**

2.1 Introduction 72
2.2 Refrigeration Systems 72
2.3 Methods of Refrigeration 73
2.3.1 Vapour Compression Refrigeration System 73
2.3.2 Vapour Absorption System 74
2.3.3 Ejector-compresion System 75
2.3.4 Electro-lux Refrigeration 75
2.3.5 Solar Refrigeration 75
2.3.6 Thermo-electric Refrigeration 76
2.3.7 Vortex Tube Refrigeration 76
2.4 Unit of Refrigeration 76
2.5 Referigeration Effect 76
2.6 Carnot Refrigeration Cycle 77
2.7 Difference Between a Heat Engine, Refrigerator and Heat Pump 78
2.8 Power Consumption of a Referigerating Machine 80
Review Questions 84
Multiple Choice Questions 84
Numerical Problems 87

CHAPTER – 3 : AIR REFRIGERATION CYCLES **90–132**

3.1 Introduction 90
3.2 Air Refrigeration Cycles 90
3.2.1 Open Air Refrigeration Cycle 90
3.2.2 Closed (or Dense) Air Refrigeration Cycle 90
3.3 Reversed Carnot Cycle 91
3.3.1 Temperature Limitations for Reversed Carnot Cycle 92

3.3.2 Vapour as a Refrigerant in Reversed Carnot Cycle 92

3.3.3 Gas as Refrigerant in Reversed Carnot Cycle 94

3.3.4 Limitations of Reversed Carnot Cycle 96

3.4 Bell-coleman Cycle (or Reversed Brayton or Joule Cycle) 96

3.4.1 Bell-Coleman Cycle with Polytropic Processes 99

Multiple – Choice Questions 128

Review Questions 130

Numerical Problems 130

CHAPTER – 4 : AIRCRAFT REFRIGERATION SYSTEMS 133–188

4.1 Aircraft Refrigeration System 133

4.2 Air Cycles for Aircraft 133

4.3 Classification of Aircraft Refrigeration Systems 134

4.3.1 Simple Air Cooling System 134

4.3.2 Simple Air Evaporative Cooling System 142

4.3.3 Boot-strap Air cooling System 146

4.3.4 Boot-strap Air Evaporative Cooling System 150

4.3.5 Reduced Ambient Air Cooling System 151

4.3.6 Regenerative Air Cooling System 153

4.4 Comparison of various Air Cooling Systems used for Aircraft 156

Review Questions 182

Multiple – Choice Quesitons 182

Numerical Problems 185

CHAPTER – 5 : SIMPLE VAPOUR COMPRESSION SYSTEMS 189–311

5.1 Introduction 189

5.2 Comparison of Vapour Compression System with Air Refrigeration System 189

5.3 Simple Vapour Compression Refrigeration System 190

5.4 Vapour Compression Refrigeration System 191

5.5 Use of T-s and p-h Charts 194

5.6 Effect of Suction Pressure 213

5.7 Effect of Discharge Pressure 214

5.8 Effect of Superheating of Refrigerant Vapour 214

5.8.1 Superheat Horn 215

5.9 Effect of Subcooling (or Undercooling) of Refrigerant Vapour 215

5.10 Actual Vapour Compression Refrigeration Cycle 220

5.11 Methods of Improving Simple Saturation Cycle 293

5.11.1 Simple Saturation Cycle with Flash Chamber 293

5.11.2 Simple Saturation Cycle with Accumulator Precooler 295

5.11.3 Simple Saturation Cycle with Sub- cooling of Liquid Refrigerant by Vapour Refrigerant 296

5.11.4 Simple Saturation Cycle with Sub-cooling of Liquid Refrigerant by Liquid Refrigerant 297

Review Questions 300
Multiple – Choice Questions 300
Numerical Problems 304

CHAPTER – 6 : MULTI–STAGE VAPOUR COMPRESSION SYSTEMS **312–358**

6.1 Multistage Vapour Compression System Requirement 312
6.2 Advantages of Multi-stage Compression over Single-stage Compression 312
6.3 Flash Gas Removal 313
6.4 Flash intercooling 314
6.5 Configurations of Multstage Systems 317
6.5.1 Two-stage Compression with Liquid Intercooler 317
6.5.2 Two-stage Compression with Water Intercooler and Liquid Sub-cooler 319
6.5.3 Two-stage Compression with Water Intercooler, Liquid Sub-cooler and Liquid Flash Chamber 321
6.5.4 Two-stage Compression with Water Intercooler, Liquid Sub-cooler and Flash Intercooler 324
6.5.5 Three-stage Compression with Water Inter Cooler 326
6.5.6 Three-stage Compression with Flash Chambers 327
6.5.7 Three-stage Compression with Flash Inter Coolers 330
6.5.8 Three-stage Compression with Multiple Expansion Valves and Flash Intercoolers 332
6.6 Multistage Compression with Multiple Evaporators 334
6.7 Cascade Refrigeration System 338
Multiple – Choice Questions 353
Review Questions 356
Numerical Problems 356

CHAPTER – 7 : VAPOUR ABSORPTION REFRIGERATION SYSTEMS **359–400**

7.1 Introduction 359
7.2 Working Principle of Vapour Absorption Refrigeration System 360
7.3 Advantages of Vapour Absorption System over Vapour Compression System 361
7.4 Coefficient of Performance of an Ideal Vapour Absorption System 362
7.5 Ammonia-water (NH_3 – H_2O) (or Practical) Vapour Absorption System 364
7.6 Lithium Bromide-water Vapour Absorption System (Libr-H_2O) 366
7.6.1 Working Principle 366
7.6.2 Lithium Bromide-water System Equipment 367
7.7 Comparison of Ammonia-water and Lithium Bromide-water Absorption Systems 369
7.8 Refrigerant Absorbent Mixtures 369
7.8.1 Properties of Aqua-ammonia Solution 369
7.9 Elementary Properties of Binary Mixtures 371
7.10 Temperature-concentration Diagram (T-C) 373
7.11 Enthalpy-concentration Diagram (H-C) 373
7.12 Elementary Steady Flow Process with Binary Mixture 375

7.13 Electrolux Refrigerator-domestic Unit 385
Review Questions 397
Multiple – Choice Questions 397
Numercial Problems 400

CHAPTER – 8 : REFRIGERANTS 401–418

8.1 Refrigerants 401
8.2 Classification of Refrigerants 401
8.3 Designation of Refrigerants 402
8.4 Desirable Properties of Refrigerants 403
8.5 Applications of Refrigerants 404
8.7 Eco-friendly Refrigerants 408
8.8 Refrigerant Selection 409
Review Questions 415
Multiple – Choice Questions 415
Theoretical Questions 418

CHAPTER – 9 : REFRIGERATION EQUIPMENT 419–457

9.1 Introduction 419
9.2 Refrigerant Compressor 419
9.2.1 Types of Compressors 419
9.2.2 Reciprocating Compressors 420
9.2.3 Definitions 420
9.2.4 Work done by a Single Stage Reciprocating Compressor 421
9.2.5 Multistage Compression 426
9.2.6 Two-stage Reciprocating Compressor 426
9.2.7 Two-stage Compression with Perfect Intercooling 426
9.2.8 Hermetically Sealed Compressor 429
9.3 Rotary Compressors 432
9.4 Centrifugal Compressors 434
9.5 Condenser 434
9.5.1 Working Principle of a Condenser 434
9.5.2 Factors Affecting Condenser Capacity 435
9.5.3 Heat Rejection Factor-HRF 436
9.5.4 Classification of Condensers 436
9.5.5 Comparison of Air-cooled and Water Cooled Condensers 439
9.5.6 Evaporative Condensers 439
9.6 Cooling Towers 440
9.7 Evaporators 443
9.8 Expansion Devices 448
Multiple – Choice Questions 454
Review Questions 456
Numerical Problems 457

CHAPTER – 10 : CRYOGENICS 458–477

10.1 Introduction 458
10.2 Limitations of Vapour Compression System for Cryogenics 458
10.3 Solid Carbon Dioxide or Dry Ice 459
10.3.1 Advantages and Disadvantages of Solid CO_2 459
10.3.2 Principle of Production of Dry Ice 460
10.4 Three Stage System with Water and Flash Intercooling 463
10.5 Liquefaction of Gases 467
10.6 Linde System for Liquefaction of Air 467
10.7 Claude System for Liquefaction of Air 470
10.8 Liquefaction of Hydrogen 472
10.9 Liquefaction of Helium 473
10.10 Applications of Low Temperature 475
Multiple – Choice Questions 476
Review Questions 476
Numerical Problems 477

CHAPTER – 11 : STEAM JET REFRIGERATION 478–494

11.1 Introduction 478
11.2 System Components 478
11.3 Analysis of Steam Jet Refrigeration System 480
11.4 Performance of Steam Jet Refrigeration System 483
11.5 Precautions for Installation of Sjrs 485
11.6 Advantages and Disadvantages of Sjrs 485
11.7 Applications of Sjrs 486
Multiple – Choice Questions 493
Multiple – Choice Questions 493
Numerical Problems 494

CHAPTER – 12 : PSYCHROMETRY 495–579

12.1 Introduction 495
12.2 Principles of Psychrometry 495
12.3 Psychrometric Relations 497
12.4 Enthalpy of Moist Air 500
12.5 Humid Specific Heat 500
12.6 Thermodynamic Bulb Temperature or Adiabatic Saturation Temperature (AST) 505
12.7 Psychrometric Chart 508
12.8 Psychrometric Processes 510
12.8.1 Sensible Heating or Cooling Process 511
12.8.2 Humidification or Dehumidification Process 512
12.8.3 Heating and Humidification 512

12.8.4 Sensible Heat Factor - SHF 513
12.8.5 Cooling and Dehumidification 514
12.8.6 Air Washer 515
12.8.7 Cooling with Adiabatic Humidification 517
12.8.8 Cooling and Humidification by Water Injection (Evaporative Cooling) 518
12.8.9 Heating and Humidification by Steam Injection 519
12.8.10 Heating and Adiabatic Chemical Dehumidification 520
12.9 Adiabatic Mixing of two Air Streams 521
Multiple – Choice Questions 571
Review Questions 577
Theoretical Questions 577
Numerical Problems 578

CHAPTER – 13 : COMFORT AIR-CONDITIONING 580–614

13.1 Introduction 580
13.2 Thermal Analysis of Human Body 580
13.2.1 Factors Affecting Human Comfort 581
13.2.2 Physiological Hazards Resulting from Heat 582
13.3 Effective Temperature 582
13.3.1 Comfort Chart 582
13.3.2 Factors Affecting Optimum Effective Temperature 585
13.4 Selection of Inside and Outside Design Conditions 586
13.4.1 Selection of Inside Design Conditions 586
13.4.2 Selection of Outside Design Conditions 587
13.5 Air Conditioning Systems 587
13.5.1 Summer Air-conditioning System with Ventilation Air and Zero By-pass Factor 587
13.5.2 Summer Air-conditioning System with Ventilation Air and By-pass Factor 589
13.5.3 Winter Air-conditioning System 591
13.5.4 Comfort Air-conditioning System 595
13.5.5 Industrial Air-conditioning System 596
Multiple – Choice Questions 611
Review Questions 612
Numerical Problems 613

CHAPTER – 14 : COOLING AND HEATING LOAD ESTIMATION 615–665

14.1 Cooling Load Estimation 615
14.1.1 Heat Transfer Through Walls and Roofs 615
14.1.2 Heat Gain from Solar Radiation 617
14.1.3 Sol Air Temperature 617
14.14 Solar Heat Gain Through Glass Areas 617
14.1.5 Heat Gain due to Infiltration 617

14.1.6 Heat Gain from Products 618
14.1.7 Heat Gain from Lights 618
14.1.8 Heat Gain from Power Equipments 618
14.1.9 Heat Gain through Ducts 619
14.2 Empirical Methods to Evaluate Heat Transfer Through Walls and Roofs 619
14.2.1 Using Decrement Factor and Time Lag 619
14.2.2 Equivalent Temperature Differential (ETD) Method 620
14.3 Heating Load Estimation 620
14.4 Room Sensible Heat Factor (RSHF) 621
14.5 Estimation of Supply Air Conditions 621
14.6 Grand Sensible Heat Factor 621
14.7 Effective Room Sensible Heat Factor 623
14.8 By-pass Factor of Heating and Cooling Coils 646
14.9 Efficiency of Heating and Cooling Coils 646

CHAPTER – 15 : DUCTS DESIGN **666–700**

15.1 Transmission and Distribution of Air 666
15.1.1 Room Air Distribution 667
15.1.2 Types of Supply Air Outlets 668
15.1.3 Considerations for Selection and Location of Outlets 669
15.1.4 Distribution Patterns of Outlets 670
15.1.5 Locating Return Air Openings 673
15.2 Processing, Transmission and Distribution of Air in Clean Rooms 674
15.3 Air-Duct Design 677
15.3.1 Duct construction 678
15.3.2 Pressure Classifications 678
15.3.3 Duct Design Procedures 679
15.3.4 Governing Equations for Ducts 680
15.4 Friction Factor for Ducts 688
15.5 Friction Chart for Circular Ducts 692
Multiple – Choice Questions 698
Review Questions 699
Theoretical Questions 699

CHAPTER – 16 : FANS **701–707**

16.1 Fans 701
16.2 Fans Performance Curves 702
16.3 Fan Similarity Laws 703
16.4 Fans in Series and Parallel 703
Multiple – Choice Questions 706
Review Questions 707
Numerical Problems 707

CHAPTER – 17 : APPLICATIONS OF REFRIGERATION AND AIR-CONDITIONING **708–727**

17.1 Food Preservation 708
17.2 Refrigerated Freezers 711
17.2.1 Types of Freezers 712
17.3 Domestic Refrigerator 714
17.4 Transport Refrigeration 715
17.5 Ice Plant 719
17.6 Cold Storage 720
17.7 Water Coolers 720
17.8 Desert Cooler 723
17.9 Room Air-conditioner 724
Multiple – Choice Questions 726
• *Review Questions* 726
Theoretical Questions 727

APPENDIX–I

Multiple – Choice Questions from GATE 729

APPENDIX–II

MCQ'S from Civil Services Examinations 742

APPENDIX–III

Table A – 1 : Properties of Saturated Steam (Temperature basis) 771
Table A – 2 : Properties of Saturated Steam (Pressure basis) 774
Table A – 3 : Moist air : Thermodynamic Properties of Saturated Air at Atmospheric Pressure of 1.01325 Bar 779
Table A – 4 : Properties of Saturated Liquid and Saturated Vapour Refrigerant – 12 781
Table A – 5 : Properties of Saturated Liquid and Saturated Vapour Refrigerant – 22 783
Table A – 6 : Properties of Saturated Liquid and Saturated Vapour Refrigerant – 134a 785
Table A – 7 : Properties of Saturated Liquid and Saturated Vapour Refrigerant – 717 (Ammonia) 787

APPENDIX–IV

Fig. A – 1 : p – h Chart for R – 11 792
Fig. A – 2 : p – h Chart for R – 12 793
Fig. A – 3 : p – h Chart for R – 13 794
Fig. A – 4 : p – h Chart for R – 22 795
Fig. A – 5 : p – h Chart for R – 134a 796
Fig. A – 6 : p – h Chart for R – 717 797
Fig. A – 7 : Psychrometric Chart for Moist Air 798

Subject Index **799**

Chapter 1

BASIC CONCEPTS OF THERMODYNAMICS

1.1 INTRODUCTION

Thermodynamics may be defined as follows:

- The science that deals with energy interactions between material systems.
- The science of energy transfer and its effect on the physical properties of substances.
- The science which deals with the transformation of energy of all kinds from one form to another.

The scope of applied thermodynamics is restricted to the study of heat and work and the conversion of one into the other.

The applications of thermodynamics are:

- Design of energy converting devices, such as, steam engines, internal combustion engines, steam and gas turbines, fuel cells, thermoelectric and thermionic generators.
- Design of refrigerators, air-conditioners, heat transfer and reaction equilibrium, etc.

In this chapter, we shall study the basic concepts of thermodynamics.

1.2 THERMODYNAMIC SYSTEM

A thermodynamic system is defined as a prescribed region or space of finite quantity of matter surrounded by an envelope called the boundary. The boundary may be a real physical surface, such as the walls of a vessel, internal combustion engine cylinder, etc. or it may be an imaginary surface enclosing some matter, such as, steam, gas, vapour, etc. The boundary may be either fixed or moving.

In a broad sense, a thermodynamic system may be defined as a specified region wherein changes due to transfer of mass or energy or both are studied.

Surroundings (or Environment). The space and matter external to the thermodynamic system is called surroundings. The system is separated from the surroundings by the system boundary. The thermodynamic system, boundary and surroundings are shown in Fig. 1.1.

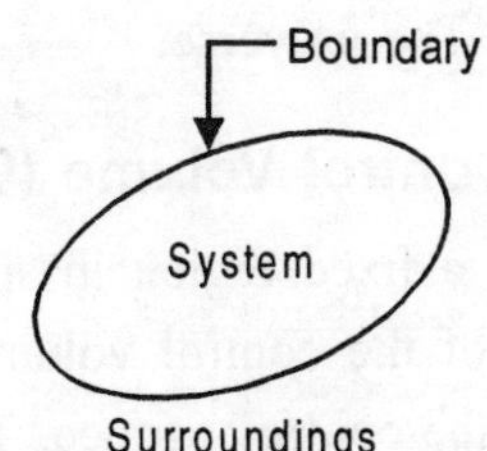

Fig. 1.1 Representation of system, boundary and surroundings.

Universe. The thermodynamic system and surroundings put together is called the universe.

1.2.1 Types of Systems

There are three types of systems: (1) Closed system, (2) open system, and (3) Isolated system.

1. Closed System: In a closed system, the mass within the boundary of the system remains constant and only the energy (heat and work) may transfer across its boundary. Such a closed system is shown in Fig 1.2. Some examples of closed system are:

(*i*) The gas confined between a piston and cylinder.

(*ii*) A rubber balloon filled with air and tightly closed.

(*iii*) Pressure cooker.

(*iv*) Dash pot.

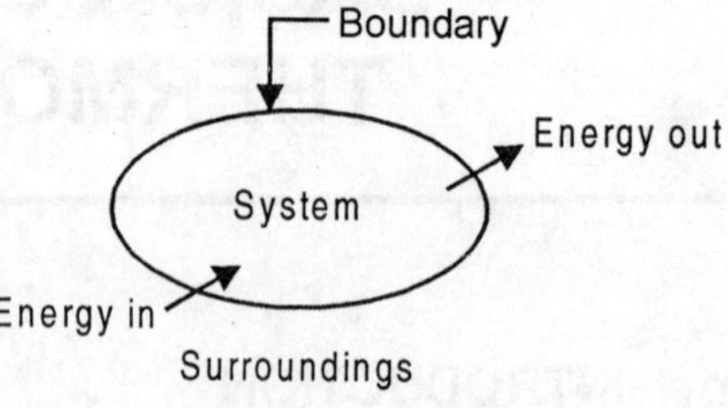

Fig. 1.2 A closed system.

2. Open System: In an open system, the mass as well as the energy transfer across its boundaries may take place. Such an open system is shown Fig. 1.3. Most of the thermodynamic systems are of the open type. Some examples of open systems are :

(*i*) Air compressors

(*ii*) Boilers

(*iii*) Turbines

(*iv*) Water pumps

(*v*) Nozzles and diffusers

(*vi*) Stem engines, and

(*vii*) Internal combustion engines, etc.

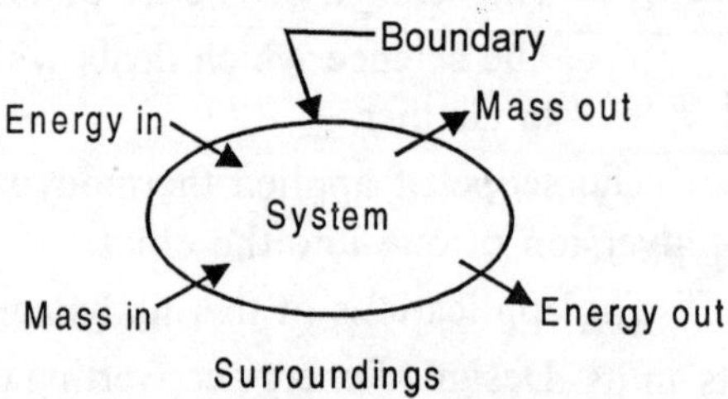

Fig. 1.3 An open system.

3. Isolated System : In an isolated system, neither mass nor energy transfer takes place across its boundaries. Such an isolated system is shown in Fig. 1.4. Some examples of isolated system are:

(*i*) Thermos flask

(*ii*) The universe.

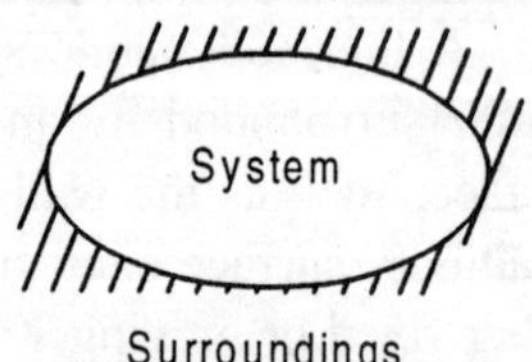

Fig. 1.4 An isolated system.

1.2.2 Control Volume (CV)

It is a fixed region in space through which mass, momentum and energy may flow. The surface of the control volume is called the **control surface (CS).** Matter as well as energy crosses the control surface. The control volume may be stationary or moving with a constant velocity. The concept of control volume and control surface is depicted with the help of an air compressor open system in Fig. 1.5.

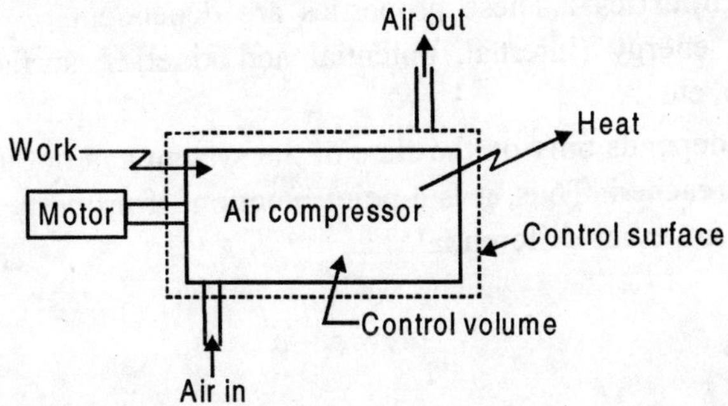

Fig. 1.5 Concept of control volume and control surface.

1.2.3 Concept of Continuum

From the continuum point of view, the matter is seen as being distributed through space and treats the substance as being continuous disregarding the action of individual molecules. The properties are taken at a point.

1.3 SOME DEFINITIONS

Density (ρ). It is the ratio of total mass of the system to its total volume. Its units are kg/m^3.

Specific weight. It is the multiplication of density (ρ) and acceleration due to gravity (g). Its units are Newton (N).

Specific volume (v): It is defined as the volume per unit mass, $v = 1/\rho$. Its units are m^3/kg.

Pressure (p): It is defined as the force exerted normal to unit area of the boundary. Its units are N/m^2 (Pa).

Acceleration due to gravity, $g = 9.80665 \cong 9.81 m/s^2$

Atmospheric pressure,
$$P_{atm} = 101.325 kN/m^2$$
$$= 1.01325 \times 10^5 Pa$$
$$= 1.01325 bar$$

Absolute pressure, $p_{abs} = p_{atm} + p_{gauge}$

1.4 THERMODYNAMIC PROPERTIES

Every system has some characteristics by which its physical condition may be described. Such characteristics are called properties of the system. A property of a system is its measurable characteristics describing the system, such as, mass, volume, physical composition, temperature, pressure, surface area, velocity, electrical potential, etc. Properties may be classified into two categories : (1) intensive, and (2) extensive.

1. **Intensive Properties** : These properties are independent of the mass of the system, e.g., temperature, pressure, density, velocity, height, viscosity, etc.

2. **Extensive Properties** : These properties are dependent on the mass of the system, e.g., volume, energy (internal, potential and kinetic), surface area, electric charge, magnetization, etc.

The property depends only on the state of the system and not upon the manner in which the state was reached. Thus, it is a point function. If property is represented by ϕ, then $d\phi$ should be an exact differential, i.e.,

$$\int_1^2 d\phi = \phi_2 - \phi_1$$

Internal (or Thermostatic) Property : It is defined as a characteristic of the matter within the equilibrium system, e.g., mass, composition, temperature, internal energy, etc.

External (or Mechanical) Property : It is defined as a characteristic of either motion or the position of the system in a gravitational field, e.g.; kinetic energy, potential energy, electric potential, etc.

Independent Properties : These properties define the state of the system.

Dependent Properties : These are those properties which become fixed when the state of the system is defined by independent properties.

1.4.1 Macroscopic and Microscopic Approach

In the macroscopic approach, a certain quantity of matter is considered without taking into account the events occurring at molecular level. This is also called classical thermodynamics point of view. The thermodynamic system is explained by measurable properties, such as, pressure, volume, etc.

In the microscopic approach, the system is considered to be made up of very large number of discrete particles called molecules, This is called statistical thermodynamics point of view.

1.4.2 Condition for a Function to be a Property

Consider the function :

$$f(x,y,z)=0$$

Let $x=x(y,z)$, $y=y(z,x)$, and $z=z(x,y)$

Now if we suppose that z is a dependent variable and x, y are independent variables, then

$$dz=\left(\frac{\partial z}{\partial x}\right)_y dx+\left(\frac{\partial z}{\partial y}\right)_x dy$$

If $\quad M=\left(\frac{\partial z}{\partial x}\right)_y dx$ and $N=\left(\frac{\partial z}{\partial y}\right)_x dy$, then

$$dz = Mdx + Ndy \qquad ...(1)$$

If Eqn. (1) has a solution, then dz is called an exact or perfect differential, and is integrable.

If $\quad z=z(x,y)$, then

$$\left(\frac{\partial}{\partial y}\right)\left(\frac{\partial z}{\partial x}\right)=\left(\frac{\partial}{\partial x}\right)\left(\frac{\partial z}{\partial y}\right)$$

or $$\left(\frac{\partial M}{\partial y}\right)_x=\left(\frac{\partial N}{\partial x}\right)_y \quad ...(2)$$

Eqn. (2) is a necessary and sufficient condition for the existence of a function of x and y. A function which satisfies Eqn. (2) is called a point function or the property of the system. Let us consider the following cases :

(*i*) $\int pdV+\int V.dp=\int(p.dV+V.dp)=\int d(pV)=pV$. Hence, it is an exact differential, and is a property of the system.

(*ii*) $\int p.dV$ is not an exact differential, hence it is not a property of the system because p and V are inter-related to each other and the integral can be evaluated only if the relationship between p and V is known.

(*iii*) $\int V.dp$ is also not an exact differential and so it is not a property.

1.5 PHASE

If a quantity of matter is homogenous or uniform throughout in physical structure and chemical composition, it is termed as a phase. The phase may be single, doublble or multiple constituents. Some examples of a single phase are :

- A piece of iron.
- Gas contained in a vessel.
- A mixture of gases.
- Pure water.
- Liquid oxygen.
- Salt dissolved in water.
- Sugar dissolved in milk.

Some examples of two phases are :

- Ice in water.
- Alcohol in water.
- Steam and water.
- Oil in water.
- Mild steel.

Some examples of three phases are :

- Mixture of solid, liquid and gas.
- Steam, water and ice at the triple point.

On the basis of phases, the system may also be classified as:

(*a*) **Homogenous System**. If a system consists of a single phase, it is called a homogenous system, e.g., mixture of air and water vapour, solution of ammonia in water (aqua ammonia), water and nitric acid, etc.

(*b*) **Hetrogenous System**. If the system consists of more than one phase, it is called a heterogenous system, e.g., water and steam, ice and water, water and oil, etc.

1.6 STATE AND PATH

State. The state is a unique condition of the system. It is the condition of a system at an instant of time described by its properties, such as, pressure, volume, temperature, density, etc.

Path. A thermodynamic system passing through a series of states constitutes a path.

In Fig.1.6, point (p_1, v_1) and point (p_2, v_2) represent the states at 1 and 2. Along curve 12, the states of pressure and volume change from point to point.

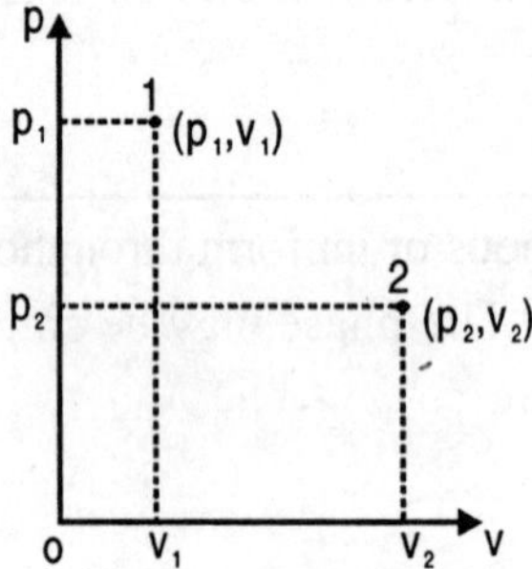

Fig. 1.6 Representation of state.

1.7 PROCESS AND CYCLE

Process. A process is defined as a transition in which a system changes from one initial state. In a process, the path is completely specified. In Fig. 1.7, the line 12 is the process that has taken place.

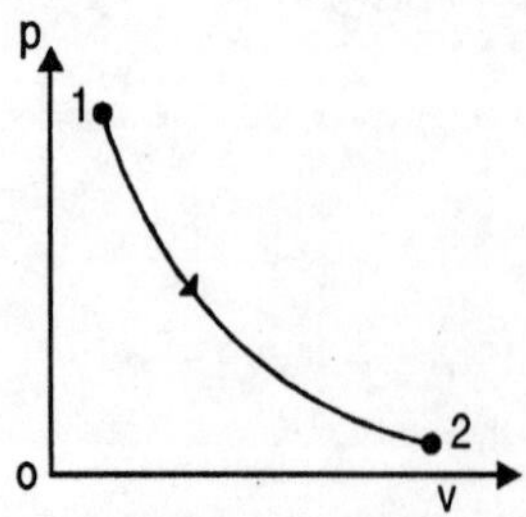

Fig. 1.7 Representation of a process.

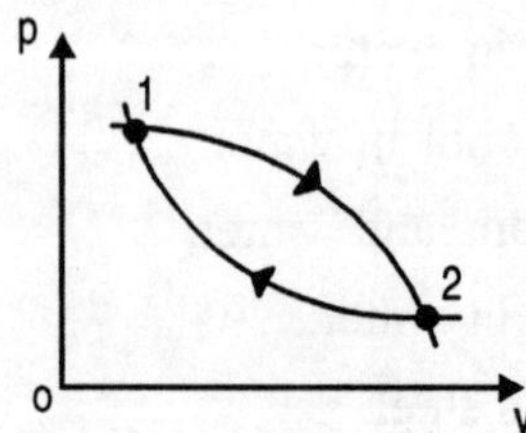

Fig. 1.8 Representation of a cycle.

Cycle. A thermodynamic cycle is defined as a series of state changes such that the final state is identical with the initial state. As shown in Fig. 1.8, 1 – 2 – 1 is the cycle.

1.7.1 Types of Processes

The various types of thermodynamic processes are :

1. Cyclic process.
2.. Quasi- static process.
3. Reversible process.
4.. Irreversible process.

1. Cyclic Process: In a cyclic process, the system undergoes through a series of processes from one state and ends with the same initial state by forming a complete cycle so that the properties of the system, at the end of the cycle are the same as at the beginning. A cyclic process is shown in Fig. 1.9 on $p - v$ diagram.

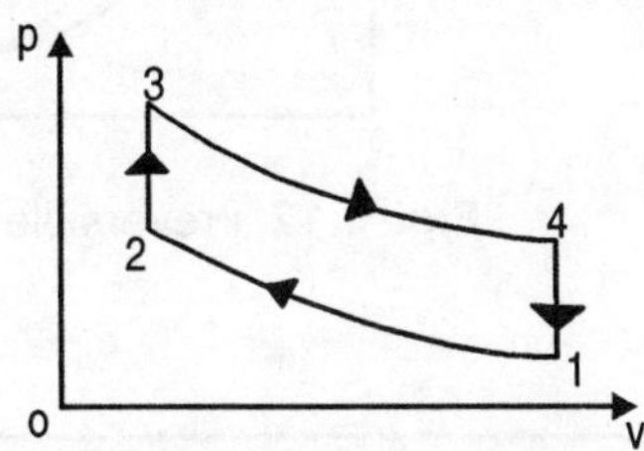

Fig. 1.9 Cyclic process.

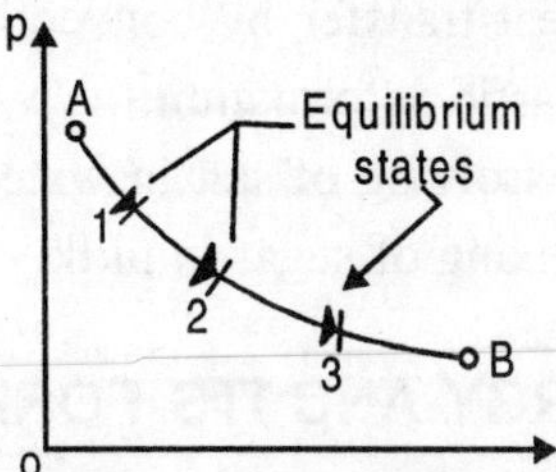

Fig. 1.10 A quasi-static process.

2. Quasi-Static Process : A quasi-static process is one in which the system deviates from one equilibrium state by only infinitesimal amounts throughout the entire process. In other words, a process closely approximating to a succession of equilibrium states is called quasi-static process, as shown in Fig. 1.10. It is a nearly reversible process.

3. Reversible Process: A process is said to be reversible if the system passes through a continuous series of equilibrium states. In other words, a process is called reversible if the initial state together with all energies transferred or transformed during the process can be completely restored in both system and environment. A process will be reversible only if there is no friction in the system. A reversible process is shown in Fig. 1.11. Some examples of reversible process are:

- Frictionless adiabatic expansion and compression of a fluid.
- Polytropic expansion and compression of a gas.
- Isothermal expansion or compression of a gas.
- Extension and compression of a spring.
- Electrolysis.
- Evaporation.

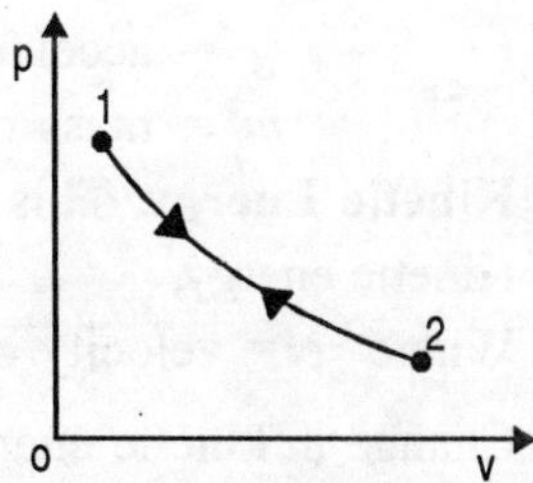

Fig. 1.11 Reversible process.

Conditions for Reversibility

The following conditions must be satisfied by a reversible process:

- There should be no heat transfer across finite temperature difference.
- The process should be carried out with absolute slowness so that the system is always in thermal, chemical and mechanical equilibrium.
- After the process, both the system and surroundings must be restored to the original state, i.e., there should be no friction, viscosity, electrical resistance, inelasticity, etc.

4. Irreversible Process: A process is said to be irreversible if a system passes through a sequence of non-equilibrium states. An irreversible process will not retrace the reverse path to restore the original state, as shown in Fig. 1.12. Some examples of irreversible process are :

- Mechanical and fluid friction.
- Combustion of air and fuel.
- Throttling.
- Free expansion.
- Adiabatic mixing.
- Heat transfer by convection.
- Plastic deformation.
- Dissolving of salt in water.
- Mixing of sugar in milk.

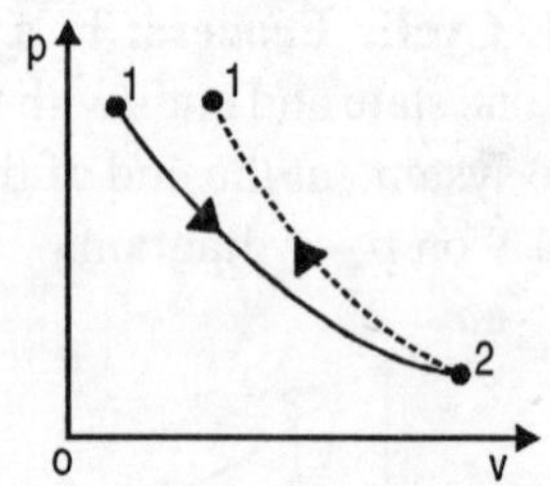

Fig. 1.12 Irreversible process.

1.8 ENERGY AND ITS FORMS

Energy is defined as the capacity to do work. The SI unit of energy are Joule (*J*). Energy may be broadly classified as : stored energy, and energy in transition.

Stored Energy. It is contained within the system boundaries, e.g., potential energy, kinetic energy, and internal energy, etc.

Energy in Transition. This type of energy crosses the system boundaries, e.g., heat, work, electricity, etc.

Potential Energy. This energy is due to the position of mass of the body with respect to the Earth.

Potential energy, $PE = mgz$

Change in potential energy, $\Delta PE = mg(z_2 - z_1)$

Where z = elevation of mass with respect to the Earth.

g = acceleration due to gravity

m = mass of the body

Kinetic Energy. This energy is due to the velocity of the mass of the body.

Kinetic energy, $KE = \frac{1}{2}\, mc^2$

Where c = velocity of mass

Change in kinetic energy, $\Delta KE = 1/2.m(c_2^2 - c_1^2)$

Internal Energy. It is a property consisting of the combined molecular kinetic and potential energies. It is determined by the intensive properties, such as, pressure and temperature. The various energies which contribute towards internal energy are:

- Translational, rotational and vibrational energy of molecules,
- Spin energy of electrons.
- Potential energy of molecules.
- Molecular binding energy.
- Magnetic or electric dipole energy.

For gases, internal energy is defined as:

$$U = mC_v \Delta T$$

Where C_v = specific heat at constant volume.

ΔT = rise in temperature.

Total energy,

$$E = U + PE + KE$$
$$= U + mgz + 1/2\ mc^2$$

High Grade Energy. It is that energy which can be completely transformed into shaft work without any loss, e.g., mechanical and electrical work; water, wind and tidal energy; kinetic energy of jets; animal and manual energy.

Low Grade Energy. This energy cannot be completely converted into mechanical work, e.g., heat energy, nuclear energy, heat from combustion of fuels like coal, wood, oil, etc.

1.9 WORK

In thermodynamics, work is defined as energy transition, i.e. a form of energy interaction between two systems. Work may be defined as follows :

- Energy transferred, without transfer of mass across the boundary of a system because of an intensive property difference other than temperature that exists between system and surroundings.
- An interaction between a system and surroundings, and is done by a system if the sole effect external to the boundaries of the system could have been the raising of a weight.

The work done is represented by the area under the p –V diagram, as shown in Fig. 1.13. Thus,

Work,

$$W_{1-2} = \int_{V_1}^{V_2} p.dV$$

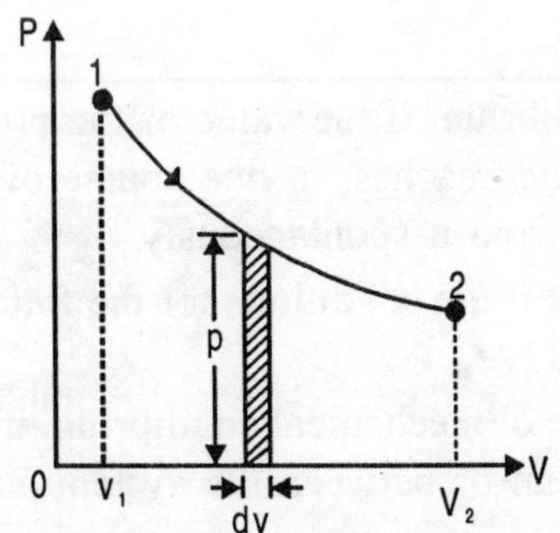

Fig. 1.13 Concept of work.

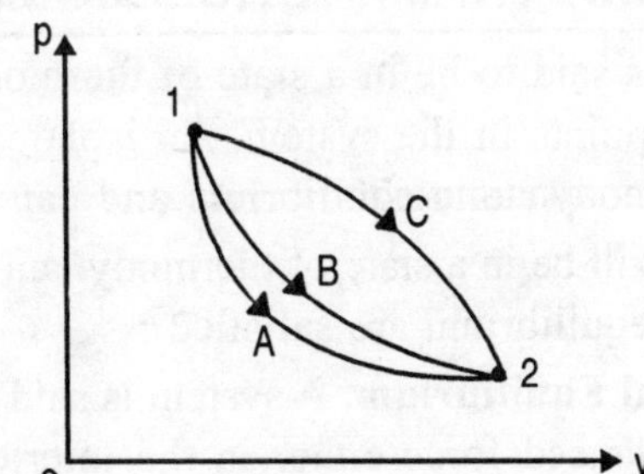

Fig. 1.14 Work a path function.

The SI units of work are N.m or Joule. Work cannot be stored in a system as it exists only during the transition. The work transferred into a system exists as stored energy. Work is a path function, as shown in Fig. 1.14.

$$W_A \neq W_B \neq W_C \text{, as}$$

$$W_{1-2} = \int_1^2 \delta W$$

SIGN CONVENTION FOR WORK. Work done by the system is considered positive and work done on the system as negative.

1.10 HEAT

Heat is an energy interaction between the system and the surroundings. Heat may be defined as the energy transferred without transfer of mass across the boundary of a system due to a temperature difference between the system and the surroundings. The SI units of heat are Joule.

A wall which does not allow the flow of heat is called an adiabatic wall where as a wall which permits the flow of heat is a **diathermal** wall. The similarities and dis-similarities between work and heat are :

Similarities :

- Energy interaction.
- Transient phenomenon.
- Boundary phenomenon.
- Not the property of the system.
- Path functions.

Dis-similarities :

- Heat transfer is the energy interactions due to temperature difference only. All other energy interactions are work interactions.
- Heat is a low grade energy while work is a high grade energy.

Sign Convention for Heat. Heat received by the system is considered positive and heat rejected by the system as negative.

1.11 THERMODYNAMIC EQULIBRIUM

A system is said to be in a state of thermodynamic equilibrium if the value of the property is same at all points in the system. An isolated system always reaches, in due course of time, a state of thermodynamic equilibrium and can never depart from it spontaneously.

A system will be in a state of thermodynamic equilibrium, if the conditions for the following three types of equilibrium are satisfied :

Mechanical Equilibrium. A system is said to be in a state of mechanical equilibrium if there exists no unbalanced force either in the interior of the system or between the system and the surroundings.

Chemical Equilibrium. A system is said to be in a state of chemical equilibrium if there exists no chemical reaction or transfer of matter from one part of the system to another, such as, diffusion or solution.

Thermal Equilibrium. A system is said to be in thermal equilibrium if there exists an uniformity of temperature throughout the system or between system and surroundings.

A system is said to be in thermodynamic equilibrium only if the system is capable of undergoing a spontaneous change in any macroscopic property when it is isolated.

1.12 POINT FUNCTION AND PATH FUNCTION

When the thermodynamic property has a definite value for a given state, it is called a point function. For example, pressure, temperature, volume, etc. are point functions. These properties depend only on the initial and final states of the system. The differentials of point functions are exact, e.g.,

$$\int_{V_1}^{V_2} dV = V_2 - V_1$$

The quantities which are not thermodynamic properties are called path functions. For example, heat and work are path functions because they depend upon the path taken from one state to the other. The work and heat are inexact differentials and are written as δW and δQ.

$$\int_1^2 \delta W = W_{1-2}$$

and

$$\int_1^2 \delta Q = Q_{1-2}$$

1.13 CONCEPT OF AN IDEAL GAS

Many gases which do not change their phase during a thermodynamic process and obey a set of common rules governing change of their properties are called ideal (or perfect) gases. For an ideal gas, the following conditions are satisfied :

1. Gas Law:

$$pv = RT$$

Where p = pressure, kPa
v = specific volume, m^3/kg
R = 0.287 kJ/kg. K, characteristic gas constant
T = absolute temperature, K

2. The internal energy is a function of temperature alone, i.e.

$$U = f(T) \text{ only}$$

None of the gases existing in nature are ideal in real sense. However, most of the gases like nitrogen, oxygen, helium, hydrogen, carbon dioxide and air obey gas laws at normal temperature and pressure. These gases may be regarded as ideal or perfect gases for all practical purposes.

1.13.1 Characteristics Gas Equation

By combining Boyle's and Charle's laws a relation between pressure, temperature and volume may be found. The resulting equation is called the equation of state for a gas. Consider one kg mass of a perfect gas undergoing changes from p_1, v_1, T to p_2, v_2, T_2 by the two processes 1 – A and A – 2, as shown in Fig. 1.15. 1 – A is a constant pressure process obeying Charles's law and A – 2 is an isothermal process obeying Boyle's law.

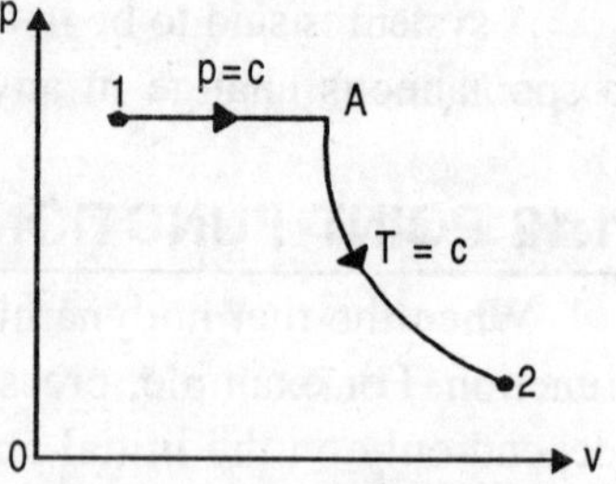

Fig. 1.15 Derivation of characteristic gas equation.

Process 1 – A:

$$\frac{v}{T_1} = \frac{v_A}{T_A}$$

or $$v_A = \frac{v_1 T_A}{T_1} = \frac{v_1 T_2}{T_1} \qquad (\because T_2 = T_A) \quad ...(1)$$

Process A – 2:

$$p_2 v_2 = p_A v_A = p_1 v_A \qquad (\because p_A = p_1) \quad ...(2)$$

or $$v_A = \frac{v_2 p_2}{p_1}$$

From Eqs. (1) and (2), we get

$$\frac{v_1 T_2}{T_1} = \frac{v_2 p_2}{p_1}$$

or $$\frac{p_1 v_1}{T_1} = \frac{p_2 v_2}{T_2}$$

or $$\frac{pv}{T} = \text{Const.} = R$$

This is called the characteristic gas equation.
For a gas having m kg mass,

$$pV = mRT$$

$$pv_{\text{mole}} = R_0 T$$

$$pV = n\, R_0 T$$

where V = total volume of gas, m^3

n = number of kg. moles of a gas = m/M

m = mass of gas, kg

M = molecular weight of a gas, kg/kg moles

v_{mole} = mole volume = $\frac{V}{n}$, m^3/kg mole

R_0 = universal gas constant

= 8.314 kJ/kg mole.K

1.14 ZEROTH LAW OF THERMODYNAMICS

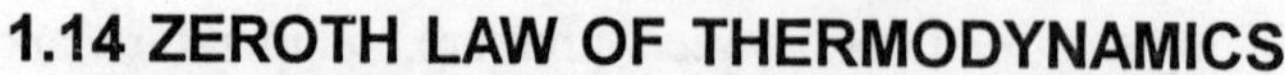

This law states that if two systems are each in thermal equilibrium with a third system, they are also in thermal equilibrium with each other.

Let us consider three bodies A, B and C. If body A and body B are in thermal equilibrium with body C, then according to this law, bodies A and B will also be in thermal equilibrium with each other, as shown in Fig. 1.16. This law is the basis of concept of temperature as well as of all temperature measurements.

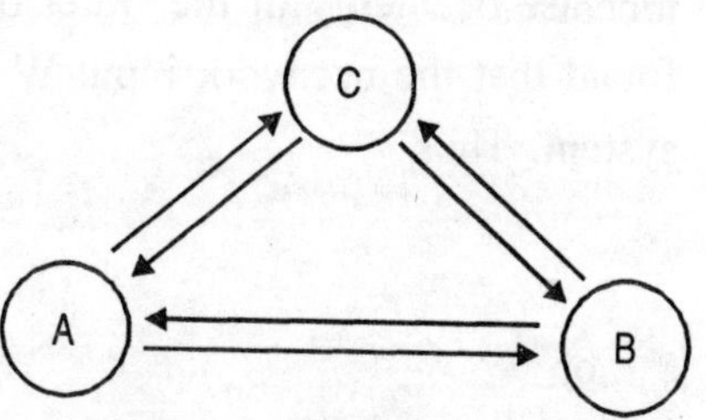

Fig. 1.16 Defining zeroth law of thermodynamics.

1.15 FIRST LAW OF THERMODYNAMICS

The first law of thermodynamics is the law of conservation of energy, i.e., energy can neither be created nor destroyed but only converted from one form to another. With reference to thermodynamics, heat and work are transferable.

This law may be stated as follows:

- For a system undergoing a cyclic change the algebraic sum of the work delivered to the surroundings is equal to the algebraic sum of heat taken from the surroundings, i.e.

$$\oint \delta Q = \oint \delta W$$

where $\oint$ = integral over a cycle

δQ = infinitesimal heat transfer

δW = infinitesimal work transfer

- Heat and work are mutually convertible one into the other.
- The total energy of an isolated system, in all its forms, remains constant.
- It is impossible to construct a perpetual motion machine of the first kind (PMM – 1).
- PMM – 1 is capable of producing energy without corresponding expenditure of the energy.

1.15.1 Verification of the First Law

The first law of thermodynamics can be verified by the Joule's experiment. The experimental set up is shown in Fig. 1.17(*a*). Work is done on the fluid kept in an insulated vessel by stirring of the paddle wheel. This work input to the fluid causes a rise in the temperature of the fluid. The amount of work on the fluid is calculated by the product of the weight and the vertical height through which the weight descends. The system is then immersed in a water bath, as shown in Fig. 1.17(*b*), after removing the insulation. The heat is transferred from the fluid to the water bath till the original temperature is reached, which will be indicated by the thermometer. Thus the system undergoes a complete cycle. The amount of heat rejected by the fluid is equal to the increase of energy of the water bath. Many such experiments were performed by Joule. It was found that the net work input W was directly proportional to the net heat transferred from the system. Thus

$$\oint \delta Q \propto \oint \delta W$$

or $$\oint \delta Q = \oint \delta W \text{, for the same units of Q and W.}$$

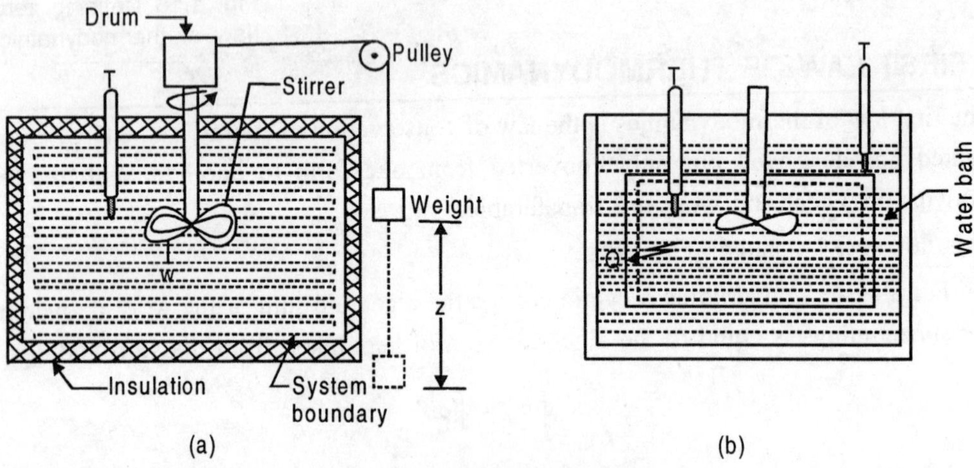

Fig 1.17 Joule's experiment.

1.15.2 Corollaries of the First Law

Corollary 1: System executing a process

For a system executing a process, change in stored energy of the system,

$$dE = dQ - dW$$

where E = energy stored in the system.

Corollary 2: Isolated system

For an isolated system,

$$dQ = 0 \quad \text{and} \quad dW = 0$$

Hence, $$dE = 0$$

or $$E = \text{const.}$$

Thus, the energy of an isolated system remains unchanged.

Corollary 3: PMM – 1

It is impossible to construct a perpetual motion machine of the first kind. The PMM – 1 violates the first law.

As per the law of conservation of energy, no engine can produce mechanical work continuously without some other form of energy disappearing simultaneously.

1.16 ENTHALPY

The summation of internal energy and flow energy is defined as enthalpy, and is given by:

Enthalpy, $H = U + p.dV$

Specific Enthalpy, $h = u + p.dv$

Where u = specific internal energy, v = specific volume

Enthalpy is an extensive property, as it depends upon mass.

Now
$$\begin{aligned} dq &= du + p.dv \\ &= du + d(pv) - v.dp \\ &= d(u + pv) - v.dp \\ &= dh - v.dp \end{aligned}$$

For a constant pressure process, $dp = 0$.

$\therefore$ $dq = dh$

Thus, enthalpy change is the heat supplied at constant pressure.

Also for a perfect gas, $u = c_v T$ and $pv = RT$

$$h = c_v T + RT$$

where c_v = specific heat at constant volume.

or $h = T(c_v + R) = c_p T$

where c_p = specific heat at constant pressure

or $H = mc_p T$

1.17 SPECIFIC HEATS

Specific heat of a substance may be defined as the amount of heat required to raise the temperature of a unit mass of the substance through one degree. The SI units of specific heat are J/kg.K. The solids and liquids have only one specific heat whereas gases have two specific heats, i.e., at constant volume, and constant pressure.

1.17.1 Specific Heat at Constant Volume, c_v

The quantity of heat required to raise the temperature of unit mass of the gas through one degree at constant volume is called specific heat at constant volume, c_v.

Consider a quantity of gas contained in afixed container shown in Fig. 1.18. When heat is supplied to the gas, the volume of the gas will not change because the container is fixed but there will be an increase in pressure and temperature of the gas.

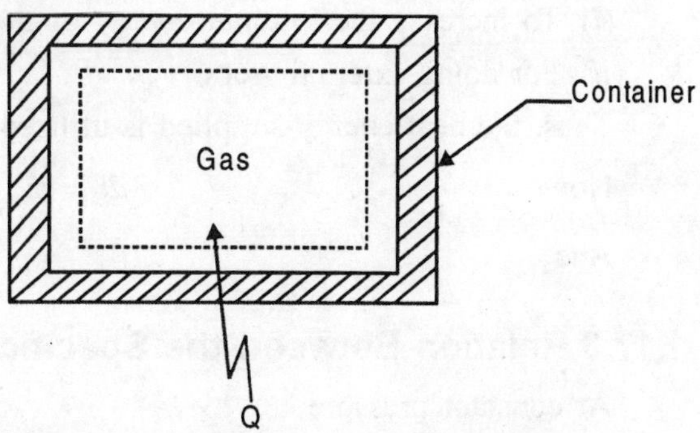

Fig 1.18 Heating at constant volume.

Let m = mass of gas

T_1 = initial temperature of gas

T_2 = final temperature of gas

Total heat supplied to gas at constant volume.

$$Q_{1-2} = mc_v(T_2 - T_1) = mc_v dT$$

Since volume of gas remains constant, therefore, no work is done by the gas and the whole heat supplied remains within the mass of gas. This represents the increase in internal energy of the gas.

$$u = c_v.dT$$

or

$$U = mc_v.dT$$

1.17.2 Specific Heat at Constant Pressure, c_p

Consider a quantity of gas contained in a cylinder tightly fitted with a frictionless piston carrying a load W, as shown in Fig. 1.19. When heat is supplied to the gas, there will be increase in temperature of the gas. But the tendency for the pressure to rise will be counter – balanced by the piston rising upward, thereby lifting the load and increasing the volume of gas. Hence the heat supplied increases the temperature and volume of gas and does work.

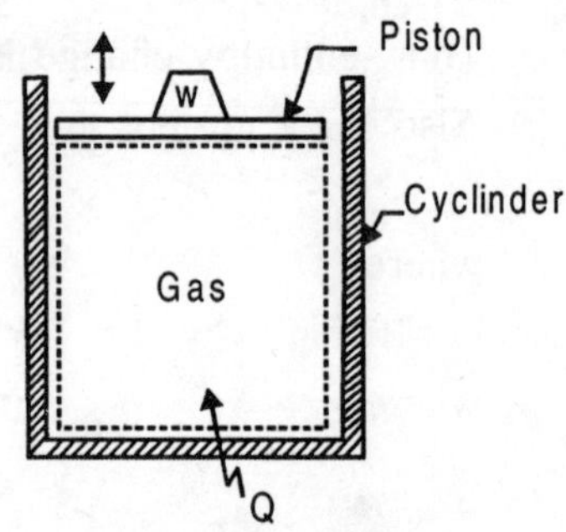

Fig. 1.19 Heating at constant pressure.

The specific heat at constant pressure, c_p, may be defined as the quantity of heat required to raise the temperature of unit mass of a gas through one degree.

Let m = mass of gas

T_1, T_2 = initial and final temperatures respectively

V_1, V_2 = initial and final volumes of gas respectively

$Q_{1-2} = mc_p(T_2 - T_1)$

The heat supplied to the gas at constant pressure is utilized for two purposes:

(*i*) To increase the internal energy of the gas.

(*ii*) For doing external work.

Thus, the heat energy supplied is utilized in increasing the enthalpy of gas.

Now $$dU = mc_v.d(T_2 - T_1)$$

And $$W_{1-2} = p(V_2 - V_1) = mR(T_2 - T_1)$$

1.17.3 Relation Between the Specific Heats

At constant pressure,

$$Q_{1-2} = mc_p.dT = dU + W_{1-2}$$

$$= mc_v.dT + p.dV = mc_v.dT + mR.dT$$

$$= m(c_v + R).dT$$

$$\therefore \quad c_p = c_v + R$$

or $$c_p - c_v = R$$

Now $\gamma = \frac{c_p}{c_v}$, is called the adiabatic constant or index.

$$\therefore \quad c_v(\gamma - 1) = R$$

or $$c_v = \frac{R}{\gamma - 1}$$

The values of c_p, c_v and γ for gases are given in Table 1.1

Table 1.1: Values of c_p, c_v and γ at 38^0 C

Gas	c_v, kJ/kg.K	c_v, kJ/kg.K	γ
Air	1.005	0.718	1.4
Carbon dioxide	0.858	0.67	1.28
Carbon monoxide	1.042	0.745	1.4
Hydrogen	14.343	10.219	1.4
Nitrogen	1.038	0.745	1.4
Oxygen	0.921	0.661	1.39

1.18 PERFECT GAS LAWS

Boyle's Law

When temperature remains constant, the volume of a given mass of a perfect gas is inversely proportional to the absolute pressure. In other words, the product of absolute pressure 'p' and volume 'V' of a given quantity of gas is constant when the temperature remains constant.

$$pV = \text{const.}$$

Charles' Law

- If the pressure remains constant, the volume of a given mass of gas varies directly as the absolute temperature 'T'.

$$\frac{V}{T} = \text{const.}$$

- If the volume remains constant, the pressure of a given mass of gas varies directly with absolute temperature.

$$\frac{p}{T} = \text{const.}$$

Where $T = 273.15 + C°$ Kelvin (K)

1.19 STEADY FLOW PROCESSES

The processes occurring in an open system which permits the transfer of mass as well as energy across its boundaries are known as flow processes. A process is said to be a steady flow process if the conditions in the control volume remain unchanged with time. The common examples of steady flow processes are : nozzles, open system turbines, compressors, pumps, etc.

The following conditions exist in a steady flow process:

- The mass flow rate remains constant within the system.
- The state of energy of the fluid at the entrance and exit of the control volume does not vary with rime.
- The state of the fluid at any point within the control volume is the same at all times.
- The rate at which heat and work transfer takes place across the control surface does not vary with time.
- There is no change in the chemical composition of the system. Therefore, there is no change in chemical energy.

1.19.1 Steady Flow Energy Equation – SFEE

Consider a control volume of a thermodynamic open system as shown in Fig. 1.20, through which a working fluid is passing at a steady rate.

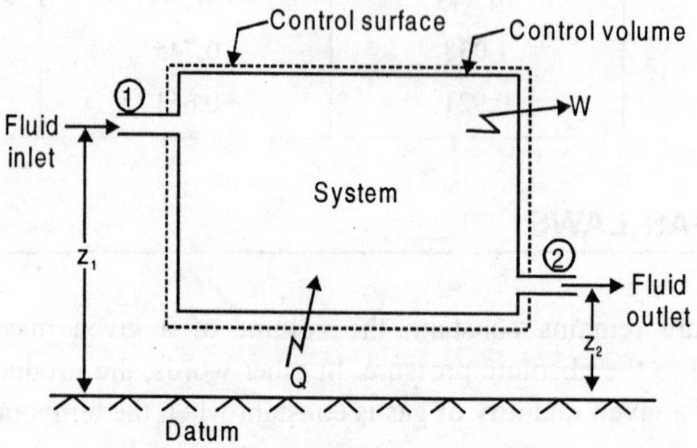

Fig. 1.20 Steady flow process.

Let $\dot{m}$ = mass flow rate, kg/s

p = pressure, Pa

v = specific volume, m^3/kg

c = velocity of fluid, m/s

u = specific internal energy, J/kg

z = height from datum, m

h = specific enthalpy of fluid, J/kg

$\dot{Q}$ = heat supplied rate, J/s

q = heat supplied per unit mass, J/kg

$\dot{W}$ = work transferred rate, J/s

w = work transferred per unit mass, J/kg

1, 2 = subscripts for inlet and outlet respectively.

Then, total energy entering the system,

$$E_1 = \dot{m}(u_1 + p_1 v_1 + \tfrac{1}{2}.c_1^2 + gz_1) + \dot{Q}$$

Total energy leaving the system,

$$E_2 = \dot{m}(u_2 + p_2 v_2 + \tfrac{1}{2}.c_2^2 + gz_2) + \dot{W}$$

According to the first law of thermodynamics,

$$E_1 = E_2$$

$$\dot{m}(u_1 + p_1 v_1 + \tfrac{1}{2}.c_1^2 + gz_1) + \dot{Q} = \dot{m}(u_2 + p_2 v_2 + \tfrac{1}{2}.c_2^2 + gz_2) + \dot{W}$$

Now $h = u + pv$

$$\therefore \quad \dot{m}(h_1 + \tfrac{1}{2}.c_1^2 + gz_1) + \dot{Q} = \dot{m}(h_2 + \tfrac{1}{2}.c_2^2 + gz_2) + \dot{W}$$

$$\dot{Q} - \dot{W} = \dot{m}[(h_2 + \tfrac{1}{2}.c_2^2 + gz_2) - (h_1 + \tfrac{1}{2}.c_1^2 + gz_1)] \quad ...(1)$$

Eqn. (1) is called the steady flow energy equation.

For a unit mass, we have

$$q - w = (h_2 - h_1) + \tfrac{1}{2}.(c_2^2 - c_1^2) + g(z_2 - z_1) \quad ...(2)$$

1.20 CLOSED SYSTEM : NON-FLOW PORCESSES

According to first law of thermodynamics, for a closed system, we have

$$Q_{1-2} - W_{1-2} = (\Delta E)_{system}$$

Where $(\Delta E)_{system} = (\Delta U + \Delta KE + \Delta PE)_{system}$

If ΔKE and ΔPE are negligible, then

$$Q_{1-2} - W_{1-2} = \Delta U$$

1.20.1 Constant Volume (or Isochoric) Process

Consider a completely closed vessel filled with a perfect gas as shown in Fig. 1.21(*a*). Let Q units of heat be supplied to the system. This results in an increase of pressure and temperature of the system at constant volume as represented by process 1-2 on the *p-v* diagram shown in Fig. 1.21(*b*). Since there is no change in volume, therefore

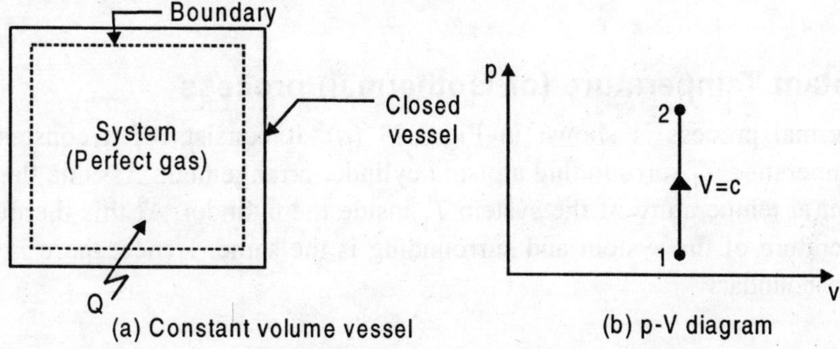

(a) Constant volume vessel (b) p-V diagram

Fig 1.21 Isochoric process.

$$\Delta W = \int p.dV = 0$$

Hence $$Q_{1-2} = dU = m\ c_v\ (T_2 - T_1)$$

1.20.2 Constant Pressure (or Isobaric) Process

Consider a cylinder with a piston carrying perfect gases shown in Fig 1.22 (*a*). When heat Q is supplied to the system, its temperature will rise and it will expand, forcing the piston to move upward. Thus a displacement work is done by the system against a constant force. The above process is shown in Fig. 1.22 (*b*) on *p-V* diagram.

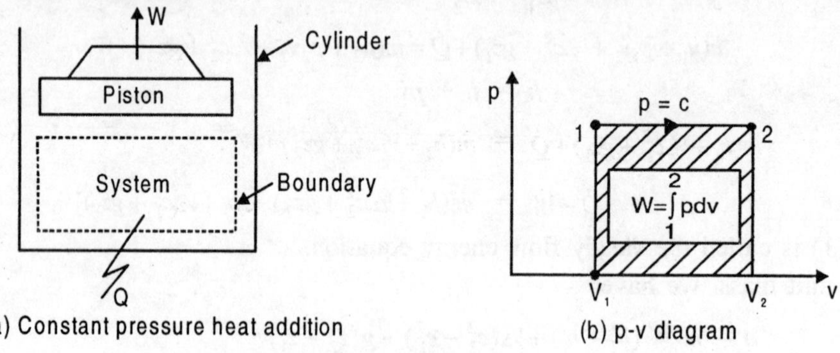

Fig 1.22 Isobaric process.

Workdone by the system, $\delta W = p.dV$

or $$W_{1-2} = \int_1^2 p.dV = p(V_2 - V_1)$$

From first law of thermodynamics, we have

$$\begin{aligned} \delta Q &= \delta W + dU \\ &= \delta(p.dV) + dU \\ &= d(pV) + dU \\ &= d(U + pV) \\ &= dH \end{aligned}$$

or $$Q_{1-2} = H_2 - H_1 = mc_p(T_2 - T_1)$$

or $$q_{1-2} = c_p(T_2 - T_1)$$

It can be seen that during isobaric process the heat transfer is equal to the change in enthalpy.

1.20.3 Constant Temperature (or Isothermal) process

An isothermal process is shown in Fig 1.23 (*a*). It consists of a constant temperature reservoir at temperature T_1 surrounding a piston cylinder arrangement. Assume that a perfect gas is at any instant at temperature of the system T_1 inside the cylinder. At this thermal equilibrium state the temperature of the system and surrounding is the same. Hence, there is no transfer of heat across the boundary.

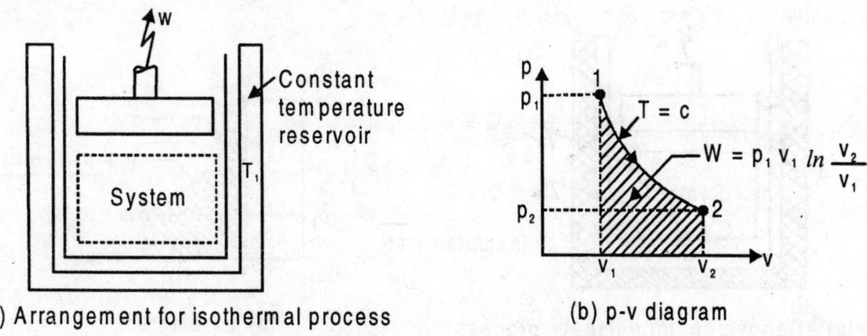

(a) Arrangement for isothermal process

(b) p-v diagram

Fig 1.23 Isothermal process.

If the piston now moves slightly downward, expansion of gas takes place increasing its volume by dV and consequently the pressure and temperature of the system drop by an amount dp and dT respectively. Therefore, heat will flow from the surrounding till the system reaches the original temperature T_1. The isothermal process will be possible only when the process is quasi-static. The isothermal expansion process is shown in Fig. 1.23(*b*) on *p-V* diagram.

For an isothermal process, from Boyle's law, we have

$$pV = \text{const.} = C$$

From first law of thermodynamics

$$\delta Q - \delta W = dU$$

But $dU = 0$ for constant temperature.

$$\delta Q = \delta W$$

$$= p \cdot dV$$

$$Q_{1-2} = W_{1-2} = \int_1^2 p.dV$$

or

$$W_{1-2} = C\int_1^2 \frac{dV}{V} = C\,|\ln V|_1^2 = C\ln\left(\frac{V_2}{V_1}\right)$$

But

$$p_1V_1 = p_2V_2 = C$$

$$W_{1-2} = p_1V_1\ln\left(\frac{V_2}{V_1}\right) = p_1V_1\ln\left(\frac{p_1}{p_2}\right) = RT_1\ln\left(\frac{V_2}{V_1}\right)$$

1.20.4 Reversible Adiabatic (or Isentropic) Process

An adiabatic process is one in which the system undergoes no heat transfer between the system and surrounding but the boundary of the system moves giving displacement work. The arrangement for the adiabatic process is shown in Fig. 1.24 (*a*). It consists of a piston cylinder arrangement where the cylinder is insulated from all sides to prevent heat transfer. Since $\delta Q = 0$, therefore, δW is only due to dU. The *p-V* diagram is shown in Fig. 1.24(*b*).

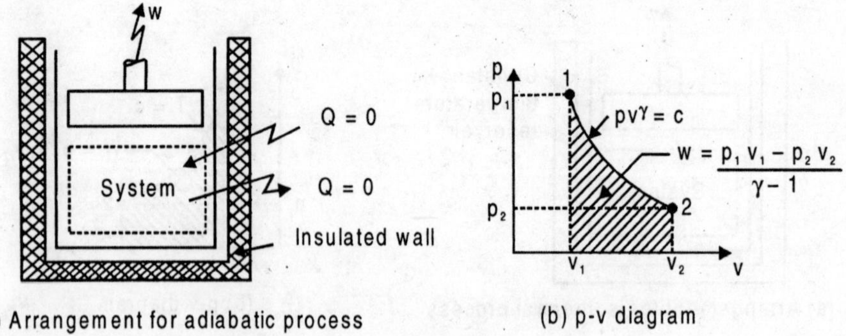

(a) Arrangement for adiabatic process (b) p-v diagram

Fig. 1.24 Adiabatic process.

Form first law of thermodynamics,

$$\delta Q - \delta W = dU$$

As $\delta Q = 0$

$\therefore$ $\delta W = -\,dU = -\,mC_v dT$

For a quasi-static (friction less) process,

$$\delta W = p.dV$$

$\therefore$ $p.dV + dU = 0$

Or $p.dV + dU = 0$

For a perfect gas, $dU = mc_v dT$

$\therefore$ $p.dV + mc_v dT = 0$

Now $pV = mRT$

So that $p.dV + V \cdot dp = mRdT$

$$dT = \frac{p.dV + V.dp}{mR}$$

$$\therefore \quad p.dV + mc_v\left[\frac{p.dV + V.dp}{mR}\right] = 0$$

or $R\,p.dV + c_v(p.dV + V.dp) = 0$

But $R = c_p - c_v$

$\therefore (c_p - c_v)\,p.dV + c_v(p.dV + V.dp) = 0$

or $c_p p.dV + c_v V.dp = 0$

Dividing by $c_p pV$, we get

$$\frac{dV}{V} + \frac{c_v}{c_p}.\frac{dp}{p} = 0$$

or

$$\frac{dp}{p} + \frac{c_p}{c_v}.\frac{dV}{V} = 0$$

Now

$$\gamma = \frac{c_p}{c_v}$$

$$\therefore \qquad \frac{dp}{p} + \frac{\gamma dV}{V} = 0$$

Integrating, we get

$$\ln p + \gamma \ln V = \ln \text{Const.}$$

or $$\ln (pV^{\gamma}) = \ln \text{Const.}$$

or $$pV^{\gamma} = \text{Const.}$$

For a unit mass, $$pv^{\gamma} = \text{Const.}$$

Now $$p_1V_1^{\gamma} = p_2V_2^{\gamma}$$

or $$\frac{V_2}{V_1} = \left(\frac{p_1}{p_2}\right)^{\frac{1}{\gamma}}$$

For a perfect gas, $$\frac{p_1V_1}{T_1} = \frac{p_2V_2}{T_2}$$

$$\frac{T_2}{T_1} = \frac{p_2}{p_1} \cdot \frac{V_2}{V_1} = \left(\frac{V_1}{V_2}\right)^{\gamma} \cdot \frac{V_2}{V_1} = \left(\frac{V_1}{V_2}\right)^{\gamma - 1}$$

or $$\frac{T_2}{T_1} = \left(\frac{p_2}{p_1}\right)^{\frac{\gamma - 1}{\gamma}}$$

Work done, $$W_{1-2} = \int_1^2 p.dV$$

Now $$pV^{\gamma} = C$$

or $$p = \frac{C}{V^{\gamma}}$$

$$W_{1-2} = C\int_1^2 \frac{dV}{V^{\gamma}} = C\int_1^2 V^{-\gamma} dV$$

$$= C\left|\frac{V^{-\gamma+1}}{-\gamma+1}\right|_1^2$$

$$= \left[\frac{C}{1-\gamma}\right]\left[V_2^{1-\gamma} - V_1^{1-\gamma}\right]$$

But $$C = p_1V_1^{\gamma} = p_2V_2^{\gamma}$$

$$W_{1-2} = \frac{p_2V_2 - p_1V_1}{1-\gamma} = \frac{p_1V_1 - p_2V_2}{\gamma - 1}$$

For a unit mass, $$W_{1-2} = \frac{p_1V_1 - p_2V_2}{\gamma - 1} = \frac{R(T_1 - T_2)}{\gamma - 1}$$

1.20.5 Polytropic Process

The polytropic law is expressed as,

Process	Isochoric	Isobaric	Isothermal	Isentropic	Polytropic
n	∞	0	1	γ	n
c_n	c_v	c_p	∞	0	$\dfrac{c_p - nc_v}{1-n}$

Work done, $$W_{1-2} = C\int_1^2 p.dV = C\int_1^2 V^{-n}dV = \frac{C\left(V_2^{1-n} - V_1^{1-n}\right)}{(1-n)}$$

But $$C = p_1V_1^n = p_2V_2^n$$

$$W_{1-2} = \frac{p_1V_1^n - p_2V_2^n}{n-1} = \frac{mR(T_1 - T_2)}{n-1}$$

For a unit mass, $$w_{1-2} = \frac{p_1V_1^n - p_2V_2^n}{n-1} = \frac{R(T_1 - T_2)}{n-1}$$

Change in internal energy, $$dU = U_2 - U_1 = mc_v(T_2 - T_1)$$

or $$du = c_v(T_2 - T_1)$$

Heat transfer, $$Q_{1-2} = dU + \int_1^2 p.dV$$

$$= mc_v(T_2 - T_1) + \frac{mR(T_2 - T_1)}{1-n}$$

$$= m\left[\frac{c_v + R}{1-n}\right](T_2 - T_1)$$

$$= m\left[\frac{c_v + \left(c_p - c_v\right)}{1-n}\right](T_2 - T_1)$$

$$= m\left[\frac{c_p - nc_v}{1-n}\right](T_2 - T_1)$$

$$= mc_v\left[\frac{\gamma - n}{1-n}\right](T_2 - T_1)$$

$$= \left[\frac{mR}{\gamma - 1}\right]\cdot\left[\frac{\gamma - n}{1-n}\right](T_2 - T_1)$$

$$= mc_n(T_2 - T_1)$$

or $$q_{1-2} = c_v\left[\frac{\gamma - n}{1-n}\right](T_2 - T_1)$$

$$= \left[\frac{R}{\gamma - 1}\right]\left[\frac{\gamma - n}{1-n}\right](T_2 - T_1)$$

The Polytropic process is shown in Fig. 1.25 on *p-v* diagram.

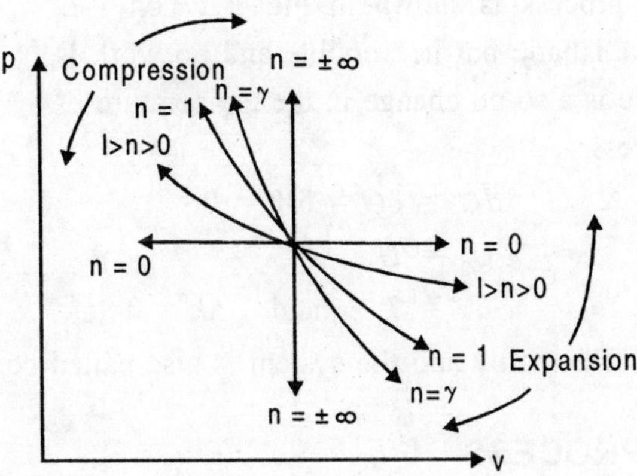

Fig. 1.25 Polytropic process on *p-V* diagram.

Now
$$p_1V_1^n = p_2V_2^n$$

or
$$\frac{p_2}{p_1} = \left(\frac{V_1}{V_2}\right)^n$$

Also
$$\frac{p_1V_1}{T_1} = \frac{p_2V_2}{T_2}$$

or
$$\frac{T_2}{T_1} = \left(\frac{p_2}{p_1}\right)\left(\frac{V_2}{V_1}\right)$$

$$= \left(\frac{V_1}{V_2}\right)^n \times \left(\frac{V_2}{V_1}\right) = \left(\frac{V_1}{V_2}\right)^{n-1} = \left(\frac{p_2}{p_1}\right)^{\frac{n-1}{n}}$$

1.20.6 Constant Internal Energy (or Free Expansion) Process

Consider a system consisting of an insulated vessel in which there are two compartments separated by a partition wall as shown in Fig 1.26 (*a*). There is a perfect gas in one compartment and the other compartment is in a state of perfect vacuum. Now if the partition is punctured the

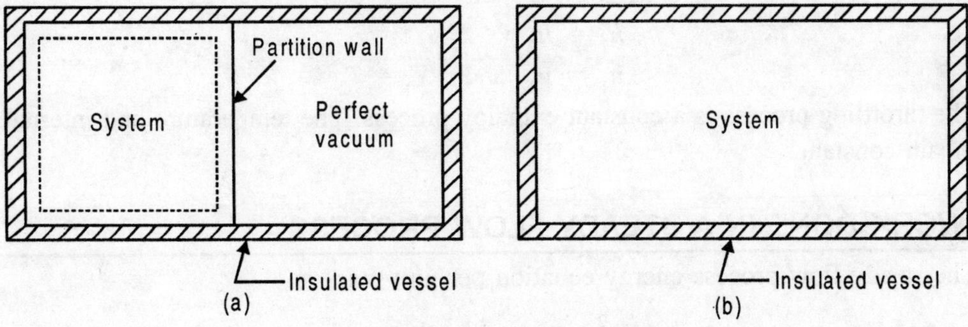

Fig. 1.26 Free expansion process.

system will expand and fill up the whole space as shown in Fig. 1.26 (*b*). The free expansion process is shown in Fig. 1.27 on *V-U* diagram. The process is adiabatic but irreversible and no work is done by the system. There is also no change in the temperature of the system. For this process,

$$dU = \delta Q - \delta W = 0$$

or $$U_2 = U_1$$

$\therefore$ $$T_2 = T_1 \quad \text{and} \quad \Delta h = 0$$

Fig. 1.27 *V-U* diagram for free expansion process.

There is no change in enthalpy and the system is also called constant enthalpy process.

1.21 THROTTLING PROCESS

Throttling process is an irreversible process in which a fluid flows across a restriction in such a manner that there is a drop of pressure without any change in enthalpy of the fluid, kinetic energy, potential energy, and there is no work or heat transfer. Such a process occurs in the flow through porous plug, a partially closed valve or a small orifice. Thus, in a throttling process (Fig. 1.28)

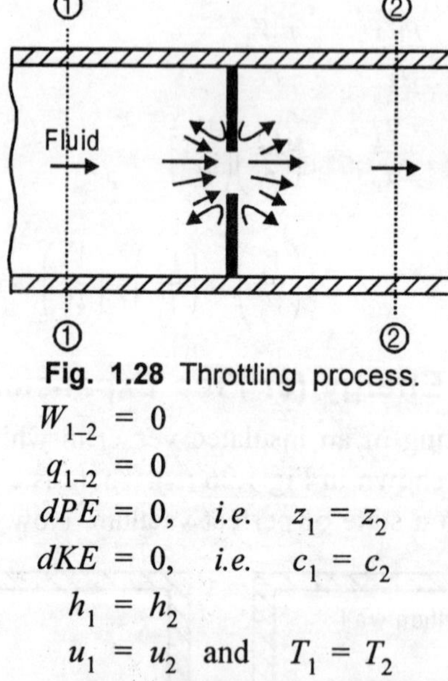

Fig. 1.28 Throttling process.

$$W_{1-2} = 0$$

$$q_{1-2} = 0$$

$$dPE = 0, \quad i.e. \quad z_1 = z_2$$

$$dKE = 0, \quad i.e. \quad c_1 = c_2$$

$$h_1 = h_2$$

$$u_1 = u_2 \text{ and } T_1 = T_2$$

The throttling process is a constant enthalpy process. The temperature and internal energy also remain constant.

1.22 WORK DONE IN A STEADY FLOW PROCESS

The steady flow process energy equation per unit mass is:

$$h_1 + ½c_1^2 + gz_1 + q = h_2 + ½c_2^2 + gz_2 + w$$

In differential form, it can be written as:

$$\delta q - \delta w = dh + d\left(\frac{1}{2}.c^2\right) + d(gz) = d(u + pv) + d\left(\frac{1}{2}.c^2\right) + d(gz)$$

$$= du + p.dv + v.dp + d\left(\frac{1}{2}.c^2\right) + d(gz)$$

For a closed system, according to first law of thermodynamic, we have

$$\delta q = du + p.dv$$

$$\therefore \quad -\delta w = v.dp + d\left(\frac{1}{2}.c^2\right) + d(gz)$$

If $dKE = 0$ and $dPE = 0$, then

$$\delta w = v.dp$$

or

$$\delta w = -v.dp$$

Integrating, we get

$$\int_1^2 \delta w = -\int_1^2 v.dp$$

or

$$w_{1-2} = -\int_1^2 v.dp$$

The work in a steady flow process is shown in Fig. 1.29.

Fig. 1.29 Work done in a steady flow process.

1.22.1 Constant Volume Process (Isochoric Process)

$$w_{1-2} = -\int_1^2 v.dp$$

$$= -v\int_1^2 dp$$

$$= -v(p_2 - p_1) = v(p_1 - p_2)$$

The work done is shown in Fig. 1.30.

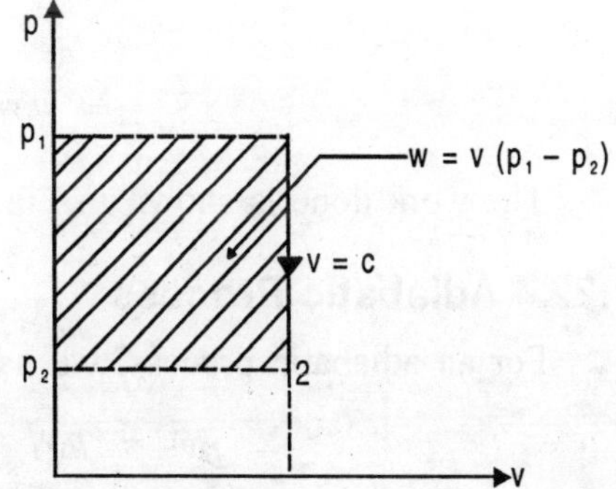

Fig. 1.30 Steady flow isochoric process.

1.22.2 Constant Pressure Process (Isobaric Process)

The $p - v$ diagram is shown in Fig. 1.31.

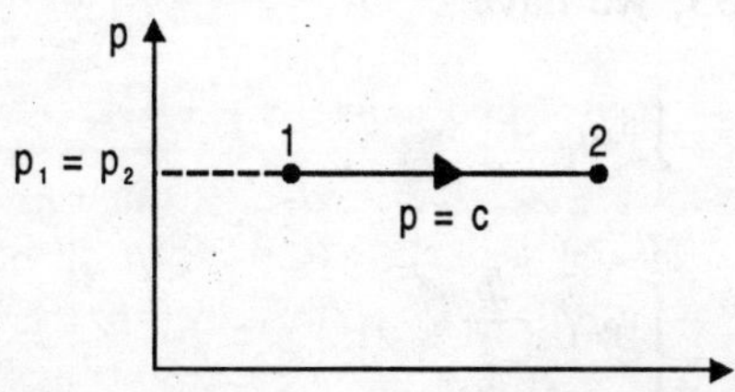

Fig. 1.31 Steady flow isobaric process.

$$w_{1-2} = -\int_1^2 v.dp$$

$$= -v(p_2 - p_1) = v(p_1 - p_2) = 0 \qquad (\because p_1 = p_2)$$

1.22.3 Constant Temperature Process (Isothermal Process)

For a perfect gas, we have

$$pv = p_1v_1 = p_2v_2 = C$$

or

$$v = \frac{p_1v_1}{p}$$

$$w_{1-2} = -\int_1^2 v.dp$$

$$= -\int_1^2 \frac{p_1v_1}{p}.dp$$

$$= -p_1v_1\int_1^2 \frac{dp}{p}$$

$$= -p_1v_1 ln\left(\frac{p_2}{p_1}\right)$$

$$= p_1v_1 ln\left(\frac{p_1}{p_2}\right)$$

$$= p_1v_1 ln\left(\frac{v_2}{v_1}\right)$$

Fig. 1.32 Steady flow isothermal process.

The work done is shown in Fig. 1.32.

1.22.4 Adiabatic Process

For an adiabatic process, we have

$$pv^\gamma = p_1v_1^\gamma = p_2v_2^\gamma = C$$

$$v = v_1\left[\frac{p_1}{p}\right]^{\frac{1}{\gamma}} = \frac{v_1 p_1^{\frac{1}{\gamma}}}{p^{\frac{1}{\gamma}}}$$

With reference to Fig. 1.33, we have

$$w_{1-2} = -\int_1^2 v.dp$$

$$= -\int_1^2 v_1 p_1^{\frac{1}{\gamma}}.\frac{dp}{p^{\frac{1}{\gamma}}}$$

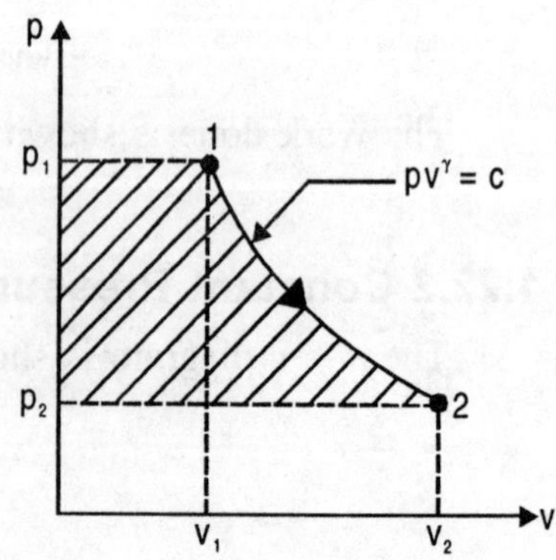

Fig. 1.33 Steady flow adiabatic process.

$$= -v_1 p_1^{\frac{1}{\gamma}} \int_1^2 \frac{dp}{p^{\frac{1}{\gamma}}}$$

$$= -v_1 p_1^{\frac{1}{\gamma}} \left| \frac{p^{-\frac{1}{\gamma}+1}}{\frac{-1}{\gamma}+1} \right|_1^2$$

$$= -v_1 p_1^{\frac{1}{\gamma}} \left(\frac{\gamma}{\gamma-1}\right)\left[\frac{\gamma-1}{p_2^{\gamma}} - \frac{\gamma-1}{p_1^{\gamma}}\right]$$

$$= \left(\frac{\gamma}{\gamma-1}\right) v_1 p_1^{\frac{1}{\gamma}} \left[p_1^{\frac{\gamma-1}{\gamma}} - p_2^{\frac{\gamma-1}{\gamma}}\right]$$

$$= \left(\frac{\gamma}{\gamma-1}\right)(p_1 v_1 - p_2 v_2)$$

1.22.5 Polytropic Process

For a Polytropic process, we have

$$pv^n = p_1 v_1^n = p_2 v_2^n = C$$

Proceeding as for adiabatic process, we get

$$w_{1-2} = \left(\frac{n}{n-1}\right)(p_1 v_1 - p_2 v_2)$$

1.23 LIMITATIONS OF FIRST LAW OF THERMODYNAMICS

The first law of thermodynamic states as follows:

- Heat and work are mutually convertible.
- It is the law of conversation of energy, i.e., energy can either be created nor destroyed.

The implications of first law are:

- No restriction is placed on the direction of flow of heat and work.
- Potential energy can be transformed into kinetic energy and vice-versa.
- Gas expands from high pressure to low pressure and vice-versa.

The actual observations are:

- Heat is not completely convertible to work.
- Kinetic energy cannot be transformed into potential energy automatically.
- Heat cannot flow automatically from cold to hot body.
- Gas cannot expand from low pressure to high pressure automatically.

The first law fails to explain these observations. The first law does not limit the extent of any energy conversion nor it gives any indication as to whether any particular process under consideration is possible or not. The first law is necessary but not a sufficient condition for a certain process to take place.

1.24 FUNDAMENTAL CONCEPTS

Thermal Reservoir: It is defined as a sufficiently large system in stable equilibrium to which and from which finite amount of heat can be transferred without any change in its temperature.

Heat Source. It is a high temperature reservoir, e.g., boiler furnace, combustion chamber, nuclear reactor, Sun etc.

Heat Engine. It is defined as a thermodynamic system operating in a cycle in which the heat is transferred and from which net work is delivered. The principle of heat engine is shown in Fig. 1.34.

Let Q_1 = Heat added to a system.

Q_2 = Heat rejected from a system.

W_{net} = net work done by a system.

Thermal efficiency, $$\eta_{th} = \frac{\text{output}}{\text{input}}$$

$$= \frac{W_{net}}{Q_1}$$

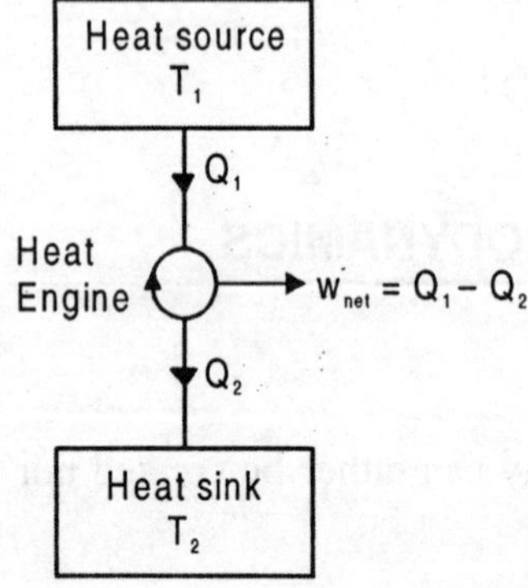

Fig. 1.34 Principle of a heat engine.

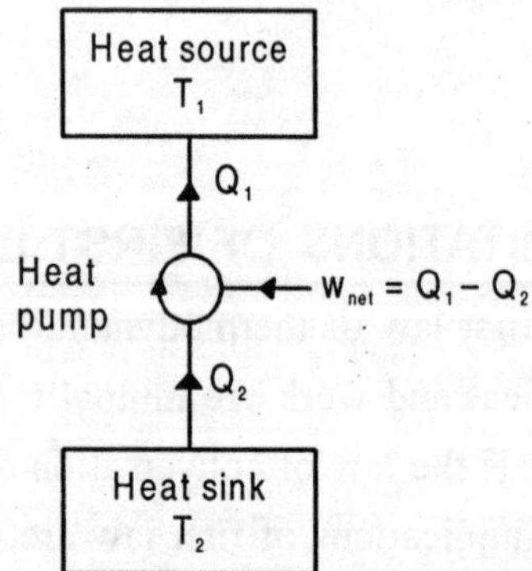

Fig. 1.35 Principle of a heat pump-refrigerator.

From first law of thermodynamics.

$$\oint \delta Q = \oint \delta W$$

or $$Q_1 - Q_2 = W_{net}$$

∴ $$\eta_{th} = \frac{Q_1 - Q_2}{Q_1}$$

$$= 1 - \frac{Q_2}{Q_1}$$

$$= 1 - \frac{T_2}{T_1}$$

HEAT PUMP. It is defined as a thermodynamic system operating in a cycle which removes heat from a low temperature body taking external energy in the form of work input. The principle of a heat pump is shown in above Fig. 1.35.

Let Q_1 = Heat rejected by a system

Q_2 = Heat supplied to a system

From first law of thermodynamics,

$$\oint \delta Q = \oint \delta W$$

or
$$Q_1 - Q_2 = W_{\text{net}}$$

Coefficient of performance (*COP*) of a refrigerator

$$= \frac{\text{Refrigerating effect}}{\text{Work input}}$$

$$= \frac{Q_2}{W_{net}} = \frac{Q_2}{Q_1 - Q_2}$$

$$= \frac{T_2}{T_1 - T_2}$$

$$COP \text{ of a heat pump} = \frac{\text{Heating effect}}{\text{Work input}} = \frac{Q_1}{W_{net}} = \frac{Q_1}{Q_1 - Q_1}$$

$$= \frac{T_1}{T_1 - T_2}$$

$$= 1 + (COP)_{\text{refrigerator}}$$

1.25 STATEMENTS OF SECOND LAW OF THERMODYNAMICS

1. Kelvin – Planck Statement

- It is impossible to construct an engine which will work in a complete cycle and produce no other effect except the raising of a weight and cooling of heat reservoir.
- No process is possible whose sole effect is the removal of heat from a single thermal reservoir at an uniform temperature and the performance of an equal amount of work.

2. Clausius Statement

- It is impossible to construct a heat pump which operating in a cycle will remove heat continuously from a lower temperature reservoir and transfer to a higher temperature without any amount of external work being done on it.
- No process is possible whose sole effect is the removal of heat from a reservoir at lower temperature and the absorption of equal amount of heat by a reservoir at a higher temperature.
- Heat cannot flow from a colder body to a hotter body.

3. Perpetual Motion Machine of Second Kind (PMM-II)

It is impossible to construct a perpetual motion machine of the second kind – which operates in a cycle and delivers an amount of work equal to heat extracted from a single reservoir at a uniform temperature.

4. **Principle of Degradation of Energy**
    - The available energy of an isolated system decreases in all real processes and is conserved in ideal processes.
    - As a result of natural processes, energy steadily becomes less available for the performance of work.
5. **Principle of Increase of Entropy**
    - The entropy of an isolated system increases in all real processes and is conserved in ideal processes.
    - As a result of natural processes, the entropy of the universe steadily increases. The statements 1 and 2 are widely used.

1.26 EQUIVALENCE OF KELVIN AND CLAUSIUS STATMENTS

The two statements for second law of thermodynamics by Kelvin-Planck and Clausius may appear to be different but they are virtually two parallel statements which are equivalent in all respects. The violation of one statement implies the violation of the second and vice-versa.

(*a*) Let us first consider a heat pump (H.P.) which transforms heat from a low temperature reservoir (T_2) to a higher temperature reservoir (T_1) with no other effect, i.e., with no expenditure of work ($W = 0$), violating Clausius statements, as shown in Fig. 1.36.

Let us assume a cyclic heat engine (H.E.) operating between the same thermal energy reservoirs, producing net work (W_{net}) in one cycle. The H.E. draws an amount of heat Q_1 from the hot reservoir equal to that discharged by H.P. Then the hot reservoir may be eliminated and the heat Q_1 discharged by the H.P is fed to the H.E. so that the H.P. acting together constitute a H.E. operating in cycles and producing net work while exchanging heat only with one body at a single fixed temperature. This violates the Kelvin-Planck statement.

(*b*) Now consider a PMM2 which produces net work in a cycle exchanging heat with only one thermal energy reservoir and thus violates the Kelvin-Planck statement, as shown in Fig. 1.37.

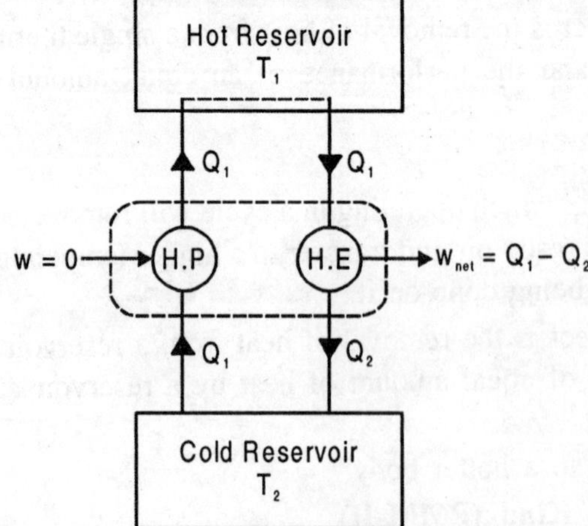

Fig. 1.36 Violation of Clausius statement.

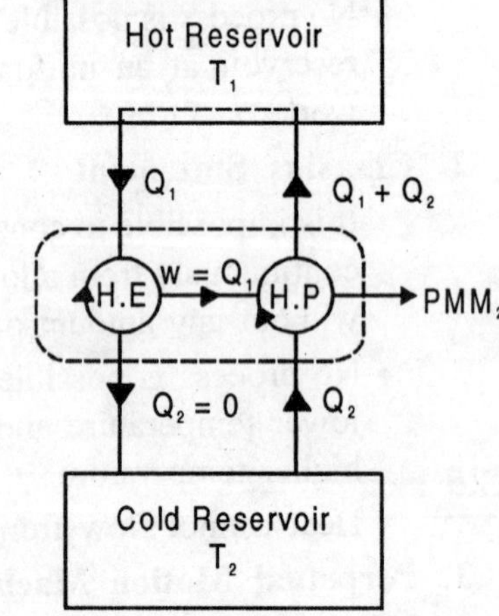

Fig. 1.37 Violation of Kelvin-Planck statement.

Let us assume a cyclic H.P. extracting heat Q_2 from a low temperature reservoir (T_2) and discharging heat to a high temperature reservoir (T_1) with a expenditure of work W equal to that delivered by PMM-2 in a complete cycle. The PMM-2 and H.P. working in cycles and producing the sole effect of transferring heat from a lower to a higher body, thus violating the Clausius statement.

1.27 CARNOT CYCLE

The Carnot cycle (Fig. 1.38) consists of an alternate series of two reversible isothermal processes and two reversible adiabatic processes.

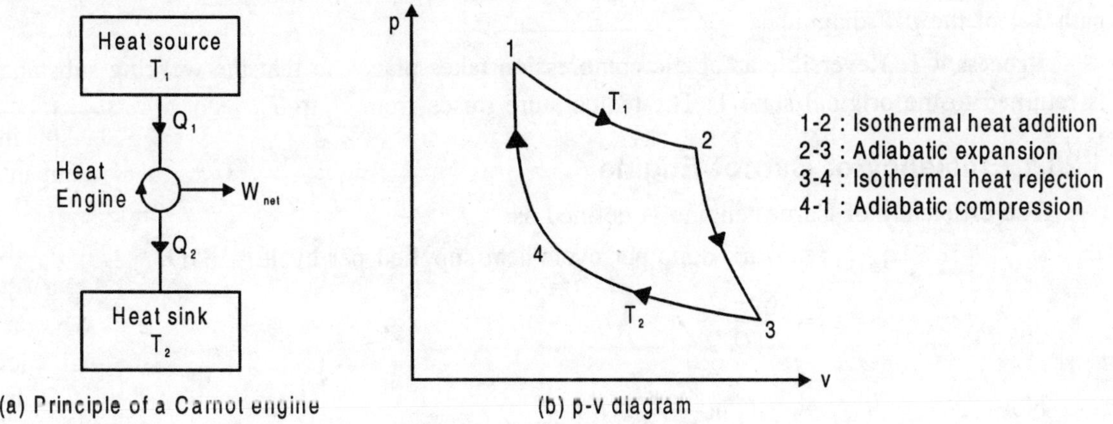

Fig. 1.38 The Carnot cycle.

The working of the Carnot cycle is shown in Fig. 1.39. The working fluid is enclosed in a cylinder having a frictionless piston. The walls of the cylinder and the piston are taken as perfect insulators of heat. The cylinder head is diathermic which permits the flow of heat. The heat source is at temperature T_1 and heat sink at temperature T_2. There is also a work reservoir.

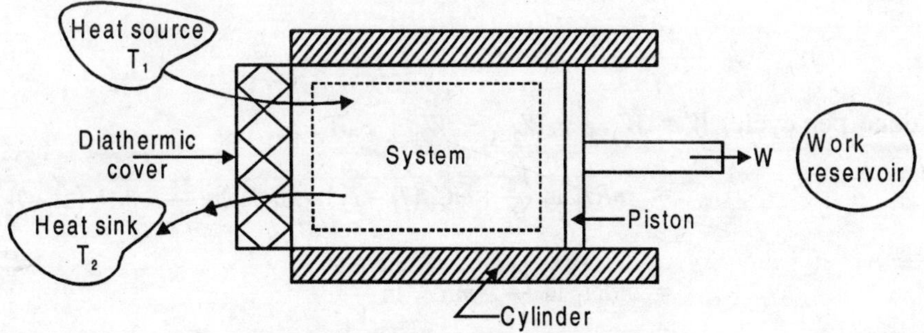

Fig. 1.39 Working of a Carnot cycle.

1.28 HEAT ENGINE

The Carnot heat engine consists of the following four processes:

Process 1-2: Heat is transferred reversibly and isothermally to the working substance from a heat source at temperature T_2. The working substance is allowed to expand slowly and the quantity of heat Q_1 is withdrawn from the heat source. The work done is given by the area under path 1-2 of the *p-V* diagram in Fig. 1.38(*b*).

Process 2-3: Reversible adiabatic expansion takes place. Since the cylinder – piston arrangement is thermally insulted, no heat transfer takes place. The temperature of working substance drops from T_1 to T_2. Work is delivered to the work reservoir and is represented by the area under path 2-3 of the *p-V* diagram.

Process 3-4: Reversible isothermal heat rejection at constant temperature T_2 takes place by slowly compressing the working substance. Work is supplied by the work reservoir while Q_1 amount of heat flows into the sink. The work done in the process is represented by the area under path 3-4 of the *p-V* diagram.

Process 4-1: Reversible adiabatic compression takes places so that the working substance is returned to the original state 1. The temperature raises from T_2 to T_1.

1.28.1 Efficiency of Carnot Engine

The efficiency of Carnot engine is defined as:

$$\eta_{\text{engine}} = \text{work done per cycle/heat supplied per cycle} = W/Q_1$$

$$= \frac{Q_1 - Q_2}{Q_1}$$

Now

$$W_{1-2} = p_1V_1 \ln \frac{V_2}{V_1} = mRT_1 \ln \frac{V_2}{V_1}$$

$$W_{2-3} = \frac{p_2V_2 - p_3V_3}{\gamma - 1} = mC_v(T_1 - T_2)$$

$$W_{3-4} = -p_3V_3 \ln \frac{V_3}{V_4} = -mRT_2 \ln \frac{V_3}{V_4}$$

$$W_{4-1} = \frac{p_4V_4 - p_1V_1}{\gamma - 1} = -mC_v(T_1 - T_2)$$

Work done per cycle, $W = W_{1-2} + W_{2-3} + W_{3-4} + W_{4-1}$

$$= mRT_1 \ln \frac{V_2}{V_1} + mC_v(T_1 - T_2) - mRT_2 \ln \frac{V_3}{V_4} - mC_v(T_1 - T_2)$$

$$= mRT_1 \ln \frac{V_2}{V_1} - mRT_2 \ln \frac{V_3}{V_4}$$

Now

$$\frac{V_2}{V_3} = \left(\frac{T_2}{T_1}\right)^{\frac{1}{(\gamma-1)}}$$

$$\frac{V_1}{V_4} = \left(\frac{T_2}{T_1}\right)^{\frac{1}{(\gamma-1)}}$$

$$\therefore \qquad \frac{V_2}{V_3} = \frac{V_1}{V_4}$$

or

$$\frac{V_2}{V_1} = \frac{V_3}{V_4}$$

$$W = mR(T_1 - T_2) \ln \frac{V_2}{V_1}$$

Also

$$Q_1 = mRT_1 \ln \frac{V_2}{V_1}$$

$$\therefore \qquad \eta_{engine} = \frac{T_1 - T_2}{T_1} = 1 - \frac{T_2}{T_1}$$

As $T_2 < T_1$, therefore, $\eta_{engine} < 100\%$.

1.29 HEAT PUMP

The Carnot cycle for the heat pump is carried out in the reverse sequence of the heat engine, as shown in Fig. 1.40. The various processes are:

Process 4-3: Heat is transferred from a low temperature reservoir to the working fluid in a reversible isothermal manner.

Process 3-2: Reversible adiabatic compression takes place and the temperature of the working fluid increases from T_2 to T_1.

Process 2-1: Heat is transferred reversibly and isothermally from the working fluid to the high temperature reservoir.

Process 1-4: The working fluid is returned to initial state 4 in a reversible adiabatic manner. The temperature of the working fluid decreases from T_1 to T_2.

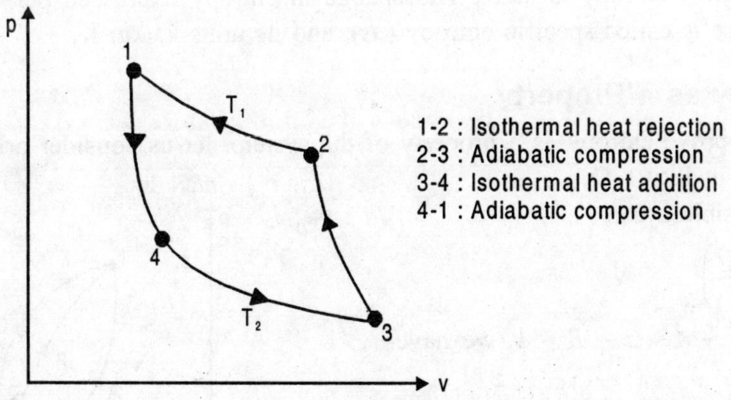

Fig. 1.40 Carnot cycle for heat pump.

Work done per cycle.

$$W_{net} = mR(T_1 - T_2) \ln \frac{V_2}{V_1}$$

$$= Q_1 - Q_2$$

$$(COP)_{\text{ref}} = \frac{Q_2}{Q_1 - Q_2} = \frac{T_2}{T_1 - T_2}$$

$$(\text{COP})_{\text{pump}} = \frac{Q_1}{Q_1 - Q_2} = \frac{T_1}{T_1 - T_2}$$

1.30 CONSEQUENCES OF CARNOT CYCLE

- All the four processes are reversible which requires that the movement of the piston should be frictionless, which is an impossible condition.
- Isothermal heat addition is not possible in practice because the temperature is bound to increase.
- During isothermal process the piston must move slowly and during adiabatic process it must move fast. The variation of speed during the same cycle is not possible.
- It is impossible to transfer the heat without finite temperature difference.

1.31 CONCEPT OF ENTROPY

The cyclic integral of $\delta Q/T$ for a reversible cycle being equal to zero suggests that the quantity $\delta Q/T$ is a point function and is therefore a property of the system. This property is called entropy. It is denoted by 'S'. Mathematically,

$$\int_1^2 \left(\frac{\delta Q}{T}\right)_{\text{rev}} = \int_1^2 dS$$

or

$$\left(\frac{\delta Q}{T}\right)_{\text{rev}} = \delta S$$

Change in entropy, δS = Heat supplied or Heat rejected, δQ/Absolute temperature, T.

The SI units of entropy is kJ/K. The change in entropy expressed per unit mass of the working substance is called specific entropy (*ds*), and its units kJ/kg. K.

1.31.1 Entropy as a Property

In order to prove entropy as a property of the system, let us consider arbitrary reversible cycles as shown in Fig. 1.41.

For a reversible cycle,

$$\oint \left(\frac{\delta Q}{T}\right)_{\text{rev}} = 0$$

For cycle $1 - A - 2 - B - 1$, we have

$$\oint \left(\frac{\delta Q}{T}\right)_{\text{rev}} = \int_1^2 \left(\frac{\delta Q}{T}\right)_A + \int_1^2 \left(\frac{\delta Q}{T}\right)_B = 0$$

and for cycle $1 - A - 2 - C - 1$, we have

$$\oint \left(\frac{\delta Q}{T}\right)_{\text{rev}} = \int_1^2 \left(\frac{\delta Q}{T}\right)_A + \int_1^2 \left(\frac{\delta Q}{T}\right)_C = 0$$

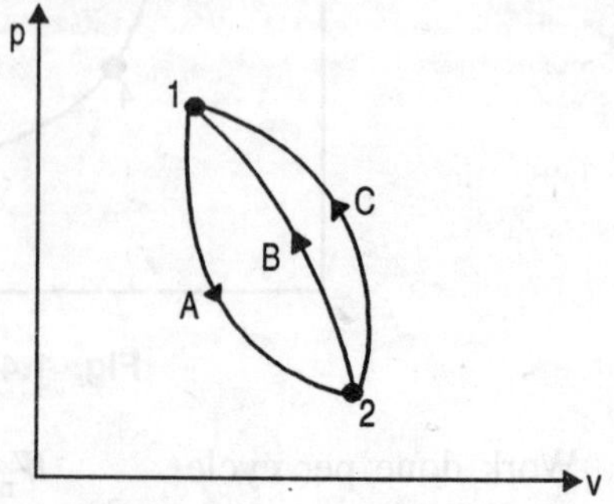

Fig. 1.41 Entropy as a point function.

Comparing the two cycles, we get

$$\int_1^2 \left(\frac{\delta Q}{T}\right)_B = \int_1^2 \left(\frac{\delta Q}{T}\right)_C = 0$$

Hence $\oint\left(\frac{\delta Q}{T}\right)$ is independent of path. It is a point function (i.e. exact differential), and is a property of the system. Thus

$$dS = \left(\frac{\delta Q}{T}\right)_{\text{rev}}$$

$$\int_1^2 dS = \oint\left(\frac{\delta Q}{T}\right)_{\text{rev}}$$

$$\Delta S = \oint\left(\frac{\delta Q}{T}\right)_{\text{rev}}$$

For a reversible adiabatic process, $\delta Q = 0$. Hence

$$\Delta S = S_2 - S_1 = 0$$

The characteristics of entropy are:

- It increases (or decreases) when heat is supplied (or removed) irrespective of the change in temperature.
- It remains constant during reversible adiabatic process.
- It increases if temperature of heat is lowered without work being done, as in a throttling process.
- It is an extensive property because it depends on the mass of the system.

1.31.2 Entropy Change in an Irreversible Process

Consider an isolated system which changes from state 1 to state 2 by an irreversible process 1-*A*-2, as shown in Fig. 1.42. Let the system return to initial state 1 by a reversible process 2-*B*-1. The processes 1-*A*-2 and 2-*B*-1 together form an irreversible cycle. From the Clausius inequality,

$$\underset{\text{Path A}}{\int_1^2 \left(\frac{\delta Q}{T}\right)_I} + \underset{\text{Path B}}{\int_1^2 \left(\frac{\delta Q}{T}\right)_R} < 0$$

But

$$\underset{\text{Path B}}{\int_1^2 \left(\frac{\delta Q}{T}\right)_R} = \underset{\text{Path B}}{\int_1^2 dS} = -\underset{\text{Path B}}{\int_1^2 dS}$$

$$\underset{\text{Path A}}{\int_1^2 \left(\frac{\delta Q}{T}\right)_I} + \underset{\text{Path B}}{\int_1^2 dS} < 0$$

or

$$\underset{\text{Path B}}{\int_1^2 dS} > \underset{\text{Path A}}{\int_1^2 \left(\frac{\delta Q}{T}\right)_I}$$

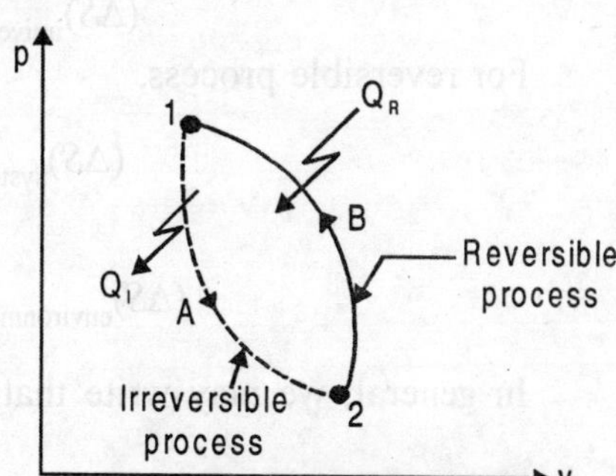

Fig. 1.42 Reversible and irreversible processes.

or $$dS > \left(\frac{\delta Q}{T}\right)_I$$

But $$dS > \left(\frac{\delta Q}{T}\right)_R$$

Hence, in general, we may write

$$dS \geq \left(\frac{\delta Q}{T}\right)$$

$$\Delta S = (S_2 - S_1) \geq \int_1^2 \left(\frac{\delta Q}{T}\right)$$

Where inequality sign holds good for irreversible process and equality sign for reversible process.

1.31.3 Entropy Change for Adiabatic Process

For adiabatic reversible or irreversible process, $\delta Q = 0$. Hence

$$\Delta S \geq 0$$

i.e, entropy change is zero for reversible adiabatic process and more than zero for irreversible adiabatic process..

1.31.4 Entropy Change for Isolated System

For an isolated system, there is no heat and work interaction with the surrounding. Thus $\delta Q = 0$. Hence

$$(dS)_{\text{isolated}} \geq 0$$

For a reversible process, $(dS)_{\text{iso}}$ = or S = const.

For an irreversible process, $(dS)_{\text{iso}} > 0$

Thus, the entropy of an isolated system can never decrease.

1.31.5 Entropy Change of the Universe

The entropy of the universe increases owing to the natural process (irreversible), such that

$$(\Delta S)_{\text{universe}} = [(\Delta S)_{\text{system}} + (\Delta S)_{\text{environmen}}] > 0$$

For reversible process,

$$(\Delta S)_{\text{system}} = \left(\frac{Q}{T}\right)_{\text{rev}}$$

$$(\Delta S)_{\text{environment}} = -\left(\frac{Q}{T}\right)_{\text{rev}}$$

In general, we may write that

$$(\Delta S)_{\text{universe}} \geq 0$$

The entropy of the universe increase in all irreversible process and is conserved in reversible process.

1.32 TEMPERATURE – ENTROPY DIAGRAM (T-S DIAGRAM)

Temperature-entropy diagram is a plot of absolute temperature on vertical axis v's entropy on horizontal axis. It is a plot od isotherms v's reversible adiabatics. During an isothermal process, temperature remains constant but entropy changes, for reversible adiabatic process entropy remains constant but temperature changes. During a reversible process,

$$(\delta Q)_{rev} = T.dS$$

The amount of heat $(\delta Q)_{rev}$ is represented by the shaded area of height T and width dS under the process path 1-2, as shown in Fig. 1.43. Total amount of heat interaction during the reversible process 1-2 is:

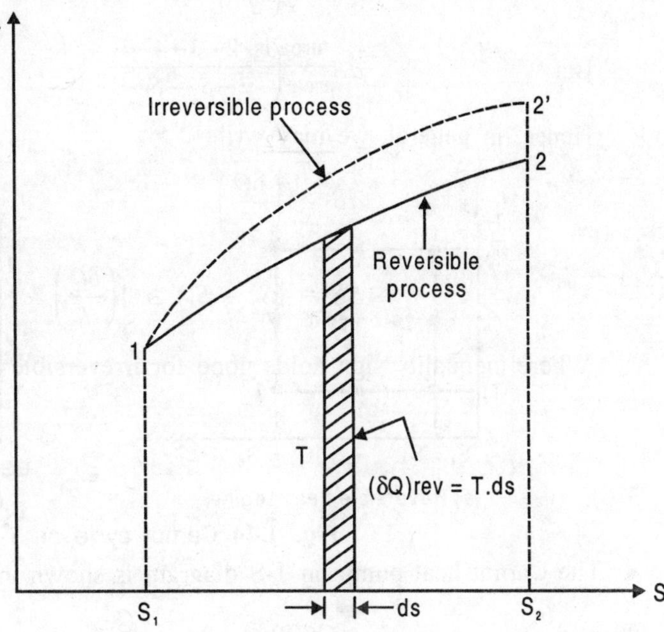

Fig. 1.43 *T-s* Diagram.

$$(Q_{1-2})_{rev} = \int_1^2 TdS$$

$$= \text{area } (1 - 2 - S_2 - S_1 - 1)$$

For an irreversible process, the area under the path is greater than the amount of heat interaction.

- Constant entropy process is called isentropic process.
- For a reversible adiabatic process, $(\delta Q)_{rev} = 0$ and since $dS = \left(\frac{\delta Q}{T}\right)_{rev}$, $dS = 0$. Hence, a reversible adiabatic process is an isentropic process. The opposite is not always true.
- An irreversible-adiabatic process requires heat flow from the system.

1.33 CARNOT CYCLE ON T-S DIAGRAM

The Carnot heat engine on *T-S* diagram is as Fig. 1.44 (*a*).

$$Q_{1-2} = T_1(S_2 - S_1), \text{ isothermal heat addition}$$

$$Q_{2-3} = 0, \text{ adiabatic expansion}$$

$$Q_{3-4} = T_2(S_1 - S_2), \text{ isothermal heat rejection}$$

$$Q_{4-1} = 0, \text{ adiabatic compression}$$

Efficiency, $\eta_{th} = \dfrac{Q_{1-2} - Q_{3-4}}{Q_{1-2}}$

$$= \frac{\text{area } 1-2-3-4-1}{\text{area } 1-2-S_2-S_1-1}$$

$$= \frac{T_1 - T_2}{T_1}$$

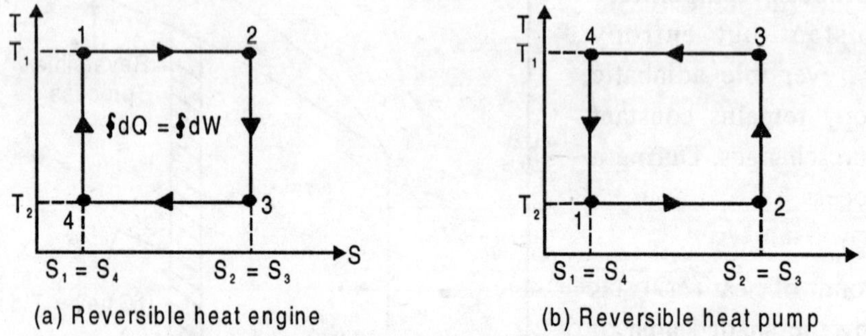
T
T1
1
2
∮dQ = ∮dW
T2
4
3
S
S1 = S4
S2 = S3
(a) Reversible heat engine
T
T1
4
3
T2
1
2
S1 = S4
S2 = S3
(b) Reversible heat pump

Fig. 1.44 Carnot cycle on *T-S* diagram.

The Carnot heat pump on T-S diagram is shown in Fig. 1.44 (*b*)

$$(COP)_{pump} = \frac{Q_{3-4}}{Q_{3-4} - Q_{1-2}}$$

$$= \frac{\text{area } 3-4-S_1-S_2-3}{\text{area } 3-4-1-2-3}$$

$$= \frac{T_1}{T_1 - T_2}$$

For the refrigerator, $(COP)_{ref} = \dfrac{Q_{1-2}}{Q_{3-4} - Q_{1-2}}$

$$= \frac{\text{area } 1-2-S_2-S_1-1}{\text{area } 1-2-3-4-1}$$

$$= \frac{T_2}{T_1 - T_2}$$

1.34 THIRD LAW OF THERMODYNAMICS

The third law of thermodynamics states that the entropy of any pure substance in thermodynamic equilibrium approaches zero as the absolute temperature approaches zero. Mathematically

$$\lim_{T \to 0} S = 0$$

This is also called the **Nernst** theorem.

The third law enables us to calculate the absolute entropy of a substance from the definition of change of entropy.

$$S - S_0 = \int_0^T \frac{(\delta Q)_{rev}}{T}$$

And since a state exists at the absolute zero, where $S_0 = 0$, hence

$$S = \int_0^T \frac{(\delta Q)_{rev}}{T}$$

The third law of thermodynamics is important in respect of the following:

- In calculating thermodynamic properties
- In analyzing chemical and phase equilibria.
- In explaining the behaviour of solids at very law temperature.

1.35 PURE SUBSTANCE

A pure substance is a substance that has a chemically homogenous and fixed chemical composition irrespective of the phase or phases in which it exists. The pure substance is sometimes also called one component system. A mixture of ice, water and steam is an example of a pure substance since the chemical composition of the three phases is the same. Steam is regarded as the most important pure substance in engineering practice.

A substance cannot be treated as pure if it undergoes a chemical reaction. A mixture of oxygen and carbon monoxide is a pure substance as long as it remains fixed in composition. If some of the CO combines with some of the O_2 to form CO_2 the system does not involve a pure substance during the process because its chemical composition is changing. The state of a pure substance at rest is specified completely by the values of two independent intensive properties.

1.36 PHASE

The phase of a substance is any homogenous part of a system that is physically distinct, Each phase in a system is separated from other phases by interfaces called phase boundaries.

In general, a pure substance exists in the following three phases:

1. Solid, 2. Liquid, and 3. Vapour or gaseous phase

Under certain conditions, all three phases may co-exist. For example, water has three phases ice, water, and steam. In a system consisting of an ice cube in liquid water, there are two phases, solid and liquid, and the phase boundary is the surface of the ice cube.

As gas mixture is always a single phase, because the molecules of the constituent gases intermingle completely throughout the entire volume of the gas and make the mixture homogenous. There could be more than one solid or liquid phase of a substance as in the case of sulphur which exists in three solid phases of monoclinic, rhombic and amorphous forms at room temperature.

A system composed of liquid water with a layer of oil on top of it would clearly involve two phases with a planar phase boundary separating the two.

1.37 TEMPERATURE-ENTROPY (T-S) DIAGRAM FOR A PURE SUBSTANCE

The *T-s* diagram for a pure substance like water is shown in Fig. 1.45. The *T-s* diagram for liquid vapour transformation only is shown in Fig. 1.46. At a particular pressure; s_f is the specific entropy of saturated water, and s_g that of saturated vapour. The entropy change of the system during the phase change from liquid to vapour is $s_{fg} = s_g - s_f$

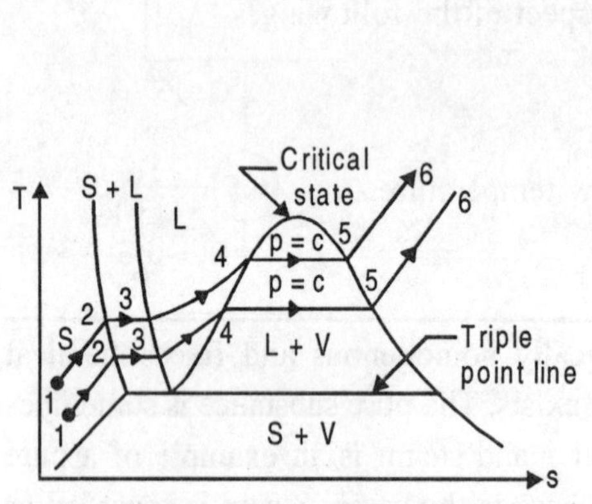

Fig 1.45 Phase equilibrium diagram on T-V-s coordinates.

Fig. 1.46 Saturation (or vapour) dome for water.

1.38 AVAILABLE AND UNAVAILABLE ENERGY

The sources of energy may be classified as:

1. High grade energy, and
2. Low grade energy.

1. High Grade Energy. It is that energy which can be completely transformed into shaft work without any loss. Its examples are : mechanical work, electrical energy, water power, wind power, tidal power and kinetic energy of a jet.

2. Low Grade Energy. This energy cannot be completely converted into mechanical work. Its examples are : thermal energy, nuclear energy, and heat energy from combustion of fossil fuels, like coal, wood, oil etc.

The bulk of the high grade energy is obtained from low grade energy, such as fuels, through the medium of a cyclic heat engine. However, the complete conversion of low grade energy into high grade energy is impossible as per the second law of thermodynamics. That part of the low grade energy which is available for conversion is called **available energy (or exergy)** and the remaining part is called **unavailable energy (or anergy)**. Therefore, **the available energy may be defined as that part of the energy which can be converted into work by ideal (reversible) processes which reduce the system to a dead state,** a state in equilibrium with the Earth and atmosphere.

The unavailable energy may be defined as that portion of energy which cannot be converted into work even by ideal (reversible) processes which reduce the system to dead state.

1.39 AVAILABLE ENERGY REFERRED TO A CYCLE

In reference to a cyclic heat engine as shown in Fig. 1.47 maximum work output obtained is called the available energy (A.E) and the minimum energy, that has to be rejected to the sink is called the unavailable energy (U.E.).

$\therefore$ Heat supplied, $\quad Q_1 = \text{A.E.} + \text{U.E.}$

or $\quad W_{max} = \text{AE.} = Q_1 - \text{U.E.}$

For the given T_1 and T_2, we have

Efficiency, $\quad \eta_{rev} = 1 - \dfrac{T_2}{T_1}$

where T_1 = higher temperature

T_2 = lower temperature

Fig. 1.47 Available and unavailable energy in a cycle.

For a given T_1, efficiency will increase with the decrease of T_2. The lowest practical temperature of rejection is the temperature of the surroundings, T_0.

$$\therefore \quad \eta_{max} = 1 - \frac{T_0}{T_1} \quad \text{and} \quad W_{max} = \left(1 - \frac{T_0}{T_1}\right) Q_1$$

Now consider a finite process 1-2, in which heat is supplied reversibly to a heat engine as shown in Fig. 1.48. For the elementary cycle, we have

$$dW_{max} = \left[\frac{T_1 - T_0}{T_1}\right] dQ_1$$

$$= \text{A.E.}$$

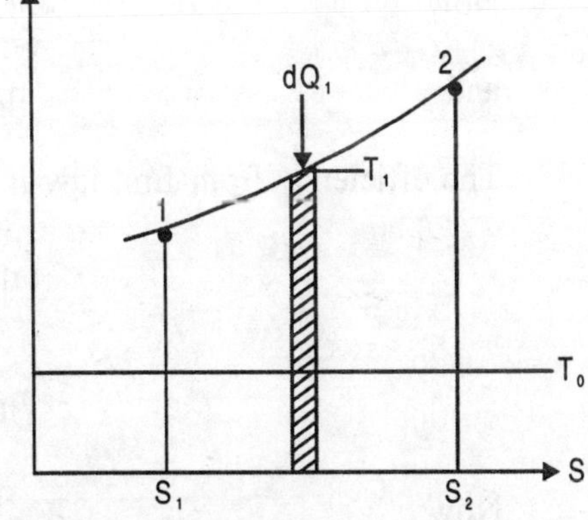

Fig. 1.48 Availability of energy.

For the whole process, we have

$$\int_1^2 dW_{max} = \int_1^2 dQ_1 - \int_1^2 \left(\frac{T_0}{T_1}\right) dQ_1 = \int_1^2 dQ_1 - \int_1^2 T_0 dS$$

$$W_{max} = \text{A.E.} = Q_{1-2} - T_0(S_2 - S_1)$$

or $$\text{U.E.} = Q_{1-2} - W_{max} = T_0(S_2 - S_1)$$

The unavailable energy is thus the product of the lowest temperature of heat rejection and the change in entropy of the system during the process of supplying heat, as shown in Fig. 1.49.

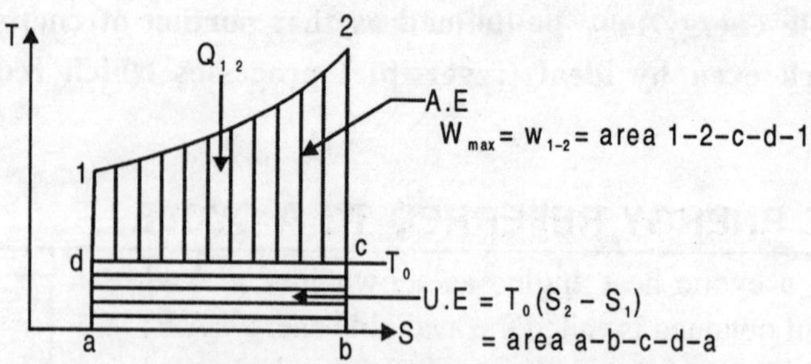

Fig. 1.49 Unavailable energy.

1.40 SECOND LAW EFFICIENCY (EFFECTIVENESS)

The second law efficiency, η_{II}, of a process is defined as the ratio of the minimum available energy which must be consumed to do a task divided by the actual amount of available energy consumed in performing a task. It is also called effectiveness. Thus

$$\eta_{II} = \frac{A_{min}}{A}$$

Where A = availability.

1. Second law efficiency of power plant. A power plant converts a fraction of available energy A or W_{max} to useful work W. For the desired output of W, $A_{min} = W$ and $A = W_{max}$.

Now $$I = W_{max} - W$$

and $$\eta_{II} = \frac{W}{W_{max}}$$

The efficiency from first law is,

$$\eta_I = \frac{W}{Q_1} = \left(\frac{W}{W_{max}}\right) \cdot \left(\frac{W_{max}}{Q_1}\right) = \eta_{II} \cdot \eta_{Carnot}$$

$\therefore$ $$\eta_{II} = \frac{\eta_I}{\eta_{Carnot}}$$

Now $$W_{max} = Q_1\left(1 - \frac{T_0}{T_1}\right)$$

$\therefore$ $$\eta_{II} = \frac{W}{Q_1\left(1 - \frac{T_0}{T_1}\right)} = \frac{\eta_I}{\eta_{Carnot}}$$

2. Second law efficiency of a turbine. The steady state energy balance of a turbine, as shown in Fig. 1.50 gives

$$\left(\frac{\dot{Q}}{\dot{m}}\right)\left[1 - \frac{T_0}{T_\sigma}\right] + a_{f1} = \frac{\dot{W}}{\dot{m}} + a_{f2} + \frac{\dot{I}}{\dot{m}}$$

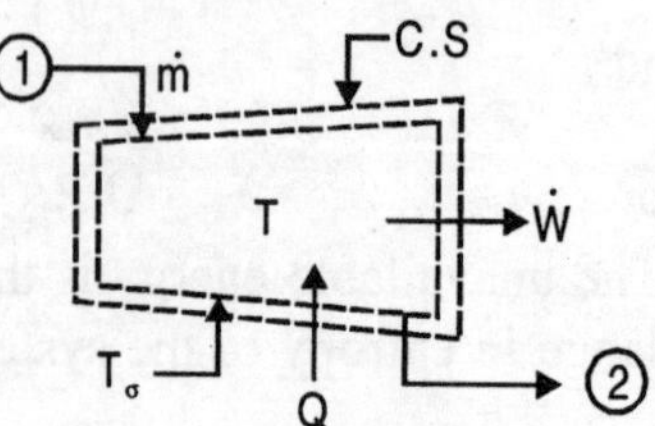

Fig. 1.50 Energy balance of a turbine.

where T_σ = temperature at he control surface of the turbine.

For no heat loss, $$a_{f1} - a_{f2} = \frac{\dot{W}}{\dot{m}} + \frac{\dot{I}}{\dot{m}}$$

Second law efficiency, $$\eta_{II} = \frac{\frac{\dot{W}}{\dot{m}}}{a_{f1} - a_{f2}}$$

3. Second law efficiency of compressor and pump. Proceeding as above for turbine,

$$\frac{\dot{W}}{\dot{m}} = a_{f2} - a_{f1} + \frac{\dot{I}}{\dot{m}}$$

$$\eta_{II} = \frac{(a_{f2} - a_{f1})}{-\frac{\dot{W}}{\dot{m}}}$$

4. Second law efficiency of heat exchanger. The exergy balance for the heat exchanger, as shown in Fig. 1.51, is

$$\sum\left[1 - \frac{T_0}{T_\sigma}\right]Q_1 - \dot{W}_{C.V.} + \left[\dot{m}_h a_{f1} + \dot{m}_c a_{f3}\right] - \left[\dot{m}_h a_{f2} + \dot{m}_c a_{f4}\right] - \dot{I}_{C.V.} = 0$$

If there is not heat transfer and work transfer, then

$$\dot{m}_h(a_{f1} + a_{f2}) = \dot{m}_c(a_{f4} - a_{f3}) - \dot{I}$$

$$\eta_{II} = \frac{\dot{m}_c(a_{f4} - a_{f3})}{\dot{m}_h(a_{f1} - a_{f2})}$$

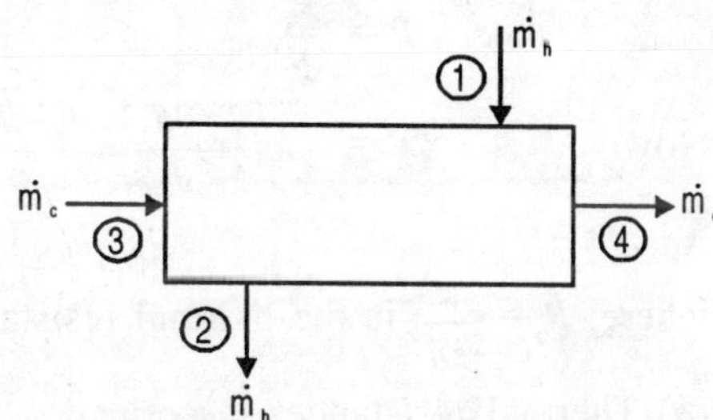

Fig. 1.51 Energy balance of heat exchanger.

5. Second law efficiency for mixing of two fluids. Exergy balance for the mixing process, as shown in Fig. 1.52, gives

$$\left[1 - \frac{T_0}{T_\sigma}\right]\dot{Q} + \dot{m}_1 a_{f1} + \dot{m}_2 a_{f2} = \dot{W}_{C.V.} + \dot{m}_3 a_{f3} + \dot{I}_{C.V.}$$

For adiabatic mixing,

$$\dot{W}_{c.v.} = 0$$

and $$\dot{m}_1 + \dot{m}_2 = \dot{m}_3$$

$$\dot{m}_1(a_{f1} - a_{f3}) + \dot{m}_2(a_{f3} - a_{f4}) + I$$

$$\eta_{II} = \frac{\dot{m}_2(a_{f3} - a_{f4})}{\dot{m}_1(a_{f1} - a_{f3})}$$

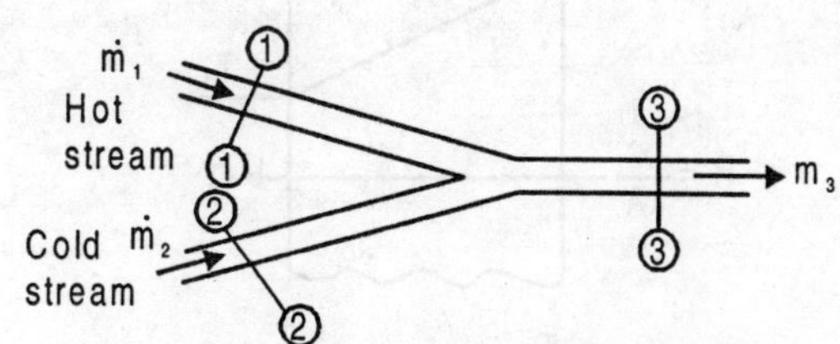

Fig. 1.52 Energy loss due to mixing.

1.41 MODES OF HEAT TRANSFER

1.41.1 Heat Conduction

1. Fourier's Law of Heat Conduction. This law states that the rate of heat flow by conduction through a body is directly proportional to the surface area (*A*) perpendicular to the direction of heat flow and the temperature gradient (*dT/dx*) through the body. As shown in Fig. 1.53, heat conduction through a wall,

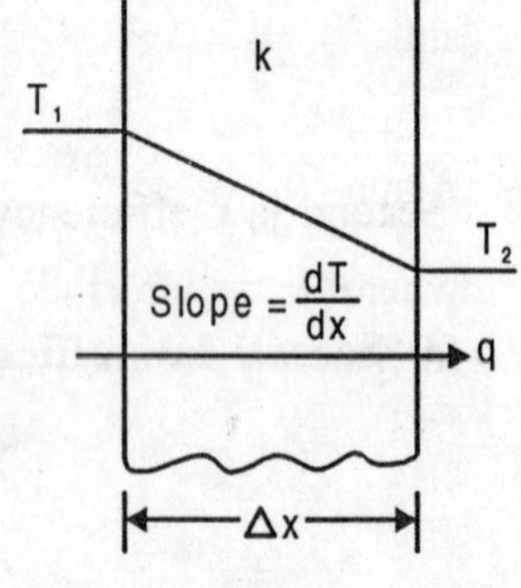

Fig. 1.53 Heat conduction through a wall.

$$Q = -\ kA\ dT/dx$$

or
$$q = Q/A = -\ k\ dT/dx$$

k is the thermal conductivity of the body in kJ/(m.h.K)

q is called the heat flux (kJ/(h.m^2)).

2. Thermal Resistance. Heat flow is analogous to flow of electricity. According to Ohm's law, current *I* flowing through a wall of resistance *R* (Fig. 1.54) is proportional to the potential difference *E*, or

Fig. 1.54 Electrical resistance concept.

$$I = \frac{E}{R}$$

Now
$$Q = -kA\frac{T_2 - T_1}{x} = \frac{T_1 - T_2}{\frac{x}{kA}} = \frac{T_1 - T_2}{R_t}$$

where, $R_t = \dfrac{x}{kA}$ is the thermal resistance to heat flow offered by the wall (Fig. 1.55).

(*a*) Thermal resistances in series

For a composite wall, as shown in Fig. 1.56, having two resistances in series, total thermal resistance,

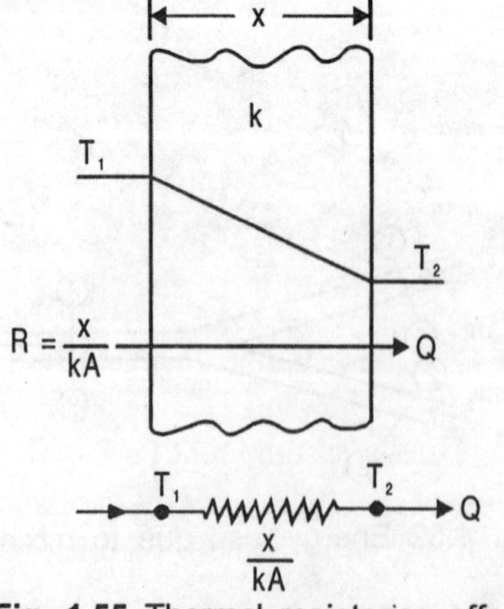

Fig. 1.55 Thermal resistance offered by a wall.

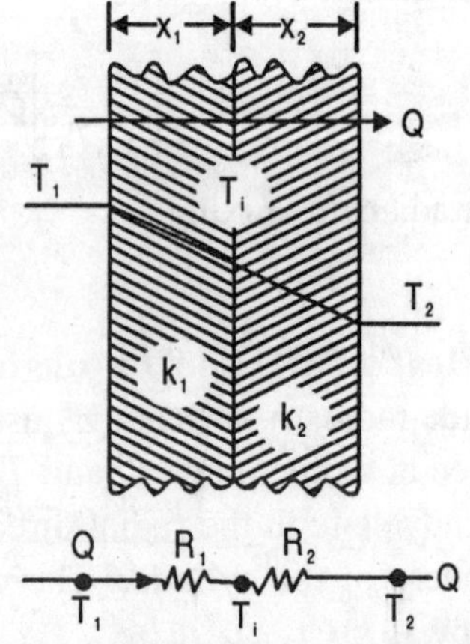

Fig. 1.56 Heat conduction through resistances in series.

$$(R_t)_c = R_1 + R_2 = \frac{x_1}{k_1 A} + \frac{x_2}{k_2 A}$$

and
$$Q = \frac{T_1 - T_2}{(R_t)_c}$$

Again
$$T_1 - T_i = QR_1 = \frac{Qx_1}{k_1 A}$$

where T_i = interface temperature.

In general, for many thermal resistances in series, we have

$$(R_t)_c = \Sigma(R_t)_i = \sum_{i=1}^{n} \frac{x_i}{k_i A_i} \text{ and } Q = \frac{T_1 - T_{n+1}}{\sum_{i=1}^{n} \frac{x_i}{k_i A_i}}$$

(*b*) Thermal resistances in parallel.

For two thermal resistances in parallel, as shown in Fig. 1.57, we have

$$\left(\frac{1}{R_t}\right)_c = \frac{1}{R_1} + \frac{1}{R_2}$$

or
$$(R_t)_c = \frac{R_1 R_2}{R_1 + R_2}$$

where
$$R_1 = \frac{x}{k_1 A_1}$$

and
$$R_2 = \frac{x}{k_2 A_2}$$

Also
$$Q = \frac{T_1 - T_2}{(R_t)_c}$$

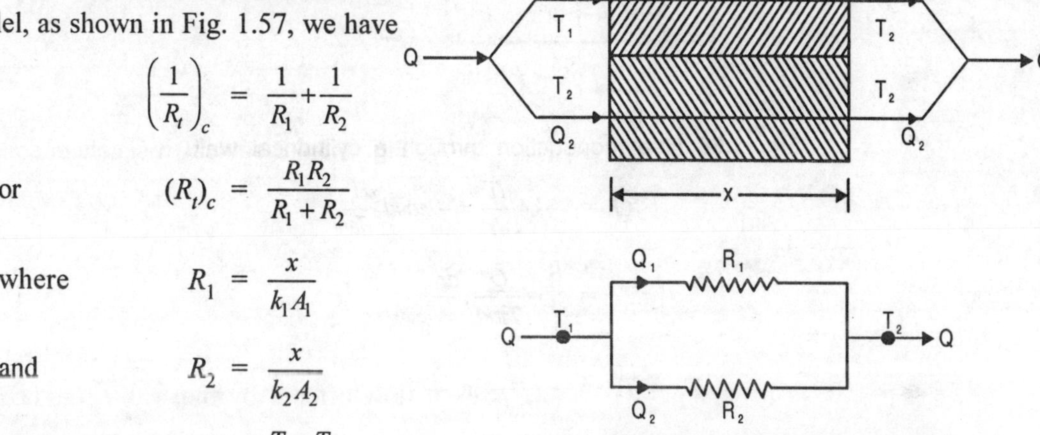

Fig. 1.57 Heat conduction through resistances in parallel.

In general, for many thermal resistances in parallel, we have

$$\frac{1}{(R_t)_c} = \sum_i \frac{1}{(R_t)_i} = \frac{1}{\sum_i \frac{x_i}{k_i A_i}}$$

and
$$Q = \frac{T_1 - T_{n+1}}{\Sigma(R_t)_i}$$

3. Heat conduction Through a Cylinder. Consider a circular cylinder of inside radius r_1 and outside radius r_2 of length l, as shown in Fig 1.58 (*a*), whose inside and outside surfaces are maintained at temperature T_1 and T_2 respectively so that $T_1 > T_2$. We assume that heat is flowing under steady state in the radial direction only and there is no heat conduction along the length or the periphery of the cylinder. The rate of heat transfer through the thin cylinder of thickness dr is given by,

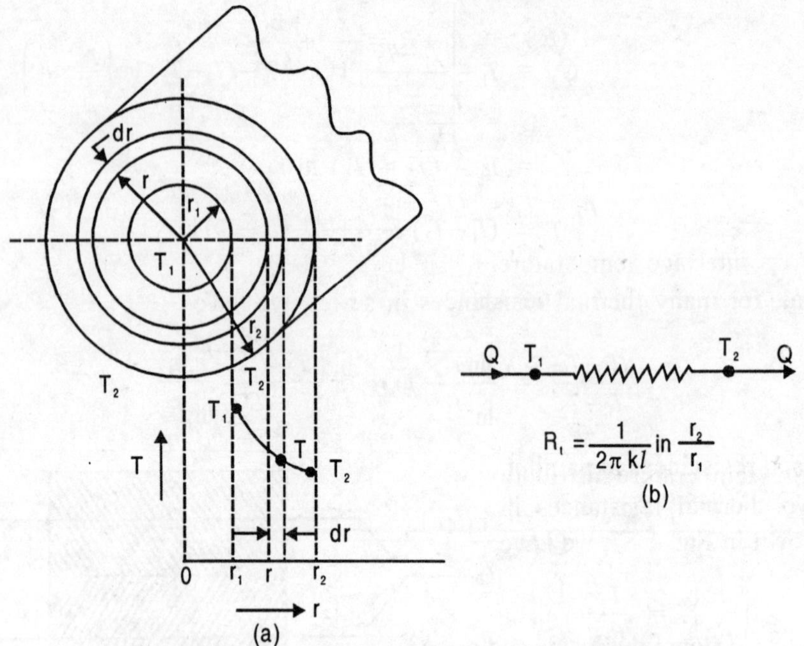

Fig. 1.58 Heat conduction through a cylindrical wall.

$$Q = -kA\frac{dT}{dr} = -2\pi rkl\frac{dT}{dr}$$

or
$$\int_{T_1}^{T_2} dT = \int_{r_1}^{r_2} -\frac{Q}{2\pi kl}.\frac{dr}{r}$$

$$T_2 - T_1 = -\frac{Q}{2\pi kl} \ln \frac{r_2}{r_1}$$

$\therefore$
$$Q = \frac{2\pi kl(T_2 - T_1)}{\ln \frac{r_2}{r_1}} = \frac{(T_2 - T_1)}{R_t}$$

where $R_t = \dfrac{1}{2\pi kl} \ln \dfrac{r_2}{r_1}$, as shown in Fig. 1.58 (*b*).

Also
$$dT = -\frac{Q}{2\pi kl}.\frac{dr}{r} = C_1\frac{dr}{r}$$

$\therefore$
$$T = C_1 \ln r + C_2$$

where C_1 and C_2 are constants to be evaluated from the temperature conditions:

When $r = r_1, T = T_1$

at $r = r_2, T = T_2$

$\therefore$
$$T_1 = C_1 \ln r_1 + C_2$$
$$T_2 = C_1 \ln r_2 + C_2$$

$$C_1 = \frac{T_1 - T_2}{\ln r_1 - \ln r_2} = \frac{T_1 - T_2}{\ln \frac{r_1}{r_2}}$$

$$C_2 = T_1 - \left[\frac{T_1 - T_2}{\ln \frac{r_1}{r_2}}\right] \ln r_1 = T_1 - (T_1 - T_2) \ln \left(\frac{r_2}{r_1} \times r_1\right)$$

$$= T_1 - (T_1 - T_2) \ln r_2$$

$$\therefore \qquad T = (T_1 - T_2)\frac{\ln r}{\ln \frac{r_1}{r_2}} + T_1 - (T_1 - T_2) \ln r_2$$

$$\frac{T - T_1}{T_1 - T_2} = \frac{\ln r}{\ln \frac{r_1}{r_2}} - \ln r_2 \text{ or } \frac{T - T_1}{T_2 - T_1} = \frac{\ln \frac{r}{r_1}}{\ln \frac{r_2}{r_1}}$$

Therefore, temperature distribution with radial conduction through a cylinder is logarithmic.

Now
$$Q = \frac{2\pi l(r_2 - r_1)k(T_1 - T_2)}{(r_2 - r_1) \ln \frac{(2\pi r_2 l)}{(2\pi r_1 l)}} = \frac{k(A_2 - A_1)(T_1 - T_2)}{(r_2 - r_1) \ln \frac{A_2}{A_1}}$$

where A_1 and A_2 are the inside and outside surface areas of cylinder.

$$Q = kA_m \left[\frac{(T_1 - T_2)}{(r_2 - r_1)}\right]$$

where $A_m = \dfrac{(A_2 - A_1)}{\ln \frac{A_2}{A_1}}$ = log-mean area

and $r_m = \dfrac{r_2 - r_1}{\ln \frac{r_2}{r_1}}$ = log-mean radius.

4. Conduction Through a Multi-layer Cylindrical Wall. Consider radial conduction of heat through a composite cylindrical wall having two layers of different materials as shown in Fig. 1.59(*a*).

For steady state conduction,

$$Q = 2\pi k_1 l\frac{(T_1 - T_2)}{\ln \frac{r_2}{r_1}} = 2\pi k_2 l\frac{(T_2 - T_3)}{\ln \frac{r_3}{r_2}}$$

$$T_1 - T_2 = \frac{Q}{2\pi k_1 l} \ln \frac{r_2}{r_1}$$

$$T_2 - T_3 = \frac{Q}{2\pi k_2 l} \ln \frac{r_3}{r_2}$$

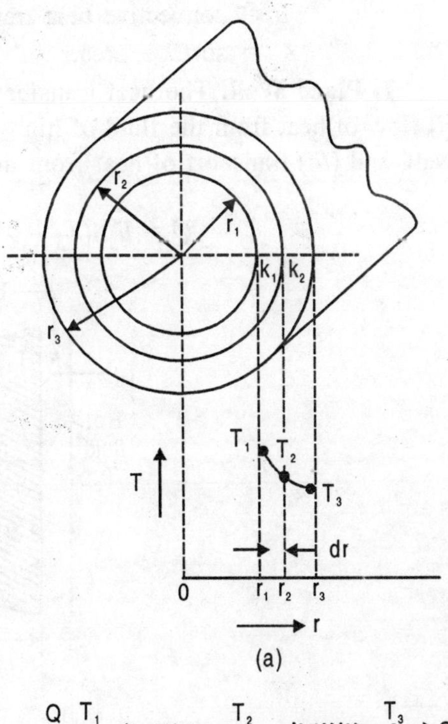

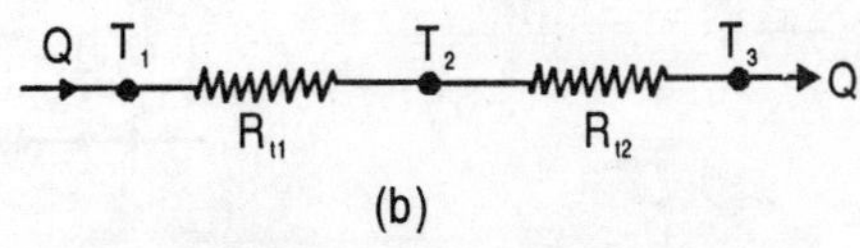

Fig. 1.59 Conduction through a composite cylindrical wall.

Adding, we get

$$T_1 - T_3 = Q\left[\frac{1}{2\pi k_1 l}\ln\frac{r_2}{r_1} + \frac{1}{2\pi k_2 l}\ln\frac{r_3}{r_2}\right]$$

$$Q = \frac{T_1 - T_3}{\frac{1}{2\pi k_1 l}\ln\frac{r_2}{r_1} + \frac{1}{2\pi k_2 l}\ln\frac{r_3}{r_2}}$$

In general for n-layered cylindrical shell,

$$Q = \frac{T_1 - T_{n+1}}{\sum_{i=1}^{n}\frac{1}{2\pi k_i l}\ln\frac{r_{i+1}}{r_i}} = \frac{T_1 - T_{n+1}}{\sum_{i=1}^{n}(R_t)_i}$$

1.41.2 Newton-Rikhman Law for Heat Convection

The Newton-Rikhman law states that the rate of convective heat transfer between solid boundary and adjacent fluid is given by :

$$Q = hA_s\,(T_w - T_\infty)$$

where T_w = wall temperature of solid
T_∞ = undisturbed fluid temperature
h = convective heat transfer coefficient, W/m^2.K
A_s = surface area.

1. Plane Wall. The heat transfer through a wall separating two moving fluids involves : (*i*) flow of heat from the fluid of high temperature to the wall, (*ii*) heat conduction through the wall, and (*iii*) transport of heat from the wall to the cold fluid (Fig. 1.60).

$$Q = h_a A(T_a - T_1) = \frac{kA}{\delta}(T_1 - T_2) = h_b A(T_2 - T_b)$$

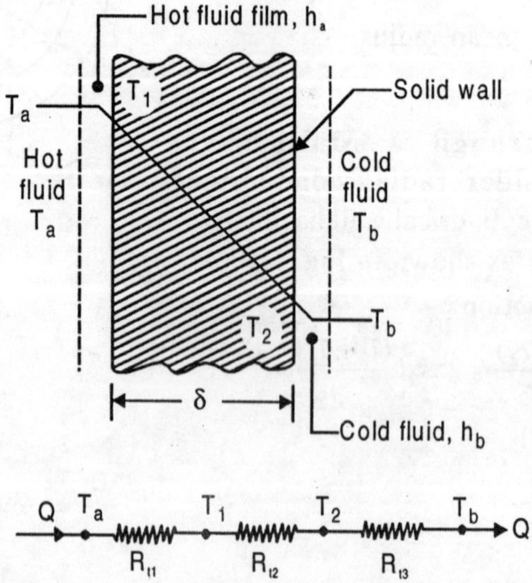

$$R_{t1} = \frac{1}{h_a A}, R_{t2} = \frac{8}{kA}, R_{t3} = \frac{1}{h_b A}$$

Fig. 1.60 Heat conduction through a wall separating two fluids.

where k_a, h_b = convective film coefficients

k = thermal conductivity of solid wall of thickness δ.

$$T_a - T_1 = \frac{Q}{h_a A}, T_1 - T_2 = \frac{Q\delta}{kA}, T_2 - T_b = \frac{Q}{h_b A}$$

Adding, we get
$$T_a - T_b = Q\left[\frac{1}{h_a A} + \frac{\delta}{kA} + \frac{1}{h_b A}\right]$$

$$Q = \frac{T_a - T_b}{\frac{1}{h_a A} + \frac{\delta}{kA} + \frac{1}{h_b A}} = \frac{A(T_a - T_b)}{\frac{1}{h_a} + \frac{\delta}{k} + \frac{1}{h_b}} = UA(T_a - T_b)$$

where $U = \dfrac{1}{\frac{1}{h_a} + \frac{\delta}{k} + \frac{1}{h_b}}$ = over all heat transfer coefficient.

2. Conduction through a composite cylindrical wall considering convective heat transfer. If the internal and external convective heat transfer coefficients for the composite cylinder shown in Fig. 1.61 are h_i and h_o respectively, then the total thermal resistance to heat flow is,

$$R_t = \frac{1}{2\pi r_1 l h_i} + \frac{1}{2\pi k_1 l} \ln \frac{r_2}{r_1} + \frac{1}{2\pi k_2 l} \ln \frac{r_3}{r_2} + \frac{1}{2\pi r_3 l h_0}$$

and
$$Q = \frac{T_i - T_0}{R_t} = UA(T_i - T_0)$$

$$= U_i A_i (T_i - T_0) = U_0 A_0 (T_i - T_0)$$

where
$$U_i = \frac{1}{\frac{1}{h_i} + \frac{r_1}{k_1} \cdot \ln \frac{r_2}{r_1} + \frac{r_1}{k_2} \ln \frac{r_3}{r_2} + \frac{r_1}{r_3} \cdot \frac{1}{h_0}}$$

$$U_0 = \frac{1}{\frac{r_3}{r_1} \cdot \frac{1}{h_i} + \frac{r_3}{k_1} \cdot \ln \frac{r_2}{r_1} + \frac{r_3}{k_2} \ln \frac{r_3}{r_2} + \frac{1}{h_0}}$$

$$A_i = 2\pi r_1 l, \; A_0 = 2\pi r_3 l$$

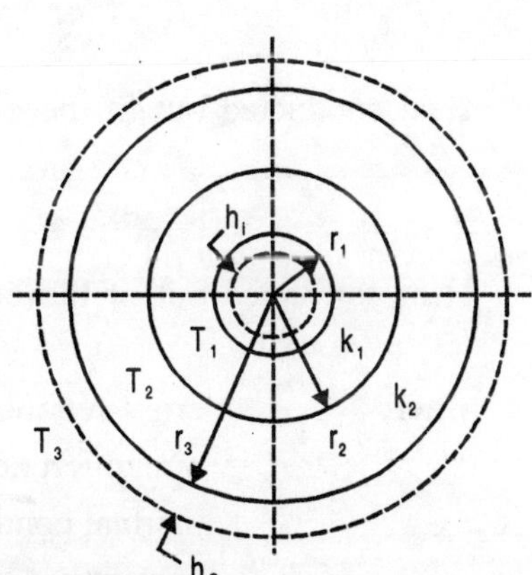

Fig. 1.61 Heat conduction through a composite cylinderical wall.

1.41.3 Heat Transfer from Fin

A fin is an extended surface used to increase the heat transfer. Common applications of finned surfaces are : air cooled cylinders of aircraft engines, I.C. engines, air compressors; electrical transformers and motors; cooling coils and condenser coils in refrigerators and air-conditioners, etc.

Consider the heat transfer from a fin as shown in Fig. 1.62. Heat from the heated wall is conducted through the fin and convected from the sides of the fin to the surroundings. Consider an infinitesimal element of the fin of thickness located at a distance x from base wall.

Heat conducted into the element at plane x,

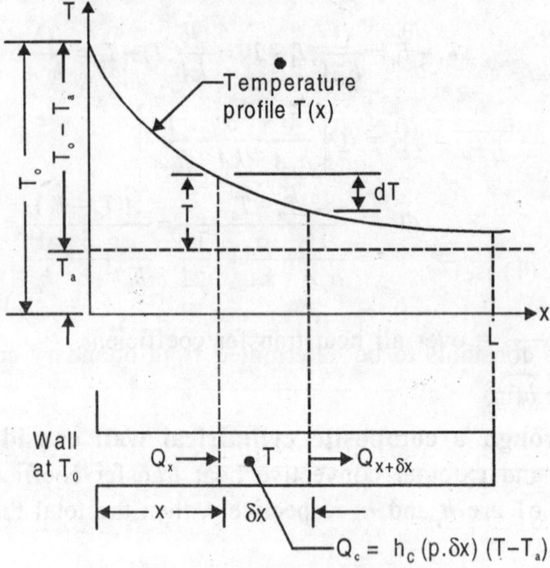

Fig. 1.62 Heat transfer through a fin.

$$Q_x = -kA\left(\frac{dT}{dx}\right)_x$$

Heat conducted out of the element at plane $(x + \delta x)$,

$$Q_{x+\delta x} = -kA\left(\frac{dT}{dx}\right)_{x+\delta x} = -kA\frac{d}{dx}\left(T+\frac{dT}{dx}\delta x\right)$$

Heat convected out of the element between the planes x and $x + \delta x$,

$$Q_c = h_c\,(P\cdot\delta x)\,(T - T_a)$$

where A = cross-sectional area of fin
P = circumferential perimeter of fin
k = thermal conductivity of fin material
h_c = convective heat transfer coefficient
T_0 = Wall temperature
T_a = ambient temperature
T = fin temperature.

Heat Balance on the element gives,

$$Q_x = Q_{x+\delta x} + Q_c$$

$$-kA\frac{dT}{dx} = -kA\frac{d}{dx}\left(T+\frac{dT}{dx}\cdot\delta x\right)+h_c P\cdot\delta x(T-T_a)$$

Simplifying, we get

$$\frac{d^2T}{dx^2}-\frac{h_c P}{kA}\cdot(T-T_a) = 0$$

Let temperature excess,

$$\theta(x) = T(x) - T_a$$

$$\therefore \quad \frac{d\theta}{dx} = \frac{dT}{dx} \text{ and } \frac{d^2\theta}{dx^2} = \frac{d^2T}{dx^2}$$

or
$$\frac{d^2\theta}{dx^2} - m^2\theta = 0 \quad ...(1)$$

where
$$m = \sqrt{\frac{h_c P}{kA}}$$

The solution of Eq. (1) is,

$$\theta = c_1 e^{mx} + c_2 e^{-mx}$$

where c_1 and c_2 are constants to be determined from boundary conditions.

For an infinitely long fin,

$$T = T_0 \text{ at } x = 0 \quad \text{or} \quad T - T_a = T_0 - T_a$$

or
$$\theta = \theta_0 \text{ at } x = 0 \quad \text{and} \quad T = T_a \text{ at } x = \infty$$

or
$$\theta = 0 \text{ at } x = \infty$$

$$\therefore \quad c_1 + c_2 = \theta_0$$

$$c_1 e^{m\infty} + c_2 e^{-m\infty} = 0$$

$$\therefore \quad c_1 = 0 \text{ and } c_2 = \theta_0$$

$$\therefore \quad \theta = \theta_0 e^{-mx} \text{ or } T - T_a = (T_0 - T_a)e^{-mx}$$

$$Q_{\text{fin}} = -kA\left(\frac{dT}{dx}\right)_{x=0}$$

$$T = T_a + (T_0 - T_a)e^{-mx}$$

$$\left(\frac{dT}{dx}\right)_{x=0} = [-m(T_0 - T_a)e^{-mx}]_{x=0} = -m(T_0 - T_a)$$

$$Q_{\text{fin}} = k\,A\,m(T_0 - T_a) = \sqrt{Ph_c kA}(T_0 - T_a)$$

1.41.4 Fin Insulated at the Tip

The boundary conditions are,

$$\theta = \theta_0 \text{ at } x = 0 \text{ and } \frac{dT}{dx} = 0 \text{ at } x = l \quad ...(1)$$

$$\therefore \quad c_1 + c_2 = \theta_0$$

$$T - T_a = c_1 e^{mx} + c_2 e^{-mx}$$

$$\frac{dT}{dx} = mc_1 e^{mx} - mc_2 e^{-mx}$$

$$\left(\frac{dT}{dx}\right)_{x=l} = mc_1 e^{ml} - mc_2 e^{-ml} = 0$$

$$c_1 e^{ml} - c_2 e^{ml} = 0 \quad ...(2)$$

Eqs. (1) and (2) give,

$$c_1 = \theta_0\left[\frac{e^{-ml}}{e^{ml}+e^{-ml}}\right]$$

$$c_2 = \theta_0\left[\frac{e^{ml}}{e^{ml}+e^{-ml}}\right]$$

$$\frac{\theta}{\theta_0} = \frac{e^{m(l-x)}+e^{-m(l-x)}}{e^{ml}+e^{-ml}}$$

$$= \frac{T-T_a}{T_0-T_a} = \frac{\cosh m(l-x)}{\cosh ml}$$

$$Q_{\text{fin}} = -kA\left(\frac{dT}{dx}\right)_{x=0}$$

Now $$T - T_a = T_a - T_a\frac{\cosh m(l-x)}{\cosh ml}$$

$$\frac{dT}{dx} = (T_0-T_a)\frac{\sinh m(l-x)}{\cosh ml}\times(-m)$$

$$\left(\frac{dT}{dx}\right)_{x=0} = -m(T_0 - T_a)\tan h\ ml$$

$$Q_{\text{fin}} = kA\ m(T_0 - T_a)\tan h\ ml$$

$$= \sqrt{Ph_c\,kA}\,(T_0 - T_a)\tan h\ ml$$

1.41.5 Fin Performance

1. Efficiency of fin. It relates the performance of an actual fin to that of an ideal or fully effective fin.

$$\text{Fin efficiency, } \eta_f = \frac{\text{Actual heat transfer rate from the fin}}{\text{Heat that would be dissipated if the whole surface of the fin were maintained at the base temperature}}$$

$$= \frac{\sqrt{Ph_c kA}(T_0-T_a)\tanh ml}{h_c Pl(T_0-T_a)} = \frac{\tanh ml}{\sqrt{\frac{Ph_c}{kA}}\,l} = \frac{\tanh ml}{ml}$$

2. Effectiveness of fin (ε_f). It represents the ratio of the fin heat transfer rate to the heat transfer rate that would exist without a fin.

$$Q_{\text{fin}} = \sqrt{Ph_c kA}(T_0-T_a)$$

$$Q_{\text{no-fin}} = h_c\ A(T_0 - T_a)$$

Fin effectiveness, $$\varepsilon_f = \frac{Q_{fin}}{Q_{no-fin}} = \frac{\sqrt{Ph_c kA}(T_0-T_a)}{h_c A(T_0-T_a)} = \sqrt{\frac{Pk}{h_c A}}$$

In case of a straight rectangular fin of thickness δ and width b,

$$\frac{P}{A} = \frac{2(b+\delta)}{b\delta} \simeq \frac{2}{\delta}$$

$$\therefore \quad \varepsilon_f = \sqrt{\frac{2k}{h_c\delta}}$$

For a fin insulated at the tip,

$$\varepsilon_f = \eta_f \frac{Pl}{A} = \eta_f \times \frac{\text{surface area of fin}}{\text{cross-sectional area of fin}}.$$

1.41.6 Radiation

Radiation intensity (I). Is the rate at which radiant energy is emitted per unit area of the emitting surface normal to a direction, per unit solid angle about the direction, per unit wave length $I = \frac{E}{\pi}$.

Diffuser emitter. Is a surface for which the emitted radiation is independent of direction.

Irradiation. Is the radiation incident on the surface coming from all directions.

Radiosity. May be defined as all the radiation leaving one of the surfaces of a translucent material.

Transmissivity (τ). Of a body is the fraction of incident energy that is transmitted through a body.

Reflectivity (ρ). Of a surface is defined as the fraction of incident energy reflected by the surface.

Absorptivity + Transmissivity + Reflectivity = 1 ...(13.75)

For an opaque surface, transmissivity is equal to zero. For a perfect reflector, reflectivity is equal to unity. For a black body, absorptivity is equal to unity.

Specular Body. Is one that reflects all the incident thermal radiations (if the reflection is regular) or an absolutely white body (if the reflection is diffused). For such a body reflectivity is unity.

Transparent (or diathermaneous) Body. Is one that allows all the incident radiation to pass through it. For such bodies transmissivity is unity.

Cavity. Is an enclosure with isothermal walls.

Kirchhoff's Law. It states that at a given temperature, the absorptivity and emissivity of any solid surface are equal when the surface emits just what it absorbs. It suggests that perfect absorber is also a perfect radiator.

Stefan-Boltzmann Law. It states that total energy emitted from a surface is proportional to the fourth power of the absolute temperature of the body.

$$E_e = \varepsilon\, \sigma\, T^4$$

where ε = emissivity of the surface, and

σ = Stefan-Boltzmann constant = 5.675×10^{-8} W/(m².K⁴).

Plankc's Law. It governs the distribution of radiant energy over wave length for a black body at a fixed temperature and states that the peak of maximum energy shifts towards shorter wave lengths with increasing temperature. For the spectral distribution of emissive power,

$$E_{b\lambda} = \frac{C_1\lambda^{-5}}{\exp\left(\frac{C_2}{\lambda T}\right)-1}$$

where $C_1 = 2\pi C_0^2 h = 0.374\times10^{-5}\ \text{J m}^2/\text{s}$

$$C_1 = \frac{C_0 h}{k} = 1.4389 \times 10^{-2}\ \text{m K.}$$

Spectral Energy Distribution. Is the variation of distribution of the monochromatic emissive power with wave length.

Wien's Law. $E_{b\lambda} = \dfrac{C_1\lambda^{-5}}{\exp\left(\frac{C_2}{\lambda T}\right)}$ when $\left(\dfrac{C_2}{\lambda T}\right) >> 1$

Rayleigh-Jean's Law. $E_{b\lambda} = \dfrac{C_1 T}{C_2\lambda^4}$ when $\left(\dfrac{C_2}{\lambda T}\right) << 1$

Wiens's Displacement Law. It states that as temperature increases, the wavelength (λ_{max}) corresponding to the maximum spectral emissive black body power decreases. In other words, the product of absolute temperature and the wavelength, at which the emissive power is maximum, is constant. Accordingly the maximum spectral intensity of radiation shifts towards the shorter wave length with rising temperature.

$$\lambda_{max}\, T = \text{constant} = 2.898 \times 10^{-3}\ \text{m. K}$$

Selective Emitters. Are those for which the emissivity is different for the various wavelengths of the emitted energy.

Lambert's Cosine Law. It states that the density of radiation in a direction making an angle θ from the normal to a black emitter is proportional to cosine of the angle θ.

$$I_\theta = I_n \cos\theta,\ I_n = \frac{\sigma T^4}{\pi} = \frac{E_b}{\pi}$$

Configuration (view or shape or geometry) Factor. It is defined as the fraction of the radiative energy that is diffused from one surface element and strikes the other surface directly with no intervening reflections. It is denoted by F_{ij}, where i refers to the emitting surface and j to the receiving surface.

Radiation energy being emitted by surface 1 and absorbed by surface 2 (Fig. 1.63) is :

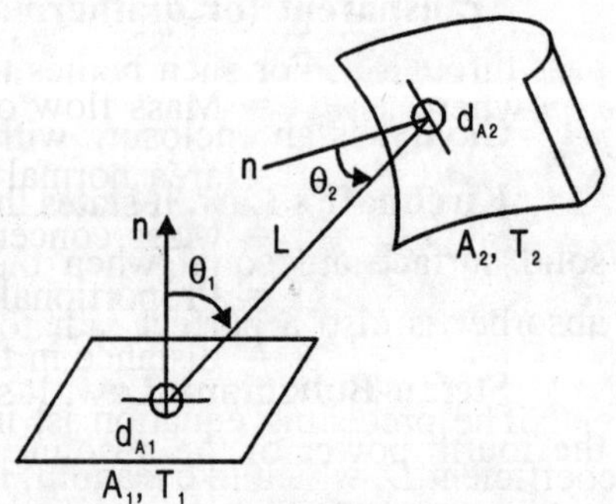

Fig. 1.63 Radiation energy exchange between two surfaces.

$$Q_{12} = \frac{\sigma_1 T^4}{\pi L^2}\iint \cos\theta_1 \cos\theta_2\, dA_1\, dA_2$$

$$F_{12} = \frac{Q_{12}}{\sigma_1 A_1 T_1^4} \text{ or } Q_{12} = A_1\, F_{12}\, \sigma_1\, T_1^{\,4}$$

1.42 MODES OF MASS TRANSFER

There are three different modes of mass transfer, as discussed below :

1. Mass transfer by molecular diffusion. Mass transfer by *molecular diffusion* as a result of the driving potential is caused by *concentration gradient*. Thus, we shall ignore the contributions, if any, to the *mass flux* as a result of other driving potentials, such as temperature difference, resulting in *thermal diffusion* or the *Soret effect*. Likewise the effect of concentration gradient on heat transfer, the so called Dufour effect, is also ignored. Both these effects are examples of coupled phenomena.

Mass transfer by molecular diffusion of a species through another stationary medium is analogous to the transfer of heat by conduction or to the transfer of momentum in laminar flow.

2. Mass transfer by convection. As in the case of heat transfer in fluids, most physical applications of mass transfer, involve bulk motion of the fluids. The diffusion rates described above, must, therefore, be modified to account for the contribution to mass transfer made by bulk motion. This gives rise to *convective mass transfer* as similar to the *transfer of heat by convection* with the additional consideration of laminar and turbulent flow situations. The evaporation of ether is an example of mass transfer by convection.

3. Mass transfer by change of phase. Mass transfer by charge of phase occurs due to simultaneous action of convection and diffusion. Some examples are :

(*i*) Hot gases escaping from the chimney rise by convection and then diffuse into the air above the chimney.

(*ii*) Mixing of water vapour with air during evaporation of water from a lake surface.

(*iii*) Boiling of water in open air.

1.43 MASS TRANSFER BY MOLECULAR DIFFUSION

1.43.1 Fick's Law of Diffusion

Fick's law relates the *diffusion rate* or *mass flux* of a species to its driving potential, *viz*., the *concentration gradient*. It is written in the form identical to the Fourier law of heat conduction as follows:

$$\frac{\dot{m}_A}{A} = -D\frac{\partial C_A}{\partial x}$$

where $\dot{m}_A$ = Mass flow or diffusion rate of component A per unit time.

A = area normal to the direction x of flow of mass

C_A = Mass concentration of component A per unit volume = ρ_A.

D = Proportionality constant called the *diffunix coefficient* or *diffusivity*

x = distance in the direction of diffusion.

The preceding equation is, in fact, not a law of nature but the definition of the diffusion coefficient D, which like k and μ, is a property of the materials present. The negative sign appears because the concentration gradient is negative in the direction of mass transfer.

It is to be noted that mass concentration C_A is the same thing as *partial molar density* ρ.

1.43.2 Diffusion Coefficient

The units of diffusion coefficient can be derived as follows:

$$[D] = \frac{[\dot{m}_A][\Delta x]}{[A][\Delta C_A]} = \frac{\left[kgs^{-1}\right][m]}{\left[m^2\right]\left[kgm^{-3}\right]} = \frac{m^2}{s}$$

Thus diffusivity or diffusion coefficient D has the dimensions of the kinematic viscosity ν, and the thermal diffusivity α, Diffusion coefficient is thus a *transport property.*

Dimensionless numbers can be formed from the ratio of any two of these properties.

Prandtl Number – It relates the velocity field with the temperature field,

$$Pr = \frac{\nu}{\alpha} = \frac{c_p \mu}{k}$$

It is an important parameter in the solution of problems involving simultaneous momentum and heat transfer, as in convection.

Lewis Number – It relates the temperature field with concentration field.

$$Le = \frac{\alpha}{D} = \frac{k}{\rho c_p D}$$

It is an important parameter in the solution of problems involving simultaneous heat and mass transfer, as in cooling towers, air washers, etc.

Schmidt Number – it relates the velocity field with the concentration field.

$$Sc = \frac{\nu}{D} = \frac{\mu}{\rho D}$$

1.43.3 Convective Mass Transfer

In molecular diffusion, bulk velocities are insignificant. Most physical applications of mass transfer, however, involve the bulk motion of fluids. This gives rise to convective mass transfer which is similar to convective heat transfer. Accordingly, we define the *mass transfer coefficient* h_M by the relation

$$\frac{\dot{m}_A}{A} = h_M (C_w - C_\infty) = \frac{k_w}{\rho} (\omega_w - \omega_\infty)$$

where C_w and C_∞ denote the concentration of the species A at the wall and free stream. We see that h_m has the units of m s^{-1}. The expression is $\rho\, h_M$ is replaced by *diffusion coefficient* k_w and C by specific humidity w in the case of diffusion of water vapour into air. It is seen that $k_w = \rho h_M$ if $\Delta C = \Delta_w$, which is a satisfactory assumption.

1.43.4 Analogy between Momentum and Heat Transfer

If $\nu = \alpha$ or $Pr = 1$, the velocity and temperature distributions in the flow will be the same. In that case, the heat transfer coefficient h can be determined from the knowledge of the friction factor f by the similarity relation called *Reynolds analogy*, viz.,

$$\text{St} = \frac{f}{2} \quad \text{...(1)}$$

where St denotes the *Stanton number* defined by

$$\text{St} = \frac{Nu}{\text{Re.Pr}} = \frac{h}{\rho u C_p} = \frac{h}{GC_p}$$

Here $G = \rho u$ is the *mass velocity.*

To account for some variation in Pr, Eq. (1) is modified to

$$j_H = \text{St}\cdot\text{Pr}^{2/3} = \frac{f}{2}\left(=\frac{h}{GC_p}\text{Pr}^{2/3}\right) \quad \text{...(2)}$$

which is called the *Colburn analogy*, and j_H is the *Colburn j-factor* for heat transfer.

1.43.5 Analogy between Momentum and Mass Transfer

The velocity and concentration profiles will have the same shape if $\nu = D$. or Sc=1. Thus, the Schmidt number plays the same role in mass transfer as does the Prandtl number in heat transfer. An equivalent of the Nusselt number in mass transfer is the *Sherwood number*, defined by

$$\text{Sh} = \frac{h_M L}{D} = \frac{k_w L}{\rho D}$$

where L is any characteristic length. The convective mass transfer correlations are, therefore, expressed in the form

$$\text{Sh} = f(\text{Re, Sc})$$

Similar to forced convection heat transfer correlations which are usually expressed in the form $\text{Nu} = f(\text{Re}\cdot\text{Pr})$.

The Stanton number can then be replaced by the number

$$\frac{\text{Sh}}{\text{Re}\cdot\text{Sc}} = \frac{h_M}{u} = \frac{\rho h_M}{G} \quad \text{...(3)}$$

in mass transfer correlations, as given in Eq. (3). And similar to Eq. (1), we can write for mass transfer factor j_M as

$$j_M = \left(\frac{\text{Sh}}{\text{Re}\cdot\text{Sc}}\right)\text{Sc}^{2/3} = \left(\frac{\rho h_M}{G}\right)\text{Sc}^{2/3} = \frac{f}{2} \quad \text{...(4)}$$

The numerical value of the Schmidt number for the diffusion of water vapour into air at 25°C and 1 atm pressure is 0.6.

1.43.6 Analogy between Heat and Mass Transfer

Combining Eqs. (2) and (4) to eliminate $f/2$, we obtain a similarity relation between heat transfer and mass transfer coefficients as follows in Eq. (5).

$$\frac{h}{h_M} = \rho C_p \left(\frac{\text{Sc}}{\text{Pr}}\right)^{2/3}$$

$$= \rho C_p \left(\frac{\alpha}{D}\right)^{2/3}$$

$$= \rho C_p \, \text{Le}^{2/3} \qquad \text{...(5)}$$

where Le = α/D is the *Lewis number*.

If Le = 1, the temperature and concentration profiles are the same and then $j_M = j_H$. Incidentally for air and water vapour mixtures at atmospheric pressure, the value of the Lewis number is approximately equal to unity. This simplifies the design of air-conditioning equipment. Writing h as f_g for heat transfer coefficient of air, and h_M as k_w/ρ for diffusion of water vapour in air, we have for Le = 1

$$k_w = \frac{f_g}{C_p}$$

Empirical relations for Sh and Nu for air over water droplets can be expressed as follows:

$$\text{Sh} = \frac{h_M L}{D} = \frac{k_w L}{\rho D} = 2 + 0.6\,\text{Re}^{0.5}\,\text{Sc}^{0.33} \qquad \text{...(6a)}$$

$$\text{Nu} = \frac{hL}{k} = 2 + 0.6\,\text{Re}^{0.5}\,\text{Pr}^{0.33} \qquad \text{...(6b)}$$

MULTIPLE – CHOICE QUESTIONS

1. In an open system, which crosses the boundary of the system?
   (a) Heat (b) Work (c) Mass (d) Heat and work
2. In an isolated system, what transfers?
   (a) mass (b) energy
   (c) both mass and energy (d) neither mass nor energy
3. Choose the extensive property:
   (a) pressure (b) volume (c) temperature (d) density
4. Which of the following is an intensive property?
   (a) volume (b) temperature (c) mass (d) energy
5. A system comprising of a single phase is called:
   (a) homogeneous (b) heterogeneous (c) isolated (d) closed
6. Consider the following statements:
   1. In an open system, mass and energy transfer takes place.
   2. In a closed system, energy transfer takes place.
   3. The universe can be considered as an isolated system.

   Of these statements:
   (a) only 1 is correct (b) only 2 is correct
   (c) 1 and 2 are correct (d) 1, 2 and 3 are correct.
7. Consider the following statements:
   1. Milk and sugar is a homogeneous system.
   2. Free expansion is an irreversible process.
   3. Thermos flask is an isolated system.

   Of these statements:
   (a) only 1 is correct (b) only 2 is correct
   (c) 1 and 2 are correct (d) 1, 2 and 3 are correct.
8. Which of the following is a path function?
   (a) heat (b) temperature (c) pressure (d) volume.
9. The following are examples of some intensive and extensive properties:
   1. Pressure 2. Temperature 3. Volume 4. Velocity
   5. Electric charge 6. Magnetisation 7. Viscosity 8. Potential energy.

   Which one of the following sets gives the correct combination of intensive and extensive properties?

	Intensive	**Extensive**
(a)	1,2,3,4	5,6,7,8
(b)	1,3,5,7	2,4,6,8
(c)	1,2,4,7	3,5,6,8
(d)	2,3,6,8	1,4,5,7

10. A mixture of gases expands from 0.03 m^3 to 0.06 m^3 at a constant pressure of 1 MPa and absorbs 84 kJ of heat during the process. The change in internal energy of the mixture is:

(a) 30 kJ (b) 54 kJ (c) 84 kJ (d) 114 kJ

11. Given that for the path 1 – 2 – 3, a system absorbs 100 kJ as heat and does 60 kJ work. While along the path 1 – 4 – 3, it does 20 kJ work (see figure below). The heat absorbed during the cycle 1 – 4 – 3 is:

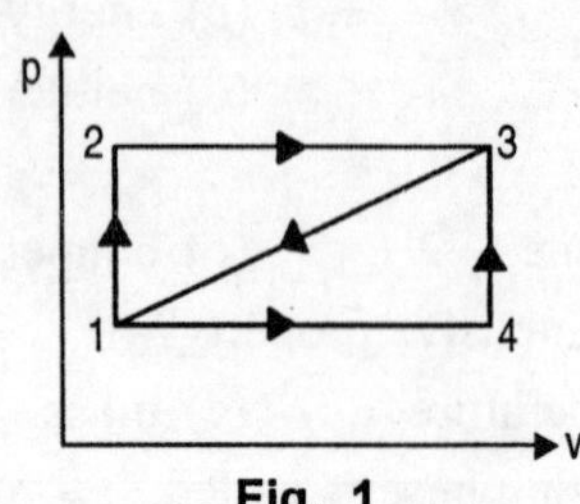

Fig. 1

(a) – 140 kJ (b) – 80 kJ (c) – 40 kJ (d) + 60 kJ

12. The given figure below shows the variation of force in an elementary system which undergoes a process during which the plunger position changes from 0 to 3 m. if the internal energy of the system at the end of the process is 2.5 kJ higher, then the heat absorbed during the process is:

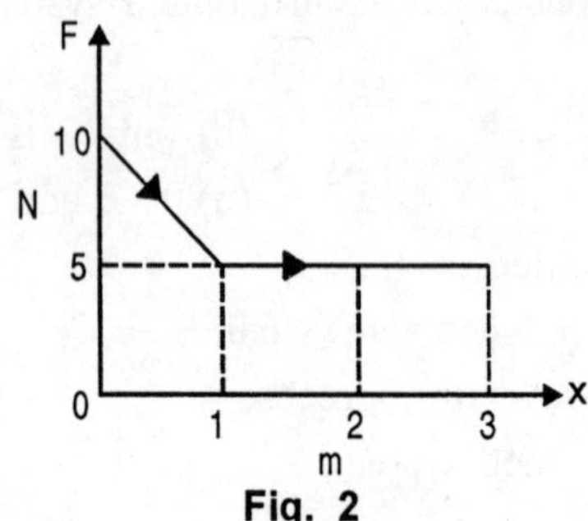

Fig. 2

(a) 15 J (b) 20 J (c) 25 J (d) 30 J

13. The net work done for the closed cycle shown in the given pressure-volume diagram below is :

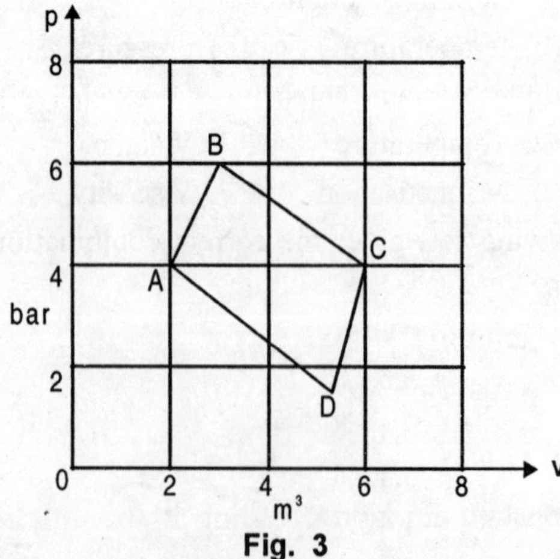

Fig. 3

(a) 600 kN.m (b) 700 kN.m (c) 900 kN.m (d) 1000 kN.m

14. The internal energy of a certain system is a function of temperature alone and is given by the formula: $E = 25 + 0.25t$ kJ. If this system executes a process for which the work done by it per degree temperature increase is 0.75 kN.m, the heat interaction per degree temperature increase is:

(a) – 1.00 kJ (b) – 0.50 kJ (c) + 0.05 kJ (d) 1.00 kJ

15. When a system is taken from state A to state B (see figure below) along the path $A - C - B$, 180 kJ of heat flows into the system and does 130 kJ of work(see figure). How much heat will flow into the system along the path $A - D - B$ if the work done by it along the path is 40 kJ?

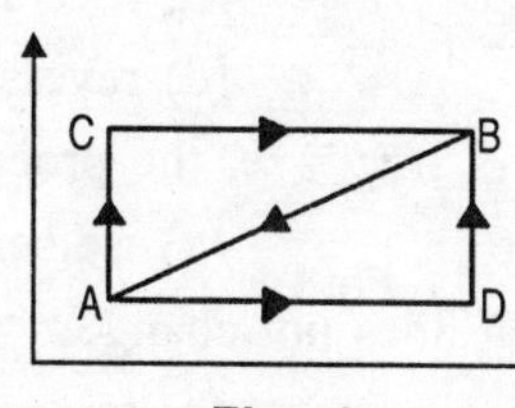

Fig. 4

(a) 40 kJ (b) 60 kJ (c) 90 kJ (d) 135 kJ

16. A control mass undergoes a process from state 1 to state 2 as shown in the given figure below. During this process, if the heat transfers to state 2 from state 1 by another process, then the work interaction during the return process (in kN.m) would be:

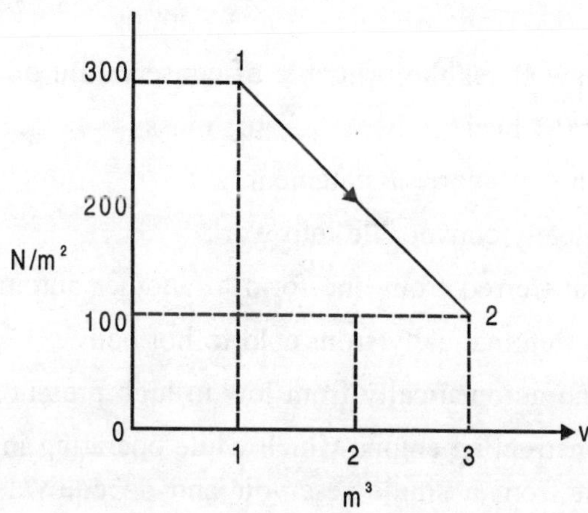

Fig. 5

(a) – 400 (b) – 200 (c) 200 (d) 400

17. A tank containing air is stirred by a paddle wheel. The work input to the paddle wheel is 9000 kJ and the heat transferred to the surroundings from the tank is 3000 kJ. The external work done by the system is:

(a) zero (b) 3000 kJ (c) 6000 kJ (d) 9000 kJ

18. The work done in a steady flow process is:

(a) $\int pdv$ (b) $\int v.dp$ (c) $-\int p\,dv$ (d) $-\int v.dp$

19. The following remains constant during a throttling process:

(a) internal energy (b) enthalpy (c) pressure (d) temperature

20. In an open system, the following can cross the boundary of the system:

(a) heat and work (b) mass

(c) mass, heat and work (d) heat only

21. In an isolated system, the following can cross the boundary of the system:

(a) energy (b) mass

(c) energy and mass (d) neither energy nor mass.

22. An isentropic process is

(a) irreversible & adiabatic (b) reversible & adiabatic

(c) irreversible & frictionless (d) reversible & isothermal

23. In an Polytropic Process $pV^n = c$, if $n = \infty$, the process is called

(a) isochoric (b) isobaric (c) isothermal (d) adiabatic

24. During isothermal process, which does not change?

(a) pressure (b) temperature

(c) volume (d) both pressure & temperature

25. Choose the incorrect relationship

(a) $c_p + c_v = R$ (b) $c_p - c_v = R$ (c) $\gamma = c_p/c_v$ (d) $c_v = R/(\gamma - 1)$

26 The second law of thermodynamics helps to define

(a) Heat (b) Work (c) Entropy (d) Internal energy

27. The Kelvin-Plancks law states the principle of conservation of

(a) Work (b) heat (c) mass (d) heat into work

28. Which of the following is incorrect statement

(a) Heat is not completely convertible into work

(b) Energy can be transferred from one form to another automatically

(c) Heat cannot flow automatically from cold to hot body

(d) Gas cannot expand automatically from low to high pressure

29. It is impossible to construct an engine which while operating in a cycle produces no other effect to extract heat from a single reservoir and do equivalent amount of work. This refers to

(a) Clausius statement (b) Kelvin-Planck's statement

(c) Carnot theorem (d) PMM-2

30. A Carnot cycle comprises of :

(a) Two isothermal and two isentropic processes

(b) Two constant volume and two isentropic processes

(c) Two constant pressure and two isentropic processes

(d) One constant volume, one constant pressure and two isentropic processes

31. A Carnot cycle operates between temperatures of 1000 K and 5000 K. Its thermal efficiency is :

(a) 50% (b) 200% (c) 100% (d) 150%

32. A Carnot cycle operates between temperatures of 800 K and 400 K. Its efficiency will be 50% if working fluid is:

(a) air (b) nitrogen (c) carbon dioxide (d) any fluid

33. A refrigerator and heat pump operate between the same temperature limits. If COP of the refrigerator is 3, the COP of pump will be :

(a) 3 (b) 4 (c) 5 (d) 6

34. A heat engine is supplied with 800 kJ/s of heat at 600 K and rejects at 300 K. For the process to be reversible, heat rejected is :

(a) 200 kJ/s (b) 400 kJ/s (c) 500 kJ/s (d) 600 kJ/s

35. For a thermodynamic cycle to be reversible :

(a) $\oint \frac{\delta Q}{T} = 0$ (b) $\oint \frac{\delta Q}{T} < 0$ (c) $\oint \frac{\delta Q}{T} > 0$ (d) $\oint \frac{\delta Q}{T} \le 0$

36. For a thermodynamic cycle to be irreversible:

(a) $\oint \frac{\delta Q}{T} = 0$ (b) $\oint \frac{\delta Q}{T} < 0$ (c) $\oint \frac{\delta Q}{T} > 0$ (d) $\oint \frac{\delta Q}{T} \le 0$

37. The more effective way to increase the thermal efficiency of a Carnot cycle is to:

(a) increase the source temperature (b) decrease the sink temperature

(c) increase the sink temperature (d) decrease the source temperature

38. Choose the correct statement

(a) Entropy is an intensive property of the system

(b) Entropy is conservative

(c) Entropy is a measure of the level of irreversibility associated with a process

(d) Entropy of the Universe is always decreasing

39. Consider the following statements :

The definition of:

1. Temperature is due to Zerorth law of Thermodynamics
2. Entropy is due to First law of Thermodynamics
3. Internal energy is due to Second law of Thermodynamics
4. reversibility is due to Kelvin- Planck's statement

Of these statements:

(a) 1, 2 and 3 are correct (b) 1, 3 and 4 are correct

(c) 1 alone is correct (d) 2 alone is correct

40. One reversible heat engine operates between 1600 K and T_2, and another reversible heat engine operates between T_2 and 400 K. If both the engines have the same heat input and output, then the temperature T_2 must be equal to :

(a) 1000 K (b) 1200 K (c) 1400 K (d) 800 K

Answers

1. (d)	**2.** (d)	**3.** (b)	**4.** (b)	**5.** (a)	**6.** (d)	**7.** (c)	**8.** (a)
9. (c)	**10.** (b)	**11.** (d)	**12.** (b)	**13.** (d)	**14.** (d)	**15.** (c)	**16.** (d)
17. (c)	**18.** (d)	**19.** (d)	**20.** (c)	**21.** (d)	**22.** (b)	**23.** (a)	**24.** (b)
25. (a)	**26.** (c)	**27.** (d)	**28.** (b)	**29.** (b)	**30.** (a)	**31.** (a)	**32.** (d)
33. (b)	**34.** (b)	**35.** (a)	**36.** (b)	**37.** (b)	**38.** (c)	**39.** (c)	**40.** (d)

SHORT ANSWER TYPE QUSETIONS

1. Define the term thermodynamic.
2. Explain the thermodynamic system, surroundings and universe.
3. Distinguish between closed, open and isolated system with examples.
4. Differentiate between homogeneous and heterogeneous systems.
5. What do you mean by phase of a system?
6. What do you mean by thermodynamic equilibrium?
7. What do you mean by property? Distinguish between intensive and extensive properties.
8. Separate the following into intensive and extensive properties :
   Mass, weight, volume, pressure, temperature, density, velocity, surface area, internal energy, potential energy.
9. Explain the terms : state, phase, process and cyclic process.
10. What do you understand by reversible and irreversible process? Give examples.
11. Explain cyclic and quasi-static process.
12. Explain the term "Energy". Discuss its various forms.
13. Define work. Show that work is a path function.
14. Define the term heat.
15. Choose the open or closed system from the following :
    Water pump, pressure cooker, automobile engine, air compressor, steam turbine, boiler system.
16. State the number of phases in the following :
    Pure water, water - alcohol mixture, coffee, water - oil mixture, aqua- ammonia, mild steel, at room temperature.
17. Explain the concept of temperature.
18. State zeroth law of thermodynamics.
19. Write characteristics gas equation.
20. What is universal gas constant?
21. What is your concept of thermal equilibrium?
22. What is a closed system?
23. Define internal energy.
24. Define thermodynamic state, path, process and cycle.

25. What is an open system?
26. What is zeroth law of thermodynamics?
27. Determine an expression for the heat transfer in a closed system isochoric process.
28. Explain the concept of a closed isobaric process and determine an expression for its heat transfer.
29. Determine an expression for the work done in a closed isothermal process.
30. What is an isentropic process? Determine an expression for the work done in a non-flow isentropic process.
31. Determine an expression for the heats transfer and work done in a non-flow Polytropic process.
32. What is a free expansion process? What are its characteristics?
33. What is a throttling process? State its characteristics?
34. What is a steady flow process? Determine the work done in such an isochoric process.
35. Determine an expression for the work done in a steady flow adiabatic process.
36. Compare the work done in a non-flow and flow type polytropic process.
37. State first law of thermodynamics.
38. What is PMM-1.
39. Show that energy of an isolated system remains unchanged.
40. Write the steady flow energy equation.
41. Write the relationship between specific heats.
42. What is adiabatic index? Give its usual value.
43. What is a steady flow process?
44. What conditions are fulfilled by a steady flow process.
45. State the limitations of First law of thermodynamics.
46. State Kelvin-Planck statement for second law of thermodynamics.
47. State Clausius statement for the second law of thermodynamic.
48. What is a PMM2?
49. Are Kelvin-Planck and Clausius statements equivalent?
50. Differentiate between a heat engine and a heat sink.
51. What is a thermal reservoir?
52. Differentiate between a heat source and a heat sink.
53. Describe the working of a Carnot cycle.
54. What are the limitations of Carnot cycle?
55. Define thermal efficiency of a heat engine.
56. Differentiate between a heat pump and a refrigerator.
57. Define COP of a heat pump and a refrigerator.
58. State Carnot theorem.
59. What is Clausius inequality?
60. State Carnot's theorem.

61. Define the term 'Entropy'.
62. Show that the entropy is a property of the system.
63. The entropy of the universe tends to be maximum Comment.
64. What are the characteristics of entropy?
65. Draw the Carnot cycle on T-S diagram.
66. State Third law of thermodynamics.
67. What is the importance of third law of thermodynamics?

EXERCISES

1.1 5 m^3 of gas at 8 ba is heated keeping the pressure same till the volume is doubled. Calculate (a), heat added, (b) external work done, and (c) change in internal energy during the process. **[Ans.** 14 MJ, 4 MJ, 10 MJ]

1.2 A jet plane is flying at a speed of 650 km/h at an altitude where the pressure is 0.5 bar and temperature – 12°C. Air passes through the diffuser of the jet plane and leaves with velocity of 100 m/s. If the flow is isentropic, calculate the temperature and pressure at the outlet of the diffuser. **[Ans.** – 0.748°C, 0.5796 bar]

1.3 A gas flows through a turbine. At entrance its pressure is 7 bar and specific volume 0.2 m^3/kg. At exit the corresponding values are 3.5 bar and 0.3 m^3/kg. The velocity at inlet and exit is 150 m/s and 300 m/s respectively. Between entrance and exit the internal energy decreases by 80 kJ/kg and 6 kJ/kg of heat is lost through radiation. Calculate the work developed per kg of gas flowing through the turbine. **[Ans.** 75.25 kJ/Kg]

1.4 An air compressor is provided with a water jacket for cooling. A test indicates that compressor requires 180 kJ/kg of work while the enthalpy of air increases by 73 kJ/kg as it passes through the compressor. The enthalpy of the circulating water increases by 60 kJ/kg of air. Calculate the heat transfer from the air to the compressor. **[Ans.** 193 kJ/kg]

1.5 A gas turbine receives gases from the combustion chamber at 7 bar and 650°C, with a velocity of 90 m/s. The gases leave the turbine at 1 bar with a velocity of 45 m/s. Calculate the work done if the expansion is isentropic. Assume gamma $\gamma = 1.33$ and $c_p = 1.11$ kJ/kg. **[Ans.** 397.45 kJ/kg]

1.6 During a steady flow process 5000 kg/h of fluid passes through an apparatus in which the exit pipe is 2 m below the level of the inlet connection. The pressure decreases from 700 to 140 kPa; velocity increases from 60 to 360 m/s; internal energy decreases by 50 kJ/kg; and specific volume increases from 0.06 to 0.36 m^3/kg. The heat radiated from the apparatus is 10 MJ/h. calculate the work exchanged with the surroundings. **[Ans.** –32.475 kW]

1.7 A blower handles 1 kg/s of air at 20°C and consumes a power of 15 kW. The inlet and outlet velocities of air are 100 m/s and 150 m/s respectively. Find the exit air temperature, assuming adiabatic conditions. Take c_p = 1.005 kJ/kg. K for air. **[Ans.** 28.70°C]

1.8 A turbine operates under steady flow conditions, receiving steam at 1.2 MPa, 188°C, enthalpy 2785 kJ/kg, 33.3 m/s and elevation 3 m. The steam leaves the turbine at 20 kPa, enthalpy 2512 kJ/kg, 100 m/s and elevation zero metre. Heat is lost to the surroundings at the rate of 0.29 kJ/s. If the rate of steam flow through the turbine is 0.42 kg/s, find the power output of the turbine in kW. **[Ans.** 112.515 kW]

1.9 At the inlet of a nozzle, the enthalpy of the passing fluid is 3000 kJ/kg. The nozzle is horizontal and there is negligible heat loss from it. Find (a) velocity at exit, and (b) mass flow rate if inlet area is 0.1 m^2 and specific volume at inlet is 0.187 m^3/kg.

[Ans. 692.5 m/s, 32.08 kg/s]

1.10 A turbo compressor delivers 2.33 m^3/s at 0.276 MPa, 43°C which is heated at this pressure to 430°C and finally expanded in a turbine which delivers 1860 kW. During expansion, there is a heat transfer of 0.09 MJ/s to the surroundings. Calculate the turbine exhaust temperature assuming changes in kinetic and potential energy to be negligible. **[Ans.** 156.33°C]

1.11 A domestic refrigerator is loaded with food and the door closed. During a certain period the machine consumes 1 kWh of energy and the internal energy of the system drops by 5000 kJ. Find the net heat transfer for the system. **[Ans.** – 8.6 MJ]

1.12 1.5 kg of liquid having a constant specific heat of 2.5 kJ/kg.K is stirred in a well-insulated chamber causing the temperature to rise by 15°C. Find the increase in energy and work done for the process. **[Ans.** 56.25 kJ, – 56.25 kJ]

1.13 A mass of 8 kg gas expands within a flexible container so that the p-v relationship is of the form $pv^{1.2}$ const. The initial pressure is 1000 kPa and the initial volume is 1 m^3. The final pressure is 5 kPa. If specific internal energy of the gas decreases by 40 kJ/kg, find the heat transfer in magnitude and direction. **[Ans.** + 2612.5 kJ]

1.14 1 kg of air at 3.5 bar and occupying 0.35 m^3 is heated at constant volume until its temperature has risen to 316°C. Find (a) heat added, and (b) gain in internal energy per kg of air. Take c_v = 0.718 kJ/kg. K. **[Ans.** 116.44 kJ/kg, 116.44 kJ/kg]

1.15 1 kg of air having an initial volume of 0.3 m^3 is heated at constant pressure of 3.2 bar until the volume is doubled. Find (a) heat added, and (b) work done. Take c_p = 1.005 kJ/kg.K, c_v = 0.718 kJ/kg.K. **[Ans.** 336.17 kJ, 96 kJ

1.16 Calculate the difference in non-flow work $\int pdv$ and flow work $-\int v.dp$, if 1 kg of air having specific volume of 0.4 m^3 at 1.35 bar is compressed to a pressure of 10 bar according to the law $pV^{1.3} = C$. **[Ans.** 31.7 kJ]

1.17 A gas undergoes the following reversible non-flow processes during its initial pressure of 15 bar and volume 1 m^3/kg to final pressure of 6 bar and volume 2 m^3/kg.

(a) Constant volume followed by constant pressure.

(b) Constant pressure followed by constant volume.

(c) Polytropic process, .

Calculate the work done by the gas during each process.

[Ans. 600 kJ/kg, 1500 kJ/kg, 1000 kJ/kg]

1.18 A certain gas occupies 0.1 m^3 at 1 bar pressure. It is compressed adiabatically to a pressure of 7 bar. Find the new volume, and change in internal energy. The density of gas at 0°C and 1 bar is 1.5 kg/m^3. Take c_v = 0.728 kJ/kg. *K*. **[Ans.** 0.0249 m^3, 18.847 kJ]

1.19 3 kg of oxygen undergoes a non-flow process during which its pressure remains constant at 7 bar from and initial volume of 1200 litres to a state where the temperature is 520 K. Find (a) changes in internal energy, (b) change in enthalpy, (c) heat transferred, and (d) work done.

For oxygen, c_v = 0.661 kJ/kg. *K*, c_p = 0.921 kJ/kg. K.

[Ans. – 1104.38 kJ, 1538.78 kJ, – 1538.78 kJ, – 434.4 kJ]

1.20 A constant volume chamber of 0.3 m^3 capacity contains 2 kg of a gas at 5°C. Heat is transferred to the gas until the temperature is 100°C. Find the work done, heat transferred, and the change in internal energy, enthalpy and the entropy. Take c_p = 1.968 kJ/kg.K and c_v = 1.507 kJ/kg.K [**Ans.** 0, 286.33 kJ, 286.33 kJ, 373.92 kJ, 0.886 kJ/K]

1.21 0.5 kg of air is compressed reversibly and adiabatically from 80 kPa, 60°C to 0.4 MPa, and is then expanded at constant pressure to the original volume. Calculate the heat transfer and work transfer for the whole path. Take R = 0.287 kJ/kg. K, c_p = 1.005 kJ/kg, K and c_v = 0.718 kJ/kg. K. [**Ans.** 571.64 kJ, 93.5 kJ]

1.22 A mass of air is initially at 260°C and 700 kPa, and occupies 0.028 m^3. The air is expanded at constant pressure to 0.084 m^3. A polytropic process, $pV^{1.5} = C$, is then carried out followed by a constant temperature process which complete the cycle. All the processes are reversible. Find (a) the heat received and rejected in the cycle, and (b) efficiency of the cycle. Take c_p = 1.005 kJ/kg. K and c_v = 0.718 kJ/kg. K.

[**Ans.** 137.13 kJ, 84.11 kJ, 38.66 %]

1.23 A mass of 8 kg gas expands within a flexible container according to the law, $pV^{1.5}$ = const. The initial pressure is 1 MPa and initial volume is 1 m^3. The final pressure is 5 kPa. If specific internal energy of the gas decreases by 40 kJ/kg. Find the heat transfer.

[**Ans.** 2612.5 kJ]

1.24 A heat engine receives 1000 kW of heat at constant temperature of 285°C and rejects heat at 5°C. The possible heat rejected are : (a) 840 kW, (b) 442 kW and (c) 300 kW. Comment on the results. [**Ans.** Not a possible cycle]

1.25 To cool water for drinking during summer, 1 kg of ice at – 2°C is mixed with 5 kg of water at 27°C in an insulated container subjected to 1 bar atmospheric pressure. Take specific heat of ice as 2.093 kJ/kg·K and the latent heat of ice as 335 kJ/kg. Specific heat of water is 4.187 kJ/kg K. [**Ans.** 9.817°C, 0.0907 kJ/K]

1.26 A heat exchanger handles 40 kg/ min of water which is heated from 20°C to 70°C by a hot gas entering the heat exchanger at 150°C and flow rate 80 kg/min.
Determine the change in entropy. Assume specific heat for water as 4.18 kJ/kg·K and for gases as 1.045 kJ/kg·K. [**Ans.** 0.0632 k W/K]

1.27 A heat pump operates between two identical bodies of specific heat C at T_1. The operation of the pump cools down one of the bodies to T_2. Show that for the operation of pump the minimum work input is given by:

$$W_{\min} = C\left[\frac{T_1^2}{T_2} + T_2 - 2T_1\right]$$

1.28 1 kg of air is compressed according to the law $pV^{1.25}$ = constant from a pressure of 1.03 bar and temperature 15°C to a pressure of 17 bar. Calculate

(a) Temperature at the end of compression.

(b) Heat rejected or received during the process, and

(c) Change in entropy.

Take $c_p = 1.005$ kJ/kg·K and $c_v = 0.715$ kJ/kg·K

[**Ans.** 504.56 K, – 96.368 kJ/kg, – 0.2495 kJ/kg·K]

1.29 4 kg of air is compressed in a reversible steady flow polytropic process, $pV^{1.25} = C$, from 1 bar and 30°C to 10 bar. Calculate the work input, heat transferred and change in entropy.

[**Ans.** 1017.24 kJ, – 304.81 kJ, – 0.793 kJ/K]

1.30 Show that the change in entropy per kg gas during Polytropic expansion is:

$$\Delta s = \left[\frac{c_p - nc_v}{n-1}\right] \ln \frac{T_1}{T_2}$$

1.31 A Carnot engine with an efficiency of 40% receives 400 kJ/h from a high temperature source and rejects heat to a sink at 27°C. What is the power output of the engine and temperature of the source in °C? [**Ans.** 0.0444 kW, 227°C]

1.32 1 kg of air is compressed according to the law $pv^{1.2}$ = Const. from 1.05 bar and 15°C to 16.8 bar. Find the change in entropy. $c_p = 1.005$ kJ/kg.k and $c_v = 0.6673$ kJ/kg.k.

[**Ans.** 0.472 kJ/K]

1.33 A reverse Carnot engine is used for the heating of a building. The building receives 209 kJ/h of heat at 20°C from a heat pump. The outside air temperature is –5°C. Calculate (a) heat taken from outside air per hour, and (b) power of motor required.

[**Ans.** 191.17 kJ/h, 4.953 W]

1.34 A piece of red hot iron is suddenly plunged into 10 kg of water at 20°C. The temperature of water rises to 55°C. Calculate the change in entropy. [**Ans.** 4.72 kJ/K]

1.35 In an air turbine, the expansion of air takes place from 6.8 bar and to 1.013 bar and 150°C. The heat loss from the turbine is negligible. Show that the process is irreversible and calculate the change of entropy per kg of air. [**Ans.** 0.0347 kJ/kg K]

1.36 Air at 15°C and 1.05 bar occupies a volume of 0.02 m³. The air is heated at constant volume until the pressure is 4.2 bar and then cooled at constant pressure back to original temperature. Calculate the net heat flow to and from the air and net entropy change.

[**Ans.** – 6.3 kJ, – 0.0101 kJ/K]

1.37 One kg of air expands reversibly in a cylinder behind a piston in such a way that the temperature remains constant at 260°C while the volume is doubled. The piston is then moved in, and heat is rejected by air reversibly at constant pressure until the volume is the same as it was initially. Determine the net heat flow and the overall change of entropy.

[**Ans.** – 161.8 kJ/ kg, – 0.49768 kJ/ kg K]

1.38 Two reversible heat engines *A* and *B* are arranged in series. A rejects heat directly to *B*. Engine *A* receives 300 kJ at a temperature of 500°C from the heat source, while engine *B* is in communication with a cold sink at a temperature of 5°C. If the work output of *A* is twice that of *B*, calculate

(a) the intermediate temperature between *A* and *B*, and

(b) thermal efficiency of each engine. [**Ans.** 170°C, 42.7%, 37.25%]

Chapter

2

FUNDAMENTALS OF REFRIGERATION

2.1 INTRODUCTION

The Process of producing cold by artificial means may be called refrigeration. The term '*refrigeration*' may be defined as the artificial withdrawal of heat, producing in a substance or within a space, a temperature lower than that which would exist under the natural influence of surroundings. In other words, the refrigeration means a continued extraction of heat from a body whose temperature is already below the temperature of its surroundings. In refrigeration, heat is virtually being pumped from a lower temperature to a higher temperature.

According to the second law of thermodynamics, the extraction of heat from a body at lower temperature to a higher temperature can only be performed with the aid of some external device. Such a device is called *refrigerator*. A refrigerator is a reversed heat engine or a heat pump which extracts heat from a cold body and delivers it to a hot body. The substance or medium which works in a heat pump to extract heat from a cold body and to deliver to a hot body is called *refrigerant*.

2.2 REFRIGERATION SYSTEMS

A refrigeration system is a device to produce the refrigeration effect. Some of the popular refrigeration systems are:

1. Ice refrigeration-used in hotels for keeping the drinks cold.
2. Evaporative refrigeration-desert bag, artificial snow.
3. Air refrigeration-aircrafts.
4. Gas throttling refrigeration-liquefaction of gases, like air, nitrogen and oxygen.
5. Vapour compression refrigeration-domestic refrigerator using Freon – 22.
6. Vapour absorption refrigeration-ice making using aqua ammonia or water plus lithium bromide as the absorbent.
7. Steam jet refrigeration-comfort cooling in air-conditioning installations or in industrial processes.
8. Liquid gases refrigeration-transportation vehicles for perishable items.
9. Dry ice (solid CO_2) refrigeration-transportation of perishable items.

2.3 METHODS OF REFRIGERATION

The methods of refrigeration are:

1. Vapour compression refrigeration.
2. Vapour absorption refrigeration.
3. Ejector-compression refrigeration.
4. Electro-Lux refrigeration.
5. Solar refrigeration.
6. Thermo-electric refrigeration.
7. Vortex-tube refrigeration.

2.3.1 Vapour Compression Refrigeration System

The schematic diagram of simple vapour compression refrigeration system is shown in Fig. 2.1. It consists of the following five essential parts:

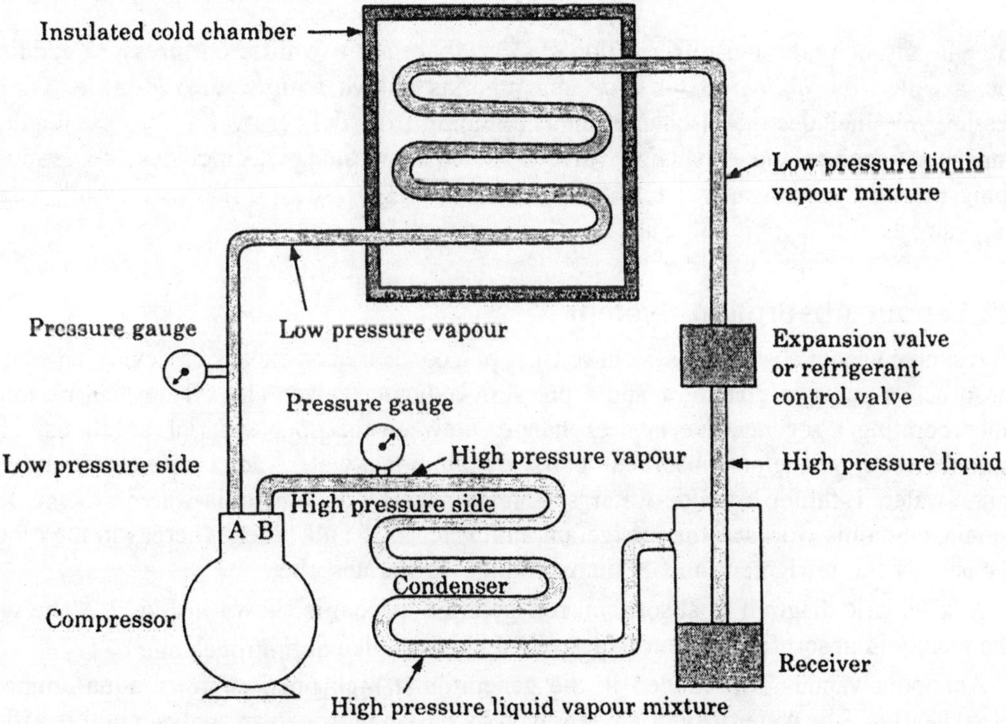

Fig. 2.1 Simple vapour compression refrigeration system.

1. ***Compressor.*** The low pressure and temperature vapour refrigerant from evaporator is drawn into the compressor though the inlet or suction valve *A*, where it is compressed to a high pressure and temperature. This high pressure and temperature vapour refrigerant is discharged into the condenser though the delivery or discharge valve *B*.

2. ***Condenser.*** The condenser or cooler consists of coils of pipe in which the high pressure and temperature vapour refrigerant is cooled and condensed. The refrigerant, while passing through the condenser, gives up its latent heat to the surrounding condensing medium which is normally air or water.
3. ***Receiver.*** The condensed liquid refrigerant from the condenser is stored in a vessel known as receiver from where it is supplied to the evaporator through the expansion valve or refrigerant control valve.
4. ***Expansion valve.*** It is also called throttle valve or refrigerant control valve. The function of the expansion valve is to allow the liquid refrigerant under high pressure and temperature to pass at a controlled rate after reducing its spressure and temperature. Some of the liquid refrigerant evaporates as it passes through the expansion valve, but the greater portion is vaporized in the evaporator at the low pressure and temperature.
5. ***Evaporator.*** An evaporator consists of coils of pipe in which the liquid-vapour refrigerant at low pressure and temperature is evaporated and changed into vapour refrigerant at low pressure and temperature. In evaporating, the liquid vapour refrigerant absorbs its latent heat of vaporizations from the medium (air, water or brine) which is to be cooled.

Note: In any compression refrigeration system, there are two different pressure conditions. One is called the *high pressure side* and other is known as *low pressure side*. The high pressure side includes the discharge line (*i.e.* piping from delivery valve *B* to the condenser, condenser, receiver and expansion valve). The low pressure side includes the evaporator, piping from the expansion valve to the evaporator and the suction line (*i.e.* piping from the evaporator to the suction valve *A*).

2.3.2 Vapour Absorption System

A simple vapour absorption system consists of a condenser, an expansion device, an evaporator, an absorber, a pump, a generator and a pressure reducing valve. The system can be improved by incorporating a regenerative heat exchanger between the poor and rich solutions. The two commonly used refrigerant absorbent pairs are ammonia-water (aqua ammonia) and lithium bromide-water. Lithium bromide-water system is superior to ammonia-water system. In aqua ammonia, ammonia works as the refrigerant and water as the absorber, whereas in the other pair, water acts as the refrigerant and lithium bromide as the absorber.

A schematic diagram of absorption refrigeration system is shown in Fig. 2.2 The working of the ammonia absorption system is described below:

Ammonia vapour is produced in the generator at high pressure from aqua ammonia by external heating. The water vapour are removed by passing through an analyser and rectifier. The dehydrated ammonia enters the condenser where vapours are condensed. The sub-cooled high pressure liquid ammonia is passed through a throttle valve to further lower its temperature. Now they enter the evaporator and leave as saturated vapour. The saturated vapour goes to absorber to become strong after absorbing ammonia vapour and the strong solution is pumped to the generator through the pump and heat exchanger. The solution in the generator becomes weak as ammonia vapours comes out of it. The weak ammonia solution from the generator is passed to the heat exchanger through the pressure reducing valve. The cycle is repeated.

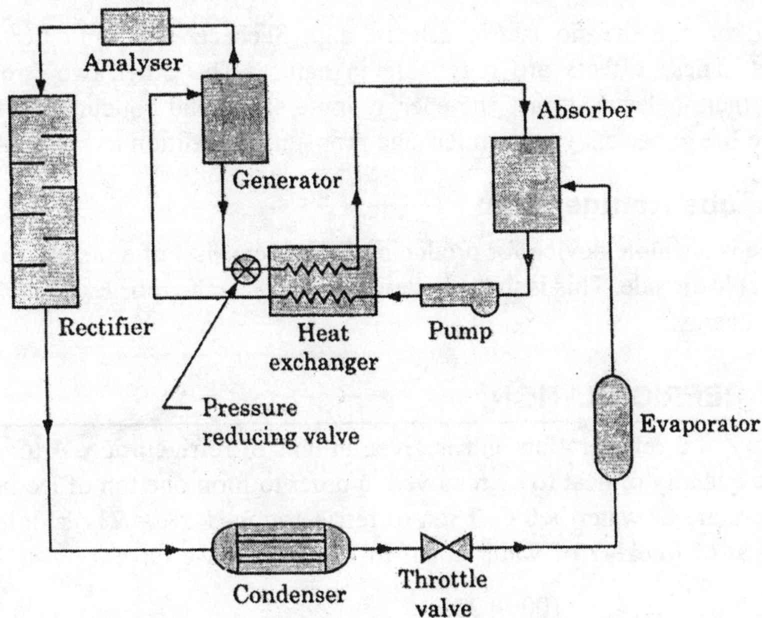

Fig. 2.2 Ammonia absorption system

2.3.3 Ejector-Compresion System

The system using water as the refrigerant in a vapour compression cycle in which compression is achieved by the principle of jet compression employing its own vapour as the motive vapour, is called *steam-ejector system.* Refrigeration effect is obtained by direct evaporation and subsequent self cooling of water.

Bubble point temperature is the temperature at which a liquid mixture begins to boil. Azeotrope is a constant temperature boiling mixture.

2.3.4 Electro-Lux Refrigeration

This vapour absorption system uses ammonia as the refrigerant and hydrogen as the inert gas medium. The inert gas is confined only to the low side of the system, *i.e.*, evaporator and absorber. In the evaporator, ammonia evaporates in the presence of hydrogen atmosphere. The partial pressure of ammonia is much lower than its pressure and thus give very low temperature. Hydrogen has the advantage of being non-corrosive and insoluble in water. This system is used for domestic refrigeration.

2.3.5 Solar Refrigeration

This uses the strong aqua ammonia varpour absorption system by utilizing the solar energy as the heat source to produce the refrigeration effect. This is an intermittent cycle. Lithium chloride-water system has also been used.

2.3.6 Thermo-electric Refrigeration

It is based on the thermoelectric effects, e.g., Seebeck effect, Peltier effect, and the Thomson effect. These effects are reversible in nature. The other two irreversible effects associated with thermoelectricity are the energy. Joule effect and conduction effect. It has low COP but infinite life expectancy and no leakage problem. In addition to many other advantages.

2.3.7 Vortex Tube Refrigeration

Vortex tube is a simple device for producing cold. It consists of a nozzle, diaphragm, valve, hot air side and cold air side. This is the only device to produce the refrigeration effect by utilizing the waste heat energy.

2.4 UNIT OF REFRIGERATION

The capacity of a refrigeration unit is given in tons of refrigeration. A ton of refrigeration is defined as the quantity of heat to be removed in order to form one ton of ice in 24 hours when the initial temperature of water is 0°C. 1 ton of refrigeration, 1 TR = 211 kJ/min or 3.5167 kW. Taking latent heat of freezing of water equal to 336 kJ/ kg, we have

$$\text{1 tonne of refrigeration} = \frac{1000 \times 336}{24} = \text{14,000 kJ/h or 3.8889kW} \quad \text{...(2.1)}$$

Coefficient of performance (COP). The COP of a refrigeration system is a ratio of the heat removed (cooling effect produced) from a system to the work supplied to achieve the heat removal.

$$\text{COP} = \frac{\text{Heat removed per unit time, } Q\text{(kJ/s)}}{\text{Work supplied per unit time, } W\text{(kJ/s)}} \quad \text{...(2.2)}$$

$$\text{COP} = \frac{211\,Tn\text{(kJ/min)}}{(60P)\text{ (kJ/min)}} = \frac{3.5167}{\left(\dfrac{P}{Tn}\right)} \quad \text{...(2.3)}$$

Where P = energy required in kW

Tn = capacity of plant in tons.

$$\text{Relative COP} = \frac{\text{Actual COP}}{\text{Theoretical COP}} \quad \text{...(2.4)}$$

2.5 REFERIGERATION EFFECT

During evaporation, the liquid-vapour refrigerant absorbs its latent heat of vaporization from the medium (air, water, brine) which is to be cooled. This heat which is absorbed by the refrigerant is called the refrigeration effect. In other words, the refrigeration effect is the amount of cooling produced.

Refrigeration effect = Enthalpy of vapour refrigerant – Enthalpy of liquid refrigerant.

2.6 CARNOT REFRIGERATION CYCLE

The reversed Carnot cycle is shown in Fig. 2.3 on *p-v* and *T-s* diagrams. The sequence of operations is given below:

(*i*) The process 1 – 2 represents the isentropic compression of the working fluid with the aid of external work. The temperature of the working fluid rises from T_1 to T_2.

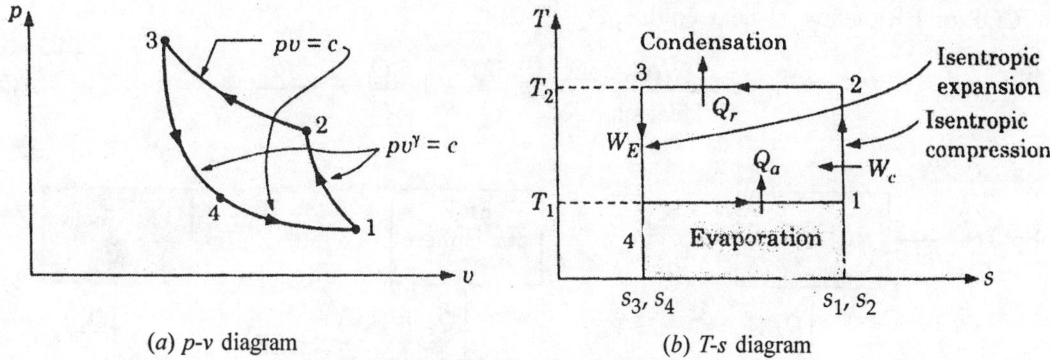

(*a*) *p-v* diagram (*b*) *T-s* diagram

Fig. 2.3 Reversed Carnot cycle.

(*ii*) The process 2-3 represents the isothermal compression of the working fluid during which heat Q_r is rejected at constant high temperature T_2.

(*iii*) The process 3-4 represents the isentropic expansion of the working fluid. The temperature or working fluid drops from T_2 to T_1.

(*iv*) The process 4-1 represents the absorption of heat Q_a by the working fluid from refrigerator at constant low temperature T_1 during isothermal expansion.

$$\text{COP} = \frac{\text{Heat absorbed in refrigerator}}{\text{Work Done}} = \frac{Q_a}{Q_r - Q_a}$$

$$= \frac{T_1(s_1 - s_4)}{T_2(s_1 - s_4) - T_1(s_1 - s_4)} = \frac{T_1}{T_2 - T_1} = \frac{h_1 - h_4}{(h_2 - h_3) - (h_1 - h_4)} \qquad ...(2.5)$$

The Carnot cycle is practically not feasible since isothermal energy rejection requires extremely slow motion followed by isentropic process during which the piston should move at extremely faster rate, which is mechanically not obtainable.

The COP of reversed Carnot cycle is the highest of all refrigeration cycles. COP increases as T_1 is increased or T_2 is lowered. The increase in COP for increased T_1 is more than for decreased T_2. Since T_2 in winter is less than T_2 in summer, therefore, COP in winter is more than that of in summer.

If the cycle is used as a heat pump, then energy performance ratio,

$$\text{EPR} = \frac{\text{Heat supplied to room}}{\text{Work Done}} = \frac{T_2}{T_2 - T_1} \qquad ...(2.6)$$

2.7 DIFFERENCE BETWEEN A HEAT ENGINE, REFRIGERATOR AND HEAT PUMP

Heat Engine : In a heat engine, as shown in Fig. 2.4 (*a*), the heat supplied to the engine is converted into useful work. If Q_2 is the heat supplied to the engine and Q_1 is the heat rejected from the engine, then the net work done by the engine is given by,

$$W_e = Q_2 - Q_1$$

COP or Efficiency of heat engine,

$$\text{COP or} \quad \eta_e = \frac{\text{Work done}}{\text{Heat supplied}} = \frac{W_e}{Q_2} = 1 - \frac{Q_1}{Q_2} = 1 - \frac{T_1}{T_2} \quad \text{...(2.7)}$$

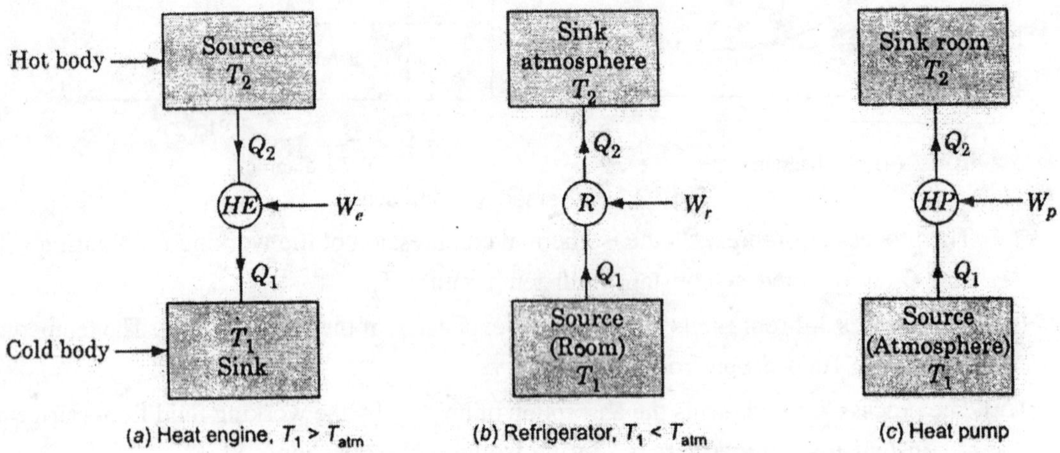

Fig. 2.4 Concept of heat engine, refrigerator and heat pump.

Refrigerator. A refrigerator is an equipment used to remove the heat continuously from the space (sink). It maintains the temperature below atmospheric temperature and reject the heat to the atmosphere (source) by increasing the temperature potential of the heat to be rejected with the help of mechanical work input (compressor).

The refrigeration system is shown in Fig. 2.4 (b). In a refrigerator system, Q_1 quantity of heat is removed from the source where its temperature T_1 is maintained below atmospheric temperature T_2. COP of refrigerator is,

$$\text{Refrigeration effect} = (\text{COP})_{\text{ref}} = \frac{Q_1}{W_{ref}} = \frac{Q_1}{Q_2 - Q_1} = \frac{Q_1/Q_2}{1-(Q_1/Q_2)} = \frac{(T_1/T_2)}{1-(T_1/T_2)} \quad \text{...(2.8)}$$

Heat Pump. It is an equipment used to supply the heat continuously to the space. It maintains the temperature above atmospheric temperature by absorbing the heat from the atmosphere and increasing its temperature potential with mechanical work input (compressor). The heat pump system is shown in Fig. 2.4 (c). In a heat pump system, Q_2 quantity of heat is supplied to the room where the temperature T_2 is maintained above atmospheric temperature T_1. The heat Q_1 is obtained from the atmosphere.

COP or Energy performance ratio of heat pump is,

$$(COP)_P \text{ or } \quad EPR = \frac{Q_2}{W_P} = \frac{Q_1 + W_P}{W_P} = 1 + \frac{Q_1}{W_P} = 1 + (COP)_{ref} \qquad ...(2.9a)$$

$$= \frac{Q_2}{Q_2 - Q_1} = 1 + \frac{Q_1}{Q_2 - Q_1} = \frac{T_2}{T_2 - T_1} = \frac{1}{1-(T_1/T_2)} \qquad ...(2.9b)$$

Example 2.1 *Find the COP of a refrigerator system if the work input is 100 kJ/kg and refrigeration effect produced is 180 kJ/kg of refrigerant flowing.*

Solution: $w_{ref} = 100$ kJ/kg, $q = 180$ kJ/kg

$$(COP)_{ref} = \frac{q}{w_{ref}} = \frac{180}{100} = 1.8$$

Example 2.2 *A machine working on a Carnot cycle operates between 310 K and 270 K. Calculate the COP when it is operated as:*

(a) a heat engine, (b) a refrigerating machine, and (c) a heat pump.

Solution: Given: $T_1 = 270$K, $T_2 = 310$ K

(*a*) Heat engine:

$$(COP)_e \text{ or } \eta_e = 1 - \frac{T_1}{T_2} = 1 - \frac{270}{310} = 0.129$$

(*b*) Refrigerating machine:

$$(COP)_{ref} = \frac{T_1}{T_2 - T_1} = \frac{270}{310-270} = \frac{270}{40} = 6.75$$

(*c*) Heat pump:

$$(COP)_P = \frac{T_1}{T_2 - T_1} = \frac{310}{310-270} = \frac{310}{40} = 7.75$$

Example 2.3 *A Carnot refrigeration cycle absorbs heat at 268 K and rejects it at 298 K.*

(a) Calculate the COP of this refrigeration cycle.

(b) If the cycle is absorbing 1150 kJ/min at 268 K, how many kJ of work is required per second?

(c) How many kJ/s will be delivered if it works as a heat pump at 298 K and absorbs 1150 kJ/min at 268 K.

Solution: Given: $T_1 = 268$ K, $T_2 = 298$ K, $Q_1 = 1150$ kJ/min

(*a*) Carnot refrigeration cycle:

$$(COP)_{ref} = \frac{T_1}{T_2 - T_1} = \frac{268}{298-268} = \frac{268}{30} = 8.933$$

Work required, $$W_{ref} = \frac{Q_1}{(COP)_{ref}} = \frac{1150}{8.933} = 128.73 \text{ kJ/min or } 2.145 \text{ kJ/s}$$

(*b*) Heat pump:

$$(\text{COP})_p = 1 + (\text{COP})_{\text{ref}} = 1 + 8.933 = 9.933$$

or
$$9.933 = \frac{Q_2}{Q_2 - Q_1} = \frac{Q_2}{Q_2 - 1150}$$

$$9.933\, Q_2 - 11422.95 = Q_2$$

$$Q_2 = \frac{11422.95}{8.933} = 1278.74 \text{ kJ/min or } 21.31 \text{ kJ/s}$$

2.8 POWER CONSUMPTION OF A REFERIGERATING MACHINE

The power consumption of a refrigerating machine is determined in terms of kW. However, the power consumption of the driving motor is sometimes rated in horsepower (H.P.). We have

$$HP = \frac{W_{\text{ref}}}{0.736} \qquad \text{...(2.10)}$$

Where W_{ref} is in kW and HP is in metric units.

Power consumption per ton refrigeration

$$= \frac{HP}{\text{kW refrigeration}}$$

$$= \frac{W_{\text{ref}}}{0.736 Q_1} \qquad \text{...(2.11)}$$

where Q_1 = refrigerating capacity in kW

Since $1\ TR = 3.5167$ kW

Therefore,
$$\frac{HP}{TR} = \frac{W_{\text{ref}} \times 3.5167}{0.736 Q_1} = \frac{4.78}{(COP)_{\text{ref}}} \qquad \text{...(2.12)}$$

Example 2.4 *The ambient air temperature during summer and winter in a particular locality are 45°C and 15°C respectively. Find the value of Carnot COP for an air-conditioner for cooling and heating, corresponding to refrigeration temperatures of 5°C for summer and heating temperature of 55°C for winter. Also find the theoretical power consumption per ton of refrigeration in each case.*

Solution: For summer: $T_1 = 273 + 5 = 278$ K, $T_2 = 45 + 273 = 318$ K

For cooling, Carnot $(\text{COP})_{\text{cooling}} = \dfrac{T_1}{T_2 - T_1} = \dfrac{278}{318 - 278} = \dfrac{278}{40} = 6.95$

Power consumption, $W_{\text{cooling}} = \dfrac{211}{6.95} = 30.36$ kJ/min or 0.506 kW

For winter: $T_1 = 273 + 15 = 288\text{K}$, $T_2 = 273 + 55 = 328$ K

For heating, $$(COP)_{heating} = \frac{T_2}{T_2 - T_1} = \frac{328}{328-288} = \frac{328}{40} = 8.2$$

Power consumption, $$W_{heating} = \frac{211}{8.2} = 25.732 \text{ kJ/min or } 0.429 \text{ kW}$$

Example 2.5 *A Carnot refrigerator has working temperature of –30°C and 35°C. The actual COP is 0.75 of the maximum. Calculate the power consumption, and heat rejected to the surroundings per ton of refrigeration.*

Solution: $$COP = \frac{T_1}{T_2 - T_1} = \frac{273-30}{35-(-30)} = \frac{243}{65} = 3.74$$

$$\text{Actual COP} = 0.75 \times 3.74 = 2.8$$

$$\text{Power consumption per ton} = \frac{3.5167}{2.8} = 1.256 \text{ kW}$$

$$\text{Heat rejected per ton} = 3.5167 + 1.256 = 4.7727 \text{ kW}$$

Example 2.6 *(a) Enumerate different methods of producing refrigeration and define the terms, 'Refrigerating effect', 'unit of refrigeration' and 'COP', as related to refrigeration systems.*

(b) A refrigeration system is working on reversed Carnot cycle between higher and lower temperatures of 40⁰C and -30⁰C respectively. Determine its COP. It is desired to increase the COP of the cycle to 4.75 by changing the temperatures of the cycle. If the increase (or decrease) in higher temperature is equal to the decrease (or increase) in lower temperature, calculate the new temperatures of the cycle.

Solution:

(*a*) Methods of refrigeration: Refer to art. 2.3

Refrigeration effect: Refer to art. 2.5

Unit of refrigeration and COP: Refer to art. 2.4

(*b*) Given : T_1 = –30°C or 273 – 30 = 243 K, T_2 = 40°C or 273 + 40 = 313 K,

COP = 4.75

For a reversed Carnot cycle,

$$COP = \frac{T_1}{T_2 - T_1} = \frac{243}{313-243} = \frac{243}{70} = 3.47$$

Let ΔT = decrease in lower temperature T_1 and increase in higher temperature T_2.

Then $$COP = \frac{T_1}{(T_2 + \Delta T) - (T_1 - \Delta T)} = \frac{T_1 - \Delta T}{T_2 - T_1 + 2\Delta T}$$

$$4.75 = \frac{243 - \Delta T}{313 - 243 + 2\Delta T} = \frac{243 - \Delta T}{70 + 2\Delta T}$$

$$332.5 + 9.50\,\Delta T = 243 - \Delta T$$

$$10.50\,\Delta T = -89.5$$
$$\Delta T = -8.524 \text{ K}$$

New temperatures are

$$T_1 = 243 - (-8.524) = 251.54 \text{ K}$$
$$T_2 = 313 - 8.524 = 304.476 \text{ K}$$

Verification:

$$\text{COP} = \frac{251.524}{304.476 - 251.524}$$
$$= \frac{251.524}{52.952} = 4.75$$

Example 2.7 *A air-conditioning unit rated at 30 TR seems to be indicating poor cooling. The heat rejection to atmosphere in its condenser was found as follows:*

Flow rate of cooling water = 3 litres/s
Inlet water temperature = 25°C
Outlet water temperature = 35°C
Power input to motor = 45 kW
Efficiency of motor = 96%

Calculate the actual refrigrating capacity of the unit.

Solution: Work input, $\dot{W} = 45 \times 0.96 = 43.2$ kW

Heat rejected in condenser, $\dot{Q}_c = \dot{m}_w c_{pw} \cdot \Delta t_w$

$$= (3 \times 10^{-3} \times 10^3) \times 4.187 \times (35 - 25)$$
$$= 125.61 \text{ kW}$$

Refrigeration capacity, $\dot{Q}_r = \dot{Q}_c - \dot{W} = 125.61 - 43.2 = 82.41$ kW or 23.4 TR

The air-conditioning unit is operating below its rated capacity of 30 TR.

Example 2.8 *The ambient air temperatures during summer and winter at a place are 40°C and 17°C respectively. Calculate the values of Carnot COP for an air-conditioner for cooling and heating, corresponding to refrigeration temperature of 5°C for summer and heating temperature of 50°C for winter. Assume 15°C temperature difference in summer and 5°C in winter in the exchanges that exchanges heat with the surroundings.*

Solution. For Summer : $t_1 = 40 + 15 = 55$°C or $T_1 = 273 + 55 = 328$ K

$t_2 = 5$°C or $T_2 = 273 + 5 = 278$ K

COP Carnot $= \dfrac{T_2}{T_1 - T_2} = \dfrac{278}{328 - 278} = \dfrac{278}{50} = 5.56$

For winter : $T_1 = 273 + 50 = 323$ K

$T_2 = (17 - 5) + 273 = 285$ K

COP Carnot $= \dfrac{T_1}{T_1 - T_2} = \dfrac{323}{323 - 285} = \dfrac{323}{38} = 8.5$

Example 2.9 *A house is to be maintained at 25°C in summer as well as in winter. For this purpose, it is proposed to use a reversible device as a refrigerator in summer and as heat pump in winter. The ambient temperature is 40°C in summer and 3°C in winter. The energy losses as heat from the roof and the walls is estimated at 5 kW per degree Celsius temperature difference between the room and the ambient conditions. Calculate the power required to operate the device in summer and in winter.* *[IES, 2005]*

Solution: For the cooled body, we have [Fig. 2.5(a)]

$$(\text{COP})_R = \frac{T_L}{T_H - T_L} = \frac{298}{313-298} = 19.867$$

$$Q_L = 5 \times (40 - 25) = 75 \text{ kW}$$

$$\text{Maximum power required} = \frac{Q_L}{(COP)_R}$$

$$\frac{75}{19.867} = 3.775 \text{ kW}$$

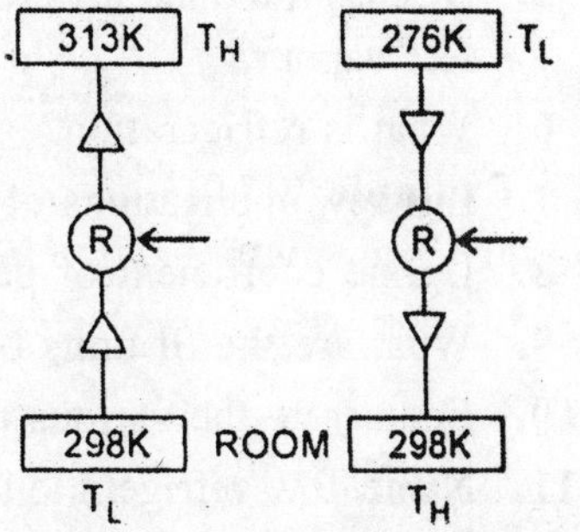

(a) Cooled body (b) Heated body

Fig. 2.5

For the heated body, we have [Fig. 2.5(b)]

$$(\text{COP})_P = \frac{T_L}{T_H - T_L}$$

$$= \frac{298}{298-276} = 13.54$$

$$Q_L = 5 \times (25 - 3) = 110 \text{ kW}$$

$$\text{Maximum power required} = \frac{Q_L}{(\text{COP})_P} = \frac{110}{113.54} = 8.12 \text{ kW}$$

REVIEW QUESTIONS

1. How is the effectiveness of a refrigeration system measured?
2. Explain the term "tonne of refrigeration".
3. Discuss the advantages of the dense air refrigerating system over an open air refrigeration system.
4. What is the difference between a refrigerator and a heat pump? Derive an expression for the performance factor for both if they are running on reversed Carnot cycle.
5. Discuss the main differences between Reversed Carnot cycle operating on perfect gas and wet vapour.
6. What is refrigeration?
7. List few applications of refrigeration.
8. Define coefficient of performance.
9. What are the SI units of refrigeration?
10. Enumerate the various methods of refrigeration.
11. Name few refrigeration systems.
12. What is refrigeration effect?

MULTIPLE CHOICE QUESTIONS

1. One tonne of refrigeration is equal to

   (a) 211 kJ/min (b) 220 kJ/min

   (c) 420 kJ/min (d) 620 kJ/min
2. One tonne refrigerating machine means that

   (a) one tonne is the total mass of the machine

   (b) one tonne of refrigerant is used

   (c) one tonne of water can be converted into ice

   (d) one tonne of ice when melts from and at 0°C in 24 hours, the refrigeration effect produced is equivalent to 211 kJ/min.
3. The ratio of heat extracted in the refrigerator to the work done of the refrigerant is called

   (a) coefficient of performance of refrigeration

   (b) coefficient of performance of heat pump

   (c) relative coefficient of performance

   (d) refrigerating efficiency
4. The relative coefficient of performance is equal to

   (a) $\frac{\text{Theoretical COP}}{\text{Actual COP}}$ (b) $\frac{\text{Actual COP}}{\text{Theoretical COP}}$

   (c) Actual COP × Theoretical COP

5. Air refrigerator works on

(a) Carnot cycle (b) Rankine cycle

(c) reversed Carnot cycle (d) Bell-Coleman cycle

6. In air-conditioning of aeroplanes, using air as a refrigerant, the cycle used is

(a) reversed Carnot cycle (b) reversed Joule cycle

(c) reversed Brayton cycle (d) reversed Otto cycle

7. Co-efficient of performance of a Reversed Carnot cycle refrigerator working between higher temperature T_2 and lower temperature T_1

(a) Will increase with increase in T_1 keeping T_2 fixed

(b) Will decrease with increase in T_1 keeping T_2 fixed

(c) Will first increase with increase in T_1 and then decrease with increase in T_1 keeping T_2 fixed

(d) None of the above.

8. A reversible refrigerator working between two fixed temperatures

(a) has the same COP whatever be the working substance

(b) has its COP increased for working substance with high enthalpy of evaporation

(c) has its COP increased for working substance with higher specific heats

(d) none of the above

9. Reversed Carnot cycle comprises

(a) two isentropic processes and two adiabatic processes

(b) two isentropic processes and two isothermal processes

(c) two isentropic processes and two constant pressure processes

(d) two isentropic processes and two constant volume processes.

10. In Reversed Carnot cycle working on perfect gas

(a) Isothermal work of compression is equal to isothermal work of expansion

(b) Isentropic work of compression is equal to isentropic work of expansion

(c) Net work of the cycle is zero

(d) Net heat transfer of the cycle is zero.

11. In a Reversed Carnot-cycle working on vapour

(a) Isentropic work of compression is equal to isentropic work of expansion

(b) Isothermal heat absorption is equal to isothermal heat rejection

(c) There is no work done during isothermal processes

(d) There is no work done during isentropic processes.

12. Reversed Carnot cycle assumes that all processes in the cycle are

(a) Non-flow only (b) Steady flow only

(c) Non-flow or steady flow (d) Transient flow.

13. Two Carnot Refrigerators are employed, one for ice making and other for comfort cooling
(a) The COP of refrigerator for ice making is higher than that for the other
(b) The COP of refrigerator for ice making is lower than that for the other
(c) The COP of refrigerator for ice making is same as that for the other
(d) The COP of Carnot refrigerator will depend on refrigerant used.

14. A reverse Carnot cycle has a COP of 4. The ratio of higher temperature to lower temperature will be
(a) 1.5 (b) 2 (c) 1.25 (d) 2.5

15. A Carnot refrigerator requires 70 kJ/min of work to produce one ton of refrigeration at –40°C. The COP of this refrigerator is
(a) 4 (b) 3
(c) 5 (d) not possible to find

16. The co-efficient of performance of refrigerator working on Reversed Carnot cycle with T_1 being lower temperature and T_2 being higher temperature is
(a) $\frac{T_2 - T_1}{T_1}$ (b) $\frac{T_1}{T_2 - T_1}$
(c) $\frac{T_2}{T_2 - T_1}$ (d) $\frac{T_2 - T_1}{T_2}$

17. The co-efficient of performance of Carnot refrigerator can be expressed: (r = Volume compression ratio for isentropic compression)
(a)) $1 - r^{\gamma-1}$ (b) $\frac{1}{1 - r^{\gamma-1}}$
(c) $\frac{1}{r^{\gamma-1} - 1}$ (d) $\frac{1}{1 + r^{\gamma-1}}$

18. The COP of Carnot Refrigerator is 3 and it produces 1ton of refrigeration. The work that will be done is equal to
(a) 210 kJ/min (b) 70 kJ/min (c) 100 kJ/min (d) 200 kJ/min

19. The ratio of high temperature to low temperature for Reversed Carnot refrigerator is 1.25. The COP will be
(a) 2 (b) 3 (c) 4 (d) 5

20. A Carnot refrigerator operates between 300.3°K and 273°K. The fraction of cooling effect required as work input is
(a) 20% (b) 10%
(c) 50% (d) not possible to find with the data.

21. The relationship between $COP_{heat\ pump}$ and $COP_{refrigerator}$ for the same range of temperature operation is
    (a) $COP_{heat\ pump} = COP_{refrigerator}$
    (b) $COP_{heat\ pump} - COP_{refrigerator} = 1$
    (c) $COP_{heat\ pump} = 2\ COP_{refrigerator}$
    (d) not possible to predict without knowing working substance.
22. Carnot refrigerator absorbs heat at –13°C and requires 1 kW for each 6.5 kW of heat absorbed, the COP and temperature of heat rejections respectively are (0°C = 273°K)
    (a) COP = 6.5, t = 27°C (b) COP = 7.5, t = 27°C
    (c) COP = 6.5, t = 30°C (d) COP = 7.5, t = 37°C

Answers

1. (a)	**2.** (d)	**3.** (a)	**4.** (b)	**5.** (d)	**6.** (c)	**7.** (a)	**8.** (a)
9. (b)	**10.** (b)	**11.** (c)	**12.** (c)	**13.** (b)	**14.** (c)	**15.** (b)	**16.** (b)
17. (c)	**18.** (c)	**19.** (c)	**20.** (b)	**21.** (b)	**22.** (a)		

NUMERICAL PROBLEMS

2.1 A Reversed Carnot Cycle operates on the following substances given below and the range of temperatures is 300°K and 250°K.

(*i*) Working substance is water between the range of dry saturated liquid to dry saturated vapour at 300°K.

(*ii*) Working substance is water in the form of super heated steam with the range of pressure from 0.01 bar to 0.05 bar.

(*iii*) Working substance is CO_2 in the super heated region with pressure range of 0.5 bar to 5 bar.

(*iv*) Working substance is air with pressure range of 1 bar to 10 bar.

For the same heat rejected from the system of 60000 kJ/hr, determine the work input, Carnot co-efficient of performance as refrigerator and as heat pump and the mass flow rate.

2.2 Carnot heat engine draws heat from a reservoir at temperature of 600 K and rejects heat to another reservoir at temperature T K. This engine drives a Carnot reverse cycle refrigerator which absorbs heat from reservoir at temperature 200 K and reject heat to reservoir at temperature T K. Determine the temperature T K such that heat supplied to Carnot heat engine is equal to heat absorbed by Carnot refrigerator.

2.3 A refrigerator using Carnot cycle requires 1.25 kW per tonne of refrigeration to maintain a temperature of –30°C. Find : (*a*) COP of the Carnot refrigerator; (*b*) Temperature at which heat is rejected, and (*c*) Heat rejected per tonne of refrigeration.

[**Ans.** 2.8; 55.4°; 284 kJ/min]

2.4 A Carnot cycle machine operates between the temperature limits of 47°C and –30°C. Determine the COP when it operates as (*a*) a refrigerating machine; (*b*) a heat pump, and (*c*) a heat engine. [**Ans.** 3.16; 4.16; 0.24]

2.5 Ten tonnes of fish is frozen to –30°C per day. The fish enters the freezing chamber at 30°C and freezing occurs at –3°C the frozen fish is Cooled to –30°C. The specific heats of fresh and frozen fish Are 3.77 kJ/kg K and 1.67 kJ/kg.K respectively while latent heat of freezing is 251.2 kJ/kg K. Find the tonnage of the plant which runs for 18 hours per day. The evaporator and condenser temperatures are –40°C and 45°C respectively. If the COP of the plant is 1.8, determine the power consumption of the plant in kW. Also find the refrigerating efficiency of the plant. [**Ans.** 18.6 TR; 36.1 kW; 65.7%]

2.6 A Carnot refrigeration system has working temperature of –30°C and 40°C. What is the maximum COP possible? If the actual COP is 75% of the maximum, calculate the actual refrigerating effect produced per kilowatt hour. [**Ans.** 3.47; 0.743 TR]

2.7 A 5 tonne refrigerating machine operating on Bell Coleman cycle has an upper limit of pressure of 12 bar. The pressure and temperature at the start of compression are 1 bar and 17°C respectively. The compressed air cooled at constant pressure to a temperature of 40°C enters the expansion cylinder. Assuming both the expansion and compression processes to be isentropic with $\gamma = 1.4$, determine: (*a*) COP; (*b*) quantity of air in circulation per minute; (*c*) piston displacement of compressor and expander; (*d*) bore of compressor and expansion cyclinders. The unit runs at 250 r.p.m. and is double acting. Stroke length is 200 mm; and (*e*) power required to drive the unit. Take $c_p = 1$ kJ/kg.K; $c_v = 0.71$ kJ/kg. K; $R = 0.287$ kJ/kg.K.

[**Ans.** 0.952; 7.65 kg/min; 6.37 m^3/min, 3.35 m^3/min; 284 mm; 18.4 kW]

2.8 An air refrigerator used for food storage provides 50 TR. The temperature of air entering the compressor is 7°C and the temperature before entering into the expander is 27°C. Assuming a 70% mechanical efficiency, find; (*a*) actual COP; and (*b*) The power required to run the compressor.

The quantity of air circulated in the system is 100 kg/min. The compression and expansion follow the law $pv^{1.3}$ = constant.

Take $\gamma = 1.4$; $c_p = 1$ kJ/kg.K for air. [**Ans.** 1.13; 110.6 kW]

2.9 A dense air refrigerating system operating between pressures of 17.5 bar and 3.5 bar is to produce 10 tonnes of refrigeration. Air leaves the refrigerating coils at –7°C and it leaves the air cooler at 15.5°C. Neglecting losses and clearance, calculate the net work done per minute and the coefficient of performance. For air $c_p = 1.005$ kJ/kg K and $\gamma = 1.4$.

[**Ans.** 1237 kJ/min; 1.7]

2.10 A refrigerating plant is required to produce 2.5 tonnes of ice per day at –4°C from water at 20°C. If the temperature range in the compressor is between 25°C and –6°C, calculate power required to drive the compressor. Latent heat of ice = 335 kJ/kg and specific heat of ice = 2.1 kJ/kg K. [**Ans.** 1.437 kW]

2.11 A refrigerator storage is supplied with 30 tonnes of fish at a temperature if 27°C. The fish has to be cooled to –9°C for preserving it for long period without deterioration. The cooling takes place in 10 hours. The specific heat of fish is 2.93 kJ/kg.K above freezing point of fish and 1.26 kJ/kg.K below freezing point of fish which is –3°. The latent heat of freezing is 232 kJ/kg. What is the capacity of the plant in tonnes of refrigeration for cooling the fish? What would be the ideal COP between this temperature range? If the actual COP is 40% of the ideal, find the power required to run the cooling plant.

[**Ans.** 78 TR; 7.33; 93.3 kW]

2.12 An air refrigeration plant working on the Carnot cycle between the temperature limits of 34°C and –1°C has a capacity of 10 ton. Determine the power input to the plant.

2.13 An air refrigeration plant working on Carnot cycle produces refrigeration capacity equivalent to production of 10 tons of ice in 24 hours at 0°C from water at 20°C. The compressor temperature limits are 30°C and –15°C. Determine the power input to the plant. The latent heat of ice is 335 kJ/kg.

2.14 Find the minimum power input to perfect reversed heat engine that will produce 1 ton of ice per hour at –5°C from water at 15°C. The temperature limits of the reversed heat engine are 15°C and –5°C. Assume specific heat of ice as 2.1 kJ/kg. K and latent heat of ice as 335 kJ/kg.

Chapter

AIR REFRIGERATION CYCLES 3

3.1 INTRODUCTION

The air refrigeration cycle uses air as a refrigerant. Air is in gaseous form and does not change its phase throughout the cycle. The heat carrying capacity per kg of air is very small. Its coefficient of performance is also very low and requires high power for operation. However, the air-refrigeration cycle is preferred because of the low weight and space occupied by equipment.

3.2 AIR REFRIGERATION CYCLES

There are two types of air refrigeration cycles.

1. Open air refrigeration cycle.
2. Closed (or dense) air refrigeration cycle.

3.2.1 Open Air Refrigeration Cycle

In an open air refrigeration cycle, the air is directly (at atmospheric pressure) led to the space to be cooled (i.e. refrigerator). The air is then allowed to circulate through the cooler and returned to the compressor to start another cycle. This requires size of the compressor and expander to be large. The moisture is regularly carried away by the air circulated through the refrigerator. This leads to the formation of frost at the end of expansion process and clog the line. Therefore, a drier is required in the open cycle system.

3.2.2 Closed (or Dense) Air Refrigeration Cycle

In a closed air refrigeration cycle, the air is passed through the pipes and component parts of the system at all times. The air is used for absorbing heat from the other fluid (say brine) which is circulated into the space to be cooled. The air does not come in contact directly with the space to be cooled.

This system has the following advantages over the open cycle:

1. The suction pressure is higher than that of atmospheric pressure. Therefore, the volume of air handled by the compressor and expander are smaller.
2. The operating pressure ratio can be reduced, which results in higher coefficient of performance.

3.3 REVERSED CARNOT CYCLE

A reversed Carnot cycle, using air as working medium is shown in Fig. 3.1 on *p-v* and T-s diagrams. The reversed Carnot cycle is represented by four processes as described below:

1. **Process 1-2: Isentropic compression :** During this process, the pressure of air increases from p_1 to p_2, specific volume decreases from v_1 to v_2 and temperature increases from T_1 to T_2. No heat is absorbed or rejected by air during this process.
2. **Process 2-3: Isothermal compression:** During this process, the pressure of air increases from p_2 to p_3 and specific volume decreases from v_2 to v_3. Temperature remains constant at $T_2 = T_3$.

   Heat rejected by air per kg of air, $q_{2\text{-}3}$ = area 2-3-3′-2′

   $$= T_3(s_2 - s_3) = T_2(s_2 - s_3)$$

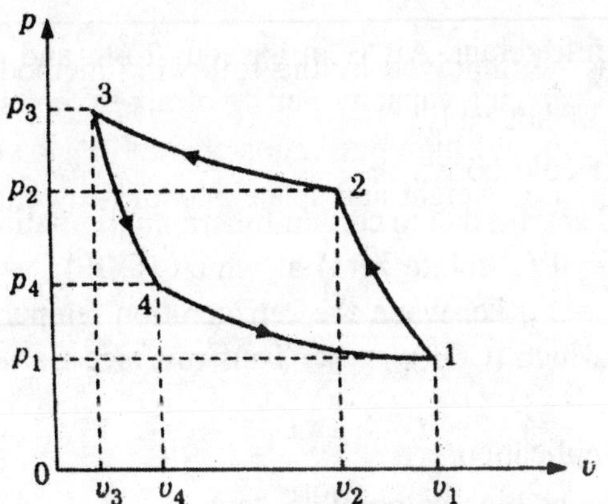

(a) p-v diagram

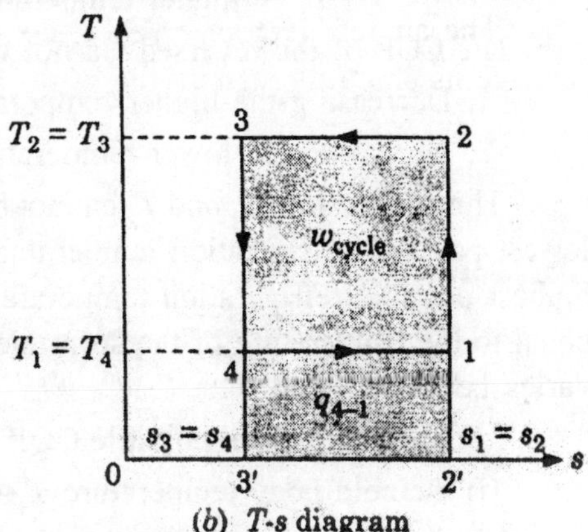

(b) T-s diagram

Fig. 3.1 Reversed Carnot Cycle.

3. **Process 3-4: Isentropic expansion:** During this process, the pressure of air decreases from p_3 to p_4, specific volume increases from v_3 to v_4 and temperature decreases from T_3 to T_4. No heat is absorbed or rejected by air.
4. **Process 4-1: Isothermal expansion:** During this process the pressure of air decreases from p_4 to p_1 and specific volume increases from v_4 to v_3. Temperature remains constant at T_4 to T_1.

   Heat absorbed by air (or heat extracted from cold body) per kg of air,

   $$q_{4\text{-}1} = \text{are } 4\text{-}1\text{-}2'\text{-}3'$$

   $$= T_4(s_1 - s_4) = T_4(s_2 - s_3) = T_1(s_2 - s_3)$$

   Work done during the cycle per kg of air, w cycle $= q_{2-3} - q_{4-1}$ = area 1–2–3–4

   $$= T_2(s_2 - s_3) - T_1(s_2 - s_3)$$

   $$= (T_2 - T_1)(s_2 - s_3)$$

   Coefficient of performance of refrigeration system,

   $$(\text{COP})_{\text{ref}} = \frac{\text{Heat absorbed}}{\text{Work done}} = \frac{q_{4-1}}{w_{\text{cycle}}}$$

$$= \frac{T_1(s_2 - s_3)}{(T_2 - T_1)(s_2 - s_3)}$$

$$= \frac{T_1}{T_2 - T_1} \qquad ...(3.1)$$

3.3.1 Temperature Limitations for Reversed Carnot Cycle

For a reversed Carnot cycle,

$$(\text{COP})_{\text{ref}} = \frac{T_1}{T_2 - T_1}$$

Where T_1 = lower temperature, and
T_2 = higher temperature.

The COP of the reversed Carnot cycle may be improved by the following methods:

1. Decreasing the higher temperature T_2 of hot body.
2. Increasing the lower temperature T_1 of cold body.

The temperature T_1 and T_2 cannot be varied at will, due to certain functional limitations. The lowest possible refrigeration temperature is $T_1 = 0$ (absolute zero) at which $(\text{COP})_{\text{ref}} = 0$. The highest possible refrigeration temperature is $T_1 = T_2$, i.e. when the refrigeration temperature is equal to the temperature of the surroundings at which $(\text{COP})_{\text{ref}} = \infty$. Thus reversed Carnot COP varies between 0 and ∞.

To obtain maximum possible COP in any application,

(*i*) the cold body temperature T_1 should be as high as possible, and
(*ii*) the hot body temperature T_2 should be as low as possible.

The lower the refrigeration temperature required and higher the temperature of heat rejection to the surroundings, the larger is the power consumption of the refrigerating machine. Also, the lower is the refrigeration temperature required, the lower is the refrigerating capacity obtained.

The temperature T_2 in winter is less than T_2 in summer. Therefore, $(\text{COP})_{\text{ref}}$ in winter will be higher than $(\text{COP})_{\text{ref}}$ in summer. In other words, the Carnot refrigerator works more efficiently in winter than in summer. Similarly, if the lower temperature fixed by the refrigeration application is high, the $(\text{COP})_{\text{ref}}$ will be high. Thus a Carnot refrigerator used for making ice at 0°C(273 K) will have less $(\text{COP})_{\text{ref}}$ than a Carnot refrigerator used for air-conditioning plant in summer at 20°C when the ambient air temperature is 40°C.

3.3.2 Vapour as a Refrigerant in Reversed Carnot Cycle

The reversed Carnot cycle can be made almost practical by operating in the liquid-vapour region of a pure substance as shown in Fig. 3.2 on *T-s* diagram.

1. **Process 1-2: Isentropic compression:** The vapours during compression are wet and becomes dry saturated at the end of the process. Such a process is called *wet compression.* The temperature rises from T_1 to T_2.

   Compression work per kg of refrigerant, $w_{1-2} = h_2 - h_1$.

2. Process 2.3: Condensation: During this process the temperature remains constant at T_2. Heat is rejected from h_2 to h_3 and refrigerant gets converted to liquid at the end of process.

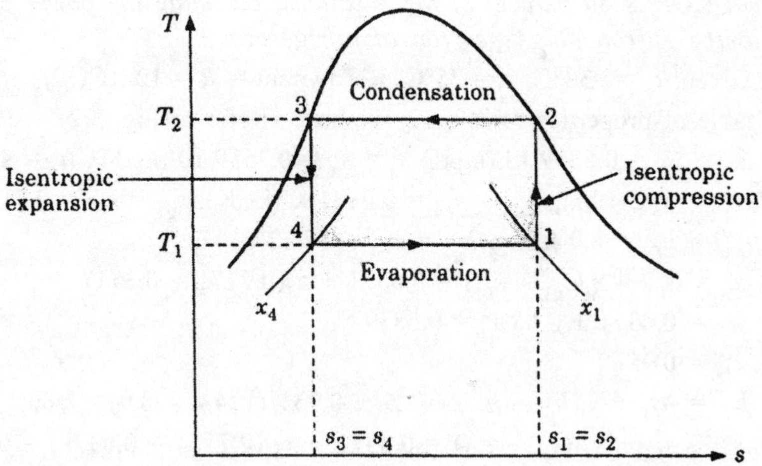

Fig. 3.2 Reversed Carnot cycle with vapour as a refrigerant.

3. Process 3-4: Isentropic expansion: During this process the flashing of the liquid refrigerant takes place with consequent temperature drop from T_2 to T_1. The refrigerant becomes wet at the end of expansion.

Work of expander, $w_{3\text{-}4} = h_3 - h_4$

4. Process 4.1: Evaporation: During this process the wet refrigerant evaporates, heat is absorbed from the medium, and refrigeration effect is produced.

Heat absorbed, $q_{4\text{-}1} = h_1 - h_4$

Net work,
$$w_{net} = w_{1-2} - w_{3-4} = (h_2 - h_1) - (h_3 - h_4)$$
$$= (h_2 - h_3) - (h_1 - h_4) = q_{2\text{-}3} - q_{4\text{-}1}$$

$$(COP)_{cooling} = \frac{\text{Refrigerating effect}}{\text{Net work}}$$

$$= \frac{h_1 - h_4}{(h_2 - h_3) - (h_1 - h_4)}$$

$$= \frac{1}{\left(\frac{h_2 - h_3}{h_1 - h_4}\right) - 1} \qquad \text{...(3.2)}$$

Example 3.1 *A Carnot refrigerator has working temperature of –30°C and 35°C. It operates with R – 12 refrigerant as a working substance. Calculate the work of isentropic compression, isentropic expansion, refrigeration effect, and COP of the cycle per kg of refrigerant.*

If the actual COP is 80 percent of the maximum, calculate the power consumption and heat rejected to the surroundings per ton of refrigeration.

Solution: Given: t_1 = –30°C, t_2 = 35°C, Refrigerant = R – 12, $(COP)_{actual} = 0.8(COP)_{max}$.

From the table of properties of R – 12, we have (Refer to Fig. 3.2)

$$s_1 = s_2 = 0.6839 \text{ kJ/(kg.K) } s_3 = s_4 = 0.2559 \text{ kJ/(kg.K)}, h_2 = 201.5 \text{ kJ/kg},$$

$$h_3 = 69.5 \text{ kJ/kg}, s_{f1} = s_{f4} = 0.0371 \text{ kJ(kg.K)}, s_{g1} = s_{g4} = 0.7171 \text{ kJ(kg/K)},$$

$$h_{f1} = h_{f4} = 8.9 \text{ kJ/kg}, h_{g1} = h_{g4} = 174.2 \text{ kJ/kg}$$

Now $s_1 = s_{f1} + x_1(s_{g1} - s_{f1}) = 0.0371 + x_1(0.7171 - 0.371)$

$$= 0.0371 + 0.68\, x_1 = 0.6839$$

$$x_1 = 0.951$$

$$h_1 = h_{f1} + x_1(h_{g1} - h_{f1}) = 8.9 + 0.951\,(174.2 - 8.9) = 166.1 \text{ kJ/kg}$$

$$s_4 = s_{f4} + x_4\,(s_{g4} - s_{f4}) = 0.0371 + x_4\,(0.7171 - 0.0371) = 0.2559$$

$$x_4 = 0.3218$$

$$h_4 = h_{f4} + x_4\,(h_{g4} - h_{f4}) = 8.9 + 0.3218\,(174.2 - 8.9) = 62.1 \text{ kJ/kg}$$

Work of compression, $w_{1-2} = h_2 - h_1 = 201.5 - 166.1 = 35.4$ kJ/kg

Work of expression, $w_{3-4} = h_3 - h_4 = 69.5 - 62.1 = 7.4$ kg

Refrigerating effect, $q_{4-1} = h_1 - h_4 = 166.1 - 62.1 = 104$ kJ/kg

Heat rejected, $q_{2-3} = h_2 - h_3 = 201.5 - 69.5 = 132$ kJ/kg

Net work, $w_{net} = w_{1-2} - w_{3-4} = 35.4 - 7.4 = 28$ kJ/kg

$$(COP)_{cooling} = \frac{q_{4-1}}{w_{net}} = \frac{104}{28} = 3.714$$

or

$$= \frac{T_1}{T_2 - T_1} = \frac{273-30}{(273+35)-(273-30)} = \frac{243}{65} = 3.738$$

$$(COP)_{act} = 0.8 \times 3.738 = 2.99$$

Power consumption per ton of refrigeration $= \dfrac{3.5167}{2.99} = 1.1761$ kW

Heat rejected per ton of refrigeration to the surroundings = 3.5167 + 1.1761 = 4.6928 kW

3.3.3 Gas as Refrigerant in Reversed Carnot Cycle

Figure 3.3 shows the *p-v* and *T-s* diagrams for the reversed Carnot cycle with a gas as a refrigerant.

1. **Process 1-2: Isentropic Compression:** During this process, the pressure increases from p_1 to p_2 and volume decreases from v_1 to v_2. Heat transfer is zero so that $q_{1-2} = 0$. Temperature of gas increases from T_1 to T_2.

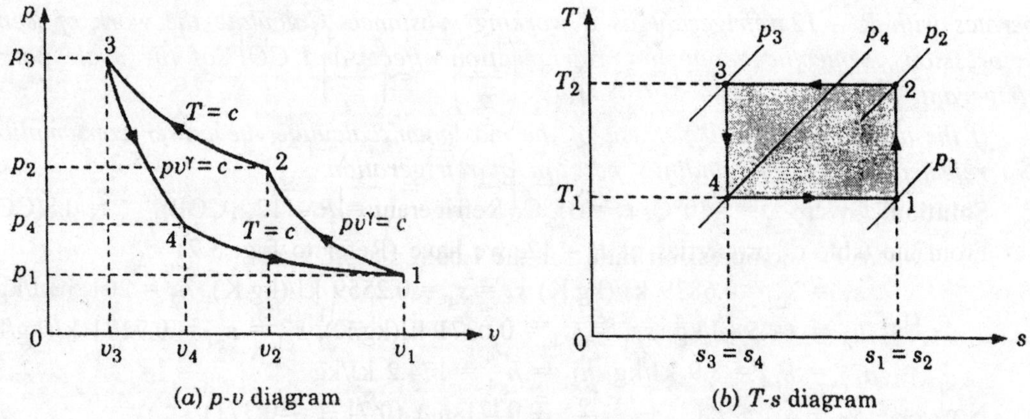

(a) p-v diagram (b) T-s diagram

Fig. 3.3 Reversed Carnot cycle with gas as a refrigerant.

Work done per kg of gas,

$$w_{1-2} = \frac{p_2v_2 - p_1v_1}{\gamma - 1} = \frac{R(T_2 - T_1)}{\gamma - 1}$$

2. **Process 2-3: Isothermal Compression:** During this process, the pressure increases from p_2 to p_3 and volume decreases from v_2 to v_3, The temperature remains constant at T_2.

Work done per kg of gas, $w_{2-3} = p_2v_2 \ln \frac{v_2}{v_3} = RT_2 \ln \frac{v_2}{v_3}$

Heat rejected, $q_{2-3} = w_{2-3}$ for a perfect gas

3. **Process 3-4: Isentropic expansion:** Pressure decreases from p_3 to p_4, volume increases from v_3 to v_4, and temperature decreases from $T_2 = T_3$ to $T_4 = T_1$.

Work done, $$w_{3-4} = \frac{p_3v_3 - p_4v_4}{\gamma - 1} = \frac{R(T_2 - T_1)}{\gamma - 1}$$

4. **Process 4.1: Isothermal expansion:** Pressure decreases from p_4 to p_1, volume increases from v_4 to v_1, and temperature remains constant at T_1.

Work done, $$w_{4-1} = p_4v_4 \ln \frac{v_1}{v_4} = RT_1 \ln \frac{v_1}{v_4}$$

Refrigerating effect, $q_{4-1} = w_{4-1}$ for a perfect gas.

Net work of the cycle, $$w_{net} = w_{2-3} - w_{4-1}$$

$$= RT_2 \ln \frac{v_2}{v_3} - RT_1 \ln \frac{v_1}{v_4}$$

Refrigerating effect, $$q_{4-1} = RT_1 \ln \frac{v_1}{v_4}$$

For the isentropic processes 1-2 and 3-4, we have

$$\frac{T_2}{T_1} = \left(\frac{p_2}{p_1}\right)^{\frac{\gamma-1}{\gamma}} = \left(\frac{p_3}{p_4}\right)^{\frac{\gamma-1}{\gamma}}$$

and
$$\frac{T_2}{T_1} = \left(\frac{v_1}{v_2}\right)^{\gamma-1} = \left(\frac{v_3}{v_4}\right)^{\gamma-1} = r^{\gamma-1}$$

where r = compression ratio for the isentropic processes.

Here
$$\frac{p_2}{p_1} = \frac{p_3}{p_4} \text{ and } r = \frac{v_1}{v_2} = \frac{u_4}{v_3}$$

or
$$\frac{v_2}{v_3} = \frac{v_1}{v_4} \text{ and } \frac{p_3}{p_2} = \frac{p_4}{p_1}$$

Since
$$p_2v_2 = p_3v_3 \text{ and } p_4v_4 = p_1v_1$$

$$(\text{COP})_{\text{cooling}} = \frac{q_{4-1}}{w_{\text{net}}} = \frac{RT_1 \ln\frac{v_1}{v_4}}{RT_2 \ln\frac{v_2}{v_3} - RT_1 \ln\frac{v_1}{v_4}}$$

$$= \frac{T_1}{T_2 - T_1} = \frac{1}{\frac{T_2}{T_1} - 1} = \frac{1}{r^{\gamma-1} - 1} \qquad \text{...(3.3)}$$

Thus, COP is a function of compression ration only.

3.3.4 Limitations of Reversed Carnot Cycle

The limitations of reversed Carnot cycle are:

1. The isentropic compression and expansion processes require high speed while the isothermal condensation and evaporation processes require an extremely low speed. This variation in speed of air of a cycle is not practicable.
2. It is difficult to design an expander to handle a mixture of largely liquid and partly vapour.
3. Because of the internal irreversibilities in the compressor and the expander, the actual COP of the reversed Carnot cycle is very low.
4. With gas as refrigerant, it is not possible to devise, in practice, isothermal processes of heat absorption and rejection.
5. The stroke volume of gas cycle cylinder is very large resulting in poor actual COP.

3.4 BELL-COLEMAN CYCLE (OR REVERSED BRAYTON OR JOULE CYCLE)

The Bell-Coleman air refrigeration cycle was developed by Bell-Coleman and Light Foot by reversing the Joule's or Brayton's air cycle. The schematic diagram of such a cycle is shown in Fig. 3.4. it consists of a compressor, a cooler, an expander, and a refrigerator. Then *p-v* and *T-s* diagram are shown in Fig. 3.5.

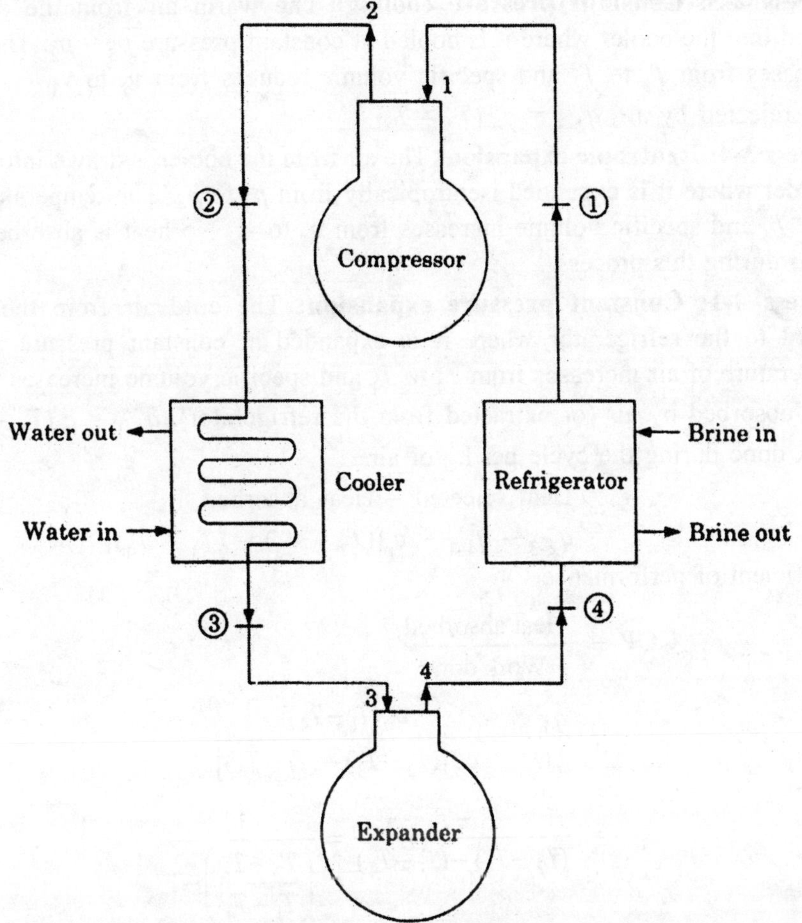

Fig. 3.4 Schematic diagram of Bell-Coleman air refrigeration cycle.

1. **Process 1-2: Isentropic compression:** The cold air from the refrigerator is drawn into the compressor cylinder where it is compressed isentropically The pressure and temperature of air increases and the specific volume decreases from v_1 to v_2. No heat is absorbed or rejected by the air.

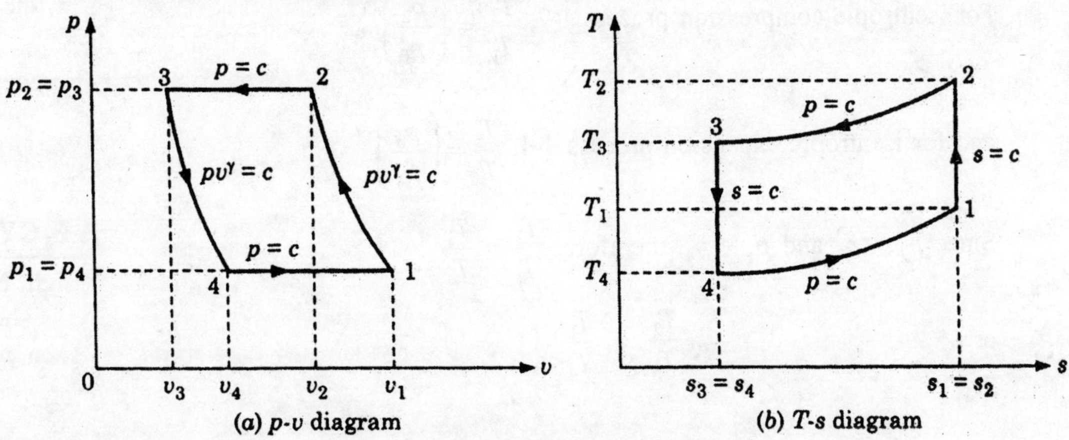

Fig. 3.5 Bell-Coleman cycle.

2. **Process 2-3: Constant pressure cooling:** The warm air from the compressor is passed into the cooler where it is cooled at constant pressure $p_3 = p_2$. The temperature decreases from T_2 to T_3 and specific volume reduces from v_2 to v_3.
   Heat rejected by air, $q_{2-3} = c_p\,(T_2 - T_3)$
3. **Process 3-4: Isentropic expansion:** The air from the cooler is drawn into the expander cylinder where it is expanded isentropically from p_3 to p_4. The temperature falls from T_3 to T_4 and specific volume increases from v_3 to v_4. No heat is absorbed or rejected by air during this process.
4. **Process 4-1: Constant pressure expansion:** The cold air from the expander is passed to the refrigerator where it is expanded at constant pressure $p_4 = p_1$. The temperature of air increases from T_4 to T_1 and specific volume increases from v_4 to v_1.
   Heat absorbed by air (or extracted from the refrigerator), $q_{4-1} = c_p(T_1 - T_4)$
   Work done during the cycle per kg of air,

$$w = \text{Heat rejected} - \text{Heat absorbed}$$

$$= q_{2-3} - q_{4-1} = c_p[(T_2 - T_3) - (T_1 - T_4)]$$

Coefficient of performance,

$$\text{COP} = \frac{\text{Heat absorbed}}{\text{Work done}}$$

$$= \frac{q_{4-1}}{w} = \frac{c_p(T_1 - T_4)}{c_p[(T_2 - T_3) - (T_1 - T_4)]}$$

$$= \frac{T_1 - T_4}{(T_2 - T_3) - (T_1 - T_4)} = \frac{1}{\left(\dfrac{T_2 - T_3}{T_1 - T_4}\right) - 1}$$

$$= \frac{1}{\dfrac{T_3}{T_4}\left[\dfrac{(T_2/T_3) - 1}{(T_1/T_4) - 1}\right] - 1} = \frac{1}{\dfrac{T_3}{T_4} - 1} \quad \text{...(3.4}a\text{)}$$

For isentropic compression process 1-2, $\dfrac{T_2}{T_1} = \left(\dfrac{p_2}{p_1}\right)^{\frac{\gamma-1}{\gamma}}$

and for isentropic expansion process 3-4, $\dfrac{T_3}{T_4} = \left(\dfrac{p_3}{p_4}\right)^{\frac{\gamma-1}{\gamma}}$

Since $p_2 = p_3$ and $p_1 = p_4$, therefore, $\dfrac{T_2}{T_1} = \dfrac{T_3}{T_4}$

or

$$\frac{T_2}{T_3} = \frac{T_1}{T_4}$$

also $$\frac{T_3}{T_4} = \left(\frac{p_3}{p_4}\right)^{\frac{\gamma-1}{\gamma}} = \left(\frac{p_2}{p_1}\right)^{\frac{\gamma-1}{\gamma}} = (r_p)^{\frac{\gamma-1}{\gamma}}$$

where $r_p = \frac{p_2}{p_1}$ = pressure ratio

$$\therefore \quad \text{COP} = \frac{1}{\frac{T_3}{T_4}-1} = \frac{1}{(r_p)^{\frac{\gamma-1}{\gamma}}-1} \quad \text{...(3.4b)}$$

3.4.1 Bell-Coleman Cycle with Polytropic Processes

Let the compression and expansion processes take place according to the polytropic law, pv^n = const.

1. **Process 1-2:** Polytropic compression

Work done per kg of air, $w_{1-2} = \left(\frac{n}{n-1}\right)(p_2v_2 - p_1v_1) = \left(\frac{n}{n-1}\right)R(T_2 - T_1)$

2. **Process 3-4:** Polytropic expansion

$$w_{3-4} = \left(\frac{n}{n-1}\right)(p_3v_3 - p_4v_4) = \left(\frac{n}{n-1}\right)R(T_3 - T_4)$$

Net work done during the cycle per kg of air,

$$w_{\text{net}} = \left(\frac{n}{n-1}\right)R[(T_2 - T_1) - (T_3 - T_4)]$$

3. Heat absorbed during constant pressure process 4-1,

$$q_{4-1} = c_p(T_1 - T_4)$$

$$\text{COP} = \frac{\text{Heat absorbed}}{\text{Work done}} = \frac{c_p(T_1 - T_4)}{\left(\frac{n}{n-1}\right)R[(T_2 - T_1) - (T_3 - T_4)]}$$

Now $$R = c_p - c_v = c_p\left(1 - \frac{c_v}{c_p}\right) = c_p\left(1 - \frac{1}{\gamma}\right) = \frac{(\gamma-1)c_p}{\gamma}$$

$$\therefore \quad \text{COP} = \frac{c_p(T_1 - T_4)}{\left(\frac{n}{n-1}\right)\left(\frac{\gamma-1}{\gamma}\right)c_p[(T_2 - T_1) - (T_3 - T_4)]}$$

$$= \frac{T_1 - T_4}{\left(\frac{n}{n-1}\right)\left(\frac{\gamma-1}{\gamma}\right)[(T_2 - T_3) - (T_1 - T_4)]}$$

$$= \frac{1}{\left(\frac{n}{n-1}\right)\left(\frac{\gamma-1}{\gamma}\right)[(r_p)^{\frac{n-1}{n}} - 1]} \quad ...(3.5)$$

(*i*) For $n = r$, Eq. (3.5) reduces to eq. (3.4b)

(*ii*) For $n < r$, $(COP)_{poly} > (COP)_{isen}$

(*iii*) for $n > r$, $(COP)_{poly} < (COP)_{isen}$

Therefore, for obtaining higher COP, the compression and expansion index should be kept less than 1.4.

Example 3.2 *A closed cycle refrigeration system working between 5 bar and 20 bar extracts 144 MJ of heat per hour. The air enters the compressor at 6°C and the expander at 22°C. The unit operates at 320 rpm. Calculate by assuming isentropic compression and expansion:*

(a) Power required to operate the unit.

(b) Bore of compressor, and

(c) Refrigerating capacity in tonnes of ice at 0°C per day.

The following data is available:

Stroke for double acting compressor and expander = 300 mm

Mechanical efficiency of compressor = 80%

Mechanical efficiency of expander = 85%

Solution: Given: $p_1 = p_4 = 5$ bar, $p_2 = p_3 = 20$ bar, $Q_{4-1} = 144$ MJ/h or 2400 kJ/min.

$T_1 = 273 + 6 = 279$ K, $T_3 = 273 + 22 = 295$ K, $N = 320$ rpm,

$L = 300$ mm, $\eta_c = 0.80$, $\eta_e = 0.85$, $\gamma = 1.4$

For *p-v* and *T-s* diagrams, refer to Fig. 3.5.

(*a*)
$$r_p = \frac{p_2}{p_1} = \frac{20}{5} = 4 = \frac{p_3}{p_4}$$

$$\frac{T_2}{T_1} = (r_p)^{\frac{\gamma-1}{\gamma}} = (4)^{\frac{0.4}{1.4}} = 1.486$$

$$T_2 = 279 \times 1.486 = 414.6 \text{ K}$$

$$\frac{T_3}{T_4} = (r_p)^{\frac{\gamma-1}{\gamma}} = 1.486$$

$$T_4 = \frac{295}{1.486} = 198.5 \text{ K}$$

Heat extracted from the refrigeration system per kg of air,

$$q_{4-1} = c_p(T_1 - T_4) = 1.005\,(279 - 198.5) = 80.9 \text{ kJ/kg}$$

Mass of air circulated,
$$\dot{m}_a = \frac{Q_{4-1}}{q_{4-1}} = \frac{2400}{80.9} = 29.67 \text{ kg/min}$$

Work done during isentropic compression process 1-2,

$$w_{1-2} = \left(\frac{\gamma}{\gamma-1}\right)R(T_2 - T_1) \times \frac{1}{\eta_c}$$

$$= \frac{1.4}{0.4} \times 0.287(414.6 - 279) \times \frac{1}{0.80} = 170.26 \text{ kJ/kg}$$

Work done during isentropic expansion process 3-4,

$$q_{3-4} = \left(\frac{\gamma}{\gamma-1}\right)R(T_3 - T_4)\eta_e$$

$$= \frac{1.4}{0.4} \times 0.287\ (295 - 198.5) \times 0.85 = 82.39 \text{ kJ/kg}$$

Net work done per kg of air supplied to the system,

$$q_{net} = w_{1-2} - w_{3-4} = 170.26 - 82.39 = 87.87 \text{ kJ/kg}$$

Power required to operate the system,

$$P = \frac{\dot{m}_a \times w_{net}}{60} = \frac{29.67 \times 87.87}{60} = 43.45 \text{ kW}$$

(*b*)

$$p_1\dot{V}_1 = \dot{m}_a R T_1$$

$$\dot{V}_1 = \frac{29.67 \times 0.287 \times 279}{5 \times 10^2} = 4.75 \text{ m}^3/\text{min}$$

Also

$$\dot{V}_1 = \frac{\pi}{4} D^2 L \times 2 \times N \quad \text{(for double acting compressor)}$$

$$4.75 = \frac{\pi}{4} D^2 \times 0.3 \times 2 \times 320$$

$$D = 0.1775 \text{ m} \quad \text{or } 177.5 \text{ mm}$$

(*c*) Refrigerating capacity of the system per day

$$= q_{4-1} \times \dot{m}_a \times 60 \times 24$$

$$= 80.9 \times 29.67 \times 60 \times 24 = 3456.436 \text{ MJ}$$

Assuming latent heat of ice = 336 kJ/kg

$$\text{Ice formation capacity of system} = \frac{3456.436 \times 10^3}{336} = 10287 \text{ kg or } 10.287 \text{ tonnes}$$

Example 3.3 *A closed cycle reversed Brayton Refrigerator is required for a capacity of 12 TR. The cooler pressure is 4.5 bar and the refrigerator pressure is 1.5 bar. The air is cooled in the cooler at a temperature of 45°C and the temperature of air at inlet to compressor is –20°C. Calculate for the ideal cycle: (a) COP, (b) mass of air circulated per minute, (c) theoretical displacement of compressor piston, (d) theoretical displacement of expander piston, and (e) net power per tonne of refrigeration. The compression and expansion are isentropic.*

Solution: Given: $Q = 12TR$, $p_1 = p_4 = 1.5$ bar, $p_2 = p_3 = 4.5$ bar, $T_1 = 273 - 20 = 253$ K, $T_3 = 273 + 45 = 318$ K.

The p-v and T-s diagrams are shown in Fig. 3.6.

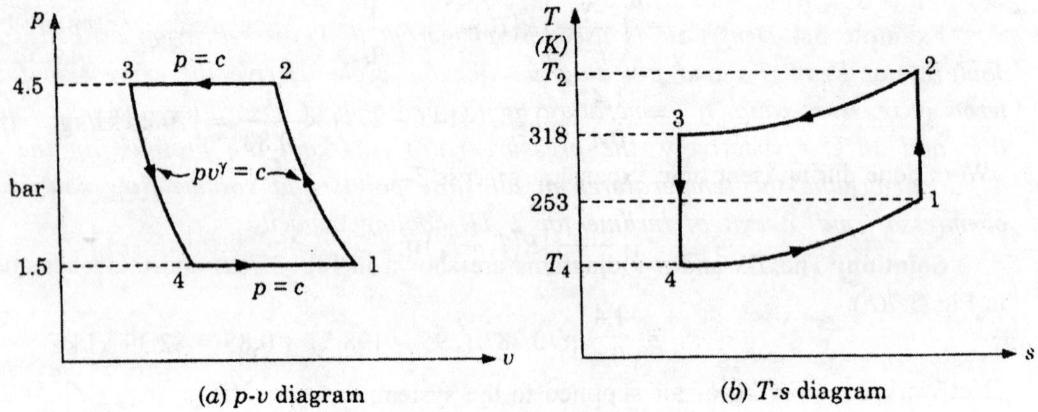

(a) p-v diagram (b) T-s diagram

Fig. 3.6 Reversed Brayton cycle.

(a)
$$\frac{T_2}{T_1} = \left(\frac{p_2}{p_1}\right)^{\frac{\gamma-1}{\gamma}} = \left(\frac{4.5}{1.5}\right)^{\frac{1.4-1}{1.4}} = 1.369$$

$$T_2 = 253 \times 1.369 = 346.3 \text{ K}$$

$$\frac{T_3}{T_4} = \left(\frac{p_3}{p_4}\right)^{\frac{\gamma-1}{\gamma}} = 1.369$$

$$T_4 = \frac{318}{1.369} = 232.3\text{K}$$

$$\text{COP} = \frac{T_1 - T_4}{(T_2 - T_3) - (T_1 - T_4)} = \frac{253 - 232.3}{(346.3 - 318) - (253 - 232.3)} = \frac{20.7}{7.6} = 2.724$$

(b) Heat extracted per minute, $\dot{Q} = 211 \times 12 = 2532$ kJ/min

Heat extracted from the refrigerator per kg of air,

$$q = c_p(T_1 - T_4) = 1.005\ (253 - 232.3) = 20.8 \text{ kJ/kg}$$

Mass of air circulated, $\dot{m}_a = \dfrac{\dot{Q}}{q} = \dfrac{2532}{20.8} = 121.73$ kg/min

(c) $\dot{V}_1 = \dfrac{\dot{m}_a R T_1}{p_1} = \dfrac{121.73 \times 0.287 \times 253}{1.5 \times 10^2} = 58.926 \text{ m}^3\text{/min}$

(d) $\dot{V}_4 = \dfrac{\dot{m}_a R T_4}{p_4} = \dfrac{121.73 \times 0.287 \times 232.3}{1.5 \times 10^2} = 54.1 \text{ m}^3\text{/min}$

(e) Work done per minute = $\dot{m}_a$ (Heat rejected – Heat extracted)

$$= \dot{m}_a c_p\ [(T_2 - T_3) - (T_1 - T_4)]$$
$$= 121.73 \times 1.005\ [(346.3 - 318) - (253 - 232.3)]$$
$$= 929.77 \text{ kJ/min}$$

$$\text{Net power per tonne of refrigeration} = \frac{929.77}{60 \times 12} = 1.29 \text{ kW/TR}$$

Example 3.4 *Dense air is used as refrigerant in Reverse-Brayton or Bell-Coleman or Joule cycle. Draw T-s and p-v diagrams for the cycle. Derive the expression for COP in terms of pressure ratio. If temperatures at the end of heat absorption and heat rejection are 0°C and 30°C respectively, the pressure ratio is 4 and the pressure in the cooler is 4 bar, determine the temperatures at all state points and volume flow rates at inlet to compressor and at exit of turbine for 1 TR cooling capacity.*

Solution: The *T-s* and *p-v* diagrams are shown in Fig. 3.7(*a*) and the schematic diagram in Fig. 3.7(*b*).

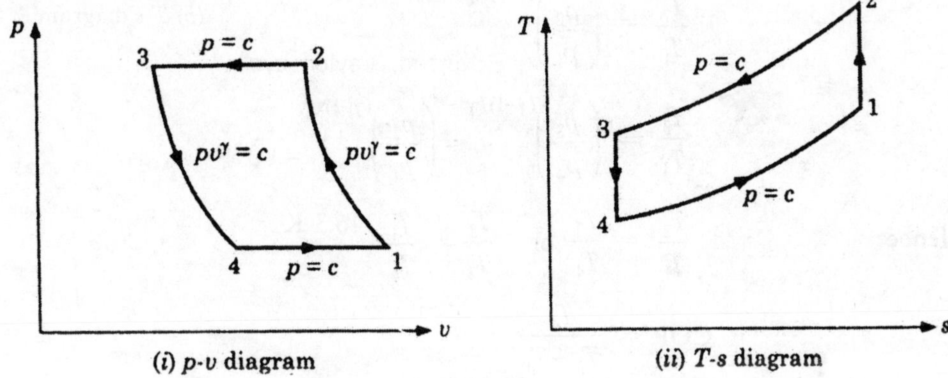

Fig. 3.7 p-v and T-s diagrams.

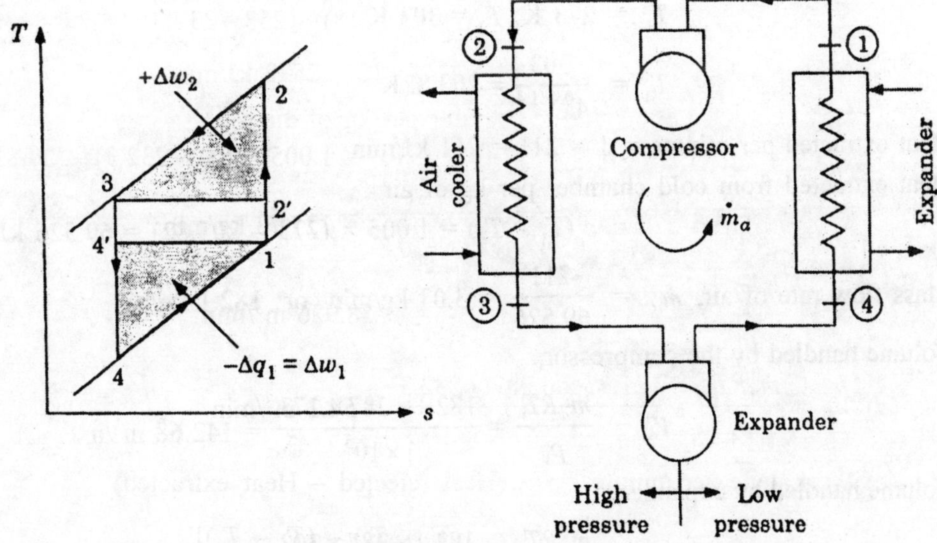

Fig. 3.7 (b) Schematic diagram of Bell-Coleman cycle.

$$q_r = q_{2-3} = c_p(T_2 - T_3)$$
$$q_a = q_{4-1} = c_{4-1}(T_1 - T_4)$$
$$\text{Work done} = q_r - q_a = c_p(T_2 - T_3) - c_p(T_1 - T_4)$$
$$\text{COP} = \frac{q_a}{\text{work done}} = \frac{c_p(T_1 - T_4)}{c_p(T_2 - T_3) - c_p(T_1 - T_4)}$$
$$= \frac{T_4\left(\frac{T_1}{T_4} - 1\right)}{T_3\left(\frac{T_2}{T_3} - 1\right) - T_4\left(\frac{T_1}{T_4} - 1\right)}$$
$$\frac{T_2}{T_1} = \left(\frac{p_2}{p_1}\right)^{(\gamma-1)/\gamma}$$
$$\frac{T_3}{T_4} = \left(\frac{p_3}{p_4}\right)^{(\gamma-1)/\gamma} = \left(\frac{p_2}{p_1}\right)^{(\gamma-1)/\gamma}$$

Hence
$$\frac{T_2}{T_1} = \frac{T_3}{T_4} \quad \text{or} \quad \frac{T_2}{T_3} = \frac{T_1}{T_4}$$
$$\text{COP} = \frac{T_4}{T_3 - T_4}$$

Let,
$$r_p = \frac{p_2}{p_1} = \frac{p_3}{p_4} = 4$$
$$T_1 = 273 \text{ K}, T_3 = 303 \text{ K}$$
$$T_4 = \frac{303}{4^{0.4/1.4}} = 203.82 \text{ K}$$

Heat extracted per minute = 1 × 211 = 211 kJ/min

Heat extracted from cold chamber per kg of air

$$= c_p(T_1 - T_4) = 1.005 \times (273 - 203.82) = 69.526 \text{ kJ/kg}$$

Mass flow rate of air, $\dot{m}_a = \dfrac{211}{69.526} = 3.03$ kg/min or 182.1 kg/h

Volume handled by the compressor,

$$V_c = \frac{\dot{m}_a R T_1}{p_1} = \frac{182.1 \times 287 \times 273}{1 \times 10^5} = 142.68 \text{ m}^3/\text{h}$$

Volume handled by expander,

$$V_e = \frac{\dot{m}_a R T_3}{p_3} = \frac{182.1 \times 287 \times 303}{4 \times 10^5} = 39.59 \text{ m}^3/\text{h}$$

Example 3.5 *An air refrigeration used for food storage provides 30 TR. The temperature of air entering the compressor is 10°C and temperature at exit of cooler is 30°C. Calculate:*

(a) COP of the cycle, and

(b) Power required per tonne of refrigeration by the compressor.

The quantity of air circulated in the system is 3600 kg/h. The compression and expansion process follow the law $pv^{1.3}$ = const. and $\gamma = 1.4$, $c_p = 1.005$ kJ/kg.K for air.

Solution: Given: $Q = 30$ TR, $T_1 = 273 + 10 = 283$ K, $T_3 = 273 + 30 = 303$ K,

$$\dot{m}_a = 3600 \text{ kg/h}, \ n = 1.3$$

(*a*) Heat extracted from the refrigerator,

$$Q = 30 \times 211 = 6330 \text{ kJ/min}$$

$$= \dot{m}_a c_p (T_1 - T_4) = \frac{3600}{60} \times 1.005(283 - T_4)$$

$$T_4 = 178 \text{ K}$$

Now,
$$\frac{T_3}{T_4} = \frac{T_2}{T_1}, \ T_2 = \frac{283 \times 303}{178} = 481.7\text{K}$$

$$\text{COP} = \frac{T_1 - T_4}{\left(\frac{n}{n-1}\right)\left(\frac{\gamma - 1}{\gamma}\right)[(T_2 - T_3) - (T_1 - T_4)]}$$

$$= \frac{283 - 187}{\frac{1.3}{0.3} \times \frac{0.4}{1.4}[(481.7 - 303) - (283 - 178)]} = 1.15$$

(*b*) Heat absorbed, $Q_{4-1} = \dot{m}_a c_p (T_1 - T_4) = \frac{3600}{60} \times 1.005\ (283 - 178) = 6331.5$ kJ/min

Work done per minute, $\dot{W} = \frac{\text{Heat absorbed}}{\text{COP}} = \frac{6331.5}{1.15} = 5505.65$ kJ/min

Power required per ton of refrigeration $= \frac{\dot{W}}{60 \times TR} = \frac{5505.65}{60 \times 30} = 3.06$ kW/TR

Example 3.6 *A dense air refrigeration cycle operates between pressures of 4 bar and 20 bar. The air temperature after heat rejection to surroundings is 40°C and air temperature at exit of refrigeration is 8°C. The isentropic efficiencies of compressor and expander are 0.82 and 0.86 respectively. Determine the compressor and expander work per TR, COP, and power required per TR. Take $\gamma = 1.4$ and $c_p = 1.005$ kJ/kg.K.*

Solution: Given: $p_1 = p_4 = 4$ bar, $p_2 = p_3 = 20$ bar, $T_3 = 273 + 40 = 313$ K,

$T_1 = 273 + 8 = 281$ K,

$\eta_c = 0.82$, $\eta_e = 0.86$, $\gamma = 1.4$, $c_p = 1.005$ kJ/kg.K

The *p-v* and *T-s* diagrams are shown in Fig. 3.8.

Process 1-2: Isentropic compression

$$\frac{T_2}{T_1} = \left(\frac{p_2}{p_1}\right)^{\frac{\gamma-1}{\gamma}} = \left(\frac{20}{4}\right)^{\frac{1.4-1}{1.4}} = 1.584$$

$$T_2 = 281 \times 1.584 = 445 \text{ K}$$

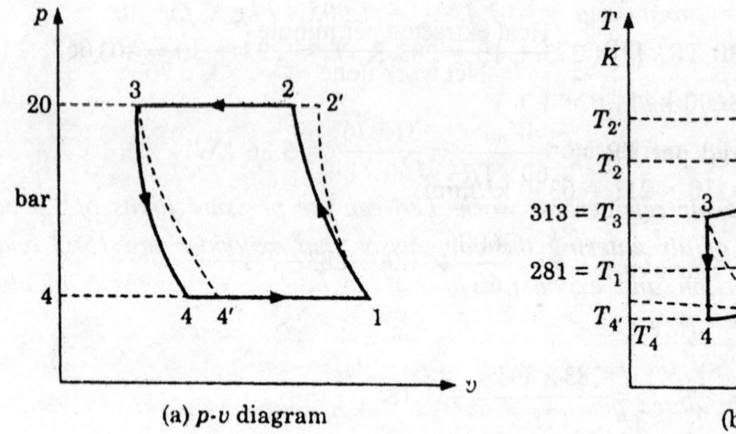

(a) *p-υ* diagram

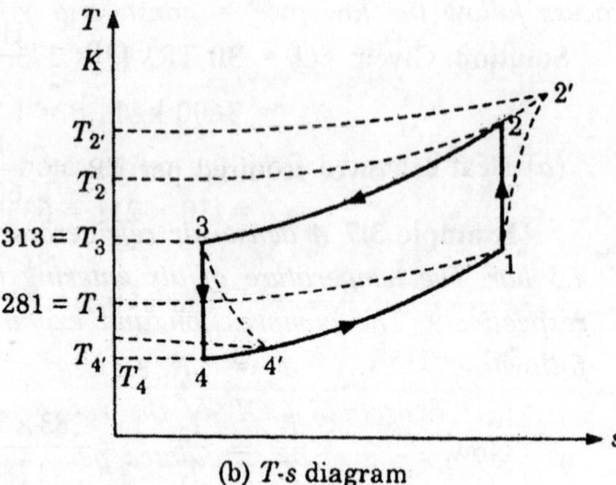

(b) *T-s* diagram

Fig. 3.8 Dense air refrigeration cycle.

Process 3-4: Isentropic expansion

$$\frac{T_3}{T_4} = \left(\frac{p_3}{p_4}\right)^{\frac{\gamma-1}{\gamma}} = \left(\frac{20}{4}\right)^{\frac{0.4}{1.4}} = 1.584$$

$$T_4 = \frac{313}{1.584} = 197.6 \text{ K}$$

Isentropic efficiency of compressor,

$$\eta_c = \frac{T_2 - T_1}{T_{2'} - T_1}$$

$$0.82 = \frac{445 - 281}{T_{2'} - 281}$$

$$T_{2'} = 481 \text{ K}$$

Isentropic efficiency of expander,

$$\eta_e = \frac{T_3 - T_{4'}}{T_3 - T_4}$$

$$0.86 = \frac{313 - T_{4'}}{313 - 197.6}$$

$$T_{4'} = 213.7 \text{ K}$$

Refrigerating effect per kg of air, $q_{4'-1} = c_p(T_1 - T_{4'}) = 1.005(281 - 213.7) = 67.6$ kJ/kg

Mass flow rate of air, $\dot{m}_a = \frac{Q}{q_{4'-1}} = \frac{211}{67.6} = 3.12$ kg/min

Compressor work per TR, $\dot{W}_c = \dot{m}_a c_p (T_{2'} - T_1) = 3.12 \times 7.005\ (481 - 281)$

$= 627.12$ kJ/min

Expander work per TR, $\dot{W}_e = \dot{m}_a c_p (T_3 - T_{4'}) = 3.12 \times 1.005\ (313 - 213.7) = 311.36$ kJ/min

Net work done, $\dot{W}_{net} = \dot{W}_c - \dot{W}_e = 627.12 - 311.36 = 315.76$ kJ/min

$$\text{COP} = \frac{\text{Heat extracted per minute}}{\text{Net work done}} = \frac{211}{315.76} = 0.668$$

$$\text{Power required per TR} = \frac{W_{net}}{60 \times TR} = \frac{315.76}{60 \times 1} = 5.26 \text{ kW}$$

Example 3.7 *A dense air refrigerator works between the pressure limits of 1.5 bar and 4.5 bar. The temperature of air entering the compressor and expander are 15°C and 30°C respectively. The compression and expansion follow the law pv'^3 = const. Calculate the following:*

(a) Theoretical COP of the refrigerating cycle.

(b) Amount of air circulated per minute for a refrigeration load of 12 TR. Actual COP is 60% of the theoretical COP.

(c) The stroke length and piston diameter of single acting compressor if it runs at 300 rpm and its volumetric efficiency is 85%.

Take $\frac{L}{D} = 1.5$, $c_p = 1.005$ kJ/kg.K and $c_v = 0.718$ kJ/kg.K.

Solution: Given: $p_1 = p_4 = 1.5$ bar, $p_2 = p_3 = 4.5$ bar, $T_1 = 273 + 15 = 288$ K,

$T_3 = 273 + 30 = 303$ K,

$n = 1.3$, $Q = 12$ TR, $(\text{COP})_{act} = 0.6\ (\text{COP})_{th}$, $N = 300$ rpm, $\eta_{vol} = 0.85$,

$\frac{L}{D} = 1.5$, $c_p = 1.005$ kJ/kg.K and $c_v = 0.718$ kJ/kg.K.

The *p-v* and *T-s* diagram are shown in Fig. 3.9.

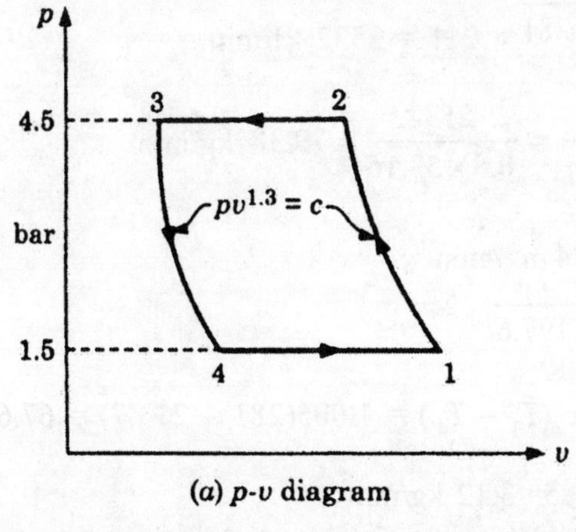

(a) *p-v* diagram

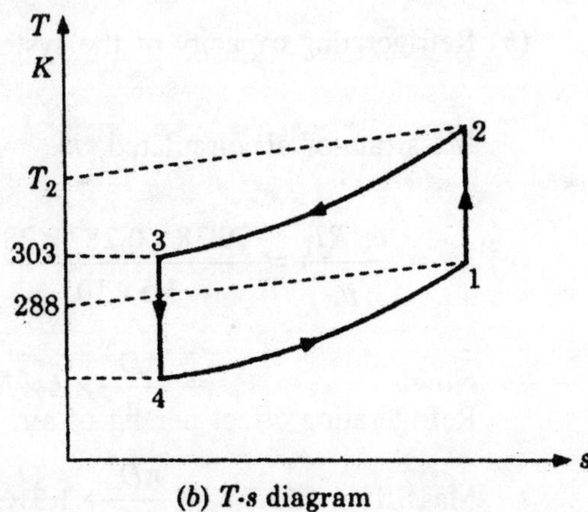

(b) *T-s* diagram

Fig. 3.9

(a) Process 1-2:

$$\frac{T_2}{T_1} = \left(\frac{p_2}{p_1}\right)^{\frac{n-1}{n}} = \left(\frac{4.5}{1.5}\right)^{\frac{1.3-1}{1.3}} = 1.2886$$

$$T_2 = 288 \times 1.2886 = 371.1 \text{ K}$$

Process 3-4:

$$\frac{T_3}{T_4} = \left(\frac{p_3}{p_4}\right)^{\frac{n-1}{n}} = 1.2886$$

$$T_4 = \frac{303}{1.2886} = 235.1 \text{ K}$$

Refrigerating effect per kg of air, $q_{4-1} = c_p(T_1 - T_4) = 1.005\ (288 - 235.1) = 53.16$ kJ/kg

Characteristic gas constant, $R = c_p - c_v = 1.005 - 0.718 = 0.287$ kJ/kg.K

Work of compression per kg of air,

$$w_e = \left(\frac{n}{n-1}\right) R(T_2 - T_1)$$

$$= \left(\frac{1.3}{1.3-1}\right) \times 0.287\ (303 - 235.1) = 84.44 \text{ kJ/kg}$$

Work of expansion per kg of air,

$$w_{net} = w_c - w_e = 103.35 - 84.44 = 18.91 \text{ kJ/kg}$$

$$(\text{COP})_{th} = \frac{\text{Refrigerating effect}}{\text{Net work done}} = \frac{53.16}{18.91} = 2.81$$

$$(\text{COP})_{net} = 0.6 \times 2.81 = 1.687$$

(*b*) Refrigerating capacity of the system, $\dot{Q} = 12 \times 211 = 2532$ kJ/min

Mass rate of air circulated, $\dot{m}_a = \dfrac{\dot{Q}}{0.6 q_{4.-1}} = \dfrac{2532}{0.6 \times 53.16} = 79.38$ kg/min

(c) $\dot{V}_1 = \dfrac{\dot{m}_a R T_1}{p_1} = \dfrac{79.38 \times 0.287 \times 288}{1.5 \times 10^2} = 43.74 \text{ m}^3/\text{min}$

Now $\dot{V}_1 = \dfrac{\pi D^2}{4} \times L \times N \times \eta_{vol}$

$$43.74 = \frac{\pi D^3}{4} \times 1.5 \times 300 \times 0.85$$

$$D = 0.526 \text{ or } 526 \text{ mm}$$

$$L = 1.5 \times 526 = 789 \text{ mm}$$

Example 3.8 *An air refrigeration plant of 20 ton capacity comprises a centrifugal compressor, a cooler heat exchanger and an air turbine. The compressor is coupled directly to the air turbine. The compressor also receives power from another prime mover. The processes in the compressor and turbines are adiabatic but not isentropic. Air at temperature 20°C and 1.0 bar enters the compressor. It leaves the compressor at 90°C. The same air enters the turbine at 40°C and 1.5 bar. The turbine exit is at 10°C. Assuming no pressure drop in the cooler and the refrigerator sections,* $c_p = 1.005$ *kJ/kg.K and* $c_v = 0.718$ *kJ/kg.K, determine:*

(a) the compressor efficiency,

(b) the turbine efficiency,

(c) the rate of air flow

(d) the power input to the plant, and

(e) COP.

Solution: Given: $T_1 = 273 + 20 = 293$ K, $p_1 = p_4 = 1.0$ bar, $T_{2'} = 273 + 90 = 363$ K,

$p_2 = p_{2'} = p_3 = 1.5$ bar,

$T_3 = 273 + 40 = 313$ K, $T_{4'} = 273 + 10 = 283$ K

The flow diagram of plant and T-s diagram are shown in Fig. 3.10.

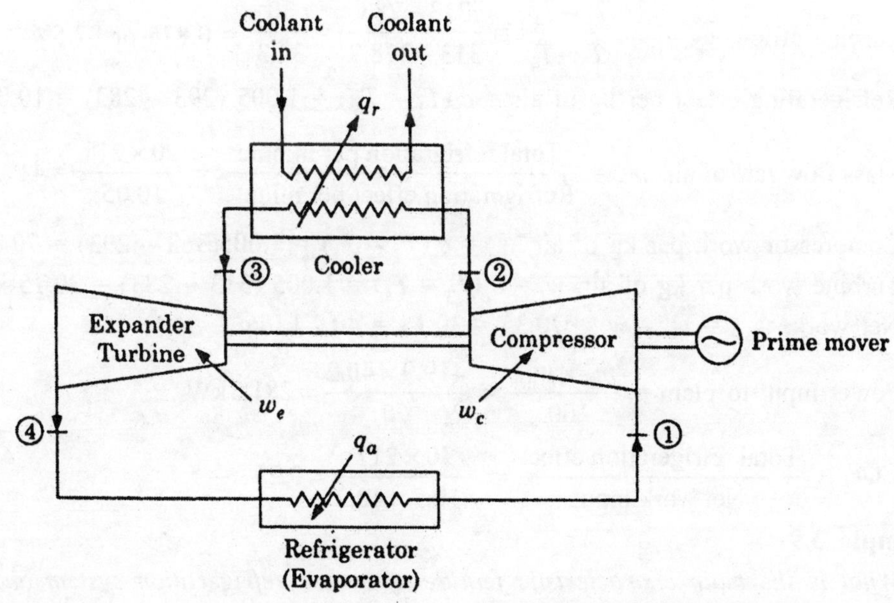

(*a*) Flow diagram of air refrigeration plant.

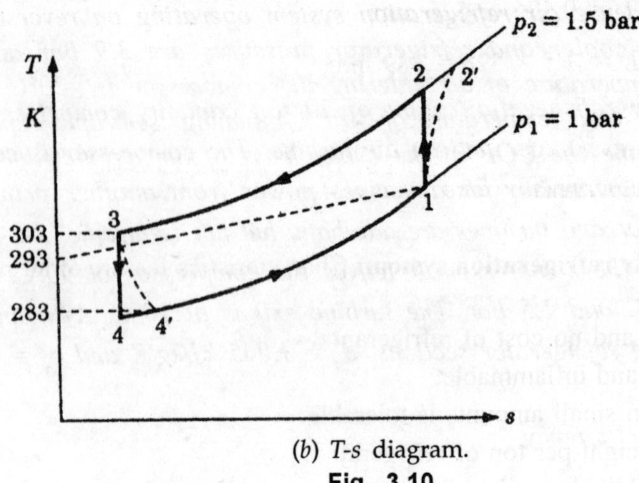

(*b*) *T-s* diagram.

Fig. 3.10

(*a*) $T_2 = T_1\left(\frac{p_2}{p_1}\right)^{\frac{\gamma-1}{\gamma}} = 293\left(\frac{1.5}{1}\right)^{\frac{1.4-1}{1.4}} = 329\text{K}$

$$\text{Compressor efficiency, } \eta_c = \frac{T_2 - T_1}{T_{2'} - T_1} = \frac{329-293}{363-293} = \frac{36}{70} = 0.514 \text{ or } 51.4\%$$

(*b*) $T_4 = T_3\left(\frac{p_4}{p_3}\right)^{\frac{\gamma-1}{\gamma}} = 313\left(\frac{1}{105}\right)^{\frac{0.4}{1.4}} = 278.7\text{K}$

$$\text{Turbine efficiency, } \eta_t = \frac{T_3 - T_{4'}}{T_3 - T_4} = \frac{313-283}{313-278.7} = \frac{30}{34.3} = 0.875 \text{ or } 87.5\%$$

(*c*) Refrigerating effect per kg of air = $c_p(T_1 - T_{4'})$ = 1.005 (293 – 283) = 10.05 kJ/kg

$$\text{Mass flow rate of air, } \dot{m}_a = \frac{\text{Total refrigration per minute}}{\text{Refrigeration effect per minute}} = \frac{20\times 211}{10.05} = 419.9 \text{ kg/min}$$

(*d*) Compressor work per kg of air, $w_c = c_p(T_2 - T_1)$ = 1.005(363 – 293) = 70.35 kJ/kg

Turbine work per kg of air, $w_t = c_p(T_3 - T_4)$ = 1.005 (313 – 283) = 30.15 kJ/kg

Net work, $w_{net} = w_c - w_t$ = 70.35 – 30.15 = 40.2 kJ/kg

$$\text{Power input to plant} = \frac{\dot{m}_a \times w_{net}}{60} = \frac{419.9\times 40.2}{60} = 281.3\text{kW}$$

(*e*) $\text{COP} = \frac{\text{Total refrigeration effect}}{\text{Net work input}} = \frac{20\times 211}{419.9\times 40.2} = 0.25$

Example 3.9

(*a*) *What is the main characteristic feature of an air refrigeration system and what is the main difference between open and closed refrigeration cycles? Derive an expression for COP of an air refrigeration system working on Bell-Coleman cycle, with the help of (p-v) and (T-s) diagrams.*

(b) The capacity of a dense air refrigeration system operating on reversed Brayton cycle is 15 T. The cooler and refrigerator pressures are 3.9 bar and 1.3 bar respectively. The temperature of air entering the compressor is –20°C and air is cooled in the cooler at a temperature of 50°C. Assuming isentropic expansion and compression, determine the COP of the cycle, mass of air circulated/min and net power of the machine. Take γ for air as 1.4.

Solution :

(*a*) **Characteristics of air refrigeration system:** Characteristics feature of air refrigeration system are:

1. Easily available and no cost of refrigerant.
2. Air is non-toxic and inflammable.
3. Leakage of air in small amounts is tolerable.
4. Air is light in weight per ton of refrigeration.
5. Low working pressure requires lighter piping, ducting etc.
6. Requires large rate of air circulation.
7. Low coefficient of performance.

Difference between open and closed refrigeration cycles: Refer to art 3.2.

Bell-Coleman cycle: Refer to art. 3.4.

(*b*) Refer to Fig. 3.5.

Given: $p_1 = p_4 = 1.3$ bar, $p_2 = p_3 = 3.9$ bar, $T_1 = -20°C$ or $273 - 20 = 253$ K.

$T_3 = 50°C$ or $273 + 50 = 323$ K, $\gamma = 1.4$

Process 1-2: Isentropic compression

$$\frac{T_2}{T_1} = \left(\frac{p_2}{p_1}\right)^{\frac{\gamma-1}{\gamma}} = \left(\frac{3.9}{1.3}\right)^{\frac{1.4-1}{1.4}} = (3)^{\frac{1}{3.5}} = 1.3687$$

$$T_2 = 253 \times 1.3687 = 346.3 \text{ K}$$

Process 3-4: Isentropic compression

$$\frac{T_3}{T_4} = \left(\frac{p_3}{p_4}\right)^{\frac{\gamma-1}{\gamma}} = \left(\frac{3.9}{1.3}\right)^{\frac{1.4-1}{1.4}} = (3)^{\frac{1}{3.5}} = 1.3687$$

$$T_4 = \frac{323}{1.3687} = 236 \text{ K}$$

Pressure ratio, $$r_p = \frac{p_2}{p_1} = \frac{3.9}{1.3} = 3$$

$$\text{COP} = \frac{1}{(r_p)^{\frac{\gamma-1}{\gamma}} - 1}$$

$$= \frac{1}{(3)^{\frac{1.4-1}{1.4}} - 1} = \frac{1}{0.36874} = 2.712$$

Heat extracted or absorbed by air per kg of air,

$$q_{4-1} = c_p(T_1 - T_4)$$
$$= 1.005\ (253 - 236) = 17.085 \text{ kJ/kg}$$

Refrigerating capacity, $Q = 15$ TR

Mass rate of air flow, $$\dot{m}_a = \frac{211Q}{c_p(T_1 - T_4)} = \frac{211Q}{q_{4-1}}$$

$$= \frac{211 \times 15}{17.085} = 185.25 \text{ kg/min or } 3.087 \text{ kg/s}$$

Net work done per kg of air,

$$w_{net} = c_p[(T_2 - T_3) - (T_1 - T_4)]$$
$$= 1.005\ [(346.3 - 323) - (253 - 236)]$$
$$= 6.3315 \text{ kJ/kg}$$

Net power of the machine, $$P = \frac{\dot{m}_a w_{net}}{60} = \frac{185.25 \times 6.3315}{60} = 19.548 \text{ kW}$$

Example 3.10 *An air refrigeration system operating on Bell-Coleman cycle draws air from a cold chamber at a pressure of 1 bar and temperature of 10°C. It is then compressed in the compressor up to a pressure of 5 bar and after being cooled at constant pressure up to a temperature of 40°C, it enters the expander. Assuming both compression and expansion processes to be isentropic, calculate the COP of the system. Take c_p = 1.005 kJ/kg.K and γ = 1.4.*

Solution: Given: p_1 = 1 bar, $T_1 = 273 + 10 = 283$ K, $p_2 = 5$ bar, $T_3 = 273 + 40 = 313$ K, $\gamma = 1.4$, c_{pa} = 1.005 kJ/kg. K

Refer to Fig. 3.11

Pressure ratio, $$r_p = \frac{p_2}{p_1} = \frac{5}{1} = 5$$

Process 1-2:

$$T_2 = T_1(r_p)^{\frac{\gamma-1}{\gamma}} = 283(5)^{\frac{1.4-1}{1.4}} = 448.2 \text{ K}$$

Process 3-4:

$$T_4 = \frac{T_3}{(r_p)^{\frac{\gamma-1}{\gamma}}} + \frac{313}{(5)^{\frac{0.4}{1.4}}} = 197.6 \text{ K}$$

$$\text{COP} = \frac{T_1 - T_4}{(T_2 - T_3) - (T_1 - T_4)}$$

$$= \frac{283 - 197.6}{(448.2 - 313) - (283 - 197.6)}$$

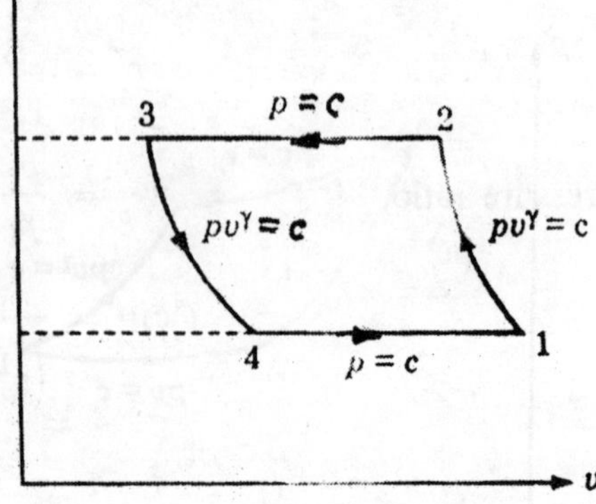

Fig. 3.11 *p-v* diagram for Bell-Coleman cycle.

$$= \frac{85.4}{49.8} = 1.715$$

$$\text{or COP} = \frac{1}{(r_p)^{\frac{\gamma-1}{\gamma}} - 1} = \frac{1}{(5)^{\frac{1.4-1}{1.4}} - 1} = 1.713$$

Example 3.11

(a) A refrigerating system operating on Carnot refrigeration cycle, is having higher and lower temperatures of 300C and –10°C respectively. If the capacity of the system is 10 tons, calculate the COP, the power required by the system in kW and the total amount of heat rejected by the system.

(b) With the help of p-v and T-s diagrams, describe the basic difference between a reversed Carnot cycle and a reversed Brayton cycle as used for air-refrigeration system.

Solution:

(*a*) Given: $T_2 = 273 + 30 = 303$ K, $T_1 = 273 - 10 = 263$ K, $TR = 10$,

For the Carnot refrigeration cycle,

$$\text{COP} = \frac{T_1}{T_2 - T_1} = \frac{263}{303-263} = \frac{263}{40} = 6.575$$

$$1TR = 211 \text{ kJ/min}$$

Power required, $$P = \frac{10 \times 211}{60} = 35.167 \text{ kW}$$

Also $$\text{COP} = \frac{\text{Heat rejected}}{\text{Work done}}$$

$$\text{Heat rejected} = \text{COP} \times \text{Work done}$$
$$= 6.575 \times 10 \times 211$$
$$= 13873.25 \text{ kJ/min}$$

(*b*) **Difference between reversed Carnot cycle and reversed Brayton Cycle:** The *p-v* and *T-s* diagram for reversed Carnot cycle and reversed Brayton cycle are shown in Fig. 3.12 and Fig. 3.13 respectively. The comparison of reversed Brayton cycle and reversed Carnot cycle is shown on T-s diagram in Fig. 3.14.

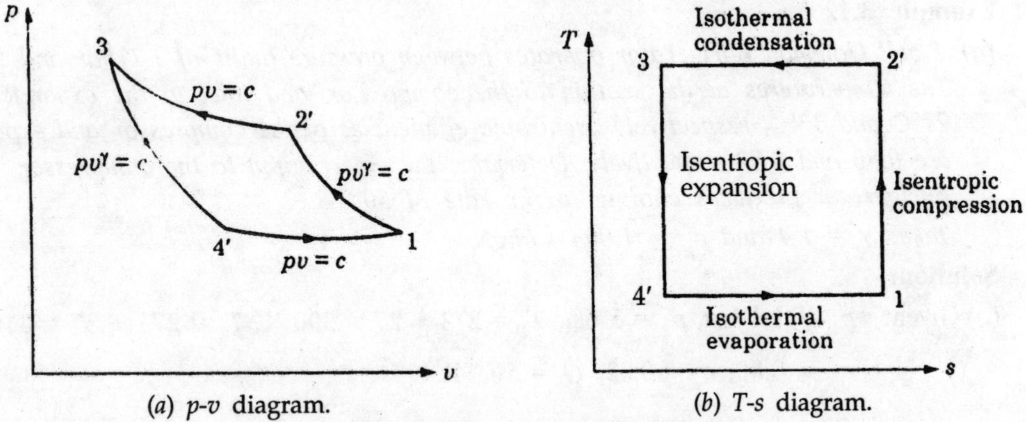

(*a*) *p-v* diagram. (*b*) *T-s* diagram.

Fig. 3.12 Reversed Carnot cycle.

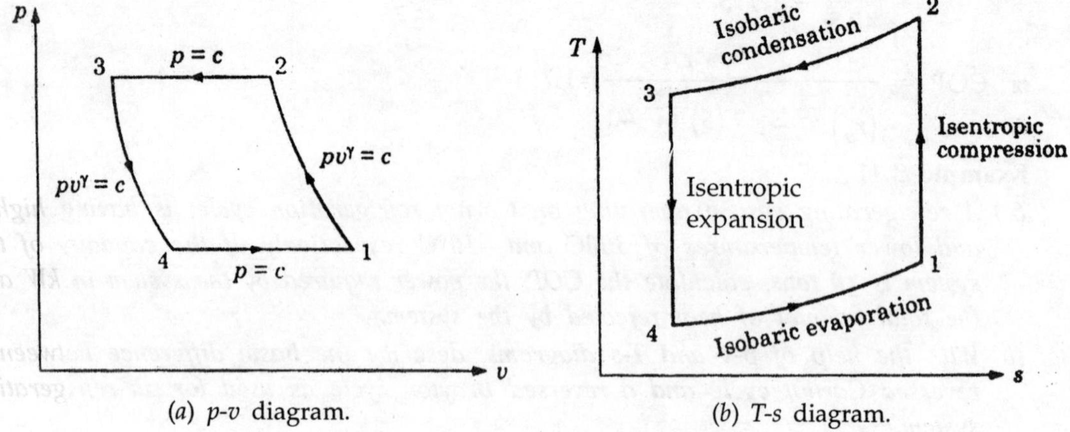

Fig 3.13 Reversed Brayton cycle.

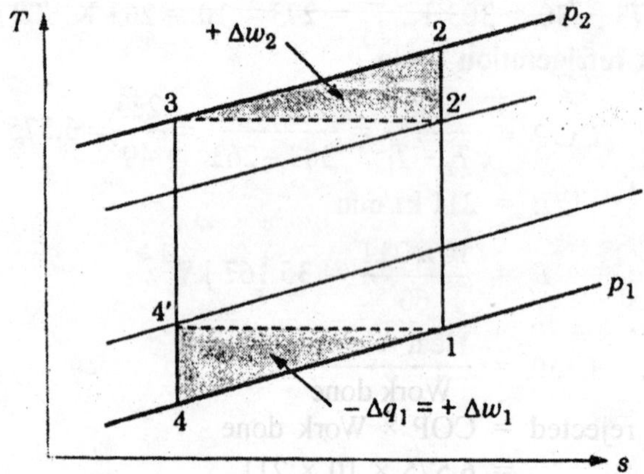

Fig. 3.14 Comparison of reversed Brayton cycle and reversed Carnot cycle.

The two isothermal processes 2′-3 and 4′-1 of the reversed Carnot cycle are replaced by the two isobaric processes 2-3 and 4-1 respectively in the reversed Brayton cycle. However, this involves a loss in the refrigerating effect Δq_1 equivalent to the area 1-4-4′, and increase in the network of the cycle, Δw_1 and Δw_2, corresponding to the areas 1-4-4′ and 3-2-2′ respectively.

Example 3.12

(a) A bell Coleman refrigerator operates between pressure limits of 1.1 bar and 5 bar. The temperatures at the suction to the compressor, and inlet to the expander are 27°C and 37°C, respectively. Isentropic efficiencies of the compressor and expander are 0.80 and 0.82 respectively. Determine the power input to the compressor, if the refrigerator produces cooling at the rate of 50 TR.

Take: $\gamma = 1.4$ *and* $c_p = 1.004$ *kJ/kgK.*

Solution:

(a) Given: $p_1 = 1.1$ bar, $p_2 = 5$ bar, $T_1 = 273 + 27 = 300$ K, $T_3 = 273 + 37 = 310$ K,

$\eta_c = 0.80, \eta_e = 0.82, \dot{Q} = 50$ TR

The *p-v* diagram for the Bell-coleman refrigeration cycle is shown in Fig. 3.15

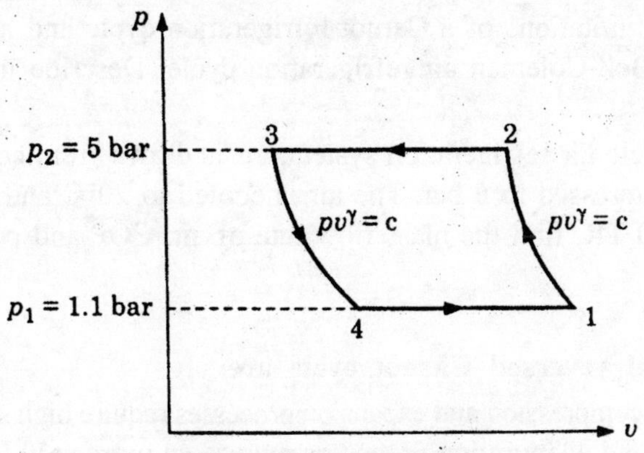

Fig. 3.15 *p-v* diagram

Process 1-2:

$$T_2 = T_1\left(\frac{P_2}{P_1}\right)^{\frac{\gamma-1}{\gamma}} = 300\left(\frac{5}{1.1}\right)^{\frac{1.4-1}{1.4}} = 462.4\,\text{K}$$

Process 3-4:

$$T_4 = T_3\left(\frac{P_1}{P_3}\right)^{\frac{\gamma-1}{\gamma}} = 310\left(\frac{1.1}{5}\right)^{\frac{1}{3.5}} = 201.1\text{ K}$$

The T-s diagram is shown in Fig 3.16.

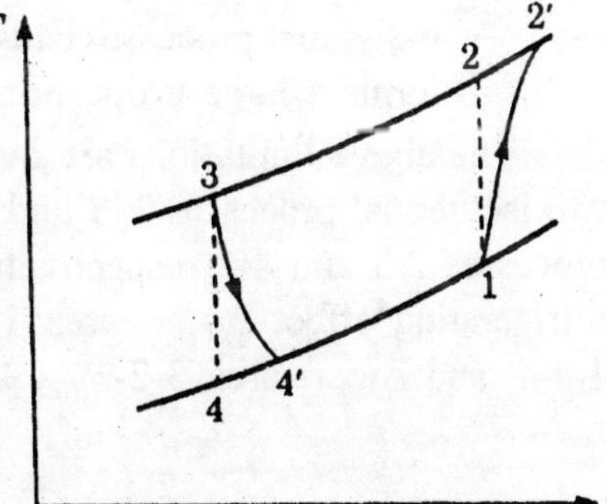

Fig. 3.16 T-s diagram.

Compressor efficiency, $\eta_c = \dfrac{T_2 - T_1}{T_{2'} - T_1}$

$$0.80 = \frac{462.4 - 300}{T_{2'} - 300}$$

$$T_{2'} = 503\text{ K}$$

Expander efficiency, $\eta_e = \dfrac{T_3 - T_{4'}}{T_3 - T_4}$

$$T_{4'} = 220.7\text{ K}$$

Refrigeration effect, $q_{4'-1} = c_{pa}(T_1 - T_{4'}) = 1.004\ (300 - 220.7) = 109.336$ kJ/kg

Mass flow rate of air, $\dot{m}_a = \dfrac{\dot{Q}}{q_{4'-1}} = \dfrac{50 \times 211}{109.336} = 96.49$ kg/min

Compressor work. $W_c = \dot{m}_a c_{pa}(T_{2'} - T_1)$

$$= 96.49 \times 1.004\ (503 - 300) = 19665.82\text{ kJ/min}$$

or $$= \frac{19665.82}{60} = 327.76\,\text{kW}$$

Example 3.13

(*a*) What are the limitations of a Carnot refrigeration cycle and how these limitations are overcome in Bell-Coleman air refrigeration cycle. Describe in brief, with the help of *T-s* diagram.

(*b*) In a closed cycle air refrigeration system, air is drawn from cold chamber at –5°C and 1 bar and compressed to 6 bar. The air is cooled to 20°C and then expanded to 1 bar. To produce 10 TR, find the mass flow rate of air, COP and power required to run the compressor.

Solution:

(*a*) Limitations of reversed Carnot cycle are:

1. The isentropic compression and expansion processes require high speed while the isothermal condensation and evaporation processes require an extremely low speed. This variation in speed of air of a cycle is not practicable.
2. It is difficult to design an expander to handle a mixture of largely liquid and partly vapour.
3. The COP is very low due to internal irreversibility's in the compressor and expander.

The reversed Carnot cycle, with gas as a refrigerant, has inherent drawbacks. These are:

1. Extreme pressures and large volumes are developed since the pressure rise takes place both during isentropic compression as well as isothermal heat rejection processes.
2. Isothermal heat transfer processes with a gas having definite specific heat are impossible to achieve in practice.
3. The *p-v* diagram of the cycle is so narrow that even small irreversibility's of the individual processes cause a significant increase in the workdone. This increase in work forms a large proportion of the network of the cycle.

The above limitations are overcome in the Bell-Coleman cycle is shown in Fig. 3.17. The two isothermal processes 2′-3 and 4′-1 of reversed Carnot cycle are replaced by the two isobaric processes 2-3 and 4-1 respectively in the Bell-Coleman cycle. However, this results in loss of refrigerating effect Δq_1 = area 1-4-4′, and increase in the network of the cycle. Δw_1= area 1-4-4′ and Δw_2 = area 3-2-2′.

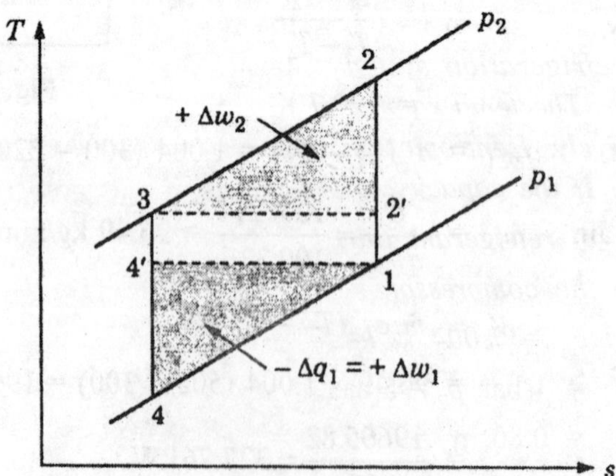

Fig. 3.17 Comparison of Bell-Coleman and reversed Carnot cycles.

(*b*) Given: $T_1 = 273 - 5 = 268$ K, $p_1 = 1$ bar, $p_2 = 6$ bar, $T_3 = 273 + 20 = 293$ K,

$p_4 = 1$ bar, $Q_{4-1} = 10\ TR$

Refer to Fig. 3.18.

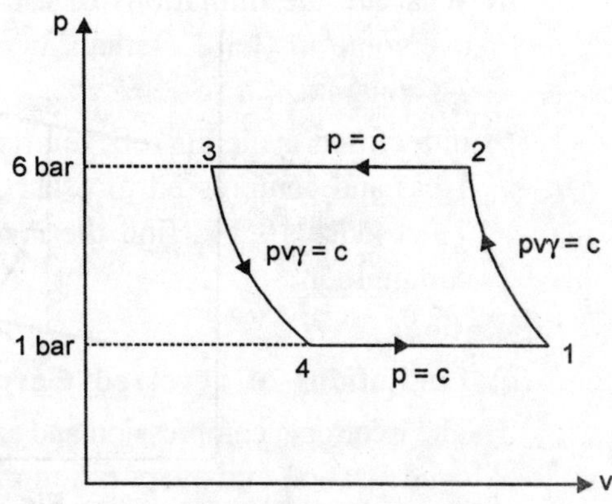

Fig. 3.18 *p-v* diagram

Process 1-2:

$$\frac{T_2}{T_1} = \left(\frac{p_2}{p_1}\right)^{\frac{\gamma-1}{\gamma}} = \left(\frac{6}{1}\right)^{\frac{1.4-1}{1.4}} = 1.6685$$

$$T_2 = 268 \times 1.6685 = 447.16 \text{ K}$$

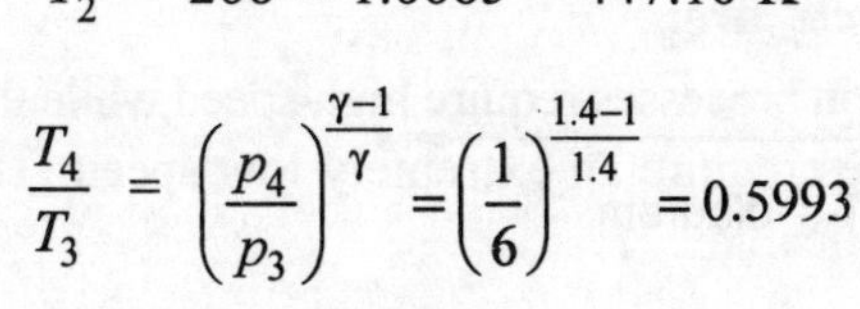

$$\frac{T_4}{T_3} = \left(\frac{p_4}{p_3}\right)^{\frac{\gamma-1}{\gamma}} = \left(\frac{1}{6}\right)^{\frac{1.4-1}{1.4}} = 0.5993$$

$$T_4 = 293 \times 0.5993 = 175.6 \text{ K}$$

Refrigeration effect, $q_{4-1} = c_{pa}\,(T_1 - T_4)$

$= 1.005\ (268 - 175.6) = 92.862$ kJ/kg air

Mass flow rate of air, $\dot{m}_a = \dfrac{\dot{Q}_{4-1}}{q_{4-1}} = \dfrac{10 \times 211}{92.862} = 22.72$ kJ/min

$$\text{COP} = \frac{1}{(r_p)^{\frac{\gamma-1}{\gamma}} - 1} = \frac{1}{(6)^{\frac{1.4-1}{1.4}} - 1} = 1.496$$

Power required to run the compressor,

$$P_c = \frac{\dot{m}_a \times c_p (T_2 - T_1)}{60}$$

$$= \frac{22.72 \times 1.005(447.16 - 268)}{60} = 68.18 \text{kW}$$

Example 3.14 *A refrigeration system working on Reversed Brayton cycle operates between 1 bar and 5 bar. The temperatures at the inlet to the compressor and expander are 30°C and 40°C, respectively. Isentropic efficiency for the compression is 0.8 and the same for the expansion is 0.9. If the capacity of the system is 5 Tons of refrigeration, determine:*

(i) mass flow rate of refrigerant (air)

(ii) power input to the compressor.

Take: $\gamma = 1.4$, *and* $c_p = 1.004$ *kJ/kg.K.*

Solution: Given: $p_1 = 1$ bar, $p_2 = 5$ bar, $T_1 = 273 + 30 = 303$ K, $T_3 = 273 + 40 = 313$ K,

$\eta_c = 0.80$, $\eta_e = 0.90$, $Q = 5TR$, $\gamma = 1.4$, $c_{pa} = 1.004$ kJ/kg.K

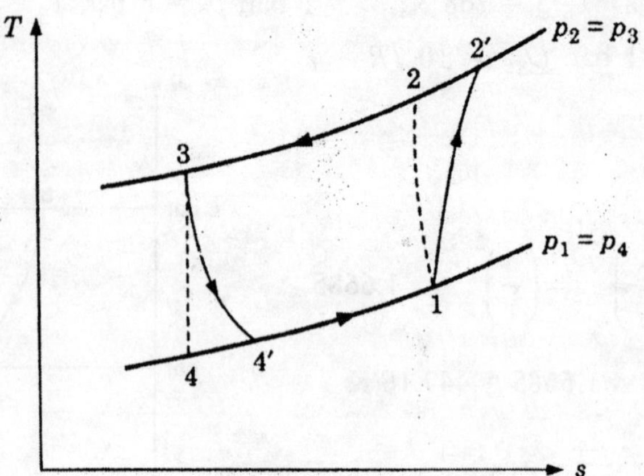

Fig. 3.19 T-s diagram.

The *T-s* diagram is shown in Fig. 3.19.

$$T_2 = T_1\left(\frac{P_2}{P_1}\right)^{\frac{\gamma-1}{\gamma}} = 303\left(\frac{5}{1}\right)^{\frac{1.4-1}{1.4}} = 479.9 \text{ K}$$

$$\eta_c = \frac{T_2 - T_1}{T_{2'} - T_1}$$

$$0.8 = \frac{479.9 - 303}{T_{2'} - 300}$$

$$T_{2'} = 524.1 \text{ K}$$

$$\frac{T_4}{T_3} = \left(\frac{P_4}{P_3}\right)^{\frac{\gamma-1}{\gamma}}, T_4 = 313\left(\frac{1}{5}\right)^{\frac{1.4-1}{1.4}} = 197.6 \text{ K}$$

$$\eta_e = \frac{T_3 - T_{4'}}{T_3 - T_4}$$

$$0.9 = \frac{313 - T_{4'}}{313 - 197.6}$$

$$T_{4'} = 209.1 \text{ K}$$

Refrigeration effect, $q_{4'-1} = c_{pq}(T_1 - T_{4'}) = 1.004\ (313 - 209.1) = 104.3$ kJ/kg

(*i*) Mass flow rate of air, $\dot{m}_a = \frac{\dot{Q}}{q_{4'-1}} = \frac{50 \times 211}{104.3} = 10.1 \text{ kg/min}$

(*ii*) Power input to the compressor, $P_c = \frac{\dot{m}_a \times c_{pa}(T_{2'} - T_1)}{60}$

$$= \frac{10.1 \times 1.004(524.1 - 303)}{60} = 37.37 \text{ kW}$$

Example 3.15

(a) What is the basic difference between open and closed Air refrigeration cycles? Describe a Bell-Coleman or Reversed Joule air refrigeration cycle with the help of a neat labelled sketch and also explain the basic difference between this cycle and a reversed Carnot air refrigeration cycle.

(b) An air refrigeration system following Bell-Coleman cycle is operating between pressure limits of 1.04 bar and 8 bar respectively. Air is drawn from the cold chamber at 9°C and after compression it is cooled to 28°C, before entering the expansion cylinder. If expansion and compression processes follow the law $p\text{-}v^{1.28}$ = constant, determine the theoretical COP of the above refrigeration system. Take $\gamma = 1.4$ and $c_p = 1.005$ kJ/kg-K for Air.

Solution:

(*a*) Difference between open a closed air refrigeration cycles: Refer to art. 3.2.

Bell-Coleman cycle: Refer to art. 3.4.

(*b*) Given: $p_1 = 1.04$ bar, $p_2 = 8$ bar, $T_1 = 273 + 9 = 282$ K, $T_3 = 273 + 28 = 301$ K

$p\text{-}v^{1.28}$ = const. $\gamma = 1.4$, $c_{pa} = 1.005$ kJ/kg.K, $n = 1.28$.

$$\text{COP} = \frac{1}{\left(\frac{n}{n-1}\right)\left(\frac{\gamma-1}{\gamma}\right)\left[(r_p)^{\frac{n-1}{n}} - 1\right]}$$

$$r_p = \frac{p_2}{p_1} = \frac{8}{1.04} = 7.6923$$

$$\text{COP} = \frac{1}{\left(\frac{1.28}{1.28-1}\right)\left(\frac{1.4-1}{1.4}\right)\left[(7.6923)^{\frac{1.28-1}{1.28}} - 1\right]}$$

$$= 1.36$$

Example 3.16 *A dense air closed Bell-Coleman refrigeration system working between 4 bar and 16 bar extracts 125 mJ/h. The air enters the compressor at 5°C and enters the expander at 23°C. The compressor is double-acting and its stroke is 0.3 m and runs at 300 rpm. Calculate:*

(a) Power required to run the unit,

(b) Bore of the compressor, and

(c) Refrigeration capacity in tons.

Assume isentropic compression and expansion.

Take $\gamma_{air} = 1.4$, $(\eta_{mech})_{comp} = 0.82$, $(\eta_{mech})_{exp} = 0.87$, $(c_p)_{air} = 1.005$ *kJ/kg.K,*

$R_{air} = 0.287$ *kJ/kg.K*

Solution: Given: $p_1 = p_4 = 4$ bar, $p_2 = p_3 = 16$ bar, $\dot{Q} = 125$ MJ/h, $T_1 = 5 + 273 = 278$ K,

$T_3 = 23 + 273 = 296$ K, $N = 300$ rpm, $L = 0.3$ m, $\gamma_a = 1.4$, $(\eta_{mech})_c = 0.82$

$(\eta_{mech})_e = 0.87$, $c_{pa} = 1.005$ kJ/kg.K, $R_a = 0.287$ kJ/kg.K (*a*) Power required to run the unit:

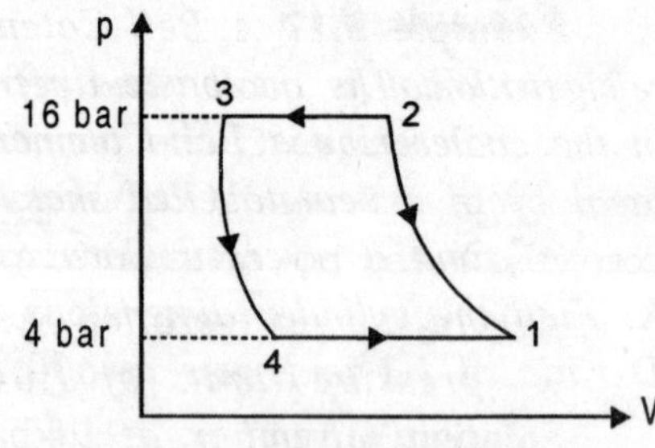

Fig. 3.20 *p-v* diagram

$$\frac{T_2}{T_1} = \left(\frac{P_2}{P_1}\right)^{\frac{r_a-1}{r_a}} = \left(\frac{16}{1}\right)^{\frac{1.4-1}{1.4}} = 1.486$$

$$T_2 = 278 \times 1.486 = 413.1 \text{ K}$$

$$\frac{T_4}{T_3} = \left(\frac{p_4}{p_3}\right)^{\frac{\gamma_a-1}{\gamma_a}} = \left(\frac{4}{16}\right)^{\frac{1.4-1}{1.4}} = 0.673$$

$$T_4 = 296.\ 0.673 = 199.2 \text{ K}$$

Refrigeration effect, $\quad RE = c_{pa}(T_1 - T_4) = 1.005\ (278 - 199.2) = 79.2$ kJ/kg

Mass flow rate of air, $\quad \dot{m}_a = \dfrac{\dot{Q}}{3600 \times RE} = \dfrac{125 \times 10^3}{3600 \times 79.2} = 0.4384$ kg/s

Compressor work, $\quad \dot{W}_c = \dfrac{\dot{m}_a \times c_{pa}(T_2 - T_1)}{(\eta_{mech})_c} = \dfrac{0.4384 \times 1.005(413.1 - 278)}{0.82}$

$$= 72.59 \text{ kW}$$

Expander work, $\quad \dot{W}_e = \dot{m}_a \times c_{pa}(T_3 - T_4)\ (\eta_{mech})$

$$= 0.4384 \times 1.005\ (296 - 199.2) \times 0.87$$

$$= 37.1 \text{ kW}$$

Power required to run the unit,

$$P = \dot{W}_c - \dot{W}_e = 72.59 - 37.1 = 35.49 \text{ kW}$$

(*b*) Bore of compressor:

$$p_1 V_1 = m_a R_a T_1$$

$$V_1 = \frac{0.4384 \times 0.287 \times 278}{4 \times 10^2} = 0.08744 \text{ m}^3/\text{s}$$

$$V_1 = \frac{\pi}{4} D^2 L \times \frac{2N}{60}$$

$$0.08744 = \frac{\pi}{4} D^2 \times 0.3 \times \frac{2 \times 300}{60}$$

$$D = 0.1926 \text{ m or } 19.26 \text{ cm}$$

(*c*) Refrigeration capacity in tons :

$$\text{Refrigeration capacity} = \frac{125 \times 10^3}{211 \times 60} = 9.874 \text{ tons}$$

Example 3.17 *A Bell-Coleman refrigeration system is used to produce 10 tons of refrigeration. The cooler and refrigerator pressure are 4.2 bar and 1.4 bar. Air is cooled in the cooler to 45°C and temperature of air at inlet of the compressor is –20°C. For an ideal cycle, calculate COP, mass of air circulated/min, theoretical piston displacement of compressor and power required per ton of refrigeration. Assume c_p for air as 1.005 kJ/kg-K. Find the cylinder dimensions if the compressor is single-acting, single-cylinder with L/D ratio of 1.2 and runs at 600 r.p.m.* **[IES, 2010]**

Solution: Given : $p_1 = 1.4$ bar, $p_2 = 4.2$ bar, $T_1 = 273 - 20 = 253$ K, $T_3 = 273 + 45 = 318$ K,

$c_p = 1.005$ kJ/kg.K

$\frac{L}{D} = 1.2$, N = 600 rpm

The T-s diagram for Bell-Coleman system is shown in Fig. 3.21.

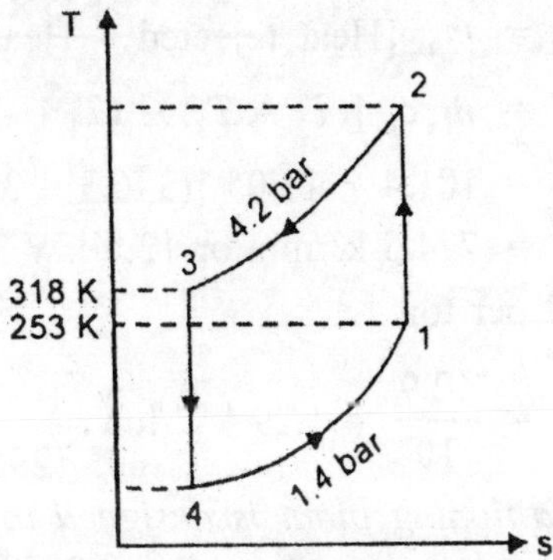

Fig. 3.21 T-s diagram.

$$\frac{T_2}{T_1} = \left(\frac{p_2}{p_1}\right)^{\frac{\gamma-1}{\gamma}}, T_2 = 253\left(\frac{4.2}{1.4}\right)^{\frac{1.4-1}{1.4}} = 346.3\text{K}$$

$$\frac{T_4}{T_3} = \left(\frac{p_1}{p_2}\right)^{\frac{\gamma-1}{\gamma}}, \; T_4 = 318 \times 0.7306 = 232.3 \text{ K}$$

$$\text{COP} = \frac{T_1 - T_4}{(T_2 - T_3) - (T_1 - T_4)} = \frac{253 - 232.3}{(346.3 - 318) - (253 - 232.3)} = 2.723$$

Refrigeration capacity = 10 TR

Heat extracted per min = 10 × 211 = 2110 kJ/min

Heat extracted from refrigerator per kg of air

$$= c_p(T_1 - T_4) = 1.005(253 - 232.3) = 20.80 \text{ kJ/kg}$$

Rate of air circulation, $\dot{m}_a = \frac{2110}{20.80} = 101.4$ kg/min

$$p_1 V_1 = \dot{m}_a R T_1$$

$$V_1 = \frac{101.4 \times 287 \times 253}{253} = 52.59 \text{ m}^3/\text{min}$$

$$V_4 = \frac{V_1 T_4}{T_1} = \frac{52.59 \times 232.3}{253} = 48.29 \text{ m}^3/\text{min}$$

$$\frac{\pi}{4} D^2 L \times N = V_1$$

$$\frac{\pi}{4} \times 1.2 \times D^3 \times 600 = 52.59$$

$$D = 0.453 \text{ m}$$

$$L = 1.2 \times 0.453 = 0.5437 \text{ m}$$

Piston displacement of compressor = 0.5437 m

Net work done on the refrigeration machine per minute

$$= \dot{m}_a \text{ [Heat rejected } - \text{ Heat extracted]}$$

$$= \dot{m}_a c_p [(T_2 - T_3) - (T_1 - T_4)]$$

$$= 101.4 \times 1.005 [(346.3 - 318) - (253 - 232.3)]$$

$$= 774.5 \text{ kJ/min or } 12.9 \text{ kW}$$

Theoretical power required per ton

$$= \frac{12.9}{10} = 1.29 \text{ kW/ton}$$

Example 3.18 *An air conditioning plant is designed to maintain a room at a condition of 20°C dry bulb temperature and specific humidity 0.0079 kg/kg dry air when the outside condition is 30°C dry bulb temperature and 40% saturation. The corresponding heat gains are 18000 W (Sensible) and 3600W (latent). The supply air contains one-third outside air by mass and the supply temperature is to be 15°C dry bulb temperature.*

The plant consists of mixing chamber for fresh and recirculated air, an air washer with chilled spray water with an efficiency of 80%, an after heater battery and a supply fan. Neglecting temperature changes in fan and ducting, calculate:

(i) the mass flow rate of supply air necessary;

(ii) the specific humidity of the supply air;

(iii) the cooling load on the washer;

(iv) the heating load on the after heater. **[CSE, 2005]**

Solution: Given: $z = 8$, $(r_p)_{st} = 1.35$, $\dot{m} = 50$ kg/s, $\eta_0 = 82\%$, $p_1 = 1.0$ bar, $T_1 = 313$ K, $\eta_{mech} = 90\%$, $c_p = 1.005$ kJ/kg.K, $\gamma = 1.4$

(i)

$$z = \frac{\ln r_p}{\ln (r_p)_{st}}$$

$$8 = \frac{\ln r_p}{\ln 1.35}, \; r_p = 11.03 = \text{overall pressure ratio}$$

Exit pressure, $p_9 = 1.0 \times 11.03 = 11.03$ bar

$$\eta_0 = \frac{T_9 - T_1}{T_{9'} - T_1}, T_9 = T_1(r_p)^{\frac{\gamma-1}{\gamma}} = 313 \times (11.03)^{\frac{0.4}{1.4}} = 621.5 \text{ K}$$

$$0.82 = \frac{621.5 - 313}{T_{9'} - T_1}, \quad T_{9'} = 689.2 \text{ K or } 416.2°\text{C}$$

(*ii*) Polytropic efficiency, $\eta_p = \dfrac{\ln\left(\dfrac{p_9}{p_1}\right)^{\frac{\gamma-1}{\gamma}}}{\ln\left(\dfrac{T_{9'}}{T_{1'}}\right)} = \dfrac{\ln(11.03)^{\frac{0.4}{1.4}}}{\ln\left(\dfrac{689.2}{313}\right)} = 0.869$ or 86.9%

(*iii*) $$\eta_p = \left(\frac{\gamma-1}{\gamma}\right)\left(\frac{n}{n-1}\right)$$

$$0.869 = \left(\frac{1.4-1}{1.4}\right)\left(\frac{n}{n-1}\right), \frac{n}{n-1} = 3.0415$$

$$3.0415\, n - 3.0415 = n$$

$$n = 1.49$$

$$\frac{T_2}{T_1} = (r_p)^{\frac{\gamma-1}{\gamma}}, T_2 = 313(1.35)^{\frac{0.4}{1.4}} = 341 \text{ K}$$

$$\frac{T_{2'}}{T_1} = (r_p)^{\frac{n-1}{n}}, T_{2'} = 313(1.35)^{\frac{1}{3.0415}} = 345.46 \text{ K}$$

Isentropic efficiency of each stage $= \dfrac{T_2 - T_1}{T_{2'} - T_1} = \dfrac{341 - 313}{345.46 - 313} = 0.8626$ or 86.26%

(*iv*) Power required to drive the compressor, $P = \dot{m} \times c_p \dfrac{T_{9'} - T_1}{\eta_{mech}}$

$$= \frac{50 \times 1.005(689.2 - 313)}{0.9} = 21004 \text{ kW}$$

Example 3.19 *A customer complained of poor cooling for an air-conditioning system of 100 TR capacity. The supplier carried out test on condenser which is water cooled and noted power input to the motor. The obeservatopns made are as under:*

Cooling water flow rate : 10 Litre/sec
Inlet water temperature : 30°C
Outlet water temperature : 41.12°C
Power input to motor : 120 kW (94.92% efficiency)

Determine the actual refrigerating capacity and state whether the cooling capacity is lower, higher or as per specifications. **[CSE, 2008]**

Solution: Given: $Q = 100\ TR$, $\dot{V}_w = 10$ litre/s or 10×10^{-3} m^3/s, $t_{wi} = 30°C$

$t_{wo} = 41.12°C$, $P_m = 120$ kW, $\eta_m = 94.92\%$

$1\ TR = 211$ kJ/min

$$\dot{m}_w = \dot{V}_w \times P = 10 \times 10^{-3} \times 10^3 = 10 \text{ kg/s}$$

$$\text{Heat transferred to cooling water} = \dot{m}_w c_{pw}(t_{wo} - t_{wi})$$
$$= 10 \times 4.2\,(41.12 - 30)$$
$$= 467 \text{ kJ/s}$$

$$\text{Capacity of air-conditioning system} = 100 \times \frac{211}{60} = 351.7 \text{ kJ/s}$$

Power output of motor = 120 × 0.9492 = 113.9 kJ/s

Actual refrigerating capacity = 467 – 113.9 = 353.1 kJ/s

Cooling capacity is lower.

Example 3.20 *Obtain an expression for the capacity of a refrigeration system (tonnage) in terms of the rate of upper and lower pressure limits, expansion index (n), v_c/v_s = (v_c = clearance volume, and v_s = swept volume), N (rpm), v_1 (specific volume at the inlet to compressor) and the refrigeration effect ($h_1 - h_4$).*

If the pressire ratio is 6.5, n = 1.1, v_c/c_s = 0.025, N = 900 rpm, v_s = 600 cc, v_1 = 0.078 m^3/kg and refrigeration effect = 150 kJ/kg, calculate capacity of the system. **[IES, 2006]**

Solution: Consider the pressure – volume diagram of a single stage reciprocating compressor with clearance as shown in Fig. 3.22.

Let, p_1 = suction pressure of compressor.

p_2 = delivery pressure of compressor.

v_c = clearance volume.

v_s = stroke (or swept) volume.

v_a = actual volume of refrigerant taken in.

v_e = volume of refrigerant after re-expansion.

v_1 = specific volume at the inlet of compressor.

n = index of compression and expansion.

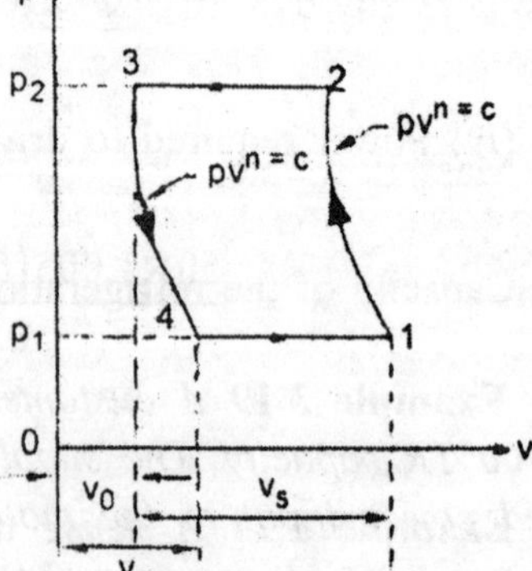

Fig. 3.22 p-v diagram for air compressor.

$$\text{Clearance ratio,} \quad c = \frac{v_c}{v_s}$$

$$\text{Volumetric efficiency,} \quad \eta_v = \frac{v_a}{v_s}$$

$$= \frac{\left[v_s - (v_e - v_c)\right]}{v_s} = \frac{\left[v_s - v_c\{(v_e / v_c) - 1\}\right]}{v_s} = 1 - \left(\frac{v_c}{v_s}\right)\left(\frac{v_e}{v_c} - 1\right)$$

Applying the law of expansion to points 3 and 4, we have

$$p_2 v_c^n = p_1 v_e^n$$

$$\frac{v_e}{v_c} = \left(\frac{p_2}{p_1}\right)^{1/n}$$

$$\eta_v = 1 - c\left[\left(\frac{p_2}{p_1}\right)^{1/n} - 1\right] = 1 + c - c\left(\frac{p_2}{p_1}\right)^{1/n}$$

Volume flow rate of piston,

$$\dot{V}_s = \left(\frac{\pi d^2}{4}\right) \times L \times N/60 \text{ m}^3/\text{s}$$

Where d = diameter of piston of compressor, m

L = length of stroke, m

N = speed of compressor, rpm

Mass flow rate, $$\dot{m}_r = \frac{\dot{V}_s \times \eta_v}{v_1} \text{ kg/s}$$

Capacity of the refrigeration system $= \dfrac{\dot{m}_r \times \text{refrigeration effect}}{3.5167}$ TR

For the given data, we have

$$\eta_v = 1 + 0.025 - 0.025\,(6.5)^{1/1.1} = 0.888$$

$$\dot{V}_s = v_s \times \text{N}/60 = 600 \times 10^{-6} \times \frac{900}{60} = 0.009 \text{ m}^3/\text{s}$$

Actual volume flow rate,

$$\dot{V}_s = \dot{V}_s\, \eta_v = 0.009 \times 0.888 = 7.992 \times 10^{-3} \text{ m}^3/\text{s}$$

Mass flow rate, $$\dot{m}_r = \frac{\dot{V}_a}{v_1} = \frac{7.992 \times 10^{-3}}{0.078} = 0.10246 \text{ kg/s}$$

Capacity of the refrigeration system = $\dot{m}_r$ × refrigeration effect/3.5167.

$$= \frac{0.10246 \times 150}{3.5167} = 4.37 \text{ TR}$$

Example 3.21 *A dense air refrigeration machine operating on Bell-Coleman cycle operates between 3.4 bar and 17 bar. The temperature of air after the cooler is 15°C and after the refrigerator is 6°C. If the refrigeration capacity is 6 tons, calculate:*

(i) Temperature after compression and expansion

(ii) Air circulation per min

(iii) Work of compressor and expander.

(iv) Theoretical COP.

(v) Rate of water circulation required in the cooler in kg/min if the rise in temperature is limited to 30°C. **[IES, 2009]**

Solution:

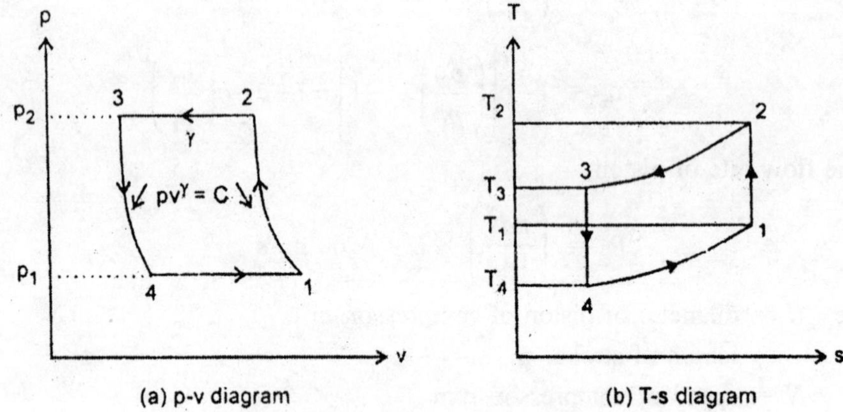

Fig. 3.23 Bell-Coleman cycle.

Refer to Fig. 3.23

Given:

$$p_1 = 3.4 \text{ bar}, p_2 = 17 \text{ bar},$$
$$T_3 = 273 + 15 = 288 \text{ K},$$
$$T_1 = 273 + 6 = 279 \text{ K},$$
$$Q = 6 \text{ TR}$$

(*i*)
$$\frac{T_2}{T_1} = \left(\frac{p_2}{p_1}\right)^{\frac{\gamma-1}{\gamma}} = \left(\frac{17}{3.4}\right)^{\frac{0.4}{1.4}} = 1.58382$$

$$T_2 = 279 \times 1.58382 = 441.9 \text{ K or } 68.9°\text{C}$$

$$\frac{T_3}{T_4} = \left(\frac{p_3}{p_4}\right)^{\frac{\gamma-1}{\gamma}} = \left(\frac{p_2}{p_1}\right)^{\frac{\gamma-1}{\gamma}} = \left(\frac{17}{34}\right)^{\frac{0.4}{1.4}} = 1.58382$$

$$T_4 = \frac{288}{1.58382} = 181.8 \text{ K or } -91.16°\text{C}$$

(*ii*) Heat extracted per min = 6 × 210 = 1260 kJ/min

$$\dot{m}_a c_p(T_1 - T_4) = 1260$$

$$\dot{m}_a = \frac{1260}{1.005(279-181.8)} = 12.9 \text{ kg/min}$$

(*iii*) Compressor work,

$$w_{1-2} = \left(\frac{\gamma}{\gamma-1}\right) R(T_2 - T_1)$$

$$= \frac{1.4}{0.4} \times 0.287\ (441.9 - 279)$$

$$= 163.63 \text{ kJ/kg}$$

$$W_{1-2} = \dot{m}_a \times W_{1-2}$$

$$= \frac{12.9}{60} \times 163.63 = 35.18 \text{ kW}$$

Work of expansion,

$$W_{3-4} = \left(\frac{r}{r-1}\right) R(T_3 - T_4)$$

$$= \frac{1.4}{0.4} \times 0.287\ (288 - 181.8) = 106.68 \text{ kJ/kg}$$

or $$W_{3-4} = \dot{m}_a \times w_{3-4} = \frac{12.9}{60} \times 106.68 = 22.94 \text{ kW}$$

(*iv*) $$\text{Theoretical COP} = \frac{\text{Heat extracted}}{\text{Network done}} = \frac{q}{w_{net}}$$

$$q = c_p(T_1 - T_4) = 1.005(279 - 181.8) = 97.686 \text{ kJ/kg}$$

$$w_{net} = w_{1-2} - w_{3-4} = 16.363 - 106.68 = 56.95 \text{ kJ/kg}$$

$$(\text{COP})_{th} = \frac{97.686}{56.95} = 1.715$$

(*v*) $$\dot{m}_w \times c_{pw} \times \Delta t = \dot{m}_a c_p(T_2 - T_3)$$

Rate of water circulated

$$\dot{m}_w = \frac{12.9 \times 1.005(441.9 - 288)}{4.18 \times 30} = 15.91 \text{ kg/min}$$

MULTIPLE – CHOICE QUESTIONS

1. The dense air system as compared to open air system for same range of temperatures using Bell Coleman system gives
   (a) Higher Power/ton of refrigeration (b) Lower Power/ton of refrigeration
   (c) Same Power/ton of refrigeration (d) Unpredictable results.
2. In Bell-Coleman refrigerator for the same temperature range
   (a) COP of dense air system is higher than COP of open air system.
   (b) COP of dense air system is lower than COP of open air system
   (c) COP of dense air system is equal to COP of open air system
   (d) COP of dense air system may be higher or lower than COP of open air system depending upon pressure ratio.
3. COP of Bell-Coleman refrigerator for pressure ratio is expressed as
   (a) $\dfrac{1}{1-r_p^{(\gamma-1)/\gamma}}$ (b) 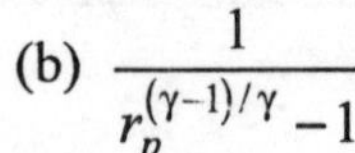 $\dfrac{1}{r_p^{(\gamma-1)/\gamma}-1}$ (c) 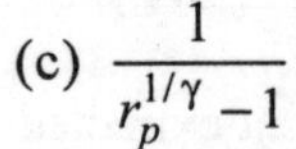 $\dfrac{1}{r_p^{1/\gamma}-1}$ (d) 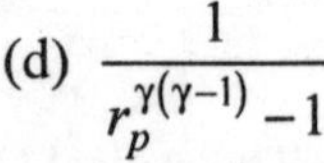 $\dfrac{1}{r_p^{\gamma(\gamma-1)}-1}$
4. For Bell-Coleman refrigeration cycle the variation of COP with pressure ratio is given by the graph. (Fig. 1).

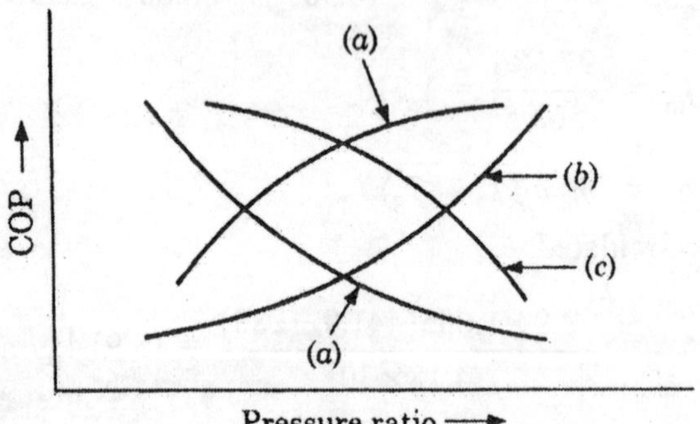

Fig. 1

5. The coefficient of performance (COP) of a refrigerator working as a heat pump is given by
   (a) $(COP)_{heat\ pump} = (COP)_{refigerator} + 2$ (b) $(COP)_{heat\ pump} = (COP)_{refigerator} + 1$
   (c) $(COP)_{heat\ pump} = (COP)_{refigerator} - 1$ (c) $(COP)_{heat\ pump} = (COP)_{refigerator}$
6. A heat pump working on a reversed Carnot cycle has a COP of 5. If it works as a refrigerator taking 1 kW of work input, the refrigerating effect will be
   (a) 1 kW (b) 2 kW (c) 3 kW (d) 4 kW
7. The coefficient of performance of heat pump $[(COP)_{HP}]$ and that of refrigerator $[(COP)_R]$
   (a) $(COP)_{HP} = (COP)_R + 1$ (b) $(COP)_{HP} = (COP)_R - 1$
   (c) $(COP)_{HP}/(COP)_{R\ =\ 1}$ (d) $(COP)_{HP} < (COP)_R$

8. For a Brayton cycle, match the following using codes given below:

List I	List II
A. Compression	1. Isothermal
B. Rejection	2. Isobaric
C. Expansion	3. Isentropic
D. Heat addition	4. Constant enthalpy

	A	B	C	D
(a)	1	2	3	4
(b)	4	3	2	1
(c)	2	3	4	1
(d)	3	4	1	2

9. In the Fig. 2, E is a heat engine with efficiency of 0.4 and R is a refrigerator. Given that $Q_2 + Q_4 = 3Q_1$, the COP of the refrigerator is

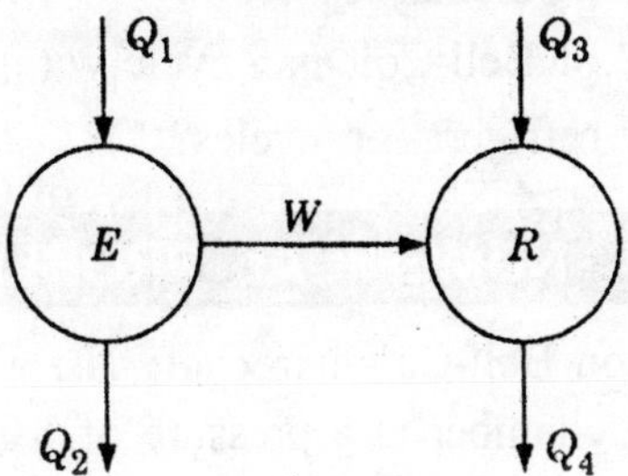

Fig. 2

(a) 2.5 (b) 3.0 (c) 4.0 (d) 5.0

10. A Bell-Coleman air refrigeration cycle works as a reversed

(a) Stirling cycle (b) Otto cycle (c) Diesel cycle (d) Brayton cycle

11. Where is an air refrigeration cycle generally employed?

(a) Domestic refrigerators (b) Commercial refrigerators
(c) Air-conditioning (d) Gas liquefaction

12. A refrigerating system operating on reversed Brayton refrigeration cycle is used for maintaining 250 K. If the temperature at the end of constant pressure cooling is 300 K and rise in the temperature of air in the refrigerator is 50 K, then the net work of compression will be (assume air as the working substance with c_p = 1 kJ per kg per °C

(a) 250 kJ/kg (b) 200 kJ/kg (c) 50 kJ/kg (d) 25 kJ/kg

13. A heat pump for domestic heating operates between a cold system at 0°C and the hot system at 60°C. What is the minimum electric power consumption if the heat rejected is 80000 kJ/hr?

(a) 2 kW (b) 3 kW (c) 4 kW (5) 5 kW

14. A Carnot refrigerator requires 1.5 kW/ton of refrigeration to maintain a region at a temperature of –30°C. The COP of the Carnot refrigerator is

(a) 1.42 (b) 2.33 (c) 2.87 (d) 3.26

Answers

1. (a) 2. (a) 3. (a) 4. (a) 5. (b) 6. (d) 7. (a) 8. (d)
9. (d) 10. (d) 11. (c) 12. (d) 13. (d) 14. (b)

REVIEW QUESTIONS

1. What is the main characteristic feature of an air-refrigeration system?
2. Differentiate between open and close air-refrigeration system.
3. State merits and demerits of an air-refrigeration system.
4. What is a dense air refrigeration cycle?
5. Define COP of reversed Carnot Cycle.
6. What are the temperature limits of reverse Carnot cycle?
7. Define COP of reversed Carnot cycle using gas as a refrigerant.
8. Draw T-S diagram for Bell-Coleman cycle.
9. Write expression for COP of Bell-Coleman cycle with polytropic processes.
10. What are the types of air refrigeration cycles?

NUMERICAL PROBLEMS

3.1 In a refrigerator working on Bell-Coleman cycle, the air is drawn into the cylinder of the compressor from the cold chamber at a pressure of 1.03 bar and temperature 12°C. After isentropic compression to 5.5 bar, the air is cooled at constant pressure to a temperature of 22°C. The polytropic expansion $pv^{1.25}$ = const. then follows and the air expanded to 1.03 bar is passed to cold chamber. Calculate (a) work done per kg of air flow, (b) Refrigerating effect per kg of air flow, (c) COP, and (d) Refrigerating capacity of the plant in tons for a mass flow rate of 90 kg/h. Take for air, $\gamma = 1.4$ and $c_p = 1.003$ kJ/kg.K

[**Ans.** 55.1 kJ/kg, 1.349, 0.528 tons]

3.2 A refrigerating system working on Bell-Coleman cycle receives air from cold chamber at –5°C and compresses it from 1 bar to 4.5 bar. The compressed air is then cooled to a temperature of 37°C before it is expanded in the expander. Calculate the COP of the system when compression and expansion are: (*i*) isentropic, and (*ii*) follow the law $pv^{1.25}$ = constant.

[**Ans.** 1.86; 1.98]

3.3 A Bell-Coleman refrigerator works between 4 bar and 1 bar pressure limits. After compression, the cooling water reduces the air temperature to 17°C. What is the lowest temperature produced by the ideal machine? Compare the coefficient of performance of this machine with that of the ideal Carnot cycle machine working between the same pressure limits, the temperature at the beginning of compression being –13°C. [**Ans.** –78°C; 2.07, 1.02]

3.4 An air refrigerator working on Bell-Coleman cycle takes air into the compressor at 1 bar and 268K. It is compressed in the compressor to 5 bar and cooled to 298 K at the same pressure. It is farther expanded in the expander to 1 bar and discharged to take the cooling load. The isentropic efficiencies of the compressor and expander are 85% and 90% respectively. Determine; (a) refrigeration capacity of the system if the air circulated is 40 kg/min; (b) power required for the compressor; and (c) COP of the system.

[**Ans.** 13.14 TR; 46 kW; 0.812]

3.5 An air refrigeration system having pressure ratio of 5 takes air at 0°C. It is compressed and then cooled to 19°C at constant pressure. If the efficiency of the compressor is 95% and that of expander is 75%, determine: (a) the refrigeration capacity of the system, if the flow of air is 75 kg/min; (b) the power of the compressor; and (c) COP of the system. Assume compression and expansion processes to be isentropic Take $\gamma = 1.4$; $c_p = 1$ kJ/kg.K and $c_v = 0.72$ kJ/kg K. [**Ans.** 31.68 TR; 106.6 kW; 1.71]

3.6 A Carnot refrigerator operates between temperatures of –50°C and 50°C. Determine COP of the refrigerator. If the COP is to be made 4 by changing the temperatures such that increase or decrease in upper temperature is equal to decrease or increase in lower temperature. Calculate the new temperatures. [**Ans.** 2.23, 203.8 K, 342.2 K]

3.7 A refrigerator has working temperatures in the evaporate and condenser coils of –30°C and 35°C respectively. Calculate the maximum possible COP of the refrigerator. [**Ans.** 4.738]

3.8 In a refrigerating plant working on the Bell-Coleman cycle the air is drawn in from the cold chamber at –5°C and 1.03 bar. The same air is compressed reversibly and adiabatically to 5.25 bar before being cooled in the heat exchanger to 15°C. After the exchanger the air is expanded adiabatically and reversibly to cold chamber state. Assuming the value of $c_p = 1.004$ kJ/kg. K and $c_v = 0.717$ kJ/kg. K for air, determine:

(*i*) Work done on the cycle per kg of air;

(*ii*) The refrigeration effect produced in cold chamber;

(*iii*) COP

3.9 In a Bell-Coleman refrigerating machine, the air is drawn from the cold chamber at a pressure of 1.03 bar and temperature of –10°C and compressed isentropically to 6.18 bar. The same is cooled to 25°C. It is then expanded in the expansion cylinder following the law pv^{13} = constant and discharged back to the cold chamber. Assume $c_p = 1.004$ kJ/kg. K, $c_v = 0.717$ kJ/kg. K for air throughout the cycle. Dteremine:

(*i*) Work input to the cycle per kg of air;

(*ii*) Refrigeration produced in cold chamber;

(*iii*) COP of the cycle.

3.10 A theoretical open air refrigerating cycle operates with a high pressure of 10 bar and a low pressure of 1 bar. The air may be cooled at 16°C and it leaves the refrigerating room at –18°C. Calculate:

(*i*) Temperature at leaving the expander and compressor;

(*ii*) kg of air be circulated per minute per ton of refrigeration;

(*iii*) Work of compression in kW per ton;

(*iv*) Work of expander in kW per ton;

(*v*) Net work in kW per ton and net power per ton;

(*vi*) COP;

(*vii*) Heat removed in air cooler in kW per ton;

(*viii*) Theoretical piston displacement of compressor in m^3/min per ton;

(*ix*) Theoretical piston displacement of expander in m^3/min per ton.

3.11 A theoretical dense air refrigerating cycle operates with high pressure of 20 bar and a low pressure of 5 bar. The air may be water cooled to 16°C and it leaves the refrigerator room at –18°C. Calculate:

(*i*) Temperature at leaving the expander and the compressor;

(*ii*) Kilograms of air to be circulated per minute per ton;

(*iii*) Work of compressor in kJ/min per ton;

(*iv*) Work of the expander in kJ/min per ton;

(*v*) Network in kJ/min per ton and net power per ton;

(*vi*) COP;

(*vii*) Heat removed in air cooler kJ/min per ton;

(*viii*) Theoretical piston displacement of compressor in m^3/min per ton;

3.12 An air refrigerating system operating on a closed cycle is required to produce 20 ton of refrigeration with a cooler pressure of 16 bar and refrigerator pressure of 4.5 bar. Leaving air temperatures are 20°C for cooler and 0°C for refrigerator. If the compressor and expander clearance is 4%, the expansion is isentropic and compression is polytropic with n = 1.3. Determine:

(*i*) Kilograms per minute of air circulated.

(*ii*) m^3/min of compressor piston displacement;

(*iii*) m^3/min of expander displacement;

(*iv*) COP;

(*v*) Net power per ton required.

Chapter

4

AIRCRAFT REFRIGERATION SYSTEMS

4.1 AIRCRAFT REFRIGERATION SYSTEM

The high-speed passenger aircraft, jet aircraft and missiles have introduced the need for compact and simple refrigeration systems. When the power requirements needed to transport the additional weight of the refrigerating system are taken into account, the air cycle systems prove to be the most efficient. The advantages of air cycle with regard to its applications in aircraft refrigeration can be listed as follows:

1. Small amounts of leakages are tolerable with air as the refrigerant.
2. Air cycle in its simplest form as an open system requires only one heat exchanger.
3. Availability of air as a refrigerant in mid-air is also an important consideration.
4. Cabin pressurization and air-conditioning can be combined into one operation.
5. Initial compression of air is obtained by the ram effect, viz, conversion of the high kinetic energy of the ambient air relative to the aircraft into enthalpy and hence pressure rise.

The coefficient of performance of air cycle refrigeration is lower than that of vapour compression cycle. The present day jet aircrafts have very high cooling loads because of their large occupancy, electronic equipment and heat generation due to high velocity and skin friction.

4.2 AIR CYCLES FOR AIRCRAFT

Two types of air cycles are used for an aircraft.

1. ***Closed air cycle system (or dense air machine)***: It has many thermodynamic advantages. It can work at a suction pressure higher than the atmospheric pressure. This reduces the volumes handled by the compressor and the expander. The operating pressure ratio can be reduced, resulting in higher coefficient of performance.
2. ***Open air cycle system***: It this system, the air after expansion is directly led to the conditioned space. It is, therefore, necessary to expand air to one atmosphere pressure. This requires larger volumes to be handled. It does not require a heat exchanger for the refrigeration process. This saves the weight and cost of the equipment. Its one disadvantage is that when the air drawn from the refrigerated space is humid, it might produce fog and ice at the end of the expansion process and clog the line. This necessitates the use of a drier in the circuit.

4.3 CLASSIFICATION OF AIRCRAFT REFRIGERATION SYSTEMS

The various methods of air refrigeration systems used for aircrafts are:

1. Simple air cooling system.
2. Simple evaporative cooling system.
3. Boot strap air cooling system.
4. Boot strap air evaporative cooling system.
5. Reduced ambient air cooling system, and
6. Regenerative air cooling system.

4.3.1 Simple Air Cooling System

The schematic diagram of a simple air cooling system for aircraft is shown in Fig. 4.1.

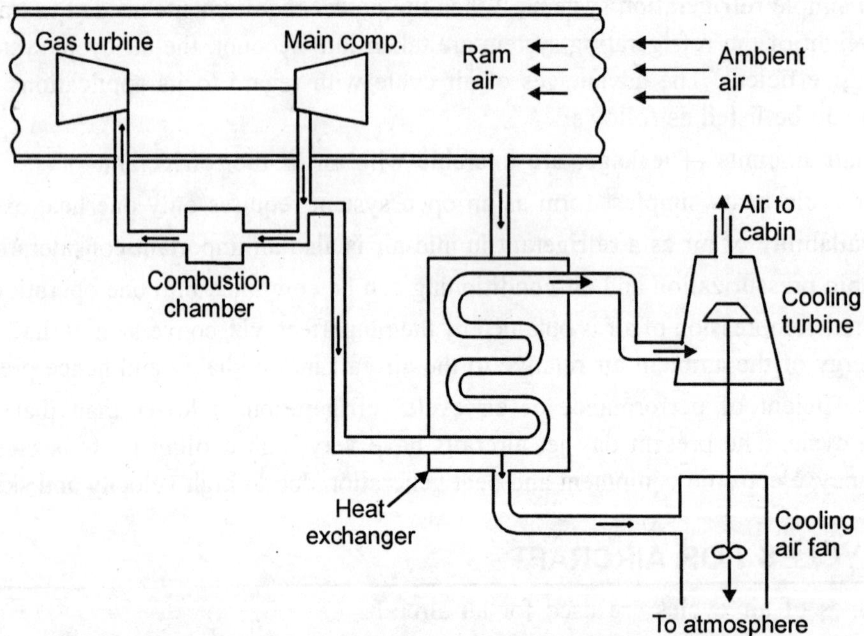

Fig. 4.1 Simple air cooling system schematic diagram.

The main components of this system are the main compressor driven by a gas turbine, a heat exchanger, a cooling turbine and a cooling air fan. The air required for refrigeration system is bled off from the main compressor. This high pressure and high temperature air is cooled initially in the heat exchanger where ram air is used for cooling. It is further cooled in the cooling turbine by the process of expansion. The work of this turbine is used to drive the cooling fan which draws cooling air through the heat exchanger. This system is good for ground surface cooling and low flight speeds.

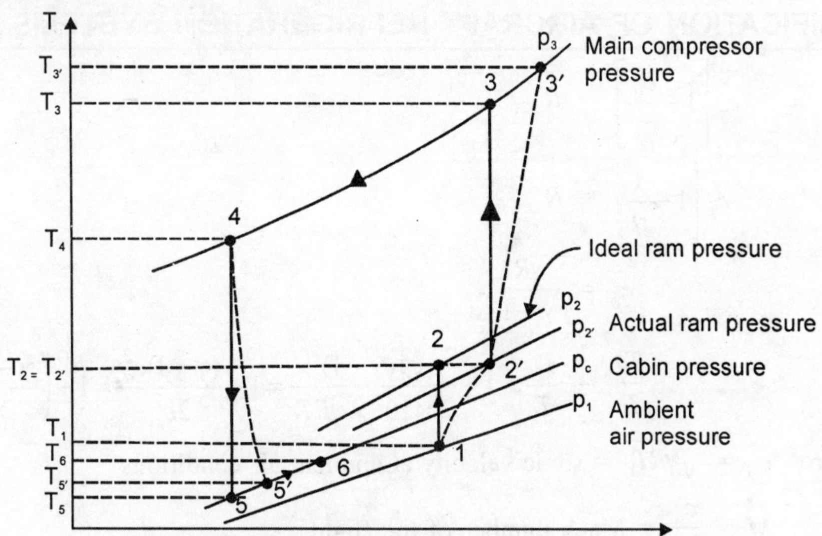

Fig. 4.2 *T-s* diagram for simple air-cycle cooling system.

The various processes are discussed below:

1. **Ramming process:** Let the pressure and temperature of ambient air be p_1 and T_1 respectively. The ambient air is rammed isentropically from pressure p_1 and temperature T_1 to the pressure p_2 and temperature T_2. This ideal ramming action is shown by the vertical line 1-2 in Fig 4.2. The actual ramming process is shown by the curve 1–2′, which is adiabatic but not isentropic due to friction. The actual pressure and temperature of rammed air are $p_{2'}$ and $T_{2'}$ respectively. If c_a is the velocity of air relative to the aircraft then kinetic energy of outside air relative to the aircraft,

$$\text{K.E.} = \frac{c_a^2}{2\times10^3} \text{ kJ/kg}$$

From the energy equation, we have

$$h_2 - h_1 = \frac{c_a^2}{2\times10^3}$$

or
$$c_p(T_2 - T_1) = \frac{c_a^2}{2\times10^3}$$

$$T_2 = T_1 + \frac{c_a^2}{2\times10^3 c_p}$$

$$\frac{T_2}{T_1} = 1 + \frac{c_a^2}{2\times10^3 c_p T_1}$$

Now
$$T_{2'} = T_2 \text{ and } h_2 = h_{2'}$$

$\therefore$
$$\frac{T_{2'}}{T_1} = 1 + \frac{c_a^2}{2\times10^3 c_p T_1}$$

Now $\quad c_p - c_v = R$

or $\quad c_p\left[1 - \dfrac{c_v}{c_p}\right] = R$

$$c_p\left[1 - \frac{1}{\gamma}\right] = R$$

$$c_p = \frac{\gamma R}{\gamma - 1}$$

$$\therefore \quad \frac{T_2}{T_1} = \frac{T_{2'}}{T_1} = 1 + \frac{c_a^2(\gamma - 1)}{2 \times 10^3 \times \gamma R T_1} = 1 + \frac{(\gamma - 1)c_a^2}{2c_s^2} = 1 + \left(\frac{\gamma - 1}{2}\right)M^2$$

where $\quad c_s = \sqrt{\gamma R T_1}$ = sonic velocity at ambient air conditions

$M = \dfrac{c_a}{c_s}$ = Mach number of the flight

$T_2 = T_{2'}$ is called the stagnation temperature of ambient air entering the main compressor.

Stagnation pressure after isentropic compression is given by,

$$\frac{p_2}{p_1} = \frac{p_2}{p_1} = \left(\frac{T_2}{T_1}\right)^{\frac{\gamma}{\gamma - 1}}$$

Ram efficiency, $\quad \eta_R = \dfrac{\text{Actual rise in pressure}}{\text{Isentropic rise in pressure}} = \dfrac{p_{2'} - p_1}{p_2 - p_1}$

2. **Compression process:** The isentropic compression of air in the compressor is represented by the line 2′-3 and the actual compression by the curve 2′-3′ because of internal friction.

   Compression work, $\quad W_c = m_a\, c_p(T_{3'} - T_{2'})$

   where m_a = mass of air bled from the main compressor for refrigeration purposes.

3. **Cooling process:** The compressed air is cooled by the ram air in the heat exchanger. This process is shown by the curve 3′-4. In actual process, there is a pressure drop in the heat exchanger (not shown). The temperature of air drops from $T_{3'}$ to T_4.

   Heat rejected in the heat exchanger during the cooling process, $Q_r = m_a\, c_p(T_{3'} - T_4)$

4. **Expansion process:** The cooled air is now expanded isentropically in the cooling turbine represented by the line 4–5 and the actual expansion by the curve 4-5′.

   Work done by the cooling turbine during expansion process, $W_t = m_a\, c_p(T_4 - T_{5'})$

   The work done by the cooling turbine is used to drive the cooling air fan which draws cooling air from the heat exchanger.

5. **Refrigeration process:** The air from the cooling turbine is sent to the cabin and cockpit where it gets heated up by the heat of equipment and occupancy. This process is shown by the curve 5′-6.

Refrigerating effect (or heat absorbed), Q_a or $R_E = m_a\, c_p(T_6 - T_{5'})$

where T_6 = inside temperature of cabin.

$$\text{COP of cycle} = \frac{\text{Refrigeration effect produced}}{\text{Compression work}} = \frac{m_a c_p (T_6 - T_{5'})}{m_a c_p (T_{3'} - T_{2'})} = \frac{T_6 - T_{5'}}{T_{3'} - T_{2'}} \quad ...(4.1)$$

Let Q = Cooling load in the cabin, tonnes of refrigeration.

$$\text{Then amount of air required, } m_a = \frac{211Q}{c_p(T_6 - T_{5'})} \text{ kg/min} \quad ...(4.2)$$

Power required for the refrigeration system,

$$P = \frac{m_a c_p (T_{3'} - T_{2'})}{60} \text{ kW} \quad ...(4.3)$$

$$\text{COP} = \frac{211Q}{m_a c_p (T_{3'} - T_{2'})} = \frac{\frac{211Q}{60}}{} = \frac{3.5167Q}{P} \quad ...(4.4)$$

Example 4.1 *A simple air cooled system is used for an aircraft having a load of 12 tonnes. The atmospheric pressure and temperature are 0.9 bar and 10°C. The pressure increases to 1.015 bar due to ramming. The temperature of air is reduced by 53°C in the heat exchanger. The pressure in the cabin is 1.01 bar and temperature of air leaving the cabin is 25°C. Calculate:*

(a) Power required, and (b) COP of the system.

Assume all compressions and expansions to be isentropic. The pressure of compressed air is 3.5 bar.

Solution: Given: Q = 12 TR, p_1 = 0.9 bar, T_1 = 273 + 10 = 283 K, p_2 = 1.015 bar, $p_5 = p_6$ = 1.01 bar, Δt_3 = 53°C, T_6 = 273 + 25 = 298 K, p_3 = 3.5 bar

The T-s diagram for the various processes is shown in Fig. 4.3

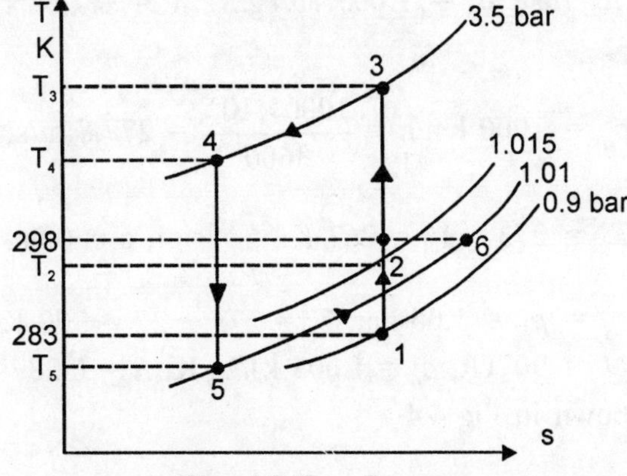

Fig. 4.3 T-s diagram.

(*a*) **Process 1-2:**

$$T_2 = T_1\left(\frac{p_2}{p_1}\right)^{\frac{\gamma-1}{\gamma}} = 283\left(\frac{1015}{0.9}\right)^{\frac{1.4-1}{1.4}} = 292.9\,\text{K}$$

Process 2-3:

$$T_3 = T_2\left(\frac{p_3}{p_2}\right)^{\frac{\gamma-1}{\gamma}} = 292.9\left(\frac{3.5}{1.015}\right)^{\frac{0.4}{1.4}} = 417.2 \text{ K or } t_3 = 144.2°\text{C}$$

Temperature of air leaving the heat exchanger,

$$T_4 = t_3 - \Delta t_3$$
$$= 144.2 - 53 = 91.2°\text{C or } 364.2 \text{ K}$$

Process 4-5:

$$T_5 = T_4\left(\frac{p_5}{p_4}\right)^{\frac{\gamma-1}{\gamma}} = 364.2\left(\frac{1.01}{3.5}\right)^{\frac{0.4}{1.4}} = 255.3 \text{ K}$$

Mass of air required, $\dot{m}_a = \dfrac{211Q}{c_p(T_6 - T_5)} = \dfrac{211\times 12}{1.005(298-255.3)} = 59$ kg/min

Power required, $P = \dfrac{m_a c_p(T_3 - T_2)}{60} = \dfrac{59\times 1.005(417.2-292.9)}{60} = 122.84$ kW

(*b*) $\text{COP} = \dfrac{211Q}{60P} = \dfrac{211\times 12}{60\times 122.84} = 0.343$

Example 4.2 *An aircraft moving with a speed of 1000 km/h uses simple gas refrigeration cycle for air conditioning. The ambient pressure and temperature are 0.35 bar and –10°C. The pressure ratio of compressor is 4. The heat exchanger effectiveness is 0.96. The isentropic efficiencies of compressor and expander are 0.85 and 0.82 respectively. The cabin pressure and temperature are 1.06 bar and 27°C. Determine the temperatures and pressures at all points of the cycle. Also calculate the volume flow rate through compressor inlet and expander outlet for 90 TR. Take* c_p *= 1.005 kJ/kg.K,* R *= 0.287 kJ/kg.K and* γ *= 1.4 for air.*

Solution: Given $c_a = 1000 \text{ km/h} = \dfrac{1000\times 10^3}{3600} = 277.8$ m/s, $p_1 = 0.35$ bar,

$T_1 = 273 - 10 = 263$ K, $\dfrac{p_3}{p_2} = 4$, $\varepsilon = 0.96$, $\eta_c = 0.85$, $\eta_e = 0.82$,

$p_5 = p_{5'} = 1.06$ bar, $T_6 = 273 + 27 = 300$ K,

$Q = 90$ TR, $c_p = 1.005$ kJ/kg.K, $R = 0.287$ kJ/kg.K, $\gamma = 1.4$

The *T-s* diagram is shown in Fig 4.4

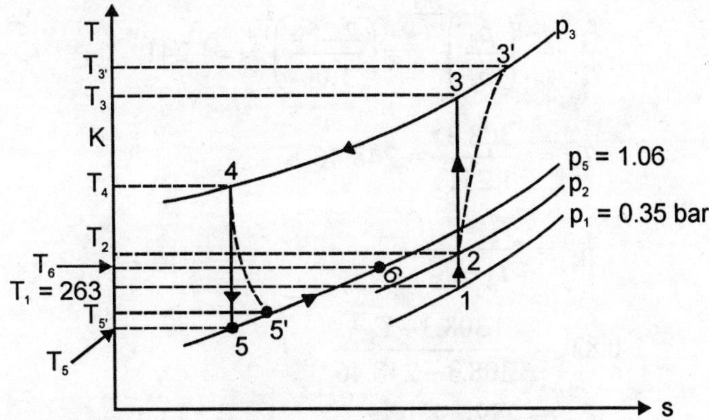

Fig. 4.4 T-s diagram.

Stagnation temperature due to ramming effect,

$$T_2 = T_1 + \frac{c_a^2}{2c_p} = 263 + \frac{(277.8)^2}{2 \times 1.005 \times 10^3} = 301.4 \text{ K}$$

$$p_2 = p_1 \left(\frac{T_2}{T_1}\right)^{\frac{\gamma}{\gamma-1}} = 0.35 \left(\frac{301.4}{263}\right)^{\frac{1.4}{0.4}} = 0.564 \text{ bar}$$

$$p_3 = 4p_2 = 4 \times 0.564 = 2.256 \text{ bar}$$

$$\frac{T_3}{T_2} = \left(\frac{p_3}{p_2}\right)^{\frac{\gamma-1}{\gamma}} = (4)^{\frac{0.4}{1.4}} = 1.486$$

$$T_3 = 301.4 \times 1.486 = 447.88 \text{ K}$$

$$\eta_c = \frac{T_3 - T_2}{T_{3'} - T_2}$$

$$0.85 = \frac{447.88 - 301.4}{T_{3'} - 301.4}$$

$$T_{3'} = 473.73 \text{ K}$$

Heat exchanger effectiveness,

$$\varepsilon = \frac{T_{3'} - T_4}{T_{3'} - T_2}$$

$$0.96 = \frac{473.73 - T_4}{473.73 - 301.4}$$

$$T_4 = 308.3 \text{ K}$$

Also $\quad p_4 = p_3 = 2.256 \text{ bar}$

$$\frac{T_4}{T_5} = \left(\frac{p_4}{p_5}\right)^{\frac{\gamma-1}{\gamma}} = \left(\frac{2.256}{1.06}\right)^{\frac{0.4}{1.4}} = 1.241$$

$$T_5 = \frac{308.3}{1.241} = 248.46 \text{ K}$$

$$\eta_e = \frac{T_4 - T_{5'}}{T_4 - T_5}$$

$$0.82 = \frac{308.3 - T_{5'}}{308.3 - 248.46}$$

$$T_{5'} = 259.2 \text{ K}$$

Mass flow rate of air, $\dot{m}_a = \dfrac{211Q}{c_p(T_6 - T_{5'})} = \dfrac{211 \times 90}{1.005(300 - 259.3)} = 463.126$ kg/min

$$\dot{V}_2 = \frac{\dot{m}_a R T_2}{p_2} = \frac{463.126 \times 0.287 \times 301.4}{0.564 \times 10^2} = 710.3 \text{ m}^3/\text{min}$$

$$\dot{V}_5 = \frac{\dot{m}_a R T_{5'}}{p_{5'}} = \frac{463.126 \times 0.287 \times 259.2}{1.06 \times 10^2} = 325.02 \text{ m}^3/\text{min}$$

Example 4.3 *The cockpit of a jet plane flying at a speed of 1500 km/h is to be cooled by a simple air cooling system. The cockpit is to be maintained at 25°C and 1 bar. The ambient air is at 30°C and 0.90 bar. The other data available is as follows:*

Cockpit cooling load = 10 TR, Compressor pressure ratio = 4.5, Ram efficiency = 0.90, Temperature of air leaving the heat exchanger and entering the cooling turbine = 65°C.

Pressure drop in heat exchanger = 0.5 bar, Pressure loss between cooler turbine and cockpit = 0.2 bar, Isentropic efficiency of compressor = 0.82, Isentropic efficiency of cooler turbine = 0.80, c_p *= 1.005 kJ/kg.K,* γ *= 1.4.*

Determine: (a) Quantity of air passed through the cooling turbine, and
(b) COP of the system.

Solution: Given: $c_a = \dfrac{1500 \times 10^3}{3600} = 416.67$ m/s, $T_6 = 25°\text{C}$ or $273 + 25 = 298$ K, $p_6 = 1$ bar,

$T_1 = 30°\text{C}$ or $273 + 30 = 303$ K, $p_1 = 0.90$ bar, $Q = 10$ *TR*, $\dfrac{p_3}{p_{2'}} = 4.5$, $\eta_R = 0.90$,

$T_4 = 65°\text{C}$ or $273 + 65 = 338$ K, $p_4 = (p_{3'} - 0.5)$ bar, $p_5 = p_{5'} = p_6 + 0.2 = 1 + 0.2 = 1.2$ bar,
$\eta_c = 0.82$, $\eta_t = 0.80$, $c_p = 1.005$ kJ/kg.K, $\gamma = 1.4$

Refer to Fig. 4.5.

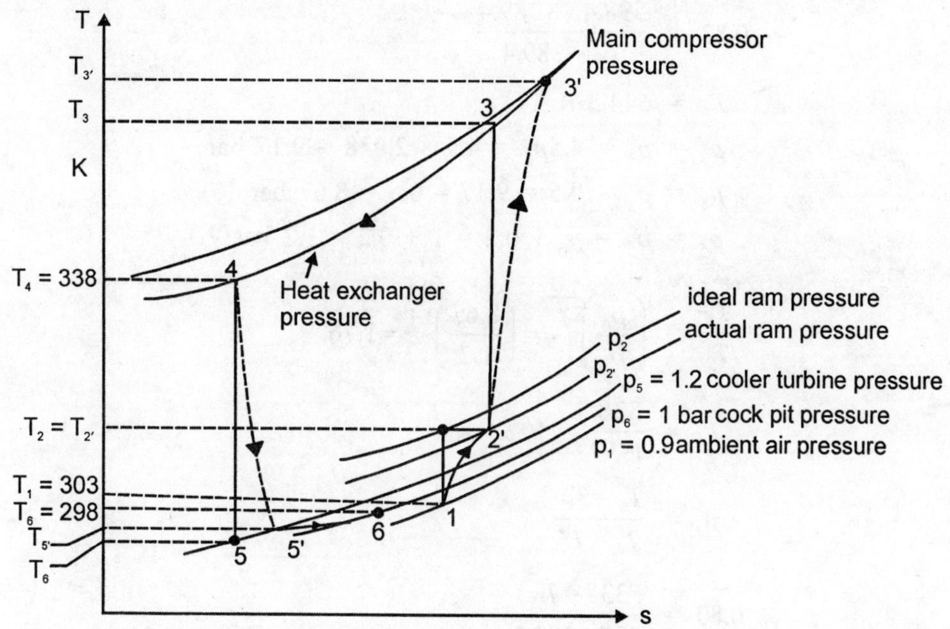

Fig. 4.5 T-s diagram for aircraft.

$$T_2 = T_{2'} = T_1 + \frac{c_u^2}{2c_p} = 303 + \frac{(416.67)^2}{2 \times 1.005 \times 10^3} = 389.4 \text{ K}$$

$$\frac{p_2}{p_1} = \left(\frac{T_2}{T_1}\right)^{\frac{\gamma-1}{\gamma}} = \left(\frac{389.4}{303}\right)^{\frac{1.4}{0.4}} = 2.046$$

$$p_2 = 0.9 \times 2.406 = 2.165 \text{ bar}$$

Ram efficiency, $\eta_R = \dfrac{p_{2'} - p_1}{p_2 - p_1}$

$$0.9 = \frac{p_{2'} - 0.9}{2.165 - 0.9}$$

$$p_{2'} = 2.038 \text{ bar}$$

$$\frac{T_3}{T_{2'}} = \left(\frac{p_3}{p_{2'}}\right)^{\frac{\gamma-1}{\gamma}} = (4.5)^{\frac{0.4}{1.4}} = 1.5368$$

$$T_3 = 389.4 \times 1.5368 = 598.45 \text{ K}$$

$$\eta_c = \frac{T_3 - T_{2'}}{T_{3'} - T_{2'}}$$

$$0.82 = \frac{598.45 - 389.4}{T_{3'} - 389.4}$$

$$T_{3'} = 644.3 \text{ K}$$

$$p_3 = p_{3'} = 4.5 p_{2'} = 4.5 \times 2.038 = 9.17 \text{ bar}$$

$$p_4 = p_{3'} - 0.5 = 9.17 - 0.5 = 8.67 \text{ bar}$$

$$p_5 = p_{5'} = p_6 + 0.2 = 1 + 0.2 = 1.2 \text{ bar}$$

$$\frac{T_4}{T_5} = \left(\frac{p_4}{p_5}\right)^{\frac{\gamma-1}{\gamma}} = \left(\frac{8.67}{1.2}\right)^{\frac{0.4}{1.4}} = 1.76$$

$$T_5 = \frac{338}{1.76} = 192.1 \text{ K}$$

$$\eta_t = \frac{T_4 - T_{5'}}{T_4 - T_5}$$

$$0.80 = \frac{338 - T_{5'}}{338 - 192.1}$$

$$T_{5'} = 221.3 \text{ K}$$

Quantity of air passed through cooling turbine,

$$\dot{m}_a = \frac{211Q}{c_p(T_6 - T_{5'})} = \frac{211 \times 10}{1.005(298 - 221.3)} = 27.37 \text{ kg/min}$$

$$\text{COP of system} = \frac{211Q}{\dot{m}_a c_p (T_{3'} - T_{2'})} = \frac{211 \times 10}{27.37 \times 1.005(644.3 - 389.4)} = 0.3$$

4.3.2 Simple Air Evaporative Cooling System

A simple air evaporative cooling system is shown in Fig. 4.6. It is similar to the simple cooling system except that the addition of an evaporator between the heat exchanger and cooling turbine. The evaporator provides an additional cooling effect through evaporation of a refrigerant such as water. At high altitudes the evaporative cooling may be obtained by using alcohol or ammonia. The water, alcohol and ammonia have different refrigerating effect at different altitudes. At 20,000 metres height, water boils at 40°C, alcohol at 9°C and ammonia at –70°C.

The *T-s* diagram for a simple air cycle evaporative cooling system is shown in Fig. 4.7.

The various processes are same as discussed in the previous article, except that the cooling process in the evaporator is shown by 4-4′ in Fig. 4.7.

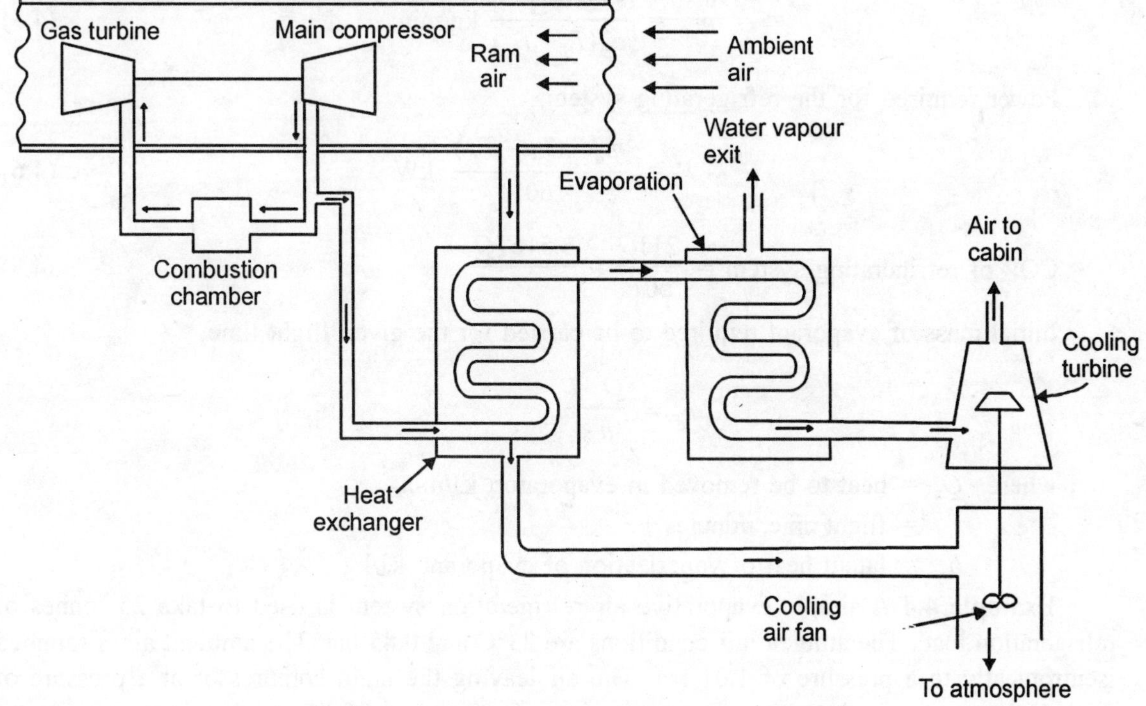

Fig. 4.6 Simple air evaporative cooling system.

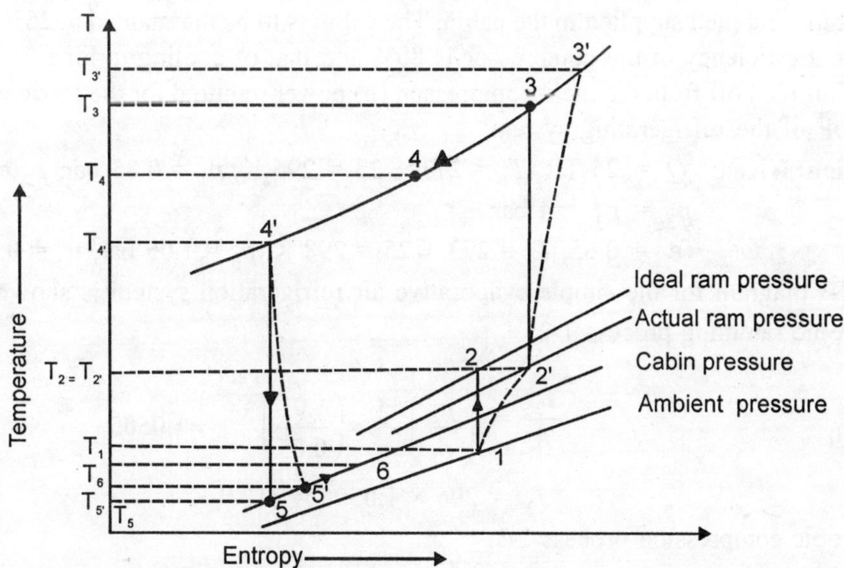

Fig. 4.7 *T-s* diagram for simple evaporative cooling system.

Let Q = Cooling load in the cabin, tonnes of refrigeration

Then air required for refrigeration,

$$\dot{m}_a = \frac{211Q}{c_p(T_6 - T_{5'})} \text{ kg/min} \quad ...(4.5)$$

Power required for the refrigerating system,

$$P = \frac{\dot{m}_a c_p (T_{3'} - T_{2'})}{60} \text{ kW} \quad ...(4.6)$$

$$\text{COP of refrigerating system} = \frac{211Q}{60P} = \frac{3.5167Q}{P} \quad ...(4.7)$$

Initial mass of evaporant required to be carried for the given flight time,

$$m_e = \frac{Q_e \cdot t}{h_{fg}}$$

where Q_e = heat to be removed in evaporator, kJ/min

t = flight time, minutes

h_{fg} = latent heat of vaporization of evaporant, kJ/kg

Example 4.4 A simple evaporative air refrigeration system is used to take 25 tonnes of refrigeration load. The ambient air conditions are 23°C and 0.85 bar. The ambient air is rammed isentropically to a pressure of 1.01 bar. The air leaving the main compressor at a pressure of 4 bar is first cooled in the heat exchanger having effectiveness of 0.65 and then in the evaporator where its temperature is reduced by 5°C. The air from the evaporator is passed through the cooling turbine and then supplied to the cabin. The cabin is to be maintained at 25°C and 1.05bar. If the internal efficiency of the compressor is 80% and that of cooling turbine 75%, determine: (a) mass of air bled off from the main compressor, (b) power required for the refrigerating system, and (c) COP of the refrigerating system.

Solution: Given: Q = 25 TR, T_1 = 273 + 23 = 296 K, p_1 = 0.85 bar, p_2 = 1.01 bar, $p_3 = p_{3'}$ = 4 bar,

ε = 0.65, T_6 = 273 + 25 = 298 K, p_6 = 1.05 bar, η_c = 0.80, η_t = 0.75

The *T-s* diagram for the simple evaporative air refrigeration system is shown in Fig. 4.8.

Isentropic ramming process 1-2:

$$\frac{T_2}{T_1} = \left(\frac{p_2}{p_1}\right)^{\frac{\gamma}{\gamma-1}} = \left(\frac{1.01}{0.85}\right)^{\frac{1.4-1}{1.4}} = 1.0505$$

$$T_2 = 296 \times 1.0505 = 310.9 \text{ K}$$

Isentropic compression process 2-3:

$$T_3 = T_2\left(\frac{p_3}{p_2}\right)^{\frac{\gamma}{\gamma-1}} = 310.9\left(\frac{4}{1.01}\right)^{\frac{0.4}{1.4}} = 460.7 \text{ K}$$

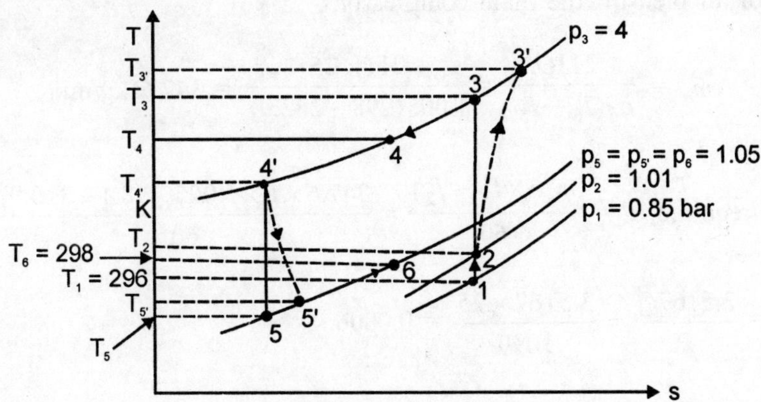

Fig. 4.8 T-s diagram.

Compressor efficiency,

$$\eta_c = \frac{T_3 - T_2}{T_{3'} - T_2}$$

$$0.8 = \frac{460.7 - 310.9}{T_{3'} - 310.9}$$

$$T_{3'} = 498.1 \text{ K}$$

Heat exchanger effectiveness,

$$\varepsilon = \frac{T_{3'} - T_4}{T_{3'} - T_2}$$

$$0.65 = \frac{498.1 - T_4}{498.1 - 310.9}$$

$$T_4 = 376.4 \text{ K or } 103.4°\text{C}$$

Temperature of air leaving the evaporator and entering the cooling turbine,

$$T_{4'} = T_4 - 5°\text{C} = 103.4 - 5 = 98.4°\text{C or } 371.4 \text{ K}$$

Isentropic expansion process 4′–5:

$$\frac{T_{4'}}{T_5} = \left(\frac{p_3}{p_6}\right)^{\frac{\gamma-1}{\gamma}} = \left(\frac{4}{1.05}\right)^{\frac{0.4}{1.4}} = 1.4654$$

$$T_5 = \frac{371.4}{1.4654} = 253.4 \text{ K}$$

Turbine efficiency,

$$\eta_t = \frac{T_{4'} - T_{5'}}{T_{4'} - T_5}$$

$$0.75 = \frac{371.4 - T_{5'}}{371.4 - 253.4}$$

$$T_{5'} = 282.9 \text{ K}$$

(*a*) Mass of air bled off the main compressor,

$$\dot{m}_a = \frac{211Q}{c_p(T_6 - T_{5'})} = \frac{211 \times 25}{1.005(298 - 282.9)} = 347.6 \text{ kg/min}$$

(*b*) Power required, $P = \frac{\dot{m}_a c_p (T_{3'} - T_2)}{60} = \frac{347.6 \times 6 \times 1.005(498.1 - 310.9)}{60} = 1090 \text{ kW}$

(*c*) $\text{COP} = \frac{3.5167Q}{P} = \frac{3.5167 \times 25}{1090} = 0.0806$

4.3.3 Boot-strap Air cooling System

A boot-strap air cooling system is shown in Fig. 4.9. This cooling system has two heat exchangers instead of one and a cooling turbine drives a secondary compressor instead of cooling fan. The air bled from the main compressor is first cooled by the ram air in the first heat exchanger. This cooled air, after compression in the secondary compressor, is led to the second heat exchanger where it is again cooled by the ram air before passing to the cooling turbine. This type of cooling system is mostly used in transport type aircraft.

The *T-s* diagram for a boot-strap air cycle system is shown in Fig. 4.10. The various processes are as follows:

1. The process 1-2 represents the isentropic ramming of ambient air from pressure p_1 and temperature T_1 to pressure p_2 and temperature T_2. The process 1–2′ represents the actual ramming process because of internal friction due to irreversibilities.
2. The process 2′-3 represents the isentropic compression of air in the main compressor and the process 2′-3′ represents the actual compression of air because of internal friction due to irreversibilities.
3. The process 3′-4 represents the cooling by ram air in the first heat exchanger. The pressure drop in the exchanger is neglected.
4. The process 4-5 represents compression of cooled air, from first heat exchanger, in the secondary compressor. The process 4-5′ represents the actual compression process because of internal friction due to irreversibilties.
5. The process 5′-6 represents the cooling by ram air in the second heat exchanger. The pressure drop in heat exchanger is neglected.
6. The process 6-7 represents isentropic expansion of the cooled air in the cooling turbine upto the cabin pressure. The process 6-7′ represents actual expansion of the cooled air in the cooling turbine.
7. The process 7′-8 represents the heating of air upto the cabin temperature T_8.

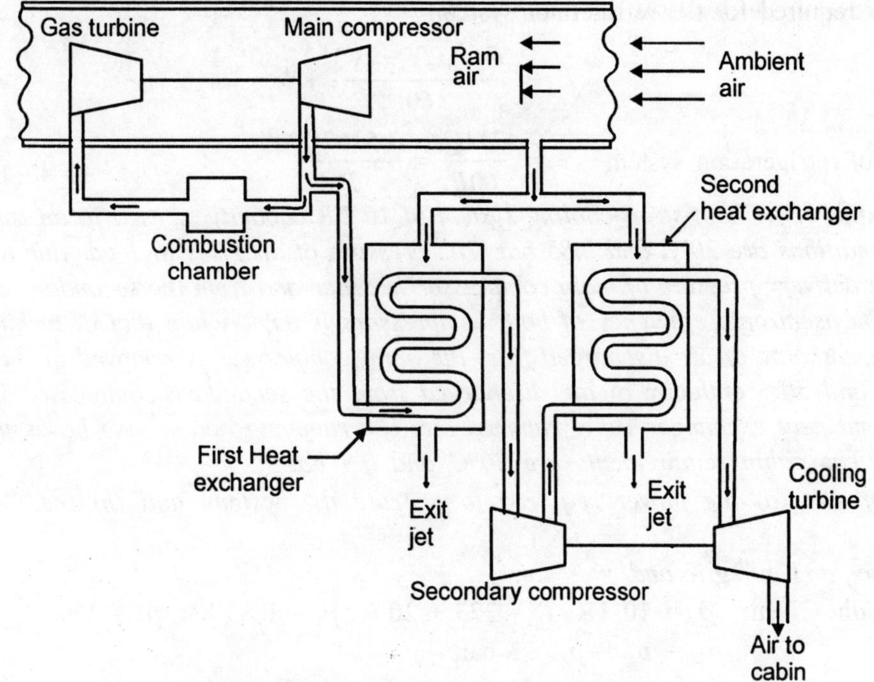

Fig 4.9 Boot-strap air cooling system.

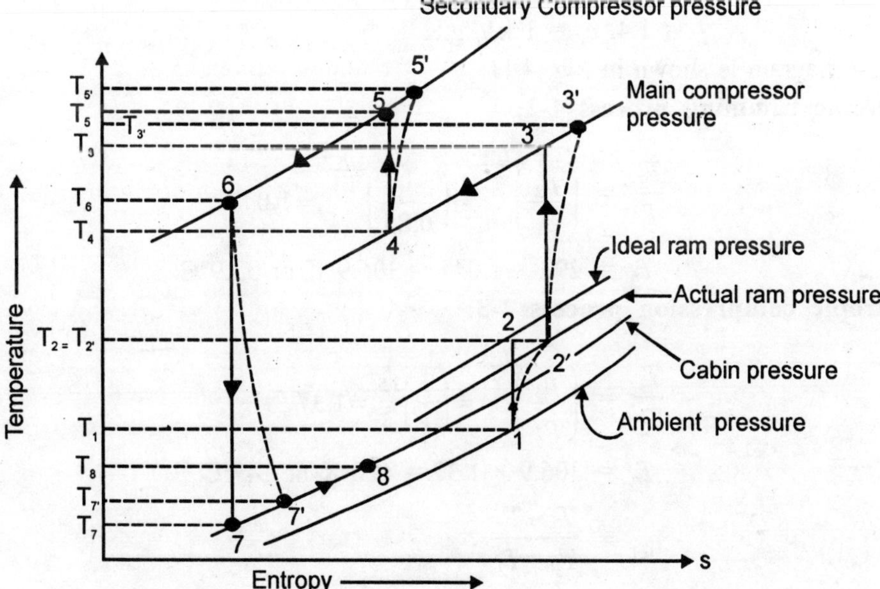

Fig. 4.10 *T-s* diagram for boot strap air cooling system.

Let Q = cooling load in the cabin, TR

Quantity of air required for refrigerator purpose,

$$\dot{m}_a = \frac{211Q}{c_p(T_8 - T_{7'})} \text{ kg/min} \quad \text{...(4.8)}$$

Power required for the refrigerator system,

$$P = \frac{\dot{m}_a c_p (T_{3'} - T_{2'})}{60} \text{ kW} \quad \text{...(4.9)}$$

$$\text{COP of refrigeration system} = \frac{211Q}{60P} = \frac{3.5167Q}{P} \quad \text{...(4.10)}$$

Example 4.5 *A boot-strap cooling system of 10 TR capacity is used in an aircraft. The ambient conditions are 20°C and 0.85 bar. The pressure of air rises to 1 bar due to ramming action. The delivery pressure of main compressor is 3 bar and from the secondary compressor is 4 bar. The isentropic efficiency of both compressors is 85%, while that of turbine is 90%. 60% of the enthalpy of air discharged from the main compressor is removed in the first heat exchanger and 30% enthalpy of air discharged from the secondary compressor is removed in the second heat exchanger using rammed air. The ramming action may be assumed to be isentropic. The cabin requirements are 20°C and 0.9 bar.*

Determine: (a) the power required to operate the system, and (b) the COP of the system.

Take c_p = 1 kJ/kg.K and γ = 1.4.

Solution: Given: Q = 10 TR, T_1 = 273 + 20 K, p_1 = 0.85 bar, p_2 = 1 bar,

$p_3 = p_{3'} = p_4$ = 3 bar,

$p_5 = p_{5'} = p_6$ = 4 bar, $\eta_{c1} = \eta_{c2}$ = 0.85, η_t = 0.90, $p_7 = p_{7'} = p_8$ = 0.9 bar,

T_8 = 273 + 20 = 293 K,

γ = 1.4, c_p = 1 kJ/kg.K

The *T-s* diagram is shown in Fig. 4.11.

Isentropic ramming process 1-2:

$$\frac{T_2}{T_1} = \left(\frac{p_2}{p_1}\right)^{\frac{\gamma-1}{\gamma}} = \left(\frac{1}{0.85}\right)^{\frac{1.4-1}{1.4}} = 1.047$$

$$T_2 = 293 \times 1.047 = 306.9 \text{ K or } 33.9°\text{C}$$

Isentropic compression process 2-3:

$$\frac{T_3}{T_2} = \left(\frac{p_3}{p_2}\right)^{\frac{\gamma-1}{\gamma}} = \left(\frac{3}{1}\right)^{\frac{1}{3.5}} = 1.37$$

$$T_3 = 306.9 \times 1.37 = 420 \text{ K or } 147°\text{C}$$

$$\eta_{c1} = \frac{T_3 - T_2}{T_{3'} - T_2}$$

$$0.85 = \frac{420 - 306.9}{T_{3'} - 306.9}$$

$$T_{3'} = 440 \text{ K or } 167°\text{C}$$

First heat exchanger 3′-4:

$$T_4 = 0.6\, t_{3'} = 0.6 \times 167 = 100.2°\text{C} \quad \text{or} \quad 373.2 \text{ K}$$

Isentropic Compression process 4-5:

$$\frac{T_5}{T_4} = \left(\frac{p_5}{p_4}\right)^{\frac{\gamma-1}{\gamma}} = \left(\frac{4}{3}\right)^{\frac{1}{3.5}} = 1.086$$

$$T_5 = 373.2 \times 1.086 = 405.2 \text{ K or } 132.2°\text{C}$$

$$\eta_{c2} = \frac{T_5 - T_4}{T_{5'} - T_4}$$

$$0.85 = \frac{405.2 - 373.2}{T_{5'} - 373.2}$$

$$T_{5'} = 410.8 \text{ K or } 137.8°\text{C}$$

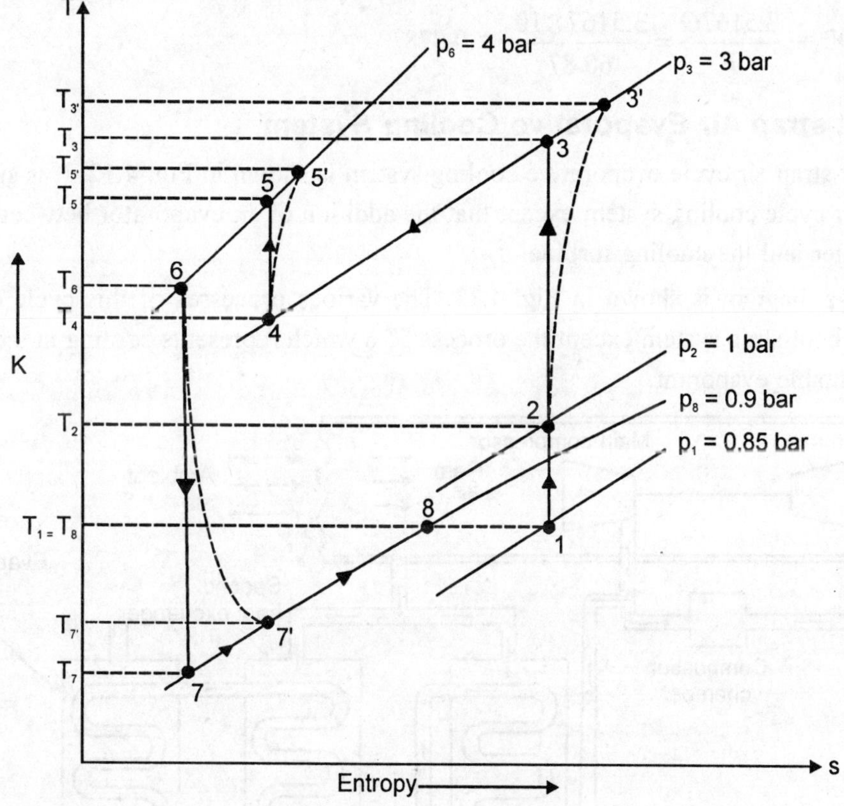

Fig. 4.11 T-s diagram

2nd heat exchanger 5′-6:

$$T_6 = 0.3t_{5'} = 0.3 \times 137.8 = 41.3°\text{C or } 314.3 \text{ K}$$

Isentropic expansion process 6-7:

$$\frac{T_7}{T_6} = \left(\frac{p_7}{p_6}\right)^{\frac{\gamma-1}{\gamma}} = \left(\frac{0.9}{4}\right)^{\frac{1}{3.5}} = 0.653$$

$$T_7 = 314.3 \times 0.653 = 205.2 \text{ K or } -67.8$$

$$\eta_t = \frac{T_6 - T_{7'}}{T_6 - T_7}$$

$$0.9 = \frac{314.3 - T_{7'}}{314.3 - 205.2}$$

$$T_{7'} = 216.1 \text{ K or } -56.9°\text{C}$$

(*a*) Air required to cool the cabin,

$$\dot{m}_a = \frac{211Q}{c_p(T_8 - T_{7'})} = \frac{211 \times 10}{1 \times (293 - 216.1)} = 27.44 \text{ kg/min}$$

Power required, $$P = \frac{\dot{m}_a c_p (T_{3'} - T_2)}{60} = \frac{27.44 \times 1 \times (440 - 306.9)}{60} = 60.87 \text{ kW}$$

(*b*) $$\text{COP} = \frac{3.5167Q}{P} = \frac{3.5167 \times 10}{60.87} = 0.578$$

4.3.4 Boot-strap Air Evaporative Cooling System

A boot-strap air cycle evaporative cooling system is shown in Fig. 4.12. It is similar to the boot strap air cycle cooling system except that the addition of an evaporator between the second heat exchanger and the cooling turbine.

The T-s diagram is shown in Fig. 4.13. The various processes of this cycle are same as for a simple boot-strap system except the process 5″-6 which represents cooling in the evaporator using any suitable evaporant.

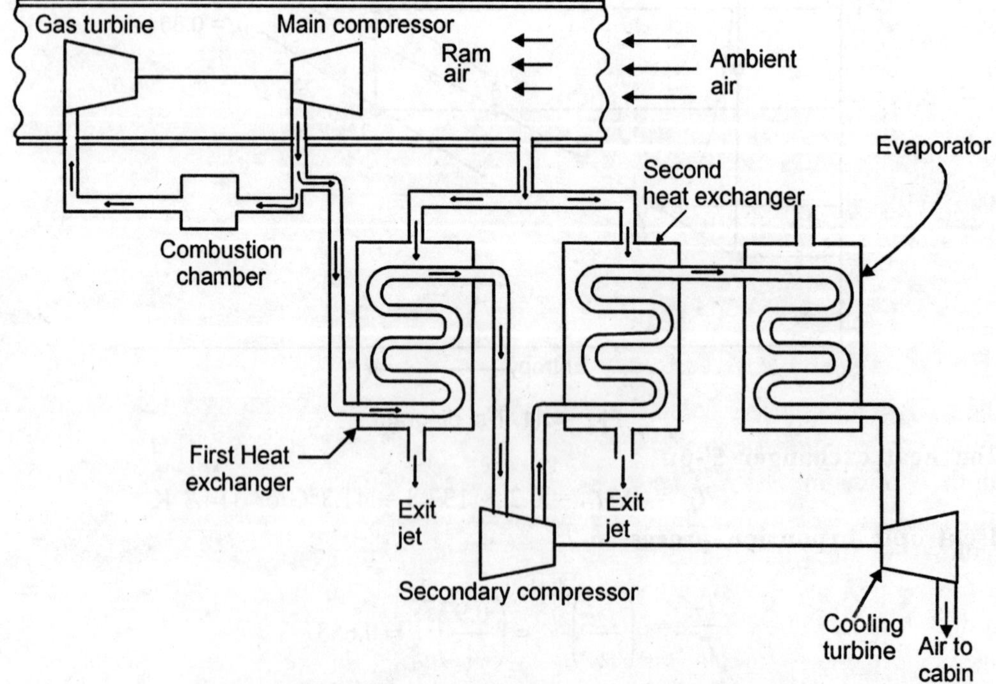

Fig. 4.12 Boot-strap air evaporative cooling system.

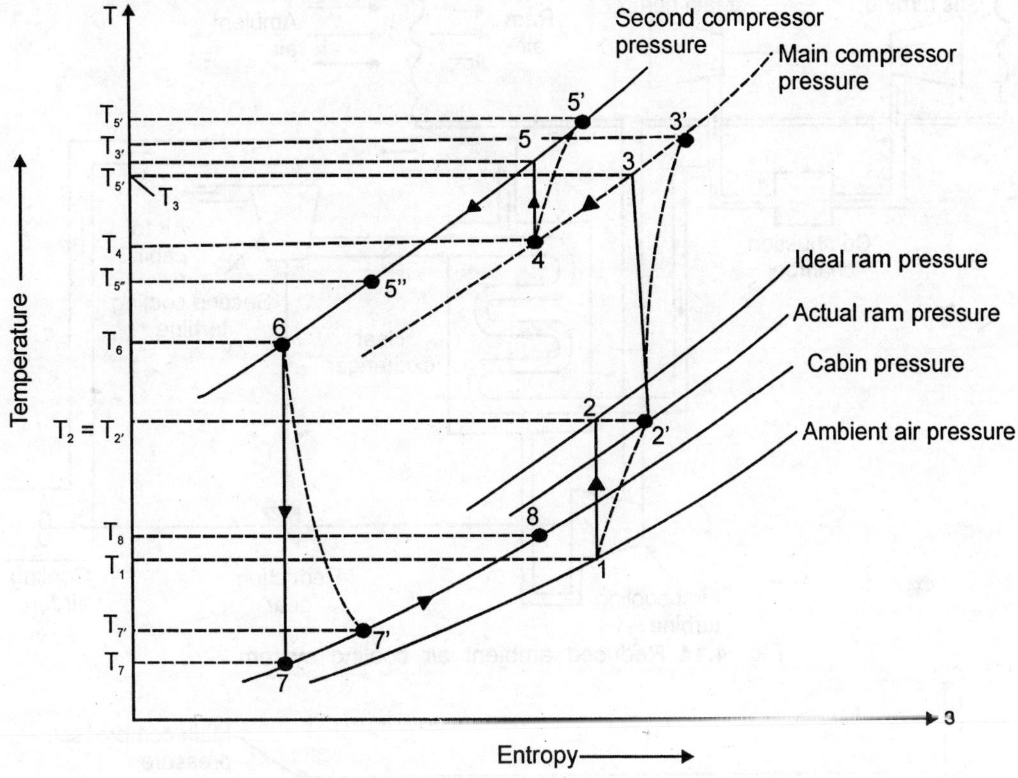

Fig. 4.13 T-s diagram for boot-strap air evaporative cooling system.

Let Q = cooling load in the cabin, TR

Quantity of air required, $$\dot{m}_a = \frac{211Q}{c_p\left(T_8 - T_{7'}\right)} \quad ...(4.11)$$

Power required, $$P = \frac{\dot{m}_a c_p \left(T_{3'} - T_{2'}\right)}{60} \text{ kW} \quad ...(4.12)$$

$$\text{COP} = \frac{3.5167Q}{P} \quad ...(4.13)$$

4.3.5 Reduced Ambient Air Cooling System

The reduced ambient air cooling system is shown in Fig, 4.14. This cooling system includes two cooling turbines and one heat exchanger. The air reduced for the refrigeration system is bled off from the main compressor. This high pressure and high temperature air is cooled initially in the heat exchanger. The air for cooling is taken from the cooling turbine which lowers the high temperature of rammed air. The cooled air from the heat exchanger is passed through the second cooling *turbine* from where the air is supplied to the cabin. The work of the cooling turbine is used to drive the cooling fan (through reduction gears) which draws cooling air from the heat exchanger. The reduced ambient air cooling system is used for very high speed aircrafts.

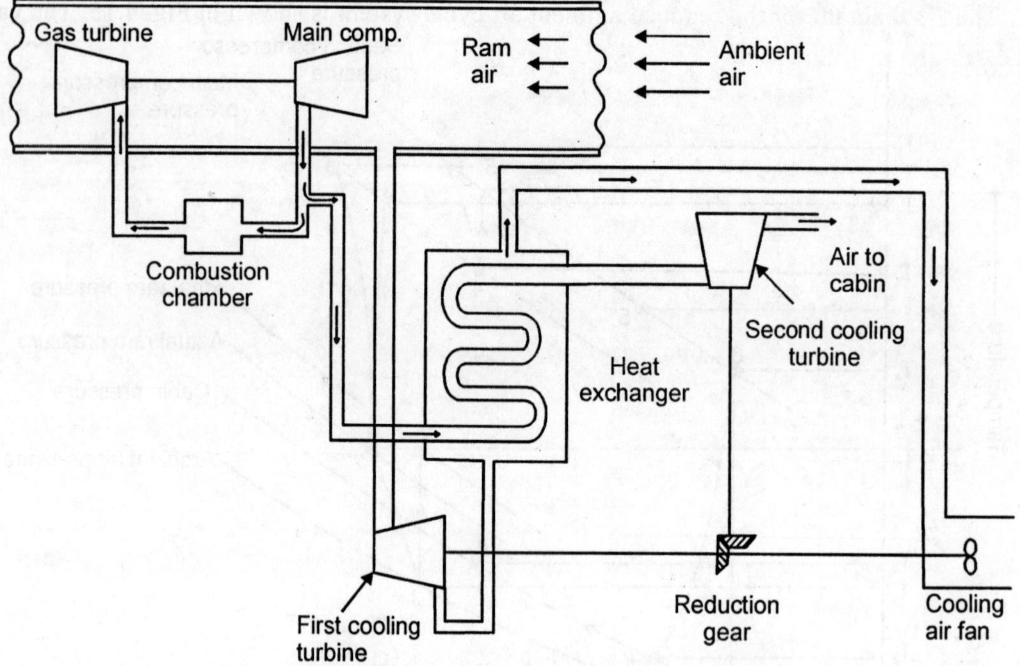

Fig. 4.14 Reduced ambient air cooling system.

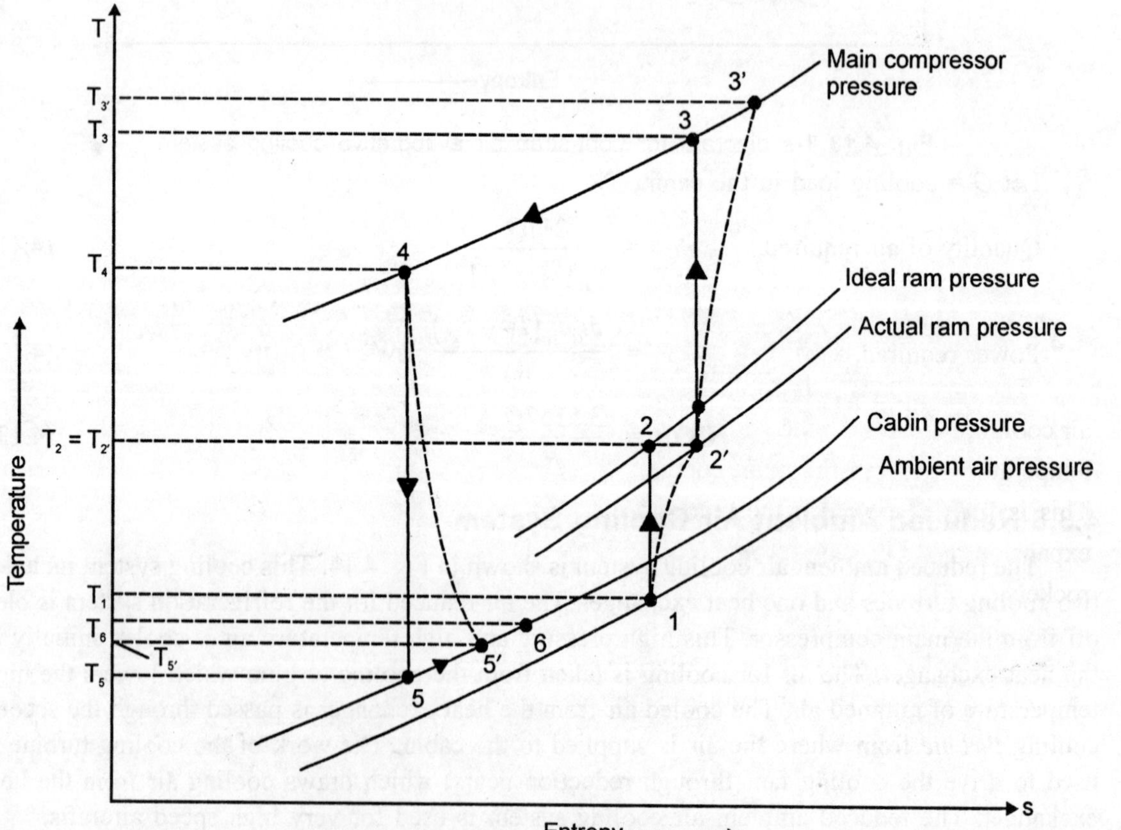

Fig. 4.15 *T-s* diagram for reduced ambient air cycle cooling system.

The *T-s* diagram for the reduced ambient air cycle system is shown in Fig. 4.15. The various processes are as follows:

1. The process 1-2 represents isentropic ramming of air and the process 1-2′ represents actual ramming of air because of internal friction due to irreversibilites.
2. The process 2′-3 represents isentropic compression in the main compressor and the process 2′-3′ represents actual compression of air, because of internal friction due to irreversibilities,
3. The process 3′-4 represents cooling of compressed air by ram air which after passing through the first cooling turbine is led to the heat exchanger. The pressure drop in the heat exchanger is neglected.
4. The process 4-5 represents isentropic expansion of air in the cooling turbine up to the cabin pressure. The actual expansion of air in the second turbine is represented by the curve 4-5′.
5. The process 5′-6 represents the heating of air upto the cabin temperature T_6.

Let Q = cooling load in the cabin, TR

Quantity of air required for refrigeration,

$$\dot{m}_a = \frac{211Q}{c_p\left(T_6 - T_{5'}\right)} \text{ kg/min} \qquad \text{...(4.14)}$$

Power required,
$$P = \frac{\dot{m}_a c_a (T_{3'} - T_{2'})}{60} \text{ kW} \qquad \text{...(4.15)}$$

COP of the system
$$= \frac{3.5167Q}{P} \qquad \text{...(4.16)}$$

4.3.6 Regenerative Air Cooling System

The regenerative air cooling system is shown in Fig. 4.16. It is a modification of a simple air cooling system with the addition of a regenerative heat exchanger. The high pressure and high temperature air from the main compressor is first cooled by the ram air in the heat exchanger. This is further cooled in the regenerative heat exchanger with a portion of the air bled after expansion in the cooling turbine. This type of cooling system is used for supersonic aircrafts and rockets.

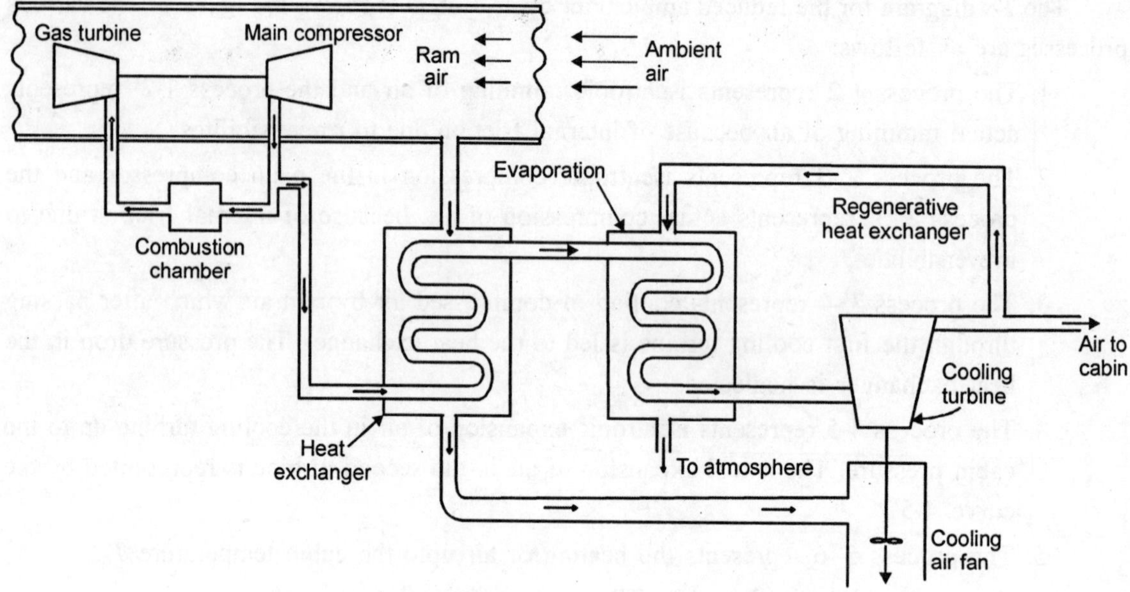

Fig. 4.16 Regenerative air cooling system.

The *T-s* diagram for the regenerative air cooling system is shown in Fig. 4.17. The various processes are as follows:

1. The process 1-2 represents isentropic ramming of air and process 1-2′ represents actual ramming of air because of internal friction due to irreversibilities.
2. The process 2′-3 represents isentropic compression of air in the main compressor and the process 2′-3′ represents actual compression of air because of internal friction due to irreversibilities.
3. The process 3′-4 represents cooling of compressed air by ram air in the heat exchanger.
4. The process 4-5 represents cooling of air in the regenerative heat exchanger.
5. The process 5-6 represents isentropic expansion of air in the cooling turbine upto the cabin pressure and the process 5-6′ represents actual expansion of air in the cooling turbine.
6. The process 6′-7 represents heating of air upto the cabin temperature T_7.

Let Q = cooling load in the cabin, TR

Quantity of air required, $$\dot{m}_a = \frac{211Q}{c_p(T_7 - T_{6'})} \text{ kg/min} \qquad ...(4.17)$$

Let m_1 = total mass of air bled from the main compressor.

m_2 = Mass of cold air bled from the cooling turbine for regenerative heat exchanger.

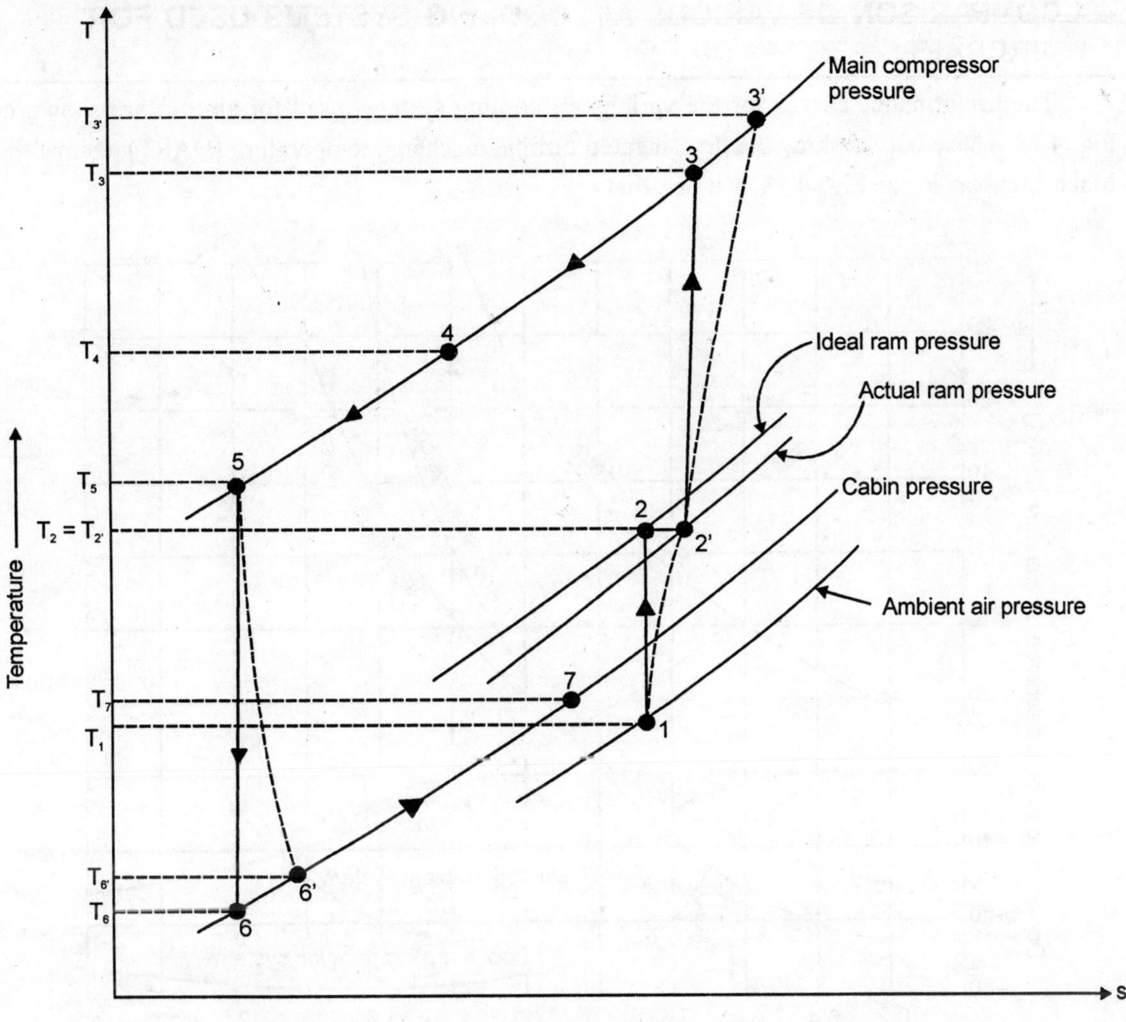

Fig. 4.17 *T-s* diagram for regenerative air cooling system.

For the energy balance of regenerative heat exchanger, we have

$$m_2 c_p \,(T_8 - T_{6'}) = m_1 \; c_p(T_4 - T_5)$$

$$m_2 = \frac{m_1(T_4 - T_5)}{(T_8 - T_{6'})} \qquad \text{...(4.18)}$$

where T_8 = temperature of air leaving to atmosphere from the regenerative heat exchangers.

Power required, $$P = \frac{m_1 c_p(T_{3'} - T_{2'})}{60} \text{ kW} \qquad \text{...(4.19)}$$

COP of refrigerating system $$= \frac{3.5167Q}{P} \qquad \text{...(4.20)}$$

4.4 COMPARISON OF VARIOUS AIR COOLING SYSTEMS USED FOR AIRCRAFT

The performance curves for the various air cooling systems used for aircraft are shown in Fig. 4.18. These curves show the dry air rated turbine discharge temperature (DART) against the Mach number. From Fig. 4.18, we see that

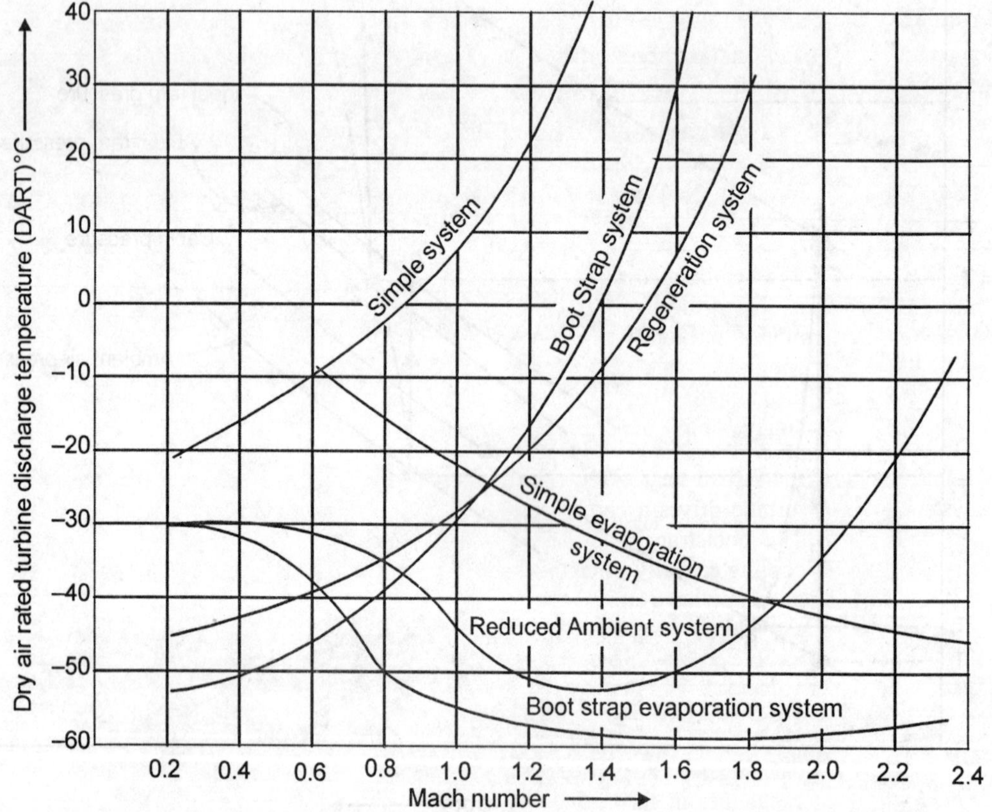

Fig. 4.18 Performance curves for various air cooling systems.

- The simple air cooling system gives maximum cooling effect on the ground surface and decreases as the speed of aircraft increases.
- The boot strap system on the other hand, requires the air plane to be in flight so that the ram air can be used for cooling in the heat exchangers.
- One method of overcoming this drawback of boot strap system is to use part of work derived from turbine to drive a fan which pulls air over the secondary heat exchanger, thus combining the features of a simple and boot strap system.
- As the speed of aircraft increases, the temperature of ram cooling air rises and the ram air becomes less effective as a coolant in the heat exchanger.

- In such cases, a suitable evaporant is used with the ram air so that the cabin temperature does not rise.
- For high speed aircrafts, the boot strap evaporative or regenerative systems are used because they give lower turbine discharge temperature than the simple cooling system.
- In some cases, aeroplanes carry an auxiliary gas turbine for cabin pressurization and air conditioning.

From Fig. 4.18, we see that the turbine discharge temperature of the air is variable. Therefore, in order to maintain the constant temperature of supply air to the cabin, it requires some control system. A comparison of air cycles for aircraft is given in Table 4.1.

Table 4.1: Comparison of air-cycle systems for aircraft refrigeration.

Name of the air cycle	Working Operation	Special features
1. Simple system	The compressed air after cooling in air-cooler is passed through a cooling turbine. The work of this turbine is to drive a fan which draws cooling air through the heat exchanger. Air is discharged from the turbine at a pressure slightly above the cabin pressure.	The system is good for ground cooling. The turbine work is not available for the compressor. It is preferable at low speeds.
2. Bootstrap system	It has two heat exchangers and the expansion turbine drives a compressor rather than a fan. The bootstrap system provides an additional cooling capacity when the primary source of air does not have a sufficiently high pressure to provide the amount of cooling required. The turbine drives the secondary compressor.	It cannot be used for ground cooling. It can be used when the aircraft is in flight. It is used at speeds above 1000 km/h.
3. Regenerative system	It has two exchangers but does not require ram air for cooling the air in the second heat exchanger. The air from the primary heat exchanger in further cooled with a portion of the refrigerated air bled after expansion in the turbine.	It is good for ground cooling. It provides lower turbine discharge temperature. It is necessary above a speed for 1000 km/h.
4. Reduced ambient system	It has two expansion turbines, one in the cabin air stream and the other in the cooling air stream. Bothe turbines are connected to the fan shaft which absorbs all the power.	It is useful supersonic aircraft and rockets.
5. Simple evaporative system	It is similar to the simple system with the addition of an evaporated between the heat exchanger and cooling turbine. The evaporated provides additional cooling effect.	It is useful for aircraft flying at high altitudes.
6. Bootstrap evaporative system	It is similar to the Bootstrap system with the addition of an evaporator between the second heat exchanger and cooling turbine.	It is useful for jet aircraft flying at high altitudes at supersonic speed.

Example 4.6 *The following data refer to a boot-strap air cycle evaporative refrigeration system used for an aeroplane to take 20 tonnes of refrigeration load:*

Ambient air-conditions = 15°C, 0.85 bar; Flight mach number = 1.3; Ram efficiency = 85%; Pressure of air bled off main and secondary compressors = 4 bar and 5 bar respectively; Isentropic efficiencies of main secondary compressors = 90% and 85% respectively; Isentropic efficiency of cooling turbine = 82%. Temperature of air leaving the first and second heat exchanger = 170°C and 160°C respectively; Temperature of air leaving the evaporator = 100°C; Cabin temperature = 25°C; Cabin Pressure = 1 bar, $\gamma = 1.4$ and $c_p = 1$ kJ/kg K.

Determine: *(a) Mass of air required to take the cabin load,*

(b) power required for the refrigeration system, and

(c) COP of the system.

Solution: Given: $Q = 20$ TR, $T_1 = 273 + 15 = 288$ K, $p_1 = 0.85$ bar, $M = 1.3$, $\eta_R = 0.85$,

$p_3 = p_{3'} = p_4 = 4$ bar

$p_5 = p_{5'} = p_{5''} = 5$ bar, $\eta_{c1} = 0.90$, $\eta_{c2} = 0.85$, $T_4 = 273 + 170 = 443$ K

$T_{5''} = 273 + 160 = 433$ K

$\eta_t = 0.82$, $T_6 = 273 + 100 = 373$ K, $T_8 = 273 + 25 = 298$ K

$p_7 = p_{7'} = p_8 = 1$ bar, $\gamma = 1.4$, $c_p = 1$ kJ/kg.K

The *T-s* diagram is shown in Fig. 4.19.

$$\frac{T_{2'}}{T_1} = 1+\frac{(\gamma-1)}{2}M^2 = 1+\left(\frac{1.4-1}{2}\right)(1.3)^2 = 1.338$$

$$T_{2'} = 288 \times 1.338 = 385.3 \text{ K} = T_2$$

Isentropic process 1-2:

$$\frac{p_2}{p_1} = \left(\frac{T_2}{T_1}\right)^{\frac{\gamma}{\gamma-1}} = \left(\frac{385.3}{288}\right)^{\frac{1.4}{0.4}} = 2.77$$

$$p_2 = 0.85 \times 2.77$$

$$= 2.36 \text{ bar}$$

Ram efficiency,

$$\eta_R = \frac{p_{2'} - p_1}{p_2 - p_1}$$

$$0.85 = \frac{p_{2'} - 0.85}{2.36 - 0.85}$$

$$p_{2'} = 2.13 \text{ bar}$$

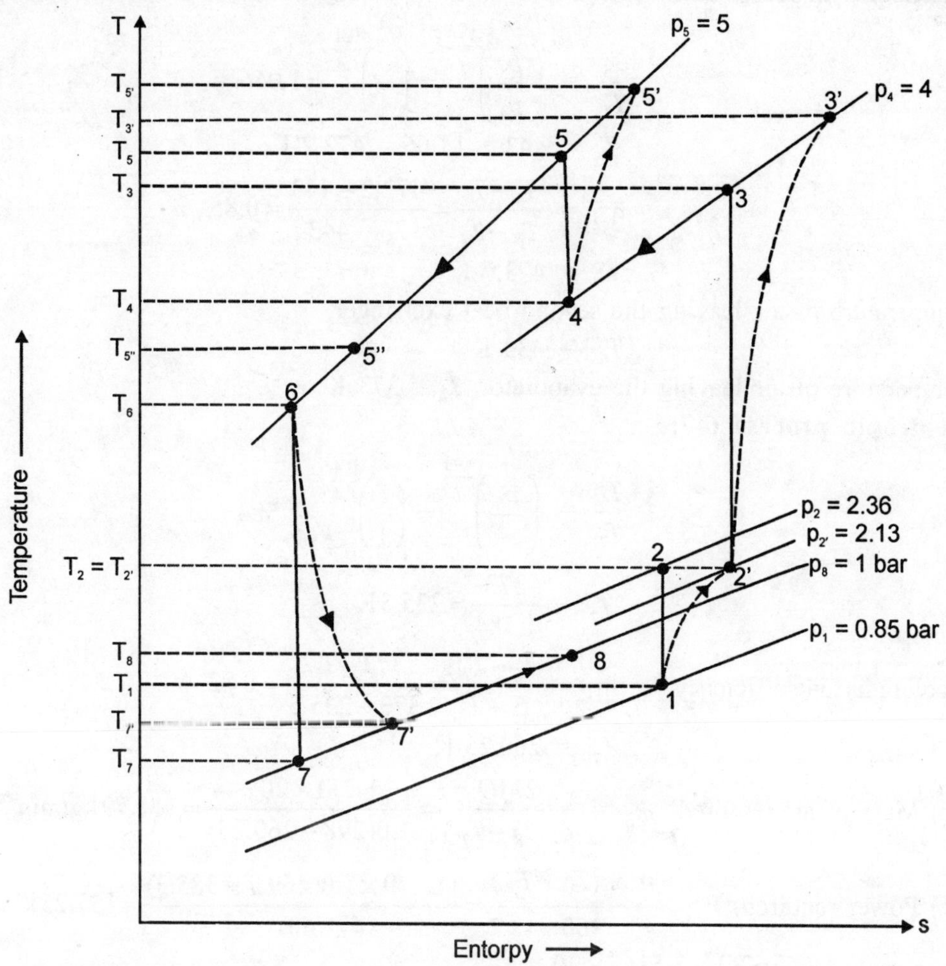

Fig. 4.19 T-s diagram.

Isentropic process 2′-3:

$$\frac{T_3}{T_{2'}} = \left(\frac{p_3}{p_{2'}}\right)^{\frac{\gamma-1}{\gamma}} = \left(\frac{4}{2.13}\right)^{\frac{0.4}{1.4}} = 1.197$$

$$T_3 = 385.3 \times 1.197 = 461.3 \text{ K}$$

$$\eta_{c1} = \frac{T_3 - T_{2'}}{T_{3'} - T_{2'}} = \frac{461.3 - 385.3}{T_{3'} - 385.3} = 0.90$$

$$T_{3'} = 469.7 \text{ K}$$

Temperature of air leaving the first heat exchanger,

$$T_4 = 443 \text{ K (given)}$$

Isentropic process 4-5:

$$\frac{T_5}{T_4} = \left(\frac{p_5}{p_4}\right)^{\frac{\gamma-1}{\gamma}} = \left(\frac{5}{4}\right)^{\frac{0.4}{1.4}} = 1.066$$

$$T_5 = 443 \times 1.066 = 472.2 \text{ K}$$

$$\eta_{c2} = \frac{T_5 - T_4}{T_{5'} - T_4} = \frac{472.2 - 443}{T_{5'} - 443} = 0.85$$

$$T_{5'} = 473.7 \text{ K}$$

Temperature of air leaving the second heat exchanger,

$$T_{5''} = 433 \text{ K}$$

Temperature of air leaving the evaporator, $T_6 = 373$ K

Isentropic process 6-7:

$$\frac{T_6}{T_7} = \left(\frac{p_6}{p_7}\right)^{\frac{\gamma-1}{\gamma}} = \left(\frac{5}{1}\right)^{\frac{0.4}{1.4}} = 1.584$$

$$T_7 = \frac{373}{1.584} = 235.5 \text{K}$$

Cooling turbine efficiency, $$\eta_t = \frac{T_6 - T_{7'}}{T_6 - T_7} = \frac{373 - T_{7'}}{373 - 235.5} = 0.82$$

$$T_{7'} = 260.25 \text{ K}$$

(*a*) Mass of air required, $\dot{m}_a = \dfrac{211Q}{c_p(T_8 - T_{7'})} = \dfrac{211 \times 20}{1.0(298 - 260.25)} = 111.79 \text{kg/min}$

(*b*) Power required, $P = \dfrac{\dot{m}_a c_p (T_{3'} - T_{2'})}{60} = \dfrac{111.79 \times 1.0(469.7 - 385.3)}{60} = 157.25 \text{kW}$

(*c*) COP $= \dfrac{3.5167Q}{P} = \dfrac{3.5167 \times 20}{157.25} = 0.4417$

Example 4.7 *The reduced ambient air refrigeration system used for an aeroplane consists of two cooling turbines, one heat exchanger and one air cooling fan. The speed of aeroplane is 1500 km/h. The ambient air conditions are 0.8 bar and 10°C. The ram efficiency is 90%. The rammed air used for cooling is expanded in the first cooling turbine and leaves at a pressure of 0.8 bar. The air bled off from the main compressor at 6 bar is cooled in the heat exchanger and leaves it at 100°C. The cabin conditions required are 20°C and 1 bar. The pressure drop between the second turbine and cabin is 0.1 bar. If isentropic efficiency for the main compressor is 85% and for both of the cooling turbines is 80%, calculate:*

(*a*) *the mass flow rate of air supplied to cabin to take load of 15 TR,*

(*b*) *the quantity of air passing through the heat exchanger if the temperature rise of ram air is limited to 80 K,*

(*c*) *the power used to drive the cooling fan, and*

(*d*) *COP of the system.*

Solution: Given: $c_a = \frac{1500 \times 10^3}{3600} = 416.67$ m/s, $p_1 = 0.8$ bar, $T_1 = 273 + 10 = 283$ K,

$\eta_R = 0.90$, $T_6 = 273 + 20 = 293$ K,

$p_3 = p_4 = 6$ bar, $T_4 = 273 + 100 = 373$ K,

$p_6 = 1$ bar, $\eta_c = 0.85$,

$\eta_{t1} = \eta_{t2} = 0.80$, $Q = 15$ TR, $\gamma = 1.4$, $c_p = 1$ kJ/kg.K.

The *T-s* diagram is shown in Fig. 4.20.

$$T_{2'} = T_1 + \frac{c_a^2}{2c_p} = 283 + \frac{(416.67)^2}{2 \times 1 \times 10^3} = 369.8 \text{ K} = T_2$$

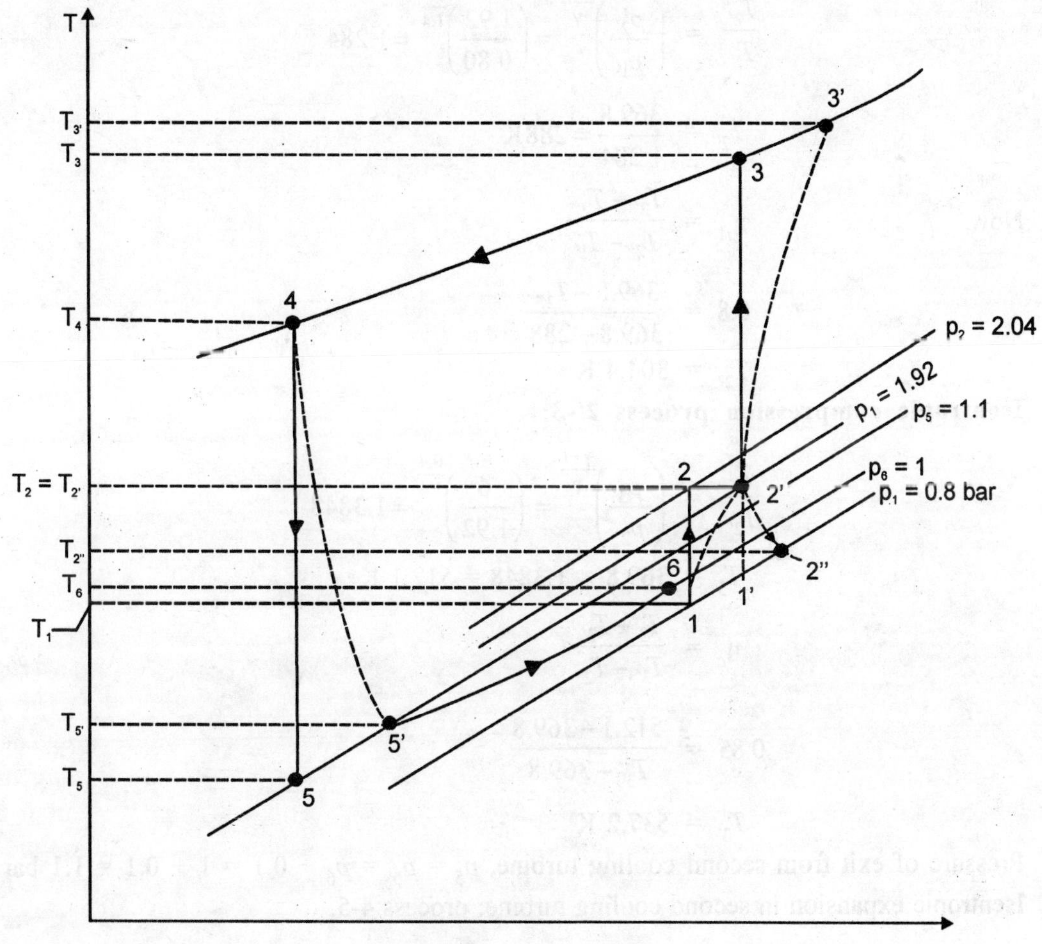

Fig. 4.20 *T-s* diagram.

Isentropic ramming process 1-2:

$$\frac{p_2}{p_1} = \left(\frac{T_2}{T_1}\right)^{\frac{\gamma}{\gamma-1}} = \left(\frac{369.8}{283}\right)^{\frac{1.4}{1.4-1}} = 2.55$$

$$p_2 = 0.8 \times 2.55 = 2.04 \text{ bar}$$

Ram efficiency, $$\eta_R = \frac{p_{2'} - p_1}{p_2 - p_1}$$

$$0.90 = \frac{p_{2'} - 0.8}{2.04 - 0.8}$$

$$p_{2'} = 1.92 \text{ bar}$$

The expression of ram air in the first cooling turbine is represented by isentropic line 2′-1′ and by actual line 2′-2″.

Isentropic cooling process 2′-1′:

$$\frac{T_{2'}}{T_{1'}} = \left(\frac{p_{2'}}{p_{1'}}\right)^{\frac{\gamma-1}{\gamma}} = \left(\frac{1.92}{0.80}\right)^{\frac{0.4}{1.4}} = 1.284$$

$$T_{2'} = \frac{369.8}{1.284} = 288\text{K}$$

Now, $$\eta_{t1} = \frac{T_{2'} - T_{2''}}{T_{2'} - T_{1'}}$$

$$0.8 = \frac{369.8 - T_{2''}}{369.8 - 288}$$

$$T_{2''} = 304.4 \text{ K}$$

Isentropic compression process 2′-3:

$$\frac{T_3}{T_{2'}} = \left(\frac{p_3}{p_{2'}}\right)^{\frac{\gamma-1}{\gamma}} = \left(\frac{6}{1.92}\right)^{\frac{0.4}{1.4}} = 1.3848$$

$$T_3 = 369.8 \times 1.3848 = 512.1 \text{ K}$$

$$\eta_c = \frac{T_3 - T_{2'}}{T_{3'} - T_{2'}}$$

$$0.85 = \frac{512.1 - 369.8}{T_{3'} - 369.8}$$

$$T_{3'} = 537.2 \text{ K}$$

Pressure of exit from second cooling turbine, $p_5 = p_{5'} = p_6 + 0.1 = 1 + 0.1 = 1.1$ bar

Isentropic expansion in second cooling turbine: process 4-5,

$$\frac{T_4}{T_5} = \left(\frac{p_4}{p_5}\right)^{\frac{\gamma-1}{\gamma}} = \left(\frac{6}{1.1}\right)^{\frac{0.4}{1.4}} = 1.624$$

$$T_5 = \frac{373}{1.624} = 229.7 \text{ K}$$

$$\eta_{t2} = \frac{T_4 - T_{5'}}{T_4 - T_5}$$

$$0.80 = \frac{373 - T_{5'}}{373 - 229.7}$$

$$T_{5'} = 258.4 \text{ K}$$

(a) $\dot{m}_a = \dfrac{211Q}{c_p(T_6 - T_{5'})} = \dfrac{211 \times 10}{1.0 \times (293 - 258.4)} = 60.98$ kg/min

(b) Let $\dot{m}_R$ = quantity of ram air passing through the heat exchanger

$$\dot{m}_R c_p \cdot \Delta T = \dot{m}_a c_p (T_{3'} - T_4)$$

$$80\dot{m}_R = 60.98\ (537.2 - 373)$$

$$\dot{m}_R = 125.16 \text{ kg/min}$$

(c) Work used to drive the cooling fan:

$$W_{t1} = \dot{m}_R c_p (T_{2'} - T_{2''})$$

$$= 125.16 \times 1.0 \times (369.8 - 304.4)$$

$$= 8185.5 \text{ kJ/min}$$

$$W_{t2} = \dot{m}_a c_p (T_4 - T_{5'})$$

$$= 60.98 \times 1.0\ (373 - 258.4)$$

$$= 6988.3 \text{ kJ/min}$$

Combined work, $W_t = W_{t1} + W_{t2} = 8185.5 + 6988.3 = 15173.8$ kJ/min

Power used, $P = \dfrac{W_t}{60} = \dfrac{15173.8}{60} = 252.9$ kW

(d) COP $= \dfrac{211Q}{\dot{m}_a c_p (T_{3'} - T_{2'})} = \dfrac{211 \times 10}{60.98 \times 1.0(537.2 - 369.8)} = 0.207$

Example 4.8 *A regenerative air cooling system is used for an aircraft to take 20 tonnes of refrigerating load. The ambient air condition is 0.8 bar and 10°C. It is rammed isentropically till the pressure rises to 1.25 bar. The air bled off the main compressor at 4.5 bar is cooled by the ram air in the heat exchanger of effectiveness 60%. The air from the heat exchanger is further cooled to 60°C in the regenerative heat exchangers. A portion of the air bled after expansion in the cooling turbine gets heated to 100°C before discharging. The conditions required in the cabin are 1 bar and 25°C. The isentropic efficiency of compressor is 90% and that of turbine 85%. Calculate:*

(a) Mass of air bled from cooling turbine,

(b) Power required, and

(c) COP of system.

Solution: Given: $Q = 20$ TR, $p_1 = 0.8$ bar, $T_1 = 273 + 10 = 283$ K, $p_2 = 1.25$ bar,
$p_3 = p_4 = p_5 = 4.5$ bar, $\varepsilon = 0.60$, $T_5 = 273 + 60 = 333$ K,
$T_7 = 273 + 25 = 298$ K,
$p_6 = p_{6'} = p_7 = 1$ bar, $\eta_c = 0.9$,
$\eta_t = 0.85$, $T_8 = 273 + 100 = 373$ K

The T-s diagram is shown in Fig. 4.21.

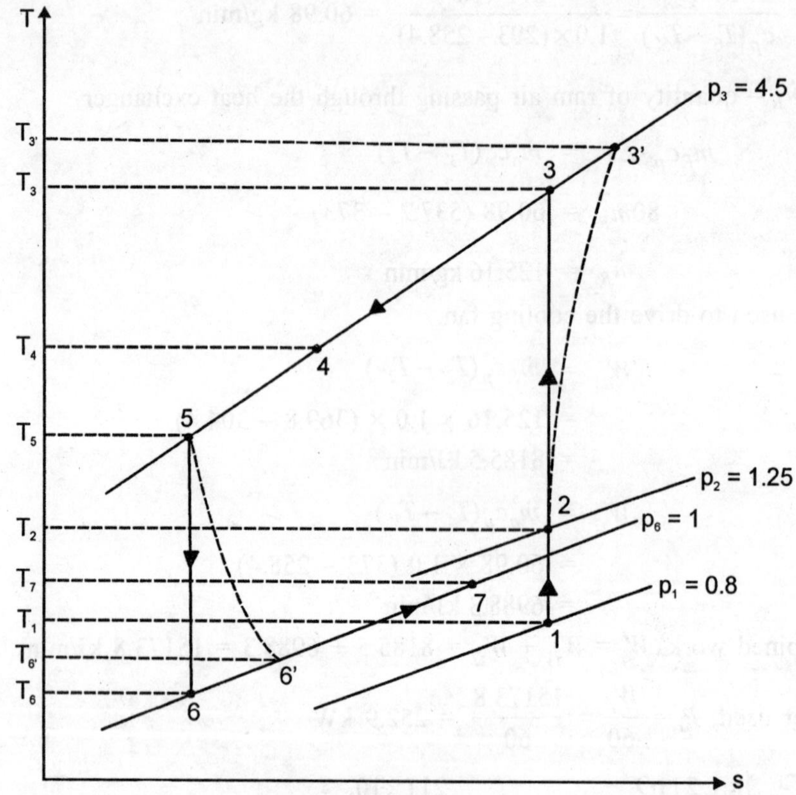

Fig. 4.21 T-s diagram.

Isentropic samming of air, process 1-2:

$$\frac{T_2}{T_1} = \left(\frac{p_2}{p_1}\right)^{\frac{\gamma-1}{\gamma}} = \left(\frac{1.25}{0.8}\right)^{\frac{0.4}{1.4}} = 1.136$$

$$T_2 = 283 \times 1.136 = 321.5 \text{ K}$$

Isentropic compression process 2-3:

$$\frac{T_3}{T_2} = \left(\frac{p_3}{p_2}\right)^{\frac{\gamma-1}{\gamma}} = \left(\frac{4.5}{1.25}\right)^{\frac{0.4}{1.4}} = 1.442$$

$$T_3 = 321.5 \times 1.442 = 463.6 \text{ K}$$

$$\eta_c = \frac{T_3 - T_2}{T_{3'} - T_2}$$

$$0.90 = \frac{463.6 - 321.5}{T_{3'} - 321.5}$$

$$T_{3'} = 479.4 \text{ K}$$

Effectiveness of heat exchanger,

$$\varepsilon = \frac{T_{3'} - T_4}{T_{3'} - T_2}$$

$$0.6 = \frac{479.4 - T_4}{479.4 - 321.5}$$

$$T_4 = 384.7 \text{ K}$$

Isentropic cooling in turbine process 5-6:

$$\frac{T_5}{T_6} = \left(\frac{p_5}{p_6}\right)^{\frac{\gamma-1}{\gamma}} = \left(\frac{4.5}{1}\right)^{\frac{0.4}{1.4}} = 1.537$$

$$T_6 = \frac{T_5}{1.537} = \frac{333}{1.537} = 216.6 \text{ K}$$

$$\eta_t = \frac{T_5 - T_{6'}}{T_5 - T_6}$$

$$0.85 = \frac{333 - T_{6'}}{333 - 216.6}$$

$$T_{6'} = 234.1 \text{ K}$$

The air at $T_{6'} = 234.1$ K gets heated to 100°C or $T_8 = 373$ K in regenerative heat exchanger.

(*a*) Let $\dot{m}_a$ = mass of air bled from cooling turbine to be used for regenerative cooling

$\dot{m}_1$ = total mass of air bled from the main compressor

$\dot{m}_2$ = mass of cold air bled from the cooling turbine for regenerative heat exchanger.

Mass of air supplied to the cabin,

$$\dot{m}_a = \dot{m}_1 - \dot{m}_2$$

$$= \frac{211Q}{c_p(T_7 - T_{6'})} = \frac{211 \times 20}{1.0 \times (298 - 234.1)} = 66.0 \text{ kg/min}$$

and $\dot{m}_2 = \dfrac{\dot{m}_1(T_4 - T_5)}{(T_8 - T_{6'})} = \dfrac{\dot{m}_1(384.7 - 333)}{(373 - 234.1)} = 0.372\dot{m}_1$

Now $\dot{m}_1 - \dot{m}_2 = \dot{m}_1(1 - 0.372) = 66$

$\dot{m}_1 = 105.1$ kg/min

and $\dot{m}_2 = 39.1$ kg/min

(*b*) Power supplied, $P = \dfrac{\dot{m}_1 c_p (T_{3'} - T_2)}{60} = \dfrac{105.1 \times 1.0(479.4 - 321.5)}{60} = 276.6$ kW

(*c*) COP $= \dfrac{3.5167Q}{P} = \dfrac{3.5167 \times 20}{276.6} = 0.254$

Example 4.9 *An air cooling system for a jet plane flying at 6000 m altitude operates on a simple air cycle. The cockpit is to be maintained at 25°C. The ambient air pressure and temperature are 0.5 bar and –0.5°C. Calculate the stagnation pressure of the air entering the first stage of the main compressor if the ram efficiency is assumed as 80% with the plane speed of 1000 km/hr. Main compressor pressure ratio is 3. Stagnation temperature of the air entering the cooling turbine is 50°C. Pressure drop through heat exchanger is 0.1 bar. Compressor and turbine adiabatic efficiencies are 80%. Pressure in the cockpit is 1 bar. The pressure of air leaving the cooling turbine is 1.05 bar. Cockpit cooling load is 20 ton.* c_p *= 1 kJ/kg.K. Neglect frictional losses and assume the air is dry and behaves as a perfect gas with* $\gamma = 1.4$.

Determine:

(*i*) *The stagnation temperature and pressure of the air entering and leaving the compressor;*

(*ii*) *the temperature of the air leaving the cooling turbine;*

(*iii*) *the air flow rate per minute;*

(*iv*) *power required for repressurisation (with and without ram work);*

(*v*) *power required for only cooling the cabin (excluding ram work and including turbine work);*

(*vi*) *COP of the cycle based on compressor work on bled off air;*

(*vii*) *COP of the cycle considering ram work and turbine work.*

Solution: Given: $c_a = \dfrac{1000 \times 10^3}{3600} = 277.8$ m/s, $p_1 = 0.5$ bar, $T_1 = 273 - 5 = 268$ K,

$T_6 = 273 + 25 = 298$ K, $\eta_R = 0.8$, $p_3/p_{2'} = 3$, $T_4 = 273 + 50 = 323$ K,

$p_4 = (p_3 - 0.1)$ bar, $\eta_c = \eta_t = 0.8$,

$p_6 = 1$ bar, $p_5 = p_{5'} = 1.05$ bar, $Q = 20$ TR, $c_p = 1$ kJ/kg.K, $\gamma = 1.4$

The schematic diagram for simple air cycle is shown in Fig. 4.22 and *T-s* diagram for the cycle in Fig. 4.23.

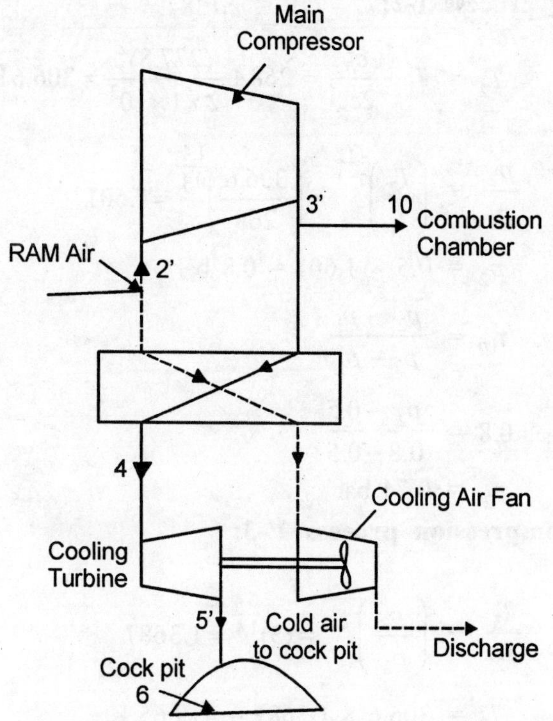

Fig. 4.22 Schematic diagram for simple air cycle.

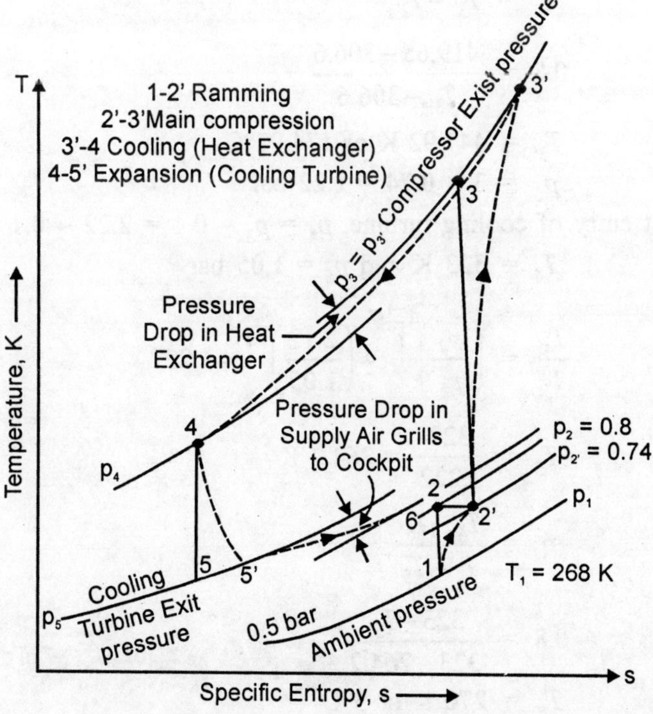

Fig. 4.23 T-s diagram for simple air cycle.

(*i*) **Air ramming process 1-2:**

$$T_2 = T_1 + \frac{c_a^2}{2c_p} = 268 + \frac{(277.8)^2}{2 \times 1 \times 10^3} = 306.6\,\text{K} = T_{2'}$$

$$\frac{p_2}{p_1} = \left(\frac{T_2}{T_1}\right)^{\frac{\gamma}{\gamma-1}} = \left(\frac{306.6}{268}\right)^{\frac{1.4}{0.4}} = 1.601$$

$$p_2 = 0.5 \times 1.601 = 0.8 \text{ bar}$$

Ram efficiency, $\eta_R = \dfrac{p_{2'} - p_1}{p_2 - p_1}$

$$0.8 = \frac{p_{2'} - 0.5}{0.8 - 0.5}$$

$$p_{2'} = 0.74 \text{ bar}$$

Isentropic compression process 2′-3:

$$\frac{T_3}{T_{2'}} = \left(\frac{p_3}{p_{2'}}\right)^{\frac{\gamma-1}{\gamma}} = (3)^{\frac{0.4}{1.4}} = 1.3687$$

$$T_3 = 306.6 \times 1.3687 = 419.65 \text{ K}$$

$$\eta_c = \frac{T_3 - T_{2'}}{T_{3'} - T_{2'}}$$

$$0.8 = \frac{419.65 - 306.6}{T_{3'} - 306.6}$$

$$T_{3'} = 447.92 \text{ K or } 174.92°\text{C}$$

$$p_3 = 3 \times 0.74 = 2.22 \text{ bar}$$

(*ii*) Pressure at entry of cooling turbine, $p_4 = p_3 - 0.1 = 2.22 - 0.1 = 2.12$ bar

Also $T_4 = 323$ K and $p_5 = 1.05$ bar

$$\frac{T_4}{T_5} = \left(\frac{p_4}{p_5}\right)^{\frac{\gamma-1}{\gamma}} = \left(\frac{2.12}{1.05}\right)^{\frac{0.4}{1.4}} = 1.222$$

$$T_5 = \frac{323}{1.222} = 264.2 \text{ K}$$

$$\eta_t = \frac{T_4 - T_{5'}}{T_4 - T_5}$$

$$0.8 = \frac{323 - T_{5'}}{323 - 264.2}$$

$$T_{5'} = 276 \text{ K or } 3°\text{C}$$

(*iii*) Mass flow rate of air supplied to cock-pit,

$$\dot{m}_a = \frac{211Q}{c_p(T_6 - T_{5'})} = \frac{211\times 20}{1.0\times(298-276)} = 191.82 \text{ kg/min}$$

(*iv*) COP of the plant considering compressor work on bled off air,

$$\text{COP} = \frac{211Q}{\dot{m}_a c_p(T_{3'} - T_{2'})} = \frac{211\times 20}{191.82\times 1.0(447.92-306.6)} = 0.1556$$

(*v*) COP of cycle considering ram work and turbine work,

$$\text{COP} = \frac{211Q}{\dot{m}_a c_p\left[(T_{2'} - T_1)+(T_{3'} - T_{2'})-(T_4 - T_{5'})\right]}$$

$$= \frac{211\times 20}{191.82\times 1.0\left[(306.6-268)+(447.92-306.6)-(323-276)\right]}$$

$$= 0.1655$$

(*vi*) Power required for pressurization with and without ram work:

With ram work, $$P = \frac{\dot{m}_a c_p T_1}{\eta_c \times 60}\left[\left(\frac{p_6}{p_1}\right)^{\frac{\gamma-1}{\gamma}} - 1\right]$$

$$= \frac{191.82\times 1.0\times 306.6}{60\times 0.8}\left[\left(\frac{1}{0.5}\right)^{\frac{0.4}{1.4}} - 1\right] = 234.56 \text{ kW}$$

Without ram work, $$P = \frac{\dot{m}_a c_p T_{2'}}{60\times\eta_c}\left[\left(\frac{p_6}{p_{2'}}\right)^{\frac{\gamma-1}{\gamma}} - 1\right]$$

$$= \frac{191.82\times 1.0\times 306.6}{60\times 0.8}\left[\left(\frac{1}{0.74}\right)^{\frac{0.4}{1.4}} - 1\right]$$

$$= 110.07 \text{ kW}$$

(*v*) Turbine work, $W_t = \dot{m}_a c_p(T_4 - T_{5'})$

Compressor work, $W_c = \dot{m}_a c_p(T_{3'} - T_{2'})$

Net work, $$W_{\text{net}} = W_c - W_t = \frac{\dot{m}_a c_p}{60}\left[(T_{3'} - T_{2'}) - (T_4 - T_{5'})\right]$$

$$= \frac{191.82\times 1.0}{60}\left[(447.92 - 306.6) - (323 - 276)\right]$$

$$= 301.54 \text{ kW}$$

Refrigeration work, W_{ref} = Net work – Cabin pressurization work

$$= 301.54 - 110.07 = 191.47 \text{ kW}$$

Example 4.10

(a) Although, the COP of an air refrigeration cycle is very low, even then it is commonly used in aircraft refrigeration. Explain the reason. Also discuss the necessity of air conditioning of aircraft at high altitudes where ambient temperatures are very low.

(b) Enumerate the classification of Aircraft refrigeration systems and differentiate between a simple aircraft refrigeration system and a Boot strap refrigeration system. What do you understand by Dry Air Rated Temperature (DART)?

Solution: (a) Reasons for using air-refrigeration cycle in aircraft refrigeration

Air has found applications as a working substance for aircraft refrigeration cycle. The coefficient of performance of air cycle is lower than that of the vapour compression cycle. Nevertheless, the air cycle continues to be favoured for aircraft refrigeration because of its many advantages.

The air cycle can work as an open cycle or as a closed cycle system. A closed air cycle system or a dense air machine has many thermodynamic advantages. It can work at a suction pressure higher than the atmospheric pressure. This reduces the volumes handled by the compressor and the expander. Also, the operating pressure ratio can be reduced, resulting in a higher coefficient of performance. On the other hand, open cycle requires only one heat exchanger. This reduces its weight and cost of equipment.

The main considerations involved in an aircraft application are weight, space and operating power. Though the power per ton of refrigeration is considerably more for air cycle refrigeration than for a vapour compression system, the bulk and weight advantages of the air cycle system result in a greater overall power saving in the aircraft.

The advantages of air cycle system for aircraft refrigeration are:

(*i*) Small amount of leakages are tolerable with air as the refrigerant.

(*ii*) Lesser weight and lower cost of equipment.

(*iii*) Availability of the refrigerant in mid air is also an important consideration.

(*iv*) Cabin pressurization and air-conditioning can be combined into one operation.

(*v*) Initial ramming effect (i.e. conversion of high kinetic energy of ambient air relative to the aircraft into enthalpy) helps in pressure rise.

Necessity of air-conditioning of aircraft at high altitudes

The present day jet aircrafts have very high cooling loads because of their large occupancy, electronic equipment and high velocity. This results in large heat generation due to skin friction, human metabolism and heat generation in the equipment. Therefore, to provide comfortable conditions in the aircraft, air-conditioning is a dire necessity.

(*b*) **Refer to art. 4.3 for classification of aircraft refrigeration systems.**

DART A comparison of the different types of air-refrigeration systems is made in terms of dry air rated temperature (DART). This is the temperature of the discharge air from the expander if there is no condensed moisture present. DART is calculated by assuming constant pressure specific heat of 1.005 kJ/kg.K for air. Thus DART will be essentially that resulting from a calculation which assumes that the humidity of air at the expander discharge is so low that it is unsaturated. The rating of aircraft units is given in terms of kg of air per unit time at the design DART. Thus the capacity of the machine giving $\dot{m}$ kg/s of air at a DART of t_o to maintain a cabin at temperature t_i is given by,

$$\dot{Q}_o = \dot{m}c_p(t_i - t_o)$$

Example 4.11 *In a simple aircraft refrigeration system, air is compressed from a temperature of 27°C and pressure of 1.2 bar to a pressure of 6 bar. The total load of the passengers of the aircraft is 30 kW, the load of control equipments is 5 kW and miscellaneous load is 3 kW. The exit pressure of the cooling turbine is equal to the cabin pressure of 1 atm. The compressor and turbine efficiencies are 80% and 75% respectively. The rise in the temperature of the cool air in the cabin is 60°C.*

Assuming c_{pair} = 1.005 kJ/kg.K, determine:

(i) the refrigeration capacity,

(ii) the mass of air-flow required,

(iii) the power required for refrigeration, and

(iv) the COP of the system.

Solution. Given: $T_1 = 273 + 27 = 300$ K, $p_1 = 1.2$ bar, $p_2 = 6$ bar

Total refrigeration load = 30 + 5 + 3 = 38 kW, $p_4 = 1$ atm, $\eta_c = 0.80$, $\eta_t = 0.75$,

$$T_1 - T_{4'} = 60°\text{C},\ C_{pa} = 1.005 \text{ kJ/kg.K}$$

Refer to Fig. 4.24.

Process 1-2:

$$\frac{T_2}{T_1} = \left(\frac{p_2}{p_1}\right)^{\frac{\gamma-1}{\gamma}} = \left(\frac{6}{1.2}\right)^{\frac{1.4-1}{1.4}} = 1.5838$$

$$T_2 = 300 \times 1.5838 = 475.15 \text{ K}$$

$$\eta_c = \frac{T_2 - T_1}{T_{2'} - T_1}$$

$$0.8 = \frac{475.15 - 300}{T_{2'} - 300}$$

$$T_{2'} = 518.9 \text{ K}$$

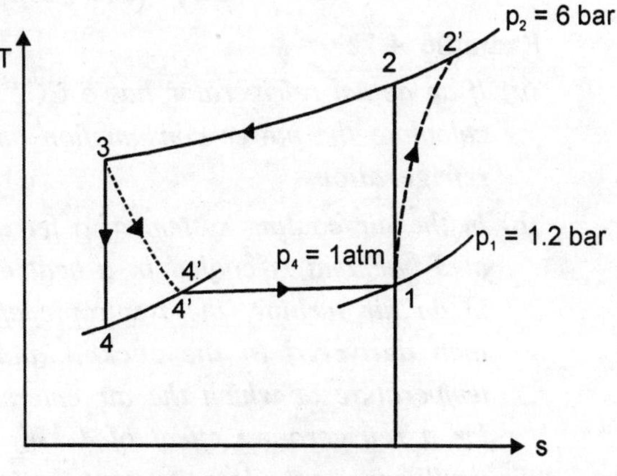

Fig. 4.24 *T-s* diagram.

Process 3-4:

$$\eta_t = \frac{T_3 - T_{4'}}{T_3 - T_4}$$

$$0.75 = \frac{T_3 - 240}{T_3 - T_4}$$

$$\frac{T_4}{T_3} = \left(\frac{p_4}{p_3}\right)^{\frac{\gamma-1}{\gamma}} = \left(\frac{1.01325}{6}\right)^{\frac{1.4-1}{1.4}} = 0.6016$$

$$T_4 = 0.6016\, T_3$$

$$0.75 = \frac{T_3 - 240}{T_3 - 0.6016T_3} = \frac{T_3 - 240}{0.3984T_3}$$

$$0.2988\, T_3 = T_3 - 240$$

$$T_3 = 342.3 \text{ K}$$

(*i*) Refrigeration capacity, $q_{4'-1} = c_{pa}(T_1 - T_{4'}) = 1.005 \times 60 = 60.3$ kJ/kg of air

or $= \dfrac{37.81 \times 60.3}{211} = 108.57$ TR

(*ii*) Now $Q_{4'-1} = 38$ kW

Mass of air circulated, $\dot{m}_a = \dfrac{Q_{4'-1}}{q_{4'-1}} = \dfrac{38}{60.3} = 0.63$ kg/s or 37.81 kg/min

(*iii*) Power required, $P = 60.3 \times 0.63 = 38$ kW

(*iv*) $\text{COP} = \dfrac{T_1 - T_{4'}}{(T_{2'} - T_3) - (T_1 - T_{4'})}$

$$= \frac{300 - 240}{(518.9 - 342.3) - (300 - 240)} = \frac{60}{116.6} = 0.515$$

Example 4.12

(a) If an actual refrigerator has a COP which is 75% of that of the ideal Carnot cycle, calculate the power consumption and heat rejected to the surroundings per ton of refrigeration.

(b) In the air cooling system of a jet aircraft, air is bled from the engine compressor at 3 bar, and is cooled in a heat exchanger to 105°C. It is expanded to 0.69 bar in an air turbine, the isentropic efficiency of the process being 85%. The air is then delivered to the cockpit and leaves the aircraft at 27°C. Calculate the temperature at which the air enters the cockpit and the mass flow of air required for a refrigerating effect of 4 kW. If the air turbine is used to help to drive the auxiliaries, calculate its contribution in power.

Solution:

(a) $$(\text{COP})_{\text{act}} = 0.75\ (\text{COP})_{\text{ideal}}$$

$$(\text{COP})_{\text{ideal}} = \frac{\text{Heat absorbed}}{\text{Work done}}$$

$$1\ \text{TR} = 211\ \text{kJ/min}$$

$$\text{Actual Work done} = \frac{211}{(\text{COP})_{\text{act}}} = \frac{211}{0.75(\text{COP})_{\text{ideal}}} = \frac{281.33}{(\text{COP})_{\text{ideal}}}\ \text{kJ/min}$$

$$\text{Power consumption} = \frac{281.33}{60(\text{COP})_{\text{ideal}}} = \frac{4.689}{(\text{COP})_{\text{ideal}}}\ \text{kW}$$

(b) The *T-s* diagram for the cooling system is shown in Fig. 4.25.

$$T_2 = 273 + 105 = 378\ \text{K}$$

$$T_4 = 273 + 27 = 300\ \text{K}$$

Air turbine isentropic process: 2-3

$$\frac{T_3}{T_2} = \left(\frac{p_3}{p_2}\right)^{\frac{\gamma-1}{\gamma}}$$

$$T_3 = 378\left(\frac{0.68}{3}\right)^{\frac{1.4-1}{1.4}} = 248.4\ \text{K}$$

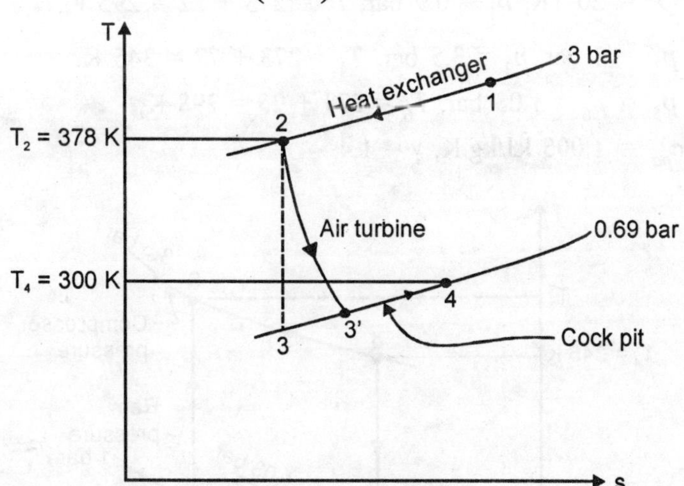

Fig. 4.25 T-s diagram.

$$\text{Turbine efficiency,}\quad \eta_t = \frac{T_2 - T_{3'}}{T_2 - T_3}$$

$$0.85 = \frac{378 - T_{3'}}{378 - 248.4}$$

$$T_{3'} = 267.8\ \text{K or } -5.2^\circ\text{C}$$

Temperature of air entering the cockpit = –5.2°C

Refrigeration effect, $q_{3'-4} = c_{pa}(T_4 - T_{3'}) = 1.005\ (300 - 267.8) = 32.33$ kJ/kg

Mass flow rate of air, $\dot{m}_a = \dfrac{\dot{Q}}{q_{3'-4}} = \dfrac{4}{32.33} = 0.1237$ kg/s or 7.423 kg/min

Power of air turbine, $W_t = \dot{m}_a\, c_{pa}(T_2 - T_{3'}) = 0.1237 \times 1.005(378 - 267.8) = 13.7$ kW

Example 4.13

(a) Discuss the different classification of Aircraft refrigeration systems and explain the working of a simple aircraft refrigeration system with the help of (T-S) diagram.

(b) A simple aircraft refrigeration system has to undertake a load of 20 TR. The ambient pressure and temperatures are 0.9 bar and 22°C respectively. The pressure of air is increased to 1bar due to isentropic ramming action. The air is further compressed in a compressor upto 3.5 bar and then cooled in a heat exchanger upto 72°C. Finally the air is passed through a cooling turbine and then supplied to the cabin at a pressure of 1.03 bar. The air leaves the cabin at 25°C. Calculate the power required and C.O.P. of the system. Take c_p = 1.005 kJ/kg-K and γ = 1.4.

Solution:

(*a*) Refer to art. 4.3 and Fig. 4.2.

(*b*) Given: $\dot{Q}$ = 20 TR, p_1 = 0.9 bar, T_1 = 273 + 22 = 295 K,

p_2 = 1 bar, p_3 = 3.5 bar, T_4 = 273 + 72 = 345 K,

$p_5 = p_6$ = 1.03 bar, T_6 = 273 + 25 = 298 K,

c_{pa} = 1.005 kJ/kg.K, γ = 1.4

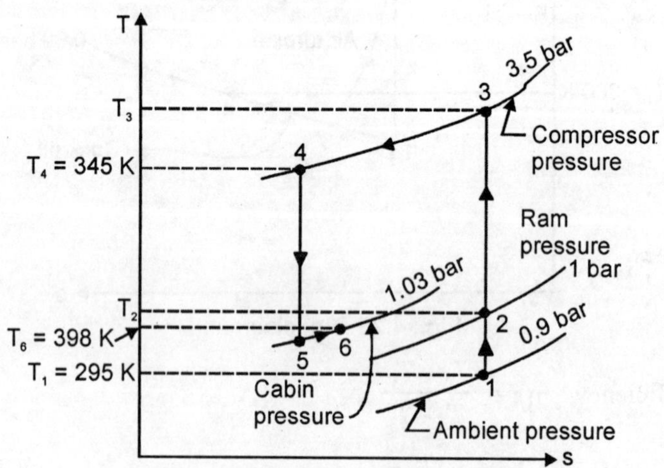

Fig. 4.26 T-s diagram.

Refer to T-s diagram of Fig. 4.26

Isentropic ramming process: 1-2

$$T_2 = T_1\left(\frac{p_2}{p_1}\right)^{\frac{\gamma-1}{\gamma}} = 295\left(\frac{1}{0.9}\right)^{\frac{1.4-1}{1.4}} = 304 \text{ K}$$

Isentropic compression process: 2-3

$$\frac{T_3}{T_2} = \left(\frac{p_3}{p_2}\right)^{\frac{\gamma-1}{\gamma}}$$

$$T_3 = 304\left(\frac{3.5}{1}\right)^{\frac{1.4-1}{1.4}} = 434.8 \text{ K}$$

Isentropic expansion process: 4-5

$$\frac{T_5}{T_4} = \left(\frac{p_5}{p_4}\right)^{\frac{\gamma-1}{\gamma}}$$

$$T_5 = 345\left(\frac{1.03}{3.5}\right)^{\frac{1.4-1}{1.4}} = 243.2 \text{ K}$$

Refrigeration effect, $q_{5-6} = c_{pa}(T_6 - T_5) = 1.005\ (298 - 243.2) = 55$ kJ/kg

Mass flow rate of air, $\dot{m}_a = \dfrac{\dot{Q}}{q_{5-6}} = \dfrac{20\times 211}{55} = 76.72$ kg/min

Compressor work done, $\dot{W}_c = \dot{m}_a c_{pa}(T_3 - T_2)$

$$= 76.72 \times 1.005\ (434.8 - 304)$$

$$= 10085.15 \text{ kJ/min}$$

$$\text{COP} = \frac{\dot{Q}}{\dot{W}_c} = \frac{20\times 211}{10085.15} = 0.418$$

Example 4.14 *In a Bootstrap air refrigeration system of an aircraft, the ram air temperature and pressure are 17°C and 1.05 bar, respectively. Air leaving the main compressor at 4 bar is cooled to 27°C using rammed air. This air is then compressed in the auxiliary compressor (driven by the cooling turbine) to 200°C and again cooled to 27°C in auxiliary heat exchanger using rammed air. Finally, expansion takes place to cabin pressure of 1.01 bar, and air leaves the cabin at 25°C. Determine the maximum pressure and COP of the system:*

Take: $\gamma = 1.4$ *and* $c_p = 1.004$ *kJ/kg K.*

Solution: Given: $p_1 = 1.05$ bar, $T_1 = 273 + 17 = 290$ K, $p_2 = p_3 = 4$ bar,

$T_3 = T_5 = 273 + 27 = 300$ K, $p_6 = p_7 = 1.01$ bar, $T_7 = 273 + 25 = 298$ K,

$T_4 = 273 + 200 = 473$ K

Process 1-2 (Fig. 427):

$$T_2 = T_1\left(\frac{p_2}{p_1}\right)^{\frac{\gamma-1}{\gamma}} = 290\left(\frac{4}{1.05}\right)^{\frac{1.4-1}{1.4}} = 424.97 \text{ K}$$

Process 3-4:

$$\frac{p_4}{p_3} = \left(\frac{T_1}{T_3}\right)^{\frac{\gamma}{\gamma-1}}$$

$$p_4 = 4\left(\frac{473}{300}\right)^{\frac{1.4}{0.4}} = 19.68 \text{ bar}$$

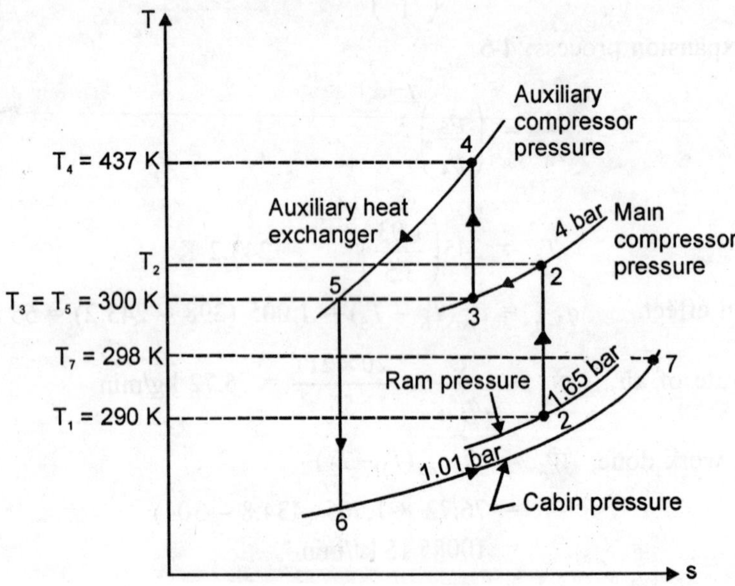

Fig. 4.27 T-s diagram.

Process 5-6:

$$\frac{T_6}{T_5} = \left(\frac{p_6}{p_5}\right)^{\frac{\gamma-1}{\gamma}}$$

$$T_6 = 300\left(\frac{1.01}{19.68}\right)^{\frac{1.4-1}{1.4}} = 128.42 \text{ K}$$

Maximum pressure in the cycle, p_4 = 19.68 bar

Refrigeration effect, $q_{6-7} = c_{pa}\,(T_7 - T_6)$ = 1.04 (298 – 128.42) = 170.258 kJ/kg

Work done, $w_c = c_{pa}(T_2 - T_1)$ = 1.004 (424.97 – 290) = 135.5 kJ/kg

$$\text{COP} = \frac{q_{6-7}}{W_c} = \frac{170.258}{135.5} = 1.256$$

Example 4.15 *A plane moving with a speed of $M = 1/3$ uses simple air refrigeration system with ambient conditions of 0.5 bar and –5°C. The pressure ratio in the main compressor is 5 and isentropic efficiency of compression is 0.85. The compressed air is cooled to a temperature which is 5°C above the rammed air temperature. This air is expanded to cabin pressure of 1 bar in a turbine having isentropic efficiency of 0.9. If the air leaves the cabin at 25°C, find*

(i) COP

(ii) Mass flow rate of air for a system of 50 TR capacity.

Take: $\gamma = 1.4$, and $c_p = 1.004$ kJ/kg.-K.

Solution:

(*a*) Given: $M = \frac{1}{3}$, $p_1 = 0.5$ bar, $T_1 = 273 - 5 = 268$ K,

$$\frac{p_3}{p_2} = 5,\ \eta_c = 0.85,\ T_4 = T_2 + 5°\text{C},\ p_5 = 1 \text{ bar},\ \eta_t = 0.9,$$

$$T_6 = 273 + 25 = 298 \text{ K},\ \dot{Q} = 50 \text{ TR},\ \gamma = 1.4,$$

$$c_{pa} = 1.004 \text{ kJ/kg.K}$$

The *T-s* diagram for the simple air refrigeration cycle is shown in Fig. 4.28.

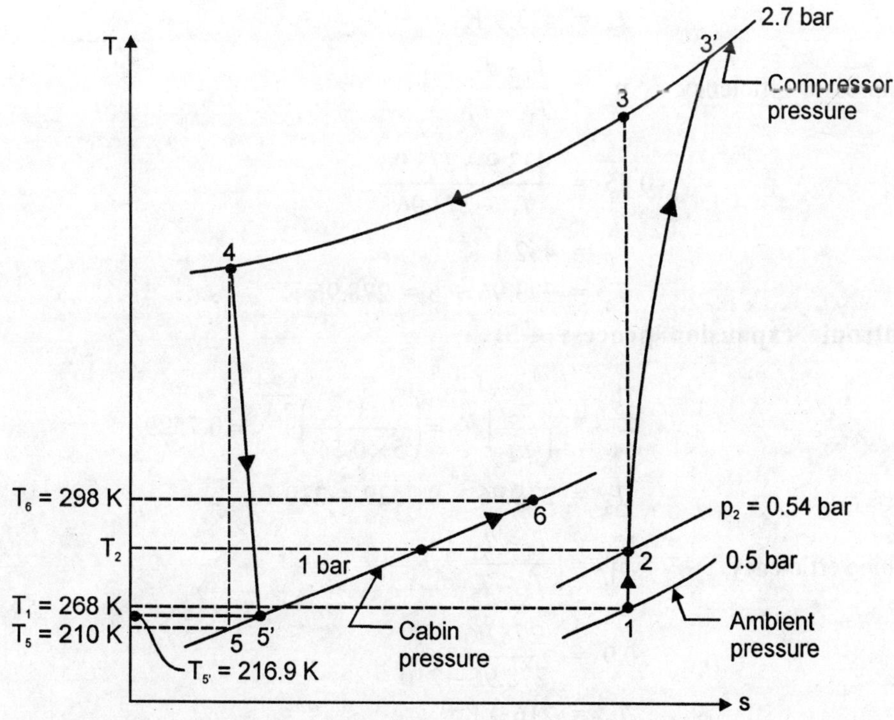

Fig. 4.28 T-s diagram.

Velocity of sound, $c = \sqrt{\gamma R T_1}$

$= \sqrt{1.4 \times 287 \times 268}$

$= 328.15$ m/s

Mach numbers, $m = \dfrac{c_1}{c} = \dfrac{1}{3}$

Velocity of plane, $c_1 = \dfrac{328.15}{3} = 109.4$ m/s

$$T_2 = T_1 + \frac{c_1^2}{2c_{pa} \times 10^3} = 268 + \frac{(109.4)^2}{2 \times 1.004 \times 10^3} = 273.96 \text{ K}$$

$$\frac{p_2}{p_1} = \left(\frac{T_2}{T_1}\right)^{\frac{\gamma}{\gamma-1}} = \left(\frac{273.96}{268}\right)^{\frac{1.4}{1.4-1}} = 1.08$$

$$p_2 = 0.5 \times 1.08 = 0.54 \text{ bar}$$

Isentropic compression process 2-3:

$$\frac{T_3}{T_2} = \left(\frac{p_3}{p_2}\right)^{\frac{\gamma-1}{\gamma}} = (5)^{\frac{1.4-1}{1.4}} = 1.5838$$

$$T_3 = 433.9 \text{ K}$$

Compressor efficiency, $\eta_c = \dfrac{T_3 - T_2}{T_{3'} - T_2}$

$$0.85 = \frac{433.9 - 273.96}{T_{3'} - 273.96}$$

$$T_{3'} = 462.1 \text{ K}$$

$$T_4 = 273.96 + 5 = 278.96 \text{ K}$$

Isentropic expansion process 4-5:

$$\frac{T_5}{T_4} = \left(\frac{p_5}{p_4}\right)^{\frac{\gamma-1}{\gamma}} = \left(\frac{1}{5 \times 0.54}\right)^{\frac{1.4-1}{1.4}} = 0.7529$$

$$T_5 = 278.96 \times 0.7529 = 210 \text{ K}$$

Turbine efficiency, $\eta_t = \dfrac{T_4 - T_{5'}}{T_4 - T_5}$

$$0.9 = \frac{278.96 - T_{5'}}{278.96 - 210}$$

$$T_{5'} = 216.9 \text{ K}$$

Refrigeration effect, $q_{5'-6} = c_{pa}(T_6 - T_{5'}) = 1.004\ (298 - 216.9) = 81.42$ kJ/kg

$$\text{Mass flow rate of air,} \quad \dot{m}_a = \frac{\dot{Q}}{q_{5'-6}} = \frac{50 \times 211}{81.42} = 129.57 \text{ kg/min}$$

$$\text{Compressor work done,} \quad \dot{W}_c = \dot{m}_a c_{pa}(T_{3'} - T_2)$$
$$= 129.57 \times 1.004\ (462.1 - 273.96)$$
$$= 24474.8 \text{ kJ/min}$$

$$\text{COP} = \frac{\dot{Q}}{\dot{W}_c} = \frac{50 \times 211}{24474.8} = 0.431$$

Example 4.16 *10 tons of refrigeration is required for a food storage space to keep the food stuff cool.*

The refrigerant used in the plant is Freon-12 and is sub-cooled by 5°C before entering to the expansion valve. The temperature of evaporator and conenser are –8°C and 30°C respectively. The vapour is superheated to 6°C before leaving the evaporator coil. The compressor is of two cylinders, single acting type having stroke equal to 1.5 time the bore. The compressor runs at 800 rpm and compression is assumed to be isentropic. Estimate the following:

(i) Refrigerating effect per kg,
(ii) Mass flow rate of the refrigerant,
(iii) Theoretical piston displacement,
(iv) Theoretical power,
(v) C.O.P., and
(vi) Theoretical bore and stroke of the compressor. **[IES, 2004]**

Solution: From *p-h* diagram for R-12, we have

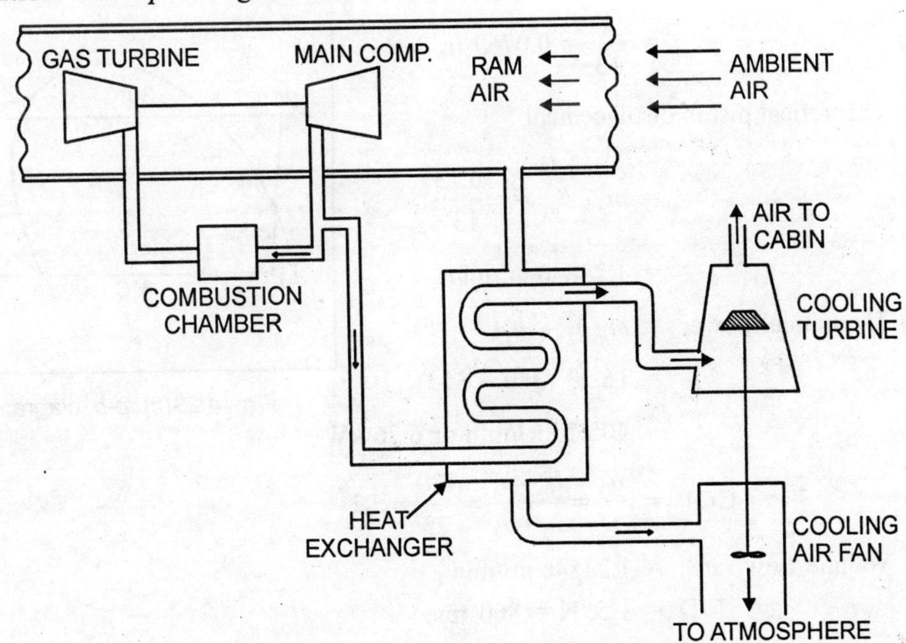

Fig. 4.29(*a*) Simple air-conditioning system.

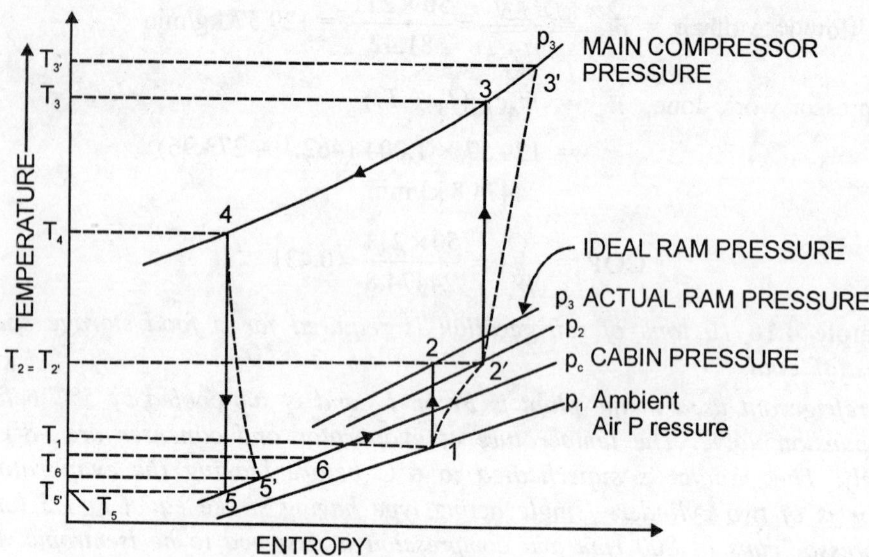

Fig. 4.29(*b*) T-s diagram.

$$h_1 = 355 \text{ kJ/kg}, h_3 = h_4 = 225 \text{ kJ/kg}, h_2 = 380 \text{ kJ/kg}$$

(*i*) Refrigerating effect, $q_0 = h_1 - h_4$

$= 355 - 225 = 130$ kJ/kg

(*ii*) Mass flow rate of refrigerant, $\dot{m}_r = \dfrac{TR}{q_0} = \dfrac{10 \times 211}{130} = 16.23$ kg/min

(*iii*) $v_{g1} = \dfrac{1}{13} = 0.0769$ m³/kg

Theoretical piston displacement

$= m_r \cdot v_{g1} = \dfrac{16.23}{13}$

$= 1.24846$ m³/min

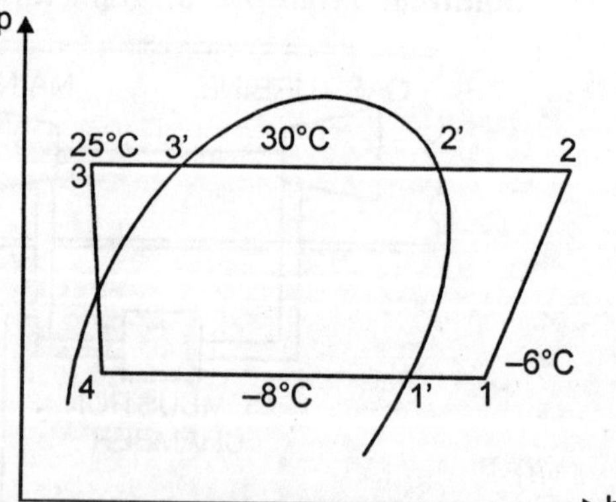

Fig. 4.29(*c*) *p-h* diagram.

(*iv*) Theoretical power $= m_r(h_2 - h_1)$

$= 16.23\ (380 - 355)$

$= 405.75$ kJ/min or 6.76 kW

(*v*) COP $= \dfrac{q_0}{w} = \dfrac{h_1 - h_4}{h_2 - h_1} = \dfrac{130}{25} = 5.2$

(*vi*) Volume flow rate $= 1.24846$ m³/min

L/D = 1.5, N = 800 rpm

$$\text{Cylinder volume} = \frac{2 \times \pi D^2 LN}{4}$$

$$D^3 = 6.6206 \times 10^{-4}$$

$$D = 0.087 \text{ m or } 87 \text{ mm}$$

$$L = 130 \text{ m}$$

Example 4.17 *The compressed air from the main compressor of an aircraft cooling system is bled off at 4.5 bar and 200°C. It is then passed through the heat exchanger in which the ram air is forced through for cooling purposes by the fan driven by the cooling turbine. The condition of the inlet to the cooling turbine is 4 bar and 30°C. The air expands in the cooling turbine is 80% and the flow rate in the turbine is 30 kg/min. By drawing the system and showing the process on T-s diagram, find*

(i) The actual exit temperature from the cooling turbine.

(ii) The power delivered to the air blow fan.

(iii) The tons of refrigeration, if the cold air is tempered by mixing with by-passed warm air and delivered to the cabin cockpit area where it warms upto 25°C before exhausting out to waste. **[IES, 2005]**

Solution: Aircraft cooling system. The T-s diagram for the given data is shown in Fig. 4.30.

(*i*)

$$\frac{T_5}{T_4} = \left(\frac{p_5}{p_4}\right)^{(\gamma-1)/\gamma}$$

$$T_5 = 303\left(\frac{0.7}{4}\right)^{0.4/1.4}$$

$$= 184.15 \text{ K}$$

$$\text{Isentropic efficiency} = \frac{T_4 - T'_5}{T_4 - T_5}$$

$$0.8 = \frac{(303 - T_5')}{303 - 184.15}$$

$$T_{5'} = 207.92 \text{ K or } -65.08°\text{C}$$

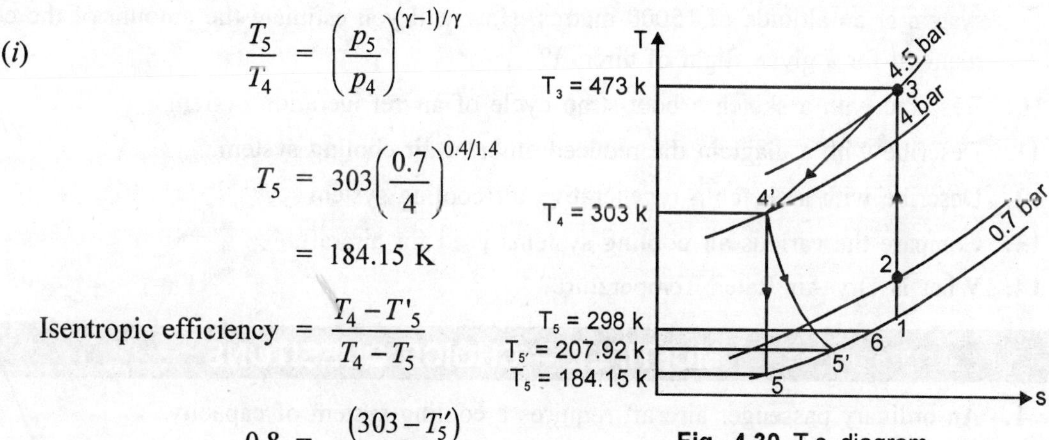

Fig. 4.30 T-s diagram.

(*ii*) Power delivered to the ram air blow fan

$$= \dot{m}_a c_p (T_6 - T_{5'}) = \left(\frac{30}{60}\right) \times 1.005 \ (303 - 207.92) = 47.78 \text{ kW}$$

(*iii*) Refrigeration effect $= m_a c_p \ (T_6 - T'_5) = 30 \times 1.005 \ (298 - 207.92) = 2715.9$ kJ/min

or

$$Q = \frac{2715.9}{211} = 12.87 \text{ TR}$$

REVIEW QUESTIONS

1. Explain refrigeration system using Brayton cycle and show the state points on Temperature-Entropy diagram considering the irreversibilities.
2. "The COP of an Air refrigeration cycle is very low, but still air refrigeration system is most common in the Air Craft", discuss the statement.
3. What is the need of air conditioning of air crafts at high altitudes where ambient temperatures are very low?
4. Discuss the different classifications of aircraft refrigeration systems.
5. Explain briefly the simple aircraft refrigeration system.
6. Explain the difference between simple air craft refrigeration and boot-strap air refrigeration system.
7. Discuss the variations in power for cabin pressurization and refrigeration with altitude. Also discuss the total power requirements with altitude.
8. Why is it necessary to pressuries the cabin in case of air refrigeration system?
9. List the names of three evaporative coolants that can be used in air-craft refrigeration system at an altitude of 15000 metres. How will you estimate the amount of the coolant required for a given flight of aircraft?
10. Describe with a sketch a boot-strap cycle of air refrigeration system.
11. Describe with a diagram the reduced ambient air cooling system.
12. Describe with a sketch a regenerative air cooling system.
13. Compare the various air cooling systems used for aircraft.
14. What is Dry Air Rated Temperature?

MULTIPLE – CHOICE QUESITONS

1. An ordinary passenger aircraft requires a cooling system of capacity
   (a) 2 TR (b) 4 TR (c) 8 TR (d) 20 TR
2. A jet fighter travelling at 950 km/h needs a cooling system of capacity
   (a) 2 to 4 TR (b) 4 to 8 TR (c) 8 to 10 TR (d) 10 to 20 TR
3. The cooling system used for supersonic aircrafts and rockets is
   (a) simple air cooling system (b) boot-strap air cooling system
   (c) reduced ambient air cooling system (d) regenerative air cooling system
4. A boot-strap air cooling system has
   (a) one heat exchanger (b) two heat exchangers
   (c) three heat exchangers (d) four heat exchangers

5. The air cooling system mostly used in transport type aircrafts is

(a) simple air cooling system (b) simple evaporative air cooling system

(c) boot-strap air cooling system (d) all of the above

6. In a boot-strap air evaporative cooling system, the evaporator is provided

(a) between the combustion chamber and the first heat exchanger

(b) between the first heat exchanger and the secondary compressor

(c) between the secondary compressor and the second heat exchanger

(d) between the second heat exchanger and the cooling turbine

7. The reduced ambient air cooling system has

(a) one cooling turbine and one heat exchanger

(b) one cooling turbine and two heat exchangers

(c) two cooling turbines and one heat exchanger

(d) two cooling turbines and two heat exchangers

8. In case of air refrigeration system for air craft the ram efficiency is expressed (p_1 = pressure before ramming, p_2 = ideal pressure after ramming, p_2 = actual pressure after ramming, T_1 = temperature before ramming, T_2 and $T_{2'}$ ideal and actual temperatures after ramming).

(a) $\eta_{Ram} = \dfrac{p_{2'} - p_1}{p_2 - p_1}$ (b) $\eta_{Ram} = \dfrac{p_2 - p_1}{p_{2'} - p_1}$

(c) $\dfrac{T_{2'} - T_1}{T_2 - T_1}$ (d) $\dfrac{T_2 - T_1}{T_{2'} - T_1}$

9. A plane is speeding at Mach No. M. The ratio of the temperature reached after ramming (T_2) to the ambient temperature (T_1) is given by

(a) $\dfrac{T_2}{T_1} = 1 + \dfrac{\gamma - 1}{2} M^2$ (b) $\dfrac{T_2}{T_1} = 1 + \dfrac{\gamma + 1}{2} M^2$

(c) $\dfrac{T_2}{T_1} = 1 - \dfrac{\gamma - 1}{2} M^2$ (d) $1 + \dfrac{1}{\gamma - 1} M^2$

10. Power for pressurization of cabin in air plane is given by (usual notations are used)

(a) $P = \dfrac{m_b c_p T_1}{\eta_c}\left[\left(\dfrac{p_{cabin}}{p_{ambient}}\right)^{(\gamma+1)/\gamma} - 1\right]$ (b) $P = \dfrac{m_b c_p T_1}{\eta_c}\left[\left(\dfrac{p_{cabin}}{p_{ambient}}\right)^{\gamma/(\gamma-1)} - 1\right]$

(c) $P = \dfrac{m_b c_p T_1}{\eta_c}\left[1 - \left(\dfrac{p_{cabin}}{p_{ambient}}\right)^{\gamma/(\gamma-1)}\right]$ (d) $P = \dfrac{m_b c_p T_1}{\eta_c}\left[\left(\dfrac{p_{cabin}}{p_{ambient}}\right)^{(\gamma-1)/\gamma} - 1\right]$

11. The Temperature-Entropy diagram for air-craft refrigeration is given by Fig. 1. This represents

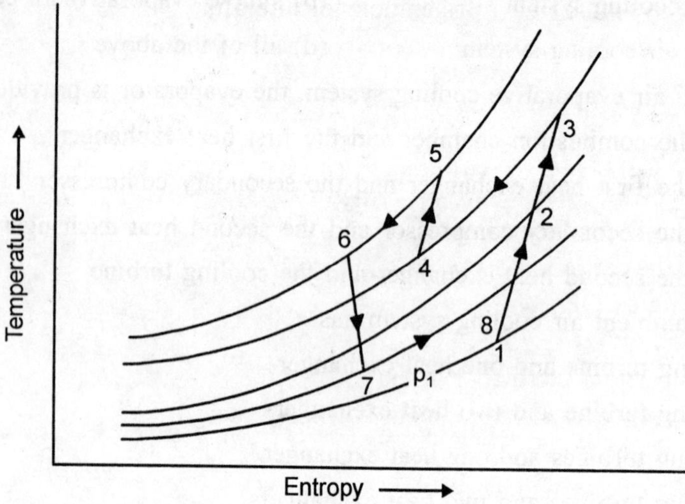

Fig. 1

(a) Reduced ambient-air system of refrigeration
(b) Regenerative air refrigeration system
(c) Boot-strap air refrigeration system
(d) Simple Evaporative type of air refrigeration system.

12. For very high speed planes cruising at Mach. No. of 2.5 and above, the air craft refrigeration system recommended is

(a) Simple evaporative system
(b) Boot-strap type
(c) Regenerative type
(d) Boot-strap evaporative type.

13. The aircraft system giving Lowest Dry Air Rated Turbine discharge temperature at supersonic cruising speeds of the plane, is

(a) Reduced ambient-system of refrigeration
(b) Boot-strap system of refrigeration
(c) Boot-strap-evaporative system of refrigeration
(d) Regenerative system of refrigeration.

14. Graph of Power required for refrigeration only with altitude at which plane is moving, in case of air refrigeration systems is given below in Fig. 2.

15. Refer above Figure 2. Power required for pressurisation of aircraft versus altitude at which plane is flying is given by the graph.

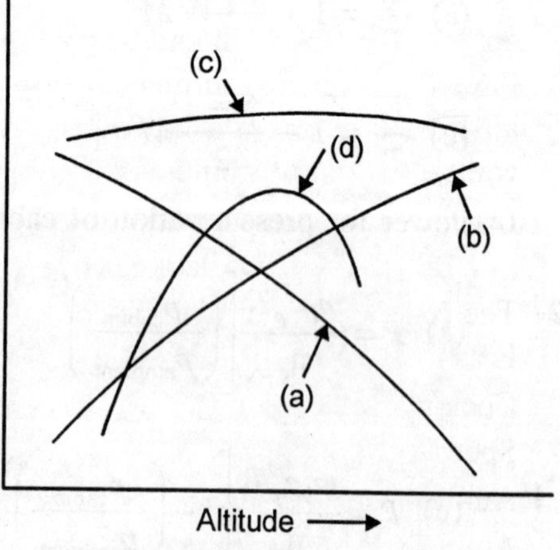

Fig. 2

16. There are two cooling turbines in

(a) Boot-strap system of air refrigeration for aircrafts

(b) Reduced ambient system of air refrigeration for aircrafts
(c) Regenerative system of air refrigeration for aircrafts
(d) Boot-strap-evaporative system of air refrigeration for aircrafts.

17. There are two stages of compression of air in
(a) Boot-strap system of air refrigeration for aircraft
(b) Reduced ambient system of air refrigeration for aircraft
(c) Regenerative system of air refrigeration for aircraft
(d) Simple-evaporative system of refrigeration for aircraft

18. While designing the refrigeration system of an aircraft prime consideration is that the
(a) system has high COP
(b) HP/ton is low
(c) weight of refrigerant circulated in the system is low
(d) weight of the refrigeration equipment is low

19. In aircraft, air refrigeration cycle is used because of
(a) low unit weight per tonne of refrigeration
(b) high heat transfer rate
(c) lower temperature at high altitudes
(d) higher coefficient of performance

Answers

1. (c)	**2.** (d)	**3.** (d)	**4.** (b)	**5.** (c)	**6.** (d)	**7.** (c)	**8.** (a)
9. (a)	**10.** (d)	**11.** (c)	**12.** (d)	**13.** (c)	**14.** (a)	**15.** (b)	**16.** (b)
17. (a)	**18.** (d)	**19.** (a)					

NUMERICAL PROBLEMS

4.1 A simple air refrigeration system is used for an aircraft to take a load of 20 TR. The ambient pressure and temperature are 0.9 bar and 22°C respectively. The pressure of air is increased to 1 bar due to isentropic ramming action. The air is further compressed in a compressor to 3.5 bar and then cooled in a heat exchanger to 72°C. Finally, the air is passed through the cooling turbine and then it is supplied to the cabin at a pressure of 1.03 bar. The air leaves the cabin at a temperature of 25°C. Assuming the isentropic efficiencies of the compressor and turbine as 80 percent and 75 percent respectively, find:
(a) Power required to take the load in the cooling cabin; and (b) COP of the system.
Take c_p = 1.005 kJ/kg K; and γ = 1.4. **[Ans.** 390 kW; 0.18]

4.2 The cockpit of a jet is to be cooled by a simple air refrigeration system. The data available is as follows:

Cock-pit cooling load	= 20 TR
Speed of the plane	= 1000 km/h
Ambient air pressure	= 0.35 bar
Ambient air temperature	= –15°C
Ram efficiency	= 90%

Pressure ratio in the main compressor	= 3
Pressure drop in the heat exchanger	= 0.1 bar
Isentropic efficiencies of main compressor and cooling turbine	= 80%
Temperature of air entering the cooling turbine	= 30°C
Pressure of air leaving the cooling turbine	= 1.06 bar
Pressure in the cock-pit	= 1 bar

If the cock-pit is to be maintained at 25°C, find:

(a) Stagnation temperature and pressure of air entering the main compressor;

(b) Mass flow rate of air to cock-pit;

(c) Power required to drive the refrigerating system; and

(d) COP of system **[Ans.** 210 kg/min; 477 kW; 0.147]

4.3 A boot-strap air refrigeration system of 20 TR capacity is used for an aeroplane flying at an altitude of 2000 m. The ambient air pressure and temperature are 0.8 bar and 0°C. The ram air pressure and temperature are 1.05 bar and 17°C. The pressure of air after isentropic compression in the main compressor is 4 bar. This air is now cooled to 27°C in another auxiliary heat exchanger and then expanded isentropically upto the cabin pressure of 1.01 bar. If the air leaves the cabin at 25°C and the efficiencies for the main compressor, auxiliary compressor and the cooling turbine are 80%; 75% and 80% respectively; find:

(a) Power required to operate the system: and (b) COP of the system.

[Ans. 106.2 kW; 0.66]

4.4 A boot-strap air refrigeration system is used for an aeroplane to take 10 tonnes of refrigeration load. The ambient air conditions are 15°C and 0.9 bar. This air is rammed isentropically to a pressure of 1.1 bar. The pressure of the air bled off the main compressor is 3.5 bar and this is further compressed in secondary compressor to a pressure of 4.5 bar. The isentropic efficiency of both the compressors is 90% and that of cooling turbine is 85%. The effectiveness of both the heat exchangers is 0.6. If the cabin is to be maintained at 25°C and the pressure in the cabin is 1 bar, find: (a) mass of air passing through the cabin; (b) power used for the refrigeration system; and (c) COP of the system.

[Ans. 55.3 kg/min; 25 kW; 0.28]

4.5 The following data refer to a reduced ambient refrigeration system:

Ambient pressure	= 0.8 bar
Pressure of ram air	= 1.1 bar
Temperature of ram air	= 20°C
Pressure at the end of main compressor	= 3.3 bar
Efficiency of main compressor	= 80%
Heat exchanger effectiveness	= 80%
Pressure at the exit of the auxiliary turbine	= 0.8 bar
Efficiency of auxiliary turbine	= 85%
Temperature of air leaving the cabin	= 25°C
Pressure in the cabin	= 1.013 bar
Flow rate of air through cabin	= 60 kg/min

Find: (a) Capacity of the cooling system required; (b) Power needed to operate the system; and (c) COP of the system. **[Ans.** 19.5 TR; 136 kW; 0.504]

4.6 The following data refer to a reduced ambient refrigeration system used for an air-craft:

Speed of aircraft	= 1500 km/h
Ambient pressure	= 0.8 bar
Ambient temperature	= 5°C
Ram efficiency	= 100%
Pressure of cooled air leaving the first cooling turbine	= 0.8 bar
Temperature of cooled air leaving the heat exchanger	= 100°C
Pressure ratio of the Main compressor	= 3
Pressure loss between the outlet of second cooling Turbine and the cabin	= 0.1 bar
Pressure in the cabin	= 1 bar
Temperature in the cabin	= 22°C
Load in the cabin	= 10 TR
Isentropic efficiency of compressor	= 85%
Isentropic efficiency of both cooling turbines	= 80%

Find: (a) mass flow of the air passing through the second cooling turbine; (b) quantity of ram air passing through the heat exchanger, if the rise in temperature is limited to 80 K; and (c) COP of the system. **[Ans.** 54 kg/min : 79.5 kg/min; 0.26]

4.7 A regenerative air refrigeration system for an aeroplane is designed to take a load of 30 TR. The temperature and pressure conditions of the atmosphere are 5°C and 0.85 bar. The pressure of the air is increased from 0.85 bar to 1.2 bar due to ramming action. The pressure of air leaving the main compressor is 4.8 bar. 60% of the total heat of the air leaving the main compressor is removed in the heat exchanger and then it is passed through the cooling turbine. The temperature of the rammed air which is used for cooling purposes in the heat exchanger is reduced to 50°C by mixing the air coming out from the cooling turbine. The isentropic efficiencies of the compressor and turbine are 90% and 80% respectively. The pressure and temperature required in the cabin are 1 bar and 25°C respectively. Assuming isentropic ramming and mass of cooled air passing through the heat exchanger equal to the mass of cooling air, find: (a) the ratio of by passed air to ram air used for cooling purposes; and (b) the power required for maintaining the cabin at required condition. **[Ans.** 41.2%; 445 kW]

4.8 Dry air is supplied to the cabin superchargers of an air plane at 1.03 bar and is discharged at 2.2 bar and 140°C. 22.5 kg per minute of this air is supplied to the primary heat exchanger of an air cycle refrigeration system. Determine:

(a) the air temperature leaving the primary heat exchanger if 1400 kJ/min is removed. From the primary heat exchanger the air enters a secondary compressor ($\gamma = 1.4$), where the pressure is increased to 3.2 bar.

(b) discharge air temperature and

(c) compressor power (mechanical efficiency = 86%). The air then flows through a secondary heat exchanger and 1500 kJ/min is removed.

(d) air temperature leaving the secondary heat exchanger. The air then enters air expansion turbine (n = 1.3), where the pressure is dropped to 1.06 bar.

(e) the final discharge, air temperature,

(f) work recovered from expansion turbine (mechanical efficiency = 68%) and

(g) ton refrigeration developed if air is heated at 25°C, in the cabin and pressure is maintained at 1 bar.

4.9 In an open cycle air refrigerating machine 500 kg of air per hour are drawn from cold chamber maintained at 5°C and compressed isentropically to 5 bar. The air is then cooled to 25°C and then expanded isentropieally to the cold chamber pressure of 1.05 bar. Determine the COP of the machine. The following data refer to air craft refrigeration system with evaporative cooling.

Ram air pressure	= 1.04 bar
Ram air temperature	= 26°C
Compressor delivery pressure	= 4 bar
Pressure drop in ram air heat exchanger	= 0.2 bar
Ram air exchanger effectiveness	= 0.8
Cabin pressure	= 1 bar
Mass flow rate of air through cabin bled from main compressor	= 30 kg/minute
Temperature of air leaving the cabin	= 25°C and pressure = 1 bar
Evaporative cooling effect	= 30 kJ/ kg of air flow

Determine :

(i) refrigeration capacity;

(ii) power for refrigeration

(iii) COP

4.10 100 kW power is used for refrigeration system working on Boot-strap air refrigeration cycle. The other data are

Ram air pressure	= 1 bar
Ram air temperature	= 288 K
Compressor delivery pressure	= 3.5 bar
Cabin pressure	= 1 bar
Maximum pressure	= 5.6 bar
Heat exchanger effectiveness for both heat exchangers	= 0.85

Efficiency of main and auxiliary compressors is 0.83 and 0.75 respectively.

Turbine efficiency	= 0.77

Cabin is maintained at 27°C and 1 bar

Determine :

(i) Refrigeration capacity in ton

(ii) COP

Also determine the percentage of turbine power used by auxiliary compressor.

Chapter

5

SIMPLE VAPOUR COMPRESSION SYSTEMS

5.1 INTRODUCTION

The vapour compression refrigeration system is based upon the fact that fluids absorb heat while changing from a liquid phase to vapour phase and reject heat while changing from a vapour phase to a liquid phase. The temperature during the phase change remains constant. However, the temperature varies with the pressure and the fluid. In vapour compression system, the refrigerant condenses and evaporates at temperatures and pressures close to the atmospheric conditions. The refrigerants generally used are ammonia, carbon dioxide and sulphur dioxide. The closed refrigeration cycle is used in which refrigerant does not leave the system. The refrigerant alternately condenses and evaporates. During evaporation the refrigerant absorbs its latent heat from the brine (salt water), which is used for circulating it around the cold camber. While condensing, it gives out its latent heat to the circulating water of the cooler. The vapour compression system is used in a small domestic refrigerator to big industrial refrigeration systems.

5.2 COMPARISON OF VAPOUR COMPRESSION SYSTEM WITH AIR REFRIGERATION SYSTEM

Following are the advantages and disadvantages of the vapour compression refrigeration system over air refrigeration system :

Advantages

1. It has smaller size for the given capacity of refrigeration.
2. It has less running cost.
3. It can be employed over a large range of temperatures.
4. The coefficient of performance is quite high.

Disadvantages

1. The initial cost is high
2. The prevention of leakage of the refrigerant is the major problem in vapour compression system.

5.3 SIMPLE VAPOUR COMPRESSION REFRIGERATION SYSTEM

The schematic diagram of a simple vapour compression refrigeration system is shown in Fig. 5.1. It consists of the following parts:

1. Compressor
2. Condenser
3. Throttle valve
4. Evaporators.

1. Compressor: The low pressure and temperature vapour refrigerant from the evaporator is drawn into the compressor through the suction valve. It is compressed to high pressure and temperature in the condenser. The high pressure and temperature in the compressor. The high pressure and temperature vapours are discharged into the condenser through the delivery valve.

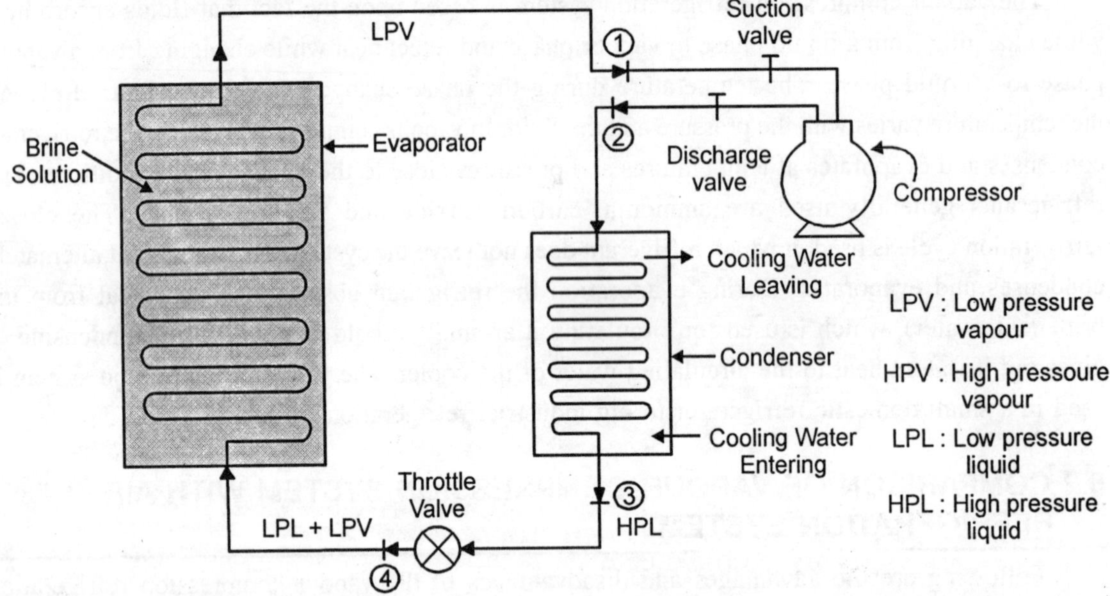

Fig. 5.1 Schematic diagram of simple vapour compression refrigeration system.

2. Condenser: The condenser or cooler consists of coils of pipes in which the high pressure and temperature vapour refrigerant is cooled and condensed. The refrigerant, while passing through the condenser coils gives up its latent heat to the condensing medium. The condensing medium is normally air or water.

3. Throttle Valve: Its function is to allow the liquid refrigerant under high pressure and temperature to pass at a controlled rate after reducing its pressure and temperature. Some of the liquid refrigerant evaporates. The enthalpy of the refrigerant remains unaltered in passing through the throttle valve.

4. Evaporator: It consists of coils of pipe in which the liquid-vapour refrigerant at low pressure and temperature is evaporated and converted into vapour refrigerant at low pressure and temperature. During evaporation, the liquid-vapour refrigerant absorbs its latent heat of vaporization from the medium (air, water or brine), which is to be cooled.

5.4 VAPOUR COMPRESSION REFRIGERATION SYSTEM

A schematic diagram of a vapour compression refrigeration system is shown in Fig. 5.2. It consists of a compressor, condenser, expansion device for throttling and an evaporator. In plants with a large amount of refrigerant charge, a receiver and a drier is installed in the liquid line. The *p-v*, *T-s*, and *p-h* diagrams for the systems are shown in Fig. 5.3. The sequence of operations is as given below:

(*i*) The process 1-2 is isentropic compression.

$s_2 = s_1$, $Q = 0$. Work done by the compressor per kg of refrigerant,

$$W_{1-2} = -\int v\,dp = -\int dh = -(h_2 - h_1)$$

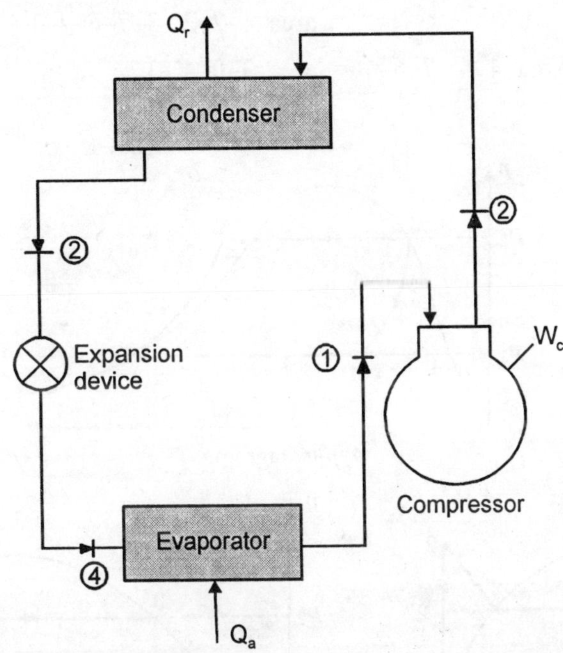

Fig. 5.2 Vapour compression refrigeration system.

(*ii*) The process 2-3 is constant pressure condensation

Heat rejected, $Q_r = h_2 - h_3$

(*iii*) The process 3-4 is isenthalpic expansion.

$h_3 = h_4 = h_{f4} + x(h_1 - h_{f4})$

or Dryness fraction, $$x = \frac{h_3 - h_{f4}}{h_1 - h_{f4}}$$

(*iv*) The process 4-1 is constant pressure evaporation. Refrigeration effect or Heat absorbed,

$Q_a = h_1 - h_4$

$$\text{COP (cooling)} = \frac{\text{Heat aborbed}}{\text{Work done}} = \frac{Q_a}{W_{1-2}} = \frac{h_1 - h_4}{h_2 - h_1} \qquad ...(5.1)$$

$$\text{COP (heating)} = \frac{\text{Heat rejected}}{\text{Work done}} = \frac{Q_r}{W_{1-2}} = \frac{h_2 - h_3}{h_2 - h_1} \qquad ...(5.2)$$

(*a*) From Fig. 5.3(*b*), we have for vapours dry and saturated at the beginning of compression and superheated at the end of compression.

Refrigerating effect, Q_a = area 1–4–6–5–1

Heat rejected, Q_r = area 2–2′–3–7–5–2

Work done, $W = Q_r - Q_a$

= area 1–2–2′–3–7–6–4–1

Area 4–4′–7–6–4 = area 3–0–4′–3

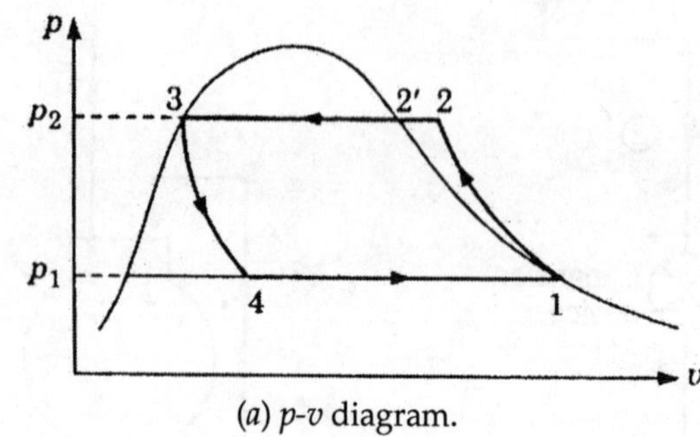

(*a*) *p-v* diagram.

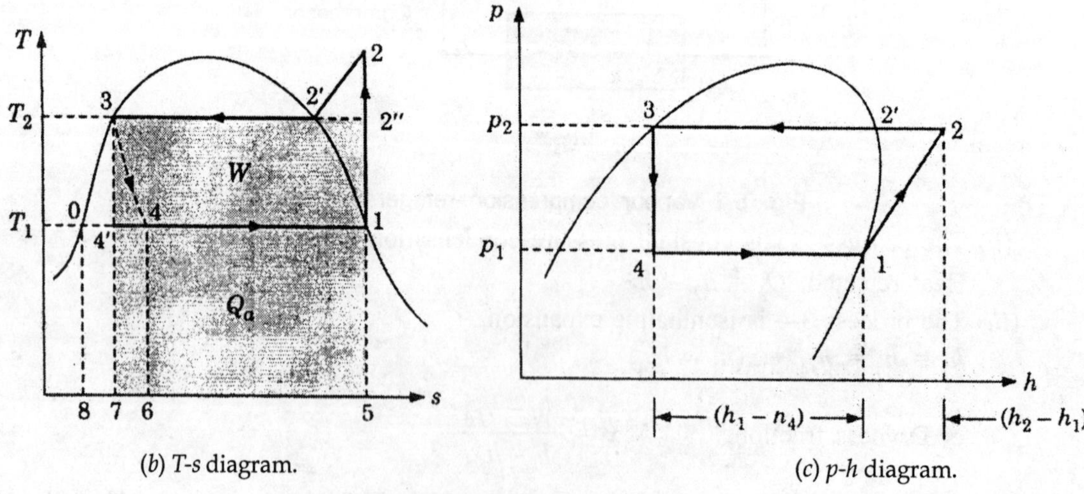

(*b*) *T-s* diagram.

(*c*) *p-h* diagram.

Fig. 5.3 Vapour compression refrigeration cycle.

(*b*) When vapours are dry and saturated at the end of compression, as shown in Fig. 5.4,

$$\text{Work done} = \text{area } 1\text{–}2\text{–}3\text{–}0\text{–}1 = \text{Heat at 2} - \text{Heat at 1} = h_2 - h_1$$

$$\text{Refrigeration effect} = \text{area } 1\text{–}4\text{–}6\text{–}5\text{–}1 = \text{Heat at 1} - \text{Heat at 4} = h_1 - h_4$$

$$\text{Area } 3\text{–}0\text{–}4'\text{–}3 = \text{area } 4\text{–}4'\text{–}7\text{–}6\text{–}4$$

$$\therefore \quad \text{Work done, } W = \text{area } 1\text{–}2\text{–}3\text{–}7\text{–}6\text{–}4\text{–}1$$

Dryness fraction at suction,

$$x_1 = \frac{s_{g2} - s_{f1}}{s_{g1} - s_{f1}}$$

$$h_1 = h_{f1} + x_1(h_{g1} - h_{f1})$$

$$v_1 = v_{f1} + x_1(v_{g1} - v_{f1})$$

$$Q_a = h_1 - h_4 = h_1 - h_3$$

$$W = h_2 - h_1$$

$$\text{COP} = \frac{Q_a}{W} = \frac{h_1 - h_3}{h_2 - h_1} \qquad \text{...(5.3)}$$

The COP of vapour compression cycle is lower than that of reversed Carnot cycle. Power per ton of refrigeration,

$$P = \frac{3.5167}{\frac{\text{COP}}{\text{TR}}} \text{kW} \qquad \text{...(5.4)}$$

$$\text{Refrigerant mass flow rate, } \dot{m} = \frac{3.5167}{Q_a} = \frac{3.5167}{h_1 - h_4} \text{kg/s} \qquad \text{...(5.5)}$$

$$\text{Motor power} = \dot{m}\left[\frac{h_2 - h_1}{1000}\right] \text{kW} \qquad \text{...(5.6)}$$

$$\text{Volume flow rate of refrigerant} = 3.5167 \times \left[\frac{v_1}{h_1 - h_4}\right] \qquad \text{...(5.7)}$$

where v_1 = specific volume of the refrigerant at the inlet of compressor

$$= \left(\frac{\pi}{4} D^2 L \frac{N}{60} n\right) \times \eta_v \qquad \text{...(5.8)}$$

D = dia of compressor piston or cylinder

L = stroke of compressor piston

N = rpm

n = number of cylinders in compressor

η_v = volumetric efficiency

$$= 1 + c - c\left(\frac{p_2}{p_1}\right)^{\frac{1}{n}}$$

c = clearance ratio

Mass flow rate of cooling medium in the condenser:

$$\dot{m}_e \cdot c_p \cdot \Delta T = \dot{m}(h_2 - h_3) \qquad \text{...(5.9)}$$

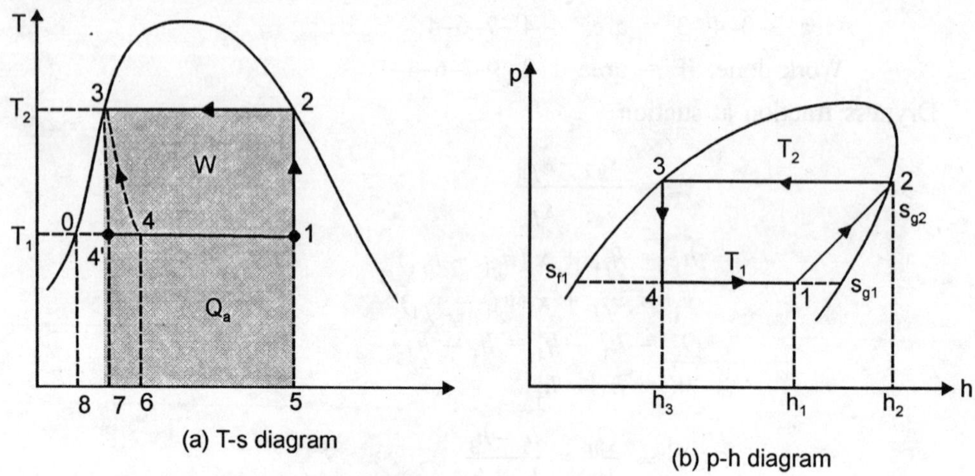

Fig. 5.4 Vapour compressions cycle.

5.5 USE OF T-s AND p-h CHARTS

The *T-s* (Temperature-entropy) diagram for the vapour compression cycle is shown in Fig. 5.5.

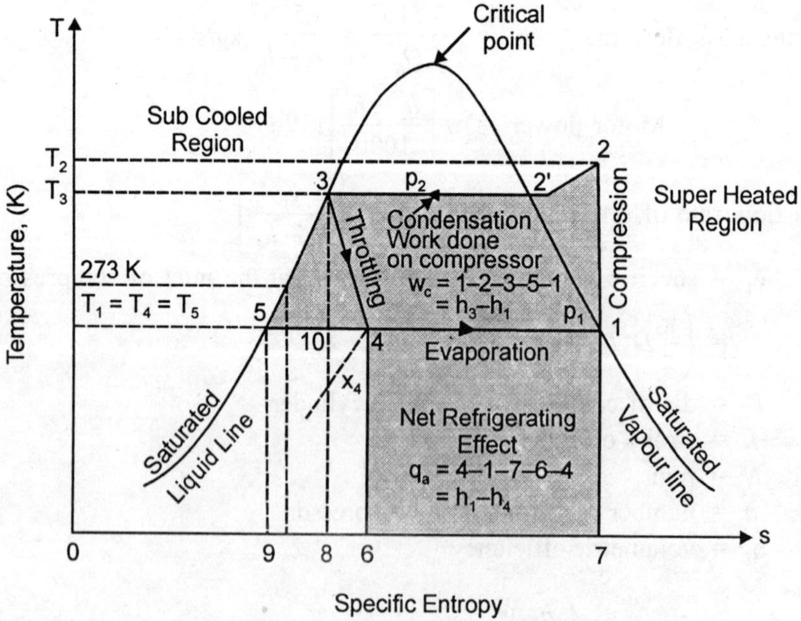

Fig. 5.5 T-s diagram for vapour compression cycle.

Critical point
$T_3 = T_{2'}$
T_2
Sub cooled Region
T_2
p_2
Super Heat Region
2'
2
3
Condensation
Throttling
Compression
$T_{4'} = T_4 = T_1$
x_4
Pressure, p
p_1
1
4'
4
Evaporation
$(h_1 - h_4) = q_a$
$(h_2 - h_1) = w_c$
$T_{2'}$ T_2
T_1
Specific Enthalpy, h

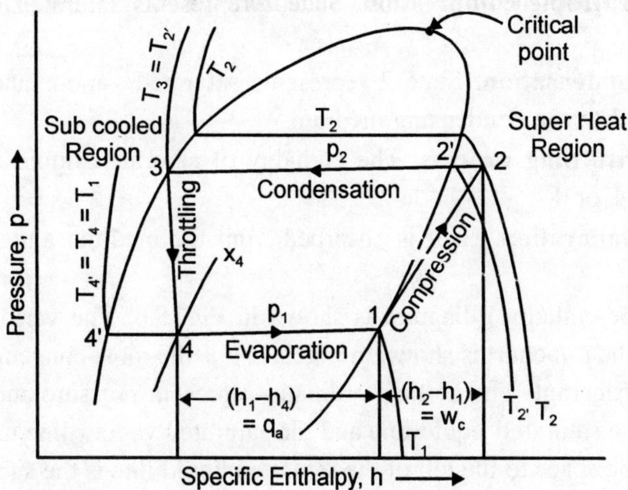

Fig. 5.6 *p-h* diagram for vapour compression cycle.

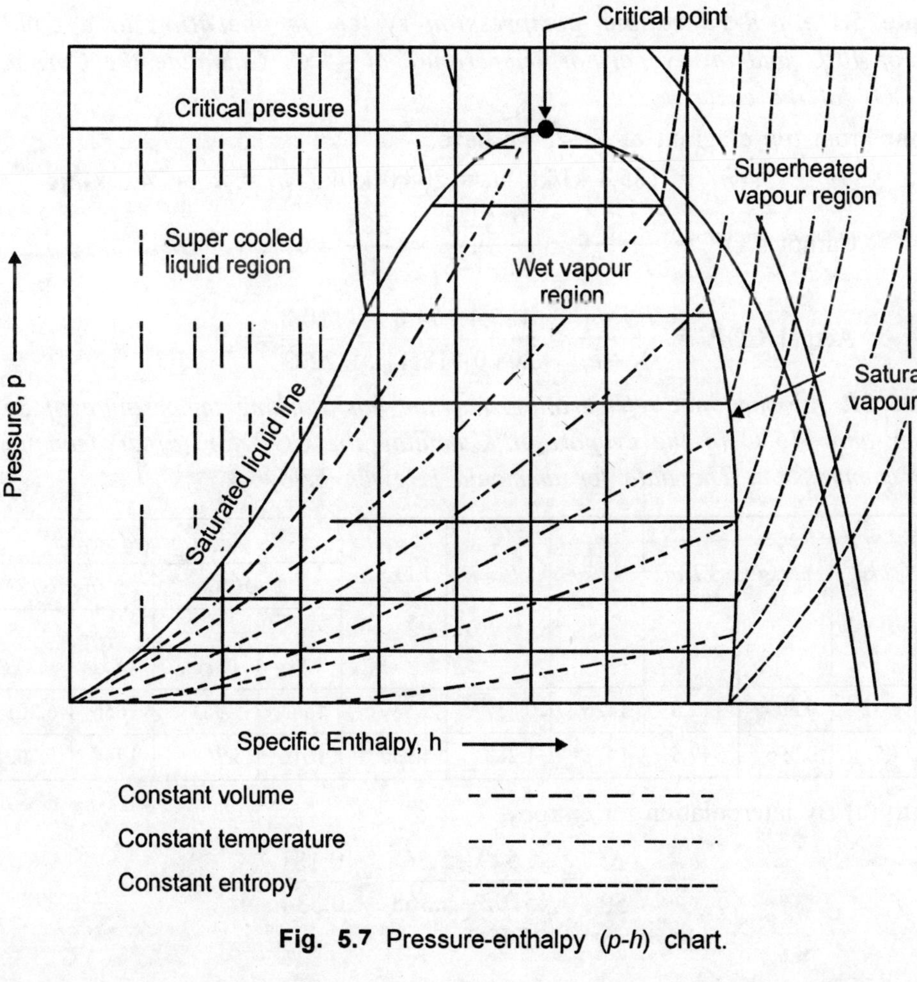

Critical point
Critical pressure
Superheated vapour region
Super cooled liquid region
Wet vapour region
Saturated vapour line
Saturated liquid line
Pressure, p
Specific Enthalpy, h
Constant volume
Constant temperature
Constant entropy

Fig. 5.7 Pressure-enthalpy (*p-h*) chart.

Process 1-2: Isentropic compression. State 1 represents saturated vapour and state 2 is super heated vapour.

Process 2-3: Condensation. State 2 represents saturated vapour and state 3 is saturated liquid. Heat is rejected to the condensing medium.

Process 3-4: Throttling process. The enthalpy of saturated liquid at state 3 is equal to the enthalpy at state 4 or 1.

Process 4-1: Evaporation. Heat is absorbed from the medium and refrigeration effect is produced.

The *p-h* (pressure-enthalpy) diagram is shown in Fig. 5.6. The various processes are also shown in the figure. The *p-h* chart is shown in Fig. 5.7. It is the most convenient chart for studying the behaviour of a refrigerant. The vertical ordinates represent pressure and horizontal ordinates represent enthalpy. The saturated liquid line and the saturated vapour line merge into one another at the critical point. The space to the left of the saturated liquid line is the sub-cooled liquid region. The space between the liquid and vapour lines is called the wet region. The space to the right of the saturated vapour line is the superheated vapour region.

Example 5.1 *An R-12 vapour compression system is operating at a condenser temperature of 40°C and an evaporator temperature of –5°C, Calculate the Carnot COP and actual COP of the cycle.*

Solution: From *p-h* diagram of *R*-12, we have

$$h_1 = 185.4 \text{ kJ/kg}, h_2 = 208.0 \text{ kJ/kg}, h_3 = h_4 = 74.6 \text{ kJ/kg}$$

$$\text{Carnot COP} = \frac{T_1}{T_2 - T_1} = \frac{273-5}{40-(-5)} = \frac{268}{45} = 6$$

$$\text{Actual COP} = \frac{h_1 - h_4}{h_2 - h_1} = \frac{185.4-74.6}{208.0-185.4} = \frac{110.8}{22.6} = 4.9$$

Example 5.2 *An ammonia refrigerating machine has working temperature of 35°C in the condenser and –15°C in the evaporator. Calculate the COP for (a) dry compression, and (b) wet compression. The data for ammonia is given below:*

Saturation Temp. °C	v_f *l/kg*	v_g *m³/kg*	h_f *kJ/kg*	h_g *kJ/kg*	s_f *kJ/kg.K*	s_g *kJ/kg.K*	*Super heated vapour*			
							by 50 °C		*by 100 °C*	
							h kJ/kg	*s kJ/kg.K*	*h kJ/kg*	*S kJ/kg.K*
–15	1.52	0.509	112.3	1426	0.457	5.549	1543	5.963	1656	6.301
35	1.7	0.096	347.5	1471	1.282	4.930	1616	5.368	1744	5.702

Solution: (a) By interpolation for entropy,

$$\frac{\Delta t}{50} = \frac{5.549-5.368}{5702-5.368} = \frac{0.181}{0.334}$$

$$\Delta t = 27.1°\text{C}$$

Degree of super heat at discharge = 50 + 27.1 = 77.1°C

Discharge temperature = 35 + 77.1 – 112.1°C

$$h_2 = 1616 + \frac{27.1}{50}(1744 - 1616) = 1685.4 \text{ kJ/kg}$$

Refrigerating effect, $Q_a = h_1 - h_4 = 1426 - 347.5 = 1078.5$ kJ/kg

Work done, $W = h_2 - h_1 = 1685.4 - 1426 = 259.4$ kJ/kg

$$\text{COP} = \frac{Q_a}{W} = \frac{1078.5}{259.4} = 4.16$$

(*b*)

$$x_1 = \frac{4.930 - 0.457}{5.549 - 0.457} = 0.88$$

$$h_1 = 112.3 + 0.88\ (1426 - 112.3) = 1268.4 \text{ kJ/kg}$$

$$v_1 = v_{f1} + x_1\ (v_{g1} - v_{f1})$$

$$= 0.00152 + 0.88\ (0.509 - 0.00152) = 0.448 \text{ m}^3\text{/kg}$$

$$Q_a = h_1 - h_4 = 1268.4 - 347.5 = 920.9 \text{ kJ/kg}$$

$$W = h_2 - h_1 = 1471 - 1268.4 = 202.6 \text{ kJ/kg}$$

$$\text{COP} = \frac{920.9}{202.6} = 4.55$$

Example 5.3 A refrigerating plant using CO_2 as refrigerant works between 25°C and –5°C, The dryness fraction of CO_2 is 0.6 at the entry of compressor. Find the ice formed per day if the relative efficiency is 50%. Ice is formed at 0°C from water at 10°C. The quantity of CO_2 circulated is 6 kg/min. For water, c_p = 4.187 kJ/kg, L = 335 kJ/kg

Properties of CO_2

Temperature °C	h_f kJ/kg	h_g kJ/kg	s_f kJ/kg.K
25	81.25	121.5	0.2513
–5	–7.53	245.8	–0.04187

Solution: The *T-s* cycle is shown in Fig. 5.8.

$$T_1 = 273 - 5 = 268 \text{ K},$$

$$T_2 = 273 + 25 = 298 \text{ K}$$

$$h_1 = h_{f1} + x_1 h_{fg1} = -7.53 + 0.6 \times 245.8$$

$$= 140 \text{ kJ/kg}$$

$$s_1 = s_2$$

$$s_{f1} + \frac{x_1 h_{fg1}}{T_1} = s_{f2} + \frac{x_2 h_{fg2}}{T_2}$$

$$-0.04187 + \frac{0.6 \times 245.8}{268} = 0.2513 + \frac{x_2 \times 121.5}{298}$$

$$x_2 = 0.63$$

Fig. 5.8

$$h_2 = h_{f2} + x_2 h_{fg2} = 81.25 + 0.63 \times 121.5 = 157 \text{ kJ/kg}$$

Work done, $w_{1-2} = h_2 - h_1 = 157 - 140 = 17$ kJ/kg

Refrigerating effect, $q_a = h_1 - h_4 = h_1 - h_3 = h_1 - h_{f3} = 140 - 81.25 = 58.75$ kJ/kg

$$\text{COP} = \frac{q_a}{w_{1-2}} = \frac{58.75}{17} = 3.45$$

$$\text{Actual COP} = 3.45 \times 0.5 = 1.725$$

$$\text{Work done per second} = \frac{17 \times 6}{60} = 1.7 \text{ kJ/s}$$

$$\text{Actual cooling effect} = 1.725 \times 1.7 = 2.94 \text{ kJ/s}$$

Heat carried to form 1 kg of ice $= c_p(t_2 - t_1) + L = 4.187 \times 10 + 335 = 376.87$ kJ

$$\text{Ice formed per hour} = \frac{2.94 \times 3600}{376.87} = 28\text{kg}$$

Example 5.4 *A refrigerator operating on a standard vapour compression cycle has a coefficient of performance of 6.5 and is driven by a 50 kW compressor. The enthalpies of saturated liquid and saturated vapour refrigerant at the operating condensing temperature of 35°C are 69.55 kJ/kg and 201.45 kJ/kg respectively. The saturated refrigerant vapour leaving the evaporator has an enthalpy of 187.53 kJ/kg. Find the refrigerant temperature at the compressor discharge. The c_p of refrigerant vapour may be taken to be 0.6155 kJ/kg.K.*

Solution: Given: COP=6.5, Compressor power = 50 kW

$$\text{Refrigerating capacity} = 50 \times 6.5 = 325 \text{ kW}$$

Heat extracted per kg of refrigerant = 187.53 – 69.55 = 117.98 kJ/kg

$$\text{Refrigerant rate, } \dot{m} = \frac{325}{117.98} = 2.7547 \text{ kg/s}$$

$$\text{Heat input} = \frac{50}{2.7547} = 18.15 \text{ kJ/kg}$$

Enthalpy of vapour after compression = 187.53 + 18.15 = 205.68 kJ/kg

$$\text{Super heat} = 205.68 - 201.45 = 4.23 \text{ kJ/kg}$$

$$\text{Degree of superheat} = \frac{4.23}{0.6155} = 6.87°\text{C}$$

Temperature of refrigerant at compressor discharge = 35° + 6.87° = 41.87°C

Example 5.5 *A refrigeration system of 15 tons capacity operates on standard simple vapour compression cycle using Refrigerant –22 to an evaporator temperature of 5°C and condensing temperature 50°C. Draw the p-h diagram for the cycle. Calculate (i) the refrigerant mass flow rate, and (ii) the compressor intake volume flow rate if the compressor volumetric efficiency is 0.72. use the refrigerant property data given in Table below.*

Temperature °C	*Pressure* bar	*Enthalpy, kJ/kg*		*Volume*	
		Sat. Liquid	*Sat. Vapour*	*Sat. Liquid* l/kg	*Sat. Vapour* m³/kg
5	5.836	205.9	407.1	0.791	0.0404
50	19.423	263.3	417.7	0.922	0.0117

Solution: Refer to Fig. 5.9.

(*i*)

$$\text{Capacity of plant} = \frac{\dot{m}(h_1 - h_4)}{3.5167}$$

$$h_1 = 407.1 \text{ kJ/kg}$$

$$h_4 = h_3 = 263.3 \text{ kJ/kg}$$

$$\dot{m} = \frac{15 \times 3.5167}{(407.1 - 263.3)} = 0.367 \text{ kg/s}$$

(*ii*)

$$\text{Volume flow rate} = \frac{\dot{m} v_g}{\eta_c} = \frac{0.367 \times 0.0117}{0.72} = 5.96 \times 10^{-3} \text{ m}^3\text{/s}$$

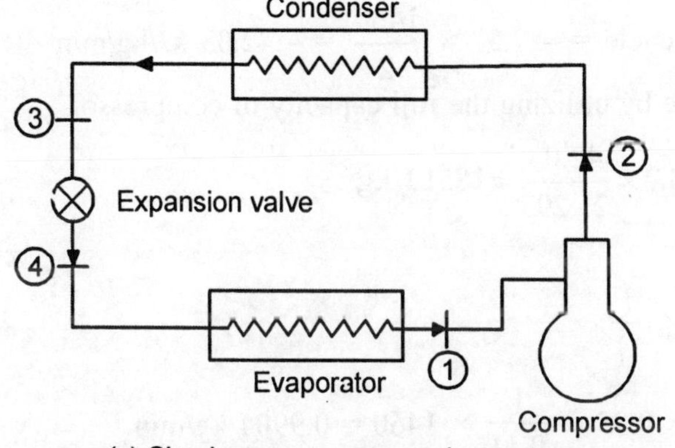

(a) Simple vapour compression system.

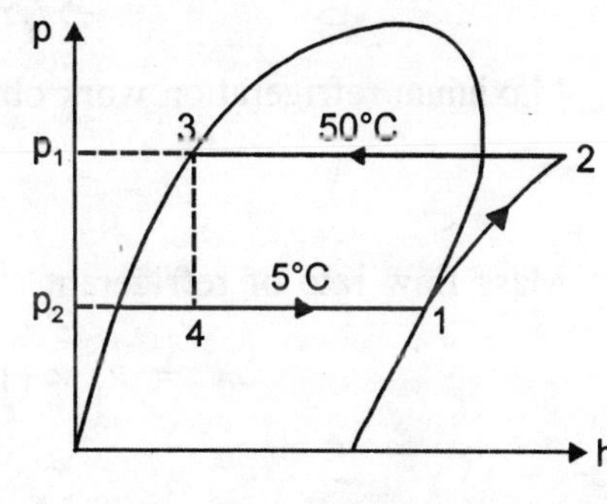

(b) p-h diagram.

Fig. 5.9

Example 5.6 A refrigeration compressor of 50 mm bore and 40 mm stroke operates at a speed of 1460 rpm between condensing and evaporating pressure of 1.219 MPa and 0.151 MPa. The clearance ratio is 5%, ratio of specific heats of refrigerant is 1.18, specific volume of refrigerant at suction is 0.11 m³/kg and enthalpy change of refrigerant in the evaporator is 93.7 kJ/kg. Calculate the refrigeration load the compressor can serve.

Solution: Given: $d = 50$ mm, $l = 40$ mm, $N = 1460$ rpm

$$p_2 = 1.219 \text{ MPa}, \ p_1 = 0.151 \text{ MPa}, \ c = 5\%, \ n = 1.18$$

$$v_1 = 0.11 \text{ m}^3\text{/kg}$$

$$V_1 = \frac{\pi}{4} d^2 l = \frac{\pi}{4} \times 50^2 \times 40 \times 10^{-9} = 78.54 \times 10^{-6} \text{ m}^3$$

$$V_2 = 0.05 \ V_1 = 3.93 \times 10^{-6} \text{ m}^3$$

$$p_2 = p_1\left(\frac{V_1}{V_2}\right)^n = 0.151 \times 20^{1.18} = 5.178 \text{ MPa (maximum)}$$

Work done by the compressor,

$$w_c = \left(\frac{n}{n-1}\right)p_1V_1\left[1-\left(\frac{p_2}{p_1}\right)^{\frac{n-1}{n}}\right]$$

$$= \frac{1.18}{0.08}\times 0.151\times 10^6 \times 78.54\times 10^{-6}\left[1-\left(\frac{1.219}{0.151}\right)^{\frac{0.18}{1.18}}\right]$$

$$= -29.17 \text{ J/kg/cycle} = -29.17 \times \frac{1460}{2} = -21.293 \text{ kJ/kg/min}$$

$$(w_c)_{max} = \frac{1.18}{0.18}\times 0.151\times 10^6 \times 78.54\times 10^6\left[1\left(\frac{5.178}{0.151}\right)^{\frac{0.18}{1.18}}\right]$$

$$= -57.57 \text{ J/kg/cycle} = -57.57 \times \frac{1460}{2} = -42.03 \text{ kJ/kg/min}$$

Maximum refrigeration work obtainable by utilizing the full capacity of compressor

$$= 93.7\times\frac{42.03}{21.293} = 185 \text{kJ/kg}$$

Mass flow rate of refrigerant,

$$\dot{m} = V_1 \times (1 - c) \times \frac{1}{v_1}$$

$$= 78.54 \times 10^{-6} \times 0.95 \times \frac{1}{0.11} \times 1460 = 0.9904 \text{ kg/min}$$

Maximum refrigeration effect obtainable from compressor

$$= 185 \times 0.9904 = 183.22 \text{ kJ/min or } 3.054 \text{ kW}$$

Example 5.7 A refrigeration cycle uses Freon –12 as the working fluid. The temperature of the refrigerant in the evaporator is –10°C. The condensing temperature of 40°C. The cooling load is 150 W and the volumetric efficiency of the compressor is 80%. The speed of compressor is 720 rpm. Calculate the mass flow rate of the refrigerant and the displacement volume of the compressor.

Properties of Freon-12

Temperature °C	*Saturation pressure MPa*	*Enthalpy, kJ/kg*		*Specific Volume, m^3/kg*
		Saturated liquid	*Saturated vapour*	*Saturated vapour*
–10	0.22	26.8	183.0	0.08
40	0.96	74.5	203.1	0.02

Solution. Cooling load, $Q = 150$ W

Refrigeration effect $= h_1 - h_4$ (Fig. 5.10)

$= 183 - 74.5$

$= 108.5$ kJ/kg

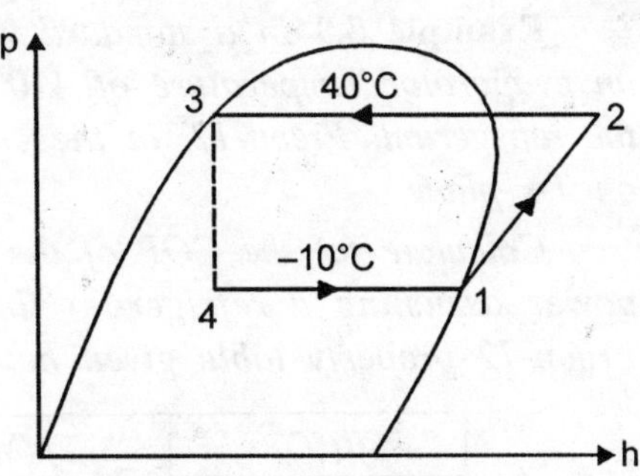

Fig. 5.10 p-h diagram.

$$\dot{m}(h_1 - h_4) = Q$$

Mass flow rate of refrigerant,

$$\dot{m} = \frac{150}{108.5 \times 10^3} = 0.001382 \text{ kg/s}$$

Displacement volume of compressor,

$$V_1 = \frac{\dot{m}v_1}{\eta_v} = \frac{0.001382 \times 0.08}{0.8} = 1.382 \times 10^{-4} \text{ m}^3/\text{s}$$

Example 5.8 *In a simple vapour compression cycle, following are the properties of the refrigerant R-12 at various points.*

Compressor inlet: $h_1 = 183.2$ kJ/kg, $v_1 = 0.0767$ m³/kg

Compressor discharge: $h_2 = 222.6$ kJ/kg, $v_2 = 0.0164$ m³/kg

Condenser exit: $h_3 = 84.9$ kJ/kg, $v_3 = 0.0083$ m³/kg

The piston displacement volume for the compressor is 1.5 litres per stroke and its volumetric efficiency is 80%. The speed of the compressor is 1600 rpm,

Find (a) power rating of the compressor (kW), (b) refrigerating effect (kW).

Solution: Piston displacement volume (Fig. 5.11),

$$V_1 = \frac{\pi}{4} d^2 l = 1.5 \times 10^{-3} \text{m}^3/\text{stroke}$$

Compressor discharge, $$q = \frac{V_1 N}{\eta_v} = \frac{1.5 \times 10^{-3} \times 1600}{0.8}$$

$= 3$ m³/min

Mass flow rate of compressor,

$$\dot{m} = \frac{q}{v_1} = \frac{3}{0.0767}$$

$= 39.12$ kg/min

Power of compressor $= \dot{m}(h_2 - h_1)$

$$= \frac{39.12}{60}(222.6 - 183.2)$$

$= 25.68$ kW

Fig. 5.11 p-h diagram.

Refrigerating effect $= \dot{m}(h_1 - h_4) = 39.12\ (183.2 - 84.9) = 3845.5$ kJ/min or 64.1 kW

Example 5.9 *In a standard vapour compression refrigeration cycle operating between an evaporator temperature of –10°C and condenser temperature of 40°C, the enthalpy of the refrigerant, Freon-12, at the end of compression is 220 kJ/kg. Show the cycle diagram on T-s plane.*

Calculate (a) the COP of the cycle, (b) the refrigerating capacity and the compressor power assuming a refrigerant flow rate of 1 kg/min. You may use the extract of the Freon-12 property table given below:

$t(°C)$	$p(MPa)$	$h_f(kJ/kg)$	$h_g(kJ/kg)$
–10	0.2191	26.85	183.1
40	0.9607	74.53	203.1

Solution: The *T-s* diagram is shown in Fig. 5.12.

$$\text{COP} = \frac{h_1 - h_4}{h_2 - h_1}$$

$$= \frac{183.1 - 26.85}{220 - 183.1} = 4.23$$

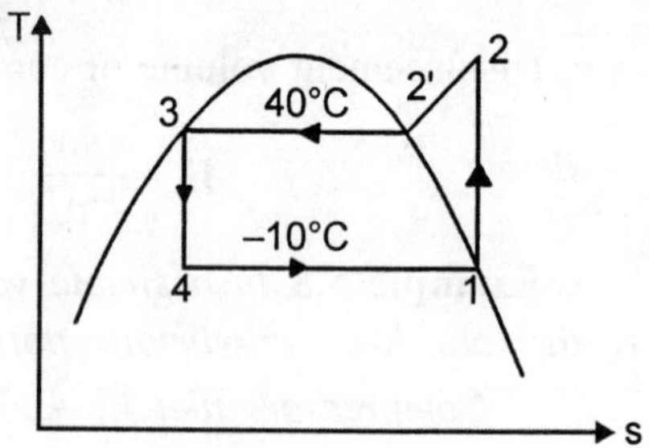

Fig. 5.12 T-s diagram.

$$\text{Refrigerating capacity} = \dot{m}(h_1 - h_4)$$

$$= 1.0\ (183.1 - 26.85)$$

$$= 156.25 \text{ kJ/min or } 2.6 \text{ kW}$$

$$\text{Compressor power} = \dot{m}(h_2 - h_1)$$

$$= 1.0\ (220 - 183.1)$$

$$= 36.9 \text{ kJ/min or } 0.615 \text{ kW}$$

Example 5.10 An R-717 (ammonia) system operates on the basic vapour compression refrigeration cycle. The evaporator and condenser pressures are 0.119 MPa and 1.389 MPa respectively. The mass flow rate of refrigerant is 0.1kg/s. The volumetric efficiency of the compressor is 84%. Determine the compressor displacement rate. If the COP of the cycle is 2, determine the power input to the compressor.

Saturation Properties of R-717 (ammonia)

Temperature °C	*Pressure MPa*	*Specific volume of vapour, m^3/kg*	*Enthalpy, kJ/kg*	
			Liquid	*Vapour*
–30	0.119	0.9638	6.9	1423.6
36	1.389	0.0930	371.4	1488.6

Solution: Compressor displacement (Fig. 5.13)

$$= \frac{\dot{m}v}{\eta_v} = \frac{0.1 \times 0.9638}{0.84} = 0.1174 \text{ m}^3/\text{s}$$

Refrigerating effect $= h_1 - h_4$

$= 1423.6 - 371.4$

$= 1052.2$ kJ/kg

$$\text{COP} = \frac{\text{Refrigerating effect}}{\text{Work done}}$$

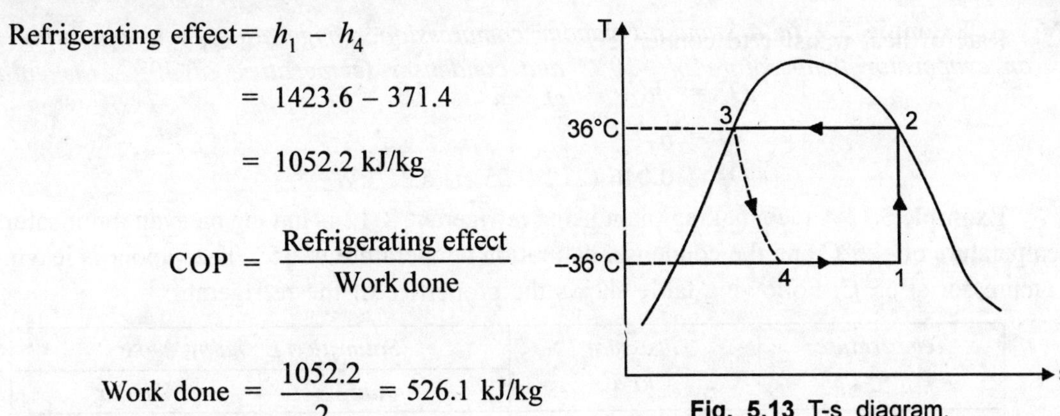

Fig. 5.13 T-s diagram.

$$\text{Work done} = \frac{1052.2}{2} = 526.1 \text{ kJ/kg}$$

Power input to compressor = 526.1 × 0.1 = 52.61 kW

Example 5.11 *In a 5 kW cooling capacity refrigeration system operating on a simple vapour compression cycle, the refrigerant enters the evaporator with an enthalpy of 75 kJ/kg and leaves with an enthalpy of 183 kJ/kg. The enthalpy of the refrigerant after compression is 210 kJ/kg. Show the cycle on T-s and p-h diagrams. Calculate the following: (a) COP, (b) power input to compressor, and (c) rate of heat transfer at the condenser.*

Solution: Refer to Fig. 5.14

$$\text{COP} = \frac{h_1 - h_4}{h_2 - h_1} = \frac{183 - 75}{210 - 183} = \frac{108}{27} = 4$$

Mass flow rate of refrigerant,

$$\dot{m} = \frac{\text{Cooling capacity}}{\text{Cooling effect}}$$

$$= \frac{5}{h_1 - h_4}$$

$$= \frac{5}{108} = 0.046 \text{ kg/s}$$

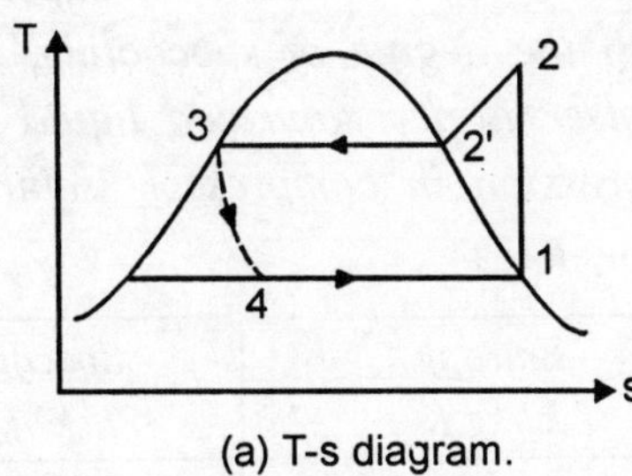

(a) T-s diagram.

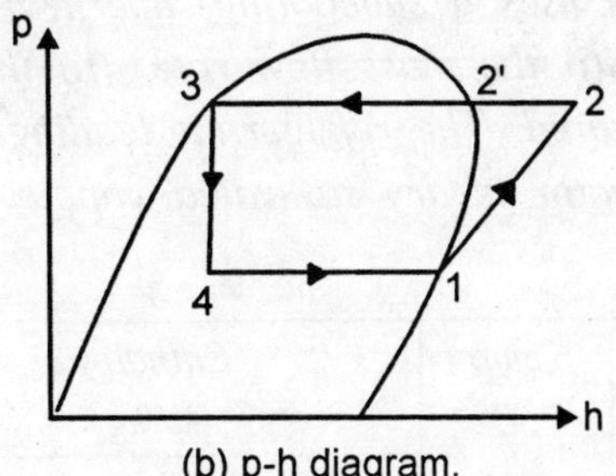

(b) p-h diagram.

Fig. 5.14

Power input to compressor

$$= \dot{m}(h_2 - h_1) = 0.046\ (210 - 183) = 1.24 \text{ kW}$$

Rate of heat transfer to condenser,

$$Q_r = \dot{m}(h_2 - h_4)$$

$$h_4 = h_3$$

$$Q_r = 0.046\ (210 - 75) = 6.21\ \text{kW}$$

Example 5.12 An ice making plant using refrigerant R-12 is having an evaporator saturation temperature of –25°C and the condenser saturation temperature of 35°. The vapour is leaving the compressor at 65°C. Following table shows the properties of the refrigerant:

Temperature °C	*Pressure kPa*	*Saturation Enthalpy, kJ/kg*	
		Liquid	*Vapour*
–25	123.7	13.3	176.6
35	850.0	69.6	201.5

Enthalpy of superheated refrigerant at 850 kPa and 65°C = 225.5 kJ/kg

(i) Calculate the coefficient of performance (COP) of this system.

(ii) If the capacity of the plant is 5 kW, calculate mass flow rate of refrigerant and power consumption.

Solution: Refer to Fig. 5.15.

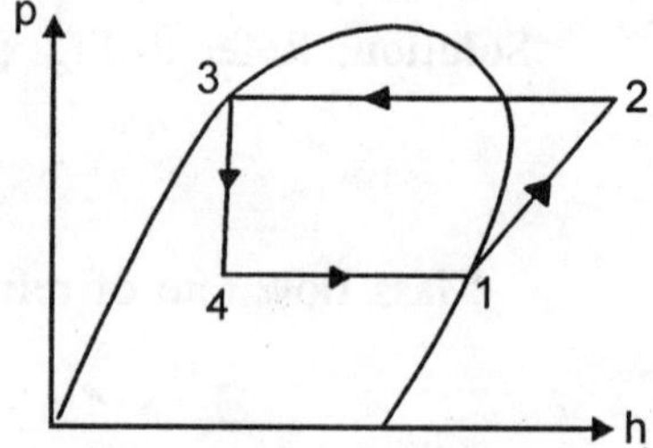

Fig. 5.15 p-h diagram.

$$(i)\ \text{COP} = \frac{h_1 - h_4}{h_2 - h_1} = \frac{176.5 - 69.6}{225.5 - 176.5} = \frac{106.9}{27} = 2.18$$

$$(ii)\ \dot{m}(h_1 - h_1) = 5$$

$$\dot{m} = \frac{5}{106.9} = 0.0467\ \text{kg/s}$$

$$\text{Power consumption} = \dot{m}(h_2 - h_1)$$

$$= 0.0467 \times 49 = 2.288\ \text{kW}$$

Example 5.13 *A 20 ton vapour compression refrigeration system using Freon-12 operates between evaporator pressure of 1.004 bar and condenser pressure of 13.663 bar. The system uses a subcooling-superheating heat exchanger with 10°C of superheat added. Determine (i) the mass flow rate, (ii) the COP, (iii) the degree of subcooling, and (iv) the power required. The refrigerant leading the condenser is dry saturated liquid and leaving the evaporator is dry saturated vapour. The compression in compressor isentropic.*

Properties for R-12

Pressure bar	*Temp °C*	*Enthalpy kJ/kg*		*Entropy kJ/kg.K*		*Specific heat kJ/kg.K*	
		Liquid	*Vapour*	*Liquid*	*Vapour*	*Liquid*	*Vapour*
1.004	–30	8.854	174.076	0.0371	0.7165	–	0.579
13.663	55	90.201	207.766	0.3194	0.6777	1.074	–

Solution : The vapour compression refrigeration cycle is shown in Fig. 5.16.

$$Q = 20 \text{ ton}$$
$$T_1 = 273 - 30 = 243 \text{ K}$$
$$T_2 = T_3 = 273 + 55 = 328 \text{ K}$$
$$T_1' = T_1 + 10 = 253 \text{ K}$$

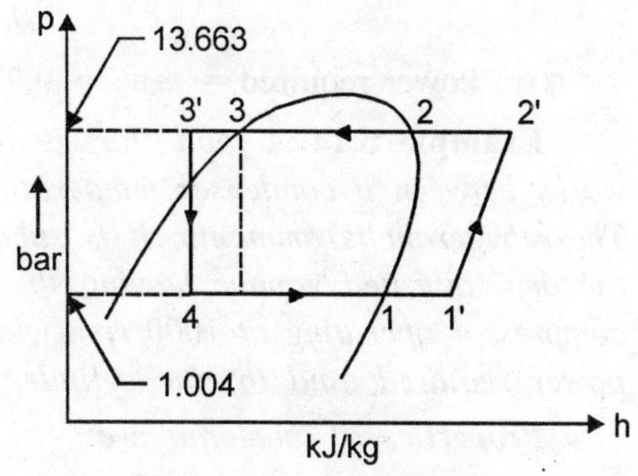

Fig. 5.16 p-h diagram.

For constant heat exchange, we have

$$h_1' = h_1 = h_3 - h_3'$$
$$c_{pv}(T_1' - T_1) = c_{pl}(T_3 - T_3')$$
$$0.579 \times 10 = 1.074\ (328 - T_3')$$
$$T_3' = 322.61 \text{ K}$$
$$s_1' = s_1 + c_{pv} \ln\left(\frac{T_1'}{T_1}\right)$$
$$= 0.7165 + 0.579 \ln\left(\frac{253}{243}\right)$$
$$= 0.7398 \text{ kJ/kg.K}$$
$$s_1' = s_2' = s_2 + c_{pv} \ln\left(\frac{T_2'}{T_2}\right)$$
$$0.7398 = 0.6777 + 0.579 \ln\left(\frac{T_2'}{328}\right)$$
$$T_2' = 365.14 \text{ K}$$
$$h_1' = h_1 + c_{pv}(T_1' - T_1) = 174.076 + 0.579 \times 10 = 179.866 \text{ kJ/kg}$$
$$h_2' = h_2 + c_{pl}(T_3' - T_3) = 90.201 + 1.074 \times (322.61 - 328) = 84.41 \text{ kJ/kg}$$
$$h_4 = h_3' = 84.41 \text{ kJ/kg}$$

(*i*) Refrigerating effect,

$$R_E = h_1' - h_4 = 179.866 - 84.41$$
$$= 95.456 \text{ kJ/kg}$$

Mass flow rate,

$$\dot{m} = \frac{3.5167Q}{R_E} = \frac{3.5167 \times 20}{95.456} = 0.737 \text{ kJ/s}$$

(*ii*) Compressor work, $w_c = h_2' - h_1' = 229.27 - 179.866 = 49.404$ kJ/kg

$$\text{COP} = \frac{R_E}{w_c} = \frac{95.456}{49.404} = 1.93$$

(*iii*) Degree of subcooling $= \dfrac{h_3 - h_3'}{c_{pl}} = \dfrac{90.201 - 84.41}{1.074} = \dfrac{5.791}{1.074} = 5.4°C$

(*iv*) Power required $= \dot{m} w_c = 0.737 \times 49.404 = 36.41$ kW

Example 5.14 *A food storage locker requires a refrigeration capacity of 50 kW. It works between a condenser temperature of 35°C and an evaporator temperature of –10°C. The refrigerant is ammonia. It is subcooled by 5°C before entering the expansion valve by the dry saturated vapour leaving the evaporator. Assuming a single cylinder, single-acting compressor operating at 1000 rpm with stroke equal to 1.2 times the bore, determine (a) the power required, and (b) the cylinder dimensions.*

Properties of ammonia are:

Saturation temperature °C	*Pressure bar*	*Enthalpy, kJ/kg*		*Entropy, kJ/kg. K*		*Specific volume, m^3/kg*		*Specific heat k.J/kg. K*	
		Liquid	*Vapour*	*Liquid*	*Vapour*	*Liquid*	*Vapour*	*Liquid*	*Vapour*
–10	2.9157	154.056	1450.22	0.82965	5.7550	–	0.417477	–	2.492
35	13.522	366.072	1488.57	1.56605	5.2086	1.7023	0.095629	4.556	2.903

Solution: $h_1 = 1450.22$ kJ/kg,

$h_2' = 1488.57$ kJ/kg

$h_{f3} = 366.072$ kJ/kg,

$h_{f3}' = h_4 = h_{f3} - c_{pf}(T_3 - T_3')$

$= 366.072 - 4.556\,(308 - 303)$

$= 343.29$ kJ/kg

Also $s_2 = s_1$

Refer to Fig. 5. 17.

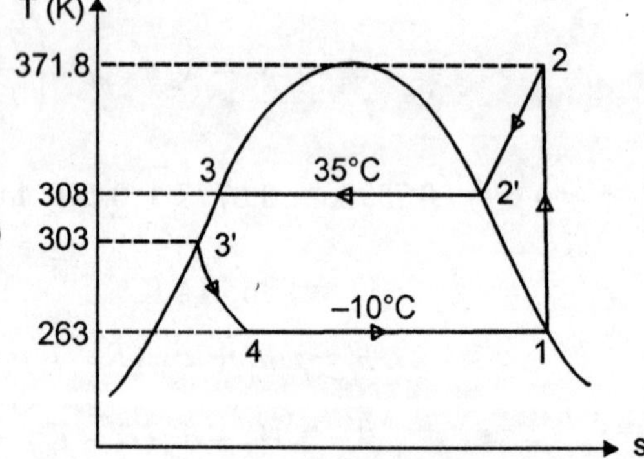

Fig. 5.17 T-s diagram.

$$s_2' + c_{pv} \ln\frac{T_2}{T_2'} = 5.7550$$

$$\ln\frac{T_2}{308} = \frac{5.7550 - 5.2086}{2.903} = 0.1882$$

$$T_2 = 371.8 \text{ K}$$

$$h_2 = h_2' + c_p(T_2 - T_2')$$

$$= 1488.57 + 2.903\,(371.8 - 308)$$

$$= 1673.8 \text{ kJ/kg}$$

Mass flow rate of refrigerant,

$$\dot{m} = \frac{50}{h_1 - h_4} = \frac{50}{1450.22 - 343.29} = 0.04517 \text{ kg/s}$$

(a) Power required = $\dot{m}(h_2 - h_1)$ = 0.04517 (1673.8 – 1450.22) = 10.1 kW

(b) Cylinder dimensions:

$$\dot{m} = \frac{\pi}{4} D^2 L \frac{N}{60} v_1$$

$$\frac{\pi}{4} D^2 \times 1.2D \times \frac{1000}{60} \times 0.417477 = 0.04517$$

$$D^3 = 0.006888$$

$$D = 0.19 \text{ m}, L = 1.2\ D = 0.228 \text{ m}$$

Examples 5.15 *An ideal vapour compression refrigerator using Freon-12 operates between temperature limits of –10°C and 40°C. The refrigerant leaves the condenser dry saturated. The rate of flow of refrigerant through the unit is 150 kg/hr.*

Calculate the refrigerating effect per kg of refrigerant, the co-efficient of performance and the capacity of the plant, if

(i) The refrigerant leaves the condenser dry saturated as stated,

(ii) The refrigerant is sub-cooled to 20°C before throttling.

Assume that the enthalpy of refrigerant before throttling is approximately equal to the enthalpy of liquid refrigerant at undercooled temperature of 20°C and take it as h_f = 50.59 kJ/kg. Give your comments on the result. Solve the problem by using tables.

Solution: Given, $\dot{m}$ = 150 kg/h. Refer to Fig. 5.18

From *p-h* chart for Freon-12, we have

$$h_1 = 183.05 \text{ kJ/kg}$$

$$s_1 = s_2 = 0.7014 \text{ kJ/kg.K}$$

$$h_3 = h_4 = 74.52 \text{ kJ/kg}$$

$$h_2 = 210 \text{ kJ/kg}$$

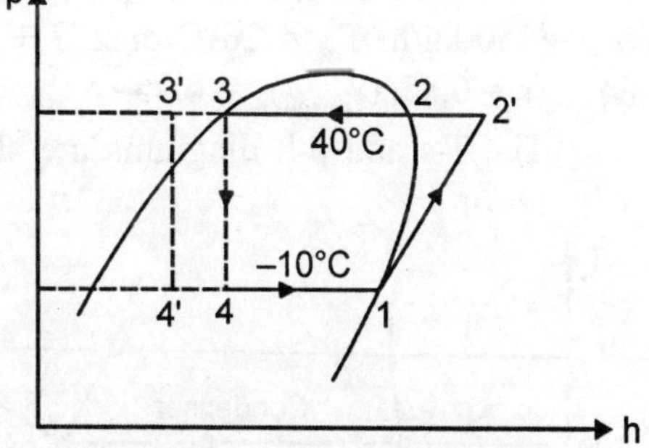

Fig. 5.18 p-h diagram.

(*i*) Refrigeration effect,

$$q_0 = h_1 - h_4$$

$$= 183.05 - 74.52$$

$$= 108.53 \text{ kJ/kg}$$

$$\text{COP} = \frac{h_1 - h_4}{h_2 - h_1} = \frac{108.53}{210 - 183.05} = 4.02$$

Plant capacity

$$= \frac{\dot{m} q_0}{3600} = \frac{150 \times 108.53}{3600}$$

$$= 4.52 \text{ kW or } \frac{4.52}{3.5167} = 1.2853 \text{ Ton}$$

(*ii*) h_1 = 183.05 kJ/kg, h_2 = 210 kJ/kg,

$h_3 = h_{f3}$ = 50.59 kJ/kg = h_4' (Given)

Refrigeration effect,

$$q'_o = h_1 - h'_4 = 183.05 - 50.59 = 132.46 \text{ kJ/kg}$$

$$\text{COP} = \frac{h_1 - h'_4}{h_2 - h_1} = \frac{132.46}{210 - 183.05} = 4.91$$

$$\text{Plant capacity} = \frac{mq'_o}{3600} = \frac{150 \times 132.46}{3600} = 5.52 \text{ kW or } 1.57 \text{ Ton}$$

With subcooling, refrigerating capacity, COP and plant capacity increase.

Examples 5.16 *A water cooler using R-12 works on the condensing and evaporating temperatures of 26°C to 2°C respectively. The vapours leave the evaporator saturated and dry. The average output of cold water is 150 kg/h cooled from 26°C to 6°C. Allowing 20% of useful heat into water cooler and the volumetric efficiency of the compressor as 80% and mechanical efficiency of compressor and electric motor as 85% and 95% respectively, calculate:*

(a) Volumetric displacement of compressor, and (b) Power of the motor.

Data for R-12 is given below:

Temp. °C	*Pressure* bar	*Enthalpy, kJ/kg*		*Entropy, kJ/kg. K*		*Specific heat, kJ/kg. K*		*Specific volume of vapor, v_g, m^3/kg*
		h_f	h_g	s_f	s_g	c_{pf}	c_{pg}	
26	6.690	60.04	198.10	0.2270	0.6865	0.996	0.674	0.026
2	3.297	37.92	188.39	0.1487	0.6956	1.067	0.620	0.052

Solution: Given: $T_{2'} = T_3 = 26°C$ or $273 + 26 = 299$ K, $T_1 = T_4 = 2°C$ or $273 + 2 = 275$ K, $m_w = 150$ kg/h, $T_{wi} = 26°C$ or $273 + 26 = 299$ K, $T_{wo} = 6°C$ or $273 + 6 = 279$ K, $\eta_{vol} = 0.80$, $(\eta_{mech})_c = 0.85$, $(\eta_{mech})_m = 0.95$.

The T-s and p-h diagrams are shown in Fig. 5.19.

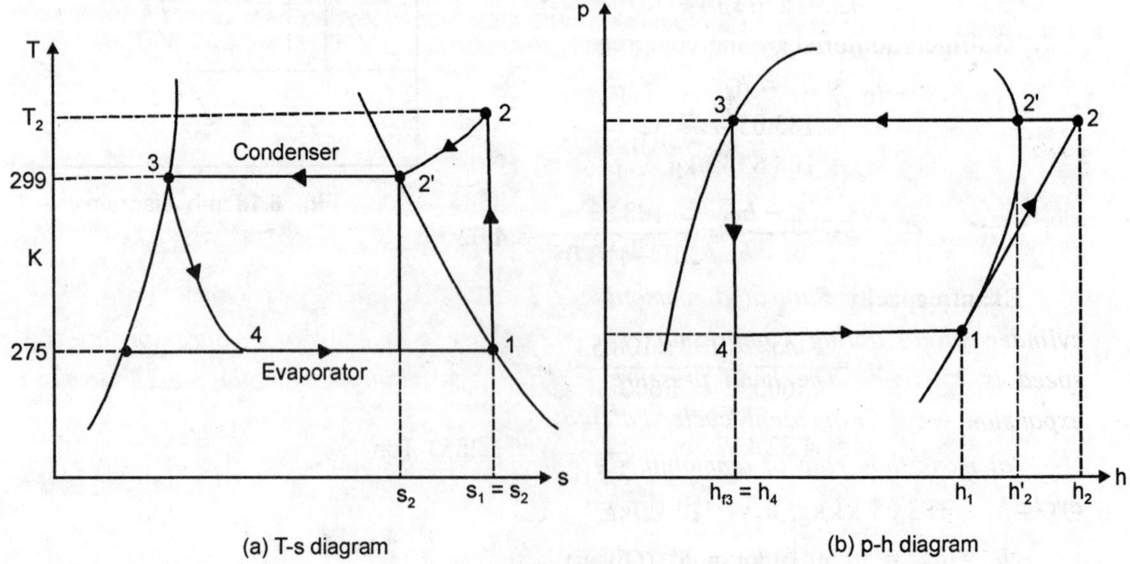

(a) T-s diagram (b) p-h diagram

Fig. 5.19

Actual heat extracted from the water cooler,

$$Q = 1.2\ m_w c_w\ (T_{wi} - T_{wo})$$
$$= 1.2 \times 150 \times 4.187\ (299 - 279) = 15073.2 \text{ kJ/k or } 251.22 \text{ kJ/min}$$

(*a*) Refrigeration effect (or heat extracted) per kg of refrigerant,

$$R_E = h_1 - h_{f3} = 188.39 - 60.64 = 127.75 \text{ kJ/kg}$$

Mass flow rate of refrigerant, $\dot{m}_r = \dfrac{Q}{R_E} = \dfrac{251.22}{127.75} = 1.9665$ kg/min

Volume displacement of compressor, $\dot{V}_1 = \dfrac{\dot{m}_r v_1}{\eta_{vol}} = \dfrac{1.9665 \times 0.052}{0.80} = 0.1278$ m^3/min

(*b*) Now $\quad s_2 = s_{2'} + c_{p_{2'}} \ln\left(\dfrac{T_2}{T_2'}\right)$

$$0.6956 = 0.6865 + 0.674\ \ln\left[\frac{T_2}{299}\right]$$

$$T_2 = 303 \text{ K}$$
$$h_2 = h_{2'} + c_{p2'}\ (T_2 - T_{2'})$$
$$= 198.01 + 0.674\ (303 - 299)$$
$$= 200.8 \text{ kJ/kg}$$

Work done by the compressor,

$$W_c = W_{1-2} = \dot{m}_r (h_2 - h_1)$$
$$= 1.9665\ (200.8 - 188.39)$$
$$= 24.4 \text{ kJ/min or } 0.4067 \text{ kW}$$

Power required by the compressor,

$$P_c = \frac{W_c}{(\eta_{mech})_c} = \frac{0.4067}{0.85} = 0.4785 \text{ kW}$$

Power of motor, $\quad P_m = \dfrac{P_c}{(\eta_{mech})_m} = \dfrac{0.4785}{0.95}$

Example 5.17 *Saturated ammonia at 2.3637 bar enters a 150 mm × 140 mm twin cylinder single acting compressor. The volumetric efficiency of compressor is 79% and speed is 250 rpm. The head pressure is 11.671 bar. Liquid ammonia at 21°C enters the expansion valve. For ideal cycle, calculate:*

(a) mass flow rate of ammonia, (b) refrigeration in tones, and (c) COP of refrigeration cycle.

Data for ammonia is given below:

Pressure bar	*Saturation Temperature °C*	*Specific volume of Vapour, V_g m^3/kg*	*Specific enthalpy, kJ/kg*		*Specific entropy, kJ/kg.K*		*Specific heat, kJ/kg. K*	
			h_f	h_g	s_f	s_g	c_{pl}	c_{pg}
2.3637	–15	0.50905	131.28	1444.19	0.7424	5.8285	4.528	2.442
11.671	+30	0.11048	342.08	1485.93	1.4892	5.2623	4.843	3.252

Solution: Given: p_1 = 2.3637 bar, T_1 = –15°C or 273 – 15 = 258 K, v_1 = 0.50905 m³/kg, η_{vol} = 0.79, N = 250 rpm, n = 2, T_2 = 30°C or 273 + 30 = 303 K, T_3 = 21°C or 273 + 21 = 294 K, D = 0.15 m, L = 0.14 m.

Isentropic compression process 1-2 (Fig. 5.20):

$$s_1 = s_2 = s_{2'} + c_{pg} \ln\frac{T_2}{T_{2'}}$$

$$5.8285 = 5.2623 + 3.252 \ln\frac{T_2}{303}$$

$$T_2 = 360.6 \text{ K}$$

$$h_2 = h_{2'} + c_{pg}(T_2 - T_{2'})$$

$$= 1485.93 + 3.252\,(360.6 - 303) = 1673.24 \text{ kJ/kg}$$

Sub-cooling process 3′–3:

$$h_3 = h_{3'} - c_{pf3'}\,(T_{3'} - T_3)$$

$$= 342.08 - 4.843\,(303 - 294)$$

$$= 298.5 \text{ kJ/kg}$$

Also during throttling process 3–4,

$$h_3 = h_4 = h_{4'} + x_4(h_1 - h_{4'})$$

$$298.5 = 131.28 + x_4\,(1444.19 - 131.28)$$

$$x_4 = 0.1274$$

(*a*) Displacement volume, $\dot{V} = \frac{\pi}{4} D^2 L N \times \eta_{vol} \times n$

$$= \frac{\pi}{4}(0.15)^2 \times 0.14 \times 250 \times 0.79 \times 2 = 0.9772 \text{ m}^3/\text{min}$$

Mass flow rate of refrigerant, $\dot{m} = \frac{\dot{V}}{v_1} = \frac{0.9772}{0.50905} = 1.92$ kg/min

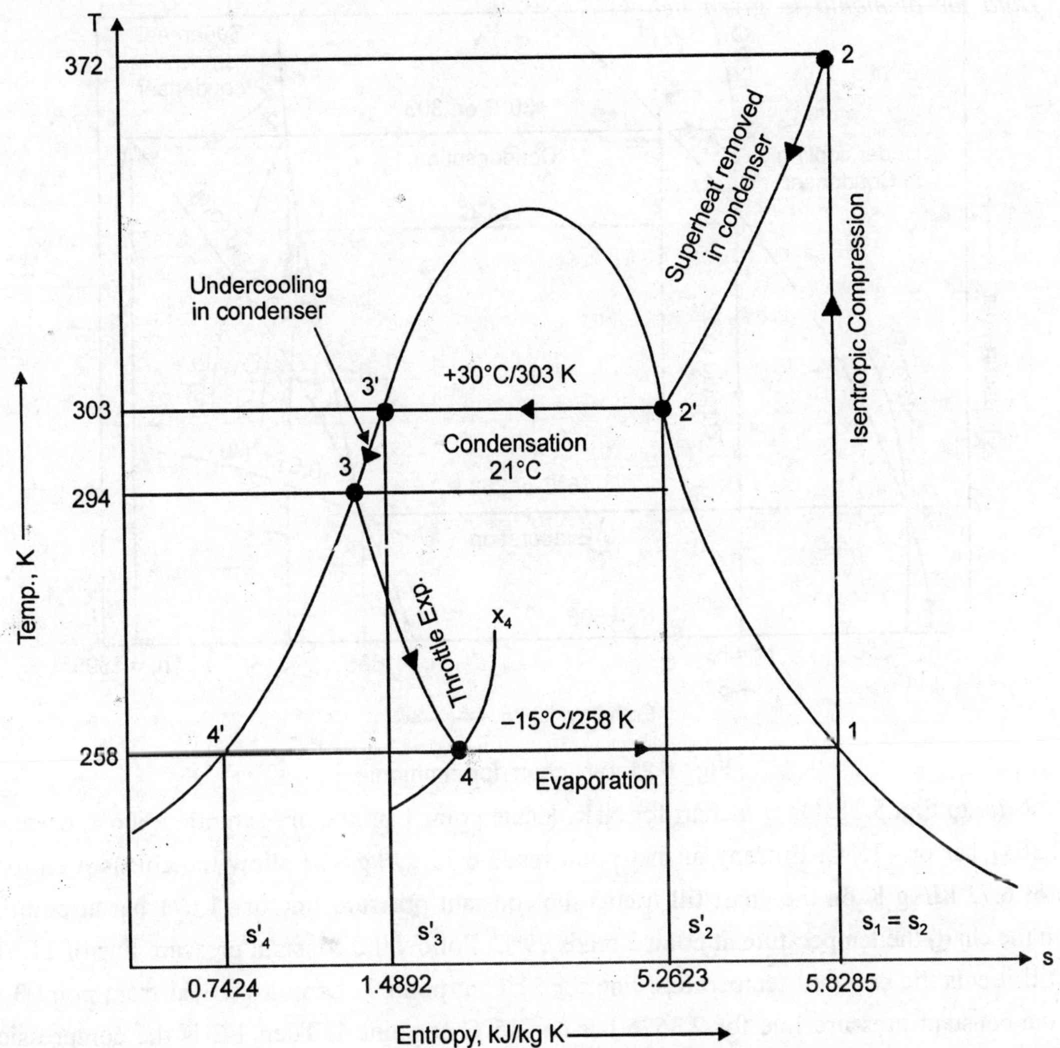

Fig. 5.20 *T-s* diagram.

(*b*) Refrigeration effect $= \dfrac{\dot{m}(h_1 - h_4)}{211} = \dfrac{1.92(1444.19 - 298.5)}{211} = 10.425$ tonnes

(*c*) Coefficient of performance, COP $= \dfrac{h_1 - h_4}{h_2 - h_1} = \dfrac{1444.19 - 298.5}{1674.55 - 1444.19} = \dfrac{1145.69}{230.36} = 4.97$

Example 5.18 *Solve example 5.17 by using p-h chart for ammonia.*

Solution: It may be noted that reference base for this chart is: At –50°C, $h_f = 200$ kJ/kg and $s_f = 1$ kJ/kg K. Though the numerical values at different state points will differ from the values by using tables because of different reference base, it is immaterial as our interest is limited to change in properties during the process which remains unaltered. Actual values that can be read have the limitations of the size of the chart.

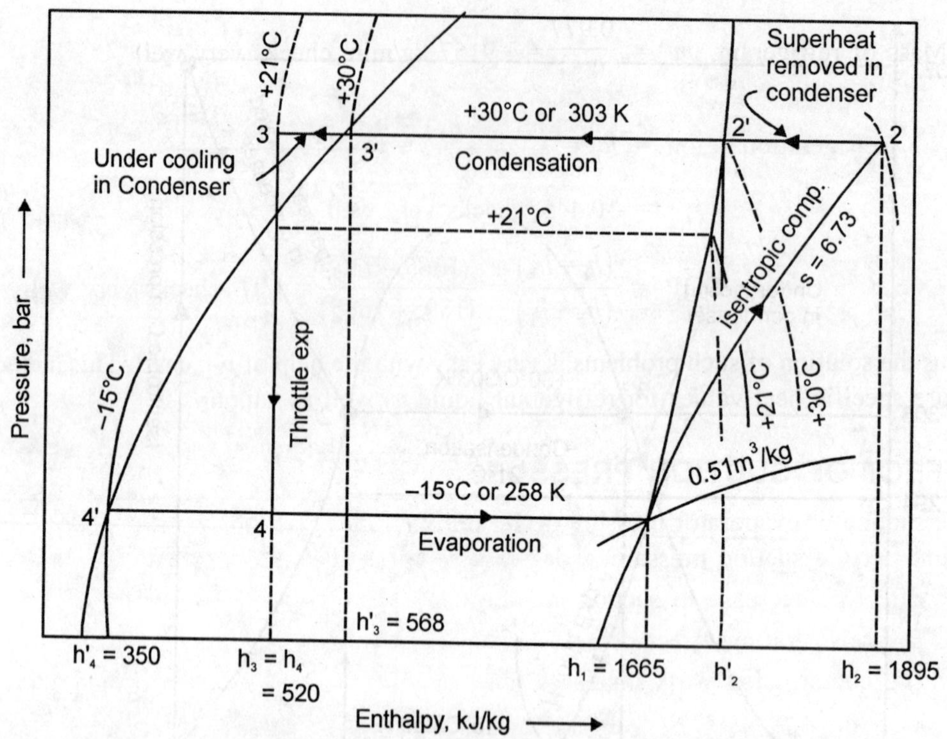

Fig. 5.21 p-h chart for ammonia.

Refer to Fig. 5.21. On *p-h* chart for NH_3, locate point 1 on the dry saturated line at pressure of 2.3637 bar or –15°C. Entropy at that point reads 6.73 kJ/kg.K. Follow the constant entropy line of 6.73 kJ/kg K on the chart till it cuts the constant pressure line of 11.671 bar at point 2. From the chart, the temperature at point 2 reads 99°C. Follow the constant pressure line of 11.671 bar till it cuts the constant temperature line for 21°C at point 3. Drop a vertical from point 3 to cut the constant pressure line for 2.3676 bar or –15°C at point 4. Then 1-2 is the compression process, 2-3 is the condensation process with under-cooling, 3-4 is the throttling process and 4-1 is the evaporation process.

Values of enthalpies read direct from the chart are as following:

$$h_1 = 1665 \text{ kJ/kg}; \ h_2 = 1895 \text{ kJ/kg}; \ h_3 = 520 \text{ kJ/kg}; \ h_4 = 520 \text{ kJ/kg}$$

Specific volume at point 1, $v_{g1} = 0.51 \text{ m}^3\text{/kg}$

Thus, having known the above five values all the results obtained with the help of tables can be calculated. This shows that the solution of such problems is very easy with the help of *p-h* chart. The slight discrepancy in the superheat and under-cooled region is due to the assumed values of the specific heat in the first method.

Displacement volume $= 0.977 \text{ m}^3\text{/min}$ as before

Mass of refrigerant, $\dot{m} = \dfrac{0.977}{0.51} = 1.9157$ kg/min, checks very well

$$\text{Refrigeration in ton} = R_E = \frac{m(h_1 - h_4)}{210} = \frac{1.9157(1665-520)}{210}$$

$$= 10.445, \text{ checks very well}$$

$$\text{COP} = \frac{(h_1 - h_4)}{(h_2 - h_1)} = \frac{(1665-520)}{(1895-1665)} = 4.97 \text{ checks very well}$$

Thus the solution of such problems is very easy with the help of *p-h* chart. This method does not require specific heat values for refrigerant liquid as well as vapour.

5.6 EFFECT OF SUCTION PRESSURE

The suction or evaporator pressure decreases due to the frictional resistance to the flow of refrigerant. Let the suction pressure p_s decrease to $p_{s'}$ as shown in *p-h* diagram of Fig. 5.22.

The effect of decrease in suction pressure are :

1. The refrigerating effect decreases from $(h_1 - h_4)$ to $(h_{1'} - h_{4'})$
2. The compression work increases from $(h_2 - h_1)$ to $(h_{2'} - h_{1'})$

Now $$\text{COP} = \frac{\text{refrigerating effect}}{\text{compression work}}$$

Therefore, the COP decreases for the same amount of refrigerant flow. Hence, the refrigerating capacity of the system decreases and the refrigeration cost increases.

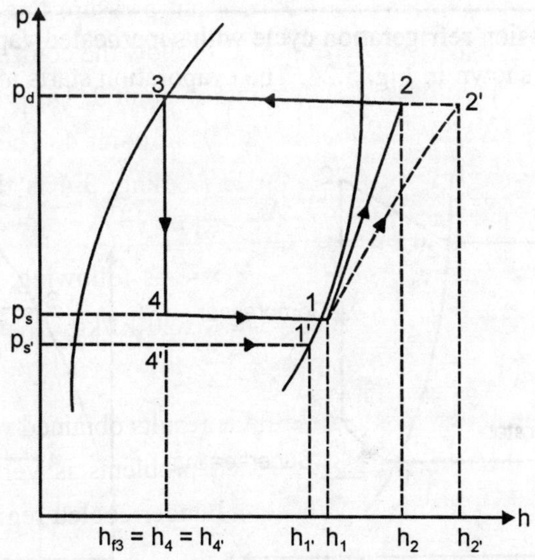

Fig. 5.22 p-h diagram.

5.7 EFFECT OF DISCHARGE PRESSURE

The discharge (or condenser) pressure p_d increases to p_d' due to frictional resistance to flow or refrigerant. The effects of increase in discharge pressure are (Fig. 5.23)

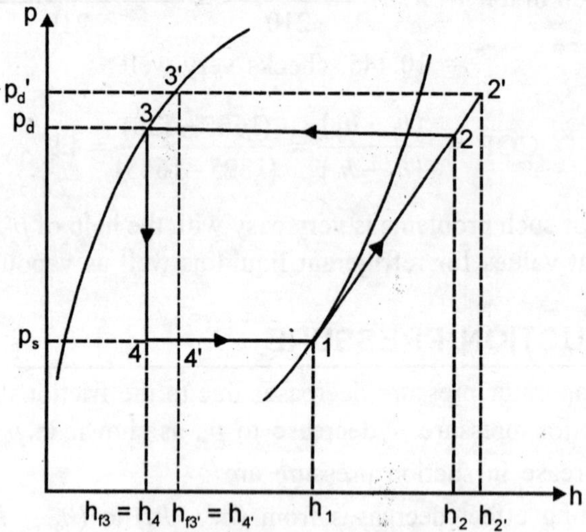

Fig. 5.23 p-h diagram.

1. The refrigerating effect decreases from $(h_1 - h_4)$ to $(h_1 - h_{4'})$
2. The compression work increases from $(h_2 - h_1)$ to $(h_{2'} - h_1)$.

Thus the COP decreases. The effect of increase in discharge pressure is similar to that of decrease in suction pressure.

5.8 EFFECT OF SUPERHEATING OF REFRIGERANT VAPOUR

The vapour compression refrigeration cycle with superheated vapour before compression on *T-s* and *p-h* diagrams are shown in Fig. 5.24. The evaporation starts at state 4 and continues upto

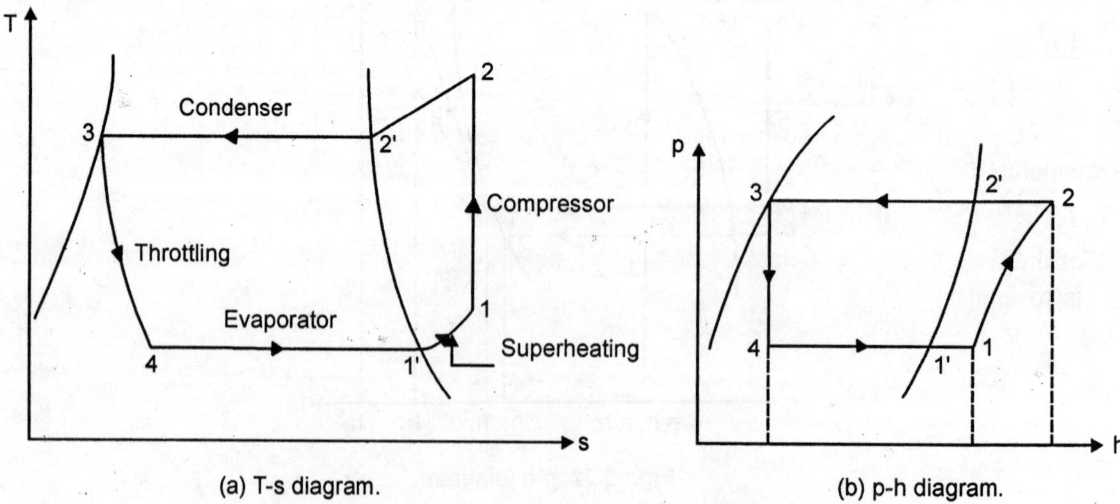

(a) T-s diagram. (b) p-h diagram.

Fig. 5.24 Superheated vapour compression cycle.

state 1′, when it is dry and saturated. The vapours are now superheated from state 1′ to state 1, where they enter compressor. Its effect is to increase the COP.

$$\text{COP} = \frac{h_1 - h_{f3}}{h_2 - h_1} = \frac{h_1 - h_4}{h_2 - h_1}$$

5.8.1 Superheat Horn

Consider the *T-s* diagram for a vapour compression cycle shown in Fig. 5.25. The vapours of refrigerant are wet at state 1′ and dry saturated at state 2′. The process 1′–2′ represents wet compression. The vapours at state 1 are dry saturated and at state 2 superheated. The process 1–2 represents dry compression. The increased work of the cycle due to the replacement of wet compression by dry compression appears as the area 2–2′–2″, generally known as superheat horn.

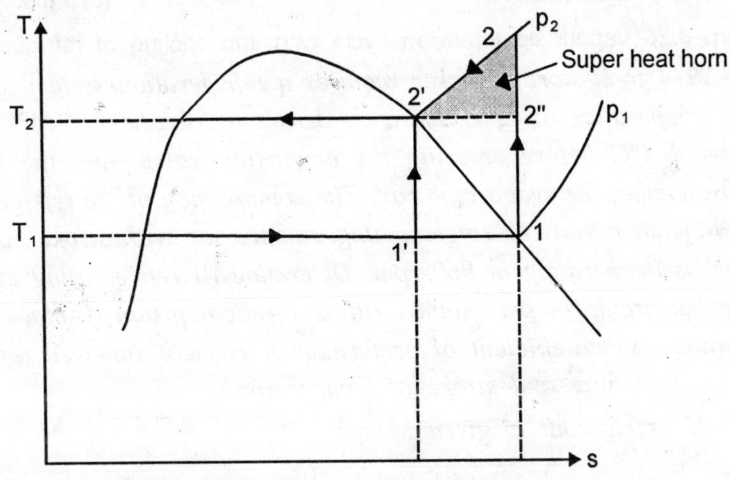

Fig. 5.25 Wet and dry compression processes.

5.9 EFFECT OF SUBCOOLING (OR UNDERCOOLING) OF REFRIGERANT VAPOUR

Consider the *T-s* and *p-h* diagrams for the vapour compression cycle shown in Fig. 5.26, in which the refrigerant after condensation process 2′–3′, is cooled below the saturation temperature T_3' before throttling process to temperature T_3. Such a process is called undercooling or subcooling of the refrigerant. It is generally done along the saturated liquid line. The effect of undercooling is to increase the COP.

Refrigerating effect, $R_E = h_1 - h_4 = h_1 - h_{f3}$, where $h_{f3} = h_{f3'} - c_{pf}(T_{3'} - T_3)$

Work done, $w = h_2 - h_1$

$$\text{COP} = \frac{R_E}{w} = \frac{h_1 - h_{f3}}{h_2 - h_1}$$

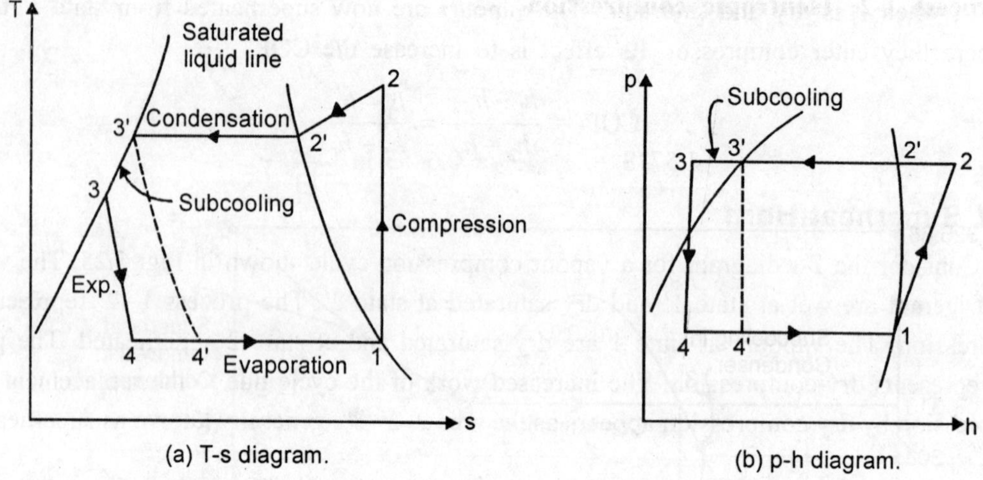

Fig. 5.26 Vapour compression cycle with sub cooling of refrigerant.

Example 5.19 *A food storage locker requires a refrigeration system of 42 kW capacity at an evaporator temperature of –5°C and a condenser temperature of 40°C. The refrigerant, R-12, is sub-cooled 5°C before entering the expansion valve, and the vapour is superheated 6°C before leaving the evaporator coil. The compression of the refrigerant is reversible adiabatic. A two-cylinder vertical single acting compressor with stroke equal to 1.5 times the bore is to be used operating at 960 rpm. Determine (i) refrigerating effect/kg. (ii) mass of refrigerant to be circulated per minute. (iii) theoretical piston displacement per minute, (iv)theoretical power, (v) co-efficient of performance, (vi)heat removed through condenser/kg, and (vii)theoretical bore and stroke of compressor.*

Data for R-12 refrigerant is given below:

Saturation temperature °C	*Absolute pressure bar*	*Specific volume of vapour m^3/kg*	*Enthalpy of liquid kJ/kg*	*Enthalpy of vapour kJ/kg*	*Entropy of liquid kJ/kg.K*	*Entropy of vapour kJ/kg.K*	*Specific heat liquid kg/kg.K*	*Specific heat vapour kJ/kg.K*
–5 °C	2.6032	0.06559	195.37	350.44	0.97805	1.5635	0.936	0.6445
+40 °C	9.5909	0.01837	239.03	368.81	1.1315	1.5459	1.030	0.762

Solution: Given: $\dot{Q}$ = 42 kW, T_1 = –5°C or 273 – 5 = 268 K, $T_{2'}$ = 40°C or 273 + 40 = 313 K, $T_{3'} - T_3$ = 5°C, $T_1 - T_{1'}$ = 6°C, n = 2, L = 1.5 D, N = 960 rpm.

The *T-s* diagram is shown in Fig. 5.27.

Process 1′-1: Superheating

$$h_1 = h_{g1'} + c_{pg1'}(T_1 - T_{1'}) = 350.44 + 0.6455 \times 6$$
$$= 354.313 \text{ kJ/kg}$$

$$s_1 = s_{g1'} + c_{pg1'} \ln \frac{T_1}{T_1'} = 1.5635 + 0.6455 \ln \frac{274}{268}$$
$$= 1.5778 \text{ kJ/kg. K}$$

Process 1-2: Isentropic compression

$$s_1 = s_2 = s_{g2'} + c_{pg2'} \ln \frac{T_2}{T_2'}$$

$$1.5778 = 1.5459 + 0.6455 \ln \frac{T_2}{313}$$

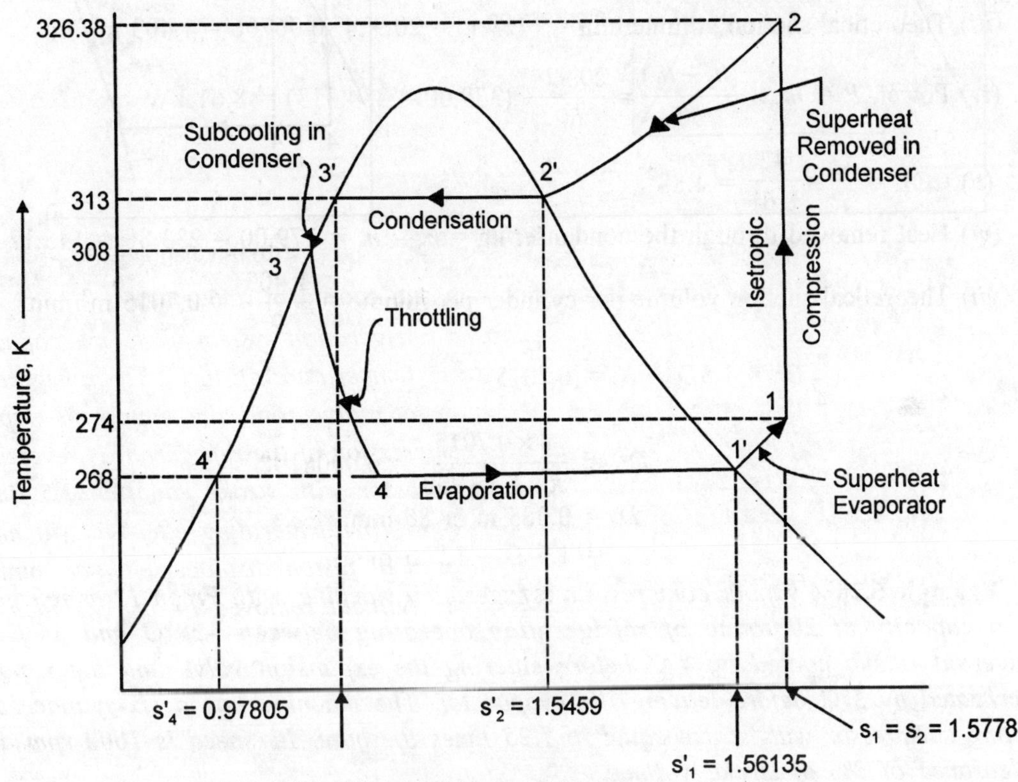

Fig. 5.27 T-s diagram.

$$T_2 = 326.38 \text{ K}$$

$$h_2 = h_{g2'} + c_{pg'2} (T_2 - T_{2'})$$

$$= 368.81 + 0.6455 \ (326.38 - 313)$$

$$= 379.00 \text{ kJ/kg}$$

Process 3' - 3: Sub cooling

$$h_3 = h_{f3'} - c_{pf3'} (T_{3'} - T_3) = 239.03 - 1.030 \times 5 = 233.88 \text{ kJ/kg}$$

Process 3 - 4: Throttling

$$h_4 = h_3 = 233.88$$

Specific volume at suction to compressor:

$$\frac{v_1}{T_1} = \frac{v_{1'}}{T_{1'}}$$

$$v_1 = \frac{274}{268} \times 0.06559 = 0.06706 \text{ m}^3/\text{kg}$$

(*i*) Refrigeration effect/kg, $q_{4-1} = h_1 - h_4 = 354.313 - 233.88 = 120.433$ kJ/kg

(*ii*) Mass of refrigerant required/min, $\dot{m}_r = \frac{\dot{Q}\times 60}{q_{4-1}} = \frac{42\times 60}{120.433} = 20.924$ kg/min

(*iii*) Theoretical suction volume/min $= \dot{m}_r \times v_1 = 20.924 \;\; 0.06706 = 1.403$ m^3/min

(*iv*) Power, $P = \dot{m}_r \times \frac{(h_2 - h_1)}{60} = \frac{20.924}{60}(379.00 - 354.313) = 8.61$ kW

(*v*) COP $= \frac{\dot{Q}}{P} = \frac{42}{8.61} = 4.88$

(*vi*) Heat removed through the condenser/kg $= h_2 - h_3 = 379.00 - 233.88 = 145.12$ kJ/kg

(*vii*) Theoretical suction volume per cylinder per minute $= \frac{1.403}{2} = 0.7015$ m^3/min

$$\frac{\pi}{4}D^2 \times 1.5D \times N = 0.7015$$

$$D^3 = \frac{4\times 0.7015}{\pi\times 1.5\times 960} = 0.0006198$$

$$D = 0.085 \text{ m or } 85 \text{ mm},$$

$$L = 1.5\,D = 127.5 \text{ mm}$$

Example 5.20 *A vapour compression refrigeration machine, with Freon-12 as refrigerant, has a capacity of 20 tonne of refrigeration operating between –28°C and 26°C. The refrigerant is sub cooled by 4°C before entering the expansion valve and the vapour is superheated by 5°C before leaving the evaporator. The machine has a six-cylinder single - acting compressor with stroke equal to 1.25 times the bore. Its speed is 1000 rpm, it has a clearance of 3% of stroke volume.*

Calculate: (a) Theoretical power required, (b) COP, (c) Volumetric efficiency, and (d) Bore and stroke of cylinder.

The properties of Freon-12 are given below:

Saturation temp °C	*Pressure bar*	*Specific volume* v_g *m³/kg*	*Enthalpy, kJ/kg*		*Entropy kJ/kg.K*		*Specific heat*	
			h_f	h_g	s_f	s_g	c_{pf}	c_{pg}
–28	1.0929	0.14817	174.58	339.76	0.9024	1.5762	0.886	0.575
26	6.6857	0.02643	224.92	363.62	1.0862	1.5498	0.987	0.707

Solution: Given: $Q = 20$ *TR*, $T_{1'} = -28$°C or $273 - 28 = 245$ K, $T_{2'} = T_{3'} = 26$°C or $273 + 26 = 299$ K, $T_{3'} - T_3 = 4$°C, $T_3 = 299 - 4 = 295$ K, $T_1 - T_{1'} = 5$°C, $T_1 = 245 + 5 = 250$ K, $L = 1.25\,D$, $c = 0.03$, $N = 1000$ rpm, $v_{1'} = 0.14817$ m^3/kg, $v_{2'} = 0.02643$ m^3/kg, $h_{f1} = 174.58$ kJ/kg $h_{f3'} = 224.92$ kJ/kg, $h_{1'} = 339.76$ kJ/kg, $h_{2'} = 363.62$ kJ/kg, $s_{f1} = 0.9024$ kJ/kg. $s_{f3} = 1.0862$ kJ/kg, $s_{1'} = 1.5762$ kJ/kg, K, $s_{2'} = 1.5498$ kJ/kg, K, L = 1.25 D

The *T-s* and *p-h* diagrams are shown in Fig. 5.28.

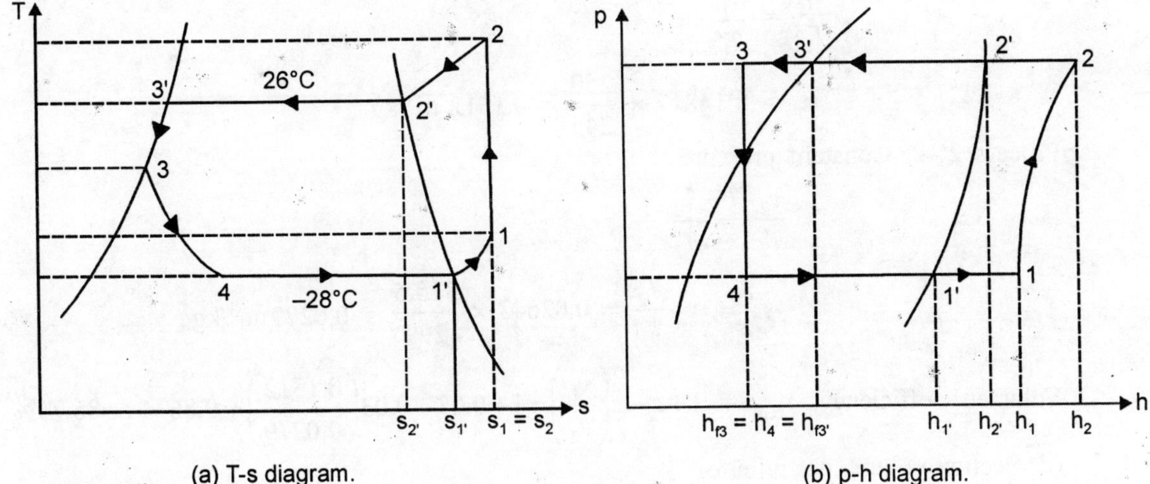

(a) T-s diagram. (b) p-h diagram.

Fig. 5.28

(a) Process 1′-1 : Superheating

$$s_1 = s_{1'} + c_{pg1'} \ln\frac{T_1}{T_{1'}} = 1.5762 + 0.575 \ln\frac{250}{245} = 1.5878 \text{kJ/kg.K}$$

$$s_2 = s_{2'} + c_{pg2'} \ln\frac{T_2}{T_{2'}} = 1.5498 + 0.707 \ln\frac{T_2}{299}$$

Process 1-2: Isentropic compression

$$s_1 = s_2$$

$$1.5878 = 1.5498 + 0.707 \ln\frac{T_2}{299}$$

$$T_2 = 315.5 \text{ K}$$

$$h_1 = h_{1'} + c_{pg1'}(T_1 - T_{1'}) = 339.76 + 0.575 \times 5 = 342.64 \text{ kJ/kg}$$

$$h_2 = h_{2'} + c_{pg2'}(T_2 - T_{2'}) = 363.62 + 0.707\,(315.5 - 299) = 375.29 \text{ kJ/kg}$$

Compressor work, $w_c = h_2 - h_1 = 375.29 - 342.64 = 32.65$ kJ/kg

Process 3¢-3: Subcooling

$$h_{f3} = h_{f3'} - c_{pf3'}(T_{3'} - T_3) = 224.92 - 0.987 \times 4 = 220.97 \text{ kJ/kg}$$

Refrigerating effect, $R_E = h_1 - h_{f3} = 342.64 - 220.97 = 121.67$ kJ/kg

Mass flow rate of refrigerant, $\dot{m}_r = \dfrac{211Q}{R_E} = \dfrac{211 \times 20}{121.67} = 34.68$ kg/min

Theoretical power required, $P = \dfrac{\dot{m}_r(h_2 - h_1)}{60} = \dfrac{34.68(375.29 - 342.64)}{60} = 18.87$ kW

(b) COP $= \dfrac{R_E}{w_c} = \dfrac{121.67}{32.65} = 3.726$

(c) Process 1′–1: Constant pressure

$$\frac{v_1}{T_1} = \frac{v_1'}{T_1'}$$

$$v_1 = 0.14817 \times \frac{250}{245} = 0.1512 \text{ m}^3/\text{kg}$$

Process 2′–2: Constant pressure

$$\frac{v_2}{T_2} = \frac{v_{2'}}{T_{2'}}$$

$$v_2 = v_{2'}\frac{T_2}{T_{2'}} = 0.02643 \times \frac{315.5}{299} = 0.0279 \text{ m}^3/\text{kg}$$

Volumetric efficiency, $\eta_{vol} = 1 + c - c\left(\frac{v_1}{v_2}\right) = 1 + 0.03 - 0.03\left(\frac{0.1512}{0.0279}\right) = 0.8674 \text{ or } 86.74\%$

(d) Suction volume per minute,

$$V_1 = \dot{m}_r \times v_1 \times \frac{1}{\eta_{vol}} = 34.68 \times 0.1512 \times \frac{1}{0.8674} = 6.045 \text{ m}^3/\text{min}$$

Suction volume per cylinder

$$= \frac{6.045}{6} = 1.0075 \text{ m}^3/\text{min}$$

$$= \frac{\pi}{4}D^2 \times 1.25D \times N = \frac{\pi}{4} \times 1.25 \times 100D^3 = 981.7477\ D^3$$

$$D = 0.1008 \text{ m or } 100.8 \text{ mm}$$

$$L = 1.25 \times 100.8 = 126 \text{ mm}$$

5.10 ACTUAL VAPOUR COMPRESSION REFRIGERATION CYCLE

The following are the main deviations to simple saturated cycle in actual practice.

(*i*) The liquid refrigerant in the condenser may be under cooled, i.e. cooled below the condensing temperature before passing through the expansion valve. This can be obtained by large quantities of circulating cooling water which should be at a temperature much lower than the condensing temperatures. The under cooling is desirable as it increases the refrigerating effect per kg of refrigerant flow.

(*ii*) Normally the gas leaves the evaporator in superheated condition before it enters the compressor. This superheating may occur in the evaporator pipes leaving the evaporator but still in the cooled state. Thus the refrigerating effect will increase per kg of refrigerant flow. But this superheating may occur in the pipes being insulated. Thus, this is a loss or an additional load on the compressor and condenser without any useful cooling effect obtained. Thus these pipes should be properly insulated from heat infiltration into the system from the surroundings.

(*iii*) Compression has been assumed isentropic, but in actual practice it may be very complex. The cylinder walls may stabilize at a temperature between the cool suction gas and the hot discharge gas. Thus during the first part of compression heat may flow from the walls to the gas and in the later part of the stroke the reversal in the direction of flow of heat may take place. Again, there may be wire drawing taking place at the suction valve as the flow of gas to take place through the valve with some finite pressure difference across the valve. Same is true at the discharge valve. Thus the compressor range of pressure is slightly different from the evaporator and condenser pressure. Since it is very difficult to evaluate these complex processes. It is a usual practice to evaluate the compression index connecting the properties at initial point of compression and final point of compression. This is however an approximate method.

(*iv*) The refrigerant has to flow through pipes in the evaporator, condenser and other connecting pipes. Thus in actual practice due to frictional resistance to flow, there is a pressure fall in the connecting pipes.

The actual vapour compression cycle is as shown in Fig. 5.29 on *T-s* and *p-h* diagrams.

1. *Process 1-2-3:* This process shows the flow of refrigerant in the evaporator. The point 1 represents the entry of refrigerant into the evaporator and the point 3 represents the exit of refrigerant from evaporator in a superheated state. The point 3 also represents the entry of refrigerant into the compressor in a superheated condition. The superheating of vapour refrigerant from point 2 to point 3 may be due to:

(*i*) Automatic control of expansion valve so that the refrigerant leaves the evaporator as the superheated vapour.

(*ii*) Picking up of larger amount of heat from the evaporator through pipes located within the cooled space.

(*iii*) Picking up of heat from the suction pipe. i.e., the pipe connecting the evaporator delivery and the compressor suction valve.

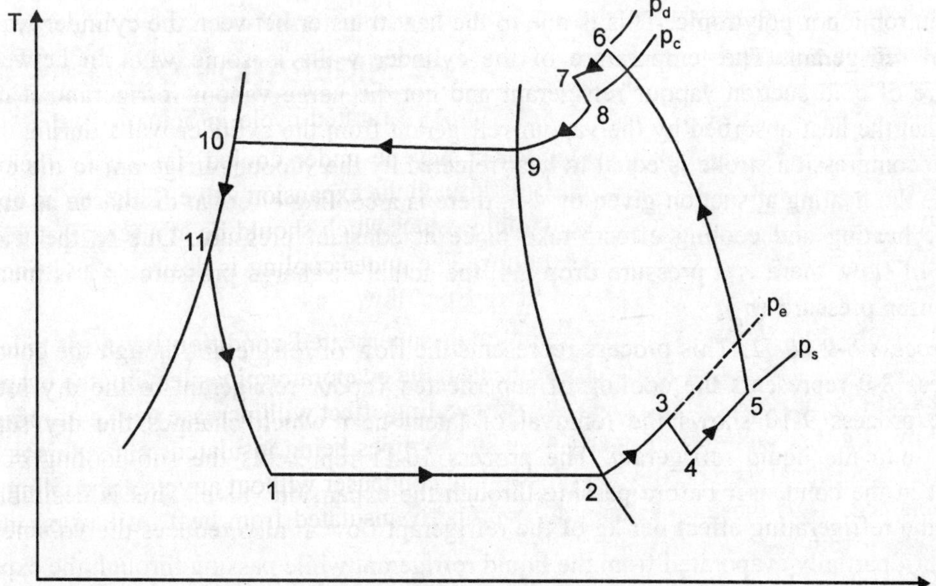

(a) T-s diagram.

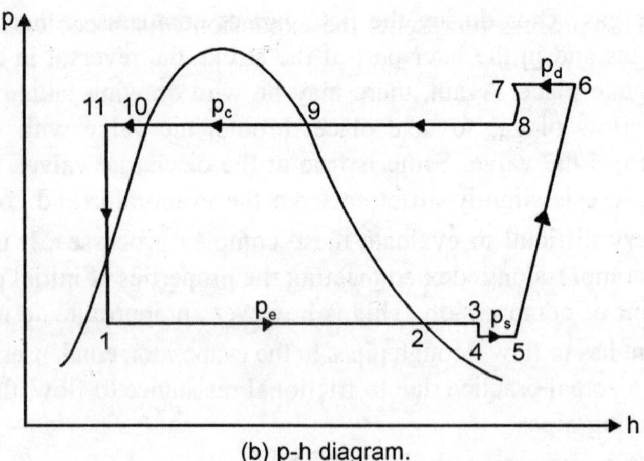

(b) p-h diagram.

Fig. 5.29 Actual vapour compression cycle.

In the first and second case of superheating the vapour refrigerant, the refrigerating effect as well as the compressor work is increased. The coefficient of performance, as compared to saturation cycle at the same suction pressure may be greater, less or unchanged.

The superheating also causes increase in the required displacement of compressor and load on the compressor and condenser. This is indicated by 2-3 on T-s diagram.

2. *Process 3-4-5-6-7-8:* This process represents the flow of refrigerant through the compressor. When the refrigerant enters the compressor through the suction valve at point 3, the pressure falls to point 4 due to frictional resistance to flow. Thus the actual suction pressure is (p_s) lower than the evaporator pressure (p_e). During suction and prior to compression, the temperature of the cold refrigerant vapour rises to point 5 when it comes in contact with the compressor cylinder walls. The actual compression of the refrigerant is shown by 5-6 which is neither isentropic nor polytropic. This is due to the heat transfer between the cylinder walls and the vapour refrigerant. The temperature of the cylinder walls is some what in between the temperature of cold suction vapour refrigerant and hot discharge vapour refrigerant. It may be assumed that the heat absorbed by the vapour refrigerant from the cylinder walls during the first part of the compression stroke is equal to heat rejected by the vapour refrigerant to the cylinder walls. Like the heating at suction given by 4-5, there is a cooling effect at discharge as given by 6-7. These heating and cooling effects take place at constant pressure. Due to the frictional resistance of flow, there is a pressure drop *i.e.*, the actual discharge pressure (p_d) is more than the condenser pressure (p_c).

3. *Process 8-9-10-11:* This process represents the flow of refrigerant though the condenser. The process 8-9 represents the cooling of superheated vapour refrigerant to the dry saturated state. The process 9-10 shows the removal of latent heat which changes the dry saturated refrigerant into the liquid refrigerant. The process 10-11 represents the sub-cooling of liquid refrigerant in the condenser before passing through the expansion valve. This is desirable as it increases the refrigerating effect per kg of the refrigerant flow. It also reduces the volume of the refrigerant of partially evaporated from the liquid refrigerant while passing through the expansion valve. The increase in refrigerating effect can be obtained by large quantities of circulating cooling water which should be at a temperature much lower than the condensing temperatures,

4. *Process 11-1:* This process represents the expansion of sub cooled liquid refrigerant by throttling from the condenser pressure to the evaporator pressure.

Example 5.21 *The condenser and evaporator temperature of a 20 T capacity, simple saturated vapour compression refrigeration system, are 40°C and –20°C respectively. The refrigerant used in the system is R-22. Draw the cycle on p-h diagram (assuming isentropic compression) and calculate (i) the mass of refrigerant to be circulated, (ii) Power required in the compressor, and (iii) the COP of the cycle. If the system employs 5°C subcooling of refrigerant liquid and a superheating of 10°C of refrigerant vapour, what will be the new COP of the cycle? Whether it will increase or decrease?*

Solution: Given: $Q = 20\ TR$, $T_{1'} = -20°C$ or $273 - 20 = 253$ K,

$$T_{2'} = 40°C \text{ or } 273 + 40 = 313 \text{ K}$$

$$T_{3'} - T_3 = 5°C,\ T_1 - T_{1'} = 10°C$$

The properties of R-22 are obtained from standard table.

Temp. °C	*Pressure* bar	*Specific volume, m³/kg*	*Enthalpy, kJ/kg*		*Entropy, kJ/kg.K*		*Specific heat, kJ/kg.K*	
			h_f	h_g	s_f	s_g	c_{pl}	c_{pg}
–20	2.4529	0.09262	177.00	396.97	0.9133	1.7822	1.125	0.677
40	15.341	0.01507	249.71	415.87	1.1667	1.6973	1.338	1.009

The T-s diagram is shown in Fig. 5.30

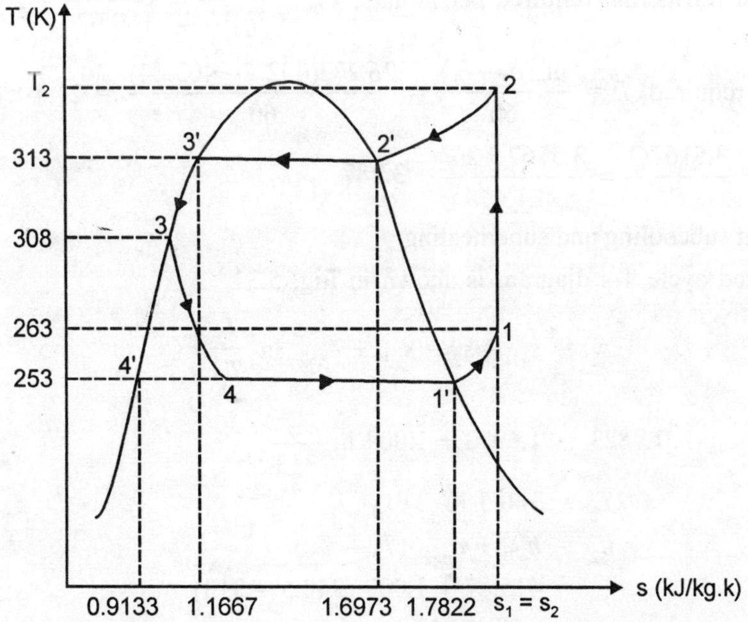

Fig. 5.30 *T-s* diagram with superheating and subcooling.

(*i*) With sub cooling and superheating $T_1 = 253 + 10 = 263$ K, $T_3 = 313 - 5 = 308$ K

Process 1′–1: Superheating

$$h_1 = h_{g1'} + c_{pg1'} + (T_1 - T_{1'}) = 396.97 + 0.667 \times 10$$
$$= 403.64 \text{ kJ/kg}$$
$$s_1 = s_{g1'} + c_{pg1'} \ln \frac{T_1}{T_{1'}} = 1.7822 + 0.667 \ln \frac{263}{253}$$
$$= 1.808 \text{ kJ/kg.K}$$

Process 1–2: Isentropic compression

$$s_1 = s_2 = s_{g2'} + c_{pg2'} \ln \frac{T_2}{T_{2'}} = 1.6973 + 1.009 \ln \frac{T_2}{313}$$
$$T_2 = 349.3 \text{ K}$$
$$h_2 = h_{g2'} + c_{pg2'} (T_2 - T_{2'}) = 415.87 + 1.009\,(349.3 - 313)$$
$$= 452.5 \text{ kJ/kg}$$

Process 3′–3: Subcooling

$$h_3 = h_{f3'} - c_{pf3'}(T_{3'} - T_3) = 249.71 - 1.338 \times 5 = 243.02 \text{ kJ/kg}$$

Process 3–4: Throttling

$$h_4 = h_{f3} = 243.02 \text{ kJ/kg}$$

Refrigerating effect, $q_{4-1} = h_1 - h_4 = 403.64 - 243.02 = 160.62$ kJ/kg

(*a*) Mass of refrigerant required per minute, $\dot{m}_r = \dfrac{211Q}{q_{4-1}} = \dfrac{211 \times Q}{160.62} = 26.273$ kg/min

(*b*) Power required, $P = \dfrac{\dot{m}_r (h_2 - h_1)}{60} = \dfrac{26.273(452.5 - 403.64)}{60} = 21.4$ kW

(*c*) COP $= \dfrac{3.5167Q}{P} = \dfrac{3.5167 \times 20}{21.4} = 3.287$

(*ii*) Without subcooling and superheating:

The saturated cycle T-s diagram is shown in Fig. 5.31

$$s_1 = s_{g1} = s_2 = s_{g2'} + c_{pg2'} \ln \frac{T_2}{T_{2'}}$$
$$1.7823 = 1.6973 + 1.009 \ln \frac{T_2}{313}$$
$$T_2 = 340.5 \text{ K}$$
$$h_2 = h_{g2'} + c_{pg2'} (T_2 - T_{2'})$$
$$= 415.87 + 1.009\,(340.5 - 313)$$
$$= 443.63 \text{ kJ/kg}$$
$$h_4 = h_{f3} = 249.71 \text{ kJ/kg}$$

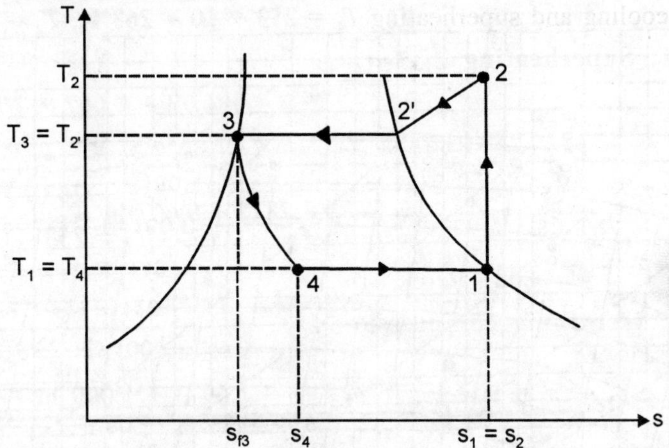

Fig. 5.31 Saturated cycle *T-s* diagram.

Refrigerating effect, $q_{4-1} = h_1 - h_4 = h_1 - h_{f3}$

(*a*) $\dot{m}_r = \dfrac{211 \times 20}{147.26} = 28.657$ kg/min

(*b*) $P = \dfrac{28.657(443.63 - 396.97)}{60} = 22.285$ kW

(*c*) $\text{COP} = \dfrac{3.5167 \times 20}{22.285} = 3.156$

We observe that COP increases with superheating and sub cooling from 3.156 to 3.287.

Example 5.22 *Solve example 5.21 by using p-h chart for R-22.*

Solution. By using *p-h* chart for saturated cycle (Fig. 5.32 and Fig. 5.33)

1. Locate point 1 on *p-h* chart for R-22 on dry saturated vapour line at 2.4529 bar or –20°C. Entropy at that point, $s_1 = 1.78$ kJ/kg.K and $h_1 = 398$ kJ/kg.
2. Follow the constant entropy line of 1.78 kJ/kg. K on the chart till it cuts the constant pressure line 15.341 bar or 40°C at point 2, $h_2 = 443$ kJ/kg.
3. Draw a horizontal line at 2 to intersect the saturated liquid line at 3. From 3, draw a vertical line to obtain $h_{f3} = h_4 = 250$ kJ/kg.

(*a*) Refrigerating effect, $q_{4-1} = h_1 - h_{f3} = 398 - 250 = 148$ kJ/kg

$$\dot{m}_r = \frac{211Q}{q_{4-1}} = \frac{211 \times 20}{148} = 28.5 \text{ kg/min}$$

(*b*) Work done, $\dot{W} = \dot{m}_r(h_2 - h_1)$

$$= 28.5\ (443 - 398) = 1282.5 \text{ kJ/min}$$

Power required, $P = \dfrac{\dot{W}}{60} = \dfrac{1282.5}{60} = 21.37$ kW

(*c*) $\text{COP} = \dfrac{R_E}{\dot{W}} = \dfrac{20 \times 211}{1282.5} = 3.29$

The results compare well with the analytical solution.

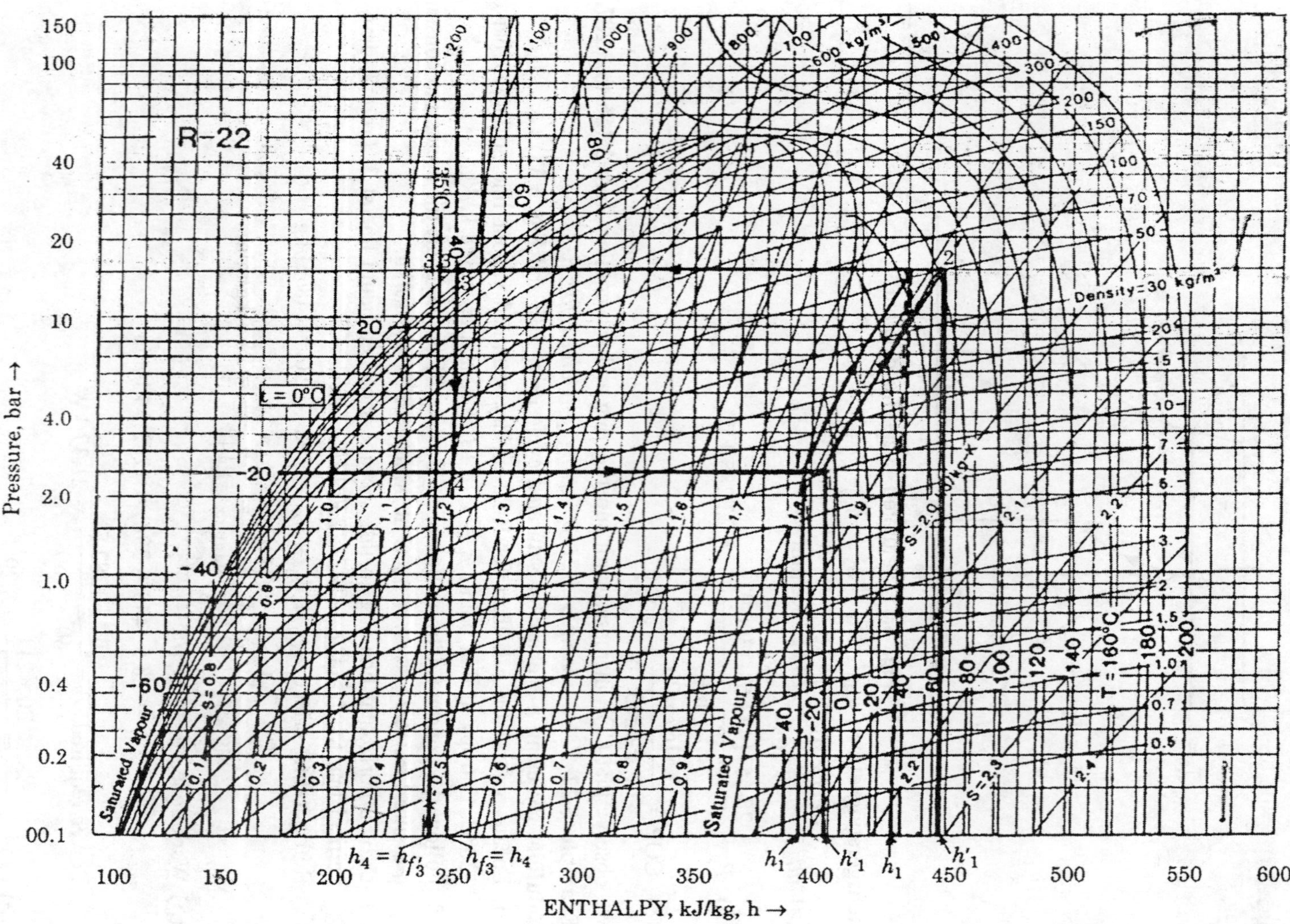
R-22
Pressure, bar →
150
100
40
20
10
4.0
2.0
1.0
0.4
0.2
00.1
100
150
200
250
300
350
400
450
500
550
600
ENTHALPY, kJ/kg, h →
$h_4 = h_{f'_3}$
$h_{f_3} = h_4$
h_1
h'_1
h_1
h'_1
Saturated Vapour
Saturated Vapour
Density=30 kg/m³
t = 0°C
35°C
T = 160°C

Fig. 5.32

Solution by p-h chart for superheated and subcooled cycle (Fig. 5.32 and 5.34)

Degree of superheat = 10°C

Degree of subcooling = 5°C

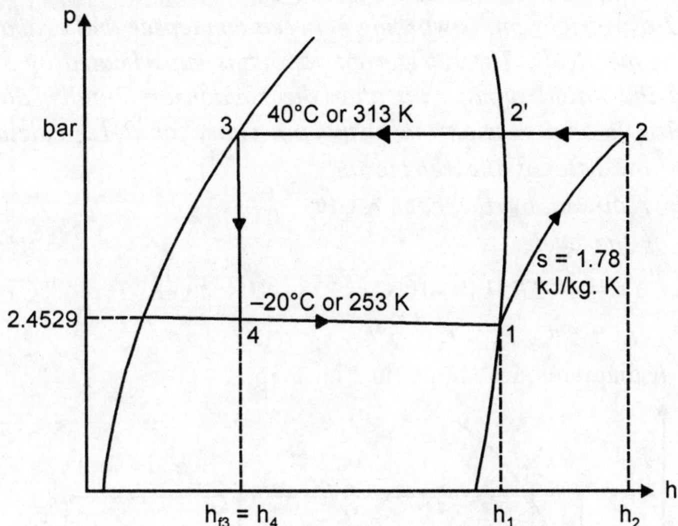

Fig. 5.33 *p-h* chart for saturated cycle

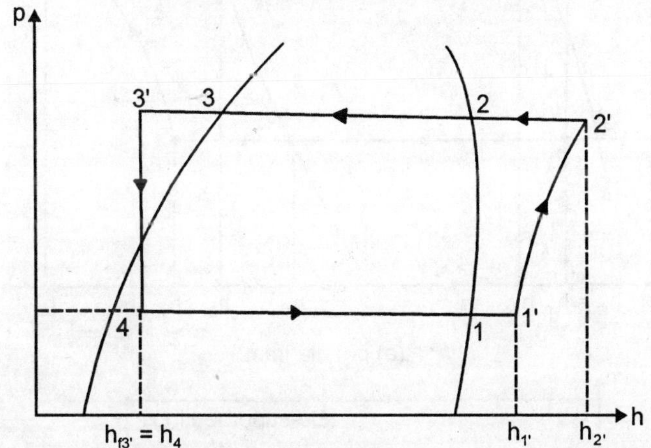

Fig. 5.34 *p-h* chart for superheated and sub cooled cycle.

1. Extend the horizontal line drawn at –20°C to the right upto –10°C to locate point 1′ so that 1–1′ = 10°C. Draw vertical line from 1′ to obtain $h_{1'}$ = 403 kJ/kg. Also $s_{1'}$ = 1.81 kJ/kg.K.
2. Follow $s_{1'}$ = 1.81 kJ/kg.K line to intersect horizontal line at 40°C at 2′. Draw vertical line at 2′ to obtain $h_{2'}$ = 452 kJ/kg.
3. Extend line 2–3 to 3′ so that 3–3′ = 50°C. Draw vertical line at 3′ to obtain $h_{f3'} = h_4$ = 247 kJ/kg.

$$\dot{m}_r = \frac{211Q}{h_{1'} - h_{4'}} = \frac{211 \times 20}{403 - 247} = 27.0 \text{ kg/min}$$

$$P = \frac{\dot{m}_r (h_{2'} - h_{1'})}{60} = \frac{27(452 - 403)}{60} = 22 \text{ kW}$$

$$\text{COP} = \frac{3.5167Q}{P} = \frac{3.5167 \times 20}{22} = 3.2$$

Example 5.23 *A 10 Ton Refrigerant System, employing a simple vapour compression cycle and using R-12 as refrigerant, is working between condenser and evaporator temperatures of 40°C and –10°C respectively. The refrigerant vapour is superheated by 5°C before entering the compressor and the liquid refrigerant after the condenser (condensate) is subcooled by 5°C. Assuming isentropic compression and using p-h chart for R-12, calculate:*

(i) The mass flow rate of the refrigerant,

(ii) The Power required by the compressor,

(iii) The COP of the cycle.

Solution: Given: $\dot{Q} = 10\ TR$, $T_1 = -10°C$ or $273 - 10 = 263$ K, $T_2 = 40°C$ or $273 + 40 = 313$ K,

$$t_1 - t_{1'} = 5°C,\ t_{3'} - t_3 = 5°C$$

The *T-s* and *p-h* diagrams are shown in Fig. 5.35.

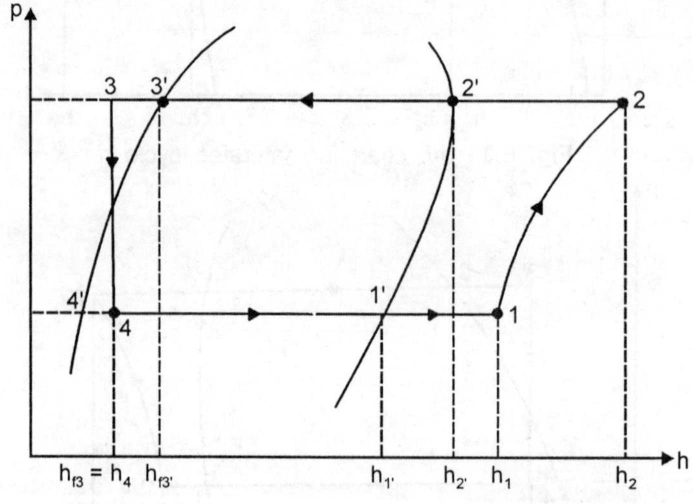

(a) p-h diagram.

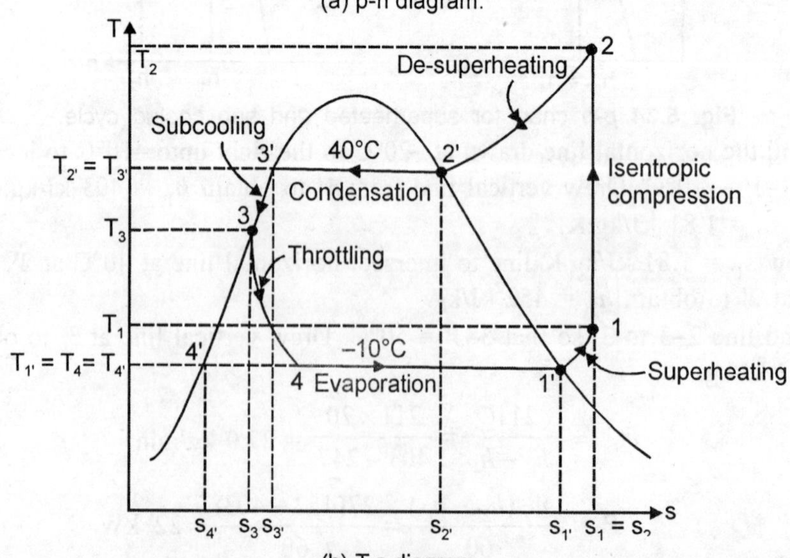

(b) T-s diagram.

Fig. 5.35

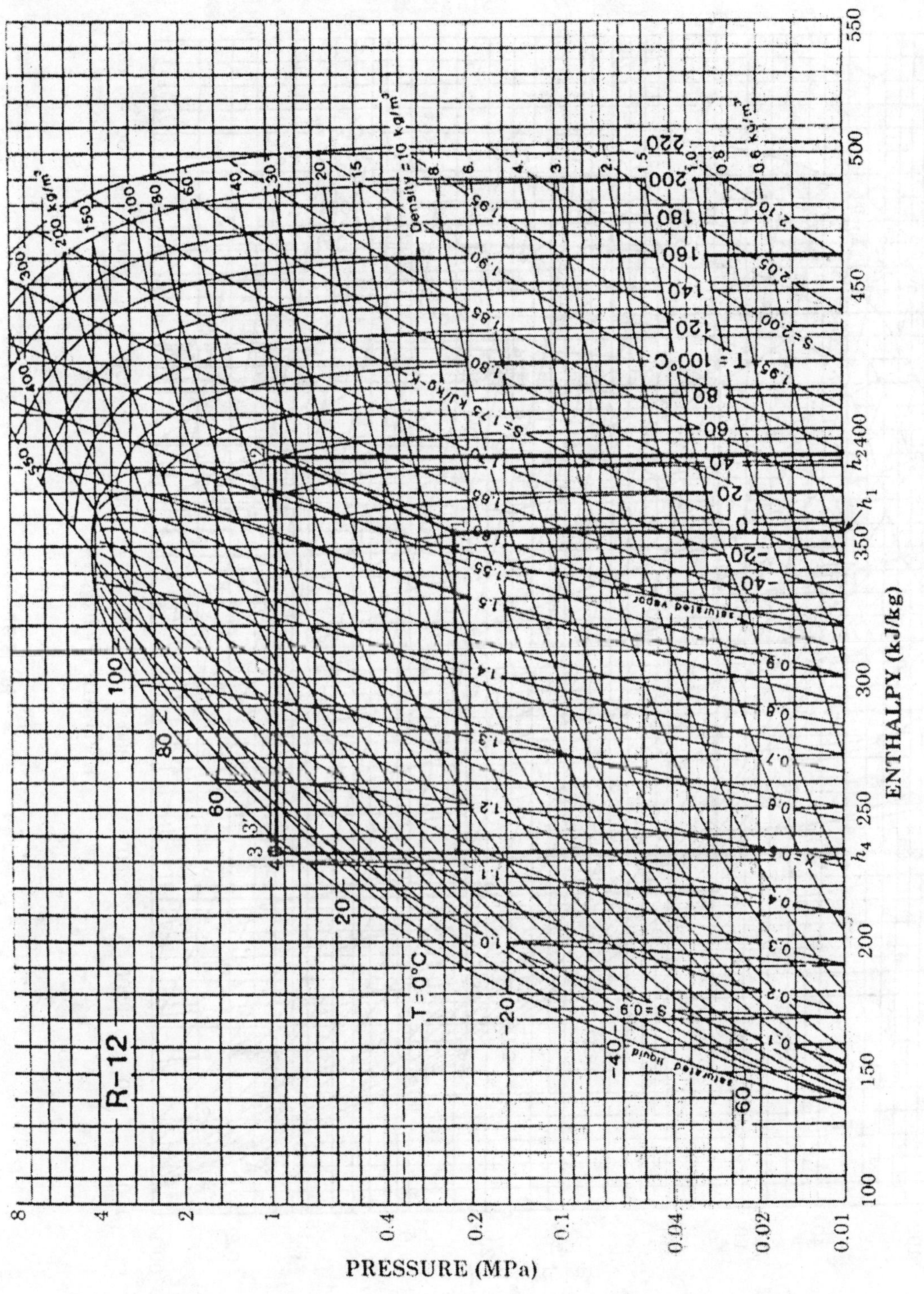
R-12
PRESSURE (MPa)
ENTHALPY (kJ/kg)
T = 0°C
T = 100°C
S = 1.75 kJ/kg-K
S = 2.00
Density = 10 kg/m³
200 kg/m³
saturated liquid
saturated vapor
X = 0.5
h_1
h_2
h_4

Fig. 5.36

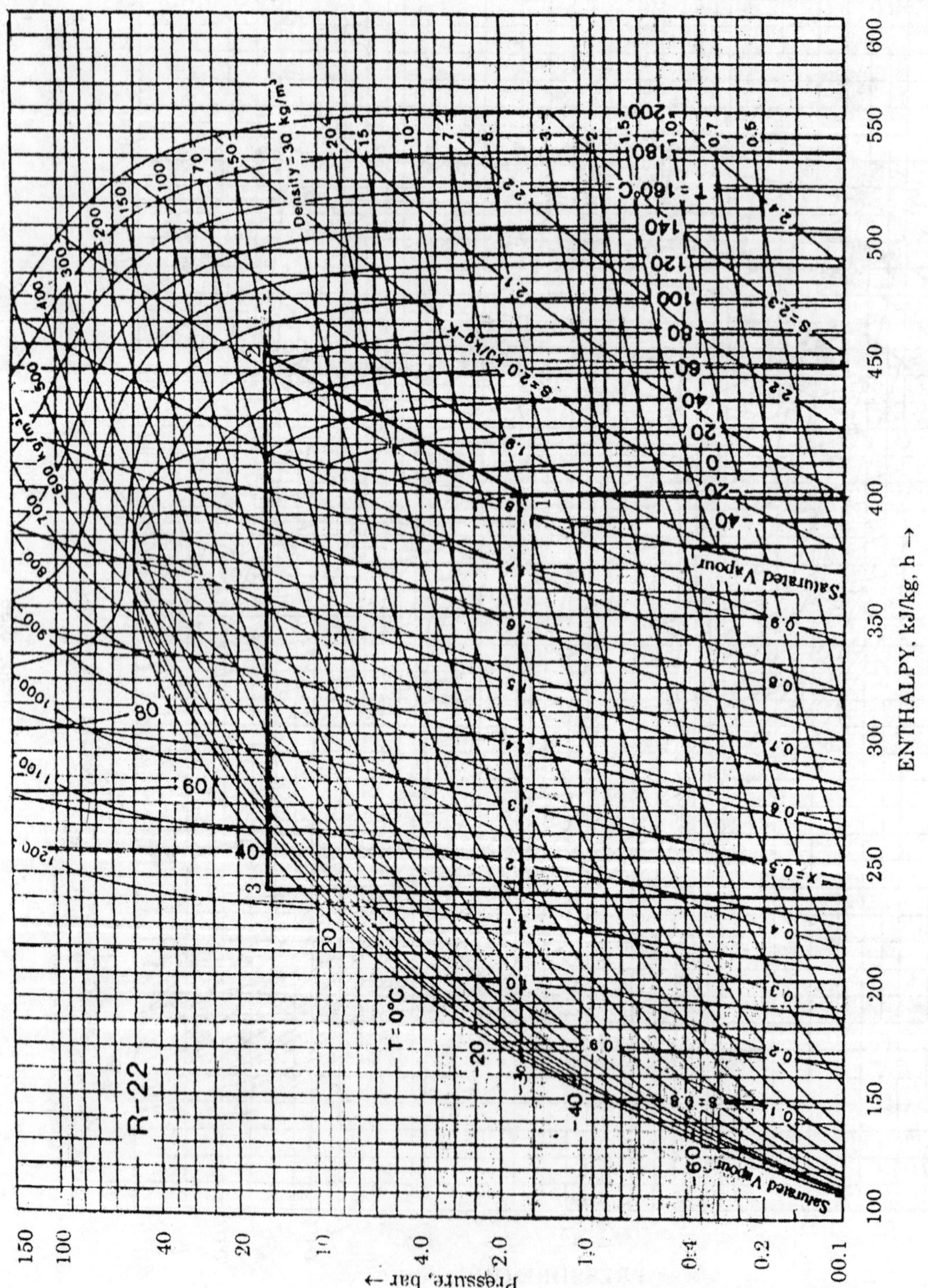
R-22
ENTHALPY, kJ/kg, h →
Pressure, bar →
Saturated Vapour
Density = 30 kg/m³
T = 0°C

Fig. 5.37

Solution by using p-h chart for R-12

Refer to Fig. 5.35 and 5.36

1. Extend the horizontal line drawn at –10°C to the right upto –5°C to locate point 1 so that 1′–1 = 50°C.
2. Draw vertical line from 1 on horizontal enthalpy scale to obtain,

$$h_1 = 355 \text{ kJ/kg}$$

Also, $$s_1 = 1.58 \text{ kJ/kg.K}$$

3. Follow s_1 = 1.58 kJ/kg.K line to intersect horizontal line at 40°C at 2.
4. Draw vertical line at 2 to obtain,

$$h_2 = 383 \text{ kJ/kg}$$

5. Draw horizontal line at 2 to locate saturation points 2′ and 3′ on vapour and liquid saturation curves respectively.
6. Extend line 2–3′ to the left upto 3 so that 3′–3 = 5°C.
7. Draw vertical line at 3 to obtain,

$$h_{f3} = h_4 = 230 \text{ kJ/kg}$$

Refrigeration effect, $$R_E = h_1 - h_4 = 355 - 230 = 125 \text{ kJ/kg}$$

(*i*) Mass flow rate of refrigerant,

$$\dot{m}_r = \frac{211Q}{h_{1'} - h_{4'}} = \frac{211 \times 20}{403 - 247} = 16.88 \text{ kg/min}$$

(*ii*) Power required by the compressor,

$$P_c = \frac{\dot{m}_r (h_2 - h_1)}{60} = \frac{16.88(383 - 355)}{60} = 7.88 \text{ kW}$$

(*iii*) $$\text{COP} = \frac{3.5167\,TR}{P_c} = \frac{3.5167 \times 10}{7.88} = 4.46$$

Example 5.24 *A simple vapour compression refrigeration system, using R-22 as refrigerant is working between evaporator and condenser temperature of –30°C and 40°C respectively. The liquid refrigerant after the condenser, is subcooled upto 30°C before it enters the expansion valve, and the refrigerant vapour is superheated up to –20°C before entering the compressor. The capacity of the system is 30 tons and the compression is isentropic.*

Draw the above cycle on p-h chart of R-22 and determine:

(i) The refrigerating effect (in kJ/kg),

(ii) The mass of refrigerant to be circulated per min.,

(iii) The power required in the compressor (in kW),

(iv) The heat removed through condenser, and

(v) The COP of the cycle.

Solution: Refer to example 5.23. Here use p-h chart for R-22, as shown in Fig. 5.37

$$h_1 = 400 \text{ kJ/kg}$$
$$s_1 = 1.82 \text{ kJ/kg.K}$$
$$h_2 = 455 \text{ kJ/kg}$$
$$h_{f3} = h_3 = 235 \text{ kJ/kg}$$

(i) $R_E = h_1 - h_4 = 400 - 235 - 165$ kJ/kg

(ii) $\dot{m}_a = \dfrac{211Q}{R_E} = \dfrac{211 \times 30}{165} = 38.36$ kg/min

(iii) $P_c = \dfrac{\dot{m}_r(h_2 - h_1)}{60} = \dfrac{38.36(455-400)}{60} = 35.17$ kW

(iv) $Q_c = \dot{m}_r(h_2 - h_{f3}) = 38.36\ (455 - 235) = 8439.2$ kJ/min

(v) $\text{COP} = \dfrac{3.5167\,TR}{P_c} = \dfrac{3.5167 \times 30}{35.17} = 3.0$

Example 5.25 *What is the effect of change of evaporator pressure and condenser pressure on the performance of simple vapour compression refrigeration system? A Freon 12 (R-12) simple vapour compression refrigeration system operating between condenser temperature of 40°C and evaporator temperature of –5°C, develops 20 tons of refrigeration. The refrigerant vapour is superheated to 0°C before entering the compressor and the liquid refrigerant after the condenser (i.e. condensate) is subcoooled to 30°C. Assuming isentropic compression, draw the cycle on p-h chart of R-12 and determine:*

(i) The mass flow rate of the refrigerant

(ii) The power required by the compressor

(iii) The theoretical piston displacement of the compressor

(iv) The heat rejected in the condenser

(v) COP of the vapour compression refrigeration cycle.

Solution: Effect of change of evaporator and condenser pressure on the performance of simple vapor compression refrigeration system:

Refer: to art 5.6 and art 5.7.

Given: R-12, condenser temperature = 40°C, Evaporator temperature = –50°C, $\dot{Q}$ = 20 TR, temperature of superheated vapour = 0°C, temperature of subcooled vapour = 30°C.

Follow the procedures as described in Example 5.23.

The solution has been obtained in Fig. 5.38.

$$h_1 = 352 \text{ kJ/kg}$$

$$v_1 = \frac{1}{15}\text{ m}^3/\text{kg}$$
$$s_1 = 1.57 \text{ kJ/kg.K}$$
$$h_2 = 375 \text{ kJ/kg}$$
$$v_2 = \frac{1}{52}\text{ m}^3/\text{kg}$$
$$h_{f3} = h_4 = 235 \text{ kJ/kg}$$
$$R_E = h_1 - h_4$$
$$= 352 - 235 = 117 \text{ kJ/kg}$$

(*i*) Mass flow rate of refrigerant, $\dot{m}_a = \frac{211\dot{Q}}{R_E} = \frac{211 \times 20}{117} = 36.07$ kg/min

(*ii*) $P_c = \frac{\dot{m}_a(h_2 - h_1)}{60} = \frac{36.07(375 - 352)}{60} = 13.82$ kW

(*iii*) Theoretical piston displacement of compressor $= v_1 - v_2 = \frac{1}{15} - \frac{1}{52} = 0.047436$ m³/kg

$= 0.047436 \times 36.07 = 1.71$ m³/min

(*iv*) Heat rejected in the condenser,

$$\dot{Q}_c = \dot{m}_r(h_2 - h_{f3}) = 36.07\ (375 - 235) = 5049.8 \text{ kJ/min or } 84.16 \text{ kW}$$

(*v*) $\text{COP} = \frac{3.5167TR}{P_c} = \frac{3.5167 \times 20}{13.82} = 5.09$

Example 5.26

(*a*) Explain the differences between throttling and isentropic expansion? Although some work can be extracted from the isentropic expansion process after the refrigerant leaves the condenser, an irreversible isenthalpic process is generally preferred. Explain the reasons.

(*b*) The pressure in the evaporator of an ammonia refrigerator is 1.902 bar and the pressure in the condenser is 12.37 bar. Calculate the refrigeration effect per unit mass of the refrigerant and the COP_{ref} for the following cycles,

(*i*) the ideal reversed Carnot cycle,

(*ii*) dry saturated vapour delivered to the condenser after isentropic compression, and no undercooling of the condensed liquid:

Given:

For ammonia at 1.902 bar,

$$T_{\text{sat}} = -20°\text{C}$$
$$s_g = 5.623 \text{ kJ/kg.K}, \ h_g = 1420.6 \text{ kJ/kg}$$
$$s_f = 0.363 \text{ kJ/kg.K}, \ h_f = 89.8 \text{ kJ/kg}$$

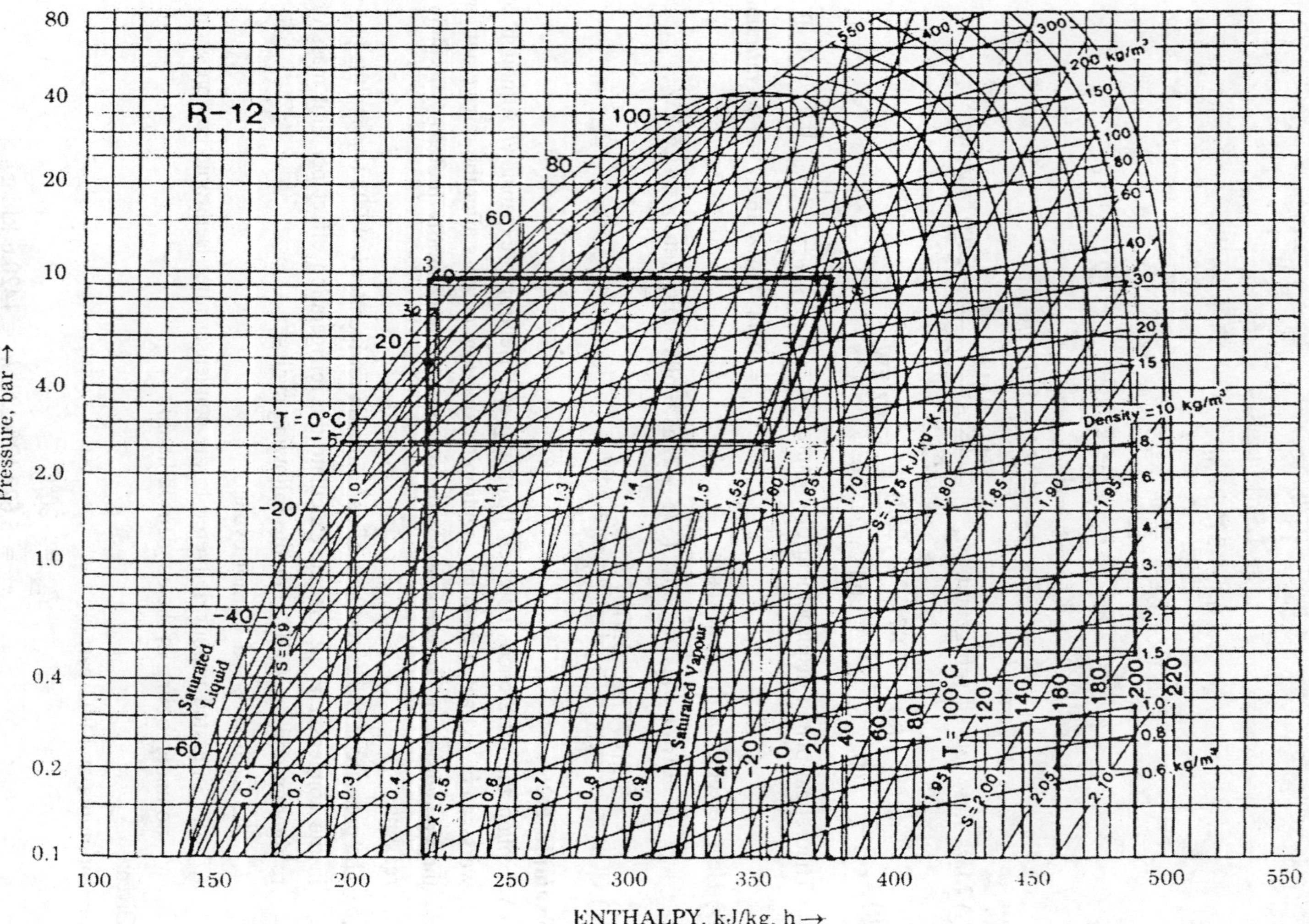
R-12
Pressure, bar →
ENTHALPY, kJ/kg, h →
Saturated Liquid
Saturated Vapour
Density = 10 kg/m³
S = 1.75 kJ/kg-K
T = 100°C
T = 0°C

Fig. 5.38

at 12.37 bar; $T_{saT} = 32°C$

$s_g = 4.962$ kJ/kg.K, $h_g = 1469.9$ kJ/kg

$s_f = 1.235$ kJ/kg.K, $h_f = 332.8$ kJ/kg

Solution:

(*a*) ***Throttling:*** During throttling process the enthalpy of the process does not change. It is an isenthalpic process.

Isentropic expansion process: During an isentropic expansion process the entropy of the process does not change. However, the enthalpy may change and some work can be extracted.

After the refrigerant leaves the condenser, an irreversible isenthalpic process is generally preferred for achieving high refrigerating effect. With isentropic expansion, the increase in work, the enthalpy of entering refrigerant to the evaporator shall increase, resulting in decrease of refrigerating effect.

(*b*) The *T-s* diagram for liquid ammonia Carnot cycle is shown in Fig. 5.39.

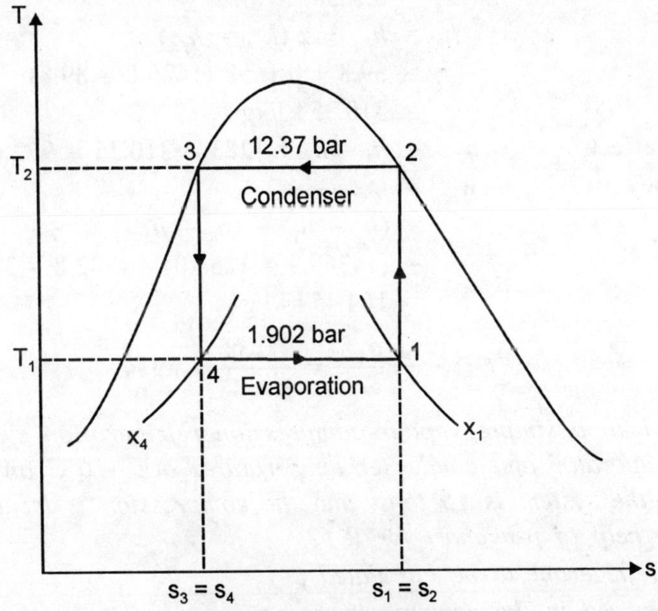

Fig. 5.39 T-s diagram.

Given: $p_1 = 1.902$ bar: $T_1 = T_s = -20°C$ or $273 - 20 = 253$.K,

$s_g = 5.623$ kJ/kg.K, $h_g = 1420.0$ kJ/kg, $s_f = 0.363$ kJ/kg.K,

$h_f = 89.8$ kJ/kg

$p_2 = 12.37$ bar: $T_2 = T_s = 32°C$ or $273 + 32 = 305$ K,

$s_g = 4.962$ kJ/kg.K, $h_g = 1469.9$ kJ/kg, $s_f = 1.235$ kJ/kg.K,

$h_f = 332.8$ kJ/kg

(*i*) Carnot $(COP)_{ideal} = \dfrac{T_1}{T_2 - T_1} = \dfrac{253}{305 - 253} = \dfrac{253}{52} = 4.865$

(*ii*) **Isentropic compression: process 1-2**

$$s_1 = s_2$$

$$s_{f1} + x_1 (s_{g1} - s_{f1}) = s_2 = s_{g2}$$

$$0.363 + x_1 (5.623 - 0.363) = 4.962$$

Dryness fraction, $x_1 = 0.874$

Enthalpy, $h_1 = h_{f1} + x_1(h_{g1} - h_{f1})$

$$= 332.8 + 0.874 (1420.0 - 332.8)$$

$$= 1283 \text{ kJ/kg}$$

$$h_2 = h_{g2} = 1469.9 \text{ kJ/kg}$$

$$h_3 = h_{f3} = 332.8 \text{ kJ/kg}$$

For isentropic expansion process 3-4:

$$s_4 = s_3$$

$$s_{f4} + x_4(s_{g4} - s_{f4}) = s_{f4}$$

$$0.363 + x_4 (5.623 - 0.363) = 1.235$$

$$x_4 = 0.1658$$

$$h_4 = h_{f4} + x_4(h_{g4} - h_{f4})$$

$$= 89.8 + 0.1658 (1420.0 - 89.8)$$

$$= 310.35 \text{ kJ/kg}$$

Refrigerating effect, $q_{4-1} = h_1 - h_4 = 1283 - 310.35 = 972.65$ kJ/kg

Net work done, $w_{net} = w_{1-2} - w_{3-4}$

$$= (h_2 - h_1) - (h_3 - h_4)$$

$$= (14369.9 - 1283.0) - (332.8 - 310.35)$$

$$= 164.45 \text{ kJ/kg}$$

$$\text{COP} = \frac{q_{4-1}}{w_{net}} = \frac{972.65}{164.45} = 5.914$$

Example 5.27 *In a simple vapour compression refrigeration system using R-12 as refrigerant, the evaporator and condenser temperatures are –10°C and 35°C respectively. If the capacity of the system is 15 tons, and the compression is isentropic, calculate the following with the help of p-h chart for R-12:*

(i) Mass of refrigerant to be circulated ,

(ii) Power required in the compressor

(iii) Total heat rejected in the condenser, and

(iv) COP of the cycle

Also calculate the Carnot COP and show the percentage difference between the two. Draw the cycle on p-h chart for *R*-12.

Solution: Given: Refrigerant: *R*-12, Evaporator temperature = –10°C,

Condenser temperature = 35°C, Q = 15 TR,

$$T_1 = 273 - 10 = 263 \text{ K}, \; T_2 = 273 + 35 = 308 \text{ K}$$

$$\text{Carnot COP} = \frac{T_1}{T_2 - T_1} = \frac{263}{308 - 236} = \frac{263}{45} = 5.84$$

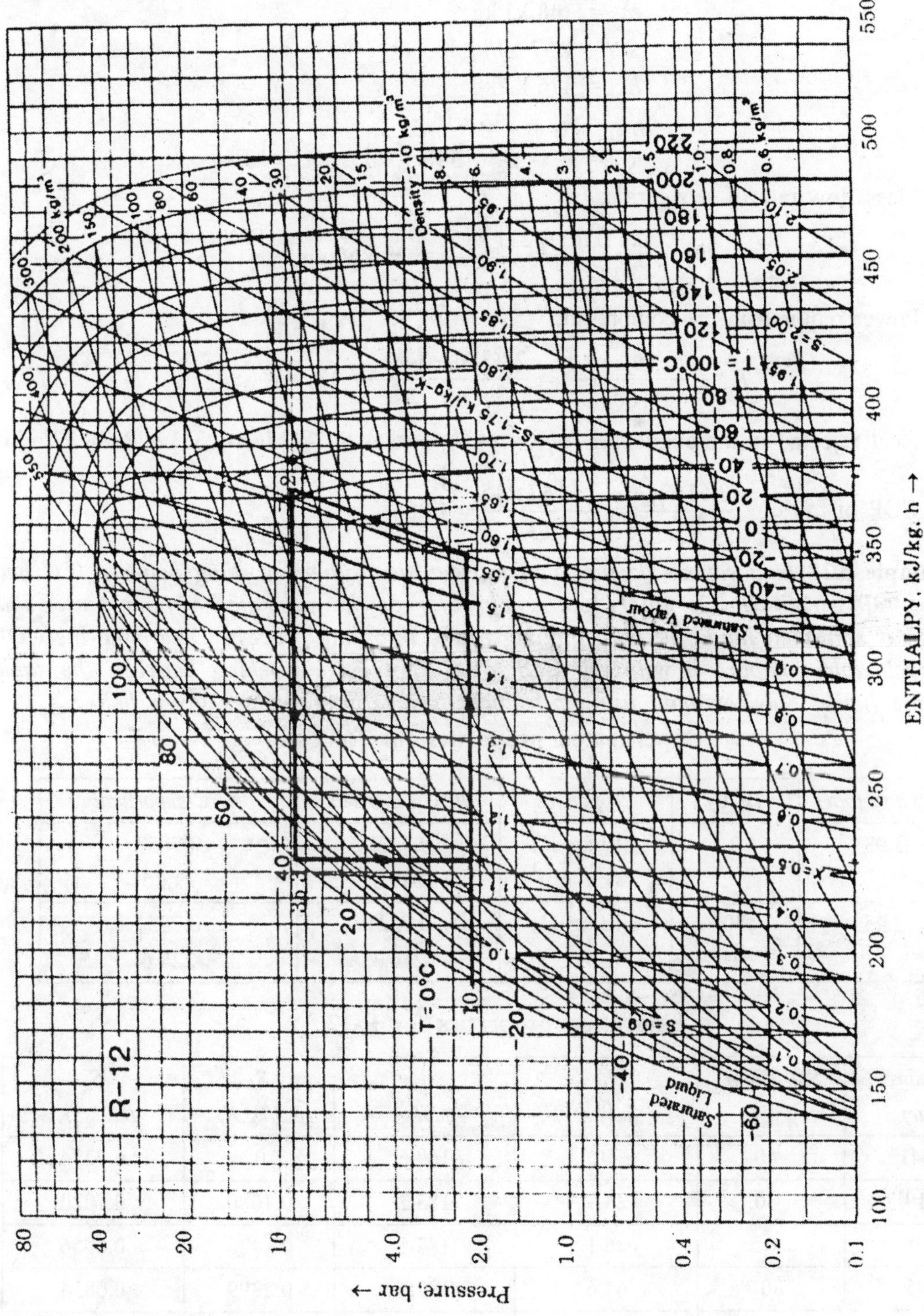
R-12
Pressure, bar →
80
40
20
10
4.0
2.0
1.0
0.4
0.2
0.1
ENTHALPY, kJ/kg. h →
100
150
200
250
300
350
400
450
500
550
Saturated Liquid
Saturated Vapour
T = 0°C
T = 100°C
Density = 10 kg/m³
S = 1.75 kJ/kg-K

Fig. 5.40

Refer to Fig. 5.38 for obtaining the solution on *p-h* chart for R-12 in Fig. 5.40.

$$h_1 = 348 \text{ kJ/kg}$$
$$s_1 = 1.57 \text{ kJ/kg.K}$$
$$h_2 = 376 \text{ kJ/kg}$$
$$h_{f3} = h_4 = 235 \text{ kJ/kg}$$
$$R_E = h_1 - h_4 = 348 - 235 = 113 \text{ kJ/kg}$$

(*i*) Mass flow rate of refrigerant,

$$\dot{m}_r = \frac{211TR}{R_E} = \frac{211\times15}{113} = 28 \text{ kg/min}$$

(*ii*) Power required in the compressor,

$$P_c = \frac{\dot{m}_r(h_2 - h_1)}{60} = \frac{28(376-348)}{60} = 13 \text{ kW}$$

(*iii*) Heat rejected in the condenser, $\dot{Q}_c = \dot{m}_r(h_2 - h_{f3}) = 28(376 - 235) = 3948$ kJ/min

(*iv*) COP of cycle $= \dfrac{3.5167\dot{Q}}{P_c} = \dfrac{3.5167\times15}{13} = 4.06$

Example 5.28 *In a vapour compression refrigeration system working between 0°C and 30°C, the refrigerant (R-12) enters the compressor at the saturated vapour state. The refrigerant is saturated liquid at the beginning of the throttling process. Determine the COP of this cycle and compare it with the COP of Carnot cycle working between the same temperature limits. Solve without using p-h chart, use property values given below :*

Superheated properties for R-12

Pressure (bar)	T_s*(°C)*	*Properties*	*20°C superheat*	*40°C superheat*
3.08	0	s(kJ/kg K)	0.7423	0.7853
		h(kJ/kg)	200.5	213.5
7.45	30	s(kJ/kg K)	0.7321	0.7751
		h(kJ/kg)	214.3	228.6

Saturation properties for R-12

Pressure (bar)	*Sat. Temp. (°C)*	h_f *(kJ/kg)*	h_g *(kJ/kg)*	S_f *(kJ/kg.K)*	S_g *(kJ/kg.K)*
0.641	–40	0	169.6	0	0.7274
2.19	–10	26.9	183.2	0.1080	0.7020
3.08	0	36.1	187.5	0.1420	0.6966
7.45	30	64.6	199.6	0.2399	0.6854
9.6	40	74.6	203.2	0.2718	0.6825

Solution: (a) Given: $h_1 = h_{g1} = 187.5$ kJ/kg, $s_1 = s_{g1} = 0.6966$ kJ/kg.K

$s_{2'} = s_{g2'} = 0.6854$ kJ/kg.K, $h_{2'} = 199.6$ kJ/kg,

$h_{f3} = 64.6$ kJ/kg

At 30°C from R-12 table, $c_{p2'} = 0.721$ kJ/kg.K

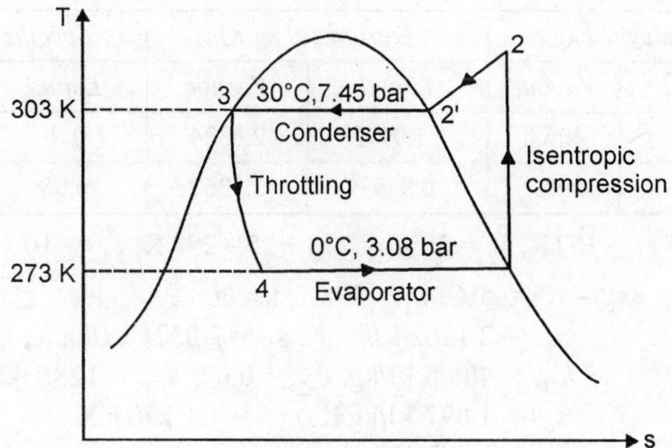

Fig. 5.41 T-s diagram.

Refer to Fig. 5.41

$$s_2 = s_{2'} + c_{p2'} \ln\left(\frac{T_2}{T_{2'}}\right)$$

$$= 0.6854 + 0.721 \ln\left(\frac{T_2}{303}\right)$$

For isentropic compression process 1–2,

$$s_1 = s_2$$

$$0.6966 = 0.6854 + 0.721 \ln\left(\frac{T_2}{303}\right)$$

$$T_2 = 307.7 \text{ K}$$

$$h_2 = h_{2'} + c_{p2'}(T_2 - T_{2'})$$

$$= 199.6 + 0.721\,(307.7 - 303) = 203.02 \text{ kJ/kg}$$

$$h_4 = h_{f3} = 64.6 \text{ kJ/kg}$$

Work done by compressor, $w_c = h_2 - h_1 = 203.02 - 187.5 = 15.52$ kJ/kg

Refrigerating effect, $q_{4-1} = h_1 - h_4 = 187.5 - 64.6 = 122.9$ kJ/kg

$$\text{COP} = \frac{q_{4-1}}{w_c} = \frac{122.9}{15.52} = 7.92$$

$$\text{Carnot COP} = \frac{T_1}{T_2 - T_1} = \frac{273}{303 - 273} = \frac{273}{30} = 9.1$$

Example 5.29 *In a 15 ton ammonia refrigerant plant, the condensing temperature is 25°C and evaporating temperature is –10°C. The refrigerant is subcooled by 5°C before reaching the inlet of the throttle valve. The vapour leaving the evaporator is dry saturated. Find the COP and mass flow rate of the refrigerant.*

Use the following table for properties of the Ammonia:

T_s(°C)	Enthalpy (kJ/kg)		Entropy (kJ/kg.K)		Specific heat(kJl/kg.K)	
	Liquid	Vapour	Liquid	Vapour	Liquid	Vapour
25	128.1	406.8	1.097	2.0324	1.1	0.67
–10	89.6	398.7	0.9593	2.1362	0.98	0.59

Solution: Given: $\dot{Q}$ = 15TR, $T_{2'}$ = 25°C or 273 + 25 = 298 K, T_1 = –10°C or 273 – 10 = 263 K

$$t_{3'} - t_3 = 5°\text{C},\ h_1 = 398.7,\ \text{kJ/kg},$$
$$s_1 = 2.1362\ \text{kJ/kg.K},\ s_{2'} = 2.0324\ \text{kJ/kg.K},$$
$$h_{2'} = 406.8\ \text{kJ/kg},\ c_{p2'} = 0.67,\ h_{f3'} = 128.1\ \text{kJ/kg},$$
$$s_g = 1.097\ \text{kJ/kg.K},\ c_{p3'} = 1.1\ \text{kJ/kg.K}$$

The *T-s* diagram is shown in Fig. 5.42.

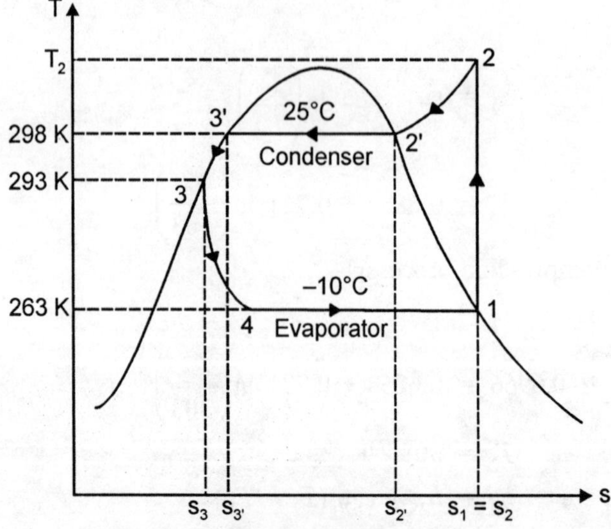

Fig. 5.42 T-s diagram.

For isentropic compression process 1–2,

$$s_1 = s_2 = s_{2'} + c_{p2'}\ \ln\left(\frac{T_2}{T_{2'}}\right)$$

$$2.1362 = 2.0324 + 0.67\ \ln\left(\frac{T_2}{298}\right)$$

$$T_2 = 455.2\ \text{K}$$

$$h_2 = h_{2'} + c_{p2'}\ (T_2 - T_{2'})$$
$$= 406.8 + 0.67\ (455.2 - 298)$$
$$= 512.12\ \text{kJ/kg}$$

For the subcooling process 3′–3,

$$h_{f3} = h_{f3'} - c_{p3'}(t_{3'} - t_3) = 128.1 - 1.1 \times 5 = 122.6 \text{ kJ/kg}$$

Now $h_3 = h_{f3} = 122.6$ kJ/kg

Refrigerating effect, $R_E = h_1 - h_4 = 398.7 - 122.6 = 276.1$ kJ/kg

Mass flow rate of refrigerant, $\dot{m}_r = \dfrac{211TR}{R_E} = \dfrac{211 \times 15}{276.1} = 11.463$ kg/min

Compressor workdone, $\dot{W}_C = \dot{m}_r(h_2 - h_1)$

$$= \frac{11.463(512.12 - 398.7)}{60} = 21.67 \text{ kW}$$

$$\text{COP} = \frac{3.5167TR}{\dot{W}_c} = \frac{3.5167 \times 15}{21.67} = 2.43$$

(*c*) Cascade refrigeration system: Refer to art. 5.14.

Example 5.30 *A refrigeration machine is required to produce ice at 0°C from water at 20°C. The machine has a condenser temperature of 298 K while the evaporator temperature is 268 K. The relative efficiency of the machine is 50% and 6 kg of Freon-12 refrigerant is circulated through the system per minute. The refrigerant enters the compressor with a dryness fraction of 0.6. Specific heat of water is 4.187 kJ/kg.K and the latent heat of ice is 335 kJ/kg. Calculate the amount of ice produced in 24 hours. The table of properties of Freon-12 is given below:* **[IES, 1992]**

Temperature, K	*Liquid heat, kJ/kg*	*Latent heat, kJ/kg*	*Entropy of liquid, kJ/kg.K*
298	59.7	138.0	0.2232
268	31.4	154.0	0.1251

Solution: Given: $x_1 = 0.6$, $T_1 = 268$ K(– 5°C), $T_2 = 298$ K(25°C), $m_h = 6$ kg/min,

$\eta_{rel} = 0.50$, $c_{pw} = 4.187$ kJ/kg.K, $(h_{fg})_{ice} = 335$ kJ/kg

The T-s diagram is shown in Fig. 5.43.

$$h_1 = h_{f1} + x_1 h_{fg1} = 31.4 + 0.6 \times 154$$
$$= 123.8 \text{ kJ/kg.K},$$

$$s_1 = s_{f1} + x_1 s_{fg1} = s_{f1} + x_1 \frac{h_{fg2}}{T_1}$$

$$= 0.1251 + \frac{0.6 \times 154}{268}$$

$$= 0.4698 \text{ kJ/kg.K}$$

$$s_1 = s_2 + s_{f2} + \frac{x_2 h_{fg2}}{T_2}$$

$$0.4698 = 0.2232 + \frac{x_2 \times 138}{298}$$

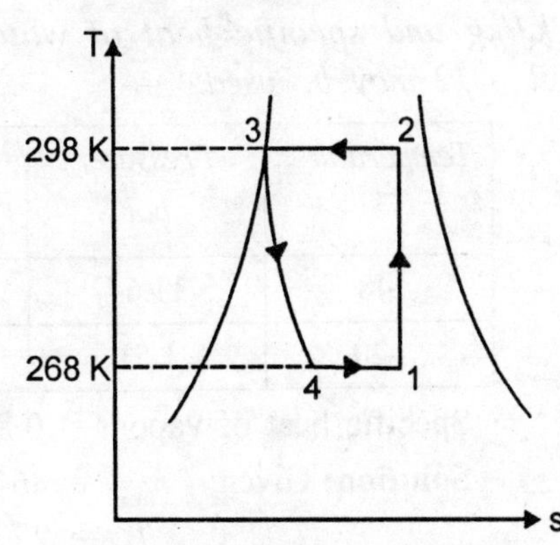

Fig. 5.43 T-s diagram

$$x_2 = 0.5325$$

$$h_2 = h_{f2} + x_2 h_{fg2} = 59.7 + 0.5325 \times 138 = 133.185 \text{ kJ/kg}$$

$$n_3 = h_4 = h_{f2} = 59.7 \text{ kJ/kg}$$

$$\text{Theoretical COP} = \frac{h_1 - h_4}{h_2 - h_1} = \frac{123.8 - 59.7}{133.185 - 123.8} = \frac{64.1}{9.385} = 6.83$$

$$(\text{COP})_{act} = \eta_{rel} \times (\text{COP})_{th} = 0.5 \times 6.83 = 3.415$$

Heat extracted from 1 kg of water at 20°C to form ice at 0°C.

$$= m_w \times [c_{pw} \times \Delta t + (h_{fg})_{ice}]$$

$$= 1 \times [4.187 \times 20 + 335] = 418.74 \text{ kJ/kg}$$

Let W = ice formed in 24 hours in tons.

$$\text{Heat taken out from water per minute, } \dot{q} = \frac{1000 \times W \times 418.74}{24 \times 60} = 290.8 \text{ W kJ/min}$$

$$\text{Work done per minute } \dot{w} = \dot{m}_r (h_2 - h_1) = 6 \times 9.385 = 56.31 \text{ kJ/min}$$

$$(\text{COP})_{act} = \frac{\dot{q}}{\dot{w}}$$

$$3.415 = \frac{290.8 \text{ W}}{56.31}$$

$$W = 0.661 \text{ tons}$$

Example 5.31 *A vapour compression refrigeration system using refrigerant R – 12 is employed to produce 8640 kg of ice per day. The condensing and evaporating temperatures of the refrigerant are 48°C and 20°C, respectively. Saturated liquid leaves the condenser and saturated vapour leaves the evaporator. Compression is isentropic. Water at 35°C is used to form ice. The temperature of ice should be –80°C. Heat flows into the brine tank from the surroundings may be taken to be 10 percent of the total heat removed from water to form ice at the specified temperature. Determine the power to form ice to drive the compressor, in kW. Take specific heat of ice = 2.26 kJ/kg.K, latent heat of ice = 334.72 kJ/kg and specific heat of water = 4.187 kJ/kg.K. The following data for refrigerant R – 12 may be used:* **[IES, 1993]**

Temperature, °C	*Pressure, bar*	*Enthalpy, kJ/kg*		*Entropy, kJ/kg. K*	
		liquid	*Vapour*	*Liquid*	*Vapour*
48	11.64	82.83	205.83	0.2973	0.6802
–20	1.51	17.82	178.74	0.0731	0.7087

Specific heat of vapour = 0.82 kJ/kg. K.

Solution: Given: $m_{ice} = 8640$ kg/day, $T_1 = 273 - 20 = 253$ K

$$T_{2'} = 273 + 48 = 321 \text{ K}$$

$$t_{wi} = 35°\text{C}, \; t_{ice} = -80°\text{C}$$

$$(c_p)_{ice} = 2.26 \text{ kJ/kg.K}$$
$$(h_{fg})_{ice} = 334.72 \text{ kJ/kg}, \; c_{pw} = 4.187 \text{ kJ/kg.K}$$
$$c_{ps} = 0.82 \text{ kJ/kg.K}$$

Refer to Fig. 5.44

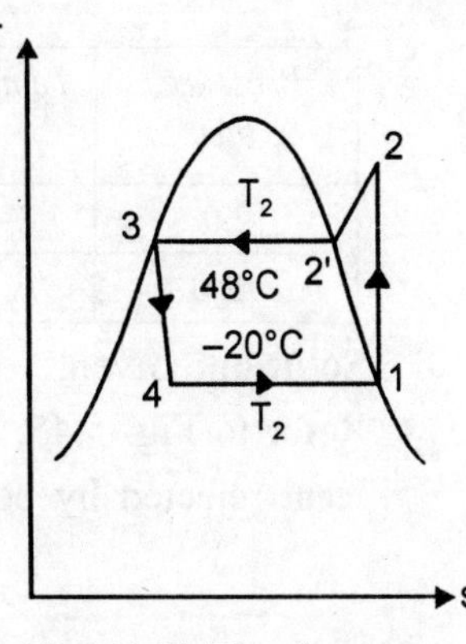

Fig. 5.44 T-s diagram.

Total heat to be removed from water at 35°C to convert it into ice at –80°C

$= m_{ice} \; [c_{pw} \; (t_{wi} - 0) + (h_{fg})_{ice} + (c_p)_{ice} \; [0 - (-8)]]$

$= 6640 \; [4.187 \times 35 + 334.72 + 2.26 \times 8]$

$= 43{,}14{,}340.8$ kJ/24 hours

Heat infiltration into brine tank from surroundings $= 0.1 \times 43{,}14{,}340.8$

$= 4{,}31{,}434.08$ kJ/24 hours

Total heat of be removed $= 43{,}14{,}340.8 + 4{,}31{,}434.08 = 47{,}45{,}774.88$ kJ/24 hours

$$= \frac{47{,}45{,}774.88}{24 \times 3600} = 54.92 \text{ kJ/s}$$

Now $h_1 = 178.74$ kJ/kg, $s_{g1} = 0.7087$ kJ/kg.K (See Fig. 5.44)

Let temperature of vapour after compression $= T_2$

Now
$$s_{g1} = s_{g2} + c_{ps} \; \ln\left(\frac{T_2}{T_{2'}}\right)$$

$$0.7087 = 0.6802 + 0.82 \; \ln\left(\frac{T_2}{321}\right)$$

$$T_2 = 332.35 \text{ K}$$
$$h_2 = h_2' + c_{ps} \; (T_2 - T_2') = 205.83 + 0.82 \; (332.35 - 321)$$
$$= 215.14 \text{ kJ/kg}$$
$$h_4 = h_{f3} = 82.83 \text{ kJ/kg}$$

Heat removed in evaporator $= h_2 - h_4 = 178.74 - 82.83 = 95.91$ kJ/kg

Flow rate of refrigerant required,

$$\dot{m}_r = \frac{\text{Total heat to be removed}}{\text{Heat removed in evaporator}} = \frac{54.92}{95.91} = 0.5726 \text{ kg/s}$$

Power required to derive the compressor $= \dot{m}_r (h_2 - h_1)$

$= 0.5726 \; (215.14 - 178.74) = 20.843$ kW

Example 5.32 *A vapour compression heat pump is driven by a power cycle having a thermal efficiency of 25%. For the heat pump, refrigerant-12 is compressed from saturated vapor at 2.0 bar to the condenser pressure of 12 bar. The isentropic efficiency of the compressor is 80%. Saturated liquid enters the expansion valve at 12 bar. For the power cycle, 80% of the heat rejected by it is transferred to the heated space which has a total heating requirement of 500 kJ/min. Determine the power input to the heat pump compressor. The following data for refrigerant-12 may be used.* **[IES, 1995]**

Pressure, bar	*Temperature, °C*	*Enthalpy, kJ/kg*		*Entropy, kJ/kg. K*	
		liquid	*Vapour*	*Liquid*	*Vapour*
2.0	–12.53	24.57	182.07	0.0992	0.7035
12.0	49.31	84.21	206.24	0.3015	0.6799

Solution: Given, $\eta_{th} = 0.25$, $h_c = 0.8$, $h_s = 0.8$, h_r, $h_s = 500$ kJ/min.

Refer to Fig. 5.45. $T_1 = 273 - 12.53 = 260.47$ K, $T_2'' = 273 + 49.31 = 322.31$ K

Heat rejected by power cycle,

$$h_r = \frac{h_s}{0.8} = \frac{500}{0.8} = 625 \text{ kJ/min}$$

For isentropic compression of refrigerant,

$$s_1 = s_2,\ h_{g1} = h_{g2} + c_{ps} \ln \frac{T_2}{T_2''}$$

$$0.7035 = 0.6799 + 0.7 \ln \frac{T_2}{322.31}$$

$$T_2 = 333.4 \text{ K}$$

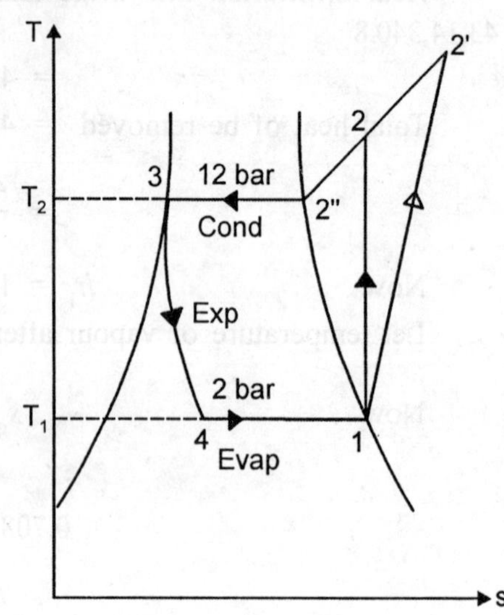

Fig. 5.45 T-s diagram.

Compressor efficiency, $\eta_c = \dfrac{T_2 - T_1}{T_2' - T_2}$

$$0.8 = \frac{333.4 - 260.47}{T_2' - 260.47}$$

$$T_2' = 351.6 \text{ K}$$

Enthalpy of superheated vapour at 12 bar,

$$h_2' = h_2'' + c_{ps}(T_2' - T_2'')$$

$$= 206.24 + 0.7\,(352.6 - 322.31) = 226.74 \text{ kJ/kg}$$

Heat rejected by cycle, $q_r = h_2' - h_{f3} = 226.74 - 84.21 = 142.53$ kJ/kg

Mass flow rate of refrigerant, $\dot{m}_r = \dfrac{h_r}{q_r} = \dfrac{625}{142.53} = 4.386$ kg/min

Compressor work, $\dot{W}_c = \dot{m}_r (h_2' - h_1) = 4.386\,(226.74 - 182.07)$

$= 195.92$ kJ/min or 3.265 kW

Example 5.33 *A Freon-12 refrigerator producing a cooling effect of 20 kJ/s operates on a simple cycle with pressure limits of 1.509 bar and 9.607 bar. The vapour leaves the evaporator dry saturated and there is no undercooling. Determine the power required by the machine.*

If the compressor operates at 3000 rpm and has a clearance volume of 3% of stroke volume determine the piston displacement of the compressor. For compressor assume that the expansion follows the law $pv^{1.13}$ = *constant.* **[IES, 1996]**

Given:

Temperature °C	p_s *bar*	v_g, m^3/kg	*Enthalpy kJ/kg*		*Entropy, kJ/kg K*		*Specific heat kJ/kg.K*
			h_f	h_g	s_f	s_g	c_{ps}
–20	1.509	0.1088	17.8	178.61	0.073	0.7082	0.747
40	9.607		74.53	203.05	0.2716	0.682	

Solution: Refer to Fig. 5.46.

Fig. 5.46 T-s diagram.

Refrigeration effect, $R = \dot{m}(h_1 - h_4) = \dot{m}(h_1 - h_{f3})$

$$\dot{m} = \frac{20}{178.61 - 74.53}$$

$$= 0.192 \text{ kg/s}$$

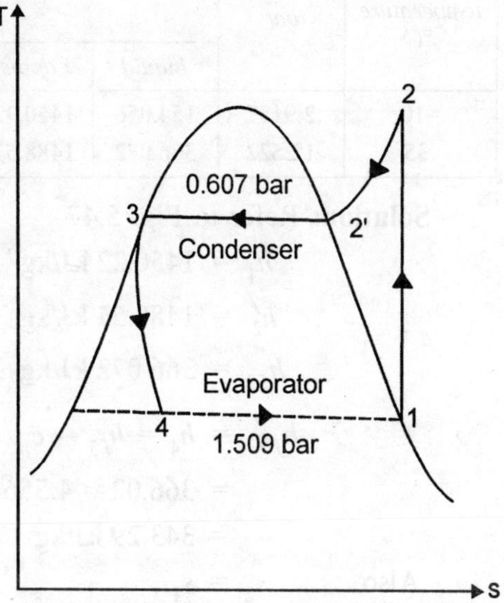

For isentropic process 1–2, we have

$$s_2 = s_1$$

$$s'_{g2} + c_{ps} \ln\left(\frac{T_2}{T'_2}\right) = s_{g1}$$

$$0.682 + 0.747 \ln\left(\frac{T_2}{273 + 40}\right) = 0.7082$$

$$T_2 = 324.17 \text{ K}$$

$$h_2 = h'_2 + c_{ps}(T_2 - T'_2)$$

$$= h'_{g2} + c_{ps}(T_2 - T'_2)$$

$$= 203.05 + 0.747\ (324.17 - 313) = 211.4 \text{ kJ/kg}$$

Work done $= \dot{m}(h_2 - h_1) = 0.192\ (211.4 - 178.61) = 6.296$ kW

Volumetri efficiency of compressor,

$$\eta_{vol} = 1 + c - c\left(\frac{p_2}{p_1}\right)^{1/n}, c = 0.03$$

$$= 1 + 0.03 - 0.03\left(\frac{9.607}{1.507}\right)^{1/1.13} = 0.8756 \text{ or } 87.56\%$$

Volume of refrigerant at the in take condition 1 is :

$$\dot{V}_1 = \dot{m} \times v_{g1} = 0.192 \times 0.1088 = 0.0209 \text{ m}^3\text{/s}$$

Swept volume, $\dot{V}_s = \frac{\dot{v}_1}{\eta_{vol}} = \frac{0.209}{0.8756} = 0.2359 \text{ m}^3\text{/s}$

Piston displacement per stroke, $V = \frac{\dot{V}_s \times 60}{N} = \frac{0.0239 \times 60}{3000} = 4.77 \times 10^{-4} \text{ m}^3$

Example 5.34 *A food storage locker requires a refrigeration capacity of 50 kW. It works between a condenser temperature of 35°C and an evaporator temperature of –10°C. The refrigerant is ammonia. It is subcooled by 5°C before entering the expansion valve by the dry saturated vapour leaving the evaporator. Assuming a single cylinder, single-acting compressor operating at 1000 rpm with stroke equal to 1.2 times the bore, determine (a) the power required, and (b) the cylinder dimensions.*

Saturation temperature °C	*Pressure bar*	*Enthalpy, kJ/kg*		*Entropy, kJ/kg K*		*Specific volume, m^3/kg*		*Specific heat kJ/kg K*	
		Liquid	*Vapour*	*Liquid*	*Vapour*	*Liquid*	*Vapour*	*Liquid*	*Vapour*
–10	2.9157	154.056	1450.22	0.82965	5.7550	–	0.417477	–	2.492
35	12.522	366.072	1488.57	1.56605	5.2086	1.7023	0.0954629	4.556	2.903

Solution: Refer to Fig. 5.47

$$h_1 = 1450.22 \text{ kJ/kg}$$

$$h_2' = 1488.57 \text{ kJ/kg}$$

$$h_{f3} = 366.072 \text{ kJ/kg}$$

$$h_{f3}' = h_4 = h_{f3} - c_{pf}\,(T_3 - T_3')$$

$$= 366.02 - 4.556\ (308 - 303)$$

$$= 343.29 \text{ kJ/kg}$$

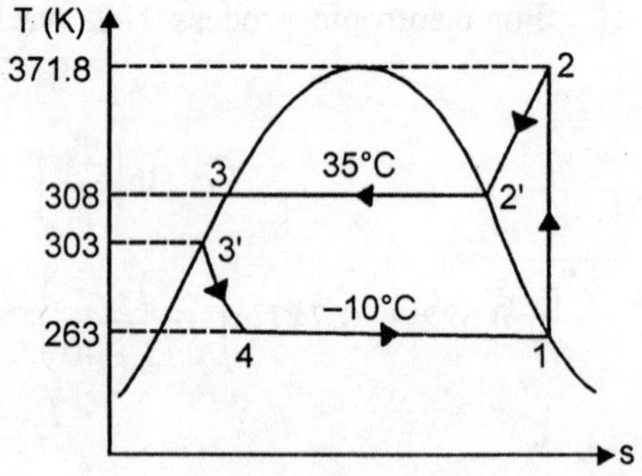

Also $s_2 = s_1$

$$s_2' + c_{ps} \ln \frac{T_2}{T_2'} = 5.7550$$

$$\ln \frac{T_2}{308} = \frac{5.7550 - 5.2086}{2.903} = 0.1882$$

$$T_2 = 371.8 \text{ K}$$

$$h_2 = h_2' + c_p(T_2 - T_2') = 1488.57 + 2.903\ (371.8 - 308)$$

$$= 1673.8 \text{ kJ/kg}$$

Fig. 5.47 T-s diagram.

Mass flow rate of refrigerant,

$$\dot{m}_r = \frac{50}{h_1 - h_4} = \frac{50}{1450.22 - 343.29} = 0.04517 \text{ kg/s}$$

(a) Power required $= \dot{m}_r(h_2 - h_1)$

$$= 0.04517\ (1673.8 - 1450.22) = 10.1 \text{ kW}$$

(b) Cylinder dimensions:

$$\dot{m} = \frac{\pi}{4} D^2 L \frac{N}{60} v_1$$

$$0.04517 = \frac{\pi}{4} \times 1.21 \times D^3 \times \frac{1000}{60} \times 0.417477$$

$$D^3 = 0.006888$$

$$D = 0.19 \text{ m}, L = 1.2 D = 0.228 \text{ m}$$

Example 5.35 *A 20 ton vapour compression system using Freon-12 operates between evaporator pressure of 1.004 bar and condenser pressure of 13.633 bar. The system uses a subcooling-superheating heat exchanger with 10°C of superheat added. Determine (i) the mass flow rate, (ii) the COP, (iii) the degree of subcooling, and (iv) the power required. The refrigerant leaving the condenser is dry saturated liquid and leaving the evaporator is dry saturated vapour. The compression in compressor is isentropic.* **[IES, 1998]**

Properties for R-12:

Pressure bar	*Temp. °C*	*Enthalpy kJ/kg*		*Entropy kJ/kg . K*		*Specific heat kJ/kg . K*	
		Liquid	*Vapour*	*Liquid*	*Vapour*	*Liquid*	*Vapour*
1.004	–30	8.854	174.076	0.0371	0.7165	0.573	0.579
13.663	55	90.201	207.766	0.3194	0.6777	1.074	–

Solution: Given: $Q = 20$ ton

$$T_1 = 273 - 30 = 243 \text{ K},$$

$$T_2 = T_3 = 273 + 35 = 328 \text{ K}$$

$$T_1' = T_1 + 10 = 253 \text{ K}$$

The vapour compression refrigeration cycle is shown in Fig. 5.48

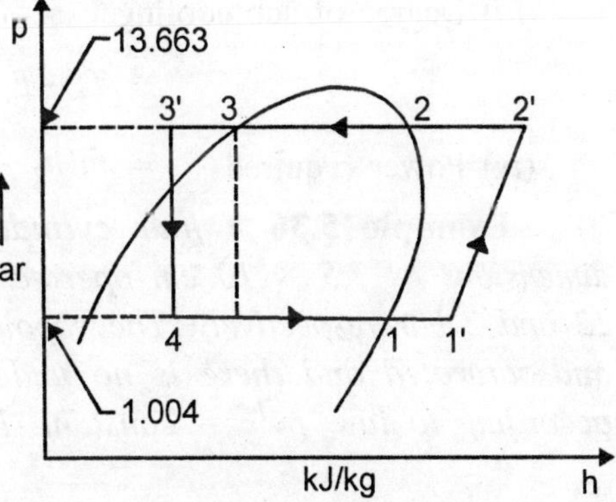

Fig. 5.48 p-h diagram.

For constant heat exchange, we have

$$h_1' - h_1 = h_3 - h_3'$$

$$c_{pv}(T_1 - T_1') = c_{pl}(T_3 - T_3')$$

$$0.579 \times 10 = 1.074\ (328 - T_3')$$

$$T_3' = 322.61 \text{ K}$$

$$s_1' = s_1 + c_{pv} \ln\left(\frac{T_1'}{T_1}\right) = 0.7165 + 0.579 \ln\left(\frac{253}{243}\right)$$

$$= 0.7398 \text{ kJ/kg. K}$$

$$s_2' = s_2 + c_{pv} \ln\left(\frac{T_2'}{T_2}\right)$$

$$T_2' = 365.14 \text{ K}$$

$$h_1' = h_1 + c_{pv}(T_1' - T_1)$$
$$= 174.076 + 0.579 \times 100 = 179.866 \text{ kJ/kg}$$

$$h_2' = h_2 + c_{pv}(T_2' - T_2)$$
$$= 207.766 + 0.579 \times (365.14 - 328)$$
$$= 229.27 \text{ kJ/kg}$$
$$h_3' = h_3 + c_{pl}(T_3' - T_3)$$
$$= 90.201 + 1.074 \times (322.61 - 328) = 84.4 \text{ kJ/kg}$$

(*i*) Refrigeration effect,

$$q = h_1' - h_3' = 179.866 - 84.41 = 95.456 \text{ kJ/kg}$$

$$Q = \dot{m}q$$

$$\dot{m} = \frac{Q}{q} = \frac{20 \times 211}{95.456} = 44.2 \text{ kg/min or } 0.737 \text{ kg/s}$$

(*ii*) $$\text{COP} = \frac{h_1' - h_3'}{h_2' - h_1'} = \frac{179.866 - 84.41}{229.87 - 179.865} = 1.932$$

(*iii*) Degree of sub-cooling

$$= T_3' - T_3 = 322.61 - 328 = -5.39°\text{C}$$

(*iv*) Power required $= \dot{m}(h_2' - h_1') = 0.737\,(229.27 - 179.866) = 36.41$ kW

Example 5.36 *A four cylinder single acting ammonia compressor with cylinder dimensions as 7.5 × 10 cm operates at 600 rpm. Condenser and evaporator pressure are 12 and 2 bar respectively. The vapour from the evaporator to suction of compressor is dry and saturated and there is no undercooling in the condenser. Compression takes place according to law, $pv^{1.2}$ = constant. If clearance is 2% of the stroke, calculate*

[IES, 2000]

(i) Refrigeration capacity in tons of refrigeration.
(ii) Power required to drive the compressor in kW.
(iii) Heat rejected in compressor water jacket.
(iv) Heat rejected to the condenser in kJ/min.

The properties of ammonia are:

t °C	p kPa	v_s, vapour m^3/kg	h_f, liquid kJ/kg	h_g, vapour kJ/kg	s_f, liquid kJ/kg. K	s_g, vapour kJ/kg. K
–19	200	0.5952	94.9	1421	–	5.6057
31	1200	–	327.6	1469	4.9724	–

Solution: Given:

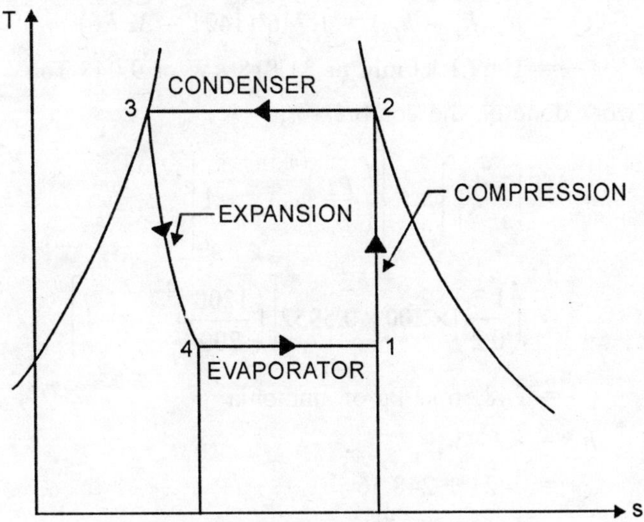

Fig. 5.49(*a*) T-s diagram.

$$L = 10 \text{ cm}, D = 7.5 \text{ cm}, N = 600 \text{ rpm},$$
$$p_1 = 2 \text{ bar}, p_2 = 12 \text{ bar}$$
$$c = 2\%, n = 1.2,$$
$$i = 4, \gamma = 1.31$$

Theoretical piston displacement,

$$V_p = \frac{\pi D^2 LNi}{4}$$
$$= \pi \times 7.5^2 \times 10 \times 10^{-6} \times 600 = 1.06 \text{ m}^3/\text{min}$$

Specific volume of refrigerant at point 1,

$$v_{g1} = 0.5952 \text{ m}^3/\text{kg}$$

Volumetric efficiency of compressor,

$$\eta_{ev} = 1 + c - c\left(\frac{p_2}{p_1}\right)^{1/n}$$

$$= 1 + 0.02 - 0.02\left(\frac{1200}{200}\right)^{1/1.2} = 0.931$$

Swept volume, $V_s = V_p \, \eta_{ev} = 1.06 \times 0.931 = 0.987 \text{ m}^3/\text{min}$

Mass flow rate of refrigerant,

$$\dot{m}_r = \frac{V_s}{v_{g1}} = \frac{0.987}{0.5952} = 1.746 \text{ kg/min}$$

The T-s and p-h diagrams are shown in Fig. 5.49.

(*i*) Refrigeration capacity

$$Q = \dot{m}_r (h_1 - h_{f3}) = 1.746\ (1421 - 327.6)$$
$$= 1909.1 \text{ kJ/min or } 31.818 \text{ kW or } 9.048 \text{ Ton}$$

Specific work done on the compressor,

$$w_c = \left[\frac{n}{n-1}\right] p_1 v_s \left[\left(\frac{p_2}{p_1}\right)^{(n-1)/n} - 1\right]$$

$$= \left(\frac{1.2}{0.2}\right) \times 200 \times 0.5952 \left[\left(\frac{1200}{200}\right)^{0.2/1.2} - 1\right]$$

$$= 248.56 \text{ kJ/kg of ammonia}$$

Enthalpy, $h_2 = h_1 + w_c$

$$= 1421 + 248.56$$
$$= 1669.56 \text{ kJ/kg}$$

(*ii*) Power required to drive the compressor

$$\dot{m}_r w_c = 1.746 \times 248.56$$
$$= 433.98 \text{ kJ/min or } 7.233 \text{ kW}$$

(*iii*) Heat rejected to compressor cooling water jacket

$= w_{\text{isentropic}} - w_{\text{actual}}$

$$w_{\text{isentropic}} = \left[\frac{\gamma}{\gamma-1}\right] p_1 v_s \left[\left(\frac{p_2}{p_1}\right)^{(n-1)/n} - 1\right]$$

$$= \left(\frac{1.31}{0.31}\right) \times 200 \times 0.5952 \left[\left(\frac{1200}{200}\right)^{0.31/1.31} - 1\right]$$

$= 265.54$ kJ/kg of ammonia

Heat rejected to compressor cooling water jacket

$$= 265.64 - 248.56$$
$$= 17.08 \text{ kJ/kg}$$
$$= 17.08 \times 1.746$$
$$= 29.82 \text{ kJ/min}$$

(*iv*) Heat rejected to condenser

$$= \dot{m}_r (h_2 - h_{f3}) = 1.746\ (1669.56 - 327.6)$$
$$= 2343.06 \text{ kJ/min or } 39.05 \text{ kW}$$

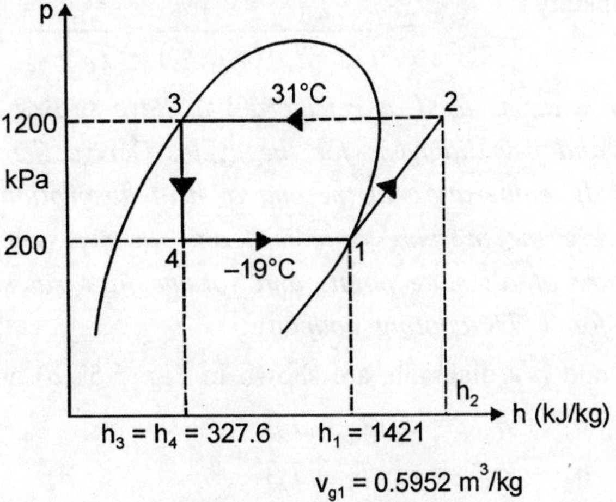

Fig. 5.49(*b*) p-h diagram.

Example 5.37 *If gas is used as refrigerant in reverse refrigeration cycle, draw T-s and p-v diagrams with heat rejection and heat absorption temperatures of T_g and T_c respectively. Show that this cycle requires two compressors and two expanders. Find their work requirement and the COP of the cycle.* **[IES, 2002]**

Solution: The *T-s* and *p-v* diagrams are shown in the Fig. 5.50. Processes 1-2 and 2-3 are isentropic compression and processes 3-4 and 4-1 are isentropic expansion. Therefore, it sequires two compressors and two expanders.

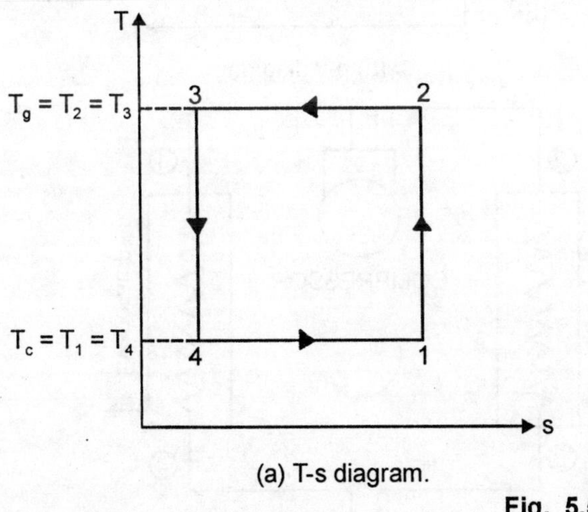

(a) T-s diagram.

(b) p-v diagram.

Fig. 5.50

Work done per cycle per kg of air,

$$q = q_r - q_a$$

$$= q_{2-3} - q_{4-1}$$

$$= (T_g - T_c)\,(s_2 - s_3)$$

$$\text{Work requirement} = mc_p\,(T_g - T_c)$$

$$(\text{COP})_{\text{ref}} = \frac{q_a}{\text{Work done}} = \frac{q_{4-1}}{q}$$

$$= \frac{T_c(s_2 - s_3)}{(T_g - T_c)(s_2 - s_3)} = \frac{T_c}{T_g - T_c}$$

Example 5.38 *Dense air is used as refrigerant in Reverse-Brayton or Bell-Coleman or Joule cycle. Draw T-s and p-v diagrams for the cycle. Derive the expression for COP in terms of pressure ratio. If temperature at the end of heat absorption and heat rejection are 0°C and 30°C respectively, the pressure ratio is 4 and the pressure in the cooler is 4 bar, determine the temperature at all state points and volume flow rates at inlet to compressor and at exit of turbine for 1 TR cooling coapcity.* **[IES, 2002]**

Solution: The *T-s* and *p-v* diagrams are shown in Fig. 5.51(*a*) and (b) respectively.

$$q_r = q_{2-3} = c_p (T_2 - T_3)$$

$$q_a = q_{4-1} = c_p (T_1 - T_4)$$

Wok done $= q_r - q_a = c_p (T_2 - T_3) - c_p (T_1 - T_4)$

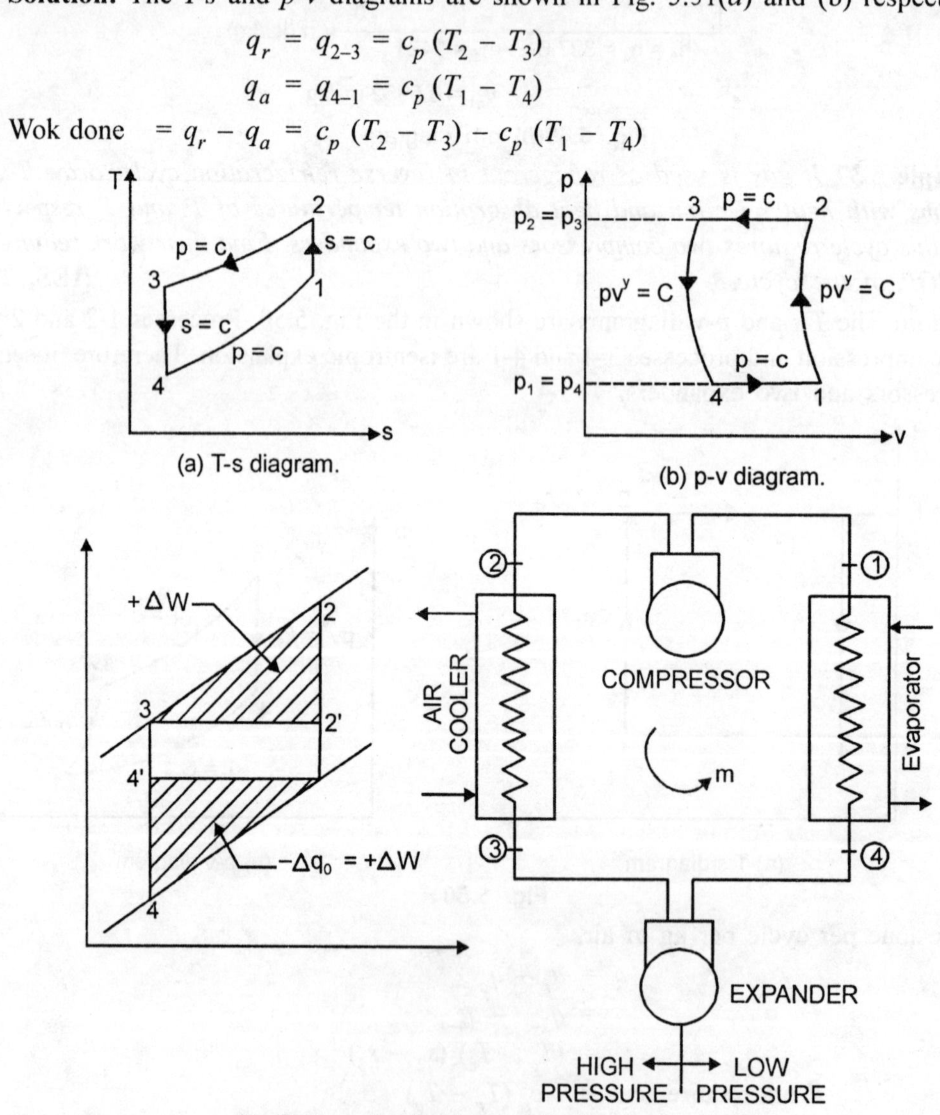

(a) T-s diagram.

(b) p-v diagram.

(*c*) Schematic diagram of Bell-Coleman Cycle.

Fig. 5.51

$$\text{COP} = \frac{q_a}{\text{Work done}} = \frac{q_{4-1}}{q}$$

$$= \frac{c_p(T_1 - T_4)}{c_p(T_2 - T_3) - c_p(T_1 - T_4)} = \frac{T_4\left(\frac{T_1}{T_4} - 1\right)}{T_3\left(\frac{T_2}{T_3}\right) - T_4\left(\frac{T_1}{T_4} - 1\right)}$$

For Isentropic compression process 1-2,

$$\frac{T_2}{T_1} = \left(\frac{p_2}{p_1}\right)^{(\gamma-1)/\gamma}$$

Similarly, $$\frac{T_3}{T_4} = \left(\frac{p_3}{p_4}\right)^{(\gamma-1)/\gamma} = \left(\frac{p_2}{p_1}\right)^{(\gamma-1)/\gamma}$$

Hence, $$\frac{T_2}{T_1} = \frac{T_3}{T_4}$$

or $$\frac{T_2}{T_3} = \frac{T_1}{T_4}$$

$$\text{COP} = \frac{T_4}{T_3 - T_4} = \frac{1}{\left(\frac{T_3}{T_4} - 1\right)} = \frac{1}{(r_p)^{\frac{\gamma-1}{\gamma}} - 1}$$

where r_p = compression or expansion ratio

$$r_p = \frac{p_2}{p_1} = \frac{p_3}{p_4} = 4$$

$$T_1 = 273 \text{ K}, T_3 = 303 \text{ K}$$

$$T_4 = \frac{303}{(4)^{0.4/1.4}} = 203.82 \text{ K}$$

Heat extracted per minute = 1 × 211 = 211 kJ/min

Heat extracted from cold chamber per kg of air

$$= c_p(T_1 - T_4) = 1.005 \times (273 - 203.82) = 69.526 \text{ kJ/kg}$$

Mass flow rate of air,

$$\dot{m}_a = \frac{211}{69.526} = 303 \text{ kg/min or } 182.1 \text{ kg/h}$$

Volume handled by the compressor,

$$V_c = \frac{m_a R T_1}{p_1} = \frac{182.1 \times 287 \times 273}{1 \times 10^5} = 142.68 \text{ m}^3\text{/h}$$

Volume handled by expander,

$$V_e = \frac{m_a R T_3}{p_3} = \frac{182.1 \times 287 \times 273}{4 \times 10^5} = 39.59 \text{ m}^3/\text{h}$$

Example 5.39 *An ideal vapour compression refrigerator using Freon-12 operates between temperature limits of –10°C and 40°C. The refrigerant leaves the condenser dry saturated. The rate of flow of refrigerant through the unit is 150 kg/hr.*

Calculate the refrigerating effect per kg of refrigerant, the co-efficient of performance and the capacity of the plant, if

(i) The refrigerant leaves the condenser dry saturated as stated,

(ii) The refrigerant is sub-cooled to 20°C before throttling. Assume that the enthalpy of refrigerant before throttling is approximately equal to the enthalpy of liquid refrigerant at undercooled temperature of 20°C and take it as h_f = 50.59 kJ/kg. Give your comments on the result. Solve the problem by using tables.

[IES, 2003]

Solution: $\dot{m}$ = 150 kg/h

The *p-h* diagram is shown in Fig. 5.52

From *p-h* chart for Freon-12, we have

$$h_1 = 183.05 \text{ kJ/kg}$$

$$s_1 = s_2 = 0.7014 \text{ kJ/kg.K}$$

$$h_3 = h_4 = 74.52 \text{ kJ/kg}$$

$$h_2 = 210 \text{ kJ/kg}$$

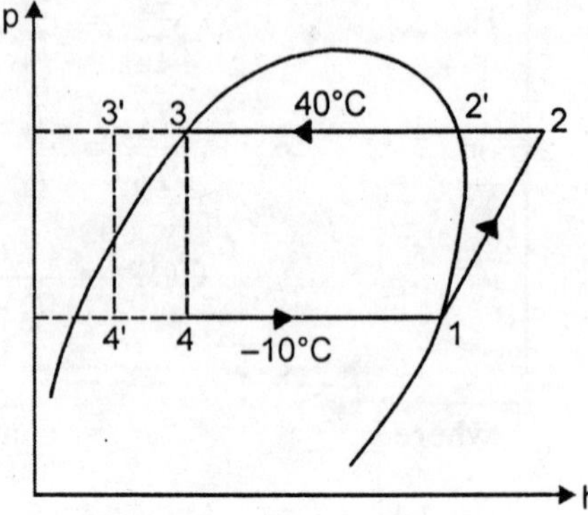

Fig. 5.52 p-h diagram.

(*i*) Refrigeration effect,

$$q_0 = h_1 - h_4 = 183.05 - 74.52$$

$$= 108.53 \text{ kJ/kg}$$

$$\text{COP} = \frac{h_1 - h_4}{h_2 - h_1} = \frac{108.53}{(210 - 183.05)} = 4.02$$

$$\text{Plant capacity} = \frac{m q_0}{3600} = \frac{150 \times 108.53}{3600} = 4.52 \text{ kW or } 1.2853 \text{ Ton}$$

(*ii*) $h_1 = 183.05$ kJ/kg, $h_2 = 210$ kJ/kg

$$h_3 = h_{f3} = 50.59 \text{ kJ/kg} = h_4' \text{ (Given)}$$

$$\text{Refrigeration effect, } q_0' = h_1 - h_4' = 183.05 - 50.59 = 132.46 \text{ kJ/kg}$$

$$\text{COP} = \frac{h_1 - h_4'}{h_2 - h_1} = \frac{132.46}{210 - 183.05} = 4.91$$

$$\text{Plant capacity} = \frac{m q_0'}{3600} = \frac{150 \times 132.46}{3600} = 5.52 \text{ kW or } 1.57 \text{ Ton}$$

With sub cooling, refrigerating capacity, COP and plant capacity increase.

Example 5.40 *What is the purpose of undercooling in vapour compression refrigeration system? Differentiate between dry and wet compression. What are the advantages of one over the other?*

[IES, 2004]

Solution: Undercooling. It is the cooling of refrigerant below the saturation temperature T_3' to T_3 before expansion by throttling, as shown in Fig. 5.53(*a*). The purpose of undercooling in vapour compression refrigeration system is to achieve the following:

1. To reduce flashing of liquid during expansion and to increase refrigeration effect and COP.
2. Piston displacement and power per ton are reduced for all refrigerants.
3. However, work of compression is increased.

Dry v's wet compression. The compression process is shown in Fig. 5.53(*b*). The wet compression involves the compression of wet refrigerant vapour at 1′ to dry saturated vapour at 1 or even slightly supersaturated is known as dry compression. The state of vapour at the end of compression is at 2 at saturation pressure p_k corresponding to the condensing temperature T_k. The refrigerant leaves the compressor superheated.

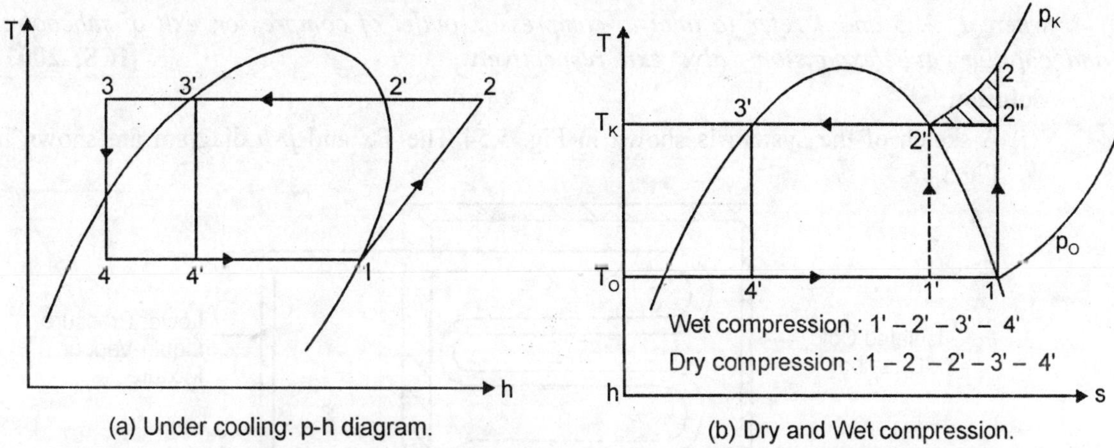

(a) Under cooling: p-h diagram. (b) Dry and Wet compression.

Fig. 5.53 Vapour Compression Refrigeration System.

The advantages of one over the other are

1. Power consumption per ton refrigeration in wet compression is less in case of ammonia as compared with dry compression.
2. In wet compression, liquid refrigerant may be trapped in the head of the cylinder and may damage the compressor valves and the cylinder, which is not so in dry compression.
3. Liquid refrigerants droplets in wet compression may wash the lubricating oil from the walls of the compressor cylinder, thus increasing wear.
4. Wet compression is suitable for screw compressors rather than reciprocating compressors.

Example 5.41 *In order to keep off foodstuff in better condition, a food storage locker requires 20 tons of refrigeration. The evaporator temperature is – 8°C and the condenser temperature is 30°C. The refrigerant used in the refrigeration plant is Fr-12. This is subcooled by 5°C before entering to the expansion valve. The vapour is superheated to 6°C before leaving the evaporator coil. The compressor employed is of two-cylinder, single acting having stroke equal to 1.5 times the bore. The speed of the compressor is 900 rpm. The compression process may be assumed to be isentropic. Work out the following:*

(i) Sketch the system and show the process on T-s and p-h diagrams;
(ii) The refrigeration effect per kg of refrigerant;
(iii) The mass flow rate of the refrigerant circulated.;
(iv) The theoretical piston displacement;
(v) The power input to compressor if mechanical efficiency is 0.9
(vi) The co-efficient of performance;
(vii) The bore and stroke if volumetric efficiency is 90%.

Use the following properties for the solution:

$$h_1 = 356 \text{ kJ/kg}$$
$$h_2 = 377 \text{ kJ/kg}$$
$$h_{f3} = h_4 = 225.5 \text{ kJ/kg}$$
$$v_1 = 0.077 \text{ m}^3\text{/kg}$$

where 1, 2, 3 and 4 refer to inlet to compressor, outlet of compressor, exit of subcoolrer and capillary tube/expansion valve exit respectively. **[IES, 2007]**

Solution:

(*i*) A sketch of the system is shown in Fig. 5.54 The *T-s* and *p-h* diagram are shown in Fig 5.55

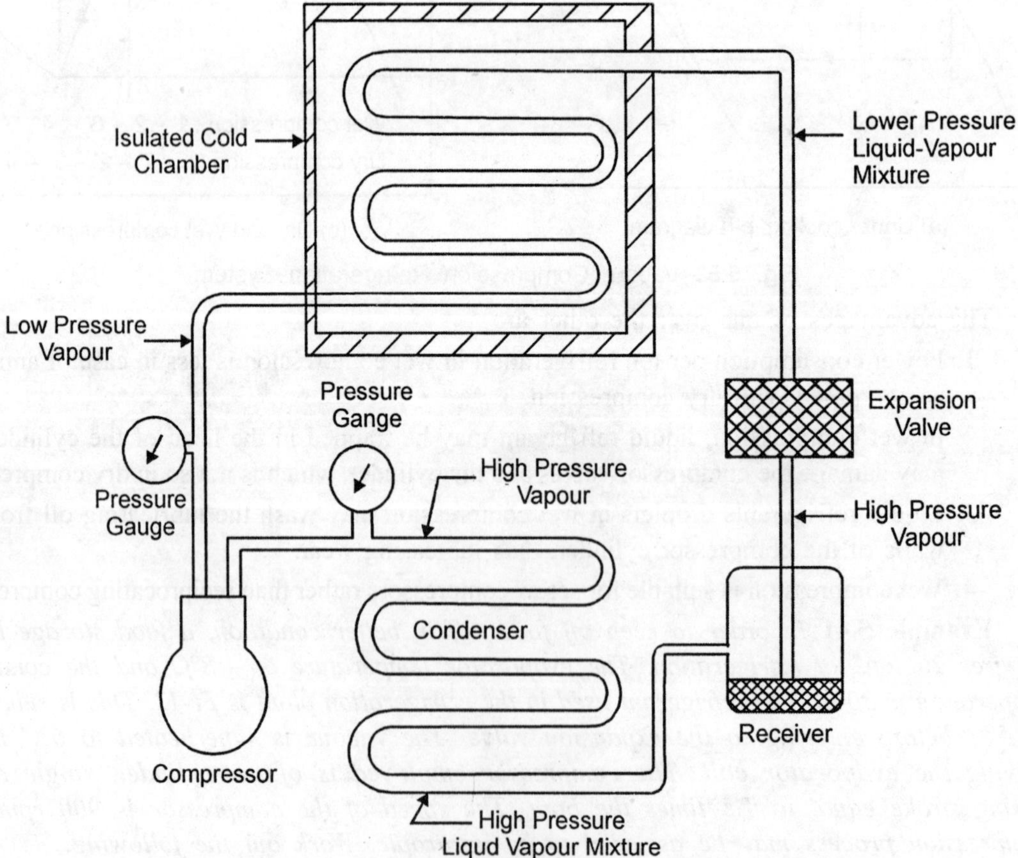

Fig. 5.54 Simple vapour compression refrigeration system.

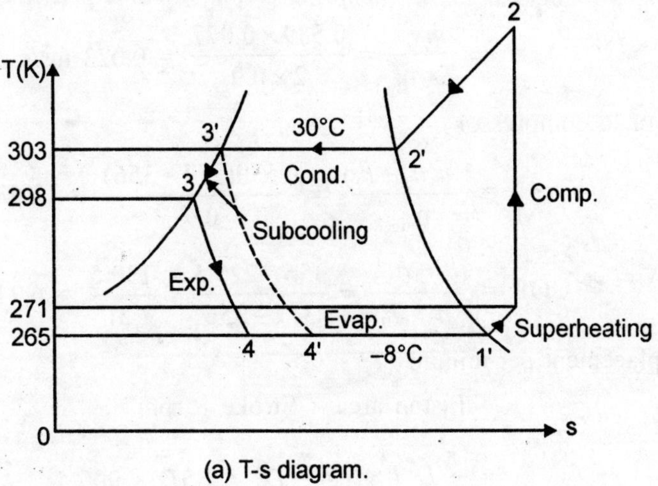

(a) T-s diagram.

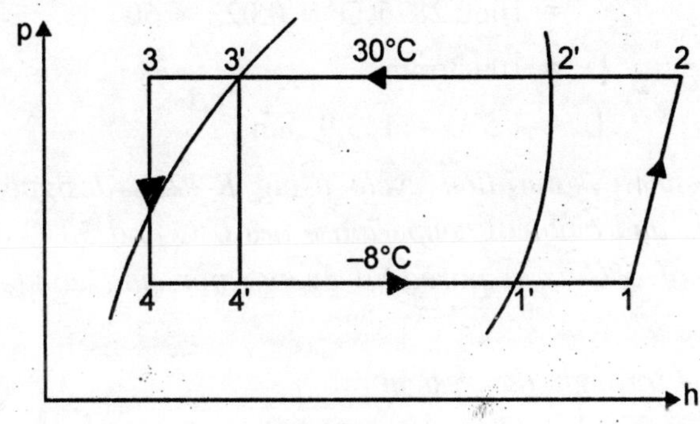

(b) p-h diagram.

Fig. 5.55

(*ii*) Refrigerating effect $= h_1 - h_4$

$= 356 - 225.5$

$= 130.5$ kJ/min

For 1 kg/min of refrigerant.

(*iii*) Refrigerating capacity,

$$\dot{m}(h_1 - h_4) = \text{TR} \times 3.5167$$

$$\dot{m}(356 - 225.5) = 20 \times 3.5167$$

$$\dot{m} = \frac{20 \times 3.5167}{130.5} = 0.539 \text{ kg/s}$$

(*iv*) Theoretical piston displacement per cylinder

$$= \frac{\dot{m} v_1}{2 \times \eta_v} = \frac{0.539 \times 0.077}{2 \times 0.9} = 0.023 \text{ m}^3/\text{s}$$

(*v*) Power input to compressor

$$= \frac{\dot{m}(h_2 - h_1)}{\eta_{\text{mech}}} = \frac{0.539(377 - 356)}{0.9} = 12.58 \text{ kW}$$

(*vi*)

$$\text{COP} = \frac{h_1 - h_{f3}}{h_2 - h_1} = \frac{356 - 225.5}{377 - 356} = \frac{130.5}{21} = 6.21$$

(*vii*) Piston displacement per minute

$$= \text{Piston area} \times \text{Stroke} \times \text{rpm}$$

$$= \frac{\pi}{4} D^2 L \times N = \frac{\pi}{4} D^2 \times 1.5D \times 900$$

$$= 1060.2875 \text{ D}^3 = 0.023 \times 60$$

$$\text{D} = 109.3 \text{ mm}$$

$$\text{L} = 1.5 \text{ D} = 163.9 \text{ mm}$$

Example 5.42 *A simple saturation cycle using R-12 is designed for taking a load of 10 tons. The refrigerants and ambient temperature are 0°C and 30°C respectively. A minimum temperature difference of 5°C is required in evaporator and condenser for heat transfer. Find the*

(*i*) *mass flow rate through the system*

(*ii*) *power required in kW*

(*iii*) *cylinder dimensions assuming L/D ratio of 1.2 for a single cylinder single acting compressor if it runs at 300 rpm with a volumetric efficiency of 0.9. The following properties are taken for R_{12}. Sp. Heat of R_{12} vap = 0.95 kJ/kg K.* **[IES, 2009]**

Sat temp °C	*Sat. pr* bar	*Sp. vol*		*Enthalpy*		*Entropy*	
		Sat. liq	*Sat. vap*	*Sat. liq*	*Sat. vap*	*Sat. liq*	*Sat. vap*
		× 10^{-3} m^3/kg		kJ/kg		kJ/kg.K	
– 5°C	2.61	0.71	0.0650	31.4	185.4	0.1251	0.6991
0°C	3.08	0.72	0.0554	36.1	187.5	0.1420	0.6966
30°C	7.45	0.77	0.0235	64.6	199.6	0.2399	0.6854
35°C	8.47	0.79	0.0206	69.5	201.5	0.2559	0.6839

Solution: Given: $Q = 10$ TR, $t_1 = -5°C$, $t_2 = 30°C$, $t_3 = 0°C$

$L/D = 1.2$, $N = 300$ rpm, $\eta_{\text{vol}} = 0.9$, $c_p = 0.95$ kJ/kg·K

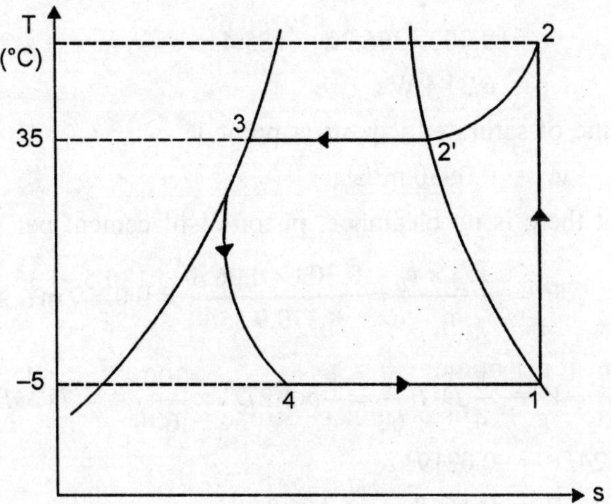

Fig. 5.56. *T-S* diagram.

As there is a difference of 5°C in both evaporator and condenser, so temperature of eneporator is – 5°C and that of condenser 35°C.

(*i*) The T-s diagram is shown in Fig 5.56

$$h_1 = 185.4 \text{ kJ/kg}, \; h_{f3} = h_4 = 69.5 \text{ kJ/kg}$$

Refrigerating effect,

$$RE = h_1 - h_{f3} = 185.4 - 69.5 = 115.9 \text{ kJ/kg}$$

Refrigerating capacity of plant,

$$\dot{Q} = \frac{10 \times 211}{60} = 35.167 \text{ kJ/s}$$

Mass flow rate of refrigerant,

$$\dot{m}_R = \frac{\dot{Q}}{RE} = \frac{35.167}{115.9} = 0.303 \text{ kg/s}$$

$$s_1 = s_2$$

$$= s_1 + c_p \ln \frac{T_2}{T_2'}$$

$$0.6991 = 0.6839 + 0.95 \ln \frac{T_2}{308}$$

$$T_2 = 312.97 \text{ K}$$

$$h_2 = h_2' + c_{pv}(T_2 - T_2')$$

$$= 201.5 + 0.95\,(312.97 - 308)$$

$$= 206.22 \text{ kJ/kg}$$

(*ii*) Power required, $P = \dot{m}_R(h_2 - h_1)$

$$= 0.303\ (206.22 - 185.4)$$

$$= 6.31\ \text{kW}$$

(*iii*) Specific volume of saturated vapour at point 1,

$$v_1 = 0.0650\ \text{m}^3/\text{kg}$$

Assuming that there is no clearance, piston displacement per second,

$$V = \frac{\dot{m}_R \times v_1}{\eta_v} = \frac{0.303 \times 0.0650}{0.9} = 0.0219\ \text{m}^3/\text{s}$$

Also $$V = \frac{\pi}{4} D^2 L \frac{N}{60} = \frac{\pi}{4} \times 1.2 D^3 \times \frac{300}{60} = 4.7124 D^3\ \text{m}^3/\text{s}$$

$$\therefore \quad 4.7124 D^3 = 0.0219$$

$$D = 0.1668\ \text{m or } 166.8\ \text{mm}$$

$$L = 1.2D = 1.2 \times 166.5 = 250\ \text{mm}$$

Example 5.43 *Saturated ammonia vapour enters a 15 cm dia × 14 cm stroke twin-cylinder single-acting compressor at 0.2365 MPa whose volumetric efficiency is 79% speed 420 r.p.m. The delivery pressure is 1.1672 Mpa. Liquid NH_3 at 21°C enters the expansion valve. For ideal cycle, find* **[IES, 2010]**

(i) the ammonia circulated in kg/min,

(ii) the refrigeration in tons and

(iii) COP of the cycle.

Assume sp. heat of NH_3 as 219 kJ/kg-K.

Properties of NH_3 are given below:

Pressure (Mpa)	*Sat. Temp. (°C)*	*Sp. volume of vap. at sat. (m³/kg)*	*Enthalpy kJ/kg-K*		*Entropy (kJ/kg-K)*		*Specific heat kJ/kg-K*	
			Sat. liq	*Sat. vap.*	*Sat. liq*	*Sat. vap.*	*Sat. liq*	*Sat. vapour*
0.2365	–15	0.5106	112.39	1426.58	0.4572	5.5497	4.396	2.303
1.1672	+30	0.1107	323.08	1468.87	1.2037	4.9842	4.60	2.19

Solution: Given : $D = 15$ cm, $L = 14$ cm, $n = 2$, $\eta_v = 0.79$, N = 420 rpm, $T_2' = T_3' = 303$ K,

$T_3 = 294$ K, $h_1 = 1426.58$ kJ/kg, $h_2' = 1468.87$ kJ/kg, $h_3' = 323.08$ kJ/kg,

$c_{pv} = 2.19$ kJ/kg·K, $c_{pl} = 4.60$ kJ/kg·K.

The *T-s* diagram is shown in Fig. 5.57.

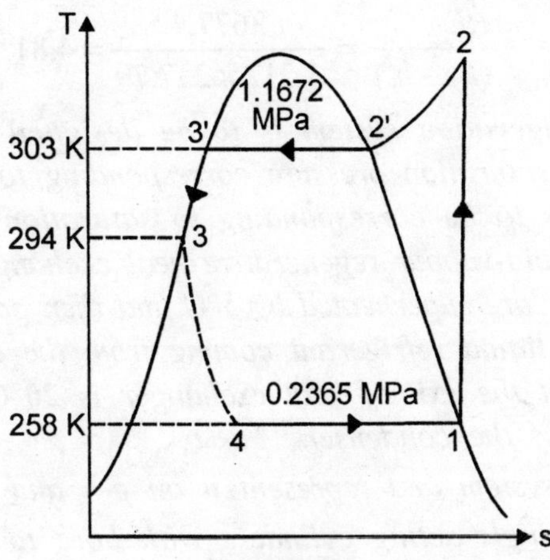

Fig. 5.57

$$s_1 = s_2 = s_2' + c_{pv} \ln \frac{T_2}{T_2'}$$

$$5.5497 = 4.9842 + 2.19 \ln \frac{T_2}{303}$$

$$T_2 = 392.27 \text{ K}$$

$$h_2 = h_2' + c_{pv}(T_2 - T_2') = 1468.87 + 2.19(392.27 - 303)$$

$$= 1664.37 \text{ kJ/kg}$$

$$h_{f3} = h_{f3}' - c_{pl}(T_3' - -T_3) = 323.08 - 4.60\ (303 - 294) = 281.68 \text{ kJ/kg}$$

(*i*) Volume of ammonia circulated,

$$V = \frac{\pi}{4} D^2 LN \times n \times \eta_v = \frac{\pi}{4}(0.15)^2 \times 0.14 \times 420 \times 2 \times 0.79$$

$$= 1.64 \text{ m}^3\text{/min}$$

Specific volume of ammonia at entry,

$$v_1 = 0.5106 \text{ m}^3\text{/kg}$$

Mass of ammonia circulated,

$$\dot{m} = \frac{V}{v_1} = \frac{1.64}{0.5106} = 3.212 \text{ kg/min}$$

(*ii*) Refrigeration effect,

$$RE = \dot{m}(h_1 - h_4) = \dot{m}(h_1 - h_{f3})$$

$$= 3.212(1426.58 - 281.68) = 3677.4 \text{ kJ/min}$$

$$= \frac{3677.4 \times 60}{14000} = 15.76 \text{ ton}$$

(*iii*) Work done $= h_2 - h_1 = 1664.37 - 1626.58 = 237.79$ kJ/kg

$$\text{COP} = \frac{RE}{\dot{m} \times (h_2 - h_1)} = \frac{3677.4}{3.212 \times 237.79} = 4.81$$

Examples 5.44 *A refrigeration system is to be designed for a cooling capacity of 7.5 tons of refrigeration at saturation pressure corresponding to −20°C. It uses refrigerant R-22. Condenser pressure is to be corresponding to saturation temperature of 40°C. The refrigerant system uses a liquid-vapour regenerative heat exchanger. The refrigerant vapours coming out of the evaporator are superheated by 5°C and then pass through the regenerative heat exchanger to cool the liquid refrigerant coming from the condenser. The temperature of the refrigerant vapours at the exit of heat exchanger is 20°C. The liquid refrigerant is subcooled to 36°C at exit of the condenser.*

Make a sketch of the system and represent it on p-h and T-S diagram.

The system uses two single-acting cylinders with bore to stroke ratio 0.8, speed of compressor is 1420 rpm, clearance factor for compressors is 0.04 and polytropic index of compression is 1.1. Mechanical Efficiency of the compressor is 0.8.

Determine:

(*i*) *temperature of liquid refrigerant at the exit of regenerative heat exchanger, specific heat of the liquid to be 1.37 kJ/kg.K.*

(*ii*) *mass flow rate of the refrigerant.*

(*iii*) *the temperature of the refrigerant vapours at the exit of compressor, assuming ideal gas behaviour in superheat region, specific heat of refrigerant vapours in the superheat region may be taken as 0.85 kJ/kg K.*

(*iv*) *Power input to the compressor.*

(*v*) *COP of the system.*

(*vi*) *Volumetric efficiency of the compressor, and*

(*vii*) *Bore and stroke of the compressor.*

The following properties of R-22 are given: **[IES, 2011]**

Sat. temp. t_s (°C)	*Sat. pressure* p_s *(bar)*	*Sp. Vol. of Sat. vapour* (m^3/kg)	*Enthalpy (kJ/kg)*	
			Sat. liquid h_f	*Sat. vapour* h_g
−20	2.448	0.0928	177.4	397.5
40	15.335	0.0151	249.08	415.95

For superheated refrigerant vapour at 2.448 bar and saturation temperature of −20°C

Temperature °C	*Specific volume*	*Enthalpy (kJ/kg)*
−15	0.0951	400.7
20	0.1107	423.9

Solution: Given: $RE = 7.5\ TR$, $i = 2$, $\dfrac{D}{L} = 0.8$, $c = 0.04$, $n = 1.1$, $\eta_{mech} = 0.8$, N = 1420 rpm

Refer to Fig 5.58.

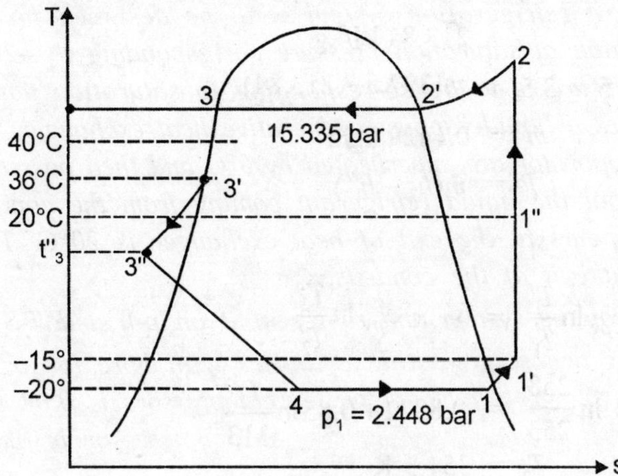

Fig. 5.58 *T-s* diagram.

1 – 1′	:	Superheating	3′ – 3″	:	Heat exchanger
1′ – 1″	:	Heat exchanger			
1″ – 2	:	Compressor	03″ – 4	:	Throttling
2′ – 3	:	Condenser	4 – 1	:	Evaporator
3 – 3′	:	Subcooling			

$$\eta_v = 1 + c - c\left(\frac{p_2}{p_1}\right)^{\frac{1}{n}} = 1 + 0.04 - 0.04\left(\frac{15.335}{2.448}\right)^{1/1.1}$$

$$= 0.8279$$

From given table, $v_1 = 0.0928\ \text{m}^3/\text{kg}$

$$\eta_v = \frac{\dot{m}v_1}{\frac{\pi}{4}D^2 L \frac{N}{60} i}$$

$$\dot{m} = 0.8279 \times \frac{\pi}{4} \times 1.25^3 D \times \frac{1420}{60} \times 2 \times \frac{1}{0.0928}$$

$$\dot{m}/D^3 = 414.57$$

(*i*) $c_{pl} = 1.37\ \text{kJ/kg·K}$, $c_{pv} = 0.85\ \text{kJ/kg·K}$

From heat balance for the heat exchanger, we have

$$c_{pl}(T_3' - T_3'') = c_{pv}(T_1'' - T_1')$$

$$1.37(309 - T_3'') = 0.85(293 - 258)$$

$$T_8'' = 287.28\ \text{K}$$

$$RE = \dot{m}(h_1 - h_4)$$

$$h_1 = 397.5 \text{ kJ/kg}$$

$$h_4 = h_3'' = h_{f3} - c_{pl}(T_3 - T_3'') = 249.08 - 1.37(313 - 287.28)$$

$$= 213.84 \text{ kJ/kg}$$

$$7.5 \times 3.5 = \dot{m}(397.5 - 213.84)$$

$$\dot{m} = 0.1429 \text{ kg/s}$$

(*ii*) Power input, $$P = \dot{m}(h_2 - h_1'')$$

$$s_1' = s_2$$

$$s_1 + c_{pv} \ln \frac{T_1'}{T_1} = s_2' + c_{pv} \ln \frac{T_2}{T_2'}$$

$$0.9638 + 0.85 \ln \frac{258}{253} = 0.8822 + 0.85 \ln \frac{T_2}{313}$$

$$T_2 = 351.3 \text{ K}$$

$$P = \dot{m} c_{pv}(T_2 - T_1'')$$

$$= 0.1429 \times 0.85\ (351.3 - 293)$$

$$= 7.08 \text{ kW}$$

(*iii*) $$\text{COP} = \frac{RE}{P} = \frac{7.5 \times 3.5}{7.08} = 3.70$$

(*iv*) $$\frac{\dot{m}}{D^3} = 414.57$$

$$D^3 = \frac{0.1429}{414.57}, \quad D = 0.070 \text{ m or } 70 \text{ mm}$$

$$L = 1.25 \times 0.070$$

$$= 0.087 \text{ m or } 87 \text{ mm}$$

Examples 5.45 *An R-12 vapour compression plant producing 10 tonnes of refrigeration operates with condensing and evaporating temperatures of 35°C and –10°C respectively. A suction line heat exchanger is used to subcool the saturated liquid leaving the condenser. Saturated vapour leaving the evaporator is superheated in the section line heat exchanger to the extent that a discharge temperature of 60°C is obtained after isentropic compression, Calculate*

(*i*) *The subcooling achieved in the heat exchanger,*

(*ii*) *The refrigerant flow rate in kg/s,*

(*iii*) *The cylinder dimensions of the two-cylinder compressor, if the speed is 900 rpm, Stroke-to-bore ratio is 1 : 1 and the volumetric efficiency is 80%,*

(*iv*) *The COP of the plant, and*

(*v*) *The power required to drive the compressor in kW.*

Draw the cycles on p-h and T-s diagrams

You may use the following table:

Temperature (°C)	*Specific volume of Vapour (v_g) (m^3/kg)*	*Specific Enthalpy (kJ/kg)*		*Specific entropy kJ/kg-K*	
		Liquid (h_f)	*Vapour ((h_g)*	*Liquid (s_f)*	*Vapour (s_g)*
–10	0.0766	190.88	347.13	0.9660	1.5600
+35	0.0206	233.50	365.92	1.1139	1.5419

Average specific heat of desuperheating vapour at condenser = 0.796 kJ/kg-K.

Average specific heat of vapour between evaporator outlet and compression suction = 0.658 kJ/kg-K. **[IES, 2012]**

Solution: Given : $Q = 10\ TR,\ T_1' = -10°C = -10 + 273 = 263$ K,

$$T_2' = T_3' = 35°C = 35 + 273 = 308 \text{ K.}$$

$$T_2 = 60°C = 60 + 273 = 333 \text{ K},\ N = 900 \text{ rpm},\ \frac{L}{D} = 1,\ \eta_{vol} = 80\%,$$

$$v_1' = 0.0766 \text{ m}^3/\text{kg},\ v_2' = 0.0206 \text{ m}^3/\text{kg},\ h_{f1}' = 190.88 \text{ kJ/kg},$$

$$h_{f3}' = 233.50 \text{ kJ/kg},\ h_1' = 347.13 \text{ kJ/kg},\ h_2' = 365.92 \text{ kJ/kg},$$

$$s_{f1}' = 0.9660 \text{ kJ/kg·K},\ s_{f3}' = 1.1139 \text{ kJ/kg·K},\ s_1' = 1.5600 \text{ kJ/kg·K},$$

$$s_2' = 1.5419 \text{ kJ/kg·K},\ c_{pl} = 0.796 \text{ kJ/kg·K},\ c_{pv} = 0.658 \text{ kJ/kg·K}$$

The *T-s* and *p-h* diagrams are shown in Fig. 5.59 (*a*) and (*b*) respectively.

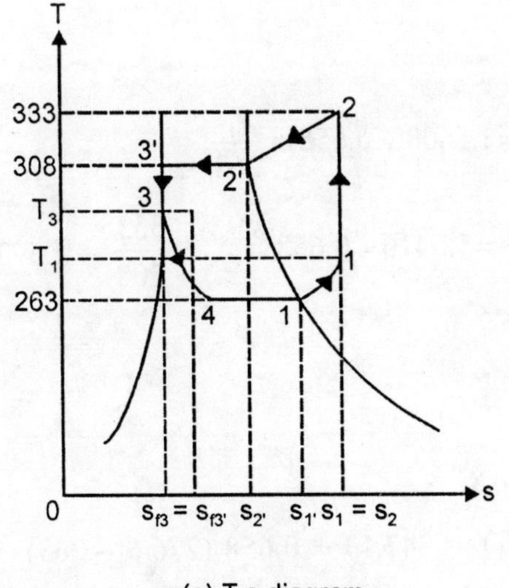

(a) T-s diagram.

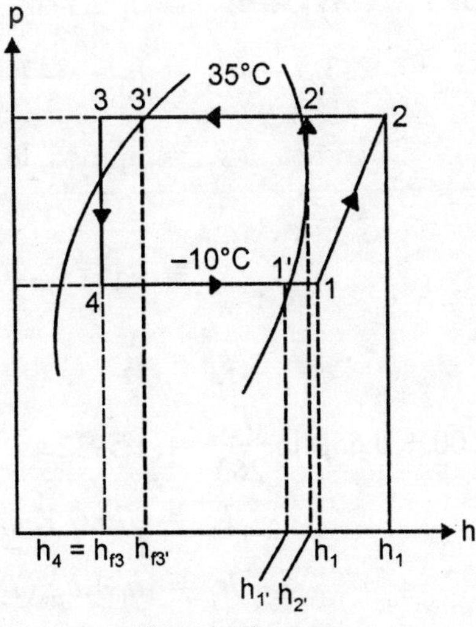

(b) p-h diagram.

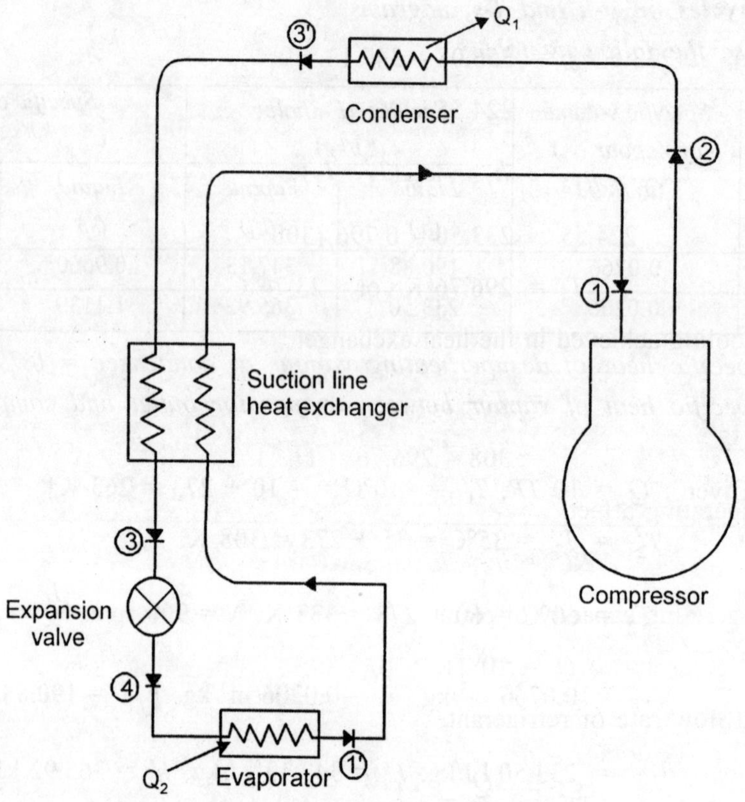

(c) Schematic diagram of vapour compression plant.

Fig. 5.59

Suction line heat exchanger: For heat balance, we have

$$h_{f3'} - h_{f3} = h_1 - h_1'$$

$$233.5 - h_{f3} = h_1 - 347.13 \quad ...(1)$$

$$s_1 = s_1' + c_{pv} \ln \frac{T_1}{T_1'} = 1.5600 + 0.658 \ln \frac{T_1}{263}$$

$$s_2 = s_2' + c_{pv} \ln \frac{T_2}{T_2'} = 1.5416 + 0.658 \ln c_{pv} \ln \frac{333}{308} = 1.5932 \text{ kJ/kg} \cdot \text{K}$$

$$s_1 = s_2$$

$$\therefore 1.5600 + 0.658 \ln \frac{T_1}{263} = 1.5932$$

$$T_1 = 276.6 \text{ K or } 3.6°\text{C}$$

$$h_1 = h_1' + c_{pv}(T_1 - T_1') = 347.13 + 0.658\ (276.6 - 263)$$

$$= 356.08 \text{ kJ/kg}$$

Eq. (1) gives,

$$233.5\, h_{f3} = 356.08 - 347.13$$

$$h_{f3} = 224.55 \text{ kJ/kg}$$

Now
$$h_{f3} = h_{f3'} - c_{pl}(T_3' - T_3)$$

$$224.55 = 233.50 - 0.796\,(308 - T_3)$$

$$T_3 = 296.76 \text{ K} \quad \text{or} \quad 23.76°\text{C}$$

(*i*) Subcooling achieved in the heat exchanger

$$= T_3' - T_3$$

$$= 308 - 296.76 = 11.24°\text{C}$$

(*ii*) Refrigerating effect,

$$RE = h_1 - h_{f3} = 356.08 - 224.55 = 131.53 \text{ kJ/kg}$$

Refrigerating capacity of plant

$$Q = 10 \text{ TR} = 10 \times 211 = 2110 \text{ kJ/kg}$$

Mass flow rate of refrigerant,

$$\dot{m}_r = \frac{Q}{RE} = \frac{2110}{131.53} = 16.04 \text{ kg/min}$$

(*iii*) Theroretical suction volume or piston displacement per minute

$$V_1 = \dot{m}_r \times v_1 \times \frac{1}{\eta_{vol}}$$

Now
$$\frac{v_1}{T_1} = \frac{v_1'}{T_1'}$$

or
$$v_1 = v_1' \times \frac{T_1}{T_1'} = 0.0766 \times \frac{276.6}{263} = 0.0806 \text{ m}^3\text{/kg}$$

$$V_1 = 16.04 \times 0.0806 \times \frac{1}{0.8} = 1.616 \text{ m}^3\text{/min}$$

Number of cylinders = 2

∴ Piston displacement per cylinder

$$= \frac{1.616}{2} = 0.808 \text{ m}^3\text{/min}$$

$$= \frac{\pi}{4} \times D^2 \times D \times N = \frac{\pi}{4} D^3 \times 900$$

$$= 706.86\ D^3\ \text{m}^3/\text{min}$$

$$\therefore \quad L = D = 0.1046\ \text{m} \quad \text{or} \quad 104.6\ \text{mm}$$

(*iv*) $$\text{COP} = \frac{h_1 - h_{f3}}{h_2 - h_1}$$

$$h_2 = h_2' + c_{pv}(T_2 - T_2') = 365.92 + 0.658\ (333 - 308)$$

$$= 382.37\ \text{kJ/kg}$$

$$\text{COP} = \frac{356.08 - 224.55}{382.37 - 356.08} = \frac{131.53}{26.29} = 5.003$$

(*v*) Work done during compression of refrigerant

$$= \dot{m}_r (h_2 - h_1) = 16.04\ (382.37 - 356.08) = 421.69\ \text{kJ/min}$$

Theoretical power required

$$= \frac{421.69}{60} = 7.028\ \text{kW}$$

Example 5.46 *A refrigerator with evaporator and condenser temperatures of –10°C and 32°C uses R-11. The vapour is superheated by 8°C at entry to the compressor and the superheating is obtained by under cooling the liquid before entry to evaporator. Assuming isentropic compression, determine the power required for a cooling load of 350 kJ/s, the overall efficiency for power use being 75%.*

Discuss about the choice of the type of compressor for this plant. Properties values are tabulated below: **[CSE, 1998]**

Temperature °C	*Specific volume of vapour* m^3/kg	*Enthalpy kJ/kg*		*Entropy of vapour kJ/kg K*	*Specific heat of vapour kJ/kg K*
		Liquid	*Vapour*		
–10	0.61141	191.44	384.64	1.7023	0.562
32	0.13104	227.96	406.23	1.6807	0.620

Solution: Given: $Q = 350\ \text{kJ/s},\ \eta_0 = 0.75$

$$T_1' = -10°\text{C or } 263\ \text{K},\ T_2' = 32° \text{ or } 305\ \text{K}$$

$$T_1 - T_1' = 8°\text{C or } T_1 = -2°\text{C or } 271\ \text{K}$$

$$c_{pv1'} = 0.562\ \text{kJ/kg} \cdot \text{K},\ c_{pv2'} = 0.620\ \text{kJ/kg} \cdot \text{K}$$

Refer to Fig. 5.60.

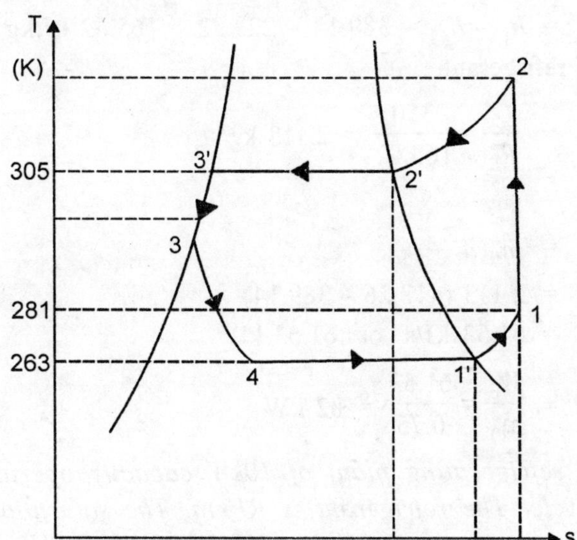

Fig. 5.60 T-s diagram.

Now $$s_1 = s_1' + c_{pv1'} \ln \frac{T_1}{T_1'}$$

$$= 1.7023 + 0.562 \ln \frac{271}{263}$$

$$= 1.719 \text{ kJ/kg} \cdot \text{K}$$

$$s_1 = s_2 \text{ for isentropic compression}$$

$$1.719 = s_2' + c_{pv2'} \ln \frac{T_2}{T_2'}$$

$$= 1.6807 + 0.620 \ln \frac{T_2}{305}$$

$$T_2 = 324.4 \text{ K}$$

$$h_1 = h_1' + c_{pv1'}(T_1 - T_1') = 384.64 + 0.562 \times 8 = 389.14 \text{ kJ/kg}$$

$$h_2 = h_2' + c_{pv2'}(T_2 - T_2') = 406.23 + 0.620\ (324.4 - 305) = 418.26 \text{ kJ/kg}$$

From properties for R-11 refrigerant,

$$c_{pl3'} = 0.888 \text{ kJ/kg} \cdot \text{K}$$

Heat added during superheating at 1′ = Heat rejected during under cooling at 3′.

$$\therefore \quad c_{pv1'}(T_1 - T_1') = c_{pl3'}(T_3' - T_3)$$

$$0.562 \times 8 = 0.888\ (305 - T_3)$$

$$T_3 = 300 \text{ K}$$

$$h_{f3} = h_{f3'} - c_{pl3'}(T_3' - T_3) = 227.96 - 0.888(305 - 300) = 223.52 \text{ kJ/kg}$$

Refrigerating effect,

$$R_E = h_1 - h_{f3} = 389.14 - 223.52 = 165.62 \text{ kJ/kg}$$

Mass flow rate of refrigerant,

$$\dot{m}_r = \frac{Q}{R_E} = \frac{350}{165.62} = 2.113 \text{ kg/s}$$

Compressor work,

$$W_c = \dot{m}_h(h_2 - h_1)$$

$$= 2.113\ (418.26 - 389.14)$$

$$= 61.53 \text{ kJ/s or } 61.53 \text{ kW}$$

$$\text{Power required, } P = \frac{W_c}{\eta_0} = \frac{61.53}{0.75} = 82 \text{ kW}$$

Example 5.47 *A refrigerating plant of 10kW capacity operates in the quasi-ideal vapour compression cycle. The refrigerant is R134a. The saturation temperatures in the evaporator and condenser are respectively – 20°C and 40°C. The plant includes a heat exchanger in which the saturated liquid leaving the condenser is subcooled to 30°C by the saturated vapour leaving the evaporator. The sub cooled liquid is then throttled to –20°C before entering the evaporator and the vapour leaving the heat exchanger is compressed isentropically to the condenser pressure. Sketch the schematic diagram and the cycle on p-h and T-s diagrams, numbering the important points. Calculate the mass flow rate of the refrigerant and power to be supplied to the compressor. The mechanical efficiency of the compressor may be assumed 80%.*

The compressor used by the plant is a single-stage, single-acting reciprocating compressor having stroke-diameter ratio of 1.2 and a speed of 240 rev/min. Assuming compressor volumetric efficiency to be 80%, determine the diameter and stroke of the compressor.

Properties of refrigerant may be obtained from table below: **[CSE, 2000]**

t °C	p_s *bar*	v_g m^3/kg	*Enthalpy, kJ/kg*		*Entropy, kJ/kg-K*		*Specific heat, kJ/kg – K*	
			Liquid	*Vapour*	*Liquid*	*Vapour*	*Liquid*	*Vapour*
–20	13268	0.1474	173.82	386.66	0.9009	1.7417	1.282	0.805
40	10.165	0.0199	256.35	419.58	1.1903	1.7115	1.5	1.12

Solution: Given:

$$Q = 10 \text{ kW},\ T_1' = -20°\text{C or } 253 \text{ K},\ T_2' = 40°\text{C or } 313 \text{ K},\ T_3 = 30°\text{C or } 303 \text{ K},$$

$$T_4 = -20°\text{C or } 253 \text{ K},\ (\eta_{mech})_c = 0.80,\ h_1' = 386.66 \text{ kJ/kg},\ h_2' = 419.58 \text{ kJ/kg},$$

$$s_1' = 1.7417 \text{ kJ/kg}\cdot\text{K},\ s_2' = 1.7115 \text{ kJ/kg}\cdot\text{K},\ h_{f3'} = 256.35 \text{ kJ/kg},$$

$$c_{pv1'} = 0.805 \text{ kJ/kg}\cdot\text{K},\ c_{pv2'} = 1.12 \text{ kJ/kg}\cdot\text{K},\ c_{pl3'} = 1.5 \text{ kJ/kg}\cdot\text{K}$$

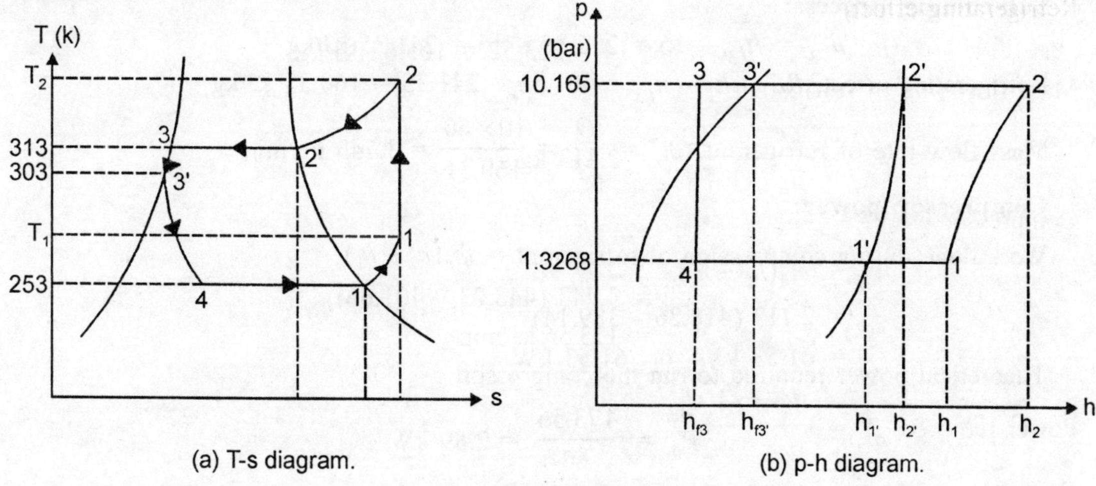

(a) T-s diagram. (b) p-h diagram.

Fig. 5.61

The *T-s* and *p-h* diagram are shown in Fig. 5.61 (*a*) and (*b*) respectively

Heat exchanger:

$$h_{f3'} - h_{f3} = h_1 - h_1'$$

$$h_{f3} = h_{f3'} - c_{pl3'}(T_3 - T_3')$$

or $$h_{f3'} - h_{f3} = c_{pl3'}(T_3 - T_{3'}) = 1.5\ (393 - 103) = 15$$

$$h_1 = h_1' + c_{pv1'}(T_1 - T_1')$$

or $$h_1 - h_1' = c_{pv1'}(T_1 - T_1') = 0.805\ (T_1 - 253)$$

$$0.805\ (T_1 - 253) = 15$$

$$T_1 = 271.63 \text{ K}$$

$$h_1 = h_1' + c_{pv1'}(T_1 - T_1') = 386.66 + 0.805\ (271.63 - 253) = 401.66 \text{ kJ/kg}$$

Isentropic compression 1-2:

$$s_1 = s_1' + c_{pv1'} \ln\frac{T_1}{T_1'} = 1.7417 + 0.805 \ln\frac{(271.63)}{253} = 1.7989 \text{ kJ/kg.K}$$

$$s_2 = s_2' + c_{pv2'} \ln\frac{T_2}{T_2'} = 1.7115 + 1.12 \ln\frac{T_2}{313}$$

$$s_1 = s_2$$

$$\therefore \quad 1.7989 = 1.7115 + 1.12\ \ln\frac{T_2}{313}$$

$$T_2 = 338.4 \text{ k}$$

$$h_2 = h_2' + c_{pv2'}(T_2 - T_2') = 419.58 + 1.12\ (338.4 - 313)$$

$$= 448.03 \text{ kJ/kg.K}$$

Sub-cooling process:

$$h_{f3} = h_{f3'} - 15 = 256.35 - 15 = 241.35 \text{ kJ/kg}$$

Refrigerating effect, $RE = h_1 - h_{f3} = 401.66 - 241.35 = 160.31$ kJ/kg

Mass flow rate of refrigerant, $\dot{m}_r = \dfrac{Q}{RE} = \dfrac{10 \times 60}{160.31} = 3.743$ kg/min

Compressor power:

Work done during compression of refrigerant $= \dot{m}_r(h_2 - h_1)$

$$= 3.743\ (448.03 - 401.66)$$

$$= 173.56 \text{ kJ/min}$$

Theoretical power required to run the compressor,

$$P_{th} = \frac{173.56}{60} = 2.89 \text{ kW}$$

Actual compressor power, $P_a = \dfrac{2.89}{0.8} = 3.616$ kW

Compressor dimensions

$$\frac{L}{D} = 1.2, \quad N = 240 \text{ rpm}, \quad \eta_{vol} = 0.8$$

Specific volume at 1′, $V_1' = 0.1474$ m^3/kg

$$\frac{V_1'}{T_1'} = \frac{V_1}{T_1}, \quad V_1 = 0.1474 \times \frac{271.63}{253} = 0.15825 \text{ m}^3/\text{kg}$$

Theoretical suction volume or piston displacement per minute

$$= \frac{\dot{m}_r \times V_1}{\eta_{vol}} = \frac{3.743 \times 0.15825}{0.8} = 0.7404 \text{ m}^3/\text{min}$$

$$\therefore \quad \frac{\pi}{4} \times 1.2D^3 N = \frac{\pi}{4} \times 1.2D^3 \times 240 = 0.7404$$

$$D = 0.1485 \text{ m}, \quad L = 0.1782 \text{ m}$$

Example 5.48 *Calculate the (i) c.o.p (ii) power requirement and (iii) cooling capacity of a CO_2 compressor working between 22.68 bar and 64.32 bar pressures. The liquid in the condenser gets cooled, by another system, to 15°C before entering the expansion valve. The system is assumed to work on wet-compression with x = 0.9. The compressor data: Stroke volume = 500 c.c., r.p.m = 500 and η_{vol} = 0.85. Property values are: sp. heat of CO_2 vapour = 2.4 kJ/kg °C.* **[CSE, 2001]**

p (bar)	*Enthalpy (kJ/kg)*		*V (m³/kg)*		*s (kJ/kg °C)*		*t (°C)*
	Liq.	*Vap.*	*Liq.*	*Vap.*	*Liq.*	*Vap.*	
22.68	49.62	322.86	0.00101	0.0166	0.1976	1.2567	–15
50.92	127.75	308.08	0.00130	0.0066	0.4697	1.0959	15
64.32	164.17	283.63	0.00147	0.0042	0.5903	0.9912	25

Solution: Given:

$$x_1 = 0.9,\ V_s = V_1 = 500 \text{ cc or } 5 \times 10^{-4} \text{ m}^3,\ N = 500 \text{ rpm}$$

$$\eta_{vol} = 0.85,\ c_{pv} = 2.4 \text{ kJ/kg.°C}$$

Refer to Fig. 5.62.

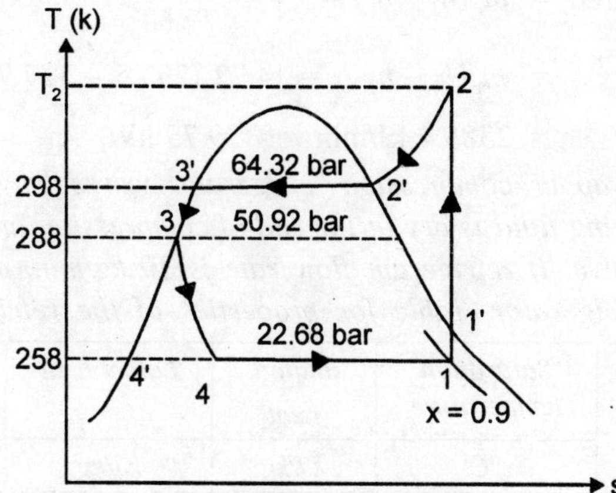

Fig. 5.62 T-s diagram.

$$s_1 = s_{f4'} + x_1 (s_{g1'} - s_{f4'})$$
$$= 0.1976 + 0.9 (1.2567 - 0.1976)$$
$$= 1.1508 \text{ kJ/kg.°C}$$

$$s_2 = s_1 = s_2' + c_{pv} \ln \frac{T_2}{T_2'}$$

$$1.1508 = 0.9912 + 2.4 \ln \frac{T_2}{298}$$

$$T_2 = 318.5 \text{ K}$$

$$h_1 = h_{f4'} + x_1 (h_{g1'} - h_{f4'}) = 49.62 + 0.9 (322.86 - 49.62) = 295.5 \text{ kJ/kg}$$

$$h_2 = h_2' + c_{pv}(T_2 - T_2') = 283.63 + 2.4 (318.5 - 298) = 332.83 \text{ kJ/kg}$$

$$h_{f3} = 127.75 \text{ kJ/kg}$$

(*i*) $$\text{COP} = \frac{h_1 - h_{f3}}{h_2 - h_1} = \frac{295.5 - 127.75}{332.83 - 295.5} = \frac{167.75}{37.33} = 4.493$$

(*ii*) Suction volume per minute

$$= V_1 \times N = 5 \times 10^{-4} \times 500 = 0.25 \text{ m}^3/\text{min}$$

Piston displacement per minute

$$= \frac{\dot{m}_r \times v_1}{\eta_{vol}} = \frac{\dot{m}_r \times 0.9 \times 0.0166}{0.85} \qquad (\because v_1 = v_{g1'}\, x_1)$$

$$= 0.017576\ \dot{m}_r$$

$$\therefore \quad 0.017576\dot{m}_r = 0.25$$

$$\dot{m}_r = 14.22 \text{ kg/min}$$

$$\text{Power required} = \dot{m}_r(h_2 - h_1) = \frac{14.22(332.83 - 295.5)}{60} = 8.847 \text{ kW}$$

(*iii*) Cooling capacity $= \dot{m}_r(h_1 - h_{f3}) = 14.22\ (295.5 - 127.75)$

$= 2385.4$ kJ/min or 39.75 kW

Example 5.49 *A vapour compression refrigerator works between pressure limits of 10 bar and 3 bar. The working fluid is dry at the end of compression and there is no undercooling before the expansion valve. If refrigerant flow rate is 10 kg/min, determine (i) COP and (ii) the capacity of the refrigerator. Table for properties of the refrigerant is an under:*

Pressure bar	*Saturation temperature*	*Liquid heat*	*Latent heat*	*Liquid entropy*
	°C	*kJ/kg*	*kJ/kg*	*kJ/kg K*
10	25	298.90	1166.94	1.1242
3	–10	135.37	1297.68	0.5443

[CSE, 2002]

Solution:

Given: $p_2 = p_3 = 10$ bar, $p_1 = p_4 = 3$ bar, $T_2 = T_3 = 25 + 273 = 298$ K

$T_1 = T_4 = -10 + 273 = 263$ K, $h_{f3} = h_{f4} = 298.9$ kJ/kg,

$h_{f'} = 135.37$ kJ/kg, $h_2 = h_{g2} = h_{f2} + h_{fg2} = 298.9 + 1166.94$

$= 1465.84$ kJ/kg, $s_{f2} = 1.1242$ kJ/kg·K,

$s_{f1} = 0.5443$ kJ/kg·K, $s_{g1} = 5.4770$ kJ/kg·K

Refer to Fig. 5.63.

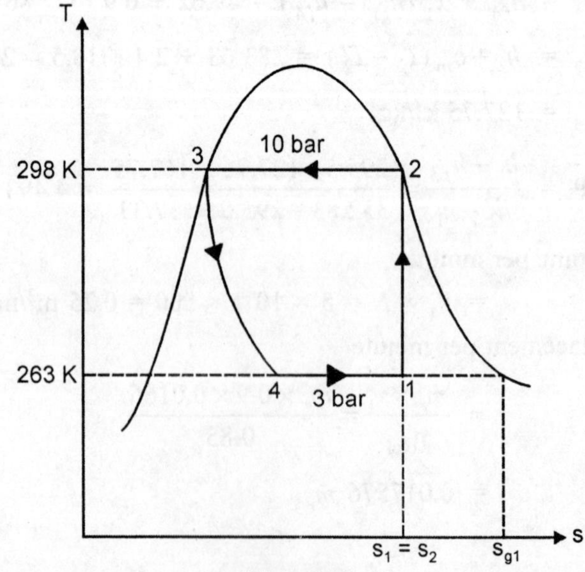

Fig. 5.63

$$s_1 = s_{f1} + \frac{x_1 h_{fg1}}{T_1} = 0.5443 + \frac{x_1 \times 1297.68}{263}$$

$$= 0.5443 + 4.9341\ x_1$$

$$s_1 = s_2 = s_{f2} + s_{fg2} = s_{f2} + \frac{h_{fg2}}{T_2} = 1.124 + \frac{1166.94}{298} = 5.0401 \text{ kJ/kg}\cdot\text{K}$$

$$0.5443 + 4.9341\ x_1 = 5.0401$$

$$x_1 = 0.9112$$

$$h_1 = h_{f1} + x_1\ h_{fg1} = 135.37 + 0.9112 \times 1297.68 = 1317.82 \text{ kJ/kg}$$

(*i*) $$\text{COP} = \frac{h_1 - h_{f3}}{h_2 - h_1} = \frac{1317.82 - 298.9}{1465.84 - 1317.82} = 6.88$$

(*ii*) Capacity of refrigerator:

Refrigerating effect,

$$RE = h_1 - h_{f3} = 1317.82 - 298.9 = 1018.92 \text{ kJ/kg}$$

$$\dot{m}_r = 10 \text{ kg/min}$$

Capacity of refrigerator

$$= \frac{\dot{m}_r \times RE}{211} = \frac{10 \times 1018.92}{211} = 48.29 \text{ TR}$$

Example 5.50 *In a refrigeration system of 10 TR cooling capacity using CHCIF2, the evaporator and condenser temperatures are –10°C and 45°C respectively. Properties of* $CHClF_2$ *at saturation are:*

Temperature °C	*Sp. Volume* m^3/kg	*Enthalpy, kJ/kg*		*Entropy kJ/kg. K*	
		h_f	h_g	s_f	s_g
–10	0.0654	34.25	247.37	0.1374	0.9473
45	0.0133	101.76	261.95	0.3662	0.8697

Consider standard vapour compression cycle with inlet to compressor as saturated vapour and inlet to expansion valve as saturated liquid. Assume that vapour may be treated as perfect gas in desuperheating process with average specific heat of 0.9335 kJ/kg-K. Show the cycle on T-s and p-h diagrams. Find (i) compressor outlet temperature and enthalpy, (ii) mass flow rate of refrigerant, (iii) work requirement, (iv) condenser heat rejection, (v) COP and (vi) swept volume of compressor in m³/s assuming 100% volumetric efficiency.

[CSE, 2003]

Solution: Given:

$$Q = 10 \text{ TR},\ T_1 = T_4 = -10°\text{C} = 263 \text{ K},\ T_2' = T_3 = 45°\text{C} = 318 \text{ K},\ c_{pv} = 0.9335 \text{ kJ/kg}\cdot\text{K},$$

$$\eta_v = 100\%,\ h_{f3} = 101.76 \text{ kJ/kg},\ h_1 = 247.37 \text{ kJ/kg},\ h_2' = 261.95 \text{ kJ/kg}$$

$$s_1 = 0.9473 \text{ kJ/kg}\cdot\text{K},$$

$$s_{2'} = 0.8697 \text{ kJ/kg}\cdot\text{K}$$

Refer to Fig. 5.64.

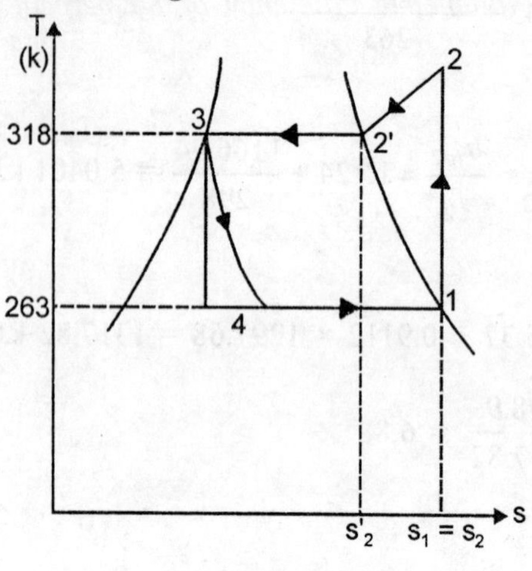

(a) T-s diagram.

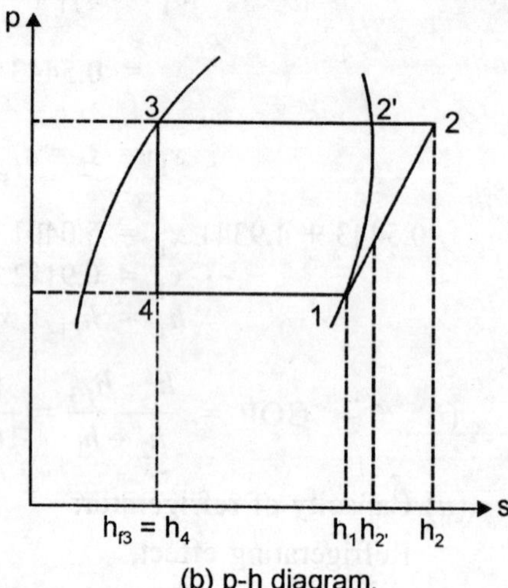

(b) p-h diagram.

Fig. 5.64

(*i*) $$s_1 = s_2 = s_{2'} + c_{pv} \ln \frac{T_2}{T_2'}$$

$$0.9473 = 0.8697 + 0.9335 \ln \frac{T_2}{318}$$

$$T_2 = 345.6 \text{ K or } t_2 = 72.6°\text{C}$$

$$h_2 = h_2' + c_{pv}(T_2 - T_2')$$

$$= 261.95 + 0.9335\ (345.6 - 318)$$

$$= 287.71 \text{ kJ/kg}$$

(*ii*) Refrigerating effect per kg of refrigerant

$$RE = h_1 - h_{f3} = 247.37 - 101.76 = 145.61 \text{ kJ/kg}$$

Mass flow rate of refrigerant,

$$\dot{m}_r = \frac{211Q}{RE} = \frac{211 \times 10}{145.61} = 14.49 \text{ kg/min}$$

(*iii*) Work requirement $= \dot{m}_r(h_2 - h_1) = \dfrac{14.44}{60}(287.71 - 247.37) = 9.742$ kW

(*iv*) Condenser heat rejection $= \dot{m}_r(h_2 - h_{f3}) = 14.49\ (287.71 - 101.76) = 2694.4$ kJ/min

(*v*) COP $= \dfrac{h_1 - h_{f3}}{h_2 - h_1} = \dfrac{145.61}{40.34} = 3.6$

(*vi*) Swept volume of compressor $= \dot{m}_r \times \dfrac{v_1}{\eta_v} = \dfrac{14.49 \times 0.0654}{1.0 \times 60} = 0.01579$ m³/s

Example 5.51 *A vapour compression refrigerator operates between the pressure limits of 462.47 kN/m² and 1785.90 kN/m². At entry to the compressor the refrigerant is dry saturated and after compression it has a temperature of 59°C. The compressor has a bore*

and stroke of 75 mm and runs at 8 rev/s with a volumetric efficiency of 80 percent. The temperature of the liquid refrigerant as it leaves the condenser is 32°C and the specific heat capacity of the superheated vapour may be assumed constant. Determine:

(i) The coefficient of performance of the refrigerator.

(ii) The mass flow of the refrigerant in kg/h.

(iii) The cooling water required by the condenser in kg/h if the cooling water temperature rise is limited to 12°C.

Take the specific heat capacity of water as 4.187 kJ/kg.K and the specific heat of liquid refrigerant as 1.32 kJ/kg.K.

The relevant properties of the refrigerant are given in the table below: **[CSE, 2004]**

Pressure	*Sat. temp.*	*Sp. enthalpy, kJ/kg*		*Sp. vol. (m^3/kg)*		*Sp. entropy, kJ/kg.K*	
kN/m²	t_f °C	h_f	h_g	v_f	v_g	s_f	s_g
462.47	–10	35.732	231.40	0.0008079	0.04573	0.1418	0.8614
1785.90	40	99.270	246.40	0.0009487	0.01105	0.3537	0.8093

Solution: Given: Refer to Fig. 5.65.

$p_1 = p_4 = 462.47$ kPa, $p_2 = p_3 = 1785.90$ kPa

$T_1 = T_4 = 263$ K, $T_2' = T_3' = 313$ K, $T_2 = 332$ K

$T_3 = 305$ K, $L = D = 75$ mm, $N = 8$ rps, $\eta_{vol} = 80\%$

$c_{pl} = c_{pv} = 1.32$ kJ/kg·K, $c_{pw} = 4.187$ kJ/kg·K

$v_1 = 0.04573$ m³/kg, $v_2' = 0.01105$ m³/kg

$h_{f3'} = 99.270$ kJ/kg, $h_1 = 231.40$ kJ/kg

$h_2' = 246.40$ kJ/kg, $s_1 = s_{g1} = 0.8614$ kJ/kg·K, $s_2' = s_{g2'} = 0.8093$ kJ/kg·K

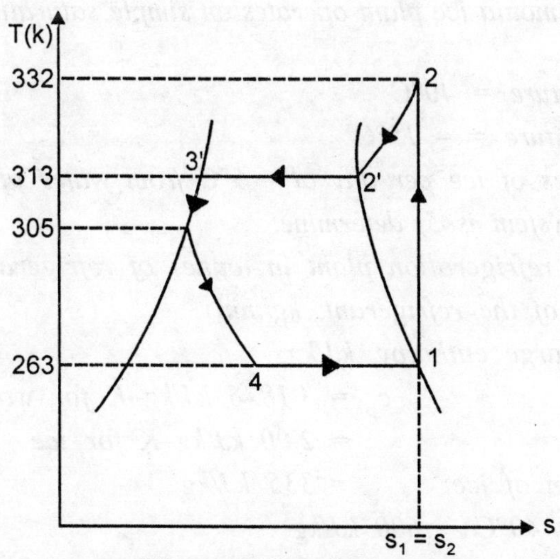

Fig. 5.65 T-s diagram.

$$h_2 = h_2' + c_{pv}(T_2 - T_2') = 246.40 + 1.32\ (332 - 313) = 271.48 \text{ kJ/kg}$$

$$h_{f3} = h_{f3'} - c_{pl}(T_3' - T_3) = 99.270 - 1.32\ (313 - 305) = 88.71 \text{ kJ/kg}$$

(*i*) $$\text{COP} = \frac{h_1 - h_{f3}}{h_2 - h_1} = \frac{231.40 - 88.71}{271.48 - 231.40} = 3.56$$

(*ii*) Suction volume $= \dfrac{\pi}{4} D^2 L \times N = \dfrac{\pi}{4}\ (0.075)^2 \times (0.075) \times 8 \times 60 = 0.159 \text{ m}^3/\text{min}$

Piston displacement per minute

$$= \frac{\dot{m}_r \times v_1}{\eta_{vol}} = \frac{\dot{m}_r \times 0.04573}{0.80} = 0.057\,\dot{m}_r$$

$$\therefore\ 0.057\,\dot{m}_r = 0.159$$

$$\dot{m}_r = 2.79 \text{ kg/min or } 167.37 \text{ kg/h}$$

(*iii*) Cooling water required by the condenser:

Let $\dot{m}_w$ = Cooling water required by the condenser, kg/min

Heat given out by the refrigerant in the condenser

$$= \dot{m}_r (h_2 - h_{f3})$$
$$= 2.79\ (271.48 - 88.71)$$
$$= 509.93 \text{ kJ/min}$$
$$= \dot{m}_w \times c_{pw} \times (\Delta t)_w$$
$$= \dot{m}_w \times 4.187 \times 12$$

$$\dot{m}_w = 10.15 \text{ kg/min or } 609 \text{ kg/h}$$

Example 5.52 *An ammonia ice plant operates on simple saturation cycle at the following temperatures:*

Condensing temperature = 40°C

Evaporation temperature = – 15°C

It produces 10 tonnes of ice per day at – 5°C from water at 30°C.

If the COP of the system is 5, determine:

(i) Capacity of the refrigeration plant in tonnes of refrigeration

(ii) Mass flow rate of the refrigerant, kg/min

(iii) Isentropic discharge enthalpy, kJ/kg

Take: c_p *= 4.1876 kJ/kg–K for water*

= 2.00 kJ/kg–K for ice

Latent heat of fusion of ice = 335 kJ/kg

Sensible enthalpy at 40°C = 600 kJ/kg

Enthalpy of saturated vapour at – 15°C = 1675 kJ/kg **[CSE, 2006]**

Solution: Given:

$$T_2 = T_3 = 273 + 40 = 313\text{ K},\ T_1 = T_4 = 273 - 15 = 258\text{ K},\ m_i = 10\text{ tonnes},$$
$$t_{wi} = 30°\text{C},\ t_{wo} = -5°\text{C},\ \text{COP} = 5,\ c_{pw} = 4.187\text{ kJ/kg·K},\ c_{pi} = 2.00\text{ kJ/kg·K},$$
$$L_i = 335\text{ kJ/kg},\ h_{f3} = 600\text{ kJ/kg},\ h_1 = 1675\text{ kJ/kg}$$

Heat rejected by water to form ice,

$$\dot{Q} = \frac{m_i}{60 \times 24}[c_{pw}(t_{wi} - 0) + L_i + c_{pi}(0 + 5)]$$
$$= \frac{10 \times 10^3}{60 \times 24}[4.187(30 - 0) + 335 + 2.0 \times 5]$$
$$= 3268.125\text{ kJ/min}$$

Refer to Fig. 5.66

Refrigeration effect,

$$R_E = h_1 - h_{f3} = 1675 - 600 = 1075\text{ kJ/kg}$$

(*i*) Capacity of refrigeration plant:

$$Q = \frac{\dot{Q}}{211} = \frac{3268.125}{211} = 15.49\text{ TR}$$

(*ii*) Mass flow rate of refrigerant,

$$\dot{m}_r = \frac{\dot{Q}}{R_E} = \frac{3268.125}{1075} = 3.04\text{ kJ/min}$$

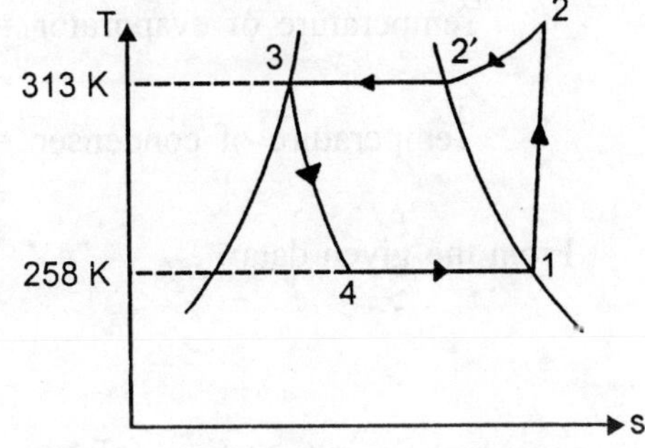

Fig. 5.66 T-s diagram.

(*iii*) Isentropic discharge enthalpy

$$\text{COP} = \frac{R_E}{h_2 - h_1}$$
$$h_2 - h_1 = \frac{1075}{5} = 215$$
$$h_2 = h_1 + 215 = 1675 + 215 = 1890\text{ kJ/kg}$$

Example 5.53 *A VCC refrigerating machine using R-12 refrigerant produces 10 tonnes of refrigeration at 10°C when the ambient is at 35°C. A temperature difference of minimum 5°C is required at the evaporator and condenser for spontaneous heat transfer. The refrigerant is dry saturated at the outlet and to the inlet of compressor. The adiabatic efficiency of the compressor is 90%. The enthalpy at the end of isentropic compression is estimated to be 370 kJ/kg. Determine:*

(i) COP

(ii) Power of the compressor

(iii) Capacity of the condenser

Represent the cycle on hand drawn T-S plane and show the refrigerating effect, compressor work and condenser capacity on the same: **[CSE, 2006]**

Properties of R-12 are:

Temperature (°C)	*Pressure (bar)*	*Specific Enthalpy, kJ/kg*	
		Sensible	*Evaporation*
5	3.62	204.64	148.97
10	4.23	209.32	146.37
15	4.91	214.10	143.69
30	7.45	228.54	135.04
35	8.47	233.50	131.90
40	9.60	238.53	128.62

Solution: Given: $Q = 10$ tonnes, $\eta_c = 0.9$, $h_2 = 370$ kJ/kg

Since there is a difference of 5°C in both evaporator and condenser, temperatures, therefore:

$$\text{Temperature of evaporator} = \text{Refrigeration temperature} - 5°\text{C}$$
$$= 10 - 5 = 5°\text{C}$$
$$\text{Temperature of condenser} = \text{Ambient temperature} + 5°\text{C}$$
$$= 35 + 5 = 40°\text{C}$$

From the given data:

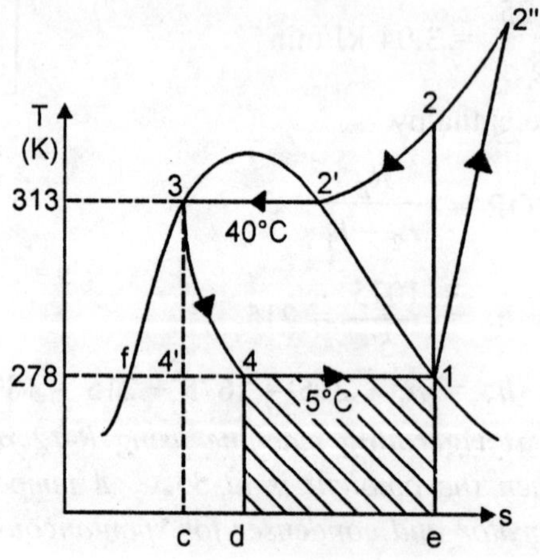

Fig. 5.67 T-s diagram.

$$h_1 = h_{f1} + h_{fg1} = 204.64 + 148.97 = 353.61 \text{ kJ/kg}$$
$$h_{f3} = h_4 = 238.53 \text{ kJ/kg}$$

Compressor efficiency,
$$\eta_c = \frac{h_2 - h_1}{h_2'' - h_1}$$
$$0.9 = \frac{370 - 353.61}{h_2'' - 353.61}$$

$$h_2'' = 371.82 \text{ kJ/kg}$$

Refrigerating effect, RE = area $1 - 4 - d - e$

Compressor work, w_c = area $1 - 2'' - 2' - 3 - c - d - 4 - 1$

Condenser capacity = area $2'' - 2' - 3 - c - e - 2''$

Refrigerating effect, $RE = h_1 - h_{f3} = 353.61 - 238.53 = 115.08$ kJ/kg

Mass flow rate of refrigerant, $\dot{m}_r = \dfrac{Q}{RE} = \dfrac{10 \times 211}{115.08} = 18.335$ kg/min

Compressor work, $w_c = h_2'' - h_1 = 371.82 - 353.61 = 18.21$ kJ/kg

Heat removed in the condenser $= h_2'' - h_{f3} = 371.82 - 238.53 = 133.29$ kJ/kg

(*i*) COP $= \dfrac{RE}{w_c} = \dfrac{115.08}{18.21} = 6.32$

(*ii*) Power of compressor $= \dfrac{\dot{m}_r \times w_c}{60} = \dfrac{18.335 \times 18.21}{60} = 5.565$ kW

(*iii*) Capacity of condenser $= \dfrac{18.335 \times 133.29}{60} = 40.73$ kW

Example 5.54 *A refrigeration system operated using simple saturated cycle with a certain refrigerant. The condensing and evaporating temperature for the refrigerant are 35°C and –15°C respectively. Determine the COP of the system. If a liquid vapour heat exchanger is installed in the system, with the temperature of the vapour leaving the heat exchanger at 15°C, what will be the change in the COP?* **[CSE, 2007]**

Use the following data for the refrigerant used:

				Superheated			
				20 K		*40 K*	
t	h_f	h_g	s_g	*h*	*s*	*h*	*S*
°C	*kJ/kg*	*kJ/kg*	*kJ/kg-K*	*kJ/kg*	*kJ/kg-K*	*kJ/kg*	*kJ/kg-K*
35	69.5	201.5	0.6839	216.4	0.731	231.0	0.7741
–15	–	181.0	0.7052	193.2	0.751	205.7	0.7942

Solution: Given:

$$t_1 = t_4 = -15°\text{C or } 258 \text{ K}, \; T_2' = t_3 = 35°\text{C or } 308 \text{ K}$$

$$h_2 = 201.5 \text{ kJ/kg}, \; s_2 = 0.6839 \text{ kJ/kg·K}$$

$$h_{f3} = h_4 = 69.5 \text{ kJ/kg}, \; h_{f1} = 22.33 \text{ kJ/kg}$$

The refrigerant is $R - 12$.

Let dryness fraction at point 1. (Fig. 5.68)

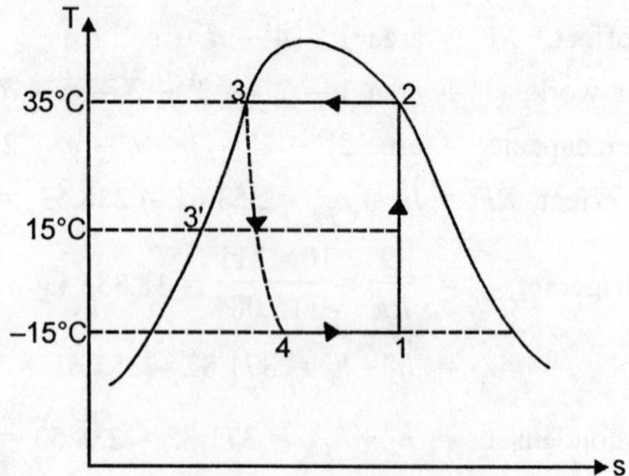

Fig. 5.68 T-s diagram.

$$s_1 = s_{f1} + x_1 \frac{h_{fg1}}{T_1} = 0.0906 + x_1 \times \frac{(181.0 - 22.33)}{258} = 0.0906 + 0.615\, x_1$$

$$s_1 = s_2$$

$$0.0906 + 0.615\, x = 0.6839$$

$$x_1 = 0.9647$$

$$h_1 = h_{f1} + x_1\, h_{fg1} = 22.33 + 0.9647\,(181.0 - 22.33)$$

$$= 175.4 \text{ kJ/kg}$$

$$\text{COP} = \frac{h_1 - h_{f3}}{h_2 - h_1} = \frac{175.4 - 69.5}{201.5 - 175.4} = \frac{105.9}{26.1} = 4.057$$

Heat exchange:

For the heat balance of heat exchanger,

$$c_{pv}(T_1' - T_1) = c_{pl}(T_3 - T_3')$$

$$0.672\ [15 - (-15)] = 0.941(35 - t_3')$$

$$t_3' = 13.6°\text{C}$$

$$h_{f3'} = h_{f3} - c_{pl}\,(T_3 - T_3') = 69.5 - 0.941\,(35 - 13.6) = 49.36 \text{ kJ/kg}$$

$$\text{COP} = \frac{h_1 - h_{f3'}}{h_2 - h_1} = \frac{175.4 - 49.36}{26.1} = 4.829$$

Increase in COP = 4.829 – 4.057 = 0.772

Example 5.55 *Discuss the importance of liquid vapour heat exchanger (LVHE) in the vapour compression refrigeration system. Clearly mention advantages and disadvantages of using such heat exchangers. Represent the cycle on T-s and p-h diagrams and illustrate the influence of LVHE.* **[CSE, 2008]**

Solution: Liquid-vapour Heat Exchanger (LVHE)

If we combine superheating of vapour with liquid subcooling, we have a liquid-vapour regenerative heat exchanger. A LVHE may be installed as shown in Fig. 5.69. In this, the

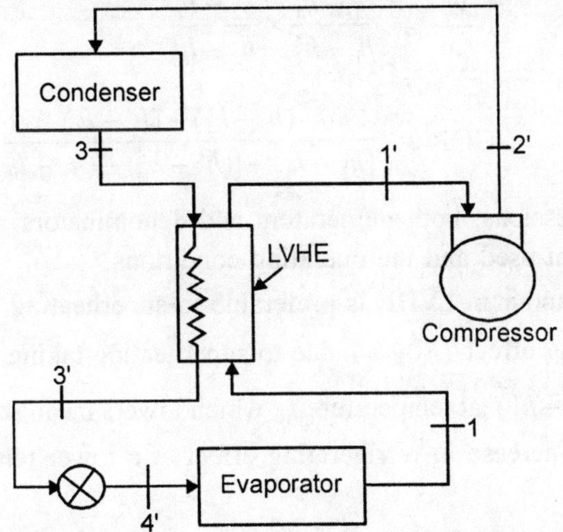

Fig. 5.69 Vapour compression system with LVHE.

refrigerant vapour from the evaporator is superheated in the regenerative heat exchanger with consequent subcooling of the liquid from the condenser. The T-s and p-h diagrams are shown is Fig. 5.70. Since the mass flow rate of liquid and vapour is the same, from energy balance of heat exchanger, we have

$$q = h_1' - h_1 = h_3 - h_3'$$

Degree of superheat $(t_1' - t_0)$ and degree of subcooling $(t_r - t_3')$ need not be the same as the specific heats of vapour and liquid phases are different.

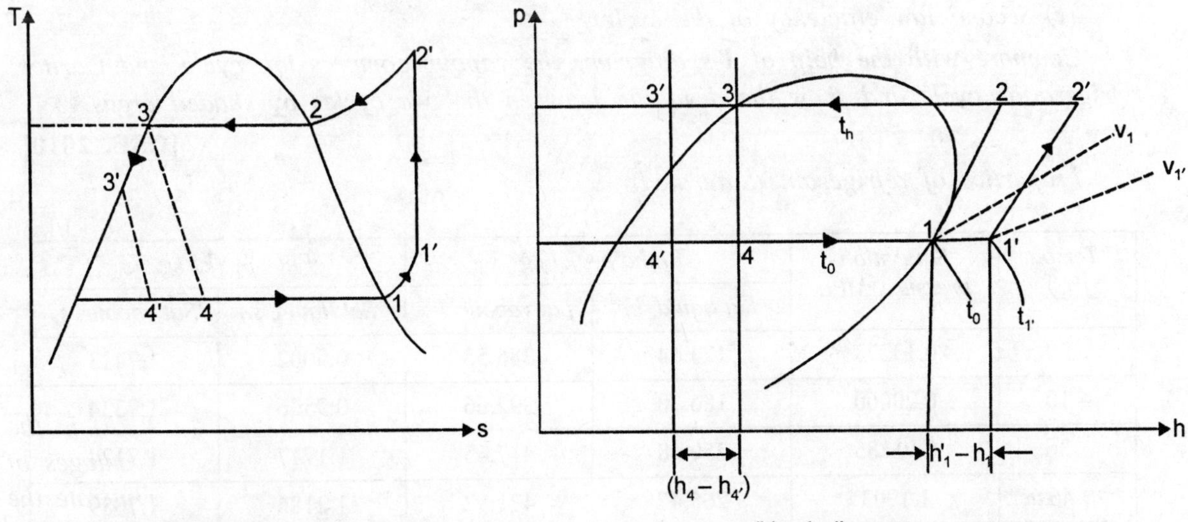

(a) T-s diagram.

(b) p-h diagram.

Fig. 5.70 Vapour compression cycle with LVRHE.

Capacity: $\dot{Q}_0' = (h_1 - h_4')/v_1'$ and $\dot{Q}_0 = (h_1 - h_4)/v_1$

Power required per unit refrigeration:

$$\frac{w'}{w} = \frac{h_1 - h_4}{h_1 - h_4'} \times \frac{h_2' - h_1'}{h_2 - h_1}$$

$$\text{COP} = \frac{(h_1 - h_4) + (h_1' - h_1)}{(h_2 - h_1) + [(h_2' - h_1') - (h_2 - h_1)]}$$

In all the above expressions, both numerators and denominators increase. The net effect depends upon the refrigerant used and the operating conditions.

Superheating of the vapour in LVHE is preferable to superheating in the evaporator itself, as the increased refrigerating effect $(h_1' - h_1)$ due to superheating taking place from temperature t_0 to t_1' is transferred as $(h_4 - h_4')$ at temperature t_0, which lowers mean refrigeration temperature. Thus, we obtain the same increase in refrigerating effect at a lower temperature by the use of LVHE.

Example 5.56 *A saturated vapour compression refrigeration system is extracting heat from a thermal reservoir at – 10°C and rejecting heat to another thermal reservoir at 36°C. The saturation temperature of evaporator is –20°C and that of condenser is 46°C. The mass flow rate of refrigerant (R – 134a) is 0.1 kg/s. Assume environment temperature equal to 36°C. Find.*

(i) Refrigerating capacity in Tons

(ii) Power input in kW

(iii) COP

(iv) COP of Carnot refrigeration cycle.

(v) Second law efficiency of the cycle.

Compare with the help of T-s diagram, the vapour compression cycle and Carnot refrigeration cycle and show the deviation between the two cycles by shaded areas.

[CSE, 2010]

Properties of refrigerant (Saturated).

Temp. (°C)	*Saturation pressure, MPa*	*Enthalpy (kJ/kg)*		*Entropy (kJ/kg. K)*	
		Sat liquid, h_f	*Sat vapour, h_g*	*Sat liquid, s_f*	*Sat vapour, s_g*
– 20	0.13273	173.64	386.55	0.9002	1.7413
– 10	0.20060	186.70	392.66	0.9506	1.7334
36	0.91185	250.48	417.65	1.1717	1.7124
46	1.1903	265.47	421.92	1.2186	1.7089

(Superheated)

Pressure (MPa)	h_g *(kJ/kg)*	s_g *(kJ/kg.K)*
1	428.91	1.7413
1.2	436.12	1.7413

Solution: Given: For Carnot cycle:

$$T_2 = 273 - 10 = 263 \text{ K}, \ T_1 = 273 + 36 = 309 \text{ K}$$

For *VCC*: $T_1 = T_4 = 273 - 20 = 253$ K, $T_2' = T_3 = 273 + 46 = 319$ K,

$$\dot{m}_r = 0.1 \text{ kg/s}, \ T_0 = 273 + 36 = 309 \text{ K}$$

Refer to Fig. 5.71

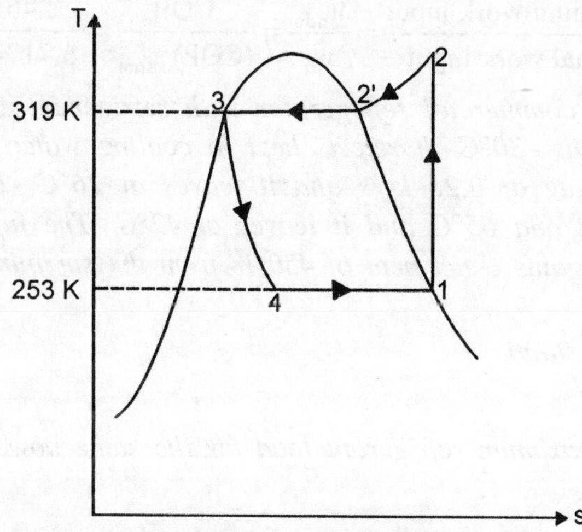

Fig. 5.71 T-s diagram.

$$h_1 = 386.55 \text{ kJ/kg}, \ h_2' = 421.92,$$

$$s_1 = 1.7413 \text{ kJ/kg}\cdot\text{K}, \ s_2' = 1.7089 \text{ kJ/kg}\cdot\text{K}$$

$$h_{f3} = h_4 = 265.47 \text{ kJ/kg}, \ c_{pv} = 1.202 \text{ kJ/kg}\cdot\text{K}$$

$$h_2 = 428.91 + \frac{(436.12 - 428.91)}{(1.2 - 1)}(1.1903 - 1)$$

$$= 435.77 \text{ kJ/kg}$$

$$s_2 = 1.7413 \text{ kJ/kg}\cdot\text{K}$$

(*i*) Refrigerating effect

$$RE = h_1 - h_{f3}$$

$$= 386.55 - 265.47$$

$$= 121.08 \text{ kJ/kg}$$

Refrigerating capacity in tons

$$= \frac{\dot{m}_r \times RE}{211} = \frac{0.1 \times 60 \times 121.08}{211} = 3.443 \text{ TR}$$

(*ii*) Power input $= \dot{m}_r(h_2 - h_1) = 0.1\ (435.77 - 386.55) = 4.922$ kW

$$(iii) \qquad \text{COP} = \frac{h_1 - h_{f3}}{h_2 - h_1} = \frac{386.55 - 265.47}{435.77 - 386.55} = \frac{121.08}{49.22} = 2.46$$

(*iv*) COP of Carnot refrigeration cycle

$$= \frac{T_2}{T_1 - T_2} = \frac{263}{309 - 263} = \frac{263}{46} = 5.7174$$

(*v*) Second law efficiency of cycle :

$$\eta_{II} = \frac{\text{Minimum work input}}{\text{Actual work input}} = \frac{w_{min}}{w_a} = \frac{\text{COP}}{(\text{COP})_{carnot}} = \frac{2.46}{5.7174} = 0.4303 = 43.03\%$$

Example 5.57 *A commercial refrigerator with refrigerant 134a as the working fluid keeps a space cooled at –30°C. It rejects heat to cooling water that enters the condenser at 18°C and at the rate of 0.25 kg/s and it leaves at 26°C. The refrigerant enters the condenser at 1.2 MPa and 65°C and it leaves at 42°C. The inlet state of compressor is 60 kPa and –34°C. It gains a net heat of 450 W from the surroundings. Sketch T-S diagram and determine.*

(i) Refrigeration load

(ii) COP

(iii) Theoretical maximum refrigerant load for the same power input to the compressor.

Given properties: **[CSE, 2012]**

$$h_{-34°C}^{60\,kPa} = 230.03 \text{ kJ/kg}$$

$$h_{65°C}^{3200\,kPa} = 295.16 \text{ kJ/kg}$$

$$h_{42°C}^{3200\,kPa} = 111.23 \text{ kJ/kg}$$

$$x_{111.23\,kJ/kg}^{60\,kPa} = 0.47$$

$$h_{water,\ 18°C} = 75.4 \text{ kJ/kg}$$

$$h_{water,\ 26°C} = 108.9 \text{ kJ/kg}$$

Solution: Given:

$$t_{wi} = 18°C,\ t_{wo} = 26°C,\ \dot{m}_w = 0.25 \text{ kg/s}$$

$$Q = +\ 450 \text{ W},\ h_1 = 230.03 \text{ kJ/kg}$$

$$h_2 = 295.16 \text{ kJ/kg},\ h_3 = 111.23 \text{ kJ/kg}$$

$$h_3 = h_4 = h_{f4} + x_4\,(h_1 - h_{f4})$$

$$x_4 = \frac{h_3 - h_{f4}}{h_1 - h_{f4}}$$

$$x_4 = 0.47,\ h_{\text{water, 18°C}} = 75.4 \text{ kJ/kg},$$

$$h_{\text{water, 26°C}} = 108.9 \text{ kJ/kg}$$

(*i*) Work done on compressor, (Fig. 5.72)

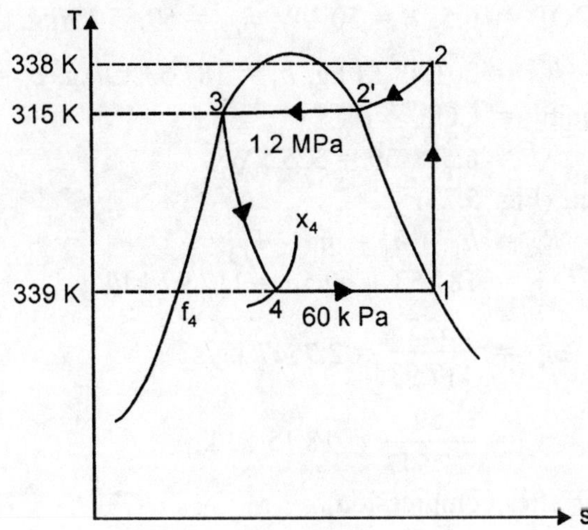

Fig. 5.72

$$w = h_2 - h_1 = 295.16 - 230.03 = 65.13 \text{ kJ/kg}$$

Refrigeration effect,

$$R_E = h_1 - h_{f3} = 230.03 - 211.23 = 118.8 \text{ kJ/kg}$$

Heat gained by water from the condenser = Heat lost by refrigerant

$$\dot{m}_w(h_{wo} - h_{wi}) = \dot{m}_r(h_2 - h_3)$$

$$0.25(108.9 - 75.4) = \dot{m}_r(295.16 - 111.23)$$

$$\dot{m}_r = 0.04553 \text{ kg/s}$$

Total refrigeration load = $\dot{m}_r \times R_E + 0.450 = 0.04553 \times 118.8 + 0.450 = 5.86$ kW

(*ii*) $$\text{COP} = \frac{\text{Total refrigeration load}}{\text{Workdone}} = \frac{5.86}{65.13 \times 0.04553} = 1.975$$

(*iii*) $$\text{Power input} = \dot{m}_r(h_2 - h_1) = 0.04553 \times 65.13 = 2.964 \text{ kW}$$

$$\text{Theoretical COP} = 118.8/65.13 = 1.524$$

Theoretical maximum refrigerant load

$$= 118.8 \times 0.04553$$

$$= 5.409 \text{ kW}$$

$$= \frac{5.409}{3.5167} = 1.538 \text{ TR}$$

Example 5.58 *A refrigerator operating on a standard vapour compression cycle has a coefficient of performance of 6.5 and is driven by a 50 kW compressor. The enthalpies of standard liquid and saturated vapour refrigerant at the operating condensing temperature of 35°C are 69.55 kJ/kg and 201.45 kJ/kg respectively. The saturated refrigerant vapour leaving the evaporator has an enthalpy of 187.53 kJ/kg. Find the refrigerant temperature at the compressor discharge. The c_p of refrigerant vapour may be taken to be 0.6155 kJ/(kg K).* **[GATE, 1992]**

Solution: Given:

$$\text{COP} = 6.5,\ P = 50 \text{ kW},\ h_{f3} = 69.55 \text{ kJ/kg},$$

$$h_2' = 201.45 \text{ kJ/kg},\ h_1 = 187.53 \text{ kJ/kg},\ c_p = 0.6155 \text{ kJ/kg·K}.$$

Refrigerating Capacity $= \text{COP} \times P$

$= 6.5 \times 50 = 325$ kW

Refrigeration effect, (Fig. 5.73)

$$R_E = h_1 - h_4 = h_1 - h_{f3}$$

$$= 187.53 - 69.55 = 117.97 \text{ kJ/kg}$$

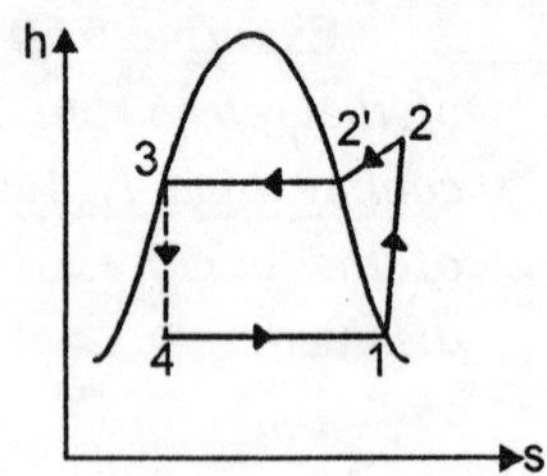

Fig. 5.73 *h-s* chart.

Refrigerant rate, $\dot{m}_r = \dfrac{325}{117.98} = 2.7547$ kg/s

Heat input per kg $= \dfrac{50}{2.7547} = 18.15$ kJ/kg

Enthalpy of vapour after compression,

$$h_2 = h_1 + 18.15 = 187.53 + 18.15 = 205.68 \text{ kJ/kg}$$

Superheat $= h_2 - h_2' = 205.68 - 201.45 = 4.23$ kJ/kg

Degree of superheat $= \dfrac{\text{superheat}}{c_p} = \dfrac{4.23}{0.6155} = 6.87$°C

Temperature of refrigerant at compressor discharge

$= 35 + 6.87 = 41.67$°C

Example 5.59 *A refrigeration system of 15 tons capacity operates on standard simple vapour compression cycle using Refrigerant-22 at an evaporator temperature 5°C and condensing temperature of 50°C. Draw the p-h diagram for the cycle. Calculate.*

(i) the refrigerant mass flow rate and

(ii) the compressor intake volume flow rate if the compressor volumetric efficiency is 0.72. **[GATE, 1993]**

Use the refrigerant property data given in Table below:

Temperature °C	*Pressure bar*	*Enthalpy*		*Volume*	
		Sat. Liquid kJ/kg	*Sat. Vapour kJ/kg*	*Sat. Liquid l/kg*	*Sat. Vapour* m^3/kg
5	5.836	205.9	407.1	0.791	0.0404
50	19.423	263.3	417.7	0.922	0.0177

Solution: Given: $h_1 = 407.1$ kJ/kg,

$$h_4 = h_3 = h_{f_3} = 26.3 \text{ kJ/kg}$$

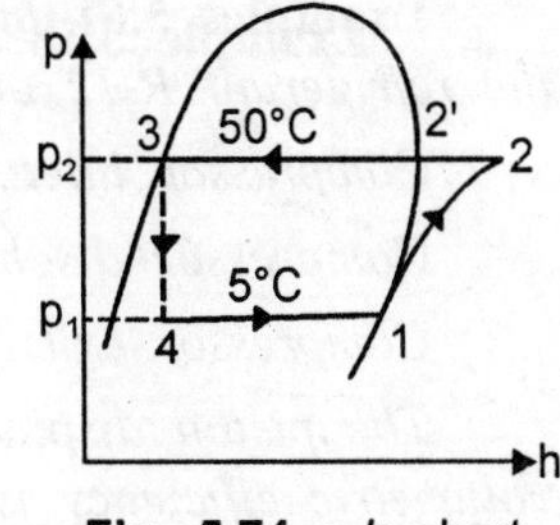

Fig. 5.74 *p-h* chart.

(*a*) Capacity of plant $= 15T_n = \dfrac{\dot{m}(h_1 - h_4)\times 3600}{14000}$

$$\dot{m} = \frac{15\times 14000}{(407.1-263.3)\times 3600}$$

$$= 0.406 \text{ kg/s}$$

(*b*) $v_1 = v_{g1} = 0.0404 \text{ m}^3\text{/kg}$

$$\dot{v}_1 = \dot{m}v_1 = 0.406 \times 0.0404 = 0.0164 \text{ m}^3\text{/s}$$

Example 5.60 *A refrigeration cycle uses Freon-12 as the working fluid. The temperature of the refrigerant in the evaporator is –10°C. The condensing temperature is 40°C. The cooling load is 150 W and the volumetric efficiency of the compressor is 80%. The speed of the compressor is 720 rpm. Calculate the mass flow rate of refrigerant and the displacdement volume of the compressor.* **[GATE, 1995]**

Properties of Freon-12

Temperature (°C)	*Saturation Pressure (MPa)*	*Enthalpy (kJ/kg) Saturated*		*Specific volume (m^3/kg) Saturated vapour*
		Liquid	*Vapour*	
–10	0.22	26.8	183.0	0.08
40	0.96	74.5	203.1	0.02

Solution: Given: $Q = 150 \text{ W}, \eta_v = 0.8, N = 720 \text{ rpm},$

$$h_4 = 183.0 \text{ kJ/kg}, v_1 = 0.08 \text{ m}^3\text{/kg},$$

$$h_{f_3} = h_4 = 74.5 \text{ kJ/kg}$$

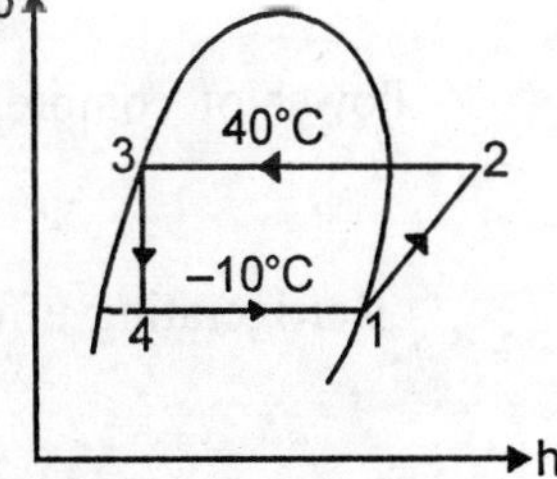

Fig. 5.75 p-h chart.

Refrigernating effect, $R_E = h_1 - h_4 = h_1 - h_{f_3}$

$$= 183.0 - 74.5 = 108.5 \text{ kJ/kg}$$

Let $\dot{m}$ = mass flow rate of refrigerant, kg/s

$$\dot{m}_r \times R_E = Q$$

$$\dot{m}_r = \frac{150}{108.5\times 10^3} = 1.382 \times 10^3 \text{ kg/s}$$

Displacement volume of compressor

$$\dot{V}_1 = \frac{\dot{m}v_1}{\eta_v} = \frac{1.382\times 10^{-13}\times 0.08}{0.8}$$

$$= 1382 \times 10^{-4} \text{ m}^3\text{/s}$$

Examples 5.61 *In a simple vapour compression cycle, following are the properties of the refrigerant R-12 at various points:*

Compressor inlet: $h_1 = 183.2$ *kJ/kg,* $v_1 = 0.767$ *m³/kg*

Compressor discharge: $h_2 = 222.6$ *kJ/kg,* $v_1 = 0.0164$ *m³/kg*

Compressor exit: $h_3 = 84.9$ *kJ/kg,* $v_3 = 0.00083$ *m³/kg*

The piston displacement volume for the compressor is 1.5 litres per stroke and its volumetric efficiency is 80%. The speed of the compressor is 1600 rpm.

Find (a) Power rating of the compressor (kW)

(b) Refrigerating effect (kW). **[GATE, 1996]**

Solution: Piston displacement, volume,

$$\frac{\pi}{4}d^2l = 1.5 \times 10^{-13} \text{ m}^3\text{/stroke}$$

$$= 1.5 \times 10^{-13} \text{ m}^3\text{/rev.}$$

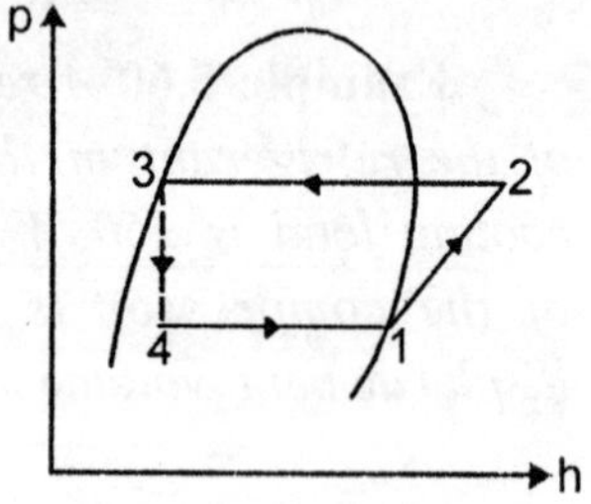

Fig. 5.76 p-h chart.

Compressor discharge rate,

$$\dot{V} = \frac{\pi^2}{4}d^2l \times \frac{N}{\eta_V}$$

$$= \frac{1.5 \times 10^{-3} \times 1600}{60 + 0.8} = 0.05 \text{ m}^3\text{/s}$$

Mass flow rate of compressor,

$$\dot{m} = \frac{\dot{V}}{v_1} = \frac{0.05}{0.0767} = 0.6519 \text{ kg/s}$$

Power of compressor, $P = \dot{m}(h_2 - h_1) = 0.6519\ (22.6 - 183.2)$

$= 25.684$ kW

Refrigerating effect, $R_E = \dot{m}(h_1 - h_{f3})$

$= 0.6519\ (183.2 - 84.9) = 64.08$ kW

Examples 5.62 *In a standard vapour compression refrigeration cycle, operating between an evaporator temperature of –10°C and a condenser temperature of 40°C, the enthalpy of the refrigerant, Feron-12, at the end of the compression is 220 kJ/kg. Show the cycle diagram on T-s plane. Calculate*

(a) The COP of the cycle.

(b) The refrigerating capacity and the compressor power assuming a refrigerant flow rate of 1 kg/min. You may use the extract of the Freon-12 property table given below:

T(°C)	p(MPa)	h_f (kJ/kg)	h_g (kJ/kg)
–10	0.2191	26.85	183.1
40	0.9607	74.53	203.1

Solution: Given: $h_2 = 220$ kJ/kg, $\dot{m} = 1$ kg/min,

$h_1 = 183.1$ kJ/kg, $h'_{f4} = 26.85$ kJ/kg,

$h'_2 = 203.1$ kJ/kg

$h_{f3} = 74.53$ kJ/kg

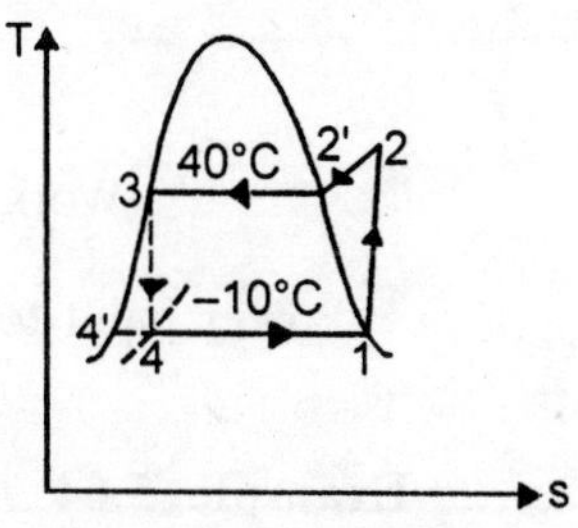

Fig. 5.77 T-s diagram.

(*a*) COP $= \dfrac{h_1 - h'_{f4}}{h_2 - h_1} = \dfrac{183.1 - 26.85}{220 - 183.1} = 4.234$

(*b*) Refrigerating capacity,

$$R_E = \dot{m}(h_1 - h'_{f4})$$
$$= 1(183.1 - 26.85) = 156.25 \text{ kJ/min}$$

Compressor power,

$$P = \dot{m}(h_2 - h_1)$$
$$= 1(183.1 - 26.85) = 156.25 \text{ kJ/min}$$
$$= \frac{1 \times (220 - 183.1)}{60} = 0.615 \text{ kW}$$

Example 5.63 *An R-717(ammonia) system operates on the basic vapour compression refrigeration cycle. The evaporator and the condenser pressure are 0.119 MPa and 1.389 MPa respectively. The mass flow rate of refrigerant is 0.1 kg/s. If the volumetric efficiency of the compressor is 84%, determine the compressor displacement rate. If the COP of the cycle is 2, determine the power input to the compressor.*

Saturation properties of R-717 (ammonia)

Temperature °C	*Pressure* *MPa*	*Specific volume of Vapour* m³/kg	*Enthalpy kJ/kg*	
			Liquid	*Vapour*
–30	0.119	0.9639	63.9	1423.6
36	1.389	0.0930	371	1488.6

Solution: Given: $\dot{m} = 0.1$ kg/s, $\eta_v = 0.84$, COP = 2

$v_1 = 0.9638$ m³/kg

$h_1 = 1423.6$ kJ/kg, $h'_{f4} = 63.9$ kJ/kg,

$h_2 = 1488.6$ kJ/kg, $h_{f3} = 371.4$ kJ/kg

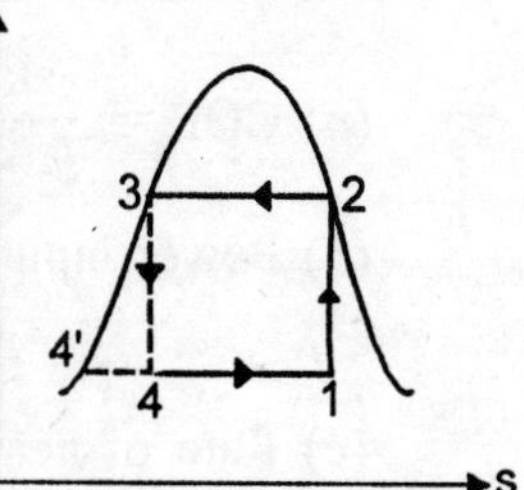

Fig. 5.78 T-s diagram.

The T-s diagram is shown in Fig. 5.78.

(*i*) Compressor displacement

$$\dot{V} = \frac{\dot{m}v_1}{\eta_v} \times \frac{0.1 \times 0.9638}{0.84} = 0.11474 \text{ m}^3/\text{s}$$

Refrigerating effect,

$$R_E = h_1 - h'_{f4} = 1423.6 - 371.4 = 1052.2 \text{ kJ/kg}$$

$$\text{COP} = \frac{R_E}{\text{Work done}}$$

$$\text{Work done} = \frac{1052.2}{2} = 526.1 \text{ kJ/kg}$$

Power input to compressor

$$= \dot{m} \times \text{Work done} = 0.1 \times 526.1 = 52.61 \text{ kW}$$

Example 5.64 *In a 5 kW cooling capacity refrigeration system operating on a simple vapour compression cycle, the refrigerant enters the evaporator with an enthalpy of 75 kJ/kg and leaves with an enthalpy of 183 kJ/kg. The enthalpy of the refrigerant after compression is 210 kJ/kg. Show the cycle on T-s or p-h diagram. Calculate the following: (a) COP (b) power input to compressor, and (c) rate of heat transfer at the condenser.*

[GATE, 2000]

Solution. Given: $P = 5 \text{ kW}, h_4 = 75 \text{ kJ/kg}, h_1 = 183 \text{ kJ/kg}$

$h_2 = 210 \text{ kJ/kg}$

Cooling effect, $R_E = h_1 - h_4 = 183 - 75 = 108 \text{ kJ/kg}$

Mass flow rate of refrigerant,

$$\dot{m} = \frac{P}{R_E} = \frac{5}{108} = 0.0463 \text{ kg/s}$$

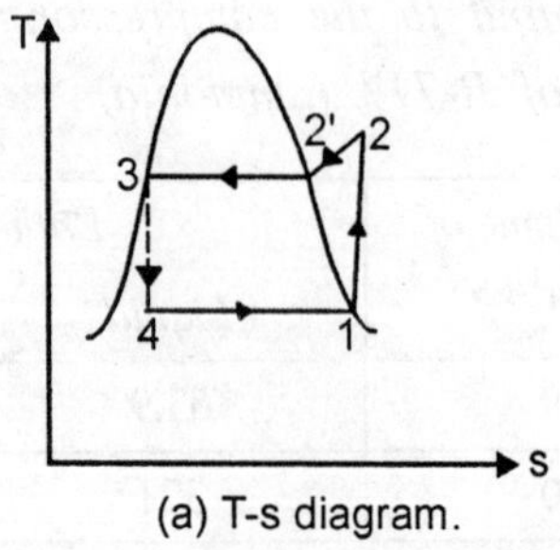

(a) T-s diagram.

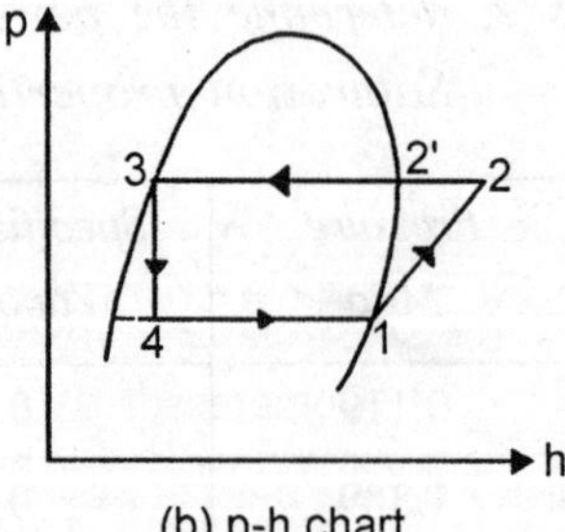

(b) p-h chart.

Fig. 5.79

(*a*) $\text{COP} = \frac{R_E}{h_2 - h_1} = \frac{108}{(210-183)} = 4$

(*b*) Power input to compressor

$$= \dot{m}(h_2 - h_1) = 0.0463\ (210 - 183) = 1.25 \text{ kW}$$

(*c*) Rate of heat transfer at the condenser

$$= \dot{m}(h_2 - h_3)$$

$$= \dot{m}(h_2 - h_4) \qquad (\because h_4 = h_3)$$

$$= 0.0463\ (210 - 75) = 6.25 \text{ kW}$$

5.11 METHODS OF IMPROVING SIMPLE SATURATION CYCLE

The simple saturation vapour compression refrigeration cycle may be improved by the following methods:

1. By introducing a flash chamber between the throttle valve and evaporator.
2. By using a precooler or accumulator.
3. By subcooling the liquid refrigerant by the vapour refrigerant.
4. By subcooling the liquid refrigerant leaving the condenser by the liquid refrigerant from the throttle valve.

5.11.1 Simple Saturation Cycle with Flash Chamber

A simple saturation cycle with flash chamber is shown in Fig. 5.80. When the high pressure liquid refrigerant from the condenser passes through the throttle valve, some of it evaporates. This partial evaporation of the liquid refrigerant is called *flash.* These formed vapours are incapable of producing any refrigeration effect. These vapours can be by-passesed around the evaporator and supplied directly to the compressor suction. This is achieved by introducing a flash chamber between the throttle valve and evaporator. The flash chamber is an insulated container which separates the liquid and vapour due to centrifugal action. Thus the mass of the refrigerant passing through the evaporator decreases. The p-h diagram is shown in Fig. 5.81

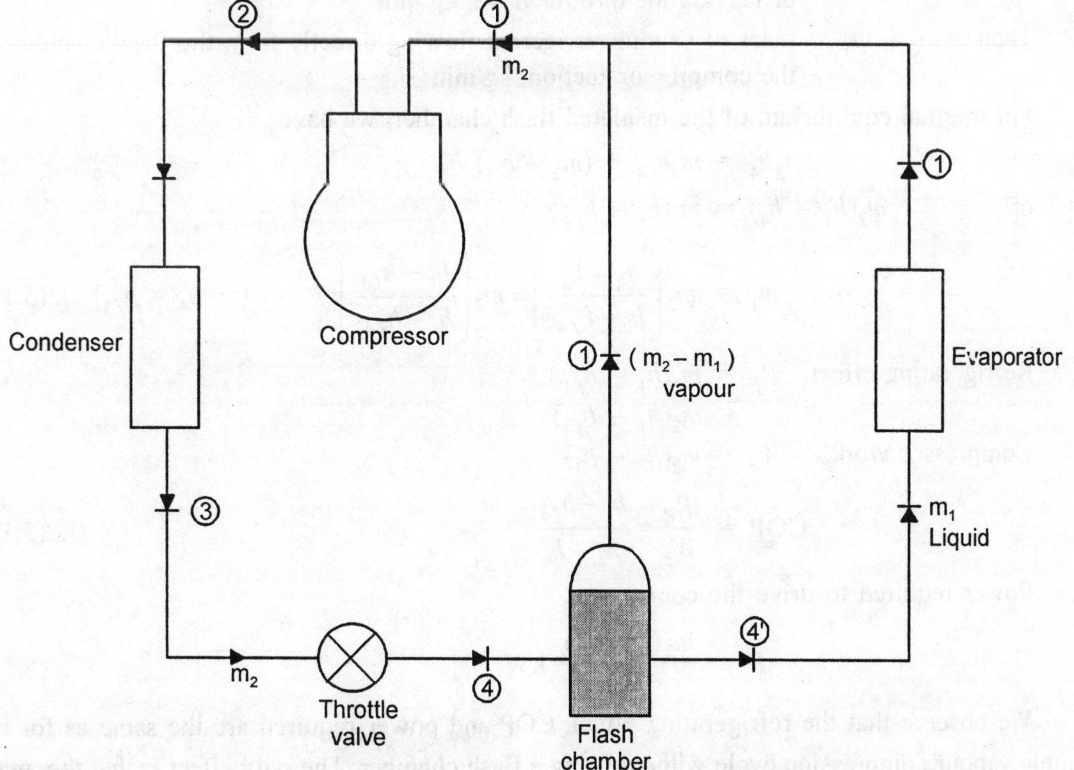

Fig. 5.80 Simple saturation cycle with flash chamber.

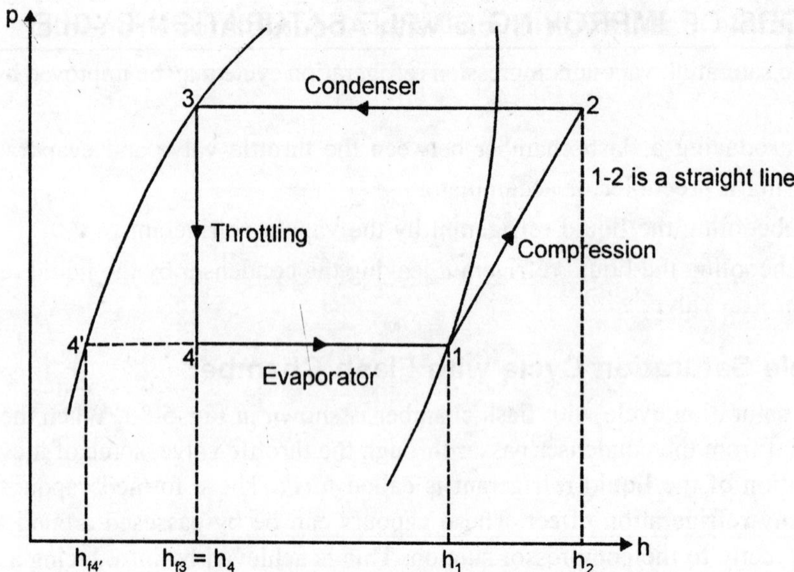

Fig. 5.81 p-h diagram of simple saturated cycle with flash chamber.

Let, m_1 = mass of liquid refrigerant supplied to the evaporator, kg/min

m_2 = mass of liquid and vapour refrigerant circulated through the condenser or leaving the throttle valve, kg/min

Then $m_2 - m_1$ = mass of vapour refrigerant flowing directly from the flash chamber to the compressor suction, kg/min

For thermal equilibrium of the insulated flash chamber, we have

$$m_2 h_4 = m_1 h_{f4'} + (m_2 - m_1)\, h_1$$

or

$$m_2(h_1 - h_4) = m_1(h_1 - h_{f4'})$$

$$m_1 = m_2\left[\frac{h_1 - h_4}{h_1 - h_{f4'}}\right] = m_2\left[\frac{h_1 - h_{f3}}{h_1 - h_{f4'}}\right] \quad [\because h_4 = h_{f3}] \quad ...(5.10)$$

Refrigerating effect, $R_E = m_1(h_1 - h_{f4'})$

$$= m_2(h_1 - h_{f3})$$

Compressor work, $W_c = m_2(h_2 - h_1)$

$$\text{COP} = \frac{R_E}{W_c} = \frac{h_1 - h_{f3}}{h_2 - h_1} \quad ...(5.11)$$

Power required to drive the compressor,

$$P = \frac{m_2(h_2 - h_1)}{60} \text{ kW} \quad ...(5.12)$$

We observe that the refrigerating effect, COP and power required are the same as for the simple vapour compression cycle without using a flash chamber. The only effect is that the, mass of refrigerant flowing through the evaporator decreases. Hence the size of evaporator reduces.

5.11.2 Simple Saturation Cycle with Accumulator Precooler

Sometimes, the liquid refrigerant passing through the evaporator is not completely evaporated. Then the compressor has to do an additional work of evaporating and raising the temperature of liquid refrigerant. In order to avoid this difficulty, an accumulator or pre-cooler, is used in the system, as shown in Fig. 5.82. The accumulator receives a mixture of liquid and vapour refrigerant from the expansion valve and supplies the liquid refrigerant only to the evaporator. The discharge from the evaporator is sent again to the accumulator which further keep off the liquid from entering the compressor. A liquid pump is provided in the system in order to maintain circulation of refrigerant in the evaporator.

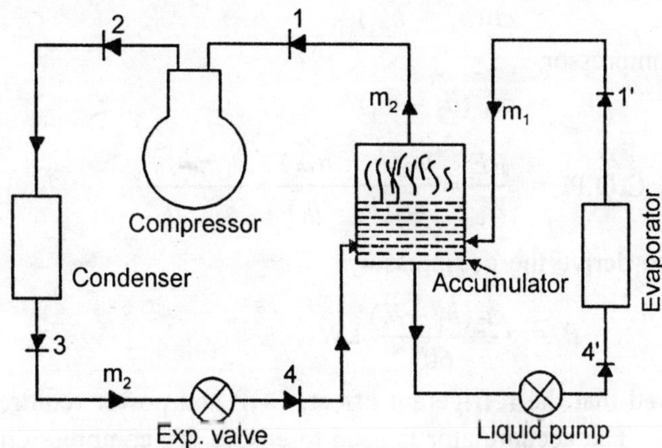

Fig. 5.82 Simple saturation cycle with accumulator or pre-cooler.

The *p-h* diagram of the system is shown in Fig. 5.83.

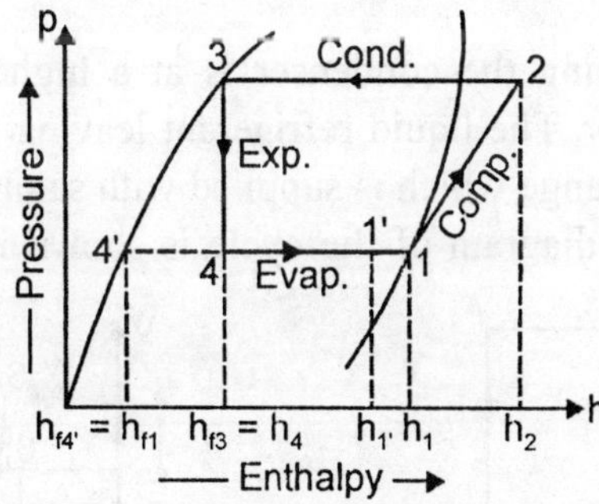

Fig. 5.83 p-h diagram of simple saturation cycle with accumulator.

Let $\quad m_1$ = mass of liquid refrigerant circulating through the evaporator, kg/min

$\quad m_2$ = mass of refrigerant flowing in the condenser, kg/min

When all the liquid refrigerant does not evaporate in the evaporator, it is represented by point 1′ on *p-h* diagram. It is assumed that mass of refrigerant at point 1′ is also m_1.

The accumulator is an insulated vessel. Therefore, there is no heat exchange between the accumulator and atmosphere. For the thermal equilibrium of accumulator, we have

Heat taken in by the accumulator = Heat given out by the accumulator

$$m_2\, h_4 + m_1 h_1' = m_2 h_1 + m_1 h_{f4'}$$

$$m_1(h_1' - h_{f4'}) = m_2(h_1 - h_4)$$

$$m_1 = m_2\left[\frac{h_1 - h_4}{h_1' - h_{f4'}}\right] = m_2\left[\frac{h_1 - h_{f3}}{h_{1'} - h_{f4'}}\right] \qquad (\because\ h_4 = h_{f3})$$

Refrigerating effect, $RE = m_1(h_1' - h_{f4'})$

$$= m_2\left[\frac{h_1 - h_{f3}}{h_{1'} - h_{f4'}}\right](h_{1'} - h_{f4'})$$

$$= m_2(h_1 - h_{f3}) \qquad \text{...(5.13)}$$

Worth done in compressor,

$$W_c = m_2\,(h_2 - h_1) \qquad \text{...(5.14)}$$

$$\text{C.O.P.} = \frac{RE}{W_c} = \frac{m_2(h_1 - h_{f3})}{m_2(h_2 - h_1)} = \frac{h_1 - h_{f3}}{h_2 - h_1} \qquad \text{...(5.15)}$$

Power required to derive the compressor,

$$P = \frac{m_2(h_2 - h_1)}{60}\ \text{kW} \qquad \text{...(5.16)}$$

It may be observed that the refrigerant effect, COP and power required is same as for the simple saturation cycle. The accumulator is used to ensure dry compression in the compression.

5.11.3 Simple Saturation Cycle with Sub- cooling of Liquid Refrigerant by Vapour Refrigerant

The liquid refrigerant leaving the condenser is at a higher temperature than the vapour refrigerant leaving the evaporator. The liquid refrigerant leaving the condenser can be subcooled by passing it through a heat exchange which is supplied with saturated vapour from the evaporator as shown in Fig. 5.84. The p-h diagram of the cycle is shown in Fig. 5.85.

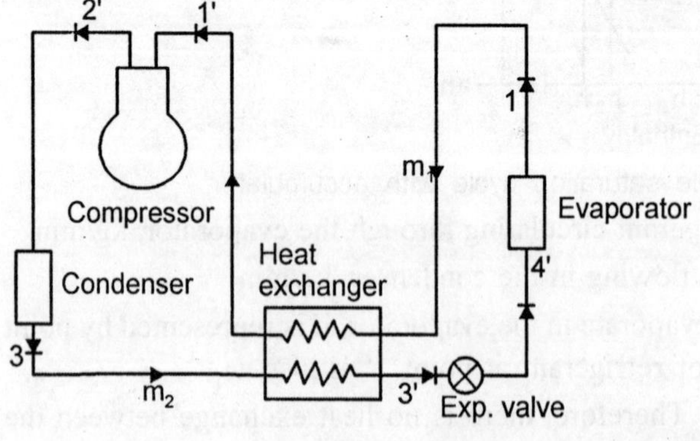

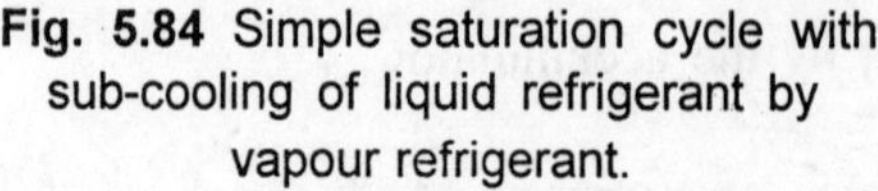

Fig. 5.84 Simple saturation cycle with sub-cooling of liquid refrigerant by vapour refrigerant.

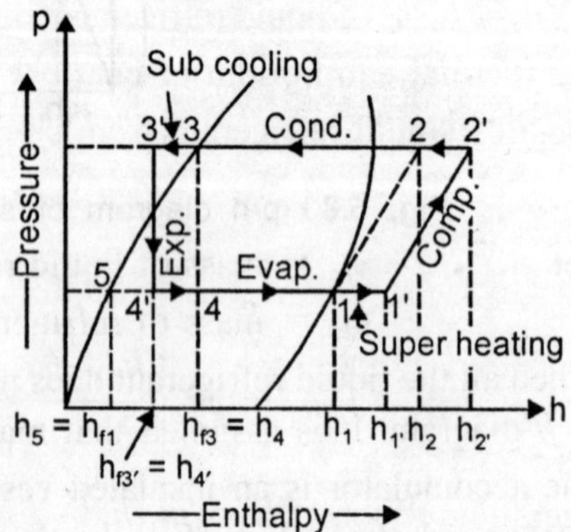

Fig. 5.85 *p-h* diagram of simple saturation cycle with sub-cooling of liquid refrigerant by vapour refrigerant.

Let m_1 = mass of vapour refrigerant

m_2 = mass of liquid refrigerant

T_3 = temperature of liquid refrigerant entering the heat exchanger

T_3' = temperature of liquid refrigerant leaving the heat exchanger

T_1 = temperature of liquid refrigerant entering the heat exchanger

T_4' = temperature of liquid refrigerant leaving the heat exchanger

5.11.4 Simple Saturation Cycle with Sub-cooling of Liquid Refrigerant by Liquid Refrigerant

The liquid refrigerant leaving the condenser is at a higher temperature than the liquid refrigerant leaving the expansion valve. Therefore, the liquid refrigerant leaving the condenser can be sub-cooled by passing it through a heat exchanger which is supplied with liquid refrigerant from the expansion valve, as shown in Fig. 5.86. The *p-h* diagram of the cycle is shown in Fig. 5.87.

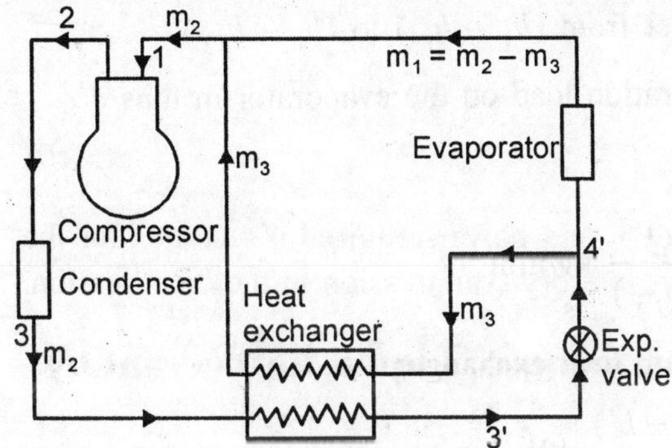

Fig. 5.86 Simple saturation cycle with sub-cooling of liquid refrigerant by liquid refrigerant.

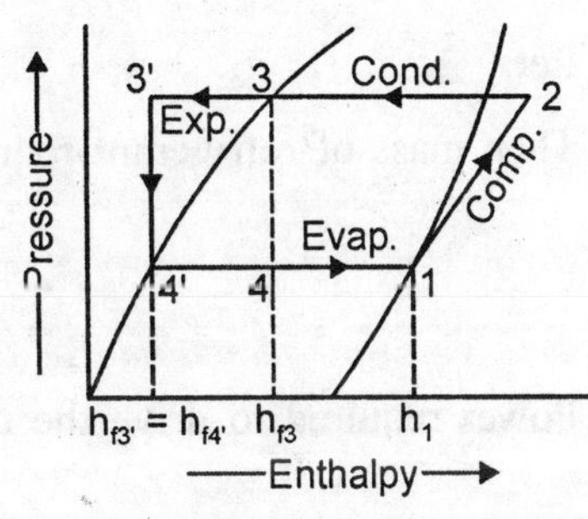

Fig. 5.87 *p-h* diagram of simple saturation cycle with sub-cooling of liquid refrigerant by liquid refrigerant.

Let m_1 = mass of refrigerant leaving the evaporator

m_3 = mass of liquid refrigerant supplied to the heat exchanger from the expansion valve

For the thermal equilibrium of heat exchanger, we have

Heat lost by liquid refrigerant from condenser = Heat gained by liquid refrigerant from expansion valve

$$m_2(h_{f3} - h_{f3'}) = m_3(h_1 - h_{f4'})$$

$$\therefore \quad m_3 = m_2\left[\frac{h_{f3} - h_{f3'}}{h_1 - h_{f4'}}\right] = m_2\left[\frac{h_{f3} - h_{f3'}}{h_1 - h_{f3'}}\right] \quad ...(a) \quad (\because h_{f4'} = h_{f3'})$$

Refrigerating effect, $RE = m_1(h_1 - h_{f4'})$

$$= (m_2 - m_3)(h_1 - h_{f4'}) \quad [\because m_1 = m_2 - m_3]$$

Substituting the value of m_3 from Eq. (*a*), use get

$$RE = \left[m_2 - m_2 \left\{ \frac{h_{f3} - h_{f3'}}{h_1 - h_{f3'}} \right\} \right] (h_1 - h_{f4'})$$

c_{pv}, c_{pl} = Specific heat of vapour and liquid refrigerant respectively

For the thermal equilibrium of heat exchanger, we have

Heat lost by liquid refrigerant = Heat gained by vapour refrigerant

$$m_2 \times c_{pl}(T_3 - T_3') = m_1 \times c_{pv}(T_1' - T_1)$$

Since $m_2 = m_1$ and $c_{pl} > c_{pv}$, therefore : $(T_3 - T_3') < (T_1' - T_1)$

For energy balance, we have

$$h_{f3} - h_{f3'} = h_1' - h_1$$

In the ideal case, $T_{1'} = T_{3'}$

Subcooling increases the refrigerating effect from $(h_1 - h_{f3})$ to $(h_1 - h_{f3'})$.

Let $\dot{Q}$ = refrigeration load on the evaporator in tons

Then mass of refrigerant required,

$$\dot{m}_r = \frac{211\dot{Q}}{(h_1 - h_{f3'})} \text{ kg/min}$$

Power required to drive the compressor with heat exchanger,

$$P_1 = \frac{\dot{m}_r (h_2' - h_1')}{60} \text{ kW}$$

$$= \frac{211\dot{Q}}{60} \left[\frac{h_2' - h_1'}{h_1 - h_{f3'}} \right] \text{ kW} \quad ...(5.17)$$

Power required to drive the compressor without heat exchanger,

$$P_2 = \frac{\dot{m}_r (h_2 - h_1)}{60} \text{ kW}$$

$$= \frac{211\dot{Q}}{60} \left[\frac{h_2 - h_1}{h_1 - h_{f3'}} \right] \text{ kW} \quad ...(5.18)$$

Excess power required to drive the compressor as compared so simple saturation cycle.

$$P_{\text{excess}} = P_1 - P_2$$

$$= \frac{211\dot{Q}}{60} \left[\frac{h_2' - h_1'}{h_1 - h_{f'3}} - \frac{h_2 - h_1}{h_1 - h_{f3}} \right] \text{ kW} \quad ...(5.19)$$

$$\text{COP} = \frac{RE}{W_c} = \frac{h_1 - h_{f3'}}{h_2' - h_1'}$$

$$= m_2(h_1 - h_{f4'}) - m_2(h_{f3} - h_{f3'}) \quad ...(5.20)$$

$$= m_2h_1 - m_2h_{f4'} - m_2h_{f3} + m_2h_{f3'}$$

$$= m_2\,(h_1 - h_{f3}) \quad ...(5.21) \quad (\because h_{f4'} = h_{f3'})$$

Work done on compressor,

$$W_c = m_2\,(h_2 - h_1)$$

$$\text{C.O.P.} = \frac{RE}{W_c} = \frac{h_1 - h_{f3}}{h_2 - h_1} \quad ...(5.22)$$

Let $\dot{Q}$ = Load on the compressor in tons.

Then
$$m_1 = \frac{211\dot{Q}}{h_1 - h_{f4'}}$$

or
$$m_2 - m_3 = \frac{211\dot{Q}}{h_1 - h_{f4'}}$$

Substituting the value of m_3 from Eq. (*a*), we get

$$m_2 - m_2\left[\frac{h_{f3} - h_{f3'}}{h_1 - h_{f4'}}\right] = \frac{211\dot{Q}}{h_1 - h_{f4'}}$$

$$\frac{m_2(h_1 - h_{f3})}{(h_1 - h_{f4'})} = \frac{211\dot{Q}}{(h_1 - h_{f4'})}$$

$$m_2 = \frac{211\dot{Q}}{(h_1 - h_{f3})}$$

Power required to drive the compressor,

$$P = \frac{m_2(h_2 - h_1)}{60}\ \text{kW}$$

$$= \frac{211\dot{Q}}{60}\left[\frac{h_2 - h_1}{h_1 - h_{f3}}\right]\ \text{kW} \quad ...(23)$$

REVIEW QUESTIONS

1. Mention the advantages of vapour compression refrigeration system over air refrigeration system.
2. Describe the mechanism of a simple vapour compression refrigeration system.
3. Sketch the *T-s* and *p-h* diagrams for the vapour compression cycles when the vapour after compression is (*i*) dry saturated, and (*ii*) wet.
4. Why in practice a throttle valve is used in vapour compression refrigerator rather than an expansion valve to reduce pressure between the condenser and the evaporator?
5. What is sub-cooling and superheating? Explain with the help of diagram. Why is superheating considered to be good in certain cases?
6. Establish how an actual cycle differs from a theoretical vapour compression cycle.
7. What is the difference between wet and dry compressions?
8. What are the effects of suction and delivery pressure on the performance of a vapour compression system?
9. What are the effects of evaporator and condenser temperatures on the performance of vapour compression system.
10. What is the function of expansion valve in vapour compression cycle?

MULTIPLE – CHOICE QUESTIONS

1. In vapour compression refrigeration system the heat is absorbed from the environment to be cooled by
   (a) evaporator (b) condenser (c) compressor (d) throttle valve.
2. During a refrigeration cycle, heat is rejected by the refrigerant in a
   (a) compressor (b) condenser (c) evaporator (d) expansion valve
3. In a vapour compression refrigeration system, the condition of refrigerant before entering the compressor is
   (a) saturated liquid (b) wet vapour
   (c) dry saturated liquid (d) Superheated vapour
4. The highest temperature during the cycle, in a vapour compression refrigeration system, occurs after
   (a) compression (b) condensation (c) expansion (d) evaporation
5. Heat is rejected by the refrigerant, during vapour compression refrigeration cycle in
   (a) condenser (b) evaporator (c) throttle valve (d) compressor
6. In a simple saturated vapour compression cycle, the refrigerant is in superheated condition
   (a) before entering compressor (b) before entering condenser
   (c) before entering throttle valve (d) before entering evaporator.

7. In a simpe saturated vapour compression refrigeration cycle the following results were obtained. Heat Rejected in condenser = 140 kJ/kg, compressor work = 28 kJ/kg. the COP of refrigeration is
(a) 5 (b) 4 (c) 6 (d) 3

8. In a vapour compression refrigeration system, the lowest temperature during the cycle occurs after
(a) compression (b) condensation (c) expansion (d) evaporation

9. In a vapour compression refrigeration system, the effect of superheating the vapour before suction to compression
(a) increases the work of compression
(b) increases the heat rejection in the condenser
(c) may increase or decrease COP depending upon the refrigerant used
(d) all of the above

10. In a domestic vapour compression refrigerator, the refrigerant commonly used is
(a) CO_2 (b) Ammonia (c) Freon-12 (d) all of these

11. A domestic refrigerator capacity may be approximately
(a) 1 ton (b) 0.1 ton (c) 1.5 ton (d) 10 ton

12. A domestic window type air conditioner capacity may be approximately
(a) 1 ton (b) 0.1 ton (c) 5 ton (d) 10 ton

13. In vapour compression refrigeration cycle for ice making the condensing temperature for better COP is desired to be
(a) near critical temperature of refrigerant
(b) Above critical temperature of refrigerant
(c) much below critical temperature of refrigerant
(d) any as it is not affected by condensing temperature.

14. COP of domestic air conditioner as compared to that of domestic refrigerator will be
(a) lower (b) higher (c) same (d) un-predictable.

15. In a domestic vapour compression refrigerator the refrigerant used is
(a) CO_2 (b) Freon-12 (c) Ammonia (d) All of the above

16. Wet compression in vapour compression refrigeration cycle means
(a) vapour compression takes place in wet region
(b) vapour compression in dry region, but evaporation in wet region
(c) vapour compression in wet region, but leaves in superheated region
(d) none of the above

17. The oil separator is incorporated in vapour compression refrigeration system
(a) between evaporator and compressor (b) between compressor and condenser
(c) between condenser and throttle valve (d) between throttle valve and evaporator.

18. The order in which main component of vapour compression refrigeration system are used is

(a) compressor-evaporator-condenser-throttle valve

(b) compressor-condenser-evaporator-throttle valve

(c) compressor-condenser-throttle valve-evaporator

(d) compressor-throttle-valve-evaporator-condenser,

19. Liquid-suction heat exchanger is used in vapour compression refrigeration system to

(a) improve COP of cycle

(b) avoid useless superheating in pipes

(c) to reduce heat rejection in condenser

(d) none of the above.

20. The COP of vapour compression refrigerating cycle using liquid suction heat exchanger as compared to simple saturated cycle

(a) is higher definitely

(b) may be higher or lower depending upon refrigerant used

(c) remains same

(d) is unpredictable

21. Sub-cooling is a process of cooling the refrigerant in vapour compression refrigeration system

(a) after compression

(b) before compression

(c) before throttling

(d) none of the above

22. The effect of liquid sub - cooling in vapour compression refrigeration system is

(a) to reduce the rate of flow or refrigerant per ton of refrigeration

(b) to reduce the volume of vapour handled by compressor per ton of refrigeration

(c) to reduce power per ton of refrigeration

(d) all the above

23. The effect of superheating the vapour before suction to compression in vapour compression refrigeration system

(a) increases the work of compression

(b) increases the heat rejection in condenser

(c) may increase COP or decrease COP depending upon the refrigerant used

(d) all of the above.

24. In vapour compression refrigeration system, if expansion valve is used in place of throttle valve, the COP will

(a) increase

(b) decrease

(c) will remain same

(d) not be predictable unless refrigerant used is known.

25. In actual large vapour compression refrigeration installations

(a) all the pipes are lagged and insulated

(b) only pipes between evaporator and compressor are lagged and insulated

(c) only pipes between compressor and condenser are lagged and insulated

(d) none of the pipes are lagged and insulated.

26. With reciprocating compressor in vapour compression refrigeration system, wet compression is not desirable because,
   (a) liquid trapped up in the head of cylinder may damage the compressor valves
   (b) COP of the cycle decreases
   (c) volumetric efficiency of compressor decreases
   (d) mass flow rate per ton of refrigerant increases,

27. In vapour compression refrigeration system, with simple saturated cycle, the isentropic expansion of Carnot cycle is replaced by throttling because
   (a) positive work of expander is very small to justify cost of expander
   (b) throttling device gives better COP
   (c) throttling device is easy to operate
   (d) throttling device increases refrigeration capacity.

28. Vapour compression cycle using R-12 gives maximum COP when
   (a) suction state to compressor is in wet region
   (b) suction state to compressor is in superheat region
   (c) suction state to compressor is dry saturated
   (d) suction state to compressor is any.

29. In vapour compression refrigerating cycle, decrease in evaporator temperature keeping condenser temperature fixed results in
   (a) decrease in refrigerating effect per kg
   (b) decrease in volumetric efficiency of compressor
   (c) increase in compressor work per kg
   (d) all of the above

30. Increase in condenser temperature keeping evaporator temperature fixed, in vapour compression refrigerating cycle results in
   (a) decrease in refrigerating capacity
   (b) increase in power consumption
   (c) decrease in volumetric efficiency of compressor
   (d) all of the above.

31. The net effect of superheating suction in vapour compression refrigeration systems using *R*-717 (Ammonia) theoretically results in
   (a) reduction in refrigeration capacity of the system
   (b) increase in refrigeration capacity of the system
   (c) keeping refrigeration capacity same of the system
   (d) none of the above

32. The effect of superheating suction vapours in vapour compression refrigeration systems using R-12 theoretically results in

(a) reduction in refrigeration capacity of the system

(b) increase in refrigeration capacity of the system

(c) keeping refrigeration capacity same of the system

(d) none of the above.

33. In vapour compression refrigeration system the following data is available. Enthalpy of refrigerant at suction to compressor = 585 kJ/kg, at discharge from compressor = 610 kJ/kg, heat rejected to cooling water = 5 kJ/kg. the work of compressor per kg is

(a) 25 kJ/kg (b) 20kJ/kg (c) 30kJ/kg (d) 35kJ/kg

Answers

1. (a)	**2.** (b)	**3.** (d)	**4.** (a)	**5.** (a)	**6.** (b)	**7.** (b)	**8.** (d)
9. (d)	**10.** (c)	**11.** (b)	**12.** (a)	**13.** (c)	**14.** (b)	**15.** (b)	**16.** (a)
17. (b)	**18.** (c)	**19.** (b)	**20.** (b)	**21.** (c)	**22.** (d)	**23.** (d)	**24.** (a)
25. (b)	**26.** (a)	**27.** (a)	**28.** (b)	**29.** (d)	**30.** (d)	**31.** (a)	**32.** (b)
33. (c)							

NUMERICAL PROBLEMS

5.1 The compressor and condenser temperature in an NH3 refrigeration system are –10°C and 40°C respectively. Determine per TR basis: (a) mass flow rate; (b) compressor work; (c) condenser heat rejection; (d) C.O.P.; and (e) refrigerating efficiency. Use only the properties of NH_3 given below:

Saturation temperature (t) °C	*Pressure (p) bar*	*Specific volume of vapour (v_g) m^3/kg*	*Enthalpy, kJ/kg*		*Entropy, kJ/kg.K*	
			Liquid	*Vapour*	*Liquid*	*Vapour*
–10	2.908	0.418	134.95	1431.41	0.5435	5.4712
40	15.55	0.09833	371.47	1472.02	1.3574	4.8728

For superheated NH3 at 15.55 bar, the following values may be taken:

Superheat K	*Specific volume (v) m^3/kg*	*Enthalpy (h) kJ/kg*	*Entropy (s) kJ/kg.K*
60	0.108	1647.9	5.3883
80	0.116	1700.3	5.5253

[**Ans.** 0.198 kg/min; 0.82 kW; 4.32 kW; 4.27; 81.2%]

5.2 In a vapour compression refrigeration system using *R*-12, the evaporator pressure is 1.4 bar and the condenser pressure is 8 bar. The refrigerant leaves the condenser sub-cooled to 30°C. The vapour leaving the evaporator is dry and saturated. The compression process is isentropic. The amount of heat rejected in the condenser is 13.42 MJ/min. Determine: (a) refrigerating effect in kJ/kg; (b) refrigerating load in TR; (c) compressor input in kW; and (d) COP.

Show the cycle on p-h diagram [**Ans.** 114 k/kg; 49 TR; 51.4 kW; 3.35]

5.3 A vapour compression refrigerator works between the temperature limits of –20°C and 25°C. The refrigerant leaves the compressor in dry saturated condition. If the liquid refrigerant is unedrcooled to 20°C before entering the throttle valve, determine:

(a) work required to drive the compressor;

(b) refrigerating effect produced per kg of the refrigerant; and

(c) theoretical COP.

Assume specific heat of the refrigerant as 4.8. The properties of the refrigerant are:

Temperature °C	*Enthalpy, kJ/kg*		*Entropy, kJ/kg.K*	
	Liquid	*Vapour*	*Liquid*	*Vapour*
–20	89.78	1420.02	0.3684	5.6244
25	298.90	1465.84	1.1242	5.0391

[**Ans.** 189.7 kJ/kg; 990.2 kJ/kg; 5.01]

5.4 A food storage chamber requires a refrigeration system of 12 TR capacity with an evaporator temperature of – 8°C and condenser temperature of 30°C. The refrigerant R-12 is subcooled by 5°C before entering the throttle valve, and the vapour is superheated by 6°C before entering the compressor. If the liquid and vapour specific heats are 1.235 and 0.733 kJ/kg.K respectively, find: (a) refrigerating effect per kg; (b) mass of refrigerant circulated per minute; and (c) coefficient of performance.

The relevant properties of the refrigerant R-12 are given below:

Saturation temperature, °C	*Enthalpy, kJ/kg*		*Entropy, kJ/kg.K*	
	Liquid	*Vapour*	*Liquid*	*Vapour*
–8	28.70	184.06	0.1148	0.7007
30	64.59	199.62	0.2400	0.6853

[**Ans.** 130.05 kJ/kg; 19.4 kg; 6.2]

5.5 An ammonia refrigerator produces 30 tonnes of ice from and at 0°C in 24 hours. The temperature range of the compressor is from 25°C to – 15°C. The vapour is dry saturated at the end of compression and an expansion valve is used. Assume a coefficient of performance to be 60% of the theoretical value. Calculate the power required to drive the compressor. Latent heat of ice = 335 kJ/kg. Properties of ammonia are:

Temperature, °C	Enthalpy, kJ/kg		Entropy, kJ/kg K	
	Liquid	Vapour	Liquid	Vapour
25	298.9	1465.84	1.1242	5.0391
–15	112.34	1426.54	0.4572	5.5490

[**Ans.** 33.24 kW]

5.6 A freezer of 20 TR capacity has evaporator and condenser temperatures of –30°C and 25°C respectively. The refrigerant *R*-12 is sub-cooled by 4°C before it enters the expansion valve and is superheated by 5°C before leaving the evporator. The compression is isentropic and the valve throttling and clearance are to be neglected. If a six cylinder, single acting compressor with stroke equal to bore running at 1000 rpm is used, determine (a) COP of the refrigerating system, (b) mass of refrigerant to be circulated per minute (c) theoretical piston displacement per minute, and (d) theoretical bore and stroke of the compressor. The specific heat of liquid *R*-12 is 1.235kJ/kg.K and of vapour *R*-12 is 0.733 kJ/kg.K. The properties of *R*-12 are given below:

Saturation temp. °C	Pressure bar	Enthalpy, kJ/kg		Entropy, kJ/kg K		Specific volume, m^3/kg	
		Liquid	Vapour	Liquid	Vapour	Liquid	Vapour
–30	1.0044	8.86	174.20	0.0371	0.7171	0.00673	0.1596
25	6.5184	59.7	197.73	0.2239	0.6868	0.00764	0.0269

[**Ans.** 3.64; 34.12 kg/min; 5.56 m/min; 0.106 m]

5.7 A refrigeration plant of 8 TR capacity has its evaporations temperature of – 8°C and condenser temperature of 30°C. The refrigerant is sub-cooled by 5°C before entering into the expansion valve and vapour is superheated by 6°C before leaving the refrigerator. The suction pressure drop is 0.2 bar in the suction valve and discharge pressure drop is 0.1 bar in the discharge valve.

If the refrigerant used is *R*-12, find out the COP, of the plant and theoretical power required for the compressor. Assume compression is isentropic. Use *p-h* chart for calculation.

5.8 An ammonia refrigerator works between – 6.7°C and 26.7°C, the vapour being dry at the end of isentropic compression. There is no under-cooling of liquid ammonia and the liquid is expanded through a throttle valve after leaving the condenser, Sketch the cycle on *T-s* and *p-h* diagram and calculate the refrigeration effect per kg of ammonia and the theoretical coefficient of performance of the unit with the help of the properties given below:

Temperature, °C	Enthalpy, kJ/kg		Entropy, kJ/kg K	
	Liquid	Vapour	Liquid	Vapour
–6.7	152.18	1437.03	0.6016	5.4308
26.7	307.18	1467.03	1.1515	5.0203

[**Ans.** 1028.3 kJ/kg; 7.2]

5.9 An ammonia refrigerating machine fitted with an expansion valve works between the temperature limits of –10°C and 30°C. The vapour is 95% dry at the end of isentropic compression and the fluid leaving the condenser is at 30°C. If the actual coefficient of performance is 60% of the theoretical, find the ice produced per kW hour at 0°C from water at 10°C. The latent heat of tice is 335 kJ/kg. the ammonia has the following properties:

Temperature, °C	*Liquid heat, kJ/kg*	*Latent heat kJ/kg*	*Entropy, kJ/kg K*	
			Liquid	*Vapour*
30	323.08	1145.79	1.2037	4.9842
–10	135.37	1297.68	0.5443	5.4770

[**Ans.** 33.24kg/kWh]

5.10 A *R*-12 refrigerating machine works on vapour-compression cycle. The temperature of refrigerant in the evaporator is –20°C. The vapour is dry saturated when it enters the compressor and leaves it in a superheated condition. The condenser temperature is 30°C. Assuming specific heat at constant pressure for *R*-12 in the superheated condition as 1,884 kJ/kg.K, determine:

(a) condition of vapour at the entrance to the condenser;

(b) condition of vapour at the entrance to the evaporator; and

(c) theoretical COP of the machine. The properties of *R*-12 are:

Temperature, °C	*Enthalpy, kJ/kg*		*Entropy, kJ/kg K*	
	Liquid	*Vapour*	*Liquid*	*Vapour*
–20	17.82	178.73	0.0731	0.7087
30	64.59	199.62	0.2400	0.6843

[**Ans.** 33.8°C; 29% dry; 4.07]

5.11 A CO_2 refrigerating plant fitted with an expansion valve, works between the pressure limits of 54.81 bar and 20.93 bar. The vapour is compressed isentropically and leaves the compressor cylinder at 32°C. The condensation takes place at 18°C in the condenser and there is no undercooling of the liquid. Determine the theoretical coefficient of performance of the plant. The properties of CO_2 are:

Pressure bar	*Saturation temperature, °C*	*Enthalpy, kJ/kg*		*Entropy, kJ/kg K*	
		Liquid	*Vapour*	*Liquid*	*Vapour*
54.81	+18	137.48	302.55	0.5065	1.0738
20.93	–18	43.27	323.06	0.1733	1.2692

[**Ans.** 4.92]

5.12 A single stage NH_3 refrigeration system has cooling capacity of 500kW. The evaporator and condenser temperatures are –10°C and 30°C respectively. Assuming saturation cycle, determine: (a) mass flow rate of refrigerant; (b) adiabatic discharge temperature; (c) compressor work in kW; (d) condenser heat rejection; (e) COP; and (f) compressor swept volume in m³/min, if volumetric efficiency is 70%.

The following values may be taken;

h_g (– 10°C) = 1431.6 kJ/kg : h_f (30°C) = 322.8 kJ/kg :

v_g (–10°C) = 0.4185 m³/kg; s_g = (–10°C) = 5.4717 kJ/kg·K.

The properties of superheated NH_3 at condenser pressure of 11.66 bar (30°C) are as follows:

At 85°C, h = 1621.8 kJ/kg; s = 5.5484 kJ/kg·K.

At 90°C, h = 1634.5 kJ/kg; s = 5.4838 kJ/kg.K;

[**Ans.** 0.45 kg/s; 88.3°C; 89.5kW; 590 kW; 5.585; 16.2 m³/min]

5.13 A 15 TR Freon-22 vapour compression system operates between a condenser temperature of 40°C and an evaporator temperature of 5°C.

(a) Determine the compressor discharge temperature:

(i) Using the *p-h* diagram for Freon-22

(ii) Using saturation properties of Freon-22 and assuming the specific heat of its vapour as 0.8 kJ/kg.K.

(iii) Using superheat tables for Freon-22.

(b) Calculate the theoretical piston displacement and power consumption of the compressor per ton of refrigeration.

5.14 A simple saturation ammonia compression system has a high pressure of 1.35 MN/m² and a low pressure of 0.19 MN/m². Find per 400,000 kJ/h of refrigerating capacity, the power consumption of the compressor and COP of the cycle.

5.15 (a) A Freon-22 refrigerating machine operates between a condenser temperature of 40°C and an evaporator temperature of 5°C. Calculate the increase (percent) in the theoretical piston displacement and the power consumption of the cycle:

(i) If the evaporator temperature is reduced to 0°C,

(ii) If the condenser temperature is increased to 45°C.

(b) Why is the performance of a vapour compression machine more sensitive to change in evaporator temperature than to an equal change in the condenser temperature?

5.16 In a vapour compression cycle saturated liquid Refrigerant –22 leaving the condenser at 40°C is required to expand to the evaporator temperature of 0°C in a cold storage plant.

(a) Determine the percentage saving in net work of the cycle per kg of the refrigerant if an isentropic expander could be used to expand the refrigerant in place of the throttling device.

(b) Also determine the percentage increases in refrigerating effect per kg of refrigerant as a result of use of the expander. Assume that compression is isentropic from saturated vapour state at 0°C to the condenser pressure.

5.17 An ammonia refrigeration system operates between saturated suction temperature of –20°C, and saturated discharge temperature of +40°C. Compare the COP of the cycle using wet compression with that of the cycle using dry compression.

Assume that the vapour leaving the compressor is saturated in the case of wet compression, and the vapour entering the compressor is saturated in the case of dry compression. The refrigerant leaves the condenser as saturated liquid.

5.18 A standard vapour compression cycle using Freon-22 operates on simple saturation cycle at the following conditions:

Refrigerating capacity = 15 TR
Condensing temperature = 40°C
Evaporating temperature = 5°C

Calculate:

(a) Refrigerant circulation rate in kg/s.
(b) Power required by the compressor in kW.
(c) Coefficient of performance.
(d) Volume flow rate of the refrigerant at compressor suction
(e) Compressor discharge temperature.
(f) Suction vapour volume and power consumption per ton of refrigeration. Refrigeration engineers assume that if this Freon-22 compressor is used with *R*-134a, its capacity would fall by about 40%. Examine this assumption by doing cycle analysis for *R*-134a with the same compressor.

5.19 An ammonia refrigeration plant operates between a condensing temperature of 40°C and an evaporating temperature of –10°C. The vapour is dry at the end of compression. Only the following property values are given:

t	h_f	h_g	s_f
°C	kJ/kg	kJ/kg	kJ/kg.K
40°C	371.5	1473	1.36
–10°C	135.4	1433	0.544

The specific heat of NH_3 vapour is 2.1897 kJ/kg.K. Calculate the theoretical coefficient of performance of the cycle.

5.20 A *R*-134a machine operates at –15°C evaporator and 35°C condenser temperatures. Assuming a simple-saturation cycle, calculate the volume of the suction vapour and power consumption per ton of refrigeration and COP of the cycle.

Calculate the same of the system has a regenerative heat exchanger with the suction vapour leaving at 20°C from the heat exchanger.

5.21 (a) What would be the necessary bore and stroke of a single acting four-cylinder, 350 rpm ammonia compressor working on simple saturation cycle between 35°C condenser and –15°C evaporator temperatures and developing 15 tons refrigeration.

Given for the compressor:

Ratio of stroke to bore = 1
Volumetric efficiency = 0.7

(b) If the index of compression is 1.15, find the error introduced in calculating the work of compression by assuming the process to be isentropic. Also, find the heat rejected to the compressor-jacket cooling water.

5.22 A simple saturation cycle using Freon-22 is designed for a load of 100 TR. The saturated suction and discharge temperatures are 5°C and 40°C respectively. Calculate:

(a) The mass flow rate of refrigerant

(b) The COP and isentropic horse power.

(c) The heat rejected in the condenser.

Use the following data:

t	p	h_f	h_g	s_f	s_g	v_g
°C	*bar*	*kJ/kg*	*kJ/kg*	*kJ/(kg.K)*	*kJ/(kg.K)*	m^3/kg
5	5.836	205.9	407.1	1.02115	1.7447	0.0404
40	15.331	249.53	416.4	1.16659	1.69953	

Specific heat of vapour is 0.65 kJ/(kg·K).

5.23 A commercial refrigerator using Isobutane operates on the simple saturation cycle with saturated suction and discharge temperature of –25°C and 55°C respectively.

(a) Calculate the COP and power required to run the compressor per ton of refrigeration.

(b) If the liquid is subcooled by 10°C in the condenser, calculate the COP and the power required per ton of refrigeration.

(c) If the liquid is further subcooled in a regenerative heat exchanger with superheating of the vapour by 30°C, what is the increase in COP and decrease in power required?

5.24 An ammonica ice plant operators on simple saturation cycle at the following temperatures:

Condensing temperature = 40°C

Evaporating temperature = –15°C

It produces 10 tons of ice per day at –5°C from water at 30°C. Determine:

(a) Capacity of the refrigerant plant.

(b) Mass flow rate of refrigerant.

(c) Isentropic discharge temperature.

(d) Compressor dimensions (bore and stroke) if its volumetric efficiency is assumed as 65%. The compressor is to run at 1400 rpm. Take stroke/bore ratio (L/D) as 1.2.

(e) Horsepower of the compressor if its adiabatic efficiency is taken as 85% and mechanical efficiency as 95%.

(f) Theoretical and actual COP.

5.25 A Refrigerant-22 vapour compression system meant for food freezing operates at 40°C condensing temperature and –35°C evaporating temperature. Its compressor is capable of pumping 30 L/s of vapour at suction.

(a) Calculate the COP of the system and its refrigerating capacity.

(b) If a regenerative heat exchanger is installed which allows suction vapour to be heated by 30°C with liquid from the condenser at 40°C to be cooled correspondingly, what is the new COP and refrigerating capacity?

5.26 A domestic refrigerator uses *R*-134a as refrigerant and operates between evaporator and condenser temperature of –25°C and 55°C in an ambient of 43°C. The liquid leaves the condenser at 43°C. It is further subcooled to 32°C in a regenerative heat exchanger in which the vapour from the evaporator is heated to 32°C. The compressor displacement volume is 4.33 cm^3. Calculate the refrigerating capacity and power consumption of the refrigerator if the volumetric efficiency of the compressor can be taken as 0.68, and its adiabatic efficiency 0.85. Also find the compressor discharge temperature. The compressor is directly coupled to a motor running at 2900 rpm.

5.27 A single compressor system using *R*-12 has two evaporates of 20 ton and 30 ton capacity, both operating at same temperature of 5°C. The condenser temperature is 40°C and the liquid is sub-cooled in the condenser subcooler section by 8°C. The discharge from the evaporators is superheated by 5°C and compression may be assumed to follow the law $pv^{1.08} = C$, Determine:

(i) Refrigerating effect in kJ/kg of flow rate

(ii) Total power, and

(iii) COP.

5.28 The following data refer to a single cylinder, single acting compressor of an ammonia refrigeration system:

Bore	=	100 mm
Stroke	=	150 mm
Speed	=	200 r.p.m.
Indicated mean effective pressure	=	3.2 bar
Condenser pressure	=	10 bar
Evaporator pressure	=	3 bar
Temperature of water at entry to condenser	=	55°C
Temperature of water at exit from condenser	=	20°C
Rate of cooling water flowing in the condenser	=	12.5 kg/min
Inlet water temperature	=	12.5°C
Outlet water temperature	=	20.5°C

If the mass of ice produced per hour from water at 15°C is 50 kg and the latent heat of ice is 335 kJ/kg, find : (a) coefficient of performance; (b) mass flow of ammonia per minute; and (c) condition of ammonia entering the compressor.

The relevant properties of ammonia are given below:

Pressure, bar	*Saturation Temperature, °C*	*Enthalpy, kJ/kg*		*Entropy, kJ/kg K*	
		Liquid	*Vapour*	*Liquid*	*Vapour*
12	31	327.9	1469.5	4.6	2.8
2.9	–10	135.4	1433	–	–

[**Ans.** 4.46; 0.326 kg; 0.86]

Chapter

MULTI–STAGE VAPOUR COMPRESSION SYSTEMS

6

6.1 MULTISTAGE VAPOUR COMPRESSION SYSTEM REQUIREMENT

The temperature achievable in single stage vapour compression refrigeration system is not very low and discharge pressure is also low. In low temperature refrigeration system the vapour refrigerant is required to be delivered at a very high pressure. In such cases, either the vapour refrigerant should be compressed to a very high pressure ratio in a single stage compressor or compress it in two or more compressors placed in series. The compression carried out in two or more compressors is called compound or multistage compression.

The coefficient of performance is the ratio of refrigerating effect to the compressure work. To achieve higher COP, either the refrigerating effect should be increased or the work of compression should be decreased. The compression work is greatly reduced if the refrigerant is compressed very close to the saturated vapour line. This can be achieved by compressing the refrigerant in more stages with intermediate intercooling. It may be remembered that the compressor work with a specified pressure ratio is lower by using multistage compression as compared to single stage compression. The single stage compression is limited to pressure ratios of 4 to 5. For higher pressure ratios, multistage compression is more economical.

The refrigerating effect can be increased by subcooling the refrigerant and removing the flashed vapour.

6.2 ADVANTAGES OF MULTI-STAGE COMPRESSION OVER SINGLE-STAGE COMPRESSION

The main advantages of multi-stage compression over single-stage compression are:

1. The work done per kg of refrigerant is reduced for the same discharge pressure.
2. The volumetric efficiency is improved for the given pressure ratio.
3. The sizes of the LP and HP cylinders may be adjusted to suit the volume and pressure of the refrigerant.

4. The leakage loss is reduced considerably.
5. It gives more uniform torque and hence a smaller size flywheel is required.
6. Because of lower temperature range, effective lubrication can be provided.
7. The cost of compressor is reduced because of lower pressure ratio.

6.3 FLASH GAS REMOVAL

In multi-stage compression, the throttling expansion of the liquid refrigerant may be done in stages as shown in Fig. 6.1 (*a*) and (*b*). Thus the liquid from the condenser at state 6 first expands into a flash chamber to state 7 at the intermediate pressure p_i. Then the liquid from the flash chamber at state 8 enters the evaporator through another expansion valve and expands to state 9.

In system without flash chamber, the liquid from the condenser expands directly to the evaporator pressure as shown by the process line 6-10 in Fig.6.1(*b*). This is wastefuel of energy as the vapour flashed at the intermediate pressure at state 3 is also throttled to state 11 at the evaporator pressure. Therefore, it is required to be again recompressed to the intermediate pressure. A flash chamber, thus, eliminates the undersiable throttling of the vapour generated at the intermediate pressure.

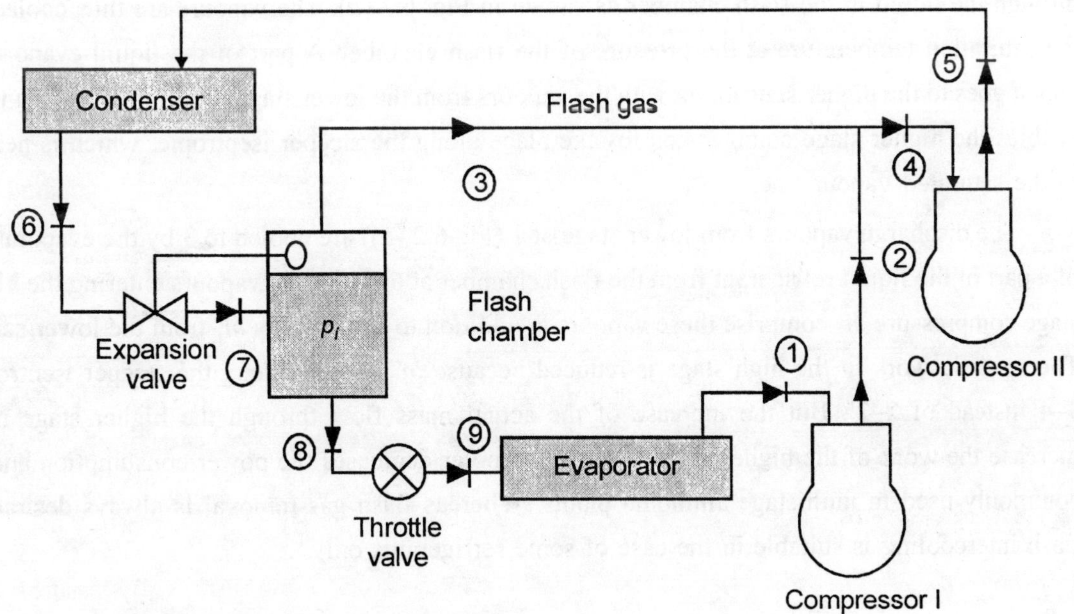

(a) Schematic diagram of two-stage system with flash gas removal.

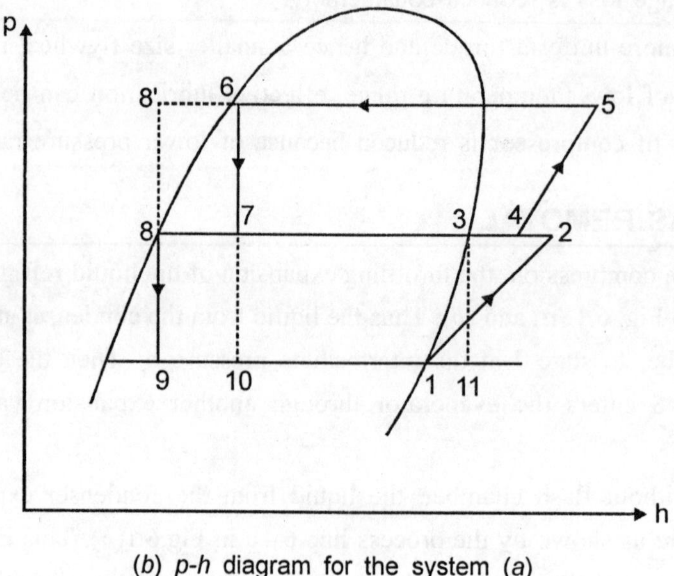

(*b*) *p-h* diagram for the system (*a*)

Fig. 6.1

6.4 FLASH INTERCOOLING

For flash intercooling, the compressed vapours from the lower stage are led and bubbled through the liquid in the flash chamber as shown in Fig. 6.2 (*a*). The vapours are thus cooled to the saturation temperature at the pressure of the flash chamber. A part of the liquid evaporates which goes to the higher state along with the vapours from the lower stage. Flash intercooling thus enables the higher stage compression to take place along the steeper isentropic, which is nearer to the saturated vapour line.

The discharge vapours from lower stage at 1 (Fig.6.2 (*b*)) are cooled to 3 by the evaporation of a part of the liquid refrigerant from the flash chamber at 6. Thus, the vapours entering the high-stage compressure m_2 comprise these vapours in addition to the vapours m_1 from the lower-stage. The specific work in the high stage is reduced because of working along the steeper isentropic 3–4 instead of 2–2′. But the increase of the actual mass flow through the higher stage may increase the work of the higher stage. Flash intercooler decreases the power consumption and is commonly used in multistage ammonia plants. Whereas flash gas removal is always desirable, flash intercooling is suitable in the case of some refrigerants only.

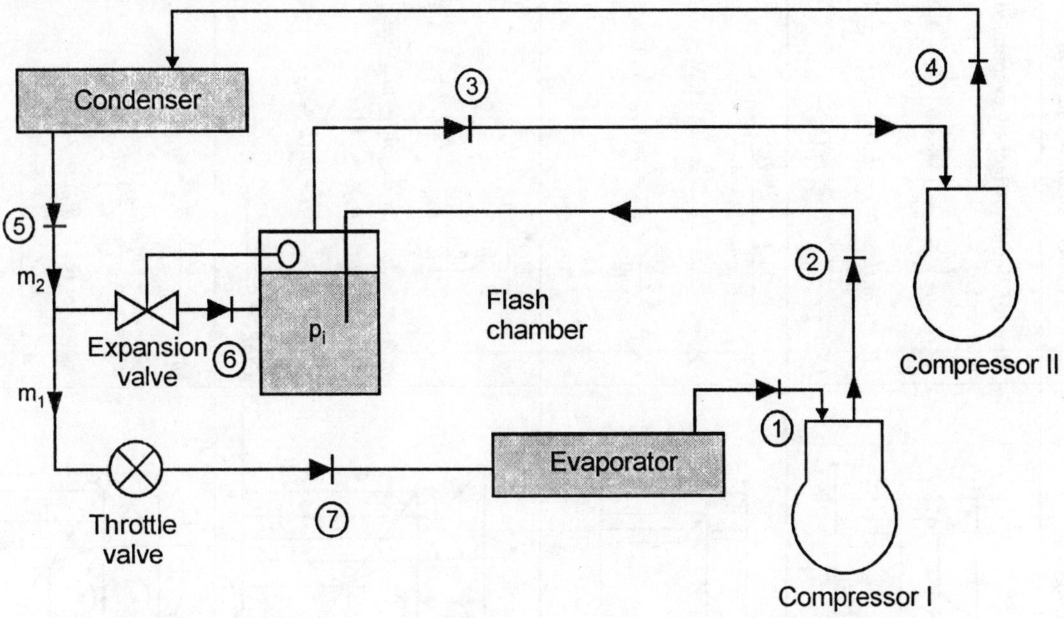

(*a*) Schematic diagram of a two-stage compression system with flash intercooling.

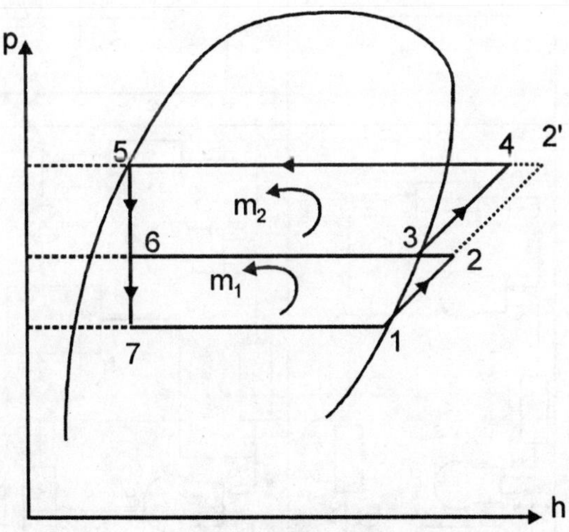

(*b*) *p-h* diagram for the system (a)

Fig. 6.2

Table 6.1: Types of intercoolers for multistage vapour compression systems

Type of intercooler	*Schematic diagram*	*p-h diagram*	*Power required and COP*
1. Two stage - liquid intercooler			Mass of liquid refrigerant evaporated in intercooler, $m_2 = m_1\left(\frac{h_2 - h_{f5}}{h_3 - h_{f5}}\right) \cdot m_3 = m_1\left(\frac{h_2 - h_9}{h_3 - h_{f5}}\right)$ $P = m_1(h_2 - h_1) + m_2(h_4 - h_3) \times \frac{1}{60}$ kW $\text{COP} = \frac{211Q}{60P}, R_E = m_1(h_1 - h_{f5})$
2. Two stage - water intercooler and liquid sub - cooler			$m = \frac{211Q}{h_1 - h_2} = \frac{211Q}{h_1 - h_{f6}}$ kg/min $P = \frac{m[(h_2 - h_1) + (h_4 - h_3)]}{60}$ kW $\text{COP} = \frac{211Q}{60P},\ R_E = m(h_1 - h_{f6})$
3. Two stage - water intercooler, liquid subcooler and liquid flash chamber			$m_1 = \frac{211Q}{h_1 - h_{11}},\ h_{11} - h_{f10}$ $m_3 = m_2\left(\frac{h_3 - h_{f10}}{h_9 - h_{f10}}\right),\ h = h_{f7},$ $m_2h_4 = m_3h_9 + m_1h_3,\ m_1 = m_2 - m_3$ $RE = m_1(h_1 - h_{11}) = 211Q$ $P = m_1(h_2 - h_1) + m_2(h_5 - h_4) \times \frac{1}{60}$ kW $\text{COP} = \frac{3.5167Q}{P}$
4. Two stage - water intercooler, liquid sub cooler and flash intercooler			$m_1 = \frac{211Q}{h_1 - h_{10}},\ h_{f9} = h_{10}$ $m_2 = m_1\left(\frac{h_3 - h_{f9}}{h_4 - h_3}\right),\ h_{f7} = h_8$ $R_E = m_1(h_1 - h_{10}) = 211Q$ $P = \frac{m_1(h_2 - h_1) + m_2(h_5 - h_4)}{60}$ $\text{COP} = \frac{211Q}{60P}$

6.5 CONFIGURATIONS OF MULTSTAGE SYSTEMS

The different configurations for multistage vapour compression systems are of the following types:

(*a*) Two-stage compression.

(*i*) Liquid intercooler.

(*ii*) Water intercooler.

(*iii*) Water intercooler, liquid intercooler, and liquid flash chamber.

(*iv*) Water intercooler, liquid subcooler and flash intercooler.

(*b*) Three-stage compression.

(i) Flash chambers.

(ii) Water intercoolers.

(iii) Flash intercoolers.

The power required and COP of various types of intercoolers are given in Table 6.1.

6.5.1 Two-stage Compression with Liquid Intercooler

The schematic diagram of a two-stage compression with liquid intercooler is shown in Fig. 6.3 (*a*). The corresponding *p-h* diagram is shown in Fig. 6.3 (*b*). The various processes are:

Process 1-2: Isentropic compression from evaporator pressure p_E to pressure p_2.

Process 2-3: Constant pressure cooling at pressure $p_2 = p_3$.

Process 3-4: Isentropic compression from pressure p_3 to condenser pressure p_c.

Process 4-5: Change of stage of refrigerant from superheated vapour to saturated liquid in the condenser.

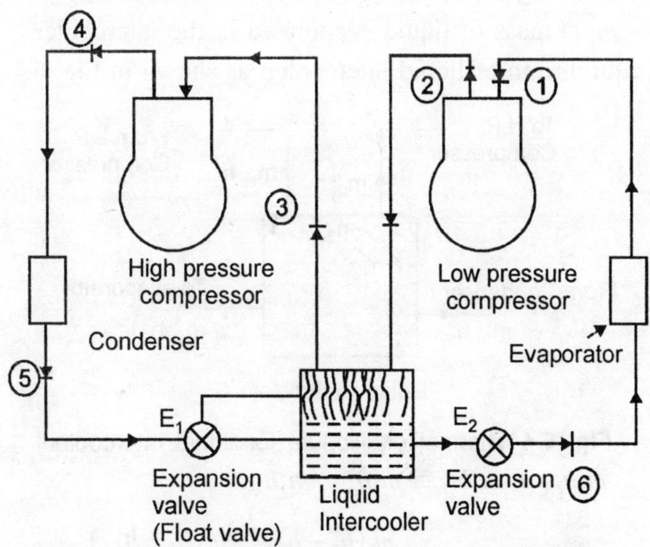

(*a*) Two stage compression with liquid intercooler

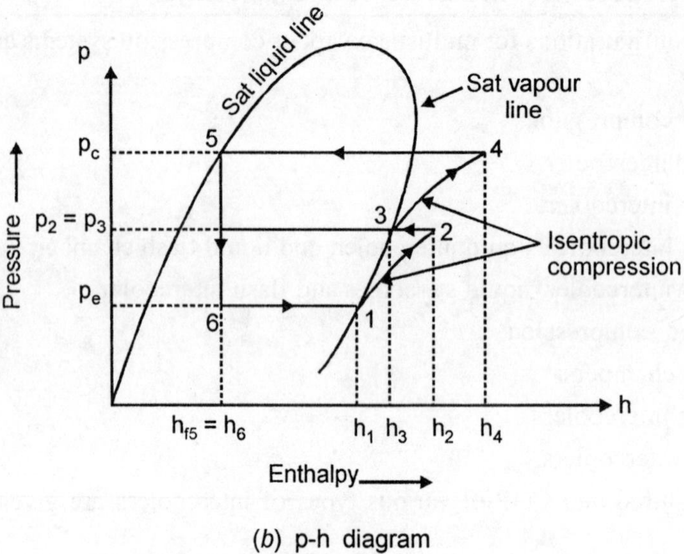

(*b*) p-h diagram

Fig. 6.3

Then the high pressure saturated liquid refrigerant from the condenser is passed to the liquid intercooler where some of the liquid refrigerant evaporates in desuperheating the superheated vapour refrigerant from the LP compressor. To maintain a constant liquid level, an expansion valve E_1, which acts as float valve, is provided. Thereafter the liquid refrigerant from the liquid intercooler is first expanded in expansion valve E_2 and then evaporated in the evaporator (process:6-1) to saturated vapour conditions.

Let, m_1 = mass of refrigerant passing through the evaporator, kg/min

m_2 = mass of refrigerant passing through the condenser, kg/min

$m_3 = m_2 - m_1$ = mass of liquid evaporated in the intercooler.

For the thermal equilibrium of liquid intercooler, as shown in Fig. 6.4, we have

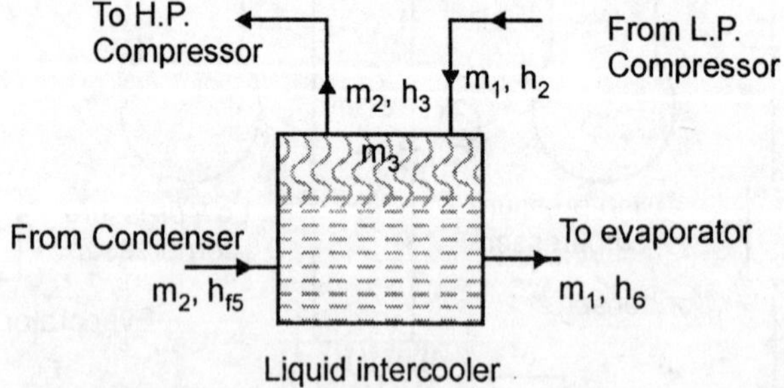

Fig. 6.4 Thermal equilibrium for liquid intercooler

$$m_2 h_{f5} + m_1 h_2 = m_1 h_6 + m_2 h_3$$

$$\therefore \quad m_2 = \frac{m_1(h_2 - h_6)}{h_3 - h_{f5}} = \frac{m_1(h_2 - h_{f5})}{h_3 - h_{f5}} \qquad (\because h_6 = h_{f5})$$

$$m_3 = m_2 - m_1 = \frac{m_1(h_2 - h_{f5})}{h_3 - h_{f5}} - m_1$$

$$= \frac{m_1(h_2 - h_3)}{h_3 - h_{f5}}$$

Refrigerating effect, $RE = m_1(h_1 - h_{f5}) = 211\dot{Q}$ kJ/min ...(6.1)

Total work done in both the compressors,

$$W = m_1(h_2 - h_1) + m_2(h_4 - h_3) \quad ...(6.2)$$

Power required to derive the system,

$$P = \frac{W}{60} \text{kW} \quad ...(6.3)$$

$$\text{COP of the system} = \frac{RE}{W} = \frac{m_1(h_1 - h_{f5})}{m_1(h_2 - h_1) + m_2(h_4 - h_3)} = \frac{211\dot{Q}}{P \times 60} \quad ...(6.4)$$

6.5.2 Two-stage Compression with Water Intercooler and Liquid Sub-cooler

The arrangement of a two-stage compression with water intercooler and liquid sub-cooler is shown in Fig. 6.5(*a*). The corresponding p-h diagram is shown in Fig. 6.5(*b*). The various processes in this system are as follows:

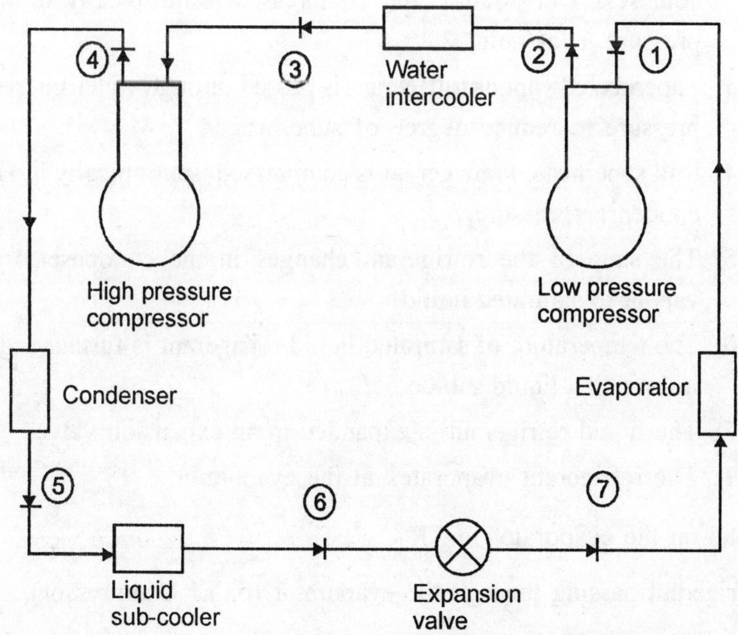

(*a*) Two stage compression with water intercooler and liquid sub-cooler

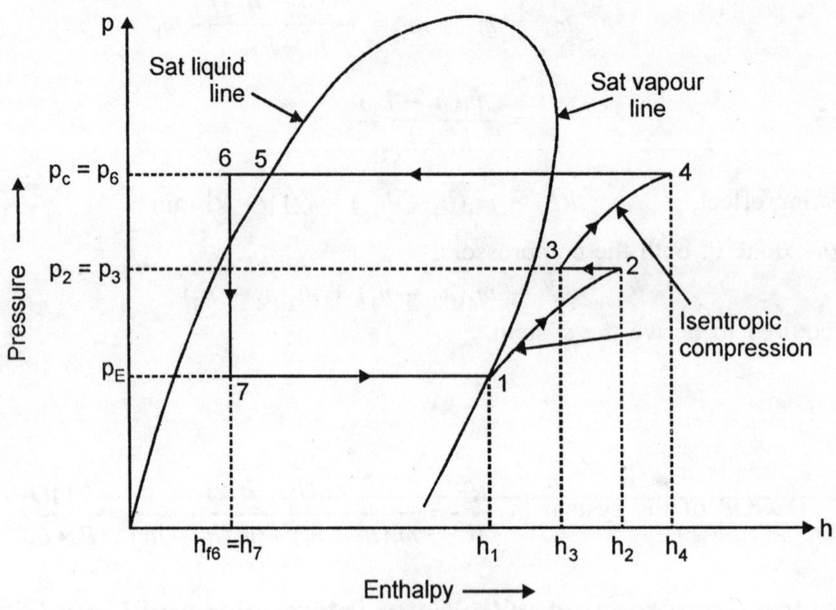

(*b*) p-h diagram

Fig. 6.5

Process 1-2: Saturated vapour refrigerant at evaporator pressure p_E is admitted to *LP* compressor at point 1 and compressed isentropically to water intercooler pressure p_2 at point 2.

Process 2-3: Superheated vapour refrigerant is passed through water intercooler at constant pressure to reduce degree of superheat.

Process 3-4: Still superheated refrigerant is compressed isentropically in HP compressor to condenser pressure p_c.

Process 4-5: The state of the refrigerant changes in the condenser from superheated vapour to saturated liquid.

Process 5-6: The temperature of saturated liquid refrigerant is further reduced by passing it through a liquid sub-cooler.

Process 6-7: The liquid refrigerant is expanded in an expansion valve.

Process 7-1: The refrigerant evaporates in the evaporator.

Let $\dot{Q}$ = load on the evaporator in TR

Mass of refrigerant passing through the evaporator (or LP compressor),

$$m = \frac{211\dot{Q}}{(h_1 - h_7)} = \frac{211\dot{Q}}{(h_1 - h_{f6})} \text{ kg/min} \qquad \text{...(6.5)}$$

The mass of refrigerant passing through LP and HP compressors is same. Therefore, total work done in both the compressors,

$$W = W_{LPC} + W_{HPC}$$
$$= [(h_2 - h_1) + (h_4 - h_3)] \text{ kJ/min} \quad ...(6.6)$$

Power required to derive the system,

$$P = \frac{W}{60} \text{ kW} \quad ...(6.7)$$

$$\text{Refrigerating effect, } RE = m(h_1 - h_{f6}) = 211\dot{Q} \text{ kJ/min} \quad ...(6.8)$$

$$\text{C.O.P. of the system} = \frac{RE}{W} = \frac{m(h_1 - h_{f6})}{m[(h_2 - h_1) + (h_4 - h_3)]} = \frac{3.5167\dot{Q}}{P} \quad ...(6.9)$$

6.5.3 Two-stage Compression with Water Intercooler, Liquid Sub-cooler and Liquid Flash Chamber

The arrangement for a two-stage compression with water intercooler, liquid sub-cooler and liquid flash chamber is shown in Fig. 6.6(*a*). The corresponding *p-h* diagram is shown in Fig. 6.6(*b*). The various processes in this system are as follows:

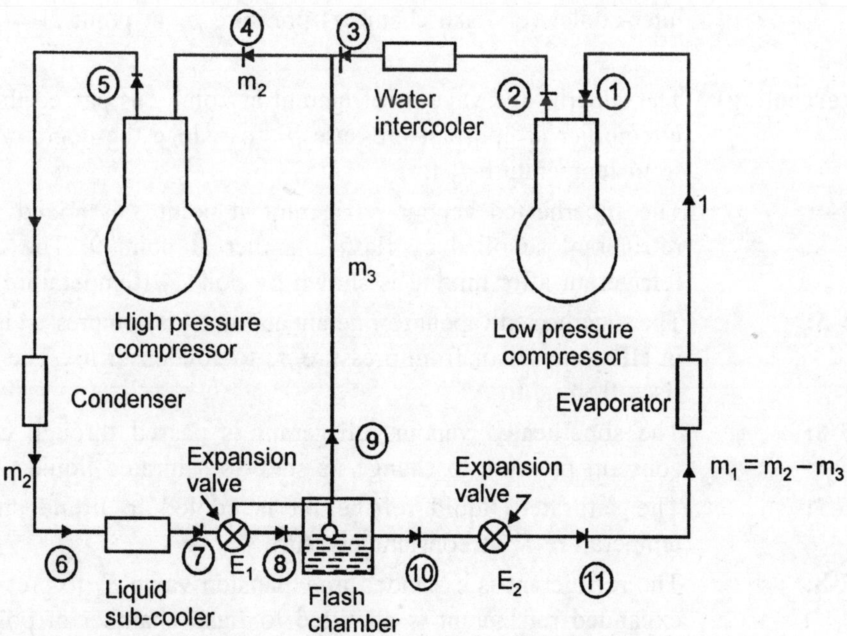

(*a*) Two stage compression with water intercooler, liquid sub-cooler and liquid flash chamber.

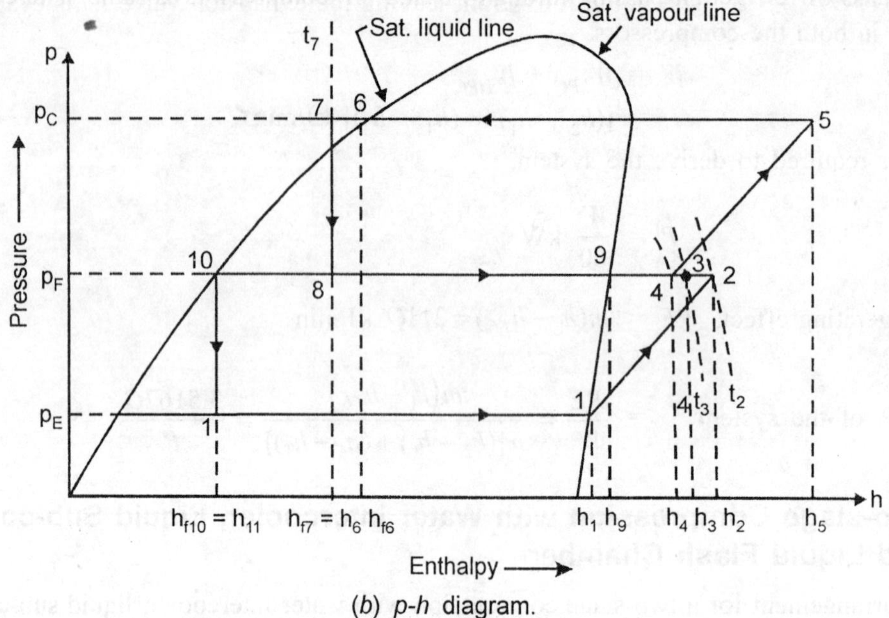

(*b*) *p-h* diagram.

Fig. 6.6

Process 1-2: The saturated vapour refrigerant at evaporator pressure p_E is admitted to *LP* compressor at point 1 and compressed isentropically to water inter-cooler (or flash chamber) pressure p_F at point 2.

Process 2-3 (water intercooling): The superheated vapour refrigerant at point 2 is passed through water intercooler at constant pressure p_F to reduce the degree of superheat from temperature t_2 to t_3.

Process 3-4: The superheated vapour refrigerant at point 3 is mixed with vapour refrigerant supplied by flash chamber at point 9. The condition of refrigerant after mixing is shown by point 4 (temperature t_4).

Process 4-5: The superheated vapour refrigerant at point 4 is compressed isentropically in HP compressor from pressure p_F to condenser p_C. The temperature rises from t_4 to t_5.

Process 5-6: The superheated vapour refrigerant is passed through condenser at constant pressure to change its state to saturated liquid.

Process 6-7: The saturated liquid refrigerant is cooled in liquid sub-cooler to temperature t_7 at constant pressure p_c.

Process 7-8: The refrigerant is expanded in expansion valve E_1 to pressure p_F. The expanded refrigerant is admitted to flash chamber at point 8, which separates the vapour and liquid refrigerants at pressure p_F. The vapour refrigerant at point 9 is mixed with the refrigerant from water inter-cooler and the liquid refrigerant at point 10 is further expanded in expansion valve E_2, as shown by process 10-11.

Process 11-9: The liquid refrigerant leaving the expansion valve E_2 is evaporated in the evaporator at pressure p_E.

Let m_2 = mass of refrigerant passing through the condenser (or HP compressor), kg/min.

m_3 = mass of vapour refrigerant formed in the flash chamber, kg/min.

Then mass of refrigerant passing through evaporator (or LP compressor),

$$m_1 = m_2 - m_3$$

$\dot{Q}$ = load on the evaporator in tons of refrigeration.

$$\therefore \quad m_1 = \frac{211\dot{Q}}{(h_1 - h_{11})} = \frac{211\dot{Q}}{(h_1 - h_{f10})} \text{ kg/min} \qquad (\because h_{11} = h_{f10})$$

For thermal equilibrium of flash chamber, we have

Heat taken by the flash chamber = Heat given by the flash chamber

$$m_2 h_8 = m_3 h_9 + m_1 h_{f10}$$

$$= m_3 h_9 + (m_2 - m_3)\, h_{f10}$$

$$m_2\,(h_8 - h_{f10}) = m_3(h_9 - h_{f10})$$

$$\therefore \quad m_3 = m_2\left[\frac{h_8 - h_{f10}}{h_4 - h_{f10}}\right] = m_2\left[\frac{h_{f7} - h_{f10}}{h_9 - h_{f10}}\right] \qquad ...(a) \qquad (\because h_8 = h_{f7})$$

The vapour refrigerant from the water intercooler (point 3) is mixed with vapor refrigerant m_3 from the flash chamber (point 9) at the same pressure before entering the HP compressor. The enthalpy of mixed refrigerant (point 4) may be calculated from the following equation:

$$m_2 h_4 = m_3 h_9 + m_1 h_3$$

$$= m_3 h_9 + (m_2 - m_3) h_3 = m_2 h_3 + m_3(h_9 - h_3)$$

Refrigerating effect of the system,

$$RE = m_1(h_1 - h_{11}) = 211\dot{Q} \text{ kJ/min} \qquad ...(6.10)$$

Work done in LP compressor,

$$W_{\text{LPC}} = m_1(h_2 - h_1)$$

Work done in HP compressor,

$$W_{\text{HPC}} = m_2(h_5 - h_4)$$

Total compressor work,

$$W_c = W_{\text{LPC}} + W_{\text{HPC}} = m_1(h_2 - h_1) + m_2(h_5 - h_4) \qquad ...(6.11)$$

Power required to drive the system,

$$P = \frac{W_c}{60} \text{ kW} \qquad ...(6.12)$$

$$\text{C.O.P. of the system} = \frac{RE}{W_c} = \frac{211\dot{Q}}{P \times 60} = \frac{3.5167\,\dot{Q}}{P} \qquad ...(6.13)$$

6.5.4 Two-stage Compression with Water Intercooler, Liquid Sub-cooler and Flash Intercooler

A two-stage compression with water intercooler, liquid sub-cooler and flash intercooler is shown in Fig. 6.7(*a*). The corresponding p-h diagram is shown in Fig. 6.7(*b*). The various processes, in this system, are as follows:

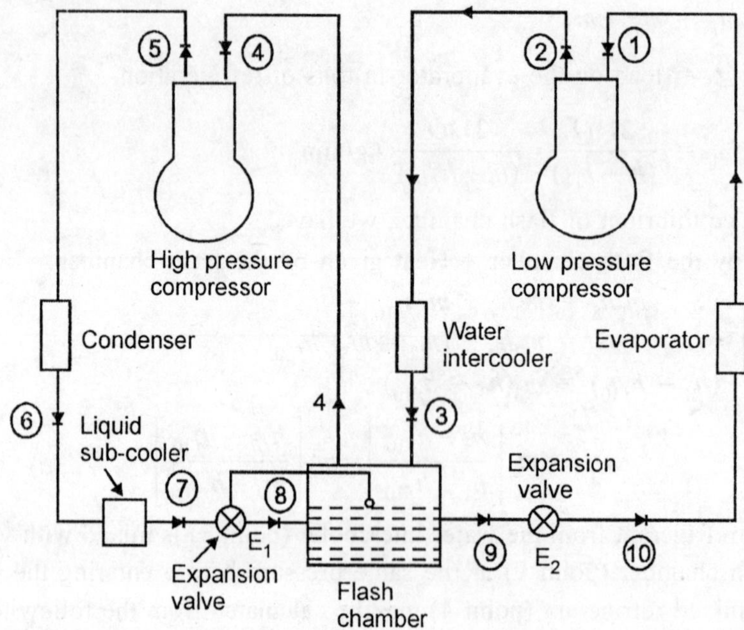

(*a*) Two stage compression with water intercooler, liquid sub-cooler and flash intercooler.

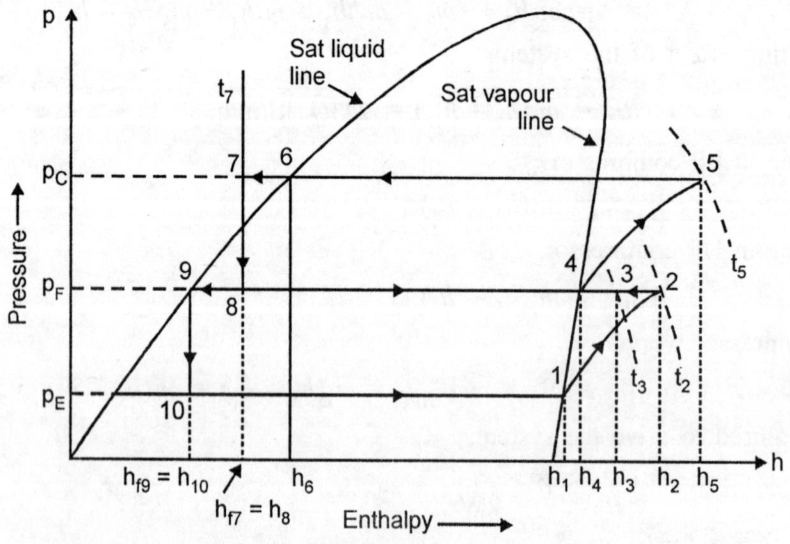

(*b*) p-h diagram

Fig. 6.7

Process 1-2: The saturated vapour refrigerant at evaporator pressure p_E is admitted to the *LP* compressor at point 1 and compressor isentropically to flash intercooler pressure p_F at point 2.

Process 2-3: The super heated vapour refrigerant at point 2 is passed through the water intercooler at constant pressure p_F to reduce the degree of superheat from temperature t_2 to t_3.

Process 3-4: The superheated vapour refrigerant at point 3 is passed through a flash inter cooler which reduces it to saturated vapour refrigerant at point 4. The cooling of refrigerant is done by the evaporation of a part of the liquid refrigerant from the flash inter cooler placed at point 8.

Process 4-5: The saturated vapour refrigerant enters the HP compressor at point 4 and is compressed isentropically from pressure p_F to condenser pressure P_c.

Process 5-6: The superheated vapor refrigerant is passed through a condenser which changes its state to saturated liquid at point 6.

Process 6-7: The saturated liquid refrigerant is cooled in the liquid sub-cooler to point 7 at temperature t_7.

Process 7-8: The liquid refrigerant is expanded in an expansion valve E_1 to pressure p_F and admitted to flash inter cooler at point 8, which also acts as a flash separator.

Process 9-10: The liquid refrigerant leaving the flash inter cooler at point 9 is passed through the second valve E_2 to point 10.

Process 10-1: The liquid refrigerant is evaporated in the evaporator.

Let m_1 = mass of refrigerant through the evaporator (or LP compressor.)

m_2 = mass of refrigerant passing through the condenser (or HP compressor)

$\dot{Q}$ = refrigeration load on the evaporator, tons

Then
$$m_1 = \frac{211\dot{Q}}{(h_1 - h_{10})} = \frac{211\dot{Q}}{(h_1 - h_{fg})} \text{ kg/min} \qquad (\because h_{10} = h_{fg})$$

For the thermal equilibrium of flash inter cooler, we have

Heat taken by the flash intercooler = Heat given by the flash intercooler

$$m_2h_8 + m_1h_3 = m_2h_4 + m_1h_{fg}$$

or
$$m_1(h_3 - h_{f4}) = m_2(h_4 - h_8)$$

$$m_2 = m_1\left[\frac{h_3 - h_{fg}}{h_4 - h_8}\right] = m_1\left[\frac{h_3 - h_{fg}}{h_4 - h_{f7}}\right] \qquad (\because h_8 = h_{f7})$$

Refrigerating effect, $RE = m_1(h_1 - h_{10}) = m_1(h_1 - h_{fg}) = 211\dot{Q}$ kJ/min ...(6.14)

Work done in both the compressors,

$$W_c = W_{\text{LPC}} + W_{\text{HPC}}$$
$$= m_1(h_2 - h_1) + m_2(h_5 - h_4) \text{ kJ/min} \qquad \text{...(6.15)}$$

Power required to derive the system,

$$P = \frac{W_c}{60} \text{ kW} \qquad \text{...(6.16)}$$

$$\text{C.O.P.} = \frac{RE}{W_c} = \frac{m_1(h_1 - h_{f9})}{m_1(h_2 - h_1) + m_2(h_5 - h_4)} = \frac{211\dot{Q}}{P \times 60} = \frac{3.5167\dot{Q}}{P} \qquad \text{...(6.17)}$$

6.5.5 Three-stage Compression with Water Inter Cooler

The arrangement of three-stage compression with water intercoolers is shown in Fig. 6.8(*a*). The corresponding p-h diagram is shown in Fig. 6.8(*b*). The following processes are involved in this system:

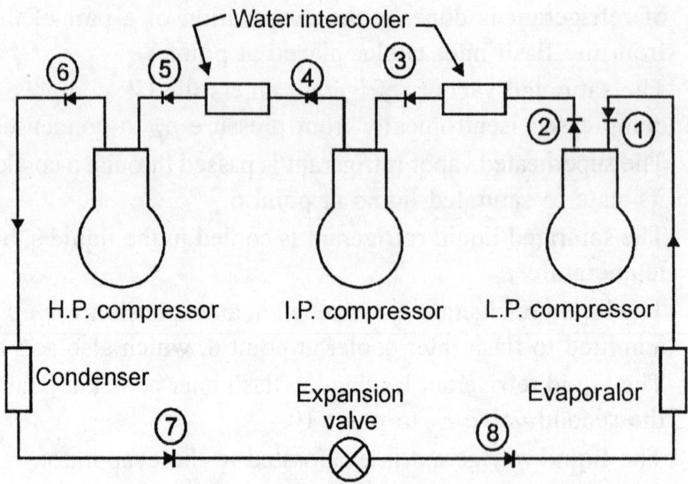

(*a*) Three stage compression with water intercoolers.

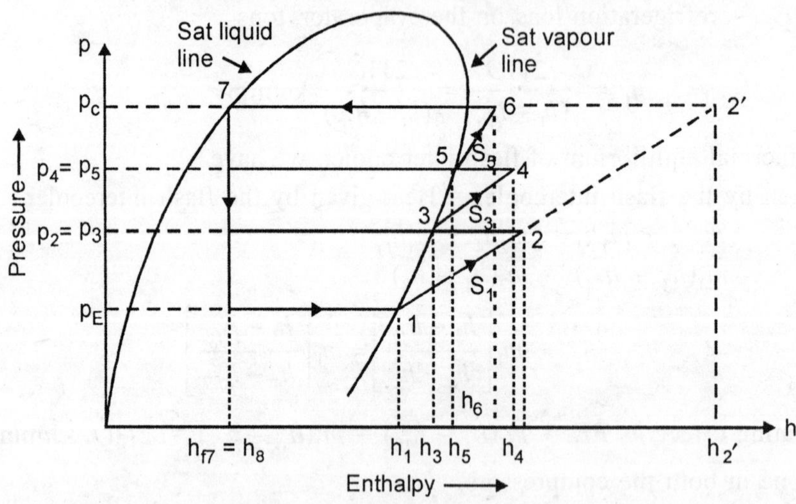

(*b*) *p-h* diagram.

Fig. 6.8

Process 1-2: The saturated vapour refrigerant at point 1 is isentropically compressed to point 2 in LP compressor to superheated state.

Process 2-3: Constant pressure cooling in the water intercooler to saturated vapour refrigerant to state point 3.

Process 3-4: Isentropic compression of saturated vapour refrigerant to superheated refrigerant to state point 4 in IP compressor.

Process 4-5: Constant pressure cooling of superheated refrigerant in water intercooler to saturated vapour refrigerant.

Process 5-6: Isentropic compression of saturated vapour refrigerant in HP compressor to superheated vapour refrigerant.

Process 6-7: Cooling of superheated vapour refrigerant to saturated liquid refrigerant in condenser at constant pressure p_c.

Process 7-8: Throttling of saturated liquid refrigerant in expansion valve.

Process 8-1: Evaporation of liquid refrigerant in evaporator.

Let $\dot{Q}$ = refrigeration load on the evaporator, tons.

Mass of refrigerant circulating through the evaporator,

$$m = \frac{211\dot{Q}}{(h_1 - h_8)} = \frac{211\dot{Q}}{(h_1 - h_{f7})} \text{ kg/min}$$

Work done in LP compressor, $W_{LPC} = m(h_2 - h_1)$ kJ/min

Work done in IP compressor, $W_{IPC} = m(h_4 - h_3)$ kJ/min

Work done in HP compressor, $W_{HPC} = m(h_6 - h_5)$ kJ/min

Total work done on three compressors, $W_c = W_{LPC} + W_{IPC} + W_{HPC}$

$$= m[(h_2 - h_1) + (h_4 - h_3) + (h_6 - h_5) \text{ kJ/min} \quad \text{...(6.18)}$$

Power required to derive the three compressors,

$$P = \frac{W_c}{60} \text{ kW} \quad \text{...(6.19)}$$

Refrigerating effect, $R_E = m(h_1 - h_{f7}) = 211\dot{Q}$ kJ/min ...(6.20)

$$\text{C.O.P.} = \frac{RE}{W_c} = \frac{(h_1 - h_{f7})}{[(h_2 - h_1) + (h_4 - h_3) + (h_6 - h_5)]} = \frac{3.5167\dot{Q}}{P} \quad \text{...(6.21)}$$

6.5.6 Three-stage Compression with Flash Chambers

The arrangement of three compressors with multiple expansion valves E_1, E_2 and E_3 and two flash chambers F_1 and F_2 is shown in Fig. 6.9(a). The corresponding p-h diagram is shown in Fig. 6.9(b).

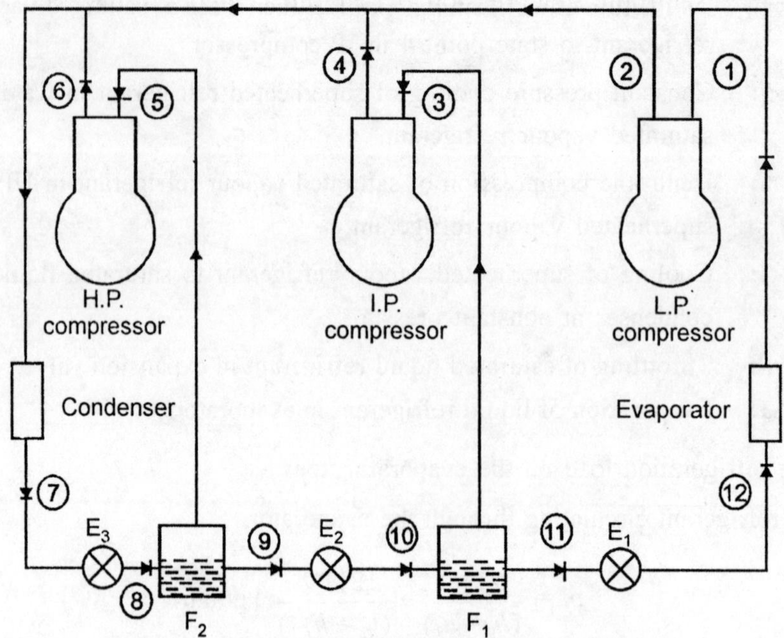

(a) Three stage compression with flash chambers.

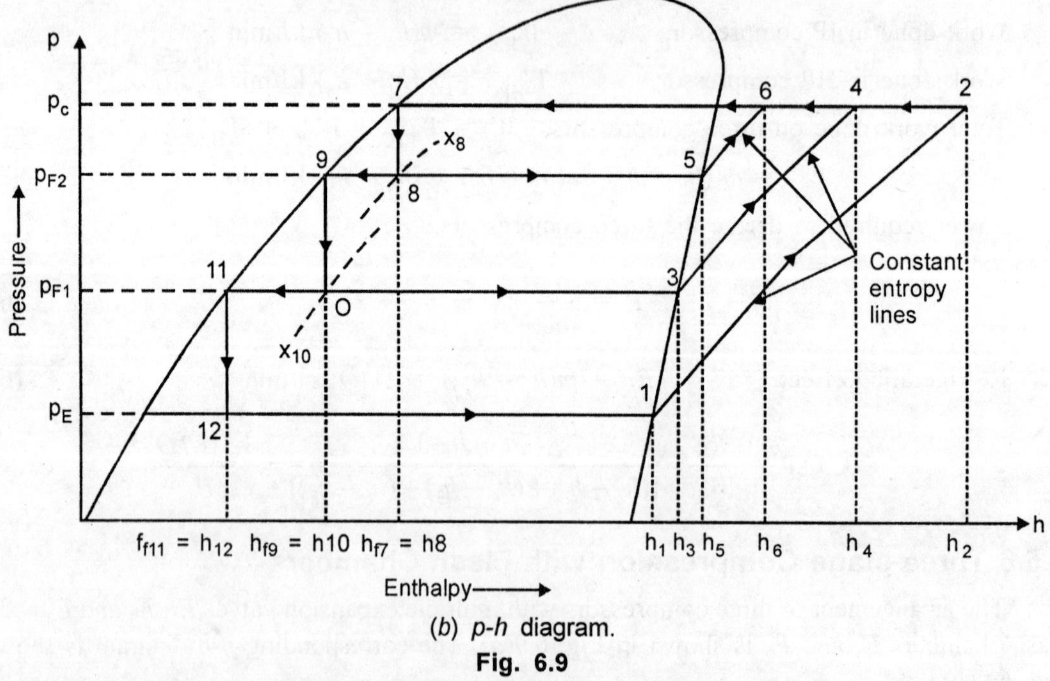

(b) p-h diagram.

Fig. 6.9

Let m = mass of refrigerant leaving the condenser at point 7.

Process 7-8: Saturated liquid refrigerant while passing through the expansion valve E_3 reduces its pressure from p_c to p_{F2}. The refrigerant leaving the expansion valve E_3 at point 8 is separated by the flash chamber F_2. If x_8 = dryness fraction of refrigerant at point 8, then mass of saturated vapour refrigerant separated at point 8 and delivered to HP compressor at point 5 is given by,

$$m_5 = m \times x_8 \text{ kg/min}$$

Mass of saturated liquid refrigerant leaving the flash chamber F_2 at point 9 is given by,

$$m_9 = m - m_5 = m - m \cdot x_8 = m(1 - x_8) \text{ kg/min}$$

Process 9-10: Saturated liquid refrigerant (m_9 kg/min) is passed through the expansion valve E_2 where its pressure reduces from p_{F2} to p_{F1} the refrigerant leaving the expansion valve E_2 is separated by the flash chamber F_1.

If x_{10} = dryness fraction of refrigerant at point 10, then mass of saturated vapour separated at point 10 and delivered to intermediate pressure compressor at point 3 is given by,

Mass of saturated liquid refrigerant leaving the flash chamber F_1 at point 11 is given by,

$$m_{11} = (m_9 - m_3) = m(1 - x_8) - m(1 - x_8) \times x_{10} \text{ kg/min}$$
$$= m(1 - x_8)(1 - x_{10}) \text{ kg/min}$$

Process 11-12: The refrigerant kg/min is passed through the expression valve E_1 where its pressure reduces from p_{F1} to P_E.

Process 12-1: The refrigerant leaving the expansion valve E_1 at point 12 is passed through the evaporator and then to LP compressor at point 1.

Mass of refrigerant passing through the evaporator or LP compressor

= Mass of saturated liquid leaving the flash chamber F_1 at point 11.

$\therefore \quad m_1 = m_{11} = m(1 - x_8)(1 - x_{10})$ kg/min

Let $\dot{Q}$ = refrigeration load on the evaporator

Then $$m_1 = \frac{211\dot{Q}}{h_1 - h_{12}} = \frac{211\dot{Q}}{h_1 - h_{f11}} \text{ kg/min} \qquad (\because h_{12} = h_{f11})$$

Mass of refrigerant passing through the condenser,

$$m = \frac{211\dot{Q}}{(1 - x_8)(1 - x_{10})(h_1 - h_{f11})} \text{ kg/min}$$

Work done in LP compressor, $W_{LPC} = m_1(h_2 - h_1) = m(1 - x_8)(1 - x_{10})(h_2 - h_1)$

Work done in IP compressor, $W_{IPC} = m_3(h_4 - h_3) = m(1 - x_8) \times x_{10} \times (h_4 - h_3)$

Work done in HP compressor, $W_{HPC} = m_5(h_6 - h_5) = m \times x_8(h_6 - h_5)$

Total work done on three compressors, $W_c = W_{LPC} + W_{IPC} + W_{HPC}$ kJ/min ...(6.22)

Power required to drive all the three compressors,

$$P = \frac{W_c}{60} \text{ kW} \qquad ...(6.23)$$

Refrigerating effect, $RE = 211\dot{Q}$ kJ/min ...(6.24)

C.O.P. of the system $= \dfrac{RE}{W_c} = \dfrac{3.5167\dot{Q}}{P}$ ...(6.25)

6.5.7 Three-stage Compression with Flash Inter Coolers

The arrangement for three-stage compression with flash inter coolers is shown in Fig. 6.10(*a*). The corresponding *p-h* diagram is shown in Fig. 6.10(*b*). Flash coolers help in superheating completely the superheated vapour at the end of each stage. The following processes are involved in this system:

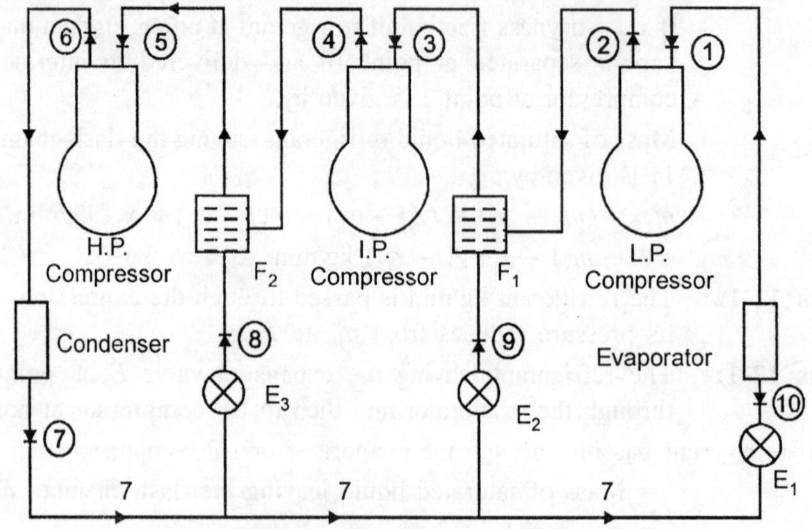

(*a*) Three stage compression with flash intercoolers.

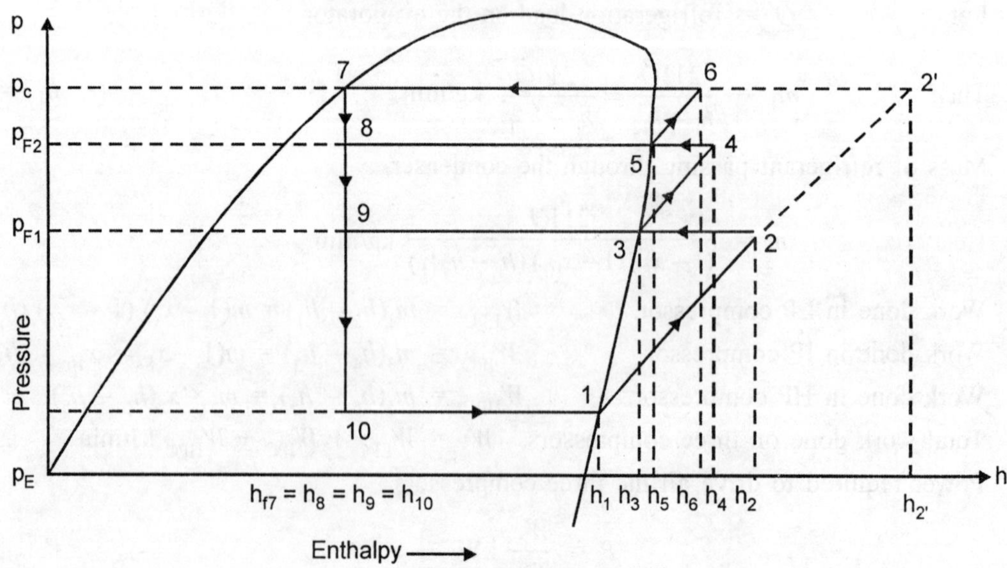

(*b*) *p-h* diagram.

Fig. 6.10

Process 1-2: Isentropic compression of saturated vapour refrigerant at point 1 to superheated state at point 2 in LP compressor.

Process 2-3: The superheated vapour is cooled to saturated vapour in flash inter cooler F_1 by the liquid refrigerant from expansion valve E_2 (as represented by point 9) at pressure p_{F1}.

Process 3-4: Isentropic compression of saturated vapour refrigerant in IP compressor to superheated state.

Process 4-5: The superheated vapour is cooled to saturated vapour in flash inter cooler F_2 by the liquid refrigerant from expansion valve E_3 (as represented by point 8) at pressure p_{F2}.

Process 5-6: Isentropic compression of saturated vapour refrigerant in HP compressor to saturated state.

Process 6-7: Cooling of superheated vapour refrigerant to saturated liquid state in the condenser at pressure p_c.

Process 7-10: Throttling of saturated liquid refrigerant in expansion valve E_1.

Process 10-1: Heating of refrigerant in the evaporator to saturated vapour state.

Let $\dot{Q}$ = refrigeration load on the evaporator in tons

Mass of refrigerant passing through the evaporator or LP compressor,

$$m_1 = \frac{211\dot{Q}}{(h_1 - h_{10})} = \frac{211\dot{Q}}{(h_1 - h_{f7})} \text{ kg/min} \qquad (\because h_{10} = h_{f7})$$

If m_3 = mass of refrigerant passing through the IP compressor, kg/min

m_5 = mass of refrigerant passing through the HP compressor, kg/min

m_2 = mass of liquid refrigerant evaporated in the flash intercooler F_1.

Then $m_3 = m_1 + m_2$

For thermal equilibrium of flash intercooler F_1, we have

Heat taken by the inter cooler F_1 = Heat given by the flash intercooler F_1

$$m_1 h_2 + m_2 h_9 = m_3 h_3$$

$$= (m_1 + m_2) h_3$$

$$\therefore \quad m_2 = \frac{m_1(h_2 - h_3)}{(h_3 - h_9)} = \frac{m_1(h_2 - h_3)}{(h_3 - h_{f7})} \text{ kg/min} \qquad (\because h_9 = h_{f7})$$

If m_4 = mass of liquid refrigerant evaporated in the flash inter cooler F_2, then

$m_5 = m_3 + m_4$

For the thermal equilibrium of flash inter cooler F_2, we have

Heat taken by the flash intercooler F_2 = Heat given by the flash intercooler F_2

$$m_3h_4 + m_4h_8 = m_5h_5$$
$$= (m_3 + m_4)\, h_5$$
$$\therefore \quad m_4 = \frac{m_3(h_4 - h_5)}{(h_5 - h_8)} = \frac{m_3(h_4 - h_5)}{(h_5 - h_{f7})} \text{ kg/min} \qquad (\because h_8 = h_{f7})$$

Work done in LP compressor, $W_{LPC} = m_1(h_2 - h_1)$

Work done in IP compressor, $W_{IPC} = m_3(h_4 - h_3)$

Work done in HP compressor, $W_{HPC} = m_5(h_6 - h_5)$

Total work done in three compressors, $W_c = W_{LPC} + W_{IPC} + W_{HPC}$ kJ/min ...(6.26)

Power required to derive the three compressors,

$$P = \frac{W_c}{60} \qquad ...(6.27)$$

Refrigeration effect, $RE = m_1(h_1 - h_{10}) = 211\dot{Q}$ kJ/min ...(6.28)

COP of the system $= \dfrac{RE}{W_c} = \dfrac{3.5167Q}{P}$ ...(6.29)

6.5.8 Three-stage Compression with Multiple Expansion Valves and Flash Intercoolers

The arrangement for three-stage compression with multiple expansion valves and flash intercooler is shown in Fig. 6.11(*a*). The corresponding p-h diagram is shown in Fig. 6.11(*b*).

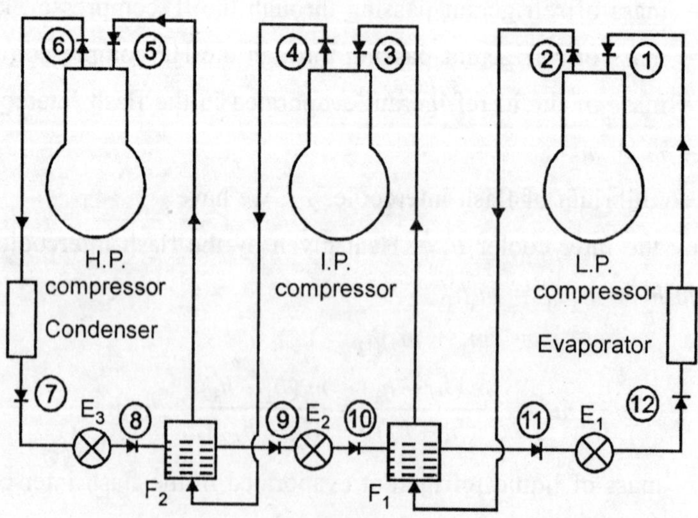

(*a*) Three stage compression with multiple expansion valves and flash intercooler

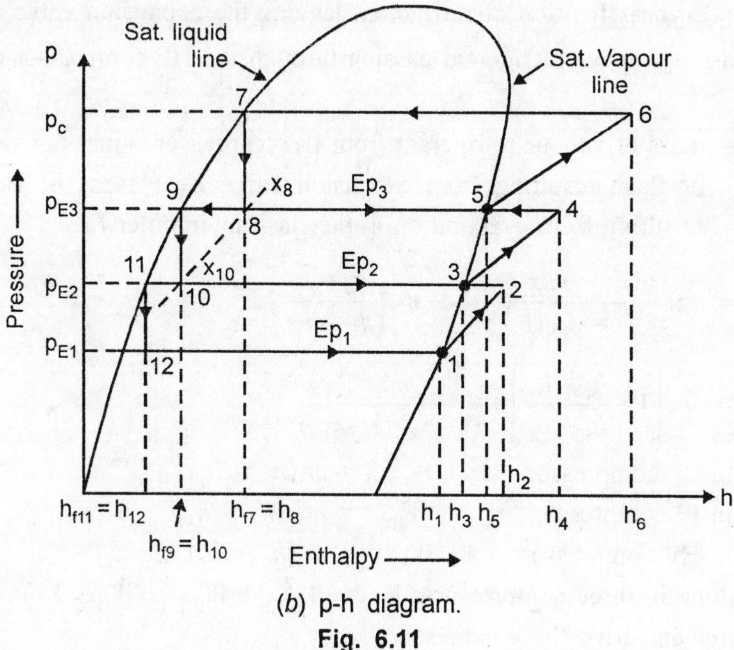

(*b*) p-h diagram.

Fig. 6.11

Let $\dot{Q}$ = refrigeration load on the evaporator, tons.

Mass of refrigerant passing through the evaporator or LP compressor,

$$m_1 = \frac{211\dot{Q}}{(h_1 - h_{12})} = \frac{211\dot{Q}}{(h_1 - h_{f11})} = \text{kg/min}$$

The following processes are involved in this system:

Process 1-2: Isentropic compression of saturated vapour refrigerant to superheat vapours from pressures p_{E1} to p_{E2} in LP compressor.

Process 2-3: Superheating in flash intercooler F_1 by the liquid refrigerant required from the expansion valve E_3 at point 10. During the process of desuperheating, some of the liquid refrigerant gets evaporated and supplied to IP compressor.

Mass of vapour refrigerant passing through the IP compressor, or second stage of compression,

m_2 = mass of vapour refrigerant from LP compressor + mass of vapour refrigerant or flash resulting from expansion valve E_2 + mass of vapour refrigerant resulting from evaporation in flash intercooler F_1.

$$= m_1 + m_1\left(\frac{x_{10}}{1 - x_{10}}\right) + m_1\left(\frac{h_2 - h_1}{h_3 - h_{10}}\right)$$

$$= m_1\left[1 + \frac{x_{10}}{1 - x_{10}} + \frac{h_2 - h_1}{h_3 - h_{10}}\right]$$

where x_{10} = dryness fraction of refrigerant leaving the expansion valve E_2 at point 10.

Similarly, mass of vapour refrigerant passing through the HP compressor or third stage of compression,

m_3 = mass of vapour refrigerant from IP compressor + mass of vapour refrigerant or flash resulting from expansion valve E_3 + mass of vapour refrigerant resulting from evaporation in the flash intercooler F_2.

$$= m_2 + \frac{m_1 x_8}{(1-x_8)(1-x_{10})} + m_2\left(\frac{h_4 - h_5}{h_5 - h_8}\right)$$

$$= m_2\left[1 + \frac{h_4 - h_5}{h_5 - h_8}\right] + \frac{m_1 x_8}{(1-x_8)(1-x_{10})}$$

Work done in LP compressor, $W_{LPC} = m_1(h_2 - h_1)$

Work done in IP compressor, $W_{IPC} = m_2(h_4 - h_3)$

Work done in HP compressor, $W_{HPC} = m_3(h_6 - h_5)$

Total work done in three compressors, $W_c = W_{LPC} + W_{IPC} + W_{HPC}$ kJ/min ....(6.30)

Power requirement drive the compressors,

$$P = \frac{W_c}{60}\text{ kW} \quad ...(6.31)$$

Refrigeration effect, $RE = m_1(h_1 - h_{f11}) = 211\dot{Q}$ kJ/min ...(6.32)

COP of the system $= \dfrac{RE}{W_c} = \dfrac{211\dot{Q}}{60P} = \dfrac{3.5167\dot{Q}}{P}$ ...(6.33)

6.6 MULTISTAGE COMPRESSION WITH MULTIPLE EVAPORATORS

In many refrigeration installations, different temperatures are required to be maintained at various points in the plant such as in hotels, large restaurants, institutions, industrial plants, and food markets. In such cases, each location is cooled by its own evaporator for better control of the system. The various types of multiple evaporator and compressor systems are:

(*i*) Single Compressor-Individual Throttle Valves

A two - evaporator single - compressor system with *individual throttle valves* (TV) for each evaporator and one compressor is shown in Fig. 6.12(*a*). Operation under these conditions means the dropping of pressure from high pressure evaporators through back pressure valves as can be seen in Fig. 6.12(*b*). This necessitates the compression of the vapour from the higher temperature evaporators through a pressure ratio greater than necessary, and hence at the expense of operating economy.

Let Q_1, Q_2 = loads on the evaporators E_1 and E_2 respectively.

Mass of refrigerant required to be circulated through E_1, $m_1 = \dfrac{Q_1}{h_7 - h_5}$

Mass of refrigerant required to be circulated through E_2, $m_2 = \dfrac{Q_2}{h_6 - h_4}$

Enthalpy of vapour mixture entering the compressor, $h_1 = \dfrac{m_1 h_7 + m_2 h_6}{m_1 + m_2}$

Net work done, $W = (m_1 + m_2)(h_2 - h_1)$

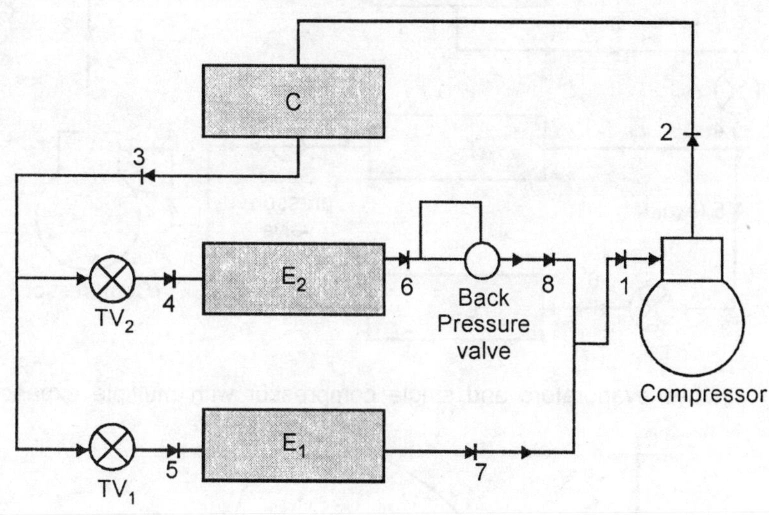

(*a*) System with two evaporators and single compressor, with individual expansion valves

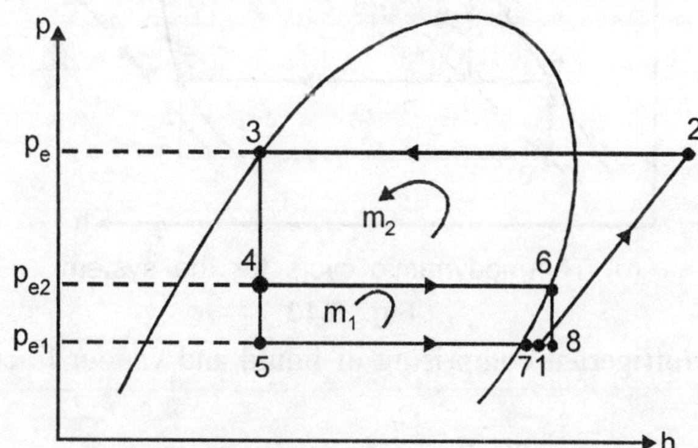

(*b*) Thermodynamic cycle for the system

Fig. 6.12

(*ii*) Single Compressor-Multiple Throttle Valves

The operation of a two-evaporator single-compressor system with *multiple arrangement* of throttle valves is shown in Fig. 6.13 (*a*) and (*b*). The only advantage of the arrangement is that the flashed vapour at the pressure of the high temperature evaporations is not allowed to go to the lower temperature evaporator, thus improving its efficiency. To gain thermodynamic advantage from this, it will be necessary to use individual compressors for each evaporator, thus eliminating the throttle losses of the back pressure valves.

The refrigerating effects for the evaporators per kg of refrigerant are :

$$R_{E1} = h_8 - h_6$$

$$R_{E2} = h_7 - h_4$$

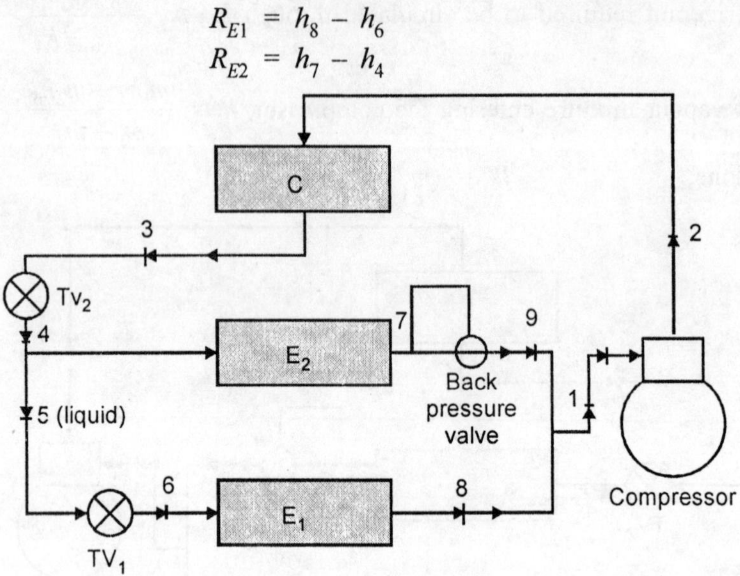

(*a*) System with two evaporators and single compressor with multiple expansion valves

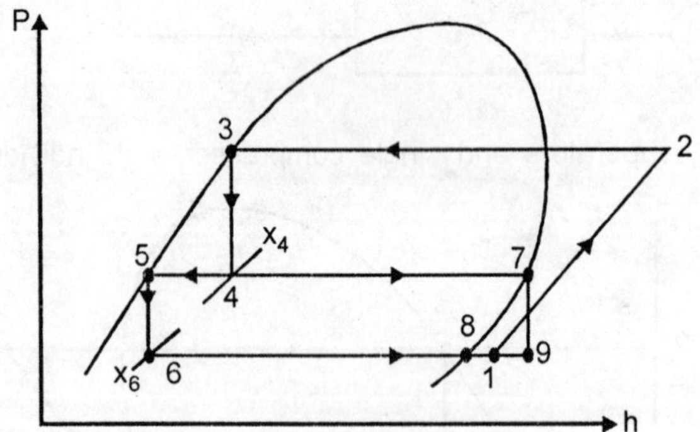

(*b*) Thermodynamic cycle for the system

Fig. 6.13

Mass flow rate of refrigerant comprising of liquid and vapour fractions are:

Evaporator 1, $$m_1 = \frac{Q_1}{R_{E1}}$$

Evaporator 2, $$m_2 = m_1 + m_1\left(\frac{x_4}{1 - x_4}\right)$$

where the second term on R.H.S represents the mass of vapour flashed at 4 corresponding to the mass of liquid going to the second evaporator.

***(iii)* Individual Compressors-Multiple Expansion Valves**

The total power requirement can be reduced by the use of individual compressor for each evaporator and by the multiple arrangement of expansion valves as shown in Fig. 6.14(*a*) and (*b*).

This amounts to parallel operation of evaporators and is called *sectionalizing*. There may be a separate condenser for each compressor or a common condenser for the whole plant.

The calculations for mass flow rates are similar to those of the preceding case.

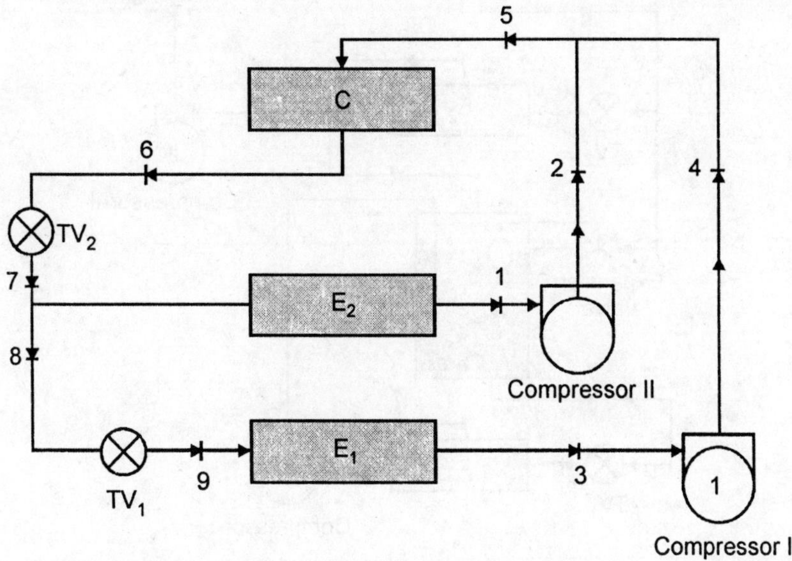

(*a*) System with two evaporators, individual compressors and multiple expansion valves

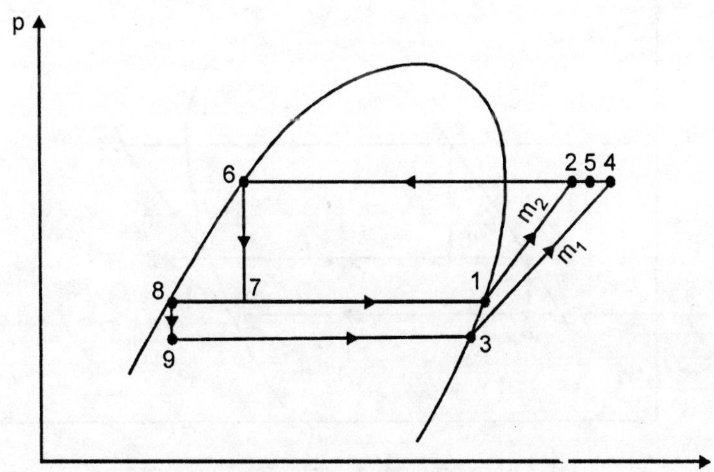

(*b*) Thermodynamic cycle for the system

Fig. 6.14

(*iv*) Individual Compressors with Compound Compression and Flash Interceding

When one or more evaporators are in operation at very low temperatures, and individual compressors are installed. Then compound compression can be used to effect power saving as shown in Figs. 6.15 (*a*) and (*b*) with multiple arrangement of expansion valves and flash intercooling. The flash chamber is maintained at the pressure of the high temperature evaporator.

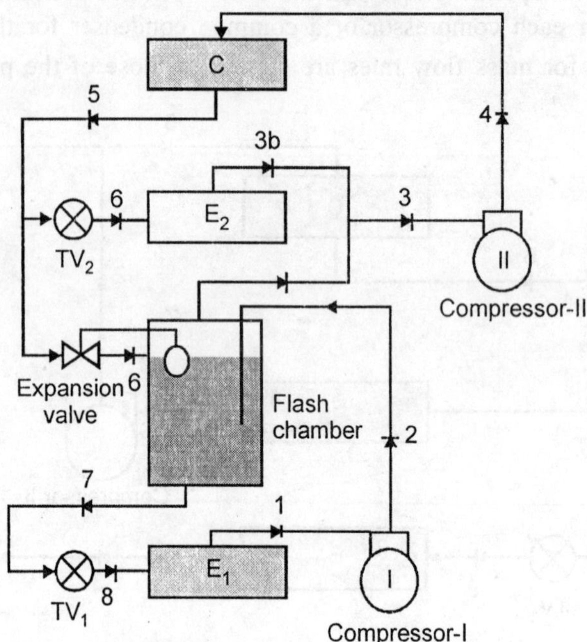

(*a*) System with two evaporators, compound compression and flash inter cooling

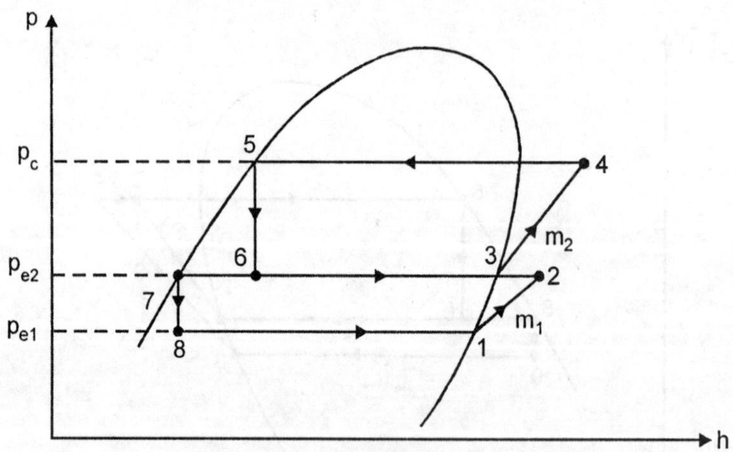

(*b*) Thermodynamic cycle for the system

Fig. 6.15

6.7 CASCADE REFRIGERATION SYSTEM

The cascade refrigeration system consists of two or more vapour compression refrigeration systems in series. Each system in the series uses refrigerants with progressively lower boiling temperatures. A two stage cascade system using two refrigerants is shown in Fig. 6.16(*a*). Its corresponding *p-h* and *T-s* diagrams are shown in Fig. 6.16(*b*) and (*c*) respectively.

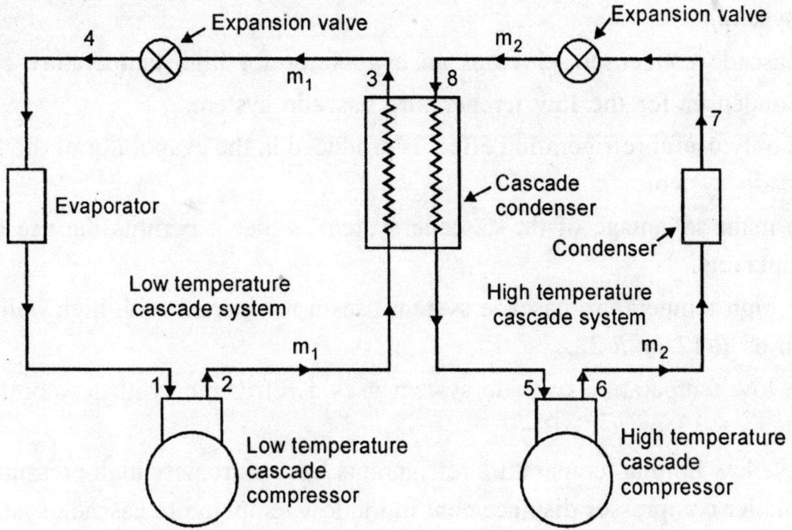

(a) Schematic diagram of a two stage cascade system.

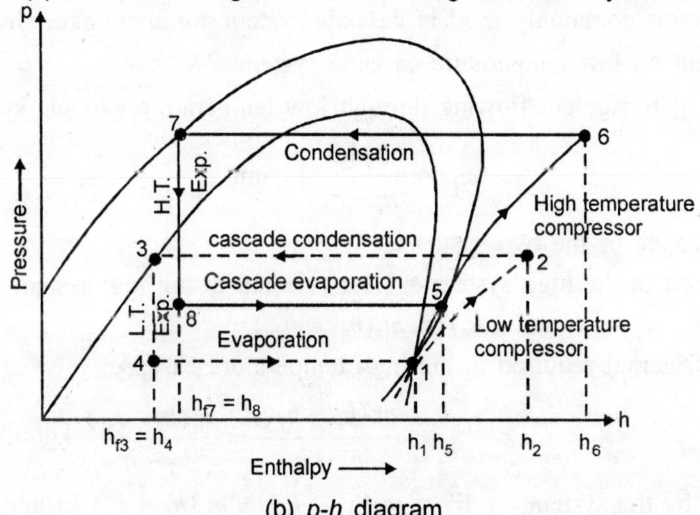

(b) *p-h* diagram

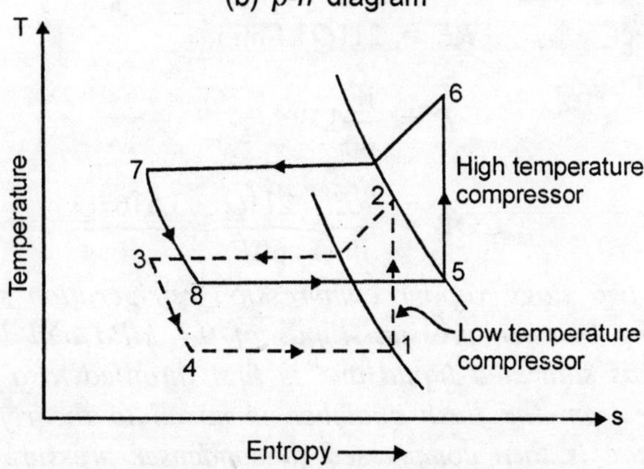

(c) *T-s* diagram

Fig. 6.16 Two stage cascade system.

In this system,

- A cascade condenser serves as an evaporator for high temperature cascade system.
- A condenser for the low temperature cascade system.
- The only useful refrigerating effect is produced in the evaporator of the low temperature cascade system.
- The main advantage of the cascade system is that it permits the use of two different refrigerants.
- The high temperature cascade system uses a refrigerant with high boiling temperature such as *R*-12 or *R*-22.
- The low temperature cascade system uses a refrigerant with low boiling temperature such as *R*-13 or *R*-13 BI.
- These low boiling temperature refrigerants have extremely high pressure which ensures a smaller compressor displacement in the low temperature cascade system and a higher COP.

The refrigerants commonly used in cascade system are ammonia, ethylene and methane.

Let Q = load on low temperature cascade system, *TR.*

Then mass of refrigerant flowing through low temperature cascade system,

$$m_1 = \frac{211Q}{h_1 - h_4} \text{ kg/min} \quad ...(6.34)$$

For heat balance of the two systems,

Heat absorbed in the high system = Heat rejected in the low system

$$m_2(h_5 - h_8) = m_1(h_2 - h_{f3})$$

Mass of refrigerant required in the high temperature cascade,

$$m_2 = \frac{m_1(h_2 - h_{f3})}{h_5 - h_8} = \frac{m_1(h_2 - h_4)}{h_5 - h_8} \text{ kg/min} \quad ...(6.35)$$

Work done by the system, $W = m_1(h_2 - h_1) + m_2(h_6 - h_5)$ kJ/min

Refrigerating effect, $RE = 211Q$ kJ/min

Power required, $$P = \frac{W}{60} \text{ kW} \quad ...(6.36)$$

$$\text{COP} = \frac{RE}{W} = \frac{211Q}{60P} = \frac{3.5167Q}{P} \quad ...(6.37)$$

Example 6.1 *A two stage vapour compression refrigeration system working on, R-12 refrigerant is operating between pressure limits of 0.1 MPa and 1.0 MPa. The refrigerant leaves the condenser as saturated liquid and is first throttled to a flash chamber operating at 0.32 MPa. Vapour from the flash chamber is mixed to the refrigerant leaving the LP compressor. The mixture is then compressed to condenser pressure by HP compressor. The liquid in the flash chamber is throttled to the evaporator pressure and vaporizes in evaporator.*

Assuming that refrigerant leaves the evaporator as a saturated vapour and both compression being isentropic, calculate:

(i) Mass fraction of vapour leaving the flash chamber

(ii) Refrigeration effect per kg of refrigerant

(iii) COP of the system.

Solution: Given : p_c = 1 MPa, p_e = 0.1 MPa, p_f = 0.32 MPa

The schematic diagram for two stage refrigerating system with the given condition is shown in Fig. 6.17(*a*) and the corresponding *p-h* diagram is shown in Fig. 6.17(*b*). The various values as read from the *p-h* diagram for *R*-12 are as follows:

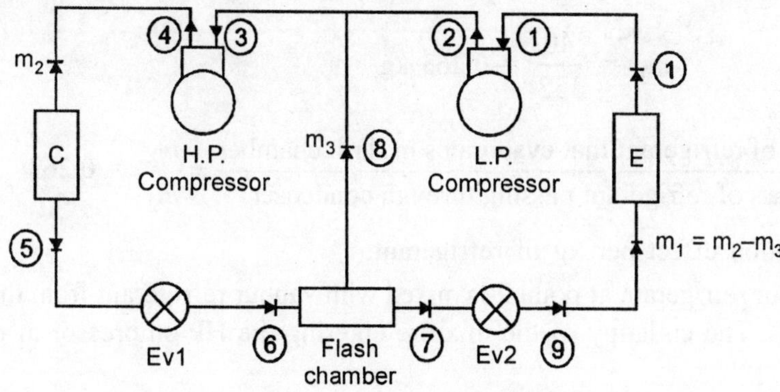

(*a*) Schematic diagram

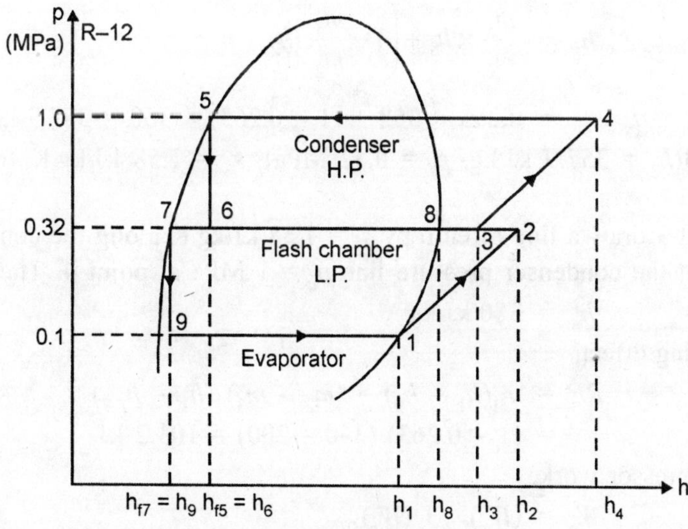

(*b*) *p-h* diagram

Fig. 6.17

$$h_1 = 340 \text{ kJ/kg}, s_1 = 1.56 \text{ kJ/kg·K}$$

$$h_2 = 360 \text{ kJ/kg}, h_8 = 352 \text{ kJ/kg}$$

$$h_{f5} = h_6 = 240 \text{ kJ/kg}, h_{fi} = h_9 = 200 \text{ kJ/kg}$$

(*i*) Mass fraction of vapour leaving the flash chamber.

Let mass of refrigerant passing through the condenser (or HP compressor) be 1 kg, i.e.

$$m_2 = 1 \text{ kg}$$

m_3 = mass of vapour refrigerant formed in the flash chamber, kg

Mass of refrigerant passing through the evaporator (or LP compressor),

$$m_1 = m_2 - m_3$$

For the thermal equilibrium of flash chamber,

$$m_2 h_6 = m_3 h_8 + m_1 h_{f7}$$

$$1 \times 240 = m_3 \times 352 + (1 - m_3) \times 200 = 152\, m_3 + 200$$

$$m_3 = \frac{40}{152} = 0.263 \text{ kg}$$

$$\therefore \quad \frac{\text{Mass of refrigerant that evaporates in flash chamber}}{\text{Mass of refrigerant passing through condenser}} = \frac{m_3}{m_2} = 0.263$$

(*ii*) Refrigeration effect per kg of refrigerant.

The vapour refrigerant at point 2 is mixed with vapour refrigerant from the flash chamber at point 8. The enthalpy of the mixture entering the HP empressor at point 3 is given by,

$$m_2 h_3 = m_3 h_8 + m_1 h_2 = m_3 h_8 + (m_2 - m_3) h_2$$

$$h_3 = \frac{m_3}{m_2} \times h_8 + \left(1 - \frac{m_3}{m_2}\right) h_2$$

$$= 0.263 \times 352 + (1 - 0.263) \times 360 = 357.9 \text{ kJ/kg}$$

At point 3(h_3 = 357.9 kJ/kg, p_f = 0.32 MPa), s_3 = 1.58 kJ/kg.K from *p-h* chart for *R*-12.

From point 3 draw a line of entropy s_3 = 1.58 kJ/kg.K along the constant entropy line to intersect the condenser pressure line p_c = 1 MPa at point 4. Then

$$h_4 = 380 \text{ kJ/kg}$$

Refrigerating effect,

$$RE = m_1(h_1 - h_9) = (m_2 - m_3)(h_1 - h_{f7})$$

$$= (1 - 0.263)(340 - 200) = 103.2 \text{ kJ}$$

(*iii*) Total compressor work,

$$W_c = (W_c)_{LP} + (W_c)_{HP}$$

$$= m_1(h_2 - h_1) + m_2(h_4 - h_3)$$

$$= (1 - 0.263)(360 - 340) + 1(380 - 357.9)$$

$$= 36.84 \text{ kJ}$$

$$\text{COP} = \frac{RE}{W_c} = \frac{103.2}{36.84} = 2.8$$

Example 6.2 *A 100 TR system using R-12 is to operate on a two stage vapour compression refrigeration system with a flash chamber. The refrigerant is evaporating at –40°C (0.641 bar), flash chamber works on an intermediate temperature of 0°C (3.08 bar), and condensation takes place at 30°C (7.45 bar). Low pressure compressor and expansion valve operate between 0.641 bar and 3.08 bar and the high pressure compressor and expansion valve operate between 3.08 bar and 7.45 bar. Saturated vapour enters both the compressors and saturated liquid enters each expansion valve. Determine:*

(1) The flow rate of refrigerant handled by each compressor

(2) The power requirement of compressors

(3) COP of the system.

May use the properties given below and Fig. 6.18

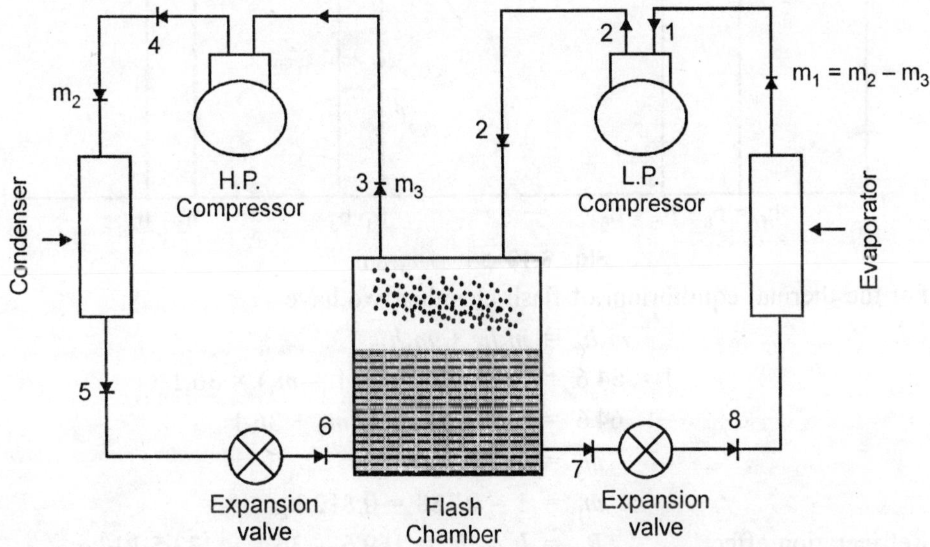

Fig. 6.18 Two stage vapour compression refrigeration system.

Solution: The schematic diagram of two-stage vapour compression refrigeration system with flash chamber is shown in Fig. 6.18. The *p-h* diagram is shown in Fig. 6.19. The various values read from the *p-h* chart for *R*-12 are as follows:

$$h_1 = 169.6 \text{ kJ/kg}, s_2 = 0.7274 \text{ kJ/kg. K}, h_3 = 187.5 \text{ kJ/kg}$$

$$h_2 = 360 \text{ kJ/kg}, h_7 = 187.5 \text{ kJ/kg}, h_4 = 364 \text{ kJ/kg}$$

$$h_{f5} = h_6 = 64.6 \text{ kJ/kg}, h_{f7} = h_8 = 36.1 \text{ kJ/kg}$$

(i) Let $m_4 = 1$ kg

m_3 = mass of vapour refrigerant formed in the flash chamber

$m_1 = m_2 - m_3$ = mass of refrigerant passing through the evaporator.

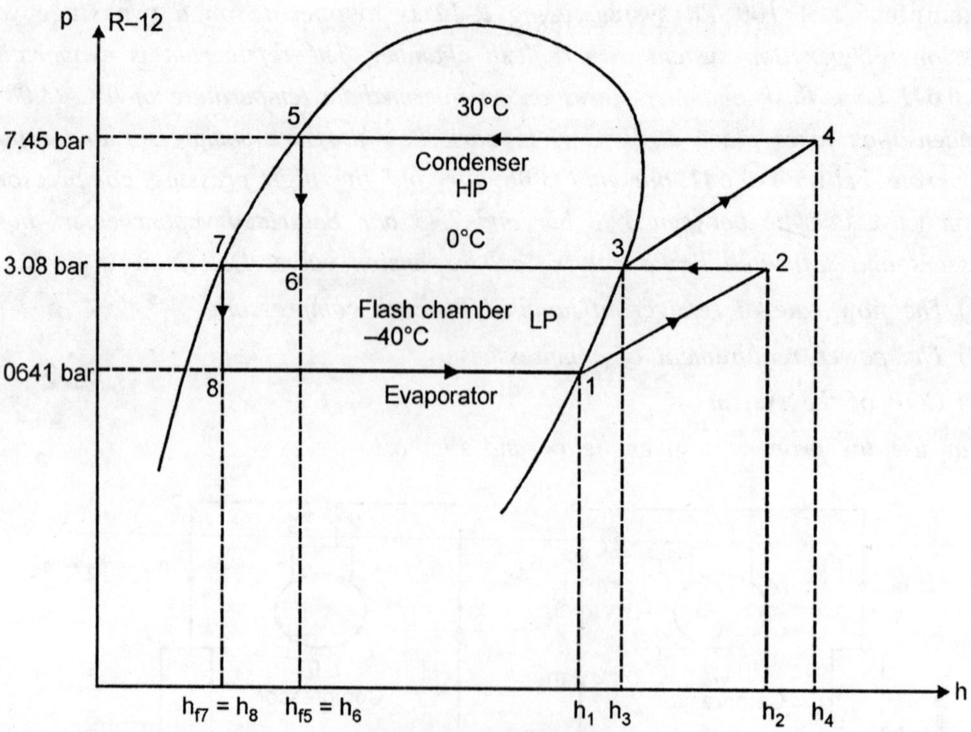

Fig. 6.19 p-h diagram.

(*ii*) For the thermal equilibrium of flash chamber, we have

$$m_2 h_6 = m_3 h_3 + m_1 h_{f7}$$

$$1 \times 64.6 = m_3 \times 187.5 + (1 - m_3) \times 36.1$$

$$64.6 = (187.5 - 36.1)\, m_3 + 36.1$$

$$m_3 = 0.188 \text{ kg}$$

$$m_1 = 1 - 0.188 = 0.812 \text{ kg}$$

Refrigeration effect, $R_E = h_1 - h_8 = 169.6 - 36.1 = 133.5$ kJ/kg

Mass flow rate of refrigerant,

$$\dot{m}_r = \frac{211TR}{R_E} = \frac{211 \times 100}{133.5} = 158 \text{ kg/min}$$

Now $$\frac{m_3}{m_2} = \frac{0.188}{1} = 0.188$$

$$m_3 = 0.188\, m_2$$

$$m_1 = m_2 - m_3 = m_2 - 0.188 m_2 = 0.812 m_2$$

$$\frac{\dot{m}_1}{\dot{m}_2} = 0.812$$

$$\dot{m}_1 + \dot{m}_2 = \dot{m}_r = 158$$

$$\dot{m}_2(0.812+1) = 158$$

$$\dot{m}_2 = 87.2 \text{ kg/min}$$

$$\dot{m}_1 = 70.8 \text{ kg/min}$$

Mass flow rate of refrigerant handled by LP compressor,

$$m_1 = 70.8 \text{ kg/min}$$

Mass flow rate of refrigerant handled by HP compressor,

$$m_2 = 87.2 \text{ kg/min}$$

(*iii*) Power required by LP compressor,

$$P_{LP} = \dot{m}_1(h_2 - h_1) = \frac{70.8(360-169.6)}{60} = 224.7 \text{ kW}$$

Power required by HP compressor,

$$P_{HP} = \dot{m}_2(h_4 - h_3) = \frac{87.2(364-187.5)}{60} = 256.5 \text{ kW}$$

Total power required,

$$P = P_{LP} + P_{HP} = 224.7 + 256.5 = 481.2 \text{ kW}$$

(*iv*) $\text{COP} = \dfrac{3.5167\,TR}{P} = \dfrac{3.5167 \times 100}{481.2} = 0.731$

Example 6.3 *A cascade refrigeration system of 100 tons (350 kW) capacity uses ammonia and carbon dioxide. The evaporating and condensing temperature of CO_2 are –40°C and 5°C, respectively. The evaporating temperature of NH_3 is –7°C. The power supplied to the ammonia compressor is 96.5 kW. In the CO_2 circuit, the liquid leaving the condenser is saturated, the vapor leaving the evaporator is dry and saturated, and compression is isentropic. Calculate the mass flow rate of CO_2 and the C.O.P of the system. The following data for CO_2 may be used.* **[IES, 1994]**

Temperature, °C	*Pressure, bar*	*Enthalpy, kJ/kg*		*Entropy, kJ/kgK*	
		Liquid	*Vapor*	*Liquid*	*Vapor*
–40	10.05	332.7	652.8	3.8531	5.2262
5	39.70	431.0	649.8	4.2231	5.0097

Specific heat of superheated vapor,

$$c_p = 0.85 \text{ kJ/kg K}$$

Solution: The schematic diagram of a two-stage cascade refrigerating system using CO_2 and NH_3 is shown in Fig. 6.20(*a*)

The corresponding *p-h* diagram of the system is shown in Fig. 6.20(*b*). The cycle 1-2-3-4 is for unit whereas the cycle 5-6-7-8 is for unit. Both these cycles are superimposed in the *p-h* diagram.

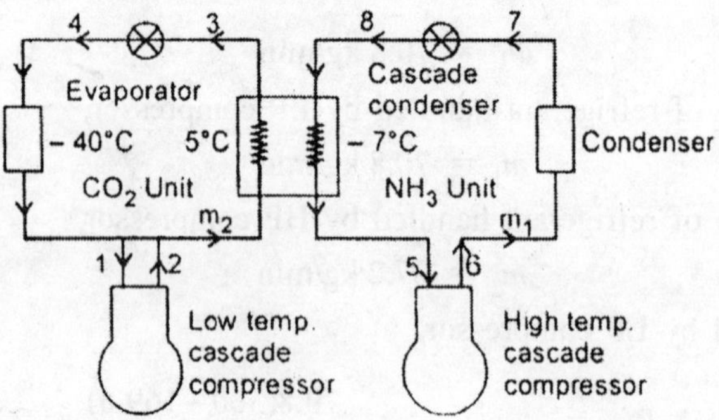

(*a*) Schematic diagram of a two stage cascade refrigerating system.

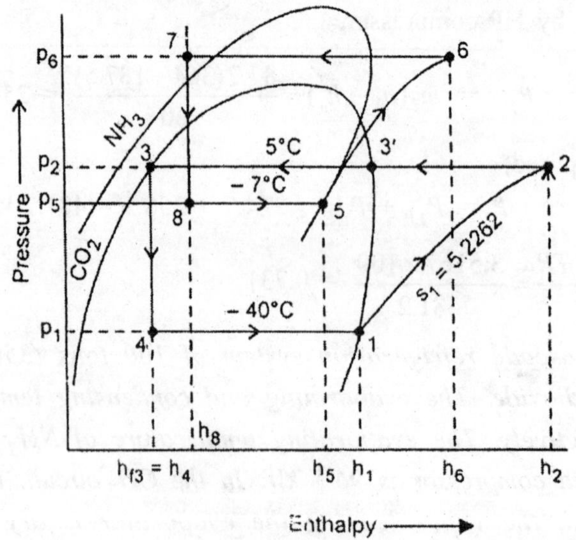

(*b*) p-h diagram

Fig. 6.20

CO_2 – unit:

Refrigerating effect $= h_1 - h_4$

$= h_1 - h_{f3}$

$= 652.8 - 431.0 = 221.8$ kJ/kg

Total refrigerating effect of plant

$= 100$ tons $= 350$ kW

Mass flow rate of CO_2, $\dot{m}_{CO_2} = \dfrac{350}{221.8}$

$= 1.578$ kg/s or 5680 kg/hr

Now $s_1 = s_2$

$$= s_{g2} + c_p \ln \frac{T_2}{(273+5)}$$

$$5.2262 = 5.0097 + 0.85 \ln\left(\frac{T_2}{278}\right)$$

$$T_2 = 358.6 \text{ K}$$

$$\therefore \quad h_2 = h_2' + c_p(T_2 - T_2')$$

$$= 649.8 + 0.85\,(358.6 - 278)$$

$$= 718.31 \text{ kJ/kg}$$

Work done in CO_2 system,

$$W_{CO_2} = \dot{m}_{CO_2} \times w_{CO_2}$$

$$= 1.578 \times 65.51$$

$$= 103.37 \text{ kJ/s}$$

$$\text{COP of system} = \frac{\text{Total refrigerating effect}}{W_{CO_2} + W_{NH_3}}$$

$$\left(W_{NH_3} = 96.5 \text{ kW}\right)$$

$$= \frac{350}{103.37 + 96.5} = \frac{350}{199.87} = 1.75$$

Example 6.4 *An ammonia vapour compression refrigeration plant has two stages of compression and one evaporator. After leaving the condenser at 11.666 bar as saturated liquid without undercooling it passes the first throttle valve from which it emerges at 4.975 bar to enter a flash chamber. The liquid from flash chamber is then throttled to 2.909 bar and passes to the evaporator while the dry saturated vapour is passed to a receiver connecting the L.P and H.P. cylinders. Determine:*

(i) The amount of vapour bled off at the flash chamber,

(ii) The coefficient of performance.

Properties of ammonia: **[CSE, 1999]**

Pressure bar	*Temperature °C*	*Enthalpy kJ/kg*		*Entropy kJ/kg. K*		c_p *value kJ/kgK*
		h_f	h_g	s_f	s_g	
11.665	30	332.9	1467.9	1.2028	4.9805	2.958
4.975	4	199.6	1447.6	0.7815	5.2852	
2.909	−10	135.5	1432.0	0.5440	5.4730	

Solution: The schematic diagram and *p-h* diagram are shown in Fig. 6.21.

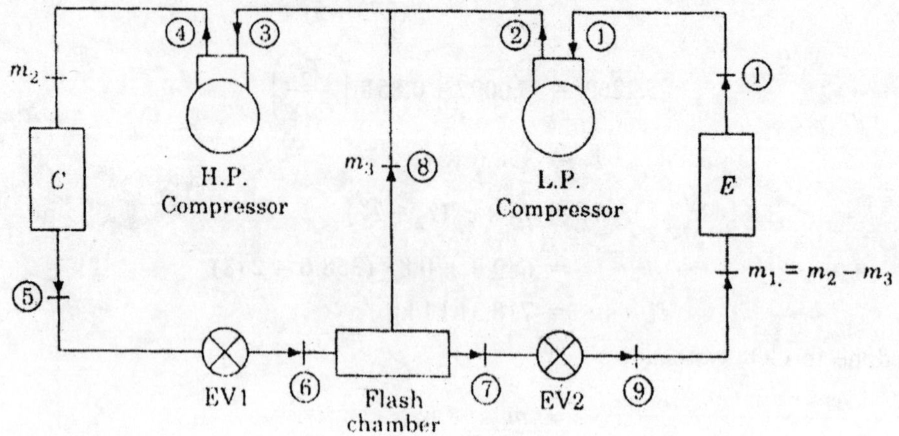

(*a*) Schematic diagram

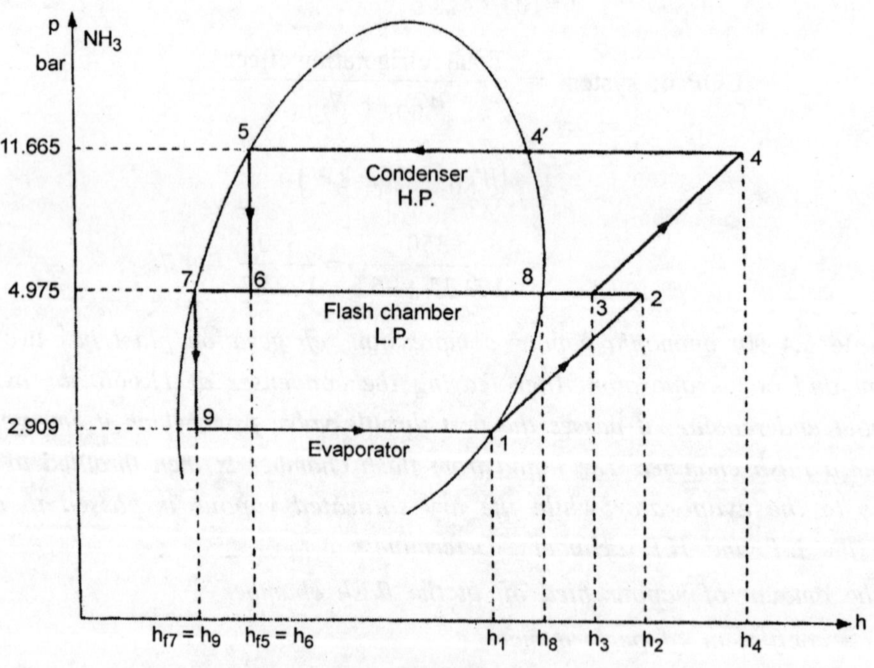

(*b*) *p-h* diagram

Fig. 6.21

$$h_1 = 1432.0 \text{ kJ/kg}$$

$$s_1 = 5.4730 \text{ kJ/kg.K}$$

$$h_8 = 1447.6 \text{ kJ/kg}$$

$$h_2 = h_8 + c_p (T_2 - T_8)$$

$$= 1447.6 + 2.958 (4 + 10) = 1489.0 \text{ kJ/kg}$$

$$h_{f5} = h_6 = 332.9 \text{ kJ/kg}$$
$$h_{f7} = h_9 = 199.6 \text{ kJ/kg}$$

(*i*) Mass fraction of vapour leaving the flash chamber.

Let mass of refrigerant passing through the condenser (or H. P. compressor),

$$m_2 = 1 \text{ kg}$$
$$m_3 = \text{mass of vapour refrigerant formed in the flash chamber.}$$

Mass of refrigerant passing through the evaporator (or L.P. compressor),

$$m_1 = m_2 - m_3 = 1 - m_3$$

For the thermal equilibrium of flash chamber,

$$m_2 h_6 = m_3 h_8 + m_1 h_{f7}$$
$$1 \times 332.9 = m_3 \times 1447.6 + (1 - m_3) \times 199.6$$
$$= 1248\, m_3 + 199.6$$
$$m_3 = 0.1068 \text{ kg}$$

Mass of refrigerant vapour bled off at the flash chamber,

$$m_3 = 0.1068 \text{ kg}$$

(*ii*) The vapour refrigerant at point 2 is mixed with vapour refrigerant from the flash chamber at point 8. The enthalpy of the mixture entering the H.P. compressor at point 3 is given by,

$$m_2 h_3 = m_3 h_8 + m_1 h_2 = m_3 h_8 + (m_2 - m_3)\, h_2$$
$$h_3 = \frac{m_3}{m_2} \times h_8 + \left(1 - \frac{m_3}{m_2}\right) h_2$$
$$= 0.1068 \times 1447.6 + (1 - 0.1068) \times 1489.0$$
$$= 1484.6 \text{ kJ/kg}$$

$$h_4 = h_4' + c_p(t_4 - t_4') = 1467.9 + 2.958\,(30 - 4) = 1544.8 \text{ kJ/kg}$$

Refrigeration effect,

$$RE = m_1\,(h_1 - h_9) = (m_2 - m_3)\,(h_1 - h_{f7})$$
$$= (1 - 0.1068) \times (1432.0 - 199.6)$$
$$= 1100.78 \text{ kJ/kg}$$

Total compressor work,

$$W_c = (W_c)_{LP} + (W_c)_{HP}$$
$$= m_1(h_2 - h_1) + m_2(h_4 - h_3) \qquad [h_3 = h_2]$$
$$= (1 - 0.1068)\,(1489.0 - 1432.0) + 1.0(1544.8 - 1489.0)$$
$$= 50.91 + 55.8 = 106.71 \text{ kJ/kg}$$

$$\text{COP} = \frac{RE}{W_c} = \frac{1100.78}{106.71} = 10.3$$

Example 6.5 *Show that the COP of a cascade refrigeration system is*

$$COP = \frac{COP_1 \times COP_2}{1 + COP_1 + COP_2}$$

where COP_1 *and* COP_2 *are COP's of low temperature and high temperature side respectively.* **[IAS, 2005]**

Solution:

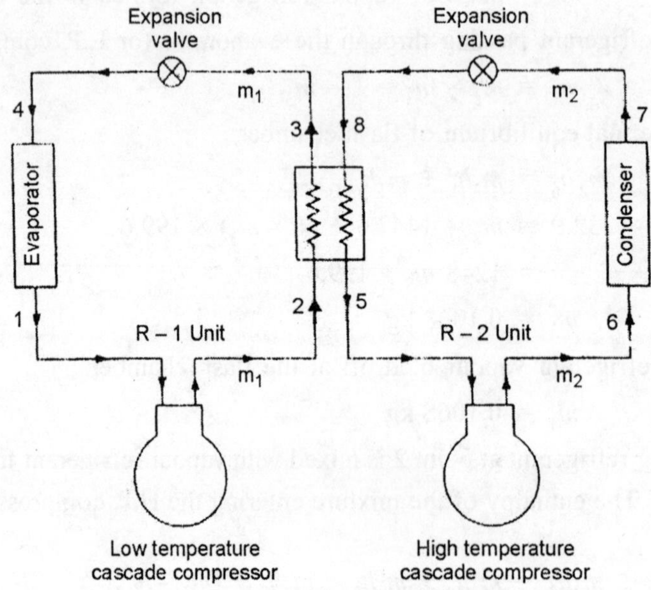

(*a*) Schematic arrangement of a two-stage cascade refrigerating system.

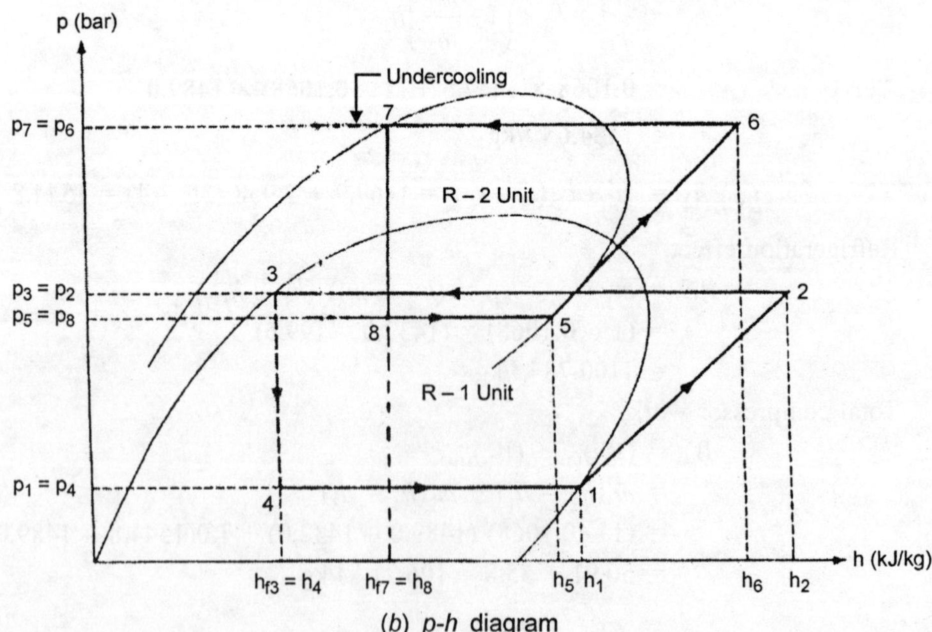

(*b*) *p-h* diagram

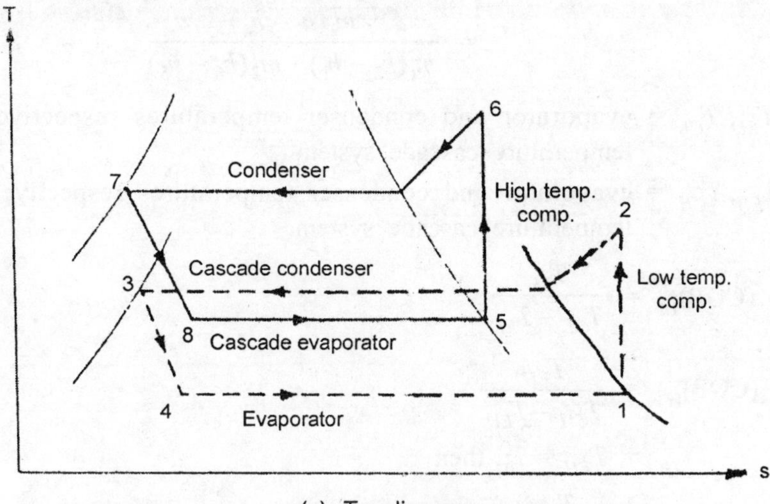

(c) T-s diagram

Fig. 6.22 Two-stage cascade system

The cascade refrigeration system consists of two or more vapour compression refrigerator systems in series which use refrigerants with progressively lower boiling temperature. A two stage cascade system using two refrigerants is shown in Fig. 6 (*a*) and its corresponding *p-h* and *T-s* diagrams are shown in Fig. 6 (*b*) and (*c*) respectively. In this system, a cascade condenser serves as an evaporator for the high temperature cascade system and a condenser for the low temperature cascade system. The only useful refrigerating effect is produced in evaporator of the low temperature cascade system.

Let Q = tonnes of refrigeration load on the low temperature cascade system.

Then the mass of refrigerant flowing through the low temperature cascade system,

$$m_1 = \frac{211Q}{h_1 - h_4} \text{ kg/min}$$

For heat balance, we have

Heat absorbed in the high temperature cascade system

= Heat rejected in the low temperature cascade system

$$\therefore \quad m_2 (h_5 - h_8) = m_1 (h_1 - h_4)$$

$$m_2 = \frac{m_1(h_1 - h_4)}{h_5 - h_8} \text{ kg/min}$$

Total work done by the system,

$$W = m_1(h_2 - h_1) + m_2(h_6 - h_5) \text{ kJ/min}$$

and refrigerating effect, $R_E = 211Q$ kJ/min

Coefficient of performance of the system,

$$\text{COP} = \frac{R_E}{W} = \frac{211Q}{m_1(h_2 - h_1) + m_2(h_6 - h_5)}$$

$$= \frac{m_1(h_1 - h_4)}{m_1(h_2 - h_1) + m_2(h_6 - h_5)}$$

Let T_{EL}, T_{CL} = evaporator and condenser temperatures respectively for the low temperature cascade system

T_{EH}, T_{CH} = evaporator and condenser temperatures respectively for the high temperature cascade system.

Then $$(\text{COP})_{\text{LT}} = \frac{T_{EL}}{T_{CL} - T_{EL}}$$

$$(\text{COP})_{\text{HT}} = \frac{T_{EH}}{T_{CH} - T_{EH}}$$

when $T_{CL} = T_{EH} = T_1$, then

$$(\text{COP})_{\text{LT}} = \frac{T_{EL}}{T_1 - T_{EL}} = (\text{COP})_1$$

and $$(\text{COP})_{\text{HT}} = \frac{T_1}{T_{CH} - T_1} = (\text{COP})_2$$

$$\text{COP} = \frac{m_1(h_1 - h_4)}{m_1(h_2 - h_1) + m_2(h_6 - h_5)}$$

$$= \frac{m_1(h_1 - h_{f3})}{m_1(h_2 - h_1) + m_2(h_6 - h_5)}$$

$$\text{COP} = \frac{(\text{COP})_1 \times (\text{COP})_2}{1 + (\text{COP})_1 + (\text{COP})_2} = \frac{\dfrac{T_{EL}}{T_1 - T_{EL}} \times \dfrac{T_1}{T_{CH} - T_1}}{1 + \dfrac{T_{EL}}{T_1 - T_{EL}} + \dfrac{T_1}{T_{CH} - T_1}}$$

$$= \frac{T_{EL} \times T_1}{(T_1 - T_{EL})(T_{CH} - T_1) + T_{EL}(T_{CH} - T_1) + T_1(T_1 - T_{EL})}$$

$$m_1 = m_2 = \frac{T_{EL}}{T_{CH} - T_{EL}}$$

$$h_2 - h_1 = c_p(T_1 - T_{EL})$$

$$h_6 - h_5 = c_p(T_{CH} - T_1)$$

$$h_1 - h_4 = c_p\, T_{EL}$$

$$\text{COP} = \frac{T_{EL}}{T_1 - T_{EL} + T_{CH} - T_1} = \frac{T_{EL}}{T_{CH} - T_{EL}}$$

Hence proved

MULTIPLE-CHOICE QUESTIONS

1. Multistage compression with intercooling
   (a) permits better lubrication as the discharge temperatures are lowered
   (b) permits better lubrication as the discharge pressures are lowered
   (c) does not affect lubrication as the discharge pressure and temperature are un-affected
   (d) does not permit any lubrication in higher stages of compression.
2. For the same refrigerant, same evaporator pressure, same condenser pressure, same intermediate pressure for two stage vapour compression system
   (a) flash intecooling gives better COP than flash cooling
   (b) flash cooling gives better COP than flash intercooling
   (c) both flash cooling and flash intercooling give same COP
   (d) none of the above
3. In compound compression system with flash intercooling the suction vapours to the second stage of compression are
   (a) superheated (b) wet
   (c) dry saturated (d) may be dry saturated or superheated
4. In compound compression system with flash cooling the suction vapour to the second stage of compression are
   (a) normally superheated (b) normally wet
   (c) normally dry saturated (d) may be dry saturated or superheated
5. During which of following process does heat rejection take place in Carnot vapour cycle
   (a) constant volume (b) constant pressure
   (c) constant temperature (d) constant entropy
6. In a vapour compression refrigeration plant, the refrigerant leaves the evaporator at 195 kJ/kg and the condenser at 65 kJ/kg. For every kg of refrigerant the plant can supply per second, a cooling load of,
   (a) 70 kW (b) 100 kW (c) 130 kW (d) 160 kW
7. A condenser of a refrigeration system rejects heat at the rate of 120 kW, while its compressor consumes a power of 30 kW. The coefficient of performance of the system would be
   (a) 1/4 (b) 1/3 (c) 3 (d) 4
8. Match items in List I with those in list II and III and select the correct answer.

List I	*List II*	*List III*
A. Reversed Carnot engine B. Subcooling C. Superheating D. Constant enthalpy	1. Condenser 2. Evaporator 3. Vortex refrigerator 4. Throttling 5. Heat pump	6. Generator 7. Increase in refrigerating-effect. 8. Highest COP 9. Adiabatic. 10. Dry compression.

Codes:	A	B	C	D
(a)	3, 10	1, 7	2, 9	4, 6
(b)	5, 8	1, 1	2, 10	4, 9
(c)	4 , 10	3, 8	3, 10	1,6
(d)	2, 1	5, 8	4, 6	1,9

9. In a vapour compression refrigeration system, throttle valve is used in place of an expander because

(a) It considerably reduces the system weight.

(b) It improves the COP, as the condenser is small.

(c) The positive work in isentropic expansion of liquid is very small.

(d) It leads to significant cost reduction.

10. Consider the following statements:

In the case of a vapour compression machine, if the condensing temperature of the refrigerant is closer to the critical temperature, then there will be

1. excessive power consumption
2. high compression
3. large volume flow

Of these statements

(a) 1, 2 and 3 are correct (b) 1 and 2 are correct

(c) 2 and 3 are correct (d) 1 and 3 are correct

11. A single-stage vapour compression system cannot be used to produce ultralow temperatures because

(a) refrigerants for ultra-low temperatures are not available

(b) lubricants for ultra-low temperatures are not available

(c) volumetric efficiency will decrease considerably

(d) heat leakage into the system will be excessive

12. In an ideal refrigeration (reversed Carnot) cycle, the condenser and evaporator temperatures are 27°C and –13°C respectively. The COP of this cycle would be

(a) 6.5 (b) 7.5 (c) 10.5 (d) 15.0

13. Consider the following statements:

A decrease in evaporator temperature of vapour compression machine lead to

1. an increase in refrigerating effect
2. an increase in specific volume of vapour
3. a decrease in volumetric efficiency compressor
4. an increase in compressor work.

Of these statements:

(a) 1, 3 and 4 are correct (b) 1, 2 and 3 are correct

(c) 2, 3 and 4 are correct (d) 2 and 4 are correct

14. A standard vapour compression refrigeration cycle consists of the following 4 thermodynamic processes in sequence:

(a) isothermal expansion, isentropic compression, isothermal compression and isentropic expansion

(b) constant pressure heat addition, isentropic compression, constant pressure heat rejection and isentropic expansion

(c) constant pressure heat addition, isentropic compression, constant pressure heat rejection and isentropic expansion

(d) isothermal expansion, constant pressure heat addition, isothermal compression and constant pressure heat rejection.

15. Subcooling heat exchanger is used in a refrigeration cycle. The enthalpies at condenser outlet and evaporator outlet are 78 kJ/kg and 182 kJ/kg respectively. The enthalpy at outlet of isentropic compressor is 230 kJ/kg and enthalpy of subcooled liquid is 68 kJ/kg. The COP of the cycle is

(a) 3.25 (b) 2.16 (c) 3.0 s (d) 3.5

16. A refrigerating machine working on reversed Carnot cycle takes out 2 kW per minute of heat from the system while working between temperature limits of 300 K and 200 K. COP and Power consumed by the cycle will be respectively.

(a) 1 and 1 kW (b) 1 and 2 kW

(c) 2 and 1 kW (d) 2 and 2 kW

17. Experimental measurements on a refrigeration system indicate that rate of heat extraction by the evaporator, rate of heat rejection in the condenser and rate of heat rejection by the compressor body to environment are 70 kW, 90 kW and 5 kW respectively. The power input (in kW) required to operate the system is

(a) 15 (b) 20 (c) 25 (d) 75

18. The working temperatures in the evaporator and condenser coils of a refrigerator are –30°C and 32°C respectively. If the actual refrigerator has a COP of 0.75 of the maximum, the required power input for a refrigerating effect of 5 kW is, nearly,

(a) 1.7 kW (b) 2.94 kW (c) 3.92 kW (d) 4.0 kW

19. In a vapour compression refrigeration plant, the refrigerant leaves the evaporator at 195 kJ/kg and the condenser at 65 kJ/kg. For 1 kg/s of refrigerant, what is the refrigeration effect?

(a) 70 kW (b) 100 kW (c) 130 kW (d) 160 kW

20. In a vapour compression refrigeration cycle for making ice, the condensing temperature for higher COP,

(a) Should be near the critical temperature of the refrigerant

(b) Should be above the critical temperature of the refrigerant

(c) Should be much below the critical temperature of the refrigerant

(d) Could be of any value as it does not affect the COP

21. In a vapour compression refrigeration plant, the enthalpy values at different points are:

(i) Enthalpy at exit of the evaporator = 350 kJ/kg

(ii) Enthalpy at exit of the compressor = 375 kJ/kg

(iii) Enthalpy at exit of the condenser = 225 kJ/kg

The refrigerating efficiency of the plant is 0.8. What is the power required per kW of cooling to be produced?

(a) 0.25kW (b) 4.0 kW (c) 12.5 kW (d) 11 kW

Answers

1. (a)	**2.** (a)	**3.** (c)	**4.** (a)	**5.** (c)	**6.** (c)	**7.** (d)	**8.** (b)
9. (c)	**10.** (c)	**11.** (a)	**12.** (a)	**13.** (a)	**14.** (b)	**15.** (b)	**16.** (c)
17. (c)	**18.** (a)	**19.** (c)	**20.** (a)	**21.** (a)			

REVIEW QUESTIONS

1. What are the advantages of multi-stage compression over single-stage compression?

2. What is the function of flash chamber ?

3. Explain the difference between compound vapour compression system with flash cooling and flash intercooling.

4. What is the function of flash intecooler in multiple vapour compression refrigerating system?

5. What is the function of liquid sub-cooler in two stage compression?

6. Discuss the relative merits and demerits of flash and water intercooling used with multiple compression.

7. Why a compound refrigeration system employed when the condenser temperature to evaporator temperature range is very large?

8. Why it is not possible to do perfect intercooling at intermediate pressure in a compound vapour compression system?

NUMERICAL PROBLEMS

6.1 Calculate the power required to compress 18 kg/min of ammonia from saturated vapour at 1.4 bar to condenser pressure of 10 bar by two stage compression with intercooling by liquid refrigerant at 4 bar. Assume saturated liquid to leave the condenser and dry saturated vapours to leave the evaporator. Also determine the power required when intercooler is not used. Use the *p-h* chart. **[Ans.** 77.865 kW, 82.8 kW]

6.2 Explain by flow diagram and on *p-h* and *T-s* diagrams the multiple evaporator at different temperatures with single compressor, multiple expansion valves and back pressure valves, vapour compression refrigeration system.

6.3 Derive an expression for COP and kW/ton of a refrigeration system consisting of two evaporators at different temperatures with multiple expansion valves and back pressure valves and single compressor. Explain by flow diagram and on *p-h* and *T-s* diagrams, the multiple evaporator at different temperatures with single compressor, individual expansion valves and back pressure valves, vapour compression refrigeration system.

6.4 A refrigeration installation using *R*-12 has one compressor, one condenser and two evaporators of capacities 20 ton and 30 ton maintained at 5°C and – 10°C respectively. Each evaporator is fitted to individual expansion valve. The condenser is maintained at 40°C. The exit condition from both the evaporators is assumed superheated by 5°C and the liquid from condenser is sub-cooled by 8°C. The compression is assumed isentropic.

Determine :

(a) refrigerating effect in kJ/kg

(b) compressor power, and

(c) COP.

6.5 A refrigeration installation using *R*-12 has one compressor, one condenser and two evaporators of capacities 20 ton and 30 ton maintained at 5°C and –10°C. The condenser temperature is 40°C. The system uses multiple expansion valves and back pressure valves. The vapour leaving the evaporators are superheated by 5°C. The liquid refrigerant is subcooled in the sub-cooler section of condenser by 8°C. Assume isentropic compression.

Determine:

(a) refrigeration capacity in kJ/kg

(b) compressor power

(c) COP of the system

6.6 A refrigeration installation plant comprises two evaporators of capacities 20 ton at 5°C and 30 ton at –10°C, with individual expansion valve and individual compressor but one condenser operating at 40°C. The discharge from each evaporator is superheated by 5°C and liquid is subcooled in the sub-cooler of condenser by 8°C. Assume isentropic compression.

Determine:

(a) compressor power

(b) heat rejected in condenser in kJ/kg

(c) COP

6.7 In a multiple expansion valve and individual compressors and evaporator system of vapour compression plant, there are two evaporators of capacities 20 ton at 5°C and 30 ton at –10°C. One condenser at 40°C handles the refrigerant *R*-12 and subcools it in sub-cooler section of 30°C. The discharge from each evaporator is superheated by 5°C. Compression may be assumed isentropic. Determine:

(a) Refrigerating effect in kJ/kg in each evaporator

(b) Compressor power in kW

(c) COP.

6.8 Calculate the power required to compress 20 kg/min of *R*-12 from saturated vapour at 1.4 bar to condenser pressure of 10 bar by two-stage compression with intercooling by liquid refrigerant at 4 bar. Assume saturated liquid to leave the condenser and dry saturated vapours to leave the evaporator. [**Ans.** 11 kW]

6.9 The following data refer to a two-stage compression ammonia refrigerating system with water intercooler:

Condenser pressure = 14 bar

Evaporator pressure = 2 bar

Intercooler pressure = 5 bar

Refrigeration load on evaporator = 12 TR

If the temperature of de-superheated vapour and sub-cooled liquid refrigerant are limited to 30°C, calculate:

(a) Power required to drive the system, and

(b) COP of the system. [**Ans.** 11.24 kW, 3.754]

6.10 A vapour compression system with ammonia as the refrigerant works between the pressure limits of 2 bar and 12 bar with three stage compression. The vapour leaving the water intercooler at pressure 4 bar and 8 bar are in a saturated state. If the load on the evaporator is 15 TR, calculate the power required to drive the three compressors and COP.

[**Ans.** 10.49 kW, 5.03]

Chapter 7

VAPOUR ABSORPTION REFRIGERATION SYSTEMS

7.1 INTRODUCTION

The vapour absorption refrigeration system uses heat energy instead of mechanical energy in order to change, the conditions of the refrigerant required for the operation of the refrigeration cycle. In the vapour absorption system, the compressor is replaced by an absorber, a pump, a generator and a pressure reducing valve. The vapour refrigerant from the evaporator is drawn into the absorber where it is absorbed by the weak solution of the refrigerant forming a strong solution. This strong solution is pumped to the generator where it is heated by some external source like steam or heating oil. During the heating process, the vapour refrigerant is driven off by the solution and enters into the condenser where it is liquefied. The liquid refrigerant then flows into the evaporator, thereby completing the cycle. A line diagram of vapour absorption refrigeration system is shown in Fig. 7.1.

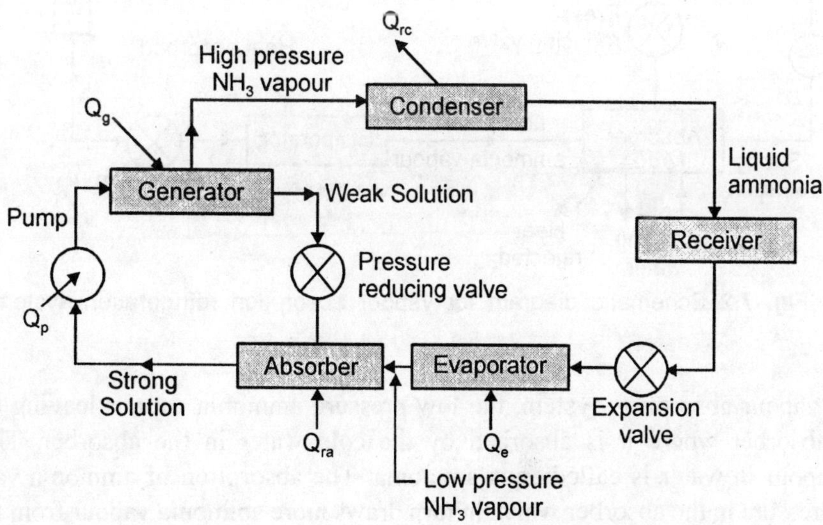

Fig. 7.1 Vapour absorption refrigeration system.

7.2 WORKING PRINCIPLE OF VAPOUR ABSORPTION REFRIGERATION SYSTEM

The schematic diagram of vapour absorption refrigeration system is shown in Fig. 7.2. It consists of:

- an absorber,
- a pump,
- a generator, and
- a pressure reducing valve.
- The function of the pressure reducing valve is to replace the compressor of vapour compression system. The other components of the system are
- condenser,
- receiver,
- expansion valve, and
- evaporator as in the vapour compression system.

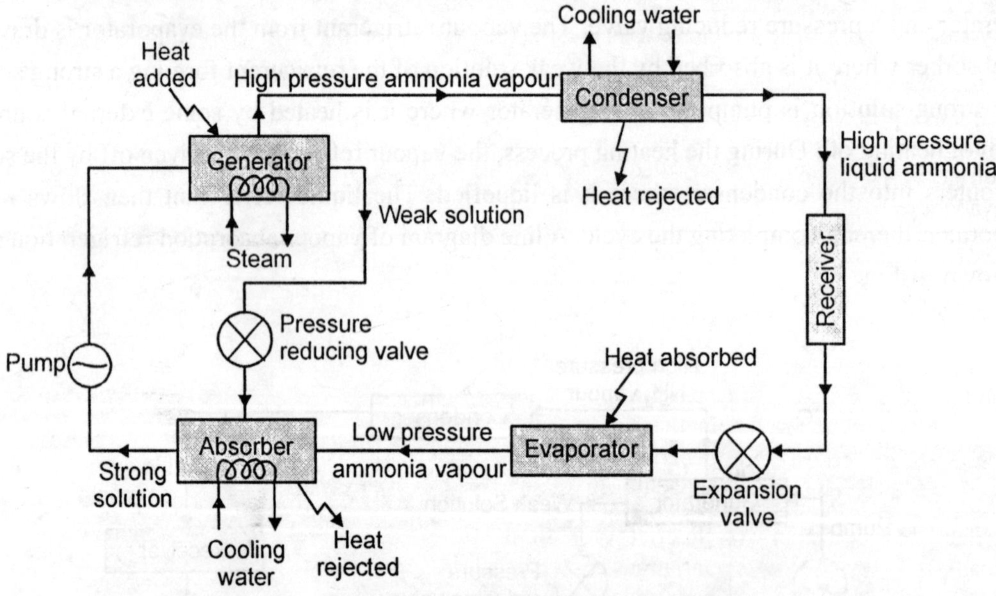

Fig. 7.2 Schematic diagram for vapour absorption refrigeration system.

Working

In the vapour absorption system, the low pressure ammonia vapour leaving the evaporator enters the absorber where it is absorbed by the cold water in the absorber. The solution of ammonia vapour in water is called aqua-ammonia. The absorption of ammonia vapour in water lowers the pressure in the absorber which in turn draws more ammonia vapour from the evaporator. As a result of this, the temperature of solution rises. Some cooling arrangement (usually water cooling) is employed in the absorber to remove the heat of solution evolved there. This increases the absorption capacity of water. The strong solution thus formed in the absorber is pumped to the generator by the liquid pump.

The strong ammonia solution in the generator is heated by some external source such as hot gas, steam or heating oil. During the heating process, the ammonia vapour is driven off the solution at high leaving behind the hot weak solution of ammonia in the generator. This weak ammonia solution flows back to the absorber at low pressure after passing through the pressure reducing valve. The high pressure ammonia vapour from the generator is condensed in the condenser to a high pressure liquid ammonia. This liquid ammonia is passed to the expansion valve through the receiver and then to the evaporator. This completes the vapour absorption cycle.

7.3 ADVANTAGES OF VAPOUR ABSORPTION SYSTEM OVER VAPOUR COMPRESSION SYSTEM

The advantages of vapour absorption system over vapour compression system are listed in Table 7.1.

Table 7.1 Comparison of vapour absorption and vapour compression systems

Feature	*Vapour absorption system*	*Vapor compression system*
1. System operation.	Pump is the only moving part. Quiet operation and little wear	Many moving parts. More wear, tear and noise.
2. Type of energy used to change the condition of refrigerant from the evaporator	Heat energy.	Mechanical energy
3. Source of energy	Steam at high or low pressure. Exhaust steam from furnaces and solar energy may be used. It is quite cheap.	Electric power, which is quite expensive.
4. Evaporative pressure and temperature.	Can operate on reduced evaporator pressure and temperature by increasing the steam pressure to generator, with little decrease in capacity.	Capacity decrease rapidly with lowered evaporator pressure.
5. Effect of variations in load	No effect on performance as the load variations are met by controlling the quantity of aqua circulated and quantity of steam supplied to generator.	The performance is poor at partial loads.
6. Effect of liquid refrigerant leaving in evaporator.	It has no bad effect on the system except that of reducing the refrigerating effect.	It is essential to superheat the vapour refrigerant leaving the evaporator so that no liquid may enter the compressor.
7. Space and automatic control requirements.	More favourable	Less favourable.

7.4 COEFFICIENT OF PERFORMANCE OF AN IDEAL VAPOUR ABSORPTION SYSTEM

The ideal vapour absorption refrigeration system is shown in Fig. 7.1.

Let Q_g = heat given to the refrigerant in the generator.

Q_{rc} = heat rejected from the condenser to the atmosphere.

Q_{ra} = heat rejected from the absorber to the atmosphere.

Q_e = heat absorbed by the refrigerant in the evaporator.

Q_p = heat added to the refrigerant due to pump work.

$Q_r = Q_{rc} + Q_{ra}$

= total heat rejected from the condenser and absorber to atmosphere

For heat balance, neglecting Q_p, according to the first law of thermodynamics, we have

$$Q_r = Q_{rc} + Q_{ra} = Q_g + Q_e \qquad ...(1)$$

Let T_g = temperature at which heat is given to the generator.

T_r = temperature at which heat is rejected to the atmosphere or cooling water from the condenser and absorber.

T_e = temperature at which heat is absorbed in the evaporator.

The vapour absorption system can be considered as a perfectly reversible system. Therefore, the initial entropy of the system must be equal to the entropy of the system after the change in its conditions.

$$\therefore \quad \frac{Q_g}{T_g} + \frac{Q_e}{T_e} = \frac{Q_{rc} + Q_{ra}}{T_r} = \frac{Q_g + Q_e}{T_r} \qquad \text{(from eq. (1)) ...(2)}$$

or
$$Q_g\left(\frac{1}{T_g} - \frac{1}{T_r}\right) = Q_e\left(\frac{1}{T_r} - \frac{1}{T_e}\right)$$

$$Q_g\left(\frac{T_r - T_g}{T_g T_r}\right) = Q_e\left(\frac{T_e - T_r}{T_r T_e}\right)$$

$$Q_g = Q_e \frac{(T_e - T_r)T_g}{(T_r - T_g)T_e} = Q_e\left(\frac{T_e - T_r}{T_e}\right)\left(\frac{T_g}{T_r - T_g}\right) \qquad ...(3)$$

Maximum coefficient of performance,

$$(\text{COP})_{\max} = \frac{\text{Heat absorbed in evaporator}}{\text{Heat supplied to generator + Work done by pump}}$$

Neglecting pump work,

$$(\text{COP})_{max} = \frac{Q_e}{Q_g} = \left(\frac{T_e}{T_e - T_r}\right)\left(\frac{T_r - T_g}{T_g}\right) \quad ...(7.1a)$$

$$= \left[\frac{T_r}{T_g} - 1\right] \Bigg/ \left[1 - \frac{T_r}{T_e}\right] \quad ...(7.1b)$$

If T_a = temperature at which heat is rejected from the absorber

T_c = temperature at which heat is rejected from the condenser.

Then

$$(\text{COP})_{max} = \left(\frac{T_e}{T_e - T_c}\right)\left(\frac{T_a - T_g}{T_g}\right) = \left(\frac{T_e}{T_c - T_c}\right)\left(\frac{T_g - T_a}{T_g}\right) \quad ...(7.2)$$

Example 7.1 *A geothermal well at 130? supplies heat at a rate of 100,500 kJ/h to an absorption refrigeration system. The environment is at 30°C and the refrigerated space is maintained at –22°C. Determine the maximum possible heat removal from the refrigerated space.*

Solution: Given: T_r = 273 + 35 = 308 K, T_g = 273 + 110 = 383 K,

T_e = 273 – 5 = 268 K

$$(\text{COP})_{max} = \frac{\left[\frac{T_r}{T_g} - 1\right]}{\left[1 - \frac{T_r}{T_e}\right]} = \frac{\left(\frac{308}{383} - 1\right)}{\left(1 - \frac{308}{268}\right)} = 1.315$$

Example 7.2 *In an absorption type refrigerator, the heat is supplied to ammonia generator by condensing steam at 2.5 bar and dryness fraction 0.9. The temperature in the refrigerator is to be maintained at –5°C. Calculate the maximum COP possible.*

If the refrigeration load is 20 tonnes and actual COP is 75% of the maximum COP, find the mass of steam required per hour. Take ambient temperature as 30°C.

Solution: Given: p = 2.5 bar, x = 0.9, T_e = 273 – 5 = 268 K, Q = 20 TR,

$(\text{COP})_{act}$ = 0.75 $(\text{COP})_{max}$, T_r = 273 + 30 = 303 K, From steam tables

h_f = 535.34 kJ/kg, h_g = 2716.9 kJ/kg

Saturation temperature of steam at

p = 2.5 bar, T_s = 273 + 127.43 = 400.43 K = T_g

$$(\text{COP})_{max} = \left[\frac{T_r}{T_g} - 1\right] \Bigg/ \left[1 - \frac{T_r}{T_e}\right]$$

$$= \left[\frac{303}{400.43} - 1\right] \Bigg/ \left[1 - \frac{303}{268}\right] = 1.863$$

$$(\text{COP})_{act} = 0.75 \times 1.863 = 1.397$$

$$\text{Actual heat supplied} = \frac{211Q}{(\text{COP})_{\text{act}}} = \frac{211 \times 20}{1.397} = 3020.76 \text{ kJ/min}$$

Latent heat of steam, $h_{fg} = h_g - h_f = 2716.9 - 535.34 = 2181.56$ kJ/kg

Mass of steam required per hour,

$$\dot{m}_s = \frac{\text{Actual heat supplied}}{x \times h_{fg}}$$

$$= \frac{3020.76 \times 60}{0.9 \times 2181.56} = 92.3 \text{ kg/h}$$

7.5 AMMONIA-WATER ($NH_3 - H_2O$) (OR PRACTICAL) VAPOUR ABSORPTION SYSTEM

In the $NH_3 - H_2O$ vapour absorption system, ammonia is the refrigerant and water is the absorber. An analyser, a rectifier and two heat exchangers are added to the simple vapour absorption system. The complete schematic diagram of ammonia-water absorption system is shown in Fig. 7.3.

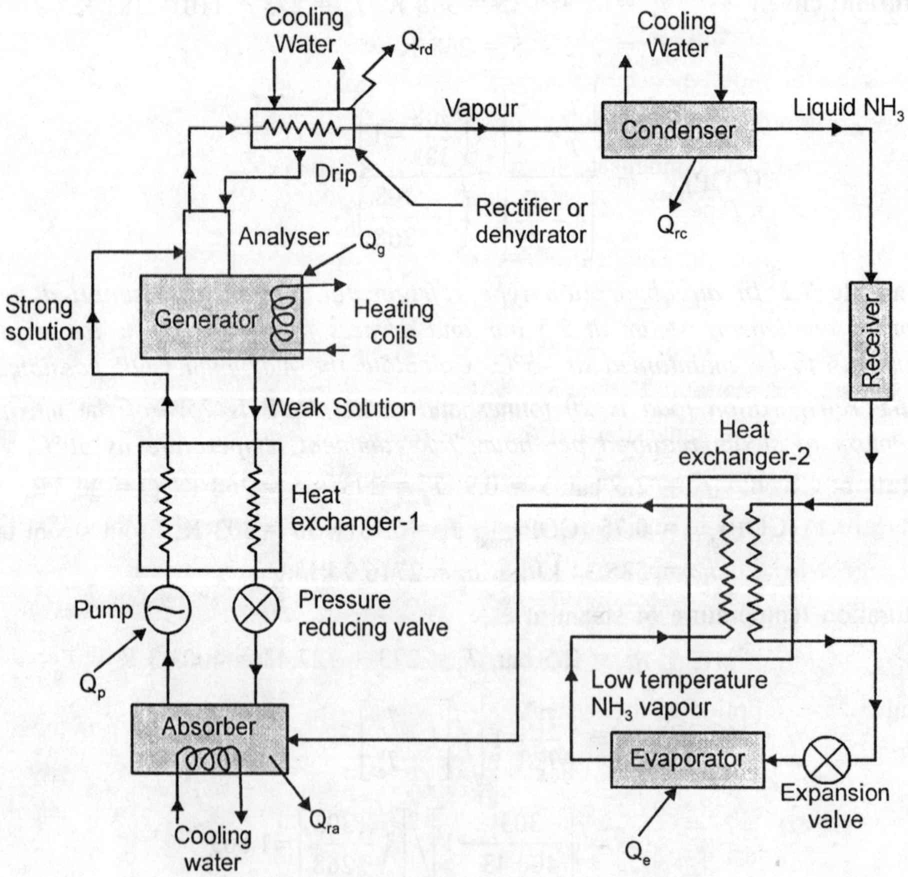

Fig. 7.3 Ammonia-water vapour absorption system.

The functions of various accessories are:

1. **Analyser:** In the simple vapour absorption system, some water is also vaporised along with ammonia and flows into the condenser. These water vapours further move and enter into the expansion valve, where they freeze and choke the pipeline. The analyser is used to remove these water vapours. The analyser may be an integral part of the generator or as a separate piece of equipment. It consists of a series of trays mounted above the generator. The strong solution from the absorber and the aqua from the rectifier (or dehydrator) are introduced at the top of the analyser and flows downward on the trays and into the generator. In this way, considerable liquid surface area is exposed to the vapour rising from the generator. The vapours are cooled and most of the water vapours condense, so that mainly ammonia vapours leave the top of the analyser. This also reduces the external heat required in the generator since aqua is heated by the vapour.
2. **Rectifier or Dehydrators:** The function of the rectifier is to completely remove the water vapours still left in the analyser. The rectifier is a closed type vapour cooler of the double pipe type, shell and coil, or shell and tube type. Its function is to further cool the ammonia vapours leaving the analyser so that the remaining water vapours are condensed. The condensate from the rectifier is returned to the top of the analyser by a drip return pipe.
3. **Heat exchangers:** The heat exchanger-1 provided between the pump and the generator is used to cool the weak hot solution returning from the generator to the absorber. The heat removed from the weak solution raises the temperature of the strong solution leaving the pump and going to analyser and generator. This operation reduces the heat supplied to the generator and the amount of cooling required for the absorber.

The heat exchanger-2 provided between the condenser and the evaporator may be called liquid sub-cooler. The liquid refrigerant leaving the condenser is sub-cooled by the low temperature ammonia vapour from the evaporator. This sub-cooled liquid is now passed to the expansion valve and then to the evaporator.

Net refrigerating effect, R_E = Heat absorbed in the evaporator

Total energy supplied to the system,

$$W_t = \text{Work done by the pump + Heat supplied in generator}$$

$$\text{COP} = \frac{R_E}{W_t} \qquad \text{...(7.3)}$$

7.6 LITHIUM BROMIDE-WATER VAPOUR ABSORPTION SYSTEM (LIBR-H_2O)

In the LiBr-water system, water is the refrigerant and LiBr the absorber. The lithium bromide solution has a strong affinity for water vapour because of its very low vapour pressure. The lithium bromide-water vapour absorption system is shown in Fig. 7.4. The evaporator and absorber are placed in one shell which operates at the same low pressure of the system. The generator and condenser are placed in another shell which operates at the same high pressure of the system.

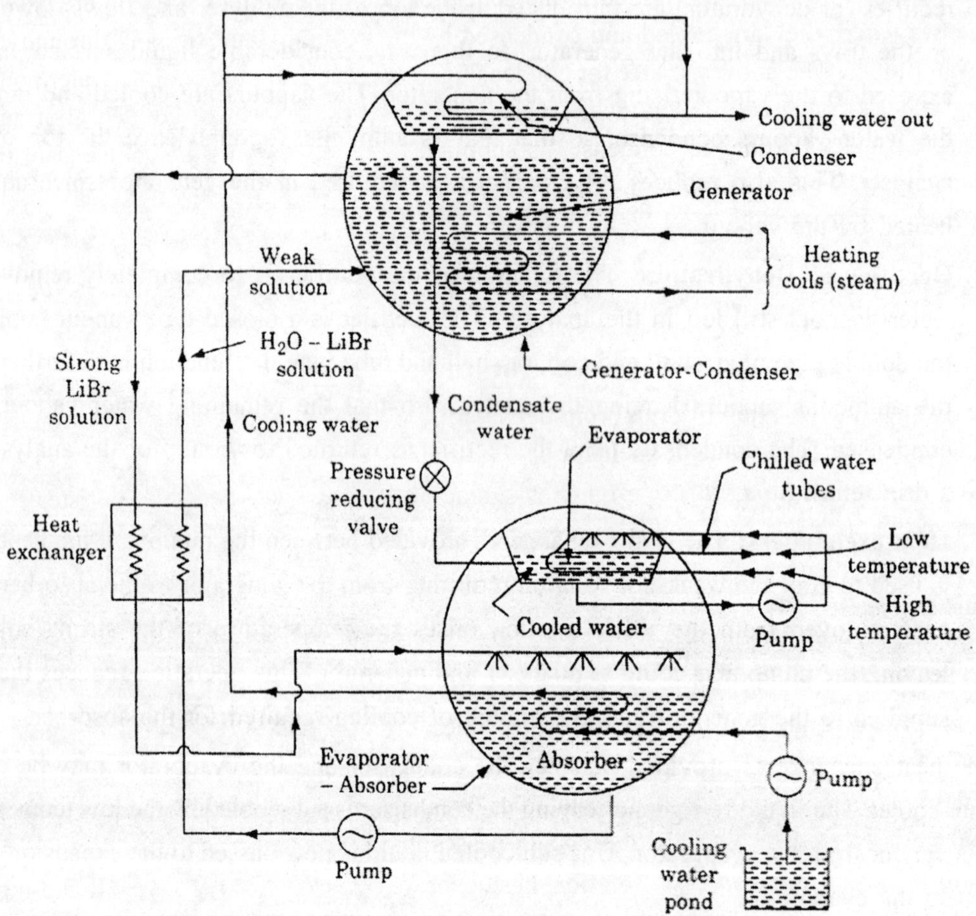

Fig. 7.4 Lithium bromide-water vapour absorption system.

7.6.1 Working Principle

The water for process requirements is chilled as it is pumped through the chilled-water tubes in the evaporator by giving up heat to the refrigerant water sprayed over the tubes. Since the pressure inside the evaporator is maintained very low, therefore, the refrigerant water evaporates. The water vapours thus formed are absorbed by the strong lithium bromide solution which is sprayed in the absorber. In absorbing the water vapour, the lithium bromide solution helps in maintaining very low pressure needed in the evaporator, and the solution becomes weak. This weak solution is pumped by a pump to the generator where it is heated up by using steam in the heating coils. A portion of water is evaporated by the heat and the solution becomes more strong. This strong solution is passed through the heat exchanger and then sprayed in the absorber. The weak solution of lithium bromide from the absorber to the generator is also passed through the heat exchanger. This weak solution gets from the strong solution in the heat exchanger, thus reducing the quantity of steam required to heat the weak solution in the generator.

The refrigerant water vapours formed in the generator due to heating of solution are passed to the condenser where they are cooled and condensed by the cooling water flowing through the condenser water tubes. The cooling water for condensing is pumped from the cooling water pond. This cooling water first enters the absorber where it takes away the heat of condensation and solution. The condensate from the condenser is supplied to the evaporator to compensate the water vapour formed in the evaporator. The pressure reducing valve reduces the pressure of condensate from the condenser pressure to the evaporator pressure. The cooled water from the evaporator is pumped and sprayed in the evaporator in order to cool the water flowing through the chilled tubes. This completes the cycle.

7.6.2 Lithium Bromide-Water System Equipment

A line diagram of two-shell lithium bromide-water vapour absorption refrigeration system is shown in Fig. 7.5. The various components are:

1. **Generator:** It consists of tube bundles submerged in the solution heated by steam or hot liquids.
2. **Condenser:** It consists of tube bundles located in the vapour space over the generator shielded from carry over of salt by eliminators. Cooling water to condenser first passes through the absorber.
3. **Absorber:** It consists of bundles over which the strong absorbent is sprayed. Refrigerant vapours are condensed into the absorbent releasing heat to the cooling water passing through it.
4. **Evaporator:** Evaporators are tube bundles over which the refrigerant water is sprayed and evaporated. The liquid to be cooled passes inside the tubes.
5. **Solution heat exchangers:** Solution heat exchangers are all of steel shell-and-tube construction,
6. **Purgers:** All units include a purger which is used to remove non-condensable gases. Non-condensable gases present in small quantities can raise the total pressure in the absorber sufficiently to significantly change the evaporator pressure. Very small pressure increases cause appreciable change in the refrigerant evaporating temperatures.
7. **Expansion Device:** Mechanical expansion valves are not used in absorption units. The flow of refrigerant liquid to evaporator is controlled by an orifice or other fixed restriction between the condenser and the evaporators.

8. Pump: Other details about pumps etc., are shown in Fig. 7.5.

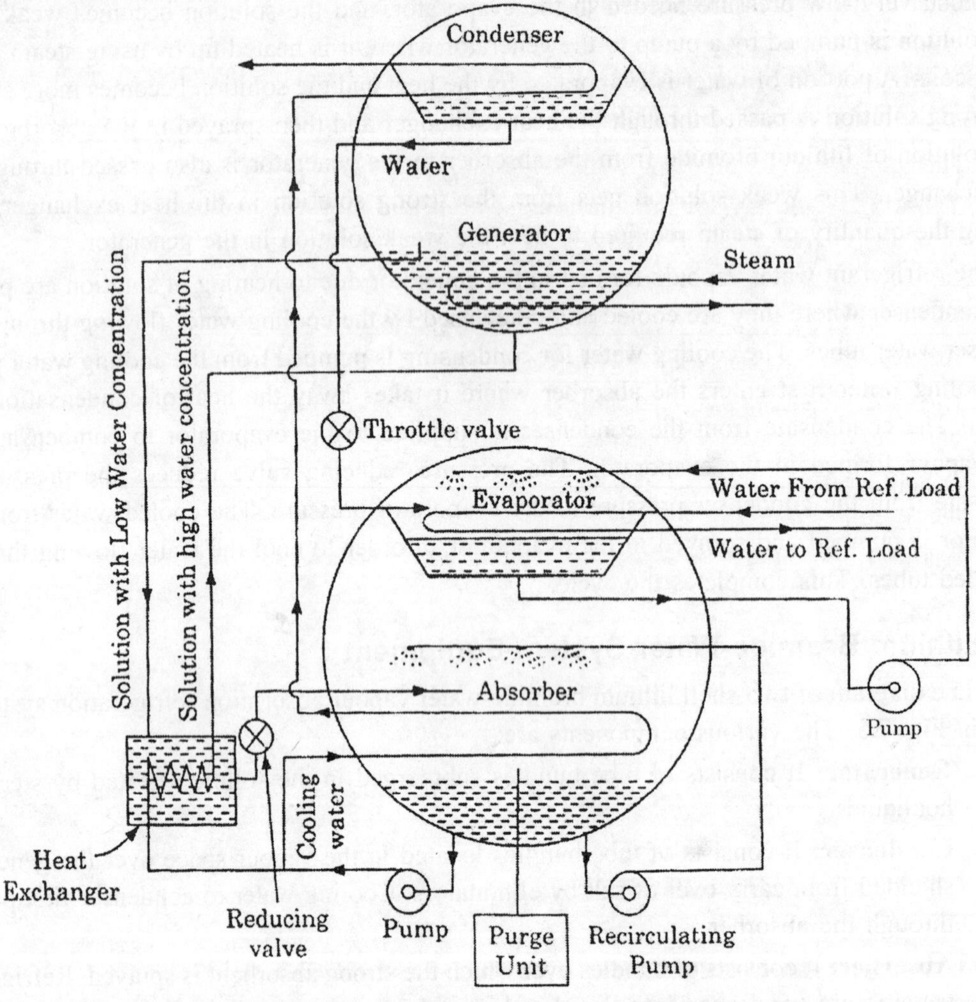

Fig. 7.5 Lithium bromide-water vapour absorption system.

Capacity control: All lithium bromide-water cycle absorption machines meet load variations and maintain chilled water temperature control by varying the rate of re-concentration of the absorbent solution.

At any given constant load, the chilled water temperature is maintained by a temperature difference between refrigerant and chilled water. The refrigerant temperature is maintained in turn by absorber being supplied with a flow rate and concentration of solution, and by absorber cooling water temperature.

Load changes are reflected by corresponding changes in chilled water temperature. A load reduction, for example, results in less temperature difference being required in the evaporator, and a reduced requirement for solution flow or concentration. The resultant chilled water temperature drop is met basically by adjusting the rate of re-concentration to match the reduced requirements of the absorber.

All units sense the chilled water temperature changes resulting from a load change with a thermostat in the leaving chilled water.

7.7 COMPARISON OF AMMONIA-WATER AND LITHIUM BROMIDE-WATER ABSORPTION SYSTEMS

The comparison of ammonia-water and lithium bromide-water systems are given in Table 7.2

Table 7.2 Comparison of ammonia-water and lithium bromide-water systems

Features	*Ammonia-Water System*	*Lithium bromide-water System*
1. Refrigerant	Ammonia solution (aqua)	Water
2. Absorbent	Water	Lithium bromide
3. Difference in boiling points of refrigerant and absorber	138°C. This system is not suitable from boiling point view.	> 200°C satisfies the boiling point requirement.
4. Quality of vapours leaving the generator	Contains some amount of water.	A pure substance
5. Solubility in water	Satisfactory	Satisfactory
6. Applications	Refrigeration systems like ice plants, ice cream plants, etc.	Air-conditioning purposes only.
7. Freezing point of refrigerant.	–78°C, does not choke the evaporator tubes at working	0°C, chokes the evaporator tubes.
8. Corrosiveness of mixture	Poisonous gas in nature and aqua solution is corrosive.	Yes

7.8 REFRIGERANT ABSORBENT MIXTURES

7.8.1 Properties of Aqua-Ammonia Solution

The main properties of aqua-ammonia system are:

1. **Concentration:** It is defined as the ratio of the mass of ammonia to total mass of solution.
2. **Equilibrium Concentration:** It is the concentration of aqua-ammonia solution at a specified temperature and pressure.
3. **Heat of absorption (or heat of solution):** It is the heat evolved when water or weak aqua solution absorbs ammonia. It is also the heat required for separating ammonia component from solution.
4. If V_w volume of water is mixed with V_{NH_3} volume of liquid ammonia, then volume of mixture, $V_m < (V_w + V_{NH_3})$.
5. The saturated vapour mixture above the aqua-ammonia solution obeys Dalton's Law of partial pressures. According to this law, the total pressure of vapours is the sum of the partial pressures of water vapour and ammonia vapour.

6. Each component in the vapour form in a mixture is in a superheated condition. Although mixture is saturated.

$$\text{Degree of superheat, } \frac{T_s}{T} = 0.66x + 0.656 \text{ for } x = 0.2 \text{ to } 0.5 \qquad ...(7.5)$$

where T_s = saturation temperature of ammonia at its partial pressure.
T = actual vapour temperature
x = concentration.

7. Mass concentration of NH_3 in aqua-ammonia solution,

$$x_m = \frac{17x}{17x + 18(1-x)} = \frac{\text{mass of ammonia}}{\text{mass of ammonia } + \text{mass of water}} = \frac{17x}{18-x} \qquad ...(7.6)$$

where x = mole-fraction of NH_3 in liquid solution.

The mass concentration of NH_3 in vapour mixture above the liquid mixture X_m may also be computed as:

$$X_m = \frac{p_{NH_3}}{p_{H_2O}} = \frac{\text{partial pressure of } NH_3 \text{ vapour in vapour mixture}}{\text{partial pressure of } H_2O \text{ vapour in vapour mixture}}$$

$$= \frac{p_t - p_{H_2O}}{p_{H_2O}} \qquad ...(7.7)$$

where p_t = total pressure.

8. **Volume of aqua-ammonia**

Let V_{NH_3} = volume of liquid ammonia
V_{H_2O} = Volume of water

Then volume of mixture,

$$V_m = 0.84\, x_m\, V_{NH_3} + (1 - x_m)\, V_{H_2O} \qquad ...(7.8)$$

Thus, there is shrinkage of volume.

9. **Enthalpy of aqua-ammonia**

If one mole of NH_3 at temperature T and pressure p_{NH_3} mixes with one mole of H_2O at pressure p_{H_2O} and temperature T, the enthalpy of mixture is given by,

$$h_m = h_{NH_3} + \left[\nu_{NH_3} - T\left(\frac{\delta \nu_{NH_3}}{\delta T}\right) \right] \times (p_m - p_{NH_3})$$

$$+ \; h_{H_2O} + \left[\nu_{H_2O} - T\left(\frac{\delta \nu_{H_2O}}{\delta T}\right)_p \right] \times (p_m - p_{H_2O}) - q_{mix} \qquad ...(7.9)$$

where q_{mix} = heat rejected due to mixing process.

if n_{NH_3} = mole of NH_3 per mole of mixture
n_{H_2O} = mole of H_2O per mole of mixture

Then enthalpy of one mole mixture is,

$$h_m = n_{NH_3} h_{NH_3} + n_{NH_3}\left[v_{NH_3} - T\left(\frac{\delta v_{NH_3}}{\delta T}\right)_p\right](p_m - p_{NH_3}) + n_{H_2O} h_{H_2O}$$

$$+ n_{H_2O}\left[v_{H_2O} - T\left(\frac{\delta v_{H_2O}}{\delta T}\right)_p\right] \times (p_m - p_{H_2O}) q_{mix} \quad ...(7.10)$$

where q_{mix} = heat rejected per mole of mixture.

10. Enthalpy of vapour mixtures

Enthalpy of one mole of vapour mixture is,

$$h_{mv} = \left[n'_{NH_3} h_{NH_3}\right]^T_{p_{NH_3}} + \left[n'_{H_2O} h_{H_2O}\right]^T_{p_{H_2O}} \quad ...(7.11)$$

where h_{NH_3} and n'_{H_2O} are mole of NH_3 and H_2O per mole of mixture.

11. Heat of mixing: The heat evolved or absorbed due to mixing is given by Moltier equation as

$Q_{absorbed}$ = 804 (1 – x_m) – 930 x_m^2 kJ/kg of NH_3 absorbed or released ...(7.12)

where x_m = mean mass concentration of aqua-ammonia solution.

$$= \frac{x_{m1} + m_{m2}}{2}$$

x_{m1}, x_{m2} are initial and final mass concentrations respectively.

Heat of aqua-ammonia solution,

$$h_{sol} = [x_2\, h_{NH_3} + (1 + x_m)\, h_{H_2O} - x_m q] \text{ kJ/kg of aqua} \quad ...(7.13)$$

where h_{NH_3} and h_{H_2O} are enthalpies of ammonia and water at solution temperature.

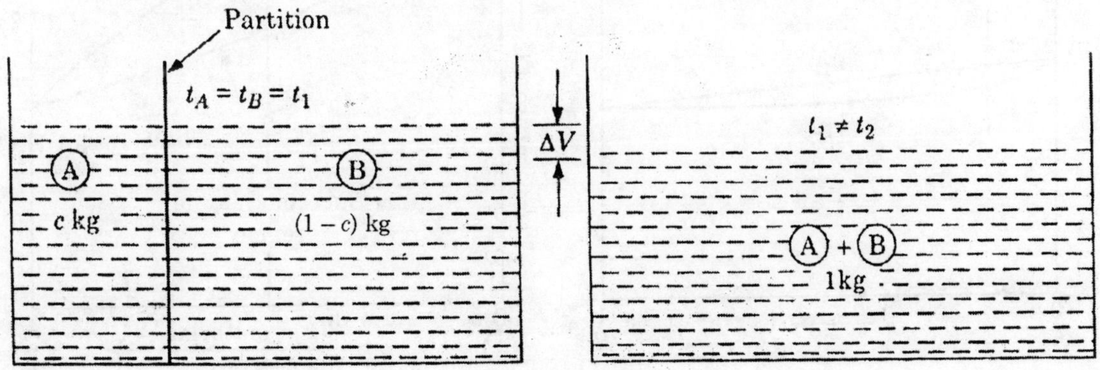

Fig. 7.6

7.9 ELEMENTARY PROPERTIES OF BINARY MIXTURES

A mixture may be homogeneous or heterogeneous. The different properties like density, pressure, temperature are uniform throughout for homogeneous mixtures. In this discussion only the homogeneous mixtures either in the liquid form or gaseous form are considered. And only the mixtures of two substances, *i.e.* binary mixtures are considered.

Unlike as in the pure substances, the state of mixtures cannot be defined only by pressure and temperature alone. In general, the quantitative composition in terms of concentration c is considered. Concentration c is the mass of one arbitrary constituent divided by the mass of the mixture. The thermodynamic state thus may be determined by pressure p, temperature t, and concentration c.

Two important phenomenon with the mixing of two liquids are:

(*i*) Change of volume, and

(*ii*) Change of temperature.

This is illustrated in Fig. 7.6. Let the divided container hold c kg of liquid A and $(1 - c)$ kg of liquid B. Each liquid is at the same temperature t_1. The volume of 1 kg of the constituents

$$v_1 = cv_A + (1 - c)v_B \qquad ...(7.14)$$

Let the dividing wall be removed and the two liquids allowed to mix completely. It is observed that the final volume v_2 is not equal to v_1. Change in volume may be positive or negative as shown in Fig. 7.7. Thus for any two constituents the experimental results are tabulated and referred later for computational work.

Also it is noted that in general final temperature t_2 is not equal to t_1. Sometimes $t_2 > t_1$ and sometimes $t_2 < t_1$, as shown in Fig. 7.8. Thus heat of solution ΔH_C may be positive or negative. Thus for original constituents,

$$h_1 = ch_A + (1 - c)h_B \qquad ...(7.15)$$

and after mixing

$$h_1 = h_1 + H_C$$
$$= ch_A + (1 - c)h_B + \Delta H_C \qquad ...(7.16)$$

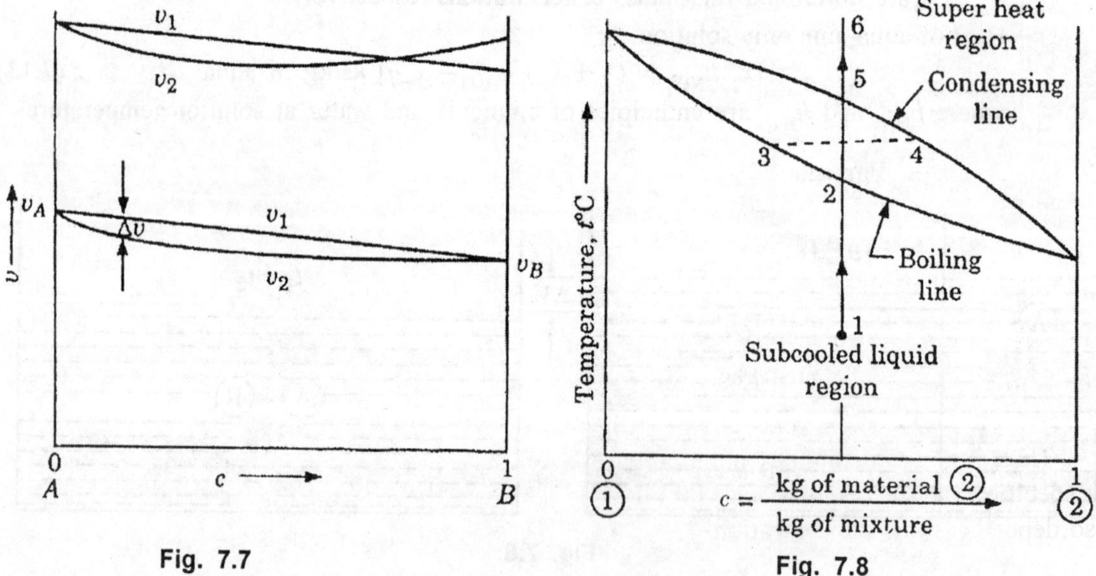

Fig. 7.7

Fig. 7.8

where ΔH_C is the heat that is either removed or added to keep the temperature constant, at constant pressure.

Thus specific enthalpy in kJ/kg of mixture for a solution of given concentration at a fixed temperature and pressure can be computed, if the enthalpies of the pure constituent and the isothermal heat of solution are known.

7.10 TEMPERATURE-CONCENTRATION DIAGRAM (*t-c*)

The relationship between temperature and pressure for binary liquid mixture when boiling and for binary vapour mixture when condensing is essential for the analysis of absorption refrigeration.

Fig. 7.9 shows schematically the experiment and the result on *t-c* co-ordinates.

At state 1, it is liquid and as heat is added maintaining pressure constant, till temperature t_2 is reached, the solution is entirely in the liquid phase. But at temperature t_2, when heat is further added, vaporisation occurs. And it is very interesting to note that concentration of liquid and vapours are different. At temperature $t_3 = t_4$ the concentration of liquid has changed from c_1 to c_3 and the concentration of vapour is now c_4. The liquid concentration $c_3 < c_1$, and vapour concentration $c_4 > c_1$.

If it is further heated, it may be noted that at point 5, there is only vapour and liquid has disappeared. At point 5, $c_1 = c_5$. And after that, the heat addition results in superheating of the vapour as shown in Fig. 7.9 follow. And the results of reversed experiment of starting with superheated vapour and then removing heat results in Fig. 7.10 at one particular pressure.

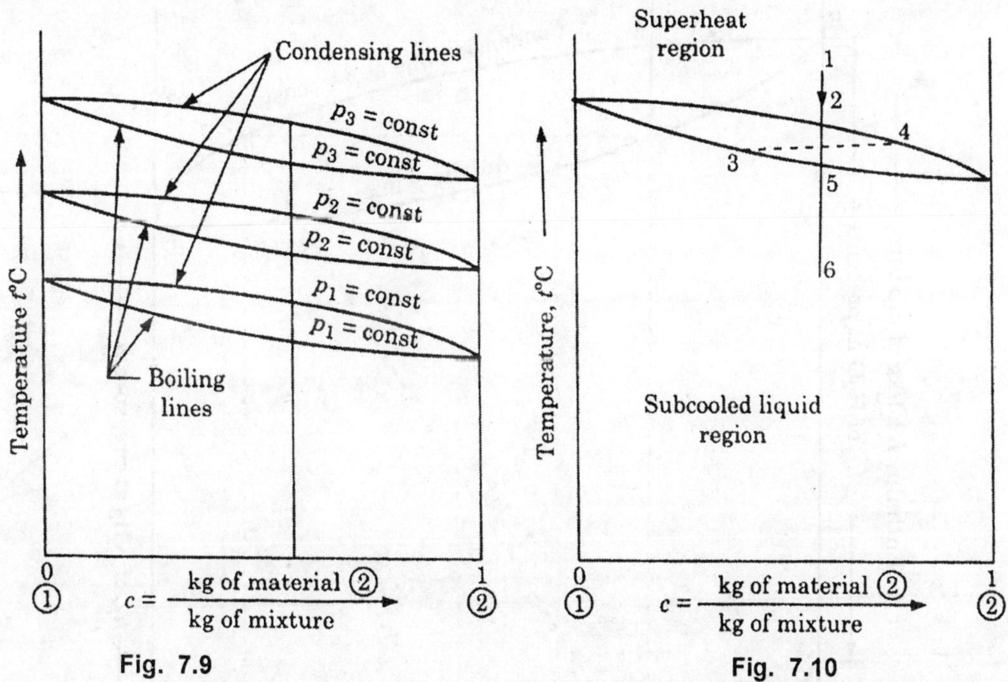

Fig. 7.9 **Fig. 7.10**

Thus, it is seen that binary in mixtures do not have a single boiling or condensing temperature for specified pressure as is the case for pure substances. The equilibrium or saturation temperature also depends on the concentration.

7.11 ENTHALPY-CONCENTRATION DIAGRAM (*h-c*)

The enthalpy concentration diagram for binary mixture is referred to as *h-c* diagram. Refer Fig. 7.11 for representative sketch for *h-c* diagram. A-2-C is the liquid saturation line or the boiling line at a particular specified pressure *p*. And B-3-D is the vapour saturation line or the condensing line at the same specified pressure *p*. The intercept *AB* is the enthalpy of evaporation at pressure

p of H_2O. Intercept is taken at $c = 0$, *i.e.* no ammonia in the mixture. Similarly at $c = 1$, there is no water in the mixture and the intercept CD gives the enthalpy of evaporation of NH_3 at pressure p. One typical isotherm, *i.e.* constant temperature t line is shown in the sub-cooled region.

If a liquid is saturated at a given pressure and temperature for example 6 bar at 40°C, the intersection of 40°C, line with 6 bar line gives the state. And, the bottom scale for that state reads the concentration. It reads 0.53 from the chart. The enthalpy reads 109 kJ/kg of mixture. (See Fig. 7.12).

In the sub-cooled region, the intersection of the temperature line and concentration line (vertical) locates the point on the *h-c* diagram. Suppose the pressure is 6 bar, temperature is 30°C and concentration 0.53, the intersection of 30°C line and 0.53 concentration line intersect at a point where enthalpy reads 65 kJ/kg mixture. And the point lies below the 6 bar lines, indicating sub-cooled liquid region (See Fig. 7.12).

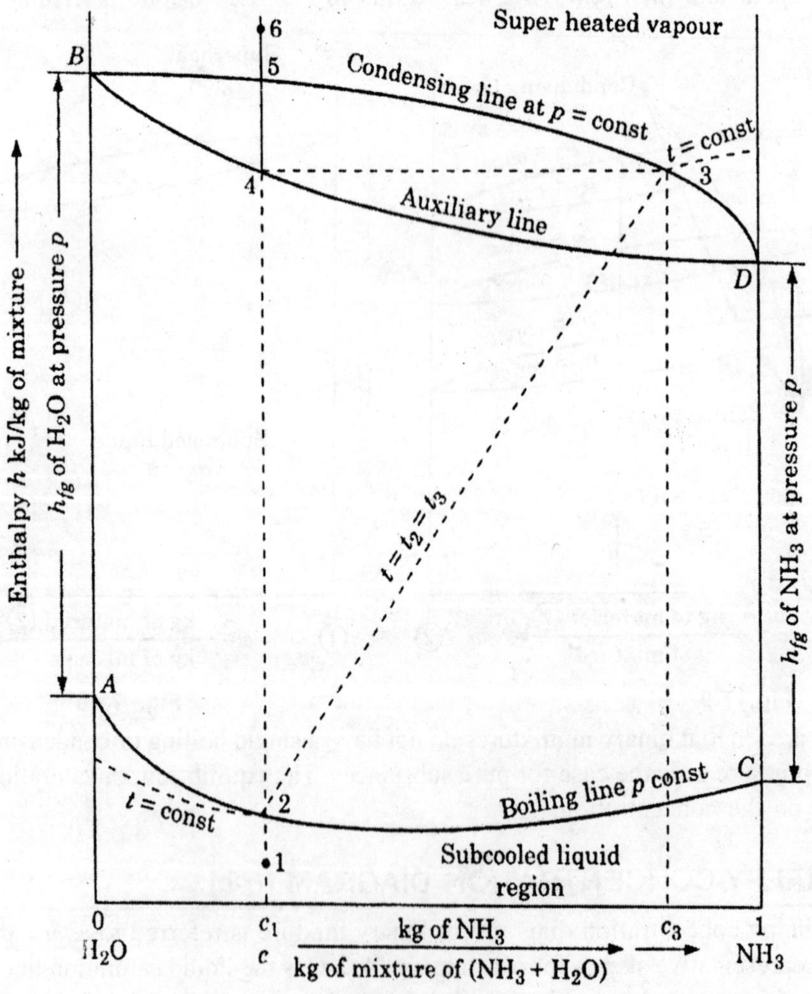

Fig. 7.11

The area between curves A-2-C and B-3-D is the two-phase region or the wet vapour region or liquid vapour region. In this region, no constant temperature lines are shown on *h-c* diagram. Thus at any given pressure and temperature, it is not directly possible to determine the concentration lines are drawn on the *h-c* chart for each pressure. Such a line is shown as B-4-D in Fig. 7.11. Suppose equilibrium vapour state at 6 bar and 40°C is to be found out. First locate the point on the intersection of boiling liquid line for 6 bar and constant temperature line for 40°C. This gives concentration of liquid as 0.53. Vertical line is drawn on the point to intersect the auxiliary line for 6 bar. From this intersection draw a horizontal line to intersect condensing line for 6 bar. This point gives the required equilibrium vapour state at 6 bar and 40°C. At this point vapour concentration reads 0.987 and enthalpy is 1710 kJ/kg of mixture from the Fig. 7.12.

7.12 ELEMENTARY STEADY FLOW PROCESS WITH BINARY MIXTURE

(*a*) ***Adiabatic mixing of two streams.*** Refer Fig. 7.13(*b*) for schematic arrangement of mixing where two fluid streams are mixed adiabatically. We have

$$m_1 h_1 + m_2 h_2 = m_3 h_3 \quad \text{...(7.17)}$$

Thus equation (7.17) shows that mixing process is a straight line on the *h-c* diagram shown in Fig. 7.13(*a*).

$$m_1 + m_2 = m_3 \quad \text{...(7.18)}$$

$$m_1 c_1 + m_2 c_2 = m_3 c_3 \quad \text{...(7.19)}$$

Thus

$$\frac{m_1}{m_2} = \frac{h_2 - h_3}{h_3 - h_1} = \frac{c_2 - c_3}{c_3 - c_1} \quad \text{...(7.20)}$$

The resulting condition is determined by

$$c_3 = c_1 + \frac{m_2}{m_3}(c_2 - c_1) \quad \text{...(7.21)}$$

and

$$h_3 = h_1 + \frac{m_2}{m_3}(h_2 - h_1) \quad \text{...(7.22)}$$

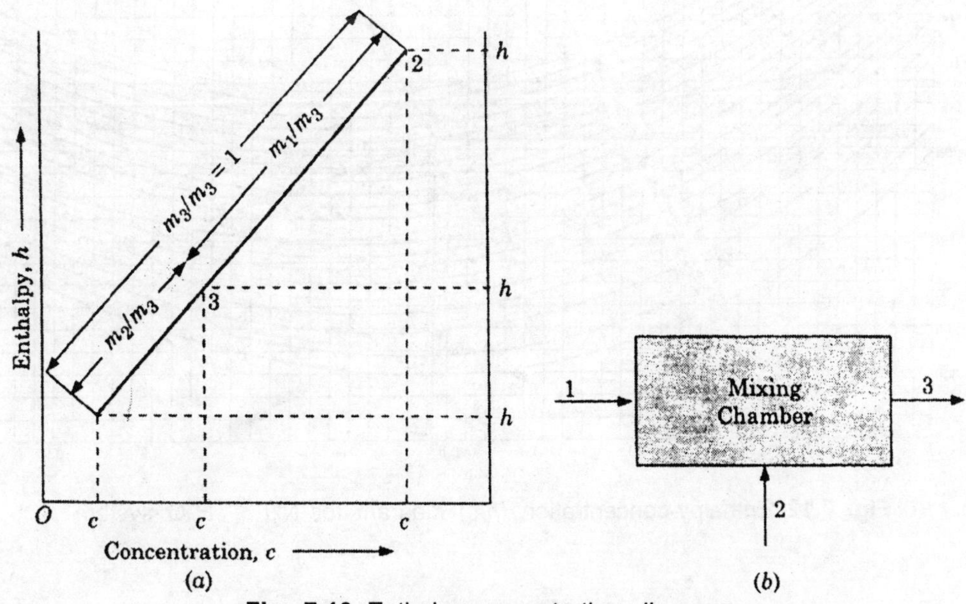

Fig. 7.13 Enthalpy-concentration diagram.

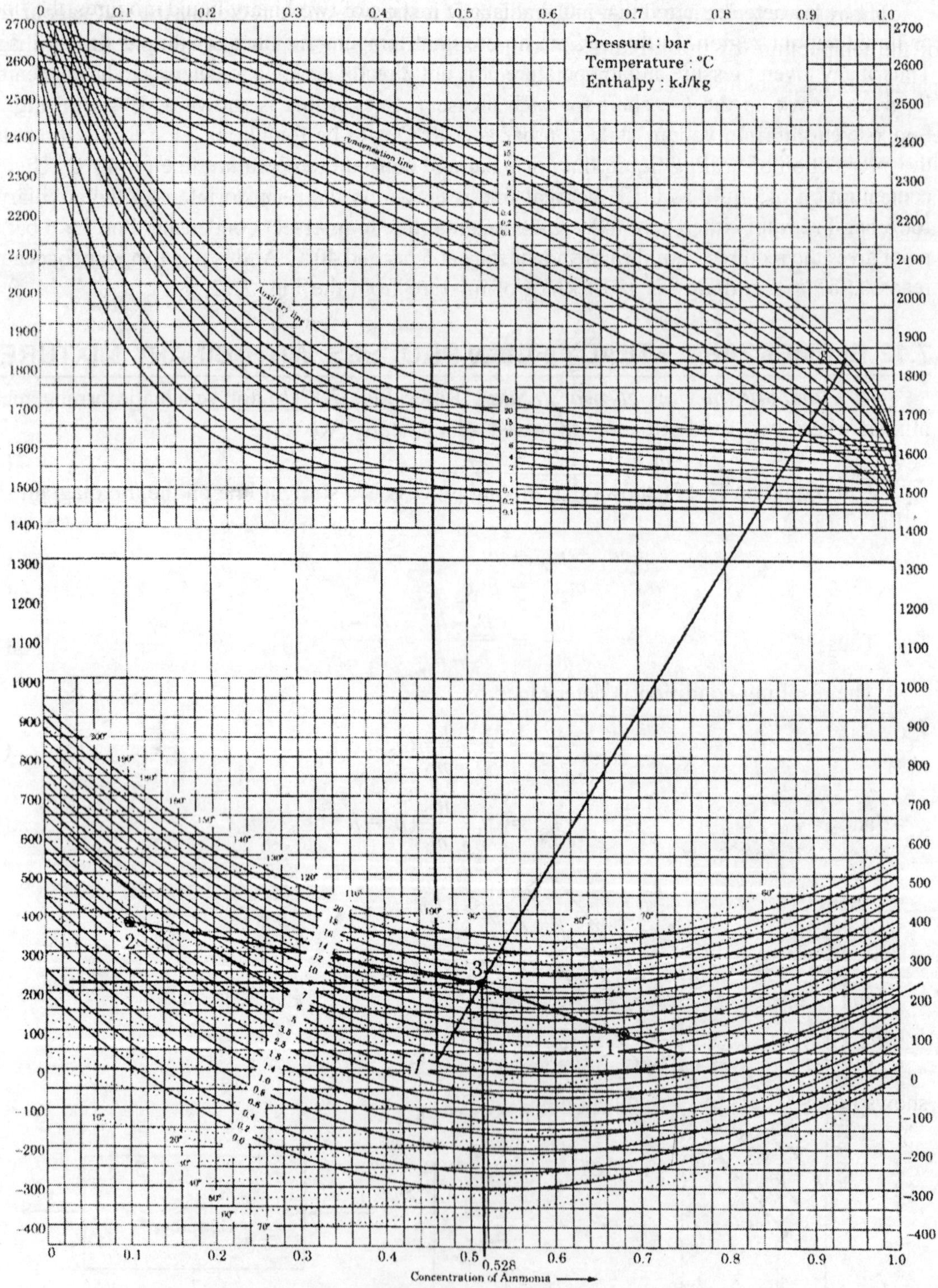

Fig. 7.12 Enthalpy-concentration (*h-c*) diagram for NH_3 – H_2O system

It may however be noted that in the adiabatic mixing of two binary liquid mixtures, the finding of the temperature of the mixture with the help of *h-c* diagram is not straight forward due to partial evaporation during mixing.

(*b*) **Mixing of two streams with heat exchange.** It is often necessary to add or reject heat while mixing two binary fluids so that the required final state is reached. This process occurs in the absorber of an absorption refrigeration system. Fig. 7.14 shows the schematic diagram of steady-flow problem in which heat is removed. Thus, in general,

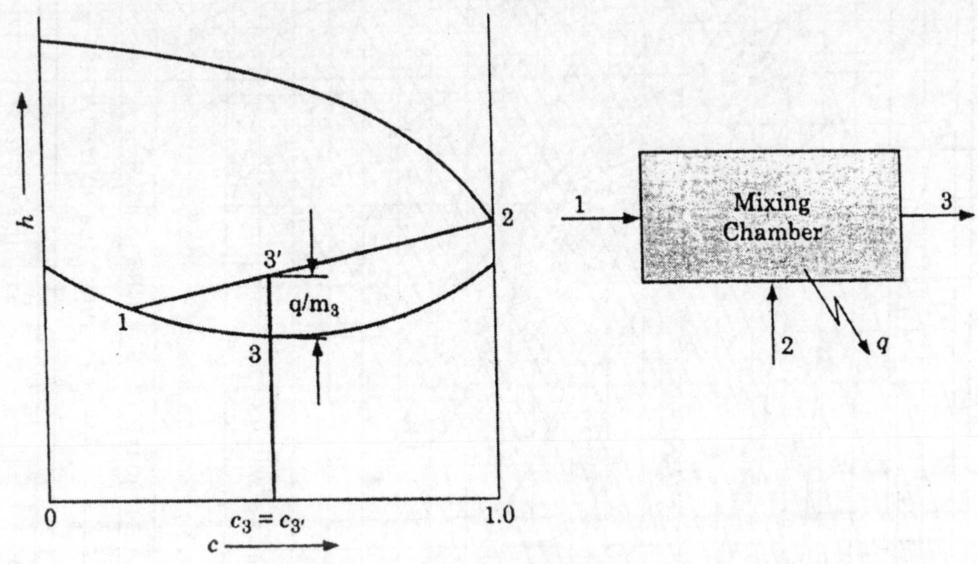

Fig. 7.14

$$m_1h_1 + m_2h_2 = m_3h_3 + q \quad \text{...(7.23)}$$

$$m_1h_2 + m_2 = m_3 \quad \text{...(7.24)}$$

$$m_1c_1 + m_2c_2 = m_3c_3 \quad \text{...(7.25)}$$

and

$$h_3 = h_1 + \frac{m_2}{m_3}(h_2 - h_1) - \frac{q}{m_3} \quad \text{...(7.26)}$$

Fig. 7.15 shows the enthalpy-concentration diagram for ammonia-water solution. Fig. 7.16 shows the *h-c* diagram for aqua-lithium bromide solution.

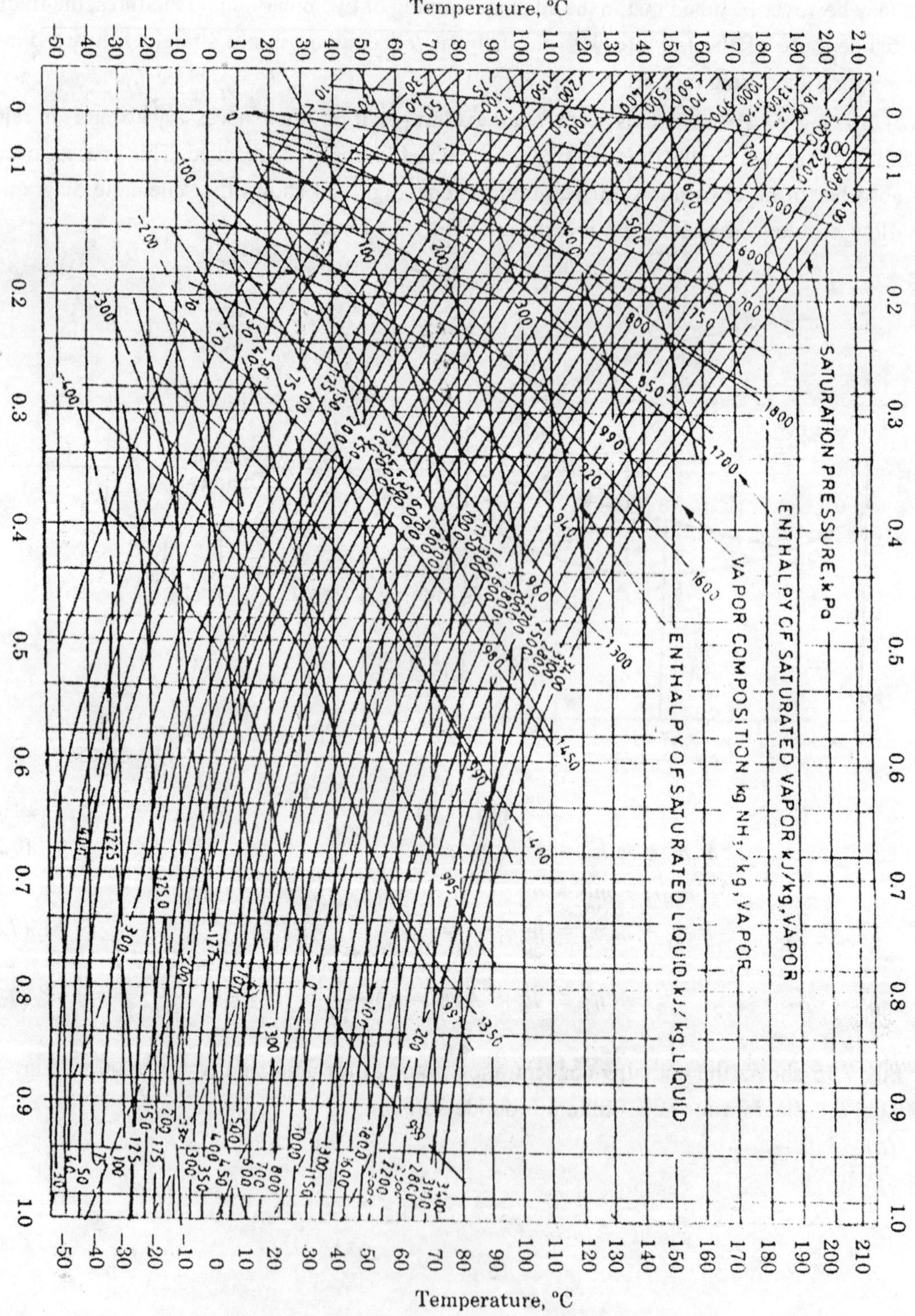

Ammonia in Saturated Liquid – kg NH_3/kg LIQUID

Fig. 7.15 Enthalpy-concentration Diagram for Ammonia water Solution

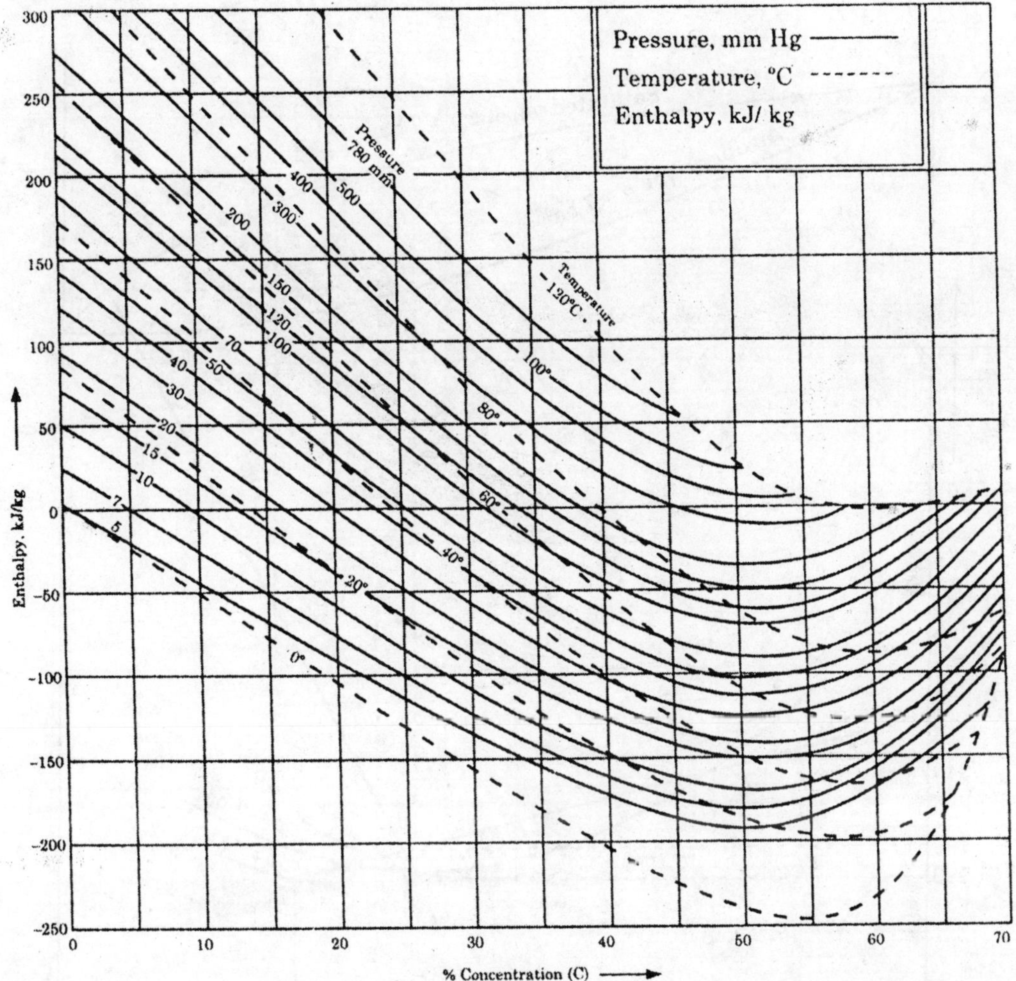

Fig. 7.16 Enthalpy-Concentration diagram for aqua-lithium bromide solution

Example 7.3 *Stream of 10 kg liquid aqua-ammonia with concentration of 0.7 kg of NH_3 per kg of mixture, at a temperature of 15°C and pressure of 7 bar is adiabatically mixed with a stream of 5 kg of saturated liquid aqua-ammonia at 100°C and 7 bar. Assuming steady flow conditions, determine:*

(a) the mixture concentration;

(b) the mixture enthalpy;

(c) the equilibrium temperature of mixture; and

(d) percentage liquid and percentage vapour composition of the mixture after equilibrium has been reached.

Solution: Chart is used for values. Refer Fig. 7.17 for schematic representation.

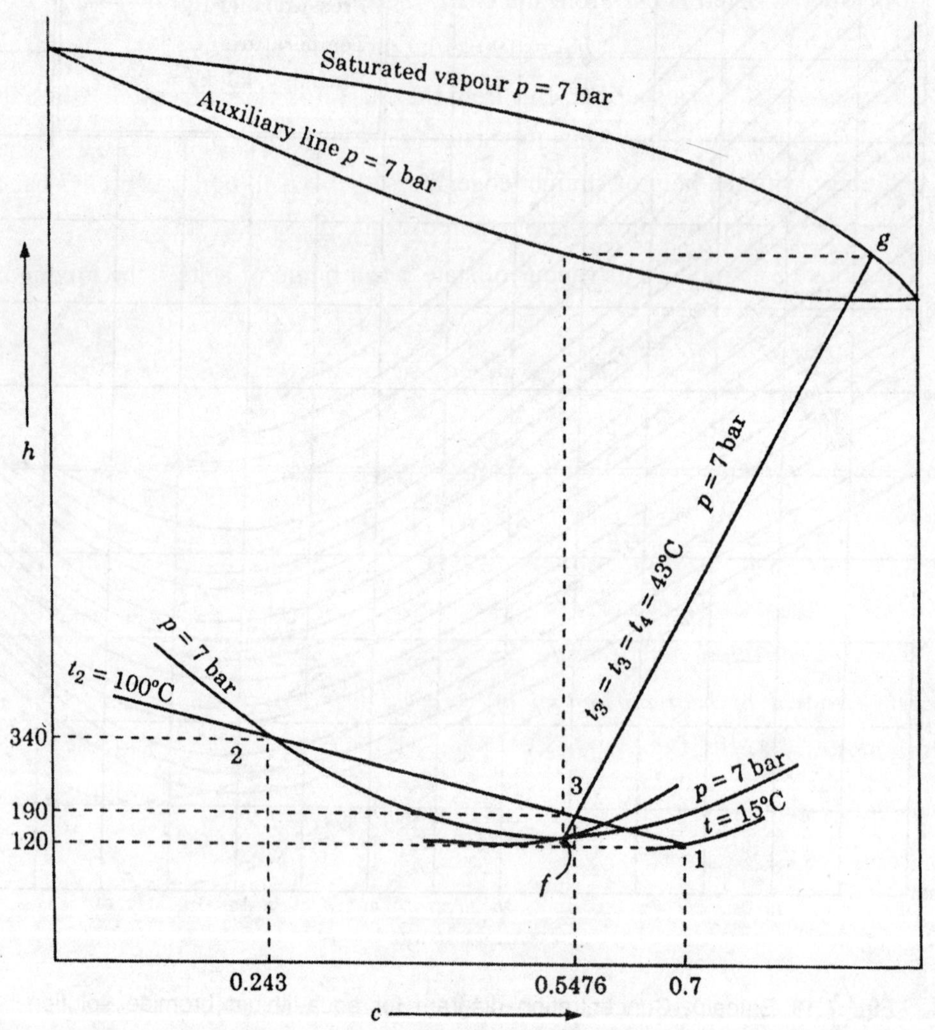

Fig. 7.17

(*a*) The state of the first stream for 10 kg flow per minute is called State 1. State 1 is a sub-cooled condition at t_1 = 15°C and c_1 = 0.7. State of second stream, *i.e.* State 2 is saturated and it is located at t_2 = 100°C and p_2 = 7 bar. Its mass is 5 kg.

Thus $\quad c_3 = c_1 + \dfrac{m_2}{m_3}(c_2 - c_1)$ and from chart $c_2 = 0.243$

Therefore, $\quad c_3 = 0.7 + \dfrac{5}{15}(0.243 - 0.7) =$ **0.5476 kg NH3 / kg mixture.**

(*b*) Join States 1 and 2 by a straight line. At the intersection of c_3 with this line, the state of mixture is determined. From the chart, the enthalpy is read as:

$$h_3 = 190 \text{ kJ/kg mix.}$$

(*c*) The pressure at 3 is also 7 bar. But from the chart it is clear that point 3 lies above the equilibrium line for 7 bar. Thus the State 3 falls in the two phase region. Thus by trial and error, with the help of straight edge, an isotherm is to be located on 7 bar and that should pass through State 3. The required isotherm *f*-3-*g* of 43°C.

(*d*) Since State 3 is mixture of vapour of state *g* and liquid of state *f*, the mixing equation gives,

$$\frac{m_g}{m_3} = \frac{c_3 - c_f}{c_g - c_f} = \frac{f_3}{f_g} = \mathbf{0.034}$$

Thus, mixture contains 96.6% liquid and 3.4% vapour.

Example 7.4 *One kg per min of saturated ammonia vapour with concentration equal to 1, i.e., c = 1 at 2 bar is mixed with 10 kg per min of saturated liquid ammonia at 2 bar and 40°C. The final desired mixture is to be saturated liquid at 2 bar. Determine:*

(a) The concentration, temperature and enthalpy of the mixture state.

(b) The removal of heat during mixing process in kJ/min.

Solution: Refer Fig. 7.18, we have

(*a*)
$$c_g = c_1 + \frac{m_2}{m_3}(c_2 - c_1)$$

$$= 0.325 + \frac{1}{10+1}(1 - 0.325)$$

$$= 0.325 + \frac{1}{11} \times 0.675$$

$$= 0.386 \text{ kg NH}_3\text{/kg Mix.}$$

Draw a line 1-2 and on it mark point 3′ such that $c_{3'} = 0.386$. From State 3′, draw a line vertically down to intersect boiling liquid line for 2 bar at 3. Thus State 3 gives the condition of saturated liquid at 2 bar.

Temperature line passing through 3 reads 30°C and enthalpy at State 3 reads h_3 = 17 kJ/kg of mix.

(*b*)
$$\frac{q}{m_3} = h_{3'} - h_3$$

$$= 200 - 17 = 183 \text{ kJ/kg of mix}$$

Thus
$$q = 183 \times 11 = 2013 \text{ kJ/min}$$

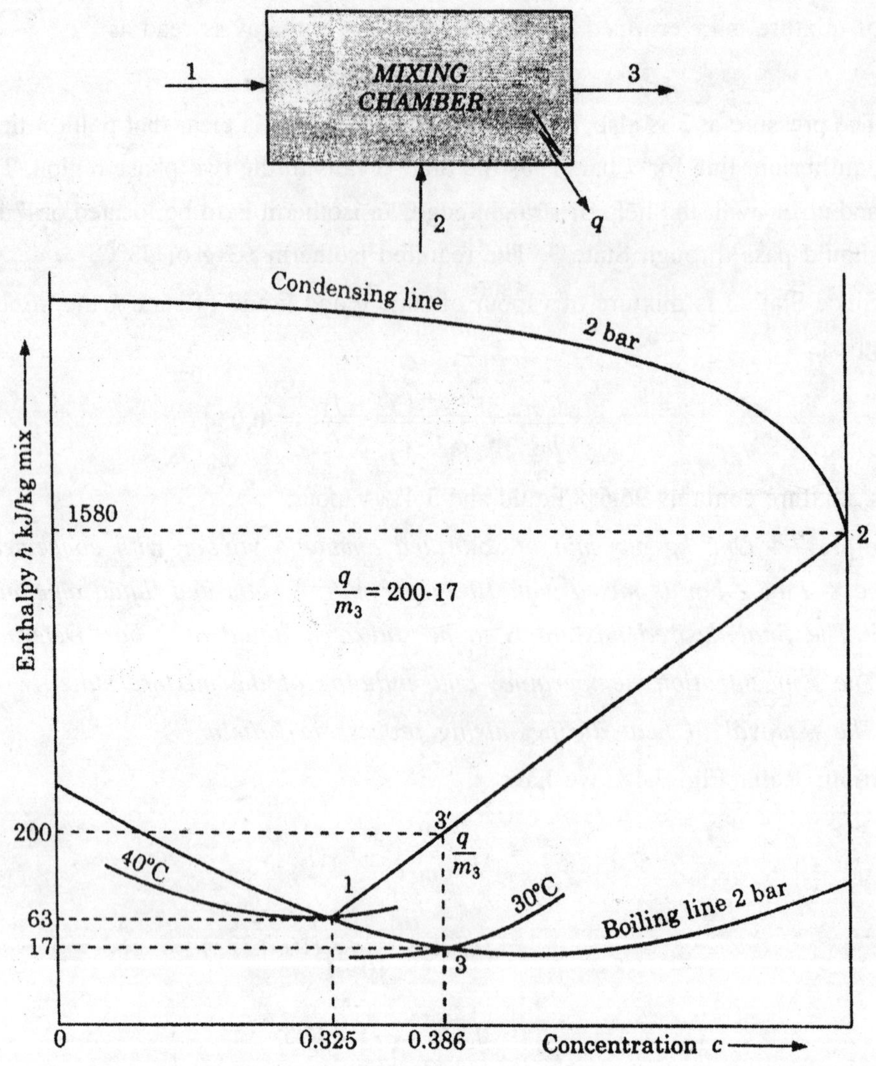

Fig. 7.18

Example 7.5 *Figure 7.19 Shows a flow diagram for the LiBr - water absorption refrigeration system. The following data apply:*

Generator Temperature	*= 100°C*
Condenser temperature	*= 40°C*
Evaporator temperature	*= 10°C*
Absorber temperature	*= 30°C*
Mass of solution handled by pump	*= 1.2 kg/s*

Determine:

(i) Heat supplied in the generator by heat source

(ii) Heat absorbed in the evaporator (refrigeration)

(iii) Heat rejected in the condenser to the coolant

(iv) Heat rejected in the absorber to the coolant.

Solution: Since a saturated condition of pure water prevails in the condenser due to simultaneous existence of liquid and vapour, the condensing temperature of 40°C fixes the pressure in the condenser and that is also the pressure in the generator. Thus p_c = 0.07375 bar 55.31 mm Hg, as read out from the steam tables. Similarly, the evaporator temperature fixes the low pressure in the evaporator = p_e = 0.01227 bar = 9.204 mm Hg from tables. Thus state points 1 (pure water vapour at condenser pressure of 55.31 mm Hg is established. Again state point 2 as pure liquid at p_c = 55.31 mm Hg is established.

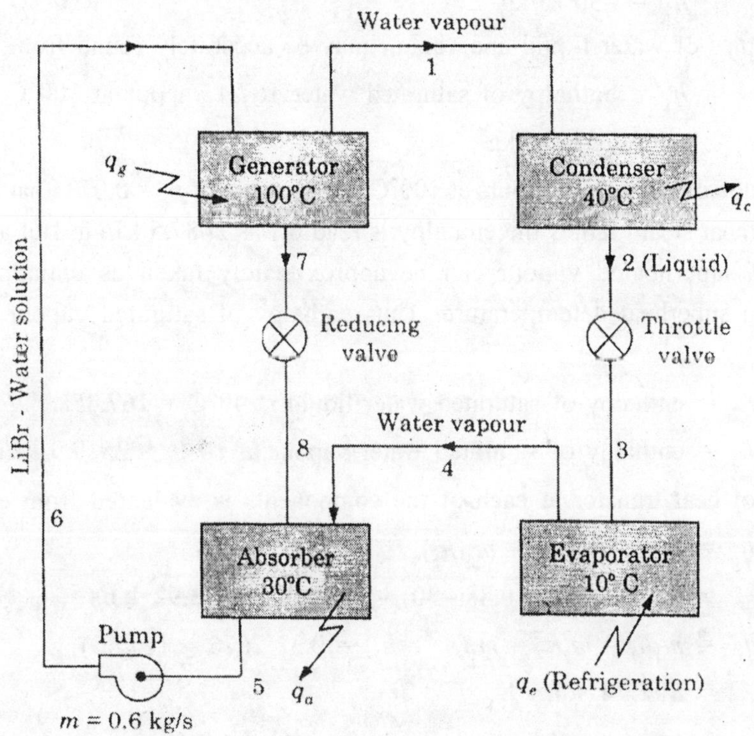

Fig. 7.19

At the intersection of generator pressure of 55.31 mm Hg and saturation temperature of 100°C, point 7 is established. The concentration as read from the *h-c* chart is c_7 = 0.68.

Leaving the absorber at solution temperature of 30°C and pressure of 9.204 mm Hg fixes state point 5 and the concentration from the *h-c* chart is read as c_7 = 0.5.

Considering the mass balance about the generator

$$m_7 + m_1 = 1.2 = m_8 + m_4 = m_5$$

and
$$m_7c_7 = (m_8 + m_4)\ c_5 = m_5c_5$$

or
$$m_7(0.68) = 1.2(0.5)$$

or
$$m_7 = \frac{1,2(0.5)}{0.68} = 0.88 \text{ kg/s}$$

and
$$m_7 = 1.2 - 0.88 = 0.32 \text{ kg/s}$$

Thus $\frac{1.2}{0.32}$ = 3.75 kg of solution is pumped for each kg of refrigerant water vapour generated.

The enthalpies of solutions can be read off the chart.

$$h_6 = h_5 = -170 \text{ kJ/kg assuming pump work as negligible.}$$

$$h_7 = -30 \text{ kJ/kg}$$

The enthalpies of water liquid and vapour may be accurately found from steam tables.

$$h_1 = \text{enthalpy of saturated water 167.4 vapour at 100°C}$$

$$= 2676 \text{ kJ/kg}$$

(It may be noted that water vapours at 100°C and pressure of p_c = 0.0735 bar are superheated. Thus from superheat steam tables the enthalpy is read off as 2687.5 kJ/kg. But at low pressures, the enthalpy of superheated vapour can be approximately taken as enthalpy at saturation corresponding to superheated temperature. Thus enthalpy of saturated vapour at 100°C reads 2676 kJ/kg.)

$$h_2 = \text{enthalpy of saturated water liquid at 40°C} = 167.45\text{kJ/kg}$$

$$h_4 = \text{enthalpy of saturated water vapour at 10°C} = 2519.9 \text{ kJ/kg.}$$

The rates of heat transfer at each of the components is evaluated from energy balance.

$$q_g = m_1h_1 + m_7h_7 - m_5(h_5)$$

$$= 0.32(2676) + 0.88(-30) - (-170) = \mathbf{1033.92\ kJ/s}$$

$$q_C = m_1h_1 - m_2h_2 - m_1(h_1 - h_2) = 0.32(2676 - 167.45)$$

$$= \mathbf{802.74\ kJ/s}$$

$$q_a = m_7h_1 - m_4h_4 - m_6h_6$$

$$= 0.88(-30) + 0.32(2519.9) - 1.2(-170) - 984 \text{ kJ/s}$$

$$q_e = m_4m_4 - m_2m_2 \qquad \text{(2-3 is throttling process)}$$

$$= 0.32(2519.9) - 0.32(167.45) = \mathbf{752.8\ kJ/s}$$

and
$$HCOP = \frac{q_e}{q_g} = \frac{752.8}{1033.92} = \mathbf{0.729}$$

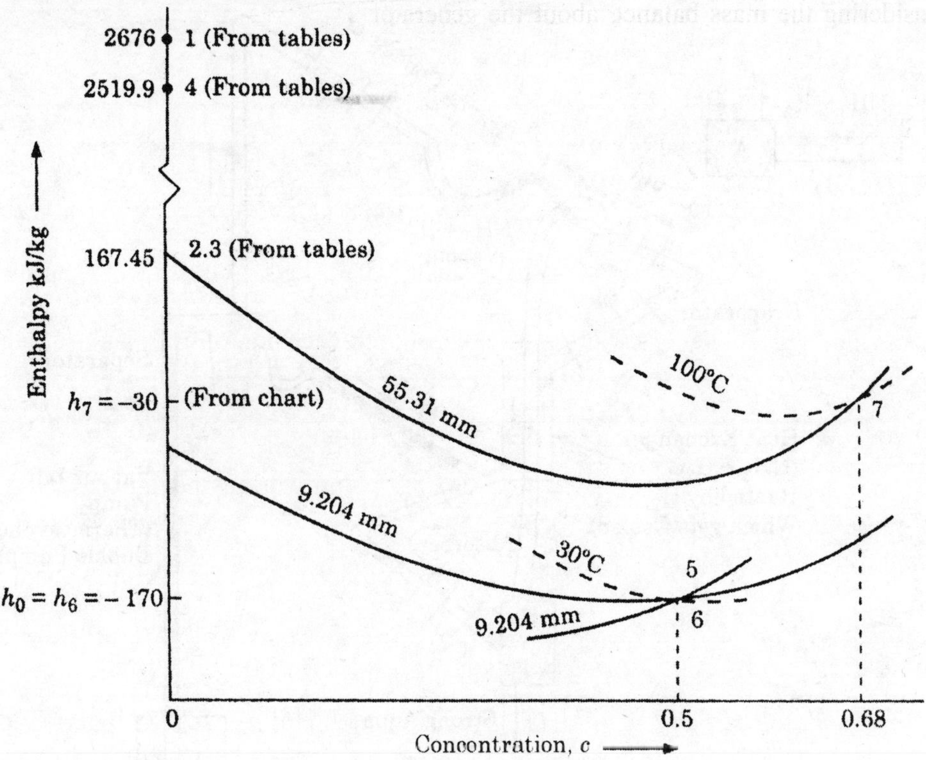

Fig. 7.20 *h-c* diagram.

7.13 ELECTROLUX REFRIGERATOR-DOMESTIC UNIT

In this system, ammonia is used as a refrigerant. The operation of this system is based on the concept of Dalton's law. A simple electrolux system is shown in Fig. 7.21. The ammonia vapour in the condenser is condensed to liquid and flows to evaporator by gravity. In the evaporator, the liquid ammonia meets an atmosphere of hydrogen where its partial pressure falls. The temperature falls to about –10°C. The vaporization of ammonia at this temperature produces the refrigerating effect. Water is used as a solvent for ammonia. Hydrogen flows in a direction opposite to that of ammonia. The ammonia vapour and hydrogen leave the top of evaporator and flow down through the gas heat exchanger getting warmed by the warmer hydrogen flowing to the evaporator.

Weak aqua ammonia solution enters at the top of absorber and absorbs ammonia gas as it passes counter flow through the absorber. The hydrogen is not soluble in weak aqua ammonia solution, gets separated and flows up to the evaporator though the heat exchanger. Strong aqua ammonia solution leaves the bottom of absorber and passes on to the generator.

Heat is supplied to the generator by a gas burner, expelling ammonia vapour out from the strong solution. Here the problem is to raise the elevation of weak solution of ammonia so that it can pass to the separator and flow back to the absorber. To achieve this the principle of bubble pump is used. The delivery tube from the generator is immersed below the liquid level in the generator. As ammonia vapours rise in the tube, they carry slugs of week solution also into the separating vessel. The weak solution flows from the separator to the absorber and ammonia vapour passes to the condenser. Thus cycle is completed.

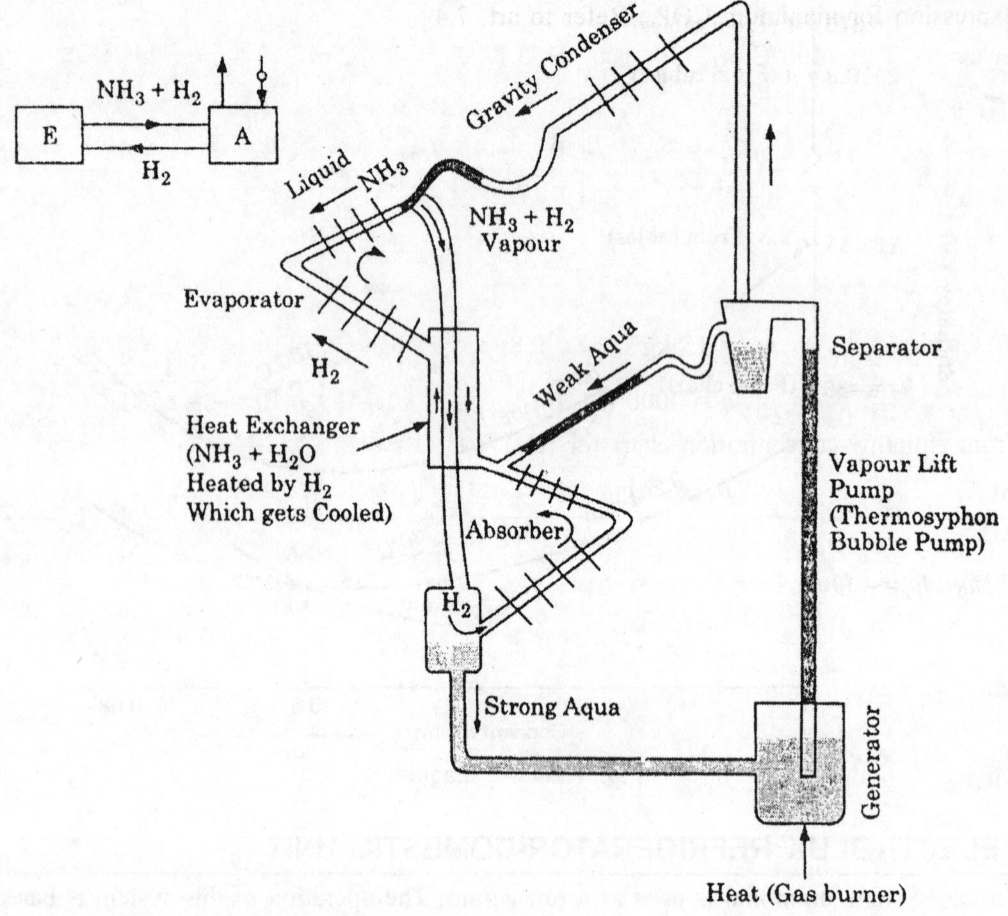

Fig. 7.21 Electrolux refrigerator schematic diagram.

Example 7.6

(a) With the help of a neat sketch, explain in brief, the working principle of a continuous Vapour Absorption Refrigeration system, obtaining an expression for maximum COP of the cycle. Also determine the COP of a Vapour Absorption system having a Generator temperature of 100°C, evaporator temperature of –15°C and absorber/ condenser temperature of 10°C.

(b) A stream of 15 kg/s of aqua ammonia having concentration of 0.8 and a pressure of 5 bar mixes with another saturated liquid stream with flow rate of 10 kg/s of 100°C at the same pressure. During mixing it receives heat at the rate of 4000 kJ/ s. Obtain mixture concentration, enthalpy, and amount of liquid and vapour after mixing.

(c) With the help of a net labeled sketch, describe the working principle of a Practical Ammonia-water vapour absorption system of refrigeration.

Solution: (*a*) Working principle of vapour absorption refrigeration system:

Refer to art. 7.2

Expression for maximum COP : Refer to art. 7.4

Given: T_g = 100°C or 273 + 100 = 373 K, T_e = –15°C or 273 – 15 = 258 K, T_r = 40°C or 273 + 40 = 313 K

$$(\text{COP})_{\max} = \frac{\left[\frac{T_r}{T_g} - 1\right]}{\left[1 - \frac{T_r}{T_e}\right]} = \frac{\left(\frac{313}{373} - 1\right)}{\left(1 - \frac{313}{258}\right)} = 0.7546$$

(*b*) Given: $m_{NH_3} = m_1 = 15$ kg/s, $c_1 = 0.8$, $p_1 = 5$ bar, $m_2 = 10$ kg/s, $t_2 = 100°C$,

$p_2 = p_1$, $q = 4000$ kJ/s, $m_3 = m_1 + m_2 = 15 + 10 = 25$ kg/s

From enthalpy-concentration chart for $NH_3 - H_2O$ system, we have

At p_1 = 5 bar and c_1 = 0.8, h_1 = 125 kJ/kg

At p_2 = 5 bar and t_2 = 100°C, c_2 = 0.12 and h_2 = 380 kJ/kg

Now $m_1c_1 + m_2c_2 = m_3c_3$

$15 \times 0.8 + 10 \times 0.12 = 25c_3$

Mixture concentration, $c_3 = \frac{13.2}{25} = 0.528$

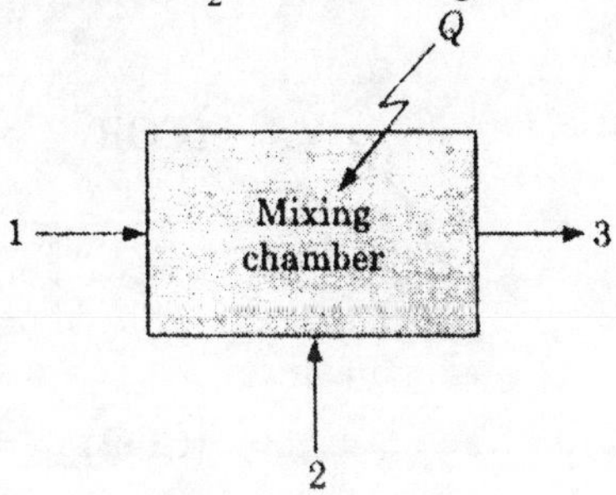

Mixture enthalpy, $h_3 = h_1 + \frac{m_2}{m_3}(h_2 - h_1) - \frac{q}{m_3}$

$= 125 + \frac{10}{25}(380 - 125) - \frac{4000}{25}$

$= 67$ kJ/kg

Fig. 7.22

Now c_3 = 0.528 and m_3 = 25 kg/s

Mass of ammonia vapour = 0.528 × 25 = 13.2 kg/s

Amount of water = 25 – 13.2 = 11.8 kg/s

(*c*) Refer to art 7.5.

Example 7.7

(a) Draw a net labelled sketch of a Practical Vapour Absorption Refrigeration Cycle and explain its working, in brief.

Also compare the relative merits and fieids of applications of vapour absorption and vapour compression systems.

(b) Derive an expression for the COP of an Ideal Vapour Absorption Refrigeration System in terms of the Generator, Evaporator, Condenser and Absorber temperatures.

In a simple ideal vapour absorption refrigeration system, following data are taken:

Generator temperature = 130°C

Condenser/Absorber temperature = *40°C*

Evaporator temperature = *–10°C*

Calculate the maximum COP

If the actual COP is 70% of the maximum COP, determine the quantity of dry-saturated steam supplied to the generator at 140°C for 50 Ton capacity system, if the condensate leaves the generator at 120°C.

(c) *Explain the concept of Adiabatic mixing of two streams, as related to Vapour Absorption Refrigeration System. Discuss the significance of 'Temperature-Concentration' and 'Enthalpy-Concentration' diagrams.*

Solution: (*a*) Practical Vapour Absorption Refrigeration cycle: Refer to art. 7.5.

(*b*) COP of an ideal vapour absorption refrigeration system : Refer to art. 7.4.

Given: T_g = 273 + 130 = 403 K, T_r = 273 + 40 313 K, T_e = 273 – 10 = 263 K, Q = 50 TR, t_s = 140°C

$$(\text{COP})_{\text{act}} = \left(\frac{T_e}{T_e - T_r}\right)\left(\frac{T_r - T_g}{T_r}\right)$$

$$= \left(\frac{263}{263-313}\right)\left(\frac{313-403}{403}\right) = 1.175$$

$$(\text{COP})_{\text{act}} = 0.70 \times 1.175 = 0.822$$

$$\text{Actual heat supplied} = \frac{211Q}{(\text{COP})_{\text{act}}} = \frac{211 \times 50}{0.822} = 12834.5 \text{ kJ/min}$$

Heat in dry saturated steam at 140°C = 2733.9 kJ/kg (from steam tables)

Heat in condensate at 120°C = 503.69 kJ/kg

Heat supplied by steam 2733.9 – 503.69 = 2230.21 kJ/kg

$$\text{Mass of steam supplied per hour} = \frac{12834.5}{2230.21} \times 60 = 345.29 \text{ kg/h}$$

(*c*) Adiabatic mixing of two streams: Refer to art. 7.8.1.

Temperature-Concentration diagram: Refer to art 7.10.

Enthalpy-Concentration diagram: Refer to art. 7.11.

Example 7.8

(*a*) *Explain, in brief, as to how the basic function of the compressor in vapour compression refrigeration cycle is achieved in a vapour absorption refrigeration system, and by which components. Also, describe the working principle of a 'continuous vapour absorption system', with the help of a neat sketch.*

(b) (i) *State the relative merits and demerits of vapour absorption and vapour compression refrigeration systems.*

(ii) *In a vapour absorption refrigeration system, the generator temperature is 100°C, evaporator temperature is –10°C and condenser/absorber temperature is 35°C. Calculate the maximum possible COP of the system.*

Also calculate the variation in COP of the system, if the generator temperature is increased to 125°C, the absorber/condenser temperature is increased to 40°C and the evaporator temperature is reduced to –15°C.

(c) *With the help of a net labeled sketch, describe the working principle of a lithium bromide-water absorption refrigeration system in brief, and also compare it with Ammonia-water-vapor absorption refrigeration system.*

Solution:

(*a*) Refer to art. 7.1 and 7.2

(*b*) (*i*) Merit and demerits of VARS: Refer to art. 7.3.

(*ii*) Given: $T_g = 273 + 100 = 373$ K, $T_e = 273 - 10 = 263$ K, $T_r = 273 + 35 = 308$ K,

$$(\text{COP})_{\max} = \left(\frac{T_e}{T_r - T_e}\right)\left(\frac{T_g - T_r}{T_g}\right)$$

$$= \left(\frac{263}{308-263}\right)\left(\frac{373-308}{373}\right)$$

$$= 1.018$$

For $T_g = 273 + 125 = 398$ K, $T_r = 273 + 40 = 313$ K, $T_e = 273 - 15 = 258$ K,

$$(\text{COP})_{\max} = \left(\frac{258}{313-258}\right)\left(\frac{398-313}{398}\right)$$

$$= 1.002$$

(c) LiBr-water absorption refrigeration system: Refer to art. 7.6.

Comparison of $NH_3 - H_2O$ and LiBr – H_2O system: Refer to art 7.7.

Example 7.9

(a) *Explain in brief, the working principle of a continuous vapour absorption refrigeration cycle. Determine the COP of a vapour absorption system working with the following operating temperatures:*

Generator temperature = 90°C

Absorber temperature = 35°C

Evaporator temperature = –12°C

If the evaporator temperature falls to –20°C, what should be the new generator temperature to operate the system with the same COP.

(b) *(i)* *Enumerate the basic difference between vapour absorption and vapour compression refrigeration system, enlisting the merits and demerits of vapour absorption system over vapour compression system.*

(ii) *A stream of 15 kg/s of aqua-ammonia [c_1= 0.8] at 0°C and 5 bar mixes adiabatically with another saturated liquid stream with the flow rate 10 kg/s and 100°C at the same pressure. Establish the state points on h-c diagram and obtain;*

(a) Mixture concentration

(b) Mixture enthalpy and

(c) Amount of liquid and vapour after mixing

(c) *Explain the concept of theory of mixtures as related to vapour absorption system. Also give the significance of Temperature-Concentration diagram explaining its importance in the analysis of vapour absorption refrigeration system.*

Solution:

(*a*) Working principle of VARS: Refer to art. 7.2.

Given: T_g = 273 + 90 = 363 K, T_r = 273 + 35 = 308 K, T_e = 273 – 12 = 261 K

$$(\text{COP})_{\max} = \left(\frac{T_e}{T_r - T_e}\right)\left(\frac{T_g - T_r}{T_g}\right)$$

$$= \left(\frac{261}{308-261}\right)\left(\frac{363-308}{363}\right) = 0.841$$

New $$T_e = 273 - 20 = 253 \text{ K}$$

Now $$0.841 = \left(\frac{253}{308-253}\right)\left(\frac{T_g - 308}{T_g}\right)$$

$$0.1828\, T_g = T_g - 308$$

$$T_g = 376.9 \text{ K or } 103.9°\text{C}$$

(*b*) (*i*) Refer to art. 7.3.

(*ii*) Aqua-ammonia: $\dot{m}_1$ = 15 kg/s, c_1 = 0.8, t_1 = 0°C, p_1 = 5 bar

Saturated liquid: $\dot{m}_2$ = 10 kg/s, t_2 = 100°C, p_2 = p_1 = 5 bar

Refer to Fig. 7.23 for schematic representation and Fig. 7.12 for obtaining the various values.

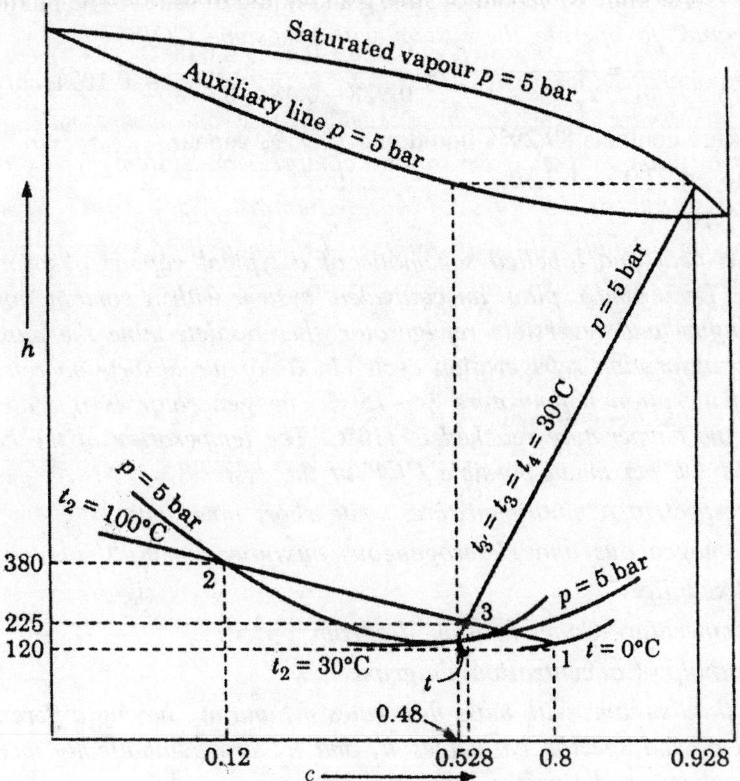

Fig. 7.23 h-c diagram.

1. Locate state 1(c_1 = 0.8, t_1 = 0°C) and p_1 = 120 kJ/kg.
2. Locate state 2(t_2 = 100°C, p_2 = 5 bar) c_2 = 0.12.
3. Join states 1 and 2 by a straight line.
4. Mass of mixture, $\dot{m}_3 + \dot{m}_2$ = 15 + 10 = 25 kg/s

   Concentration of mixture,

$$c_3 = c_1 + \frac{\dot{m}_2}{\dot{m}_3}(c_2 - c_1)$$

$$= 0.8 + \frac{10}{25}(0.12 - 0.8) = 0.528 \text{ kg NH}_3\text{/kg mixture}$$

5. With c_3 = 0.528, locate state 3 on line 1-2.
   Enthalpy of mixture, h_3 = 225 kJ/kg
6. The pressure at 3 is also 5 bar. But point 3 lies above the equilibrium line for 5 bar. Thus the state 3 lies in the two phase region. Thus by trial and error, with the help of a straight edge, an isotherm is to be located on 5 bar which should pass through state 3. The required isotherm is *f*-3-*g* of 30°C. c_f = 0.48 and c_g = 0.928.

7. State 3 is a mixture of vapour of state g and liquid of state f. The mixing equation gives,

$$\frac{\dot{m}_g}{\dot{m}_3} = \frac{c_3 - c_f}{c_g - c_f} = \frac{f_3}{f_g} = \frac{0.528 - 0.48}{0.928 - 0.48} = \frac{0.048}{0.448} = 0.1071$$

∴ Mixture contains 89.29% liquid and 10.71% vapour.

(*c*) Refer to art. 7.9 and 7.10.

Example 7.10

(a) Draw a neat and labelled schematic of a typical vapour absorption refrigeration system. Draw and explain an equivalent system with a combination of a reversible heat engine and reversible refrigerator, thereby determine the maximum COP of a vapour absorption refrigeration cycle. In a vapour absorption refrigeration system, the refrigeration temperature is –15°C The generator is operated by solar heat where the temperature reached is 110°C. The temperature of the heat sink is 55°C. What is the maximum possible COP of the system?

(b) With respect to a binary mixture, write short notes on:

(i) Homogeneous and Heterogeneous mixtures

(ii) Mixcibility

(iii) Temperature-Concentration diagram

(iv) Enthalpy-Concentration diagram

If two fluid streams with mass flow rates m_1 and m_2, having different concentrations, c_1 and c_2 and specific enthalpies h_1 and h_2, are adiabatically mixed together in a mixing chamber, derive the expressions for the resultant concentration and enthalpy, explaining the process with an Enthalpy-Concentration diagram.

(c) Describe the operation of a lithium Bromide-Water absorption refrigeration system with the help of a neat labelled diagram. Comment on the possibility of utilizing Solar Energy for the purpose of refrigeration coupled with this system.

Solution: (*a*) Vapour absorption refrigeration system: Refer to art. 7.2.

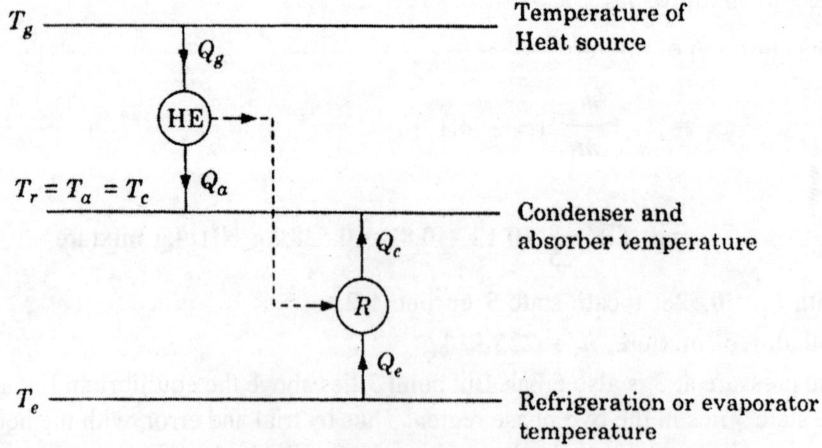

Fig. 7.24 Representation of vapour absorption system as a combination of heat engine and a refrigerator

Equivalent System

The vapour-absorption system is a heat-operated refrigerating machine. It may be considered as a combination of a heat engine and a refrigerating machine as shown in Fig. 7.24. The energy supplied to the system is in the form of heat Q_g at temperature T_g. The thermodynamic cycle is considered to comprise of a heat engine H.E. cycle, operating between the heat source temperature T_g and the temperature of heat rejection T_r (= T_a = T_c), and a refrigerator R cycle operating between the refrigeration temperature T_e and temperature of heat rejection T_a. The work done in the heat engine part of the cycle is equal to the work requirement of the refrigeration part of the cycle. Thus COP of the cycle can be written as,

$$\text{COP} = \frac{Q_e}{Q_g} = \frac{W}{Q_g} \times \frac{Q_e}{W} = (\eta_{th})_{\text{H.E.}} \times (\text{COP})_{\text{cooling}}$$

which implies that the COP of heat-operated refrigerating machine is equal to the product of the thermal efficiency of the heat engine part of the cycle and the coefficient of performance for cooling of the refrigeration part of the cycle.

The COP of a heat-operated refrigerating machine shall be maximum when each of the two terms has a maximum value. This shall be so when both are equal to their respective Carnot values.

Thus $\quad (\text{COP})_{\max} = (\eta_{th})_{\text{Carnot}} \times [(\text{COP})_{\text{cooling}}]_{\text{Carnot}}$

$$= \left[\frac{T_g - T_r}{T_g}\right]\left[\frac{T_e}{T_r - T_e}\right] = \left[1 - \frac{T_r}{T_g}\right]\left[\frac{1}{\frac{T_r}{T_e} - 1}\right]$$

When condenser and absorber temperature are not the same, then

$$(\text{COP})_{\max} = \left(1 - \frac{T_a}{T_g}\right)\left(\frac{1}{\frac{T_c}{T_e} - 1}\right)$$

Given: $T_e = 273 - 15 = 258$ K, $T_g = 273 + 110 = 383$ K, $T_r = 273 + 55 = 328$ K

$$(\text{COP})_{\max} = \left(1 - \frac{T_r}{T_g}\right)\left(\frac{1}{\frac{T_r}{T_e} - 1}\right)$$

$$= \left(1 - \frac{328}{383}\right)\left[\frac{1}{\frac{328}{258} - 1}\right]$$

$$= 0.529$$

(b) (*i*) Refer to art. 7.9.

(*ii*) Refer to art. 7.9.

(*iii*) Refer to art. 7.10.

(*iv*) Refer to art. 7.11

Adiabatic mixing of two fluids: Refer to art 7.12(*a*).

(*c*) Refer to art. 7.6.

Example 7.11

(a) What is the basic difference between a vapour absorption refrigeration system and a vapour compression system? Explain how the function of the compressor in vapour compression system is achieved in a vapour absorption system, and by which components? Draw a neat sketch of a practical vapour absorption system and describe its working in brief.

(b) What do you understand by maximum COP of a vapour Absorption refrigeration system and how do you calculate its value in terms of the Evaporator, Generator, Condenser and Absorber temperatures?

In a vapour absorption refrigeration system, the generator is operated by solar heat where the temperature achieved is 100°C. If the evaporator temperature is –10°C and the condenser/absorber temperature are 35°C, what is the maximum possible COP of the system?

(c) (i) Explain in brief, the concept of Adiabatic mixing of two streams, as related to vapour absorption Refrigeration system. Also discuss the significance of Temperature - Concentration diagram.

Solution:

(*a*) Refer to art. 7.1 and 7.5

(*b*) Refer to art. 7.4.

Given: $T_g = 273 + 100 = 373$ K, $T_e = 273 - 10 = 263$ K, $T_r = 273 + 35 = 308$ K

$$(\text{COP})_{\max} = \left(1 - \frac{T_r}{T_g}\right)\left(\frac{1}{\dfrac{T_r}{T_e} - 1}\right)$$

$$= \left[1 - \frac{308}{373}\right]\left[\frac{1}{\dfrac{308}{263} - 1}\right] = 1.1018$$

(c) Refer to art 7.12(*a*) and art. 7.10.

Example 7.12

(a) Answer the following:

(i) Explain the method of obtaining an isotherm (in two phase region) on enthalpy-concentration (h-c) diagram for a mixture.

(ii) Compare aqua ammonia vapour absorption refrigeration system with Li-Br water absorption refrigeration system.

(*b*) *Two aqua ammonia mixtures at saturated liquid state and pressure of 20 bar each are adiabatically mixed. Mixture A has a mass 6 kg and a concentration of 0.7, while mixture B has a mass of 3 kg and concentration of 0.1. Find the temperature, concentration, and specific enthalpy of mixture after adiabatic mixing. Also determine the concentration of liquid and vapour phases, and the masses of liquid and vapour phases.*

Solution:

(*a*) (*i*) Method of obtaining an isotherm on *h-c* diagram

Refer to Fig. 7.11

1. Locate state 1 on *h-c* diagram for given values of p_1, c_1 and t_1. Draw a horizontal line from state 1 to the left to read the value of h_1 on *h*-scale.
2. Locate state 2 on $p_2 = p_1$ and given value of t_2. Draw a vertical line downward to read the value of c_2 on *c*-scale.
3. Join states 1 and 2 by a straight line.
4. Calculate the concentration of mixture c_3 for the given values of m_1, m_2, c_1 and then read value of c_2 from the *h-c* chart.

$$c_3 = c_1 + \frac{\dot{m}_2}{\dot{m}_3}(c_2 - c_1)$$

5. Locate state 3 on line 1-2. Draw a horizontal line at state 3 to the left and read the value of h_3 on *h*-scale. State 3 shall lie on the two-phase region.
6. Locate the isotherm on p_1 with the help of a straight edge, by trial and error, which should pass through state 3. Mark points *f* and *g* on the boiling line corresponding to p_1 and saturated vapour line corresponding to p_1 respectively.
7. Draw vertical and horizontal lines at *f* and *g* to read the values of *c* and *h* respectively.

(*a*) Refer to art. 7.7.

(*b*) Given : $p_1 = p_2 = 20$ bar, $m_A = 6$ kg, $c_A = 0.7$, $m_B = 3$ kg, $c_B = 0.1$

A → Mixing chamber → C; B ↑

Fig. 7.25 Adiabatic mixing of two streams

For adiabatic mixing of two streams, we have (see Fig. 7.25)

$$m_A + m_B = m_C$$

$$\therefore \quad m_C = 6 + 3 = 9 \text{ kg}$$

$$m_A c_A + m_B c_B = m_C c_C \qquad ...(2)$$

$$6 \times 0.7 + 3 \times 0.1 = 9 c_C$$

Mixture concentration, $$c_C = \frac{4.5}{9} = 0.5$$

$$m_A h_A + m_B h_B = m_C h_C \qquad ...(3)$$

For saturated aqua ammonia solution at 20 bar pressure,

$$h_f = 437 \text{ kJ/kg}$$

Mixture enthalpy, $$h_c = \frac{9 \times 437}{9} = 437 \text{ kJ/kg}$$

Temperature of mixture, $$t_c = 49.3°\text{C}$$

To determine the concentration and masses of liquid and vapour phases, follow the procedure of Example 7.9(*b*)-(*ii*).

Example 7.13

(a) With the help of schematic diagram, explain the working of a practical aqua ammonia vapour absorption refrigeration system. Highlight the advantage of using Analyser, Rectifier, and Heat Exchangers. Derive an Expression for the maximum COP of a simple vapour absorption refrigeration system.

(b) Two aqua ammonia streams at a pressure of 20 bar with a mass flow rate of 2 kg/s each are adiabatically mixed. First stream is saturated liquid at concentration of 0.8 while second stream is at a temperature of 150°C and concentration of 0.2. Find the temperature, concentration, and specific enthalpy of mixed stream after adiabatic mixing. Also determine the concentration of liquid and vapour phases.

Solution:

(*a*) Refer to art. 7.5, art 7.1 and art 7.4

(*b*) Given: p = 20 bar, $\dot{m}_1 = \dot{m}_2$ = 2 kg/s, c_1 = 0.8, c_2 = 0.2, t_2 = 150°C,

$$\dot{m}_3 = \dot{m}_1 + \dot{m}_2 = 2 + 2 = 4 \text{ kg/s}$$

$$\dot{m}_1 c_1 + \dot{m}_2 c_2 = \dot{m}_3 c_3$$

$$2(0.8 + 0.2) = 4c_3$$

$$c_3 = 0.5$$

Follow the procedure of Example 7.9(*b*)-(*ii*) to get the answers.

REVIEW QUESTIONS

1. How the function of compressor in vapour compression system is achieved in vapour absorption system?
2. Explain the function of generator and absorber in absorption system.
3. What is the function of rectifier and heat exchanger in absorption system?
4. What is the role of analyser and rectifier in absorption system?
5. List the advantages of vapour absorption system over vapour compression system.
6. List the applications of aqua-ammonia vapour absorption system.
7. Give a comparison of aqua ammonia and lithium bromide absorption systems.
8. Write the expression for maximum COP of vapour absorption system.
9. Define the following:

   (*a*) Concentration (*b*) Degree of superheat

   (*c*) Mass concentration

MULTIPLE – CHOICE QUESTIONS

1. The refrigerant commonly used in vapour absorption system is

   (a) water (b) ammonia (c) aqua ammonia (d) Freon-12
2. In aqua-ammonia absorption refrigeration system, incomplete rectification leads to accumulation of water in

   (a) Generator (b) Condenser (c) Evaporator (d) Absorber.
3. In aqua-ammonia and lithium bromide-water absorption refrigeration systems, the refrigerants are respectively

   (a) Water, water (b) Water, LiBr (c) NH_3, LiBr (d) NH_3, H_2O
4. In lithium bromide absorption refrigeration system

   (a) lithium bromide is used as refrigerant and water as absorbent.

   (b) water is used as refrigerant and lithium bromide as absorbent

   (c) ammonia is used as refrigerant and lithium bromide as absorbent

   (d) lithium bromide is used as refrigerant and ammonia as absorbent.
5. COP of vapour absorption system as compared to vapour compression system is

   (a) less (b) more

   (c) may be less or more (d) depends upon working temperature
6. The function of compressor in vapour absorption system is performed by

   (a) generator (b) absorber

   (c) absorber and pump (d) generator, absorber and pump

7. Maximum COP of vapour absorption system is given by

(a) $\frac{T_e(T_g - T_r)}{T_g(T_r - T_e)}$ (b) $\frac{T_g(T_r - T_e)}{T_e(T_g - T_r)}$ (c) $\frac{T_r(T_g - T_e)}{T_g(T_r - T_e)}$ (d) $\frac{T_g(T_r - T_e)}{T_r(T_g - T_e)}$

where T_g = generator temperature, T_r = environment temperature,
T_e = evaporator temperature.

8. In vapour absorption system, heating in generator is done at 100°C, refrigeration in evaporator at –10°C and cooling by cooling water at 30°C. The maximum COP is

(a) 0.81 (b) 1.234 (c) 2.234 (d) 0.45

9. If V_w = volume of water, V_{NH_3} = volume of ammonia, and V_m = volume of mixture, then

(a) $V_m = V_w + V_{NH_3}$ (b) $V_m < (V_w + V_{NH_3})$

(c) $V_m > (V_w + V_{NH_3})$ (d) Unpredictable.

10. Hydrogen is essential in an Electrolux refrigeration system, because

(a) it acts as a catalyst in the evaporator

(b) the reaction between hydrogen and ammonia is endothermic in evaporator and exothermic in absorber

(c) the cooled hydrogen leaving the heat exchanger cools the refrigerant entering the evaporator

(d) it helps in maintaining a low partial pressure for the evaporating ammonia

11. The refrigerant used for absorption refrigerators working heat from solar collectors is mixture of water and

(a) carbon dioxide (b) sulphur dioxide

(c) Lithium bromide (d) Freon-12

12. Consider the following statements:

In ammonia refrigeration systems, oil separator is provided because

1. Oil Separation in evaporator would lead to reduction in heat transfer coefficient.
2. Oil accumulation in the evaporator causes choking of evaporator.
3. Oil is partially miscible in the refrigerant.
4. Oil causes choking of expansion device

Of these statements

(a) 1 and 2 are correct (b) 2 and 4 are correct

(c) 2, 3 and 4 are correct (d) 1, 3 and 4 are correct.

13. Waste heat can be effectively used in which one the following refrigeration system?

(a) Vapour compression cycle (b) Vapour absorption cycle

(c) Air refrigeration cycle (d) Vortex refrigeration system

14. Vapour absorption refrigeration system works using the
(a) ability of a substance to get easily condensed or evaporated
(b) ability of a vapour to get compressed or expanded
(c) affinity of a substance for another substance
(d) absorptivity of a substance

15. Which one of the following statements regarding ammonia absorption system is correct ?
(a) a function of the temperature and pressure of a solution
(b) a function of the pressure of the solution irrespective of the temperature
(c) a function of the temperature of the solution
(d) independent of the temperature and pressure of the solution

16. Match List I (Basic components of Aqua-ammonia refrigeration system) with List II (functions of the components in the system) and select the correct answer using the codes given below the lists:

List I	*List II*
A. Generator	1. Dehydration
B. Analyser	2. Removal of vapour from strong aqua-ammonia solution
C. Rectifier	3. Producing dry ammonia vapour by removing traces of water particles completely
D. Receiver	4. Storage of high pressure liquid ammonia
	5. Formation of liquid ammonia from high pressure vapours.

Codes:	**A**	**B**	**C**	**D**
(a)	3	1	2	5
(b)	5	3	4	2
(c)	1	3	2	5
(d)	2	1	3	4

17. Maximum possible COP of a solar absorption refrigeration system with generator temperature of 360 K, absorber temperature of 300 K, condenser temperature of 300 K and evaporator temperature of 270 K is
(a) 9 (b) 6 (c) 3 (d) 1.5

18. In a vapour absorption refrigerator, heat is rejected in
(a) Condenser only (b) Generator only
(c) Absorber only (d) Condenser and absorber

Answers

1. (c) **2.** (b) **3.** (d) **4.** (b) **5.** (a) **6.** (d) **7.** (a) **8.** (b)
9. (b) **10.** (d) **11.** (c) **12.** (b) **13.** (b) **14.** (c) **15.** (a) **16.** (d)
17. (d) **18.** (d)

NUMERCIAL PROBLEMS

7.1 In a vapour absorption system, heat is supplied to the generator at a temperature of 90°C. The cooling in the condenser and evaporator takes place at 25°C and –10°C respectively. Calculate the maximum COP of the system. [**Ans.** 1.328]

7.2 In a vapour absorption system, the heat is supplied to the generator by condensing steam at 3 bar and dryness fraction 0.85. The temperature in the evaporator is to be maintained at –10°C. If the cooling water rejects heat at 30°C in the condenser, calculate the maximum COP of the system. Also calculate the mass of steam required per hour if the refrigeration load is 15 tonnes and the actual COP is 40 percent of the maximum COP.

[**Ans.** 1.674, 153.5 kg/h]

7.3 In a vapour absorption refrigeration system, the refrigeration temperature is –15°C. The generator is operated by a heat source at 110°C. The temperature of heat sink is 55°C. Find the maximum possible COP of the system. [**Ans.** 0.529]

7.4 Steam at 3 bar and 0.85 dry is used in the generator of vapour absorption system. The evaporator of the refrigeration system is maintained at –10°C. The circulating cooling water rejects heat at 30°C in the condenser. Determine maximum COP for the system. If the steam leaves the generator as saturated liquid, determine the stream consumption per hour for 10 TR plant. Assume relative COP as 0.4. [**Ans.** 1.6745, 102.3 kg/h]

Chapter

8

REFRIGERANTS

8.1 REFRIGERANTS

A refrigerant may be defined as a substance which absorbs heat through expansion or vaporization and loses it through condensation in a refrigeration system.

8.2 CLASSIFICATION OF REFRIGERANTS

1. **Primary refrigerants:** They take part directly in refrigeration and cool the substance on absorption of their latent heat. Some of the primary refrigerants are NH_3, CO_2CH_3, Cl_2, CH_2Cl_2, SO_2, C_2H_5Cl, and Freon group.
2. **Secondary refrigerants:** They are first cooled with the help of primary refrigerant and then used for cooling other substances by absorption of their sensible heat only, e.g. ice, solid CO_2.

Primary refrigerants may be grouped as follows:

(*i*) **Halo-Carbon compounds:** They contain one or more of three halogens, fluorine, chlorine and bromine. Commercially they are sold under the names as: Freon, Genetron, Irontron and Areton, They are used in domestic, commercial and industrial refrigerating devices. Some of their important examples are:

Carbon tetrachloride – R10 : CCl_4

Trichloromono-fluro methane – R11 : CCl_3F

Dichloro-difluro methane – R12 : CCl_2F_2

Monobromotri-fluro methane – R13 : $CBrF_3$

(*ii*) **Azeotropes:** They are the mixtures of different refrigerants. They do not undergo any separation with changes in temperature and pressure, e.g.

R-500(73.8% R12 + 26.2% R-152) : $CCl_2F_2 + CH_3CHF_2$

R-502 (48.8% R22 + 51.2% R-115) : $CHClF_2 + CCl\ F_2CF_3$

(*iii*) **Hydrocarbons : They are hydrocarbons, e.g.**

Methane – R50 : CH_4

Ethane – R170 : C_2H_6

Propane – R200 : C_2H_8

(*iv*) **Inorganic compounds**

Ammonia – R717 : NH_3

Water – R718 : H_2O

Air – R720 : –

Carbon dioxide – R720 : CO_2

(*v*) **Unsaturated organic compounds:** They contain ethylene or propylene as main constituents, e.g

Ethylene – R1150 : C_2H_4

Propylene – R1720 : C_3H_6

Trichloroethylene – R1120 : $C_2H_4Cl_3$

8.3 DESIGNATION OF REFRIGERANTS

1. For a compound derived from a saturated hydrocarbon denoted by the chemical formula

$$C_mH_nF_pCl_q$$

In which $(n + p + q) = 2m + 2$, the complete designation of a refrigerant is

$$R(m - 1)\ (n + 1)\ (p)$$

2. The brominated refrigerants are denoted by putting an additional *B* and a number to denote as to how many chlorine atoms are replaced by bromine atoms.
3. In the case of isomers, i.e., compounds with the same chemical formula but different molecular structure, subscripts *a*, *b*, etc are used after the designations.
4. Unsaturated compounds, for which $(n + p + q) = 2m$, are distinguished by putting the digit 1 before $(m - 1)$.
5. For inorganic refrigerants, numerical designations have been given according to their molecular weight added to 700.
6. In the case of butane, C_4H_{10} and higher hydrocarbons, arbitrary designation R 600 is used.

8.4 DESIRABLE PROPERTIES OF REFRIGERANTS

A good refrigerant should have the following properties:

(*a*) *Thermodynamic properties*

1. Low boiling point
2. Low freezing point
3. High saturation temperature
4. High latent heat of vaporization
5. High critical pressure and temperature,

(*b*) *Chemical properties*

1. Non-toxic
2. Non-flammable
3. Non-explosive
4. Non-corrosive
5. Chemically stable
6. Odourless

(*c*) *Physical properties*

1. Low specific volume
2. High thermal conductivity
3. High electrical insulation
4. Low specific heat
5. Miscibility with lubricating oil.

(*d*) Other properties

1. Ease of leakage detection
2. Ease of handling
3. High COP
4. Ease of availability and of low cost.
5. Low power consumption per ton of refrigeration.

8.5 APPLICATIONS OF REFRIGERANTS

The refrigerants and their applications are given in table 8.1.

Table 8.1 Refrigerants and their applications

Refrigerant	*Applications*
1. Inorganic	
(i) Ammonia (NH_3)	Reciprocating and screw compressors, in cold storages, ice plants, food refrigeration, etc.
(ii) Water (H_2O)	Steam-ejector system for air-conditioning, lithium bromide-water absorption system
(iii) Carbon dioxide (CO_2)	Dry ice (solid CO_2) in transport refrigeration.
2. Organic:	
Freon-11	Centrifugal compressors in large capacity central air-conditioning plants.
Freon-12	Reciprocating compressors in small units, especially domestic refrigerators, water coolers, etc.
Refrigerant-22	Reciprocating compressors large units such as package units and central air-conditioning plants, cold storage, food freezing and storage, etc.
Refrigerant-502	Large supermarket frozen food cabinets.
Refrigerant-13	Low temperature refrigeration
Refrigerant-113	Centrifugal compressors for air-conditioning.
Refrigerant-114	Rotary compressors.
3. Secondary refrigerants: Brines ($CaCl_2$, Industrial ice plants, cold storages, skating rinks, etc. $NaCl_2$ in water) Inhibited glycols Freezing point depressants, (ethylene glycol, propylene glycol in water)	

The primary-refrigerants are further classified into the following four groups:

1. Halo-carbon refrigerants
2. Azeotrope refrigerants.
3. Inorganic refrigerants, and
4. Hydro-carbon refrigerants.

1. Halo-carbon Refrigerants

The commonly used halo-carbons refrigerants, their applications and method of leak testing are given in Table 8.2.

Table 8.2 : Commonly used halo-carbon refrigerants

Refrigerant number	*Chemical formula*	*Chemical name*	*Applications*	*Method of leak testing*
R-11	CCl_3F	Trichloromono-fluoromethane	Large centrifugal compressors. Flushing agent for cleaning compressor parts.	Soap solution, halide torch, electronic detector
R-12	CCl_2F_2	Dichlorofluoro-methane	Refrigerators, freezers, water coolers, room and window air-conditioners, Large centrifugal compressors.	Soap solution, halide torch, electronic leak detector
R-13	$CClF_3$	Monochlorotri-fluoromethane	Cascade system, reciprocating compressors	–
R-14	CF_4	Carbontetra-fluoride	Cascade system	–
R-21	$CHCl_2F$	Dichloromono-fluoromethane	Centrifugal compressors.	–
R-22	$CHClF_2$	Monochloro-difluoromethane	Household refri-gerators, air-conditioners, reciprocating and centrifugal compressors.	Soap solution, halide torch, electronic leak detector
R-30	CH_2Cl_2	Methylene Chloride	Theatre air-conditioners, auditoriums, office buildings	Halide torch
R-40	CH_3Cl	Methyl chloride	Domestic units	Soap solution, electronic leak detector
R-100	C_2H_5Cl	Ethyl chloride	Not used	–
R-118	CCl_2FCClF_2 or $C_2Cl_3F_3$	Trichlorotri-fluoroethane	Commercial and industrial air-conditioning	–
R-114	$CClF_2CClF_2$ or $C_2Cl_2F_4$	Dichlorotetra-fluoroethane	Household refrigerating system, water coolers.	–
R-115	$CClF_2CF_3$ or C_2ClF_5	Monochloro-peantafluoroethane	–	–

2. Azeotrope Refrigerants

The term 'azeotrope' refers to a stable mixture of refrigerants whose vapour and liquid phases retain identical compositions over a wide range of temperatures.

Some of the azeotropic refrigerants, their applications and method of leak testing are given in Table 8.3.

Table 8.3 Azeotrope refrigerants

Refrigerant number	*Chemical formula*	*Applications*	*Leak testing method*
R-500	CCl_2F_2 / CH_3CHF_2	Industrial and commercial reciprocating compressor systems	Soap solution, halide torch, coloured tracing agent
R-502	$CHClF_2$ / $CClF_2CF_3$	Frozen food lockers, frozen food processing plants, display cases, storage units and ice cream	Soap solution, halide torch, electronic leak detector
R-503	CHF_3 / $CClF_3$	Cascade systems	Soap solution, halide torch, electronic leak detector
R-504	CH_2F_2 / $CClF_2CF_3$	Industrial processes	Soap solution, halide torch, electronic leak detector

3. Inorganic Refrigerants

The various inorganic refrigerants, their applications and leak testing method are given in Table 8.4.

Table 8.4 Inorganic refrigerants

Refrigerant number	*Chemical formula*	*Chemical name*	*Applications*	*Method of leak testing*
R-717	NH_3	Ammonia	Large and commercial reciprocating compression systems, absorption systems. Cold storage, warehouse plants, ice cream manufacture, beer manufacture, food freezing plants.	Burning sulphur candle gives white fumes
R-729	–	Air	Aircraft air-conditioning	–
R-744	CO_2	Carbon di oxide (dry ice)	Industrial unit and aboard ships	–
R-764	SO_2	Sulphur dioxide	Ripener and preservative of foods	Soap solution, ammonia swab white giving smoke
R-118	H_2O	Water	Absorption system	–

4. Hydro-Carbon Refrigerants

The various hydro-carbons are given in Table 8.5. They are mostly used for industrial and commercial installations.

Table 8.5 : Hydrocarbon refrigerants

Refrigerant number	*Chemical formula*	*Chemical name*
R-170	C_2H_6	Ethane
R-290	C_3H_3	Propane
R-600	C_4H_{10}	Butane
R-600a		Isobutane
R-1120	$C_2H_4Cl_3$	Trichloroethylene
R-1130	$C_2H_4Cl_2$	Dichloroethylene
R-1150	C_2H_4	Ethylene
R-1270	C_3H_6	Propylene

8.6. Ozone Layer Depletion

Stratosphere is about 25-30 km away from the Earth's surface. It contains ozone (O_3). Ozone is formed in the stratosphere by the reaction that occurs when sunlight interacts with oxygen. This ozone shelters the Earth from ultraviolet (UV) radiation. This protection is essential for human health. This ozone is easily destroyed by the UV radiation. The Sun's radiation will break down the O_3 molecule into a standard O_2 molecule and an elemental free oxygen (O) atom. More O_3 is produced through photosynthesis and the bonding of O_2 with free oxygen O. Thus ozone is constantly being formed and destroyed in the Stratosphere.

In the stratosphere, UV radiation will break off a chlorine atom (Cl) from the CFC and HCFC molecule. The chlorine atom destroys many O_3 molecules. The depleted O_3 molecules in the stratosphere let more UV radiation reach the Earth. This causes an increase in skin cancers and frequency of cataracts in humans; weakens human immune system and decreases plant and marine life. Ozone depletion potential (ODP) has been used as an index to show the effect of a refrigerant on O_3. Lower the ODP, it will be better. The ODP of some of the refrigerants is given in Table 8.6.

Table 8.6 Ozone Depletion Potential (ODP) of Refrigerants

Sub Group	*Refrigerant Designation*	*Chemical Formula*	*Refrigerant Name*	*ODP*
CFCs	R-11 R-12 R-113 R-114	CCl_3F CCl_2F_2 $CF_2Cl-CFCL_2$ CF_2Cl-CF_2CL	Trichloro-fluoro methane Dichlorodi-fluoro methane Trifluro-trichloro ethane Tetrafluro-dichloro ethane	1.0 1.0 0.8 1.0
HCFCs	R-22 R-123	$CHClF_2$ $CHCl_2F_3$	Chlorodi-fluoro methane Dichlororri-fluoro methane	0.05 0.02
HFC	R-134a	CH_2FCF_3	Tetra-fluoro ethane	0.0
Azeotropes	R-500 R-717	$R-12/R-152$ NH_3	– Ammonia	0.74 0.0

8.7 ECO-FRIENDLY REFRIGERANTS

With the increasing awareness of environmental degradation, the production, use and disposal of Chloro Fluoro Carbons (CFCs) and Hydro Chloro Fluoro Carbons (HCFCs) as refrigerants in mechanical refrigeration systems has become a subject of great concern. However, such systems are being developed using more eco-friendly refrigerants, such as

1. Hydrocarbons (HG-290, HC-600a).
2. Ammonia (NH_3)
3. Carbon Dioxide (CO_2)
4. Water (R-718)
5. Air

Eco-friendly refrigerants are also known as natural refrigerants. The important features and applications of eco-friendly refrigerants are given in Table 8.7.

Table 8.7 Important features and applications of eco-friendly refrigerants

Refrigerant	*Important Features*	*Applications*
1. Hydrocarbons	1. Zero Global Warming Potential (GWP) 2. Zero Ozone Depletion Potential (ODP) 3. COP comparable to CFC-12. 4. Flammable 5. Compatible with mineral oils	1. Domestic refrigerator (HC-270, HC-600a) 2. Commercial refrigeration (HC-290 and HC-600a) 3. Transport refrigeration (HC-290/HC-600a or HC-290) 4. Cold storage and food processing (HC-290) 5. Industrial refrigeration (HC-290) 6. Air-conditioning (HC-290)

2. Ammonia	1. Excellent thermodynamic and thermophysical properties 2. Zero ODP 3. Zero GWP 4. High efficiency 5. Low cost 6. Flammable	1. In small and larger plants. 2.Commercial and cold storage. 3. Super markets 4. Water chillers 5. Centralised air-conditioning 6. Water
3. Water	1. Thermodynamically attractive 2. Low volumetric refrigeration capacity 3. High COP (7-10) 4. Can be used both as primary and secondary refrigerant 5.Less corrosive	1. Water cooled chillers 2. Eco-chiller
4. Carbon dioxide	1. Zero ODP 2. Low critical temperature 3. High operating pressure 4. Non-flammable, non-toxic and cheap 5. Excellent performance 6. Low COP	1. Commercial air-conditioning system 2. Mobile air-conditioning system
5. Air	1. Simple and compact 2. Non-toxic and non-flammable. 3. Easy repair and maintenance 4. Low COP 5. Requires less space and lighter in weight	1. Used in aircrafts. 2. Used in railways and bus air-conditioning

8.8 REFRIGERANT SELECTION

The choice of refrigerant depends upon the following factors:

1. Refrigeration temperature required.
2. Refrigerating capacity of plant.
3. Thermodynamic requirements: Normal boiling point, condensing and evaporating pressure, critical temperature and pressure , freezing point, volume of suction vapour per ton, COP, power consumption per ton etc.
4. Chemical requirements : Flammability, toxicity, action with water and oil (mixibility), and action with materials of construction.
5. Physical requirements: Dielectric strength, thermal conductivity, viscosity, heat capacity, surface tension, leak tendency and cost of refrigerant.

6. Economics.
7. Equipment type and size.
8. Application.

Example 8.1 *Give the refrigerant number for the following: $CHClF_2$, NH_3, CH_2F-CF_3, and $CH_3CH_2CH_2CH_3$.*

Solution: Designation of refrigerants:

Sl. No.	*Refrigerant formula*	*C (m)*	*H (n)*	*Cl (q)*	*F (p)*
1.	$CHClF_2$	1	1	1	2
2.	CH_2F-CF_3	2	2	–	4
3.	$CH_3CH_2CH_2CH_3$	4	10	–	–

Chemical formula of hydrocarbons : $C_mH_nF_pCl_q$

$$n + p + q = 2m + 2$$

Designation: $R(m - 1)\ (n + 1)\ (p)$

1. $$1 + 2 + 1 = 2m + 2$$
$$m = 1$$
$$R - 22$$

2. $$2 + 4 = 2m + 2$$
$$m = 2$$
$$R - 134$$

3. $$10 = 2m + 2$$
$$m = 4$$
$$R - 311$$

4. NH_3: $R - [700 + (14 + 3)] = 717$

$\therefore$ $$R - 717$$

Example 8.2 *What factors are considered in selecting a refrigerant for the following mentioning the name of the refrigerant in each case*

(i) Household refrigerator

(ii) An ice plant

(iii) Air conditioning plant for a cinema hall. **[IES, 2009]**

Solution: The factors to be considered in selecting a refrigerant are :

(*i*) Household refrigerator

1. It should be non-toxic, non-flammable, and non-explosive.
2. It should be fully oil miscible.
3. It should have high COP.
4. It should condense at moderate pressure and under atmospheric conditions.

5. It should have no effect on quality of stored food.
6. It should be odourless.
7. Ease of leak detection.
8. Availability and low cost.
9. Ease of handling.
10. High electrical insulation.

Refrigerant : R-12: Dichloro-difluoro methane (CCl_2F_2)
R-22: Monochloro-difluoro ethane ________ ($CHClF_2$)
R-114: Dichloro-tetafluoro ethane ($C_2Cl_2F_4$)

(*ii*) Ice plant

1. Excellent thermal properties.
2. High refrigerating effect per kg of refrigerant.
3. Low volumetric displacement.
4. Low cost.
5. High efficiency.
6. Ease of leak defection.
7. Low weight of liquid circulated per ton of refrigeration.
8. Low boiling point and high heat of vaporization.

Refrigerants : R-717: Ammonia (NH_3)

(*iii*) Cinema hall air-conditioning

1. Colourless and odourless.
2. Non-toxic and non-flammable
3. Non-explosive and non-corrosive.
4. High operating pressure.
5. Moderate refrigerating effect.
6. Low cost.
7. Easy handling.

Refrigerants : R-744 : Carbon dioxide (CO_2)
R-113 : Trichloro-trifluoromethane ($C_2Cl_3F_3$)
R-22 : Monochloro-difluoromethane ($CHClF_2$)

Example 8.3 *What are the harmful environmental effects of conventional refrigerants? Name any one eco-friendly refrigerant which is being commercially used and discuss its important characteristics.* **[CSE, 2000]**

Solution: Harmful Environmental Effects of conventional Refrigerants

The two major environmental effects of conventional refrigerants which are harmful to human health are: ozone layer deplection and global warming. The Earth's ozone layer in the upper atmosphere is needed for the absorptionof harmful ultraviolet rays from the Sun. These rays can cause skin cancer. CFC'S (Chloro-Flurocarbons) have been linked to the depletion of ozone layer. They have varying degrees of ozone depletion potential (ODP). They also act as greenhouse gases. Hence, they have global warming-potential (GWP) as well.

Eco-friendly Refrigerants

Eco-friendly refrigerants fall into four categories: Those based on N_2, HFCs and HCFCs, HCs, and inert gases. Hydrocarbons (HCs) and hydroflurocarbons (HFCs) provide an alternative to fully halogenated CFC refrigerants. They contain no chlorine atom at all and, therefore, have zero ODP. Even hydrochloroflurocarbons (HCFCs) Like R-22 and R-123 which contain chlorine atoms, have much reduced ODP due to the presence of the H-atoms,

Eco-frineldy refrigerants are:

HFC R - 134a ($CH_2F - CF_3$), Tetrafluro - ethane

HFC R - 143a ($CH_3 - CF_3$), Trifluro - ethane

HFC R - 152a ($CH_3 - CHF_2$), Difluro - ethane

R - 717 (NH_3), Ammonia

R - 744 (CO_2), Carbon dioxide

R - 134a is being used commercially in domestic refrigerators. It has been observed that it has zero ODP and GWP. It is non-flammable, normal boiling point –26.07°C, COP = 1.91, h_{fg} = 221.83 kJ/kg.

Example 8.4 *What are the commonly used drying agents and antifreeze solution?*

Solution: Antifreeze solutions are those compounds which are soluble in water and it lowers the freezing point of water. e.g.,

(*i*) If 70% by weight of glycrine is added in water, it reduces the freezing point of water to about – 40°C.

(*ii*) Methyl alcohol.

(*iii*) Ethylene glycol and propylene glycol are some of the commonly used antifreeze solutions.

Drying Agents of Dessicants are used to remove moisture from the refrigerant. e.g. Silica gel, aluminium oxide, anhydrous calcium sulphate.

Example 8.5 *Compare field of application of various refrigerants.*

Solution:

No.	*Refrigerants*	*Applications*
1.	R-11, R-113, $CH_2 Cl_2$	For air-conditioning theatres, offices, factories, auditoriums etc.
2.	R-12, SO_2, R-500, CH_3Cl and isobutance	For domestic purposes as house hold refrigerators, water coolers and small air-conditioners.
3.	NH_3, R-12, R-22, and CO_2	For domestic and commercial freezers, industrial purposes as cold storage plants and ice cream factories.
4.	R-13, R-14, C_2H_6 and Ethylene	For liquefaction of gases, metallurgical operations and in aerodynamic wind tunnel.

Example 8.6 *Write the types of refrigerant used under the following applications :*

(i) Domestic refrigerator.

(ii) Air craft.

(iii) Room Air conditioner.

(iv) Ice plants

(v) Cold storages

Solution:

(*i*) Domestic refrigerator.	R–12, R–22, R–134a
(*ii*) Air craft.	Air
(*iii*) Room Air conditioner.	R–12, R–22
(*iv*) Ice plants	NH_3
(*v*) Cold storages	NH_3

Example 8.7 *Write a short note on leak detection of Refrigerants.*

Solution: A refrigerating machine in use may lose its charge of refrigerant due to leaks in the system in due course of time. Due to the shortage of refrigerant inside the system the working capacity of the system is reduced, therefore it becomes necessary to detect the leaks immediately and repair the leaking spots or joints at once. Leaks can develop any time during the operation of the machine due to normal mechanical wear, but poor workmanship, use of wrong techniques and improper materials are major contributors. The first sign of a leaking joint may well be the appearance of an oil film at that place.

Methods of leak Detection :

Following are the methods of leak detection:

(*a*) Detection of NH_3 : (*i*) Sulphur test : The best way to detect leaks in NH_3 piping is by using sulphur candle which is prepared by melting sulphur powder, dipping 10 cm long thick threads in it and then allowing it to dry. This sulphur candle is lighted and held near all the joints. If any leak is there, NH_3 comes in contact with the sulphur fumes, a dense white smoke will result.

(*ii*) Litmus paper: Litmus paper will turn blue if NH_3 present or leaking when it is brought to the suspected leak point.

(*b*) Freon – 12 : Halide torch may be used to detect the leak of Freon vapour. This halide torch detector produces nearly colourless flame. The air tube is held near the suspected leak and if any Freon is present, the flame changes to brilliant green. This type of torch is able to detect Freon concentration as low as 0.01%.

Example 8.8 *What is a drier or dehydrator ?*

Solution: Drier or Dehydrator is a device employed to remove moisture from a refrigerant. It is generally placed in a liquid line. The drying material, used is called desiccant which is a solid substance capable of removing moisture from a gas, liquid or solid.

The drier is in the shape of shell made of steel or brass. It is provided with inlet and outlet openings. A strainer is provided at the outlet of the shell and is provided with moisture absorbing chemicals like silica gel, activated alumina, calcium sulphate, and calcium chloride.

There are 2 types of driers.

(*i*) Sealed type, (*ii*) Refillable type.

Sealed type : In sealed type of drier, the shell charged with moisture absorbing chemical is permanently sealed and the chemical cannot be replaced. This type of drier is used on small sealed units.

Refillable type : In this drier there is a provision for filling the moisture absorbing chemical as and when required.

Example 8.9 *State the effect and causes of moisture in a refrigeration system. How can it be removed from the system?*

Solution: Effects and causes of moisture in refrigeration system : The following are the detrimental effects of moisture in a refrigeration system :

(*i*) The moisture may freeze and clog the expansion valve and stop the circulation of refrigerant.

(*ii*) The moisture deteriorate the quality of lubricating oils.

(*iii*) In the hermatically sealed compressor, the moisture may cause short circuiting of motor winding.

Causes of moisture:

(*i*) The equipment may not be completely dried before charging.

(*ii*) As air always contains some quantity of water vapour in it, so the moisture may enter the system with air leakage on low pressure side.

(*iii*) Water leakage may be there in the system with water cooled condensers.

(*iv*) Formation of water due to oxidation of certain hydrocarbons in lubricating oil.

Removal of moisture from the system:

(*i*) By heating the entire assembly.

(*ii*) By circulating the dehumidified air through the system.

(*iii*) By creating vacuum within the system using vacuum pump and removing moisture by the combined use of alcohol and CO_2.

(*iv*) By installing a suitable drier or dehydrator.

REVIEW QUESTIONS

1. What are the desirable properties of an ideal refrigerant?
2. What are the factors considered for the selection of a refrigerant for a system?
3. What is a azeotrope? Give some examples to indicate its importance.
4. Assign number to the following refrigerants:
   (a) Methyl chloride – CH_3CI (b) Tetrachloroethane – $C_2H_4Cl_4$
   (c) Ammonia – NH_3 (d) Dischlorodifluoromethane – CCl_2F_2
5. Name the refrigerants commonly used for the following systems:
   (a) Domestic refrigerator (b) Cold storage – 100 TR
   (c) Ice cream plant (d) Room air conditioner
6. Differentiate between primary and secondary refrigerants
7. Write the chemical formula for the following refrigerants:
   (a) R-12 (b) R-22 (c) R-717 (d) R-764
   (e) R-134a

MULTIPLE – CHOICE QUESTIONS

1. The Freon group of refrigerants are of the type
   (a) halo-carbon (b) azeotrpe (c) inorganic (d) hydro-carbon
2. In case of leakage of Freon refrigerant, the colour of the flame of halide torch will change to
   (a) bright green (b) yellow (c) red (d) orange.
3. The refrigerant having lowest boiling point is
   (a) ammonia (b) carbon dioxide (c) sulphur dioxide (d) R-12
4. The refrigerant having highest freezing point is
   (a) R-12 (b) sulphur dioxide (c) ammonia (d) carbon dioxide
5. Which refrigerant is highly toxic and flammable?
   (a) NH_3 (b) CO_2 (c) SO_2 (d) R-12
6. The refrigerant widely used in refrigerators is
   (a) ammonia (b) sulphur dioxide (c) carbon dioxide (d) R-12
7. The refrigerant used in ice plants is
   (a) R-12 (b) R-22 (c) ammonia (d) carbon dioxide.
8. The chloro-fluoro-carbon refrigerants are prohibited for use to guard against environmental protection because these react with
   (a) water vapour and cause acid rain (b) plants and cause green house effect
   (c) oxygen and cause its depletion (d) ozone layer
9. The leaks in a refrigeration system using Freon are detected by
   (a) A halide torch, which on detection produces greenish flame lighting
   (b) Sulphur sticks, which on detection give white smoke
   (c) Using reagents
   (d) Sensing reduction in pressure

10. Which of the following statements are true for Ammonia as a refrigerant?

1. It has higher compressor discharge temperature compared to fluorocarbons.
2. It is toxic to mucous membranes.
3. It requires larger displacement per TR compared to fluorocarbons.
4. It reacts with copper and its alloys.

Select the correct answer using the codes given below:

(a) 1 and 2 (b) 1, 2 and 3 (c) 1, 2 and 4 (d) 2, 3 and 4

11. March List I with List II and select the correct answer using the codes given below the lists:

List I	*List II*
A. Freon 12	1. Centrifugal systems
B. Freon 22	2. Low temperature cold storage
C. Freon 11	3. Window type units
D. Ammonia	4. Ice plants

Codes:	**A**	**B**	**C**	**D**
(a)	3	2	1	4
(b)	3	1	2	4
(c)	1	2	4	3
(d)	1	3	4	2

12. Consider the following statements: Moisture Should be removed from refrigerants to avoid.

1. compressor seal failure.
2. freezing at the expansion valve.
3. restriction to refrigerant flow.
4. corrosion of steel parts.

On these statements:

(a) 1, 2, 3 and 4 are correct

(b) 1 and 2 are correct

(c) 2, 3 and 2 are correct.

(d) 1, 3 and 4 are correct.

13. A good refrigerant should have:

(a) High latent heat of vaporization and low freezing point

(b) High operating pressure and low freezing point

(c) High specific volume and high latent heat of vaporization

(d) Low COP and low freezing point.

14. Which of the following refrigerant has the maximum ozone depletion in the stratosphere?
(a) Ammonia (b) Carbon dioxide (c) Sulphur dioxide (d) Fluorine

15. Consider the following statements:

1. Practically all common refrigerants have approximately the same COP and power requirement.
2. Ammonia mixes freely with lubricating oil and this helps lubrication of compressors.
3. Dielectric strength of refrigerants is an important property in hermetically sealed compressor units.
4. Leakage of ammonia can be detected by halide torch method.

Of these statements:
(a) 1, 2 and 4 are correct (b) 1 and 4 are correct
(c) 1, 3 and 4 are correct (d) 1 and 3 are correct.

16. Ozone depletion by CFCs occurs by breakdown of
(a) chlorine atoms from refrigerant by UV radiation and reaction with ozone in troposphere
(b) fluorine atoms from refrigerant by UV radiation and reaction with ozone in troposphere
(c) chlonne atoms from refrigerant by UV radiation and reaction with ozone in stratosphere
(d) fluorine atoms from refrigerant by UV radiation and reaction with ozone in stratosphere

17. Match List I with List II and select the correct answer using the codes given below the lists:

List I (Refrigerant)	*List II (Principal application)*
A. Air	1. Direct contact freezing of food
B. Ammonia	2. Centrifugal compressor system
C. Carbon dioxide	3. Large industrial temperature installation
D. Refrigerant-II	4. Automotive air-conditioners.
	5. Aircraft refrigeration

Codes:	**A**	**B**	**C**	**D**
(a)	4	3	1	2
(b)	5	3	1	2
(c)	2	4	3	5
(d)	5	3	2	1

Answers

1. (a) **2.** (a) **3.** (b) **4.** (d) **5.** (a) **6.** (d) **7.** (c) **8.** (d)
9. (a) **10.** (c) **11.** (a) **12.** (a) **13.** (a) **14.** (d) **15.** (d) **16.** (c)
17. (b)

THEORETICAL QUESTIONS

8.1 What is the basic function of refrigerant in a refrigeration cycle and how they are classified? Write some desirable properties of refrigerants. What is the basic difference between primary, and secondary refrigerants?

8.2 How are refrigerants classified? What are the desirable properties of refrigerants? Name some common refrigerants generally used in refrigeration systems.

8.3 What is the importance of refrigerants in a refrigeration system, and what is the basic difference between Primary and Secondary refrigerants? Enumerate the desired properties of refrigerants giving two examples of common refrigerants.

8.4 What is a refrigerant? Enumerate the desired chemical, physical and thermodynamic properties of refrigerants used in vapour compression refrigeration cycle. Write a short note on the effects of release of CFC class of refrigerants into the atmosphere.

8.5 Enumerate the classification of Refrigerants. What are the desirable properties of refrigerants? Name some common refrigerants generally used in refrigeration systems. What do you understand by CFC free refrigerants?

8.6 Give classification and nomenclature of refrigerants in details. Discuss the effects of CFC refrigerants on Ozone layer briefly.

8.7 Give azeotropic mixing refrigerants for the following refrigerants and their chemical formula:
(a) R-500 (b) R-502 (c) R-503 (d) R-504.

8.8 Name the refrigerants and their operating pressures for the following applications:
(a) Domestic refrigerator (b) Cold storage – 100 TR
(c) Ice cream plant (d) Room air – conditioner.

8.9 What are eco-friendly refrigerants? Name these refrigerants and give their chemical formulae.

Chapter

9

REFRIGERATION EQUIPMENT

9.1 INTRODUCTION

The major refrigeration equipment in vapour compression system is:

1. Compressor 2. Condensers 3. Evaporators, and 4. Refrigeration controls.

9.2 REFRIGERANT COMPRESSOR

A refrigerant compressor is a machine used to compress the vapour refrigerant from the evaporator. This process raises the vapour pressure so that the corresponding saturation temperature is higher than that of the cooling medium. A compressor is driven by some prime mover to do the desired work, e.g., an electric motor or I.C. engine.

9.2.1 Types of Compressors

The compressors are of the following types:

1. Based on the method of compression.
   - (*i*) Reciprocating compressors.
   - (*ii*) Rotary compressors.
   - (*iii*) Centrifugal compressors.
2. Based on the number of working strokes.
   - (*i*) Single acting compressors.
   - (*ii*) Double acting compressors
3. Based on the number of stages.
   - (*i*) Single -stage compressors.
   - (*ii*) Multi-stage compressors.
4. Based on the method of drive used.
   - (*i*) Direct drive compressors.
   - (*ii*) Belt drive compressors.
5. Based on the location of prime mover.
   - (*i*) Semi-hermetic compressors.
   - (*ii*) Hermetic compressors.

9.2.2 Reciprocating Compressors

In reciprocating compressors, the vapour refrigerant is compressed by the reciprocating motion of the piston in a cylinder. These compressors are used for refrigerants which have comparatively low specific volume and a large differential pressure. The commonly handled refrigerants by these compressors are: R-12, R-22, R-40, and R-717. They are available in sizes from about 0.1 kW to 150 kW.

(*i*) **Single acting reciprocating compressor:** A line diagram of single-acting reciprocating compressor is shown Fig. 9.1. The principle parts of a reciprocating compressor are: Prime mover (an electric motor), crankshaft, crank, connecting rod, piston, cylinder, cylinder head, suction and delivery valves, and suction and discharge ports. The working principle of a reciprocating compressor is similar to that of a two-stroke internal combustion engine. The one cycle is completed in one revolution.

(*ii*) **Double acting reciprocating compressor:** In a double acting reciprocating compressor, the suction and compression takes place on both sides of the piston. Such a compressor delivers double the volume of refrigerant as compared to the single acting reciprocating compressor.

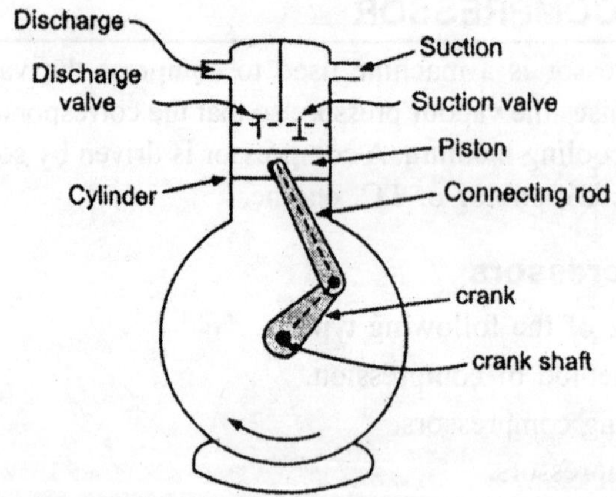

Fig. 9.1 Single-acting reciprocating compressor

9.2.3 Definitions

Some of the important terms used are:

Suction pressure: It is the absolute pressure of refrigerant at the compressor inlet.

Discharge pressure: It is the absolute pressure of refrigerant at the compressor outlet.

Compression or pressure ratio: It is the ratio of absolute discharge pressure to absolute suction pressure. $r = p_2/p_1$.

Suction volume, V_1: It is the volume of refrigerant sucked by the compressor during its suction stroke.

Piston displacement (or stroke or swept) volume, V_s: It is the volume swept by the piston during one stroke.

$$V_s = \frac{\pi}{4} D^2 L$$

where D, L = piston diameter and stroke length respectively.

Clearance ratio, c: It is the ratio of clearance volume V_c to swept volume.

$$c = \frac{V_c}{V_s}$$

Compressor capacity: It is the volume of the actual amount of refrigerant passing through the compressor in a unit time. It is the suction volume expressed in m^3/s.

Volumetric efficiency, η_v: It is the ratio of suction volume to swept volume.

$$\eta_V = \frac{V_1}{V_s}$$

9.2.4 Work done by a Single Stage Reciprocating Compressor

(*a*) Without clearance volume

The *p-v* and *T-s* diagrams for a single stage reciprocating compressor without clearance are shown in Fig. 9.2.

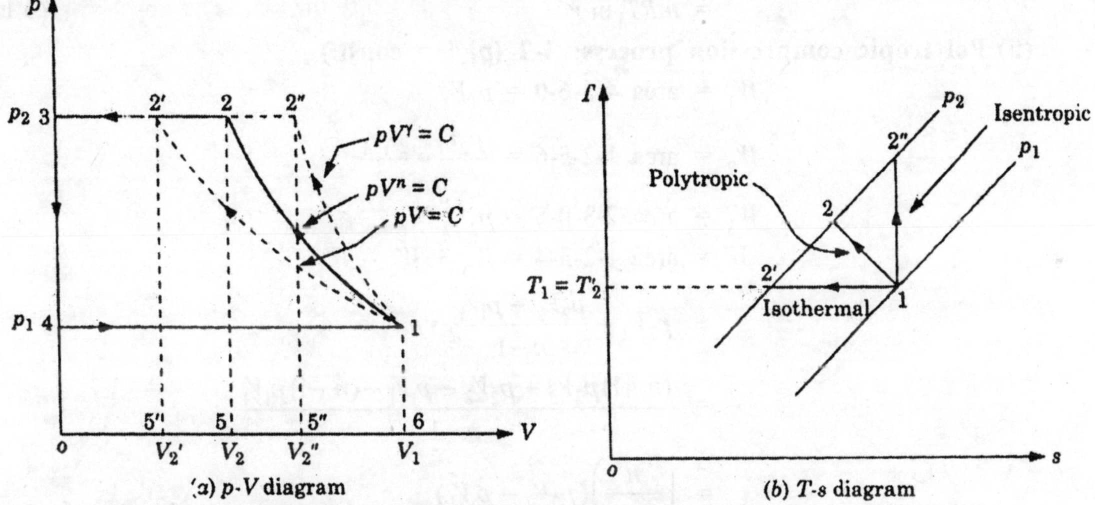

Fig. 9.2 p-v and T-s diagrams for a single stage reciprocating compressor.

Let p_1 = suction pressure of refrigerant

V_1 = suction volume of refrigerant

T_1 = suction temperature of refrigerant

p_2, V_2, T_2 = corresponding values for the refrigerant at discharge

r = compression or pressure ratio

$= p_2/p_1$

(*i*) Isothermal compression process: 1-2′ (pV = const.)

Work done during suction stroke 4-1,

$$W_1 = \text{area 4-1-6-0} = p_1 V_1$$

Work done during compression stroke 1-2′,

$$W_2 = \text{area 1-2'-5''-6} = p_1 V_1 \ln\left(\frac{V_1}{V_{2'}}\right)$$

Work done during discharge stroke 2′-3,

$$W_3 = \text{area } 2'\text{-}3\text{-}0\text{-}5'' = p_2 V_{2'}$$

Work done by the compressor per cycle,

$$\begin{aligned} W &= \text{area } 1\text{-}2'\text{-}3\text{-}4 \\ &= W_3 + W_2 - W_1 \\ &= p_2 V_{2'} + p_1 V_1 \ln\left(\frac{V_1}{V_{2'}}\right) - p_1 V_1 \\ &= p_1 V_1 \ln\left(\frac{V_1}{V_{2'}}\right) \qquad [\because p_1 V_1 = p_2 V_{2'}] \\ &= p_1 V_1 \ln r \qquad \left[\because \frac{V_1}{V_{2'}} = \frac{p_2}{p_1} = r\right] \qquad \text{...(9.1}a\text{)} \\ &= mRT_1 \ln r \qquad \text{...(9.1}b\text{)} \end{aligned}$$

(ii) Polytropic compression process: 1-2 (pV^n = const.)

$$W_1 = \text{area } 4\text{-}1\text{-}6\text{-}0 = p_1 V_1$$

$$W_2 = \text{area } 1\text{-}2\text{-}5\text{-}6 = \frac{p_2 V_1 - p_1 V_1}{n-1}$$

$$W_3 = \text{area } 2\text{-}3\text{-}0\text{-}5 = p_2 V_2$$

$$\begin{aligned} W &= \text{area } 1\text{-}2\text{-}3\text{-}4 = W_3 + W_2 - W_1 \\ &= p_2 V_2 + \frac{p_2 V_2 - p_1 V_1}{n-1} - p_1 V_1 \\ &= \frac{(n-1) p_2 V_2 + p_2 V_2 - p_1 V_1 - (n-1) p_1 V_1}{n-1} \\ &= \left(\frac{n}{n-1}\right)(p_2 V_2 - p_1 V_1) \\ &= \left(\frac{n}{n-1}\right) p_1 V_1 \left[\frac{p_2 V_2}{p_1 V_1} - 1\right] \end{aligned}$$

Now for polytropic compression,

$$p_1 V_1^n = p_2 V_2^n$$

$$\therefore \quad \frac{V_2}{V_1} = \left(\frac{p_1}{p_2}\right)^{\frac{1}{n}}$$

$$\therefore \quad W = \left(\frac{n}{n-1}\right) p_1 V_1 \left[\frac{p_2}{p_1}\left(\frac{p_1}{p_2}\right)^{\frac{1}{n}} - 1\right]$$

$$= \left(\frac{n}{n-1}\right) p_1 V_1 \left[\left(\frac{p_2}{p_1}\right)^{\frac{n-1}{n}} - 1\right]$$

$$= \left(\frac{n}{n-1}\right) p_1 V_1 \left[(r)^{\frac{n-1}{n}} - 1\right] \quad \text{...(9.2}a\text{)}$$

$$= \left(\frac{n}{n-1}\right) mRT_1 \left[r^{\frac{n-1}{n}} - 1\right] \quad \text{...(9.2}b\text{)}$$

Now $\left(\frac{p_2}{p_1}\right)^{\frac{n-1}{n}} = \frac{T_2}{T_1}$

$$\therefore \quad W = \left(\frac{n}{n-1}\right) mRT_1 \left(\frac{T_2}{T_1} - 1\right) = \left(\frac{n}{n-1}\right) mR(T_2 - T_1) \quad \text{...(9.2}c\text{)}$$

where m = mass of the refrigerant sucked by the compressor per cycle.

(*iii*) Isentropic compression process: 1-2″ (pv^γ = const.)

Changing polytropic index n to isentropic index γ in Eqs. (9.2), we have

$$W = \left(\frac{\gamma}{\gamma-1}\right) p_1 V_1 \left[r^{\frac{\gamma-1}{\gamma}} - 1\right] \quad \text{...(9.3}a\text{)}$$

$$= \left(\frac{\gamma}{\gamma-1}\right) mRT_1 \left[r^{\frac{\gamma-1}{\gamma}} - 1\right] \quad \text{...(9.3}b\text{)}$$

$$= \left(\frac{\gamma}{\gamma-1}\right) mR[T_2 - T_1] \quad \text{...(9.3}c\text{)}$$

Also $\frac{c_p}{c_v} = \gamma$ and $R = c_p - c_v = c_p\left(\frac{\gamma-1}{\gamma}\right)$

$$\therefore \quad W = \left(\frac{\gamma}{\gamma-1}\right) mc_p \left(\frac{\gamma-1}{\gamma}\right)(T_2 - T_1) = mc_p(T_2 - T_1) \quad \text{...(9.4)}$$

Power required, $P = \frac{W \times n}{60 \times 10^3}$ kW ...(9.5)

where n = number of working strokes per minute

= N for single acting compressor

= $2N$ for double acting compressor

N = speed of compressor, rpm

(b) With Clearance Volume

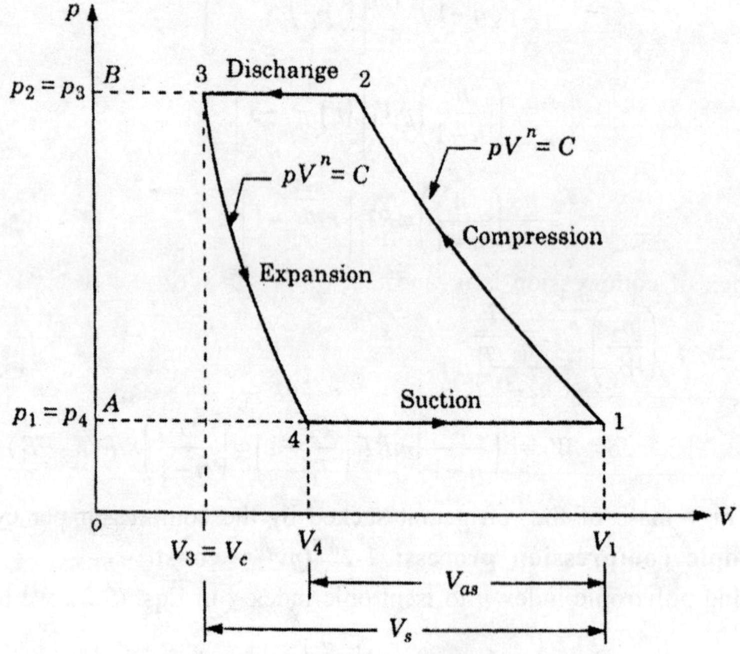

Fig. 9.3 *p-V* diagram with clearance volume

The *p-V* diagram for single-stage reciprocating compressor with clearance volume is shown in Fig. 9.3.

Let
$$V_3 = V_c = \text{clearance volume}$$
$$V_s = V_1 - V_c = \text{swept (or stroke) volume}$$
$$V_{as} = V_1 - V_4 = \text{actual swept volume}$$
$$r = \frac{p_2}{p_1} = \text{compression ratio}$$
$$n = \text{compression index.}$$
$$c = \frac{V_c}{V_s} \text{ clearance ratio}$$

Work done by the compressor,
$$W = \text{area 1-2-3-4}$$
$$= \text{area } A\text{-1-2-}B - \text{area } A\text{-4-3-}B$$
$$= \left(\frac{n}{n-1}\right) p_1 V_1 \left[\left(\frac{p_2}{p_1}\right)^{\frac{n-1}{n}} - 1\right] - \left(\frac{n}{n-1}\right) p_4 V_4 \left[\left(\frac{p_2}{p_1}\right)^{\frac{n-1}{n}} - 1\right]$$

$$= \left(\frac{n}{n-1}\right) p_1 (V_1 - V_{4)} \left[\left(\frac{p_2}{p_1}\right)^{\frac{n-1}{n}} - 1\right] \qquad [\because p_1 = p_4]$$

$$= \left(\frac{n}{n-1}\right) p_1 V_{as} (r^{\frac{n-1}{n}} - 1) \qquad ...(9.6a)$$

$$= \left(\frac{n}{n-1}\right) mRT_1 (r^{\frac{n-1}{n}} - 1) \qquad ...(9.6b)$$

If the index of compression is n_1 and that of expansion n_2, then

$$W = \left(\frac{n_1}{n_1 - 1}\right) p_1 V_1 \left[r^{\frac{n_1 - 1}{n_1}} - 1\right] - \left(\frac{n_2}{n_2 - 1}\right) p_1 V_4 \left[r^{\frac{n_2 - 1}{n_2}} - 1\right] \qquad ...(9.7)$$

Volumetric efficiency,

$$\eta_v = \frac{\text{Actual Swept Volume}}{\text{Swept Volume}} = \frac{V_{as}}{V_s} = \frac{V_1 - V_4}{V_1 - V_c}$$

For process 3-4, we have

$$p_3 V_3^n = p_4 V_4^n$$

or

$$p_2 V_c^n = p_1 V_4^n \qquad [\because V_4 = V_e]$$

$$V_4 = V_c \left(\frac{p_2}{p_1}\right)^{\frac{1}{n}}$$

$$\therefore \quad \eta_v = \frac{V_1 - V_c \left(\frac{p_2}{p_1}\right)^{\frac{1}{n}}}{V_1 - V_c} = \frac{V_1 + V_c - V_c - V_c \left(\frac{p_2}{p_1}\right)^{\frac{1}{n}}}{V_1 - V_c}$$

$$= \frac{V_c - V_c}{V_1 - V_c} + \frac{V_c}{V_1 - V_c} - \left(\frac{V_c}{V_1 - V_c}\right) \left[\frac{p_2}{p_1}\right]^{\frac{1}{n}}$$

$$= 1 + \frac{V_c}{V_s} - \frac{V_c}{V_s} \left(\frac{p_2}{p_1}\right)^{\frac{1}{n}} = 1 + c - c \left(\frac{p_2}{p_1}\right)^{\frac{1}{n}} \qquad ...(9.8a)$$

$$= 1 + c \left(1 - \left(\frac{p_2}{p_1}\right)^{\frac{1}{n}}\right) \qquad ...(9.8b)$$

The pressure ratio p_2/p_1 is more than unity. Therefore, the quantity within the brackets is less than one. Hence the volumetric efficiency decreases with increase in pressure ratio.

9.2.5 Multistage Compression

The drawbacks of single stage compression for producing very high pressure refrigerant are:

1. Too large cylinder size.
2. Rise in temperature of refrigerant.
3. Danger of burning of lubricating oil due to heat.
4. High friction losses.
5. High running cost.
6. Low volumetric efficiency.

The main advantages of multistage compression are:

1. Workdone per kg refrigerant is reduced with intercooling between stages.
2. The Volumetric efficiency improves.
3. The leakage losses reduces considerably.
4. Due to uniform torque the flywheel size decreases.
5. Lubrication becomes more effective due to lower temperature range.
6. The cost of compressor reduces.

9.2.6 Two-stage Reciprocating Compressor

The schematic arrangement for a two-stage reciprocating compressor with intercooling is shown in Fig. 9.4. First of all the vapour refrigerant from the evaporator is sucked by the LP compressor at point 1 during its suction stroke. The compressed refrigerant at point 2 is passed through the intercooler. This cooled refrigerant is now sucked by the HP compressor at point 3 during its suction stroke. Finally, the refrigerant after compression in HP compressor is delivered at point 4 to the condenser.

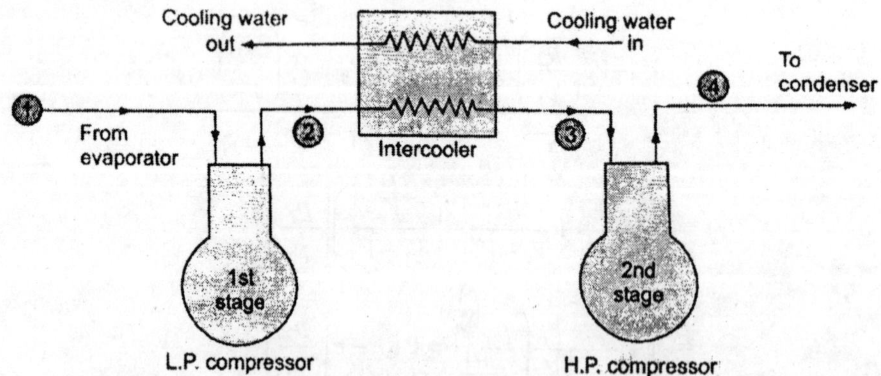

Fig. 9.4 Two stage reciprocating compressor with intercooler.

9.2.7 Two-stage Compression with Perfect Intercooling

Consider the two-stage compression with perfect intercooling shown in Fig. 9.5 on *p-V* and *T-s* diagrams.

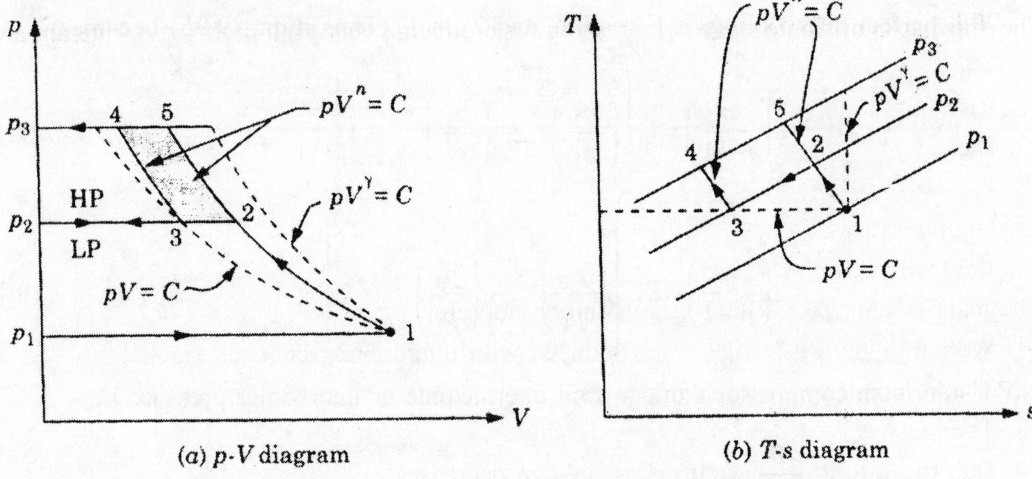

Fig. 9.5 Two stage compression with perfect intercooling

(a) Imperfect intercooling

LP cylinder work,
$$W_{LP} = \left(\frac{n}{n-1}\right) p_1 V_1 \left[\left(\frac{p_2}{p_1}\right)^{\frac{n-1}{n}} - 1\right]$$

$$= \left(\frac{n}{n-1}\right) mRT_1 \left[\left(\frac{p_2}{p_1}\right)^{\frac{n-1}{n}} - 1\right]$$

HP cylinder work,
$$W_{HP} = \left(\frac{n}{n-1}\right) p_2 V_3 \left[\left(\frac{p_3}{p_2}\right)^{\frac{n-1}{n}} - 1\right]$$

$$= \left(\frac{n}{n-1}\right) mRT_3 \left[\left(\frac{p_3}{p_2}\right)^{\frac{n-1}{n}} - 1\right]$$

Total work done per cycle,

$$W = W_{LP} + W_{HP}$$

$$= \left(\frac{n}{n-1}\right) p_1 V_1 \left[\left(\frac{p_2}{p_1}\right)^{\frac{n-1}{n}} - 1\right] + \left(\frac{n}{n-1}\right) p_2 V_3 \left[\left(\frac{p_3}{p_2}\right)^{\frac{n-1}{n}} - 1\right] \quad \text{...(9.9a)}$$

$$= \left(\frac{n}{n-1}\right) mR \left[T_1 \left(\frac{p_2}{p_1}\right)^{\frac{n-1}{n}} + T_3 \left(\frac{p_3}{p_2}\right)^{\frac{n-1}{n}} - T_1 - T_3\right] \quad \text{...(9.9b)}$$

(*b*) Perfect intercooling

With perfect intercooling, $p_1V_1 = p_2V_3$ and $T_1 = T_3$. Thus, Eqn. (9.9*b*) becomes,

$$W = \left(\frac{n}{n-1}\right)p_1V_1\left[\left(\frac{p_2}{p_1}\right)^{\frac{n-1}{n}} + \left(\frac{p_2}{p_1}\right)^{\frac{n-1}{n}} - 2\right] \quad ...(9.10a)$$

$$= \left(\frac{n}{n-1}\right)mRT_1\left[\left(\frac{p_2}{p_1}\right)^{\frac{n-1}{n}} + \left(\frac{p_3}{p_1}\right)^{\frac{n-1}{n}} - 2\right] \quad ...(9.10b)$$

For minimum compressor work to find intermediate or intercooler pressure p_2,

$$\frac{dW}{dp_2} = 0$$

Differentiating Eq. (7.9*b*) w.r.t. p_2, we have

$$\frac{d}{dp_2}\left[\left(\frac{n}{n-1}\right)^{\frac{n-1}{n}} mR\left\{T_1\left(\frac{p_2}{p_1}\right)^{\frac{n-1}{n}} + T_3\left(\left(\frac{p_3}{p_1}\right)^{\frac{n-1}{n}}\right) - T_1 - T_3\right\}\right] = 0$$

Let $\left(\frac{n-1}{n}\right) = x$, then

$$\frac{d}{dp_2}\left[\frac{mR}{x}\left\{T_1\left(\frac{p_2}{p_1}\right)^x + T_3\left(\frac{p_3}{p_2}\right)^x - T_1 - T_3\right\}\right] = 0$$

$$T_1xp_1^{-x}p_2^{x-1} + T_3p_3^x(-x)p_2^{-x-1} - 0 - 0 = 0$$

$$\frac{T_1p_2^{x-1}}{p_1^x} - \frac{T_3p_3^x}{p_1^{x+1}} = 0$$

or $\quad p_2^{2x} = \frac{T_3}{T_1}(p_1p_3)^x$

$$p_2 = p_2 = \left(\frac{T_3}{T_1}\right)^{\frac{1}{2x}}(p_1p_3)^{\frac{1}{2}} = \left(\frac{T_3}{T_1}\right)^{\frac{n}{2(n-1)}}(p_1p_3)^{\frac{1}{2}}$$

$$= \sqrt{\left(\frac{T_3}{T_1}\right)^{\frac{n}{(n-1)}} p_1p_3}$$

$$W_{min} = \left(\frac{n}{n-1}\right) mR\left[T_1 \times \frac{1}{p_1^{\frac{n-1}{n}}} \times \left\{\left(\frac{T_3}{T_1}\right)^{\frac{n}{2(n-1)}} (p_1\, p_3)^{\frac{1}{2}}\right\}^{\frac{n-1}{n}} + T_3 p_3^{\frac{n-1}{n}} \left\{\left(\frac{T_3}{T_1}\right)^{\frac{n}{2(n-1)}} (p_1\, p_3)^{\frac{1}{2}}\right\}^{\frac{1-n}{n}} - T_1 - T_3\right]$$

$$= \left(\frac{n}{n-1}\right) mR\left[\frac{T_1}{p_1^{\frac{n-1}{n}}} \times \left\{\left(\frac{T_3}{T_1}\right)^{\frac{1}{2}} (p_1\, p_3)^{\frac{n-1}{2n}}\right\} + T_3 p_3^{\frac{n-1}{n}} \left\{\left(\frac{T_3}{T_1}\right)^{\frac{-1}{2}} (p_1\, p_3)^{\frac{1-n}{2n}}\right\} - T_1 - T_3\right]$$

$$= \left(\frac{n}{n-1}\right) mR\left[\sqrt{T_1 T_3}\left(\frac{p_3}{p_1}\right)^{\frac{n-1}{2n}} + \sqrt{T_1 T_3}\left(\frac{p_3}{p_1}\right)^{\frac{n-1}{2n}} - T_1 - T_3\right] \qquad \text{...(9.12)}$$

For perfect intercooling, $T_3 = T_1$, so that

$$p_2 = \sqrt{p_1 p_3} \qquad \text{...(9.13)}$$

Therefore, intermediate pressure with perfect intercooling is equal to the geometric mean of the suction and discharge pressures.

and
$$W_{min} = \left(\frac{2n}{n-1}\right) mRT_1\left[\left(\frac{p_3}{p_1}\right)^{\frac{n-1}{2n}} - 1\right] \qquad \text{...(9.14)}$$

9.2.8 Hermetically Sealed Compressor

When the compressor and motor operate on the same shaft and are enclosed in a common casing, they are known as hermetically sealed compressor. The hermetic sealed units are widely used for small capacity refrigerating systems such as in domestic refrigerators, home freezers and window air-conditioners.

The advantages of hermetic sealed compressors are:

1. Use of crankshaft seal is eliminated.
2. The leakage of refrigerant is completely prevented.
3. It is less noisy.
4. It requires less space due to compactness.
5. The lubrication is simple as the motor and compressor operate in a sealed space with the lubricating oil.

Its disadvantages are:

1. The maintenance is difficult as the moving parts are inaccessible.
2. It requires a separate pump for evacuation and charging of refrigerant.

Example 9.1 *A single stage reciprocating compressor is required to compress 2m³/min of vapour refrigerant from 1 bar to 8 bar. Calculate the power required to drive the compressor when the compression process is (a) isothermal, (b) isentropic with index 1.4, and (c) Polytropic with compression index 1.30.*

Solution: Given: $V_1 = 2\text{m}^3/\text{min}$, $p_1 = 1$ bar, $p_2 = 8$ bar, $\gamma = 1.4$, $n = 1.30$.

(a) Isothermal compression:

Work done by the compressor,

$$W = p_1 V_1 \ln\left(\frac{p_2}{p_1}\right)$$

$$= 1\times10^5 \times \frac{2}{60} \times \ln\left(\frac{8}{1}\right) = 6.93 \text{ kW}$$

(b) Isentropic process:

$$\text{Work done by compressor, } W = \left(\frac{\gamma}{\gamma-1}\right) p_1 V_1 \left[\left(\frac{p_2}{p_1}\right)^{\frac{\gamma-1}{\gamma}} - 1\right]$$

$$= \left(\frac{1.4}{1.4-1}\right) \times 1\times10^5 \times \frac{2}{60}\left[\left(\frac{8}{1}\right)^{\frac{1.4-1}{1.4}} - 1\right]$$

$$= 9.467 \text{ kW}$$

(c) Polytropic process:

$$\text{Work done by compressor, } W = \left(\frac{n}{n-1}\right) p_1 V_1 \left[\left(\frac{p_2}{p_1}\right)^{\frac{n-1}{n}} - 1\right]$$

$$= \left(\frac{1.30}{1.30-1}\right) \times 1\times10^5 \times \frac{2}{60}\left[\left(\frac{8}{1}\right)^{\frac{1.3-1}{1.3}} - 1\right]$$

$$= 8.896 \text{ kW}$$

Example 9.2 *A single-stage, single acting reciprocating compressor has a bore of 150 mm and a stroke of 200 mm. It receives vapour refrigerant at 1 bar and delivers it at 6 bar. The compression and expansion follow the law* $pV^{1.35}$ = *const. and the clearance volume is 5 percent of stroke volume. Calculate:*

(a) The power required to drive the compressor, if it runs at 600 rpm, and

(b) The volumetric efficiency of the compressor.

Solution: Given: $D = 150$ mm, $L = 200$ mm, $p_1 = 1$ bar, $p_2 = 6$ bar, $n = 1.35$, $V_c = 0.05\ V_s$

or $c = 0.05$, $N = 600$ rpm

Refer to Fig. 9.2.

Swept volume, $V_s = \dfrac{\pi}{4} D^2 L = \dfrac{\pi}{4}\ (0.15)^2 \times 0.2 = 3.5343 \times 10^{-3}\ \text{m}^3$

Clearance volume, $V_c = 0.05 \times 3.534 \times 10^{-3} = 0.1767 \times 10^{-3}\ \text{m}^3$

Total cylinder volume,

$$V_1 = V_c + V_s = (0.1767 + 8.5343) \times 10^{-3} = 3.711 \times 10^{-3}\ \text{m}^3$$

Expanded clearance volume,

$$V_4 = V_c\left(\frac{p_2}{p_1}\right)^{\frac{1}{n}} = 0.1767\times10^{-3}\left(\frac{6}{1}\right)^{\frac{1}{1.35}} = 0.6663 \times 10^{-3}\ \text{m}^3$$

Actual swept volume,

$$V_{as} = V_1 - V_4 = (3.711 - 0.6663) \times 10^{-3} = 3.0447 \times 10^{-3}\ \text{m}^3$$

(*a*) Work done by the compressor,

$$W = \left(\frac{n}{n-1}\right)p_1 V_{as}\left[\left(\frac{p_2}{p_1}\right)^{\frac{n-1}{n}} - 1\right]$$

$$= \left[\frac{1.35}{1.35-1}\right]\times1\times10^5\times3.0447\times10^{-3}\left[\left(\frac{6}{1}\right)^{\frac{1.35-1}{1.35}} - 1\right]$$

$$= 694.374\ \text{N.m}$$

Power required to drive the compressor,

$$P = \frac{WN}{60\times10^3} = \frac{694.34\times600}{60\times10^3} = 6.944\ \text{kW}$$

(*b*) Volumetric efficiency of compressor,

$$\eta_v = \frac{V_{as}}{V_s} = \frac{3.0447\times10^{-3}}{3.5343\times10^{-3}} = 0.8615 \text{ or } 86.15\%$$

or

$$\eta_v = 1 + c - c\left(\frac{p_2}{p_1}\right)^{\frac{1}{n}}$$

$$= 1 + 0.05 - 0.05\left(\frac{6}{1}\right)^{\frac{1}{1.35}}$$

$$= 0.8615 \text{ or } 86.15\%$$

9.3 ROTARY COMPRESSORS

In rotary compressors, the vapour refrigerant from the evaporator is compressed due to the movement of blades (or vanes). They have high volumetric efficiency as the clearance in them is negligible. They are used with R-12, R-22, R-114 and R-717 refrigerants. The rotary compressors are of the following types:

1. Rotating impeller type.
2. Rotating vane type.
3. Screw type.

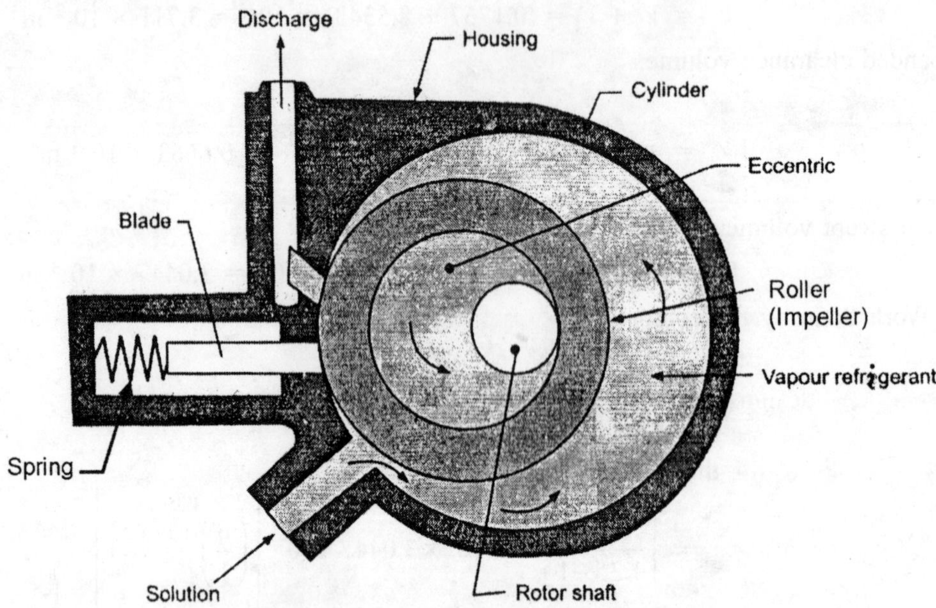

Fig. 9.6 Rotating impeller type rotary compressor.

1. Rotating Impeller Type Rotary Compressor

The rotating impeller type rotary compressor, shown in Fig. 9.6, consists of a stationary cylinder, roller or impeller and a roller shaft. The shaft has an eccentric on which the impeller is mounted. A blade is set into the slot of a cylinder in such a way that it always maintains contact with the impeller by means of a spring. The blade moves in and out of the slot to follow the impeller when it rotates. The blade separates the suction and the discharge ports. When the shaft rotates, the impeller also rotates so that it always touches the cylinder wall. Fig. 9.6 shows the impeller position at the completion of suction stroke and beginning of compression stroke. As the impeller rotates the new intake stroke is continued till the compressed vapours of refrigerant are discharged.

2. Rotating Vane Type Rotary Compressor

In the rotating vane type compressor, shown in Fig. 9.7, there are four vanes. The rotor is concentric with the shaft. The vanes slide within the rotor but keep contact with the cylinder. The assembly of rotor and vanes is off-centre with respect to the cylinder. The vanes are forced against the cylinder wall by the centrifugal action rotation of the motor.

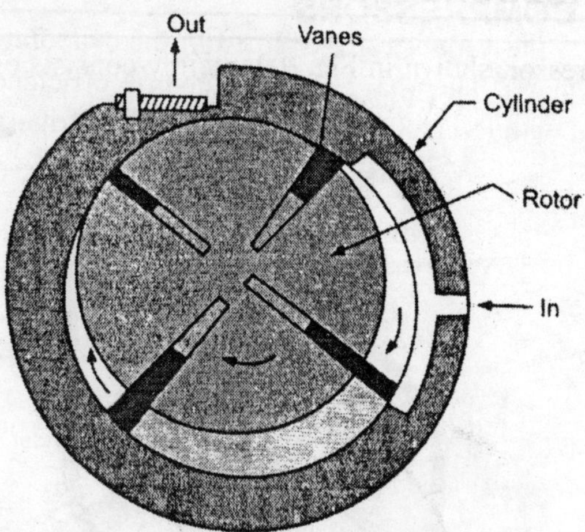

Fig. 9.7 Rotating Vane Type Compressor.

The low pressure and temperature vapour refrigerant from the evaporator is drawn in through the suction port. As the rotor turns, the sucked in vapour refrigerant entrapped between the two adjacent vanes is compressed. The compressed refrigerant at high pressure and temperature is discharged through the discharge port to the condenser.

In both the designs of rotating impeller and rotating vane type compressors, the whole assembly is enclosed in a housing (not shown in Fig. 9.7), filled with oil and remains submerged in oil. An oil film forms the seal between the high-pressure and low-pressure sides. When the compressor stops, this seal is lost and the pressure equalizes.

3. Screw Compressors

A screw compressor, shown in Fig. 9.8, essentially consists of two helically-grooved rotors which rotate in a housing. The male rotor consists of lobes and is normally the driving rotor. The female rotor has gullies and is normally the driven rotor. When the male rotor turns clockwise, an interlope space between a pair and housing nearest to the suction end, opens and is filled with the vapour refrigerant. The vapour refrigerant is moved helically and compressed until the trapped volume reaches the discharge end.

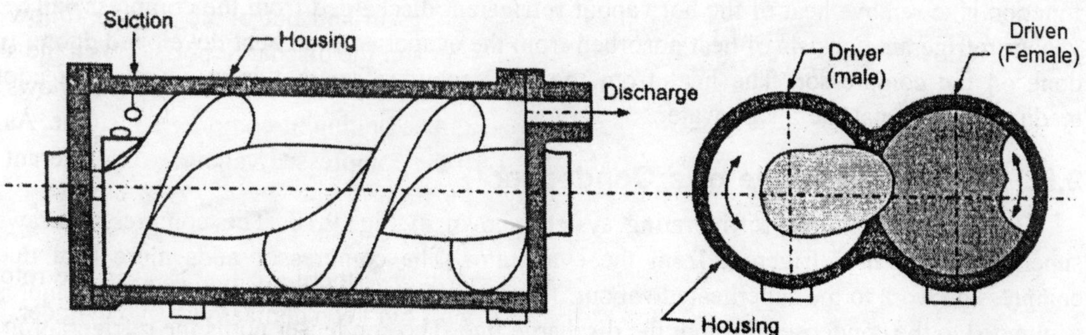

Fig. 9.8 Screw compressor

9.4 CENTRIFUGAL COMPRESSORS

A centrifugal compressor, shown in Fig. 9.9, mainly consists of the following components:

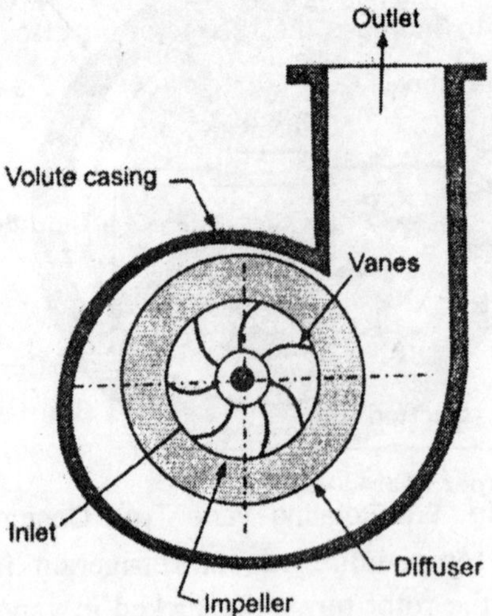

Fig. 9.9 Centrifugal compressor.

1. An inlet casing to accelerate the vapour refrigerant to the impeller inlet.
2. An impeller to transfer energy to the vapours in the form of increased static pressure and kinetic energy.
3. A diffuser to convert the kinetic energy at the impeller outlet pressure energy.
4. A volute casing to collect the vapours and to further convert the kinetic energy into pressure rise.

Thus, the centrifugal compressor increases the pressure of low pressure vapour refrigerant to a high pressure by centrifugal force. It is generally used for refrigerants that require large displacement and low condensing pressure, such as R-11, R-113 and R-12.

9.5 CONDENSER

The condenser is a device used in the high pressure side of a refrigeration system. Its main function is to remove heat of the hot vapour refrigerant discharged from the compressor. The hot vapour refrigerant consists of heat absorbed from the evaporator and heat developed due to work done on the compressor. The heat from the hot vapour refrigerant is removed by a cooling medium, which may be air or water.

9.5.1 Working Principle of a Condenser

Consider the simple refrigerating system shown in Fig. 9.10. The compressor draws in superheated vapour refrigerant from the evaporator. The compressor adds more heat due to compression work to the superheated vapour. This highly superheated vapour from the compressor is pumped to the condenser through the discharge line. The condenser cools the refrigerant in the following way:

1. The superheated vapour is cooled to saturation temperature (called desuperheating) corresponding to the pressure of the refrigerant.

2. The saturated vapour refrigerant gives up its latent heat and is condensed to a saturated liquid refrigerant. This process is called condensation.
3. The temperature of the liquid refrigerant is reduced below its saturation temperature (i.e. sub-cooled) in order to increase the refrigeration effect.

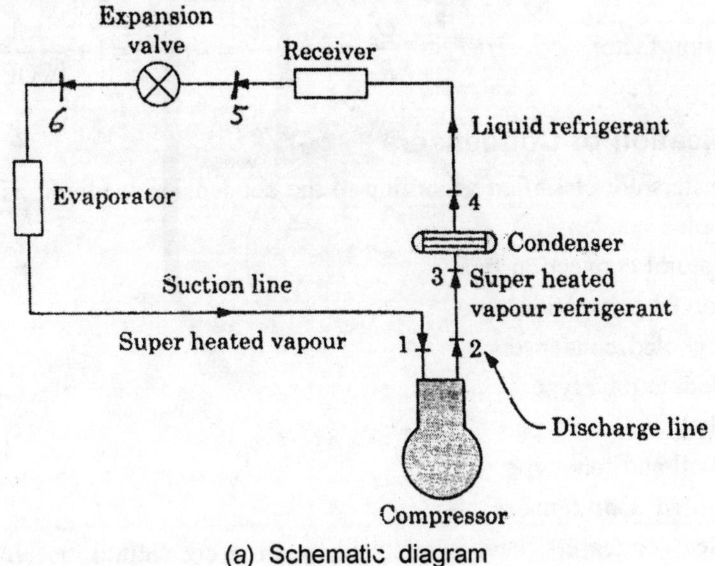

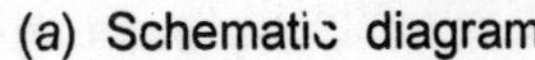
(a) Schematic diagram

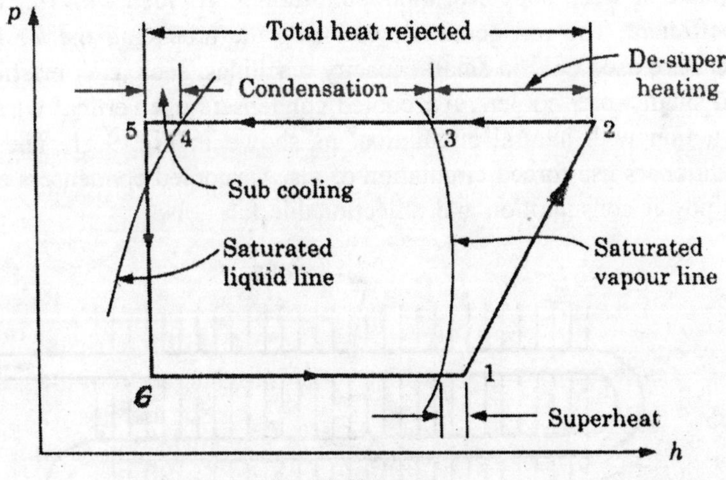

(b) p-h diagram

Fig. 9.10 Simple refrigeration system.

9.5.2 Factors Affecting Condenser Capacity

The condenser capacity is the ability of the condenser to transfer heat from the hot vapour refrigerant to the cooling medium. It depends upon the following factors:

1. Material of condenser.
2. Amount of contact area between the condenser surface and cooling medium.
3. Rate of flow of condensing medium over the condenser surface.
4. Temperature difference between the condensing medium and the vapour refrigerant.

9.5.3 Heat Rejection Factor-HRF

Heat rejection factor is defined as the load on the condenser per unit of refrigeration capacity.

Load on the condenser, Q_c = Refrigeration capacity (R_E) + Compressor work (W)

Heat rejection factor,
$$HRF = \frac{Q_c}{R_E} = \frac{R_E + W}{R_E} = 1 + \frac{W}{R_E} = 1 + \frac{1}{(\text{COP})_{\text{ref}}} \quad ...(9.15)$$

9.5.4 Classification of Condensers

The condensers are classified according to the condensing medium, as given below:

1. Air cooled condensers
   (*a*) Natural convection type
   (*b*) Forced convection type
2. Water cooled condensers
   (*a*) Double tube type
   (*b*) Shell and coil type
   (*c*) Shell and tube type

1. Air Cooled Condensers

In air cooled condensers, heat is removed by air using natural or forced convection. The condensers are made of steel, copper or aluminum tubing provided with fins to improve air side heat transfer coefficient. The refrigerant flows inside the tubes and the air flows outside. Air-cooled condensers are used only in small capacity machines, such as domestic refrigerators, air-conditioners and small water coolers. Air cooled condensers use vertical wire and tube or plate and tube construction with natural circulation, as shown in Fig. 9.11. The window type and package air-conditioners use forced circulation of air. Air cooled condensers are used upto 5 TR because of high power consumption and objectionable fan noise.

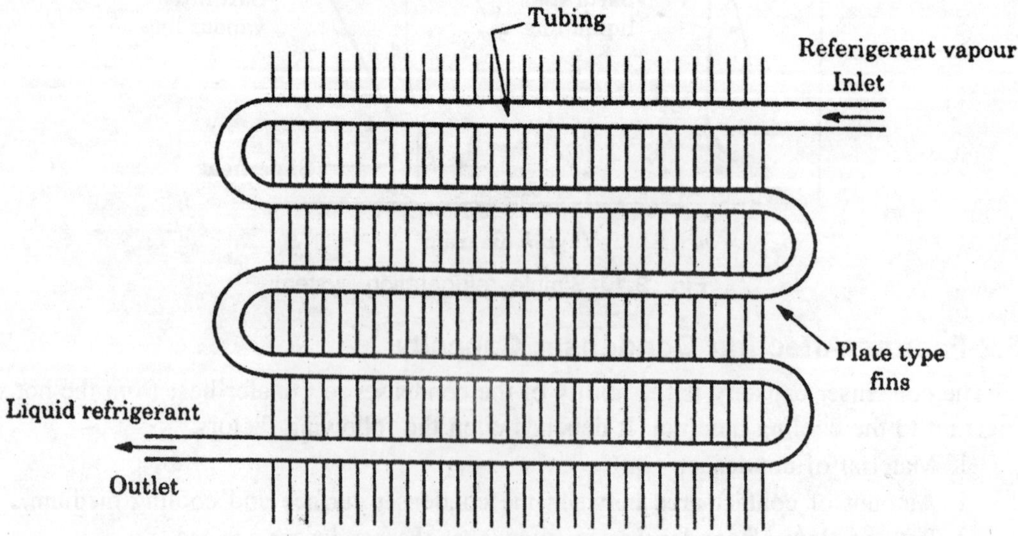

Fig. 9.11 Air cooled condenser.

2. Water Cooled Condensers

In water-cooled condensers, water is used as the condensing medium. These condensers are used commonly in commercial and industrial refrigerating units. A water cooled condenser with recirculating water system is shown in Fig. 9.12.

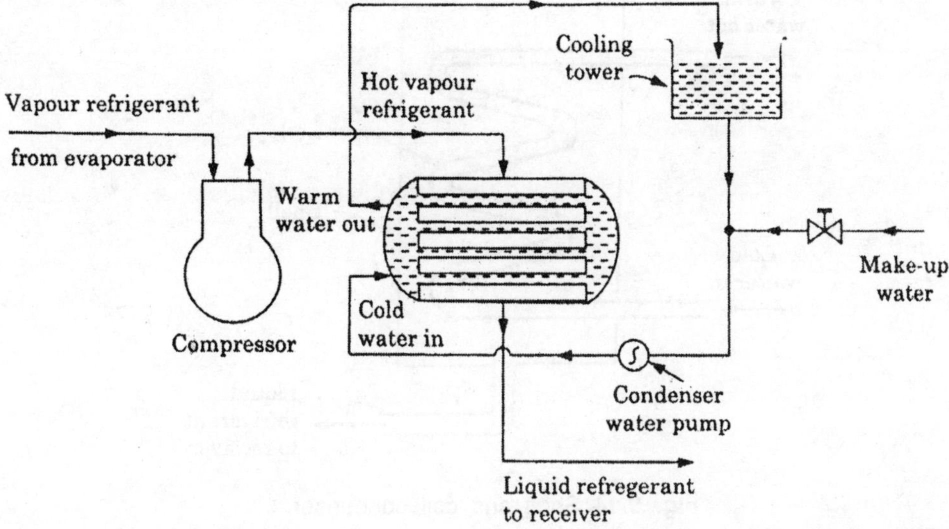

Fig. 9.12 Water cooled condenser with recirculating water system.

Water cooled condensers can be of three types: shell and tube, shell and coil and double tube.

(*a*) Fig. 9.13 shows the arrangement for a two-pass water cooled shell and tube condenser. In this system, water flows inside tubes and the refrigerant condenses in the shell. This type of condenser also serves as a receiver for pumping down the refrigerant. The shell is made of steel. Copper tubes are used for fluorocarbons and steel tubes for ammonia.

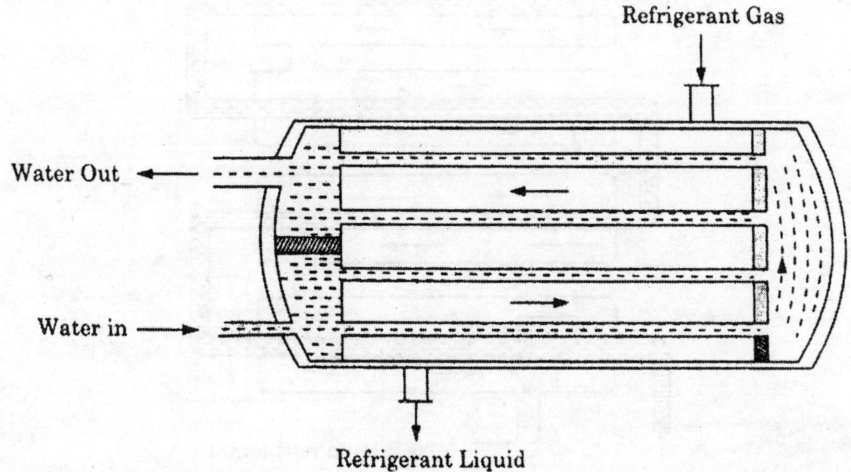

Fig. 9.13 Schematic respresentation of a two-pass water-cooled shell and tube condenser.

(*b*) The shell-and-coil condenser arrangement, shown in Fig. 9.14, consists of an electrically welded closed shell containing a water coil, sometimes of finned tubing.

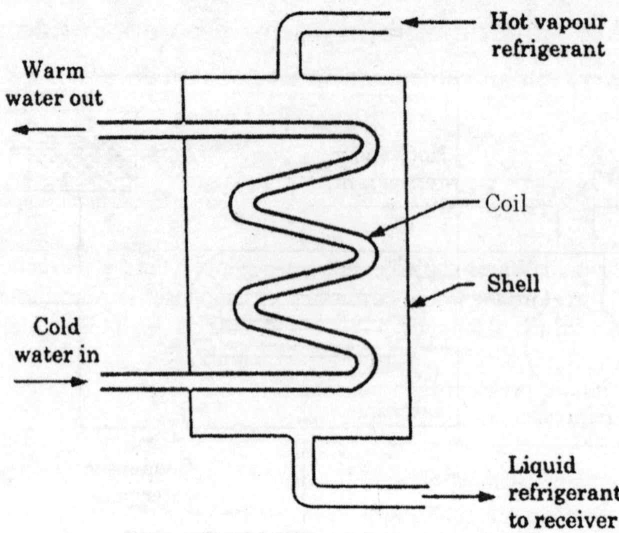

Fig. 9.14 Shell and coil condenser.

(*c*) In the double tube arrangement, shown in Fig. 9.15, the refrigerant condenses in the outer tube and water flows through the inner tube in the opposite direction.

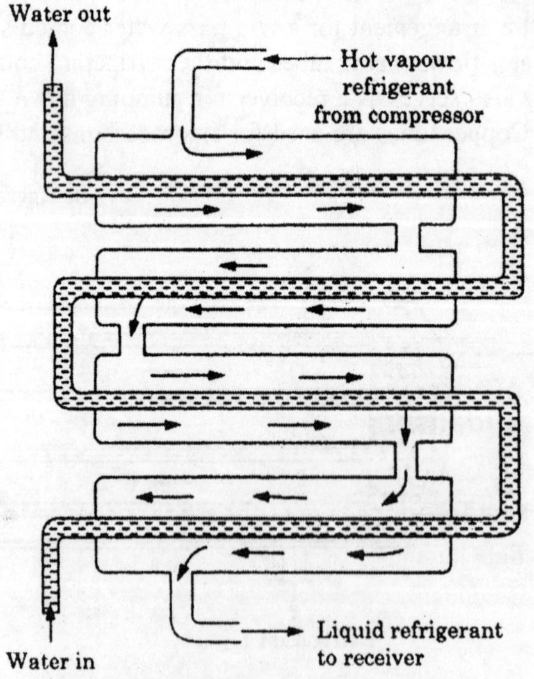

Fig. 9.15. Tube-in-tube condenser.

Water cooled condensers are invariably used in conjuction with cooling towers, spray ponds, etc. Heated water from the condenser is led to the cooling tower where it is cooled by self-evaporation into a stream of air. After cooling, the water is pumped back to the condenser.

9.5.5 Comparison of Air-Cooled and Water Cooled Condensers

The comparison between air-cooled and water cooled condensers is given in Table 9.1.

Table 9.1

Sl. No.	*Air Cooled condenser*	*Water cooled condenser*
1.	Since the construction of air cooled condenser is very simple, therefore the initial cost is less. The maintenance cost is also low.	Since the construction of water cooled condenser is complicated, therefore the initial cost is high. The maintenance cost is also high.
2.	There is no handling problem with air cooled condensers.	The water cooled condensers are difficult to handle.
3.	The air cooled condensers do not require piping arrangement for carrying the air.	The pipes are required to take water to and from the condenser.
4.	There is no problem in disposing of used air.	There is a problem of disposing the used water unless a recirculation system is provided.
5.	Since there is no corrosion, therefore fouling effect is low.	Since corrosion occurs inside the tubes carrying the water, therefore fouling effects are high.
6.	The air-cooled condensers have low heat transfer capacity due to low thermal conductivity of air.	The water cooled condensers have high heat transfer capacity due to high thermal conductivity of water.
7.	These condensers are used for low capacity plants (less than 5 TR).	These condensers are used for large capacity plants.
8.	Since the power required to drive the fan is excessive, therefore, the fan noise becomes objectionable.	There is no fan noise.
9.	The distribution of air on condenser surface is not uniform.	There is even distribution of water on the condensing surface.
10.	The air-cooled condensers have high flexibility.	The water cooled condensers have low flexibility.

9.5.6 Evaporative Condensers

Figure 9.16 shows the schematic diagram of an evaporative condenser. The refrigerant first rejects its heat to water and then water rejects its heat to air, mainly in the form of evaporated water. Air leaves with high humidity as in a cooling tower. Thus an evaporative condenser combines the functions of condenser and cooling tower.

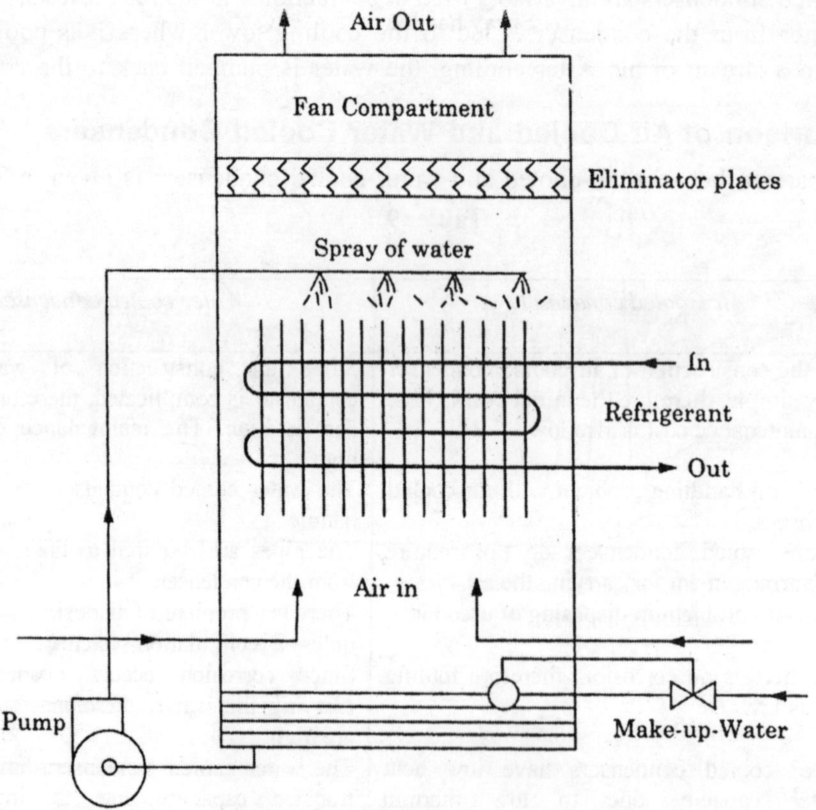

Fig. 9.16 Evaporative condenser

Evaporative condensers are commonly used on large ammonia plants as they are found to be cheaper. Such condensers require a larger amount of the refrigerant charge due to the longer length of the refrigerant piping. But in the case of ammonia systems this is immaterial since the refrigerant is quite cheap.

9.6 COOLING TOWERS

There is a necessity to recycle the cooling water through the condenser when cooling water supply is limited. Cooling tower is an artificial device used to cool hot cooling water coming out from the condenser.

The cooling towers are mainly divided, according to their method of air circulation, into the following two groups:

1. Natural draft cooling towers, and 2. Mechanical draft cooling towers.

In *natural draft cooling towers*, the air circulates through the tower by natural convection whereas in *mechanical draft cooling towers*, the air is forced through the tower by means of fans or blowers,

1. Natural Draft Cooling Towers

Since the air circulating through the natural draft cooling towers is atmospheric air, therefore these cooling towers are also known as *atmospheric natural draft cooling towers* or simply *atmospheric cooling towers*.

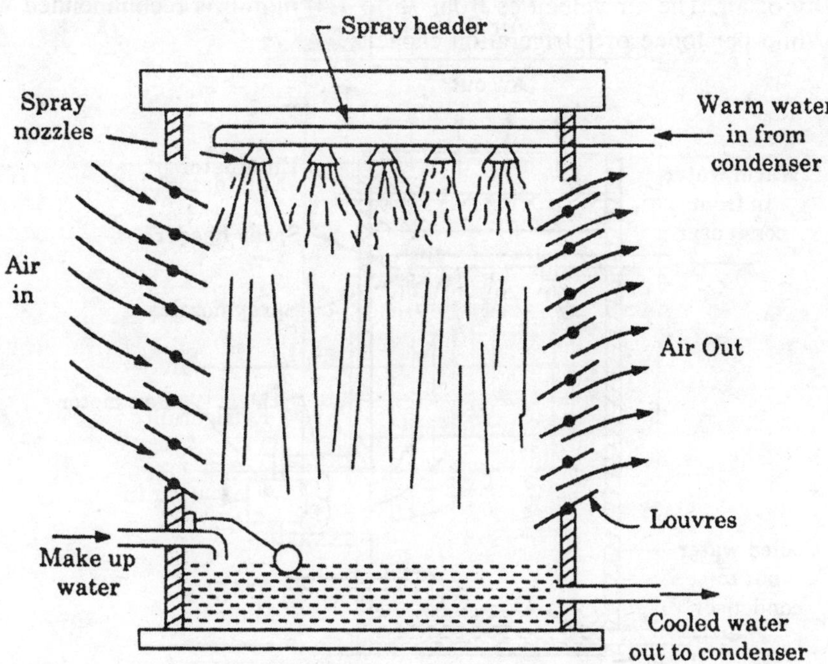

Fig. 9.17 Atmospheric natural draft (spray type) cooling tower.

The atmospheric natural draft (spray type) cooling tower, as shown in Fig. 9.17, consists of a box-shaped structure with louvers. The louvers allow the atmospheric air to pass through the tower, but slant down towards the inside of the tower to retain water in it. The frame work and louvers are usually made of steel. The size of the cooling tower depends upon the capacity of the unit. The atmospheric natural draft (spray type) cooling towers should be located in the open space or on the roof of a building where the air can blow freely through them.

In this type of cooling tower, warm water from the condenser is pumped to a spray header provided at the top of a tower. It is sprayed down into the tower through the nozzles. Since the heat transfer from water to air is dependent upon the surface of water exposed to the air stream, therefore a spray nozzle having finer spray pattern is essential for good performance of the cooling tower. It may be noted that the finer spray exposes more water surface to air. However, if the spray is too fine, too much water is blown away. The water spray blown away by the air is called drift.

2. Mechanical Draft Cooling Towers

In these towers, fans are used to force the air through them. These towers are either of the forced draft or induced draft type.

(*a*) Forced Draft Cooling Tower

In the forced draft cooling tower, as shown in Fig. 9.18, a fan forces the air through the tower. In its operation, the warm water from the condenser is sprayed at the top of the tower through the spray nozzles. The air is forced upward through the tower by the propeller fan provided on the side near the bottom of the tower as shown in the figure. The condenser warm water is cooled by means of evaporation as discussed earlier. The effectiveness of the cooling tower may be improved by increasing the height of the tower, area of water surface exposed to air or the velocity of air. The air velocities from 75 to 120 m/min is recommended with a flow of 90 to 120 m^3/min per tonne of refrigeration capacity.

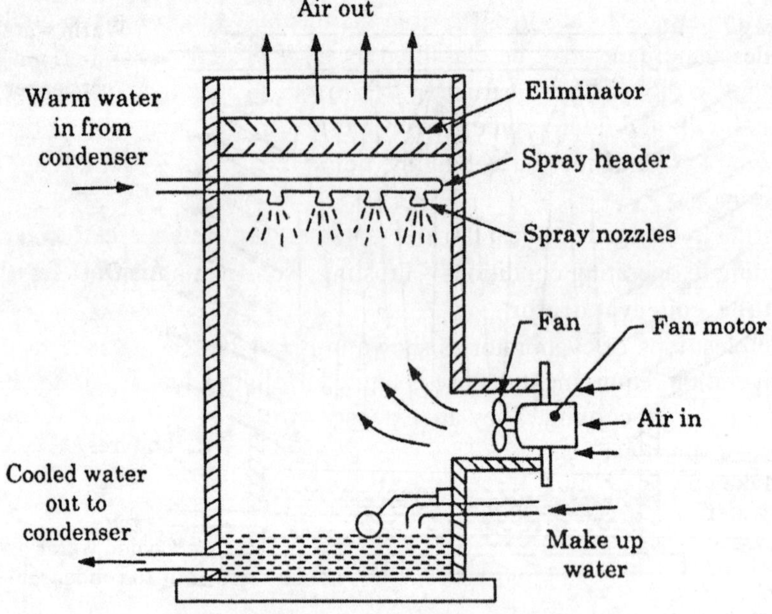

Fig. 9.18 Forced draft cooling tower.

(*b*) Induced Draft Cooling Towers

In the induced draft cooling tower, as shown in Fig. 9.19, the fan sucks the air through the tower. The induced draft cooling towers are similar to forced draft cooling towers except that the fans are located at the top instead of at the bottom and draw the air upward through the tower.

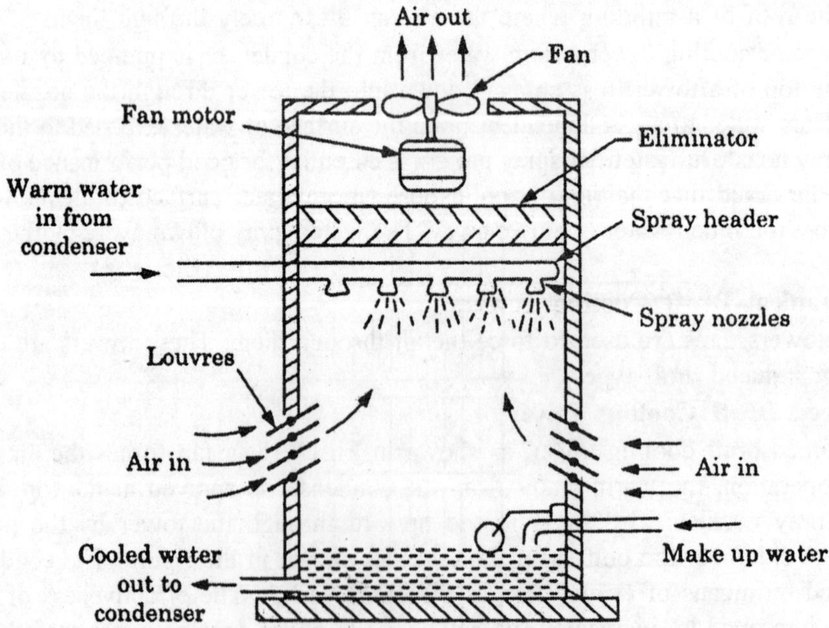

Fig. 9.19 Induced draft cooling tower.

9.7 EVAPORATORS

The evaporator is a device used in the low pressure side of a refrigeration system. The function of an evaporator is to absorb heat from the medium which is to be cooled by means of a refrigerant. The evaporators may be classified as follows:

1. According to the type of construction – Bare tube, finned tube, plate type, shell and tube, shell and coil, and double tube type evaporators.
2. According to the manner of feeding the liquid refrigerant – Flooded and Dry Expansion evaporator.
3. According to the mode of heat transfer – Natural convection and forced convection type.
4. According to operating conditions – Frosting, Non-frosting, and Defrosting evaporators.

(*i*) Bare tube coil evaporator

It is the simplest type of evaporator as shown in Fig. 9.20. This evaporator may be used for any type of refrigeration requirement. Its use is limited to applications where the box temperatures are below °C and in liquid cooling. They are extensively used in household refrigerators because they are easier to keep clean.

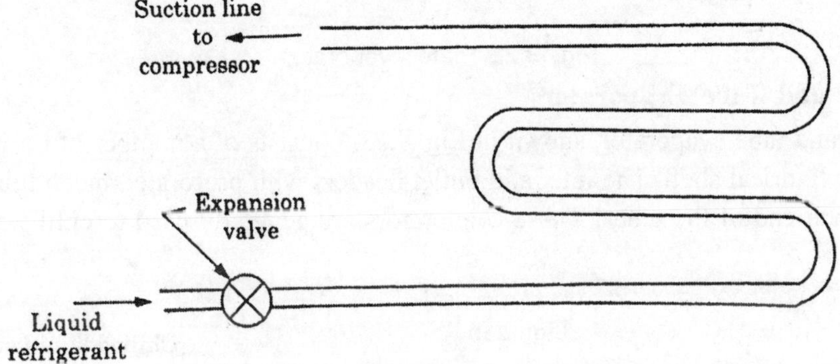

Fig. 9.20 Bare tube coil evaporator

(*ii*) Finned Evaporators

The finned evaporators, as shown in Fig. 9.21, consist of bare tubes or coils over which the metal plates or fins are fastened. They are also called extended surface evaporators. They are primarily designed for air-conditioning applications where the refrigerator temperature is above 0°C.

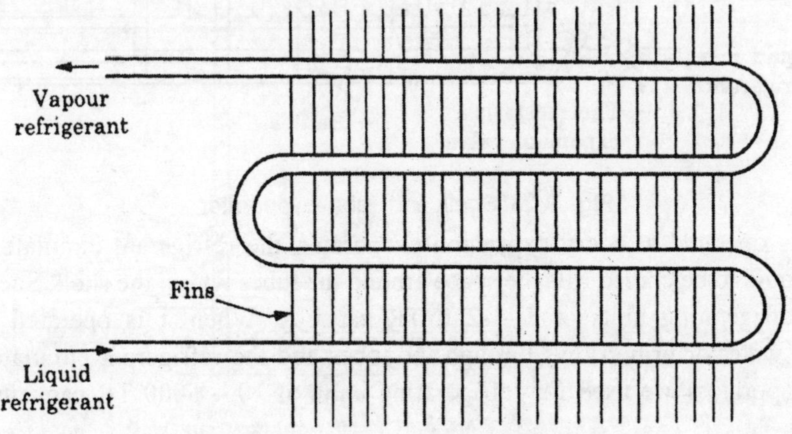

Fig. 9.21 Finned evaporator

(*iii*) Plate Evaporators

A common type of plate evaporator, shown in Fig. 9.22, consists of coils either welded on one side of a plate or between the two plates welded at the edges. These are generally used in household refrigerators, home freezers, beverage coolers, ice cream cabinets, etc.

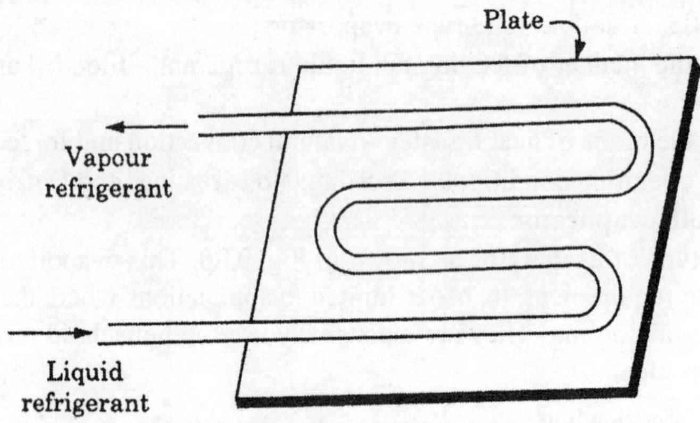

Fig. 9.22 Plate evaporator

(*iv*) Shell and Tube Evaporators

The shell and tube evaporator, shown in Fig. 9.23, consists of a number of horizontal tubes enclosed in a cylindrical shell. The inlet and outlet headers with proforated metal tube sheets are connected at each end of the tubes. These evaporators are generally used to chill water or brine solutions.

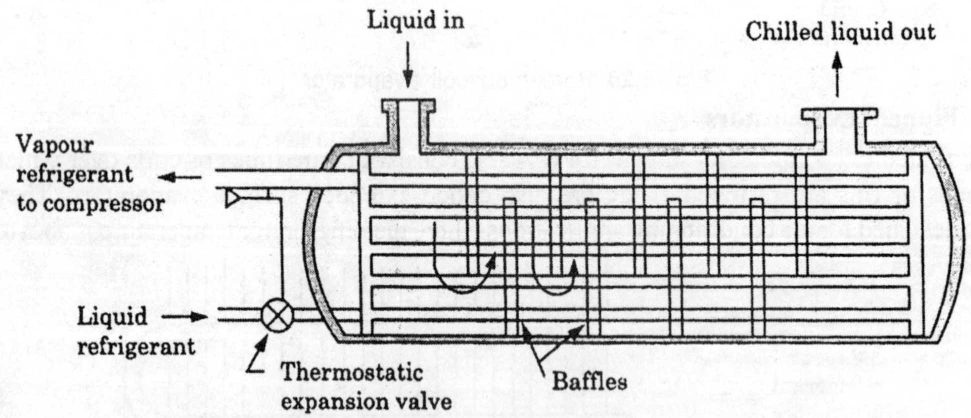

Fig. 9.23 Shell and tube evaporator

When it is operated as a dry expansion evaporator, the refrigerant circulates through the tubes and the liquid to be cooled fills the space around the tubes within the shell. Such evaporators are used for refrigerating units of 2 – 250 TR capacity. When it is operated as a flooded evaporator, the water or brine flows through the tubes and the refrigerant circulates around the tubes. Such evaporators are used for refrigerating units of 10 – 5000 TR capacity.

(*v*) Shell and Coil Evaporators

The shell and coil evaporator, as shown in Fig. 9.24, is generally dry expansion evaporator to chill water. The cooling coil is a continuous tube that can be in the form of a single or double spiral. The shell may be sealed or open. These evaporators are used to cool drinking water.

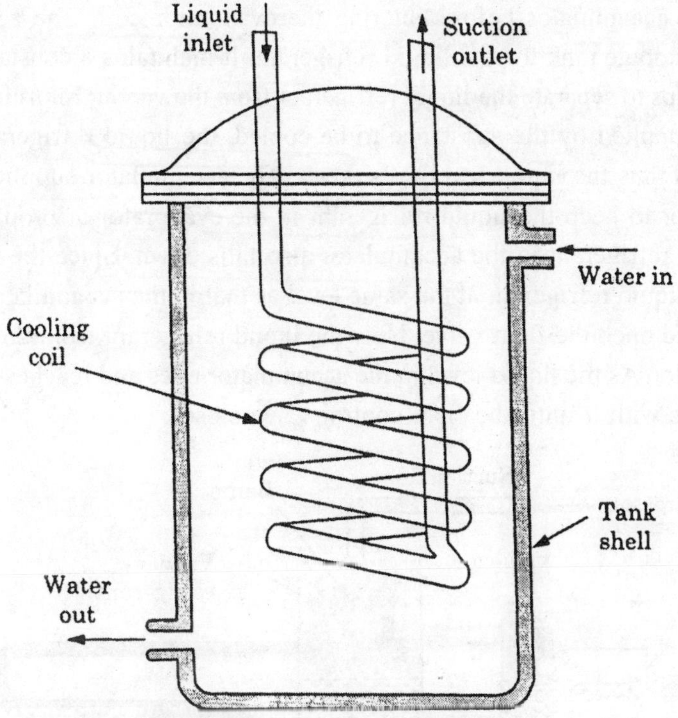

Fig. 9.24 Shell and coil evaporator

(*vi*) Double Tube Evaporators

The double tube evaporator, as shown in Fig. 9.25, consists of one tube inside another tube. The liquid to be cooled flow through the inner tube while the primary refrigerant or secondary refrigerant circulates in the space between the two tubes. These evaporators provide high transfer rates. These evaporators are used wine cooling and in petroleum industry for chilling of oil.

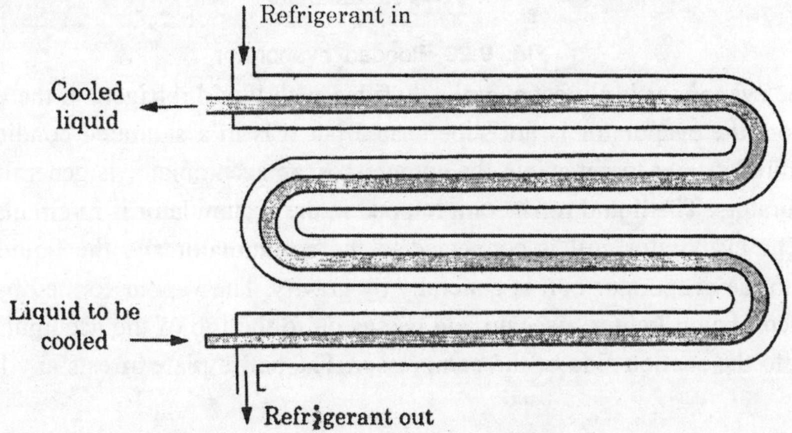

Fig. 9.25 Tube in tube or double tube evaporator

(*vii*) Flooded Evaporators

In a flooded evaporator, as shown in Fig. 9.26, a constant liquid refrigerant level is always maintained. A float control valve is used as an expansion device which maintains constant liquid level in the evaporator. The liquid refrigerant from the receiver passes through a low side float control valve and accumulator before entering the evaporator coil. The accumulator (or surge tank) serves as a storage tank for the liquid refrigerant. It maintains a constant liquid level in the evaporator and helps to separate the liquid refrigerant from the vapour returning to the compressor. Due to the heat supplied by the substance to be cooled, the liquid refrigerant in the evaporator coil vaporizes and thus the liquid level falls down. The accumulator supplies more liquid to the evaporator in order to keep the liquid refrigerant in the evaporator at proper level. In this way, the level of liquid refrigerant in the accumulator also falls down. Since the float within the float chamber rests on liquid refrigerant at the same level as that in the accumulator, therefore the float also falls down and open the float valve. Now the liquid refrigerant from the receiver is admitted into the accumulator. As the liquid level in the accumulator rises and reaches to the constant level, the float also rises with it until the float control valve closes.

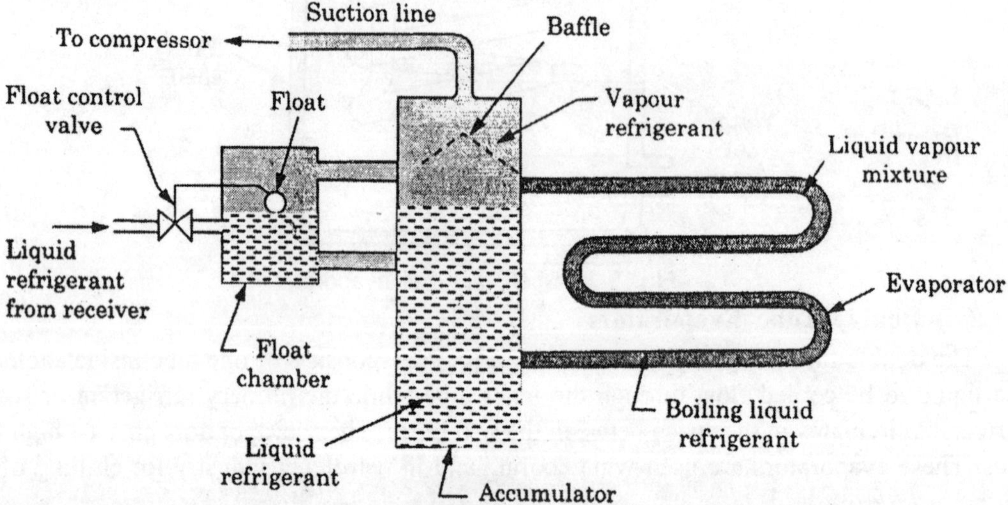

Fig. 9.26 Flooded evaporator

Since the evaporator is almost completely filled with liquid refrigerant, therefore the vapour refrigerant from the evaporator is not superheated but it is in a saturated condition. In order to prevent liquid refrigerant to enter into the compressor, an accumulator is generally used with the flooded evaporators. The liquid refrigerant trapped in the accumulator is re-circulated through the evaporator. The evaporator coil is connected to the accumulator and the liquid flow from the accumulator to the evaporator coil is generally by gravity. The vapour formed by vaporizing the liquid in the coil being lighter, rises up and passes on to the top of the accumulator from where it is supplied to the suction side of the compressor. The baffle plate arrests any liquid present in the vapour.

The advantage of the flooded evaporator is that the whole surface of the evaporator coil is in contact with the liquid refrigerant under all the load conditions. Thus, it gives high heat transfer rates (i.e., more efficient cooling) than a dry expansion evaporator of the same size. However, the flooded evaporator is more expensive to operate because it requires more refrigerant charge.

The flooded evaporators have many industrial applications, especially in the chemical and food processing industries. These evaporators used in comfort and process air cooling installations may be the bare coil type or finned type.

(viii) Dry Expansion Evaporators

The dry expansion evaporators are not really dry at all. They simply use relatively little refrigerant as compared to flooded evaporators having the same coil volume. The dry expansion evaporators are usually only one-fourth to one-third filled with liquid refrigerant. A simple bare tube dry expansion evaporator is shown in Fig. 9.27. The finned coil dry expansion evaporators are also available.

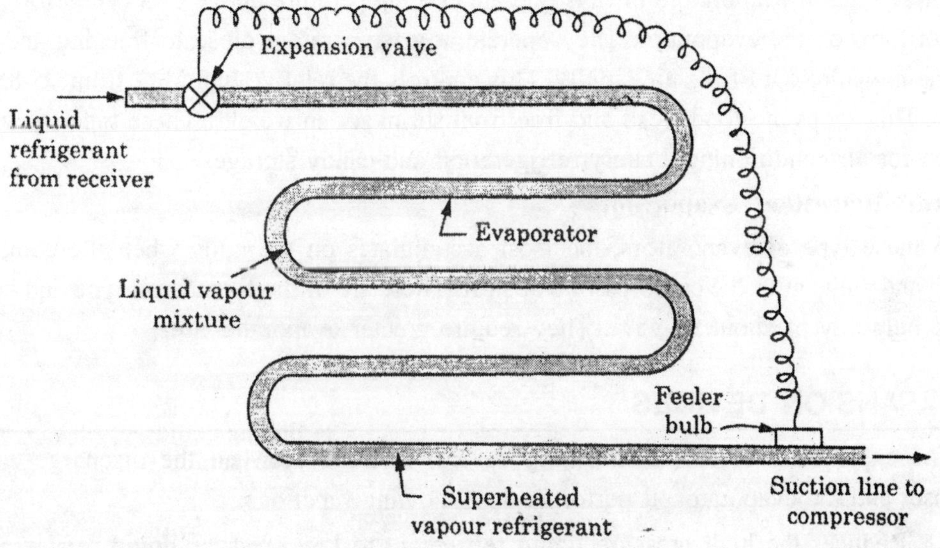

Fig. 9.27 Dry expansion evaporator

In dry expansion evaporators, the liquid refrigerant from the receiver is fed by the expansion valve to the evaporator coil. The expansion valve controls the rate of flow of liquid refrigerant in such a way that all the liquid refrigerant is vaporized by the time it reaches at the end of the evaporator coils or the suction line to the compressor. The vapour is also superheated to a limited extent. It may be noted that in a dry expansion system, the refrigerant does not recirculate within the evaporator as in flooded type evaporator.

(*ix*) Natural Convection Evaporators

The natural convection evaporators are used where low velocity and minimum hydration of the product is desired. The domestic refrigerators, water coolers and small freezers have natural convection evaporators.

(*x*) Forced Convection Evaporators

In these evaporators, the air is forced over the refrigerant cooled coils and fins. This is done by a fan driven by an electric motor. These type of evaporators are suited for air cooling units as well as for refrigerator cabinets used to store bottled beverages or foods in sealed containers.

(*xi*) Frosting Evaporators

These type of evaporators always operate at temperatures below 0°C. The frost on the evaporator comes from the moisture in the air. As the frost grows in thickness, the cooling efficiency decreases until the ice and frost is removed. The evaporators in household refrigerators and bare pipe coils in storage boxes are of the frosting type.

(*xii*) Non-frosting Evaporators

These type of evaporators always operate at a temperature above 0°C. Therefore, the frost does not form on the evaporator. They operate at a temperature close to freezing and does not draw the moisture out of the air rapidly. This controls the relative humidity from 75-85% in the cabinet. This keeps the food fresh and free from shrinkage in weight. These type of evaporators are used for air-conditioning, bakery refrigeration and candy storage.

(*xiii*) Defrosting Evaporators

In these type of evaporators, the frost accumulates on the coils when the compressor is running and melts after the compressor shuts off. These are of the finned coil type and keeps high relative humidity of about 90-95%. They require greater evaporator area.

9.8 EXPANSION DEVICES

The expansion device (or throttling device) is fitted between the discharge end of the condenser and the evaporator. It performs the following functions:

1. Reduces the high pressure liquid refrigerant to low pressure liquid refrigerant before being fed to the evaporator.
2. Supplies the liquid refrigerant to the evaporator at the rate at which it is evaporated.

An expansion device is essentially a restriction offering resistance to flow so that the pressure drops, resulting in a throttling process.

Basically these are two types of expansion devices. These are:

1. Constant-restriction type.

The capillary tube is such a device, which is merely a long tube with a narrow diameter bore.

2. Variable-restriction type.

In these devices, the area of flow keeps on changing depending on the type of control. There are three common types of such control devices.

(*i*) Automatic or constant pressure expansion valve

(*ii*) Thermostatic expansion valve.

(*iii*) Float valves

(*a*) *High-side float:* It maintains constant level in the condenser.

(*b*) *Low-side float:* It maintains constant level in the evaporator.

1. Capillary Tube

It is copper tube of small bore and of varying length depending upon the application, as shown in Fig. 9.28. The bore varies from 0.5 to 2.25 mm and the length varies from 0.5 to 5 m. it is installed in the liquid line between condenser and evaporator. It is used in domestic refrigerators, water coolers, room air - conditioners and freezers. The pressure drops due to frictional resistance offered by a small bore tube. The frictional resistance is directly proportional to tube length and inversely proportional to its bore.

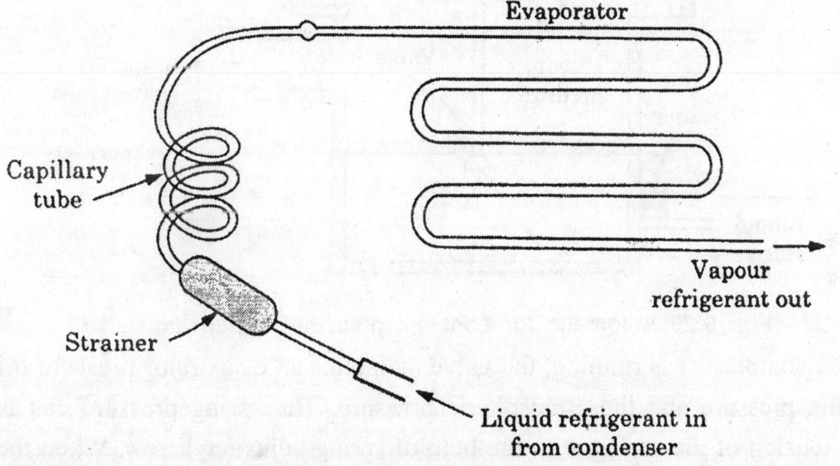

Fig. 9.28 Capillary tube

The advantages of capillary tube are:

(*i*) Lower cost

(*ii*) When the compressor stops, the refrigerant continues to flow into the evaporator so that the pressure equalizes between the high and low sides of the system. This decreases the load on the compressor requiring a low cost motor.

(*iii*) No receiver is required as the refrigerant charge in capillary tube is critical.

2. Automatic Expansion Valve

It maintains constant evaporator pressure regardless of the load on the evaporator. It is used with dry expansion evaporators where the load is relatively constant.

The automatic expansion valve, as shown in Fig. 9.29, consists of a needle valve and a seat (which forms an orifice), a metallic diaphragm or bellows, spring and an adjusting screw. The opening and closing of the valve with respect to the seat depends upon the following two opposing forces acting on the diaphragm:

1. The spring pressure and atmospheric pressure acting on the top of the diaphragm, and
2. The evaporator pressure acting below the diaphragm.

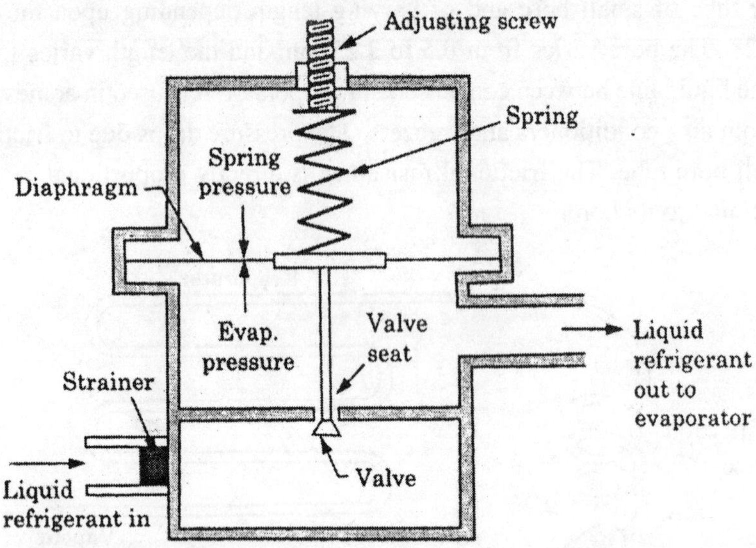

Fig. 9.29 Automatic (or constant pressure) expansion valve

When the compressor is running, the valve maintains an evaporator pressure in equilibrium with the spring pressure and the atmospheric pressure. The spring pressure can be varied by adjusting the tension of the spring with the help of spring adjusting screw. When the evaporator pressure falls down, the diaphragm moves downwards to open the valve. This allows more liquid refrigerant to enter into the evaporator thereby increasing the evaporator pressure to the desired level. The reverse phenomena occurs on the increase of evaporator pressure. When the compressor stops, the liquid refrigerant continues to flow into the evaporator, which causes the evaporator pressure to increase. This increased pressure causes the diaphragm to move upwards and the valve is closed. The valve opens when the compressor starts again.

3. Thermostatic Expansion Valve

This valve maintains constant superheat of the vapour refrigerant at the end of the evaporator coil by controlling the flow of liquid refrigerant. The thermostatic expansion valve shown in Fig. 9.30, consist of a needle valve and a seat, a metallic diaphragm, spring and an adjusting screw.

It has also a feeler (or thermal) bulb mounted on the suction line near the outlet of the evaporator coil. The feeler bulb is filled partly with the same liquid refrigerant as used in the refrigeration system. The opening and closing of the valve depends upon the spring pressure (p_s), evaporator pressure (p_E) and feeler bulb pressure (p_B) acting on the diaphragm. It is in the most commonly used expansion device in commercial and industrial system.

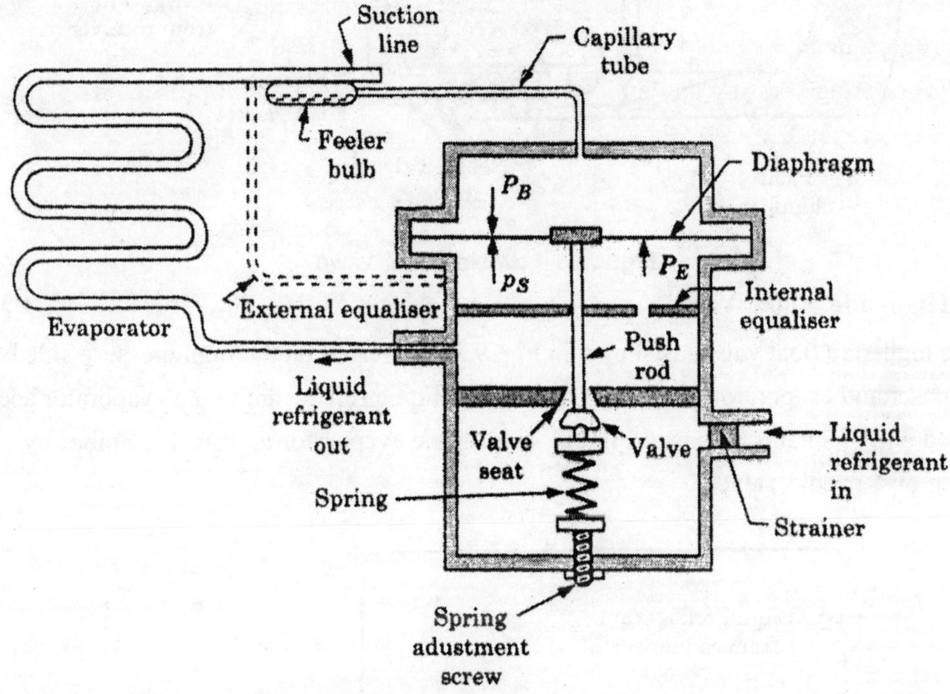

Fig. 9.30 Thermostatic expansion valve

4. Low-side Float Valve

The low side float valve, as shown in Fig. 9.31 is located in the low side (i.e. between the evaporator and compressor suction line) of the refrigeration system. It maintains a constant level of liquid refrigerant in the evaporator and float chamber by opening and closing a needle valve.

The float valve is a hollow ball attached to one end of a float arm. The other end of the arm is connected to a needle valve. The movement of the float ball (rise or (fall)) is transmitted to the needle valve by the float arm which closes or opens the flow of liquid refrigerant. Since the float valve is hollow, therefore it floats on the liquid refrigerant, in the float chamber.

When the liquid refrigerant in the evaporator vaporizes, its level falls down This causes the float to drop and thus opens the needle valve, thereby allowing liquid refrigerant from the liquid line to the float chamber and then to the evaporator to make up for the amount of vaporization. When the desired liquid level is reached, the float rises and closes the needle valve. The major advantage of the low-side float valve is that it maintains a constant liquid level in the evaporator under all loading conditions.

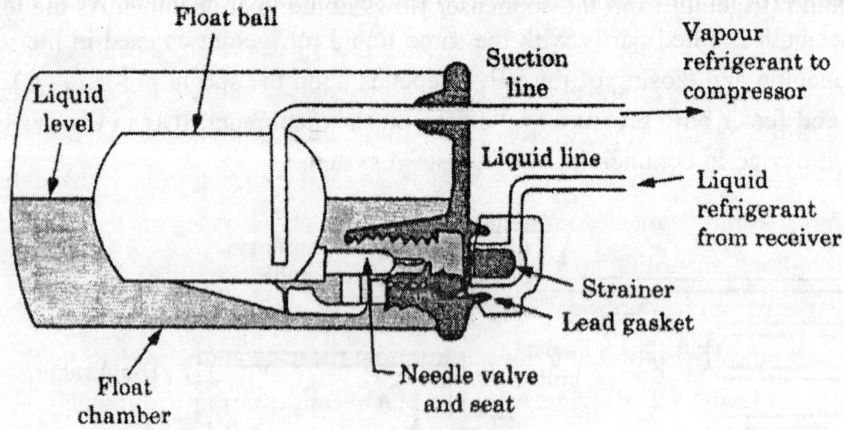

Fig. 9.31 Low-side float valve

5. High-side Float Valve

The high-side float valve, as shown in Fig. 9.32 is located on the high pressure side between the condenser and evaporator. It controls the flow of liquid refrigerant to the evaporator according to the load and maintains a constant liquid level in the evaporator and float chamber by opening or closing of a needle valve.

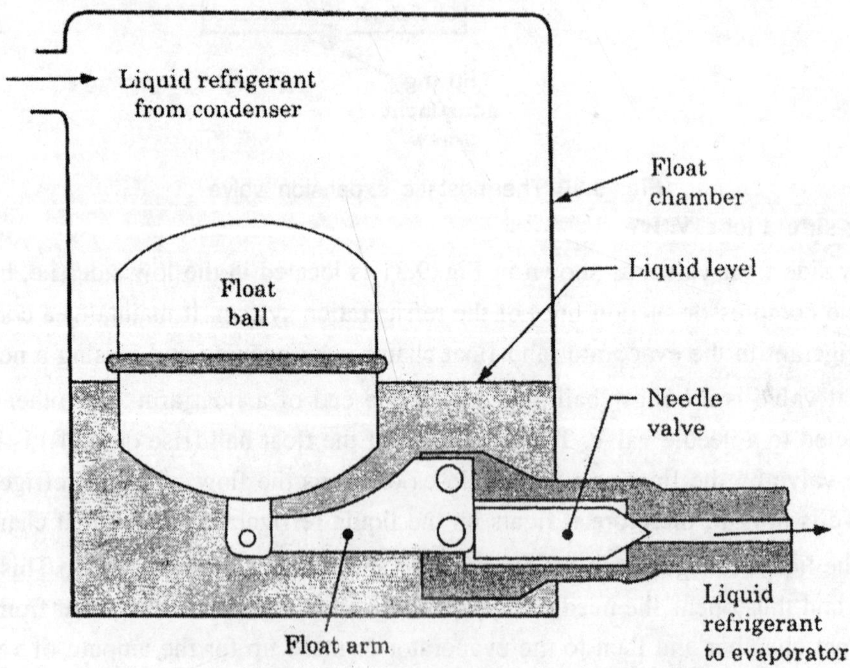

Fig. 9.32 High-side float valve

The liquid refrigerant from the condenser flows to the float chamber. As the level of liquid refrigerant in the float chamber rises, the float ball also rises, thereby opening the needle valve. This allows the liquid refrigerant to flow into the evaporator. When the liquid level in the float chamber falls down, the float ball also drops, thereby closing the needle valve. It may be noted that the condenser supplies the liquid refrigerant at the same rate as it evaporates in the evaporator. Since the rate of vaporization depends upon the load on the evaporator, therefore the high side float valve functions according to the load. The high side float valve may be used with dry expansion evaporators.

A summary of refrigeration and air-conditioning controls is given in Table 9.2.

Table 9.2 Refrigeration and Air-conditioning Controls

S. No.	*RAC control*	*Functions*
1.	Expansion devices (i) Automatic expansion valve (ii) Thermostatic expansion valve	(i) It serves the thermodynamic function of expanding the liquid refrigerant from the condenser pressure to the evaporator pressure. (ii) To control the supply of the liquid to the evaporator at the rate at which it is evaporated.
2.	Cut – outs for refrigeration units (i) High pressure cut – out (ii) Low pressure cut – out	To check the rise and fall of pressure in a refrigeration system.
3.	Back pressure valves	To control the back pressure in the compressor.
4.	Liquid level regulating devices (i) High – side float valve (ii) Low – side float valve	To check the liquid level on the high and low sides of the compressor.
5.	Flow – regulating devices (i) Solenoid valves (ii) Check valves (iii) Water and brine valves	To regulate the flow of water and brine in the refrigeration system.
6.	Thermostats (i) Evaporator thermostat (ii) Room thermostat	To control the temperature.
7.	Humidity controls	To control the moisture content
8.	Motor controls	Act as a safety device for the electric motor by controlling its speed.
9.	Modulating motors	For proportional or floating control

MULTIPLE – CHOICE QUESTIONS

1. For minimum work of a reciprocating compressor, the compression process should be
(a) isentropic (b) polytropic (c) isothermal (d) adiabatic
2. The clearance ratio is defined as:
(a) clearance volume/swept volume (b) swept volume/clearance volume
(c) clearance volume/total volume (d) total volume/clearance volume
3. In a two-stage compression process with perfect intercooling, the intermediate pressure is given by
(a) $p_2 = \frac{p_1 + p_3}{2}$ (b) $p_2 = \frac{p_3 - p_1}{2}$ (c) $p_2 = \sqrt{p_1 p_3}$ (d) $p_2 = p_1 p_3$
4. The heat rejection factor (HRF) is defined as:
(a) 1 + COP (b) 1 – COP (c) $1 + \frac{1}{\text{COP}}$ (d) $1 - \frac{1}{\text{COP}}$
5. The natural convection air-cooled condensers are used in
(a) water cooler (b) domestic refrigerators
(c) room air-conditioners (d) all of these
6. For ammonia refrigerating system, the tubes of a shell and tube condenser are made of
(a) aluminium (b) copper (c) steel (d) brass
7. An evaporator is also called
(a) Freezing coil (b) cooling coil (c) chilling coil (d) all of the above
8. The main function of the evaporator is to change the low pressure liquid refrigerant from the expansion valve into
(a) high pressure liquid refrigerant (b) low pressure liquid and vapour mixture
(c) low pressure vapour refrigerant (d) high pressure vapour refrigerant
9. The evaporator generally used in household refrigerators is of the following type:
(a) plate (b) finned (c) double tube (d) shell and coil
10. The evaporator generally used for wine cooling is of the following type:
(a) plate (b) finned (c) double tube (d) shell and tube
11. The evaporator used in household refrigerators is of the following type:
(a) frosting (b) none-frosting (c) defrosting (d) all of the above
12. In a refrigeration system, the expansion device is connected between the
(a) compressor and condenser (c) condenser and evaporator
(c) evaporator and compressor (d) receiver and evaporator

13. The capillary tube is used as an expansion device in
(a) water cooler (b) domestic refrigerator
(c) room air-conditioner (d) all of the above

14. The thermostatic expansion valve is also called
(a) constant pressure valve (b) constant temperature valve
(c) constant superheat valve (d) constant volume valve

15. Which type of valves is generally used in reciprocating refrigerant compressors?
(a) Mushroom valve (b) Puppet valve (c) Plate valve (d) Throttle valve

16. In a domestic refrigerator, a capillary tube controls the flow of refrigerant from the
(a) expansion valve of the evaporator (b) evaporator to the thermostat
(c) condenser to the expansion valve (d) condenser to the evaporator

17. A pressure gauge on discharge side of a refrigerant compressor reads too high. The reason could be:
1. Lack of cooling water 2. Water temperature being high
3. Dirty condenser surface 4. Refrigerant temperature being too high

Of these reasons
(a) 1, 2, and 4 are valid (b) 1, 2 and 3 are valid
(c) 2, 3 and 4 are valid (d) 1, 2 and 4 are valid.

18. A valve maintains a constant degree of superheat at the end of the evaporator coil, is called
(a) automatic expansion valve (b) high side float valve
(c) thermostatic expansion valve (d) low side float valve

19. Air cooling is used for Freon compressors whereas water jacketing is adopted for cooling ammonia compressors. This is because
(a) latent heat of ammonia is higher than that of Freon
(b) thermal conductivity of water is higher than that of air
(c) specific heat of water is higher than that of air
(d) of the larger superheat horn of ammonia compression cycle

20. For small installations of refrigeration systems (up to 35 kW), which type of condenser is used?
(a) Shell and tube type (b) Shell and coil type
(c) Double tube type (d) Air cooled type

21. The ratio of the clearance volume to the displacement volume of a R-12 reciprocating compressor is 0.05. Specific volume at inlet and outlet of compressor are 0.04 and 0.02 m^3/kg respectively. Volumetric efficiency of the compressor is
(a) 95.0% (b) 47.5% (b) 38.0% (d) 19.0%

22. Match List-I (Refrigeration equipment) with List-II (Characteristic) and select the correct answer using the codes given below the lists:

List – I (Refrigeration equipment)	List – II (Characteristic)
A. Hermetically sealed compressor B. Semi – hermetically sealed compressor C. Open type compressor D. Expansion device	1. Capillary tube 2. Both compressor and motor enclosed in a shell or casing 3. Both compressor and motor enclosed in a shell or casing with a removable cylinder cover 4. Driving motor not enclosed in a shell or casing and connected to the shaft driving the compressor.

Codes:	**A**	**B**	**C**	**D**
(a)	1	4	3	2
(b)	2	3	4	1
(c)	1	3	4	2
(d)	2	4	3	1

Answers

1. (c) **2.** (a) **3.** (c) **4.** (c) **5.** (b) **6.** (c) **7.** (d) **8.** (c)
9. (a) **10.** (a) **11.** (a) **12.** (d) **13.** (d) **14.** (c) **15.** (d) **16.** (d)
17. (b) **18.** (c) **19.** (a) **20.** (c) **21.** (a) **22.** (b)

REVIEW QUESTIONS

1. What is the effect of clearance volume in a reciprocating compressor?
2. Define volumetric efficiency of a reciprocating compressor.
3. What is the effect of pressure ratio on volumetric efficiency of a reciprocating compressor?
4. What are the advantages of multistage compression?
5. What is the condition for minimum work done by a two - stage reciprocating compressor with perfect intercooling?
6. What is the hermetically sealed compressor? What are its advantages?
7. What is a condenser? What is its function in a refrigeration system?
8. Explain the principle of working of a condenser.
9. Define heat rejection factor of a condenser.
10. What are the various types of condensers?
11. How an evaporative condenser works?
12. What is the function of a cooling tower?
13. Differentiate between flooded and non-flooded shell and tube type evaporators.

14. What is the main function of an evaporator?

15. What are frosting and defrosting evaporators?

16. What is a capillary tube? What is its use?

17. Why a capillary tube is preferred as a throttling device in household refrigerators?

18. Differentiate between low-side and high-side float valve.

NUMERICAL PROBLEMS

9.1 A reciprocating compressor compresses 1.2 m^3/min of a refrigerant vapour from 1 bar to 5 bar. The clearance is 5% and compression index is 1.35. Calculate the percentage change in compression work if the compression index is changed to 1.25.

9.2 A reciprocating compressor compresses a refrigerant from 1 bar to 5 bar with a compression index 1.35 and 5% clearance. Calculate the volumetric efficiency.

9.3 A single cylinder, single acting reciprocating compressor using R-12 as refrigerant has a base 100 mm and stroke 80 mm. The compressor runs at 1440 rpm. If the condensing temperature is 35°C, calculate the mass of refrigerant circulated per minute and the refrigerating capacity of the compressor when the suction temperature is (a) 10°C, and (b) –10°C. Assume simple cycle with no clearance.

9.4 Repeat problem 9.3 when clearance factor is 5% and index of isentropic compression is 1.135.

9.5 A two cylinder R-12 reciprocating compressor has a base 100 mm and stroke 80 mm. The compressor runs at 1440 rpm. The condensing temperature is 40°C. Calculate the mass of the refrigerant circulated per minute and refrigerant capacity of the compressor when the evaporating temperature is (a) 5°C, and (b) –5°C. Neglect clearance.

9.6 What are the important equipments being used in refrigeration and air-conditioning systems? Describe the basic function and significance of each equipment in brief.

9.7 Explain the importance of Expansion Devices in refrigeration Systems. What are the different types of expansion devices generally used in a refrigeration system? What are the advantages of a capillary tube over the other expansion devices in a refrigeration system?

9.8 Enumerate the functions of expansion devices in a vapour-refrigeration system and explain the working principle of a thermostatic expansion valve, with the help of a neat sketch, in brief.

9.9 What are the different types of condensers used in the refrigeration systems?

9.10 What do you understand by hermetically sealed compressors and what are its main advantages over ordinary coupled units?

9.11 With the help of a neat diagram, explain the working of an automatic expansion valve used for refrigerant flow control.

Chapter 10

CRYOGENICS

10.1 INTRODUCTION

The production of very low temperatures from –100°C to –273°C or absolute zero may be treated as cryogenics. The term "Cryogenics" is derived from the Greek word Kayos which means cold or frost.

The various activities associated with cryogenics are:

1. Cold treatment of metals.
2. Liquefaction and separation of gases.
3. Environmental testing of processes and products.

The various methods used for the production of low temperature are:

1. Vapour compression cascade systems. These methods have been discussed in chapter 6.
2. Solid carbon dioxide or dry ice production.
3. Liquefaction of gases by using Joule- Thomson principle, Linde system, Claude System.
4. Liquefaction of hydrogen.

10.2 LIMITATIONS OF VAPOUR COMPRESSION SYSTEM FOR CRYOGENICS

The evaporator temperature is limited to –40°C for different refrigerants with single compressor vapour compression system. Below this temperature, it has the following drawbacks:

1. The refrigerant must have a freezing point much below the required temperature to be attained. The freezing temperature of some of the refrigerants commonly used for vapour compression systems are given in Table 10.1.

Table 10.1 Freezing temperatures of commonly used refrigerants

Refrigerant	R-11	R-12	R-13	R-21	R-22	R-113	R-114	R-717 (NH_3)	R-744 (CO_2)	R-764 (SO_2)
Freezing temperature at atmospheric pressure, °C	–111	–157.5	–181	–135	–160	–35	–94	–77.8	–56-7	–75.6

Thus R-717, R-744, R-764 and R-113 cannot be used for cryogenics.

2. For high boiling point refrigerants, the pressure in the evaporator will be very low and suction volumes very large. For low boiling point refrigerants the condenser pressure will be very excessive. For some vapour compression refrigerants the suction pressure and specific volume at –40°C and condenser pressure at 40°C are given in Table 10.2.

Table 10.2 Suction pressure, specific volume and condenser pressure for typical refrigerants

Refrigerant	*R-11*	*R-12*	*R-13*	*R-21*	*R-22*	*R-113*	*R-114*	*R-717 (NH_3)*	*R-744 (CO_2)*
Pressure at -40°C, bar	0.05088	0.64152	6.0699	0.0956	1.0527	0.02858 (–30°C)	0.1303	0.7154	10.05
Specific volume at -40°C, m^3/kg	2.7642	0.24264	0.02609	2.004	0.2048	3.7614	0.86072	1.5567	0.03826
Pressure at +40°C, bar	1.7483	9.5944	38.7 (at+28.9 °C)	2.95867	15.34	0.78278	3.3941	15.555	73.825 (at 31.06°C)
Pressure ratio	34.36	14.9557	6.3757	30.948	14.57	27.389	26.048	21.74	7.346

Thus, except for R-13, R-744 and R-22, the suction pressure is below atmospheric even at –40°C. Also except for R-13, R-22 and R-12, the specific volumes at suction (–40°C) are very high.

3. Due to very high pressure ratios, the COP is very low.
4. At very low temperatures the difficulties are encountered in the operation of mechanical equipment.
5. Vapour compression system can operate upto –100°C with multi-stage or cascade systems.

10.3 SOLID CARBON DIOXIDE OR DRY ICE

The solid behavior under the influence of heat absorption depends upon the relation of pressure at which it exists to the pressure of triple point. In case the pressure is below the triple point the solid will be saturated and any heat addition will evaporate it directly from solid phase to gaseous phase without passing through the liquid phase. This process is called sublimation.

For CO_2, the triple point is –56.66°C at 5.178 bar. At atmospheric pressure (1.01325 bar) it exists as solid at –78.33°C.

10.3.1 Advantages and Disadvantages of Solid CO_2

Advantages

1. It is non-corrosive and non-toxic.
2. It can be easily cut and handled.
3. It has lower temperature than water ice and occupies smaller space for a required cooling capacity.
4. It can be used to maintain a wide range of temperatures required for the preservation of frozen perishables and foods.

Disadvantages

1. It has high cost of production.
2. It requires careful control due to its very low temperature.

10.3.2 Principle of Production of Dry Ice

Carbon dioxide can be liquefied by compressor to pressure of 60 to 70 bar and then condensed by using water. Solidification of liquid CO_2 can be done by using the following methods:

1. By removing the latent heat of solidification using auxiliary refrigeration. This method is uneconomical because of very low temperature requirement of the auxiliary refrigeration system.
2. By removing the heat through evaporation of a portion of liquid carbon dioxide. This method is not a practical means of production.
3. By throttling the liquid carbon dioxide to pressure below that of triple point. This method is widely used for production of dry ice.

The Schematic arrangement for production of dry ice is shown in Fig. 10.1. The corresponding *p-h* diagram is shown in Fig. 10.2 and the pressure temperature diagram in Fig. 10.3. The CO_2 in the form of gas is compressed to a sufficiently high pressure so that it can be condensed by cooling water available. This liquid which may even be subcooled as such as possible with

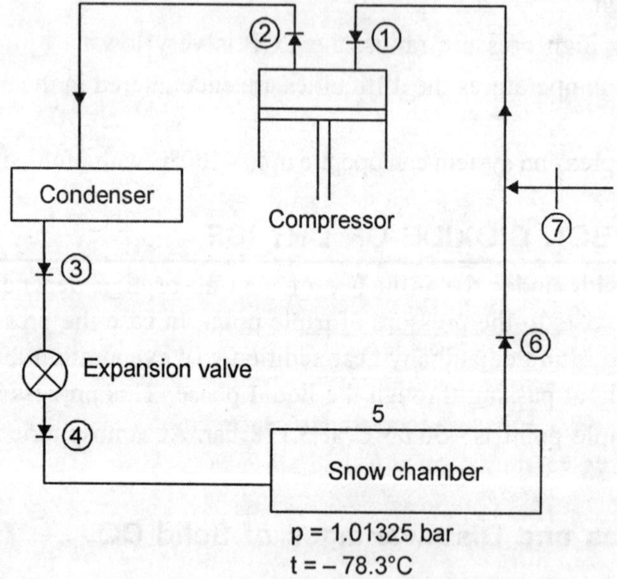

Fig. 10.1 Schematic diagram for production of dry ice.

available water, is then throttled through expansion valve into a snow chamber maintained at atmospheric pressure which is below the triple point pressure of CO_2. Thus after throttle expansion the CO_2 will exit as two phase mixture of solid carbon dioxide and gaseous carbon dioxide. Solid CO_2 can be removed and pressed mechanically in blocks as Dry ice. Make up CO_2 gas equal in weight to dry ice removed is fed in the cycle.

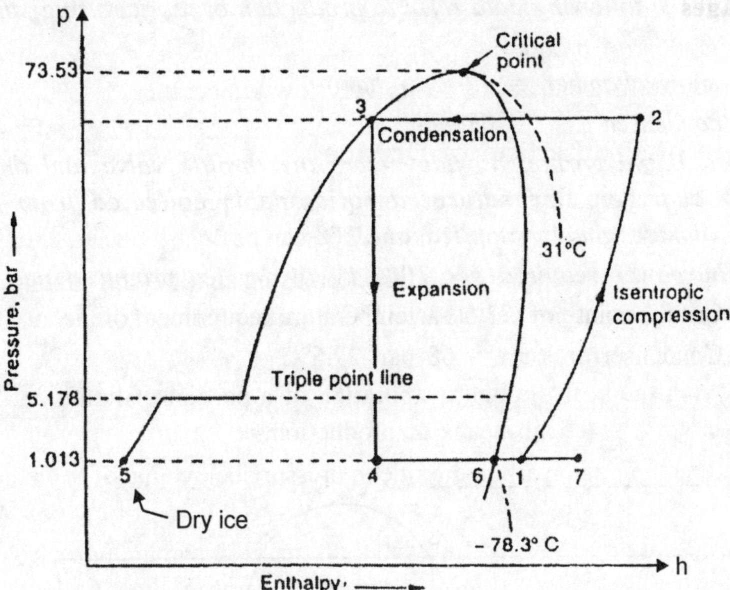

Fig. 10.2 *p-h* diagram for production of solid carbon dioxide or dry ice.

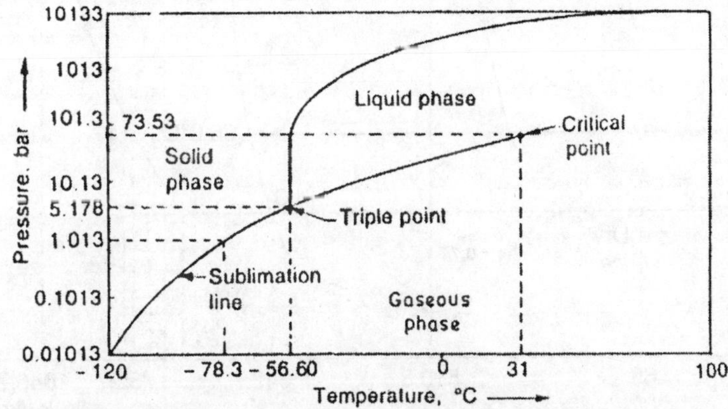

Fig. 10.3 Pressure-temperature diagram for carbon dioxide.

The various processes shown on *p-h* diagram are:

Process 1-2: Isentropic compression of CO_2 in a compressor.

Process 2-3: Condensation of CO_2 in the condenser by cooling water.

Process 3-4: Throttling of high pressure CO_2 in an expansion valve to pressure below the triple point pressure. The CO_2 at point 4 is a mixture of solid CO_2 and gaseous CO_2.

Point 5 : Solid CO_2 or dry ice can be removed at point 5.

Points 6 & 7 CO_2 gas removed at point 6 is mixed with make-up CO_2 gas at point 7 before being compressed at point 1.

Example 10.1 *The following data refer to production of dry ice using single compression system:*

Pressure in snow chamber = 1.01325 bar

Pressure in condenser = 68 bar

Dry saturated liquid carbon dioxide enters the throttle valve and dry saturated gas leaves the snow chamber. Dry saturated solid ice is collected from snow chamber. Make-up carbon dioxide gas is supplied at 28°C.

Determine the power required per 1000 kg of dry ice produced per hour.

Solution: Critical point = 73.5 bar, 31°C

Condenser pressure = 68 bar, 27.5°C

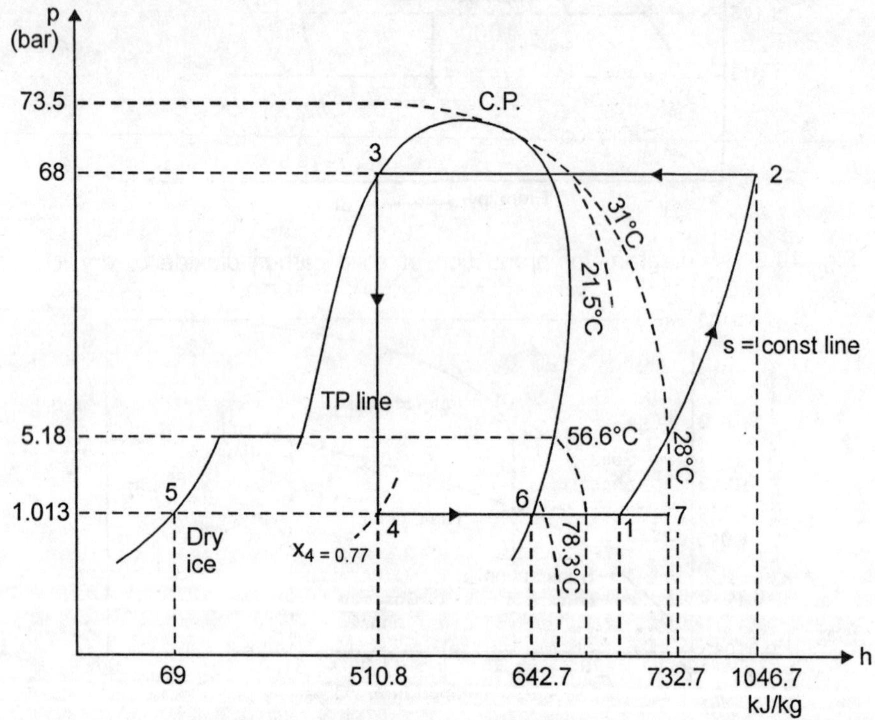

Fig. 10.4 *p-h* diagram

$$\text{Fraction of solid } CO_2 = \frac{4-6}{5-6}$$

$$\text{Fraction of gaseous } CO_2 = \frac{5-4}{5-6}$$

From the *p-h* chart for CO_2,

$$x_4 = 0.77, \text{ i.e. } 0.77 \text{ kg of gas}$$

compressed and 0.23 kg of solid/kg of gas compressed.

∴ Make-up CO_2 gas at 28°C = 0.23 kg per kg of gas compressed.

$$m_6h_6 + m_7h_7 = m_1h_1$$

$$0.77 \times 642.7 + 0.23 \times 732.7 = 1 \times h_1$$

$$h_1 = 663.4 \text{ kJ/kg}$$

$$h_2 = 1046.7 \text{ kJ/kg}$$

Work done/kg gas = $h_2 - h_1$ = 1046.7 – 663.4 = 383.3 kJ/kg of gas

= 383 kJ/0.23 kg of dry ice

For 1000 kg dry ice, power required

$$= \frac{383.3 \times 1000}{0.23 \times 3600} = 462.9 \text{ kW}$$

Refrigeration effect, $RE = 1000(h_7 - h_5) = 1000(732.7 - 69) = 663.7$ kJ

$$\text{COP} = \frac{663.7 \times 0.23}{383.3} = 0.398$$

10.4 THREE STAGE SYSTEM WITH WATER AND FLASH INTERCOOLING

The single-stage compression system for the production of dry ice is very uneconomical due to high consumption of power, low COP and high pressure ratio (~ 70). The production cost of solid CO_2 can be reduced by using the following modifications:

1. Usc of multistage compression with intercoolers to reduce cost of compressions
2. Snow chamber may be operated at a pressure only slightly below that corresponding to the triple point instead of atmospheric pressure.
3. Using flash chambers for sub-cooling the liquid and de-superheating the vapour.

A diagrammatic layout of three-stage compression system with water and flash intercooler for production of solid CO_2 is shown in Fig. 10.5 and the corresponding *p-h* diagram in Fig. 10.6.

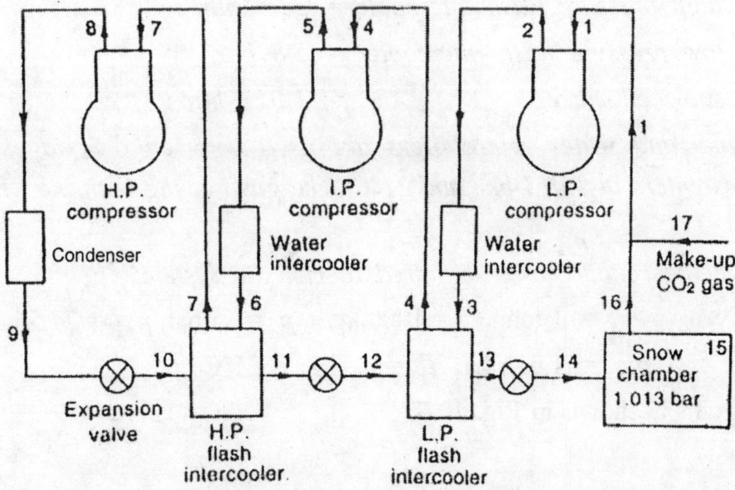

Fig. 10.5 Three stage compression system with water and flash intercooler for production of solid carbon dioxide or dry ice.

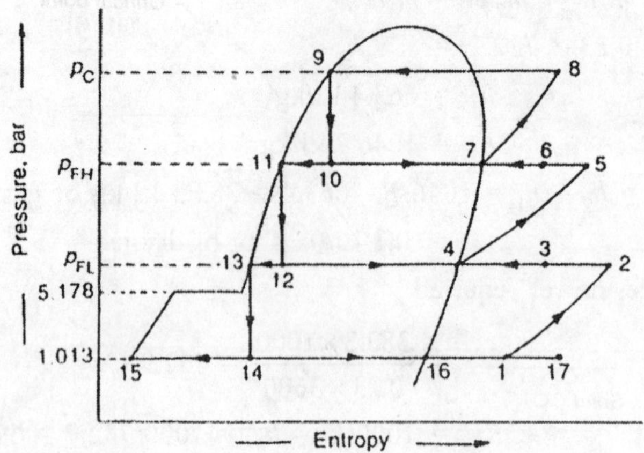

Fig. 10.6 p-h diagram of three stage compression system with water and flash intercooler for production of solid carbon dioxide or dry ice.

Another problem is the formation of solid snow in the expansion valve which will block the flow to the snow chamber. This may be eliminated by first producing liquid CO_2 in the low pressure flash intercooler at a pressure (say PFC) above the triple point pressure of 5.178 bar, and then reducing its pressure to one atmosphere (1:013 bar) in the snow chamber, as shown in Fig. 10.6. Thus the pressure ratio in the low pressure stage has to be increased. This increase in pressure ratio is compensated by the larger mass flow rate in higher pressure stages at lower pressure ratios.

Example 10.2 *The following data refer to a three-stage compression system for the manufacture of one tonne of solid carbon dioxide per hour:*

Condenser pressure = *65 bar*

Pressure in high pressure flash intercooler= 27.5 bar

Pressure in low pressure flash intercooler = *5.8 bar*

Pressure in snow chamber = *1.013 bar*

Both the flash and water intercoolers are used between the stages. The make-up carbon dioxide gas enters at 1.013 bar and 21°C. The gas leaving both the water intercoolers is at 21°C.

Calculate the theoretical power required to run the system.

Solution: Given: m_a = 1 tonne/h = 1000 kg/h, p_c = 65 bar, p_{FH} = 27.5 bar, p_{FL} = 5.8 bar,

p_{SC} = 1.013 bar, $t_{17} = t_3 = t_6$ = 21°C

The *p-h* diagram is shown in Fig. 10.7.

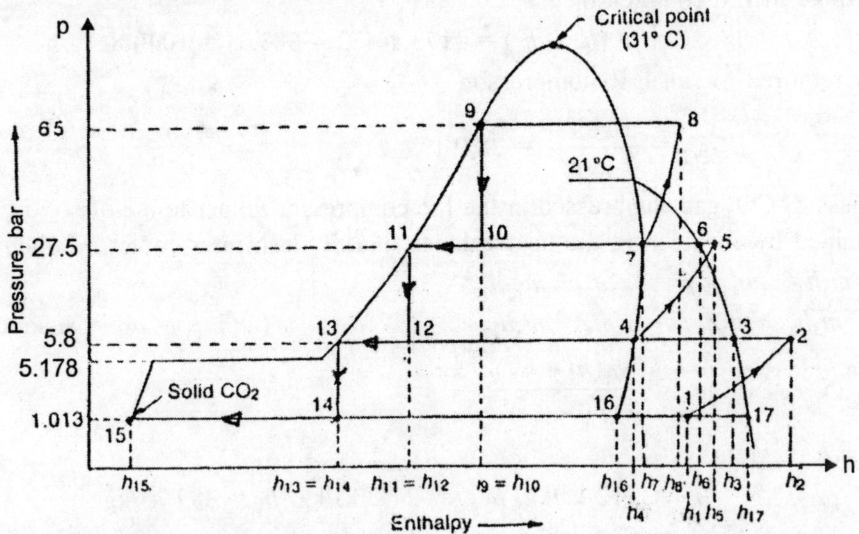

Fig. 10.7 *p-h* diagram for a three-stage system with flash and water intercoolers.

At point 14, the CO_2 is a mixture of solid CO_2 and gaseous CO_2. The solid CO_2 is removed at point 15 and the gaseous CO_2 at point 16.

Let m_{14} = mass of solid CO_2 + mass of gaseous CO_2

$$= m_{15} + m_{16}$$

$$= 1000 \text{ kg/h} + m_{16}$$

or $m_{16} = m_{14} - 1000$

For the heat balance of snow chamber, we have

$$m_{14}h_{14} = m_{15}h_{15} + m_{16}h_{16}$$

From *p-h* diagram of CO_2,

$$h_{14} = -24.5 \text{ kJ/kg}, \; h_{15} = -82.5 \text{ kJ/kg}, \; h_{16} = 310 \text{ kJ/kg}$$

$\therefore \quad m_{14} \times (-24.5) = 1000 \times (-82.5) + (m_{14} - 1000) \times 310$

or $\quad -24.5m_{14} = -82500 + 310m_{14} - 310{,}000$

$$m_4 = 1173.4 \text{ kJ/h}$$

and $\quad m_6 = 1173.4 - 1000 = 173.4 \text{ kg/h}$

To find the enthalpy at point 1, consider the heat balance between points 1, 16 and 17.

$$m_1h_1 = m_{16}h_{16} + m_{17}h_{17}$$

Now $\quad m_{17} = m_{15} = 1000 \text{ kg/h}$ and $m_1 = m_{14} = 1173.4$ kg/h

Also from *p-h* diagram,

$$h_{17} = 396 \text{ kJ/kg}$$

$\therefore \quad 1173.4h_1 = 173.4 \times 310 \times 1000 \times 396$

$$h_1 = 383.3 \text{ kJ/kg}$$

From *p-h* diagram,

$$h_2 = 472 \text{ kJ/kg}$$

Work done in L.P compressor,

$$W_{LPC} = m_1(h_2 - h_1) = 1173.4(472 - 383.3) = 104080 \text{ kJ/h}$$

Power required to run L.P. compressor,

$$P_{LPC} = \frac{104080}{3600} = 28.9 \text{ kW}$$

The mass of CO_2 gas compressed in the I.P. compressor m_4 per tonne of solid CO_2 per hour may be obtained by considering the heat balance of L.P. flash inter cooler. Mathematically,

$$m_4h_4 + m_{13}h_{13} = m_1h_3 + m_{12}h_{12}$$

or $$m_4h_4 + m_1h_{13} = m_1h_3 + m_4h_{12} \qquad (\because m_{13} = m_3 = m_1 \text{ and } m_{12} = m_4)$$

$$m_4 = \frac{m_1(h_3 - h_{13})}{(h_4 - h_{12})}$$

From *p-h* diagram,

$$h_4 = 312 \text{ kJ/kg}, h_{13} = -24.2 \text{ kJ/kg}, h_3 = 387 \text{ kJ/kg}$$

$$h_{12} = 63.5 \text{ kJ/kg}, h_5 = 382 \text{ kJ/kg}$$

$$m_4 = \frac{1173.4(387 - 24.2)}{(312 - 63.5)} = 1713.1 \text{ kg/h}$$

Work done in I.P. compressor,

$$W_{IPC} = m_4(h_5 - h_4)$$

$$= 1713.1\,(382 - 312) = 119917 \text{ kJ/h}$$

Power required to run I.P. compressor,

$$P_{IPC} = \frac{119917}{3600} = 33.3 \text{ kW}$$

To find the mass m_7 of CO_2 compressed in the H.P. compressor per tonne of solid CO_2 per hour, consider the heat balance of high pressure flash intercooler. Mathematically,

$$m_7h_7 + m_{11}h_{11} = m_6h_6 + m_{10}h_{10}$$

or $$m_7h_7 + m_4h_{11} = m_4h_6 + m_7h_{10} \qquad (\because m_{11} = m_6 = m_4 \text{ and } m_{10} = m_7)$$

$$\therefore \quad m_7 = \frac{m_4(h_6 - h_{11})}{(h_7 - h_{10})}$$

From p-h diagram, $h_6 = 360$ kJ/kg, $h_{11} = 63$ kJ/kg, $h_7 = 318$ kJ/kg,

$$h_{10} = 158 \text{ kJ/kg}, h_9 = 354 \text{ kJ/kg}$$

$$m_7 = \frac{1713.1(360 - 63)}{(318 - 158)} = 3180 \text{ kg/h}$$

Work done in H.P. compressor,

$$W_{HPC} = m_7(h_8 - h_7) = 3180\,(354 - 318) = 114480 \text{ kJ/h}$$

Power required to sum the H.P. compressor,

$$P_{HPC} = \frac{114480}{3600} = 31.8 \text{ kW}$$

Total required to sum the system,

$$P = P_{LPC} + P_{IPC} + P_{HPC}$$
$$= 28.9 + 33.3 + 31.8 = 94 \text{ kW}$$

10.5 LIQUEFACTION OF GASES

The gas must be liquefied for producing low temperature for refrigeration purposes. The gases may be expanded by the following two methods to produce refrigeration:

1. **Isentropic expansion:** This method is used in air refrigeration cycle for aircraft refrigeration systems. Its use is limited to high temperatures since very low temperature of the gas may be obtained.
2. **Free, irreversible expansion:** The free irreversible expansion of a perfect gas is an isothermal process. But no real gas is truly perfect and there is actual drop in temperature experienced upon free expansion between two pressures. This may be used as a measure of the degree of imperfection of that gas and as a means of imperfection.

The change in temperature with drop in pressure at constant enthalpy is termed as Joule-Thomson coefficient (μ).

Mathematically, $$\mu = \left(\frac{dT}{dp}\right)_h$$

μ varies with both the temperature and pressure of the gas.

The magnitude of μ is a measure of the imperfection of a gas as its deviation from perfect gas behaviour. μ for real gases may be either positive or negative.

(*i*) The temperature at which $\mu = 0$, is called the inversion temperature for a given pressure.

(*ii*) If $\mu > 0$, the temperature of gas decreases with throttling.

(*iii*) If $\mu < 0$, the temperature of gas increases with throttling.

Thus, in cooling of a gas by throttling, gas should show a large positive value of μ.

The value of μ is negative for all gases under high temperatures and pressures. Under these conditions, the throttling results in warming of the gas. The inversion temperature for most common gases are above those of temperature normally encountered. Therefore, if throttling occurs at ordinary temperatures and from moderately high pressures, then cooling effect results. However, hydrogen and helium behave at ordinary temperatures as most gases do at high temperatures, since their temperatures are –77.8°C and –250°C respectively. Therefore, if throttling of these gases is to be used to accomplish refrigeration, it is first necessary to cool them, by some other means, below their inversion temperature.

10.6 LINDE SYSTEM FOR LIQUEFACTION OF AIR

The schematic arrangement of Linde system and its corresponding T-s diagram are shown in Fig. 10.8 (*a*) and (*b*) respectively. The equipment includes a *compressor*, a *heat exchanger* and a *separator*. The working of the system is as follows:

- **Process 1-2:** The atmospheric air is compressed isothermally to pressure of 100 to 200 atmosphere.

- **Process 2-3:** The high pressure air is cooled to about – 106.7°C in the heat exchanger.
- **Process 3-4:** The cooled air from the heat exchanger is throttled to atmospheric pressure and a temperature of – 190°C.
- **At point 5:** A portion of the air which is liquefied is removed from the separator. The remainder of the cold air leaves the separator at **point 6** and returns to the compressor through the heat-exchanger where it cools the incoming high pressure air.
- **At point 7** The make-up air from the atmosphere (equal to the amount of air liquefied) is supplied to the compressor along with the air removed from the heat exchanger.

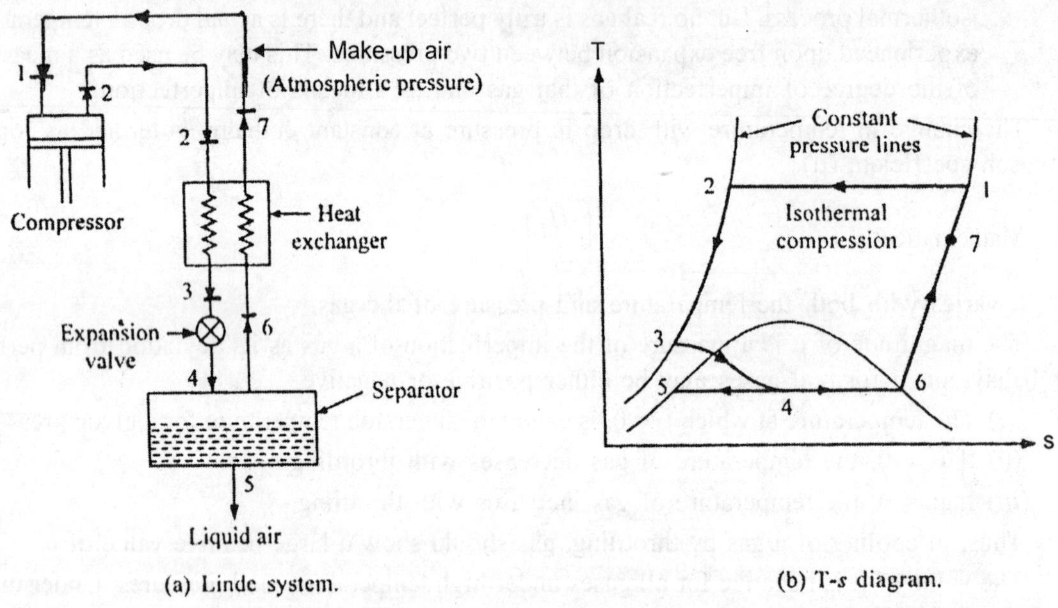

Fig. 10.8 Linde system for liquefaction of air.

Let m_2 = Mass of air compressed in the compressor (or mass of high pressure air passing through the heat exchanger, between points 2 and 3),

m_5 = Mass of air liquefied in the separator, and

m_6 = Mass of low pressure air passing through the heat exchanger between points 6 and 1

$= (m_2 - m_5)$

Considering *heat balance of heat exchanger,* we have:

$$m_2(h_2 - h_3) = m_6(h_1 - h_6) = (m_2 - m_5)(h_1 - h_6) \qquad (\because m_6 = m_2 - m_5) \quad ...(1)$$

Considering heat balance of the separator, we get

$$m_2 h_4 = m_5 h_5 + m_6 h_6 = m_5 h_5 + (m_2 - m_2)\, h_6 \quad ...(2)$$

For throttling process 3-4, we have:

$$h_3 = h_4 \qquad ...(3)$$

The mass of air liquefied by the system may be obtained from the eqns. (1) and (2).

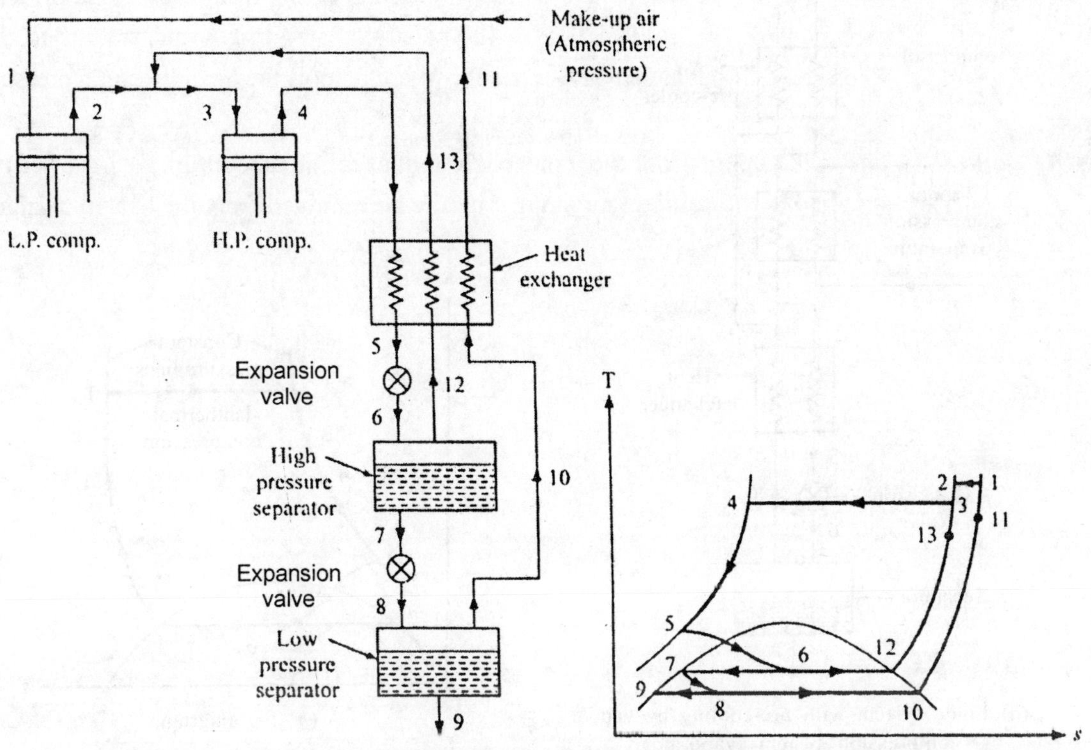

(*a*) Schematic arrangement for dual pressure Linde system. (*b*) T-*s* diagram for dual pressure Linde system.

Fig. 10.9 Dual pressure Linde system.

The simple Linde system is comparatively *inefficient* and is employed only when *quantity of liquid air need is small*. In order to get an increased efficiency, a dual pressure Linde system is employed as shown in Fig. 10.9 (a). Its corresponding T-s diagram is shown in Fig. 10.9 (b). In this system, a portion of air after the first throttling process from point 5 to point 6 is *bled back* through a heat exchanger into the discharge of the first (low pressure) stage of compression. This system is more economical since the majority of the air undergoes only one expansion to the intermediate pressure and one compression from this intermediate pressure to the high pressure.

Both the systems desctibed above may be improved if the *high pressure compressed air is precooled by an evaporator of a vapour compression system* as shown in Fig. 10.10. The best performance is obtained when the compressed air is precooled to a relatively low temperature.

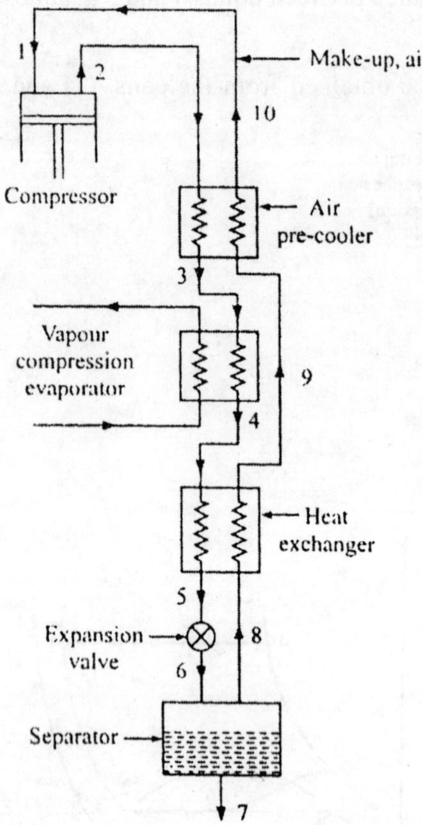

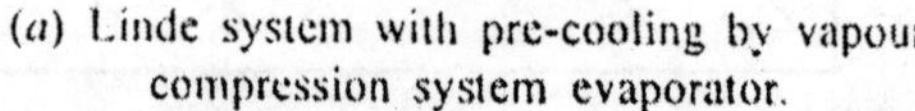
(a) Linde system with pre-cooling by vapour compression system evaporator.

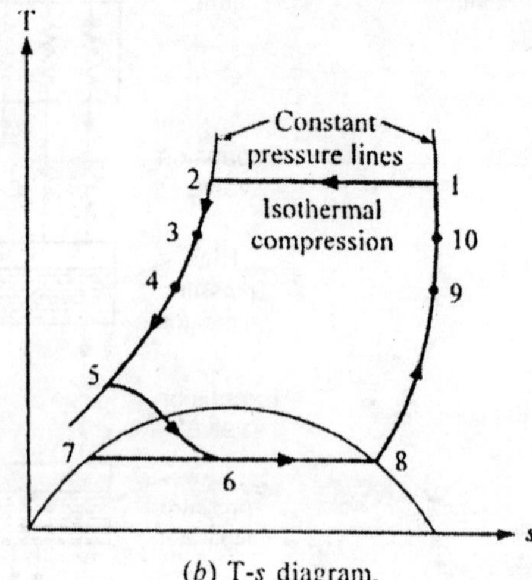

(b) T-s diagram.

Fig. 10.10

10.7 CLAUDE SYSTEM FOR LIQUEFACTION OF AIR

The Claude system of liquefying air *utilizes the principle that a gas which performs external work as it expands experiences a greater drop in temperature than when it expands through throttling valve.* In this process, a large part of the air is expanded in an engine which transfers the energy produced to the compressor to compress the incoming air. Thus energy removed from expanding air *represents extra refrigeration produced in the cycle.*

The schematic diagram of the system is shown in Fig. 10.11(*a*) and its T-s diagram is shown in Fig. 10.11(*b*).

The *working of the system* is as follows:

- At point 3 about 80% of air is *bled* and cooled by expansion between 3 and 8. The remaining 20% portion of air passes through second heat exchanger between point 3 and 4.

- The air from the second heat exchanger is throttled between points 4 and 5 at atmospheric pressure.
- At point 6, the liquid air is removed from the separator.
- The unliquefied air at point 7 is mixed with the low temperature air from the expander at point 8, giving increased mass flow of air at point 9. This air passes back through the two heat exchangers to the compressor.

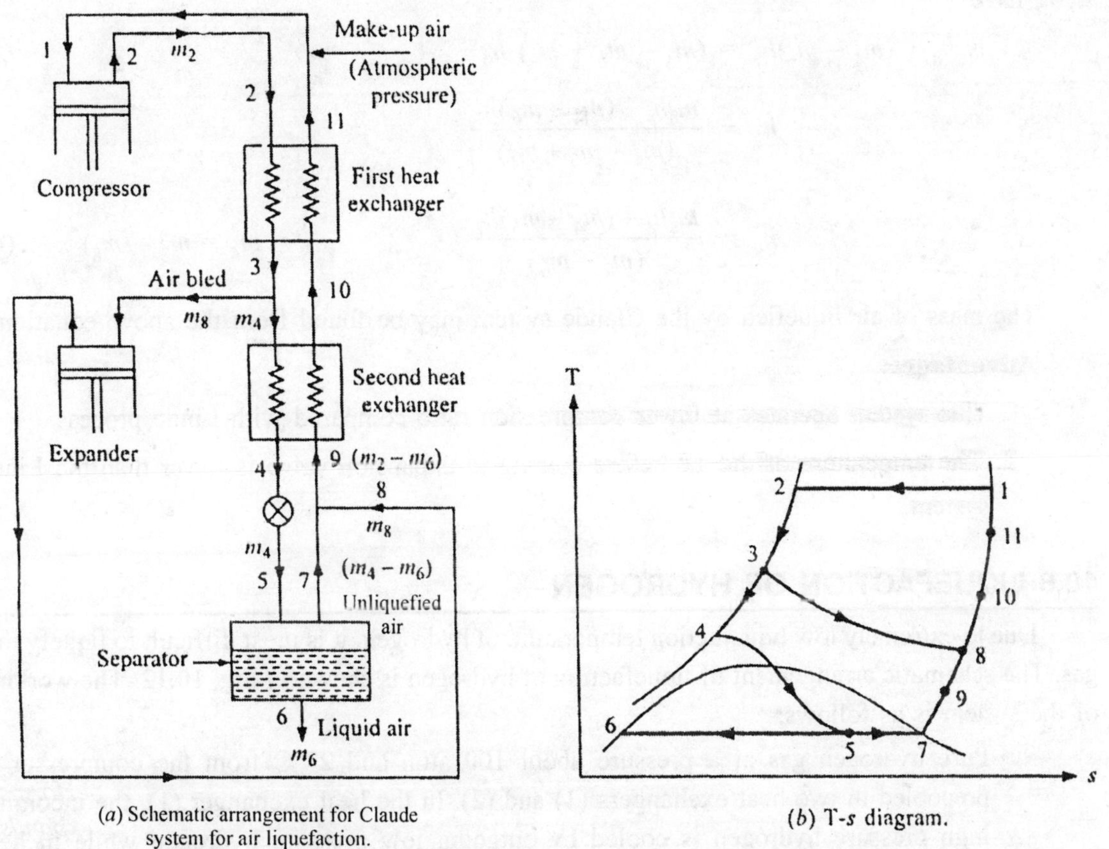

(*a*) Schematic arrangement for Claude system for air liquefaction.

(*b*) T-*s* diagram.

Fig. 10.11 Claude system for liquefaction of air.

Let m_2 = Mass of compressed air,

m_8 = Mass of air bled/passed to expander,

m_4 = Mass of air passing through second heat exchanger, and

$= (m_2 - m_8)$, and

m_6 = Mass of air liquefied in the separator.

Considering the heat balance of first heat exchanger, we have:

$$m_2 = (h_2 - h_3) = (m_2 - m_6)(h_1 - h_{10}) \quad ...(1)$$

Considering the heat balance of the second heat exchanger, we get:

$$m_4(h_3 - h_4) = (m_2 - m_6)(h_{10} - h_9) \quad ...(2)$$

Considering the heat balance of the separator, we have:

$$m_4 m_5 = m_6 h_6 + (m_4 - m_6)\, h_7 \quad ...(3)$$

Considering the heat balance resulting from mixing of air from the expander at point 8 with unliquefied air from the separator at point 7, the combined condition of air represented by point 9, we have:

$$m_8 h_8 + (m_4 - m_6)h_7 = (m_4 - m_6 + m_8)\, h_9$$

$$\therefore \quad h_9 = \frac{m_8 h_8 + (m_4 - m_6)h_7}{(m_4 - m_6 + m_8)}$$

or

$$h_9 = \frac{m_8 h_8 + (m_4 - m_6)h_7}{(m_2 - m_6)} \quad (\because\ m_4 = m_2 - m_8) \quad ...(4)$$

The mass of air liquefied by the Claude system may be found from the above equations.

Advantages:

1. This system operates at lower compression ratio compared with Linde process.
2. The temperature of the air before coming to expansion valve is lower than the Linde system.

10.8 LIQUEFACTION OF HYDROGEN

Due to extremely low liquefaction temperature of hydrogen, it is most difficult to liquefy this gas. The schematic arrangement of liquefaction of hydrogen is shown in Fig. 10.12. The working of the system is as follows:

- Pure hydrogen gas at a pressure about 100 atm. and 27°C from the compressor is precooled in two heat exchangers (1) and (2). In the heat exchanger (1), the incoming high pressure hydrogen is cooled by outgoing low pressure hydrogen while in heat exchanger (2), it is cooled by nitrogen
- The high pressure hydrogen gas from both the heat exchangers is passed through heat exchanger (3) where the hydrogen is further cooled to about –209°C *by nitrogen boiling under reduced pressure.*
- The hydrogen gas is further cooled to about – 230°C in the heat exchanger (4) by the low pressure hydrogen gas returning from the separator.
- The liquid hydrogen is produced by throttling the hydrogen gas from the heat exchanger (4) to atmospheric pressure.

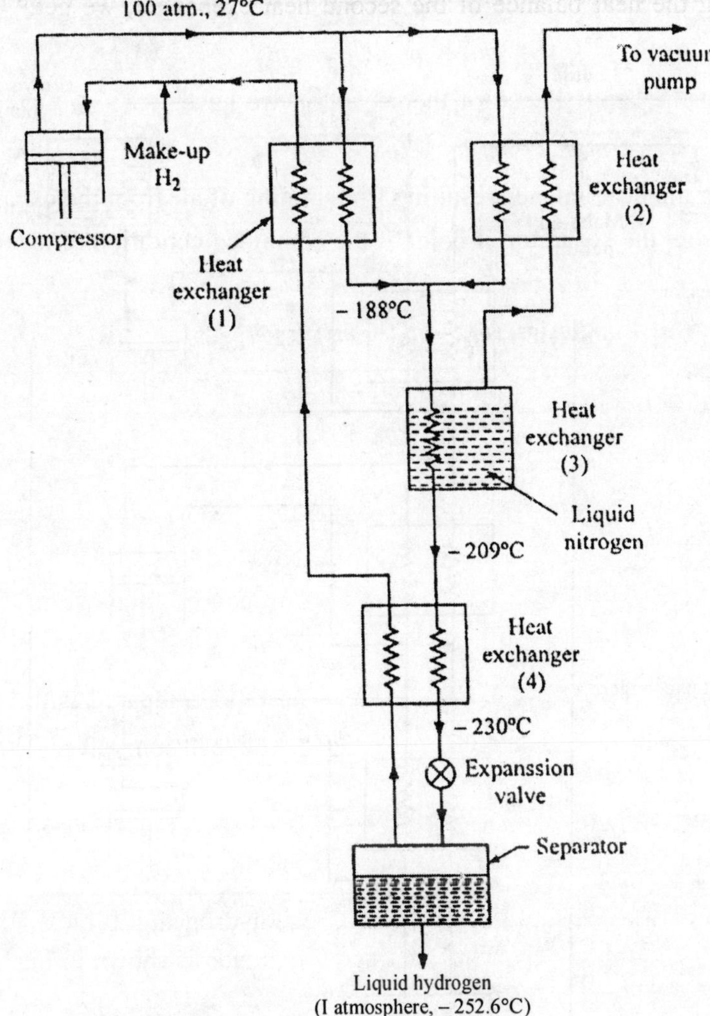

Fig. 10.12. Hydrogen liquefier.

10.9 LIQUEFACTION OF HELIUM

Helium is most difficult among all gases to liquefy. It boils at –268.8°C at atmospheric pressure and its maximum inversion temperature is –233°C. Fig. 10.13 shows S.C. Collins helium liquefier. The working of this system is as follows:

- Helium gas at a pressure of 12 atm. (approx) is supplied to the liquefier by a four-stage compressor.
- A part of helium is *precooled by liquid nitrogen*.
- The high pressure helium gas may be cooled to –257°C by the use of *heat exchangers*. (HE-1, HE-2, HE-3, HE-4) and expanders (EP-1, EP-2, EP-3 and EP-4) as per arrangement shown in Fig. 10.14).
- The helium gas obtained above is then *throttled* to atmospheric pressure to produce liquid helium at –268.8°C.

It is found that one such unit may provide a liquefaction rate of 22 to 28 litres per hour with 40 kW power.

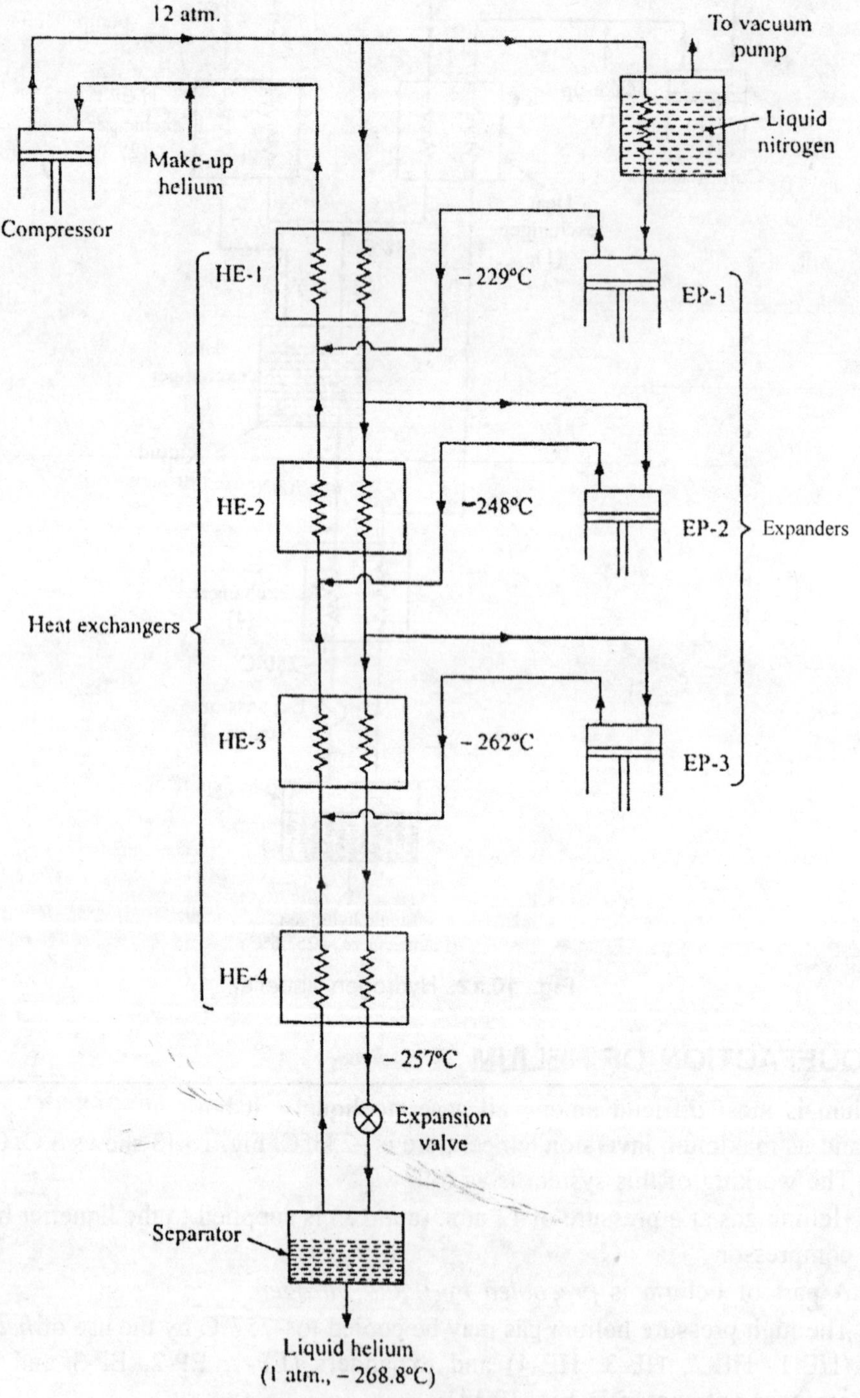

Fig. 10.13 S.C. Collins' helium liquefier

10.10 APPLICATIONS OF LOW TEMPERATURE

Following are a *few applications of low temperatures*:

1. Expansion fitting;
2. *Cryobiology* (This term relates to the effects of low temperatures on materials of biological origin);
3. *Semen preservation*;
4. *Cryosurgery*;
5. *Space research*;
6. *Computers*;
7. *Underground power lines*;
8. *Miscellaneous uses*:
   - *Infrared detectors;*
   - *Frictionless superconductive magnetic bearings* etc.

MULTIPLE - CHOICE QUESTIONS

1. Liquid oxygen boils at

   (a) 20.1 K (b) 51.1 K (c) 70.4 K (d) 90.2 K

2. The triple point for carbon dioxide at 5.178 bar is

   (a) –20.4°C (b) –40.6°C (c) –56.6°C (d) –76.4°C

3. In the simple Linde system, the high pressure air from the compressor is cooled in a heat exchanger to a temperature of

   (a) –53.6°C (b) –72.8°C (c) –106.7°C (d) –138.5°C

4. The following temperature falls in the field of cryogenics

   (a) 123 K (b) 133 K (c) 143 K (d) all of the above

5. The Joule-Thomson principle states that real gas enthalpy is a function of

   (a) temperature alone (b) pressure alone

   (c) both pressure and temperature (d) internal energy alone

6. In the Linde system of liquefaction of gas the compression process of gas is

   (a) adiabatic (b) polytropic (c) isentropic (d) isothermal

Answers

1. (d) 2. (c) 3. (c) 4. (a) 5. (c) 6. (d)

REVIEW QUESTIONS

1. What do you understand by cryogenics?
2. What are the limitations of vapour compression refrigeration system for production of low temperatures?
3. What is dry ice? What are its advantages?
4. Explain the working of a system for the production of dry ice.
5. Explain the two methods used for liquefaction of gases.
6. Explain the Linde system for liquefaction of gases.
7. Describe the Claude system for liquefaction of gases.
8. Describe briefly the methods used for the liquefaction of (a) hydrogen gas, and (b) helium gas.
9. What are the applications of low temperature refrigeration?

NUMERICAL PROBLEMS

10.1 The following data refer to a three-stage compression system for the manufacture of 1000 kg/h of solid CO_2:

Condenser pressure = 65.3 bar

Pressure in the high pressure flash intercooler = 28 bar

Pressure in the low pressure flash intercooler = 5 bar

Pressure in the snow chamber = 1.013 bar

Both the flash and water intercoolers are used between the stages. The gas leaving both the water intercoolers is at 21°C.

Calculate the theoretical power required to run the system. [**Ans.** 102.8 kW]

10.2 Dry air at 20°C and 1 bar is to be liquified by the simple Linde method. The air is isothermally compressed at 20°C to 170 bar. The make-up air is supplied to the system at 20°C and 1 bar. Calculate the yield of liquid air in kg per kg of air compressed and the temperature of air before throttling. [**Ans.** 0.08 kg, –98°C]

10.3 Dry air at 20°C and 1 bar is to be liquified by the Claude system. The air is compressed isothermaly at 20°C to 170 bar. Assume that 80 percent of the total mass of air compressed passes through the expander. The temperature of air entering the expander is –80°C while the temperature of air leaving the expander is –140°C. The make-up air is supplied also at 20°C and 1 bar. Determine the yield of liquid air in kg per kg of air compressed and the temperature of air before throttling. [**Ans.** 0.014 kg, –100°C]

Chapter

11

STEAM JET REFRIGERATION

11.1 INTRODUCTION

In the steam jet refrigeration cycle, water is used as the refrigerant and the cooling effect is produced in the stream jet refrigeration continuous cycle by evaporation of a part of water in the evaporator at a low absolute pressure level. This is also known as ejector refrigeration system.

11.2 SYSTEM COMPONENTS

A steam-jet refrigeration system consists of an evaporator (or flash tank), a steam jet ejector (or booster ejector), a condenser, and two-stage ejectors non-condensable pump. Auxiliary components include a chilled water circulating pump, a cooling water circulating pump, a condensate pump, and necessary valves and controls. Fig. 11.1 shows a commonly used steam jet refrigeration system.

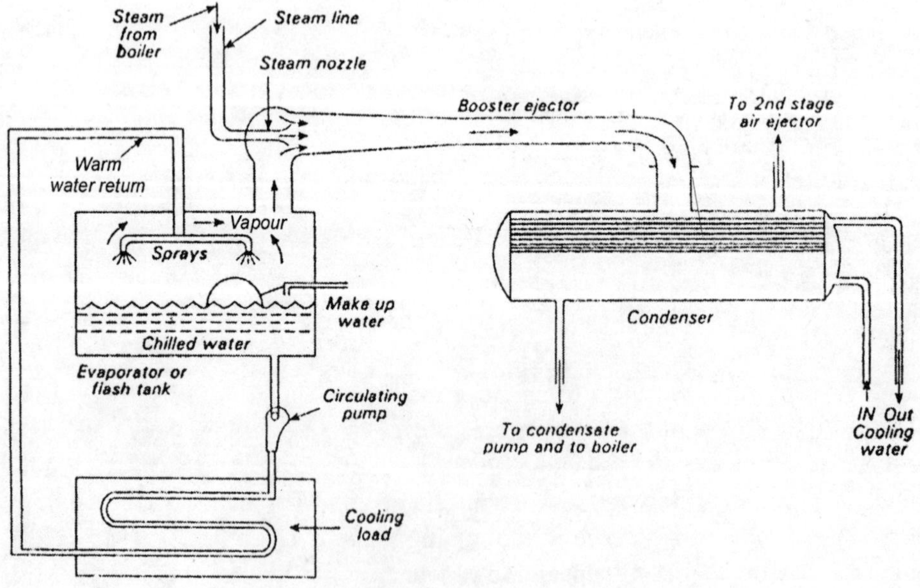

Fig. 11.1 Steam-jet refrigeration system.

1. Evaporator : The evaporator is a large volume vessel which must provide a large water surface area for efficient evaporative cooling action. The design of an evaporator requires water to be distributed in small droplets or thin sheets so that the main body of water can attain equilibrium temperature by the time it reaches the bottom of the evaporator. Evaporator can be classified into two types.

(*i*) **Open**—Chilled water system – in which refrigerant water from the evaporator is circulated directly to the cooling load, such as heat exchangers.

(*ii*) **Closed**—Chilled water system – in which the refrigerant water is recirculated within the evaporator over a oil or other heat transfer device through which is circulated another fluid (water or brine) which never mixes with the refrigerant water, and which in turn is pumped to the cooling pond.

The open-chilled water system is more efficient since one less stage of heat transfer is required.

2. Steam-Jet Ejector: A stream-jet ejector is shown in Fig. 11.2. Its various parts are:

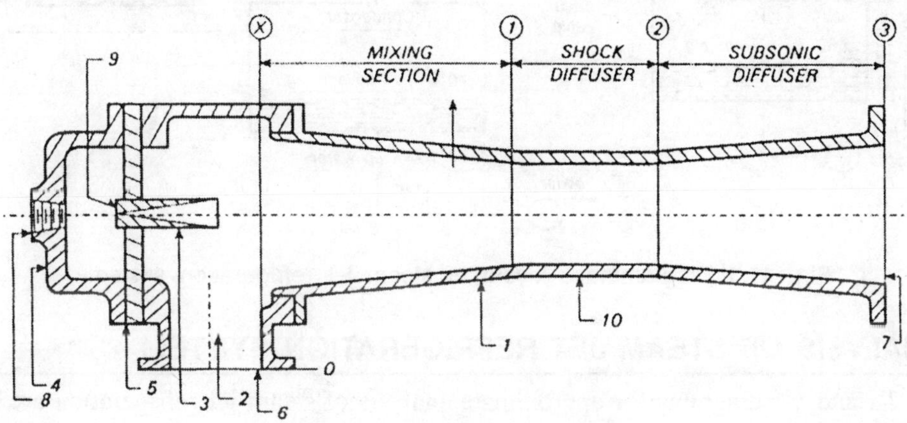

Fig. 11.2 Steam jet ejector

Legend:

1. Diffuser
2. Suction chamber
3. Steam nozzle
4. Steam chest
5. Nozzle plate
6. Suction
7. Discharge
8. Seam inlet
9. Nozzle throat
10. Diffuser throat

The high pressure steam used to energize the ejector is called the primary fluid or motive steam. The induced steam raised to high pressure is called the secondary fluid or vapour. The motive steam is expanded through a convergent-divergent nozzle to high velocities up to 1200 m/s. The high velocity steam issuing from the nozzle enterain the water vapour leaving the suction chamber, and the two streams merge in the mixing section. The mean velocity of the mixture is supersonic. The kinetic energy of supersonic vapour steam is recovered in the constant area throat section, followed by the subsonic diffuser.

3. Condenser: A conventional surface condenser is employed because of its compact size. A cooling tower is used when natural supply of cooling water is not available.

4. Air Ejector: The air ejector employs two small steam powered ejectors in series. An inter and after condenser is often used to improve economy.

A complete diagrammatic layout of steam jet refrigeration system is shown in Fig. 11.3.

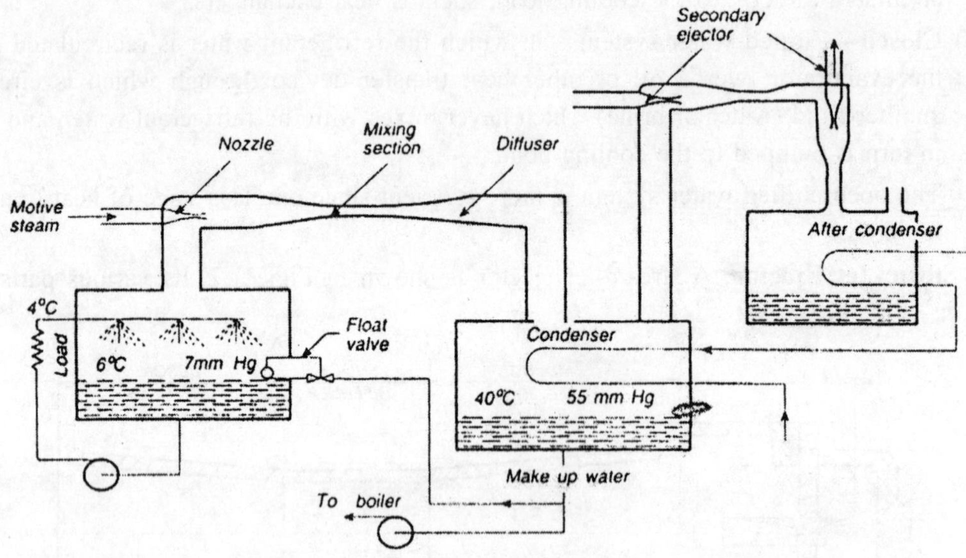

Fig. 11.3 Diagrammatic layout of steam jet refrigeration system.

11.3 ANALYSIS OF STEAM JET REFRIGERATION SYSTEM

The *T-s* and *h-s* diagrams for approximate analysis of steam jet refrigeration are shown in Fig. 11.4.

Let p_B = Pressure of motive steam from boiler.

p_E = Flash tank or evaporator pressure.

p_C = Condenser pressure.

1–2 = isentropic expansion of motive steam from boiler pressure to evaporator pressure.

1–2′ = actual irreversible adiabatic expansion of motive steam from p_B to p_E.

3 = state of dry saturated vapour formed in the evaporator.

1 = state of motive steam assumed dry saturated at boiler pressure.

2′ = state of steam leaving the nozzle.

4 = state of steam after the steam from nozzle and water vapour from evaporator have mixed at p_E.

4–5 = reversible adiabatic compression in steam jet ejector or booster.

4–5′ = actual irreversible adiabatic compression in steam jet ejector or booster.

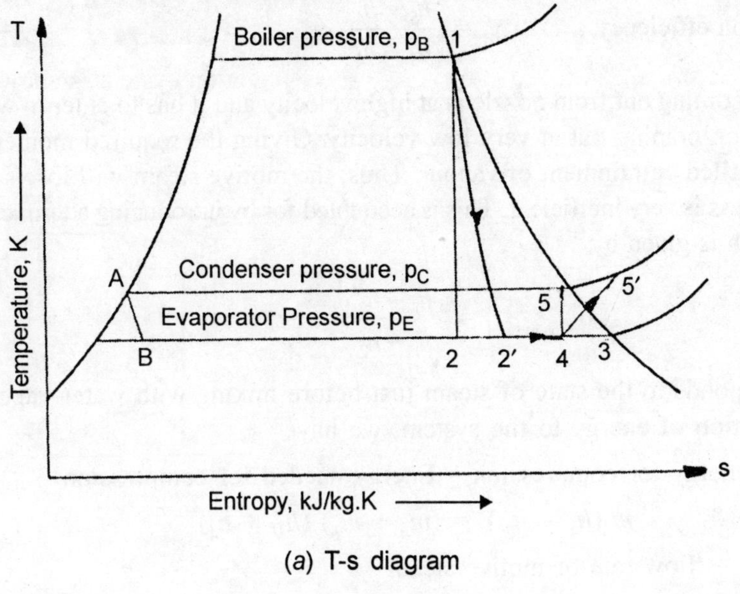

(*a*) T-s diagram

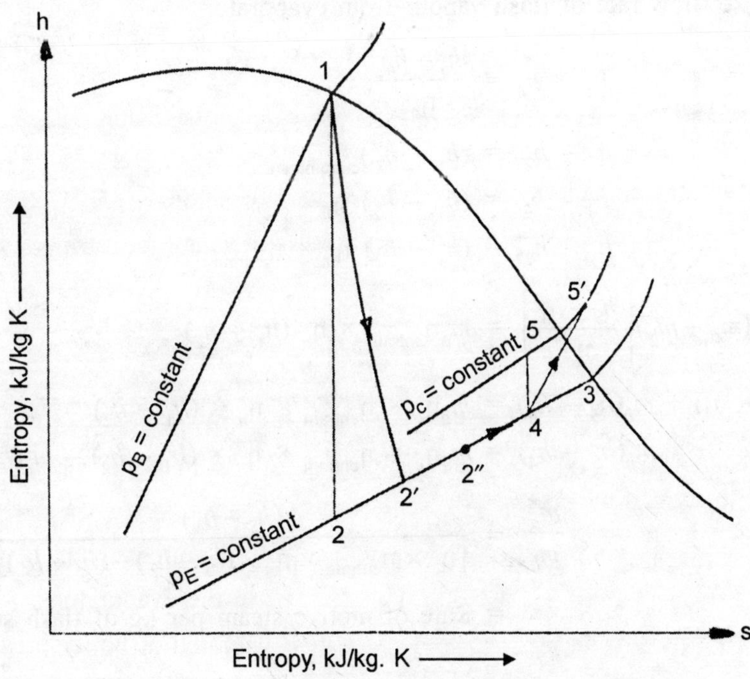

(*b*) h-s diagram

Fig. 11.4 Approximate analysis for steam jet refrigeration system.

Nozzle efficiency, $$\eta_n = \frac{h_1 - h_2'}{h_1 - h_2} \quad ...(11.1)$$

Compression efficiency, $$\eta_c = \frac{h_5 - h_4}{h_5' - h_4} \quad ...(11.2)$$

The steam coming out from nozzle is at high velocity and it has to enter in water vapour flash tank or evaporator, coming out at very low velocity. Giving the required momentum to the flash tank vapour is called entrainment of vapour. Thus, the motive steam will lose some energy. This entrainment process is very inefficient. This is accounted for by introducing a term called Entrainment Efficiency, which is given by,

$$\eta_{entrainment} = \frac{h_1 - h_2''}{h_1 - h_2'} \quad ...(11.3)$$

where 2″ corresponds to the state of steam just before mixing with water vapour. Applying the law of conservation of energy to the system, we have

Available energy for compression = Energy needed for compression

$$m_a(h_1 - h_{2''}) = (m_a + m_b)(h_{5'} + h_4) \quad ...(11.4)$$

where m_a = flow rate of motive steam

m_b = flow rate of flash vapour from evaporator

Also $$h_{5'} - h_4 = \frac{h_5 - h_4}{\eta_c} \quad ...(11.5)$$

and $$h_1 - h_{2''} = (h_1 - h_{2'})_{entrainment} \quad ...(11.6)$$

But $$h_1 - h_{2'} = (h_1 - h_2)\,\eta_n \quad ...(11.7)$$

∴ $$h_1 - h_{2''} = (h_1 - h_2)\,\eta_n \times \eta_{entrain}$$

Thus $$(m_a + m_b)\left[\frac{h_5 - h_4}{\eta_c}\right] = m_a\eta_{entrain} \times \eta_n\,(h_1 - h_2) \quad ...(11.8)$$

∴ $$m_a(h_5 - h_4) + m_b(h_5 - h_4) = m_a\eta_c \times \eta_{entrain} \times \eta_n \times (h_1 - h_2)$$

or $$m_b(h_5 - h_4) = m_a\eta_c \times \eta_{entrain} \times \eta_n \times (h_1 - h_2) - m_a(h_5 - h_4)$$

or $$\frac{m_a}{m_b} = \frac{(h_5 - h_4)}{[\eta_c \times \eta_{entrain} \times \eta_n \times (h_1 - h_2) - (h_5 - h_4)]} \quad ...(11.9)$$

= Rate of motive steam per kg of flash steam produced ...(11.10)

The make up water is available at condenser pressure as shown by state point A on Fig. 11.4(a). It throttles to evaporator pressure in the float valve and enters evaporator at state point B on Fig. 11.4(a).

Enthalpy of make up water, $h_A = h_B$

Enthalpy of flash steam from evaporator = h_3

Heat absorbed per kg of vapour = $h_3 - h_A$

Vapour formed in kg per ton of refrigeration = $\dfrac{211}{(h_3 - h_A)}$ kg vapour/ton minute ...(11.11)

Motive steam required in kg per ton of refrigeration

$$= \frac{m_a}{m_b} \times \frac{211}{(h_3 - h_A)} \text{ kg(motive steam/ ton minute)} \quad ...(11.12)$$

11.4 PERFORMANCE OF STEAM JET REFRIGERATION SYSTEM

The performance of steam jet refrigeration system is defined as the mass of steam required per ton of refrigeration with the decrease in temperature of chilled water (evaporating temperature) or increase in pressure of condensing vapour (condensing temperature).

The steam rate in steam jet refrigeration is a function of steam pressure, condenser temperature and evaporator temperature. The performance curves are shown in Fig. 11.5 to 11.8.

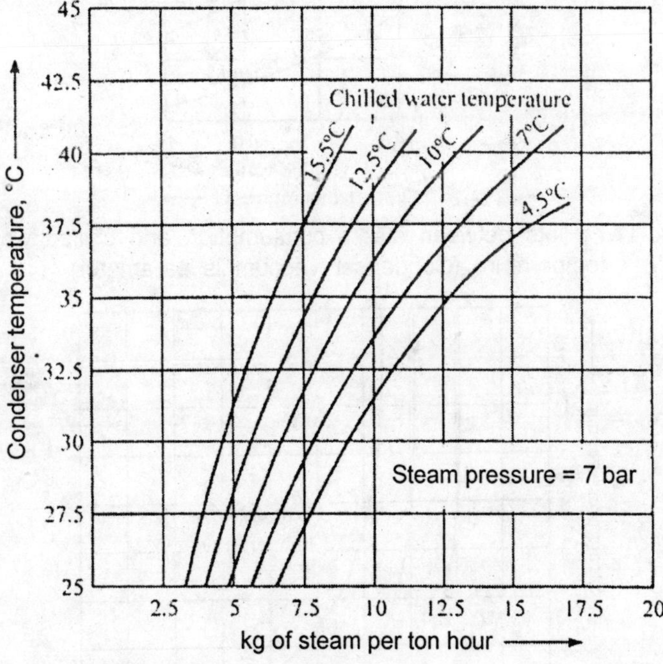

Fig. 11.5 Plots between condensing temperature and steam consumption (Chilled water temperature as parameter).

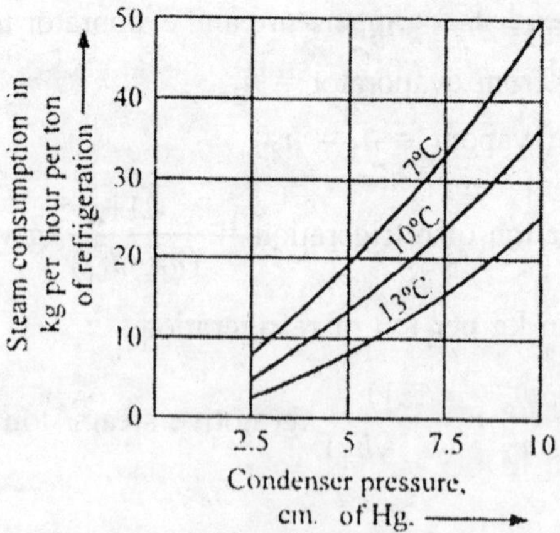

Fig. 11.6 Plots between steam consumption and condenser pressure (Evaporation temperature as parameter).

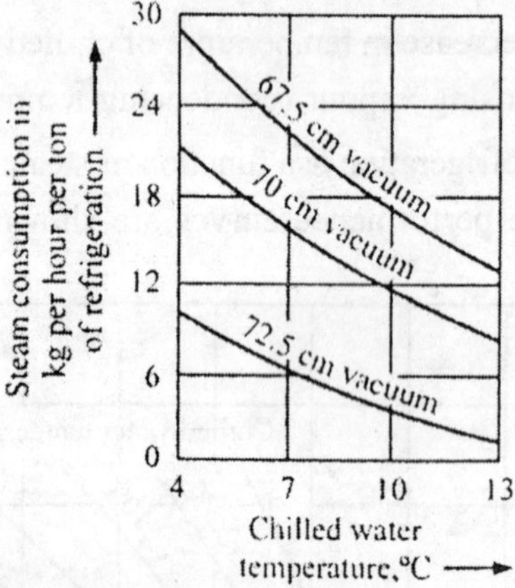

Fig. 11.7 Plots between steam consumption and chilled water temperature (Condenser vaccum is parameter).

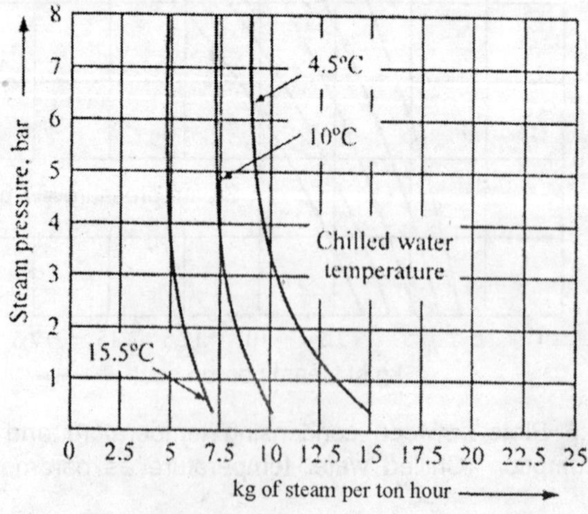

Fig. 11.8 Plots between steam pressure and steam consumption (Chilled water temperature as parameter).

1. The effect of both condenser temperature and evaporator temperature at constant steam pressure (7 bar) is shown in Fig. 11.5. The actual steam consumption will vary within 5% of the curves shown.
2. The amount of steam required to produce one ton of cooling effect decreases with decreasing condenser temperature and with increasing chilled water temperature.
3. It is more economical to operate at the highest steam pressure available. The optimum pressure of operation is about 7 bar.
4. The influence of steam pressure is more pronounced with chilled water temperature below 100°C than at higher chilled water temperature.
5. Low pressure steam below 2 bar are not economical to use for steam jet refrigeration cooling.

11.5 PRECAUTIONS FOR INSTALLATION OF SJRS

1. Chilled water circulating pump should have an adequate net positive suction head (NPSH) to prevent cavitations.
2. Any throttling valves in the return line on the evaporator should be located as close as feasible to the evaporator to minimize cavitations induced noise.
3. The steam supply should be free from suspended or dissolved solids carry over from the boiler, otherwise nozzle may plug.
4. All piping should be leak tight, clean and free from dirt.
5. The chilled liquid pump seat should be leak free and the chilled liquid pump vented to flash tank to prevent loss of capacity due to accumulation in casing.

11.6 ADVANTAGES AND DISADVANTAGES OF SJRS

Advantages

1. Simple and rugged in construction.
2. Highly reliable and vibration free system.
3. Low production and maintenance cost.
4. Relatively less plant mass (kg/ton of refrigeration)
5. Water as refrigerant is perfectly safe.
6. Low running cost.
7. Quickly adjustable to load variations.
8. It can be installed where electricity is not available.

Disadvantages

1. Not suitable for water temperature below 4°C.
2. The mäintenance of high vaccum required in the evaporator is achieved by direct vaporization to produce chilled water which is usually limited as a very large volume of vapours need to be handled.
3. A large size condenser is required to handle ejected vapour and motive steam.
4. The COP is very low (≃ 0.4 to 0.6).
5. Either centrifugal or ejector devices are required to handle high volume of vapours.

11.7 APPLICATIONS OF SJRS

1. Food processing plants for precooling of vegetables and concentrating fruit juices.
2. Gas plants.
3. Breweries.
4. Rubber and vulcanizing industries.
5. Paper and pulp industries.

Example 11.1 The following data refer to a steam-jet refrigeration system:

Motive steam pressure = 7 bar, flash chamber (evaporator) water temperature = 4°C, temperature of make up water to flash chamber = 18°C, pressure in the condenser = 0.05 bar, nozzle efficiency = 90%, entrainment efficiency = 60%, compression efficiency = 70%. Calculate:

(a) mass flow of motive steam per kg of flash vapour generated.

(b) refrigerating capacity in ton per kg of flash vapour.

(c) mass flow rate of motive steam per ton of refrigeration.

(d) volume flow of vapour leaving the flash chamber per ton of refrigeration.

(e) dryness fraction of steam from flash chamber.

Assume condition of motive steam at entry to nozzle to be dry saturated and at entry to thermo-compressor as 0.92.

Solution: At 7 bar dry saturated motive steam, from steam tables, we have

$$h_1 = h_g = 2762 \text{ kJ/kg}, \ t = 165°\text{C}, \ s_1 = s_{g1} = 6.705 \text{ kJ/kg}\cdot\text{K}$$

At condenser pressure of 0.05 bar,

$$t = 32.98°\text{C}, \ h_f = 137.8 \text{ kJ/kg}, \ h_{fg} = 2423.8 \text{ kJ/kg},$$

$$h_g = 2561.6 \text{ kJ/kg}, \ s_f = 0.476 \text{ kJ/kg}, \ s_g = 8.396 \text{ kJ/kg}\cdot\text{K}$$

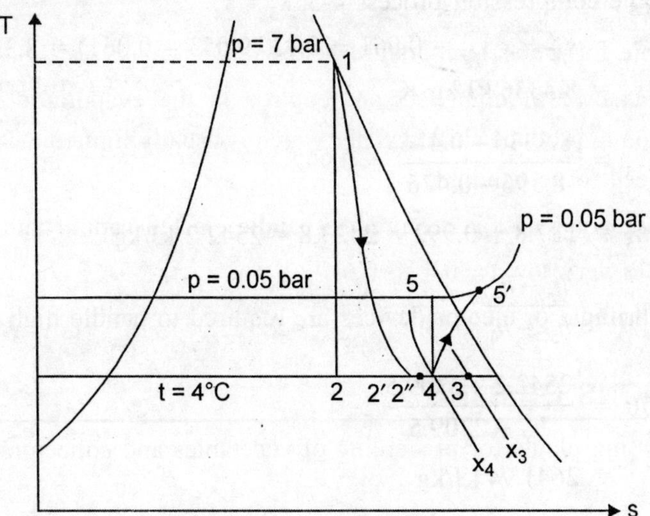

Fig. 11.9 T-s diagram.

At 4°C for flash chamber,

$$p_e = 0.00813 \text{ bar}, h_f = 16.8 \text{ kJ/kg}, h_{fg} = 2492.1 \text{ kJ/kg}, s_f = 0.061 \text{ kJ/kg}\cdot\text{K},$$
$$s_g = 9.053 \text{ kJ/kg}\cdot\text{K}, v_g = 157.27 \text{ m}^3\text{/kg}$$

For isentropic expansion process 1–2, $s_1 = s_2 = s_{f2} + x_2 s_{fg2}$

$$6.705 = 0.061 + x_2(9.053 - 0.061)$$
$$x_2 = 0.739$$

$$\therefore \quad h_2 = h_{f2} + x_2 h_{fg2} = 16.8 + 0.739 \times 2492.1 = 1858.8 \text{ kJ/kg}$$

Nozzle efficiency, $$\eta_n = \frac{h_1 - h_2'}{h_1 - h_2}$$

$$0.90 = \frac{2762 - h_2'}{2762 - 1858.5}$$

$$h_2' = 1948.8 \text{ kJ/kg}$$

Entrainment efficiency, $$\eta_{ent} = \frac{h_1 - h_{2''}}{h_1 - h_{2'}} = \frac{2762 - h_{2''}}{2762 - 1948.8} = 0.6$$

$$h_{2''} = 2274.1 \text{ kJ/kg}$$

and $$x_{2''} = \frac{h_{2''} - h_f}{h_{fg}} = \frac{2274.1 - 16.8}{2492.1} = 0.905$$

From the given data, at entry to thermo-compressor, $x_4 = 0.92$

$$\therefore \quad h_4 = h_f + x_4 h_{fg} = 16.8 + 0.92 \times 2492.1 = 2309.5 \text{ kJ/kg}$$

For the isentropic compression process 4-5, $s_5 = s_4$

$$s_4 = s_{f4} + x_4 s_{fg} = 0.061 + 0.92\,(9.053 - 0.061) = 8.3336 \text{ kJ/kg·K}$$

$$\therefore \quad s_5 = 8.3336 \text{ kJ/kg·K}$$

and
$$x_5 = \frac{8.3336 - 0.476}{8.396 - 0.476} = 0.992$$

$$h_5 = 137.8 + 0.992 \times 2423.8 = 2542.2 \text{ kJ/kg}$$

$$\eta_c = \frac{h_5 - h_4}{h_{5'} - h_4}$$

$$0.70 = \frac{2542.2 - 2309.5}{h_{5'} - 2309.5}$$

$$h_{5'} = 2641.94 \text{ kJ/kg}$$

$$\frac{m_a}{m_b} = \frac{(h_5 - h_4)}{[\eta_c \times \eta_{ent} \times \eta_n \times (h_1 - h_2) - (h_5 - h_4)]}$$

$$= \frac{(2542.2 - 2309.5)}{[0.70 \times 0.6 \times 0.9 \times (2762 - 1858.5) - (2542.2 - 2309.5)]} = \frac{232.7}{108.82}$$

$$= 2.138 \text{ kJ/kg of vapour}$$

For mass and enthalpy balance at entry to thermo- compressor, we have

$$h_3 + \frac{m_a}{m_b} \times h_{3''} = \left(\frac{m_a + m_b}{m_b}\right) \times h_4$$

$$m_b h_3 + m_a h_{2''} = (m_a + m_b) h_4$$

$$h_3 = \frac{(m_a + m_b)\,h_4 - m_a h_{2''}}{m_b}$$

$$= \left(\frac{m_a}{m_b} + 1\right) h_4 - \frac{m_a}{m_b} h_{2''}$$

$$= (2.138 + 1) \times 2309.5 - 2.138 \times 2274.1$$

$$= 2385.18 \text{ kJ/kg}$$

Dryness fraction of steam from flash chamber is given by,

$$x_3 = \frac{2385.18 - 16.8}{2492.1} = 0.95$$

Refrigerating effect per kg of flashed vapour from the flash chamber

$$= h_3 - \text{enthalpy of make up water at 18°C}$$

$$= 2385.8 - 75.5 = 2310.3 \text{ kJ/kg of flash vapour}$$

Motive steam required per ton of refrigeration per hour

$$= \frac{211 \times 60}{2310.3} \times 2.138 = 11.716 \text{ kg/h.ton}$$

Volume of vapour removed from flash chamber per hour-ton

$$= \text{Volume of 1 kg of vapour at condition 3} \times \frac{211 \times 60}{2310.3}$$

$$= x_3 v_g \times \frac{211 \times 60}{2310.3}$$

$$= 0.95 \times 157.27 \times \frac{211 \times 60}{2310.3} = 820.95 \text{ m}^3/\text{h} \cdot \text{ton}$$

Example 11.2 *A steam power plant operates with steam at 30 bar and 400°C from the boiler and develops 2100 kW. The condenser pressure is maintained at 0.04 bar. Steam is bled off the turbine at 5 bar for meeting a refrigeration load of 150 ton by steam-jet refrigeration system, at flash chamber pressure of 0.01 bar (approx. 7°C). The various efficiencies are:* $\eta_n = 0.9$, $\eta_t = 0.9$, $\eta_{ent} = 0.65$, $\eta_c = 0.65$.

Steam enters the thermo-compressor at 0.01 bar with dryness fraction of 0.94. Make up water enters the flash chamber at 18°C.

Calculate using Mollier diagram:

(a) State of steam at all salient points, (b) quality of steam leaving the flash chamber, (c) quality of motive steam bled-off the power turbine, (d) quality of steam generated in the boiler for power and refrigeration, (e) additional quantity of steam generated in boiler for refrigeration load, and (f) COP of the steam-jet refrigeration system based on the heat input of the motive steam.

Assume same condenser for power turbine and steam-jet refrigeration system.

Solution: The schematic diagram of steam jet refrigeration system is shown in Fig. 11.10(*a*) and the Mollier diagram in Fig. 11.10(b). Point B is located on Mollier diagram for boiler condition or entry to power turbine, i.e. for 30 bar and 400°C to 5 bar. B-1 represents isentropic process and B–1′ the actual process. From Mollier diagram, we have

$$h_B = 3232.5 \text{ kJ/kg}, \; h_1 = 2792 \text{ kJ/kg}$$

Turbine efficiency,

$$\eta_t = \frac{h_B - h_1'}{h_B - h_1} = \frac{3232.5 - h_1'}{3232.5 - 2792} = 0.9$$

$$h_1' = 2836 \text{ kJ/kg}$$

This is the motive steam for steam-jet refrigeration nozzle. It expands from point 1′ to 2′, i.e., to flash chamber pressure of 0.01 bar. 1′-2 is isentropic expansion in nozzle, giving

$$h_2 = 1968 \text{ kJ/kg}$$

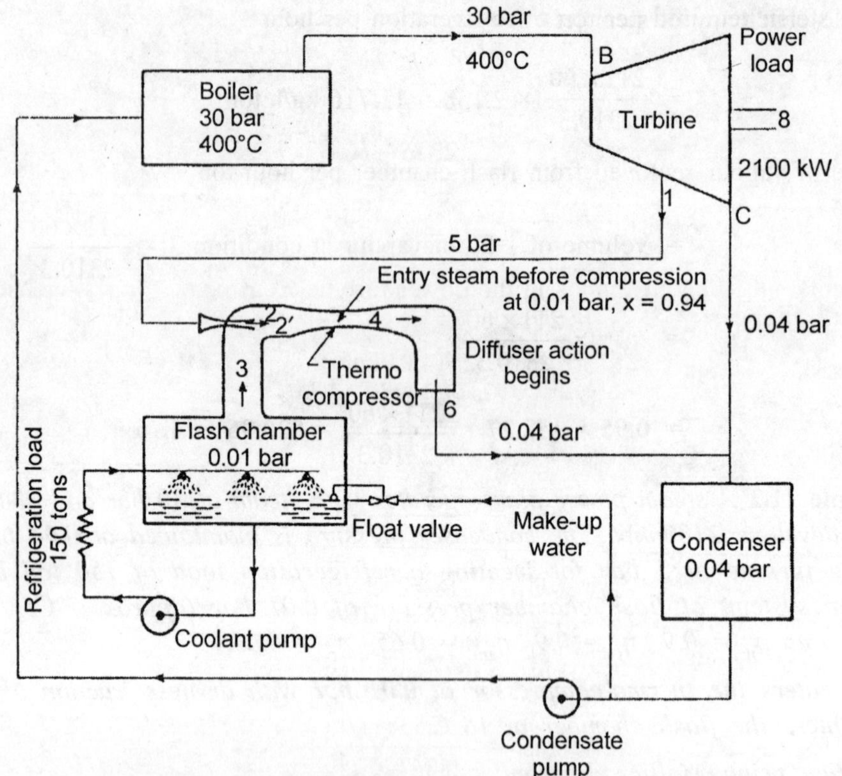

(a) Schematic diagram of steam-jet refrigeration system.

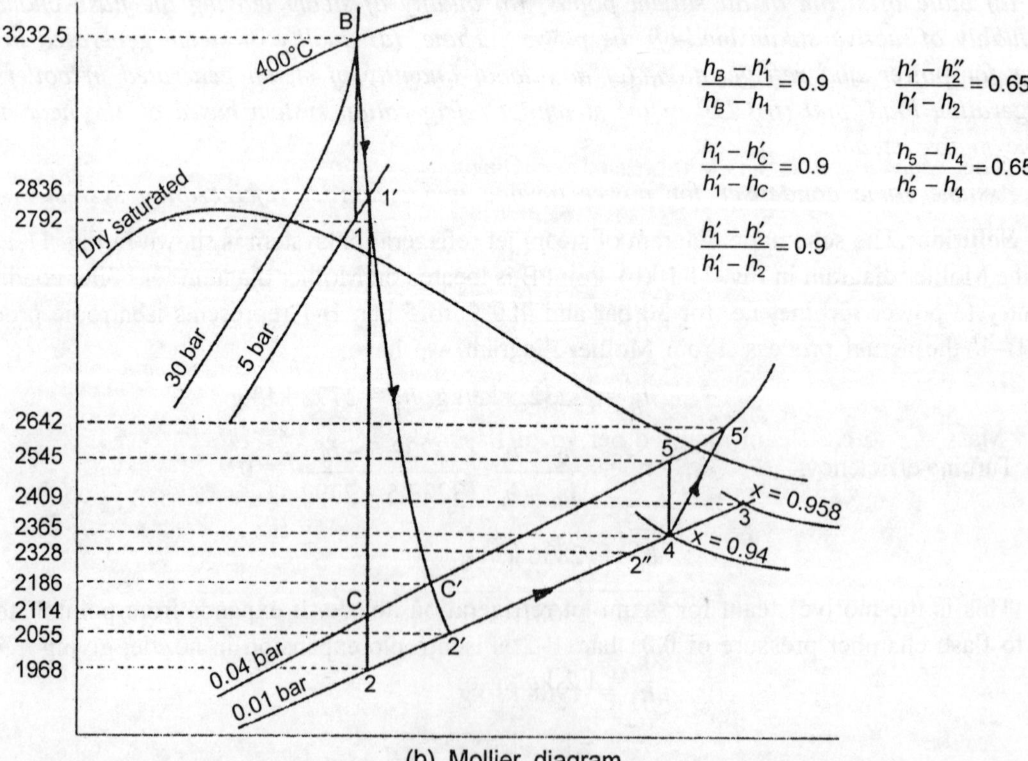

(b) Mollier diagram.

Fig. 11.10

Nozzle efficiency of stem jet,

$$\eta_n = \frac{h_1' - h_2'}{h_1' - h_2}$$

$$0.9 = \frac{2336 - h_2'}{2836 - 1968}$$

$$h_2' = 2054.8 \text{ kJ/kg}$$

The remaining steam expands in the power turbine from point 1′ to condenser pressure of 0.04 bar to point C.

$$h_c = 2114 \text{ kJ/kg}$$

$$\eta_t = \frac{h_1' - h_c'}{h_1' - h_c}$$

$$0.9 = \frac{2836 - h_c'}{2836 - 2114}$$

$$h_c' = 2186.2 \text{ kJ/kg}$$

Entrainment efficiency,

$$\eta_{ent} = \frac{h_1' - h_2''}{h_1' - h_2'}$$

$$0.65 = \frac{2836 - h_2''}{2836 - 2054.8}$$

$$h_{2''} = 2328.2 \text{ kJ/kg}$$

This steam at condition 2″ mixes with flash chamber vapour at condition 3. The conditions at the beginning of thermo-compression, i.e., point 4 are 0.01 bar and $x_4 - 0.94$. Thus $h_4 = 2365$ kJ/kg.

The steam at point 4 is compressed to 0.04 bar, giving $h_5 = 2545$ kJ/kg.

Now

$$\eta_c = \frac{h_5 - h_4}{h_5' - h_4}$$

$$0.65 = \frac{2545 - 2365}{h_5' - 2365}$$

$$h_5' = 2642 \text{ kJ/kg}$$

Mass of motive steam required per kg of flash vapour is given by,

$$\frac{m_a}{m_b} = \frac{h_5 - h_4}{(h_1' - h_2)\eta_n \times \eta_{ent} \times \eta_e - (h_5 - h_4)}$$

$$= \frac{2545 - 2365}{(2836 - 1968) \times 0.9 \times 0.65 \times 0.65 - (2545 - 2365)}$$

$$= \frac{180}{150} = 1.2 \text{ kg/kg of flash vapour.}$$

For enthalpy balance,

$$m_b h_3 + m_a h_{2''} = (m_a + m_b)\, h_4$$

$$h_3 + \frac{m_a}{m_b} h_{2''} = \left(\frac{m_a}{m_b} + 1\right) h_4$$

$$h_3 = \left(\frac{m_a}{m_b} + 1\right) h_4 - \frac{m_a}{m_b} h_{2''}$$

$$= (1.2 + 1) \times 2365 - 1.2 \times 2328.2$$

$$= 2409 \text{ kJ/kg}$$

Thus point 3 can be marked on Mollier chart on 0.01 line giving $x_3 = 0.958$

Refrigeration effect per kg of flash vapour

$= h_3$ – enthalpy of make up water

$= 2409 - 4.187 \times 18 = 2333.64$ kJ of refrigeration /kg of flash vapour.

Amount of water flashed out per hour for 150 ton refrigeration

$$= \frac{150 \times 211 \times 60}{2333.64} = 813.75 \text{ kg/h}$$

Motive steam bled-off the turbine at 5 bar

$$= \frac{m_a}{m_b} \times 813.75 = 1.2 \times 813.75 = 976.5 \text{ kg/h}$$

Let m_p = total steam entering the power turbine.

Then, $m_p[(h_B - h_{1'})] + (m_p - 976.5)(h_{1'} - h_{c'})$ = Power developed in kW × 3600

or $m_p[(h_B - h_{1'}) + (h_{1'} - h_{c'})] - 976.5\,(h_{1'} - h_c) = 2100 \times 3600$

or $m_p[(h_B - h_{c'})] - 976.5\,(h_{1'} - h_c)] = 2100 \times 3600$

$m_p\,[3232.5 - 2186.2] - 976.5\,(2836 - 2186.2) = 2100 \times 3600$

$m_p = 7800$ kg/h

If refrigeration load is not there, the quantity of steam required for 2100 kW is,

$$m_{p1} = \frac{2100 \times 3600}{h_B - h_c'} = \frac{2100 \times 3600}{3232.5 - 2186.2} = 7225.5 \text{ kg/h}$$

Additional steam required to be generated by the boiler at 30 bar and 400°C

$= m_p - m_{p1} = 7800 - 7225.5 = 574.5$ kg/h

Heat per kg of motive steam required for refrigeration purposes

$= h_{1'}$ – enthalpy of water at 0.04 bar

$= 2836 - 121.41 = 2714.59$ kJ/kg

$$\text{COP} = \frac{150 \times 211 \times 60}{\text{Motive steam bled per hour} \times 2714.59}$$

$$= \frac{150 \times 211 \times 60}{976.5 \times 2714.59} = 0.716$$

MULTIPLE – CHOICE QUESTIONS

1. In a steam jet refrigeration system, the motive steam expands in

(a) convergent nozzle (b) divergent nozzle

(c) convergent-divergent nozzle (d) any of the above

2. The velocity of steam at the exit of the nozzle is

(a) supersonic (b) sonic (c) subsonic (d) any of the above.

3. The compression device used in a steam jet refrigeration system is a

(a) vapour compressor (b) steam ejector

(c) diffuser (d) liquid pump.

4. The COP of steam jet refrigeration system varies from

(a) 0.2 – 0.3 (b) 0.4 – 0.6 (c) 2 – 4 (d) 5 – 8

5. The ratio of isentropic enthalpy increase to the actual enthalpy increase required for the compression of motive steam and water vapour is known as

(a) boiler efficiency (b) compression efficiency

(c) nozzle efficiency (d) entrainment efficiency

Answers

1. (c) **2.** (a) **3.** (b) **4.** (b) **5.** (b)

REVIEW QUESTIONS

1. Explain the principle of steam-jet refrigeration system.
2. What are the main components of steam-jet refrigeration system?
3. List the advantages and disadvantages of steam jet refrigeration system.
4. What are the applications of steam jet refrigeration system?
5. Draw the T-s and h-s diagrams for steam jet refrigeration system.
6. Define the following:

   (a) Motive steam (b) Entrainment efficiency

   (c) Compression efficiency (d) Nozzle efficiency.

NUMERICAL PROBLEMS

11.1 The following data refer to a steam jet refrigeration system:

Condition of motive steam = 8 bar, dry saturated temperature of water in the flash chamber = 5°C, temperature of make-up water = 20°C, condenser pressure = 0.06 bar, nozzle efficiency = 0.86, entraining efficiency = 0.64, compression efficiency = 0.80, quality of steam and flash vapour at the beginning of compression = 0.9. Calculate the following:

(a) Mass of motive steam required per kg of flash vapour, (b) Quality of vapour flashed from the flash chamber, (c) Refrigerating effect per kg of flash vapour, (d) Mass of motive steam required per hour per ton of refrigeration, (e) Volume of vapour removed from the flash chamber per hour per ton of refrigeration, and (f) COP of the system.

[**Ans.**1.463 kg/kg of flash vapour, 0.903, 2185.6 kJ/kg, 8.47 kg/h/ton of refrigeration, 770 m^3/h/ton of refrigeration, 0.57]

11.2 The following data refer to a steam jet refrigeration system:

Capacity of thermal power plant = 2 MW, Condition of steam supplied from the boiler = 25 bar, 350°C, Refrigeration load = 110 ton, pressure at which steam is bled-off the steam turbine = 4 bar, vaccum maintained in the condenser = 730 mm of Hg, pressure and condition at which steam enters the thermo-compressor = 0.01 bar, 0.92 dry; temperature at which make-up water enters the flash chamber = 20°C, efficiency of steam turbine = 0.88, nozzle efficiency = 0.88, entrainment efficiency = 0.64, thermo-compressor efficiency = 0.64. Using Mollier chart, determine the following:

(a) Dryness fraction of steam leaving the flash chamber, (b) Mass of motive steam bled-off from the steam turbine, (c) Mass of steam generated in the boiler, (d) Mass of additional steam to be generated by the boiler due to refrigeration load, and (e) COP of the system.

[**Ans.** 0.927, 1044.2 kg/h, 8334.5 kg/h, 656.1 kg/h, 0.502]

Chapter 12

PSYCHROMETRY

12.1 INTRODUCTION

The air-conditioning is that branch of engineering which deals with the study of supplying and maintaining desirable internal room atmospheric conditions of air for human comfort. The four important factors for human comfort are: temperature, humidity, purity, and motion of air.

The working substance for air-conditioning is moist air, which is a mixture of two gases. One of these is dry air, which itself is a mixture of a number of gases and the other is water vapour which may exist in a saturated or superheated state. Moist air is not a pure substance in any process in which condensation or evaporation of moisture occurs. Both dry air and water vapour can be considered as perfect gases since both exist in the atmosphere at low pressures. Hence, perfect gas laws to both and Dalton's law of partial pressures for non-reactive mixture of gases can be applied to dry air part only.

12.2 PRINCIPLES OF PSYCHROMETRY

The psychrometry is that branch of engineering which deals with the study of moist air i.e. dry air mixed with water vapour.

Dry air: The dry air consists of O_2 = 0.2099, N_2 = 0.7803, Ar = 0.0094, CO_2 = 0.0003 and H_2 = 0.0001 parts by volume.

Molecular weight of air = 28.966

Gas constant, R_a = 0.287 kJ/kg.K

Density of air = 1.293 kg/m^3 at 1.01325 bar and 0°C

Specific heat of air at constant pressure = 1.005 kJ/kg.K

Molecular weight of water (vapour) = 18.02

Gas constant for water vapour, R_v = 0.461 kJ/kg.K

Barometric pressure = 1.01325 bar or 760 mm Hg

Dalton's law of partial pressures

The Dalton's law of partial pressures states that for a mixture of ideal gases, the total pressure is equal to the sum of the partial pressures, which each constituent would exert, if it occupied the whole space alone. Thus,

Barometric pressure of mixture of dry air and water vapour is given by

$$p_b = p_a + p_v \qquad ...(12.1)$$

Where p_a = partial pressure of dry air

p_v = partial pressure of water vapour

Moist air: It is a mixture of dry air and water vapour.

Saturated air: It is a mixture of dry air and water vapour when air has absorbed maximum amount of water vapour. At saturation, $p_v = p_s$, the saturation pressure.

Degree of saturation (μ)

$$= \frac{\text{actual mass of water vapour in a unit mass of dry air}}{\text{mass of water vapour in the same mass of dry air when it is saturated at the same temperature}}$$

Humidity (or specific humidity or humidity ratio) : It is the mass of water vapour present in one kg of dry air. It is expressed in terms of gram per kg of dry air (g/kg d.a.)

Absolute humidity: It is the mass of water vapour present in 1 m^3 of dry air. It is expressed in terms of gram per cubic meter of dry air (g/m^3 d.a.)

Relative humidity (RH or ϕ): It is the ratio of actual mass of water vapour in a given volume of moist air to the mass of water vapour in the same volume of saturated air at the same temperature and pressure.

Dry Bulb Temperature (DBT, t_d): It is the temperature of air recorded by a thermometer, when it is not affected by the moisture present in the air.

Wet Bulb Temperature (WBT, t_w): It is the temperature of air recorded by a thermometer, when its bulb is surrounded by a wet cloth exposed to the air.

The DBT and WBT are measured by a Sling Psychrometer, as shown in Fig. 12.1.

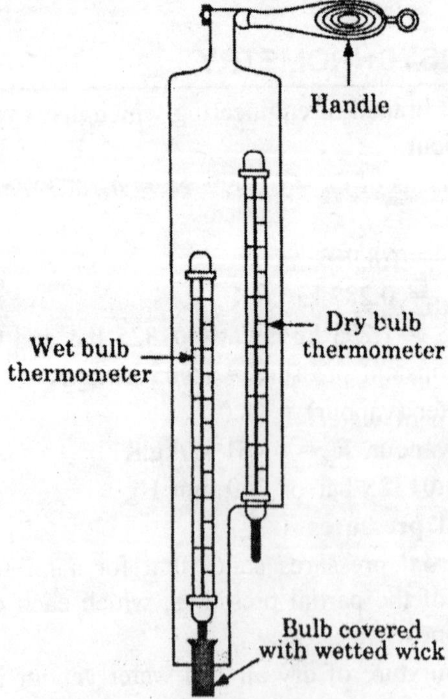

Fig. 12.1 Sling psychrometer

Wet Bulb Depression, WBT = DBT – WBT

Dew Point Temperature (DPT, t_{dp}): It is the temperature of air recorded by a thermometer, when the water vapour present in it begins to condense. Thus, DPT is the saturation temperature (t_{sat}) corresponding to the partial pressure of water vapour (p_v).

Dew Point Depression, DPD = DBT – DPT

12.3 PSYCHROMETRIC RELATIONS

1. Specific (or absolute) humidity, humidity ratio or moisture content, w.

It is defined as the ratio of the mass of water vapour to the mass of dry air in a given volume of the mixture.

Let, p, V, T, m = pressure, volume, absolute temperature and mass respectively

$R, M, v, \bar{R}$ = gas constant, molecular mass, specific volume, and universal gas constant respectively

Subscripts a, v = air and vapour respectively.

Specific humidity,
$$w = \frac{m_v}{m_a} = \frac{\frac{V}{v_v}}{\frac{V}{v_a}} = \frac{v_a}{v_v} \quad \text{...(12.2)}$$

For air,
$$p_a v_a = \frac{\bar{R}}{M_a} T \quad \text{...(1)}$$

and
$$p_a V = m_a \frac{\bar{R}}{M_a} T \quad \text{...(2)}$$

For water vapour,
$$p_v v_v = \frac{\bar{R}}{M_v} T \quad \text{...(3)}$$

and
$$p_a V = m_v \frac{\bar{R}}{M_v} T \quad \text{...(4)}$$

Substituting for m_v and m_a from these equations in Eq. (12.2), we get

$$w = \frac{p_v M_v}{p_a M_a} = \frac{18.02 p_v}{28.966 p_a} = 0.622 \frac{p_v}{p_a} \text{ (kg of water vapour per kg of d.a.)} \quad \text{...(12.3}a\text{)}$$

$$= 622 \frac{p_v}{p_a} \text{ (gram of water vapour/kg d.a.)} \quad \text{...(12.3}b\text{)}$$

Now actual total pressure, $p_b = p_a + p_v$

$$\therefore \quad w = \frac{0.622 p_v}{p_b - p_v} \quad \text{...(12.4)}$$

Now $p_v = 0$ so that $p_a = p_b$. Thus

$$w = \frac{0.622 p_v}{p_b} \quad \text{...(12.5)}$$

For $m_a = 1$ kg, $m_v = w$ kg and total mass of moist air,

$$m = (1 + w) \text{ kg}$$

For saturated air, $p_v = p_s$ and maximum humidity ratio,

$$w_s = w_{max} = \frac{0.622\,p_s}{p_b - p_s} \qquad ...(12.6)$$

where p_s = partial pressure of air corresponding to saturation temperature i.e. t_d.

2. Degree of saturation or percentage humidity, μ

The ratio of actual specific humidity w to the specific humidity ω_s of saturated air at the same temperature is termed as the degree of saturation.

$$\mu = \frac{w}{w_s} = \frac{\left[\dfrac{0.622\,p_v}{p_b - p_v}\right]}{\left[\dfrac{0.622\,p_s}{p_b - p_s}\right]} = \left(\frac{p_v}{p_s}\right)\left[\frac{p_b - p_s}{p_b - p_v}\right] = \left(\frac{p_v}{p_s}\right)\left[\frac{1-(p_s/p_b)}{1-(p_v/p_b)}\right] \qquad ...(12.7)$$

The degree of saturation is a measure of the capacity of air to absorb moisture.

3. Relative Humidity (RH or ϕ)

It is defined as the ratio of mass of water vapour m_v in a certain volume of moist air at a given temperature to the mass of water vapour m_{vs} in the same volume of saturated air at the same temperature.

$$\phi = \frac{m_v}{m_{vs}} = \frac{\left(\dfrac{p_v V}{\bar{R}T}\right)}{\left(\dfrac{p_s V}{\bar{R}T}\right)} = \frac{p_v}{p_s} \qquad ...(12.8)$$

For a perfect gas, $p_v v_v = p_s v_s$

$$\therefore \qquad \frac{p_v}{p_s} = \frac{v_s}{v_v}$$

or

$$\phi = \frac{p_v}{p_s} = \frac{v_s}{v_v} \qquad ...(12.9)$$

Now

$$\mu = \frac{p_v}{p_s}\left[\frac{1-\left(\dfrac{p_s}{p_b}\right)}{1-\left(\dfrac{p_v}{p_b}\right)}\right] = \phi\left[\frac{1-\left(\dfrac{p_s}{p_b}\right)}{1-\phi\left(\dfrac{p_s}{p_b}\right)}\right]$$

$$\phi = \frac{\mu}{1-(1-\mu)\dfrac{p_s}{p_b}} \qquad ...(12.10)$$

4. Pressure of water vapour

(*i*) The partial pressure of water vapour, according to Carrier's equation, is given by

$$p_v = p_w - \frac{(p_b - p_w)(t_d - t_w)}{1544 - 1.44 t_w} \quad ...(12.11)$$

where p_w = saturation pressure corresponding to WBT

p_b = barometric pressure

t_d = DBT

t_w = WBT

(*ii*) Modified Apjohn equation:

$$p_v = p_w - \frac{p_b(t_d - t_w)}{1500} \quad ...(12.12)$$

(*iii*) Modified Ferrel equation:

$$p_v = p_w - 0.00066 p_b\,(t_d - t_w)\left[1 + \frac{t_d}{872.78}\right] \quad ...(12.13)$$

5. Vapour density or Absolute humidity

Let ρ_v = density of water vapour corresponding to its partial pressure and DBT, t_d.

ρ_a = density of dry air.

Then $$m_v = V_v \rho_v \text{ and } m_a = V_a \rho_a$$

$$\frac{m_v}{m_a} = \frac{V_v \rho_v}{V_a \rho_a}$$

Since $V_a = V_v$, therefore,

$$w = \frac{m_v}{m_a} = \frac{\rho_v}{\rho_a}$$

or $$\rho_v = w \rho_a$$

Now $$p_a V_a = m R_a T_d$$

For $m_a = 1$ kg, $$V_a = v_a = \frac{1}{\rho_a}$$

$$\therefore \quad p_a \times \frac{1}{\rho_a} = R_a T_a$$

or $$\rho_a = \frac{p_a}{R_a T_a}$$

$$\therefore \quad \rho_a = \frac{w p_a}{R_a T_d} = \frac{w(p_b - p_v)}{R_a T_a} \quad ...(12.14)$$

12.4 ENTHALPY OF MOIST AIR

The temperature-entropy diagram is shown in Fig. 12.2.

Enthalpy of moist air = Enthalpy of dry air + Enthalpy of water vapour associated with dry air

$$h = h_a + wh_v \text{ per kg of dry air}$$

where $h_a = c_{pa} \cdot t_d = 1.005\, t_d$ kJ/kg

$c_{pa} = 1.005$ kJ/kg·K

t_d = DBT of air in °C.

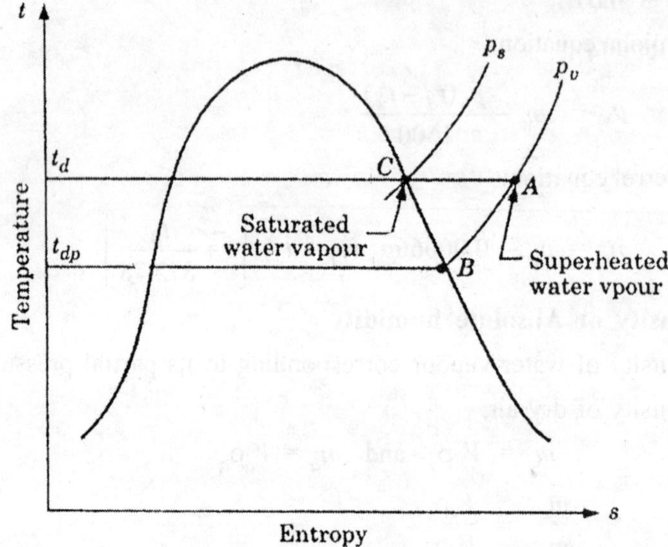

Fig. 12.2 Temperature-entropy diagram

$$h_v = c_{pw} t_{dp} + [(h_{fg})_{dp} + c_{pv}(t_d - t_{dp})] \text{ kJ/kg}$$
$$= 4.1868\, t_{dp} + (h_{fg})_{dp} + 1.88\,(t_d - t_{dp})$$

where c_{pw} = 4.1868 kJ/kg.K is the specific heat of liquid water

$(h_{fg})_{dp}$ = latent heat of vaporisation at DPT

c_{pv} = 1.88 kJ/kg.K is the specific heat of superheated water vapour

t_{dp} = DPT

w = specific humidity

$$h = 1.005 t_d + w[4.1868 t_{dp} + (h_{fg})_{dp} + 1.88\,(t_d - t_{dp})]$$
$$= (1.005 + 1.88w)\, t_d + w\,[(h_{fg})_{dp} + 2.3068 t_{dp}] \text{ kJ/kg} \quad \text{...(12.15)}$$
$$= 1.005 t_d + w(2500 + 188 t_d) \text{ kJ/kg } d.a. \quad \text{...(12.16)}$$

12.5 HUMID SPECIFIC HEAT

Eq. (12.16) can be written as,

$$h = (c_{pa} + wc_{pv})\, t_d + w(h_{fg})_{0°C} \quad \text{...(12.17)}$$
$$= c_p t_d + w(h_{fg})_{0°C} \quad \text{...(12.18)}$$

where $c_p = c_{pa} + wc_{pv} = 1.005 + 1.88w$ kJ/(kg d.a.K)
is termed as the humid specific heat. It is the specific heat of moist air $(1 + w)$ kg per kg of dry air.

At low temperatures of air-conditioning the value of w is very small. An approximate value of c_p of 1.0216 kJ/(kg d.a.).K may be taken for all practical purposes.

$$h = 1.0216\, t_d + w\, (h_{fg})_{dp} + 2.3068 t_{dp}] \quad ...(12.19)$$

Example 12.1 *A mixture of dry air and water vapour is at a temperature of 20°C under a total pressure of 735 mm Hg. The dew point temperature is 15°C. Calculate:*

(a) Partial pressure of water vapour,

(b) Relative humidity,

(c) Specific humidity

(d) Specific enthalpy of water vapour

(e) Enthalpy of air per kg of dry air, and

(f) Specific volume of air per kg of dry air.

Solution:

(*a*) From steam tables, the partial pressure of water vapour at t_{dp} = 15°C,

$$p_v = 12.79 \text{ mm Hg}$$

Now $$760 \text{ mm Hg} = 1.01325 \times 10^5 \text{ N/m}^2$$

$$1 \text{ mm Hg} = 133.32 \text{ N/m}^2$$

$$\therefore \quad p_v = 12.79 \times 133.32 = 1705.16 \text{ N/m}^2$$

(*b*) Saturation pressure of water vapour at 20°C *DBT*,

$$p_s = 17.54 \text{ mm Hg} = 17.54 \times 133.32 = 2338.4 \text{ N/m}^2$$

$$\text{Relative humidity, } \phi = \frac{p_v}{p_s} \times 100 = \frac{12.79}{17.54} \times 100 = 72.92\%$$

(*c*) Specific humidity, $$w = \frac{0.622 p_v}{p_b - p_v} = \frac{0.622 \times 12.79}{735 - 12.79} = 0.011 \text{ kg water vapour/kg.d.a.}$$

(*d*) Latent heat of vaporisation of water at t_d = 20°C, $(h_{fg})_{20°C}$ = 2454.1 kJ/kg

Latent heat of vaporisation of water at t_{dp} = 15°C, $(h_{fg})_{15°C}$ = 2465.9 kJ/kg

Specific enthalpy of water vapour,

$$h_v = [4.1868\, t_{dp} + (h_{fg})_{dp} + 1.88\, (t_d - t_{dp})]$$
$$= 4.1868 \times 15 + 2465.9 + 1.88(20 - 15)]$$
$$= 2538.1 \text{ kJ/kg w.v}$$

(*e*) Enthalpy of air per kg d.a.,

$$h = 1.005 t_d + w h_v$$
$$= 1.005 \times 20 + 0.011 \times 2538.1$$
$$= 48.02 \text{ kJ/kg d.a.}$$

(*f*) Specific volume of air per kg of d.a.

$$v = v_a = \frac{R_a T}{p_a}$$

$$= \frac{0.287 \times 10^3 \times (273+20)}{(735-12.79) \times 133.32}$$

$$= 0.873 \text{ m}^3\text{/kg d.a.}$$

Example 12.2 *A sling psychrometer gave the following readings: DBT = 30°C, WBT = 20°C, Barometer reading = 740 mm of Hg.*

Determine: (a) Dew point temperature, (b) Relative humidity, (c) Specific humidity, (d) Degree of saturation, (e) Vapour density, and (f) Enthalpy of mixture per kg of dry air.

Solution: Given: t_d = 30°C, t_w = 20°C, p_b = 740 mm Hg.

(*a*) Corresponding to t_w = 20°C, from steam tables, saturation pressure, p_w = 0.023385 bar

Barometric pressure, $p_b = 740 \times 133.32 \times 10^{-5} = 0.986568$ bar

Partial pressure of water vapour,

$$p_v = p_w - \frac{(p_b - p_w)(t_d - t_w)}{1544 - 1.44 t_w} \quad \text{(Carrier's eqn.)}$$

$$= 0.023385 - \frac{(0.986568 - 0.023385)(30-20)}{(1544 - 1.44 \times 20)}$$

$$= 0.01703 \text{ bar}$$

Corresponding to p_v = 0.01703 pressure, from steam tables, dew point temperature,

$$t_{dp} = 15°\text{C}$$

(*b*) Corresponding to t_d = 30°C, from steam tables, saturation pressure of water vapour,

$$p_s = 0.04246 \text{ bar}$$

$$\text{RH, } \phi = \frac{p_v}{p_s} = \frac{0.01703}{0.04246} = 0.401 \text{ or } 40.1\%$$

(*c*) Specific humidity,

$$w = \frac{0.622 p_v}{p_b - p_v} = \frac{0.622 \times 0.01703}{0.986568 - 0.01703}$$

$$= 0.010925 \text{ kg/kg d.a. or } 10.925 \text{ g/kg d.a.}$$

(*d*) Specific humidity of saturated air,

$$w_s = \frac{0.622 p_s}{p_b - p_s} = \frac{0.622 \times 0.04246}{0.986568 - 0.04246} = 0.02797 \text{ kg/kg d.a.}$$

$$\text{Degree of saturation, } \mu = \frac{w}{w_s} = \frac{0.010925}{0.02797} = 0.3905 \text{ or } 39.05\%$$

(*e*) Vapour density,

$$\rho_v = \frac{w(p_b - p_v)}{R_a T_d} = \frac{0.010925(0.986568 - 0.01703) \times 10^5}{287(273+30)}$$

$$= 0.01218 \text{ kg/m}^3 \text{ of d.a.}$$

(f) From steam tables, $(h_{fg})_{dp\,=\,15°C}$ = 2465.9 kJ/kg

Enthalpy of mixture per kg d.a.,

$$h = 1.0216\, t_d + w[(h_{fg})_{dp} + 2.3068\, t_{dp}]$$

$$= 1.0216 \times 30 + 0.010925\, [2465.9 + 2.3068 \times 15] = 57.966 \text{ kJ/kg d.a.}$$

Example 12.3 *The atmospheric air was found to have a DBT = 35°C and WBT = 25°C and barometric pressure, p_b = 750 mm Hg. Calculate:*

(a) Relative humidity, (b) Specific humidity, (c) Dew point temperature, (d) Enthalpy of air per kg of dry air, and (e) The volume of mixture per kg of dry air.

Solution: Given: t_d = 35°C, t_w = 25°C, p_b = 750 mm Hg

(a) Saturation pressure corresponding to t_w = 25°C from steam tables is,

$$p_w = 0.03169 \text{ bar or } 3169 \text{ N/m}^2$$

$$= \frac{3169}{133.32} = 23.77 \text{ mm Hg}$$

$$p_v = p_w - \frac{(p_b - p_w)(t_d - t_w)}{1544 - 1.44 t_w} \quad \text{(Carrier's eqn.)}$$

$$= 23.77 - \frac{(750 - 23.77)(35 - 25)}{(1544 - 1.44 \times 25)} = 18.95 \text{ mm Hg}$$

Saturation pressure corresponding to t_d = 35°C from steam tables is,

$$p_s = 0.05628 \text{ bar or } 5628 \text{ N/m}^2$$

$$= \frac{5628}{133.32} = 42.21 \text{ mm Hg}$$

$$\text{Relative humidity, } \phi = \frac{p_v}{p_s} = \frac{18.95}{42.21} = 0.4489 \text{ or } 44.89\%$$

(b) Specific humidity, $$w = \frac{0.622 p_v}{p_b - p_v} = \frac{0.622 \times 18.95}{750 - 18.95} = 0.016 \text{ kg/kg d.a. or } 16 \text{ g/kg.d.a.}$$

(c) DPT = Saturation temperature corresponding to p_v = 18.95 mm Hg or 18.95 × 133.32 = 0.02526 bar

From steam tables corresponding to p_v = 0.02526 bar, t_{dp} = 21.2°C (by interpolation)

(d) Latent Heat of vaporisation of water at t_{dp} = 21.2°C, from steam tables is,

$$(h_{fg})_{dp} = 2451.2 \text{ kJ/kg}$$

Enthalpy of air per kg of dry air,

$$h = 1.0216\, t_d + w[(h_{fg})_{dp} + 2.3068\, t_{dp}]$$

$$= 1.0216 \times 35 + 0.016\, [2451.2 + 2.3068 \times 21.2]$$

$$= 75.76 \text{ kJ/kg d.a.}$$

(e) Specific volume of dry air at 760 mm Hg and t_d = 35°C, from psychrometric tables is, v_a = 0.8728 m³/kg d.a.

Volume of mixture at t_d = 35°C and p_v = 18.95 mm Hg is,

$$v = v_a \times \frac{760}{p_b - p_v} = \frac{0.8728 \times 760}{(750 - 18.95)} = 0.9074 \text{ m}^3\text{/kg d.a.}$$

or
$$v = v_a = \frac{R_a T_d}{p_a} = \frac{287 \times 308}{(750 - 18.95) \times 133.32} = 0.907 \text{ m}^3\text{/kg d.a.}$$

Example 12.4 *A room 5 m × 4m × 3m contains moist air at 40°C. The barometric pressure is 1 bar and relative humidity is 70%. Determine: (a) Humidity ratio, (b) Dew point, (c) Mass of dry air, and (d) Mass of water vapour.*

If the moist air is further cooled at constant pressure until the temperature is 15°C, find the amount of water vapour condensed.

Solution: Given: $V = 5 \times 4 \times 3 = 60$ m³, t_d = 40°C, p_b = 1 bar, ϕ = 0.70.

(*a*) Saturation pressure of vapour corresponding to t_d = 40°C, from steam tables is,

$$p_e = 7.384 \text{ kPa or } 0.07384 \text{ bar}$$

$$RH, \phi = \frac{p_v}{p_s}$$

$$p_v = 0.7 \times 0.07384 = 0.05169 \text{ bar}$$

$$\text{Humidity ratio, } w = \frac{0.622 p_v}{p_b - p_v} = \frac{0.622 \times 0.05169}{1.0 - 0.05169}$$

$$= 0.0339 \text{ kg/kg d.a. or } 33.9 \text{ g/kg d.a.}$$

(*b*) DPT, t_{dp} = Saturation temperature corresponding to p_v = 0.05169 bar or 5.169 kPa

= 33.46°C from steam tables

(*c*)
$$p_a = p_b - p_v = 1.0 - 0.5169 = 0.94831 \text{ bar}$$

$$\text{Mass of water vapour, } m_a = \frac{p_a V}{R_a T_d} = \frac{0.94831 \times 10^2 \times 60}{0.287 \times (273 + 40)} = 63.34 \text{ kg}$$

(*d*) Now $w = \dfrac{m_v}{m_a}$

$$\text{Mass of water vapour, } m_v = 0.0339 \times 63.34 = 2.147 \text{ kg}$$

Saturation pressure corresponding to 15°C, from steam tables, is,

$$p_s = p_v = 1.707 \text{ kPa or } 0.01707 \text{ bar}$$

$$w = \frac{0.622 p_v}{p_b - p_v} = \frac{0.622 \times 0.01707}{1.0 - 0.01707}$$

$$= 0.0108 \text{ kg/kg d.a. or } 10.8 \text{ g/kg d.a.}$$

Now
$$p_a = p_b - p_v = 1 - 0.01707 = 0.98293 \text{ bar or } 98.293 \text{ kPa}$$

Mass of dry air, $m_a = \frac{p_a V}{R_a T_a} = \frac{98.293 \times 60}{0.287 \times (273 + 15)} = 71.35$ kg

Mass of water vapour, $m_v = wm_a = 0.0108 \times 71.35 = 0.77$ kg

Amount of water vapour condensed = 2.147 – 0.77 = 1.377 kg

12.6 THERMODYNAMIC BULB TEMPERATURE OR ADIABATIC SATURATION TEMPERATURE (AST)

The thermodynamic wet bulb temperature or AST is the temperature at which the air can be brought to saturation state adiabatically by the evaporation of water in the flowing air. A schematic representation of this process, called the adiabatic saturation process, is shown in Fig. 12.3. It consists of an adiabatic enclosure containing adequate quantity of water. There is also an arrangement for feed water from the top.

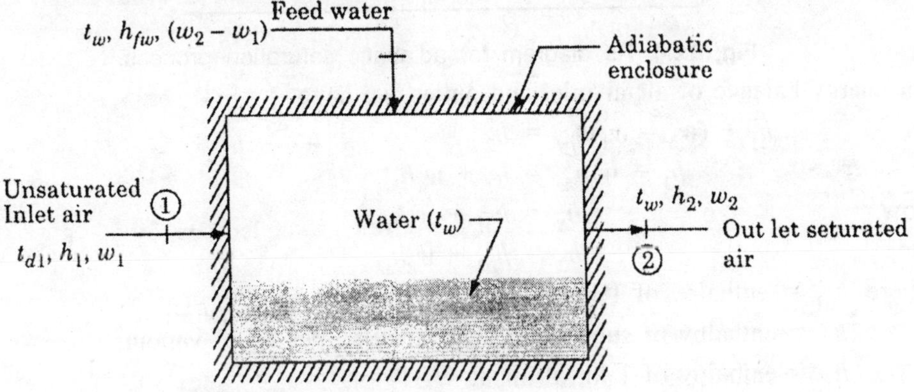

Fig. 12.3 Adiabatic saturator

The unsaturated air enters the enclosure at section 1. As the air flows through the enclosure, water evaporates which is carried by the flowing stream of air. Thus the specific humidity of air increases. The water level in the enclosure is maintained by the feed water. Both the air and water are cooled as the evaporation takes place. The process continues until the energy transferred from air to water is equal to the energy required to vaporise water. When steady conditions are reached, the air flowing at section 2 becomes saturated with water vapour. The temperature of the saturated air at section 2 is known as thermodynamic wet bulb temperature or adiabatic saturation temperature. The adiabatic saturation process is represented on *T-s* diagram as shown by the curve 1-2 in Fig. 12.4. The adiabatic saturation temperature is taken equal to the wet bulb temperature.

Let t_{d1} = DBT of initial unsaturated air

t_{d2} = DBT of saturated air

= adiabatic saturation temperature, t_w

h_1, h_2 = enthalpy of unsaturated and saturated air respectively

w_1, w_2 = specific humidity of air at sections 1 and 2 respectively

h_{fw} = sensible heat of water at adiabatic saturation temperature

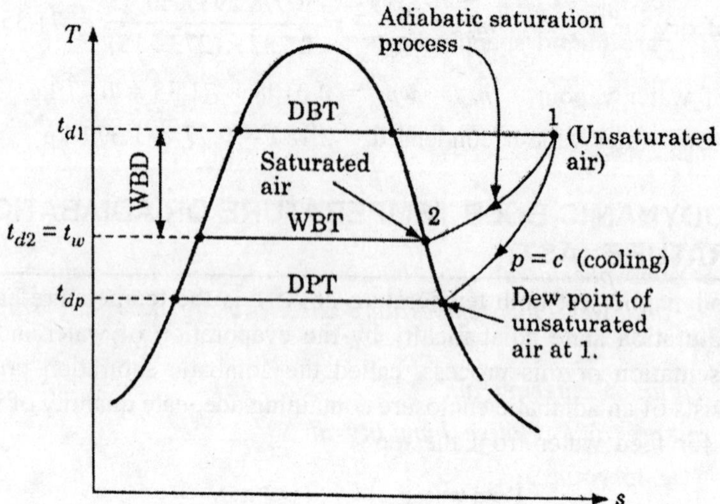

Fig. 12.4 *T-s* diagram for adiabatic saturation process.

For energy balance of air at inlet and outlet, we have

$$h_1 + (w_2 - w_1)h_{fw} = h_2 \quad ...(1)$$

or

$$h_1 + w_1 h_{fw} = h_2 + w_1 h_{fw} \quad ...(2)$$

Now

$$h_1 = h_{a1} + w_1 h_{s1}$$

and

$$h_2 = h_{a2} + w_2 h_{s2}$$

where h_{a1} = enthalpy of 1 kg of dry air at *DBT* of $t_{d1} = c_{pa}\, t_{d1}$

h_{s1} = enthalpy of superheated vapour at t_{d1} per kg of vapour

h_{a2} = enthalpy of 1 kg of air at WBT = $t_w = c_{pa} t_{d2}$

h_{s2} = enthalpy of saturated vapour at WBT = t_w per kg of vapour

Eq. (2) then becomes,

$$(h_{a1} + w_1 h_{s1}) - w_1 h_{fw} = (h_{a2} + w_2 h_{s2}) - w_2 h_{fw}$$

$$w_1(h_{s1} + h_{fw}) = w_2(h_{s2} + h_{fw}) + (h_{a2} - h_{a1})$$

$$w_1 = \frac{w_2(h_{s2} - h_{fw}) + (h_{a2} - h_{a1})}{(h_{s1} - h_{fw})} \quad ...(12.20)$$

Also

$$h_1 = c_{pa} t_{d1} + w_1 h_{s1}$$

$$h_2 = c_{pa} t_{d2} + w_2 h_{s2}$$

where

$$h_{s2} = h_{f2} + h_{fg2}$$

$$h_{s1} = h_{s2} + c_{pv}(t_{d1} - t_{d2}) = h_{f2} + h_{fg2} + c_{pv}(t_{d1} - t_{d2})$$

Eq. (1) becomes,

$$c_{pa} t_{d1} + w_1[h_{f2} + h_{fg2} + c_{pv}\,(t_{d1} - t_{d2})] + (w_2 - w_1)\, h_{f2}$$

$$= c_{pa} t_{d2} + w_2(h_{f2} + h_{fg2})$$

or $(c_{pa} + w_1\, c_{pv})\, t_{d1} - (c_{pa} + w_2\, c_{pv})\, t_{d2} = h_{fg2}\, (w_2 - w_1)$

or $$c_{p1} t_{d1} - c_{p2} t_{d2} = h_{fg2} (w_2 - w_1) \quad ...(12.21)$$

where c_{p1} and c_{p2} are humid specific heats

If $c_{p1} = c_{p2} = c_p$, then

$$c_p(t_{d1} - t_{d2}) = h_{fg2}(w_2 - w_1)$$

$$\therefore \quad t_{d2} = t_{d1} - \frac{h_{fg2}}{c_p}(w_2 - w_1) \quad ...(12.22)$$

Example 12.5 *Atmospheric air at 1 bar enters the adiabatic saturator. The dry bulb temperature is 30°C and wet bulb temperature 20°C during adiabatic saturation process. Calculate:*

(a) Humidity ratio of entering air,

(b) Vapour pressure and relative humidity at 30°C, and

(c) Dew point temperature.

Solution: Given: p_b = 1 bar, t_d = 30°C, t_w = 20°C

(*a*) Saturation pressure of vapour at 20°C from steam tables is,

$$p_{v2} = 2.34 \text{ kPa or } 0.0234 \text{ bar}$$

Enthalpy of saturated vapour at 20°C, $h_{s2} = h_{g2}$ = 2538.2 kJ/kg

Sensible heat of water at 20°C, h_{fw} = 83.9 kJ/kg

Enthalpy of saturated vapour at 30°C, $h_{s1} = h_{g1}$ = 2556.4 kJ/kg

Enthalpy of 1 kg of unsaturated air at t_d = 30°C,

$$h_{a1} = mc_{pa}t_d = 1 \times 1.005 \times 30 = 30.15 \text{ kJ/kg}$$

Enthalpy of 1 kg of saturated air at t_w = 20°C,

$$h_{a2} = mc_{pa}t_w = 1 \times 1.005 \times 20 = 20.10 \text{ kJ/kg}$$

$$w_2 = \frac{0.622\,p_{v2}}{p_b - p_{v2}} = \frac{0.622 \times 0.0234}{1.0 - 0.0234} = 0.0149 \text{ kg/kg d.a.}$$

$$w_1 = \frac{w_2(h_{s2} - h_{fw}) + (h_{a2} - h_{a1})}{(h_{s1} - h_{fw})}$$

$$= \frac{0.0149(2538.2 - 83.9) + (20.10 - 30.15)}{(2556.4 - 83.9)}$$

$$= 0.0107 \text{ kg/kg d.a.}$$

(*b*) Now $$w_1 = \frac{0.622\,p_{v1}}{p_b - p_{v1}}$$

$$0.0107 = \frac{0.622\,p_{v1}}{1.0 - p_{v1}}$$

$$0.0107 - 0.0107\,pv_1 = 0.622\,pv_1$$

$$p_{v1} = 0.0169 \text{ bar}$$

Saturation pressure corresponding to 30°C from steam tables is,

$$p_s = 4.246 \text{ kPa or } 0.04246 \text{ bar}$$

$$\text{Relative humidity, } \phi = \frac{p_{v1}}{p_s} = \frac{0.0169}{0.04246} = 0.3983 \text{ or } 39.83\%$$

(*c*) Dew point temperature is the saturation temperature corresponding to the partial pressure of water vapour, p_{v1} = 0.0169 bar or 1.69 kPa

From steam tables, we have

$$t_{dp} = 14.85°\text{C}$$

12.7 PSYCHROMETRIC CHART

The psychrometric chart is a graphical representation of the complete thermodynamic properties of moist air and psychrometric analysis of air-conditioning processes. The chart most commonly used is the w-t_d chart, which specifies specific humidity (w) or water vapour pressure (p_v) along the ordinate and the dry bulb temperature (t_d) along the abscissa. The chart is normally constructed for a standard atmospheric pressure of 760 mm Hg or 1.01325 bar, corresponding to the pressure at the mean sea level. A typical layout of this chart is shown in Fig. 12.5.

Some of the important lines on the psychrometic chart are explained below:

DBT lines: These lines are vertical, i.e. parallel to the ordinate and are uniformly spaced. These lines are drawn at an interval of 5°C and upto the saturation line.

Specific humidity or moisture content lines: These lines are horizontal, i.e. parallel to the abscissa and are uniformly spaced.

DTP lines: These lines are horizontal, i.e. parallel to the abscissa and are non-uniformly spaced. At any point on the saturation line, the DBT and DPT are equal.

WBT lines: These are inclined straight lines and are non-uniformly spaced. At any point on the saturation line DBT and WBT are equal.

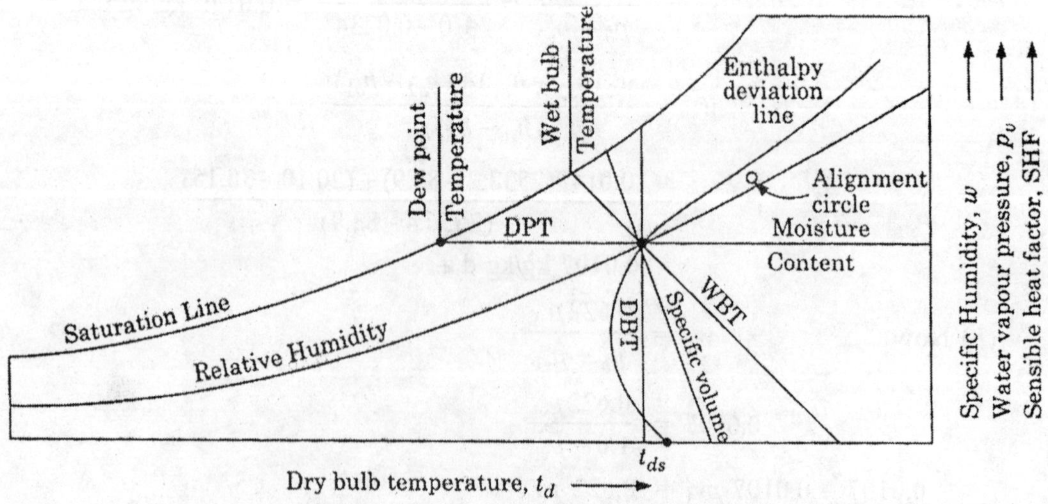

Fig. 12.5 Layout of Psychrometric chart.

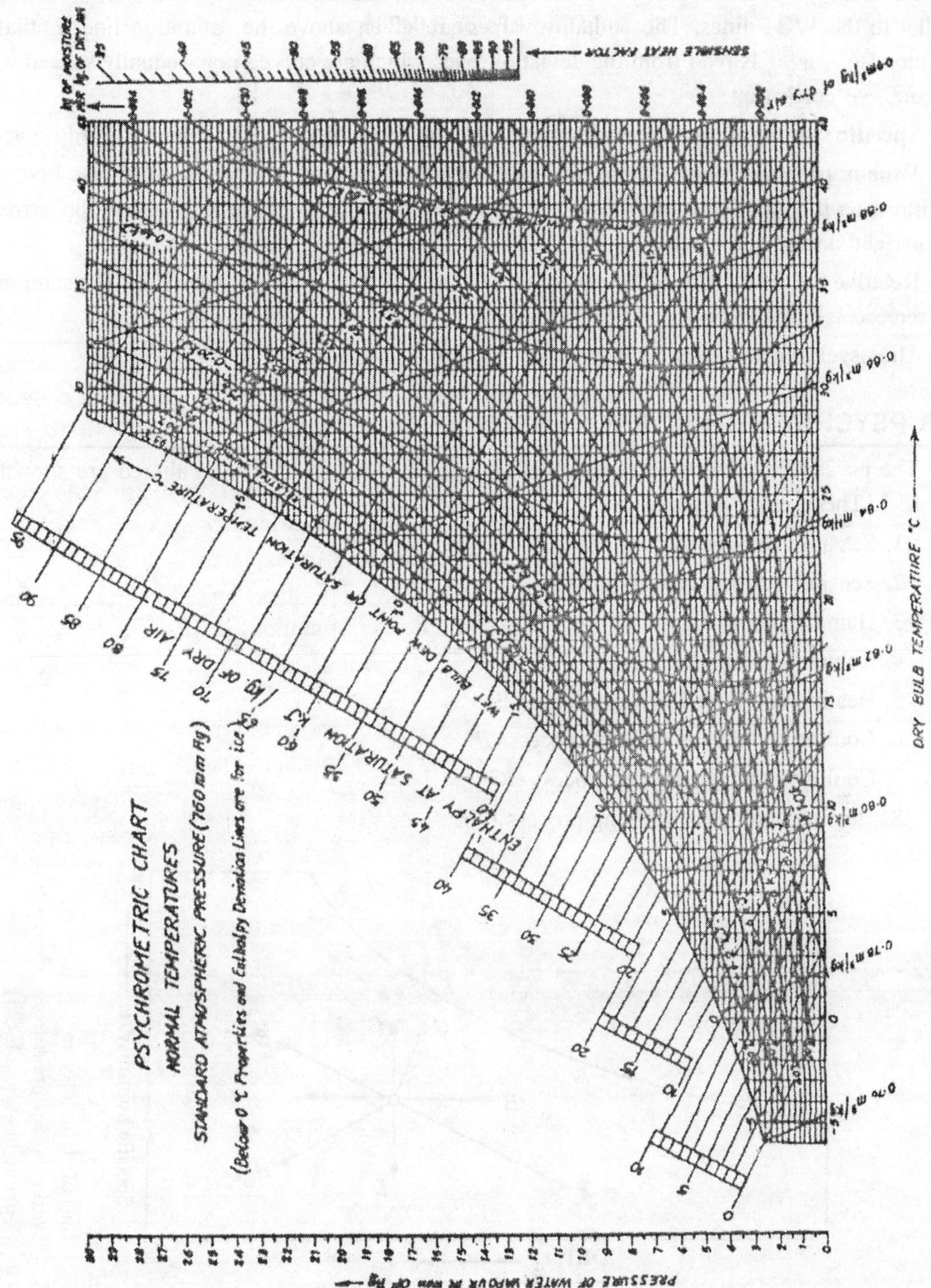

Fig. 12.6 Psychrometric chart.

Enthalpy line: These are inclined straight lines and uniformly spaced. These lines are parallel to the WBT lines. The enthalpy values are given above the saturation line. Enthalpy deviation $(w_s - w)h_w$ is read from the deviation lines which are curved non - equally spaced with +ve and –ve correction.

Specific volume lines: These are obliquely inclined straight lines and are uniformly spaced.

Vapour pressure lines: These lines are horizontal and uniformly spaced. These lines are not drawn on the chart. A scale showing vapour pressure in mm of Hg is given on the extreme left or right side of the chart.

Relative humidity lines: These are curved lines and follow the saturated line. The saturation line represents 100% RH.

The psychrometric chart is shown in Fig. 12.6.

12.8 PSYCHROMETRIC PROCESSES

The psychrometric processes by which the state of moist air can be altered are shown in Fig. 12.7. These processes are:

1. Sensible heating-process *OA*
2. Sensible cooling-process *OB*
3. Humidifying-process *OC*
4. Dehumidifying-process *OD*
5. Heating and humidifying-process *OE*
6. Cooling and dehumidifying-process *OF*
7. Cooling and humidifying-process *OG*
8. Heating and dehumidifying-process *OH*

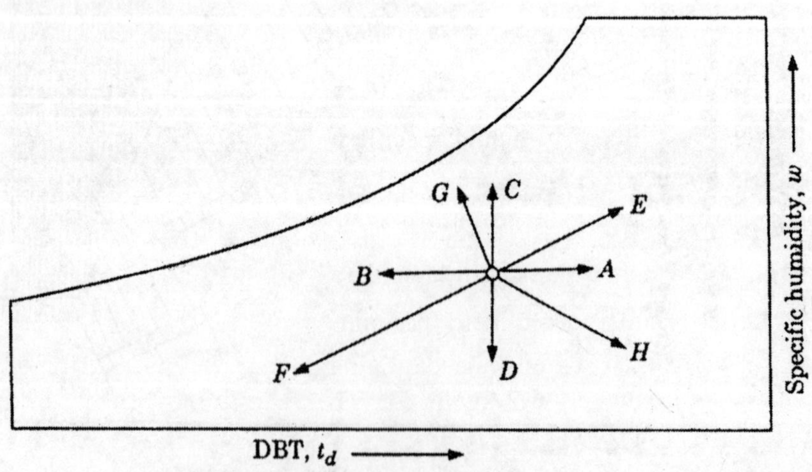

Fig. 12.7 Basic psychrometric processes.

12.8.1 Sensible Heating or Cooling Process

The heating/cooling of air without any change in its specific humidity is known as sensible heating/cooling. The process of sensible heating (line 1-2) or cooling (line 2-1) is shown by a horizontal line in Fig. 12.8. In this process, the specific humidity of air remains constant, i.e. $w_1 = w_2$. The heat has to be transferred which goes to change the temperature of air. The sensible heat transfer rate is given by,

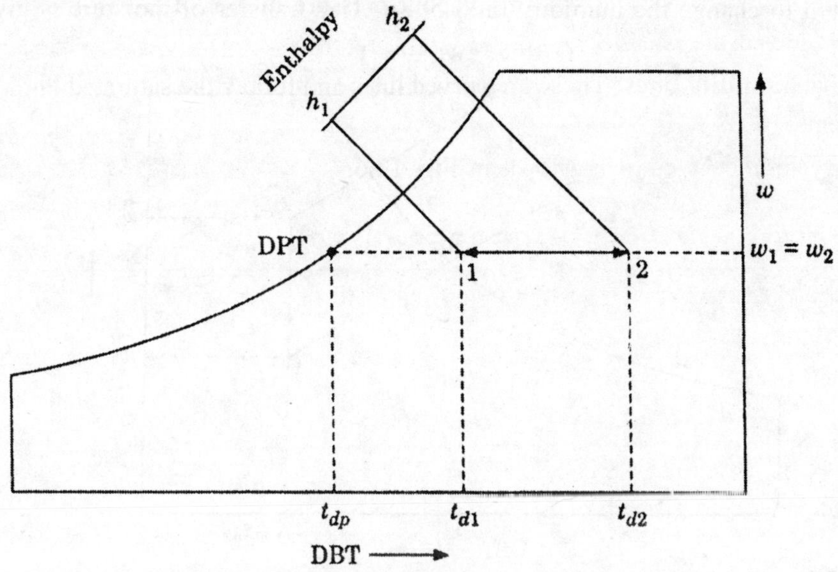

Fig. 12.8 Sensible heating or cooling process.

$$\dot{Q}_s = \dot{m}_a(h_2 - h_1)$$

$$= \dot{m}_a[c_{pa}(t_2 - t_1) + wc_{pv}(t_2 - t_1)]$$

$$= \dot{m}_a[1.005 + 1.88w](t_2 - t_1) \quad \text{...(12.23)}$$

$$= \dot{m}_a c_p(t_2 - t_1)$$

where $c_p = c_{pa} + wc_{pv}$ = humid specific heat

$\dot{m}_a$ = mass flow rate of dry air, kg d.a./s

$= \dfrac{\dot{V}\rho}{60}$ = Volume flow rate of air $\dot{V}$ in cmm × density of air ρ in kg/m³.

(cmm = cubic meter per minute)

For the purpose of calculations, standard air is taken at 20°C and 50% RH. The density of standard air is approximately 1.2 kg/m³ d.a. The value of humid specific heat is taken as 1.0216 kJ/(kgd.a.) (K).

$$\therefore \quad \dot{Q}_s = \frac{(\text{cmm})\,(1.2)\,(1.0216)\cdot\Delta t}{60} = 0.0204\,(\text{cmm}).\cdot\Delta t \text{ kW} \quad \text{....(12.24)}$$

The sensible heating of moist air can be done to any desired temperature. But the sensible cooling can be done only upto the dew point temperature t_{dp}, as shown in Fig. 12.8.

12.8.2 Humidification or Dehumidification Process

Humidification is the addition of moisture to air without change in its DBT. Similarly dehumidification is the removal of moisture without change in DBT. When the state of air is altered along the constant DBT line, such as 1-2 in Fig. 12.9, moisture in the form of vapour has to be transferred to change the humidity ratio of air. This transfer of moisture is given by,

$$G = m_a(w_2 - w_1) \quad ...(12.25)$$

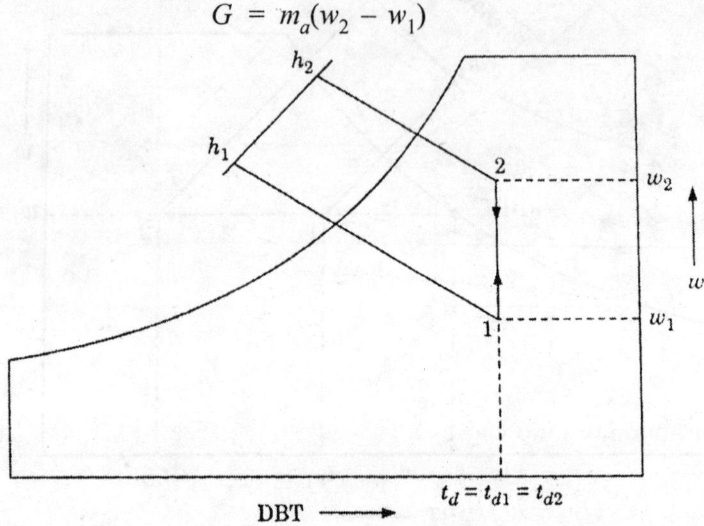

Fig. 12.9 Humidification or dehumidification process.

Because of this change in humidity ratio, there is also a change in the specific enthalpy of air, as shown in Fig. 12.9. This change in enthalpy due to the change in humidity ratio is considered to cause a latent heat transfer, given by

$$\dot{Q}_L = \dot{m}_a(h_2 - h_1) = m_a[(c_p t_{d2} + h_{fg} w_2) - (c_p t_{d1} + h_{fg} w_1)]$$

Since $t_{d1} = t_{d2}$

$$\therefore \quad \dot{Q}_L = \dot{m}_a h_{fg}(w_2 - w_1) = 2500\ G \quad ...(12.26)$$

where h_{fg} = latent heat of vaporization at t_{d1}

$$\dot{Q}_L = \frac{(\text{cmm})(1.2)(2500)}{60} \times \Delta w \text{ kW}$$

$$= 50\ (\text{cmm}).\ \Delta w \text{ kW} \quad ...(12.27)$$

12.8.3 Heating and Humidification

This process involves both a change in temperature as well as in the humidity ratio as shown in Fig. 12.10. The change in temperature causes a sensible heat load, given by

$$Q_S = m_a(h_A - h_1) = m_a c_p(t_{d2} - t_{d1}) \quad ...(1)$$

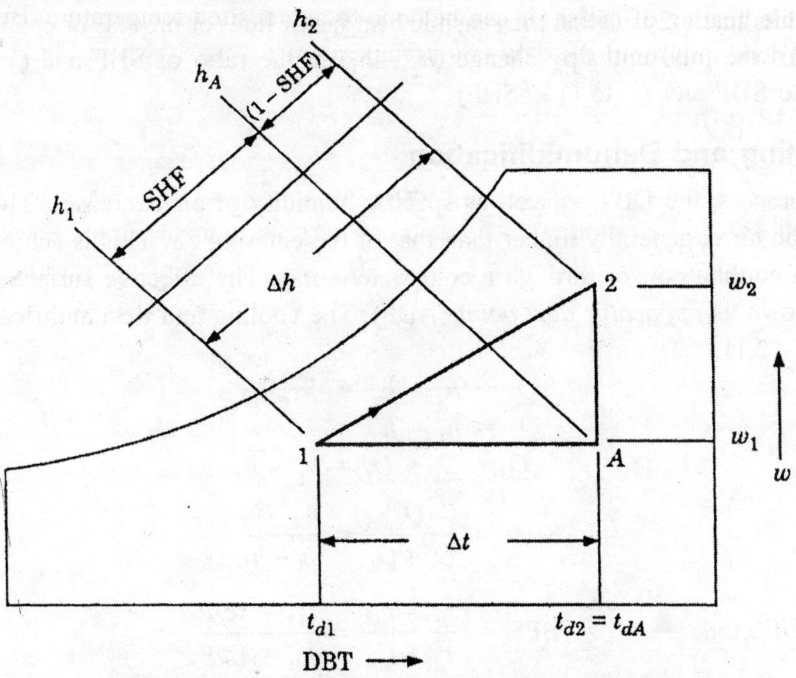

Fig. 12.10 Heating and humidification process.

The change in humidity ratio causes a moisture transfer, given by

$$G = m_a(w_2 - w_A) = m_a(w_2 - w_1)$$

Latent heat load, $\quad Q_L = m_a(h_2 - h_A) = m_a h_{fg}(w_2 - w_1) \quad$ ...(2)

Adding Eqs. (1) and (2), we get

Total heat load, $\quad Q = Q_S + Q_L = m_a[c_p(t_{d2} - t_{d1}) + h_{fg}(w_2 - w_1)] \quad$ ...(12.28)

Now
$$\dot{m}_a = \frac{\dot{V}\rho}{60}$$

$\therefore$
$$\dot{Q} = \frac{(\text{cmm})(1.2)}{60} \cdot \Delta h = 0.2\ (\text{cmm}) \cdot \Delta h$$
$$= (\text{cmm})\ [0.0204\ \Delta h + 50\ \Delta w]\ \text{kW} \quad \text{...(12.29)}$$

12.8.4 Sensible Heat Factor - SHF

The ratio of the sensible heat transfer to the total heat transfer is termed as the sensible heat factor. Thus

$$\text{SHF} = \frac{Q_S}{Q} = \frac{Q_S}{Q_S + Q_L}$$
$$= \frac{h_A - h_1}{(h_A - h_1) + (h_2 - h_A)} = \frac{h_A - h_1}{h_2 - h_1}$$
$$= \frac{0.0204\Delta t}{0.0204\Delta t + 50\Delta w} = \frac{0.0204\Delta t}{0.02\Delta h} \quad \text{...(12.30)}$$

The process line 1-2 is called the sensible heat factor line (or process or condition line). The point A divides the total enthalpy change $(h_2 - h_1)$ in the ratio of SHF and (1 – SHF). Q_S is proportional to SHF and Q_L to (1 – SHF).

12.8.5 Cooling and Dehumidification

In this process, the DBT as well as specific humidity of air decreases. The final relative humidity of the air is generally higher than that of the entering air. This is achieved by passing the air over a cooling coil or through a cold water spray. The effective surface temperature of the coil is known as *apparatus dew point* (ADP). The cooling and dehumidification process is shown in Fig. 12.11.

$$Q_s = h_A - h_2$$

$$Q_L = h_1 - h_A$$

$$Q = Q_s + Q_L = h_1 - h_2$$

$$\text{SHF} = \frac{Q_s}{Q_s + Q_L} = \frac{h_A - h_2}{h_1 - h_2} \quad \text{...(12.31)}$$

By Pass Factor, $$\text{BPF} = \frac{t_{d2} - t_{d4}}{t_{d1} - t_{d4}} = \frac{t_{d2} - ADP}{t_{d1} - ADP} \quad \text{...(12.32a)}$$

Also $$\text{BPF} = \frac{w_2 - w_4}{w_1 - w_4} = \frac{h_2 - h_4}{h_1 - h_4} \quad \text{...(12.32b)}$$

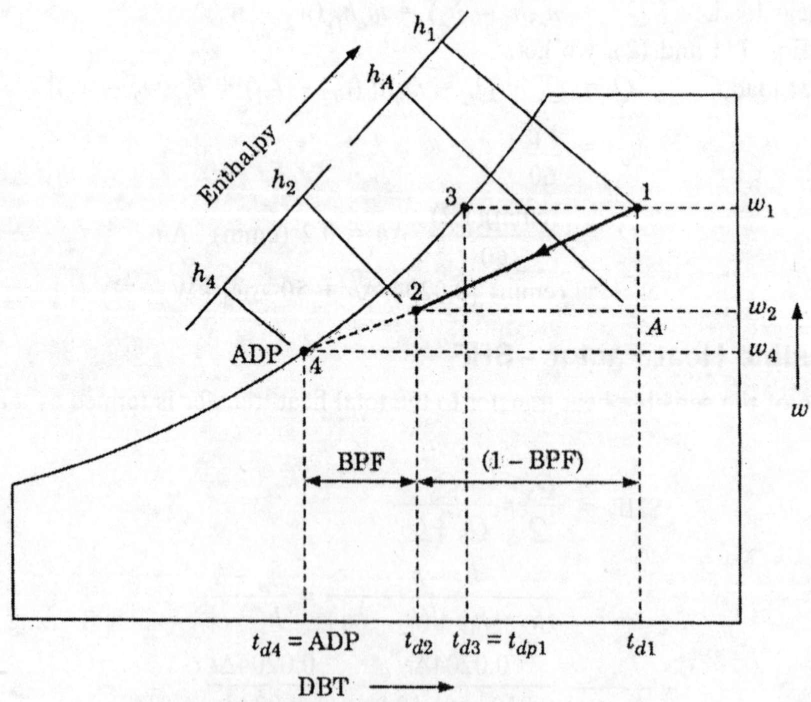

Fig. 12.11 Cooling and dehumidification process.

12.8.6 Air Washer

The schematic representation of an air washer is shown in Fig. 12.12. It involves the flow of air through a spray of water. During the course of flow, the air may be cooled or heated, humidified or dehumidified, or simply adiabatically saturated, depending on the mean surface temperature of water. The water is, accordingly cooled or externally cooled or heated or simply circulated by a pump. Make-up water is added for any loss in the case of humidification of air. Eliminator plates are provided to minimize the loss of water droplets.

The thermodynamic changes of state of air along path 1-2 in an air washer, depending on the mean surface temperature of water droplets (equal to), is shown in Fig. 12.13. The following processes are possible:

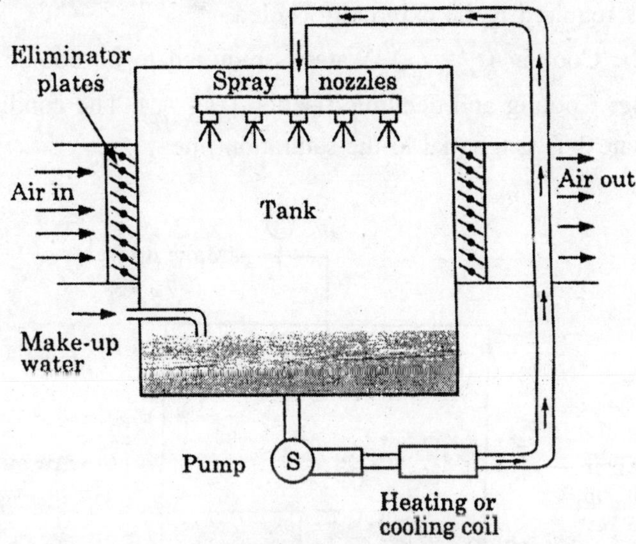

Fig. 12.12 Air-washer

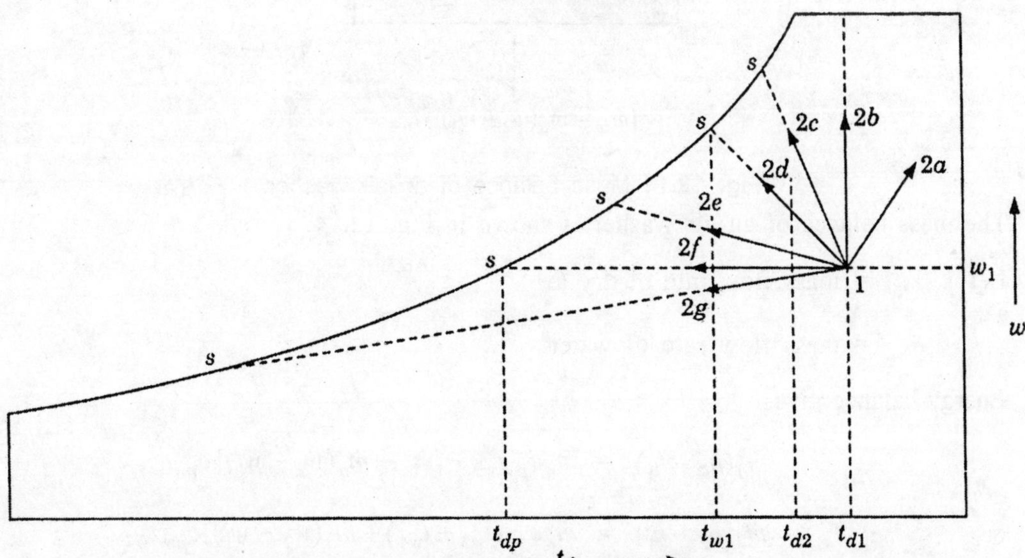

Fig. 12.13 Range of psychrometric processes with an air washer.

1. **Process 1-2*a*:** Heating and humidification ($t_s > t_{d1}$). The water is externally heated.
2. **Process 1-2*b*:** Humidification ($t_s = t_{d1}$). The enthalpy of air increases. The water is externally heated.
3. **Process 1-2*c*:** Cooling and humidification ($t_{w1} < t_s < t_{d1}$). The mean surface temperature of water is greater than the wet bulb temperature of air, t_{w1}. The enthalpy of air increases as a result of humidification. The water has to be externally heated.
4. **Process 1-2*d*:** Adiabatic saturation ($t_{w1} = t_s$). The water is simply recirculated without any external heating or cooling.
5. **Process 1-2*e*:** Cooling and humidification ($t_{dp} < t_s < t_{w1}$). The enthalpy of air decreases. The water is required to be externally cooled.
6. **Process 1-2*f*:** Cooling ($t_s = t_{dp}$). Water is required to be cooled.
7. **Process 1-2*g*:** Cooling and dehumidification ($t_s < t_{dp}$). The condition line drawn from the initial state 1 is tangential to the saturation line.

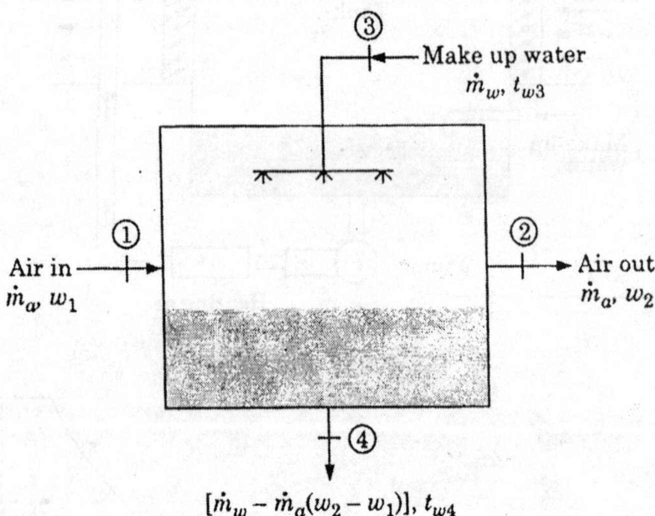

Fig. 12.14 Mass balance of an air washer.

The mass balance of an air washer is shown in Fig. 12.14.

Let $\dot{m}_a$ = mass flow rate of dry air

$\dot{m}_w$ = mass flow rate of water

Energy balance gives,

$$\dot{m}_a(h_2 - h_1) = \dot{m}_w c_{pw} t_{w3} - [\dot{m}_w - \dot{m}_a(w_2 - w_1)]\, c_{pw} t_{w4}$$

or

$$\dot{m}_a(h_2 - h_1) = \dot{m}_w c_{pw}(t_{w3} - t_{w4}) + \dot{m}_a(w_2 - w_1)]c_{pw}\, t_{w4}$$

Neglecting the effect of temperature of water in the last term, we have

$$\dot{m}_a(h_2 - h_1) = \dot{m}_w c_{pw}(t_{w3} - t_{w4}) \quad ...(12.33a)$$

or

$$\dot{m}_a dh = \dot{m}_w c_{pw} dt_w \quad ...(12.33b)$$

For adiabatic saturation case, $dh = 0$

$$\therefore \quad dt_w = 0 \text{ and } t_{w3} = t_{w4}$$

Humidifying efficiency of air washer,

$$\eta_H = \frac{h_2 - h_1}{h_s - h_1} = \frac{w_2 - w_1}{w_s - w_1} \quad ...(12.34)$$

By Pass Factor,

$$\text{BPF} = \frac{w_s - w_2}{w_s - w_1} = 1 - \frac{w_2 - w_1}{w_s - w_1} = 1 - \eta_H \quad ...(12.35)$$

12.8.7 Cooling with Adiabatic Humidification

When the air is passed through an insulated chamber, as shown in Fig. 12.15(*a*), having water sprays maintained at a temperature (t_f) higher than the DPT of entering air (t_{dp1}), but lower than its DBT (t_{d1}) of entering air equal to the WBT of entering air (t_{w1}), then the air is said to be cooled and humidified. Since the chamber is insulated and same water is circulated again and again, therefore, a condition of adiabatic saturation is reached. Ultimately $t_f = t_{w1}$. This process is shown by line 1-3 in Fig. 12.15(*b*). This line follows the path along constant WBT line or constant enthalpy line.

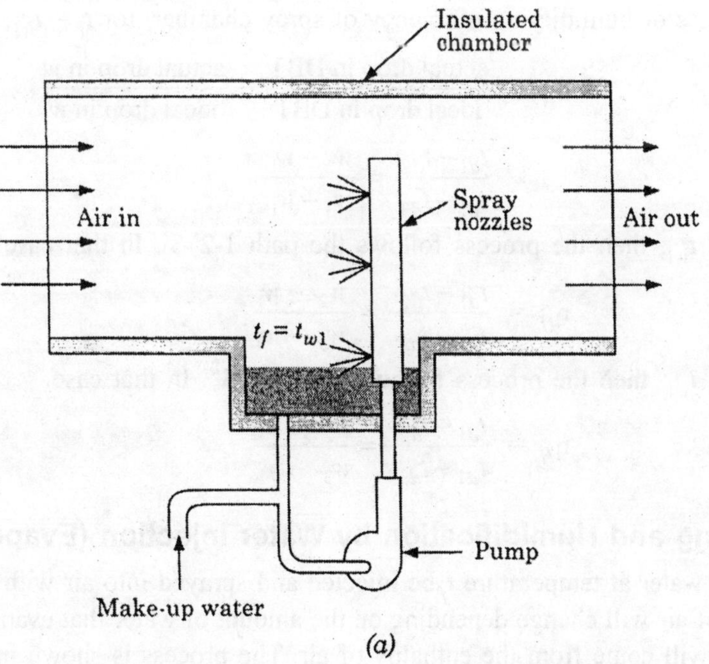

(*a*)

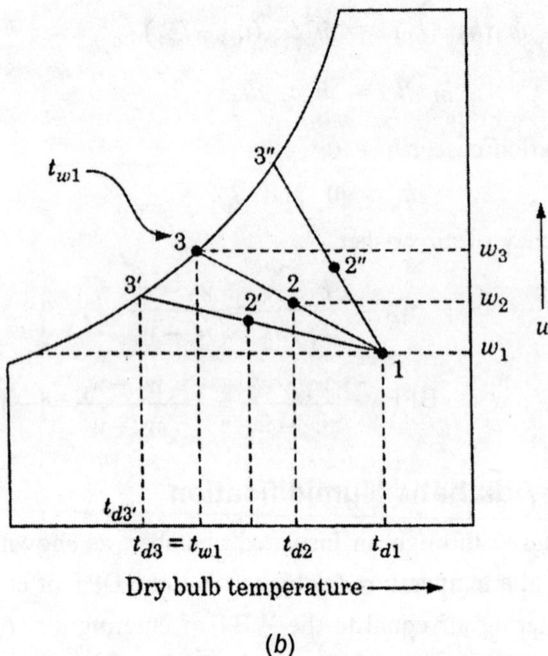

(*b*)

Fig. 12.15 Cooling with adiabatic humidification.

For perfect humidification, the final condition of air will be at point 3, where $t_{d3} = t_{w1}$. In actual practice the humidification is not perfect, therefore, the final condition of air at outlet will be represented by point 2 on line 1-3.

Effectiveness or humidifying efficiency of spray chamber, for $t_f = t_{w1}$, is given by

$$\eta_H = \frac{\text{actual drop in DBT}}{\text{ideal drop in DBT}} = \frac{\text{actual drop in } w}{\text{ideal drop in } w}$$

$$= \frac{t_{d1} - t_{d2}}{t_{d1} - t_{d3}} = \frac{w_2 - w_1}{w_3 - w_1} \quad \text{...(12.36}a\text{)}$$

When $t_f < t_{w1}$, then the process follows the path 1-2′-3′. In that case

$$\eta_H = \frac{t_{d1} - t_{d2'}}{t_{d1} - t_{d3'}} = \frac{w_{2'} - w_1}{w_{3'} - w_1} \quad \text{...(12.36}b\text{)}$$

When $t_f > t_{w1}$, then the process follows line 1-2″-3″. In that case

$$\eta_H = \frac{t_{d1} - t_{d2''}}{t_{d1} - t_{d3''}} = \frac{w_{2''} - w_1}{w_{3''} - w_1} \quad \text{...(12.36}c\text{)}$$

12.8.8 Cooling and Humidification by Water Injection (Evaporative Cooling)

Let liquid water at temperature t_f be injected and sprayed into air with the help of nozzles. The condition of air will change depending on the amount of water that evaporates. The enthalpy of evaporation will come from the enthalpy of air. The process is shown in Fig. 12.16.

Let $\dot{m}_a$ = mass flow rate of air

$\dot{m}_v$ = mass flow rate of water evaporated

= mass flow rate of water supplied.

h_f = enthalpy of liquid water

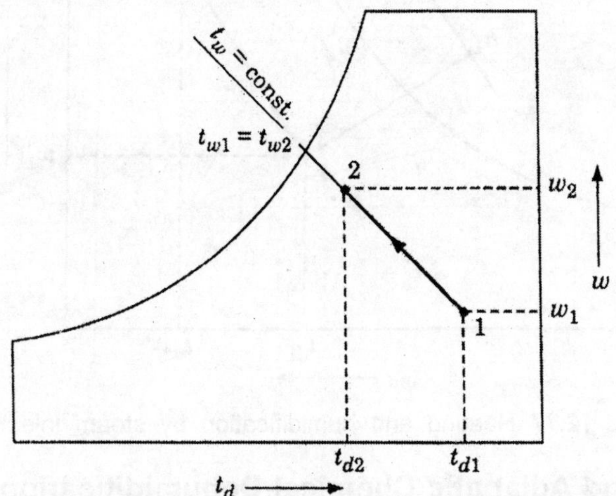

Fig. 12.16 Cooling and humidification by water injection.

Then mass and enthalpy balances give

$$w_2 = w_1 + \frac{\dot{m}_v}{\dot{m}_a} \qquad ...(12.37)$$

$$h_2 = h_1 + \frac{\dot{m}_v}{\dot{m}_a} h_f = h_1 + (w_2 - w_1) h_f \qquad ...(12.38)$$

The term $(w_2 - w_1)h_f$ is extremely small as compared to h_1 and h_2.

Hence $h_1 = h_2$. Thus the water injection process is a constant enthalpy process, irrespective of the temperature of injected water.

12.8.9 Heating and Humidification by Steam injection

Steam is normally injected into fresh outdoor air which is then supplied for the conditioning of textile mills where high humidity needs to be maintained. The process is shown in Fig. 12.17. If m_s is the mass of steam supplied with enthalpy h_s and m_a the mass of air, then for the mass balance,

$$w = w_1 + \frac{m_s}{m_a} \qquad ...(12.39)$$

and for the enthalpy balance,

$$h_2 = h_1 + \frac{m_s}{m_a} h_s = h_1 + (w_2 - w_1) h_s \qquad ...(12.40)$$

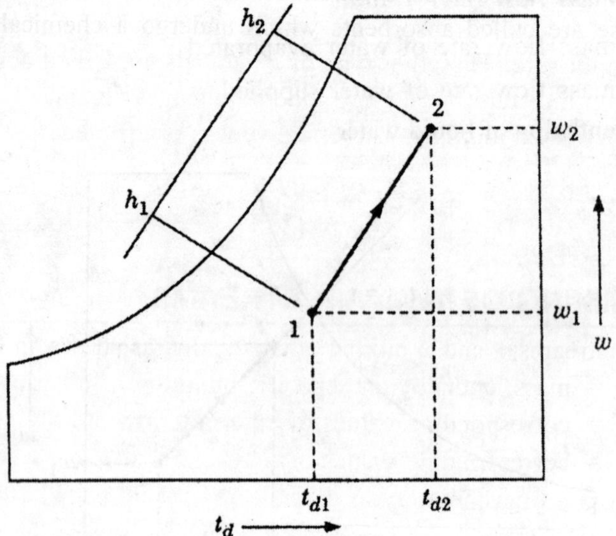

Fig. 12.17 Heating and humidification by steam injection.

12.8.10 Heating and Adiabatic Chemical Dehumidification

This process is mainly used in industrial air-conditioning. It can also be used for some comfort air-conditioning requiring low relative humidity or low dew point temperature in the room. In this process, the air is passed over chemicals which have infinity for moisture. As the air comes in contact with these chemicals, the moisture gets condensed out of air and gives up its latent heat. Due to condensation, the specific humidity decreases and the heat of condensation supplies sensible heat for heating of air. As a result of this DBT increases. The process is shown in Fig. 12.18 by line 1-2. The path followed during the process is along the constant WBT line or constant enthalpy line.

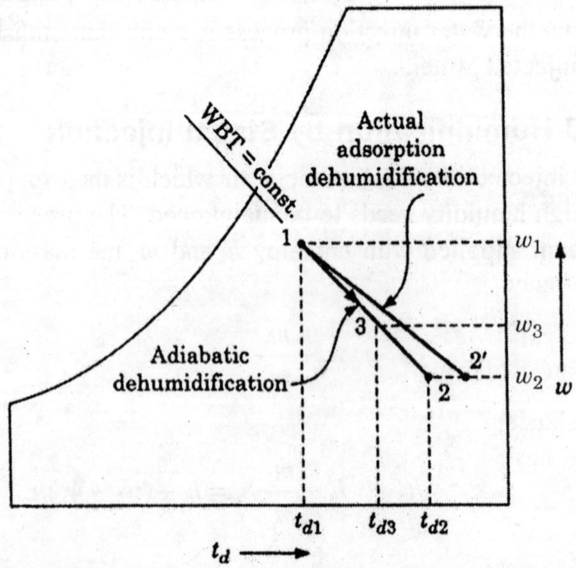

Fig. 12.18 Heating and adiabatic chemical dehumidification.

The chemicals used are hygroscopic solutions or brines of calcium chloride, lithium chloride and ethyl glycol. These are called absorbents which undergo a chemical or physical or both changes during taking moisture. The other type of chemicals are called absorbents which are in the solid state to take up moisture but do not undergo changes chemically or physically. These include silica gel and activated alumina. The effectiveness or efficiency of the dehumidifier,

$$\eta_H = \frac{t_{d3} - t_{d1}}{t_{d2} - t_{d1}} \qquad \text{...(12.41)}$$

12.9 ADIABATIC MIXING OF TWO AIR STREAMS

Consider two air streams 1 and 2 mixing adiabatically, as shown in Fig. 12.19 (*a*).

Let m_{a1}, h_1, w_1 = mass, enthalpy and specific humidity of entering air at 1.

m_{a2}, h_2, w_2 = corresponding values of entering air at 2.

m_{a3}, h_3, w_3 = corresponding values of leaving air at 3.

The mixing process on psychrometic process is shown in Fig. 12.19(*b*).

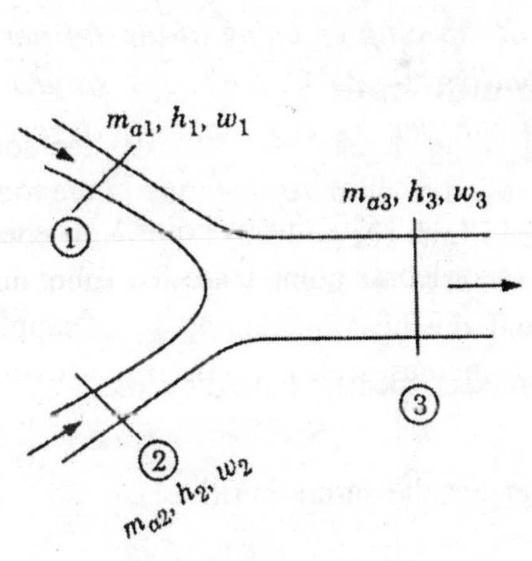

(a) Adiabatic mixing of two air streams.

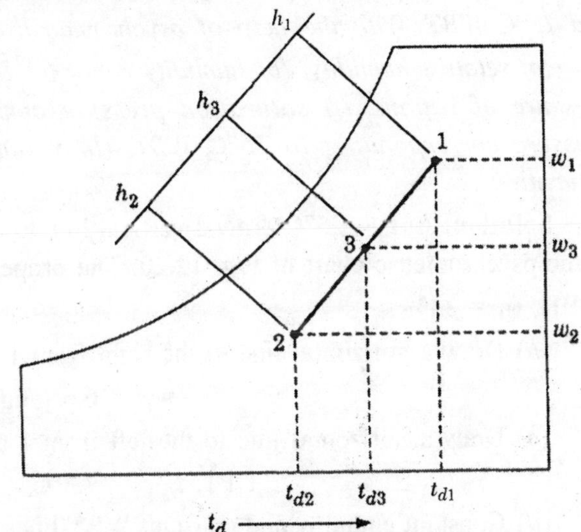

(b) Mixing process on psychrometric chart.

Fig. 12.19

For the mass balance,

$$m_{a1} + m_{a2} = m_{a3} \qquad \text{...(1)}$$

For the energy balance,

$$m_{a1}h_1 + m_{a2}h_2 = m_{a3}h_3 \qquad \text{...(2)}$$

For the mass balance of water vapour,

$$m_{a1}w_1 + m_{a2}w_2 = m_{a3}w_3 \qquad \text{...(3)}$$

From Eq. (2), we have

$$h_3 = \frac{m_{a1}h_1 + m_{a2}h_2}{m_{a3}} \qquad \text{...(12.42)}$$

and from Eq. (3),

$$w_3 = \frac{m_{a1}w_1 + m_{a2}w_2}{m_{a3}} \quad ...(12.43)$$

Eq. (12.42) can be written as,

$$c_p t_{d3} + h_{fg} w_3 = \frac{m_{a1}}{m_{a3}}(c_p t_{d1} + h_{fg} w_1) + \frac{m_{a1}}{m_{a3}}(c_p t_{d2} + h_{fg} w_2)$$

Simplifying, we get

$$t_{d3} = \frac{m_{a1}t_{d1} + m_{a2}t_{d2}}{m_{a3}} + \frac{h_{fg}}{c_p}\left[\frac{m_{a1}}{m_{a3}}w_1 + \frac{m_{a2}}{m_{a3}}w_2 - w_3\right]$$

The second term in the above expression being negligible, we get

$$t_{d3} = \frac{m_{a1}t_{d1} + m_{a2}t_{d2}}{m_{a3}} \quad ...(12.44)$$

Example 12.6 *Atmospheric air at 760 mm of Hg barometric pressure has 25°C DBT and 15°C WBT. With the help of psychrometric chart, determine:*

(a) relative humidity, (b) humidity ratio, (c) DPT, (d) specific enthalpy of air, (e) partial pressure of vapour, (f) saturation pressure corresponding to 25°C DBT, (g) saturation pressure corresponding to 15°C WBT, (h) volume of air per kg d.a., and (i) degree of saturation.

Solution: At t_d = 25°C draw a vertical line to intersect t_w = 15°C line at point 1, as shown in the psychrometric chart of Fig. 12.20. The properties recorded at point 1 are the following:

(*a*)RH, ϕ_1 = 34%

(*b*) Draw a horizontal line to the right from 1 to intersect specific humidity line.

$$w_1 = 6.6 \text{ g/kg d.a.}$$

(*c*) Draw a horizontal line to the left from 1 to intersect the saturation line

$$\text{DPT, } t_{dp1} = 7.8°\text{C}$$

(*d*) Constant enthalpy and constant WBT lines are the same. Produce the constant enthalpy line through 1 to cut enthalpy scale at h_1 = 42 kJ/kg d.a.

(*e*) Draw a horizontal line from 1 to the left to intersect the pressure scale at p_{v1} = 7.9 mm of Hg.

(*f*) Draw a vertical line at t_d = 25°C to intersect the saturation line at *A*. At point *A*, draw a horizontal line to the left to the vapour pressure scale at p_{s1} = 23.7 mm of Hg.

(*g*) Let the WBT line, t_{w1} = 15°C intersect the saturation line at *B*. At point *B*, draw a horizontal line to the left to meet the vapour pressure scale at p_{sw1} = 12.5 mm of Hg.

(*h*) At point 1, v_1 = 0.8542 m³/kg

(*i*) At A draw a horizontal line to intersect specific humidity scale at w_s = 20.2 g/kg d.a

$$\text{Degree of saturation, } \mu = \frac{w_1}{w_s} = \frac{6.6}{20.2} = 0.326 \text{ or } 32.6\%$$

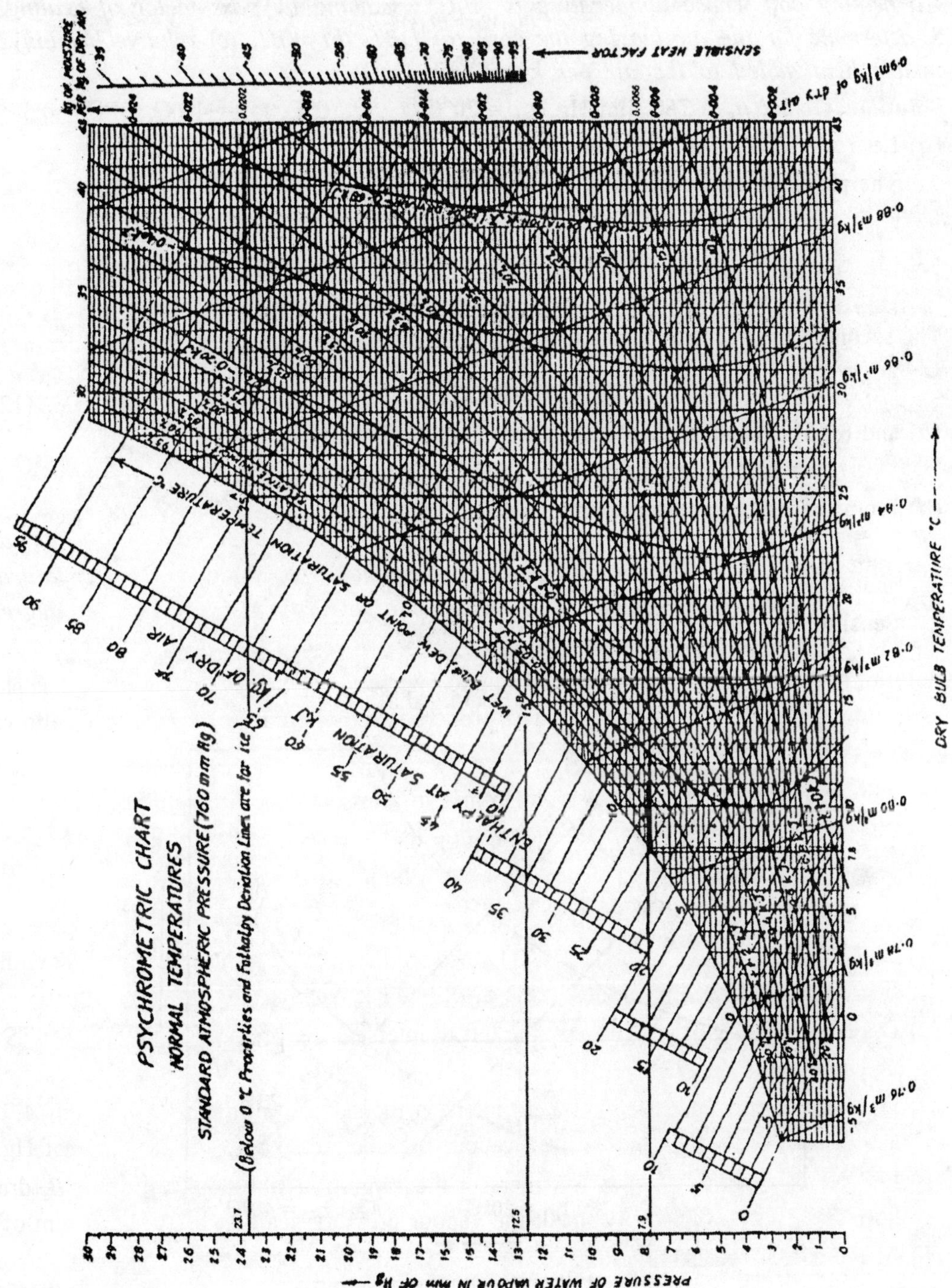

Fig. 12.20 Psychrometric chart.

Example 12.7 *The atmospheric air at 760 mm of Hg, DBT = 20°C and WBT = 10°C enters a heating coil whose temperature is 40°C. Assuming by pass factor of heating coil as 0.5, determine for the air leaving the coil: (a) DBT, (b) WBT, (c) relative humidity, and (d) sensible heat added to the air per kg of dry air.*

Solution: Given: p_b = 760 mm Hg, t_{d1} = 20°C, t_{w1} = 10°C, t_{d3} = 40°C, BPF = 0.5

(*a*) Let t_{d2} = DBT of air leaving the coil.

The psychrometic process is shown in Fig. 12.21.

$$\text{BPF} = \frac{t_{d3} - t_{d2}}{t_{d3} - t_{d1}}$$

$$0.5 = \frac{40 - t_{d2}}{40 - 20}$$

$$t_{d2} = 30°\text{C}$$

(*b*) and (*c*) From the psychrometric chart, we find that

$$t_{w2} = 14.4°\text{C and } RH,\ \phi = 15\%$$

(*d*) From the psychrometric chart,

$$h_1 = 30 \text{ kJ/kg. d.a.}$$

and $$h_2 = 40 \text{ kJ/kg. d.a.}$$

Sensible heat added to air $= h_2 - h_1$

$= 40 - 30$

$= 10$ kJ/kg d.a.

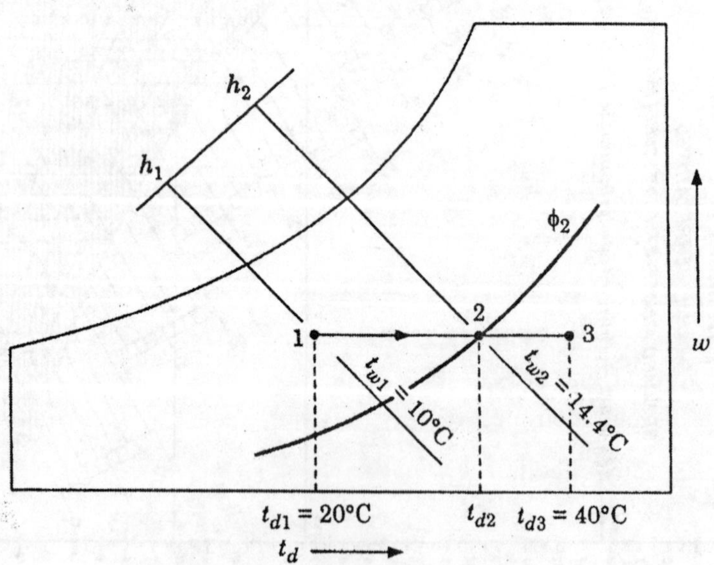

Fig. 12.21 Psychrometric chart.

Example 12.8 *Moist air enters a refrigeration coil at the rate of 120 kg d.a./min at 40°C and 50% RH. The apparatus dew point of coil is 10°C and by-pass factor is 0.20. Determine the outlet state of moist air and capacity of coil in TR.*

Solution: Given: $\dot{m}_a$ = 120 kg d.a./min, t_{d1} = 40°C, ϕ_1 = 50%, ADP = 10°C,

BPF = 0.20

(*a*) Let t_{d2} and ϕ_2 be the DBT and RH of air leaving the cooling coil respectively.
The psychrometric process is shown in Fig. 12.22.
Locate point 1 (t_{d1} = 40°C, ϕ_1 = 50%) on the psychrometric chart. Join point 1 with ADP = 10°C point on the saturation line.

Now
$$\text{BPF} = \frac{t_{d2} - t_{d3}}{t_{d1} - t_{d3}} = \frac{t_{d2} - ADP}{t_{d1} - ADP}$$

$$0.2 = \frac{t_{d2} - 10}{40 - 10}$$

$$t_{d2} = 16°\text{C}$$

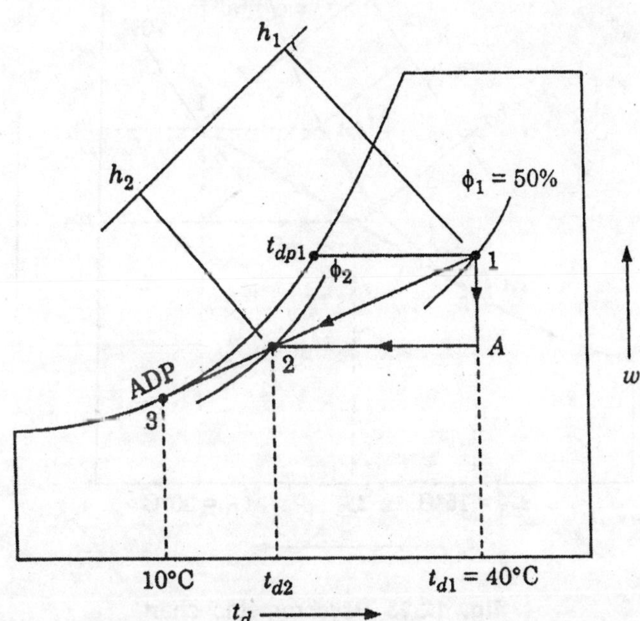

Fig. 12.22 Psychrometric chart.

(*b*) Draw a vertical line at t_{d2} = 16°C to intersect line 1-3 at point 2. Then ϕ_2 = 93%

(*c*) From the psychrometric chart, h_1 = 101 kJ/kg d.a, and h_2 = 43 kJ/kg d.a.

Cooling capacity of coil, $\dot{Q} = \dot{m}_a(h_1 - h_2)$

$$= 120(101 - 43) \times \frac{1}{211} = 32.986 \text{ TR}$$

Example 12.9 *The atmospheric air at 30°C DBT and 70% RH enters a cooling coil at the rate of 220 m³/min. The coil DPT = 15°C and its BPF = 0.15. Determine*

(*a*) *the temperature of air leaving the cooling coil,*

(*b*) *the capacity of the cooling coil in TR,*

(c) the amount of water vapour removed per minute, and

(d) Sensitive heat factor for the process.

Solution: Given: t_{d1} = 30°C, ϕ_1 = 70%, $\dot{V}_1$ = 220 m³/min (cmm), ADP, t_{d3} = 15°C, BPF = 0.15

(*a*) Let t_{d2} = temperature of air leaving the cooling coil

The psychromatric process is shown in Fig. 12.23.

At t_{d1} = 30°C and ϕ_1 = 70%, t_{dp1} = 24°C from psychrometric chart.

$$\text{BPF} = \frac{t_{d2} - t_{d3}}{t_{d1} - t_{d3}}$$

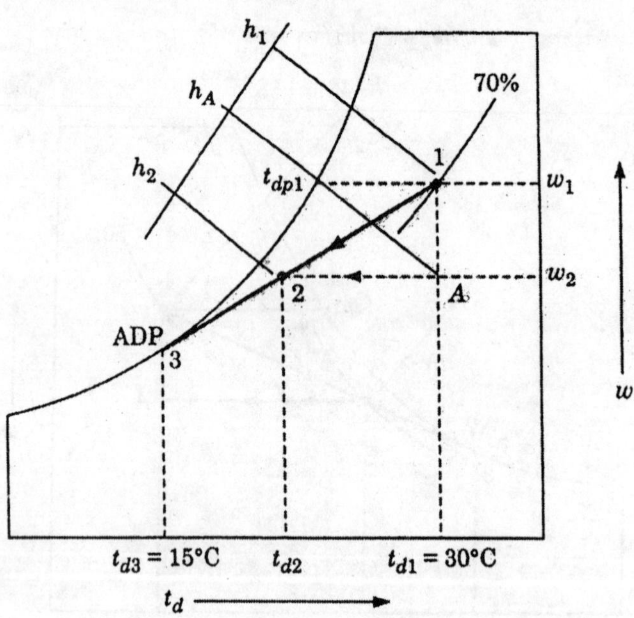

Fig. 12.23 Psychrometric chart.

$$0.15 = \frac{t_{d2} - 15}{30 - 15}$$

$$t_{d2} = 17.25°\text{C}$$

From psychrometric chart, we have

$$w_1 = 18.8 \text{ g/kg, d.a.}$$

$$w_2 = 11.8 \text{ g/kg. d.a.}$$

$$v_1 = 0.885 \text{ m}^3\text{/kg}$$

$$h_1 = 78.5 \text{ kJ/kg d.a.}$$

$$h_2 = 47.8 \text{ kJ/kg d.a.}$$

$$h_A = 61.8 \text{ kJ/kg d.a.}$$

Mass of air flowing through cooling coil,

$$\dot{m}_a = \frac{\dot{V}_1}{v_1} = \frac{220}{0.885} = 248.6 \text{ kg/min}$$

(*b*) Capacity of cooling coil $= \dot{m}_a(h_1 - h_2)\text{d} = 248.6(78.5 - 47.8) = 7632 \text{ kJ/min}$

$$= \frac{7632}{211} = 36.17 \text{ TR}$$

$$= \frac{7632}{60} = 127.2 \text{ kW}$$

(*c*) Amount of water vapour removed per minute,

$$\dot{m}_v = \dot{m}_a(w_1 - w_2)$$

$$= 248.6\ (18.8 - 11.8) \times 10^{-3} = 1.74 \text{ kg/min}$$

(*d*) Sensible heat factor, $\text{SHF} = \frac{h_A - h_2}{h_1 - h_2} = \frac{61.8 - 47.8}{78.8 - 47.8} = 0.45$

Example 12.10 *In a certain environment, the DBT of air is 25°C and RH is 40%. Determine the specific humidity, dew point and WBT of air. If this air is cooled in the air washer using recirculated spray water and having a humidifying efficiency of 0.85, find the DBT and DPT of air leaving the air washer.*

Solution: Given: $t_{d1} = 25°\text{C}$, $\phi_1 = 40\%$, $\eta_H = 0.85$

The psychrometric process is shown in Fig. 12.24.

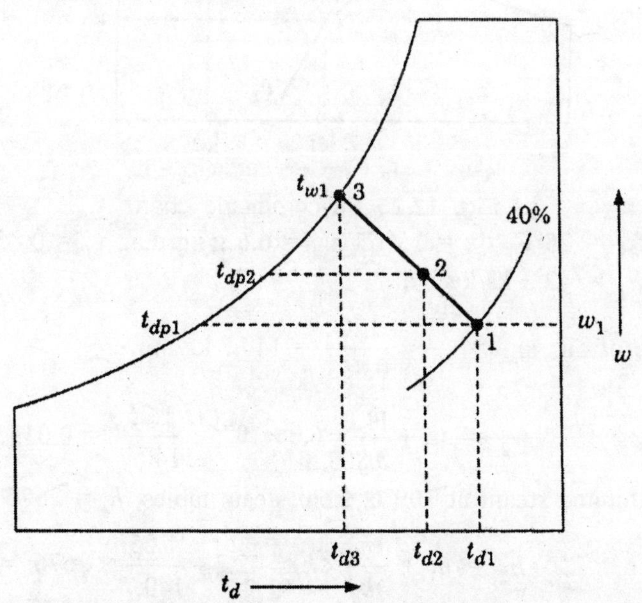

Fig. 12.24 Psychrometric chart.

At point 1 $(t_{d1} = 25°C,\ \phi_1 = 40\%)$, $w_1 = 7.6$ g/kg d.a., $t_{dp1} = 10.4°C$

and $t_{w1} = 16°C = t_{d3}$

Let t_{d2} = DBT of air leaving the air washer

$$\eta_H = \frac{t_{d1} - t_{d2}}{t_{d1} - t_{d3}} = \frac{25 - t_{d2}}{25 - 16} = 0.85$$

$$t_{d2} = 17.35°C$$

Then $t_{dp2} = 15.2°C$

Example 12.11 *The atmospheric air at 25°C DBT and 15°C WBT is flowing at the rate of 120 cmm through a duct. The dry saturated steam at 100°C is injected into the air stream at the rate of 75 kg/h. Calculate the specific humidity, enthalpy, DBT, WBT, and RH of leaving air.*

Solution: Given: $t_{d1} = 25°C$, $t_{w1} = 15°C$, $\dot{V}_1 = 120$ cmm, $t_s = 100°C$,

$\dot{m}_s = 75$ kg/h or 1.25 kg/min

The psychrometric process is shown in Fig. 12.25.

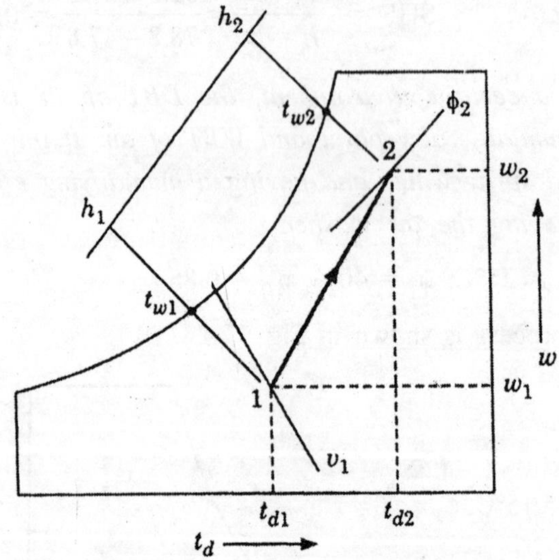

Fig. 12.25 Psychrometric chart.

At state 1 $(t_{d1} = 25°C,\ t_{w1} = 15°C)$, $w_1 = 6.6$ g/kg d.a., $v_1 = 0.853$ m³/kg,

$h_1 = 42.2$ kJ/kg d.a.

$$\text{Mass flow rate of air, } \dot{m}_a = \frac{\dot{V}_1}{v_1} = \frac{120}{0.853} = 140.7 \text{ kg/min}$$

$$w_2 = w_1 + \frac{\dot{m}_s}{\dot{m}_a} = 6.6 \times 10^{-3} + \frac{1.25}{140.7} = 0.0155 \text{ kg/kg d.a.}$$

Enthalpy of saturated steam at 100°C from steam tables, $h_s = 2676$ kJ/kg

$$h_2 = h_1 + \frac{\dot{m}_s}{\dot{m}_a} \times h_s = 42.2 + \frac{1.25}{140.7} \times 2676 = 65.97 \text{ kJ/kg d.a.}$$

Locate point 2 (w_2 = 0.0155 kg/kg d.a., h_2 = 65.97 kJ/kg) on the psychrometric chart, Then t_{d2} = 260°C, *WBT*, t_{w2} = 22.4°C, ϕ_2 = 73%.

Example 12.12 *Air at 15°C DBT and 90% RH is to be heated and humidified to 40°C DBT and 25°C WBT. The air is preheated sensibly before passing to the air washer in which water is recirculated. The RH of air coming out of air washer is 90%. This air is again reheated sensibly to obtain the final desired condition. Determine: (a) the temperature to which air should be preheated, (b) the total heating required, (c) the make up water required in the air washer, and (d) the humidifying efficiency of air washer.*

Solution: Given: t_{d1} = 15°C, ϕ_1 = 90%, t_{d2} = 40°C, t_{w2} = 25°C

The psychrometric process is shown in Fig. 12.26

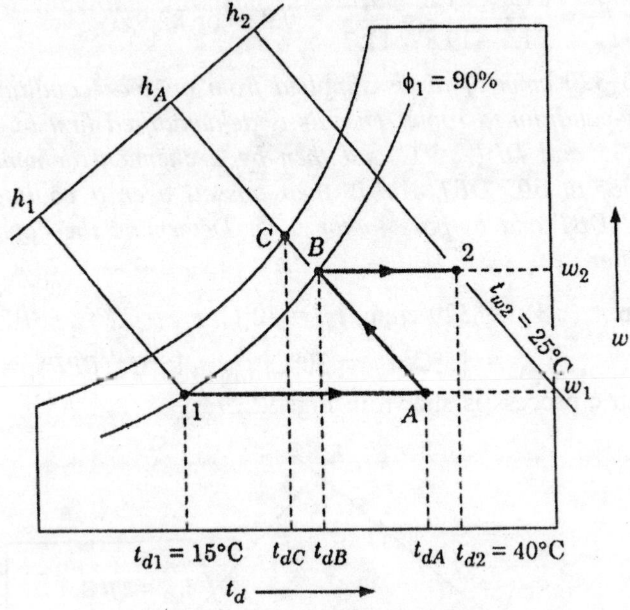

Fig. 12.26 Psychrometric chart.

Locate point 1(t_{d1} = 15°C, ϕ_1 = 90%) and point 2(t_{d2} = 40°C, t_{w2} = 25°C) on the psychrometric chart.

From point 1 draw a horizontal line to the right to represent sensible heating.

From point 2 draw a horizontal line to the left to intersect 90% *RH* line at point *B*. Then from point *B* draw a constant WBT line to intersect horizontal line at point 1 at point *A*. The line 1-*A* represents preheating of air, *AB* represents humidification and *B*-2 represents reheating to final condition.

From the psychrometric chart, we have

t_{dA} = 31.5°C	h_1 = 39 kJ/kg d.a.
h_2 = 76.8 kJ/kg d.a.	h_A = 56.4 kJ/kg d.a.
w_1 = 0.0096 kg/kg d.a.,	w_2 = 0.0138 kg/kg d.a.
t_{dB} = 21.8°C	t_{dC} = 19.8°C

(*b*) Heat required for preheating of air,

$$Q_p = h_A - h_1 = 56.4 - 39 = 17.4 \text{ kJ/kg d.a.}$$

Heat required for reheating of air,

$$Q_r = h_2 - h_B = 76.8 - 56.4 = 20.4 \text{ kJ/kg d.a.} \qquad (\because h_B = h_A)$$

Total heat required, $Q = Q_p + Q_r = 17.4 + 20.4 = 37.8$ kJ/kg.d.a.

(*c*) Make up water required in air washer

$$= w_B - w_A = w_2 - w_1$$

$$= 0.0138 - 0.0096 = 0.0042 \text{ kg/kg.d.a.}$$

(*d*) $\eta_H = \dfrac{t_{dA} - t_{dB}}{t_{dA} - t_{dC}} = \dfrac{31.5 - 21.8}{31.5 - 19.8} = \dfrac{9.7}{11.7} = 0.829$ or 82.9%

Example 12.13 *320 cmm of air is supplied from outdoor conditions of 40°C DBT and 27°C WBT to an air-conditioned room. The air is dehumidified first by a cooling coil having by-pass factor of 0.35 and DPT 15°C and then by a chemical dehumidifier. Air leaves the chemical dehumidifier at 30° DBT. Air is then passed over a cooling coil whose surface temperature is 15°C DBT and by-pass factor 0.25. Determine the capacities of two cooling coils and dehumidifier.*

Solution: Given: $\dot{V}_1 = 320$ cmm, $t_{d1} = 40°C$, $t_{w1} = 27°C$, $(BPF)_1 = 0.35$, ADP, $t_{dp3} = 15°C$, $t_{d4} = 30°C$, $t_{d5} = 15°C$, $(BPF)_2 = 0.25$.

The psychrometric process is shown in Fig. 12.27.

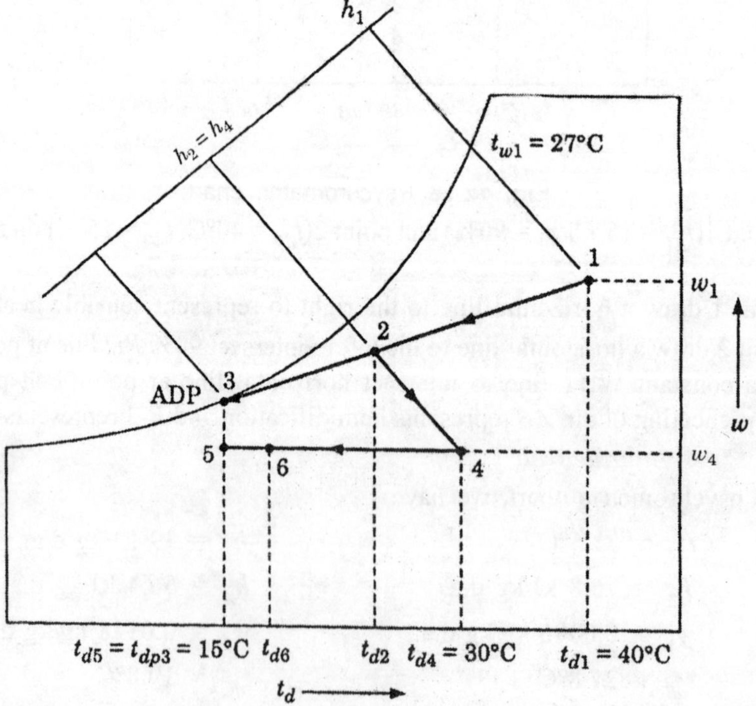

Fig. 12.27 Psychrometric chart.

Locate point 1 (t_{d1} = 40°C, t_{w1} = 27°C) on the psychrometric chart. Similarly locate point 3(t_{dp3} = 15°C) on the saturation curve.

For the first cooling coil,

$$(BPF)_1 = \frac{t_{d2} - t_{d3}}{t_{d1} - t_{d3}}$$

$$0.35 = \frac{t_{d2} - 15}{40 - 15}$$

$$t_{d2} = 23.75°C$$

Point 2 represents the state of air leaving the first cooling coil.

From point 2, draw constant enthalpy line to intersect the vertical line at t_{d4} = 30°C. Process 2-4 represents the chemical dehumidification of air. At state 4, the air now enters the second cooling coil having surface temperature of t_{d5} = 15°C.

For the second cooling coil,

$$(BPF)_2 = \frac{t_{d6} - t_{d5}}{t_{d4} - t_{d5}}$$

Point 6 represents the condition of air leaving the second cooling coil.

From the psychromatic chart, we have

$$w_1 = 0.0172 \text{ kg/kg d.a.}, \; h_1 = 85.2 \text{ kJ/kg d.a.}, \; v_1 = 0.91 \text{ m}^3\text{/kg}$$

$$h_2 = 57 \text{ kJ/kg}, \; h_4 = h_2 = 57 \text{ kJ/kg}, \; w_4 = 0.014 \text{ kg/kg d.a.}$$

$$h_6 = 45.5 \text{ kJ/kg}$$

$$\text{Mass of air supplied, } \dot{m}_a = \frac{\dot{V}_1}{v_1} = \frac{320}{0.91} = 351.6 \text{ kg/min}$$

Capacity of first cooling coil,

$$\dot{Q}_1 = \dot{m}_a(h_1 - h_2) = 351.6\,(85.2 - 57) = 9915 \text{ kJ/min}$$

Capacity of second cooling coil,

$$\dot{Q}_2 = \dot{m}_a(h_4 - h_6) = 351.6\,(57 - 45.5) = 4043 \text{ kJ/min}$$

Capacity of chemical dehumidifier,

$$\dot{m}_d = \dot{m}_a(w_1 - w_4) = 351.6\,(0.0172 - 0.014) = 1.125 \text{ kg/min}$$

Example 12.14 *Saturated air leaving the cooling section of an air-conditioning system at 15°C at the rate of 60 cmm is mixed adiabatically with outside air at 30°C and 60% RH at the rate of 20 cmm. Find the specific humidity, relative humidity, DBT and the volume flow rate of the mixture.*

Solution: Given: t_{d1} = 30°C, ϕ_1 = 60°, $\dot{V}_1$ = 20 cmm, t_{d2} = 15°C, $\dot{V}_2$ = 60 cmm

The psychrometric process for mixing is shown in Fig. 12.28.

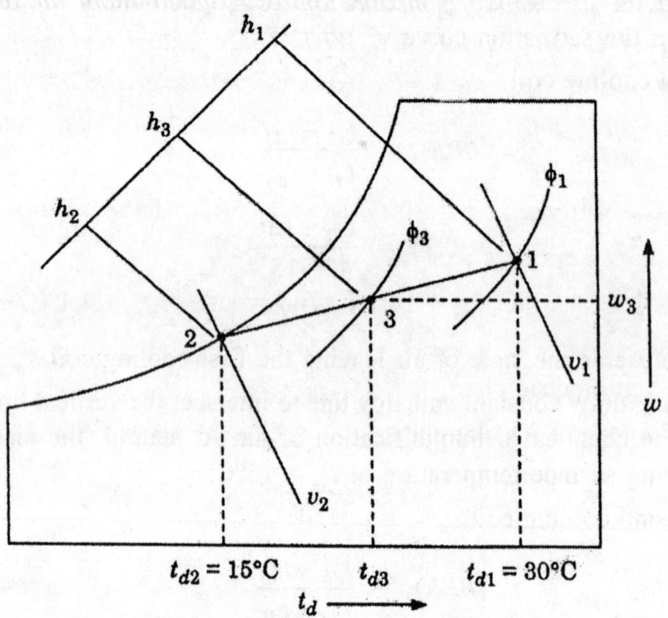

Fig. 12.28 Psychrometric chart.

From the psychrometric chart, we have

$$h_1 = 71.6 \text{ kJ/kg.d.a.}, \; h_2 = 42.1 \text{ kJ/kg.d.a.},$$

$$v_1 = 0.88 \text{ m}^3\text{/kg.d.a.}, \; v_2 = 0.83 \text{ m}^3\text{/kg.d.a.}$$

$$\dot{m}_{a1} = \frac{\dot{V}_1}{v_1} = \frac{20}{0.88} = 22.73 \text{ kg/min}$$

$$\dot{m}_{a2} = \frac{\dot{V}_2}{v_2} = \frac{60}{0.83} = 72.29 \text{ kg/min}$$

Now

$$\frac{\dot{m}_{a1}}{\dot{m}_{a2}} = \frac{h_3 - h_2}{h_1 - h_3}$$

$$\frac{22.73}{72.29} = \frac{h_3 - 42.1}{71.6 - h_3}$$

$$0.3144(71.6 - h_3) = h_3 - 42.1$$

$$h_3 = 49.16 \text{ kJ/kg.d.a.}$$

$$w_3 = 0.012 \text{ kg/kg.d.a.}, \; \phi_3 = 88\% \text{ and } t_{d3} = 18.8°\text{C}$$

$$v_3 = 0.843 \text{ m}^3\text{/kg.d.a.}$$

$$\dot{V}_3 = (\dot{m}_{a1} + \dot{m}_{a2})\, v_3 = (22.73 + 72.29) \times 0.843 = 80.1 \text{ m}^3\text{/min.}$$

Example 12.15 *40 cmm of a mixture of recirculated room air and outdoor air enters a cooling coil at 31°C DBT and 18.5°C WBT. The effective surface temperature of the coil is 4.4°C. The surface area of the coil is such as would give 12.5 kW of refrigeration with the given entering air state. Determine the dry and wet bulb temperatures of the air leaving the coil and by-pass factor.*

Solution: $\dot{V}_1$ = 40 cmm, t_{d1} = 30°C, t_{w1} = 18.5°C, ADP = t_{w3} = 4.4°C, $\dot{Q}$ = 12.5 kW

The psychrometric process is shown in Fig. 12.29.

Locate point 1(t_{d1} = 31°C, t_{w1} = 18.5°C) and point 3 (t_{dp3} = 4.4°C) on the saturation curve. Joint points 1 and 3.

From the psychrometric chart,

$$h_1 = 52.5 \text{ kJ/kg d.a.}, \qquad h_3 = 17.7 \text{ kJ/kg d.a.}$$

$$w_1 = 0.0082 \text{ kg/kg d.a.}, \qquad w_3 = 0.00525 \text{ kg/kg d.a.}$$

$$v_1 = 0.872 \text{ m}^3\text{/kg.}$$

Mass flow rate of air, $$\dot{m}_a = \frac{\dot{V}_1}{v_1} = \frac{40}{0.872} = 45.87 \text{ kg/min}$$

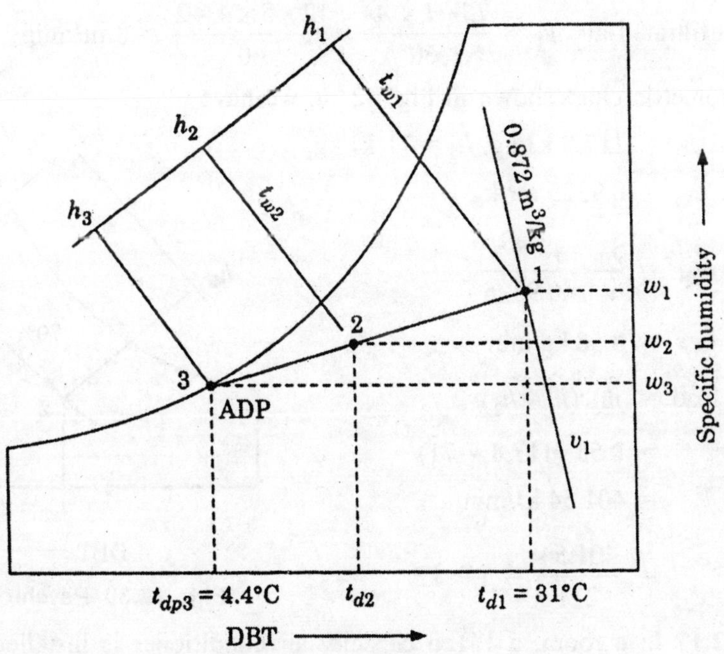

Fig. 12.29 Psychrometric chart.

Cooling capacity of coil, $$\dot{Q} = \dot{m}_a(h_1 - h_2)$$

$$h_1 - h_2 = \frac{12.5 \times 60}{45.87} = 16.35 \text{ kJ/kg d.a.}$$

$$h_2 = 52.5 - 16.35 = 36.15 \text{ kJ/kg d.a.}$$

Now $$\text{BPF} = \frac{h_2 - h_3}{h_1 - h_3} = \frac{w_2 - w_3}{w_1 - w_3}$$

$$\frac{w_2 - 0.00525}{0.0082 - 0.00525} = \frac{36.15 - 17.7}{52.5 - 17.7}$$

$$w_2 = 0.00681 \text{ kg/kg d.a.}$$

Now locate point 2 on line 1-3 such that h_2 = 36.15 kJ/kg d.a. and w_2 = 0.00681 kg/kg d.a.

Then $$t_{d2} = 18°\text{C},\ t_{w2} = 12.6°\text{C}$$

$$\text{BPF} = \frac{h_2 - h_3}{h_1 - h_3} = \frac{36.15 - 17.7}{52.5 - 17.7} = 0.53$$

Example 12.16 *A room of size 12 m × 5 m × 4 m high is maintained at DBT = 28°C and RH = 70%. The ambient air is at DBT = 41°C and RH = 60%. The infiltration amounts to two air changes per hour. Determine the infiltration load.*

Solution: Given: Room size 12 m × 5 m × 4 m, t_{d2} = 28°C, ϕ_2 = 70%, t_{d1} = 41°C, ϕ_1 = 60%, infiltration = air changes per hour, A_C = 2 for a room having four walls exposed.

Amount of infiltrated air, $$\dot{V}_1 = \frac{LWH \times A_C}{60} = \frac{12 \times 5 \times 4 \times 2}{60} = 8 \text{ m}^3/\text{min}$$

From psychrometric chart shown in Fig. 12.30, we have

$$h_1 = 117.8 \text{ kJ/kg},\ h_2 = 71 \text{ kJ/kg},$$

$$v_1 = 0.9325 \text{ m}^3/\text{kg}$$

$$\dot{m}_{a1} = \frac{\dot{V}_1}{v_1} = \frac{8}{0.9325}$$

$$= 8.58 \text{ kg/min}$$

$$\text{Infiltration load} = \dot{m}_{a1}(h_1 - h_2)$$

$$= 8.58\,(117.8 - 71)$$

$$= 401.54 \text{ kJ/min}$$

$$= \frac{401.54}{211} = 1.9 \text{ TR}$$

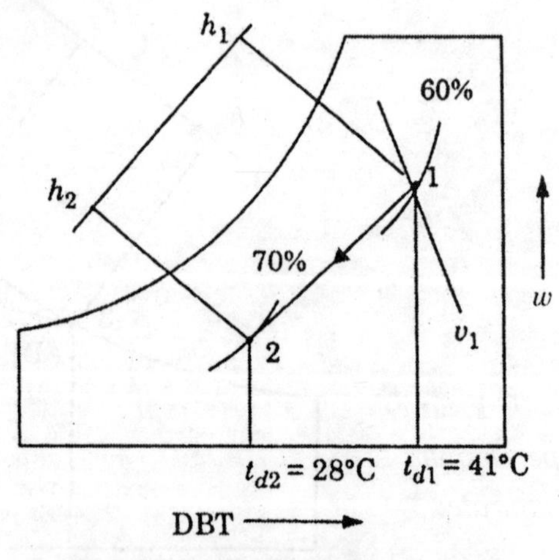

Fig. 12.30 Psychrometric chart.

Example 12.17 In a room, a 1 Ton capacity air conditioner is installed which supplies 12 m³/min air at dry bulb temperature of 18°C and wet bulb temperature of 15°C. If it sucks air at DBT at 28°C and relative humidity of 60%, determine the actual capacity of air conditioner, with the help of psychrometric chart.

Solution: Given: Q = 1 TR, $\dot{V}_1$ = 12 m³/min, t_{d2} = 18°C, t_{w2} = 15°C, t_{d1} = 28°C, ϕ_1 = 60%

The psychrometric process is shown in Fig. 12.31.

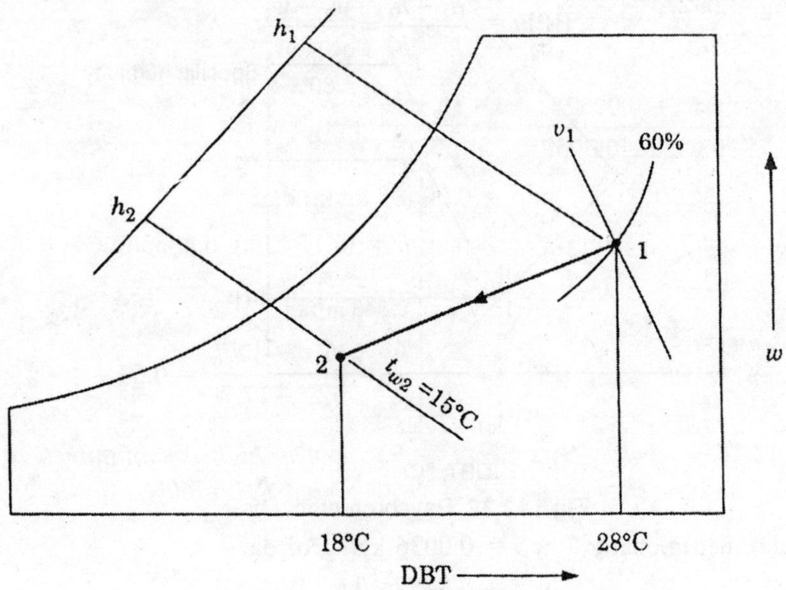

Fig. 12.31 Psychrometric chart.

From the chart, we have (Fig. 12.31)

$$h_1 = 57.5 \text{ kJ/kg}, v_1 = 0.87 \text{ m}^3\text{/kg}, h_2 = 42.1 \text{ kJ/kg}$$

$$\dot{m}_{a1} = \frac{\dot{V}_1}{v_1} = \frac{12}{0.87} = 13.79 \text{ kg/min}$$

$$\text{Capacity of air-conditioner} = \frac{\dot{m}_{a1}(h_1 - h_2)}{211}$$

$$= \frac{13.79(57.5 - 42.1)}{211}$$

$$= 1.006 \text{ TR}$$

Example 12.18 *100 cmm of atmospheric air at 25°C DBT and 12°C WBT is flowing per minute through a duct. Dry saturated steam at 100°C is injected through the duct. Dry saturated steam at 100°C is injected into the airstream with a rate of 1.2 kg/min.*

(i) What is the temperature of air after mixing the steam?

(ii) What is the relative humidity of air after mixing the steam ? **[CSE, 2007]**

Solution: $\dot{V}_1$ = 100 m³/min, t_{d1} = 25°C, t_{w1} = 12°C, $\dot{m}_s$ = 1.2 kg/min, t_s = 100°C

Refer to Fig. 12.32.

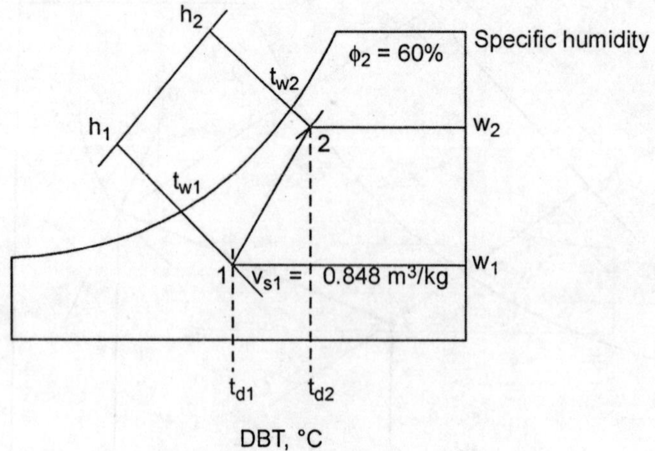

Fig. 12.32 Psychrometric chart.

From psychrometric chart, $w_1 = 0.0036$ kg/kg of d.a.

$v_{s1} = 0.848$ m³/kg

$h_1 = 34$ kJ/kg of d.a.

Mass of air flowing,
$$\dot{m}_a = \frac{\dot{V}_1}{v_{s1}} = \frac{100}{0.848} = 117.9 \text{ kg/min}$$

$$w_2 = w_1 + \frac{\dot{m}_s}{\dot{m}_a} = 0.0036 + \frac{1.2}{117.9} = 0.0138 \text{ kg/kg of d.a.}$$

Enthalpy of dry saturated steam at 100°C,

$h_s = 2676$ kJ/kg

$$h_2 = h_1 + \frac{\dot{m}_s}{\dot{m}_a} \times h_s = 34 + \frac{1.2}{117.9} \times 2676 = 61.24 \text{ kJ/kg of d.a.}$$

Locate point 2 corresponding to

$w_2 = 0.0138$ kg/kg d.a. and $h_2 = 61.24$ kJ/kg d.a.

$\phi_2 = 60\%$ and $t_{d2} = 27.2$°C, $t_{w2} = 21.5$°C

Example 12.19 *Moist air enters a chamber at 5°C DBT and 2.5°C WBT at a rate of 90 m³/min. The barometric pressure is 1.01325 bar. While passing through the chamber, the air absorbs sensible heat at the rate of 40.7 kW and picks up 40 kg/hr of saturated steam at 110°C. Determine the dry and wet bulb temperatures of leaving air.*

Properties of steam are given below :

Enthalpy of saturated steam at 110°C is 2691.3kJ/kg. **[CSE, 2008]**

Solution: Given: $t_{d1} = 5$°C, $t_{w1} = 2.5$°C, $\dot{V}_1 = 90$ m³/min, $p_b = 1.01325$ bar, $\dot{Q}_s = 40.7$ kW,

$\dot{m}_s = 40$ kg/h, $t_s = 110$°C, $h_s = 2691.3$ kJ/kg

Refer to Fig. 12.33.

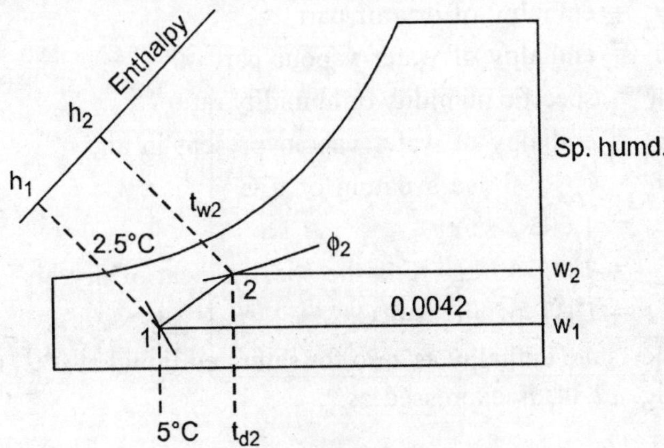

Fig. 12.33 Psychrometric chart.

Mark point 1 (t_{d1} = 5°C, t_{w1} = 2.5°C) on the psychrometric chart. We find that:

Sp. volume, v_{s1} = 0.79 m³/kg of d.a.

Sp. humidity, w_1 = 0.0042 kg/kg of d.a.

Enthalpy of air, h_1 = 15.5 kJ/kg

Mass of air flowing, $\dot{m}_a = \dfrac{\dot{V}_1}{v_{s1}} = \dfrac{90}{0.79} = 113.9$ kg/min

$$w_2 = w_1 + \frac{\dot{m}_s}{\dot{m}_a} = 0.0042 + \frac{40}{60} \times \frac{1}{113.9} = 0.0100 \text{ kg/kg of d.a.}$$

Let h_2 = Enthalpy of leaving air

$\dot{Q}_s$ = 40.7 kJ/s

$$q_s = \frac{\dot{Q}_s}{\dot{m}_a} = \frac{40.7 \times 60}{113.9} = 21.44 \text{ kJ/kg}$$

$$h_2 = h_1 + \frac{\dot{m}_s}{\dot{m}_a} \times h_s + q_s = 15.5 + \frac{40}{60} \times \frac{2691.3}{113.9} + 21.44 = 52.69 \text{ kJ/kg}$$

Corresponding to w_2 = 0.0100 kg/kg of d.a. and h_2 = 52.69 kJ/kg, from psychrometric chart, we have

$$t_{d2} = 26.5°\text{C}, \quad t_{w2} = 18.5°\text{C}$$

Example 12.20 *Develop an expression for enthalpy of moist air per unit mass of dry air as under:*

h = 1.005 t + w(2500 + 1.88 t) kJ/kg where t is DBT and w is humidity ratio.

[CSE, 2008]

Solution: Enthalpy of Moist Air

According to Gibb's law, the enthalpy of a mixture of perfect gases is obtained by the summation of the enthalpies of the constituents. Thus the enthalpy of moist air *h* is equal to the sum of enthalpies of dry air and associated water vapour, i.e.,

$$h = h_a + wh_v \text{ per kg of dry air}$$

where h_a = enthalpy of dry air part

wh_v = enthalpy of water vapour part

w = specific humidity or humidity ratio

h_v = enthalpy of water vapour present in air.

Now $h_a = c_{pa} \cdot t$ above a datum of 0°C

$= 1.005\ t$ kJ/kg

where c_{pa} = 1.005 kJ/kg · K is the specific heat of dry air

t = DBT of air in 0°C.

Taking the reference state enthalpy as zero for saturated liquid at 0°C, the enthalpy of water vapour at point A in Fig. 12.34, is expressed as:

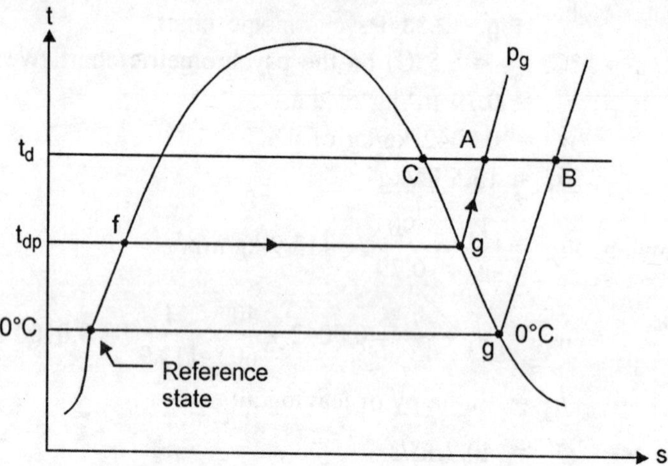

Fig. 12.34 Evaluation of enthalpy of water vapour part (t-s diagram).

$$h_v = h_A = c_{pw} t_{dp} + (h_{fg})_{dp} + c_{pv}\ (t_d - t_{dp}) \text{ kJ/kg}$$

where c_{pw} = specific heat of liquid water

$\simeq$ 4.1868 kJ/kg·K

t_{dp} = dew point temperature, t_d = DBT

$(h_{fg})_{dp}$ = Latent heat of vaporization at DPT

c_{pv} = Specific heat of superheated vapour

$\simeq$ 1.88 kJ/kg · K

$$\therefore \quad h_v = 4.1868\ t_{dp} + (h_{fg})_{dp} + 1.88\ (t - t_{dp})$$

At low pressures for an ideal gas,

$$h = f(t) \text{ only.}$$

$$h_A = h_C = (h_g)_{td}$$

and

$$h_A = h_B = (h_g)_{0°C} + c_{pv}\ (t_d - 0)$$

Taking $(h_{fg})_{0°C} = 2501$ kJ/kg

$\therefore$ $h_v = 2501 + 1.88\ t_d$ kJ/kg

Thus $h = 1.005\ t_d + w\ (2500 + 1.88\ t_d)$ kJ/kg of d.a.

Example 12.21 *20 m³ of air per minute at 30°C DBT and 60% RH is sensibly cooled to 22°C DBT. Take saturation pressure of water vapour at 30°C and 22°C temperatures to be 0.425 bar and 0.0265 bar respectively.*

Find heat removed from air.

Take atmospheric air pressure p_b = 1 bar. **[CSE, 2009]**

Solution: Given: $\dot{V}_1 = 20$ m³/min, $t_{d1} = 30°C$, $\phi_1 = 60\%$, $t_{d2} = 22°C$, $p_{sv1} = 0.425$ bar,

$p_{sv2} = 0.0265$ bar, $p_b = 1$ bar

In sensible cooling of air, there is no change in specific humidity. (Fig. 12.35)

From psychrometric chart,

$$h_1 = 72.2 \text{ kJ/kg}, \ h_2 = 64.5 \text{ kJ/kg}$$

Heat removed from air $= h_1 - h_2 = 72.2 - 64.5 = 7.7$ kJ/kg

Partial pressure of air at 30°C

$$p_{a1} = p_b - p_{sv1} = 1 - 0.425 = 0.575 \text{ bar}$$

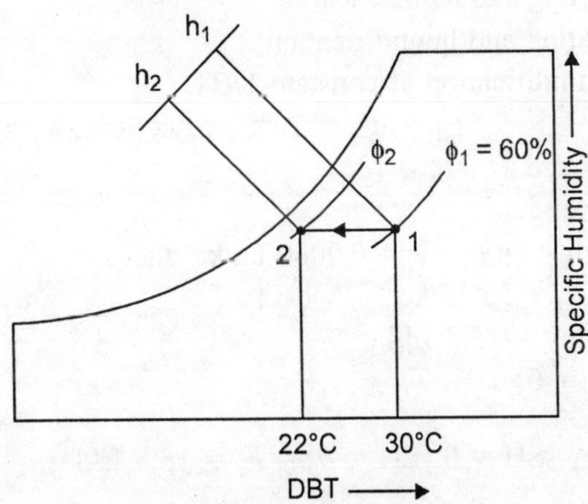

Fig. 12.35 Psychrometric chart.

Mass of air, $$m_a = \frac{p_{a1} V_1}{R_a T_{d1}} = \frac{0.575 \times 10^5 \times 20}{60 \times 287 \times 303} = 0.22 \text{ kg/s}$$

Total heat removed $= 0.22 \times 7.7 = 1.7$ kW

Example 12.22 *Illustrate the following processes on psychrometric chart with initial state of moist air as dry bulb temperature equal to 20°C and relative humidity of 50%*

(i) Cooling and humidification

(ii) Heating and humidification

(iii) Humidification at constant dry bulb temperature

Mention one application each for above humidification processes.

If the moist air leaves the system for the case (iii) above, at 90% relative humidity, determine per unit mass of dry air :

(i) increase in humidity ratio, Δw

(ii) increase in enthalpy, Δh

(iii) increase in dry bulb temperature, Δt

(iv) Determine Sensible Heat Factor (SHF) for this process. **[CSE, 2010]**

Solution: Refer to Fig. 12.36

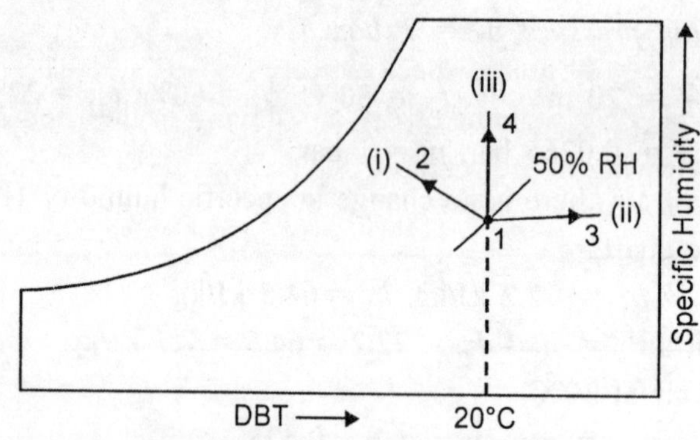

Fig. 12.36 Psychrometric chart.

(*i*) Process 1-2 : Cooling and humidification

(*ii*) Process 1-3 : Heating and humidification

(*iii*) Process 1-4 : Humidification at constant DBT.

$$t_{d1} = t_{d2} = 20°C, \ \phi_1 = 50\%, \ \phi_2 = 90\%$$

(*i*) $w_1 = 0.0076$ kg/kg. d.a.

$w_2 = 0.0136$ kg/kg. d.a.

Increase in humidity ratio, $\Delta w = 0.0060$ kg/kg d.a.

(*ii*) $h_1 = 39.2$ kJ/kg, $h_2 = 55$ kJ/kg

Increase in enthalpy = 15.8 kJ/kg

(*iii*) Increase in DBT = 0

(*iv*) $\text{SHF} = \dfrac{\text{SH}}{\text{SH} + \text{LH}}$, SH = 0, LH = $h_2 - h_1$ = 15.8 kJ/kg

SHL = 0

Example 12.23 *It is desired to maintain a room at temperature of 20°C when outside temperature is 30°C. The volume of this room is 300 m³. The pressure in the room and outside is 1 bar.*

(i) The air in the room is renewed completely in 1 hour. Calculate the mass of air that the air conditioning system pumps into the room.

(ii) Compute maximum possible C.O.P. of this A/C system.

(iii) When this A/C system is switched-off, the temperature inside the room reaches 21°C in 20 minutes. Calculate the amount of heat transferred from the surroundings to the room.

(iv) Calculate power required by this system. **[CSE, 2011]**

Solution: Given: t_{d1} = 30°C, t_{d2} = 20°C, p = 1 bar, V = 300 m³

(*i*) Mass of air in the room:

$$m_r = \frac{pv}{RT_2} = \frac{1 \times 10^5 \times 300}{287 \times 293} = 356.76 \text{ kg}$$

Heat rejected, $Q = m_r \, c_{pm} \, (t_{d1} - t_{d2})$

$= 356.76 \times 1.022 \times (30 - 20)$

$= 3646$ kJ

where c_{pm} = humid specific heat

Let V_a = volume of dry air flowing in the room, m³

Density of air flowing in the room,

$$\rho_a = \frac{p}{RT_1} = \frac{1 \times 10^5}{287 \times 303} = 1.15 \text{ kg/m}^3$$

Sensible heat, $SH = V_a \rho_a c_{pm} \cdot \Delta t = V_a \times 1.15 \times 1.022 \times 10 = 11.753\, V_a$ kJ

$SH = Q$

$11.753\, V_a = 3646$

$V_a = 310$ m³

$m_a = V_a \rho_a = 310 \times 1.15 = 356.7$ kg/h or 5.945 kg/min

(*ii*) Maximum possible COP $= \dfrac{T_1}{T_1 - T_2} = \dfrac{303}{303 - 293} = \dfrac{303}{10} = 30.3$

(iii) Amount of heat transferred

$$= \frac{m_r \times c_{pm} \times \Delta t}{t}$$

$$= \frac{356.76 \times 1.022 \times (30 - 21)}{20}$$

$= 164$ kJ/min

(*iv*) Power required $= \dfrac{164}{60} = 2.73$ kW

Example 12.24 *A thermostatic expansion valve using R-134a in the power assembly is designed to produce 10°C superheat at an evaporator temperature of 0°C. What will be the superheat that it will maintain when the evaporator temperature is -30°C? Given for R-134a, the saturation pressure temperature data:*

Temp (°C)	*–30*	*–25*	*–20*	*–15*	*–10*
Pressure (bar)	0.8438	1.065	1.327	1.64	2.006
Temp (°C)	–5	0	5	10	
Pressure (bar)	2.43	2.928	3.5	4.146	

Solution: Given: $T_{1'} = 0°C$ or 273 K, $T_1 = 10°C$ or 283 K,

$$p_{01} \text{ (at } 0°C) = 2.928 \text{ bar}$$

$$p'_0 \text{ (at } 10°C) = 4.146 \text{ bar}$$

Refer to Fig. 12.37.

Required pressure difference or initial compression of follow - up spring per unit area of diaphragm,

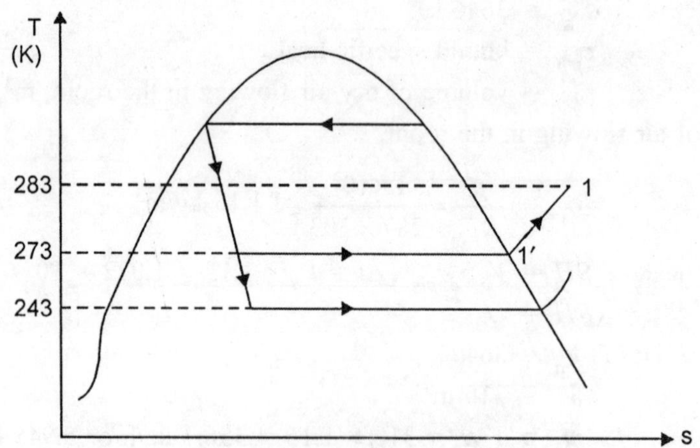

Fig. 12.37 T-s diagram.

$$f_{14} = p'_0 - p_{01} = 4.146 - 2.928 = 1.218 \text{ bar}$$

When the evaporator temperature is –30°C, then pressure at evaporator inlet,

$$p_{01} \text{ (at } -30°C) = 0.8438 \text{ bar}$$

For the same adjustment of follow-up spring compression, the equilibrium pressure of power fluid,

$$p'_0 = p_{01} + f_{14} = 0.8438 + 1.218 = 2.0618 \text{ bar}$$

The corresponding saturation temperature of power fluid in bulb,

$$t'_0 \text{ (at 2.0618 bar)} = -10 + \frac{(-5+10)}{(2.43 - 2.006)} = -9.34°C$$

∴ Superheat of suction gas = –9.34°C

Example 12.25 *Cold water bottles kept in a room often start condensing atmospheric moisture. What is the minimum temperature to which water bottle can be cooled without any dripping moisture from its surface when kept in a room at 25°C DBT and 60% RH? If it is desired that a water bottle at 10°C should also not condense moisture on its surface what should be the RH in the room, keeping the DBT the same?* **[CSE, 2011]**

Solution: Refer to Fig 12.38.

From psychrometric chart, for

$$t_d = 25°C \text{ and } \phi = 60\%,$$

Dew point temperature, $t_{dp} = 17°C$

For $t_d = 25°C$ and $t_w = 10°C$,

$\phi = 8\%$

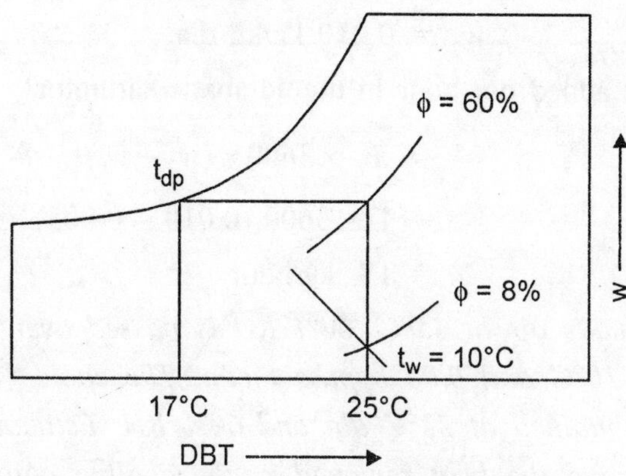

Fig. 12.38 Psychrometric chart.

Example 12.26 *In a winter heating air-conditioning system 1 kg/s of ambient air at 7°C DBT, 80% RH is sensibly heated, then adiabatically saturated to a high humidity of RH = 90%, and then further sensibly heated to the desired outlet state of 25°C DBT and 50% RH.*

Determine the temperature of air at various state points and show all the processes on a psychrometric chart. Also determine the amount of moisture added per hour in the adiabatic saturator. **[CSE, 2011]**

Solution: Given: $\dot{m}_a = 1$ kg/s, $t_{d1} = 7°C$, $\phi_1 = 80\%$, $\phi_2 = 90\%$, $t_{d3} = 25°C$, $\phi_3 = 50\%$

Refer to Fig 12.39.

State 1: (7° DBT, 80% RH)

Process 1-A: Sensible heating

Process A-2: Adiabatic saturation to 90% RH

Process 2-3: Sensible heating to (25°C DBT, 50% RH)

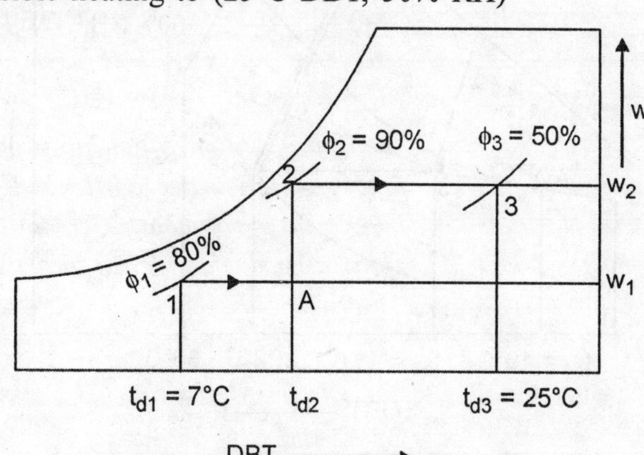

Fig. 12.39 Psychrometric chart.

From the psychrometric chart:

$$t_{d2} = 15°\text{C}$$

$$w_1 = 0.005 \text{ kg/kg d.a.}$$

$$w_2 = 0.010 \text{ kg/kg d.a.}$$

Amount of moisture added per hour in the adiabatic saturator

$$= \dot{m}_a \times 3600 \times (w_2 - w_1)$$

$$= 1 \times 3600\ (0.010 - 0.005)$$

$$= 18 \text{ kg/hour}$$

Example 12.27 *Outside air at 35°C, 60% RH is passed over a cooling coil having an apparatus dew point of 10°C and 0.06 bypass factor. The cooled air is then supplied to a room which is to be maintained at 22°C dbt and 60% RH. Estimate the sensible and latent heat loads of the room and the heat removed in the cooling coil in kW.* **[IES, 1993]**

Solution: Given: $t_{d1} = 35°\text{C}$, $(\text{RH})_1 = 60\%$, $t_{d4} = 10°\text{C}$,

$t_{d2} = 22°\text{C}$, $x = 0.06$, $\dot{m}_a = 4500 \text{ kg/h}$

$$\text{BPF} = x = \frac{t_{d3} - t_{d4}}{t_{d1} - t_{d4}}$$

$$0.06 = \frac{t_{d3} - 10}{35 - 10}$$

$$t_{d3} = 11.5°\text{C}$$

From psychrometric chart, we have (Fig. 12.40)

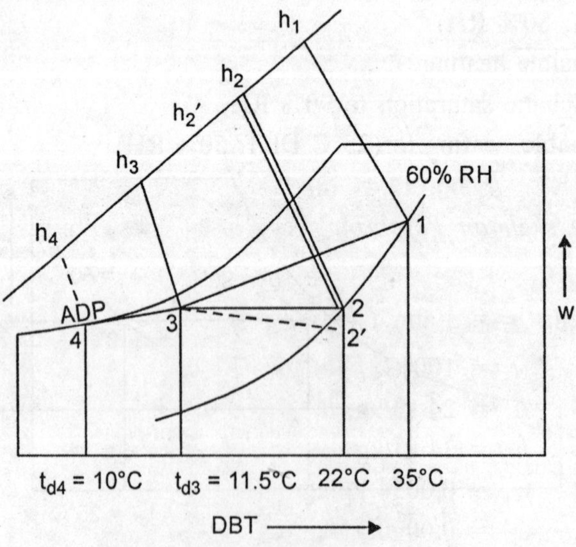

Fig. 12.40 Psychrometric chart.

$$v_{s1} = 0.91 \text{ m}^3/\text{kg}$$

$$\dot{m}_a = \frac{4500}{3600} = 1.25 \text{ kg/s}$$

$$h_3 = 35.5 \text{ kJ/kg}$$

$$h_2 = 48.5 \text{ kJ/kg}$$

$$h_{2'} = 44.5 \text{ kJ/kg}$$

Sensible heat, $\text{RSH} = \dot{m}_a(h_2' - h_3) = 1.25\ (44.5 - 33.5) = 13.75 \text{ kW}$

$$\text{RLH} = \dot{m}_a(h_2 - h_2') = 1.25\ (48.5 - 44.5) = 5.0 \text{ kW}$$

Example 12.28 *An air conditioning system (see Fig. 12.41(a)) operates at a total pressure of 1 atm. It consists of a heating section and a humidifier that supplies wet steam (saturated) at 100°C. Air enters the heating section at 10°C and 70 percent relative humidity at the rate of 35 m³/min. It leaves at 20°C and 60% relative humidity. Determine*

(i) temperature and relative humidity of air when it leaves the heating section.

(ii) the rate of heat transfer to the heating section, and

(iii) the rate at which water is added to the air in the humidifying section.

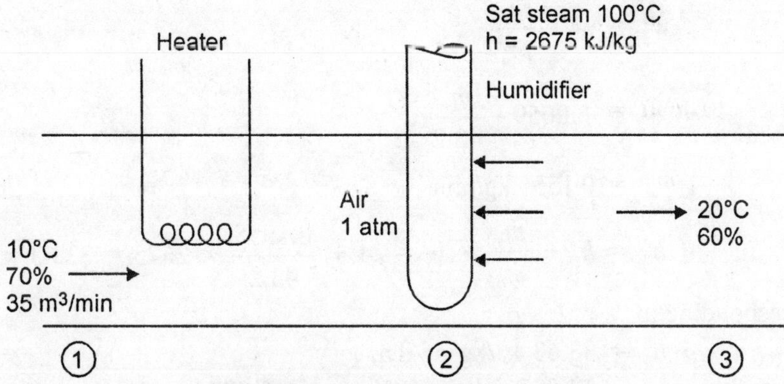

Fig. 12.41 (a) Air conditioning system.

Also draw the skeleton Psychrometric chart representation showing the process.

[CSE, 2012]

Solution: Given: $p_t = 1$ atm, $t_{d1} = 10°\text{C}$, $\phi_1 = 70\%$, $\dot{V}_a = 35 \text{ m}^3/\text{min}$, $t_{d3} = 20°\text{C}$, $\phi_3 = 60\%$,

$$t_s = 100°\text{C},\ h_s = 2675 \text{ kJ/kg}$$

$$h_1 = 24 \text{ kJ/kg},\ h_3 = 43.5 \text{ kJ/kg}$$

$$h_A = 34 \text{ kJ/kg}$$

$$w_1 = 0.0054 \text{ kg/kg d.a.}$$

$$w_3 = 0.0090 \text{ kg/kg d.a.}$$

$$v_{s1} = 0.81 \text{ m}^3/\text{kg of d.a.}$$

(*i*) Mass of air flowing,

$$\dot{m}_a = \frac{\dot{V}}{v_{s1}} = \frac{35}{0.81}$$

$$= 43.2 \text{ kg/min}$$

Refer to Fig. 12.41 (b).

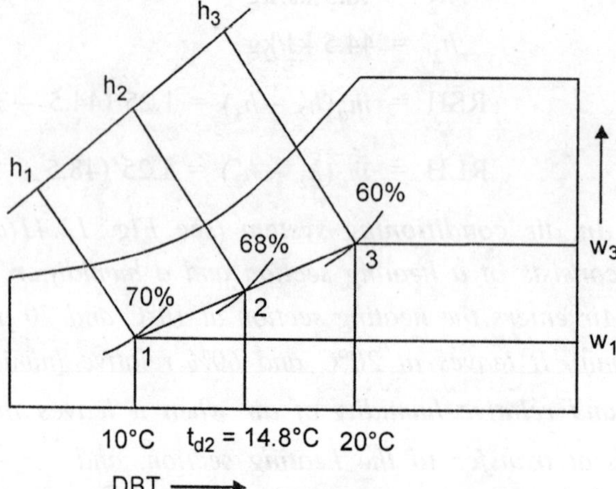

Fig. 12.41 (b) Psychrometric chart.

$$w_3 = w_1 + \frac{\dot{m}_s}{\dot{m}_a}$$

$$0.0090 = 0.0059 + \frac{\dot{m}_s}{43.2}$$

$$\dot{m}_s = 0.1555 \text{ kg/min}$$

$$h_2 = h_1 + \frac{\dot{m}_s}{\dot{m}_a} \times h_s = 24 + \frac{0.1555}{43.2} \times 2675 = 33.63 \text{ kJ/kg}$$

Corresponding to

$$h_2 = 33.63 \text{ kJ/kg of d.a.}$$

$$t_{d2} = 14.8°\text{C}, \ \phi_2 = 68\%, \ w_2 = 0.0074 \text{ kg/kg d.a.}$$

(*ii*) Rate of heat transfer to the heating section

$$= h_2 - h_1 = 33.63 - 24 = 9.63 \text{ kJ/kg of d.a.}$$

(*iii*) Moisture added to air in the humidifying section

$$= w_3 - w_2 = 0.0090 - 0.0074 = 0.0016 \text{ kg/kg of d.a.}$$

or $$43.2 \times 0.0016 = 0.069 \text{ kg/min}$$

Example 12.29 *A small auditorium is required to be maintained at 22°C dry bulb temperature and 70% relative humidity. The ambient conditions are at 30°C dry bulb tempeture and 75% relative humidity. The amount of free air circulated is 200 m³/min. The required conditions are achieved by first cooling coil having apparatus dew point of 14°C and then by heating. With the help of psychrometric chart find:*

(i) The cooling capacity of cooling coil in tons of refrigeration and its bypass factor.

(ii) The amount of water vapour removed by the cooling coil in kg/h.

(iii) Assuming the by pass factor as 0.2, determine the capacity of the heating coil in kW and its surface temperature. **[IES, 1992]**

Solution: Given: $t_{d2} = 22°C$, $(RH)_2 = 70\%$, $t_{d1} = 30°C$, $(RH)_1 = 75\%$,

$\dot{V}_a = 200 \text{ m}^3/\text{min}$, $t_{ADP} = 14°C$, BPF = 0.2

From psaychrometric chart (Fig. 12.42), we have

At 30°C DBT and RH = 75%,

$$v_1 = 0.886 \text{ m}^3/\text{kg}$$
$$v_4 = 0.838 \text{ m}^3/\text{kg}$$
$$h_1 = 82 \text{ kJ/kg}$$

At 22°C DBT and 70% RH,

$$h_2 = 51 \text{ kJ/kg}$$
$$h_4 = 47 \text{ kJ/kg}$$
$$t_4 = 16.5°C$$
$$w_1 = 20.5 \text{ gm/kg d.a.}$$
$$w_2 = 11.8 \text{ gm/kg d.a.}$$

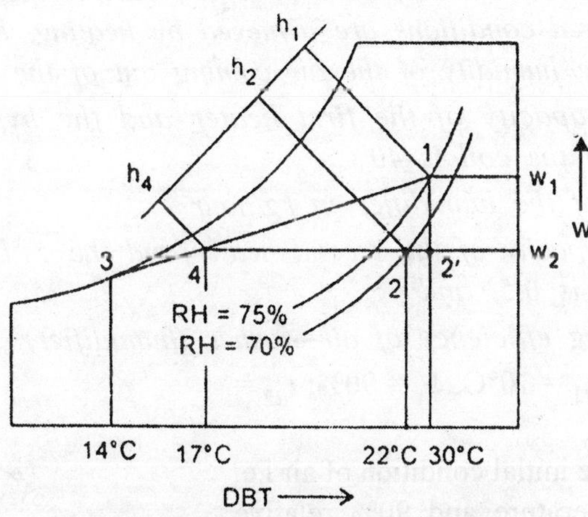

Fig. 12.42. Psychrometric chart

(*i*) Cooling capacity of coil

$$= \frac{\dot{V}_a}{v_1} \times \frac{h_1 - h_4}{211}$$

$$= \frac{200}{0.886} \times \frac{(82-47)}{211} = 37.47 \text{ ton}$$

(*ii*) Amount of water vapour removed by cooling coil,

$$\dot{m}_w = \frac{\dot{V}_a}{v_1}(w_1 - w_2)$$

$$= \frac{200}{0.886} (20.5 - 11.8) \times 60 \times 10^{-3} = 117.8 \text{ kg/hr.}$$

(*iii*)
$$\text{BPF} = \frac{t_{d2} - t_{d2}'}{t_{d4} - t_{d2}'}$$

$$0.2 = \frac{22 - t_{d2}'}{17 - t_{d2}'}$$

$$3.4 - 0.2\, t_{d2}' = 22 - t_{d2}'$$

$$t_{d2}' = \frac{18.6}{0.8} = 23.25°\text{C}$$

Capacity of heating coil

$$= \frac{\dot{V}_a}{v_4} (h_2 - h_4)$$

$$= \frac{200}{0.838} (51 - 47) \times \frac{1}{60} = 15.91 \text{ kW}$$

Example 12.30 *300 m³/min of air at 10°C dry bulb temperature and 90% relative humidity is to be heated and humidified to 35°C dry bulb temperature and 22.5°C wet bulb temperature. The required conditions are achieved by heating, humidifying and then again by heating. The relative humidity of the air coming out of the humidifier is 90%. Find*

(i) *the heating capacity of the first heater and the bypass factor if the surface temperature of the coil is 40°C.*

(ii) *the capacity of the humidifier in kg/hour*

(iii) *the heating capacity of the second heater and the coil surface temperature if the bypass factor is 0.5, and*

(iv) *the humidifying efficiency of air washer (humidifier).* **[IES,1997]**

Solution: Given: $t_{d1} = 10°\text{C}$; $\phi_1 = 90\%$; $t_{d2} = 35°\text{C}$; $t_{w2} = 22.5°\text{C}$

First of all, mark the initial condition of air i.e. at 10°C dry bulb temperature and 90% relative humidity, on the psychrometric chart at point 1, as shown in Fig. 12.43. Now mark the final condition of air at 35°C dry bulb temperature and 22.5°C wet bulb temperature at point 2.

From point 1, draw a horizontal line to represent sensible heating and from point 2 draw horizontal line to intersect 90% relative humidity curve at point *B*. Now from point *B*, draw a constant wet bulb temperature line which intersects the horizontal line drawn through point 1 at point *A*. The line 1-*A* represents reheating to final conditions.

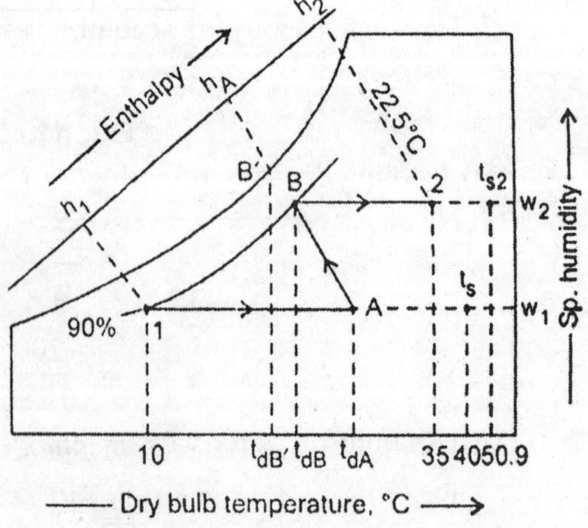

Fig. 12.43 Psychrometric chart.

(*i*) Heating capacity of first heater.

From the psychrometric chart, the temperature to which the air should be preheated (corresponding to point A) is

$$t_{dA} = 32.6°C$$

$$v_1 = 0.82 \text{ m}^3/\text{kg}$$

$$\dot{m}_a = \frac{300}{60 \times 0.82} = 6.09 \text{ kg/s}$$

From the psychrometric chart, we find that enthalpy of air at point 1,

$$h_1 = 27.2 \text{ kJ/kg of dry air}$$

Enthalpy of air at point A,

$$h_A = 51 \text{ kJ/kg of dry air}$$

Heating capacity of first heater,

$$Q_1 = \dot{m}_a(h_A - h_1) = 6.09\ (51 - 27.2) = 144.942 \text{ kW}$$

$$\text{By pass factor} = \frac{t_s - t_{dA}}{t_s - t_1} = \frac{40 - 32.6}{40 - 10} = 0.247$$

(*ii*) Capacity of humidifier.

From the psychrometric chart, we find that specific humidity of entering air

$$w_1 = 0.0068 \text{ kg/kg of dry air}$$

and specific humidity of leaving air,

$$w_2 = 0.0122 \text{ kg/kg or dry air}$$

$$\text{Capacity of humidifier} = \dot{m}_a(w_2 - w_1) = 6.09\ (0.0122 - 0.0068) \times 3600 = 118.39 \text{ kg/kr}$$

(*iii*) Heating capacity of second heater.

Enthalpy of air at point 2, h_2 = 68 kJ/kg of dry air

Also $t_{dB} = 19.1°C$

and $h_B = 51$ kJ/kg of ry air

Heating capacity of second heater,

$$Q_2 = \dot{m}_a(h_2 - h_B)$$

$$= 6.09\ (68 - 51) = 103.53 \text{ kW}$$

Now

$$\frac{t_{s2} - t_2}{t_{s2} - t_{dB}} = 0.5$$

$$\frac{t_{s2} - 35}{t_{s2} - 19.1} = 0.5$$

$$t_{s2} = 50.9°C$$

(*iv*) Humidifying efficiency of the air washer.

From the psychrometric chart, we find that

$$t_{dB} = 19.1°C \text{ and } t_{dB'} = 18°C$$

We know that humidifying efficiency of the air washer,

$$\eta_H = \frac{\text{Actual drop in DBT}}{\text{Ideal drop in DBT}} = \frac{t_{dA} - t_{dB}}{t_{dA} - t_{dB'}}$$

$$= \frac{32.6 - 19.1}{32.6 - 18} = \frac{13.5}{14.6} = 0.924 \text{ or } 92.4\% \textbf{ Ans.}$$

Example 12.31 *The properties of air in a room are: temperature 35°C and relative humidity 70 per cent. Determine (i) the partial pressure of air, (ii) the humidity ratio, (iii) the saturation ratio, (iv) dew point temperature, (v) density of the air-vapour mixture, (vi) the enthalpy of mixture per kg of dry air. Assume room pressure to be 1.0132 bar and the specific heat for the super-heated vapour as 1.863 kJ/kg K. Solve the problem with the help of steam tables.* **[IES, 1996]**

Solution: Given: Room conditions:

$$t = 35°\text{C}, \phi = 0.70, p_t = 1.0132 \text{ bar},$$

$$(c_p)_{sup} = 1.863 \text{ kJ/kg}\cdot\text{K}$$

From steam tables, saturation pressure at 35°C,

$$p_s = 0.05628 \text{ bar}$$

Partial pressure of vapour,

$$p_v = \phi p_s = 0.7 \times 0.05628 = 0.0394 \text{ bar}$$

Humidity ratio, w = weight of water vapour/kg of d.a.

$$= \frac{0.622\, p_v}{p_t - p_v} = \frac{0.622 \times 0.0394}{1.0132 - 0.0394} = 0.02517 \text{ kg/kg of d.a.}$$

$$\text{Saturation ratio} = \frac{\text{mass of water vapour/kg of d.a.}}{\text{mass of water vapour/kg of dry air at saturation condition}}$$

$$= \phi\left[\frac{p_t - p_s}{p_t - \phi\, p_s}\right] = 0.7\left[\frac{1.0132 - 0.05628}{1.0132 - 0.7 \times 0.05628}\right] = 0.6879$$

Dew point temperature,

DPT = Saturation temperature corresponding to partial pressure of water vapour, p_v

= 28°C

Partial pressure of air, $p_a = p_t - p_v = 1.0132 - 0.0394 = 0.9738$ bar

Density of air, $$\rho_a = \frac{p_a}{R_a T} = \frac{0.9738}{0.287 \times (273 + 35)} = 1.102 \text{ kg/m}^3$$

Density of vapour, $$\rho_v = \frac{p_v}{R_v T} = \frac{0.394 \times 10^5}{461 \times 308} = 0.028 \text{ kg/m}^3$$

Density of mixture, $\rho = \rho_a + \rho_v = 1.102 + 0.028 = 1.13 \text{ kg/m}^3$

Enthalpy of mixture, $h = h_a + h_w$

$$= c_{pa} \times \text{DBT} + w[c_w \times \text{DPT} + h_{fg} + (c_p)_{\text{sup}} (\text{DBT} - \text{DPT})]$$
$$= 1.005 \times 35 + 0.02517\ [4.18 \times 28 + 2450 + 1.863\ (35 - 28)]$$
$$= 100.195 \text{ kJ/kg of d.a.}$$

Example 12.32 *Atmospheric air at 40°C dry bulb temperature and 55% relative humidity is passed over a cooling coil at 600 m³/min. At exit from the coil the air is saturated and its temperature is 8°C. The condensate also leaves at 8°C. Determine the amount of condensate leaving the coil per minute and the refrigeration required in kW. The specific heat of superheated steam may be assumed to be 1.88 kJ/kg. K. Solve the problem without the use of Psychrometric chart.* **[IES, 1995]**

Solution: Given: $t_{d1} = 40°\text{C}$, $\phi_1 = 55\%$, $V_1 = 600 \text{ m}^3/\text{min}$, $t_{d2} = 8°\text{C}$, $\phi_2 = 100\%$

Saturation pressure corresponding to 40°C, from steam tables,

$$p_{s1} = 0.07384 \text{ bar}$$
$$p_{v1} = \phi_1\, p_{s1} = 0.55 \times 0.07384$$
$$= 0.04061 \text{ bar}$$

Humidity ratio,
$$w_1 = \frac{0.622\, p_{v1}}{p_b - p_{v1}}$$
$$= \frac{0.622 \times 0.04061}{1.01325 - 0.04061}$$
$$= 0.02597 \text{ kg/kg of dry air}$$

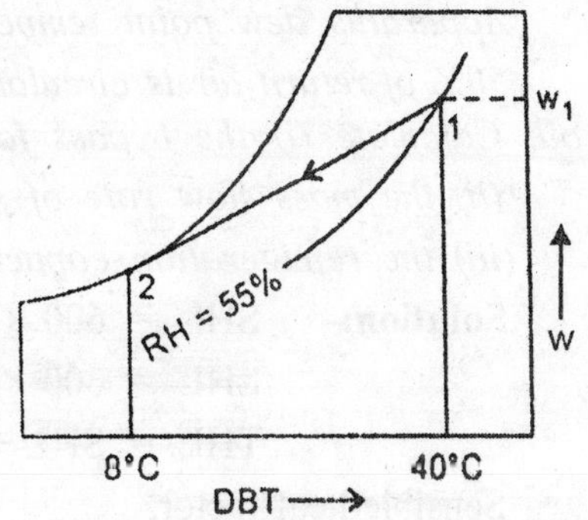

Fig. 12.44 Psychrometric chart.

Refer to Fig. 12.44

At 8°C,
$$p_{s2} = 0.010724 \text{ bar}$$

$$w_2 = \frac{0.622 \times 0.010724}{1.01325 - 0.010724} = 0.00665 \text{ kg/kg of dry air}$$
$$h_1 = 1.005\, t_{d1} + w_1\, (2500 + 1.88 \text{ kg})$$
$$= 1.005 \times 40 + 0.2597\ (2500 + 1.88 \times 40) = 107.08 \text{ kJ/kg}$$
$$h_2 = 1.005 \times 8 + 0.00665\ (2500 + 1.88 \times 8)$$
$$= 24.76 \text{ kJ/kg}$$

Mass of air,
$$\dot{m}_a = \frac{p_{a1}\dot{V}_1}{R_a T_1}$$

Now
$$p_{a1} = p_b - p_{v1} = 1.01325 - 0.04061 = 0.97264 \text{ bar}$$
$$\dot{m}_a = \frac{0.97264 \times 10^2 \times 600}{0.287 \times 313} = 649.65 \text{ kg/min}$$

Mass of water condensed
$$= \dot{m}_a (w_1 - w_2)$$
$$= 649.64\ (0.02597 - 0.00665) = 12.64 \text{ kg/min}$$

Capacity of cooling coil $= \dot{m}_a(h_1 - h_2)$

$$= \frac{649.65}{60}(107.08 - 24.76) = 891.32 \text{ kW}$$

Example 12.33 *The following data refers to design of an air-conditioning system of a theatre :*

Total seating capacity = 600 persons
Sensible heat per person = 200 kJ/h
Latent heat per person = 130 kJ/h
Solar heat gain though glass, doors and walls = 2,50,000 kJ/h
Outside design condition = 40°C DBT, 27°C WBT
Inside design condition = 24°C DBT, 50% RH
Apparatus dew point temperature of cooling coil = 4°C

50% of return air is circulated and it is mixed with fresh air before entering the cooling coil. Calculate (i) the bypass factor of the cooling coil,

(ii) the mass flow rate of supply air to the theatre, and

(iii) the refrigeration capacity of the coil in tons of refrigeration. **[IES, 1998]**

Solution: SHL = 600 × 200 + 2,50,000 = 3,70,000 kJ/h

LHL = 600 × 130 = 78,000 kJ/h

THL = SHL + LHL = 448,000 kJ/h

Sensible heat factor,

$$\text{SHF} = \frac{\text{SHL}}{\text{THL}} = \frac{3,70,000}{4,48,000} = 0.8259$$

ADP = 4°C

Locate point '*a*' at 40°C DBT and 27°C WBT, point "*b*" at 24°C and 50% RH and point '*c*' on the saturation line of the psychrometric chart. Locate point 1, the mid-point *ab* for 50% return air recirculation. Join 1 and '*c*'. Draw the SHF line to intersect the saturation line at '*d*'. Join *bd* to intersect 1*c* at point 2, as shown in Fig. 12.45.

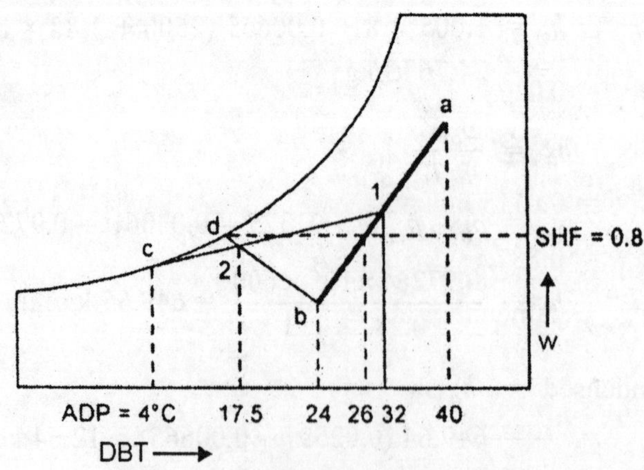

Fig. 12.45 Psychrometric chart.

Then

$$t_1 = 32°C$$

and

$$t_2 = 17.5°C$$

(*i*) $\text{BPF} = \dfrac{t_2 - \text{ADP}}{t_1 - \text{ADP}} = \dfrac{17.5° - 4}{32 - 4} = 0.48$

(*ii*) Mass flow rate of supply air,

$$\dot{m}_a = \frac{\text{THL}}{h_1 - h_2} = \frac{4,48,000}{49 - 41} = 56,000 \text{ kg/h}$$

(*iii*) Refrigerating capacity of cooling coil

$$= \dot{m}_a \times \frac{h_a - h_2}{211 \times 60} = 56,000\left[\frac{67 - 41}{211 \times 60}\right] = 115 \text{ TR}$$

Example 12.34 *An air conditioning system is designed under the following conditions:*

Outdoor conditions = 30°C DBT, 75% RH

Required indoor conditions = 22°C DBT, 20% RH

Amount of free air circulated = 3.33 m³/s

Coil dew point temperature = 14°C

The required conditions achieved first by cooling and dehumidification, and then by heating. Estimate

(i) The capacity of the cooling coil in tonnes,

(ii) The capacity of the heating coil in kW, and

(iii) The amount of water vapour removed in kg/s. **[IES, 1999]**

Solution: Refer to Fig. 12.46.

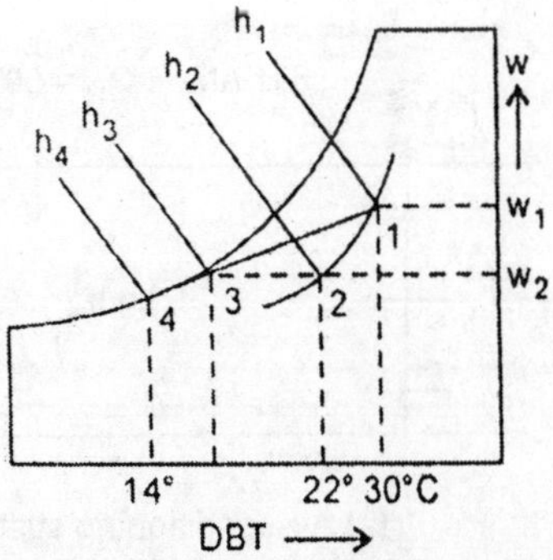

Given:

$$h_1 = 82 \text{ kJ/kg da},$$

$$h_2 = 52 \text{ kJ/kg da},$$

$$h_3 = 47 \text{ kJ/kg da}$$

$$h_4 = 40 \text{ kJ/kg da}$$

$$w_1 = 0.020,$$

$$w_2 = w_3 = 0.0115 \text{ kg/kg da},$$

$$v_{s1} = 0.887 \text{ m}^3\text{/kg da (See Fig. 12.46)}$$

Fig. 12.46 Psychrometic chart.

Dry air flow,

$$G = \frac{\text{Amount of free air circulated}}{v_{s1}}$$

$$= \frac{3.33}{0.887} = 3.754 \text{ kg da/s}$$

(*i*) Capacity of cooling coil

$$= G(h_1 - h_3) = \frac{3.754\,(82-47)}{3.5167} = 37.362 \text{ tonnes}$$

(*ii*) Capacity of heating coil

$$= G(h_2 - h_3) = 3.753\ (52 - 47) = 18.77 \text{ kW}$$

(*iii*) Rate of water vapour removed

$$= G(w_1 - w_2) = 3.754\ (0.020 - 0.0115) = 0.0319 \text{ kg/s}$$

Example 12.35 *Air enters a chamber at 5°C dry bulb temperature (DBT) and 2.5°C wet bulb temperature at the rate of 100 m³/min at the pressure of 1 bar. While passing through the chamber, the air absorbs 50 kW heat and picks up 35 kg/hr of saturated steam at 110°C. Show the process on a psychrometric chart and find the dry and wet bulb temperatures of the leaving air. At 110°C, enthalpy of saturated steam is 2691.3 kJ/kg.*

[IES, 2001]

Solution: Given: DBT, $t_1 = 5°C$, WBT $t_1' = 2.5°C$, $\dot{V}_1 = 100 \text{ m}^3/\text{min}$

$$v_{s1} = 0.792 \text{ m}^3/\text{kg d.a.}$$

$$\dot{m}_a = \dot{m}_{a1} = \dot{m}_{a2} = \frac{\dot{V}_1}{v_{s1}} = \frac{100}{0.792} = 126.26 \text{ kg/min}$$

Refer to Fig. 12.47.

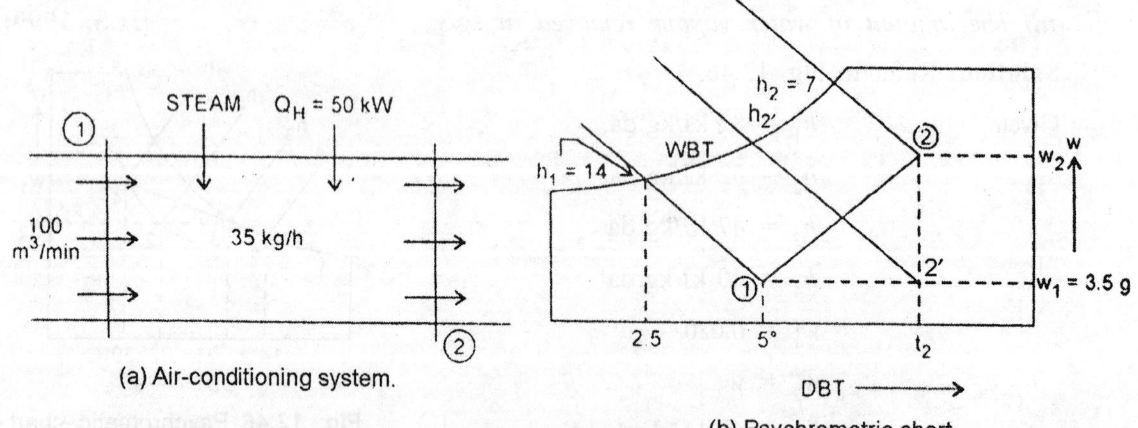

(a) Air-conditioning system.

(b) Psychrometric chart.

Fig. 12.47

From energy balance:

$$\dot{m}_{a1}.h_1 + \dot{m}_s.h_s + Q_H = \dot{m}_{a2}.h_2$$

$$\frac{126.26 \times 14 + 35 \times 2691.3}{60} + 50 \times 60 = 126.26 \times h_2$$

$$h_2 = 50.2 \text{ kJ/kg d.a.}$$

From sensible heat balance:

$$\dot{m}_{a1}.h_1 + Q_H = \dot{m}_a \,.\, h_2'$$
$$126.26 \times 14 + 50 \times 60 = 126.26\; h_2'$$
$$h_2' = 37.76 \text{ kJ/kg.d.a.}$$

From moisture balance:

$$\dot{m}_{a1}.w_1 + w = \dot{m}_{a2}.w_2$$

Knowing h_2' and w_2, point 2 can be located on the psychrometric chart for leaving air.

$$\text{DBT} = 25.5°\text{C}, \text{ WBT} = 17.5°\text{C}$$

DBT at point 2 can also be found out by sensible heat balance as:

$$\dot{m}_{a1}.c_p.t_1 + 50 \text{ kJ/s} = \dot{m}_{a2}.c_p.t_2$$

Dry air mass balance gives,

$$m_{a1} = m_{a2} = m_a$$
$$126.26 \times 1.005 \times 5 + 50 \times 60 = 126.26 \times 1.005 \times t_2$$
$$t_2 = 28.64°\text{C}$$

Example 12.36 *Explain adiabatic saturation and thermodynamic wet-bulb temperature. When are the adiabatic saturation and wet-bulb temperatures equivalent for atmospheric air?* **[IES, 2006]**

Solution: Adiabatic Saturation Temperature, t^*. For any state of unsaturated moist air, there exists a temperature at which the air becomes adiabatically saturated by the evaporation of water into air, at exactly the same temperature, called the *adiabatic saturation temperature.*

Thermodynamic Wet Bulb Temperature, t′. The temperature measured by the wick-covered bulb of a thermometer indicates the temperature of liquid water in the wick and is called the wet bulb temperature.

$t^* = t'$ when the dimensionless quantity $\dfrac{f_g}{k_w c_p}$, called the

Lewis number, is unity;

where f_g = heat transfer co-efficient,

k_w = diffusion co-efficient for mass transfer, and

c_p = specific heat at constant pressure

Example 12.37 *What is metabolism? What is the range of metabolic rate for an average man? Why are we interested in the metabolic rate of the occupants of a building when we deal with heating and air-conditioning?* **[IES, 2006]**

Solution: Metabolism. It is the process of generation of heat within the body of human beings on account of food consumed and respiration. The metabolism rate for an average man varies from 115 W to 265 W.

The occupants in a conditioned space give out heat at a metabolic rate that more or less depends on their rate of working. The relative proportion of the sensible and latent heats given out depends on the ambient DBT. The lower the DBT, the greater the heat given out as sensible heat. The metabolic rate adds cooling and heating loads for the building to be air-conditioned and therefore is of interest to the air-conditioning designer.

Example 12.38 *Calculate all the psychrometric properties of air at 1 bar and 25°C dbt and 15°C wbt. The following properties of water may be assumed:*

Temp (°C)	*Sat. pressure (bar)*	*Sp. Volume of vapour (m^3/kg)*	*Enthalpy (kJ/kg)*	
			Sat. liq	*Sat. vap.*
25	0.03166	43.40	–	2547.3
15	0.01703	77.98	62.94	–
10	0.01078	–	–	–
0	0.01002	–	–	–

The following expression may be used, if necessary: **[IES, 2010]**

$$p_v = (p_{vs})_{wb} = \frac{[p_t - (p_{vs})_{wb}](t_{db} - t_{wb})}{1527.4 - 1.3\, t_{wb}}$$

Solution: Given:

$$t_{db} = 25°\text{C},\ t_{wb} = 15°\text{C},\ p_t = 1 \text{ bar}$$

$$p_v = (p_{vs})_{wb} - \frac{[p_t - (p_{vs})_{wb}](t_{db} - t_{wb})}{1527.4 - 1.3 t_{wb}}$$

Corresponding to 15°C, $(p_{vs})_{wb} = 0.01703$ bar

$$p_v = 0.01703 - \frac{[1 - 0.01703](25° - 15°)}{1527.4 - 1.3 \times 15} = 0.0105 \text{ bar}$$

(*i*) Specific humidity, $w = \dfrac{0.622\, p_t}{p_t - p_v} = \dfrac{0.0622 \times 0.0105}{(1 - 0.0105)} = 6.6 \times 10^{-8}$ kg/kg d.a.

(*ii*) Relative humidity, $\phi = \dfrac{p_v}{p_w} = \dfrac{0.0105}{0.01703} = 0.6165$ or 61.65%

(*iii*) Vapour density, $\phi_v = \dfrac{p_v}{R_v T_v} = \dfrac{0.0105 \times 10^5}{\dfrac{8314.3}{18} \times (273 + 25)} = 7.628 \times 10^{-3}$ kg/m^3

(*iv*) Dew point temperature, t_{dp} = 10.5°, corresponding to p_v = 0.0105 bar from steam tables.

(*v*) Enthalpy of moisture, $h = 1.005\, t_{db} + w\,(2500 + 1.88\, t_{db})$

$= 1.005 \times 25 + 6.6 \times 10^{-3}\,(2500 + 1.88 \times 25)$

$= 41.93$ kJ/kg d.a.

Example 12.38

(*i*) *Discuss the measurement of wet bulb temperature by a means of a wick-covered bulb of a thermometer. How is it related to dry-bulb temperature and specific humidity of moist air and on what other factors does it depend?*

(*ii*) *Define thermodynamic wet bulb temperature. How is it determined? How does it differ from measured wet bulb temperature?* **[IES, 2011]**

Solution:

(*i*) WBT is measured by a thermometer whose bulb is covered by a cotton wick saturated with water. As air stream flows past the wick, some water evaporates taking away the latent heat from the wick, thus decreasing its temperature. In turn, energy is transferred to the wick from the surrounding air. When equilibrium condition is reached, there is a balance between energy removed from water film of wick by evaporation and energy supplied to the wick by heat transfer from surrounding air. The temperature indicated by the thermometer at this instant is called the wet bulb temperature.

The WBT is always less than the DBT except when air is saturated. For saturated air, WBT = DBT.

The other factors affecting WBT are:

1. Dry bulb temperature
2. Dew point temperature
3. Barometric pressure.

(*ii*) **Thermodynamic WBT :** The thermodynamic WBT (or adiabatic saturation temperature) is the temperature at which air can be brought to saturation state adiabatically by the evaporation of water into the flowing air. It can be measured by adiabatic saturation process. The equipment (see Fig. 12.48) used for adiabatic saturation of air consists of an insulated chamber containing adequate quantity of water. The make-up water is sprayed from the top of chamber. The unsaturated air enters from one side of the chamber, passes over a long sheet of water. The water evaporates and the specific humidity of air increases. The water level is kept constant by make-up water spray. Both the air and water cool as evaporation takes place. This process continues until the

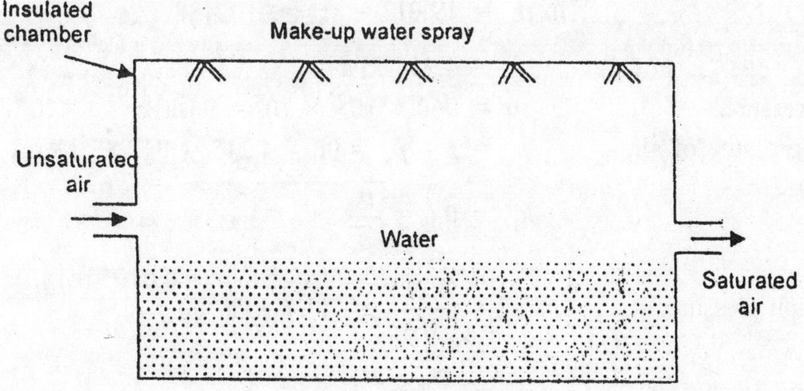

Fig. 12.48 Adiabatic saturation of air.

energy transferred from the air to water is equal to the energy required to vaporize the water. When steady state conditions are reached the leaving air from the chamber is saturated with water vapour. The temperature of saturated air at exit from chamber is known as thermodynamic WBT. The thermodynamic WBT is taken equal to the WBT for all practical purposes.

Example 12.39 *At a location where the atmospheric pressure is 960 m bar, the temperature is 30°C and the relative humidity 60%. Calculate the humidity ratio, degree of saturation, enthalpy and dew point. Assume that air and water vapour may be treated as perfect gases.*

The following data may be used:

$$M_{air} = 29; \quad M_{water} = 18$$

Universal gas constant $= 8.314$ *kJ/kg mole.K*

$$c_{p\ air} = 1 \text{ kJ/kg.K}$$

$$c_{p\ water} = 4.1856 \text{ kJ/kg.K}$$

Heat of vaporization of water $= 2514$ *kJ/kg*

Vapour pressure of water follows the equation

$$\ln p = 19.013 - \frac{5325}{T}$$

where p is in kPa and T in K

[GATE, 1994]

Solution: Given: $p = 960$ m bar, $T = 273 + 30 = 303$ K, $w = 60\%$, $M_{air} = 29$, $M_{water} = 18$,

$R_u = 8.314$ kJ/kg mol.K, $c_{pa} = 1$ kJ/kg.K, $c_{pw} = 4.186$ kJ/kg·K,

$h_{fg} = 2514$ kJ/kg,

$$\ln p_w = 19.013 - \frac{5325}{T}$$

where p in kPa and T in K.

Vapour pressure of water

$$\ln p_w = 19.013 = \frac{5325}{303} = 1.438$$

$$p_w = 4.215 \text{ kPa}$$

Total pressure, $p = 960 \times 10^{-3} \times 10^5 = 96$ kPa

Partial pressure of air, $p_a = p - p_w = 96 - 4.215 = 91.785$ kPa

$$R_H = 0.6 = \frac{p_w}{p_s}$$

Saturation pressure, $p_s = \dfrac{4.215}{0.6} = 7.026$ kPa

(*a*) humidity ratio, $w = \dfrac{m_w}{m_a} = \dfrac{R_a}{R_w} \cdot \dfrac{p_w}{p_a}$

where, $R_a = \frac{R_u}{M_a} = \frac{8.314}{29} = 0.2867 \text{ kJ/kg} \cdot \text{K}$

$$R_w = \frac{R_u}{M_w} = \frac{8.314}{18} = 0.4619 \text{ kJ/kg} \cdot \text{K}$$

$$w = \frac{0.2867}{0.4617} \times \frac{4.215}{91.785} = 0.0285 = 2.85\%$$

(*b*) Degree of saturation, $\mu = \frac{w}{w_s} = \frac{p_w}{p_s}\left(\frac{p - p_s}{p - p_w}\right)$

$$= \frac{4.215}{7.026} \times \left[\frac{96 - 7.026}{96 - 4.215}\right]$$

$$= 0.5815$$

(*c*) Enthalpy $h = c_{pa} T + (c_{pw} \cdot T + h_{fg})\, w$

$$= 1 \times 303 + (4.186 \times 303 + 2514) \frac{2.85}{100}$$

$$= 407.975 \text{ kJ/kg}$$

(*d*) $\ln p_s = 19.013 - \frac{5325}{\text{DPT}}$

$$\ln 7.026 = 19.013 - \frac{5325}{\text{DPT}}$$

$$DPT = 312.07 \text{ K or } \mathbf{39.07°C}$$

Example 12.40 *Atmospheric air at a pressure of 1.0132 bar, has a dry bulb temperature of 30°C and relative humidity of 70%. Using STEAM TABLES only, calculate:*

(i) partial pressure of Water Vapour and air in moist air,

(ii) the specific humidity,

(iii) the degree of saturation, and

(iv) the dew Point temperature.

Given:

(*i*) Corresponding to t_d = 30°C, saturation pressure of water vapour, from steam tables, is

$$p_s = 0.04246 \text{ bar},$$

$$\phi = \frac{p_v}{p_s},\ p_v = 0.7 \times 0.04246 = 0.02972 \text{ bar}$$

Now $p_b = p_a + p_v$

Partial pressure of air, p_a = 1.0132 – 0.02972 = 0.98348 bar

(*ii*) Specific humidity, $w = \frac{0.622\, p_v}{p_b - p_v} = \frac{0.622\, p_v}{p_a} = \frac{0.622 \times 0.02972}{0.98348} = 0.01879$ kg/kg d.a.

$$\text{Degree of saturation, } \mu = \frac{w}{w_s} = \frac{0.01879}{0.02721} = 0.6906 \text{ or } 69.05\%$$

(*iv*) DPT = saturation temperature corresponding to p_v = 0.02972 bar.

From steam tables, corresponding to p_v = 0.02972 bar

$$\text{or} \quad \frac{0.02972 \times 10^5}{133.32} = 22.29 \text{ mm Hg,}$$

$$\text{DPT, } t_{dp} = 23.72 + \left(\frac{24.46 - 23.72}{23 - 22}\right)(22.29 - 22)$$

$$= 23.72 + 0.21 = 23.93°\text{C}$$

Example 12.41 *In a laboratory test, on a particular day, a psychrometer recorded the dry bulb and wet bulb temperatures for atmospheric air as 35°C and 25°C respectively. The atmospheric pressure is 1.013 bar. Using STEAM TABLES only (psychrometric chart not allowed), calculate:*

(i) The relative humidity

(ii) The specific humidity

(iii) The dew point temperature, and

(iv) The degree of saturation.

The partial pressure of water vapour (p_v) can be calculated with the help of Carrier's equation, given as;

$$p_v = \left[p_{wb} - \frac{(p - p_{wb})(t_{db} - t_{wb})}{1547 - 1.44\, t_{wb}}\right]$$

Where p_{wb} = *saturation pressure corresponding to wet bulb temperature,*

p = *Barometric pressure, and*

t_{db} *and* t_{wb} = *are the dry and wet bulb temperatures of air (in 0°C).*

Gas constant of air, R_a = *0.287 kJ/kg-K.*

Solution: Psychrometric definitions: Refer to art. 12.2.

Given: t_d = 35°C, t_w = 25°C, p_b = 1.013 bar,

(*i*) Corresponding to t_d = 35°C, p_s = 5622 bar, and for t_w = 25°C, p_w = 0.03166 bar

Partical pressure of water vapour,

$$p_v = p_w - \frac{(p - p_w)(t_d - t_w)}{1547 - 1.44\, t_w}$$

$$= 0.03166 - \frac{(1.013 - 0.03166)(35 - 25)}{1547 - 1.44 \times 25} = 0.02516 \text{ bar}$$

$$\text{Relative humidity,} \quad \phi = \frac{p_v}{p_s} = \frac{0.0516}{0.05622} = 0.4476 \text{ or } 44.76\%$$

(*ii*) Specific humidity, $w = \dfrac{0.622\,p_v}{p_b - p_v} = \dfrac{0.622 \times 0.02516}{1.013 - 0.02516} = 0.01584$ kg/kg d.a.

(*iii*) Corresponding to $p_v = 0.02516$ bar,

$$\text{DPT, } t_{dp} = 21 + \frac{(22-21)}{(0.02643 - 0.02486)} \cdot (0.02516 - 0.02486) = 21 + 0.191 = 21.191°\text{C}$$

(*iv*) Specific humidity of saturated air,

$$w_s = \frac{0.622\,p_s}{p_b - p_s}$$

$$= \frac{0.622 \times 0.05622}{1.013 - 0.05622} = 0.03655 \text{ kg/kg d.a.}$$

Degree of saturation, $\mu = \dfrac{w}{w_s} = \dfrac{0.01584}{0.03655} = 0.4333$ or 43.33%

Example 12.42 *Air at 20°C, 40% RH is mixed adiabatically with air at 40°C and 40% RH in the ratio of 1 kg of former to 2 kg of latter (dry basis). Find the final condition of air*

Solution: Given: $t_1 = 20°\text{C}, \phi_1 = 40\%, m_1 = 1$ kg; $t_2 = 40°\text{C}, \phi_2 = 40\%, m_2 = 2$ kg

$$m_{a3} = m_{a1} + m_{a2} = 1 + 2 = 3 \text{ kg}$$

$$m_{a1}h_1 + m_{a2}h_2 = m_{a3}h_3$$

For dry air, we have

$$m_{a1}c_p t_{d1} + m_{a2}c_p t_{d2} = m_{a3}c_p t_{d3}$$

$$\therefore \quad 1 \times 20 + 2 \times 40 = 3 \times t_{d3}$$

$$\therefore \quad t_{d3} = \frac{100}{3} = 33.3°\text{C}$$

Now $\phi = \dfrac{p_v}{p_s}$

and $w = \dfrac{0.622\,p_v}{p_b - p_v}$

$$w = \frac{0.622\,p_v}{p_s} = 0.622\,\phi$$

Now $m_{a1}w_1 + m_{a2}w_2 = m_{a3}w_3$

or $m_{a1}\phi_1 + m_{a2}\phi_2 = m_{a3}\phi_3$

$$1 \times 0.4 + 2 \times 0.4 = 3\phi_3$$

$$\phi_3 = \frac{1.2}{3} = 0.4$$

Example 12.43 *What do you understand by Adiabatic saturation and Thermodynamic Wet Bulb temperature? In an adiabatic saturator, moist air enters at a temperature of 32°C and leaves at 22°C which is also the adiabatic saturation temperature. If the pressure remains constant at 1 bar, determine the Specific humidity and Relative humidity of the inlet mixture. Take* c_{pa} *for air as 1.005 kJ/kg.K.*

Solution: Adiabatic saturation and thermodynamic WBT: Refer to art. 12.6.

Given: $t_d = 32°C,\ t_w = 22°C,\ p_b = 1\ \text{bar},\ c_{pa} = 1.005\ \text{kJ/kg.K °C}$

Saturation pressure of vapour at $t_w = 22°C$ from steam tables,

$$p_{v2} = 0.02643\ \text{bar}$$

Enthalpy of saturated vapour at 22°C, $h_{s2} = h_{g2} = 2541.47$ kJ/kg

Sensible heat of water at 22°C, $h_{fw} = 92.27$ kJ/kg

Enthalpy of 1 kg of unsaturated air at $t_d = 32°C$,

$$h_{a1} = m_a c_{pa} t_d = 1 \times 1.005 \times 32 = 32.16\ \text{kJ/kg}$$

Enthalpy of 1 kg of saturated air at $t_w = 22°C$,

$$h_{a2} = 1 \times 1.005 \times 22 = 22.11\ \text{kJ/kg}$$

$$w_2 = \frac{0.622\, p_{v2}}{p_b - p_{v2}} = \frac{0.622 \times 0.02643}{1.0 - 0.02643} = 0.01688\ \text{kg/kg d.a.}$$

$$w_1 = \frac{w_2(h_{s2} - h_{fw}) + (h_{a2} - h_{a1})}{h_{s1} - h_{fw}}$$

$$= \frac{0.01688(2541.47 - 92.27) + (22.11 - 32.16)}{2559.07 - 92.27}$$

$$= 0.01268\ \text{kg/kg d.a.}$$

Now $$w_1 = \frac{0.622\, p_{v1}}{p_b - p_{v1}}$$

$$0.01268 = \frac{0.622\, p_{v1}}{1.0 - p_{v1}}$$

$$0.01268 - 0.01268\, p_{v1} = 0.622\, p_{v1}$$

$$p_{v1} = 0.01998\ \text{bar}$$

Saturation pressure corresponding to 32°C from steam tables is,

$$p_s = 0.04753\ \text{bar}$$

Relative humidity, $$\phi = \frac{p_{v1}}{p_s} = \frac{0.01998}{0.04753} = 0.4203 \text{ or } 42.03\%$$

Example 12.44 *Calculate (i) relative humidity, (ii) humidity ratio, (iii) dew point temperature (iv) density and enthalpy of atmospheric air when the DBT is 35°C, WBT is 23°C and the barometer reads 750 mm of mercury (Use Carrier Equation for determining the partial pressure of vapour). Use properties of water from steam tables.* **[June,2006]**

Solution: Given: $t_d = 35°C$, $t_w = 23°C$, $p_b = 750$ mm Hg or $750 \times 133.32 \times 10^{-5} = 0.9999$ bar

Corresponding to $t_w = 23°C$, $p_w = 0.02808$ bar

According to Carrier's equation,

$$p_v = p_w - \frac{(t_d - t_w)(p_b - p_w)}{1544 - 1.44 t_w}$$

$$= 0.02808 - \frac{(35-23)(0.9999 - 0.02808)}{1544 - 1.44 \times 23}$$

$$= 0.02036 \text{ bar}$$

(*i*) For $t_d = 35°C$, $p_s = 0.05622$ bar

$$RH, \phi = \frac{p_v}{p_s} = \frac{0.02036}{0.05622}$$

$$= 0.3621 \text{ or } 36.21\%$$

(*ii*) Humidity ratio, $w = \frac{0.622\, p_v}{p_b - p_v} = \frac{0.622 \times 0.02036}{0.9999 - 0.02036} = 0.01293$ kg/kg d.a.

(*iii*) Corresponding to $p_v = 0.02036$ bar, DPT from steam tables is,

$$t_{dp} = 17 + \frac{(18-17)(0.02036 - 0.019364)}{(0.020620 - 0.019364)}$$

$$= 17.79°C$$

(*iv*) Density of air, $\rho_a = \frac{p_a}{R_a T_a}$

$$p_a = p_b - p_v = 0.9999 - 0.02036 = 0.97954 \text{ bar}$$

$$R_a = 0.287 \text{ kJ/kg.K}, \; T_a = T_d = 273 + 35 = 308 \text{ K}$$

$$\rho_a = \frac{0.97954 \times 10^5}{287 \times 308} = 1.108 \text{ kg/m}^3$$

Enthalpy of air, $h = 1.216\, t_d + w[(h_{fg})_{dpt} + 2.3068 t_{dp}]$

At $t_{dp} = 17.79°C$, $(h_{fg})_{dpt} = 2461 + \left[\frac{2458 - 2461}{18 - 17}\right][17.79 - 17] = 2458.63$ kJ/kg

$$h = 1.0216 \times 35 + 0.01293\,[2458.63 + 2.3068 \times 17.79]$$

$$= 68.08 \text{ kJ/kg d.a.}$$

Example 12.45 *On a particular day, the atmospheric air at a pressure of 1.0132 bar, was found to have a dry bulb temperature of 35°C and relative humidity of 65%. Using STEAM TABLES only, calculate: the partial pressure of water vapour and air in moist air, the specific humidity, the degree of saturation, and the dew point temperature (psychrometric chart not allowed).*

Solution: Given: $p_b = 1.0132$ bar, $t_d = 35°C$, $\phi = 65\%$

Corresponding to $t_d = 35°C$, $p_s = 0.05622$ bar

Relative humidity, $\phi = \dfrac{p_v}{p_s}$

Partial pressure of water vapour, $p_v = 0.65 \times 0.05622 = 0.03654$ bar

Partial pressure of air, $p_a = p_b - p_v = 1.0132 - 0.03654 = 0.97668$ bar

Specific humidity, $w = \dfrac{0.622\, p_v}{p_a} = \dfrac{0.622 \times 0.03654}{0.97668} = 0.02327$ kg/kg d.a.

Corresponding to $p_v = 0.03654$, DPT,

$$t_{dp} = 27 + \left[\frac{28 - 27}{0.03779 - 0.03564}\right] (0.03654 - 0.03564)$$

$$= 27.42°C$$

Specific humidity of saturated air,

$$w_s = \frac{0.622\, p_s}{p_b - p_s}$$

$$= \frac{0.622 \times 0.05622}{1.0132 - 0.05622} = 0.03654 \text{ kg/kg d.a.}$$

Degree of saturation, $\mu = \dfrac{w}{w_s} = \dfrac{0.02327}{0.3654} = 0.6368$ or 63.68%

Example 12.46 *The out door air (at 39°C DBT and 45% RH) is mixed with return (from room) air (at 27°C DBT and 50% RH) in the ratio of 1:1 before entering the cooling coil. The by-pass factor for the cooling coil is 0.2 and the ADP is 12°C. Air flow rate (total) through the cooling coil is 8 kg/s.*

Determine:

(i) Condition of air at inlet and exit of the cooling coil

(ii) RSHF (room sensible heat factor)

(iii) Tonnage of the plant, and

(iv) Rate of condensation.

Solution: Given: $t_{d1} = 39°C$, $\phi_1 = 45\%$, $t_{d2} = 27°C$, $\phi_2 = 50\%$, $\dot{m}_{a1} : \dot{m}_{a2} :: 1 : 1$,

BPF $= 0.2$ ADP $= 12°C$, $\dot{m}_{a3} = 8$ kg/s

The flow diagram for air-conditioning is shown in Fig. 12.49. The various processes are shown on the psychrometric chart in Fig. 12.50.

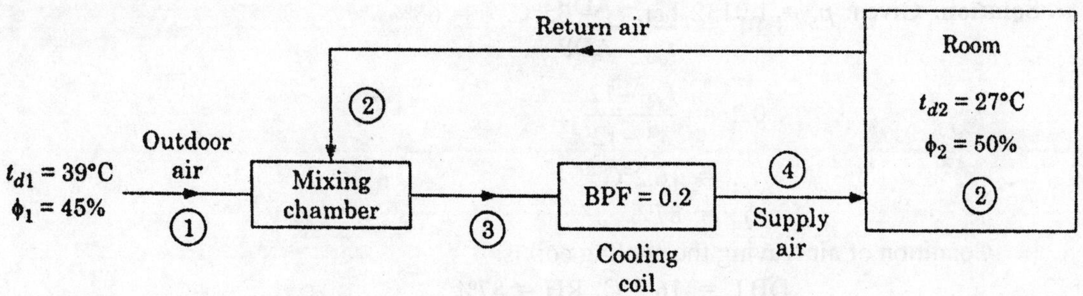

Fig. 12.49 Flow diagram for air conditioning

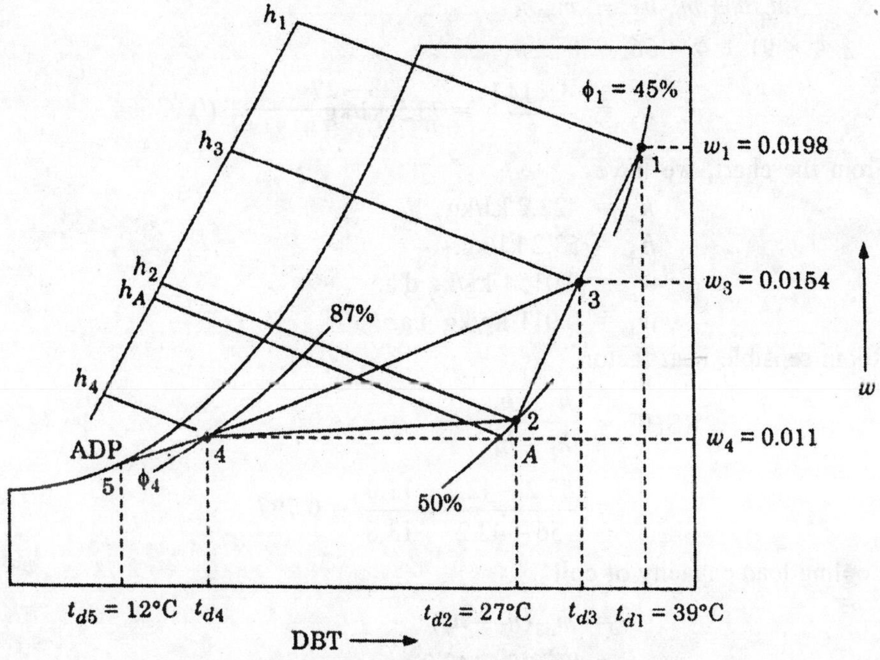

Fig. 12.50 Psychromatric chart.

(*i*) Locate point 1 (t_{d1} = 39°C, ϕ_1 = 45%) and point 2 (t_{d2} = 27°C, ϕ_2 = 50%)
Join points 1 and 2 by a straight line.

Since $\dot{m}_{a1} : \dot{m}_{a2} :: 1:1$, therefore

Length 2-3 = length 3-1.

Point 3 represents the condition of air entering the cooling coil after mixing process.

Location of point 3 is t_{d3} = 33°C, ϕ_3 = 49%

From the psychrometric chart, we have

$$h_1 = 91 \text{ kJ/kg},\ h_2 = 56 \text{ kJ/kg},\ h_3 = 73 \text{ kJ/kg}$$

Locate point 5 on the saturation curve such that ADP = 12°C. Join point 3 with point 5.

Now BPF = 0.2

$$\text{BPF} = \frac{t_{d4} - \text{ADP}}{t_{d3} - \text{ADP}}$$

$$0.2 = \frac{t_{d4} - 12}{33 - 12}$$

$$t_{d4} = 16.2°\text{C}$$

$$\phi_4 = 87\%$$

Condition of air leaving the cooling coil is:

DBT = 16.2°C, RH = 87%.

By calculations:

$$\dot{m}_{a1}h_1 + \dot{m}_{a2}h_2 = \dot{m}_{a3}h_3$$

$$4 \times 91 + 4 \times 56 = 8 \times h_3$$

$$h_3 = \frac{4 \times 147}{8} = 73.5 \text{ kJ/kg}$$

(*ii*) From the chart, we have

$$h_4 = 422.2 \text{ kJ/kg}$$

$$h_A = 53.2 \text{ kJ/kg},$$

$$w_3 = 0.0154 \text{ kg/kg d.a.}$$

$$w_4 = 0.011 \text{ kg/kg d.a.}$$

Room sensible heat factor,

$$\text{RSHF} = \frac{h_A - h_4}{h_2 - h_4}$$

$$= \frac{53.2 - 42.2}{56 - 42.2} = \frac{11.0}{13.8} = 0.797$$

(*iii*) Cooling load capacity of coil

$$= \dot{m}_{a3}(h_3 - h_4)$$

$$= 8(73.5 - 42.2)$$

$$= 250.4 \text{ kJ/s}$$

$$= \frac{250.4 \times 60}{211} = 71.2 \text{ TR}$$

(*iv*) Rate of condensation $= \dot{m}_a(w_3 - w_4)$

$$= 8(0.0154 - 0.011)$$

$$= 0.0352 \text{ kg/s or } 2.1 \text{ kg/min}$$

Example 12.47 *In an air conditioner the out door air (at 40°C DBT and 43% RH) is mixed with return (from room) air (at 25°C DBT and 50% RH) in the ratio of 3:2 before entering the cooling coil. The by-pass factor for the cooling coil is 0.25 and the room sensible heat factor (RSHF) is 0.8. air flow rate (total) through the cooling coil is 8 kg/s. determine:*

(i) ADP,

(ii) Condition of air at inlet and exit of the cooling coil,

(iii) Tonnage of the plant, and

(iv) Rate of condensation.

Solution: Given: $t_{d1} = 40°C, \phi_1 = 43\%, t_{d2} = 25°C, \phi_2 = 50\%, \dot{m}_{a1} : \dot{m}_{a2} :: 3 : 2, \dot{m}_{a3} = 8$ kg/s, BPF = 0.25, RSHF = 0.8

The flow diagram for the room is shown in Fig. 12.51. The various processes are shown on the psychrometric chart in Fig. 12.52.

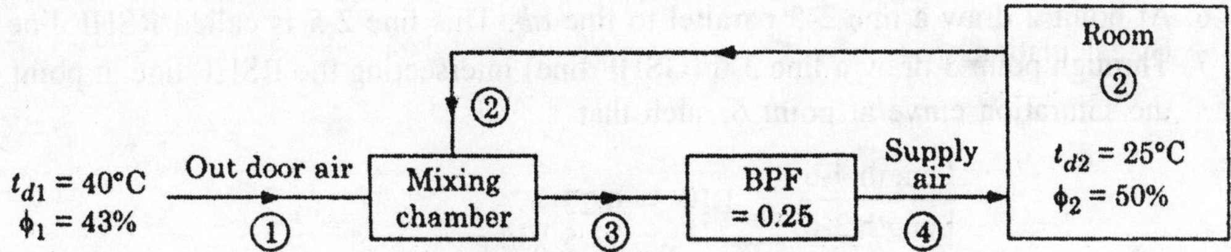

Fig. 12.51 Air-conditioning system.

1. Locate point 1 (t_{d1} = 40°C, ϕ_1 = 43%) and point 2 (t_{d2} = 25°C, ϕ_2 = 50%).
2. Joint points 1 and 2 by a straight line.
3. $\dot{m}_{a1} = \frac{3}{5}\dot{m}_{a3} = \frac{3}{5} \times 8 = 4.8$ kg/s

$\dot{m}_{a2} = 8 - 4.8 = 3.2$ kg/s

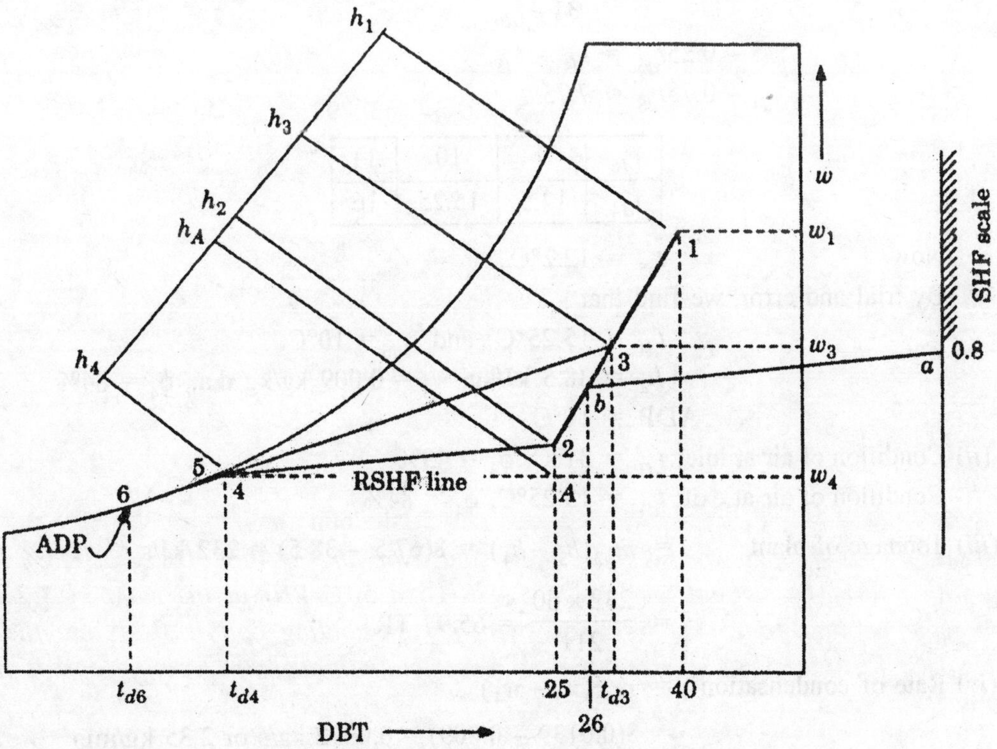

Fig. 12.52 Psychrometric chart.

4. Divide line 1-2 in the ratio of 3 : 2 to locate point 3. Length 2-3 = (Length 1-2) $\times \frac{2}{5}$

From the chart,

$$h_1 = 92 \text{ kJ/kg}, \quad h_2 = 50.5 \text{ kJ/kg}$$

Point 3 represents the condition of air at inlet to the cooling coil.

$$t_{d3} = 31°\text{C}, \phi_3 = 49\%, h_3 = 67.5 \text{ kJ/kg}, w_3 = 0.0139$$

5. Locate the alignment circle (i.e. 26° DBT and 50% RH) point *b* and point '*a*' at RSHF = 0.8 on the SHF scale. Joint points '*a*' and *b*. $t_{d5} = 12.2°\text{C}$.
6. At point 2 draw a line 2-5 parallel to line *ab*. This line 2-5 is called RSHF line.
7. Through point 3 draw a line 3-6 (GSHF line) intersecting the RSHF line at point 4 and the saturation curve at point 6, such that

$$\frac{\text{Length 4-6}}{\text{Length 3-6}} = \text{BPF} = 0.25$$

(*i*) Apparatus dew point.

The point 6 on the psychrometric chart is the ADP.

$$\text{BPF} = \frac{t_{d4} - t_{d6}}{t_{d3} - t_{d6}} = \frac{h_4 - h_6}{h_3 - h_6} = \frac{w_4 - w_6}{w_3 - w_6}$$

From the chart, $t_{d3} = 31°\text{C}$

$$\therefore \quad 0.25 = \frac{t_{d4} - t_{d6}}{31 - t_{d6}}$$

$$7.75 - 0.25t_{d6} = t_{d4} - t_{d6}$$

$$t_{d4} - 0.75t_{d6} = 7.75$$

t_{d6}	9	10	11
t_{d4}	14.5	15.25	16

Now $t_{d5} = 12.2°\text{C}$

By trial and error, we find that

$$t_{d4} = 15.25°\text{C} \text{ and } t_{d6} = 10°\text{C}.$$

$$h_4 = 38.5 \text{ kJ/kg}, w_4 = 0.009 \text{ kg/kg d.a.}, \phi_4 = 84\%$$

$$\therefore \quad \text{ADP} = 10°\text{C}$$

(*ii*) Condition of air at inlet: $t_{d3} = 31°\text{C}, \phi_3 = 49\%$

Condition of air at exit: $t_{d4} = 15.25°\text{C}, \phi_4 = 84\%$

(*iii*) Tonnage of plant $= \dot{m}_{a3}(h_3 - h_4) = 8(67.5 - 38.5) = 232 \text{ kJ/s}$

$$= \frac{232 \times 60}{211} = 65.97 \text{ TR}$$

(*iv*) Rate of condensation $= \dot{m}_{a3}(w_3 - w_4)$

$$= 8(0.0139 - 0.009) = 0.0392 \text{ kg/s or } 2.35 \text{ kg/min}$$

Example 12.48 *What do you understand by Psychrometrics and Psychrometric properties? Define 'Dew Point temperature', 'specific humidity', 'relative humidity' and 'degree of saturation', as related to psycrometrics. Calculate the (i) relative humidity, (ii) specific humidity (iii) dew point temperature and (iv) degree of saturation; for atmospheric air having dry bulb temperature of 40°C, wet bulb temperature of 25°C and atmospheric pressure of 1.013 bar, using steam tables only. The partial pressure of water vapour (p_v) can be calculated by Carrier' s equation,*

Given as:

$$p_v = \left[p_{wb} - \frac{(p - p_{wb})(t_{db} - t_{wb})}{1547 - 1.44 t_{wb}} \right]$$

Where, p_{wb} = *saturation pressure corresponding to wet bulb temperature.*

p = *barometric pressure*

t_{db} *and* t_{wb} = *dry bulb and wet bulb temperatures of air (in °C) and,*

Gas constant for air, R_a = *0.287 kJ/kg.K.*

Solution: Refer to art. 12.2 and art. 12.8.

Given: t_d = 40°C, t_w = 25°C, p_b = 1.013 bar, R_a = 0.287 kJ/kg.K.

Corresponding to t_d = 40°C, p_s = 0.07375 bar and for t_w = 25°C, p_w = 1.01366 bar

$$p_v = p_w - \frac{(p_b - p_w)(t_d - t_w)}{1547 - 1.44 t_w}$$

$$= 0.03166 - \frac{(1.013 - 0.03166)(40 - 25)}{1547 - 1.44 \times 25}$$

$$= 0.02192 \text{ bar}$$

Relative humidity, $$\phi = \frac{p_v}{p_s} = \frac{0.02192}{0.07375} = 0.2972 \text{ or } 29.72\%$$

Specific humidity, $$w = \frac{0.622\, p_v}{p_b - p_v} = \frac{0.622 \times 0.02192}{1.013 - 0.02192} = 0.01376 \text{ kg/kg d.a.}$$

Corresponding to p_v = 0.02192 bar, dew point temperature,

$$t_{dp} = 18 + \left[\frac{19 - 18}{0.02196 - 0.02062} \right] (0.02192 - 0.02062) = 18.97°\text{C}$$

Specific humidity of saturated air,

$$w_s = \frac{0.622\, p_s}{p_b - p_s}$$

$$= \frac{0.0622 \times 0.07375}{1.013 - 0.07375} = 0.04884 \text{ kg/kg d.a.}$$

Degree of saturation, $$\mu = \frac{w}{w_s} = \frac{0.01376}{0.04884} = 0.2817 \text{ or } 28.17\%$$

Example 12.49 *Room Air having Dry bulb temperature of 40°C and wet bulb temperature of 25°C is cooled through sensible cooling process upto a temperature of 25°C show it on a Psychrometric chart and determine the amount of heat removed (in kJ/kg of d.a.)*

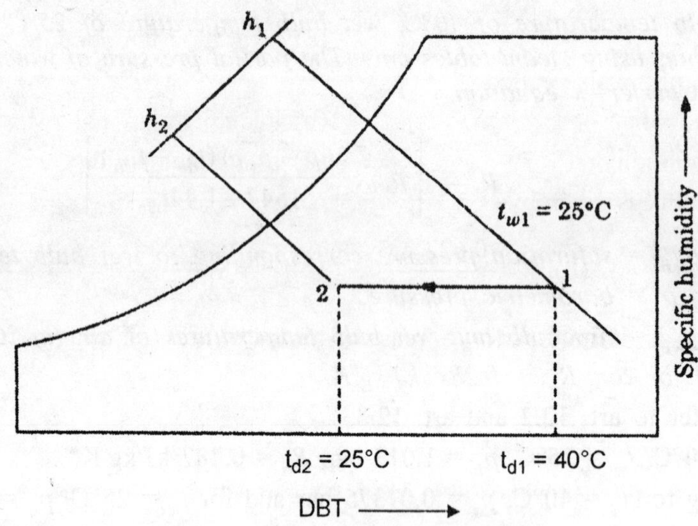

Fig. 12.53 Psychrometric chart.

Solution: Given: t_{d1} = 40°C, t_{w1} = 25°C, t_{d2} = 25°C.

From psychromietric chart, (Fig. 12.53)

$$h_1 = 76.8 \text{ kJ/kg}$$

$$h_2 = 60.8 \text{ kJ/kg}$$

Amount of heat removed $= h_1 - h_2$

$= 76.8 - 60.8 = 16$ kJ/kg d.a.

MULTIPLE – CHOICE QUESTIONS

1. The ambient air temperature as recorded by ordinary thermometer is called
(a) WBT (b) DPT
(c) DBT (c) saturation temperature

2. Specific humidity is also called
(a) relative humidity (b) humidity ratio
(c) absolute humidity (d) saturation humidity

3. Specific humidity is given by
(a) $w = \frac{0.622\,p_v}{p_b}$ (b) $w = \frac{0.622\,p_v}{p_a}$ (c) $w = \frac{0.622\,p_v}{p_b - p_a}$ (d) $w = \frac{0.622\,p_v}{p_b - p_v}$

4. Humidity ratio is defined as (m_a = mass of air, m_v = mass of water vapour in mixture)
(a) $\frac{m_v}{m_a}$ (b) $\frac{m_v}{m_v + m_a}$ (c) $\frac{m_a}{m_a + m_v}$ (d) $\frac{m_v}{m_a - m_v}$

5. The relative humidity ϕ and degree of saturation μ are related to each other as
(a) $\mu = \phi\frac{[p_b - p_s]}{p_b - p_v}$ (b) $\mu = \phi\frac{[p_b - p_v]}{p_b - p_s}$ (c) $\mu = \phi\frac{p_v}{p_b}$ (d) $\mu = \phi\frac{p_s}{p_v}$

6. RSHF is defined as:
(a) $\frac{\text{RSH}}{\text{RTH}}$ (b) $\frac{\text{RLH}}{\text{RTH}}$ (c) $\frac{\text{RSH}}{\text{RLH}}$ (d) $\frac{\text{RSH} - \text{RLH}}{\text{RSH}}$

7. Effective temperature depends on:
(a) DBT only (b) DBT and WBT
(c) DBT and RH (d) DBT, WBT and air motion.

8. In summer air-conditioning, the air is
(a) cooled and humidified (b) cooled and dehumidified
(c) heated and humidified (d) heated and dehumidified

9. In winter air-conditioning, the air is
(a) cooled and humidified (b) cooled and dehumidified
(c) heated and humidified (d) heated and dehumidified

10. The alignment circle is marked on the psychrometric chart at
(a) 20°C DBT and 50% RH (b) 26°C DBT and 50% RH
(c) 20°C DBT and 40% RH (d) 26°C DBT and 40% RH

11. Consider the following statements:
During sensible heating.
1. moisture content increases.
2. dry bulb temperature and wet bulb temperature increase.

3. dew point remains constant.
4. relative humidity increases.

Of these statements

(a) 1, 2 and 3 are correct (b) 2, 3 and 4 are correct
(c) 2 and 3 are correct (d) 1 and 2 are correct

12. When the air is passed through an insulated chamber having sprays of water maintained at a temperature higher than the dew point temperature of entering air but lower than its dry bulb temperature, then the air is said to be

(a) cooled and humidified (b) cooled and dehumidified
(d) heated and humidified (d) heated and dehumidified

13. The atmosphere air at dry bulb temperature of 15°C enters a heating coil maintained at 40°C. The air leaves the heating coil at 25°C. The bypass factor of the heating coil is

(a) 0.376 (b) 0.4 (c) 0.6 (d) 0.67

14. If the specific heats of dry air and water vapour are 1.00 kJ/kg-K. and 1.88 kJ/kg-K respectively and the humidity ratio is 0.011, then the specific heat of moist air at 25°C and 50% relative humidity will be

(a) 1.0207 kJ/kg-K (b) 1.869 kJ/kg-K (c) 1.891 kJ/kg-K (d) 0.9793 kJ/kg-K

15. In case of sensible cooling of air, the coil efficiency is given by (BPF=Bypass factor)

(a) BPF – 1 (b) 1 – BPF (c) $\frac{1}{BPF}$ (d) 1+ BPF

16. In a spray washing system if the temperature of water is higher than the dry bulb temperature of entering air, then the air is

(a) heated and dehumidified (b) heated and humidified
(c) cooled and humidified (d) cooled and dehumidified

17. In a psychrometric process, the sensible heat added is 30 kJ/s and the latent heat added is 20 kJ/s. The sensible heat factor for the process will be

(a) 0.3 (b) 0.6 (c) 0.67 (d) 1.5

18. During sensible cooling

(a) Relative humidity remains constant
(b) Wet bulb temperature increases
(c) Specific humidity increases
(d) Partial pressure of vapour remains constant.

19. By pass factor for a cooling coil

(a) increases with increase in velocity of air passing through it
(b) decreases with increase in velocity of air passing through it
(c) remains unchanged with increase in velocity of air passing through it
(d) may increase or decrease with velocity of air passing through it depending upon the condition of air entering.

20. If air is passed through a solid chemical absorbent, the psychrometic process followed is
(a) heating and dehumidification with the bulb temperature remaining fairly constant
(b) cooling and dehumidification
(c) dehumidification with sharp rise in wet bulb temperature
(d) dehumidification at constant dry bulb temperature.

21. Consider the following statements:
A psychrometer measures
1. wet bulb temperature
2. Dew point temperature
3. dry bulb temperature
On these statements:
(a) 1 alone is correct (b) 2 and 3 are correct
(c) 1 and 3 are correct (d) 1, 2 and 3 are correct

22. During the adiabatic cooling of moist air
(a) DBT remains constant (b) specific humidity remains constant
(c) relative humidity remains constant (d) WBT remains constant.

23. When a stream of moist air is passed over a cold and dry cooling coil such that no condensation takes place, then the air stream will get cooled along the line of
(a) constant wet bulb temperature (b) constant dew point temperature
(c) constant relative humidity (d) constant enthalpy.

24. Consider the following statements:
In chemical dehumidification process

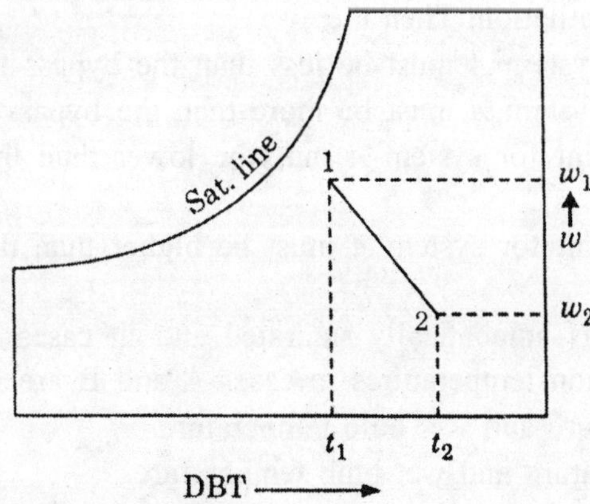

1. dew point temperature decreases.
2. wet bulb temperature decreases.
3. dry bulb temperature increases.
On these statements
(a) 1, 2 and 3 are correct (b) 1 and 2 are correct
(c) 2 and 3 are correct (d) 1 and 3 are correct

25. Moist air is a mixture of dry air and water vapour. Hence three independent intrinsic thermodynamic properties are required to fix its thermo-dynamic state. While using psychrometric chart, however, only two thermodynamic properties are needed since, psychrometric chart
   (a) is an aprroximination to actual properties
   (b) assumes that both water vapour and dry air behave like perfect gases
   (c) is drawn for actual properties of water vapour and dry air
   (d) is drawn for a fixed pressure
26. The deposition of frost on evaporator tubes of an air conditioner will result in
   (a) decrease in heat transfer (b) increase in heat transfer
   (c) no change in heat transfer (d) increase in capacity of evaporator
27. Consider the following statements:
   In psychrometric chart, •
   1. the constant enthalpy lines are also constant wet bulb temperature lines
   2. the wet bulb and dry bulb temperature are same at any condition
   3. the wet bulb and dry bulb temperature are equal at saturation condition.
   Of these statements.
   (a) 1 alone is correct (b) 1 and 2 are correct
   (c) 1 and 3 are correct (d) 2 and 3 are correct
28. At 100% relative humidity, the wet bulb temperature is
   (a) more than dew point temperature (b) same as dew point temperature
   (c) less than dew point temperature (d) equal to ambient temperature
29. Two summer air-conditioning systems with nonzero bypass factor are proposed for a room with a known sensible and latent heat load. System *A* operates with ventilation but system *B* operates without ventilation. Then the
   (a) bypass factor of system *A* must be less than the bypass factor of system *B*
   (b) bypass factor of system *A* must be more than the bypass factor of system *B*
   (c) apparatus dew point for system *A* must be lower than the apparatus dew point for system *B*
   (d) apparatus dew point for system *A* must be higher than the apparatus dew point for system *B*.
30. In case *A*, moist air is adiabatically saturated and in cases *B*, moist air is isotarically saturated. The saturation temperatures in cases A and B are respectively
   (a) dry bulb temperature and wet bulb temperature
   (b) dew point temperature and wet bulb temperature
   (c) wet bulb temperature and dew point temperature
   (d) wet bulb temperature and dry bulb temperature
31. If p_a and p_v denote repectively the partial pressure of dry air and that of water vapour in moist air, the specific humidity of air is given by

   (a) $\dfrac{p_v}{p_a + p_v}$ (b) $\dfrac{p_v}{p_a}$ (c) $\dfrac{0.622\, p_v}{p_a}$ (d) $\dfrac{0.622\, p_v}{p_a + p_v}$

32. In a saturated air-water vapour mixture, the
 (a) dry bulb temperature is higher than the wet bulb temperature
 (b) dew point temperature is lower than the wet bulb temperature
 (c) dry bulb, wet bulb and dew point temperatures are the same
 (d) dry bulb temperature is higher than the dew point temperature
33. What is the sensible heat factor during the heating and humidification process equal to?
 (a) $\frac{H_1+H_2}{H_3-H_1}$ (b) $\frac{H_2-H_1}{H_3-H_1}$ (c) $\frac{H_1+H_2}{H_1-H_2}$ (d) $\frac{H_3-H_1}{H_2-H_1}$
 Where,
 H_1 = Total heat of air entering the heating coil
 H_2 = Total heat of air leaving the heating coil
 H_3 = Total heat of air at the end of the humidification
34. For cooling and dehumidifying of unsaturated moist air, it must be passed over a coil at a temperature:
 (a) of adiabatic saturation of incoming stream
 (b) which is lower than the dew point of incoming stream
 (c) which lies between dry bulb and wet bulb temperature
 (d) which lies between wet bulb and dew point temperature of incoming stream.
35. The latent heat load in an auditorium is 25% of the sensible heat load. The value of sensible heat factor (SHF) is equal to
 (a) 0.25 (b) 0.5 (c) 0.8 (d) 1.0
36. Evaporative air cooler is used effectively when
 (a) dry bulb temperature is very close to the wet bulb temperature
 (b) dry bulb temperature is high and humidity is high
 (c) dry bulb temperature is low and relative humidity is high
 (d) dry bulb temperature is high and the relative humidity is low
37. Which one of the following is correct for the process 1-2 shown below?

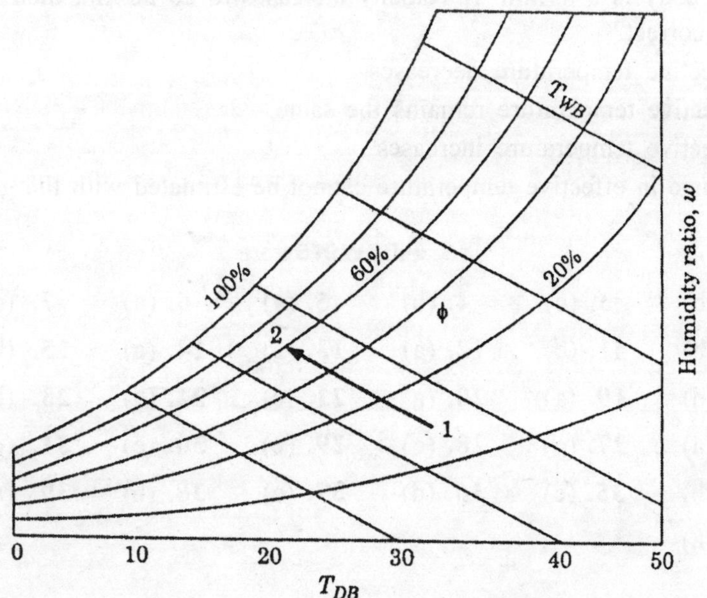

Which one of the following is correct for the process 1-2 shown above?
(a) The partial pressure of water vapour in air remains constant
(b) Specific humidity of air remains constant
(c) Enthalpy of air remains constant
(d) Dry bulb temperature of air remains constant

38. For an office building the outdoor design conditions are 45°C DBT and humidity ratio of 0.015. The indoor design conditions are 25°C DBT and 0.01 humidity ratio. The supply air state is 15°C DBT and 0.007 humidity ratio. If the supply air flow rate is 1000 m^3/per min and fresh air flow rate is 100 m^3/min, room sensible and room latent heat loads are, respectively,
(a) 408 kW and 400 kW (b) 408kW and 150 kW
(c) 204 kW and 400 kW (d) 204 kW and 150 kW

39. The wet bulb depression is zero, when relative humidity is equal to
(a) 100% (b) 60% (c) 40% (d) zero

40. Atmospheric air at 35°C and 60% RH can be brought to 20°C and 60% RH by
(a) Cooling and dehumidification process (b) Cooling and humidification process
(c) Adiabatic saturation process (d) Sensible cooling process

41. Consider the following statements:
Air washer can work as
1. Humidifier only
2. Dehumidifier only
3. Filter only

Which of the statements given above is/are correct?
(a) Only 1 (b) Only 2 and 3 (c) Only 1 and 3 (d) 1,2 and 3

42. Dry bulb temperature and wet bulb temperature is 25°C each, and velocity of air passing over human body is 6 m/min. If velocity increases to 20 m/min, then which one of the following is correct?
(a) The effective temperature decreases
(b) The effective temperature remains the same.
(c) The effective temperature increases
(d) The change in effective temperature cannot be estimated with the given information,

Answers

1. (c)	**2.** (b)	**3.** (c)	**4.** (b)	**5.** (a)	**6.** (a)	**7.** (d)	**8.** (b)
9. (c)	**10.** (b)	**11.** (c)	**12.** (a)	**13.** (c)	**14.** (a)	**15.** (b)	**16.** (b)
17. (b)	**18.** (d)	**19.** (a)	**20.** (a)	**21.** (c)	**22.** (d)	**23.** (b)	**24.** (b)
25. (d)	**26.** (a)	**27.** (c)	**28.** (c)	**29.** (c)	**30.** (c)	**31.** (d)	**32.** (c)
33. (b)	**34.** (b)	**35.** (c)	**36.** (d)	**37.** (c)	**38.** (b)	**39.** (a)	**40.** (a)
41. (d)	**42.** (a)						

REVIEW QUESTIONS

1. Explain the following terms:
   (a) Degree of saturation (b) Specific humidity
   (c) Relative humidity (d) Dew point temperature
   (e) Thermodynamic wet bulb temperature
2. Represent the following processes on psychrometric chart:
   (a) Sensible heating and cooling (b) Adiabatic heating and dehumidification
   (c) Adiabatic cooling and humidification (d) Cooling and dehumidification
3. What is the difference between WBT and thermodynamic WBT?
4. Define SHF and LHF
5. What is by-pass factor?

THEORETICAL QUESTIONS

12.1 Define the following terms, with the help of Temperature Entropy diagram:
   (i) Dew point temperature of air
   (ii) Degree of saturation of air
   (iii) Wet bulb temperature

12.2 If the total atmospheric pressure remains constant at a location, prove that the specific humidity is approximately a linear function of the partial pressure of the vapour in the atmosphere.

12.3 What do you understand by the followings? Explain by plotting psychrometric chart(s) on your answer book:
   (i) Enthalpy deviation lines
   (ii) Grand sensible heat factor and room sensible heat factor
   (iii) Adiabatic saturation of air.

12.4 With the help of a neat sketch, explain the importance of a 'Psychrometric Chart' in solving the problems of air-conditioning.

12.5 Draw the following processes on psychrometric chart and describe them in brief:
   (i) Sensible heating and cooling
   (ii) Latent heating and cooling, humidification and dehumidification
   (iii) Heating and humidification
   (iv) Cooling and dehumidification

12.6 What is the basic difference between refrigeration and air conditioning? What is psychrometrics? How do you define the terms: Dew Point Temperature, Specific humidity, Relative humidity and Degree of saturation, as related to psychrometric chart describing them in brief:

12.7 What is the importance of Psychrometric chart and how different Psychrometric processes are represented on it? Draw the following process on a Psychrometric chart describing them in brief:

(i) Sensible heating and cooling

(ii) Latent heating and cooling i.e. humidification and dehumidification.

(iii) Cooling and dehumidification

(iv) heating and humidification

12.8 What is 'air conditioning' and what is basic difference between refrigeration and air conditioning? Define the terms: 'specific humidity', 'relative humidity', 'dew point temperature' and 'degree of saturation', as related to psychrometrics.

12.9 Why psychrometric chart is most commonly used in solving the problems of air conditioning? Which are different psychrometric process commonly used in air conditioning? Show them on psychrometric chart.

NUMERICAL PROBLEMS

12.1 An air conditioned space is to be supplied with 300 cmm of air at 20°C dry bulb and 40% relative humidity. Unconditioned air from atmosphere at 32°C dry bulb and 28°C dew point is supplied to the dehumidifier coils, from where the same passes on to a heating coil before entering the conditioned space. Determine the cooling required. Solve the problem both from first principles and with Psychrometric chart.

12.2 Room air at 26°C dry bulb and 50% relative humidity is mixed with outdoor air at 43°C dry bulb and 40% relative humidity in the ratio of 4:1. The mixture is passed through a cooling coil maintained at 5°C with a by-pass factor of 0.15. The air from the cooling coil is mixed with room air in the ratio of 4:1. The mixture is then reheated to 20°C dry bulb and supplied to the conditioned space.

(i) Show the diagram schematically.

(ii) Show the different processes on Psychrometric chart.

(iii) For 450 kg of supply air per minute, determine the quantity of fresh air needed, the refrigeration load and the heat supplied to the reheated coils.

12.3 The air 35°C dry bulb temperature and 25°C wet bulb temperature is passed through a cooling coil at the rate of 280 m^3/min. The air leaves the cooling coil at 26.5°C dry bulb temperature and 50% relative humidity. Find:

(a) Capacity of the cooling coil in tonnes of refrigeration;

(b) Wet bulb temperature of the leaving air;

(c) Water vapour removed per minute; and

(d) Sensible heat factor.

[**Ans.** 31.98 TR; 19.2°C; 1.56 kg/kg of dry air; 0.39]

12.4 Air at 40°C dry bulb temperature and 15% relative humidity is passed through the adiabatic humidifier at the rate of 200 m^3/min. The outlet conditions of air are 25°C dry bulb temperature and 20°C wet bulb temperature. Find (a) dew point temperature; (b) relative humidity of exit air; (c) amount of water vapour added to the air per minute.

[**Ans.** 17.8°C; 65%; 1.26 kg/min]

12.5 Air at 0°C and 95% relative humidity has to be heated and humidified to 25°C and 40% relative humidity by the following three processes:

(a) preheating

(b) adiabatic saturation in a recirculated water air-washer; and

(c) reheating to final state.

Calculate:

(a) The heating required in two heaters.

(b) The make-up water required in washer and temperature of washer.

Assume effectiveness of washer as 80 per cent.

[**Ans.** 27.3 kJ/kg of dry air; 9 kJ/kg of dry air; 4.1 g/kg of dry air; 12.6°C]

12.6 A sling psychrometer reads 40°C DBT and 28°C WBT. Calculate the following:

(a) Specific humidity
(b) Relative humidity
(c) Dew Point Temperature
(d) Vapour density
(e) Enthalpy of mixture per kg of dry air.

Assume atmospheric pressure = 1.03 bar.

[**Ans.** 0.0189 kg/kg d.a., 40.4%, 24°C, 0.0338 kg/m^3, 88.8 kJ/kg d.a.]

12.7 The atmospheric conditions are 25°C DBT and specific humidity of 10 g/kg of d.a. Determine:

(a) Partial pressure of vapour,

(b) Relative humidity, and

(c) Dew point temperature.

Take atmospheric pressure = 1 bar.

[**Ans.** 0.0158 bar, 49%]

12.8 Air enters an evaporative cooler at a pressure of 1 atm, DBT of 38°C and 25% RH. The air leaves the cooler at 1 atm and 18°C DBT. Assuming adiabatic process, determine:

(a) relative humidity of cool air, and

(b) quality of water that must be supplied to cooler for each kg of dry air

[**Ans.** 63%, 0.00808 kg/kg d.a.]

Chapter 13

COMFORT AIR-CONDITIONING

13.1 INTRODUCTION

The human comfort depends upon that condition of mind which expresses satisfaction with the thermal environment. Comfort air-conditioning deals with to create those living conditions of environment for human beings which satisfy their mind. We shall study these conditions of comfort, in this chapter.

13.2 THERMAL ANALYSIS OF HUMAN BODY

The rate at which body produces heat is called the metabolic rate. The heat produced by a normal healthy person while sleeping is called the basal metabolic rate, which is of the order of 60 W. The maximum value may be 10 times as much as this for a person engaged in sustained hard work. Human comfort is influenced by physiological factors determined by the rate of heat generation within the body and the rate of heat dissipation to the environment.

The body loses heat to the environment mainly by convection (Q_{conv}), radiation (Q_{rad}) and evaporation of moisture (Q_{evap}). In addition, there is heat loss by respiration having sensible component (Q_{Sr}) and latent component (Q_{Lr}). The total heat loss from the body is thus,

$$\begin{aligned} Q &= Q_{conv} + Q_{rad} + Q_{evap} + (Q_{Sr} + Q_{Lr}) \\ &= (Q_{conv} + Q_{rad} + Q_{Sr}) + (Q_{evap} + Q_{Lr}) \\ &= Q_{ST} + Q_{LT} \end{aligned}$$

where Q_{ST} = total sensible heat loss

Q_{LT} = total latent heat loss.

The total heat component depends on the temperature difference between the surface of the body and the surroundings. The latent heat component depends on the difference in the water vapour pressures.

In summer, the temperature difference available for sensible heat transfer is less. Thus, the convective and radiative heat losses are reduced. To maintain thermal equilibrium, the body starts perspiring to increase the evaporative loss. On the other hand, in winter, the sensible heat transfer is increased so that the evaporative losses tend towards zero.

The heat exchange between human body and environment can be expressed by the following energy balance equation:

$$Q_M - W = Q \pm Q_S$$

where Q_M = metabolic heat produced within the body

W = useful work done by human being

$Q_M - W$ = heat to be dissipated to environment

$Q = \pm Q_{conv} + Q_{rad} + Q_{evap} + (Q_{Sr} + Q_{Lr})$

Q_{Sr} = heat stored in the body

1. The metabolic heat produced depends upon the rate of food energy consumption in the body.
2. The heat loss by evaporation is always positive. It depends upon the vapour pressure difference between the skin surface and the surrounding air.

$$Q_{evap} = C_d A(p_s - p_v) h_{fg} C_c \quad ...(13.1)$$

where C_d = diffusion coefficient, kg of water evaporated per unit surface area and pressure difference per hour.

A = skin surface area ≃ 1.8 m^2 for a normal human being

p_s = saturation vapour pressure corresponding to skin temperature

p_v = vapour pressure of surrounding air

h_{fg} = latent heat of vaporisation = 2450 kJ/kg

C_c = factor which accounts for clothing worn.

3. The heat loss or gain by radiation from the body to the surroundings depends upon the mean radiant temperature. It is the average surface temperature of the surrounding objects. Q_{rad} is positive when the mean radiant temperature is lower than the DBT of room air, i.e. the human body will undergo a radiant heat loss. When Q_{rad} is negative, the body will undergo a radiant heat gain.
4. The heat loss by convection from the body to the surroundings is given by

$$Q_{conv} = UA\,(t_b - t_s) \quad ...(13.2)$$

where U = body film coefficient of heat transfer

A = body surface area = 1.8 m^2 for normal human being

t_b = body temperature

t_s = surroundings temperature.

The heat will be gained by body when $t_s > t_b$ and Q_{conv} is negative.

5. Q_S is negative when $Q > (Q_M - W)$, i.e. body temperature falls down. Q_S is positive when $Q < (Q_M - W)$, i.e. a human being gets fever.

A human body feels comfortable when there is no change in the body temperature, i.e. when $Q_S = 0$. Any variation in the body temperature acts as a stress on the brain which results in either perspiration or shivering.

13.2.1 Factors Affecting Human Comfort

The following factors affect human comfort:

1. Effective temperature.
2. Heat production and regulation in human body.

3. Heat and moisture losses from the human body.
4. Moisture content of air.
5. Quality and quantity of air.
6. Air velocity.
7. Hot and cold surfaces.
8. Air stratification.

13.2.2 Physiological Hazards Resulting from Heat

The physiological hazards resulting from rise in body temperature are:

1. **Heat exhaustion:** It is due to the failure of normal blood circulation. The symptoms of heat exhaustion include fatigue, headache, dizziness, vomiting and abnormal mental reactions such as irritation. Severe heat exhaustion may cause fainting. It does not cause permanent injury to the body and recovery is usually rapid when the person is shifted to a cool place.
2. **Heat Cramp:** It results from loss of salt due to an excessive rate of body perspiration. It causes severe pain in thigh muscles. This may be avoided by using salt tablets.
3. **Heat stroke:** It occurs when a person is exposed to excessive heat and work. If the body temperature rises rapidly to 40.5°C or 105°F, sweating ceases and a person may enter a coma, with imminent death. It may lead to permanent damage to brain. Heat stroke can be avoided by drinking sufficient water at frequent intervals.

13.3 EFFECTIVE TEMPERATURE

Effective temperature (*ET*) is defined as the temperature of saturated air (*RH* = 100%) at which the human beings would experience the same feeling of comfort as experienced in the actual saturated environment. It may also be defined as that index which correlates the combined effects of air temperature, relative humidity and air velocity on the human body.

Corresponding, another effective temperature (*ET**) can also be defined as the temperature at 50% *RH* at which the human body would experience exactly same feeling of comfort as at *ET* at 100% *RH*, and as experienced in the actual environment.

The effective temperature is defined to evaluate the combined effect of DBT, *RH*, and air velocity. The numerical value of effective temperature is made equal to the temperature of still (i.e. 5 to 8 m/min air velocity) saturated air, which produces the same sensation of warmth or coolness as produced under the given conditions.

13.3.1 Comfort Chart

The comfort chart is defined as that index which correlates the combined effects of temperature, relative humidity and air velocity on the human body. The numerical value of effective temperature is made equal to the temperature of still (i.e. 5 to 8 m/min air velocity) saturated air, which produces the same sensation of warmth or coolness as produced under the given conditions.

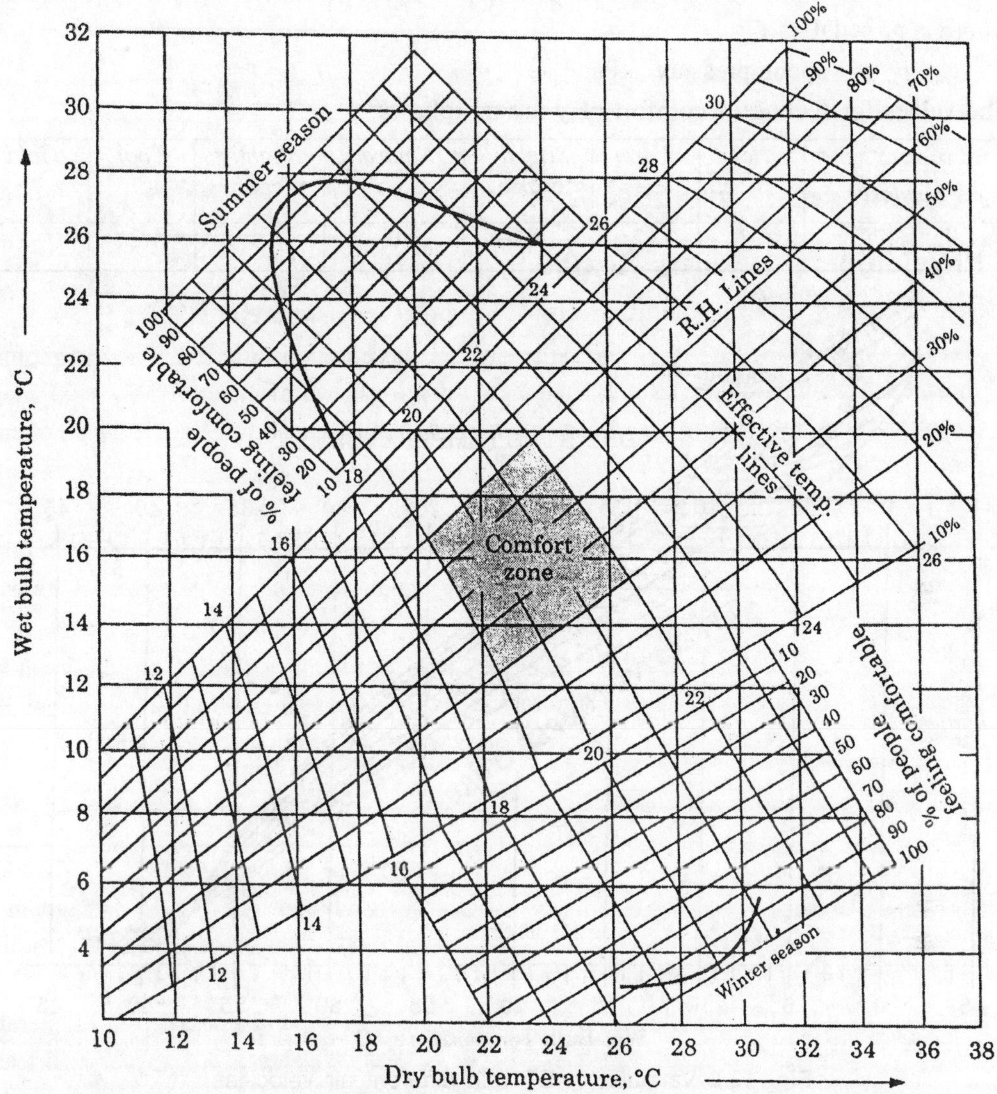

Fig. 13.1 Comfort chart for still air (air velocities from 5 to 8 m/min).

The comfort chart, shown in Fig. 13.1, represents the concept of effective temperature. In the comfort chart, DBT is taken as the abscissa and WBT as ordinates. The *RH* lines are replotted from the psychrometric chart. The statistically prepared graphs corresponding to summer and winter season are also superimposed. These graphs have effective temperature scale as abscissa and percentage of people feeling comfortable as ordinate.

There is a correlation between comfort level, temperature, humidity, sex, length of exposure, etc. According to ASHRAE (American Society for Heating, Refrigeration and Air-conditioning Engineering) thermal sensation scale for exposure for a period of one hour, the value of index y for comfort for men, women and both sexes separately, are expressed as follows:

Men: $$y = 0.22t_d + 0.233p_v - 5.673 \quad ...(13.3a)$$

Women: $$y = 0.272t_d + 0.248p_v - 7.245 \quad ...(13.3b)$$

Both: $y = 0.245t_d + 0.248p_v - 6.475$ ...(13.3c)

where t_d = DBT, C°

p_v = vapour pressure, kPa

The values for feeling of comfort of y are as follows:

Feeling of comfort	*Hot*	*Warm*	*Slightly warm*	*Neutral*	*Slightly cool*	*Cool*	*Cold*
Value of *y*	+3	+2	+1	0	–1	–2	–3

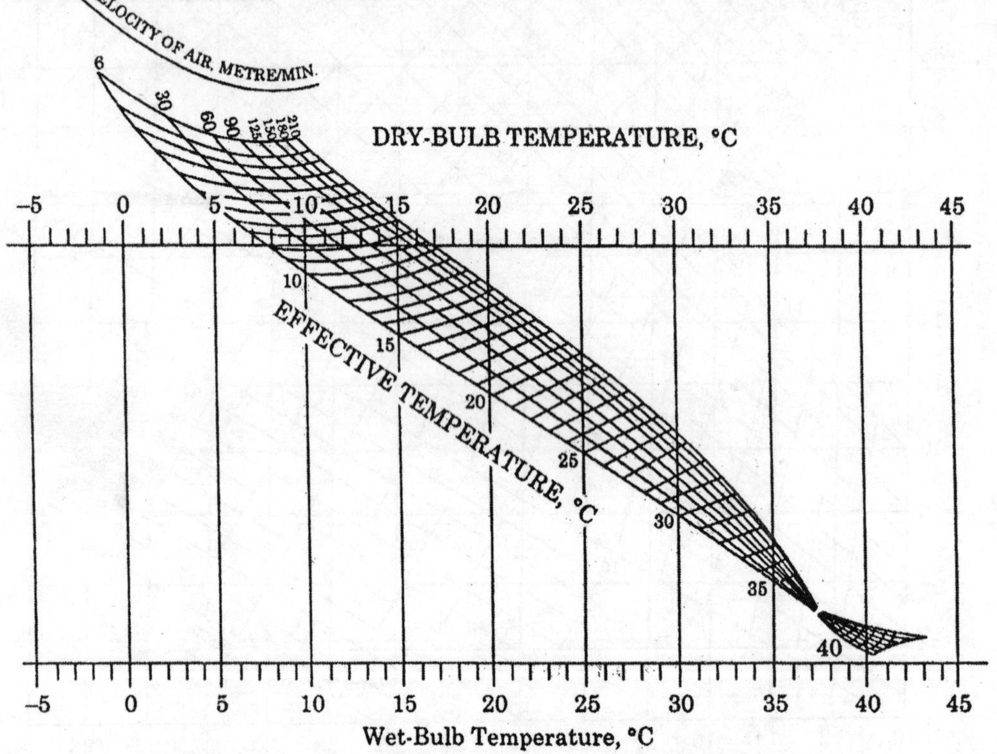

Fig. 13.2 Variation in ET with different air velocities.

The comfort chart shown in Fig. 13.1 represent the range for both summer and winter conditions within which a condition of comfort exists for most people. The general practice is to recommend the following optimum inside design conditions for comfort:

Summer air-conditioning:

ET = 22°C

DBT = 25° ± 1°C

RH = 50 ± 5%

Room air velocity = 0.4 m/s

Winter air-conditioning:

ET = 19°C

DBT = 21°C

RH = 50%

Room air velocity = 0.15 – 0.20 m/s

The comfort chart does not take into account the variations in comfort conditions when there are wide variations in the mean radiant temperature. The variation in effective temperature with different air velocities is shown in Fig. 13.2.

The comfort chart shown in Fig. 13.1 has become obsolete due to its short comings of over exaggeration of humidity at lower temperature and under estimation of humidity at heat tolerance level. The modified comfort chart shown in Fig. 13.3 is commonly used now.

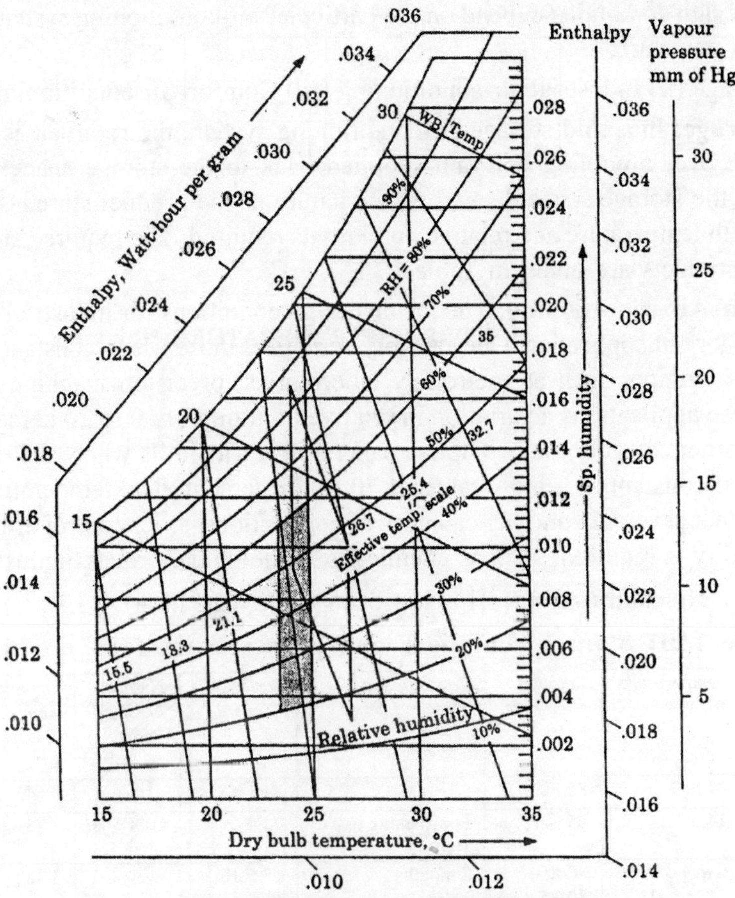

Fig. 13.3 Modified comfort chart.

13.3.2 Factors Affecting Optimum Effective Temperature

The factors affecting optimum effective temperature are:

1. **Climate and seasonal differences.** The *ET* changes with season. During summer it is 22°C and during winter 19°C.
2. **Clothing.** Persons with light clothings need lower optimum *ET* than a person with heavy clothings.
3. **Age and sex.** Women require higher *ET* (about 0.5°C) than men. Children also need higher *ET* than adults.
4. **Duration of stay.** For shorter stay, higher *ET* is required than that needed for longer stay.
5. **Kind of activity.** Low *ET* is required for people doing heavy activity, like factory workers, dancers, etc.
6. **Density of occupants.** Lower *ET* is needed when density of persons in a place is high.

13.4 SELECTION OF INSIDE AND OUTSIDE DESIGN CONDITIONS

13.4.1 Selection of Inside Design Conditions

The inside design conditions depend on the particular air-conditioning system. The important air-conditioning systems are:

(*i*) Cold Storage (*ii*) Industrial air-conditioning (*iii*) Comfort air conditioning.

(*i*) Cold Storage: In a cold storage air-conditioning system, the room air is cooled to much lower temperature over a cooling coil and supplied back to the storage space. The conditions maintained inside the storage space depend on the nature of the product stored. In cold storages, strict control of both temperature and relative humidity is required. The required storage conditions for a number of products are given in Table 13.1.

(*ii*) Industrial Air-conditioning: The inside design conditions for industrial air-conditioning depend on the category of application. One category comprises those where constancy of temperature is the prime consideration, such as metrology laboratories, precision machine tools, computer centres, etc. In these applications a variation in relative humidity of 10 to 20 percent will not have much effect. The other category may comprise paper and textile mills where the relative humidity is to be maintained constant at a high value of 70 to 75 percent. The temperature requirements of such spaces are not severe. In another category of applications, strict control of both temperature and relative humidity is required, such as chemical and biological process industries.

(*iii*) Comfort air-conditioning: This has been dealt with in art. 13.3.

Table 13.1: Storage conditions and properties of food products

Product	*Temperature °C*	*Relative Humidity %*	*Approximate Storage Life*	*Water Content %*	*Highest Freezing Point °C*	*Sp. Heat Above Freezing kJ/kg.K.*	*Sp. Heat Below Freezing kJ/kg. K.*
1. Apples	–1 to 0	85-90		84.1	–1.5	3.643	1.884
2. Bananas	14.5	95	For ripening in 8-10 days	74.8	–0.8	3.35	1.76
3. Butter	0 to 4.4 –18 to –23	80-85 80-85	2 months 1 year	15.5-16.5 15.5-16.5	–	1.382	–
4. Milk, Pasteurized	0.5	–	7 days	87	–0.6	3.77	1.93
5. Eggs	1.5 to –0.5	80-85	6-9 months	66	–2.2	3.06	1.68
6. Fish-fresh frozen	0.5 to 1.5 –23.5 to –18	90-96 90-95	5-15 days 8-10 months	62-85 62.85	–2.2 –	2.93-3.6 –	– 1.59-1.884
7. Grapes	-0.5 –	85-90 –	3-8 weeks (American) 3-6 months (European)	81.9 81.6	–1.3 –2.2	3.6 3.6	1.842 1.842
8. Beef-fresh Frozen	0 to 1 –23.5 to –18	88-92 90-95	1-6 weeks 9-12 months	62-77 62-77	–2.2 –	2.93-3.52 –	– 1.59-1.8
9. Mangoes	10	85-90	2-3 weeks	81.4	–0.9	3.56	1.842
10. Potatoes, late crop	3 to 4.5	85-90	5-8 months	77.8	–0.6	3.433	1.8
11. Tomatoes-green ripe	14 to 21 7 to 10	85-90 85-90	2-4 weeks 2-7 weeks	94.7 94.1	–0.6. –0.5	3.98 3.98	2.01 2.01

13.4.2 Selection of Outside Design Conditions

For the outside design conditions in summer, it is recommended to use the mean monthly maximum dry bulb temperature and its corresponding wet bulb temperature.

During winter it has been observed that the fuel consumption for heating of buildings varies also directly as the difference between the outside temperature and a reasonably comfortable inside temperature of 18.5°C. Thus the fuel consumption would be practically nil if the outside temperature is 18.5°C. On the other hand, the fuel consumption would double if the outside temperature drops from 18.5°C to 8.5°C. A degree-day is obtained for every degree when the mean outside temperature is below 18.5°C during the 24 hour period. If in a given locality the outside temperature average of 30 days is 10°C, then the degree days (D) for the period would be

$$D = (18.5 - 10) \times 30 = 255$$

The outside design temperature may be calculated as follows:

$$\text{Number of days in a month} \times (18.5 - t_{do}) = D \qquad \text{...(13.4)}$$

where t_{do} = outside design temperature.

13.5 AIR CONDITIONING SYSTEMS

13.5.1 Summer Air-conditioning System with Ventilation Air and Zero By-pass Factor

The schematic arrangement of summer air-conditioning system with ventilation without by-pass factor is shown in Fig. 13.4. The outside air flows through the damper and mixes with the recirculated air. The mixed air then passes through the filter to remove dirt and dust and then through cooling coil whose temperature is kept much below the DBT of air in the conditioned space. The cooled air then passes through a perforated membrane where it loses its moisture in the condensed form. The moisture is collected in a sump. After that, the air is made to pass through a heating coil which heats the air slightly as per the requirement of designed air DBT and RH. Now the conditioned air is supplied to the room by a supply air fan. From the room, a part of the used air is exhausted to the atmosphere by ventilation. The remaining part of the used air (recirculated air) is mixed with fresh air for conditioning.

Fig. 13.4 shows the air-conditioning processes.

o, i = outside and inside air states respectively

1 = state of air after mixing recirculated room air with ventilation air.

$\dot{m}_{ai}, \dot{m}_{ao}$ = mass flow rate of recirculated room air and ventilation air respectively

2 = supply air state for a minimum rate of supply air

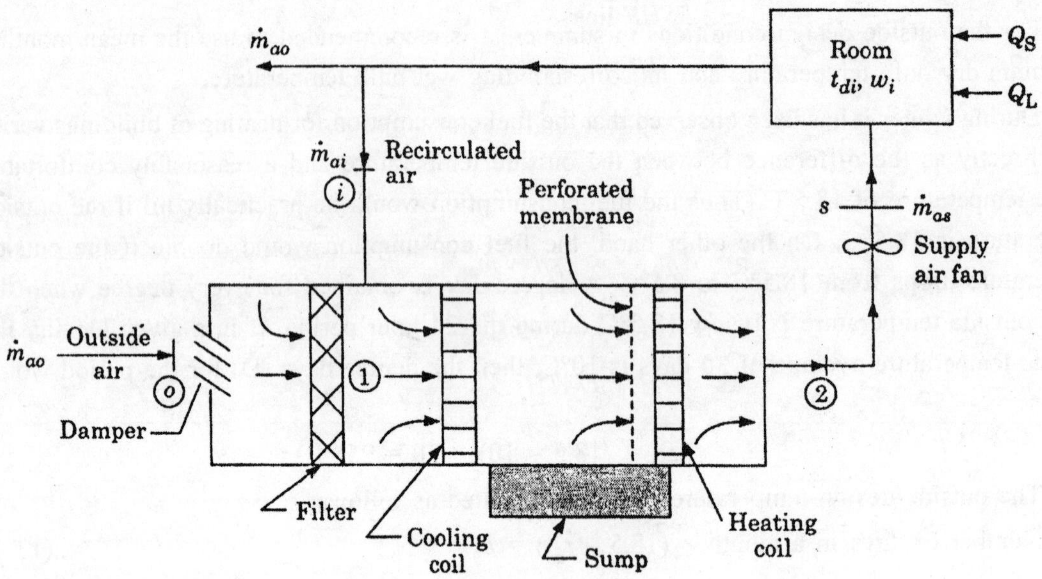

Fig. 13.4 Schematic diagram of system with ventilation air without by-pass factor.

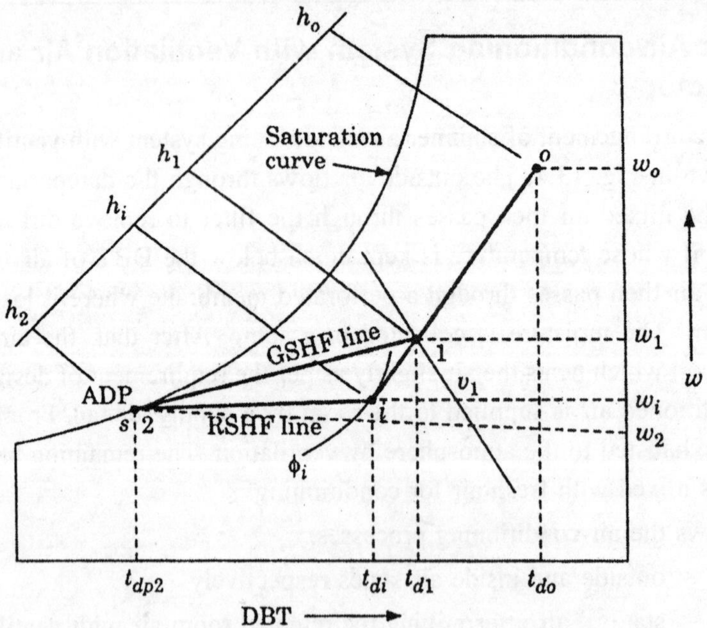

Fig. 13.5 Summer air conditioning processes with ventilation without BPF.

$$\text{line } i\text{-}2 = ADP, t_{dp2}$$
$$= RSHF \text{ line}$$
$$\text{line } 1\text{–}2 = GSHF \text{ line}$$
$$\dot{m}_{as} = \text{mass flow rate of mixture air}$$
$$\dot{Q} = \dot{m}_{as}(h_1 - h_2) = (\dot{m}_{ai}h_i + \dot{m}_{ao}h_o) - \dot{m}_{as}h_2$$
$$= (\dot{m}_{as} - \dot{m}_{ao})\, h_i + \dot{m}_{ao}h_o - \dot{m}_{as}h_2$$
$$= \dot{m}_{as}(h_i - h_2) + \dot{m}_{ao}(h_o - h_i) \quad \text{...(13.5)}$$
$$= RTH + OATH$$

(*a*) **Ventilation load**

Let $(\text{cmm})_o$ = outside ventilation air volume flow rate = $\dot{m}_{as} \times v_o$.

Then $OASH = 0.0204\ (\text{cmm})_o\ (t_{do} - t_{di})$ ...(13.6)

$OALH = 50\ (\text{cmm})_o\ (w_o - w_i)$ ...(13.7)

and $OATH = OASH + OALH = 0.02\ (\text{cmm})_o\ (h_o - h_i)$ ...(13.8)

(b) **Room load**

Let $(\text{cmm})_1$ = mixture air volume flow rate

$$\dot{m}_{a1} = \dot{m}_{ai} + \dot{m}_{ao} = \dot{m}_{as}$$

Now
$$\dot{m}_{a1} = \frac{(\text{cmm})_1}{v_1} \quad \text{...(13.9)}$$

$$RTH = RSH + RLH$$

(*c*) Air-conditioning equipment load.

$$TSH = RSH + OASH$$
$$TLH = RLH + OALH$$
$$GTH = TSH + TLH$$

Grand sensible heat factor,

$$GSHF = \frac{TSH}{TSH + TLH} = \frac{TSH}{GTH} \quad \text{...(13.10)}$$

13.5.2 Summer Air-conditioning System with Ventilation Air and By-pass Factor

The summer air-conditioning process with ventilation air and BPF is shown in Fig. 13.6.

$$BPF = \frac{t_{d2} - (t_{dp})_{\text{coil}}}{t_{d1} - (t_{dp})_{\text{coil}}} = \frac{w_2 - w_{S'}}{w_1 - w_{S'}} = \frac{h_2 - h_{S'}}{h_1 - h_{S'}} \quad \text{...(13.11)}$$

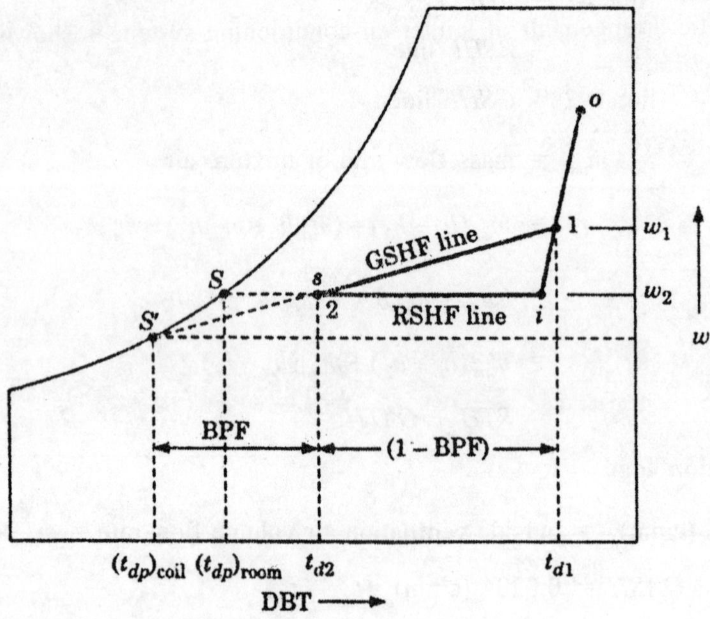

Fig. 13.6 Summer air-conditioning process with ventilation air and BPF.

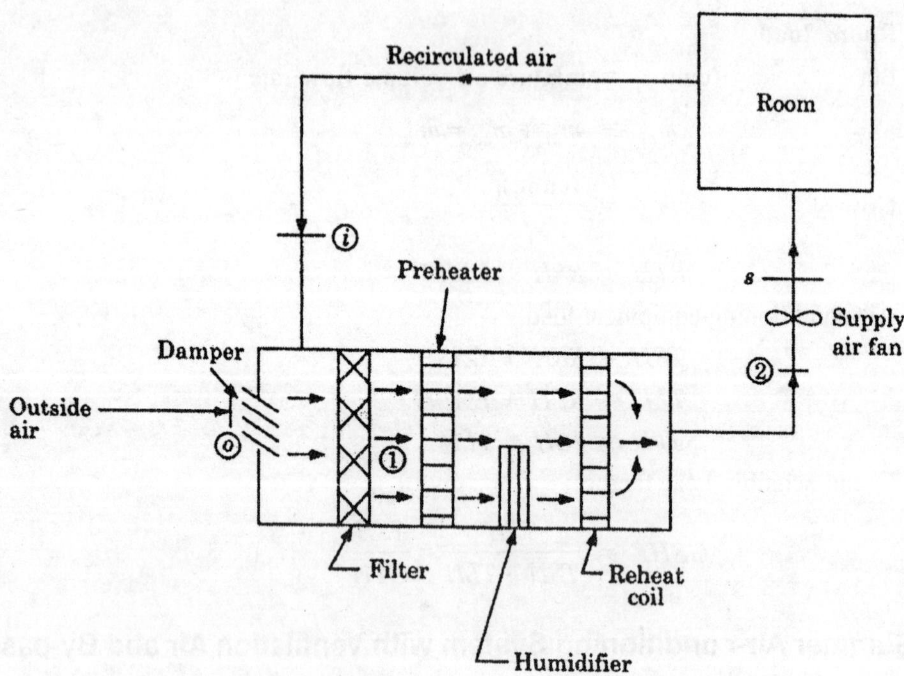

Fig. 13.7 Winter air-conditioning system.

13.5.3 Winter Air-conditioning System

The schematic arrangement of winter air-conditioning system is shown in Fig. 13.7. The outside air flows through a damper and mixes up with the recirculated air. The mixed air passes

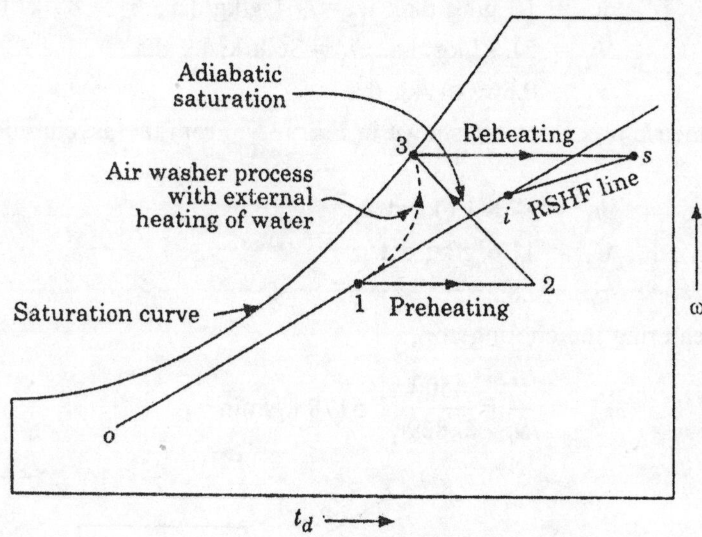

Fig. 13.8 Winter air-conditioning processes.

through a filter to remove dirt, dust and other impurities. The air now passes through a preheating coil to prevent the possible freezing of water and to control the evaporation of water in the humidifier. Then the air is passed through a reheating coil to bring the air to the designed DBT. This conditioned air is supplied to the room by means of a supply air fan. From the room, a part of the used air is exhausted to the atmosphere and the remaining part is recirculated again.

The air-conditioning processes are shown in Fig. 13.8.

Process 1-2: Preheating

Process 2-3: Adiabatic saturation.

Process 1-3: Heating and humidifying air in air washer with pumped recirculation and external heating of water.

Process 3-*s*: Reheating

Example 13.1 *An air handling unit in an air-conditioning plant supplies 4500 cmm of dry air which comprises by mass 20% of fresh air at 40°C DBT and 27°C WBT and 80% recirculated air at 25°C DBT and 50% RH. The air leaves the cooling coil at 13°C saturated. Calculate the total cooling load and room heat gain. The following data is given.*

Condition	*DBT* °C	*WBT* °C	*RH* %	*Sp. Humidity* g/kg d.a.	*Enthalpy* kJ/kg d.a.
Outside	40	27	–	17.2	8.5
Inside	25	–	50	10.0	5.1
ADP	13	–	100	9.4	36.8

Specific volume of air entering the cooling coil is 0.869 m³/kg of d.a.

Solution: Given: $\dot{V}_1 = 4500$ cmm, $t_{do} = 40°C$, $t_{wo} = 27°C$, $t_{di} = 25°C$.

$\phi_i = 50\%$, $t_{dp2} = ADP = 13°C$, $w_o = 17.2$ g/kg d.a.,

$w_i = 10$ g/kg d.a., $w_2 = 9.4$ g/kg/d.a., $h_o = 85$ kJ/kg d.a.,

$h_i = 51$ kJ/kg d.a., $h_2 = 36.8$ kJ/kg d.a.,

$v_1 = 0.869$ m³/kg d.a.

The psychrometric processes are shown in Fig. 13.9. From the psychrometric chart, we find that,

$$h_1 = 57.8 \text{ kJ/kg d.a.}$$

$$w_1 = 11.6 \text{ g/kg d.a.}$$

$$t_{d1} = 28.3°C$$

Mass of air entering the cooling coil,

$$\dot{m}_{a1} = \frac{\dot{V}_1}{v_1} = \frac{4500}{0.869} = 5178 \text{ kg/min}$$

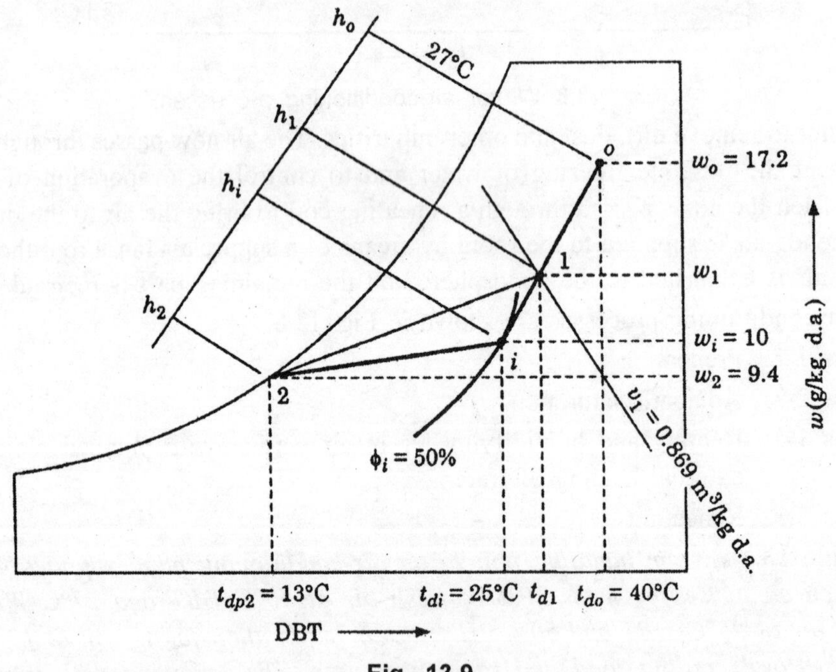

Fig. 13.9

$$\text{Total cooling load} = \dot{m}_{a1}(h_1 - h_2) = 5178\ (57.8 - 36.8) = 108738 \text{ kJ/min}$$

$$= \frac{108738}{211} = 515.3 \text{ TR}$$

Mass of fresh air supplied at 0,

$$\dot{m}_{ao} = 0.2 \times \dot{m}_{a1} = 0.2 \times 5178 = 1035.6 \text{ kg/min}$$

$$\text{Fresh air load} = \dot{m}_{ao}(h_o - h_i) = 1035.6\ (85 - 51) = 35210 \text{ kJ/min}$$

$$= \frac{35210}{211} = 166.9 \text{ TR}$$

Room heat gain = total cooling load – fresh air load

$= 515.3 - 166.9 = 348.4$ TR

Example 13.2 *Air at 10°C DBT and 90% RH is to be brought to 35°C DBT and 22.5°C WBT with the help of winter air-conditioner. If the humidified air comes out of the humidifier at 90% RH, determine: (a) the temperature to which the air should be preheated, and (b) the efficiency of the air washer.*

Solution: Given: $t_{d1} = 10°C$, $\phi_1 = 90\%$, $t_{ds} = 35°C$, $t_{ws} = 22.5°C$

The psychrometric processes are shown in Fig. 13.10.

Locate point $1(t_{d1}, \phi_1)$ and point $s(t_{ds}, t_{ws})$.

Draw horizontal lines from points 1 and *s*. From point 3 draw WBT = const. line to locate point 2.

From psychrometric chart, the temperature to which air should be preheated,

$$t_{d2} = 31.2°C$$

$$t_{d3} = 18.5°C \text{ and } t_{d3'} = 17.5°C$$

$$\text{Efficiency of air-washer} = \frac{t_{d2} - t_{d3}}{t_{d2} - t_{d3'}} = \frac{31.2 - 18.5}{31.2 - 17.5} = 0.927 \text{ or } 92.7\%$$

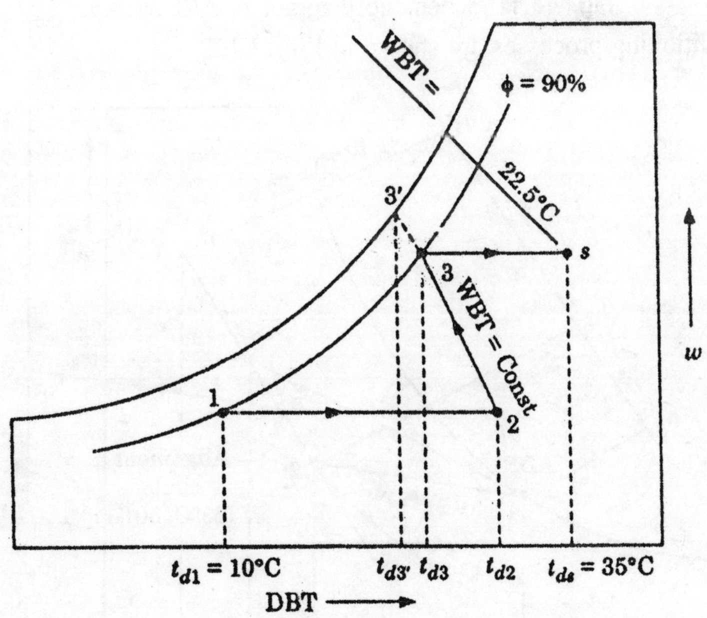

Fig. 13.10

Example 13.3 *In an air-conditioning system, the inside and outside conditions are DBT 25°C. RH 50% and DBT 40°C, WBT 27°C respectively. The room sensible heat factor is 0.8. 50% of the room air is rejected to atmosphere and an equal amount of fresh air added* *before air enters the air conditioning apparatus. If the fresh air added is 100 m³/min, determine:*

(a) Room sensible and latent heat load, (b) Sensible and latent heat load due to fresh air, (c) Apparatus dew point, (d) Humidity ratio and DBT of air entering air conditioning apparatus.

Assume by-pass factor as zero, density of air as 1.2 kg/m³ at a total pressure of 1.01325 bar.

Solution: Given: $t_{d1} = 40°C$, $t_{w1} = 27°C$, $t_{d2} = 25°C$, $\phi_2 = 50\%$, RSHF = 0.8,

$\dot{V}_1 = 100 \text{ m}^3\text{/min}$, $\rho_a = 1.2 \text{ kg/m}^3$.

The schematic diagram of air-conditioning system is shown in Fig. 13.11.

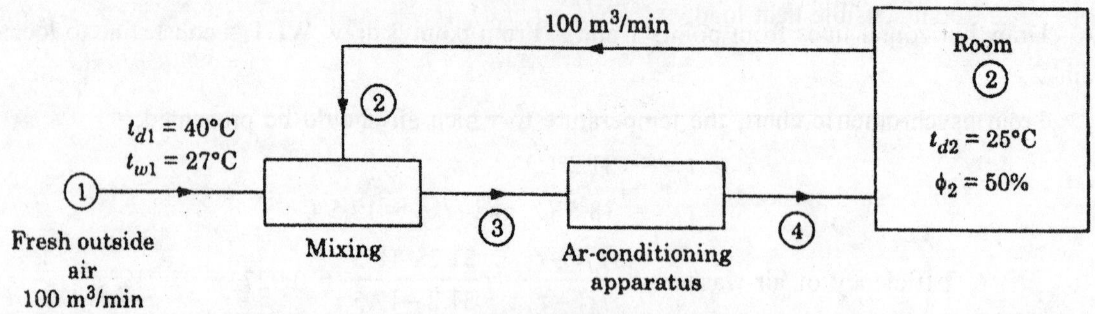

Fig. 13.11 Schematic diagram of A/C system

The air-conditioning processes are shown in Fig.13.12.

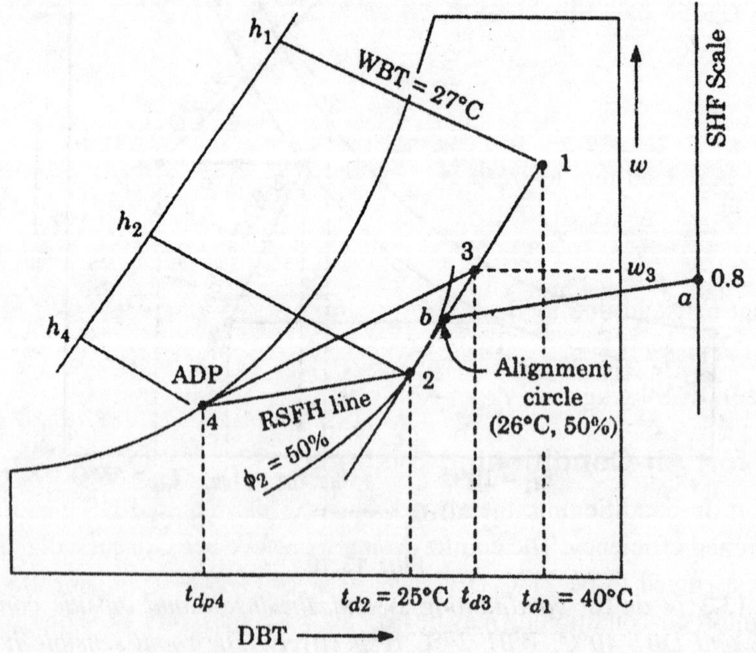

Fig. 13.12 Air-conditioning processes.

Locate point 1(t_{d1} = 40°C, t_{w1} = 27°C) and point 2(t_{d2} = 25°C, ϕ_2 = 50%) on psychrometric chart. Mark point 3 on line 1-2 such that

$$\text{Length 2-3} = \frac{1}{2}\text{(Length 1-2)}$$

Now mark *RSHF* = 0.8 value (point '*a*') on the *SHF* scale on the right of chart and join this with the alignment circle (point *b*, 26°C *DBT* and 50% *RH*).

From point 2, draw a line 2-4 parallel to line *ab*. The point 4 represents the *ADP*.

$$t_{dp4} = 11.8°\text{C}$$

Now h_1 = 82.5 kJ/kg d.a., h_2 = 50 kJ/kg d.a., h_4 = 33 kJ/kg d.a.

(*a*) Mass of air supplied to room; $\dot{m}_a = \dot{V}_3\rho_a$ = (100 + 100) × 1.2 = 240 kg/min

Room sensible heat load, $RSH = \dot{m}_a c_{pm}(t_{d2} - t_{dp4})$

$= 240 \times 1.022\ (25 - 11.8) = 3238$ kJ/min

$= \dfrac{3238}{60} = 53.96$ kW

Room total heat load, $RTH = \dot{m}_a(h_2 - h_4) = 240(50 - 33) = 4080$ kJ/min

$= \dfrac{4080}{60} = 68$ kW

Room latent heat load, $RLH = RTH - RSH = 68 - 53.96 = 14.04$ kW

(*b*) Mass of fresh air supplied, $\dot{m}_f = \dot{V}_1\rho_a = 100 \times 1.2 = 120$ kg/min

Sensible het load due to fresh air $= m_f\, c_{pm}\,(t_{d1} - t_{d2})$

$= 120 \times 1.022\ (40 - 25) = 1840$ kJ/min

$= \dfrac{1840}{60} = 30.67$ kW

Total heat load due to fresh air $= \dot{m}_f(h_1 - h_2) = 120(82.5 - 50) = 4224$ kJ/min

$= \dfrac{4224}{60} = 70.4$ kW

Latent het load due to fresh air $= 70.4 - 30.67 = 39.73$ kW

(*c*) *ADP* = 11.8°C

(*d*) w_3 = 0.0138 kg/kg d.a., t_{d3} = 32.5°C

13.5.4 Comfort Air-Conditioning System

In comfort air-conditioning, the air is brought to the required DBT and *RH* for the human health, comfort and efficiency. The comfort conditions have been discussed in detail in art. 13.3.1. Normally it is assumed to be 21°C DBT and 50% *RH*. The SHF is generally kept between 0.7 to 0.9. The comfort air-conditioning may be adopted for homes, offices, shops, restaurants, theatres, hospitals, schools, etc.

13.5.5 Industrial Air-Conditioning System

In an industry, the inside DBT and *RH* of air have to be kept constant for proper working of the machines and manufacturing processes. Special type of air-conditioning systems have to be provided for textile mills, paper mills, machine-parts manufacturing plants, tool rooms, CNC machining centres, photo-processing plants, etc.

Example 13.4 *A mixture of air and water vapour at 1.01325 bar 16°C has a dew point of 5°C. Calculate the relative humidity and specific humidity.* **[CSE, 2000]**

Solution:

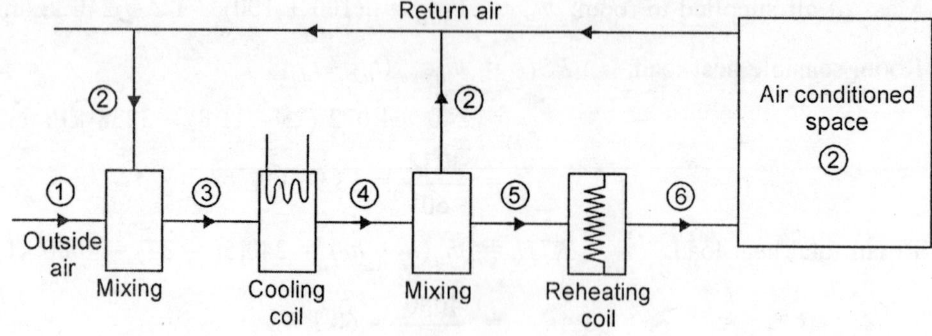

(*a*) Flow diagram

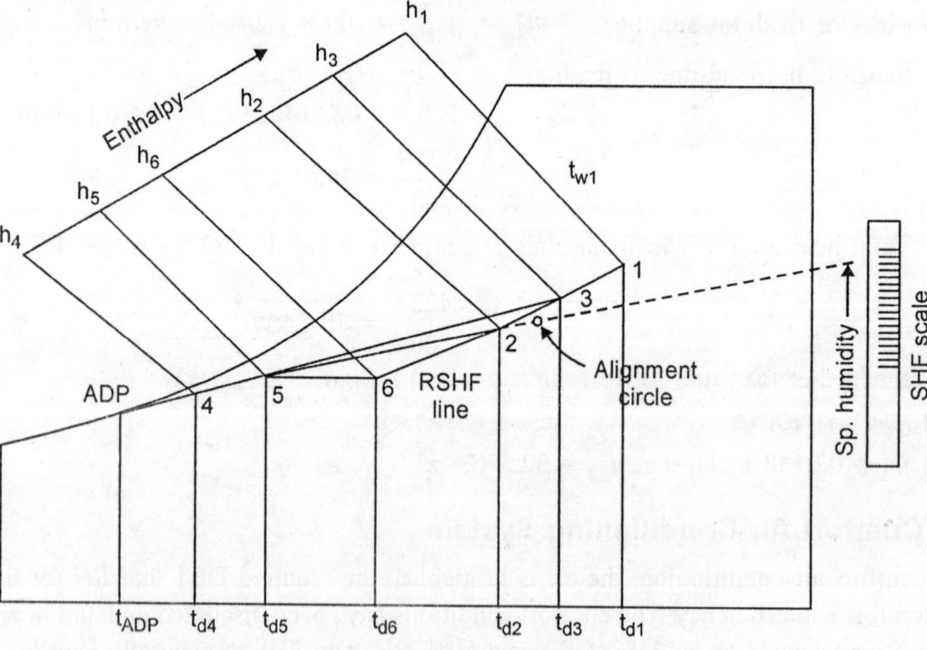

(*b*) Psychometric processes

Fig. 13.13 Summer air-conditioning plant

Refer to Fig. 13.13

Point 1 : Outside air condition (t_{d1}, t_{w1})

Point 2 : Conditioned space air (t_{d2}, % RH)

Point 3 : Condition of air after mixing return air from the conditioned space and fresh outside air in some ratio.

$$\text{Length } 2-3 = \frac{\text{Length } 2-1}{\text{Sum of ratio}}$$

Point 4: DBT of air leaving the cooling coil.

$$\text{BPF of cooling coil} = \frac{t_{d4} - \text{ADP}}{t_{d3} - \text{ADP}}$$

Locate point 4 on the line joining point 3 and t_{ADP}.

Since the return air from the conditioned space is also mixed with the air leaving the cooling coil in some ratio.

Point 5: Length $$4-5 = \frac{\text{Length } 4-2}{\text{Sum of ratio}}$$

$$RSHF = \frac{RSH}{RSH + RLH}$$

Draw *RSHF* line from the SHF scale. Then draw a horizontal line from point 5 to intersect RSHF line at point 6, which represents the condition of supply air to the conditioned space.

Example 13.5 *In an air-conditioning plant equipment exists for re-heating the air before supply to a room to be maintained at 25° dbt with 55% R.H. The RSHL = 150,000 W and RLHL = 122,000 W. The discharge from cooling coil is 12°C dbt, 95% R.H. Calculate:*

(i) The supply air state and rate

(ii) reheating, if any. **[CSE, 2001]**

Solution: Given: t_{d2} = 25°C, ϕ_2 = 55%, *RSHL* = 150,000 W, *RLHL* = 122,000 W,

t_{d1} = 12°C, ϕ_1 = 95%

Room sensible heat factor,

$$RSHF = \frac{RSHL}{RSHL + RLHL} = \frac{150{,}000}{150{,}000 + 122{,}000} = 0.55$$

Refer to Fig. 13.14.

1. Locate points 1 and 2 on the psychrometric chart.
2. Join points 1 and 2.
3. Mark point '*a*' at *SHF* = 0.55 on *SHF* scale.
4. Join alignment circle point '*b*' with '*a*'.
5. Produce line 12 to intersect line *abdc*. Then

   (*i*) t_{dc} = 33°C, ϕ_c = 40%

   which is the supply air state.

(*ii*) Heating is required.

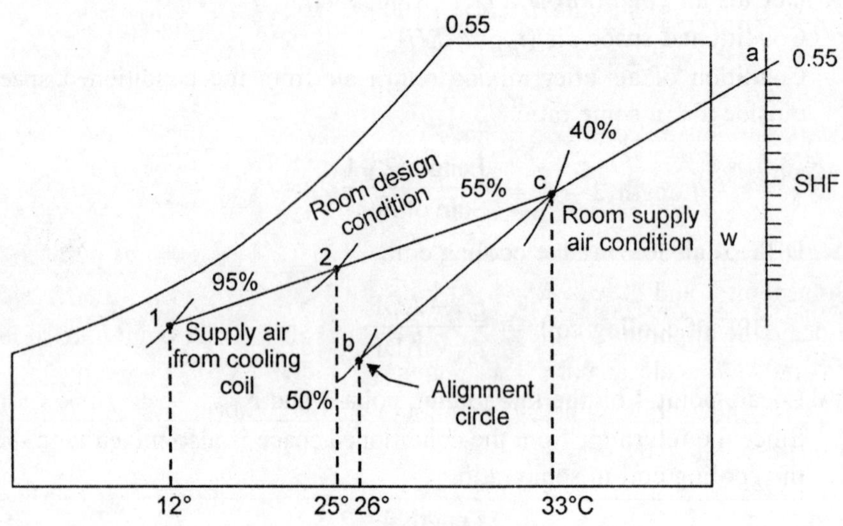

Fig. 13.14

Example 13.6 *In an air-conditioning system 30 m³/min of fresh outdoor air is introduced at 43°C dry-bulb temperature and 30% relative humidity. The remaining air is re-circulated from the room maintained at 25°C dry-bulb temperature and 50% RH. The by-pass factor of the cooling coil is 0.15 and apparatus dew point is 11.8°C. RSH = 100 kW and RLH = 15 kW. Determine (i) humidity ratios for outdoor and room condition, (ii) OASH and OALH, (iii) ERSH and ERLH, (iv) supply air temperature, (v) supply air volume flow rate using ERSH and (vi) temperature at inlet to cooling coil by assuming the density of outdoor and recirculated air to be the same.*

Given that saturation pressures of water at 25°C and 43°C are 3.1693 kPa and 8.6495 kPa respectively. Show the process on psychrometric chart schematically. Standard atmospheric pressure = 1.01325 bar. **[CSE, 2003]**

Solution: Given: $\dot{V}_F$ = 30 m³/min, t_{d1} = 43°C, ϕ_1 = 30%, t_{d2} = 25°C, ϕ_2 = 50%, *BPF* = 0.15,

T_{ADP} = 11.8°C, *RSH* = 100 kW, *RLH* = 15 kW, p_{s1} = 3.1693 kPa,

p_{s2} = 8.6495 kPa, p_b = 1.01325 bar or 101.325 kPa

(*i*) Humidity ratios:

$$w = \frac{0.622\, p_s}{p_b - p_s}$$

$$w_1 = \frac{0.622 \times 3.1693}{101.325 - 3.1693} = 0.020 \text{ kJ/kg d.a.}$$

$$w_2 = \frac{0.622 \times 8.6495}{101.325 - 8.6495} = 0.058 \text{ kJ/kg d.a.}$$

(*ii*) Room sensible heat factor,

$$RSHF = \frac{RSH}{RSH + RLH} = \frac{100}{100+15} = \frac{100}{115} = 0.8696 \simeq 0.87$$

Refer to Fig. 13.15.

1. Mark the outside condition of air, i.e. t_{d1} = 43°C, ϕ_1 = 30% on psychrometric chart at point 1.
$$v_{s1} = 0.92 \text{ m}^3/\text{kg}$$
2. Mark the inside condition of air, i.e. t_{d2} = 25°C, ϕ_2 = 50% as point 2.
3. Join point 1 and 2.
4. Locate the alignment circle at 26°C DBT and 50% *RH* at point *b*. Also locate point '*a*' on *SHF* scale at 0.87. Join points '*a*' and '*b*'.
5. From point 2, draw a line 2–5 parallel to the line *ab*, which is called the *RSHF* line.
6. Locate point *A* by drawing vertical and horizontal lines through points 1 and 2 respectively.

$$h_1 = 86 \text{ kJ/kg}, \; h_2 = 50 \text{ kJ/kg}, \; h_A = 70 \text{ kJ/kg}$$

$$\text{Supply air quantity, } \dot{V} = \frac{RSH}{0.0204(t_{d2} - t_{ADP})} = \frac{100}{0.0204(25-11.8)}$$

$$= 371.36 \text{ m}^3/\text{min}$$

$$\text{Recirculated room air} = \dot{V} - \dot{V}_F = 371.36 - 30 = 341.36 \text{ m}^3/\text{min}$$

$$w_1 = 0.0162 \text{ kJ/kg d.a.}, \; w_2 = 0.010 \text{ kJ/kg d.a.}$$

$$OASH = 0.0204\,\dot{V}_F(t_{d1} - t_{d2}) = 0.0204 \times 30\,(43-25) = 11.016 \text{ kW}$$

$$OALH = 50\,\dot{V}_F(w_1 - w_2) = 50 \times 30\;(0.0162 - 0.0100) = 9.3 \text{ kW}$$

(*iii*) $EASH = RSH + OASH \times BPF$ = 100 + 11.016 × 0.15 = 101.65 kW

$ERLH = RLH + OALH \times BPF$ = 15 + 9.3 × 0.15 = 16.395 kW

$$\frac{\text{Fresh air}}{\text{Supply air}} = \frac{30}{371.36} = \frac{1}{12.37}$$

$$\text{Length } 2-3 = (\text{Length } 2-1) \times \frac{1}{12.37} = \frac{65}{12.37} = 5.25 \text{ mm}$$

$$t_{ds} = 26.5°\text{C}$$

$$t_6 = ADP = 11.8°\text{C}$$

$$BPF = \frac{t_{d4} = t_{d6}}{t_{d3} - t_{d6}}$$

$$0.15 = \frac{t_{d4} - 11.8}{26.5 - 11.8}$$

$$t_{d4} = 14°\text{C}$$

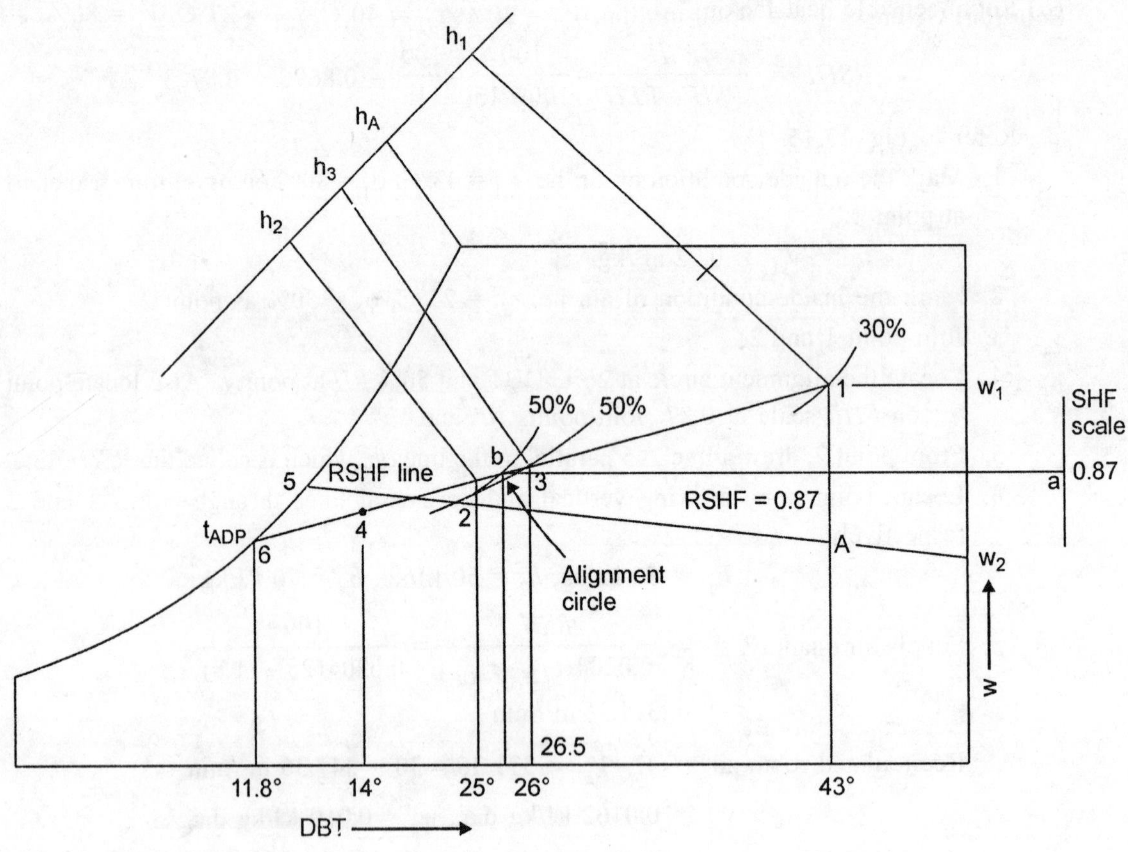

Fig. 13.15

Example 13.7 *The air handling unit of an air conditioning plant supplies a total 4500 cubic metre/min of. dry air which comprises 20 per cent fresh air to 40°C DBT and 27°C WBT and 80 per cent of recirculated air at 25°C DBT and 50% RH. The air leaves the cooling coil at 13°C saturated state. Determine the total cooling load and the room heat gain. Take specific volume of air entering the cooling coil as 0.869 m³/kg da.*

Condition	*DBT* °C	*WBT* °C	*RH* %	*Sp. Humidity* gwv/kg da	*Enthalpy* kJ/kg da
Outside	40	27	–	17.2	85.0
Inside	25	–	50	10.0	50.8
ADP	13	–	100	9.4	37.0

Note: *ADP – Apparatus dew point*
wv – water vapour
da – dry air

[CSE, 2004]

Solution: Given: $\dot{V}_a = 4500$ m³/min, $V_F = 20\%$, $t_{d1} = 40°C$, $t_{w1} = 27°C$, $V_r = 80\%$,

$t_{d2} = 25°C$, $\phi_2 = 50\%$, $ADP = 13°C$, $V_s = 0.869$ m³/kg d.a.

Refer to Fig. 13.16.

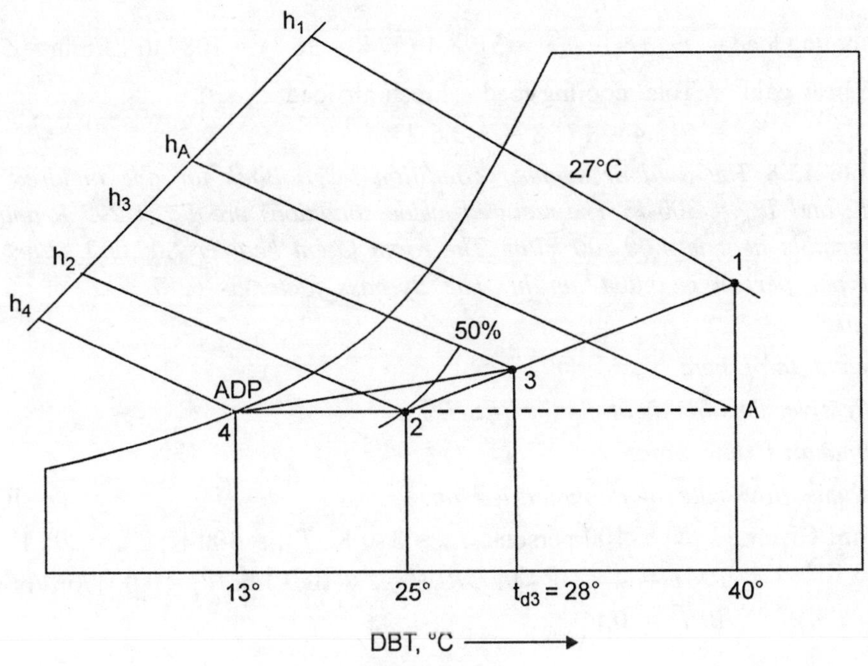

Fig. 13.16

1. Locate point 1(t_{d1} = 40°C, t_{w1} = 27°C) and point 2(t_{d2} = 25°C, ϕ_2 = 50%) on the psychrometric chart as shown in Fig. 13.16.
2. Join points 1 and 2.
3. By measurement, length of line 1-2 is 6 cm.
4. For V_F = 20%, locate point 3 on line 1-2 such that length 2 – 3 = 0.2 × Length 1-2, i.e. 0.2 × 6 = 1.2 cm.
5. Locate point 4 (ADP = 13°C) on the saturation line.
6. Join points 2 and 4 and 3 and 4.

Mass of Fresh air supplied at 3,

$$\dot{m}_{a3} = \frac{\dot{V}_a}{v_s} = \frac{4500}{0.869} = 5178.4 \text{ kg/min}$$

Mass of fresh air supplied at point 1,

$$\dot{m}_{a1} = 0.2 \times 5178.4 = 1035.68 \text{ kg/min}$$

Enthalpies:

$$h_1 = 85 \text{ kJ/kg air},\ h_2 = 50 \text{ kJ/kg air},\ h_3 = 57.8 \text{ kJ/kg air},\ h_4 = 36.8 \text{ kJ/kg air}$$

$$\text{Fresh air load} = \dot{m}_{a1}(h_1 - h_2) = 1035.68\ (85 - 50) = 36248.8 \text{ kJ/min}$$

$$= \frac{36248.8}{211} = 1711.8 \text{ TR}$$

$$\text{Total cooling load} = \dot{m}_{a3}(h_3 - h_4) = 5178.4\ (57.8 - 36.8) = 108746 \text{ kJ/min} = 515.4 \text{ TR}$$

$$\text{Room heat gain} = \text{Total cooling load} - \text{Fresh air load}$$

$$= 515.4 - 171.8 = 343.6 \text{ TR}$$

Example 13.8 *The outdoor summer condition for a Bank for one hundred persons is T_{db} = 310 K, and T_{wb} = 300 K. The required inside conditions are T_{db} = 295 K and ϕ = 60%. The room sensible heat is 4,00,000 kJ/hr. The room latent heat is 2,00,000 kJ/hr. Ventilation requirement per person is 0.0047 m³/hr. The By-pass factor is 0.15.*

Evaluate:

(i) Grand total heat

(ii) Effective sensible heat factor (ESHF)

(iii) Apparatus dew point

(iv) Volume flow rate of dehumidified air. [CSE, 2009]

Solution: Given: N = 100 persons, T_{d1} = 310 K, T_{w1} = 300 K, T_{d2} = 295 K, ϕ_2 = 60%, RSH = 4×10^5 kJ/hr, RLH = 2×10^5 kJ/hr, V_1 = 0.0047 m³/hr per person, BPF = 0.15

$$Q_s = \frac{4 \times 10^5}{360} = 111.11 \text{ kW},\ Q_L = \frac{2 \times 10^5}{3600} = 55.55 \text{ kW}$$

$$V_1 = 0.0047 \times 100 \times 60 = 28.2 \text{ m}^3/\text{min}$$

$$RSHF = \frac{Q_S}{Q_S + Q_L} = \frac{4}{4+2} = \frac{4}{6} = 0.667$$

Outside air sensible heat,

$$OASH = 0.02044\ V_1\ (T_{d1} - T_{d2}) \text{ kW}$$

$$= 0.02044 \times 28.2\ (310 - 295) = 8.64 \text{ kW}$$

Now w_1 = 0.0206 kJ/kg d.a., w_2 = 0.0100 kJ/kg d.a.

Outside air latent heat,

$$OALH = 50V_1\ (w_1 - w_2) \text{ kW}$$

$$= 50 \times 28.2 \times (0.0206 - 0.0100) = 14.95 \text{ kW}$$

Effective room sensible heat,

$$ERSH = RSH + OASH \times BPF = 111.11 + 8.64 \times 0.15 = 112.41 \text{ kW}$$

Effective room latent heat,

$$ERLH = RLH + OALH \times BP = 55.55 + 14.95 \times 0.15 = 57.79 \text{ kW}$$

(*i*) Grand total heat, $GTH = RSH + RLH + OASH + OALH$

$$= 111.11 + 55.55 + 8.64 + 14.95 = 190.25 \text{ kW}$$

(*ii*)
$$ERSHF = \frac{ERSH}{ERSH + ERLH} = \frac{112.41}{112.41 + 57.79} = 0.660$$

(*iii*) Apparatus dew point:

Locate point 1(t_{d1} = 37°C, t_{w1} = 27°C) and

Refer to Fig. 13.17.

Point 2 (t_{d2} = 22°C, ϕ_2 = 60%)

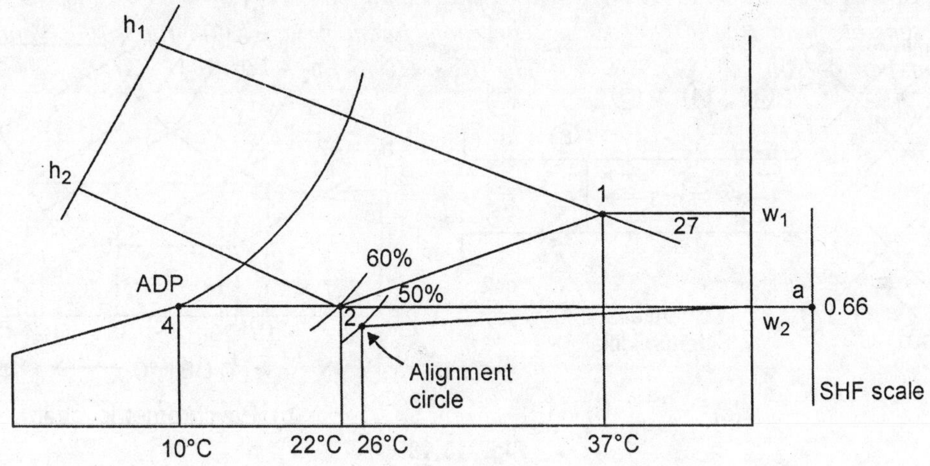

Fig. 13.17

(*a*) Join points 1 and 2

(*b*) Locate the alignment circle point *b* (26°C, 50% *RH*) and *ERSHF* = 0.66 point '*a*' on *SHF* scale. Join points '*a*' and *b* to get *ERSHF* line.

(*c*) Draw a line 2-4 from point 2 parallel to line *ab* to intersect saturation line at point 4. Then Apparatus Dew Point,

$$ADP = t_{d4} = 10°\text{C}$$

$$w_1 = 0.0206 \text{ kJ/kg d.a., } w_2 = 0.0100 \text{ kJ/kg d.a., } h_1 = 85 \text{ kJ/kg, } h_2 = 46 \text{ kJ/kg}$$

(*iv*) Dehumidified air volume flow rate

$$(\text{cmm})_{da} = \frac{ERSH}{0.02044\,(1 - BF)\,(t_{d2} - t_{d4})}$$

$$= \frac{112.41}{0.02044\,(1 - 0.15)\,(22 - 10)}$$

$$= 539.2 \text{ m}^3\text{/min}$$

Example 13.9 *100 m³ per minute outdoor air at 43.3°C d.b.t. and 37% relative humidity is mixed with 200 m³ per minute of air at 38.2°C d.b.t. and 24.5°C wet bulb temperature. The mixed air is dehumidified first by a cooling coil having bypass factor of 0.32 and* *apparatus dew point of 15°C and then by a chemical dehumidifier at 30°C dry bulb temperature. Air is then passed over a cooling coil whose surface temperature is 15°C and bypass factor 0.26. Calculate:*

(i) Capacity of two cooling coils in tons of refrigeration.

(ii) Capacity of coil dehumidifier in kg/min.

(iii) Capacity of chemical dehumidifier in kg/min.

(iv) Total dehumidifying capacity of the system in kg/min. **[IES, 2000]**

Solution: Refer to Fig. 13.18 (*a*)

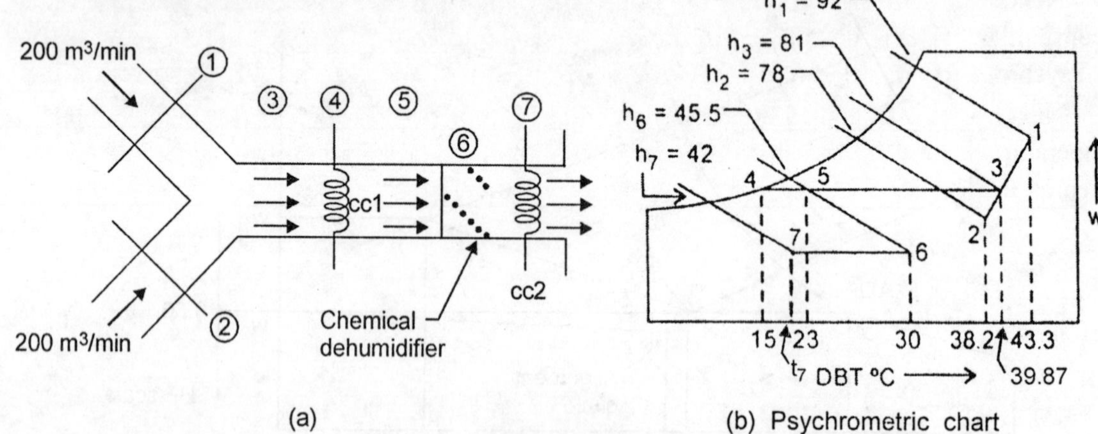

Fig. 13.18

1st stream:

$$\text{Volume flow} = 100 \text{ m}^3\text{/min}$$

$$\text{DBT} = 43.3°\text{C}$$

$$\text{RH} = 37\%$$

Specific volume, $v_{s1} = 0.92 \text{ m}^3\text{/kg d.a.}$

2nd stream:

$$\text{Volume flow} = 200 \text{ m}^3\text{/min}$$

$$\text{DBT} = 38.2°\text{C}$$

$$\text{WBT} = 24.5°\text{C}$$

Specific volume, $v_{s2} = 0.90 \text{ m}^2\text{/kg d.a}$

Determination of mixture condition:

From energy balance, we have

$$\dot{m}_{a1} \cdot c_p \cdot t_1 + \dot{m}_{a2} \cdot c_p \cdot t_2 = (\dot{m}_{a1} + \dot{m}_{a2}) \cdot c_p \cdot t_3$$

$$\left(\frac{100}{0.92}\right) \times 43.3 + \left(\frac{200}{0.90}\right) \times 38.2 = \left(\frac{100}{0.92} + \frac{200}{0.90}\right) t_3$$

$$t_3 = 39.87°\text{C}$$

Next calculate temperature of air leaving cooling coil from by-pass factor of coil (Fig. 13.18 *b*).

$$\frac{t_5 - t_4}{t_3 - t_4} = 0.32$$

$$\frac{t_5 - 15}{39.87 - 15} = 0.32$$

$$t_5 = 23°C$$

Next dehumidification takes place from 5-6. The process can be assumed to take place along a constant WBT line. It is given that air leaves dehumidifier at 30°C. So point 6 can be located on Psychrometric chart.

Next air is again cooled by a cooling coil at 15°C. The DPT of air is less than cooling coil temperature. Hence process shall be sensible cooling only. Since b.p.f. of second cooling coil is given as 0.26, point 7 can be located on Psychrometric chart, as

$$\frac{t_7 - t_4}{t_6 - t_4} = 0.26$$

$$\frac{t_7 - 15}{30 - 15} = 0.26$$

$$t_7 = 18.9°C = 19°C$$

(*i*) Total capacity of two cooling coils

$$= \dot{m}_{a3}[(h_3 - h_5) + (h_6 - h_7)]$$

$$= 330.92\ [(81 - 54.5) + (54.5 - 42)]$$

$$= 12909 \text{ kJ/min or } 61.18 \text{ Ton.}$$

(*ii*) Moisture removed in cooling coil

$$= \dot{m}_{a3}[(h_3 - h_5) + (h_6 - h_7)]$$

$$= 1.489 \text{ kg/min.}$$

(*iii*) Capacity of chemical dehumidifier

$$= \dot{m}_{a3}(w_5 - w_6)$$

$$= 330.92\ (12.5 - 9.2)\ 10^3 = 1.092 \text{ kg/min}$$

(*iv*) Total dehumidification capacity

$$= 1.489 + 1.092 = 2.581 \text{ kg/min}$$

Example 13.10 *In an air-conditioned space, 50 kga/s of fresh air at 45°C dry-bulb temperature and 30°C relative humidity is introduced. The room air at 25°C dry-bulb temperature and 50% relative humidity is recriculated at 450 kga/s. The mixed air flows over a cooling coil which has apparatus dew point of 12°C and bypass factor of 0.15. Determine the conditions at outlet of cooling coil, RSH, RLH, the cooling load of coil and the condensate rate. The saturation pressure of water at required temperature are*

t°C	p_{ws} (bar)
12	0.014016
25	0.03166
45	0.09584

$$h = 1.005t + w\ (2500 + 1.88t)\ kJ/kga$$

t is in °C and kga refers to kg of dry air. **[IES, 2002]**

Solution: From heat balance, we get

$$\dot{m}_{a1}.c_p t_1 + \dot{m}_{a2}.c_p.t_2 = (\dot{m}_{a1} + \dot{m}_{a2}) \cdot c_p.t_3$$

$$50 \times 45 + 450 \times 25 = (50 + 450) \times t_3$$

$$t_3 = \frac{13500}{500} = 27°\text{C}$$

Refer to Fig. 13.19.

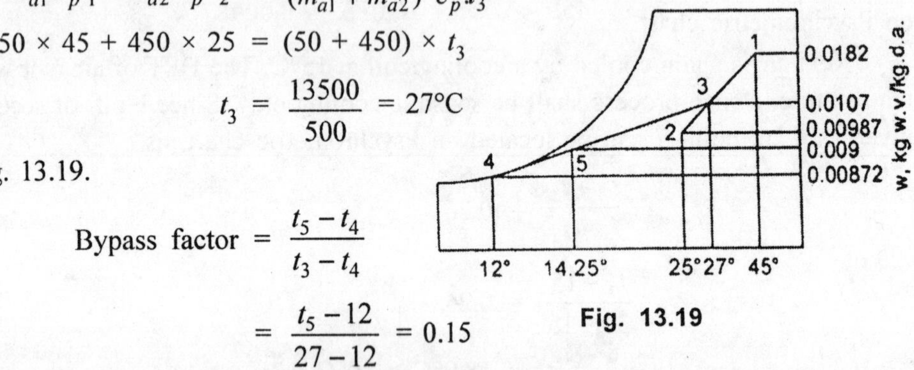

Fig. 13.19

$$\text{Bypass factor} = \frac{t_5 - t_4}{t_3 - t_4}$$

$$= \frac{t_5 - 12}{27 - 12} = 0.15$$

$$t_5 = 14.25°\text{C}$$

Therefore, DBT at outlet of cooling coil = 14.25°C

From Psychrometric chart for air, WBT = 13.25°C and RH = 88%

Now, $p_v = p_{ws} \times \text{RH}$ and $w = \dfrac{0.622}{p - p_v}$

where p = 1.01325 bar

$$w_1 = \frac{0.622 \times 0.09584 \times 0.30}{1.01325 \times 0.09584 \times 0.30} = 0.0182 \text{ kg w.v./kg. d.a.}$$

$$w_2 = \frac{0.622 \times 0.03166 \times 0.50}{1.01325 \times 0.03166 \times 0.50} = 0.00987 \text{ kg w.v./kg d.a}$$

$$w_4 = \frac{0.622 \times 0.04016 \times 1.00}{1.01325 \times 0.014016 \times 1.00} = 0.00872 \text{ kg w.v./kg d.a}$$

By interpolation,

$$w_3 = w_2 + \left[\frac{w_1 + w_2}{t_1 - t_2}\right](t_3 - t_2)$$

$$= 0.00987 + \left[\frac{0.0182 - 0.00987}{45 - 26}\right](27 - 25)$$

$$= 0.00987 + 0.000833 = 0.0107 \text{ kg.w.v./kg d.a.}$$

$$w_5 = \omega_4 + \left[\frac{w_3 - w_4}{t_3 - t_4}\right](t_5 - t_4)$$

$$= 0.00872 + \left[\frac{0.0107 - 0.00872}{27 - 12}\right](14.25 - 12)$$

$$= 0.00872 + 0.000297 = 0.009 \text{ kg.w.v./kg d.a.}$$

$$\text{RSH} = m_{a2}.c_p.(t_1 - t_2) = 450 \times 1.005 \times (45 - 25) = 9045 \text{ kJ/s}$$

$$\text{RLH} = m_{a2}.(h_{fg})_0 \, (w_1 - w_2)$$

$$= 450 \times 2500 \times (0.0182 - 0.00987) = 9371.25 \text{ kJ/s}$$

$$h_1 = 1.005 \times 45 + 0.0182 \, (2500 + 1.88 \times 25)$$

$$= 92.265 \text{ kJ/kga}$$

$$h_2 = 1.005 \times 25 + 0.00987 \, (2500 + 1.88 \times 25)$$

$$= 50.265 \text{ kJ/kga}$$

$$h_3 = 1.005 \times 12 + 0.0107 \, (2500 + 1.88 \times 27)$$

$$= 54.428 \text{ kJ/kga}$$

$$h_4 = 1.005 \times 12 + 0.00872 \, (2500 + 1.88 \times 12) = 34.056 \text{ kJ/kga}$$

$$h_5 = 1.005 \times 14.25 + 0.009 \, (2500 + 1.88 \times 14.25)$$

$$= 37.062 \text{ kJ/kga}$$

Cooling load of coil

$$= h_3 - h_5 = 54.428 - 37.062 = 17.366 \text{ kJ/kga}$$

Condensate rate

$$= \dot{m}_a(w_3 - w_5) = 500 \, (0.0107 - 0.009) = 0.85 \text{ kg/s}$$

Example 13.11 *Define human comfort. Give the factors governing optimum effective temperature. Outline the main points considered for cooling load estimate and heating load estimate.* **[IES, 2003]**

Solution: Human comfort: It is defined as the state of human body when the heat produced by metabolism of human body is equal to the sum of the heat dissipated to the surroundings and the heat stored in the human body is raising the temperature of body tissues.

During summer, for comfort air-conditioning:

DBT = 25 ± 1°C, RH = 50 ± 5%, Air Velocity = 0.4 m/s

During winter, DBT = 21°C, RH = 50%, Air velocity = 0.15 – 0.20 m/s

Factor governing optimum effective temperature:

1. Climate and seasonal differences.
2. Clothing
3. Age and sex.

4. Duration of stay.
5. Kind of activity.
6. Density of occupants.

Main points for Cooling and Heating load estimate:

1. **Internal heat gains:**
   (*a*) Occupancy load.
   (*b*) Lighting load.
   (*c*) Appliances load.
   (*d*) Piping, tanks and steam load.
   (*e*) Heat of respiration.
   (*f*) Process load.
2. **System heat gains:**
   (*a*) Supply air duct gain and leakage loss.
   (*b*) Air-conditioning fan load.
   (*c*) Return air duct heat and leakage gain load.
   (*d*) Heat gain from dehumidifier pump and piping.
3. **Safety factor : 5% additional load.**

Example 13.12

(i) What are the factors affecting comfort air conditioning?

(ii) Explain the air conditioning system used in winter season.

(iii) Show how air washer can be used for year round air conditioning?

(iv) A spray cooling coil is used to operate under the following conditions-Air inlet 28° DBT/21° WBT, Air outlet WBT. Total air flow rate 2000 m³/mt. Chilled water inlet and outlet temperatures are 7°C and 12°C. Find

(a) Cooling load on the coil

(b) Water flow rate through the coil. **[IES, 2009]**

Properties of moist air from psychometric chart are as follows:

DBT	*WBT*	*Sp. Humidity gm/kg*	*Sp. Vol. m³/kg*	*Enthalpy kJ/kg*
28	21.0	12.95	0.87	61.0
10	6.0	4.2	0.81	21.0

Solution: (*i*) Factors affecting comfort air-conditioning: The factors affecting comfort air-conditioning are:

1. **Temperature of air:** It is the temperature maintained within an enclosed space. This is accomplished either by the addition or removal of heat from the enclosed space. Human beings feel comfortable when the air is at 21°C with 56% relative humility.

2. **Humidity of air:** The humidity of air is controlled by either decreasing or increasing the moisture content of air. In general, for summer air-conditioning, the relative humidity should not be less than 60%, whereas for winter air-conditioning, it should not be more than 40%

3. **Purity of air:** To feel comfortable while breathing, air should be contaminated. Purity in air can be achieved by proper filtration and cleaning to make it free from dust and other impurities.

4. **Motion of air:** The motion or circulation of air is essential to keep constant temperature throughout the conditioned space. Recommended air velocity for comfort air-conditioning is 0.4 m/s during summer and 0.15 to 0.20 m/s during winter.

(*ii*) **Winter Air Conditioning System:** In winter air conditioning, the air is heated, which is generally accompanied by humidification. The schematic arrangement of the system is shown in Fig. 12.

The outside air flows through a damper and mixes up with the *recirculated air* (which is obtained from the conditioned space). The mixed air passes through a filter to remove dirt, dust and other impurities. The air now passes through a preheat coil in order to prevent the possible freezing of water and to control the evaporation of water in the humidifier. After that, the air is made to pass through a reheat coil to bring the air to the designed dry bulb temperature. Now, the conditioned air is supplied to the conditioned space by a fan. From the conditioned space, a part of the used air is exhausted to the atmosphere by the exhaust fans or ventilators. The remaining part of the used air (known as recirculated air) is again conditioned as shown in Fig. 13.20.

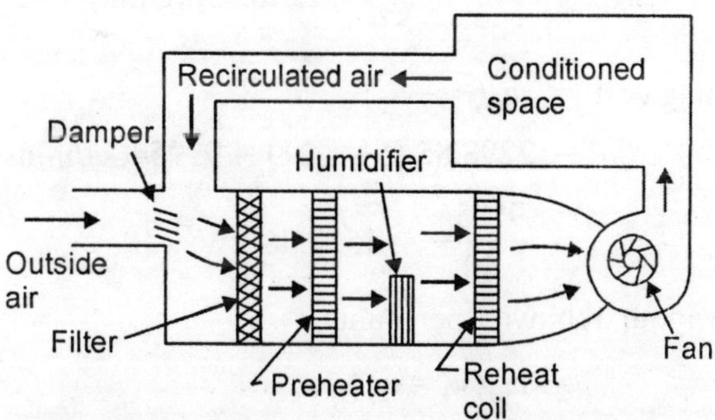

Fig. 13.20 Winter air conditioning system.

The outside air is sucked and made to mix with recirculated air, in order to make up for the loss conditioned (or used) air through exhaust fans or ventilation from the conditioned space.

(*iii*) **Year-Round Air Conditioning System Using air washer:** An air washer involves the flow of air through a spray of water. During the of flow the air may be cooled or heated, humidified or dehumidified, or simply adiabatically saturated, depending upon the mean surface temperature of water.

The year-round air conditioning system should have equipment for both the summer and winter air conditioning. The schematic arrangement of a modern summer year round air conditioning system is shown in Fig. 13.21

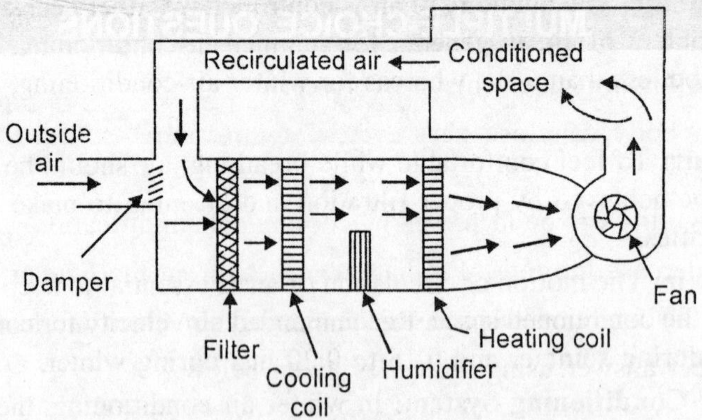

Fig. 13.21 Year-round air conditioning system.

The outside air flows through the damper and mixes up with the recirculated air (which is obtained from the conditioned space). The mixed air passes through a filter to remove dirt, dust and other impurities. In summer air conditioning, the cooling coil operate to cool the air to the desired value. The dehumidification is obtained by operating the cooling coil at a temperature lower than the dew point temperature (apparatus dew point). In winter, the cooling coil is made inoperative and the heating coil operates to heat the air. The spray type humidifier is also made use in the dry season to humidify the air.

(*iv*) (*a*) Mass of air flowing through the cooling coil,

$$\dot{m}_a = \frac{V_1}{v_{s1}} = \frac{2000}{0.87} = 2298.85 \text{ kg/min}$$

$$\text{Capacity of cooling coil} = \dot{m}_a(h_1 - h_2)$$

$$= 2298.85\ (61 - 21) = 91954 \text{ kJ/min}$$

$$= \frac{91954}{60} = 1532.56 \text{ kW}$$

Amount of water vapour removed per minute

$$= \dot{m}_a(w_1 - w_2)$$

$$= 2298.85\ (12.95 - 4.2) \times 10^{-3} = 20.1 \text{ kg/min}$$

(*b*) Let $\dot{m}_w$ = water flow rate through the coil, kg/min

$$91954 = \dot{m}_w \times c_{pw} \times \Delta t = \dot{m}_w \times 4.18 \times (12 - 7)$$

$$\dot{m}_w = 4399.7 \text{ kg/min or } 73.33 \text{ kg/s}$$

MULTIPLE-CHOICE QUESTIONS

1. Which of the following statements are correct?

   (i) The human body can lose heat even if temperature is less than the atmospheric temperature.

   (ii) Relative humidity can be increased by cooling dehumidification.

   (iii) Warm air increase the rate of radiation of heat from the human body.

   (iv) Increase in air movement increases the evaporation from the human body.

   Select the correct answer using the codes given below:

   Codes:

   (a) 1 and 4 (b) 2 and 4 (c) 1 and 3 (d) 2 and 3

2. Consider the following factors:

   (i) Wind velocity

   (ii) Type of activity

   (iii) Indoor design conditions

   (iv) Door openings

   Occupancy load in cooling load calculations depends upon

   (a) 1 and 2 (b) 1 and 3 (c) 1 and 4 (d) 2 and 3

3. Effective temperature is that temperature of saturated air which gives the same degree of comfort as the air at given

   (a) DBT, WBT and incidental solar radiation

   (b) WBT, incidental solar radiation and air flow rate

   (c) DBT, sol-air temperature and air flow rate

   (d) DBT; WBT and air flow rate.

4. A human body feels comfortable when the heat produced by the metabolism of human body is equal to the

   (a) heat dissipated to the surroundings (b) heat stored in the human body

   (c) sum of (a) and (b) (d) difference of (a) and (b)

5. In system: Metabolic rate = M, work done by man = W, rate of convective, radiative and evaporative heat losses = Q and rate of heat storage = S. Then heal exchange between man and his environment is given by

   (a) $M + W = Q + S$ (b) $M - W = Q - S$

   (c) $M + W = Q - S$ (d) $M - W = Q + S$

6. Fresh air intake (air change per hour) recommended for ventilation purposes in the air-conditioning system of an office building is

   (a) 1/2 (b) 3/2 (c) 9/2 (d) 25/2

7. What are the general comfort conditions in an air-conditioning system?
   (a) 20°C DBT, 80% RH
   (b) 24°C DBT, 60% RH
   (c) 25°C DBT, 40% RH
   (d) 25°C DBT, 100% RH
8. What is Sol-air temperature?
   (a) It is equal to the sum of outdoor air temperature, and absorbed total radiation divided by convective heat transfer coefficient outer surface
   (b) It is equal to the absorbed total radiation divided by convective heat transfer coefficient at outer surface
   (c) It is equal to the total incident radiation divided by convective heat transfer coefficient at outer surface
   (d) It is equal to the sum of indoor air temperature and absorbed total radiation divided by convective heat transfer coefficient at outer surface.
9. Effective temperature depends on dry bulb temperature, and
   (a) Wet bulb temperature only
   (b) Relative humidity
   (c) Specific humidity
   (d) Wet bulb temperature, and air motion
10. A human body feels comfortable when the heat produced by the metabolism of human body is equal to
   (a) Heat dissipated to the surroundings
   (b) Heat stored in the human body
   (c) Sum of (a) and (b)
   (d) Difference of (a) and (b)

Answers

1. (a) **2.** (d) **3.** (d) **4.** (a) **5.** (a) **6.** (c) **7.** (c) **8.** (a)
9. (b) **10.** (a)

REVIEW QUESTIONS

1. What are the factors that determine human comfort?
2. What is effective temperature? List the factors that affect effective temperature.
3. Differentiate between heat stroke and heat cramp.
4. What is the difference between summer and winter air-conditioning?
5. What is alignment circle?
6. Define human comfort.
7. What is metabolic rate?
8. What is a comfort chart?
9. List the factors which affect effective temperature
10. Define BPF, GSHF and ESHF

NUMERICAL PROBLEMS

13.1 An air-conditioned space is maintained at 27°C dry bulb temperature and 50% relative humidity. The ambient conditions are 40°C dry bulb temperature and 27°C wet bulb temperature. The space has a sensible heat gain of 14 kW. The air is supplied to the space at 7°C saturated. Calculate.

(a) Mass of moist air supplied to the space in kg/h;

(b) Latent heat gain of space in kW; and

(c) Cooling load of the air washer in kW if 30 per cent of the air supplied to the space is fresh the remaining being recirculated. [**Ans.** 2480 kg/h; 8.6 kW; 28.7 kW]

13.2 A cinema hall of seating capacity 1500 persons has been provided with an air conditioned plant with the following data:

Outdoor conditions = 40°C *DBT* and 20°C *WBT*

Required indoor conditions = 20°C *DBT* and 60% *RH*

Amount of outdoor air supplied = 0.3 m^3/min/person

If the required condition is achieved first by adiabatic humidifying and then by cooling, find: (a) the capacity of cooling coil and surface temperature of the coil if by-pass factor is 0.25; and (b) the capacity of the humidifier and its efficiency.

[**Ans.** 118.8 kW; 15.27°C; 72 kg/h, 29%]

13.3 The following data refers for a space to be air conditioned:

Inside design conditions = 25°C *DBT*, 50% *RH*

Outdoor air conditions = 43°C *DBT*, 27.5°C *WBT*

Room sensible heat gain = 20 kW

Room latent heat gain = 5 kW

By-pass factor of the cooling coil = 0.1

The return air from the space is mixed with the outside air before entering the cooling coil in the ratio of 4 : 1 by mass. Determine (a) apparatus dew point; (b) condition of air entering and leaving the cooling coil; (c) dehumidified air quantity; (d) fresh air mass flow and volume flow rate; and (e) total refrigeration toad on the air-conditioning plant.

[**Ans.** 11.8°C; 28.6°C, 13.5°C; 6265 kg/h; 1253 kg/h, 19.2 m^3/min; 40.7 kW]

13.4 Given: Summer Design = 43°C DBT, 24°C WB

Inside Design = 25°C DBT, 50% RH

Room Sensible Heat = 300000 kJ/hr

Room Latent Heat = Nil

Ventilation = 30 cmm.

Determine:

(a) Effective sensible heat factor

(b) Dehumidified air-quantity

(c) Supply air temperature

Assume by-pass factor = 0.05.

13.5 In an industrial Application:

Summer Design	= 40°C DBT; 24°C WBT
Inside Room Design	= 60% RH
Room Sensible Heat	= 200000 kJ/hr (55.556 kW)
Room Sensible Heat Factor	= 0.9.

Use all outdoor air and evaporative cooling process, with washer efficiency 80%. Determine:

(a) Room DBT maintained

(b) Supply quantity in cmm.

13.6 The following data refer to a laboratory

Summer Design	= 40°C DBT; 24°C DBT
Inside Design	= 25°C DBT, 50% RH
Room Sensible Heat	= 60000 kJ/kg
Room Latent Heat	= 12000 kJ/hr
Ventilation	= 40 cmm.

Also all out side fresh air to be supplied through coil, and no return air. Determine:

(a) Effective sensible heat factor

(b) ADP

(c) Actual quantity of outdoor air to be supplied to take care of the loads

(d) Dehumidified air quantity supplied.

Assume by-pass factor = 0.05

13.7 The outside air at 30°C dry bulb temperature and 18.5°C wet bulb temperature enters a cooling coil at the rate of 40 m^3/min. The effective surface temperature of the cooling coil is 4.5°C and its cooling capacity is 12.5 kW of refrigeration. Find:

(a) Dry bulb and wet bulb temperatures of the air leaving the coil,

(b) Enthalpy of air leaving the coil, and

(c) By pass factor of the coil.

[**Ans.** 18.8°C; 12.7°C; 35.6 kJ/kg of dry air; 0.52]

Chapter 14

COOLING AND HEATING LOAD ESTIMATION

14.1 COOLING LOAD ESTIMATION

The components of a cooling load are:

1. Sensible heat gain

This is the direct addition of heat to the space to be conditioned. It may occur due to any one or all of the following sources of heat transfer:

(*a*) The heat flowing into the building by conduction through exterior walls, floors, ceilings, doors and windows due to the temperature difference on their two sides.

(*b*) The heat received from solar radiation. It consists of the following:

(*i*) The heat transmitted directly through glass of windows, ventilators, and doors.

(*ii*) The heat absorbed by walls and roofs exposed to solar radiation and later on transferred to the space to be conditioned by conduction.

(*c*) The heat conducted from adjoining un-conditioned rooms.

(*d*) The heat generated by fans, lights, machinery, cooking operations, industrial processes, etc.

(*e*) Occupancy load.

(*f*) Air infiltration load through doors, windows, and their frequent opening.

(g) Heat gain from walls of ducts, etc.

2. Latent heat gain

The latent heat load occurs due to the presence of water vapour in the air conditioned space. This load occurs due to the following sources:

(*a*) Moisture in the air entering by infiltration.

(*b*) Condensation of moisture from occupants, food cooking process.

(*c*) Moisture entering directly into the conditioned space through permeable walls, partitions, etc.

Total cooling load = Sensible heat load + Latent heat load.

14.1.1 Heat Transfer Through Walls and Roofs

The heat transferred through a plane wall by conduction is given by Fourier law. For a single wall shown in Fig. 14.1 (*a*), we have

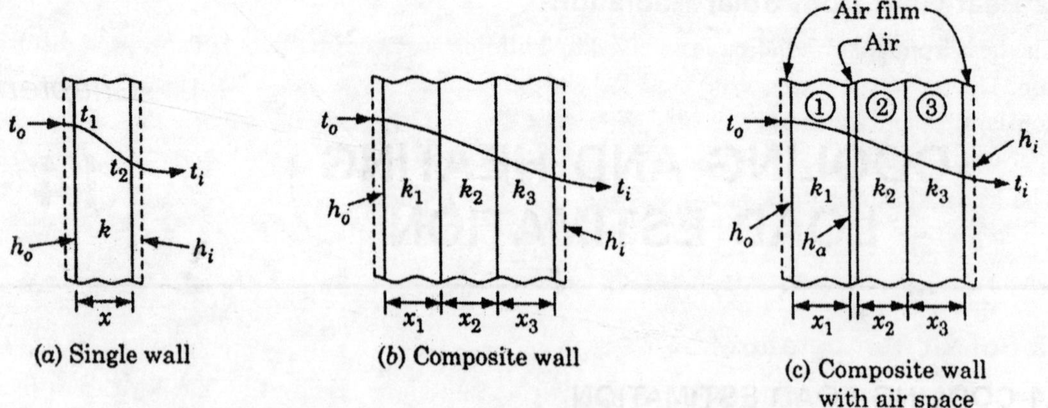

Fig. 14.1 Heat transfer by conduction through a plane wall.

$$Q_{\text{cond}} = \frac{kA_c \cdot \Delta t}{\Delta x} \qquad \text{...(14.1)}$$

where k = thermal conductivity of wall, W/(m. °C)

A_c = area of wall cross-section, m^2

Δt = temperature difference on two sides of wall, °C

Δx = wall thickness, m

The heat transfer by convection from a plane wall is given by Newton's law.

$$Q_{\text{conv}} = hA_s \cdot \Delta t \qquad \text{...(14.2)}$$

where h = film or surface conductance, W/(m$^2\cdot$°C)

A_s = surface area of wall, m^2

Δt = temperature difference between wall and surroundings, °C

Heat gained through a wall by the combined effect of conduction and convention becomes,

$$Q = h_o\,(t_o - t_1)\,A_c + \frac{k}{x}(t_1 - t_2)\,A_c + h_i(t_2 - t_i)\,A_c = UA_c(t_o - t_i) \qquad \text{...(14.3)}$$

Where subscripts o, i refer to outside and inside wall conditions respectively and U = overall coefficient of heat transfer of wall

$$= \frac{1}{\dfrac{1}{h_o} + \dfrac{x}{k} + \dfrac{1}{h_i}} \text{ for a single wall (Fig. 14.1}a\text{)} \qquad \text{...(14.4)}$$

$$= \frac{1}{\dfrac{1}{h_o} + \sum_{m=1}^{3}\dfrac{x_m}{k_m} + \dfrac{1}{h_i}} \text{ for a composite wall (Fig. 14.1 }b\text{)} \qquad \text{...(14.5)}$$

$$= \frac{1}{\dfrac{1}{h_o} + \dfrac{1}{k_a} + \sum_{m=1}^{3}\dfrac{x_m}{k_m} + \dfrac{1}{h_i}} \text{ for a composite wall with air gap (Fig. 14.1 }c\text{)} \qquad \text{...(14.6)}$$

where k_a = thermal conductance of air space

14.1.2 Heat Gain from Solar Radiation

The heat from solar radiation is received by building surfaces by direct radiation and diffuse radiation. The direct radiation is the impingement of the Sun's rays upon the surface. The diffuse radiation is received from moisture and dust particles in atmosphere which absorbs part of the energy of the Sun's rays. The diffuse radiation is received by surfaces which do not face the Sun.

The heat gain through outside walls and roofs is given by,

$$Q = UA.\Delta t_e \quad \text{...(14.7)}$$

where Δt_e = equivalent temperature differential.

14.1.3 Sol Air Temperature

It is a hypothetical temperature used to calculate the heat received by the outside surface of a building wall by the combined effect of convection and radiation.

Total heat received by the outside surface of the wall per unit area,

$$q_{os} = q_{\text{conv}} + q_{\text{rad}} = h_o(t_o - t_{os}) + I \propto$$

$$= h_o t_o - h_o t_{os} + I\alpha = h_o\left(t_o + \frac{I\alpha}{h_o}\right) - h_o t_{os}$$

$$= h_o t_e - h_o t_{os} = h_o(t_e - t_{os}) \quad \text{...(14.8)}$$

where $t_e = t_o + \dfrac{I_a}{h_o}$ = sol air temperature.

t_o = temperature of outside air

t_{os} = temperature of outside surface of wall

h_o = outside film coefficient

I = total radiation intensity

α = absorptivity of the surface

14.14 Solar Heat Gain Through Glass Areas

Heat gain through glass area = Absorbed solar radiation convection and radiation heat exchanges between glass and outdoor surfaces. ...(14.9)

14.1.5 Heat Gain Due to Infiltration

There are two methods of estimating the infiltrated air:

(*i*) Crack length method, and (*ii*) Air change method.

The crack length method is usually used where greater accuracy is required. In most cases the air change method is used for calculating the quantity of infiltrated air. According to this method, the quantity of infiltrated air through windows and walls is,

$$\dot{V} = \frac{1}{2}\left[\frac{LWH \times A_c}{60}\right] \text{ m}^3\text{/min} \quad \text{...(14.10)}$$

where L, W, H = length, width and height of room respectively, m

A_c = air changes per hour

Factor $\frac{1}{2}$ is used because infiltration takes place on the windward side of building only.

14.1.6 Heat Gain From Products

This heat gain is very important in case of cold storages.

(*i*) Chilling load above freezing,

$$Q_{ch} = \frac{mc_{pm}(T_1 - T_2)}{t_{ch}} \quad ...(14.11)$$

where m = mass of product

c_{pm} = mean specific heat of product

T_1, T_2 = initial and final temperatures of products respectively

t_{ch} = chilling time.

(*ii*) Freezing load,

$$Q_f = \frac{m \times h_{fg}}{t_f} \quad ...(14.12)$$

where h_{fg} = latent heat of freezing

t_f = freezing time

(*iii*) Cooling load below freezing,

$$Q_c = \frac{mc'_{pm}(T_{1'} - T_{2'})}{t_c} \quad ...(14.13)$$

where c'_{pm} = mean specific heat of freezing product

T'_1, T'_2 = actual storage and freezing temperatures of product

T_c = cooling time

(*iv*) Product respiration heat, $Q_r = m \times$ rate of heat evolution of food ...(14.14)

14.1.7 Heat Gain From Lights

Heat gain from lights, Q_l = total wattage of lights × use factor × allowances factor

where use factor = 1.0 for residences and commercial stores ...(14.15)

= 0.5 for industrial workshops

Allowance factor = 1.25 for fluorescent tubes.

14.1.8 Heat Gain From Power Equipments

Heat gain from electric motor, $Q_{em} = \frac{\text{kW rating}}{\eta_{em}} \times \text{Load factor}$ ...(14.16)

where η_{em} = efficiency of electric motor.

14.1.9 Heat Gain Through Ducts

Heat gain through the duct, $Q_D = UA_D\,(t_a - t_s)$ ...(14.17)

where U = overall heat transfer coefficient

A_D = surface area of duct

t_a, t_s = temperatures of ambient and supply airs respectively

14.2 EMPIRICAL METHODS TO EVALUATE HEAT TRANSFER THROUGH WALLS AND ROOFS

There are two approaches to empirical calculations of heat transfer through walls and roofs. They are:

(*i*) The decrement factor and time lag method.

(*ii*) The equivalent temperature differential method.

Both the methods use analytical-experimental results for their formulations. The equivalent temperature differential method is more commonly used by the air-conditioning engineers as it is also applicable to sunlit walls and roofs.

14.2.1 Using Decrement Factor And Time Lag

If the thermal capacity of the wall is ignored, then the instantaneous rate of heat transfer through the wall at any time is given by

$$Q_\tau = UA(t_e - t_i) \quad ...(14.18)$$

where t_e = sol-air temperature.

and on an average basis, the mean heat flow is given by

$$Q_m = UA(t_{em} - t_i) \quad ...(14.19)$$

where t_{em} = mean sol-air temperature.

For the sake of simplicity, the *dot* above Q has been dropped to indicate the *rate*.

But most building materials have a finite thermal capacity which is expressed as

$$mc = \rho c V = \rho c\,(A\Delta x) \quad ...(14.20)$$

where m = Mass of wall, V = volume of wall

ρ, c = Density and specific heat of wall material respectively.

A = Cross-sectional area of wall

Δx = wall thickness.

It has been seen that there is a two-fold effect of thermal capacity on heat transfer.

(*i*) There is a time lag between the heat transfer at the outside surface q_o and the heat transfer at the inside surface q_i, defined by ϕ.

(*ii*) There is a decrement in the heat transfer due to the absorption of heat by the wall and subsequent transfer of a part of this heat back to the outside air when its temperature is lower, defined by λ.

The rigorous analytical method to determine the time lag ϕ and decerement factor is quite complicated. In the limiting case, when the wall thickness approaches zero, $\lambda \to 1$ and $\phi \to 0$.

Then $$Q_\tau = UA(t_e - t_i)$$

i.e. the heat transfer through the wall is equal to its instantaneous value.

Consider the effect of thermal capacity, the actual heat transfer at any time τ is,

$$Q_\tau = UA\,(t_{e_m} - t_i) + UA\lambda(t_{e_{\tau-\phi}} - t_{e_m}) \quad \text{...(14.21)}$$

where $t_{e_{\tau-\phi}}$ = sol-air temperature at time $\tau - \phi$, i.e., ϕ hours before the heat transfer is to be calculated.

In a locality where the daily range of variation of the outside air temperature is small, it is immaterial what thickness of wall is provided. But in a locality when the daily range of temperature is large, it is desirable to have thick wall so as to cut the cooling load in summer and the heating in winter. Such walls also do not allow the inside temperature to rise very much during the day and drop at night.

14.2.2 Equivalent Temperature Differential (ETD) Method

Eq. (14.16) can be written in terms of an equivalent temperature differential defined by the equation,

$$Q = UA\Delta t_E \quad \text{...(14.22)}$$

where $$\Delta t_E = (t_{e_m} - t_i) + \lambda(t_{e_{\tau-\phi}} - t_{e_m}) \quad \text{...(14.23)}$$

Δt_E depends on the following:

(*i*) λ and ϕ, which in turn depend on the thermo-physical properties of construction.

(*ii*) Then outside air temperature t_a and solar radiation I.

(*iii*) Room temperature t_i.

Thus, the equivalent temperature differential approach takes care of the exposure of the wall or roof to the Sun.

14.3 HEATING LOAD ESTIMATION

An estimate of the heating load is made on the basis of maximum probable heat loss of the conditioned heated space. The heating loads are:

1. **Transmission heat loss:** The transmission heat loss from walls, roofs, etc. is given by,

$$Q = UA(t_i - t_o) \quad \text{...(14.24)}$$

where t_i, t_o = inside and outside design temperature respectively.

2. **Solar radiation:** The solar radiation heat gain is neglected as peak heating load occurs in the early hours of morning.

3. **Internal heat gains:** The internal heat gains from occupants, lights, appliances, etc. decrease the heating load. These negative loads should be accounted for in applications, such as theatres assembly halls, stores, office buildings, etc.

14.4 ROOM SENSIBLE HEAT FACTOR (RSHF)

It is defined as the ratio of room sensible heat to the total heat (Fig. 14.2)

$$RSHF = \frac{RSH}{RTH} = \frac{RSH}{RSH + RLH} \quad ...(14.25)$$

where *RSH, RLH, RTH* = room sensible, latent and total heats respectively.

14.5 ESTIMATION OF SUPPLY AIR CONDITIONS

When the supply air conditions are not known, then the *RSHF* line may be drawn from the *RSHF* calculated value, as described below (Fig. 14.3)

1. Mark point '*a*' on the sensible heat factor scale drawn on the right hand corner of psychrometric chart. The point '*a*' represents the calculated value of *RSHF.*
2. Joint point '*a*' with the alignment circle or reference point $b(t_d = 26°C, \phi = 50\%)$. The line *ab* is called the base line.
3. Mark point *R* on the psychrometric chart to represent the room design conditions.
4. Draw a line *RR′* through point *R* parallel to the base line *ab*. This line is the required RSHF line.

14.6 GRAND SENSIBLE HEAT FACTOR

It is defined as the ratio of total sensible heat to the grand total heat which the cooling coil or conditioning apparatus is required to handle (Fig. 14.4)

$$GSHF = \frac{TSH}{GTH} = \frac{TSH}{TSH + TLH} = \frac{RSH + OASH}{(RSH + ORSH) + (RLH + OALH)} \quad ...(14.26)$$

Where TSH = total sensible heat = $RSH + OASH$

TLH = total latent heat = $RLH + OALH$

GTH = grand total heat = $TSH + TLH = (RSH + OASH) + (RLH + OALH)$

Let $\dot{V}_1$ = volume rate of outside air, m³/min

t_{d1}, t_{d2} = DBT of outside and room air respectively, °C

w_1, w_2 = specific humidity of outside and room air respectively, kg/kg d.a.

h_1, h_2 = enthalpy of outside and room air respectively, kJ/kg d.a.

Then outside air sensible heat, $OASH = 0.0204\ \dot{V}_1 (t_{d1} - t_{d2})$ kW

Outside air latent heat, $OALH = 50\dot{V}_1 (w_1 - w_2)$ kW

Outside air total heat, $OATH = OASH + OALH$

$= 0.02\ \dot{V}_1 (h_1 - h_2)$ kW

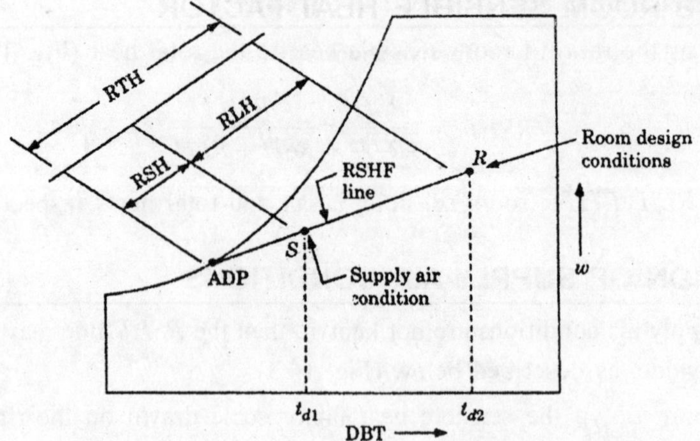

Fig. 14.2 Room and supply air conditions and ADP

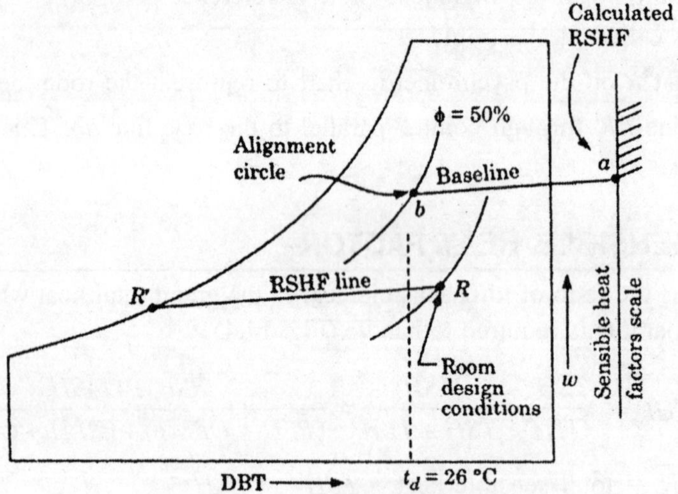

Fig. 14.3 RSHF line estimation

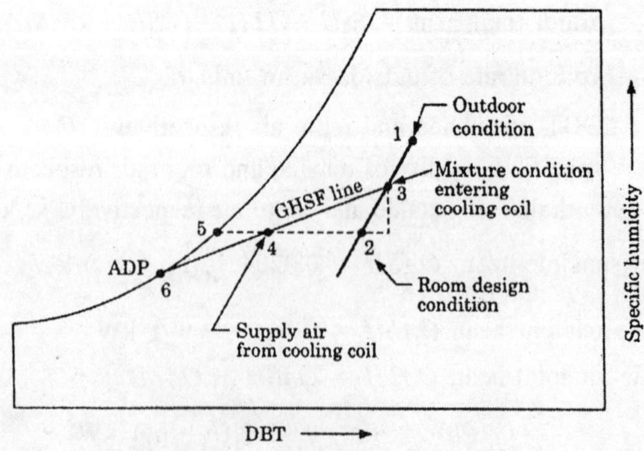

Fig. 14.4 GHSF line estimation

14.7 EFFECTIVE ROOM SENSIBLE HEAT FACTOR

It is defined as the ratio of the effective room sensible heat to the effective room total heat.

$$ERSHF = \frac{ERSH}{ERTH} = \frac{ERSH}{ERSH + ERLH} \quad ...(14.30)$$

where ERSH = Effective room sensible heat

$= RSH + OASH \times BPF$

$= RSH + 0.0204\,\dot{V}_1(t_{d1} - t_{d2}) \times BPT$

$ERLH$ = effective room latent heat

$= RLH + OALH \times BPF$

$= RLH + 50\dot{V}_1(w_1 - w_2) \times BPF$

ERTH = effective room total heat

$= ERSH + ERLH$

BPF = By-pass factor

The line joining the point 2 and 6, i.e. ADP, as shown in Fig. 14.5, gives the effective room sensible heat factor line (*ERSHF* line). From point 4, draw line 4-4′ parallel to line 3-2. From similar triangles 6-4-4′ and 6-3-2, we have

$$BPF = \frac{\text{length 4-6}}{\text{length 3-6}} = \frac{\text{length 4}'\text{-6}}{\text{length 2-6}} \quad ...(14.31a)$$

Also

$$BPF = \frac{t_{d4} - ADP}{t_{d3} - ADP} = \frac{t_{d4'} - ADP}{t_{d2} - ADP} \quad ...(14.31b)$$

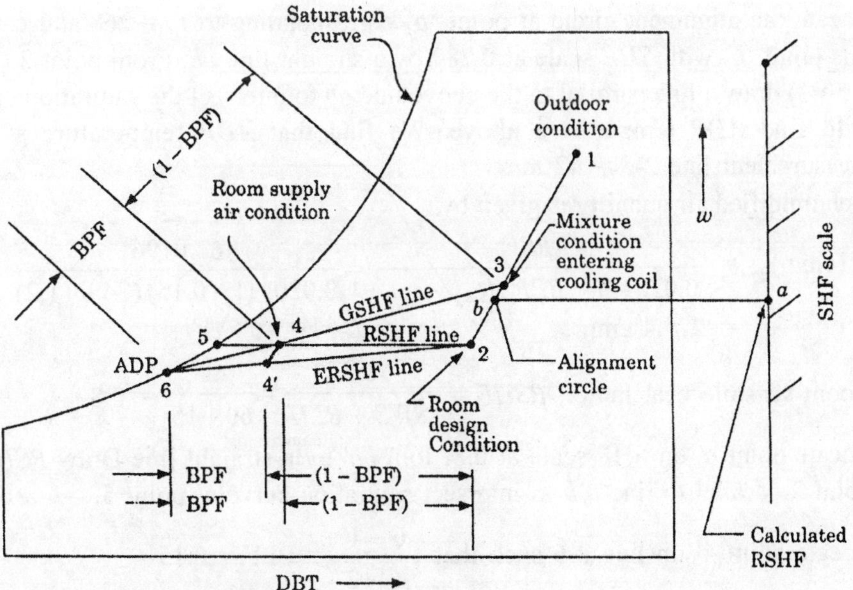

Fig. 14.5 Effective room sensible heat factor (ERSHF)

Example 14.1 *A room is to be maintained at 24°C DBT and 50% RH when the outside conditions are 35° DBT and 24° WBT. The room sensible and latent heat gains are 60 kW and 15 kW respectively. If the ventilation is 60 m³/min, calculate: (a) Outdoor air load (b) Grand total heat (c) Effective sensible heat factor (d) Apparatus dew point temperature (e) Dehumidified air quantity, and (f) Entering and leaving conditions at the apparatus. Assume by pass factor = 0.15.*

Solution: Given: $t_{d1} = 35°C, t_{w1} = 24°C, \phi_2 = 0.5, RSH = 60 \text{ kW}, RLH = 15 \text{ kW},$

$\text{cmm} = 60 \text{ m}^3/\text{min}, BPF = 0.15$

From psychrometric chart, we have (Refer to Fig. 14.5)

$$h_1 = 72.5 \text{ kJ/kg}, h_2 = 48 \text{ kJ/kg}$$

(*a*) Outdoor air load,

$$OATH = 0.0204 \text{ (cmm)} (h_1 - h_2) = 0.0204 \times 60 \times (72.5 - 48) = 29.988 \text{ kW}$$

(*b*) Grand total heat,

$$GTH = RSH + RLH + OATH = 60 + 15 + 29.988 = 104.988 \text{ kW}$$

(*c*) Effective room sensible heat,

$$ERSH = RSH + 0.0204 \text{ (cmm)} \times BPF \times (t_{d1} - t_{d2})$$

$$= 60 + 0.0204 \times 60 \times 0.15 \times (35 - 24) = 62.0196 \text{ kW}$$

Effective room total heat,

$$ERTH = RSH + RLH + BPF \times OATH$$

$$= 60 + 15 + 0.15 \times 29.988 = 79.4982 \text{ kW}$$

Effective room sensible heat factor,

$$ERSHF = \frac{ERSH}{ERTH} = \frac{62.0196}{79.4982} = 0.780$$

Refer to Fig. 14.5

(*d*) Locate the alignment circle at point '*b*' corresponding to $t_d = 26°$ and $\phi = 50\%$. Join this point '*b*' with *SHF* scale at 0.780 by a straight line *ab*. From point 2 ($t_{d2} = 24°$, $\phi_2 = 50\%$) draw a line parallel to the above line *ab* to intersect the saturation curve at point 6 to find *ADP* temperature above. We find that *ADP* temperature = 11.2°C. By measurement line 2-6 = 42 mm.

(*e*) Dehumidified air quantity is given by,

$$(\text{cmm})_{da} = \frac{ERSH}{0.0204\,(1 - BPF)\,(t_{d2} - t_{adp})} = \frac{62.0196}{0.0204\,(1 - 0.15)\,(24 - 11.2)}$$

$$= 279.4 \text{ cmm}$$

(*f*) Room sensible heat factor, $RSHF = \dfrac{RSH}{RWS + RLH} = \dfrac{60}{60 + 15} = 0.8$

Locate point *a*′ on SHF scale at 0.8. Join *ba*′ by a straight line Draw *RSHF* line from point 2 parallel to line *a*′*b* to intersect saturation curve at point 5.

Locate point 4′ on line 2-6 such that $\dfrac{6 - 4'}{6 - 2} = \text{BPF} = 0.15$

Then 6-4′ = 6.3 mm

From point 4′ draw a line 4′4 parallel to line 1-2. Then 4 is the room supply point. Join 6 and 4 and produce it to intersect line 1-2 at point 3. Then point 3 is the entering condition of air to the apparatus.

Thus supply condition or leaving air condition from apparatus,

$$t_{d4} = 12°C, \quad RH = 85\%$$

Entersing air conditions to apparatus,

$$t_{d3} = 27.5°C, \quad RH = 48\%$$

Example 14.2 *A laboratory requires 24°C DBT and 50% RH to be maintained when the outdoor conditions are 32°C DBT and 23°C WBT. The room sensible and latent heat gains are 36 kW and 20 kW respectively. The ventilation is 65 m³/min. Calculate: (a) outdoor air load (b) Grand total heat load (c) Effective sensible heat factor (d) Apparatus dew point temperature (e) Reheat required (f) Supply air quantity (g) Entering and leaving conditions at the apparatus (h) Supply air conditions to the space. Assume by pass factor = 0.05. ADP should not be less than 8°C.*

Solution:

$$\text{RSHF} = \frac{RSH}{RSH + RLH} = \frac{36}{36+20} = \frac{36}{56} = 0.643$$

$$ESHF = \frac{RSH + BPF \times OASH}{RTH + BPF \times OATH}$$

$$OASH = 0.0204\ (\text{cmm})_v\ (t_{d1} - t_{d2}) = 0.0204 \times 65 \times (32 - 24) = 10.608 \text{ kW}$$

$$OALH = 50 \times (\text{cmm})_v \times (w_1 - w_2) = 50 \times 65 \times (0.0142 - 0.0092) = 16.25 \text{ kW}$$

$$ERSHF = \frac{36 + 0.05 \times 10.608}{(36+02) + 0.05\,(10.608 + 16.25)} = \frac{36.53}{57.34} = 0.637$$

1. Locate the alignment circle point b (26°C DBT, 50% RH) and point 'a' at $SHF = 0.637$. Join points 'a' and b by a straight line
2. Locate outside air point 1 (32°C DBT and 23°C WBT) and inside room point 2 (24°C DBT and 50% RH). Join points 1 and 2 by a straight line.
3. At point 2 draw a line parallel to line ab to intersect the saturation curve at point C. This gives $t_{ADP} = 5.5°C$, which is less than 8°C. So we choose $t_{ADP} = 8°C$ and solve the problem.
4. corresponding to $t_{ADP} = 8°C$, locate point 6. Join points 2 and 6 by a straight line
5. Form point b draw a line parallel to line 26 to intersect SHF scale at a'. Point a' corresponds to $ERSHF = 0.685$.

(*e*) This $ERSHF = 0.685$ can be obtained with reheat.

Thus

$$ERSHF = \frac{RSH + BPF \times OASH + \text{Reheat}}{RTH + BPF \times OATH + \text{Reheat}}$$

$$0.685 = \frac{36.53 + \text{Reheat}}{57.34 + \text{Reheat}}$$

$$39.278 + 0.685 \text{ Reheat} = 36.53 + \text{Reheat}$$

$$\text{Reheat} = 8.72 \text{ kW}$$

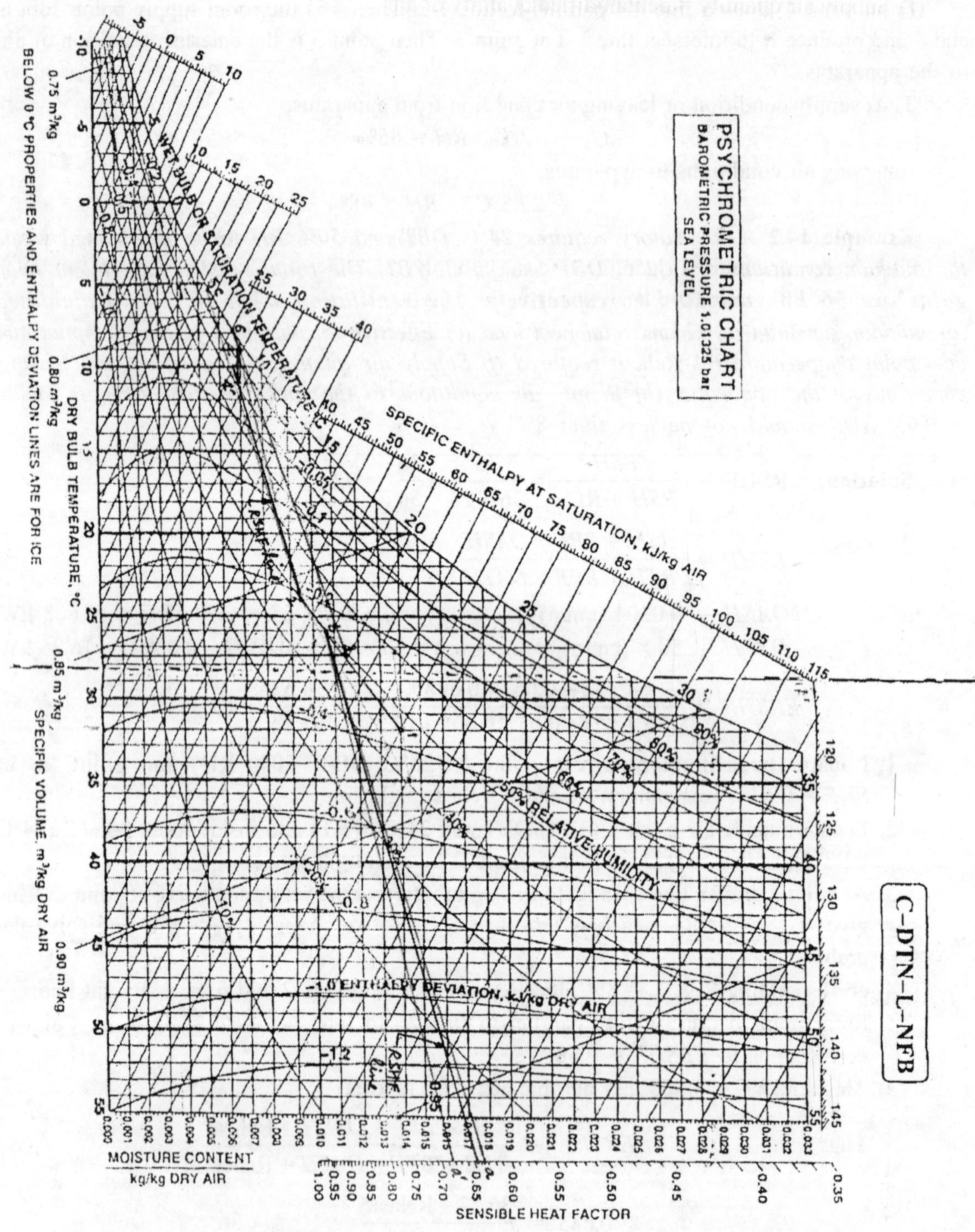
PSYCHROMETRIC CHART
BAROMETRIC PRESSURE 1.01325 bar
SEA LEVEL
C–DTN–L–NFB
WET BULB OR SATURATION TEMPERATURE
SPECIFIC ENTHALPY AT SATURATION, kJ/kg AIR
DRY BULB TEMPERATURE, °C
SPECIFIC VOLUME, m³/kg DRY AIR
BELOW 0 °C PROPERTIES AND ENTHALPY DEVIATION LINES ARE FOR ICE
RELATIVE HUMIDITY
ENTHALPY DEVIATION, kJ/kg DRY AIR
MOISTURE CONTENT
kg/kg DRY AIR
SENSIBLE HEAT FACTOR
Ref. Point for S.H.F. is 25°C, 50% R.H.

Fig. 14.6

(*f*) Supply air quantity = dehumidified quantity of air

$$(\text{cmm})_{da} = \frac{ESH}{0.0204 \times (1 - BPF)\,(t_{d2} - t_{ADP})} = \frac{36 + 8.72}{0.0204 \times (1 - 0.05)\,(24 - 8)}$$

$$= \frac{44.72}{0.31008} = 144.22 \text{ cmm}$$

(*g*) Return air quantity mixing before cooling coil, $(\text{cmm})_r = (\text{cmm})_{da} - (\text{cmm})_r$

$$= 144.22 - 65 = 79.22 \text{ cmm}$$

To determine entry air condition to the cooling coil, either divide line 1-2 in the ratio of 65 : 79.22 or calculate analytically as:

$$t_{d3} \times (\text{cmm})_{da} = t_{d1} \times (\text{cmm})_v + t_{d2} \times (\text{cmm})_r$$

$$t_{d3} = \frac{32 \times 65 + 24 \times 79.22}{144.22} = 27.6°\text{C}$$

Corresponding to $t_{d3} = 27.6°\text{C}$, point 3 on line 1-2 gives, $t_{w3} = 19.8°\text{C}$ for the entry air.

Leaving air conditions from cooling coil:

Join points 3 and 6 by a straight line. Locate point 4 on line 3-6 such that

$$(6 - 4)/(6 - 3) = 0.05$$

By measurement from the psychrometric chart, (see Fig. 14.7)

$$6 - 3 = 67 \text{ mm}$$

$$\therefore \quad 6 - 4 = 0.05 \times 67 = 3.35 \text{ mm}$$

At point 4, $t_{d4} = 9°\text{C}$ and $t_{w4} = 8.7°\text{C}$

(*h*) Supply air condition to the space:

Draw a horizontal line from point 4 to intersect *RSHF* line at 4′, which gives supply air condition to the space.

$$t_{d4'} = 11.2°\text{C},\ t_{w4'} = 9.8°\text{C}$$

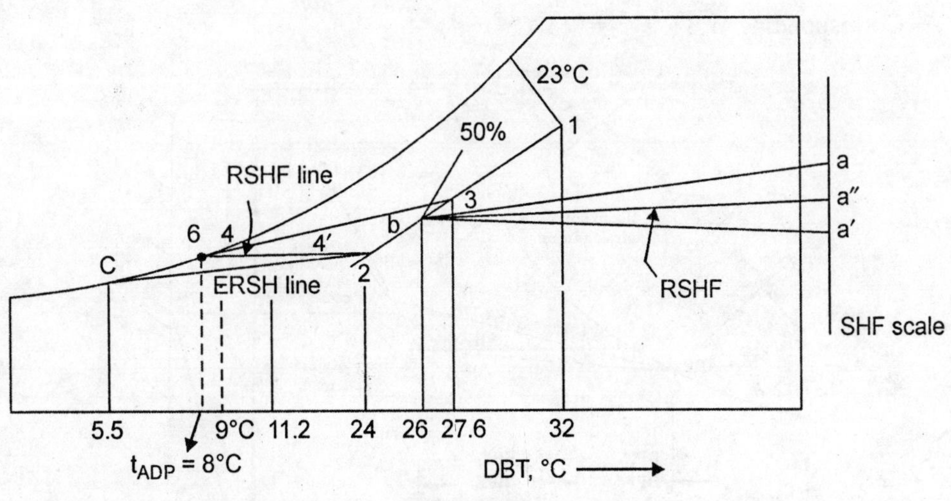

Fig. 14.7

Example 14.3 *The following date refer to a high humidity chamber:*

Summer design = 35°C DBT, 25°C WBT

Inside design = 21°C DBT, 70% RH

RSH = 45 kW, RLH = 3 kW, Ventilation = 110 cmm, BPF = 0.05

Find : (a) Outside air load (b) Grand total heat (c) ESHF (d) ADP (e) Dehumidified air quantity (f) Entering and leaving conditions at the apparatus.

Solution:

(*a*) $OASH = 0.0204 \times (\text{cmm})_v \times (t_{d1} - t_{d2}) = 0.0204 \times 110 \times (35 - 21) = 31.416 \text{ kW}$

(*b*) $OALH = 50\ (\text{cmm})_v\ (w_1 - w_2) = 50 \times 110 \times (0.016 - 0.0108) = 28.6 \text{ kW}$

$OATH = OASH + OALH = 31.416 + 28.6 = 60.016 \text{ kW}$

$GTH = OATH + RSH + RLH = 60.016 + 45 + 3 = 108.016 \text{ kW}$

(*c*) $$ERSHF = \frac{RSH + BPF \times OASH}{RTH + BPF \times OATH} = \frac{45 + 0.05 \times 31.416}{(45+3) + 0.05 \times 60.016} = \frac{46.5708}{51.0008} - 0.913$$

Also $$RSHF = \frac{RSH}{RTH} = \frac{45}{45+3} = 0.9375$$

(*d*) Corresponding t_{ADP} from the psychrometric chart, t_{ADP} = 15°C (see Fig. 14.9)

(*e*) Dehumidified air quantity,

$$(\text{cmm})_{da} = \frac{RSH + BPF \times OASH}{0.0204 \times (1 - BPF)\,(t_{d2} - t_{ADP})} = \frac{46.5708}{0.0204 \times 0.95\,(21 - 15)} = 400.5 \text{ cmm}$$

Return air, $(\text{cmm})_r = (\text{cmm})_{da} - (\text{cmm})_v = 400.5 - 110 = 290.5$ cmm

(*f*) Entering air DBT : $$t_{d3} = \frac{(\text{cmm})_v \times t_{d1} + (\text{cmm})_r \times t_{d2}}{(\text{cmm})_{da}} = \frac{110 \times 35 + 290.5 \times 21}{400.5} = 24.8°\text{C}$$

Corresponding WBT, t_{w3} = 19.7°C

Leaving air DBT from the cooling coil,

$$t_{d4} = t_{ADP} + BPF \times (t_{d3} - t_{ADP}) = 15 + 0.05\ (24.8 - 15) = 15.5°\text{C}$$

Corresponding WBT, t_{w4} = 15°C

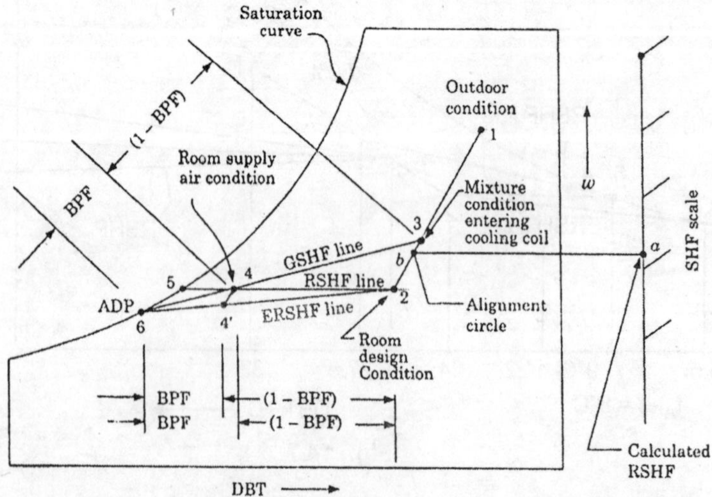

Fig. 14.8 Effective room sensible heat factor (ERSHF)

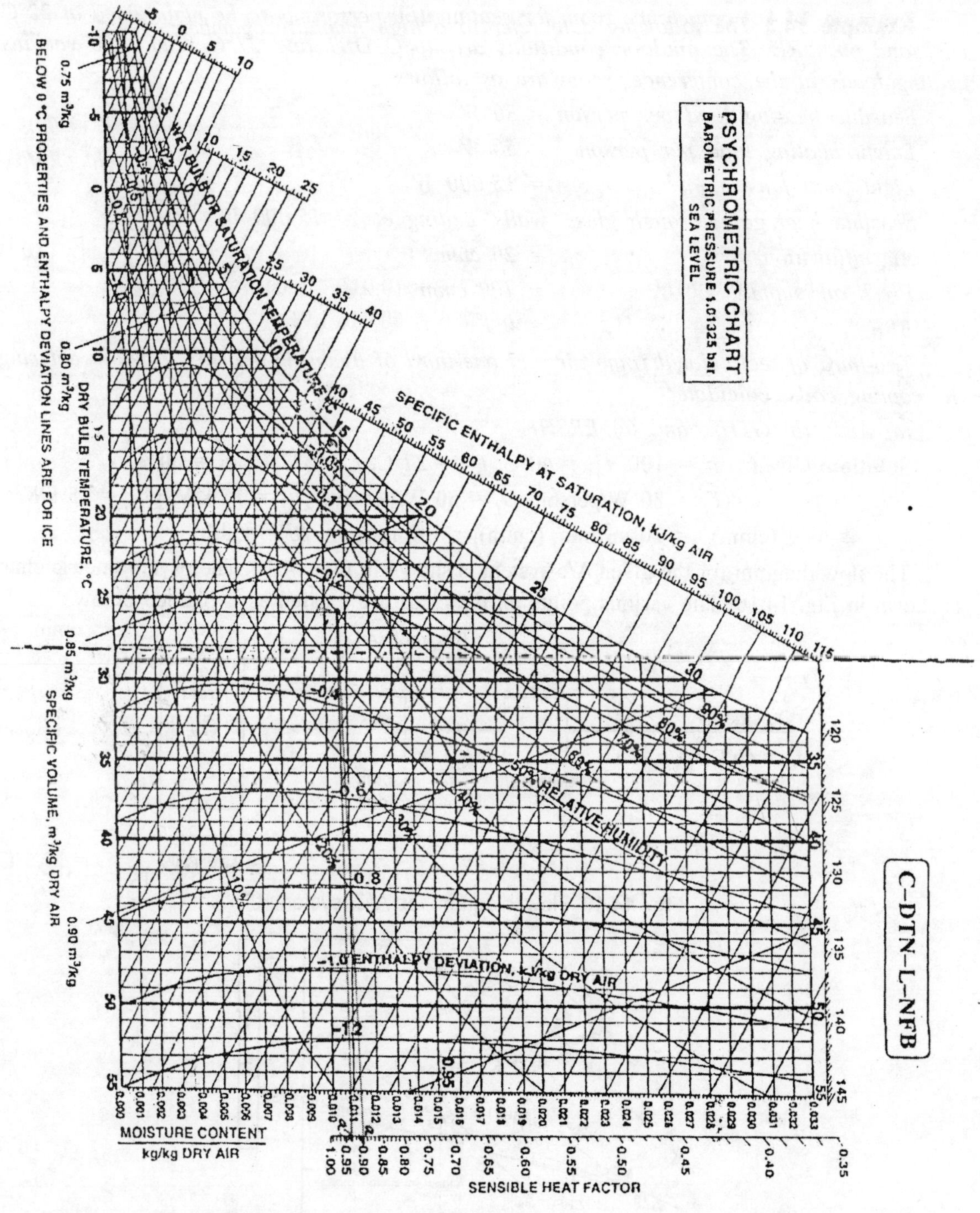
PSYCHROMETRIC CHART
BAROMETRIC PRESSURE 1.01325 bar
SEA LEVEL
WET BULB OR SATURATION TEMPERATURE
SPECIFIC ENTHALPY AT SATURATION, kJ/kg AIR
RELATIVE HUMIDITY
ENTHALPY DEVIATION, kJ/kg DRY AIR
DRY BULB TEMPERATURE, °C
BELOW 0 °C PROPERTIES AND ENTHALPY DEVIATION LINES ARE FOR ICE
SPECIFIC VOLUME, m³/kg DRY AIR
MOISTURE CONTENT
kg/kg DRY AIR
SENSIBLE HEAT FACTOR
C–DTN–L–NFB
Rel. Point for S.H.F. is 25°C, 50% R.H.

Fig. 14.9

Example 14.4 *A conference room for seating 100 persons is to be maintained at 22°C DBT and 60% RH. The outdoor conditions are 40°C DBT and 27°C WBT. The various heating loads in the conference, room are as follows:*

Sensible heating load per person = 80 W
Latent heating load per person = 50 W
Lights and fans = 15,000 W
Sensible heat gain through glass, walls, ceiling etc = 15,000 W
Air infiltration = 20 cmm
Fresh air supply = 100 cmm
BPF = 0.1

Two-third of recirculated room air and one-third of fresh air are mixed before entering the cooling coils, calculate:

(a) ADP, (b) GTHL, and (c) ERSHF.

Solution: Given: $n = 100$, $t_{d1} = 40°C$, $t_{w1} = 27°C$, $t_{d2} = 22°C$, $\phi_2 = 60\%$

$Q_S = 80$ W/person, $Q_L = 50$ W/person, $Q_{SL} = 15$ kW, $Q_{SG} = 15$ kW,

$(\text{cmm})_r = 20\ \text{m}^3/\text{min}$, $(\text{cmm})_f = 100\ \text{m}^3/\text{min}$, $BPF = 0.1$

The flow diagram for the given A/c system is shown in Fig. 14.10. The psychrometric chart is shown in Fig. 14.10. The various points on the chart are located as explained below.

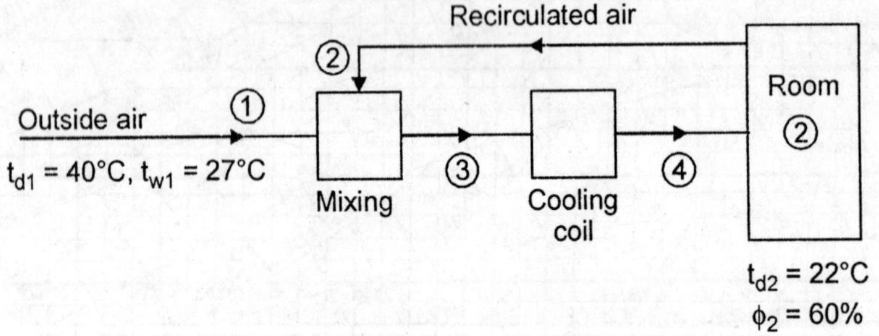

Fig. 14.10 Flow diagram for A/C system.

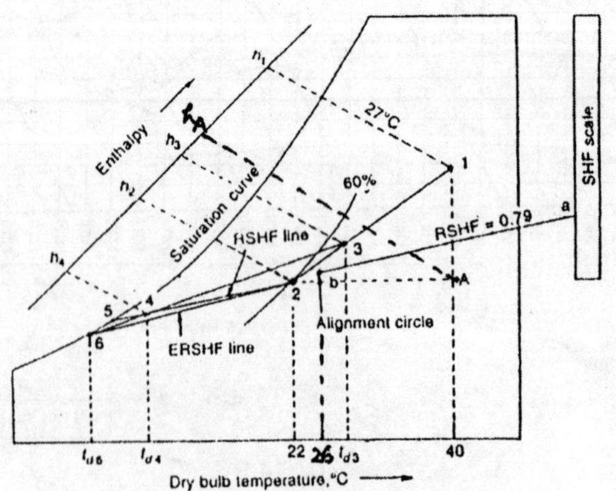

Fig. 14.10 Psychrometric chart

1. Mark point 1 as the outside condition of air at t_{d1} = 40°C, t_{w1} = 27°C and point 2 as the inside room condition at t_{d2} = 22°C and ϕ_2 = 60%. Join points 1 and 2 by a straight line. Also locate point A by drawing vertical line at 1 and horizontal line at 2.

   From the psychrometric chart, we get

$$v_{s1} = 0.911 \text{ m}^3\text{/kg of d.a., } h_1 = 81 \text{ kJ/kg of d.a., } h_2 = 47.5 \text{ kJ/kg of d.a.,}$$

$$h_A = 66 \text{ kJ/kg of d.a.}$$

   Mass of infiltrated air at point 1,

$$m_1 = \frac{(\text{cmm})_1}{v_{s1}} = \frac{20}{0.911} = 21.95 \text{ kg/min}$$

   Sensible heat gain due to air infiltration,

$$Q_{Si} = m_1(h_A - h_2) = 21.95\ (66 - 47.5) = 406 \text{ kJ/min}$$

$$= \frac{406}{60} = 6.77 \text{ kW}$$

   Latent het gain due to air infiltration,

$$Q_{Li} = m_1(h_1 - h_A) = 21.95\ (85 - 66) = 417 \text{ kJ/min}$$

$$= \frac{417}{60} = 6.95 \text{ kW}$$

   Total sensible heat gain from persons,

$$Q_{Sp} = Q_s \times n = 80 \times 100 = 8000 \text{ W or } 8 \text{ kW}$$

   Total latent heat gain from persons,

$$Q_{LP} = Q_L \times n = 50 \times 100 = 5000 \text{ W or } 5 \text{ kW}$$

   Total sensible heat gain in room,

$$RSH = Q_{Si} + Q_{Sp} + Q_{SL} + Q_{SG}$$

$$= 6.77 + 8 + 15 + 15 = 44.77 \text{ kW}$$

   Total latent heat gain in room,

$$RLH = Q_{Li} + Q_{Lp} = 6.95 + 5 = 11.95 \text{ kW}$$

   Room sensible heat factor,

$$RSH = \frac{RSH}{RSH + RLH} = \frac{44.77}{44.77 + 11.95} = 0.79$$

2. Locate the alignment circle point b at t_d = 26° DBT and RH = 50% and point 'a' on *SHF* scale at 0.79. Join points 'a' and b by a straight line.
3. From point 2, draw a line 2-5 parallel to line ab to intersect saturation curve at point 5. This line 2-5 is called the RSHF line.
4. Mark the mixing point 3 on the line 1-2, such that

$$\text{Length 2-3} = \frac{1}{3} \times \text{Length 1-2}$$

   At point 3, t_{d3} = 28°C

(*a*) ADP:

Let t_{d6} = ADP

Now $$BPF = \frac{t_{d4} - t_{d6}}{t_{d3} - t_{d6}}$$

$$0.1 = \frac{t_{d4} - t_{d6}}{28 - t_{d6}}$$

By hit and trial method, we find that

$$t_{d4} = 13.15°C \text{ and } t_{d6} = 11.5°C$$

(*b*) Grand total heat load:

$$OATH = 0.0204 \times (\text{cmm})_1 \times (h_1 - hd_2) = 0.0204 \times 20 \times (85 - 47.5) = 15.3 \text{ kW}$$

$$GTH = RSH + RLH + OATH = 44.77 + 11.95 + 15.3 = 72.02 \text{ kW}$$

(*c*) ERSHF:

Joint points 2 and 6 by a straight line and produce upto the SHF scale.

ERSHF = 0.77.

Example 14.5 *The following data relates to the office air conditioning plant having maximum seating capacity of 25 persons:*

Outside design conditions	= *34°C DBT, 28°C WBT*
Inside design conditions	= *24°C DBT, 50% RH*
Solar heat gain	= *9000 W*
Latent heat gain per person	= *100 W*
Sensible heat gain per person	= *95 W*
Lighting Load	= *2300 W*
Sensible heat load from other sources	= *11600 W*
Infiltration load	= *15 m³/min*

Assuring 40% fresh air and 60% of recirculated air passing through the evaporator coil and the by-pass factor of 0.15, find the dew point temperature of the coil and the capacity of the plant.

Solution: Given: $n = 25$, $Q_{SO} = 900$ W, $Q_L = 100$ W/person, $Q_S = 95$ W/person, $Q_{SL} = 2300$ W, $Q_{SS} = 11600$ W, $v_1 = 15$ m³/min, $BPF = 0.15$

The flow diagram for the A/C system is shown in Fig. 14.11 and the psychrometric chart in Fig. 14.12. Follow the following procedure to locate various points on the psychrometric chart.

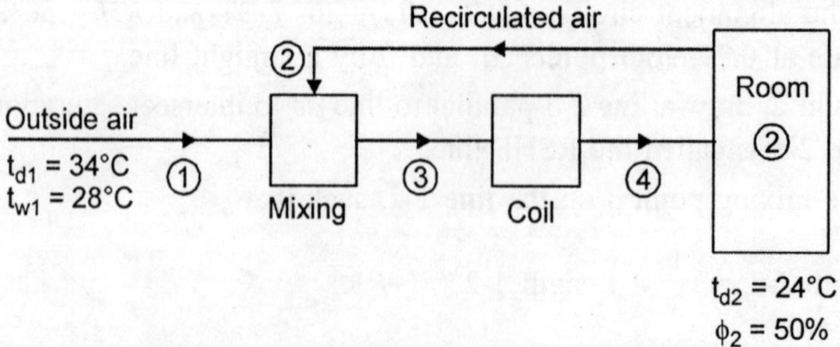

Fig. 14.11 Flow diagram for A/C system.

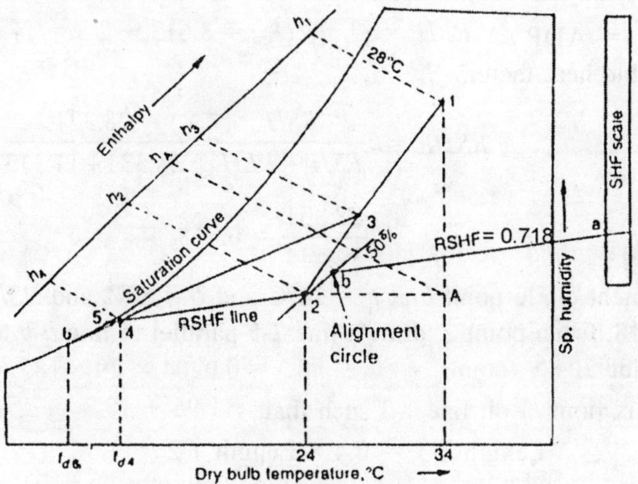

Fig. 14.11 Psychrometric chart.

1. Locate point 1 corresponding to outside design conditions, $t_{d1} = 34°C$ and $t_{w1} = 28°C$ and and point 2 corresponding to inside room conditions $t_{d2} = 24°C$ and $\phi_2 = 50\%$. Join points 1 and 2 by a straight line. Locate point A by drawing vertical and horizontal lines at points 1 and 2 respectively.

   Specific volume at point 1,

$$v_{s1} = 0.9 \text{ m}^3\text{/kg of d.a.}$$
$$h_1 = 90 \text{ kJ/kg of d.a.}$$
$$h_2 = 48 \text{ kJ/kg of d.a.}$$
$$h_A = 59 \text{ kJ/kg of d.a.}$$

   Mass of infiltrated air at point 1,

$$m_1 = \frac{V_1}{v_{s1}} = \frac{15}{0.9} = 16.67 \text{ kg/min}$$

   Sensible heat gain due to infiltration air,

$$Q_{Si} = m_1(h_A - h_2) = 16.67\,(59 - 48) = 183.37 \text{ kJ/min or } 3.056 \text{ kW}$$

   Latent heat gain due to infiltration air,

$$Q_{Li} = m_1\,(h_1 - h_1) = 16.67\,(90 - 59) = 516.77 \text{ kJ/min or } 8.613 \text{ kW}$$

   Total sensible heat gain from persons,

$$Q_{Sp} = Q_S \times n = 95 \times 25 = 2375 \text{ W or } 2.375 \text{ kW}$$

   Total latent heat gain from persons,

$$Q_{LP} = Q_L \times n = 100 \times 25 = 2500 \text{ W or } 2.5 \text{ kW}$$

   Total Sensible heat gain in the room,

$$RSH = Q_{SO} + Q_{Si} + Q_{Sp} + Q_{SL} + Q_{SS}$$
$$= 9 + 3.056 + 2.375 + 2.3 + 11.6 = 28.331 \text{ kW}$$

Total latent heat gain in the room,

$$RLH = Q_{Li} + Q_{LP} = 8.613 + 2.5 = 11.13 \text{ kW}$$

Room sensible heat factor,

$$RSHF = \frac{RSH}{RSH + RLH} = \frac{28.331}{28.331 + 11.113}$$

$$= \frac{28.331}{39.444} = 0.718$$

2. Mark alignment circle point *b* at $t_d = 26°C$ and $\phi = 50\%$ and *RSHF* point '*a*' on *SHF* scale at 0.718. From point 2, draw a line 2-5 parallel to line *a-b* to intersect saturation curve at point 5.
3. Mark the mix point 3 on line 1-2 such that,

$$\text{Length 2-3} = 0.4 \times \text{Length 1-2}$$

DPT of the coil

Let

$$t_{d6} = \text{DPT of coil}$$

$$BPF = \frac{t_{d4} - t_{d6}}{t_{d3} - t_{d6}} = \frac{t_{d4} - t_{d6}}{28 - t_{d6}}$$

By trial and error method, we find that

$$t_{d4} = 10.5°C \text{ and } t_{d6} = 7.4°C$$

Capacity of the plant:

$$h_4 = 29 \text{ kJ/kg d.a.,} \quad h_3 = 64.8 \text{ kJ/kg d.a.}$$

Mass of air entering the room,

$$m_a = \frac{RSH + RLH}{h_2 - h_4} = \frac{\text{Total room heat}}{\text{Total heat removed}} = \frac{28.331 + 11.113}{48 - 29}$$

$$= \frac{39.444}{19} = 2.076 \text{ kg/s or } 124.56 \text{ kg/min}$$

$$\text{capacity of the plant} = m_a(h_3 - h_4) = 124.56\,(64.8 - 29) = 4459.248 \text{ kJ/min}$$

$$= \frac{4459.248}{211} = 21.134 \text{ TR}$$

Example 14.6 *An air-conditioning system is designed for a restaurant with the following data:*

Total heat flow through the walls:

Roof and floor = 6 kW, Solar heat gain through glass = 2 kW, Equipment sensible heat gain = 3 kW, Equipment latent heat gain = 0.6 kW.

Total infiltration air = 400 m³/h, outdoor condition = 35°C DBT, 26°C WBT,

Inside designed conditions = 27°C DBT, 55% RH, Minimum temperature of air supplied to room = 17°C DBT, Total amount of fresh air supplied = 1600 m³/h, Seating chairs for dining = 50, Employees serving the meals = 5, Sensible heat gain per person = 60 W, Latent heat gain per P sitting person = 45 W, Latent heat gain per employee = 75 W, Sensible heat added from meals = 0.15 kW, Latent heat added from meals = 0.4 kW, Motor power connected to fan = 7.5 kW.

The fan is situated before the conditioner.

Calculate the following:

(a) Amount of air delivered to the room in m³/h,

(b) Percentage of recirculated air,

(c) Refrigeration load on the will in tons of refrigeration, and

(d) Dew point temperature of cooling coil and by-pass factors.

Solution: Given: $Q_{SW} = 6$ kW, $Q_{SG} = 2$ kW, $Q_{SE} = 3$ kW, $Q_{LE} = 0.6$ kW, $V_1 = 400$ m³/h,

$t_{d1} = 35°C$, $t_{w1} = 26°C$, $t_{d2} = 27°C$, $\phi_2 = 55\%$, $t_{d4} = 17°C$,

$V_F = 1600$ m³/h,

$n_{DC} = 50$, $n_E = 5$, $Q_{Sp} = 60$ W/person, $Q_{Lp} = 45$ W/person,

$Q_{LE} = 75$ W/employee, $Q_{Sm} = 0.16$ kW,

$Q_{LM} = 0.4$ kW, $Q_M = 7.5$

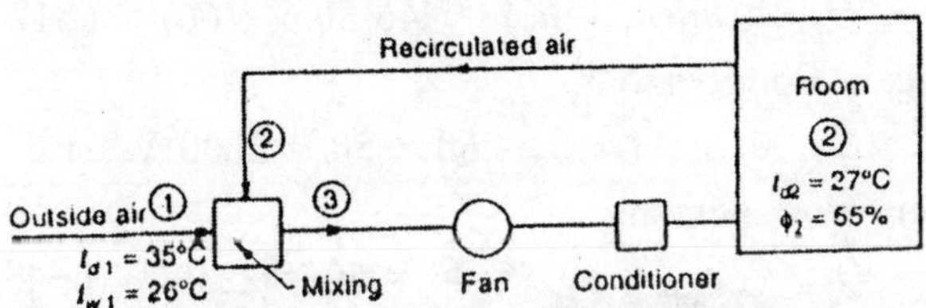

Fig. 14.12 Flow diagram for A/C system.

The flow diagram for the air-conditioning system is shown in Fig. 14.12. and the psychrometric chart is shown in Fig. 14.13.

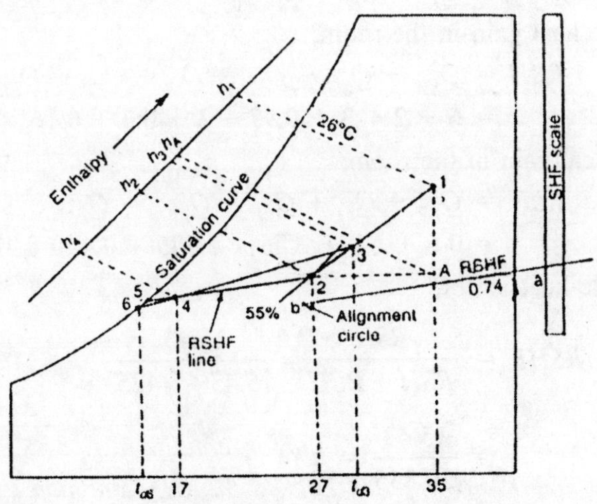

Fig. 14.13 Psychrometric chart

1. Mark the outside conditions of air at point 1 as t_{d1} = 35°C, t_{w1} = 26°C, as shown in Fig. 14.13. Then mark the inside designed conditions at point 2 as t_{d2} = 27°C and ϕ_2 = 55%. Join points 1 and 2 by a straight line.
2. Locate point *A* by drawing vertical and horizontal lines at points 1 and 2 respectively.
3. From the psychrometric chart, we have

$$v_{s1} = 0.897 \text{ m}^3\text{/kg of d.a.}$$
$$h_1 = 80.6 \text{ kJ/kg of d.a.}$$
$$h_2 = 58.2 \text{ kJ/kg of d.a.}$$
$$h_A = 66 \text{ kJ/kg of d.a.}$$

Mass of infiltration air,

$$m_1 = \frac{V_1}{v_{s1}} = \frac{400}{0.897} = 446 \text{ kg/h}$$

Sensible heat gain due to infiltration air,

$$Q_{SI} = m_1(h_A - h_2) = 446\ (66 - 58.2) = 3480 \text{ kJ/h or } 0.97 \text{ kW}$$

Latent heat gain due to infiltration air,

$$Q_{LI} = m_1(h_1 - h_A) = 446(80.6 - 66) = 6512 \text{ kJ/hr or } 1.8 \text{ kW}$$

Sensible heat gain from persons,

$$Q_{Sp} = q_{sp} \times n_{DC} = 60 \times 50 = 3000 \text{ W or } 3 \text{ kW}$$

Latent heat gain from persons,

$$Q_{LP} = q_{LP} \times n_{DC} = 45 \times 50 = 2250 \text{ W or } 2.25 \text{ kW}$$

Sensible heat from employees serving the meals,

$$Q_{SEM} = q_{sp} \times n_E = 60 \times 5 = 300 \text{ W or } 0.3 \text{ kW} \qquad (\because q_{SE} = q_{Sp})$$

Latent heat gain from employees serving the meals,

$$Q_{LEM} = Q_{LE} \times n_E = 75 \times 5 = 375 \text{ W or } 0.375 \text{ kW}$$

Total sensible heat gain in the room,

$$RSH = Q_{SW} + Q_{SG} + Q_{SE} + Q_{SI} + Q_{SP} + Q_{SEM} + Q_{SM}$$
$$= 6 + 2 + 3 + 0.97 + 3 + 0.3 + 0.16 = 15.43 \text{ kW}$$

Total latent heat gain in the room,

$$RLH = Q_{LE} + Q_{LI} + Q_{LP} + Q_{LEM} + Q_{LM}$$
$$= 0.6 + 1.8 + 2.25 + 0.375 + 0.4 = 5.425 \text{ kW}$$

Room sensible heat factor,

$$RSHF = \frac{RSH}{RSH + RLH} = \frac{15.43}{15.43 + 5.425}$$
$$= \frac{15.43}{20.855} = 0.74$$

4. Locate alignment circle point *b* at t_d = 26°C and ϕ = 50%. Also mark point '*a*' on SHF scale at RSHF = 0.74. Join points '*a*' and *b* by a straight line.

5. From point 2, draw a line 2-5 parallel to line *ab* to intersect saturation curve at point 5. Line 2-5 is the RSHF line.
6. Draw a vertical line at t_d = 17°C (minimum temperature of air supplied to the room) to intersect line 2-5 at point 4.

$$v_{s4} = 0.836 \text{ m}^3\text{/kg of d.a.}$$
$$h_4 = 45 \text{ kJ/kg of d.a.}$$

(*a*) Amount of air delivered to the room,

$$m_a = \frac{\text{Total room heat}}{\text{Total heat removed}} = \frac{RSH + RLH}{h_2 - h_4} = \frac{20.855}{58.2 - 45}$$
$$= 1.58 \text{ kg/s or } 5687.7 \text{ m}^3\text{/h}$$

Amount of air delivered o the room,

$$V_a = m_a \times v_{s4} = 5687.7 \times 0.836 = 4755 \text{ m}^3\text{/h}$$

(*b*) Mass of fresh air supplied,

$$m_F = \frac{V_F}{v_{s1}} = \frac{1600}{0.897} = 1784 \text{ kg/h}$$

Mass of recirculated air,

$$m_R = m_a - m_F = 5687.7 - 1784 = 39037 \text{ kg/h}$$

Percentage of recirculated air

$$= \frac{3903.7}{5687.7} = 0.6863 \text{ or } 68.63\%$$

% of fresh air supplied at 1 = 1 – 68.63 = 31.37%

(*c*) The mixing of recirculated air and fresh air before entering to the cooling coil is marked at point 3 on the psychrometric chart, such that

$$\text{Length 2-3} = \text{Length 1-2} \times 0.03137$$

Then

$$h_3 = 64.6 \text{ kJ/kg of d.a.}$$
$$\text{Refrigeration load on coil} = m_a(h_3 - h_4) + Q_M$$
$$= 1.58\,(64.6 - 45) + 7.5 = 38.468 \text{ kW}$$
$$= \frac{38.468}{3.5167} = 10.939 \text{ TR}$$

(*d*) Join point 3 with point 4 and produce the line to intersect the saturation curve at point 6. Then

$$t_{dp} = t_{d6} = 14.6°\text{C}$$
$$BPF = \frac{t_{d4} - t_{d6}}{t_{d3} - t_{d6}} = \frac{17 - 14.6}{29.5 - 14.6} = 0.161$$

Example 14.7 *A hall is to be maintained at 24°C DBT and 60% RH under the following conditions:*

Outdoor conditions = 38°C DBT, 28°C WBT, Room sensible heat load = 46 kW, Room latent heat load = 12 kW, Total infiltration air = 1200 m³/h, ADP temperature = 10°C, Quantity of recirculated air from the hall = 60%.

If the recirculated air is mixed with the conditioned air after the cooling coil, find the following:

(a) The condition of air leaving the conditioner coil and before mixing with the recirculated air, (b) The condition of air before entering the hall, (c) The mass of air entering the cooler, (d) The mass of total air passing through the hall, (e) The BPF of cooling coil, and (f) The refrigeration load on the cooling coil in tons of refrigeration.

Solution: Given: t_{d2} = 24°C, ϕ_2 = 60%, t_{d1} = 38°C, t_{w1} = 28°C, Q_S = 46 kW, Q_L = 12 kW,

V_1 = 1200 m³/h, t_{ADP} = 10°C, V_R = 60%

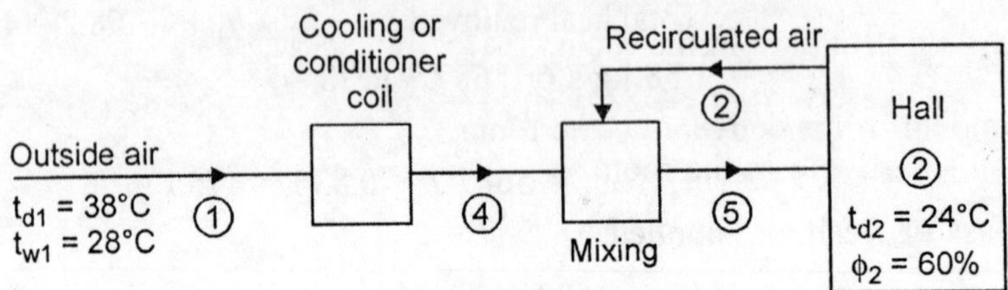

Fig. 14.14 Flow Diagram.

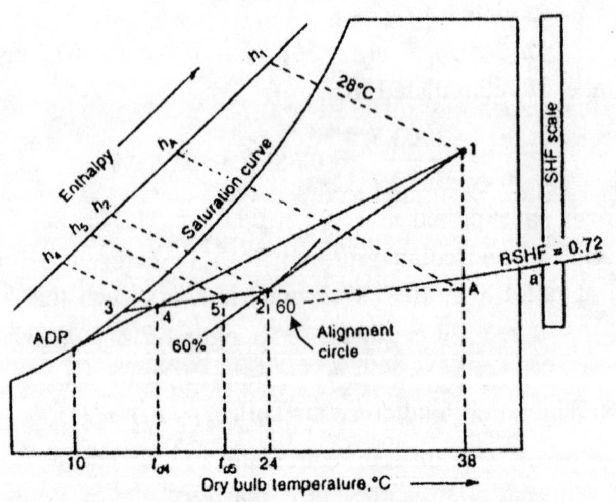

Fig. 14.15 Psychrometric chart.

The flow diagram is shown in Fig. 14.14 and the psychrometric chart in Fig. 14.15.

1. Mark the outdoor condition as point 1 at t_{d1} = 38°C and t_{w1} = 28°C, and the room inside design condition as point 2 at t_{d2} = 24°C and ϕ_2 = 60%. Join points 1 and 2 by a straight line.
2. Locate point A by drawing vertical line at point 1 and horizontal line at point 2.
3. From psychrometric chart, we find that

$$v_{s1} = 0.907 \text{ m}^3\text{/kg of d.a.}, \; h_1 = 89.6 \text{ kJ/kg of d.a.}$$

$$h_2 = 53 \text{ kJ/kg of d.a., and } h_A = 68 \text{ kJ/kg of d.a.}$$

Mass of infiltrated air,

$$m_1 = \frac{V_1}{v_{s1}} = \frac{1200}{0.907} = 1323 \text{ kg/h}$$

Sensible heat gain due to infiltrated air,

$$Q_{SI} = m_1(h_A - h_2) = 1323\ (68 - 53) = 19845 \text{ kJ/h or } 5.5 \text{ kW}$$

Latent heat gain due to infiltrated air,

$$Q_{LI} = m_1(h_1 - h_A) = 1323(89.6 - 68) = 28577 \text{ kJ/h or } 7.94 \text{ kW}$$

Total sensible heat gain in the room,

$$RSH = Q_S + Q_{SI} = 46 + 5.5 = 51.5 \text{ kW}$$

Total latent heat gain in the room,

$$RLH = Q_L + Q_{LI} = 12 + 7.94 = 19.94 \text{ kW}$$

Room sensible heat factor,

$$RSHF = \frac{RSH}{RSH + RLH} = \frac{51.5}{51.5 + 19.94} = \frac{51.5}{71.44} = 0.72$$

4. Locate the alignment circle point *b* at $t_d = 26°\text{C}$ and $\phi = 50\%$. Also locate point '*a*' on the SHF scale corresponding to RSHF = 0.72. Join points '*a*' and *b* by a straight line. Line *ab* is known as RSHF line.
5. From point 2 draw a line parallel to line *ab* to intersect the saturation curve at point 3.
6. Draw a line joining the points 1 and $t_{ADP} = 10°\text{C}$, intersecting the line 2-3 at point 4. This point 4 represents the condition of air leaving the conditioner coil and before mixing with the recircultated air.

(*a*) $t_{d4} = 16°\text{C}$ and $\phi_4 = 90\%$

(*b*) Condition of air before entering the hall.

Since 60% of air at point 2 is recirculated and mixed with 40% of air leaving the conditioner coil at point 4, therefore, mark the mixing point 5, such that

Length 2-5 = 0.4 × Length 2-4

Point 5 represents the condition of air before entering the hall.

$$t_{d5} = 20.8°\text{C} \text{ and } \phi_5 = 70\%$$

(*c*) Mass of air entering the cooler,

$$m_c = \frac{\text{Total room heat}}{\text{Total heat removed}} = \frac{RSH + RLH}{h_2 - h_4}$$

$$= \frac{71.44}{53 - 42} = 6.5 \text{ kg/s or } 390 \text{ kg/min}$$

(*d*) Mass of total air passing through the hall.

Let m_H = mass of total air passing through the hall

$$390 + 0.6\,m_H = m_H$$

$$m_H = \frac{390}{0.4} = 975 \text{ kg/min}$$

(*e*) BPF of cooling coil.

$$BPF = \frac{t_{d4} - t_{ADP}}{t_{d1} - t_{ADP}} = \frac{16-10}{38-10} = \frac{6}{28} = 0.214$$

(*f*) Refrigeration load on cooling coil

$$= m_c(h_1 - h_4) = 390(29.6 - 42)$$

$$= 18564 \text{ kJ/min or } \frac{18564}{211} = 87.98 \text{ TR}$$

Example 14.8 *An air-conditioning system is to be designed for a restaurant with the following data:*

Outside design conditions = 40°C DBT, 28°C WBT, Inside design conditions = 25°C DBT, 50% RH, Solar heat gain through walls, roof and floor = 6 kW, Solar heat gain through glass = 5.5 kW. Occupants = 25, Sensible heat gain per person = 60 W, Latent heat gain per person = 60 W, Inter lighting load = 15 lamps of 100 W, 10 CFL of 80 W, Sensible heat gain from other sources = 12 kW, Infiltrated air = 15 m³/min.

If 25% fresh air and 75% recirculated air is mixed and passed through the conditioner coil, find:

(a) Amount of total air required in m³/h, (b) DPT of coil, (c) Condition of supply air to the room, and (d) capacity of the conditioning plant.

Assume $BPF = 0.2$

Solution: Given:

t_{d1} = 40°C, t_{w1} = 28°C, t_{d2} = 25°C, ϕ_2 = 50%, Q_{SW} = 6 kW, Q_{SG} = 5.5 kW,
n = 25, q_{Sp} = 60 W/person, q_{Lp} = 60 W/person, Q_{SL} = 15 × 100 + 10 × 80 = 2300 W

or 2.3 kW, Q_{SS} = 12 kW, V_1 = 15 m³/min, BPF = 0.2

The flow diagram is shown in Fig. 14.16 and psychrometric chart in Fig. 14.17.

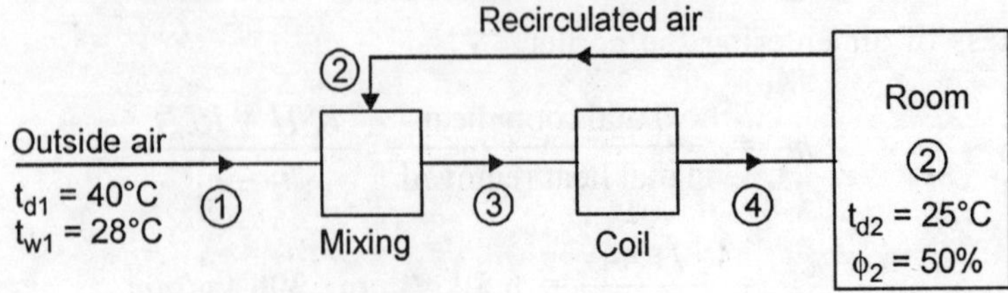

Fig. 14.16 Flow diagram for A/C system.

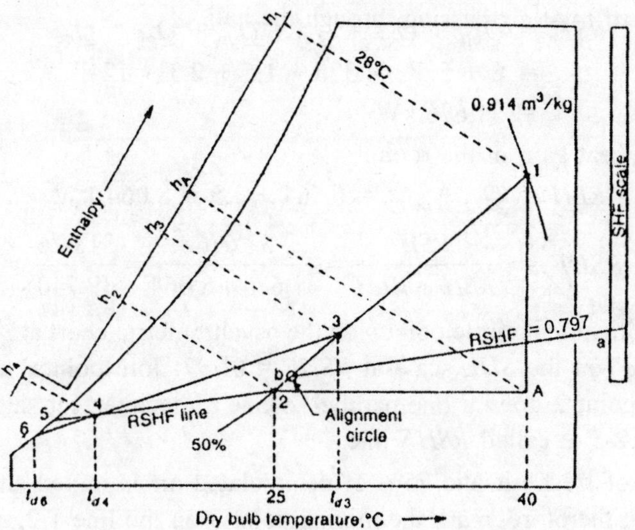

Fig. 14.17 Psychrometric chart.

1. Mark point 1 on the Fig. 14.16 representing the outside conditions of air at t_{d1} = 40°C, t_{w1} = 28°C and point 2 representing the inside design conditions of air at t_{d2} = 25°C and ϕ_2 = 50%.
2. Locate point A by drawing vertical line at point 1 and horizontal line at point 2.

Then v_{s1} = 0.914 m³/kg of d.a.

h_1 = 90 kJ/kg of d.a.

h_2 = 50 kJ/kg of d.a.

h_A = 66 kJ/kg of d.a.

Mass of infiltrated air at point 1,

$$m_1 = \frac{V_1}{v_{s1}} = \frac{15}{0.914} = 16.41 \text{ kg/min}$$

Sensible heat gain due to infiltrated air,

$$Q_{SI} = m_1(h_A - h_2) = 16.41\ (66 - 50) = 262.56 \text{ kJ/min or } 4.376 \text{ kW}$$

Latent heat gain due to infiltrated air,

$$Q_{LI} = m_1(h_1 - h_A) = 16.41\ (90 - 66) = 393.84 \text{ kJ/min or } 6.564 \text{ kW}$$

Total sensible heat gain from occupants,

$$Q_{Sp} = q_{Sp \times n} \times n = 60 \times 25 = 1500 \text{ W or } 1.5 \text{ kW}$$

Total latent heat gain from occupants,

$$Q_{Lp} = q_{Lp} \times n = 60 \times 25 = 1500 \text{ W or } 1.5 \text{ kW}$$

Sensible heat gain due to internal lighting, Q_{SL} = 2.3 kW

Sensible heat gain from other sources, Q_{SS} = 12 kW

Total sensible heat gain in the room,

$$RSH = Q_{SW} + Q_{SG} + Q_{SI} + Q_{SP} + Q_{SL} + Q_{SS}$$
$$= 6 + 5.5 + 4.376 + 1.5 + 2.3 + 12$$
$$= 31.676 \text{ kW}$$

Total latent heat gain in the room,

$$RLH = Q_{LI} + Q_{LP} = 6.564 + 1.5 = 8.064 \text{ kW}$$

$$RSHF = \frac{RSH}{RSH + RLH} = \frac{31.676}{31.676 + 8.064} = \frac{31.676}{39.740} = 0.797$$

3. Locate the alignment circle point b on the psychrometric chart at $t_d = 25°\text{C}$ and $\phi = 50\%$ and point 'a' on the *SHF* scale at *RSHF* = 0.797. Join points 'a' and b by a straight line. From point 2 draw a line parallel to line ab to intersect saturation curve at point 5. The line 2-5 is called *RSHF* line.
4. Since 25% of fresh air and 75% of decirculated air is mixed and passed through the cooling coil, therefore, mark the mixing point 3 on the line 1-2, such that,

$$\text{Length 2-3} = 0.25 \times \text{Length 1-2.}$$

Corresponding to point 3, $t_{d3} = 28.8°\text{C}$ and $h_3 = 60$ kJ/kg of d.a.

5. Now $$BPF = \frac{t_{d4} - t_{d6}}{t_{d3} - t_{d6}} = \frac{t_{d4} - t_{d6}}{28.8 - t_{d6}}$$

By trial and error method, we find that

$$t_{d4} = 14.4°\text{C and } t_{d6} = 10.8°\text{C}$$

Point 4 represents the condition of air leaving the coil and entering the room.

The point 6 represents the dew point temperature of the coil.

$$h_4 = 36.5 \text{ kJ/kg of d.a.}$$
$$v_{s4} = 0.826 \text{ m}^3\text{/kg of d.a.}$$

(*a*) Mass of total air required,

$$m_a = \frac{\text{Total room heat}}{\text{Total heat removed}} = \frac{RSH + RLH}{h_2 - h_4} = \frac{39.74}{(50 - 36.5)}$$

$$= \frac{39.74}{13.5} = 2.943 \text{ kg/s or } 10597 \text{ kg/h}$$

$$v_a = m_a \times v_{s4} = 10597 \times 0.826 = 8753.4 \text{ m}^3\text{/h}$$

(*b*) Dew Point temperature, $t_{d6} = 10.8°\text{C}$

(*c*) Condition of supply air to room, $t_{d4} = 14.40°\text{C}$ and

(*d*) Capacity of the conditioning plant

$$= m_a(h_3 - h_4) = 2.943\ (60 - 36.5) = 69.16 \text{ kW}$$

$$= \frac{69.16}{3.5167} = 19.67 \text{ TR}$$

Example 14.9 *A cold storage is to be designed to store 500 tons of vegetables when the following data is available:*

Outdoor conditions = 35°C DBT and 28° WBT, Required indoor conditions = 20°C DBT and 60% RH, water content of vegetables = 76%, Loss of water content = 0.01% per hour, Persons working in the cold storage = 25, Fresh air supplied from outside = 4500 m³/h, Sensible heat gain through glass = 6 kW, sensible heat gain through walls and ceilings = 12 kW, Heat from equipment and respiration heat of vegetables = 3.5 kW, Infiltrated air = 200 m³/h. The air-conditioning is achieved by first cooling and dehumidifying and then heating. Find: (a) Amount of recirculated air, if the recirculated air is mixed with fresh air before entering the cooling coil, (b) Capacity of the cooling coil in tons of refrigeration and its BPF, if DPT of coil is 5°C, and (c) Capacity of the heating coil in kW.

The temperature of air entering the room is not to exceed 15°C

Solution: Given:

$$m_v = 500 \text{ tons or } 500 \times 10^3 \text{ kg}, t_{d1} = 35°\text{C}, t_{w1} = 28°\text{C}, t_{d2} = 20°\text{C},$$
$$\phi_2 = 60\%, W_v = 76\%, W_{loss} = 0.01\% \text{ per hour}, n = 25, V_F = 4500 \text{ m}^3/\text{h},$$
$$Q_{SG} = 6 \text{ kW}, Q_{SW} = 12 \text{ kW}, Q_{SR} = 3.5 \text{ kW}, V_I = 200 \text{ m}^3/\text{h}, t_{DPT} = 5°\text{C}$$

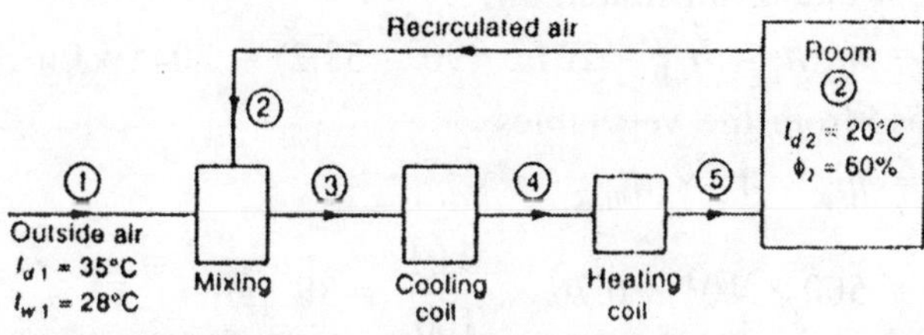

Fig. 14.18 Flow diagram

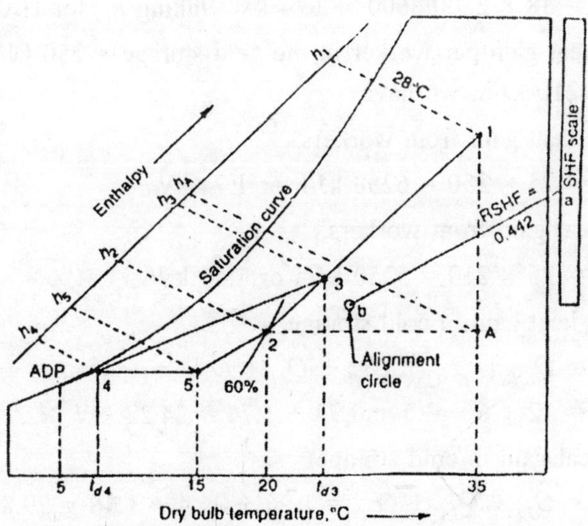

Fig. 14.19 Psychrometric chart.

The flow diagram is shown in Fig. 14.18 and the psychrometric chart in Fig. 14.19.

1. Mark outdoor conditions of air at t_{d1} = 35°C and t_{w1} = 28°C on the psychrometric chart as point 1 in Fig. 14.19. Now mark the required indoor conditions of air at t_{d2} = 20°C and as point 2. Also locate point A by drawing vertical line at point 1 horizontal line at point 2.

From psychrometric chart, we have

Specific volume, v_{s1} = 0.903 m³/kg of d.a.

Emthalpies, h_1 = 90 kJ/kg of d.a.

h_2 = 42.3 kJ/kg of d.a.

h_A = 58.2 kJ/kg of d.a.

Mass of infiltrated air supplied at point 1,

$$m_1 = \frac{V_1}{v_{s1}} = \frac{200}{0.903} = 221.5 \text{ kg/h}$$

Sensible heat gain due to infiltrated air,

$$Q_{SI} = m_1(h_A - h_2) = 221.5\ (58.2 - 42.3) = 3522 \text{ kJ/h or } 0.98 \text{ kW}$$

Latent heat gain due to infiltrated air,

$$Q_{LI} = m_1(h_1 - h_A) = 221.5\ (90 - 58.2) = 7044 \text{ kJ/h or } 1.96 \text{ kW}$$

Total water loss from the vegetables

$$= m_v \times W_v \times W_{\text{loss}}$$

$$= 500 \times 10^3 \times 0.76 \times \frac{0.01}{100} = 38 \text{ kg/h}$$

Latent heat gain due to evaporation of water from vegetables

$$Q_{LV} = 38 \times 2500/3600 = 26.4 \text{ kW (taking } h_{fg} \text{ for } H_2O = 2500 \text{ kJ/kg)}$$

Let sensible heat gain per worker in the cold storage = 250 kJ/h, and latent heat gain per worker = 210 kJ/h, we have

Total sensible heat gain from workers,

$$Q_{SP} = 25 \times 250 = 6250 \text{ kJ/h or } 1.74 \text{ kW}$$

Total latent heat gain from workers,

$$Q_{LP} = 25 \times 210 = 5250 \text{ kJ/h or } 1.46 \text{ kW}$$

Total sensible heat gain in cold storage,

$$RSH = Q_{SN} + Q_{SG} + Q_{SR} + Q_{SI} + Q_{SP}$$

$$= 12 + 6 + 3.5 + 0.98 + 1.74 = 24.22 \text{ kW}$$

Total latent heat gain in cold storage,

$$RLH = Q_{LI} + Q_{LV} + Q_{LP} = 1.96 + 26.4 + 1.46 = 29.8 \text{ kW}$$

$$RSHF = \frac{RSH}{RSH\ \ RLH} = \frac{24.22}{24.22 + 29.82} = 0.448$$

2. Locate the alignment circle point b at t_d = 26°C and ϕ = 50% on the psychrometric chart. Also mark the *RSHF* value of 0.448 on the *SHF* scale at point '*a*'. Join points '*a*' and *b* by a straight line.
3. Draw a vertical line at t_d = 15°C (i.e. temperature of air entering the room) from point 2, draw a line parallel to *ab* line to intersect above vertical line at point 5. Line 2-5 is the *RSHF* line. The point represents the condition of air entering the room.

$$h_5 = 31.6 \text{ kJ/kg d.a.}$$

(*a*) Mass of circulated air.

Total mass of air passing through the room,

$$m_a = \frac{\text{Total room heat}}{\text{Total heat removed}} = \frac{RSH + RLH}{h_2 - h_5} = \frac{24.22 + 29.82}{42.3 - 31.6}$$

$$= \frac{54.04}{10.7} = 5 \text{ kg/s or } 18000 \text{ kg/h}$$

Mass fresh air supplied from outside,

$$m_F = \frac{V_F}{v_{s1}} = \frac{4500}{0.903} = 4983.4 \text{ kg/h}$$

Mass of recirculated air, $m_R = m_a - m_F$ = 18000 – 4983.4 = 13,016.6 kg/h

(*b*) Mark the air mixing point 3 such that,

$$\text{Length 2-3} = (1 - 0.723) \times \text{Length 1-2} = 0.277 \times \text{Length 1-2}$$

Now draw a line joining point 3 and t_{ADP} = 5°C intersecting the constant specific humidity line drawn through point 5, at point 4. Point 4 represents the condition of air leaving the cooling coil. The line 4-5 represents the heating of air. At point 3, t_{d3} = 24.3°C, and at point 4, t_{d4} = 7.8°C.

Also h_3 = 56 kJ/kg of d.a and h_4 = 24 kJ/kg of d.a.

$$\text{Capacity of cooling coil} = m_a(h_3 - h_4) = 5(56 - 24) = 160 \text{ kW or } \frac{160}{3.5167} = 45.5 \text{ TR}$$

$$BPF = \frac{t_{d4} - t_{ADP}}{t_{d3} - t_{ADP}} = \frac{7.8 - 5}{24.3 - 5} = 0.145$$

(*c*) capacity of heating coil $= m_a(h_5 - h_4)$

$$= 5(31.6 - 24)$$

$$= 38 \text{ kW}$$

14.8 BY-PASS FACTOR OF HEATING AND COOLING COILS

Let 1 kg of air at temperature t_{d1} is passed over the coil having its temperature t_{d3} as shown in Fig. Fig. 14.20 when air passes over a coil, some of it (say x kg) just by-passes unaffected while the remaining $(1 - x)$ kg comes in direct contact with the coil. This by-pass process of air is measured in terms of a by-pass factor.

Balancing the enthalpies, we have

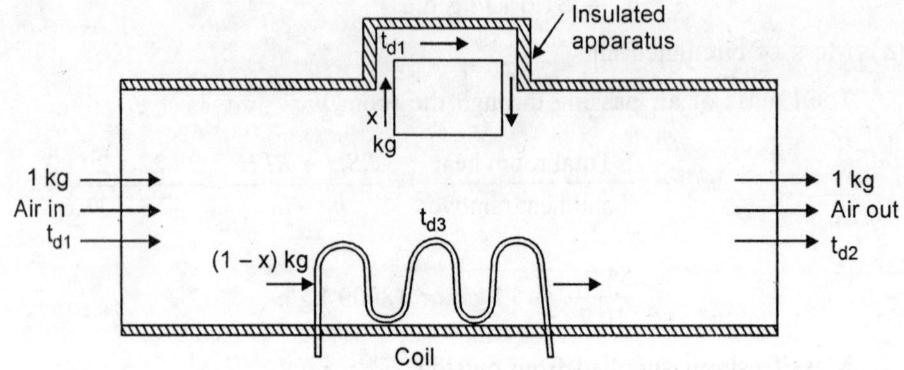

Fig. 14.20 By-pass factor (BPF)

$$xc_{pm}t_{d1} + (1 - x)\, c_{pm}t_{d3} = 1 \times c_{pm}t_{d2}$$

$$x(t_{d3} - t_{d1}) = t_{d3} - t_{d2}$$

$$x = \frac{t_{d3} - t_{d2}}{t_{d3} - t_{d1}}$$

where $x = BPF$

For heating coil, $$BPF = \frac{t_{d3} - t_{d2}}{t_{d3} - t_{d1}} \quad ...(14.32a)$$

For cooling coil, $$BPF = \frac{t_{d2} - t_{d3}}{t_{d1} - t_{d3}} \quad ...(14.32b)$$

14.9 EFFICIENCY OF HEATING AND COOLING COILS

Efficiency of coil or contact factor = $1 - BPF$

Efficiency of heating coil, $$\eta_H = 1 - \frac{t_{d3} - t_{d2}}{t_{d3} - t_{d1}} = \frac{t_{d2} - t_{d1}}{t_{d3} - t_{d1}} \quad ...(14.33a)$$

Efficiency of cooling coil, $$\eta_C = 1 - \frac{t_{d2} - t_{d3}}{t_{d1} - t_{d3}} = \frac{t_{d1} - t_{d2}}{t_{d1} - t_{d3}} \quad ...(14.33b)$$

Example 14.10 *A laboratory 30 m × 20 m × 4 m high is to be air conditioned. The 30 m wall faces north. The north wall has two doors of 2.5 m × 3 m each. The south wall has four glass windows of 2 m × 1.5 m each. The east and west walls also have four windows of the same size. The lighting load is 15 W fluorescent per m² floor area. The infiltration is one air change. The solar heat gain factors (SHGF) for south, east and west glass are 150, 50 and 350 W/m² respectively. The overall heat transfer coefficients for walls, roof, floor, door and windows are 2.5, 2, 3, 1.5 and 6 W/m² K respectively. The correctly equivalent temperature differences for north, south, east, west walls, roof and floor are 12, 15, 12, 17, 20 and 2.5°C respectively. There are 100 persons with sensible and latent heat loads of 75 W and 55 W each respectively. The ventilation requirement is 0.3 m³/min per person. The outdoor condition is 43°C dry bulb temperature and 0.0277 kg/kg of dry air of humidity ratio. The indoor condition is 25°C dry bulb temperature and 0.01 kg/kg of dry air of humidity ratio. Use a factor of 1.25 for fluorescent light.*

Determine room sensible heat load and room latent heat load with 5% safety factor, 5% for fan power, 1% leakage of supply air and 0.5% heat leakage to supply air duct.

Solution: Given: L = 30 m; B = 20 m; H = 4 m; Door size = 2.5 m × 3 m; Window size = 2.5 m × 3 m; Q_{SL} = 15 W/m² floor area; No. of air changes (A_C) = 1; *SHGF* for south glass = 150 W/m²; *SHGF* for east glass = 50 W/m²; *SHGF* for west glass = 350 W/m²; U_{walls} = 2.5 W/m² K; U_{roof} = 2 W/m² K; U_{floor} = 3 W/m² K; U_{door} = 1.5 W/m² K; U_{windows} = 6 W/m² K; t_e for north wall = 12°C; t_e for south wall = 15°C; t_e for east wall 12°C; t_e for west wall = 17°C; t_e for roof = 20°C; t_e for floor = 2.5°C : No. of persons = 100; Q_S per person = 75 W; Q_L per person = 55 W; v_1 = 0.3 m³/min per person; t_{d1} = 43°C; W_1 = 0.0277 kg/kg of dry air; t_{d2} = 25°C; W_2 = 0.01 kg/kg of dry air; Allowance factor for fluorescent light = 1.25; Safety factor = 5%; Factor for fan power = 5%; Factor for leakage of supply air = 1%; Factor for heat leakage to supply air duct = 0.5%.

The plan of a laboratory room is shown in Fig. 14.21. The sensible heat gain from various sources is given in the following Table 14.1.

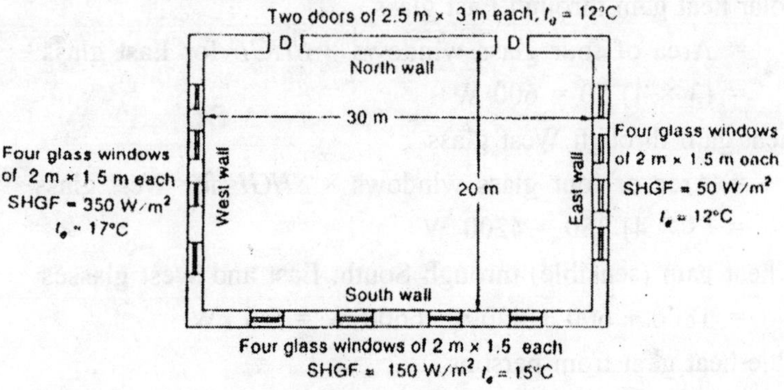

Fig. 14.21

Table 14.1: Estimation of sensible heat gain

Source	*Overall heat transfer coefficient (U) W/m² K*	*Area (A) m²*	*Equivalent temperature difference (t_e)°C*	*Sensible heat gain = U.A. t_e W*
1. North wall	2.5	30 × 4 – 2 × 7.5 = 105	12	2.5 × 105 × 12 = 3150
2. South wall	2.5	30 × 4 – 4 × 3 = 108	15	2.5 × 108 × 15 = 4050
3. East wall	2.5	20 × 4 – 4 × 3 = 68	12	2.5 × 68 × 12 = 2040
4. West wall	2.5	20 × 4 – 4 × 3 = 68	17	2.5 × 68 × 17 = 2890
5. Roof	2	30 × 20 = 600	20	2 × 600 × 20 = 24000
6. Floor	3	30 × 20 = 600	2.5	3 × 600 × 2.5 = 4500
7. Doors (2 Nos.) in North wall	1.5	2 × 7.5 = 15	12 (same as north wall)	1.5 × 15 × 12 = 270
8. Windows				
(a) South wall (4 Nos)	6	3 × 4 = 12	*18	6 × 12 × 18 = 1296
(b) East wall (4 Nos)	6	3 × 4 = 12	18	6 × 12 × 18 = 1296
(c) West wall (4 Nos)	6	3 × 4 = 12	18	6 × 12 × 18 = 1296
				Total = 44 788 W = 44.788 kW

* For glass windows, the equivalent temperature difference is given as

$$t_e = t_{d1} - t_{d2} = 43 - 25 = 18°C$$

Solar heat gain through South glass

= Area of four glass windows × *SHGF* for South glass

= (3 × 4) 150 = 1800 W

Similarly solar heat gain through East glass

= Area of four glass windows × *SHGF* for East glass

= (3 × 4) 50 = 600 W

and solar heat gain through West glass

= Area of four glass windows × *SHGF* for West glass

= (3 × 4) 350 = 4200 W

Total solar heat gain (sensible) through South, East and West glasses

= 1800 + 600 + 4200 = 6600 W = 6.6 kW

Total sensible heat gain from persons

= Q_S per person × No. of persons = 75 × 100 = 7500 W = 7.5 kW

Total latent heat gain from persons

$= Q_L$ per person × No. of persons = 55 × 100 = 5500 W = 5.5 kW

We know that the amount of infiltrated air,

$$v_1 = \frac{L \times W \times H \times A_C}{60} = \frac{30 \times 20 \times 4 \times 1}{60} = 40 \text{ m}^3/\text{min} \quad ...(\because A_C = 1)$$

∴ * Sensible heat gain due to infiltration air

$= 0.02044\ v_1\ (t_{d1} - t_{d2}) = 0.02044 \times 40\ (43 - 25) = 14.7$ kW

and *latent heat gain due to infiltration air

$= 50 v_1\ (W_1 - W_2) = 50 \times 40\ (0.0277 - 0.01) = 35.4$ kW

We know that volume of ventilation or outside air,

$v = 0.3$ m^3/min/person $= 0.3 \times 100 = 30$ m^3/min

∴ **Outside air sensible heat,

$OASH = 0.02044\ v\ (t_{d1} - t_{d2}) = 0.02044 \times 30\ (43 - 25) = 11.04$ kW

and **outside air latent heat,

$OALH = 50\ v\ (W_1 - W_2) = 50 \times 30\ (0.0277 - 0.01) = 26.6$ kW

Sensible heat gain due to lighting

= Total wattage of lights × Use factor × Allowance factor

= 15(30 × 20) × 1 × 1.25 = 11250 W = 11.25 kW ...(∵ Use factor = 1)

Assuming that the fan is placed before the conditioner (*i.e.*, neglecting factor of 5% for fan power), the total room sensible heat (*RSH*) is to be increased by 6.5% (*i.e.* 5% for safety factor; 1% for leakage of supply air and 0.5% for heat leakage to duct).

∴ Total room sensible heat,

RSH = 1.065 [Heat gain from walls, roof, floor and windows (From Table 14.1) + Solar heat gain through glasses + Sensible heat gain from persons + Sensible heat gain due to infiltration air + Sensible heat gain due to ventilation (*OASH*) + Sensible heat gain due to lighting]

= 1.065 [44.788 + 6.6 + 7.5 + 14.7 + 11.04 + 11.25] = 102 kW

The total room latent heat (*RLH*) is to be increased by 6% (*i.e.* 5% for safety factor and 1% for leakage of supply air).

∴ Total room latent heat,

RLH = 1.06 [Latent heat gain from persons + Latent heat gain due to infiltration air + Latent heat due to ventilation (*OALH*)]

= 1.06 [5.5 + 35.4 + 26.6] = 71.6 kW

If the fan is located after the conditioner, then the 5% for fan power should be added to *RSH*. The *RLH* will remain unaffected.

∴ *RSH* = (1.065 + 0.05) (44.788 + 6.6 + 7.5 + 14.7 + 11.04 + 11.25) kW

= 107 kW

EXERCISES

14.1 The following data is available for the design of air conditioning of a small theatre:

Outdoor design conditions	= 34°C *DBT* and 70% *RH*
Comfort conditions required	= 22°C *DBT* and 50% *RH*
Total seating capacity	= 350 persons
Sensible heat gain per person	= 90 W
Latent heat gain per person	= 30 W
Sensible heat due to solar gain and infiltrated air	= 46.6 kW
Latent heat gain due to infiltrated air	= 23.3 kW
Fresh air supplied	= 0.4 m^3/min/person
Desirable temperature rise in the theatre	= 8°C

Assume that the recirculated air is mixed with the fresh air after leaving the conditioner. Find : (*a*) Room sensible heat factor : (*b*) The percentage of total air circulated; (*c*) The refrigeration capacity of the conditioner coil.

Assume that the air leaves the conditioner coil with 100% *RH*.

[**Ans.** 0.72; 76%; 50 TR]

14.2 The design data for an air-conditioning plant of a restaurant is given below:

Outdoor design conditions	= 35°C *DBT* and 24°C *WBT*
Indoor design conditions	= 27°C *DBT* and 55% *RH*
Seating capacity of the restaurant	= 50
Latent heat gain per person	= 44 W
Latent heat gain from meals per person	= 6 W
Sensible heat gain per person	= 58 W
Sensible heat gain from meals per person	= 3.5 W
Number of service employees	= 5
Latent heat gain per employee	= 75 W
Sensible heat gain per employee	= 58 W
Sensible heat gain from outside	= 8.14 kW
Sensible heat gain from inside equipment	= 2.9 kW
Latent heat gain from inside equipment	= 0.7 kW
Rate of infiltrated air	= 400 m^3/h
Rate of fresh air supply	= 1600 m^3/h
Minimum temperature of air supplied to room	= 17°C *DBT*

The fan is situated before the conditioner and has a motor of 11 kW. Calculate : (*a*) volume of air passing through the room in m^3/h; (*b*) percentage of recirculated air; (*c*) apparatus dew point temperature and by-pass factor; and (*d*) cooling capacity in tones of refrigeration.

[**Ans.** 4745 m^3/h; 68.6%; 14.6°C, 0.161; 11.9 TR]

14.3 The following are the design data for an air conditioning system proposed for a restautant:

Outdoor conditions	= DBT and WBT
Indoor conditions	= 24°C *DBT* and 50% *RH*
Solar heat gain through walls, roof and floor	= 4.7 kW
Solar heat gain through glass areas	= 4.4 kW
Occupants	= 25
Sensible heat gain per person	= 85 W
Latent heat gain per person	= 105 W
Internal lighting load	= 15 lamps of 100 watts capacity each plus 10 fluorescent fixtures of 80 watts each
Sensible heat gain from other sources	= 11.6 kW
Infiltration air	= 14 m^3/min
Coil by-pass factor	= 0.15

If the return and out-door air are adiabatically mixed in the ratio of 3 : 2 (by mass) and then passed through the conditioner, determine:
(*a*) Dry bulb and wet bulb temperature of supply air; (*b*) apparatus dew point; and (*c*) capacity of the air-conditioning plant. [**Ans.** 10.4°C; 10°C; 7.3°C; 72 kW]

14.4 A space to be air conditioned for comfort at 25°C, 50% *RH* has the following peak load data:

Room size : 20 m wide, 20 m long and 4 m high.

North glass : 4 windows of 2 m × 1.5 m each with solar heat gain factor (*SHGF*) = 50 W/m^2.

East glass : 4 windows of 1 m × 0.75 m each with = 60 W/m^2.

West glass : 4 windows of 2 m × 1.5 m each with *SHGF* = 350 W/m^2.

East wall and North wall have two doors each 2.5 m × 3 m.

Overall heat transfer coefficient (U_o) in W/m^2 K are : – Roof = 2.0; walls = 2.5; Glass = 6.0; Door = 1.5; and Floor = 3.0.

Corrected equivalent temperature difference (ΔT_{eq}) : East wall = 10 K : West wall = 15 K; North wall = 10 K; South wall = 13 K; Roof = 18 K and Floor = 2.5 K.

Infiltration	: One air change
Occupancy	: 100 persons dissipating 75 W sensible heat and 65 W latent heat
Lighting	: Fluorescent 10 W/m^2 of floor area
Ventilation	: 0.3 m^3/min per person
Out-door conditions	: 43°C and 30% *RH*
Indoor conditions	: 25°C and 50% *RH*

Determine room sensible heat (*RSH*) load; room latent heat (*RLH*) load; overall sensible heat load (*OASH*) : overall latent heat load (*OALH*).

Given apparatus dew point (*ADP*) = 7°C, find effective sensible heat factor (*ESHF*) and supply air flow rate. Humidity ratio of outdoor and indoor air may be taken as 0.017 and 0.01 kg/kg of dry air

[**Ans.** 69.3 kW; 14.95 kW; 79.2 kW; 24.2 kW; 0.8; 404.6 kg/min]

14.5 The following data relates to a cold storage for storing 450 tonnes of vegetables

Inside design conditions	19°C *DBT* 60% RH
Outside design conditions	36°C DBT WBT
Infiltrated air	180 m^3/hr
Freash air supply	4500 m^3/hr
No. of persons working in cold store	20
Sensible heat gain through glass, walls, ceiling etc.	16.3 kW
Water content of vegetables	74%
Loss of water content	0.01% per hour
Heat from equipment and reaction heat of vegetables	3.1 kW

If the air conditioning is achieved by first cooling and dehumidifying and then heating and the temperature of air entering the room is not to exceed 16°C, determine

(*a*) Amount of recirculated air, if the recirculated air is mixed with fresh air before entering the cooling coil.

(*b*) Capacity of the cooling coil in tones of refrigeration and its by pass factor if the dew point temperature of the coil is 6°C and

(*c*) Capacity of heating coil. [**Ans.** 3.384 kg/s, 46.75 TR, 34.82 kW.]

14.6 The following data relates to a conference room having a seating capacity of 80 persons;

Inside design conditions	DBT 55% RH
Outside design conditions	DBT WBT
Sensible and latent heat loads per person	75 W and 45W respectively
Lights and fan loads	12000W
Sensible heat gain through glass, walls, ceiling etc.	12000W
Air infiltration	18 m^3/min
Fresh air supply	80 m^3/min
By pass factors of the coils	0.1

If two third of recirculated room air and one third of fresh air are mixed before entering the cooling coils determine.

(a) Apparatus Dew point (b) Grand total heat load

(c) Effective room sensible heat factor.

Table 14.2: Overall coefficient of heat transmission (U) for structures with a wind velocity of 24 km/h outside.

Structure	*Overall coefficient of heat transmission (U) in W/m² K*
Brick wall, 20 cm, bare	2.84
Brick wall, 20 cm, plaster one side on brick	2.61
Brick wall, 20 cm plaster one side on metal lath-furred	1.82
Brick wall, 40 cm, bare	1.59
Brick wall, 40 cm, plaster one side on brick	1.53
Brick wall, 40 cm, plaster one side on metal lath-furred	1.20
Hollow tile, stucco exterior, 20 cm, bare	2.27
Hollow tile, stucco exterior, 20 cm plaster on metal lath furred	1.59
Hollow tile, stucco exterior, 30 cm bare	1.70
Hollow tile, stucco exterior 30 cm, plaster on metal lath furred	1.25
Cinder blocks, 24 cm bare	2.38
Cinder blocks 20 cm, plaster one side on metal lath-furred	1.59
Concrete blocks, 20 cm bare	3.18
Clap-board frame construction, plaster on wood lath	1.42
Wood shingle frame construction, plaster on wood lath	1.42
Stucco frame construction, plaster on wood lath	1.70
Brick veneer frame construction, plaster on wood lath	1.53

Table 14.3 Thermal conductivity (k) of building materials.

Material	*Thermal conductivity (k) in W / m K*
Brick, low density	8.66
Brick, high density	15.93
Cement mortar	20.82
Cement plaster, typical	20.82
Concrete, typical	20.82
Concrete, typical fibre gypsum, 87.5% gypsum, 12.5% wood chips	1.61
Sand and gravel	21.86
Lime stone	18.72
Cinder	8.50
Cinder, boiler (12.5 mm to 19 mm)	2.13
Title or terrazzo, typical	20.82
Asbestos building board	4.71
Gypsum, between layers or heavy paper	2.44
Gypsum, plaster, typical	5.72
Wood across grain, typical	1.73
Balsa	0.657
Maple, hard, 16% moisture	2.0
Pine, yellow short leaf, 16% moisture	1.45

Table 14.4 Thermal conductivity (k) of insulating materials.

Material	*Thermal conductivity (k) in W / m K*
Asbestos packed	2.812
Asbestos loose	1.852
Asbestos, paper, thin layers organic binder	0.85
Asbestos mill board	1.454
Asbestos wood	4.85
Asphalt roofing (felt)	1.212
Balsa	0.607
Cement wood (saw dust and Portland cement)	1.683
Charcoal (hard woods) coarse	0.617
Cork board, typical	0.468 to 0.486
Cork regranulated, coarse	0.536
Cork regranulated	0.468
Cotton	0.676
Diatomaceous earth	0.460
Eel grass	0.412
Glass wool, high grade	0.388 to 0.460
Glass wool, commercial grade	0.444
Hair felt, not compressed	0.444
Insulation boards, various fibres	0.553 to 0.657
Kapok, loosely packed	0.416
Planer shavings, various woods	0.692
Rock wool	0.468 to 0.486
Rubber, expanded	0.364
Saw dust, various woods	0.710

Table 14.5: Conductance of building materials and construction for thickness indicated

Material	*Conductance in $W/m^2\ K$*
Plaster board 9 mm thick	21.17
Plaster board 12.5 mm thick	16.14
Roofs:	
Asphalt, composition or prepared roofing	36.85
Shingles, asbestos	34.05
Shingles, asphalt	36.65
Shingles, slate	58.96
Metal lath and plaster, total thickness 12.5 mm	24.96
Wood lath and plaster, total thickness 18 mm	14.20
For shealthing 25 mm and building paper	4.88
For shealthing 25 mm building paper and yellow pine lap siding	2.83
For shealthing 25 mm building paper and stucco	4.65
Air space over 18 mm faced ordinary building materials	6.15
Typical hollow clay tile, 10 cm	5.67
Typical hollow clay tile, 15 cm	3.64
Typical hollow clay tile, 20 cm	3.40
Typical hollow clay tile, 25 cm	3.29
Typical hollow clay tile, 30 cm	2.277
Typical hollow clay tile, 40 cm	1.762
Concrete block, sand and gravel, typical, 20 cm	5.675
Concrete block, sand and gravel, typical, 30 cm	4.547
Concrete block, cinder aggregate, typical, 20 cm	3.404
Concrete block, cinder aggregate, typical, 30 cm	3.006
Surface coefficient, outside for 24 km. p.h. wind	34.05

Table 14.6: Film or surface conductance for air film and air spaces.

Material	*Surface position*	*Thermal conductance (f) in (f) in W/m^2 K*
Air film (Surface)		
Still air (f_i) heat flowing up	Horizontal	9.25
	Sloping, 45°	9.1
Still air, heat flowing down	Horizontal	6.13
	Sloping 45°	7.5
Still air, heat flowing horizontal	Vertical	8.3
Wind, 24 km/h (f_a)	Any position	34.0
Wind, 12 km/h (f_a)	Any position	22.7
Air spaces bounded by		
Structural material		
Heat flowing up		
	Horizontal, 19 to 100 mm	6.7
Heat flowing horizontal	Sloping 45°, 19 to 100 mm	6.3
Heat flowing down	Vertical, 19 to 100 mm	5.85
	Sloping 45°, 19 to 100 mm	5.5
	Horizontal, 19 mm	5.56
	Horizontal, 38 mm	4.94
	Horizontal, 100 mm	4.6

Note: The above values of thermal conductance are for a mean temperature of 10°C to 32°C.

Table 14.7 Total equivalent temperature differentials for calculating heat gain through sunlit shaded roofs.

Description of roof construction[1]	Sun-time								
	A.M.			P.M.					
	8	10	12	2	4	6	8	10	12
Light construction roofs-exposed to sun									
2.5 cm wood[2] or 2.5 cm wood + 2.5 cm or 5 cm insulation or 5 cm wood	12	38	54	62	50	26	10	4	0

Medium construction roofs-exposed to sun

5 cm concrete or 5 cm concrete + 2.5 cm or 5 cm insulation or 5 cm wood	6	30	48	58	50	32	14	6	2
5 cm gypsum or 5 cm gypsum + 2.5 cm insulation 2.5 cm wood[2] / 5 cm wood[2] or / 5 cm concrete or / 5 cm gypsum + 10 cm rock wool in furred ceiling	0	20	40	52	54	42	20	10	6
10 cm concrete or 10 cm concrete + 5 cm insulation	0	20	38	50	52	40	22	12	6

Heavy construction roofs-exposed to sun

15 cm concrete	4	6	24	38	46	44	32	18	12
5 cm concrete + 5 cm insulation	6	6	20	34	42	44	34	20	14

Roofs covered with water-exposed to sun

Light construction roof with 2.5 cm water	0	4	16	22	18	14	10	2	0
Heavy construction roof with 2.5 cm water	–2	–2	–4	10	14	16	14	10	6
Any roof with 15 cm water	–2	0	0	6	10	10	8	4	0

Roofs with roof sprays-exposed to sun

Light construction	0	4	12	18	16	14	10	2	0
Heavy construction	–2	–2	2	8	12	14	12	10	6

Roofs in shade

Light construction	– 4	0	5	12	14	12	8	2	0
Medium construction	– 4	– 2	2	8	12	12	10	6	2
Heavy construction	– 2	– 2	7	4	8	10	10	8	4

1. Includes 9 mm felt, roofing with or without slag. May also be used for single roof.
2. Nominal thickness of wood.

Table 14.8 Total equivalent temperature differentials for calculating heat gain through sun lit and shaded walls.

North Latitude Wall Facing	Sun – Time																		South Latitude Wall Facing
	A.M.						P.M.												
	8		10		12		2		4		6		8		10		12		
	Exterior colour of wall; D-dark, L-light																		
	D	L	D	L	D	L	D	L	D	L	D	L	D	L	D	L	D	L	
	Frame																		
E	22	13	24	12	14	10	12	10	14	14	14	14	10	10	6	4	2	2	SE
	30	14	36	18	32	16	12	12	14	14	14	14	10	10	6	6	2	2	E
	13	6	26	16	28	18	24	16	16	14	14	14	10	10	6	4	2	2	NE
	–4	–4	4	6	22	12	30	20	26	20	16	14	10	10	6	6	2	2	N
V	–4	–4	0	–2	6	4	26	22	40	28	42	28	24	20	6	4	2	2	NW
	–4	–4	0	6	6	6	20	12	40	28	48	34	22	22	8	8	2	2	W
V	–4	–4	0	–2	6	4	12	10	24	20	40	26	34	24	6	4	2	2	SW
ade)	–4	–4	–2	–2	4	4	10	10	14	14	2	12	8	8	4	4	0	2	S (shade)
	10 cm brick or stone veneer frame																		
NE	–2	–4	24	12	20	10	10	6	12	10	14	14	12	12	10	10	5	4	SE
E	2	0	30	14	31	17	14	14	12	12	14	14	12	12	10	8	6	6	E
SE	2	–2	20	10	28	16	26	16	18	14	14	14	12	12	10	8	6	6	NE
S	–4	–4	–2	–2	12	6	24	16	26	18	20	16	12	12	8	8	4	4	N
SW	0	–2	0	–2	2	2	12	8	32	22	36	26	34	24	24	10	8	6	NW
W	0	–2	0	0	4	2	10	8	26	18	40	28	42	28	28	16	14	6	W
NW	–4	–4	–2	–2	2	2	8	6	12	12	30	22	34	24	24	12	10	6	SW
N (shade)	–4	–4	–2	–2	0	0	6	6	10	10	12	12	12	12	12	8	8	4	S (shade)
	20 cm hollow tile or 20 cm cinder block																		
NE	0	0	0	0	20	10	16	10	10	6	12	10	14	12	12	10	8	8	SE
E	4	2	12	4	24	12	26	14	20	12	12	10	14	12	14	10	10	8	E
SE	2	0	2	0	16	8	20	12	20	14	14	12	14	12	12	10	8	6	NE
S	0	0	0	0	2	0	12	6	24	14	26	16	20	14	12	10	8	9	N
SW	2	0	2	0	2	0	6	4	12	10	26	18	30	20	26	18	8	6	NW
W	4	2	4	2	4	2	6	4	10	8	18	14	30	22	32	22	18	11	W
NW	0	0	0	0	2	0	4	2	8	6	12	10	22	18	30	22	10	14	SW
N (shade)	–2	–2	–2	–2	–2	–2	0	0	6	6	10	11	10	10	10	10	6	6	S (shade)

20 cm brick or 30 cm hollow tile or 30 cm cinder block

NE	2	2	2	2	10	2	16	8	14	8	10	6	10	8	10	10	10	8	SE
E	8	6	8	6	14	8	18	10	18	10	14	8	14	10	14	10	12	10	E
SE	8	4	6	4	6	4	14	10	18	12	16	12	12	10	12	10	12	10	NE
S	4	2	4	2	4	2	4	2	10	6	16	10	16	12	12	10	12	8	N
SW	8	4	6	4	6	14	8	4	10	6	12	8	20	12	24	16	20	14	NW
W	8	4	6	4	6	6	8	6	10	6	14	8	20	16	24	16	24	16	W
NW	2	2	2	2	2	2	4	2	6	4	8	6	10	8	16	14	18	14	SW
N (shade)	0	0	0	0	0	0	0	0	2	2	6	6	8	8	8	8	6	6	S (shade)

30 cm brick

NE	8	6	8	6	8	4	8	4	10	4	12	6	12	6	10	6	10	6	SE
E	12	8	12	8	12	8	10	6	12	8	14	10	14	10	14	8	14	8	E
SE	10	6	10	6	10	6	10	6	10	6	12	8	14	10	14	10	12	8	NE
S	8	6	8	6	6	4	6	4	6	4	8	4	10	6	12	8	12	8	N
SW	10	6	10	6	10	6	10	10	10	6	10	6	10	8	12	8	14	10	NW
W	12	8	12	8	12	8	10	10	10	6	10	6	10	6	12	8	10	10	W
NW	8	6	8	6	8	4	8	4	8	4	8	4	8	6	10	6	10	6	SW
N (shade)	4	4	2	2	2	2	2	2	2	2	2	2	2	2	4	4	4	6	S (shade)

20 cm concrete or Stone or 15 cm or 20 cm concrete block

NE	4	2	4	0	16	8	14	8	10	6	12	8	12	10	10	8	8	6	SE
E	6	4	14	8	24	12	24	12	18	10	14	10	14	10	12	10	10	8	E
SE	6	2	6	4	16	10	18	12	18	12	14	12	12	10	12	10	10	8	NE
S	2	1	2	1	4	1	12	6	16	12	18	12	14	12	10	8	1	6	N
SW	6	2	4	2	6	2	8	4	14	10	22	16	24	16	22	16	10	3	NW
W	6	4	6	4	6	4	8	6	12	8	20	14	28	18	26	13	14	10	W
NW	4	2	4	0	4	2	4	4	6	6	12	10	20	14	22	16	8	6	SW
N (shade)	0	0	0	0	0	0	2	2	4	4	6	6	8	8	6	6	4	4	S (shade)

100 cm concrete or stone

NE	6	4	6	2	6	2	14	8	14	8	10	8	10	8	12	10	10	8	SE
E	10	6	8	6	10	6	18	10	18	12	16	10	12	10	14	10	14	10	E
SE	8	4	8	4	6	4	14	8	16	10	16	10	14	10	12	10	12	10	NE
S	6	4	4	2	4	2	4	2	10	6	14	10	16	12	14	10	10	8	N
SW	8	4	8	4	6	4	6	4	8	6	10	8	8	14	20	14	18	16	NW
W	10	6	8	6	8	6	10	6	10	6	12	8	16	10	24	14	22	16	W
NW	6	4	6	2	6	2	6	4	6	4	8	6	16	8	18	12	20	14	SW
N (shade)	0	0	0	0	0	0	0	0	2	2	4	4	0	6	8	8	6	8	S (shade)

Table 14.9: Number of air changes per hour

Kind of room or building	*Number of air changes per hour (A_C)*
Rooms with no windows or outside doors	0.5 to 0.75
Rooms, one wall exposed	1
Rooms, two wall exposed	1.5
Rooms, three wall exposed	2
Rooms, four wall exposed	2
Entrance halls	2 to 3
Reception halls	2
Bath rooms	2

Table 14.10: Infiltrated air

Usage of door	*Infiltrated air for 1.8 metres revolving door in m^3 per person per passage*	
	Freely revolving door	*Door equipped with brake*
Infrequent	2.5	2
Average	2	1.75
Heavy	1.5	1.25

Notes : 1. For rooms with weather stripped windows or storm sash, use 1/2 of the values given in Table 19.8 but it should never be less than 0.5.

2. The figures in Table 19.9 are based on the assumption that there is no wind pressure and the swinging doors are in use in one wall only. Any swinging door in other walls should be kept closed to ensure air conditioning in accordance with the recommended standards.

3. The door infiltration for a 0.9 m swinging door may be taken as 3 m^3/min.

Table 14.11: Outside air required per person

Application	*Smoking*	*Outside air in m^3/min/person*	
		Recommended	*Minimum*
(1)	*(2)*	*(3)*	*(4)*
Apartment	Some	0.6	0.45
Banking space	Occasional	0.3	0.23
Barber shops	Considerable	0.45	0.3
Beauty Parlour	Occasional	0.3	0.23
Cocktail bars	Heavy	0.9	0.75
Departmental stores	None	0.23	0.15

Drug stores	Considerable	0.3	0.23
Factories	None	0.3	0.23
Funeral parlours	None	0.3	0.23
Hospitals, private rooms	None	0.9	0.75
Hospitals, wards	None	0.6	0.45
Hotel rooms	Heavy	0.9	0.75
Meeting rooms	Very heavy	1.5	0.9
Offices, general	Some	0.45	0.3
Offices, private	None	0.75	0.45
Restaurants	Considerable	0.45	0.36
Cafeterias	Considerable	0.36	0.3
Theatres	None	0.23	0.15

Table 14.12: Heat gain from occupants (in wants)

Degree of activity (1)	*Typical application (2)*	*Total heat Adults-Male (3)*	*Total heat adjusted (4)*	*Sensible heat (5)*	*Latent heat (6)*
1. Seated- at rest	Theatre-Matinee	114	97	53	44
	Theatre- Evening	114	102	57	45
2. Seated- very light work	Offices, hotels, Apartment	132	117	57	67
3. Moderately active office work	Offices, Hotels, Apartments	139	132	59	73
4. Standing, light work, walking slowly	Department store, Retail store	161	132	59	73
5. Walking, seated, standing, walking slowly	Drug store, Bank	161	146	58	88
6. Sedentary work	Restaurant	144	161	64	97
7. Light bench work	Factory	234	220	64	155
8. Moderate dancing	Dance hall	264	249	72	177
9. Walking 4.8 km / h, moderately heavy work	Factory	293	293	88	205
10. Bowling	Bowling alley	440	425	136	288
11. Heavy work	Factory	440	425	136	288

Notes :

1. The above values are based on 27°C dry bulb temperature. For room dry bulb, the total heat remains the same, but the sensible heat values should be increased approximately 10 per cent and the latent heat values decreased accordingly.
2. The adjusted total heat gain is based on normal percentage of men, women and children for the application listed, with the postulate that the gain from an adult female is 85 per cent of that for an adult male and that the gain from a child is 75 per cent of that for an adult male.
3. The adjusted total heat value for sedentary work, restaurant includes 17.5 W for food per individual (8.75 W sensible and 8.75 W latent).
4. For bowling, figure one person per alley actually bowling and all other as sitting (117 W) and standing (161 W).

Table 14.13: Heat gain from appliances without hoods (in watts)

Appliances (1)	Electrical		Gas	
	Sensible (2)	Latent (3)	Sensible (4)	Latent (5)
1. Coffee brewer, $2\frac{1}{4}$ litres	264	64	396	103
2. Coffee warmer, $2\frac{1}{4}$ litres	68	18	117	29
3. Coffee brewer with tank, 20 litres	1406	35	2110	528
4. Coffee urn, $13\frac{1}{2}$ litres	645	436	733	733
5. Coffee urn, $22\frac{1}{2}$ litres	996	674	1143	1143
6. Egg boiler, 2 cups	352	234	–	–
7. Food warmer with plate warmer, per sq m of top surface	1140	1140	–	–
8. Food warmer only per sq m of top surface	650	1140	2756	1401
9. Fry kettle per sq m of fry area	1026	1465	1758	1172
10. Griddle, per sq m of fry area	880	470	–	–
11. Grill, meat, per sq m of fry area	1378	733	2930	733
12. Grill, sand witch, per sq m of fry area	790	205	–	–
13. Hair dryer, blower type	674	117	–	–
14. Hair dryer, helmet type	547	98	–	–
15. Toaster, 360 slices per hour	1495	380	2257	967
16. Stoves, sort order				
Open top, per sq m top	–	–	1230	1230
Closed top, per sq m top	–	–	967	967
Fry top, per sq m top	–	–	1055	1055
17. Permanent wave machine	250	44	–	–
18. Sterilizer instrument for physicians	190	352	–	–

Table 14.14(a): Rate of heat production for different type of foods

Commodity (1)	*Temperature in °C* (2)	*Heat evolved per tonne per 24 hours in kJ* (3)
1. Apples	0	690 – 1045
	4.5	1170 – 1865
	15.5	4650 – 6950
	30	6950 – 16 250
2. Bananas, green and ripe	20	8835
3. Beets	0	2800
	4.5	4270
	15.5	7640
4. Cabbage	0	1240
	4.5	1770
	15.5	4260
5. Carrots	0	2240
	4.5	3660
	15.5	8520
6. Cauliflower	4.5	4650
	15.5	10500
7. Cherries (sour)	0	1380 – 1840
	15.5	11600 – 13920
8. Cucumbers	15.5	2250 – 6850
9. Grape fruit	0	480
	4.5	1130
	15.5	2930
10. Lemons	0	610
	4.5	860
	15.5	2180
11. Melons	0	1350
	4.5	2050
	15.5	8840
12. Mushrooms (cultivated)	0	6490
	10	23215
13. Onions	0	690 – 1170
	10	1860 – 2090
14. Oranges	0	730 – 965
	4.5	1465
	15.5	5275

15.	Peaches	0	900 – 1450
		4.5	1450 – 2050
		15.5	7560 – 9820
16.	Pears	0	690 – 920
		15.5	9295 – 13920
17.	Peppers	0	2870
		4.5	4960
		15.5	8940
18.	Peas	0	8500 – 8700
		4.5	13500 – 16600
		15.5	40800 - 46200
19.	Potatoes	0	460 – 920
		4.5	1170 – 1860
		20	2320 – 3725
20.	Raspberries	2.2	4650 – 6970
		15.5	16 245 – 18 570
21.	Straw berries	0	2890 – 4020
		4.5	5400 – 6970
		15.5	16 500 – 20 200
22.	Tomatoes (green)	0	610
		4.5	1130
		15.5	6570
23.	Tomatoes (ripe)	0	1090
		4.5	1340
		15.5	5950
24.	Turnips	0	1960
		4.5	2290
		15.5	5500

Table 14.14(b): Rate of heat production for different type of foods

Commodity	*Temperature (°C)*	*Humidity (%)*	*Period for loss*	*Loss of moisture (%)*
Apples	0	90	6 – 8 months	3 -4
	2.8	95		0.03 – 1.1
	2.8	75		0.08 – 3.5
	15	95	1 -50 days	0.06 – 2.5
	15	75		0.18 – 8.2
Beef : 0.9 kg steak	–	90 and 80	1 day	0.9 and 1.8
18 kg roast	–	90 and 80	1 day	0.4 and 8
Beef	Normal storage	Normal storage	1 month	3
Beets	0	98	1 month	2.5
Cabbage: curled	-0.5	85	2 – 7 months	10 – 15
red			6 – 7 months	6 – 8
white			6 – 7 months	8 - 10
Carrots	0	98	1 month	1.5
Cauliflower	0	98	1 month	11
Cheese	15	75	–	10
Cucumber	0	98	1 month	6
Currants	–0.5	85	2 months	6
Eggs	0	99	3 months	2
	0	88	3 months	7
Goose berries	–1.1	85	2 months	6
Onions	–0.5	85	6 – 8 months	6 – 8
Peaches	–1.1	90	1 – 2 months	8 – 12
Pears	0	90	4 – 6 months	3 – 6
Peppers	0	98	1 month	3
Plums	–1.1	85	2 months	4 – 6
Spinach	–1.1	90	6 – 7 months	4 -5
Squash	0	98	1 month	1
Straw berries	1.1	90	1 month	4

Chapter

15

DUCTS DESIGN

15.1 TRANSMISSION AND DISTRIBUTION OF AIR

The air-handling system consists of:

(*i*) The air distribution system comprising various inlets for recirculated air and outlets for the supply air.

(*ii*) The duct system including the return duct, supply duct, and air-conditioning apparatus comprising of dampers, filters, coil/air washer, etc.

(*iii*) The fan which provides the necessary energy to move the air.

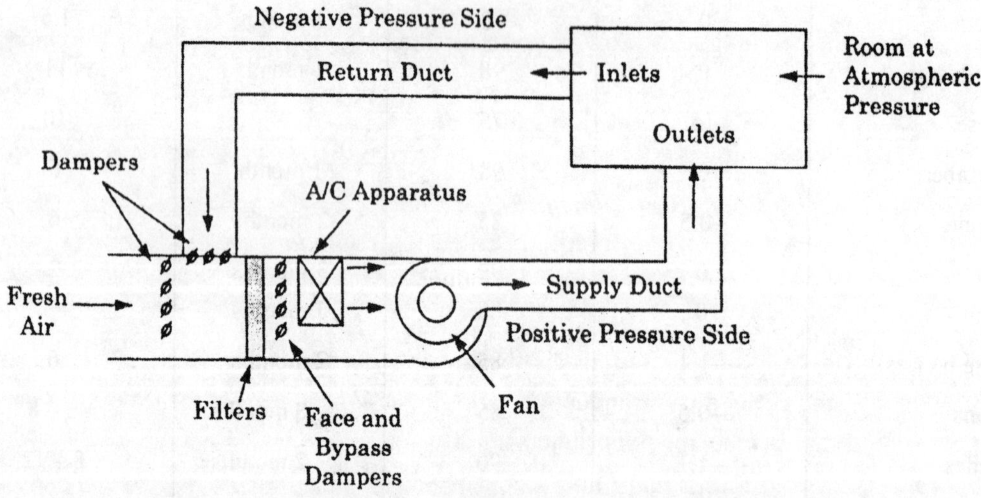

Fig. 15.1 Schematic air-flow diagram for an air-conditioning system

Fig. 15.1 shows how air is handled in a simple air-conditioning system. It is seen that a closed loop is formed for the circulation of air. The reference point in this loop is the room itself which can be considered at atmospheric pressure. The air enters the return duct through the inlets from the room and continues to drop in pressure until it reaches the fan. The fan raises the pressure. Thereafter, the pressure starts dropping in the supply duct until the air is released to the space. Therefore, the pressure on the suction side of the fan is negative, and on the discharge side positive. The pressure in the room is atmospheric, i.e., zero gauge.

15.1.1 Room Air Distribution

The requirement of good room air distribution is to create a proper combination of temperature, humidity and air motion in the occupied zone which is normally at 1.8 m above the floor level. The maximum variation in temperature in a single room should not be more than 1°C. The desirable air velocity is 9.1 mpm at the occupancy level.

Draft is defined as any localized feeling of coolness or warmth of any portion of the body due to both air movement and air temperature, with humidity and radiation considered constant. The warmth or coolness of a draft is measured above or below the controlled room condition of 24.4°C DBT at the centre of the room and air moving at approximately 9.1 mpm velocity.

The most desirable direction of cooled supply air is from the front side towards the face of a person. Cold air directed towards the feet or from the back is undesirable while warm air directed towards the feet is comfortable and directed towards the face is undesirable.

The principles of air distribution involve the following factors:

1. ***Blow or Throw:*** Blow or throw L is the distance travelled by the supply air stream in the horizontal direction on leaving the air outlet and reaching a velocity of 15 mpm. The velocity is measured in the occupied zone at 1.8 m above the floor level. The localized condition just at the end-point of blow will not be comfortable due to lower temperature than 24.4°C, and higher velocity of 15 mpm compared to a velocity of 9 mpm after diffusion of supply air in room.

   The desirable length of blow is up to three-fourths of the distance to the opposite side of wall.

2. ***Drop:*** Drop is the vertical distance the air moves after it leaves the outlet and reaches the end of the blow.

3. ***Induction Ratio or Entrainment Ratio:*** The air leaving the outlet is *primary* etc. This supply air entrains some room air, called the *secondary air*. The sum of primary and secondary air is called *total air*. The induction ratio R is defined as the ratio of total air to primary air.

   Blow depends on the initial supply air velocity, temperature difference between the supply and room air and induction ratio. Induction, in turn, depends on the primary air velocity and perimeter of primary air stream.

   In addition to the primary and secondary air streams, air motion also develops in the room as a result of natural convection due to the density difference between the room air and the colder or hotter supply air.

4. ***Spread:*** Spread is the angle of divergence of an air stream after it leaves the outlet. The spread can be both horizontal and vertical.

Air outlets can have three types of vanes, viz., straight, converging or divergent vanes. Outlets with straight vanes produce a spread of 14 to 24° (approximately 19°) in both the vertical and horizontal planes. Outlets with converging vanes have approximately the same spread at straight vanes but the blow is about 15 per cent longer. Outlets with diverging vanes give a fanning effect and have a marked effect on the direction and blow. Fig. 15.2 shows an outlet with vertical vanes. The end vanes are turned at an angle of 45° and intermediate vanes are turned by progressively smaller angles. Such an outlet has a horizontal spread of 60°. The blow is reduced by abot 50 per cent as compared to that of a straight-vane outlet.

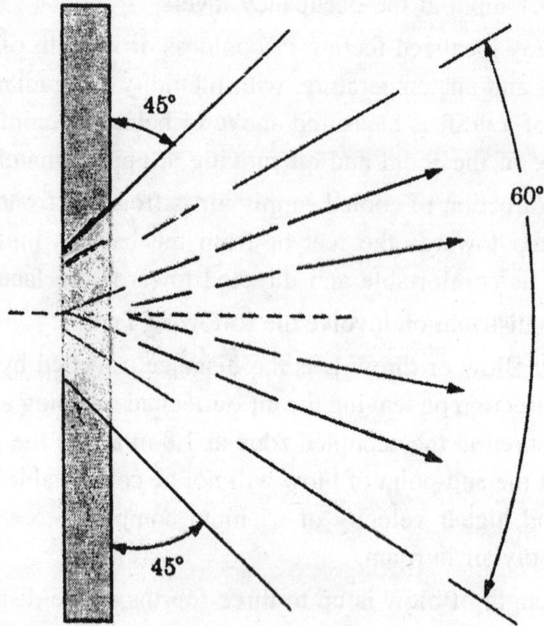

Fig. 15.2 Spread in a 45° outlet

15.1.2 Types of supply Air Outlets

The outlets may be classified as high sidewall, low sidewall, ceiling and floor outlets according to their location. There are also numerous outlet designs which are described by their construction feature. Accordingly, there are four basic types of outlets:

(*i*) Grille outlets,

(*ii*) Slot diffuser outlets,

(*iii*) Ceiling diffuser outlets,

(*iv*) Perforated ceiling panels.

Grille outlets may have *adjustable bar grilles* which are the most common types with vertical and horizontal vanes. They may have fixed bar grilles in which case vanes are not adjustable, but they are fixed either straight or at an angle. The application of grille outlets is in a high sidewall location. They are not acceptable in ceiling location for comfort application as they may cause draft conditions. The accessories to grille outlets are shown in Fig. 15.3. These help in the uniform distribution of air over the whole grille area. A combination of a grille and a damper is called a *register*.

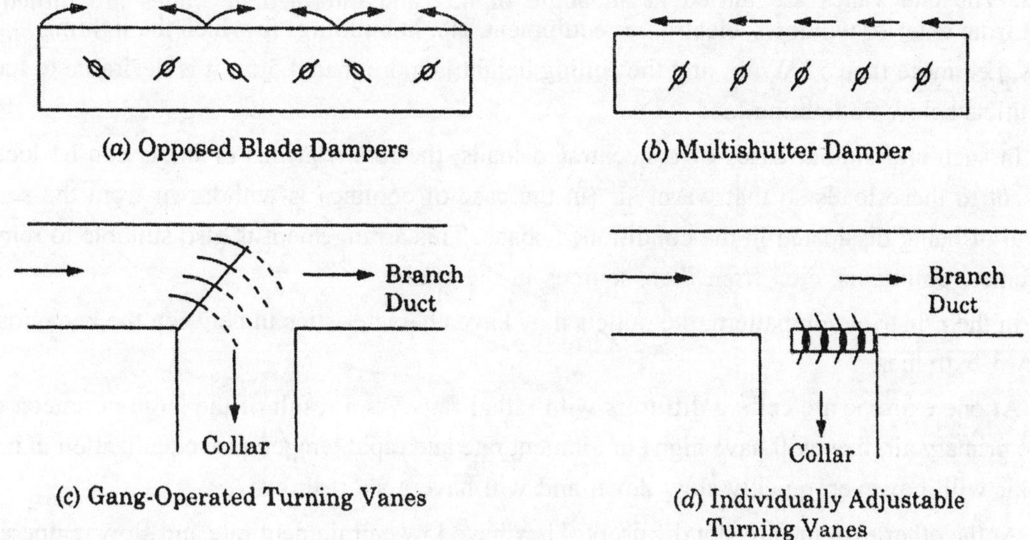

Fig. 15.3 Accessories to grill outlets

A slot diffuser is an elongated outlet with an aspect ratio of 25 : 1 and a maximum width of 7.5 cm. Their application is in high sidewall locations or perimeter installations in floors. They should not be installed in ceilings. Accessories to slot diffusers are dampers normally available as integral equipment.

Ceiling diffusers are mounted in the ceilings. Multi-passage round, square or rectangular are the most common types. They consist of a series of flaring rings or louvers, which may be of adjustable pattern or variable area type. They may also be fitted with dampers.

Perforated ceiling panels use the confined space above the ceiling as a supply plenum. Air is delivered to the room through holes or slots. The upper and lower limit of the plenum pressure is 3.5 mm and 0.025 mm H_2O respectively. The recommended supply air rates are 0.3 to 4.5 cmm/m^2 of the floor area. Perforated ceiling panels are suited to large zones of uniform room temperature.

15.1.3 Considerations for Selection and Location of Outlets

The selection and location of supply air outlets is governed by the following considerations:

(*i*) The amount of air to be delivered by the outlet should be proportional to the load of the part of the space for which it is installed.

(*ii*) The selection of the type of outlet is governed by the ceiling height, nature of room occupancy, etc.

(*iii*) The location of the outlet should be governed by the condition of uniform air distribution and *rapid temperature equalization*.

(*iv*) The selection of size of the outlet can be made from the manufacturer's catalogue data according to the air delivery, core velocity, distribution pattern, sound levels, throw, drop, spread, etc.

The outlets should be located so as to neutralize the *concentrated loads*, such as those that result from exterior windows, electronics equipment, etc. in buildings in which the lighting load is heavy, i.e., more than 55 W/m^2, and the ceiling height is more than 4.5 m, it is desirable to locate the outlets below tie lighting load.

In such and similar cases of concentrated loads, the return grilles or inlets can be located adjacent to these loads so that warm air (in the case of cooling) is withdrawn from the source instead of being dissipated in the conditioned space. This arrangement is also suitable to remove the fumes, pollutants, etc., from their sources in the space.

In their distribution pattern, the outlets may have characteristics in between the behaviour at the two extremes.

At one extreme are ceiling diffusers with radial flow. As a result of the large perimeter area of the primary air, they will have high entrainment rate and rapid temperature equalization in room. The air will, however, quickly slow down and will have a short throw.

At the other extreme are slot diffusers. They have low entrainment rate and slow temperature equalization. But they have a long throw.

Thus, generally speaking, ceiling diffusers can deliver more air to a space than grilles and slot diffusers. Because of their high entrainment, ceiling diffusers may also be used in systems with low supply air temperatures. In spite of the low supply air temperatures, induction will result in rapid-temperature equalization.

The same cannot be done in the case of slot diffusers and grilles. In their case, this temperature difference may not exceed 11°C. They are used only when the throw required is very long.

15.1.4 Distribution Patterns of Outlets

The general distribution patterns of various types of supply air outlets will now be described. The representation for primary air, total air, natural convection air and stagnant zone will be as shown in Fig. 15.4. It will be seen that the distribution patterns follow differently for cooling and heating. The best distribution pattern is one in which the whole room air is set in motion, and there are neither any stagnation zones nor zones of draft at the occupancy level.

(*a*) Primary Air

(*b*) Total Air

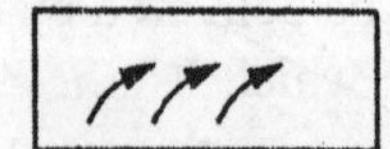

(c) Natural Convection

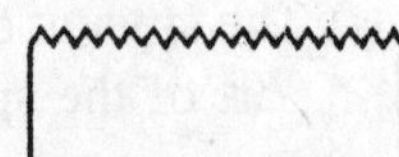

(*d*) Stagnant Zone

Fig. 15.4 Representation of primary, total and natural convection air and stagnation zone

1. ***High Sidewall Grilles/Ceiling Diffusers Discharging Air Horizontally.*** Fig. 15.5 (*a*) and (*b*) shows the distribution patterns for cooling and heating respectively for a high sidewall grille. The variation of the vane setting may affect the flow to some extent but the general pattern will be the same.

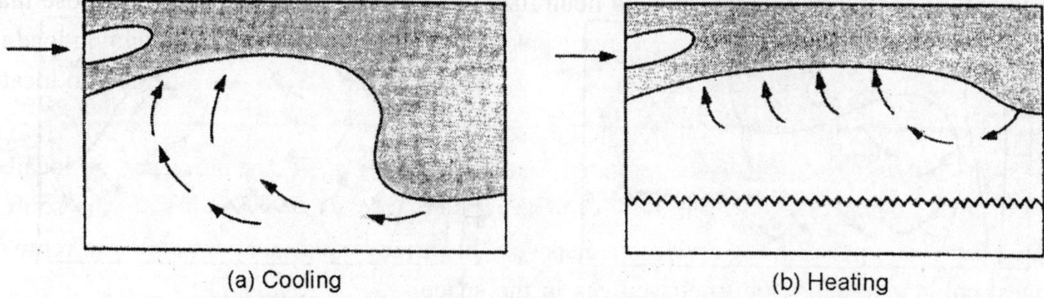

Fig. 15.5 Distribution patterns for high sidewall grille

During heating, the total warm air tends to rise. This results in a large stagnation degree zone. A degree of over-blow may be helpful in minimizing the stagnation zone.

The general pattern for ceiling diffusers projecting air horizontally is similar though symmetrical on the two sides as shown in Fig. 15.6 (*a*) and (*b*). There is hardly any stagnation zone for cooling application though the same cannot be said for the case of heating. During heating, cold air from the walls tends to drop but warm air tends to remain near the ceiling. A large stagnation zone results. An attempt must be made to direct the air towards the cold walls.

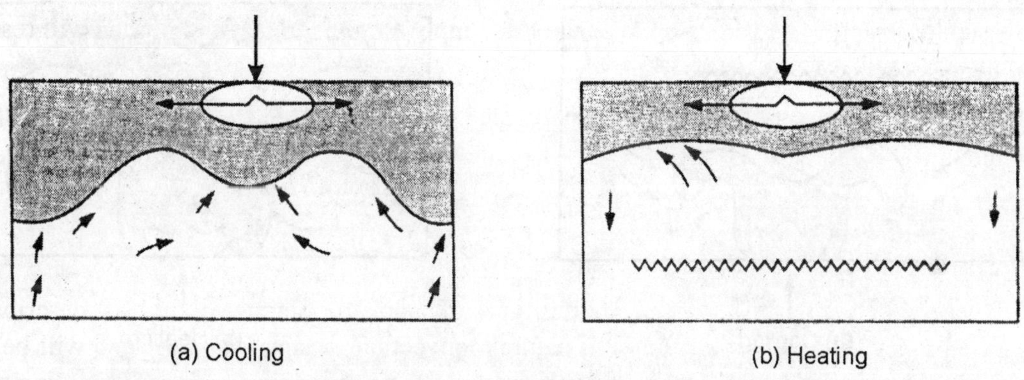

Fig. 15.6 Distribution patterns for ceiling diffusers projecting air horizontally

These outlets with high location are practically suited for cooling with a large dehumidified rise, of the order of 13 to 17°C. They are selected on the basis of drop and throw.

2. ***Floor Registers Discharging Air Vertically.*** Floor registers normally discharge primary air in a straight vertical jet as shown in Fig. 15.7. Ultimately the total air, after reaching the ceiling, fans out. In the case of cooling, it falls out soon after travelling short distance.

The cooling diagram shows stagnation region above the terminal point of the total air. In a large space, this stagnation zone may extend much farther and to a lower level. In the case of heating, the total air follows the ceiling and then descends down if flowing along the cold exterior walls.

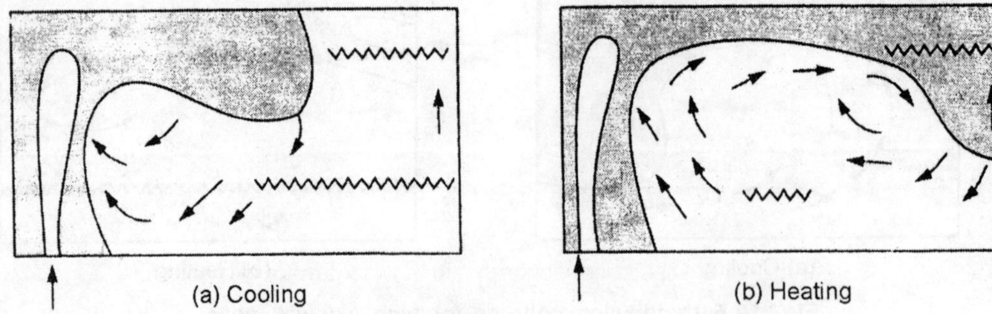

Fig. 15.7 Distribution patterns for floor registers

There is a better temperature equalization for heating than for cooling. In these outlets, generally, an increase of the supply air velocity will improve the air distribution. These outlets are more suited for heating only.

3. ***Floor Diffusers Discharging Air in a Spreading Jet.*** Floor diffusers are similar to floor registers. The only difference is in the nature of the jet which is spreading in this case, instead of being nonspreading as seen from Fig. 15.8. Although the characteristics are similar, the stagnation zone is much larger during cooling but smaller during heating.

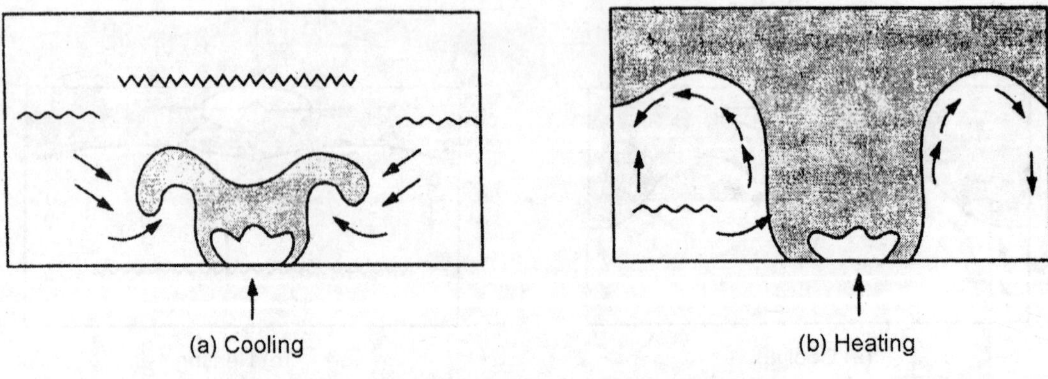

Fig. 15.8 Distribution patterns for floor diffusers

These outlets are suitable when the heating requirement is severe and primary, and the cooling requirement moderate and secondary.

Floor outlets are not permissible when people are seated such as in theatres. But where people are moving, as in stores, they are quite permissible. However, a very low dehumidified rise, say not more than 8°C should be used. This will require a large volume flow. One disadvantage of floor outlets is that they become dust collectors.

4. ***Low Sidewall Outlets Discharging Air Horizontally.*** As is seen from Fig. 15.9. the total air during cooling remains near the floor level resulting in low temperature in the occupied zone and a large stagnation zone above. During heating the warm air rises and temperature equalization takes place except in the region of total air.

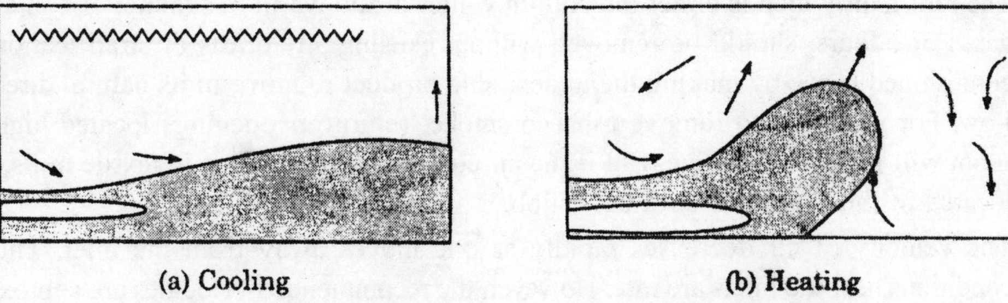

Fig. 15.9 Distribution patterns for low sidewall diffusers

These outlets discharge air directly into the occupied zone with high velocity. They are not recommended for comfort air conditioning.

5. ***Ceiling Diffusers Discharging Air Vertically.*** These are ceiling diffusers which do not project air horizontally, but vertically as shown in Fig. 15.10. During cooling, the total air drops to the floor and then fans out, finally rising along the walls. The stagnation region is near the ceiling. During heating, the total air, after reaching the floor, returns back towards the ceiling. There is no stagnation zone.

   These outlets have completely different distribution patterns for cooling and heating because of the different throws obtained. They are, therefore, used either for cooling or for heating, but seldom for both. For cooling, we require low values of supply air volume, velocity and temperature difference, whereas for heating, the same should be high to get proper throw.

   Nevertheless, ceiling diffusers can be conveniently applied to ducts or plenums in the ceiling in large spaces and halls/auditoriums.

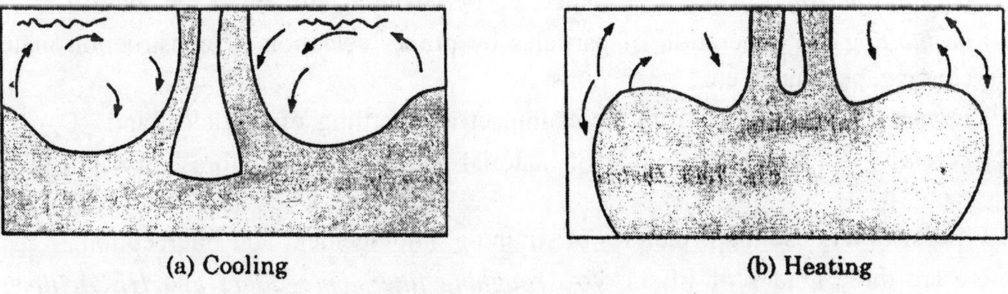

Fig. 15.10 Distribution patterns for ceiling diffusers projecting air vertically.

15.1.5 Locating Return Air Openings

While a great deal has been written about supply air outlets, not enough is known about locating return air openings. The most important considerations involved in the location of inlets are the following:

1. There should not be any short-circuiting of air between the supply outlet and return inlet. The return air opening should be complementary to the flow pattern generated by the supply air.

2. The return air which is either too cool in winter or too warm in summer or contains dirt, gases or odours, should be removed without causing any drafts or stratification in the conditioned space by making the undesirable product to move in its natural direction of flow. For example, to remove tobacco smoke, return air openings located high in the room will be most effective. But if the air contains cotton lint as in textile mills, returns located in the floor will do a better job.
3. The velocity of air decreases rapidly as one moves away from the inlet. Thus draft conditions near the inlets are rare. However, the recommended velocities are approximately 240 mpm for inlets above the occupied zone and 180 mpm within the occupied zone.
4. The consideration of noise is a little more severe in the case of return inlets as they are normally located near the ear-level. In that case the velocity in the return inlet should not exceed 75 per cent of the velocity in the supply air outlet. From the point of view of location, a wall return near the floor is the best. A ceiling return may short-circuit the cold air. A wall return near the ceiling is as undesirable as a ceiling return.

Floor returns should be avoided because they act as dust collectors. Whenever floor returns are used, it is necessary to provide a low velocity settling chamber.

Also, as stated earlier, in the case of concentrated load, the return opening should if/located near the load.

15.2 PROCESSING, TRANSMISSION AND DISTRIBUTION OF AIR IN CLEAN ROOMS

In a clean room, particles within the room are removed and controlled by

(*i*) *preventing* the entry of particles by using HEPA filters,

(*ii*) *purging* the room of particles generated by changing the air in the room,

(*iii*) *prohibiting* the generation of particles by proper selection of construction materials, clothing, personnel, etc.,

(*iv*) *protecting the product* from direct impact and settling of particles, and

(*v*) *providing* support for cleaning of material and personnel before entry into the clean room.

1. **Filters:** Filters use the principles of straining, impingement and interception.

There are three grades of filters, viz,. *roughing filters, pre-filters* and *HEPA filters.*

Roughing filters use the principle of straining. Roughing filters used in clean rooms have a number of layers of nets, with a layer provided with resin impregnated synthetic media, all stitched together. Pre-filters use the method of impingement which depends on the inertia force of the filtered particles. HEPA filters use a combination of impingment and interception.

The filtration media in HEPA filters is made of submicronic glass fibres. It is a continuous sheet formed into accordion pleats. The glass fibres are heat resistant retaining their strength upto 320°C. Aluminium separators are inserted between glass fibre sheet folds to form a *filter pack.* The filter pack is then sealed into an aluminium frame using suitable adhesive, and also provided with rubber gaskets on the sealing faces.

HEPA filters are tested both before installation, and once again, at site after installation to ensure that there are no leakages. The filter efficiency is computed as below:

$$\eta = \frac{\text{No. of particles before filter} - \text{No. of particles after filter}}{\text{No. of particles before filter}}$$

The tests are conducted with a uniform velocity of 0.46 m/s. Disctyl phthalate (DOP) smoke is introduced upstream of the filters. The entire filter bank is then scanned using a properly sized aerosol photometer probe.

HEPA filters should have a filtration efficiency of 99.97% down to 0.3 micron size particles. They can handle their rated capacities at an initial pressure drop of 25 mm H_2O, and are normally operated to a final pressure drop of 75 mm H_2O *before being discarded*. The life of HEPA filters can be enhanced by using roughing filters, and microfine pre-filters, particularly in the *make-up/ ventilation* air circuit, prior to them.

2. **Conventional Flow Clean Rooms:** Conditioned air in the clean rooms should be directed to obtain the *cleanest air at the most critical work areas*. Thereafter, as the particles are entrained, they are led to the less critical areas. These criteria require introduction of large quantities of air at low velocities.

Further, all rooms should be maintained at positive pressure, i.e.-at a static pressure sufficiently above atmospheric pressure-so as to prevent infiltration of contaminated outside air and permit only exfiltration.

A conventional flow clean room A/c system (cooling and dehumidification) is illustrated in Fig. 15.11.

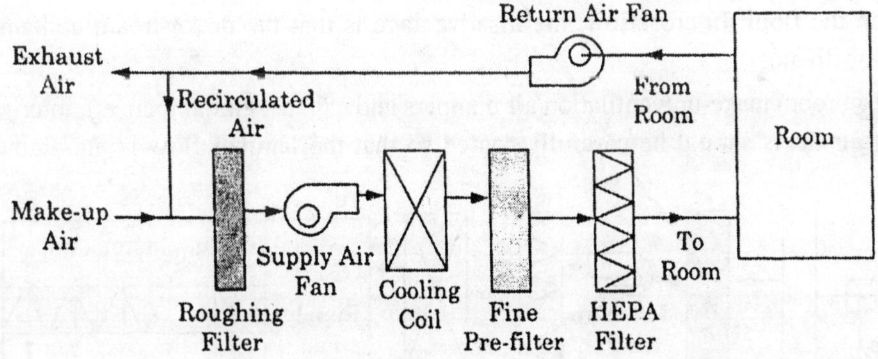

Fig. 15.11 Conventional flow clean room A/C systems

Major design features of the conventional flow clean rooms are 20 air changes per hour, positive pressure, final filters to be HEPA filters, installed only on discharge side of the fan, and to be preferably *terminally mounted*, people particle generating activities near air exit, etc. the building materials used should not chip off or flake. All surfaces should be flush finished. Surface finish is done by epoxy paint, etc. special lint free garments and clothing are to be used. Further, support areas are provided for *air showers* and *air locks*, restricting entry of personnel and introducing clean room discipline.

3. **Laminar Flow Clean Rooms:** It was found that even after taking all the measures described above, cleanliness levels could not be improved any further in conventional flow rooms. Major drawback is the build-up of smaller size particles. This method also does not prevent *cross-contamination* from one work area to another.

There was a major breakthrough in this respect with the application of *laminar* flow to clean rooms. It was realized that, instead of providing multiple outlets, if the entire ceiling or one wall were to made of HEPA filters and used as supply panel, and the entire flooring or opposite wall were to made up of return air grilles with prefilters installed, then the air distribution system would be the most effective. The advantages of laminar flow could be listed as follows:

(*i*) Air flows in parallel stream lines.

(*ii*) No cross -contamination of stream lines.

(*iii*) Predictable air path.

(*iv*) Independent of operation/activity/personnel.

(*v*) HEPA filters give lowest contamination.

(*vi*) No build-up of particles.

(*vii*) Isolating operations possible.

There are two versions of laminar flow; *downflow* in which the air is supplied from the ceiling and return is taken from the floor, and crossflow in which the air is supplied from one wall and return is taken from the opposite wall. The processing, transmission and distribution of the two is shown in Figs.15.12 and 15.13 respectivley.

In downflow, since the floor is made of open grilles, it is not possible to place heavy machinery on the floor. In crossflow, the disadvantage is that the downstream contamination is higher than upstream.

The clean room make-up/ventilation air dampers and exhaust air dampers are inter-connected. The exhaust air ducts should be carefully routed so that the laminar flow is maintained.

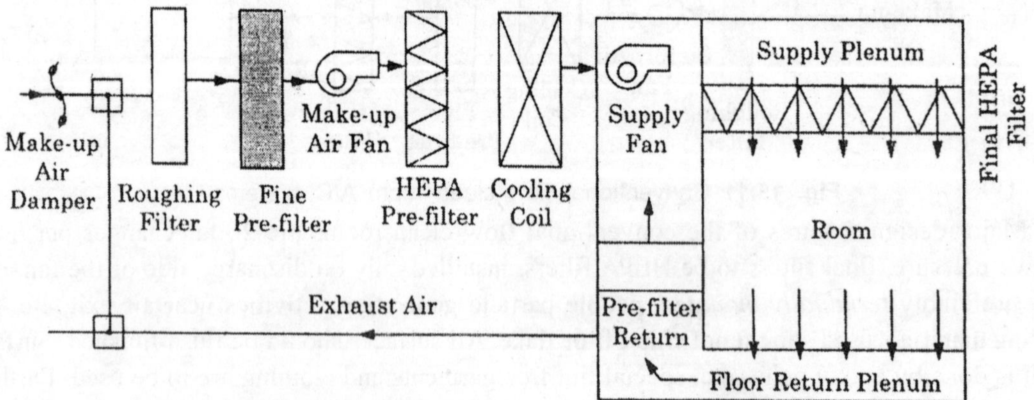

Fig. 15.12 Downflow laminar flow clean room A/C system

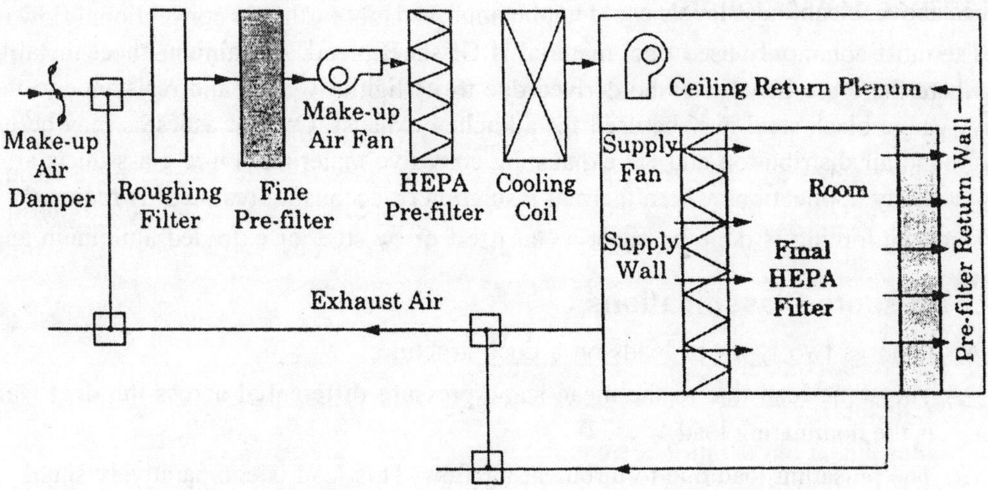

Fig. 15.13 Crossflow laminar flow clean room A/C system

15.3 AIR-DUCT DESIGN

The essential economics of an air transmission system is achieved by a proper balance between the initial or first cost and operating cost for the given flow rate of air. The first cost is determined by the cost of the duct system which depends on duct sizes. The operating cost is determined by the fan power consumption which depends on the pressure drop in the air-handling equipment and the duct system. The pressure drop can be reduced by increasing the size of the ducts but it will increase the first cost. Hence the need for a proper balance.

A few general rules are stated below which should be followed in the design of ducts.

(*i*) Air should be conveyed as directly as possible to economise on power, material and space,

(*ii*) Sudden changes in direction should be avoided. When bends are essential, turning vanes should be used to minimise the pressure loss,

(*iii*) Air velocities in ducts should be within permissible limits to minimise noise.

(*iv*) Diverging sections should be made gradual. The angle of divergence should not exceed 20°.

(*v*) Rectangular ducts should be made as nearly square as possible. This will ensure minimum duct surface, and hence cost, for the same air-carrying capacity. An aspect ratio of less than 4:1 should be maintained.

(*vi*) Ducts should be made of smooth materials such as galvanized iron (GI) or aluminium sheet metals. Whenever other materials are used, allowance should be made for the roughness of the material.

(*vii*) Dampers should be provided in each branch outlet for balancing the system,

(*viii*) Avoid duct obstructions.

15.3.1 Duct construction

The most commonly used duct material is GI sheet metal. Aluminium sheet metal may be used where there are benefits to be derived due to its lighter weight and resistance to moisture Heavy gauge black-steel may be used for a kitchen exhaust. Cement asbestos may be used for underground air distribution and for exhausting corrosive materials. Fibre glass ducts are in use in low velocity applications where thermal insulation or accoustic treatment is required.

Duct reinforcing is done by sheet metal itself or by steel or extruded aluminum angles.

15.3.2 Pressure Classifications

Air imposes two types of loads on a duct structure:

(*i*) The static load due to the mean static pressure differential across the duct wall. This is the dominating load.

(*ii*) The pulsating load due to turbulent air flow. This load is comparatively small.

Accordingly, the ducts are classified as follows:

(*i*) *Low-pressure system*

Velocities < 600 mpm and static pressure < 5 cm H_2O gauge,

(*ii*) *Medium-pressure system*

Velocities < 600 mpm or static pressure up to 15 cm H_2O gauge,

(*iii*) *High-pressure system*

Velocities < 600 mpm or static pressure over 15 cm and up to 25 cm H_2O gauge.

The low pressure system permits the use of simple forming methods. A majority of duct construction belongs to the low-pressure type. The gauge of the sheet metal depends on the dimension of the longest side as given in Table 15.1.

The recommended maximum duct velocities for the low-velocity are given in Table 15.2.

Medium and high pressure rectangular ducts may be used to meet the space limitations. It is generally a more expensive construction. Tie rods as well as transverse reinforcements are used. Recommended GI gauges for the two systems are also given in table 15.1

Table 15.1: Recommended GI gauges

Longest side in cm			
Low pressure	*Medium Pressure*	*High Pressure*	*GI Sheet Metal Gauge*
Upto 30	–	–	26
30-75	Upto 45	–	24
75-135	45-120	Upto 120	22
135-210	120-180	120-180	20
Above 210	Above 180	180-240	18
–	–	Above 240	16

15.3.3 Duct Design Procedures

There are three common methods for the sizing of ducts. They are:

(*i*) Equal friction method

(*ii*) Velocity reduction method

(*iii*) Static regain method

(*i*) ***Equal Friction (EF) Method:*** In the equal friction method, the fractional pressure drop per unit length of the duct is maintained constant throughout the duct system. The procedure is to select a suitable velocity in the main duct from sound level considerations as given in Table 15.2. Knowing the air flow rate and the velocity in the main duct, the size and friction loss are determined from the chart. The remaining ducts are then sized, maintaining the friction loss per unit length at this value for respective air-flow rates.

This method of sizing ducts, automatically, reduces the air velocity in the direction of flow. The method is generally recommended because of its simplicity. Moreover, the use of a calculator, called the *ductulator*, speeds up the design work.

If an equal friction design has a mixture of short and long runs of duct, the shortest will need a considerable amount of dampering. This is a drawback of the equal friction design.

To determine the total friction loss in the duct system, it is necessary to calculate the loss in the duct run that has the highest resistance.

(*ii*) ***Velocity Reduction Method:*** In this method, the main duct is designed in the same manner as in the equal friction method. Thereafter, arbitrary reductions are made in the air velocity as we go down the duct run. Equivalent diameters are found, as before, from the friction chart.

Through the method allows safe velocities, it is not normally adopted unless the person using it has considerable practical experience and knowledge to design within reasonable accuracy.

The starting velocities should not exceed those recommended in Table 15.2

Note: *High velocity duct/pipe system, with velocity approaching 1800 rpm, are used for air conditioning in ships and aircraft to economise on spaces.*

Table 15.2: Recommend maximum duct velocity for low velocity system (mpm)

Application	*Controlling factor: Noise*	*Controlling Factor :Friction*			
		Main Ducts		*Branch Ducts*	
	Main Ducts	*Supply*	*Return*	*Supply*	*Return*
Residences	180	300	240	180	180
Apartments					
Hotel bed rooms					
Hospital bed rooms	300	450	390	360	300
Offices Libraries	360	600	450	480	360
Threatres	240	390	330	300	240
Stores, Banks					
Restaurents	550	600	450	480	360
Cafeterias	540	600	450	480	360

(*iii*) ***Static Regain (SR) Method:*** The principle of the static regain method is to maintain a constant static pressure before each terminal and each branch. This is achieved by sizing the duct in such a manner that after each branch or outlet, the static pressure gain—due to reduction in velocity—exactly balances the pressure drop in the succeeding duct section.

The method meets the essential requirement of maintaining uniform static pressure at all branches and outlets. Thus it is a balanced system and the duct does not require the use of dampers anywhere.

15.3.4 Governing Equations for Ducts

1. **Continuity equation**

Quantity of air, $Q_1 = Q_2$

or $A_1c_1 = A_2c_2$ ...(15.1)

where A = cross-sectional area of duct

c = air velocity

Eq. (15.1) is the continuity equation.

Mass flow rate, $\dot{m}_1 = \dot{m}_2$

or $\rho_1 A_1 c_1 = \rho_2 A_2 c_2$

2. **Bernoulli's equation**

$$\frac{p}{\rho} + \frac{1}{2}c^2 + gz = \text{const.} \qquad ...(15.2)$$

If $\rho_1 = \rho_2$ and $z_1 = z_2$, then

$$p_{s1} + \frac{1}{2}\rho_1 c_1^2 = p_{s2} + \frac{1}{2}\rho_1 c_2^2$$

where p_s = static gauge pressure of air

or $p_{s1} + p_{v1} = p_{s2} + p_{v2}$

where p_v = dynamic or velocity pressure

Total pressure, $p_T = p_{s1} + p_{v1} = p_{s2} + p_{v2}$

Let P_{TF} = = fan total pressure

and P_L = total pressure drop

Then $p_{s1} + p_{v1} + p_{TF} = p_{s2} + p_{v2} + P_L$ ...(15.3)

3. **Pressure losses in ducts**

(*a*) Pressure loss due to friction,

$$p_f = \frac{fL\rho_a c^2}{2\ \text{m}} \qquad ...(15.4)$$

where f = friction factor

L = length of duct, m

c = mean velocity of air flowing through the duct, m/s

m = hydraulic mean depth, m

$= \dfrac{\text{Cross-section area of duct, } A}{\text{Wetted perimeter of duct, } P}$

ρ_a = density of air, kg/m^3

also $$p_f = \frac{fL}{m}\left(\frac{\rho_a c^2}{2}\right) = \frac{fL}{m} \times P_v \text{ mm of } H_2O \qquad ...(15.5)$$

$$p_v = \frac{1}{2}\rho_a c^2 = \left(\frac{c}{4.04}\right)^2 \text{ mm of } H_2O = \text{velocity pressure} \qquad ...(15.6)$$

For a circular duct of diameter D,

$$m = \frac{A}{P} = \frac{\frac{\pi}{4}D^2}{\pi D} = \frac{D}{4} \qquad ...(15.7)$$

For a rectangular direct of sides 'a' and b,

$$m = \frac{A}{P} = \frac{ab}{2(a+b)} \qquad ...(15.8)$$

Frictional factor, $$f = \frac{64}{Rc} \text{ for laminar flow} \qquad ...(15.9a)$$

$$= \frac{0.3164}{(Re)^{1/4}} \text{ for turbulent flow.} \qquad ...(15.9b)$$

where $Re = \dfrac{cD}{\nu}$ = Reynolds number

ν = kinematic viscosity, m^2/s

For rough pipes, $$f = \frac{1}{\left[1.74 - 2\log\left(\frac{2e}{D}\right)\right]^2} \qquad ...(15.10)$$

where e = absolute roughness of pipe surface, $\dfrac{e}{D}$ = roughness factor.

4. **Equivalent diameter of a circular duct for a rectangular duct**

$$D = \left[\frac{32a^3b^3}{\pi^2(a+b)}\right]^{1/5} = 1.265\left[\frac{a^3b^2}{a+b}\right]^{1/5} \text{ for same discharge of air} \qquad ...(15.11a)$$

$$= \frac{2ab}{a+b}, \text{ for same air velocity} \qquad ...(15.11b)$$

5. **Dynamic pressure loss**

$$p_d = \frac{fL_e}{m}\left(\frac{c}{4.04}\right)^2 \text{ mm of } H_2O \qquad ...(15.12)$$

where L_e = equivalent additional length

$$= \frac{C_d m}{f}$$

C_d = dynamic loss coefficient.

6. **Pressure losses**

(i) $p_L = \left(1-\frac{A_1}{A_2}\right)^2\left(\frac{c_1}{4.04}\right)^2 = \left(1-\frac{A_2}{A_1}\right)^2\left(\frac{c_2}{4.04}\right)^2$ for sudden enlargement or contraction ...(15.13)

(ii) $p_L = C_L\left(\frac{c_1}{4.04}\right)^2$ mm of H_2O, at suction of duct ...(15.14a)

$= \left(\frac{c_1}{4.04}\right)^2$ mm of H_2O, at discharge of duct ...(15.14b)

where C_L = loss coefficient

7. **Standard air:**

Temperature = 20°C
Pressure = 1.013 bar
Density = 1.2 kg/m^3
Dynamic viscocity = 18.18×10^{-6} N.s/m^2

Example 15.1 *A circular duct of 40 cm diameter is selected to carry air in an air - conditioned space at a velocity of 440 m/min. If this duct is to be replaced by a rectangular duct of aspect ratio 1.25, find the size of rectangular duct for equal friction method when:*

(i) Velocity of air in two ducts is same

(ii) Discharge rate of air in two ducts is same.

Solution: Given: d = 40 cm, v = 440 m/min, aspect ratio, $\frac{a}{b}$ = 1.25,

(i) For same air velocity in two ducts,

Equivalent diameter of circular duct,

$$d = \frac{2ab}{a+b}$$

$$40 = \frac{2\times 1.25b^2}{1.25b+b}$$

$$= \frac{2.5b^2}{2.25b} = \frac{2.5}{2.25}b$$

$$b = \frac{40 \times 2.25}{2.5} = 36 \text{ cm}$$

$$a = 1.25 \times 36 = 45 \text{ cm}$$

(*ii*) For same discharge in two ducts,

$$d = 1.265\left[\frac{a^3b^3}{a+b}\right]^{1/5}$$

$$40 = 1.265\left[\frac{(1.25)^3 b^6}{2.25b}\right]^{1/5}$$

$$= 1.2297b$$

$$b = 32.53 \text{ cm}$$

$$a = 40.66 \text{ cm}$$

Example 15.2 *With respect to the passive heating and cooling of buildings, explain the following concepts:*

(*i*) Direct gain principle

(*ii*) Indirect gain principle

(*iii*) Isolated gain principle

Solution: Passive Heating and Cooling of Buildings

A *passive solar building* may be defined as that in which thermal energy flows by natural means, enabling the system to function without external power. The operation of a passive system involves the control of the flow of thermal energy, and includes the ability to completely stop energy from escaping or entering the building (*e.g.*, by insulation) and the ability to vary the timing or location of energy flow inside the building (*e.g.*, by opening or closing spaces to each other).

A space-heating system comprises the space, a solar collector and a thermal storage. A space-cooling system comprises the space, an environmental sink (sky and ground) and a thermal or cold storage. The heat exchange between the three elements falls into two categories, viz., forced exchange (using fans, pumps, etc.) and natural exchange (involving conduction, natural convection and radiation). If all the significant exchanges linking the three elements of a heating or cooling system involve forced flow, the system is classified as active, and if they involve purely natural flow, the system is classified as *passive*. A mixture of the two may be called-hybrid system.

Three basic passive solar concepts have been defined; each involving differential relationships relating the Sun, storage mass and living space. They are *direct gain*, *indirect gain* and *isolated gain*. Each concept will now be dealt with, separately, from the viewpoints of principles, requirements, variations and controls.

It must be stated here that without the controls, the addition of the passive system with large glass areas and storage masses can cause much discomfort due to winter heat losses and summer overheating.

1. Direct Gain: This is the most common passive solar gain building. The living space is directly heated by the Sun and serves as a live-in collector. The basic requirements for direct-gain building are:

(*i*) A large south-facing glazing (collector area) with the living space exposed directly behind.

(*ii*) A floor and/or wall storage mass of significant dimension for solar exposure and for thermal capacity.

(*iii*) A method for isolating the storage from the exterior climatic conditions by using insulation between the storage mass and outdoors.

The best location of the storage mass is often decided by the physical laws governing natural heat flow by radiation and convection. Typical variations of this mass are given below:

(*a*) External building walls

(*b*) Internal walls

(*c*) Floor

(*d*) Free-standing masses inside the space.

In addition to storage location, there may be significant variation in storage materials providing thermal capacities and different time-lag properties. Storage materials may vary from concrete, brick, sand and ceramics, to water and other liquids either singly or in various combinations.

To add to the efficiency and usefulness of direct gain, several controls must be considered. To prevent unwanted heat gain, sunshading is required for the large expanse of south-facing glass. Due to the high location of the southern summer sun overhangs can provide adequate protection for vertical southern glazing, but some other solutions must be found for titled glazing, or those with east-west orientations faced with low sun angles.

To prevent unwanted heat loss, for example on sunless winter days and nights, movable insulation panels, shutters, etc., may be used. These will also prevent heat gain on hot summer days. Other controls include exhaust and ventilation arrangements.

2. Indirect Gain: In an indirect gain building, a storage mass collects heat directly from the sun, stores it, and then transfer it to the living space. The storage wall is placed just behind the glass as shown in Fig. 15.14. The basic requirements of such a wall, called the mass *Trombe wall*, is a large glass area with a large storage mass directly behind it to intercept the sun's rays.

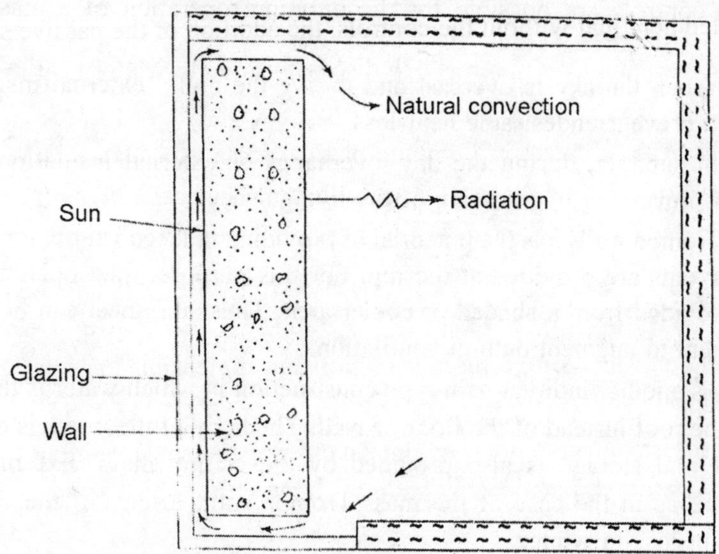

Fig. 15.14 Indirect gain principle illustrated by a mass trombe wall.

Again, variations are possible in the material used and the thickness of wall, depending on which the radiant distribution from the storage mass to the living space can be either instantaneous or delayed. It is possible to have distribution by natural convection through openings at the top of the wall as shown in Fig. 15.15 in addition to radiation and convection dirctly from the wall. If the openings are controllable, natural convection can be stopped and started as and when desired. Also, insulation placed behind the storage mass can eliminate radiant distribution.

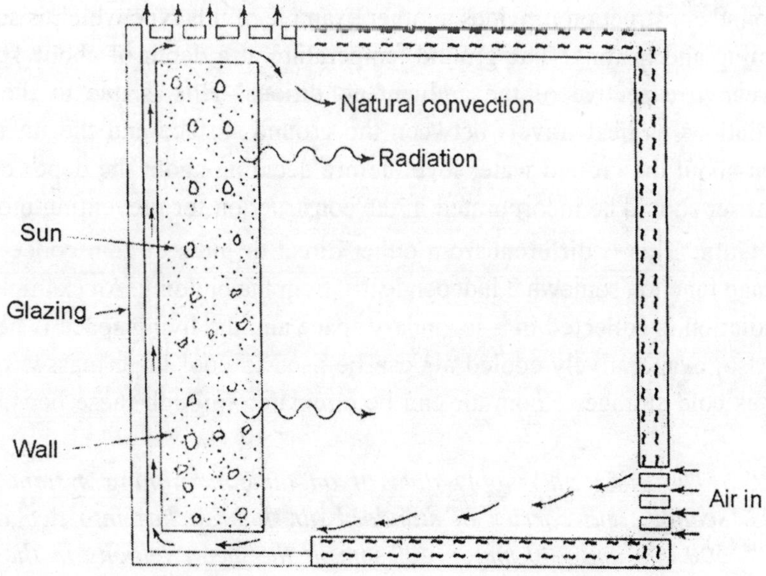

Fig. 15.15 Mass trombe wall with induced ventilation.

The following controls are possible for the optimum operation of a mass Trombe wall building.

(*i*) In winter, when the sky is overcast and during the night, external insulation must be provided to prevent undesirable heat loss.

(*ii*) Similarly in summer, during the day, overhangs or external insulation will serve the purpose of preventing overheating of the living space.

(*iii*) The mass Trombe walls has the potential of providing induced ventilation during summer if exhaust vents are provided at the top. For this purpose, inlet openings for fresh air must be provided from a shaded or cooler area. Thus sun's heat can be used to induce air movement to augment natural ventilation.

The roof pond is another indirect gain-type construction in which water is the storage mass, and it is located at the roof instead of the floor or wall. The body of the water is directly exposed to the sun. The thermal storage is also provided by the ceiling mass. External insulation is necessary for control, as in the case of the mass Trombe wall, to cut off the day heat gain in summer and the night loss in winter.

A clear night sky acts as a large heat sink. Thus, the heat collected by water during the day and stored in its mass is dissipated to the sky during the night. During the day, therefore, the water mass acts as a cold storage, drawing the heat away from the living space. It is obvious that such a system will be effective in a climate with a large swing in day and night temperatures.

Direct evaporative cooling and *Indirect* evaporative cooling are the other methods. For example, a shaded roof pond also acts as an indirect evaporative cooler. The water is cooled to near about the wet bulb temperature of air. Thus, if the roof is completely shaded, there will be a loss of heat from the living space which is at the dry bulb temperature of air—to the outside.

An underground construction provides another example of this type which is suitable both for passive solar cooling and heating. The ground temperature at a depth of about 10 m is uniform throughout the year, irrespective of the ambient conditions. This is due to the decrement in temperature oscillations as heat travels between the ground surface and the inside. Care must, however, be taken about the ground water level before deciding about the depth of construction. Also, a vapour barrier should be incorporated in the construction for preventing moisture transfer.

3. Isolated Gain: This is different from either direct or indirect gain concepts; In this, the collector and storage function somewhat independently from the building. An example is a solarium, in which solar radiation is collected in a secondary space and the living space is heated or cooled by convection. Also, evaporatively-cooled air can be used to cool large masses of rock beds at night, which act as cold storages Room air can be circulated through these beds during the day for cooling.

Example 15.3 *The main air supply duct of an air-conditioning system is 900 mm × 700 mm in cross- section, and carrier of standard air. It branches into two ducts of cross-section 700 mm × 500 mm and 700 mm × 400 mm. If the mean velocity in the larger branch is 540 m/min Calculate : (a) mean velocity in the main duct and smaller branch, and (b) mean velocity pressure in each duct.*

Solution: Given: $a_1 = 900$ mm, $b_1 = 700$ mm, $Q_1 = 30$ m^3/min or 6 m^3/s, $a_2 = 700$ mm, $b_2 = 500$ mm, $a_3 = 700$ mm, $b_3 = 400$ mm, $c_2 = 540$ m/min or 9 m/s

The cross - section of the duct is shown in Fig. 15.16.

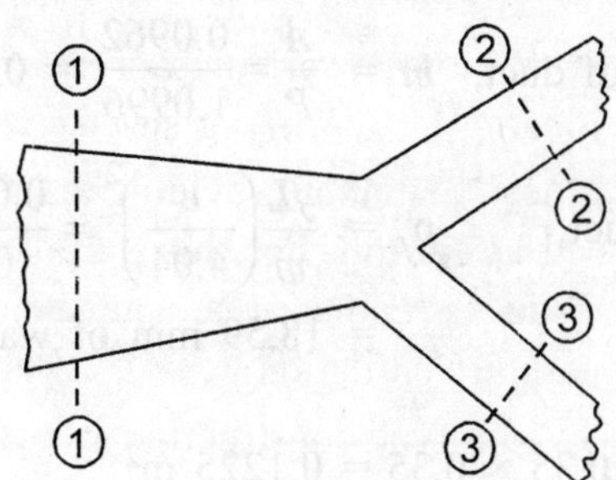

Fig. 15.16

$$A_1 = a_1 b_1 = 0.9 \times 0.7 = 0.63 \text{ m}^2$$
$$A_2 = a_2 b_2 = 0.7 \times 0.5 \text{ m} = 0.35 \text{ m}^2$$
$$A_3 = a_3 b_3 = 0.7 \times 0.4 = 0.28 \text{ mm}^2$$

$$(a)\ c_1 = \frac{Q_1}{A_1} = \frac{6}{0.63} = 9.836 \text{ m/s}$$

$$Q_2 = A_2 c_2 = 0.35 \times 9 = 3.15 \text{ m}^3\text{/s}$$
$$Q_3 = Q_1 - Q_2 = 6 - 3.15 = 2.85 \text{ m}^3\text{/s}$$

$$C_3 = \frac{Q_3}{A_3} = \frac{2.85}{0.28} = 10.18 \text{ m/s}$$

$$(b)\ p_{v1} = \left(\frac{c_1}{4.04}\right)^2 = \left(\frac{9.836}{4.04}\right)^2 = 5.927 \text{ mm of water}$$

$$p_{v2} = \left(\frac{c_2}{4.04}\right)^2 = \left(\frac{9}{4.04}\right)^2 = 4.963 \text{ mm of water}$$

$$p_{v3} = \left(\frac{c_3}{4.04}\right)^2 = \left(\frac{10.18}{4.04}\right)^2 = 6.349 \text{ mm of water}$$

Example 15.4 *A duct of 16 m length passes air at the rate of 96 m^3/min. Assuming the friction factor as 0.006, calculate the pressure drop in duct in mm of water when (a) the duct is circulated of diameter 0.35 m, and (b) the duct is of 0.35 m square cross-section.*

Solution: Given: $L = 16$ m, $Q = 96$ m^3/min or 1.6 m^3/s, $f = 0.006$, $D = 0.35$ m, $a = b = 0.35$ m

(*a*) Circular duct:

$$A = \frac{\pi}{4} D^2 = \frac{\pi}{4}(0.35)^2 = 0.0962 \text{ m}^2$$

$$C = \frac{Q}{A} = \frac{1.6}{0.0962} = 16.63 \text{ m/s}$$

Wetted perimeter, $p = \pi D = \pi \times 0.35 = 1.0996$ m

Hydraulic mean depth of duct, $m = \frac{A}{P} = \frac{0.0962}{1.0996} = 0.0875$

Pressure drop in the duct, $p_f = \frac{fL}{m}\left(\frac{c}{4.04}\right)^2 = \frac{0.006 \times 16}{0.0875}\left(\frac{16.63}{4.04}\right)^2$

$= 18.59$ mm of water

(*b*) Square duct:

$$A = a \times b = 0.35 \times 0.35 = 0.1225 \text{ m}^2$$

$$c = \frac{Q}{A} = \frac{1.6}{0.1225} = 13.06 \text{ m/s}$$

$$p = 2(a + b) = 2 \times 2 \times 0.35 = 1.4 \text{ m}$$

$$m = \frac{A}{p} = \frac{0.1225}{1.4} = 0.0875$$

$$p_f = \frac{fL}{m}\left(\frac{c}{4.04}\right)^2 = \frac{0.006 \times 16}{0.0875}\left(\frac{13.06}{4.04}\right)^2 = 11.46 \text{ mm of water}$$

15.4 FRICTION FACTOR FOR DUCTS

(*a*) Smooth Ducts

1. Friction Factor for laminar flow,

$$f = \frac{64}{\text{Re}} \quad ...(15.15)$$

2. Friction factor for turbulent flow,

$$f = \frac{0.3164}{(\text{Re})^{0.25}} \quad ...(15.16)$$

where R_e = Reynold's number

$$= \frac{CD}{\nu} \quad ...(15.17)$$

c = velocity of air, m/s

D = diameter of duct, *m*

ν = kinematic viscosity in m²/s $= \frac{\mu}{\rho_a}$

μ = absolute or dynamic viscosity, N.s/m²

ρ_a = mass density of air, kg/m³

(*b*) Rough ducts

In case of rough ducts, the friction factor f depends upon the roughness factor $\frac{e}{D}$, where e = absolute roughness of the surface and D is the diameter of the duct.

e = 0.15 mm for galvanized iron ducts

For R_e < 2000, the flow is laminar

= 2000 to 4000, the flow is transitional

> 4000, the flow is turbulent.

$$f = \frac{1}{\left[1.74 - \ln\left(\frac{2e}{D}\right)\right]^2} \qquad ...(15.18)$$

The values of friction factor f for different Reynold's numbers Re and roughness factors (e/D) may be read directly from Moody chart as shown in Fig. 15.17. Then the pressure loss due to friction in a circular duct is given by,

$$p_f = \frac{fL}{D} \times p_v = \frac{fL}{D}\left(\frac{c}{4.04}\right)^2 \text{ mm of water} \qquad ...(15.19)$$

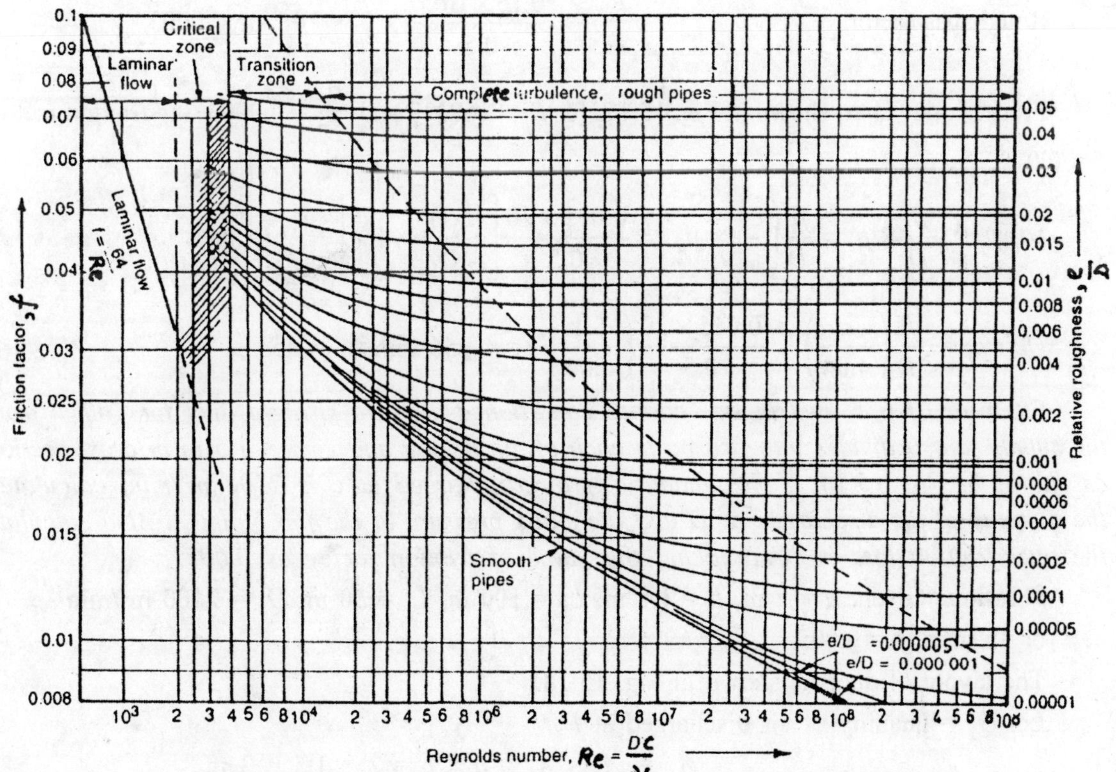

Fig. 15.17 Moody chart for friction factor in circular ducts.

Example 15.5 *A galvanized steel duct of 0.5 m diameter and 20 m long carriers air at 20°C and 1.013 bar. If the flow rate of air through the duct is 60 m³/min, calculate the pressure loss due to friction in (a) Pa, and (b) mm of water*

Solution: Given: $D = 0.5$, $L = 20$ m, $t = 20°C$, $p = 1.013$ bar, $Q = 60$ m³/min or 1 m³/s

Velocity of air, $$c = \frac{Q}{A} = \frac{Q}{\frac{\pi}{4}D^2} = \frac{1 \times 4}{\pi (0.5)^2} = 5.1 \text{ m/s}$$

Now $$pv = RT$$

Density of air, $$\rho_a = \frac{p}{RT} = \frac{1.013 \times 10^5}{287 \times 293} = 1.2 \text{ kg/m}^3$$

Dynamic viscosity of air at 20°C, $\mu = 18.18 \times 10^{-6}$ N.s/m²

Kinematic viscosity of air at 20°C, $$\nu = \frac{\mu}{\rho_a} = \frac{18.18 \times 10^{-6}}{1.2} = 1.515 \times 10^{-5} \text{ m}^2\text{/s}$$

Reynold's number, $$R_e = \frac{cD}{\nu} = \frac{5.1 \times 0.5}{1.515 \times 10^{-5}} = 1.68317 \times 10^5 \text{ or } 168317$$

For galvanized steel, $e = 0.15$ mm

Roughness factor, $$\frac{e}{D} = \frac{0.15 \times 10^{-3}}{0.5} = 3 \times 10^{-4} \text{ or } 0.0003$$

From Moody chart shown in Fig. 15.17, for $R_e = 168317$ and $\frac{e}{D} = 0.0003$, the friction factor, $f = 0.015$.

$$(a)\ p_f = \frac{fL}{D} \times p_v = \frac{fL}{D}\left(\frac{c^2}{2} \times \rho_a\right) = \frac{0.015 \times 20}{0.5}\left[\frac{(5.1)^2}{2} \times 1.2\right] = 9.36 \text{ pa}$$

$$(b)\ p_f = \frac{fL}{D}\left(\frac{c}{4.04}\right)^2 = \frac{0.015 \times 20}{0.5}\left(\frac{5.1}{4.04}\right)^2 = 0.956 \text{ mm of water}$$

Example 15.6 *A duct in size carries conditioned air in a straight duct for 100 m from the supply fan. It divides into two parts each of 50 m long and 2 m × 1.5 m in cross-section as shown in Fig. 15.16. If the quantity of air discharged at c is 2100 m³/min, calculated the quantity of air discharged at D and the static pressure at the fan outlet A. Also calculate the duct friction loss in N/m² taking the value of friction factor as 0.005.*

Solution: Given: $a = 2$ m, $b = 1.5$ m, $L_1 = 100$ m, $L_2 = 50$ m, $Q_3 = 2100$ m³/min or 35 m³/s, $f = 0.005$

The layout of ducts is shown in Fig. 15.18.

Let Q_2 = quantity of air discharged at D.

$$A_1 = A_2 = A_3 = a \times b = 2 \times 1.5 = 3 \text{ m}^2$$

$$C_3 = \frac{Q_3}{A_3} = \frac{35}{3} = 11.67 \text{ m/s}$$

Velocity pressure, $$p_v = \frac{c^2}{2} \times p_a = \frac{c^2}{2} \times 1.2 = 0.6 \text{ C}^2 \text{ N/m}^2$$

$$p_{v3} = 0.6c_3^2 = 0.6 \times \left(\frac{35}{3}\right)^2 = 81.67 \text{ N/m}^2$$

Hydraulic mean depth, $$m_1 = m_2 = m_3 = \frac{ab}{2(a+b)}$$

$$= \frac{2 \times 1.5}{2(2+1.5)} = 0.4286$$

Pressure loss due to friction in BC duct,

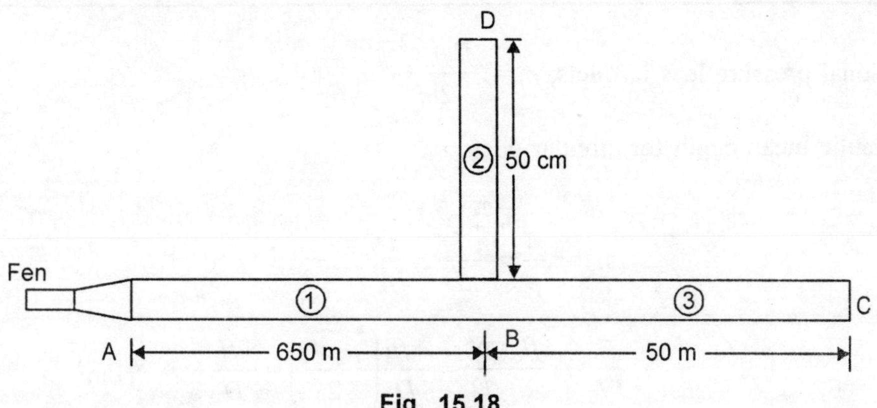

Fig. 15.18

$$p_{f3} = \frac{fL_3}{m_3} \times p_{v3} = \frac{0.005 \times 50}{0.4286} \times 81.67 = 47.63 \text{ N/m}^2$$

Total pressure at B, $$p_{tB} = p_{f3} + p_{v3} = 47.63 + 81.67 = 129.3 \text{ N/m}^2 \quad ...(1)$$

Let p_{v2} = velocity pressure in duct BD

c_2 = velocity of air in duct BD

$$\therefore \quad p_{f2} = \frac{fL_2}{m_2} \times p_{v2} = \frac{0.005 \times 50}{0.4286} \times p_{v2} = 0.5833\, p_{v2} \text{ N/m}^2 \quad ...(2)$$

Comparing Eqs. (1) and (2), we have

$$0.5833\, p_{v2} = 129.3$$

$$p_{v2} = 221.67 \text{ N/m}^2$$

Also $$p_{v2} = 0.6c_2^2 = 221.67$$

$$c_2 = 19.22 \text{ m/s}$$

$$Q_2 = A_2 c_2 = 3 \times 19.22 = 57.66 \text{ m}^3\text{/s or } 3460 \text{ m}^3\text{/min}$$

Static pressure at fan outlet:

$$Q_1 = Q_2 + Q_3 = 2100 + 3460 = 5560 \text{ m}^3/\text{min}$$

$$C_1 = \frac{Q_1}{A_1} = \frac{5560}{3} = 1853.3 \text{ m/min or } 30.9 \text{ m/s}$$

$$p_{v1} = 0.6c_1^2 = 0.6 \times (30.9)^2 = 572.9 \text{ N/m}^2$$

$$p_{f1} = \frac{fL_1}{m_1} \times p_{v1} = \frac{0.005 \times 100}{0.4286} = 572.9 = 668.3 \text{ N/m}^2$$

$$p_{tA} = p_{tB} + p_{f1} = 129.3 \; 668.3 = 797.6 \text{ N/m}^2$$

Static pressure at fan outlet, $p_{sA} = p_{tA} - p_{v1} = 797.6 - 572.9 = 224.7 \text{ N/m}^2$

15.5 FRICTION CHART FOR CIRCULAR DUCTS

Frictional pressure loss in ducts, $p_f = \dfrac{fLp_a c^2}{2m}$

Hydraulic mean depth for circular ducts,

$$m = \frac{A}{p} = \frac{\frac{\pi}{4}D^2}{\pi D} = \frac{D}{4}$$

$$\therefore \quad p_f = \frac{fLp_a c^2}{2 \times \frac{D}{4}} = \frac{4fL}{D}\left(\frac{p_a c^2}{2}\right) = \frac{4fL}{D} \times p_v \text{ N/m}^2 \qquad \text{...(15.20)}$$

where velocity pressure, $p_v = \dfrac{1}{2} p_a c^2$

According to first sche, $p_f = \dfrac{0.01422\, L(c)^{1.852}}{D^{1.269}}$ ...(15.21)

Where c is in m/s and L and D in m.

Mean velocity of air flowing through the duct,

$$c = \frac{\text{Volume flow rate}}{\text{Area of cross-section}} = \frac{4Q}{\pi D^2}$$

Eq. (15.21) becomes,

$$p_f = \frac{0.22243\, L\, Q^{1.852}}{D^{4.973}} \text{ N/m}^2 \qquad \text{...(15.22}a\text{)}$$

$$= \frac{0.002268\,LQ^{1.852}}{D^{4.973}} \text{ mm of water} \qquad \text{...(15.22}b\text{)}$$

$$= \frac{0.012199\,LC^{2.4865}}{Q^{0.06343}} \text{ N/m}^2 \qquad \text{...(15.22}c\text{)}$$

The frictional pressure loss for circular ducts in mm of water for various velocities (in m/s) and duct diameters (in mm) may be obtained directly from the friction charts as shown in Fig. 15.19 (*a*) and (*b*) for low and high flow rates respectively. There charts are valid for 20°C and 1.013 bar (p_a = 1.2 kg/m³) and lean galvanized iron ducts of circular cross-section. The following corrections may be made for other parameters:

1. For other materials than galvanized iron, correction factors should be applied.
2. For variation in density of air, $p_f \propto p_a$.
3. At other temperatures, $p_f \propto \dfrac{1}{T^{0.857}}$
4. For rectangular duct, equivalent circular diameter should be used.

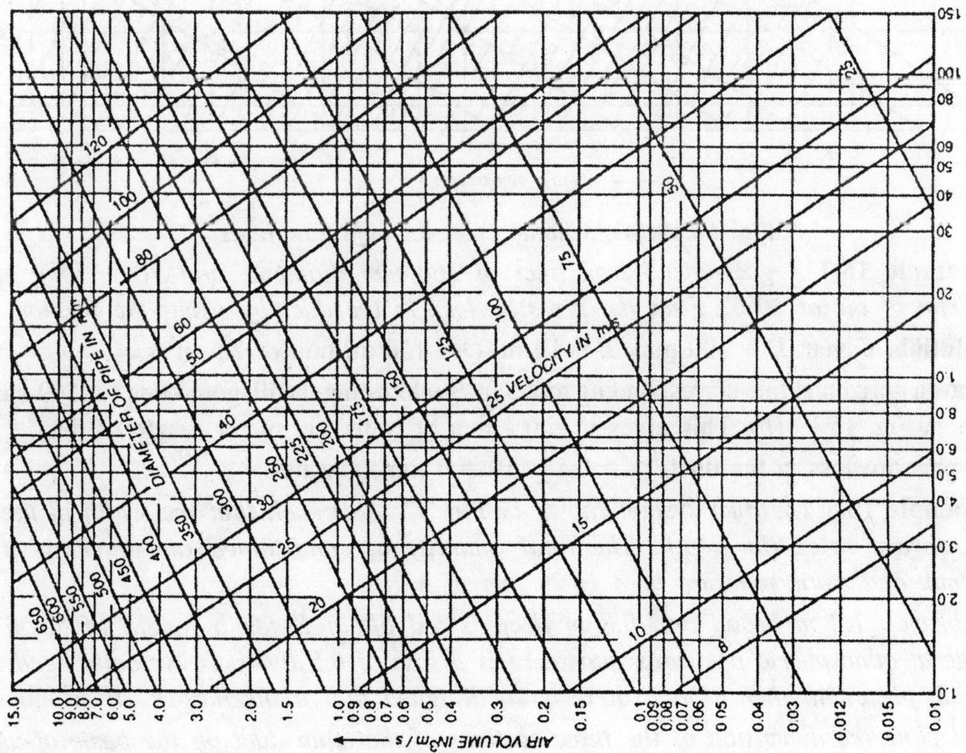

Fig. 15.19(*a*) Frictional chart for low flow rates

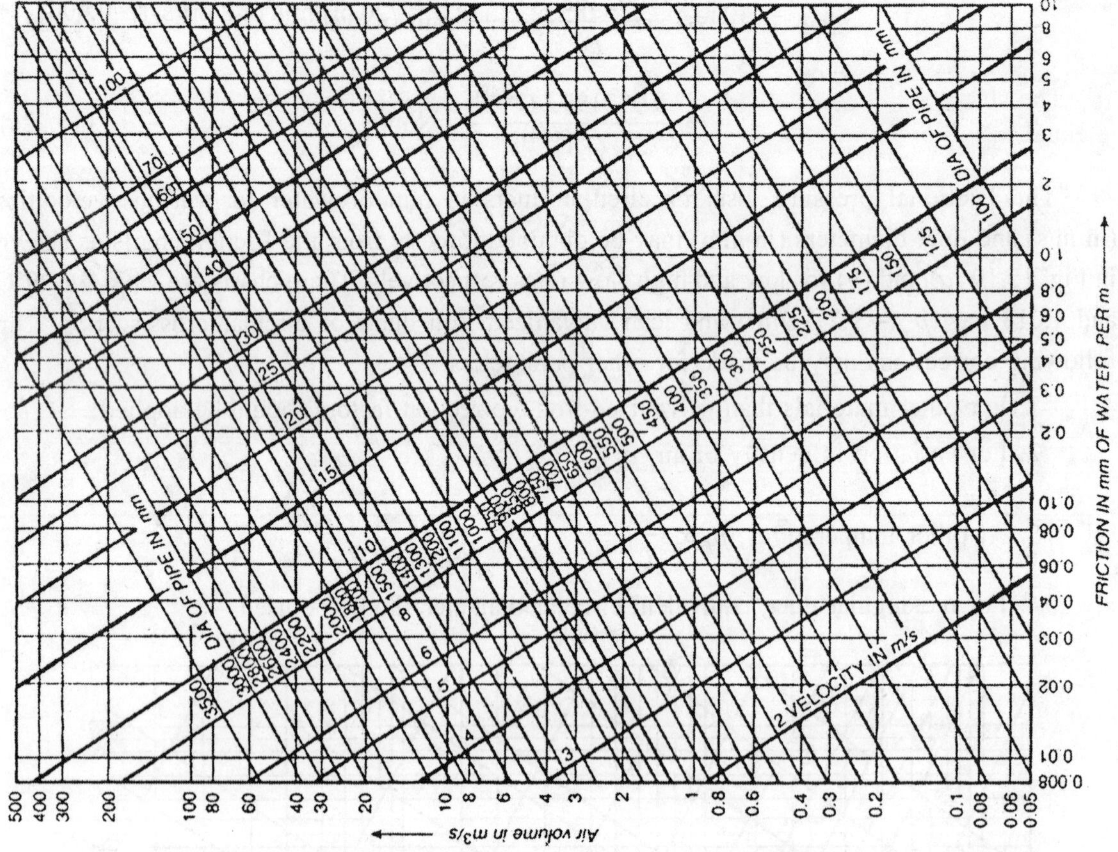

Fig. 15.19(*b*) Frictional chart for high flow rates

Example 15.7 *A galvanized iron duct of 500 mm diameter and 10 m long carries 150 m³/min of air at 20°C. Find the pressure loss in the duct by using the friction chart.*

Solution: Given: D = 500 mm, L = 10 m, Q = 150 m^3/min or 2.5 m^3/s

Draw a horizontal line corresponding to 2.5 m^3/s interesting the diagonal line for 500 mm duct diameter in Fig. 15.19 (*b*). This gives p_f = 0.4 mm of water per meter length of duct.

$\therefore$ pressure loss in the duct, $p_f = 0.4 \times 10 = 4$ mm of water.

Example 15.8 *The duct system shown in Fig. 15.18 is such that the width of the main duct is always twice the depth. The total quantity of conditioned air entering at A is 720 m³/min and static pressure at A is 20 mm of water.*

A branch 1.2 m broad and 0.8 m deep is led off at B for a length of 50 m. This discharge at atmospheric pressures and there is a loss of 0.5 times the velocity head at the conversion piece. Another branch at C is so designed that it discharges 20 m³/min.

Calculate the dimension of the three sections of the main duct on the basis of uniform pressure drop. Neglect velocity changes in the main duct and assume that discharge pressure at D atmospheric. The value of friction factor may be taken as 0.006.

Solution: Given: $Q_{AB} = 720 \text{ m}^3/\text{min or } 12 \text{ m}^3/\text{s}, p_{sA} = 20 \text{ mm of water}, a = 1.2 \text{ m}, b = 0.8 \text{ m},$
$L_{BE} = 50 \text{ m}, p_{l(BE)} = 0.5\, p_{v(BE)}, Q_{CF} = 720 \text{ m}^3/\text{min or } 2 \text{ m}^3/\text{s}, f = 0.006,$
$L_{AB} = 30 \text{ m}, L_{BC} = 40 \text{ m}, L_{CD} = 30 \text{ m}$

The velocity changes in main duct *AD* are neglect. Therefore the static pressure at *A* must be equal to the pressure loss due to friction between *A* and *D*.

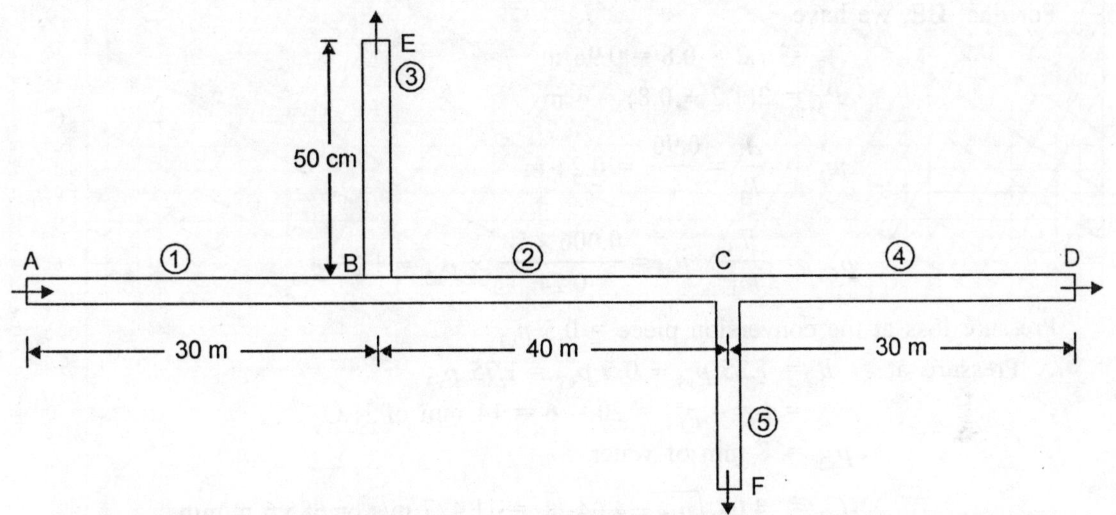

Fig. 15.20

$$\therefore \quad p_{f(AB)} + p_{f(BC)} + p_{f(CD)} = p_{sA} = 20 \text{ mm of } H_2O.$$

Total length of duct *AD*, $L = 30 + 40 + 30 = 100$ m

Pressure drop per m length of duct

$$AD = \frac{20}{100} = 0.2 \text{ mm of } H_2O$$

Dimensions of duct AB:

$$a_1 = 2b_1$$

$$A_1 = 2b_1^2$$

$$P_1 = 2(a_1 + b_1) = 2(2b_1 + b_1) = 6b_1$$

$$m_1 = \frac{A_1}{P_1} = \frac{2b_1^2}{6b_1} = \frac{b_1}{3}$$

$$c_1 = \frac{Q_{AB}}{A_1} = \frac{12}{2b_1^2} = \frac{6}{b_1^2}$$

$$p_{v1} = \left(\frac{c_1}{4.04}\right)^2 = \left(\frac{6}{b_1^2 \times 4.04}\right)^2 = \frac{2.2}{b_1^4}$$

$$p_{f1} = \frac{fL_1}{m_1} \times p_{v1} = \frac{0.006 \times 30 \times 3}{b_1} \times \frac{2.2}{b_1^4} = \frac{1.188}{b_1^5}$$

$$= 0.2 \times 30 = 6$$

$$b_1 = 0.723 \text{ m}, a_1 = 2b_1 = 1.446 \text{ m}$$

Dimensions of duct BC:

For duct BE, we have

$$A_3 = 1.2 \times 0.8 = 0.96 \text{ m}^2$$

$$P_3 = 2(1.2 + 0.8) = 4 \text{ m}$$

$$m_3 = \frac{A_3}{P_3} = \frac{0.96}{4} = 0.24 \text{ m}$$

$$p_{f3} = \frac{fL_3}{m_3} \times p_{v3} = \frac{0.006 \times 50}{0.24} \times p_{v3} = 1.25\, p_{v3}$$

Pressure loss at the conversion piece = $0.5\, p_{v3}$

$\therefore$ Pressure at $B = 1.25\, p_{v3} + 0.5\, p_{v3} = 1.75\, p_{v3}$

$$= p_{sA} - p_{f1} = 20 - 6 = 14 \text{ mm of } H_2O$$

$\therefore$ $p_{v3} = 8$ mm of water

$$C_3 = 4.04\sqrt{p_{v3}} = 4.04\sqrt{8} = 11.427 \text{ m/s or } 685.6 \text{ m/min}$$

$$Q_3 = A_3C_3 = 0.96 \times 685.6 = 658.2 \text{ m}^3\text{/min}$$

Then $Q_2 = Q_1 - Q_3 = 720 - 658.2 = 61.8 \text{ m}^3\text{/min}$

$$A_2 = a_2b_2 = 2b_2^2$$

$$p_2 = 6b_2$$

$$m_2 = \frac{A_2}{p_2} = \frac{b_2}{3}$$

$$c_2 = \frac{Q_2}{A_2} = \frac{61.8}{2b_2^2} = \frac{30.9}{b_2^2} \text{ m/min or } \frac{0.515}{b_2^4} \text{ m/s}$$

$$p_{v2} = \left(\frac{c_2}{4.04}\right)^2 = \left(\frac{0.515}{4.04\, b_2^2}\right)^2 = \frac{0.01625}{b_2^4}$$

$$p_{f2} = \frac{fL_2}{m_2} \times p_{v2} = \frac{0.006 \times 40 \times 3}{b_2} \times \frac{0.01625}{b_2^4} = \frac{0.0117}{b_2^5}$$

Now $p_{f2} = 0.2 \times 40 = 8 \text{ mm of } H_2O$

$\therefore$ $\frac{0.0117}{b_2^5} = 8$

$$b_2 = 0.271 \text{ m}, a_2 = 0.542 \text{ m}$$

Dimensions of duct CD:

$$A_4 = a_4 b_4 = 2b_4^2$$

$$p_4 = 6b_4$$

$$m_4 = \frac{A_4}{p_4} = \frac{b_4}{3}$$

$$Q_4 = Q_2 - Q_5 = 61.8 - 20 = 41.8 \text{ m}^3\text{/min}$$

$$C_4 = \frac{Q_4}{A_4} = \frac{41.8}{2b_4^2 \times 60} = \frac{0.3483}{b_4^2} \text{ m/s}$$

$$p_{v4} = \left(\frac{c_4}{4.04}\right)^2 = \left(\frac{0.3483}{4.04 \times b_4^2}\right)^2 = \frac{7.434 \times 10^{-3}}{b_4^4} \text{ mm of H}_2\text{O}$$

$$p_{f4} = \frac{fL_4}{m_4} \times p_{v4} = \frac{0.006 \times 30 \times 3}{b_4} \times \frac{7.434 \times 10^{-3}}{b_4^4} = \frac{4.014 \times 10^{-3}}{b_4^5}$$

Now $p_{f4} = 0.2 \times 30 = 6$ mm of H_2O

$$\therefore \quad \frac{4.014 \times 10^{-3}}{b_4^5} = 6$$

$$b_4 = 0.232 \text{ m}, \; a_4 = 0.464 \text{ m}$$

MULTIPLE-CHOICE QUESTIONS

1. If the velocity of air flowing through a duct is v m/min, then the velocity pressure p_v in mm of water is given by

(a) $\left(\frac{v}{4.04}\right)^2$ (b) $\left(\frac{v}{40.4}\right)^2$ (c) $\left(\frac{v}{242.4}\right)^2$ (d) $\left(\frac{v}{2424}\right)^2$

2. The equivalent diameter of a circular duct for a rectangular duct for the same pressure loss per unit length is

(a) $\frac{a+b}{ab}$ (b) $\frac{ab}{a+b}$ (c) $\frac{2ab}{a+b}$ (d) $\frac{a+b}{2ab}$

3. Equal friction method for duct design is ideal for

(a) return ducts (b) balanced system
(c) non-balanced system (d) minimum cost

4. The Velocity of air in a low velocity duct is upto

(a) 10 m/s (b) 15 m/s (c) 20 m/s (d) 25 m/s

5. A duct is made of

(a) galvanized iron sheet (b) aluminium sheet
(c) fiber glass (d) cement asbestos

6. The equivalent diameter D of a circular disc for a rectangular duct of size for the same pressure loss per unit length and same rate of air flow is given by

(a) $D = 1.265\left(\frac{a^3b^3}{a+b}\right)^{1/5}$ (b) $D = 1.265\left(\frac{a^4b^4}{a+b}\right)$

(c) $D = 1.265\left(\frac{a+b}{a^3+b^3}\right)^{1/5}$ (d) $D = 1.265\left(\frac{a+b}{a^4+b^4}\right)^{1/5}$

7. The aspect ratio for a rectangular duct is

(a) sum of longer and shorter sides (b) difference of longer and shorter sides
(c) product of longer and shorter sides (d) ratio of linger and shorter sides.

8. The static regain method of designing the ducts as compared to equal friction method

(a) increases balancing problems
(b) increases the cost of sheet metal for the duct
(c) decreases the cost of sheet metal for the duct
(d) has no effect on the cost of sheet metal for the duct.

Answers

1. (a) **2.** (c) **3.** (b) **4.** (a) **5.** (a) **6.** (a) **7.** (d) **8.** (b)

REVIEW QUESTIONS

1. What is the role of ducts in air-conditioning?
2. Which material is generally used for making ducts?
3. Explain static and velocity pressure in a duct?
4. What is the use of dampers?
5. Name the different methods used for duct design.

THEORETICAL QUESTIONS

15.1 What are the various methods of air distribution?

15.2 What are the general rules for duct design?

15.3 Describe the procedures for duct design.

15.4 Why are 'Ducts' used in an air-conditioning system and how do they help in proper air distribution? Describe, in brief, the concept of air-flow through ducts, explaining the different factors, which lead to pressure drop in ducts.

15.5 What is the importance of Ducts in an air conditioning system and what are the different methods generally used for designing of ducts? What do you understand by passive heating or cooling of buildings? What effect does infiltration air have on cooling loads?

15.6 With the help of example, explain any one of the following methods of duct design:

(1) Equal Friction method.

(2) Static Regain method.

15.7 Write short notes on frictional losses and dynamic losses in a flow through a duct. Write the expression for frictional pressure drop in ducts with proper nomenclature.

15.8 What is the importance of Ducts and Fans in transmission and distribution of air in a central air conditioning system? Describe in brief, the concept of air flow through ducts, explaining the different factors which leads to pressure drop in ducts.

15.9 With the help of example any one of the following methods of duct design

(i) Equal friction method

(ii) Velocity Reduction method

15.10 A duct of 25 m length passes air at the rate of 120 m3/min. Assuming friction factor as 0.006, calculate the pressure drop in the duct in mm of water when (a) the duct is circular of diameter 0.45m, and (b) the duct is rectangular of sides 0.6 m and 0.45m.

[**Ans.** 13 mm of water; 3.93 mm of water]

15.11 A duct of galvanized steel is of 300 mm diameter and 15 m long. Find the pressure drop in the duct in N/m^2 when air at 30°C is flowing with a mean velocity of 480 m/min. Use Moody chart to find the value of friction factor '*f*'. [**Ans.** 33.4 N/m^2]

15.12 A 0.3 m diameter circular duct carries standard air at a velocity of 360 m/min. it is replaced by a rectangular duct having the same pressure loss per unit length due to friction. Determine the dimensions of the rectangular duct if the aspect ratio is to be 1.5 for (a) the same volume flow through the two ducts, and (b) the same velocity in the two ducts. [**Ans.** 0.336 m, 0.224 m; 0.375 m, 0.25 m]

15.13 A main duct 0.5 m × 0.5 m carries air at a floe rate of 300 m^3/min. If branches into two ducts 0.5 m × 0.25 m and 0.5 m × 0.3 m. If the mean velocity in the larger branch is 540 m/min, calculate the mean velocity in the main duct and the other branch. Also calculate the mean velocity pressure in each duct assuming standard density.

15.14 The velocity of air chosen through a circular duct which carries standard air is 360 m/min. the size of the duct is 0.3 m diameter. If this duct is replaced by a rectangular duct of aspect ratio 1.5, calculate the size of the rectangular duct for equal friction method when:

(a) The velocity of air in two ducts is same

(b) The discharge rate through the line duct is same

[**Ans.** 0.375 m, 0.250 m; 0.336 m, 0.224 m]

15.15 A ventilation duct 1.5 m × 1 m runs straight for 50 m from the supply fan. It divides then into two parts *BC* and *BD* each 70 m long and of size 0.5 m × 1 m. The branch *BD* is lined with noise absorbent and divided into three parts. If the air discharged at C is 1600 m^3/min, calculate the quantity discharged at *D* and the static pressure at fan outlet. Take friction factor = 0.005 for sheet metal and 0.006 for absorbent material.

[**Ans.** 825 m^3/min, 8.52 mm of water]

Chapter

16

FANS

16.1 FANS

A fan is a kind of pump which is used for pumping or circulating the air through the entire duct system and the conditioned space. It consists of a rotating wheel (impeller) which is surrounded by a stationary member (housing). As the impeller fitted with blades rotate, energy is transmitted to air and a pressure difference is created to provide flow of air. The air fed into a fan is called induced draft and the exhaust from the fan is called forced draft.

There are two types of fans used for transmission of air: centrifugal or radial flows fans, and axial flow fans.

Total pressure developed by a fan,

$$p_t = p_{t2} - p_{t1}$$

where p_{t2} = total pressure at fan outlet

$= p_{s2} + p_{v2}$

p_{t1} = total pressure at fan inlet

$= p_{s1} + p_{v1}$

p_{s1}, p_{s2} = static pressure at inlet and outlet respectively.

p_{v1}, p_{v2} = velocity pressure at inlet and outlet respectively.

Total fan air power = work done by the fan

$$P_{at} = \frac{9.81\, Q p_t\, K_p}{60} \quad ...(16.1)$$

where Q = air flow rate, m^3/min

p_t = fan total pressure, mm of H_2O

K_p = compressibility coefficient.

Static fan air power, $$P_{as} = \frac{9.81\, Q p_s\, K_p}{60} \text{ watts} \quad ...(16.2)$$

Total fan efficiency or mechanical efficiency,

$$\eta_t = \frac{\text{Total fan air power}}{\text{Input or brake power}} = \frac{P_{at}}{B.P.} \quad ...(16.3)$$

Static fan efficiency, $$\eta_s = \frac{\text{Static fan air power}}{\text{Input or brake power}} = \frac{P_{as}}{B.P.} \quad ...(16.4)$$

16.2 FANS PERFORMANCE CURVES

The fans performance curves are shown in Fig. 16.1.

$$\text{Specific speed of a fan,} \quad N_s = \frac{N\sqrt{Q}}{H^{3/4}} \quad \ldots(16.5)$$

where N = impeller speed, rpm

Q = fan discharge, m^3/s

H = fan head developed, m

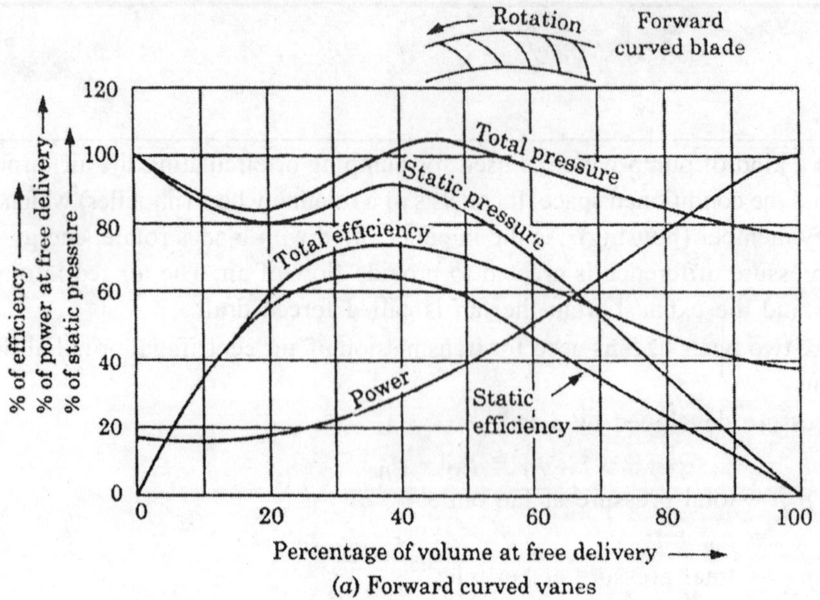

(a) Forward curved vanes

Percentage of volume at free delivery

(b) Backward curved vanes.

Fig. 16.1 Performance curves for a centrifugal fan

16.3 FAN SIMILARITY LAWS

Fan similarity laws are:

1. $Q \propto N \propto D^3$

$$\frac{Q_1}{Q_2} = \frac{N_1}{N_2} = \left(\frac{D_1}{D_2}\right)^3 \quad ...(16.6)$$

2. $p_t \propto N^2 \propto D^2 \propto \rho_a$

$$\frac{p_{t1}}{p_{t2}} = \left(\frac{N_1}{N_2}\right)^2 = \left(\frac{D_1}{D_2}\right)^2 = \frac{\rho_{a1}}{\rho_{a2}} \quad ...(16.7)$$

3. $$\frac{H_1}{H_2} = \left(\frac{N_1 D_1}{N_2 D_2}\right)^2 \quad ...(16.8)$$

4. $P_a \propto N^3 \propto D^5 \propto \rho_a$

$$\frac{P_{a1}}{P_{a2}} = \left(\frac{N_1}{N_2}\right)^3 = \left(\frac{D_1}{D_2}\right)^5 = \frac{\rho_{a1}}{\rho_{a2}} \quad ...(16.9)$$

5. $\eta_1 = \eta_2$ ...(16.10)

16.4 FANS IN SERIES AND PARALLEL

1. Fans in series:

$$Q = Q_1 = Q_2 = Q_3 \quad ...(16.11)$$

$$p_t = p_{t1} + p_{t2} + p_{t3} \quad ...(16.12)$$

2. Fans in parallel:

$$Q = Q_1 + Q_2 + Q_3 \quad ...(16.13)$$

$$p_{t1} = p_{t2} = p_{t3} \quad ...(16.14)$$

Example 16.1 *The following data refer to a centrifugal fan:*

Diameter of inlet duct = *0.6 m*
Size of rectangular duct = *0.6 m × 0.5 m*
Static pressure at fan inlet = *–12 mm of water*
Static pressure at fan outlet = *25 mm of water*
Quantity of air delivered = *140 m³/min*
Power absorbed = *1.3 kW*

Calculate: (a) Total pressure at fan inlet and outlet.
(b) Fan total pressure and fan static pressure
(c) Fan total efficiency and fan static efficiency.

Take compressibility coefficient = *1*

Solution: Given: $D = 0.6$ m, $a = 0.6$ m, $b = 0.5$ m, $p_{s1} = -12$ mm of water, $p_{s2} = 25$ mm of water, $Q = 140$ m^3/min, $BP = 1.3$ kW, $K_p = 1$

(*a*)
$$A_1 = \frac{\pi}{4}D^2 = \frac{\pi}{4}(0.6)^2 = 0.2827 \text{ m}^2$$

$$C_1 = \frac{Q}{A_1} = \frac{140}{0.2827 \times 60} = 8.25 \text{ m/s}$$

$$p_{v1} = \left(\frac{C_1}{4.04}\right)^2 = \left(\frac{8.25}{4.04}\right)^2 = 4.17 \text{ mm of water}$$

$$p_{t1} = p_{s1} + p_{v1} = -12 + 4.17 = -7.83 \text{ mm of water}$$

$$A_2 = a \times b = 0.6 \times 0.5 = 0.3 \text{ m}^2$$

$$C_2 = \frac{Q}{A_2} = \frac{140}{0.3 \times 60} = 7.78 \text{ m/s}$$

$$p_{v2} = \left(\frac{C_2}{4.04}\right)^2 = \left(\frac{7.78}{4.04}\right)^2 = 3.7 \text{ mm of water}$$

$$p_{t2} = p_{s2} + p_{v2} = 25 + 3.7 = 28.7 \text{ mm of water}$$

(*b*)
$$(p_t)_{fan} = p_{t2} - p_{t1} = 28.7 - (-7.83) = 36.53 \text{ mm of water}$$

$$(p_s)_{fan} = (p_t)_{fan} - p_{v2} = 36.53 - 3.7 = 32.83 \text{ mm of water}$$

(*c*)
$$(P_{fan})_{total} = \frac{9.81 Q (p_t)_{fan} \times K_p}{60 \times 10^3} \text{ kW}$$

$$= \frac{9.81 \times 140 \times 36.53 \times 1}{60 \times 10^3} = 0.836 \text{ kW}$$

$$(P_{fan}) = \frac{9.81 Q (p_s)_{fan} \times K_p}{60 \times 10^3} \text{ kW}$$

$$= \frac{9.81 \times 140 \times 32.83 \times 1}{60 \times 10^3} = 0.751 \text{ kW}$$

$$(\eta_{fan})_{total} = \frac{(P_{fan})_{total}}{BP} = \frac{0.836}{1.3} = 0.6430 \text{ or } 64.3\%$$

$$(\eta_{fan})_{static} = \frac{(P_{fan})_{static}}{BP} = \frac{0.751}{1.3} = 0.5777 \text{ or } 57.77\%$$

Example 16.2 *A fan 600 mm is diameter is running at 100 rpm, delivering 100 m^3/min of air at 15°C against 72 mm of water of total pressure. Its total efficiency is 80%. Calculate the volume of air delivered, total pressure developed and power consumed, if (a) air temperature is 42°C (b) fan speed is increased to 1250 rpm at 42°C, (c) conditions are same in (b) but a 450 mm diameter, geometrically similar fan is used.*

Solution: Given: $D_1 = 600$ mm, $N_1 = 100$ rpm, $Q_1 = 100$ m^3/min, $t_1 = 15$°C, $(p_{t1})_{fan} = 72$ mm of water, $(\eta_{fan})_{total} = 80\%$, $N_3 = 1250$ rpm, $D_4 = 450$ mm, $t_2 = 42$°C

$$\frac{\rho_{a1}}{\rho_{a2}} = \frac{273+t_2}{273+t_1} = \frac{273+42}{273+15} = 1.094$$

(*a*)
$$\frac{Q_1}{N_1 D_1^3} = \frac{Q_2}{N_2 D_2^3}$$

$$D_1 = D_2 \text{ and } N_1 = N_2$$

$$\therefore \quad Q_1 = Q_2 = 100 \text{ m}^3\text{/min}$$

$$(p_{t2})_{\text{fan}} = (p_{t1})_{\text{fan}} \times \frac{\rho_{a2}}{\rho_{a1}} = 72 \times \frac{1}{1.094} = 65.83 \text{ mm of water}$$

$$= 65.83 \times 9.81 = 645.8 \text{ N/m}^2$$

$$P_2 = \frac{Q_2 \times (p_{t2})_{\text{fan}}}{\eta_{\text{fan}}} = \frac{100 \times 645.8}{60 \times 0.8 \times 10^3} = 1.345 \text{ kW}$$

(*b*)
$$\frac{Q_3}{N_3} = \frac{Q_2}{N_2}$$

$$Q_3 = \frac{1250}{1000} \times 100 = 125 \text{ m}^3\text{/min}$$

$$\frac{(p_{t3})_{\text{fan}}}{N_3^2} = \frac{(p_{t2})_{\text{fan}}}{N_2^2} \qquad (\because D_2 = D_3 \text{ and } \rho_{a3} = \rho_{a2})$$

$$(p_{t3})_{\text{fan}} = 65.83 \times \left(\frac{1250}{1000}\right)^2 = 102.86 \text{ mm of water}$$

$$\frac{P_3}{N_3^3} = \frac{P_2}{N_2^3} \qquad (\because D_2 = D_3 \text{ and } \rho_{a2} = \rho_{a3})$$

$$P_3 = 1.345\left(\frac{1250}{1000}\right)^3 = 2.627 \text{ kW}$$

(*c*)
$$\frac{Q_4}{D_4^3} = \frac{Q_3}{D_3^3}$$

$$Q_4 = 125 \times \left(\frac{450}{600}\right)^3 = 52.73 \text{ m}^3\text{/min}$$

$$\frac{(p_{t4})_{\text{fan}}}{D_4^2} = \frac{(p_{t3})_{\text{fan}}}{D_3^2} \qquad (N_3 = N_4,\ \rho_{a3} = \rho_{a4})$$

$$(p_{t4})_{\text{fan}} = 102.86 \times \left(\frac{450}{600}\right)^3 = 43.4 \text{ mm of water}$$

$$\frac{P_4}{D_4^5} = \frac{P_3}{D_3^5} \qquad (\because N_3 = N_4,\ \rho_{a3} = \rho_{a4})$$

$$P_4 = 2.672 \times \left(\frac{450}{600}\right)^5 = 0.623 \text{ kW}$$

MULTIPLE-CHOICE QUESTIONS

1. The power of a fan is directly proportional to

(a) N (b) N^2 (c) N^3 (d) N^4

Where N = fan speed

2. The power of a fan is directly proportional to

(a) D^2 (b) D^3 (c) D^4 (d) D^5

Where D = impeller diameter

3. Two fans that are of different sizes, but have the same basic shape to their performance curves, are called

(a) base fans (b) tube axial fans

(c) geometrically similar fans (d) multi-stage fans.

4. The capacity of a fan is directly proportional to

(a) D^2 (b) D^3 (c) D^4 (d) D^5

Where D = impeller diameter

5. The air at inlet of a fan is

(a) above atmospheric pressure (b) below atmospheric pressure

(c) at atmospheric pressure (d) at any of the above.

6. The specific speed of a fan is

(a) $\frac{N\sqrt{Q}}{H^{2/3}}$ (b) $\frac{N\sqrt{Q}}{H^{3/4}}$ (c) $\frac{N\sqrt{Q}}{H^{5/4}}$ (d) $\frac{N\sqrt{Q}}{H^{4/3}}$

7. For fans in series

(a) $Q = \Sigma Q_i,\ p_t = \Sigma p_{ti}$ (b) $Q = \Sigma Q_i,\ p_t = p_{t1} = p_{t2} = ...$

(c) $Q = Q_1 = Q_2 = ...,\ p_t = \Sigma p_{ti}$ (d) $Q = Q_1 = Q_2 = ...,\ p_t = p_{t1} = p_{t2} = ...$

8. For fans in parallel

(a) $Q = \Sigma Q_i,\ p_t = \Sigma p_{ti}$ (b) $Q = \Sigma Q_i,\ p_t = p_{t1} = p_{t2} = ...$

(c) $Q = Q_1 = Q_2 = ...,\ p_t = \Sigma p_{ti}$ (d) $Q = Q_1 = Q_2 = ...,\ p_t = p_{t1} = p_{t2} = ...$

Answers

1. (c) **2.** (d) **3.** (c) **4.** (b) **5.** (b) **6.** (b) **7.** (c) **8.** (b)

REVIEW QUESTIONS

1. Define a fan.
2. What is the function of fan in an air-conditioning system?
3. Define fan static pressure, velocity pressure, and total fan pressure.
4. Explain static fan air power and total fan air power.
5. Define static fan efficiency and total fan efficiency.
6. Define specific speed of a fan and write its expression.
7. What do you understand by a geometrically similar fan?

NUMERICAL PROBLEMS

16.1 A fan draws in air freely and discharges through a duct of cross-section 0.08 m^2 in which the static pressure is 25 percent of the velocity pressure. If the total efficiency of the fan is 66 percent and the input power is 1kW, find the quantity of air being delivered in m^3/min.

16.2 A fan is to deliver 600 m^3/min at a static pressure of 25 mm of water when running at 240 rpm and requiring 6 kW. If the fan speed is changed to 300 rpm, find the capacity, static pressure and the power required.

APPLICATIONS OF REFRIGERATION AND AIR-CONDITIONING

Chapter 17

17.1 FOODF PRESERVATION

The preservation of perishable foods by refrigeration involves the use of low temperature as a means of eliminating or retarding the activity of spoilage agents. The degree of low temperature required for adequate preservation varies with the type of product stored and length of time for which the product is to be kept in storage. The food conservation is common in the following:

(*i*) Domestic refrigerators

(*ii*) Commercial refrigerators, and

(*iii*) Cold storages.

1. Domestic refrigerators for food preservation

The main purpose of a domestic refrigerator is to provide a low temperature for food storage and drinks. A temperature of about 4°C is enough for preserving fruits, vegetables, fish, meat, milk etc. The storage in domestic refrigerator is generally short term or temporary storage. Their capacity varies from 100 to 250 litres. Generally R-12 or R-134a refrigerant is used in hermetically sealed units.

2. Commercial refrigerators for food preservation

The commercial refrigerators are usually small ready built, refrigerated fixtures used for processing, storing, displaying and dispersing of perishable commodities. These are generally used by retail stores, markets, hotels, and restaurants.

3. Cold storages for food preservation

The storages which are used for short term storage purposes are known as cold storages. The period for short term storage is not more than fifteen days. The following points are to be kept in mind while storing foods in cold storages:

(*i*) Storage temperature

(*ii*) Relative humidity and air motion

(*iii*) Mixed storage

(*iv*) Condition of products at the time of entering storage

(*v*) Product chilling.

Table 17.1 gives the storage conditions and properties of perishable products.

Products	Range of storage (°C)		Optimum rel-humidity %	Freezing temp (°C)	Composition (1% of water	Sp. Heat (kJ/kg °C)		Latent heat of fusion kJ/kg	Heat of respiration		Max. storage period
	Short time storage	Ware-house storage				Above freezing	Below freezing		T °C	kJ/24h-ton	
1	2	3	4	5	6	7	8	9	10	11	12
(a) FRUITS											
Apples	2–4.5	–1.0	85	2.0	8.5	3.76	2.05	283.5	–	1576	8months
										2725	
Apricots	–	–	–	–2.2	85.4	3.68	1.92	28.5	–	–	–
Avocadoes	–	–	–	2.7	94	3.80	2.05	316.1	–	15876	–
Bananas (Green)	12–13	12–13	80	–2.2	75	3.35	1.76	251.0	12	3475	10 days
Blackberries	–	–	–	–1.7	85	3.68	1.92	283.3	–	–	–
Cantaloupes	–	–	–	–1.7	93	3.93	2.00	306.8		6590	
									15	15344	
Cherries	–	–	–	–3.3	83	3.64	1.88	278.9	–	–	–
Cranberries	–	–	–	–2.6	87	3.76	1.92	288.2	–	–	–
Currants	–	–	–	–1.0	85	3.68	1.88	278.9	–	–	–
Dates (drycured)	–	–	–	–2.0	20	1.506	1.09	67.4	–	–	–
Dates (dry-non-cured)	–	–	–	–2.0	20	1.506	1.09	67.4	–	–	–
Dates (fresh)	–	–	–	–2.7	78	3.43	1.79	260.3	–	–	–
Figs (fresh)	–	–	–	–2.7	78	3.43	1.79	260	–	–	–
Figs (dried)	–	–	–	–	24	1.63	1.13	79.0	–	–	–
Grapes	2–4.5	–1–0	80	–2.21	77	3.76	2.55	260.4	2	1576	1 to 6 months
Lemons	4.15	10–13	80	–2.2	89	3.93	2.09	302.4	5	1538	–
									16	5640	90 days
Oranges	5–7	0–1	80	–2.2	86	3.76	1.97	288.3	0	1510	2 months
Peaches	2–4.5	–1.1	80	–1	88	3.85	2.00	297.5	0	2108	30 days
										3295	
Pears	2–4.5	–1-0	85	–1.9	64	3.81	2.05	283	0	1462	1 to 7 months

Table 17.1 Continued.

(b) VEGETABLES											
Strawberries	2–4.5	–	80	–1.1	90.5	3.85	2.00	304.4	–	–	10 days
Asparagus	5–7	0–1	90	–1.1	94	3.81	2.05	316.1	4.5	21841	30 days
Bears (string)	5–7	0.1	85	–1.1	68.5	3.35	1.92	228.9	0	9000	30 days
									4.5	12800	
Beet	5–7	0–1	85	–2.8	88.5	3.60	2.00	297.5	0	2210	7 to 90 days
									4.5	3460	
Cabbage	2–4.5	0–1	90	–0.6	91.5	3.89	1.97	306.8	–	–	4
Carrots	2–4.5	0–1	90	–1.4	88	3.60	1.88	2928	0	4040	2 to 4 months
									4.5	6590	
Celery	2–4.5	–10.1	90	–1.1	94.5	3.81	1.92	316.1	4.5	5350	2 to 4 months
									4.5	8620	
Corn (green)	2–4.5	–1.1	85	–1.7	755	3.60	1.59	251.0	0	11180	10 days
									4.5	8620	
Corn (dried)	10–15	2–4.5	60		10.5	1.21	1.00	62.7	–	–	12 days
Lettuce	2–4.5	0–1	95	–0.6	94.5	3.76	1.92	316.1	0	1211	20 days
									4.5	14040	
Potatoes •	2–10	3–6	85	–1.7	78.5	3.60	1.97	262.6	0	1250	6
									4.	2720	
Tomatoes	10–13	10–13	80	–0.8	94.5	3.85	1.92	306.8	4.5	2390	10
(c) MEAT & FISH											
Bacon	5–7	–1– (–1)	80	–	20	2.09	1.25	67.40	–	–	15 days
Beef (fresh)	2–4.5	–1.0	84	–2.78	68	3.14	1.67	227.75	–	–	3 weeks
Fish (frozen)	–12– (–1)	–2– (–1)	80	–2.22	70	3.18	1.71	234.72	–	–	6 to 8
Fish (iced)	1-3	–1.0	85	–	70	3.18	1.71	234.72	–	–	15

17.2 REFRIGERATED FREEZERS

Since properly frozen foods have a much longer storage life than cold-stored foods and also retain better quality, the emphasis is on the development of frozen food storages and techniques of quick freezing. Among the common products that are preserved by freezing are fish, meat, vegetables, such as beans, sprouts, carrots, cauliflower, peas and spinach, various tissues and organs, etc.

Freezing requires the removal of heat, and the product from which heat is removed falls in temperature in the manner shown in Fig. 17.1. The first stage involves sensible cooling. The temperature drops rapidly to the freezing temperature true below 0°C. The second stage involves the extraction of latent heat of fusion during which the temperature changes very little. This stage is known as the period of *thermal arrest*. When about 75 per cent of the water is converted into ice, the temperature begins to fall again in the third stage in which the remaining water is frozen. The unfrozen water has a higher concentration of salts and other compounds, the effect of which is to depress the freezing point. Even at temperatures as low as –30°C, a proportion of water in the product tissues remains in the unfrozen state.

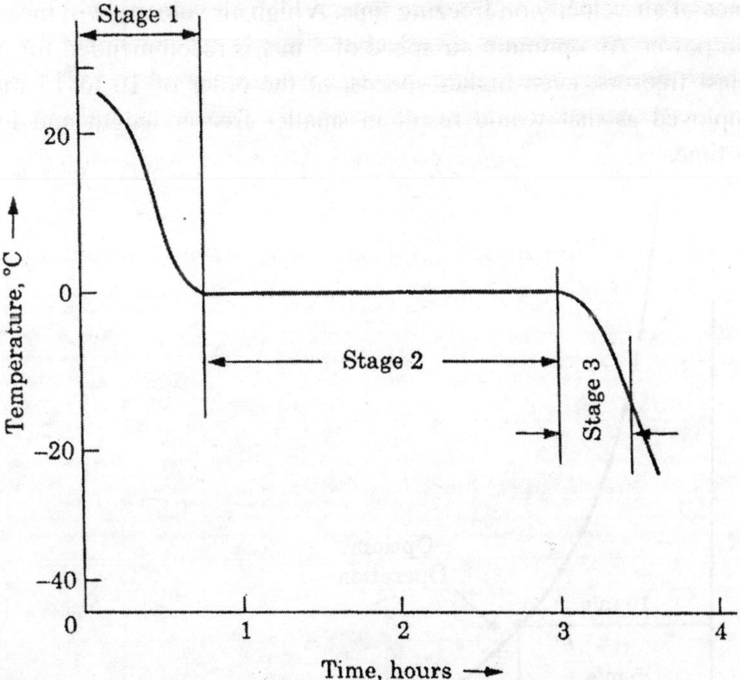

Fig. 17.1 Temperature-time graph of a product during freezing

In order to produce a good product, the freezing must be accomplished quickly so that no change takes place in the product salt concentrations. Changes take place during freezing as the salt tend to separate out. The product is altered from its *native state*. The process is termed *denaturation*. Slow freezing is, therefore, detrimental as it means a longer time being spent in the zone of denaturation activity which is in the region of –1 to –2°C.

17.2.1 Types of Freezers

There are three basic types of freezers:

(*i*) *Air-blast freezers* in which cold air is blown over the product.

(*ii*) *Contact or plate freezers* in which the product is brought in direct contact with a refrigerated surface.

(*iii*) *Immersion or spray freezers* in which the product is immersed in or sprayed with a liquid refrigerant.

Air blast and contact-plate freezers are the two most common types used in industry. Contact-plate freezers do not have the versatility of the air-blast freezers. The former can only be used to freeze regularly shaped blocks. The latter can be used for a variety of irregular shaped as well as small sized products such as peas, french fries, shrimp, etc., which results in *individual quick freezing* (IQF) of each piece. Air blast freezers can be freezing tunnel type, belt freezers or fluidized bed type freezers. Immersion freezers use *liquid freezants* at atmospheric pressure, the corresponding temperatures being –196°C with N_2 and –30°C with R-12.

To achieve quick freezing in air blast freezers, the air flow rates should be high. Fig. 17.2 shows the influence of air velocity on freezing time. A high air velocity will mean shorter freezing time but more fan power. An optimum air speed of 5 m/s is recommended for most purposes. In *continuous* air-blast freezers, even higher speeds, of the order of 10 to 15 m/s, could also be economically employed as that would result in smaller freezer length and lower cost due to reduced freezing time.

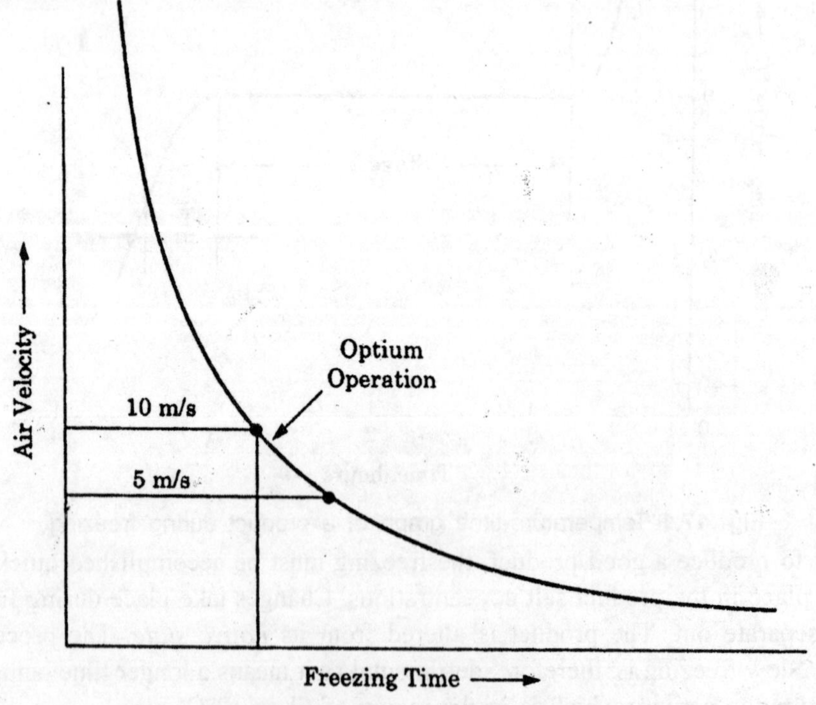

Fig. 17.2 Influence of air velocity on freezing time

Fig. 17.3 shows the schematic diagram of an air-blast freezer in which a moving belt carries the product. The belt can also be replaced by trolleys and the product stacked in multilayered trays. The air is handled by axial-flow fans developing a pressure of about 14 cm H_2O gauge. The air is in cross-flow to the product. Likewise, there may be three or more fan-coil units along the length of the belt. To minimize dehydration, the temperature of the air is kept close to that of the product. The temperature rise of air is thus limited to about 2°C only. Accordingly, the fan handles a large quantity of air. Also, the cooling coil is 12-16 rows deep to achieve BPF $\cong$ 0, in order to obtain ADP as high as possible for higher COP in this low refrigeration temperature application. As the frost formation is more in the front rows, their spacing must be larger so as to maintain a certain free-flow area with continuous operation without defrosting for at least 8 hours and without affecting the fan-air flow.

In well-designed freezers, the fan power is of the order of 25 to 30 per cent of the Refrigeration load. If the air-flow rate is very large, the fan power may be greater. As a result of small dehumidified rise and high air flow the temperature rise of air over the product will be very small. An average rise in air temperature of about 1.5-2°C is permissible.

Air blast freezers with *batch operation* use trolleys or shelves for loading the product. Those with *continuous operation* make use of conveyors or belts. The latter are used only if the product can be frozen quickly, the freezing time not exceeding 30 minutes, otherwise the length of the freezer will become very large. To reduce the length, sometimes double, triple or spiral belts are used. The speed of the belt should be kept variable to increase its versatility to accommodate the varying freezing time requirements of various products and their thicknesses.

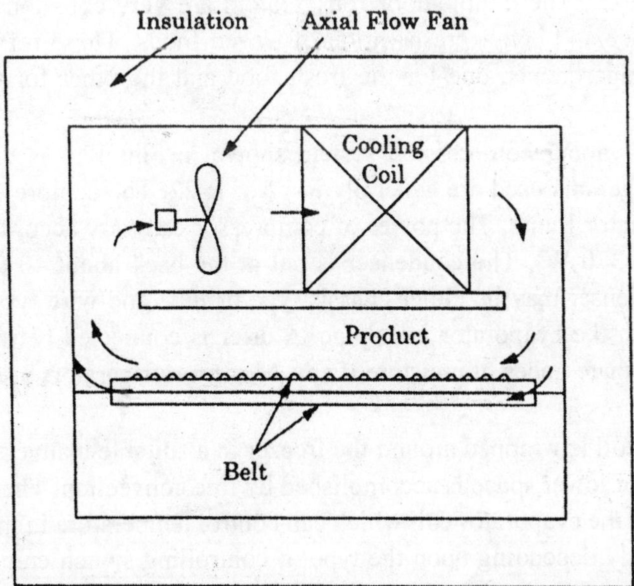

Fig. 17.3 A continuous belt air blast freezer with cross flows air

17.3 DOMESTIC REFRIGERATOR

The primary function of a domestic refrigerator is to provide food storage space maintained for low temperature for the preservation of food. Its essential secondary function is the formation of ice cubes for domestic consumption. They are usually specified by the internal gross volume of storage.

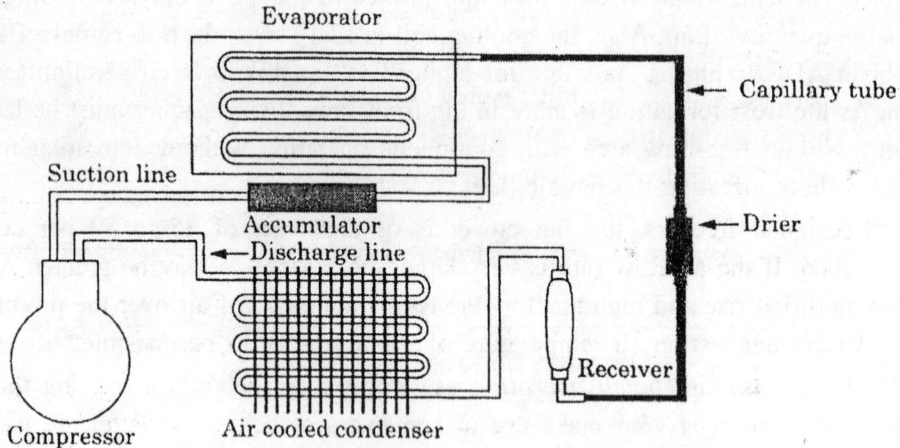

Fig. 17.4 Refrigeration system for a domestic refrigerator

A temperature of to 0°C is satisfactory for the preservation of most of the fresh foods. The freezers are generally provided at the top portion of the refrigerators space.

The refrigerators may be single door, double-door top freezer, double-door button freezer, and side-by-side door freezer. The double-door refrigerators are very common because of the need for larger storage space and better preservation of frozen foods, These refrigerators are divided into two separate compartments, one for the fresh food and the other for the storage of frozen foods.

The mechanical vapour compression system shown in Fig. 17.4 is used for the domestic refrigerator. The refrigerants used are generally R-12 or R-22. The compressor is mounted at the bottom of the refrigerator frame. The power of compressor can vary according to the size of the refrigerator (75 W to 370 W). The condenser is put at the back about 40 to 60 mm away from the cabinet. The condenser may be either chassis type or tube and wire type. The capillary tube is kept in contact with the evaporator inlet pipe. A drier is connected between the receiver and the evaporator to eliminate traces of moisture if any. Modern refrigerators use R-134a refrigerants, which is eco-friendly.

The evaporator coil is wrapped around the freezer in a suitable manner to give efficient heat transfer. The cooling of lower space is accomplished by free convection. The thermostatic sensing element is provided to the evaporator coil which can control temperature in the freezer upto –15°C in steps or continuously depending upon the type of controlling switch employed.

The refrigerator body is provided with good quality insulation in order to prevent heat transfer into the system. Usually 60 to 100 mm thick glass wool or thermocole is used.

Precaution for Use of Refrigerator

1. Refrigerator should be placed at least at a distance of 15 to 20 cm from the wall to allow adequate circulation of atmospheric air over the condenser tubes.
2. Keep the condenser tubes clean for adequate heat transfer.
3. Refrigerator should not be kept in a hot room like kitchen.
4. Never store hot food in the refrigerator.
5. Frequent opening of refrigerator should be avoided.
6. All dishes/containers should be stored covered to minimize moisture lose. This would help in decreasing frost formation on evaporator.
7. Use stabilized electric supply to avoid overloading of the compressor motor during low voltage.
8. Place the refrigerator on a level surface to ensure that magnetic rubber gaskets on the door seals the refrigerator.

17.4 TRANSPORT REFRIGERATION

Refrigerated Trucks and Trailers

These vehicles are refrigerated to maintain temperatures of either 1.5 to 4°C foods or –1.8°C for frozen foods. The types of refrigeration systems used are given below.

1. ***Product Subcooling:*** With the use of improved insulting materials, it is possible to drastically cut the transmission load of vehicles. In that case, use may be made of the storage capacity of the product itself for cold by subcooling it to as low a temperature as possible before transporting for short distances. Thus even after reaching the destination, the temperature of the product is below the storage requirement. Examples are tankers for milk, orange juice, etc.
2. ***Using Water Ice:*** The top of the product can be suitably iced. Again, it is a satisfactory method for short distances and for some products only. The refrigerating effect produced by the melting of ice is 335.4 kJ/kg.
3. ***Water Ice in Bunker with Forced Air Circulation:*** Fig. 17.5 shows the sketch of an ice bunker that is fitted in front of an insulated vehicle. Air is sucked over ice by the blower taking its drive from the engine. A ½ HP blower will add a heat equivalent of 0.37 kW. A mixture of ice and salt can also be used for lower temperature up to –9°C.
4. ***Using dry Ice:*** Dry ice is used in many small retail tracks for the delivery of frozen food, such as ice-cream. The usual positioning of the dry-ice blocks is in the ceiling. The cooling is by natural convection. When forced convection is employed, dry ice may be placed in bunkers just like water ice. The refrigerating effect produced by the sublimation of dry ice, which takes place at a temperature of –78.5°C, is 605.5 kJ/kg. In the use of dry ice, care must be taken to avoid burns due to low temperature, and suffocation due to lack of oxygen.

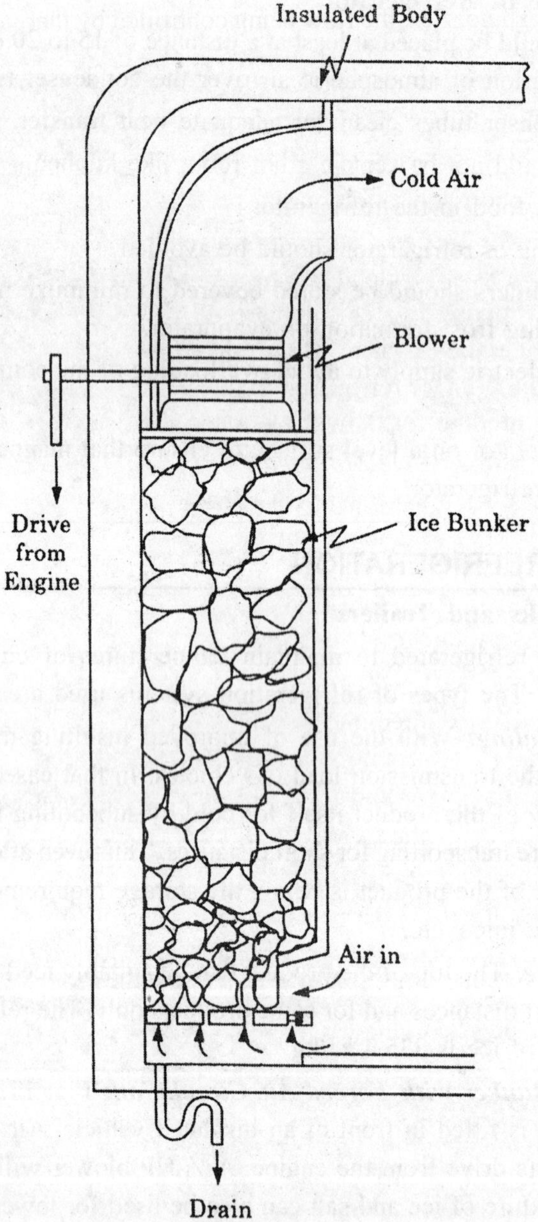

Fig. 17.5 Ice bunker in transport refrigeration

5. ***Using Liquid Nitrogen or Liquid Carbon Dioxide Spray:*** In recent times, due to the expansion of the industry for the manufacture of oxygen, by liquefaction of atmospheric air, liquid nitrogen has become available almost as a by-product. It has, therefore, come in use for the transport of frozen food. It may, however, be noted that liquid nitrogen has a normal boiling point of – 195.6°C. This is an extremely low temperature from the point of view of COP and refrigeration economy. Liquid nitrogen is, therefore, not recommended if it is not available as a by-product.

In both liquid nitrogen and liquid carbon dioxide cooling, the refrigerant is introduced into the vehicle through nozzles, the flow being controlled by thermostats and solenoid valves. The storage vessels for the refrigerants are kept within the refrigerated space. Whereas a liquid nitrogen vessel need not be under pressure. A liquid carbon dioxide vessel has to be under a pressure of about 5.2 bar as at atmospheric pressure it cannot exist as a liquid. Spray nozzles for liquid carbon dioxide should, therefore, be designed to avoid blockage due to snow formation during pressure drop.

6. ***Eutectic Plates with Station Charging:*** Eutectic plates forming channels are placed all round the body of the vehicle. They contain an eutectic solution which can be frozen by a refrigerant flowing in a coil immersed in the solution. The coil is connected to a mechanical refrigeration system at the charging station. Cooling is produced by the evaporation of the primary refrigerant in the coil. When the solution is frozen, the truck loaded with the product for delivery is ready for departure.
7. ***Eutectic Plates with Vehicle-Mounted Condensing Unit:*** This system is increasingly popular because of economy as well as reliability for the transport of frozen foods. Eutectic plates are used for maintaining the product at the required temperature. However, when the frozen eutectic has melted, the vehicle-mounted condensing unit can be started. Further, the condensing unit has an auxiliary drive mounted on the vehicle when it is in use and an electric motor drive for use at the charging station.
8. ***Mechanical Refrigeration with Independent Engine or Electric Motor:*** The mechanical refrigeration system mounted on the vehicle has an independent engine. Some are also equipped with an electric motor for stand-by operation at the charging station. These units are available in two ranges:

   (*i*) 5.86 to 10.3 kW capacity at 1.5°C.

   (*ii*) 1.76 to 5.28 kW capacity at – 18°C in a 40°C environment.

   By far, this is one of the most common methods.
9. ***Mechanical Refrigeration Deriving Power from Vehicle Engine or Transmission*** There are many practical designs available. The refrigeration compressor-may be either directly coupled to the engine, or may take off the power through its transmission. In one design, the engine runs an alternator. The A.C. current is rectified and is used to run D.C. motors for compressor and fans. The refrigeration is, however, produced only when the vehicle engine is in operation.

Example 17.1 *Calculate the heat that must be removed to lower the temperature of 800 kg of Rahu fish from 10°C to –20°C. The following data is given:*

Freezing temperature = –2°C

Specific heat above freezing = 1.63 kJ/kg°C

Latent heat of freezing = 214 kJ/kg

Solution: Heat removed,

$$Q = [c_{p1}(t_1 - t_2) + h_{fg} + c_{p2}(t_2 - t_3)]$$
$$= 800[2.97\ \{10 - (-2)\} + 214 + 1.63\ \{(-2) - (-20)\}]$$
$$= 2,23,184 \text{ kJ}$$

Example 17.2 *32,400 kg/day of cold fish is to be frozen to –30°C in 10 cm thick blocks each weighing 45 kg. The secondary refrigerant temperature is –40°C. The evaporating refrigerant temperature is –47°C: The fish enters at 30°C. The freezing cycle time may be taken as 4 hours.*

Given:

Specific heat of thawed fish = 3.77 kJ/kg

Latent heat of fusion of fish = 251.2 kJ/kg

Specific heat of frozen fish = 1.67 kJ/kg

Calculate the number of blocks frozen per cycle and refrigeration duty of the plant for 18 hours running time.

Solution:

$$\text{Number of blocks per day} = \frac{32{,}400}{45} = 720$$

$$\text{Number of freezing cycles per day} = \frac{24}{4} = 6$$

$$\text{Number of blocks frozen per cycle} = \frac{720}{6} = 120$$

$$\text{Fish loading,} \quad \dot{m} = \frac{32{,}400}{24 \times 3600} = 0.375 \text{ kg/s}$$

Sensible cooling to a freezing temperature of 0°C,

$$\dot{Q}_1 = \dot{m} c_p \cdot \Delta t = 0.375 \times 3.77\ [30 - 0] = 42.4 \text{ kW}$$

$$\text{Latent heat of fusion,} \quad \dot{Q}_2 = \dot{m} \cdot L = 0.375 \times 251.2 = 94.2 \text{ kW}$$

$$\text{Sensible cooling to} - 30°\text{C}, \quad \dot{Q}_3 = \dot{m} \cdot c_p \cdot \Delta t = 0.375 \times 1.67 \times 30 = 18.8 \text{ kW}$$

$$\text{Total product load,} \quad \dot{Q}_P = \dot{Q}_1 + \dot{Q}_2 + \dot{Q}_3 = 42.4 + 94.2 + 18.8 = 155.4 \text{ kW}$$

Product load for 18 hours running time,

$$\dot{Q} = 155.4 \times \frac{24}{18} = 207.2 \text{ kW}$$

Example 17.3 *30 tons of 'Fish' are required to be stored in a cold storage plant at a temperature of – 10°C. The temperature of the fish when supplied is 30°C. The specific heat of the fish above freezing point is 2.94 kJ/kg°C and below freezing point is 1.25 kJ/kg°C. The freezing point of the fish is –3°C and its latent heat is 233 kJ/kg. If the cooling is achieved in 10 hours, calculate the capacity of the refrigerating plant and the Carnot cycle C.O.P. between this temperature range.*

Solution: Given: m = 30 tons, t_2 = –3°C, t_1 = 3°C, t_3 = –10°C, c_{p1} = 2.94 kJ/kg·°C, c_{p2} = 1.25 kJ/kg.°C, h_{fg} = 233 kJ/kg, cooling time = 10h

Heat removed from fish, $Q = m[c_{p1}(t_1 - t_2) + h_{fg} + c_{p2}(t_2 - t_3)]$

$= 30 \times 10^3\ [2.94\{30 - (-3)\} + 233 + 1.25\ \{(-3) - (-10)\}]$

$= 10163.1 \times 10^3$ kJ

$$\text{Capacity of refrigerating plant} = \frac{Q}{10 \times 60 \times 211} = \frac{10163.1 \times 10^3}{10 \times 60 \times 211} = 288998 \text{ TR}$$

$$T_1 = 273 - 10 = 263 \text{ K}$$

$$T_2 = 273 + 30 = 303 \text{ K}$$

$$\text{Carnot COP} = \frac{T_1}{T_2 - T_1} = \frac{263}{303 - 263} = \frac{263}{40} = 6.575$$

Example 17.4 *200 kg/h of fish are to be frozen, the product requiring a residence time of 20 minutes, in an air-blast freezer with a 1 m wide stainless steel belt. The possible belt loading density is 3 kg/m². Find the length of the freezer.*

Solution:

$$\text{Belt loading} = 200 \times \frac{20}{60} = 66.6 \text{ kg}$$

$$\text{Belt loading per unit length of belt} = \frac{3}{1} = 3 \text{ kg/m}$$

$$\text{Length of belt required} = \frac{66.6}{3} = 22.2 \text{ m}$$

Note : To allow for the loading and unloading of fish outside the freezer, a total length of about 25 m would be required.

17.5 ICE PLANT

The layout of an ice manufacturing plant is shown in Fig. 17.6. The ammonia gas is used as the refrigerant. The high pressure and high temperature ammonia vapours from the compressor are condensed in a condenser which may be of shell and tube type or evaporative type. The condensed liquid ammonia is collected in the receiver and then expanded through the expansion valve. The pressure of liquid ammonia is considerably reduced after expansion. It is then passed through the evaporator coils surrounding a brine tank containing brine solution. The low pressure liquid ammonia absorbs heat from the brine solution. It gets converted into vapour state and fed to the compressor to complete the cycle.

The commercial ice is produced by freezing potable water in standard cans placed in rectangular tanks. The tanks are filled with chilled brine. The brine solution is kept in constant motion by agitators to increase the heat transfer. The agitators are operated by means of electric motors. The brine temperature is maintained at about – 10°C. Generally sodium chloride is used as brine.

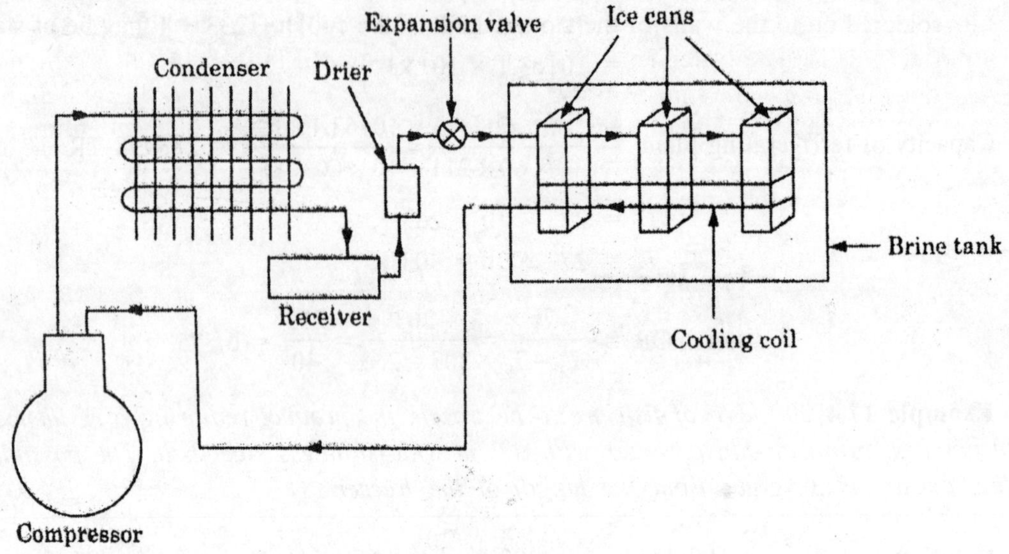

Fig. 17.6 Ice plant layout

17.6 COLD STORAGE

The cold storage is a building designed to store food stuffs, fruits, vegetables, and dairy products within specified range and relative humidity. The temperature and humidity conditions depend upon the type of product to be stored. The conditions required for storage can be divided into two categories:

1. For products to be stored at temperatures of 0°C and above.
2. For products to be stored at temperature below 0°C.

The fresh vegetables and fruits produce heat of respiration during storage. The refrigeration plant must be designed to take care of this load in addition to the usual heat loads.

The types of cooling plants for cold storage are as follows:

1. Brine coils placed parallel to and near the centre of ceiling.
2. Unit conditioner with the condensing unit outside the store.
3. Small ceiling mounted units.

17.7 WATER COOLERS

A water cooler is a device to make water available at a constant temperature irrespective of ambient temperature. A water cooler is used to produce cold water at about 7°C to 15°C. The water cooler is mainly used during summer season. There are two types of water coolers:

1. Storage type, and
2. Instantaneous type.

1. **Storage type water coolers:** In this type of water coolers, the evaporator coils are soldered on to the walls of the storage tank of the cooler. The tank may be of stainless steel sheets. The water level in the tank is maintained by a float valve. The machine runs for quite some time to bring down the temperature. Once the temperature touches the set point of the thermostat, the machine cycle is stopped. The machine starts again when the water is drawn from the cooler. There is always a reservoir of cold water in the cooler all the time.

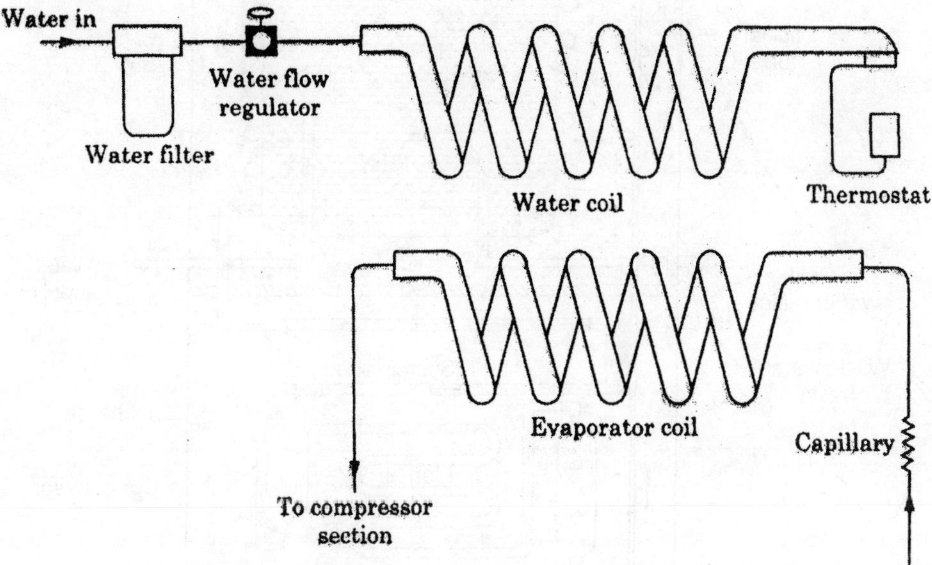

Fig. 17.7 Cooling coil of an instantaneous type cooler.

2. **Instantaneous type Water cooler:** In such a water cooler, the evaporator, as shown in Fig. 17.7, consists of two separate cylindrically wound coils made of copper or stainless steel tube. The water and evaporator coils are bonded together by soldering. The water is cooled by the refrigerant in evaporator by conduction. These water coolers are classified as; bottle type, and self - contained remote type.

In bottle type cooler, water to be cooled is stored in a bottle or reservoir. For filling glass tumblers or containers faucet or similar means are provided. The dripping water from the faucet is collected in the waste water basin or water drip. Its usual size is 25 litres and is suitable for places where plumbing installations is expensive and drains are not available. Fig. 17.8 shows the outside view of bottle type water cooler and Fig. 17.9 of a storage type water cooler.

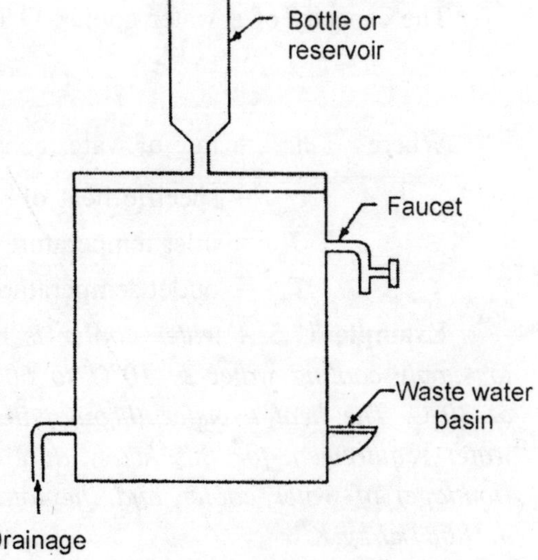

Fig. 17.8 Bottle type water cooler

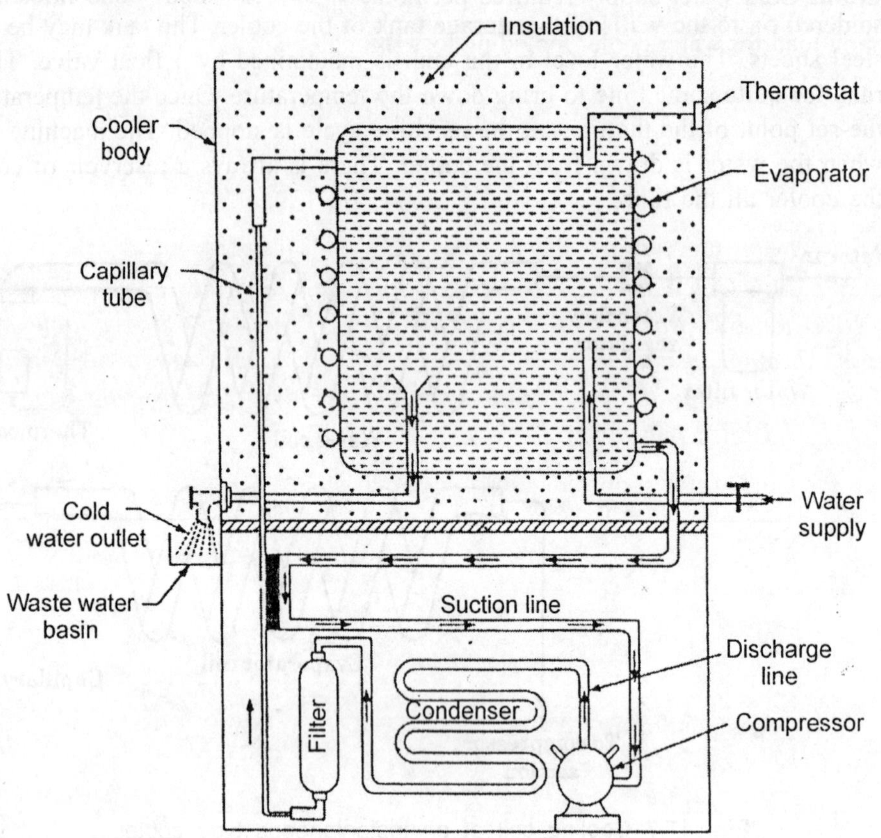

Fig. 17.9 Storage type water cooler

Capacity of water cooler:

The capacity of a water cooler, Q is given by,

$$\dot{Q} = \dot{m}_w \times c_{pw}(T_{iw} - T_{ow})$$

Where $\dot{m}_w$ = rate of water consumption, litres/hour

c_{pw} = specific heat of water = 4.187 kJ/kg.K

T_{iw} = inlet temperature of water, °C

T_{ow} = outlet temperature of water, °C

Example 17.5 *A water cooler is required to be installed in an industrial organization to supply cooling water at 10°C to 600 workers for 8 hours duty time. Water is available at 30°C. The heat transfer through insulation is 5% of the total heat load. The drinking water requirement for this heavy duty cooler is 1.1 litre per hour. Determine the capacity (tonnage) of water cooler and the total water consumption per day. Take* c_p *for water as 4.1868 kJ/kg.K.*

Solution: Cold water supply required per hour, $\dot{m}_w$ = 1.1 × 600 = 660 litres/h = 660 kg/h

Cooling load for water cooler including loss through insulation,

$$\dot{Q} = 1.05\ \dot{m}_w c_p \Delta t = 1.05 \times 660 \times 4.1868\ (30 - 10) = 58029\ \text{kJ/h}$$

$$\text{Capacity (in tonnage) of water cooler} = \frac{\dot{Q}}{60 \times 211} = \frac{58029}{60 \times 211} = 4.584\ \text{TR}$$

Total water consumption per day = 660 × 8 = 5280 litres/day

Example 17.6 *Calculate the refrigeration tonnage of a water cooler meant for drinking water at 10°C for 600 workers of 8 hours duty time. The water is available at 30°C. The heat transfer through insulation is 5% of the total heat load. Suggest the tonnage of water cooler. Also calculate the total water consumption per day. The drinking water requirement per worker is 1.1 litre per hour.*

Solution: Cold water supply required per hour, $\dot{m}_w$ = 1.1 × 600 = 660 kg/h

Cooling load for water cooler, $\dot{Q} = \dot{m}_w \cdot c_{pw} \cdot \Delta t$ = 660 × 4.187 × (30 – 10) = 55268.4 kJ/h

Total cooling load, $\dot{Q}_t$ = $1.05Q$ = 1.05 × 55268.4 = 58031.82 kJ/h

$$\text{Refrigeration tonnage of water cooler} = \frac{58031.82}{211 \times 60} = 4.584\ \text{TR}$$

Water consumption per day = $8\dot{m}_w$ = 8 × 660 = 5280 litres/day

Example 17.7 *What is the function of cold storages and what is their importance in today's life? 5000 tons of potato are available at a temperature of 30°C. It has to be preserved in cold storage at a temperature of 2°C. How much refrigeration is necessary. If this refrigeration is to be obtained in 3 days, what should be the capacity (tonnage) of the plant?*

Solution: Cold storage: Refer to art. 17.6.

The required data for potatoes from Table 17.1 is,

Specific heat above freezing = 3.60 kJ/kg.°C

Total cooling load, $Q = mc_p\Delta t = 5000 \times 10^3 \times 3.60\ (30 - 2) = 504 \times 10^6$ kJ

$$\text{Capacity in tonnage} = \frac{504 \times 10^6}{3 \times 24 \times 60 \times 211} = 552.92\ \text{TR}$$

17.8 DESERT COOLER

Principle of Working: A desert cooler is suitable for a place when the air humidity is quite low and temperature quite high, as in a desert. A desert cooler works on the principle of evaporative cooling in which sensible heat is removed and moisture added to the air. This is achieved by passing air through a spray of water. In this process, air gives up heat to water, some of the water evaporates and picks up heat from the air equivalent to its latent heat. In this way, air is cooled and humidified.

Construction and working: A desert cooler consists of the following parts:

1. Fan
2. Water recirculating pump
3. Water wetted pads
4. Water tank
5. Float valve
6. Three water traps.

Two views of a desert cooler are shown in Fig. 17.10. Water is filled in the tank (or sump) from water supply mains, whose level is controlled by a float valve. A reciscolating pump lifts the water from the tank and sprays on the pads through water distribution system. The water which drops back from the pads is recirculated. The fan pulls the air through the water pads and delivers it to be the space to be cooled through an opening on the side of the fan in the cabinet of desert cooler.

Desert cooler efficiency, $\eta_{desert} = \dfrac{t_{da} - t_{dga}}{t_{da} - t_{wa}}$

where t_{da} = DBT of ambient air

t_{dga} = temperature of air leaving the grills

t_{wa} = WBT of ambient air.

η_{desert} should be around 75% and air velocity around 40 m/min.

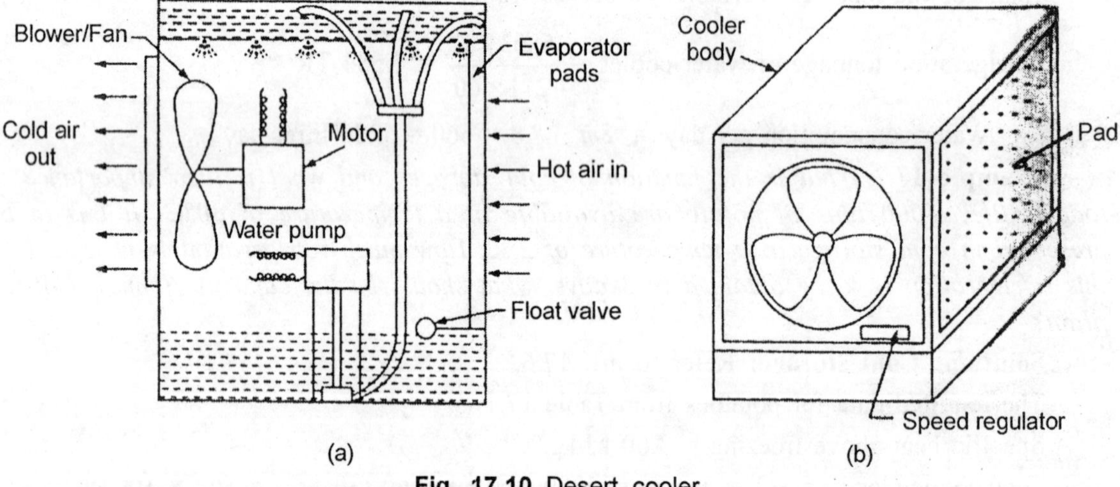

Fig. 17.10 Desert cooler

17.9 ROOM AIR-CONDITIONER

A room air-conditioner is a compact, self contained air-conditioning unit which is normally installed in window or wall opening of the room. It is widely known as window type air-conditioner. It works on vapour compression cycle. A complete unit of a room air conditioner consists of the following:

1. Refrigeration system
2. Control system (thermostat and selector switch)
3. Electrical protection system (motor overload switches and thermostat)
4. Air circulation system (fan motor, centrifugal evaporator blower)
5. Ventilation (fresh air damper)
6. Exhaust system.

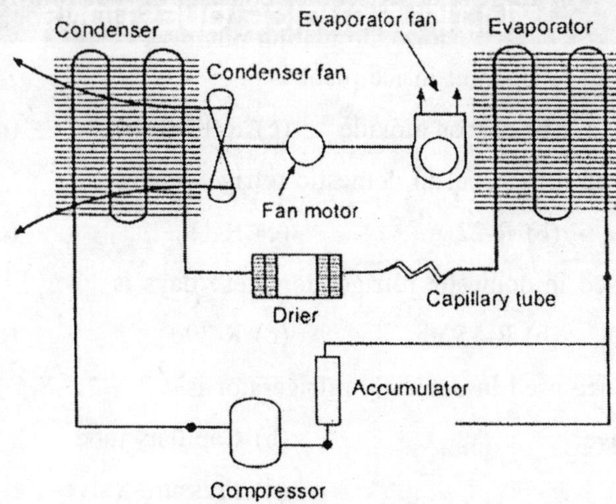

Fig. 17.11 Layout of a room air conditioner

The layout of a window type room air-conditioner is shown in Fig. 17.11.

1. Refrigeration System. It consists of a hermetically sealed compressor, forced air-cooled finned condenser coil, finned cooling coil, capillary tube as the throttling device and a refrigerant drier. The refrigerant used is R-12 or R-22. In hermetically sealed compressor, a winding thermostat is embedded in the compressor motor windings. It puts off the compressor if the winding temperature exceeds the safe limits, thus protecting the winding against high temperature.

The condenser is continuous coil made of copper tubing with aluminium fins attached to it to increase the heat transfer rate. A propeller type fan provides the necessary air to cool the refrigerant in the condenser and also exhausts air from the air- conditioned space when the exhaust damper is opened.

The evaporator is a cooling coil also made of copper with aluminium fins to increase the heat transfer rate. A centrifugal blower is installed behind the cooling coil which sends cool air in the room. The quantity of fresh air may be varied by adjusting the dampers.

2. Control system: A thermostat element is located in the return air passage of the unit. It controls the operation of the compressor based on the return air temperature. When the required temperature is obtained, the compressor is stopped. A selector switch controls the compressor motor, condenser fan motor, and evaporator fan motor. When the control switch is in "Ventilate" position, only evaporator blower motor operates and outside fresh air is supplied in the room. In the "Exhaust" position, the condenser fan motor operates and all the room air is exhausted to the atmosphere. In the "Cool" position, all the motors are in working state and cool air is supplied in the room.

MULTIPLE-CHOICE QUESTIONS

1. The commonly used refrigerant in ice plant is

   (a) ammonia (b) carbon dioxide (c) R-12 (d) R-22

2. The commonly used refrigerant in domestic refrigerator is

   (a) NH_3 (b) R-22 (c) R-11 (d) R-12

3. The refrigerant used in domestic refrigerator these days is

   (a) R-134 a (b) R-152 a (c) R-290 (d) R-600 a

4. The throttling device used in domestic refrigerator is

   (a) Expansion valve (b) Capillary tube

   (c) Float valve (d) Pressure valve

5. The temperature maintained in the brine tank of ice plant is

   (a) –10°C (b) –20°C (c) –15°C (d) –18°C

6. The milk is stored at a temperature of

   (a) 4°C (b) –5°C (c) 10°C (d) 12°C

7. Meat products are stored at a temperature of

   (a) –10°C (b) –20°C (c) –30°C (d) –40°C

8. Poultry products are frozen at a temperature range of

   (a) –5°C to –10°C (b) –15°C to –20°C (c) –25°C to –40°C (d) below –40°C

Answers

1. (a) 2. (d) 3. (a) 4. (b) 5. (a) 6. (a) 7. (c) 8. (c)

REVIEW QUESTIONS

1. Which refrigerant is generally used in an ice plant?
2. What are the various types of water coolers?
3. What are the needs for food preservation?
4. What is the difference between a refrigerator and a freezer?
5. What are the advantages of food preservation?
6. What are the various types of freezers?
7. Which throttling device is used in domestic refrigerator?
8. Name various types of water coolers.

THEORETICAL QUESTIONS

17.1 What is 'food preservation' and what are its advantages? Enumerate some common methods being used for food preservation, in brief.

17.2 What is the importance of cold storages in today's life? Describe a cold storage, in brief, What is the importance of storage period and preservation conditions in the design of a cold storage?

17.3 What is the usual temperature, of ice formation? What would be the nature of ice if it is formed quickly. If you want to increase the rate of ice formation from the same plant, how can it be done? How does this ice differ from 'Dry Ice'?

17.4 What are the advantages of 'Food Preservation'? Explain in brief, how refrigerators and freezers are used for food preservation.

17.5 Describe a cold storage, in brief. What factors are considered in the design of a cold storage?

17.6 What is the function of cold storages and what is their importance in today's life? What do you understand by 'Short Term' and 'Long Term' storages? What factors are considered in the design of a cold storage?

17.7 Explain the application of refrigeration for food preservation in 'Cold storages', discussing the various factors to be kept in mind while storing food products in cold storages. What do you understand by long-term and short-term storages?.

17.8 What do you understand by food preservation and what are its advantages? Explain, in brief, as to how refrigerators and freezers are used for food preservation.

17.9 What is the design procedure of an 'Ice Plant'? With the help of a net sketch, describe the working of an 'Ice Plant', in brief.

17.10 Explain two types of refrigeration systems used in trucks/trailers during road transport?

17.11 Explain the construction and working of a domestic refrigerators with a neat sketch.

17.12 Explain briefly different types of water coolers.

17.13 Describe with a neat sketch the construction and working of a storage type water cooler.

17.14 Describe with a neat sketch the construction and working of a desert cooler.

17.15 Describe briefly various types of freezers.

17.16 What are the precautions to be taken for the use of a domestic refrigerators?

17.17 Describe the construction and working of a window type room air-conditioner.

APPENDIX–I

MULTIPLE CHOICE QUESTIONS FROM GATE

1. A condenser of a refrigeration system rejects heat at a rate of 120 kW, while its compressor consumes a power of 30 kW. The coefficient of performance of the system would be :

(a) 1/4 (b) 4 (c) 1/3 (d) 3

2. Round the clock cooling of an apartment having a load of 300 MT/day requires an air-conditioning plant of capacity about

(a) 1 ton (b) 5 tons (c) 10 tons (d) 100 tons

3. The use of Refrigerant-22 (R-22) for temperatures below –300°C is not recommended due to its

(a) good miscibility with lubricating oil (b) poor miscibility with lubricating oil

(c) low evaporating pressure (d) high compressor discharge temperature

4. Match list-I with List-II

List I	*List II*
(A) Liquid to suction heat exchanger	1. Vapor absorption refrigeration
(B) Constant volume heat addition	2. Vapor compression refrigeration
(C) Normal shock	3. Diesel cycle
(D) Ammonia water	4. Otto cycle
	5. Converging nozzle
	6. Converging diverging nozzle.

5. A Carnot cycle refrigerator operates between 250 K and 300 K. Its coefficient of performance is

(a) 6.0 (b) 5.0 (c) 1.2 (d) 0.8

6. Ambient air dry-bulb temperature is 45°C and wet bulb temperature is 27°C. Select the lowest possible condensing temperature from the following for an evaporatively cooled condenser.

(a) 25°C (b) 30°C (c) 42°C (d) 48°C

7. Select statements from List II matching the processes in List I. Enter your answer as D, C if the correct choice for (1) is (D) and that for (2) is (C)

List I	*List II*
(1) Cooling and dehumidification	(A) Dry bulb temperature increases, but dew-point temperature decreases.
(2) Chemical dehumidification	(B) Dew-point temperature increases and dry-bulb temperature remains uncharged
	(C) Dry-bulb and wet-bulb temperatures decrease.
	(D) Dry-bulb temperature decreases, but dew-point temperature increases

8. In a vapour compression refrigeration system, liquid to section heat exchanger is used to
(a) keep the COP constant
(b) prevent the liquid refrigerant from entering the compressor
(c) subcool the liquid refrigerant leaving the condenser
(d) subcool the vapour refrigerant from the evaporator

9. Global warming is caused by
(a) ozone (b) carbon dioxide (c) nitrogen (d) carbon monoxide

10. When atmospheric air is heated at constant pressure, its
(a) humidity ratio does not change (b) relative humidity increases
(c) dew point temperature does not change (d) wet bulb temperature increases

11. For air at a given temperature, as the relative humidity is increased isothermally,
(a) the wet bulb temperature and specific enthalpy increase
(b) the wet bulb temperature and specific enthalpy decrease
(c) the wet bulb temperature increases and specific enthalpy decreases
(d) the wet bulb temperature decreases and specific enthalpy increases

12. For air with a relative humidity of 80%
(a) the dry bulb temperature is less than the wet bulb temperature
(b) the dew point temperature is less than the wet bulb temperatue
(c) the dew point and wet bulb temperatures are equal
(d) the dry bulb and dew point temperatures are equal

Data for Q. 13-14 are given below. Solve the problems and choose correct answers.
A refrigerator based on ideal vapour compression cycle operates between the temperature limits of –20°C and 40°C. The refrigerant enters the condenser as saturated vapour and leaves as saturated liquid and vapour at these temperatures are given in the table below.

T(°C)	h_f(kJ/kg)	h_g(kJ/kg)	s_f(kJ/kg K)	s_g(kJ/kg K)
–20	20	180	0.07	0.7366
40	80	200	0.3	0.67

13. If refrigerant circulation rate is 0.025 kg/s the refrigeration effect is equal to

(a) 2.1 kW (b) 2.5 kW (c) 3.0 kW (d) 4.0 kW

14. The COP of the refrigerator is

(a) 2.0 (b) 2.33 (c) 5.0 (d) 6.0

15. In the window air conditioner, the expansion device used is

(a) capillary tube (b) thermostatic expansion valve

(c) automatic expansion valve (d) float valve

16. Environment friendly refrigerant R134 a is used in the new generation domestic refrigerators. Its chemical formula is

(a) $CHClF_2$ (b) $C_2Cl_2F_3$ (c) $C_2Cl_2F_4$ (d) $C_2H_2F_4$

17. During chemical dehumidification process of air

(a) dry bulb temperature and specific humidity decrease

(b) dry bulb temperature increases and specific humidity decreases

(c) dry bulb temperature decreases and specific humidity increases

(d) dry bulb temperature and specific humidity increase

18. Dew point temperature of air at one atmospheric pressure (1.013 bar) is 18°C. The air dry bulb temperature is 30°C. The saturation pressure of water at 18°C and 30°C are 0.02062 bar and 0.04241 bar respectively. The specific heat of air and water vapour respectively are specific 1.005 and 1.88 kJ/kg. K and the latent heat of vaporization of water at 0°C is 2500 kJ/kg. The specific humidity (kJ/kg of dry air) and enthalpy (kJ/kg of dry air) of this moist air respectively, are

(a) 0.01051, 52.64 (b) 0.01291, 63.15 (c) 0.01481, 78.60 (d) 0.01532, 8140

19. A R-12 refrigerant reciprocating compressor operates between the condensing temperature of 30°C and evaporator temperature of –20°C. The clearance volume ratio of the compressor is 0.03. Specific heat ratio of the vapour is 5 and the specific volume at the section is 0.1089 m^3/kg. Other properties at various states are given in the Fig. 1 To realize 2 Tons of refrigeration, the actual volume displacement rate considering the effect of clearance is

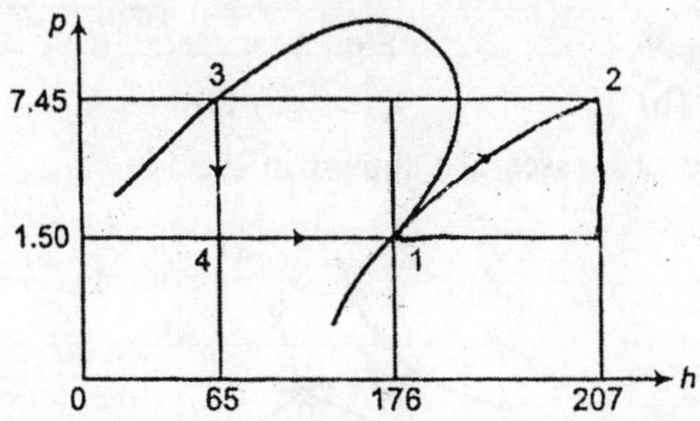

Fig. 1

(a) 6.35×10^{-3} m/s (b) 63.5×10^{-3} m^3/s (c) 635×10^{-3} m^3/s (d) 4.88×10^{-3} m^3/s

20. For a typical sample of ambient air (at 35°C, 75% relative humidity and standard atmospheric pressure), the amount of moisture in kg per kg of dry air will be

(a) 0.002 (b) 0.027 (c) 0.25 (d) 0.75

21. Water at 42°C is sprayed into a stream of air at atmospheric pressure, dry bulb temperature of 40°C and a wet bulb temperature of 20°C. The air leaving the spray humidifier is not saturated. Which of the following statements is true?

(a) Air gets cooled and humidified (b) Air gets heated and dehumidified

(c) Air gets heated and humidified (d) Air gets cooled and dehumidified

22. The vapour compression refrigeration cycle is represented as shown in the Fig. 2, with state I being the exit of the evaporator. The coordinate system used in this figure is

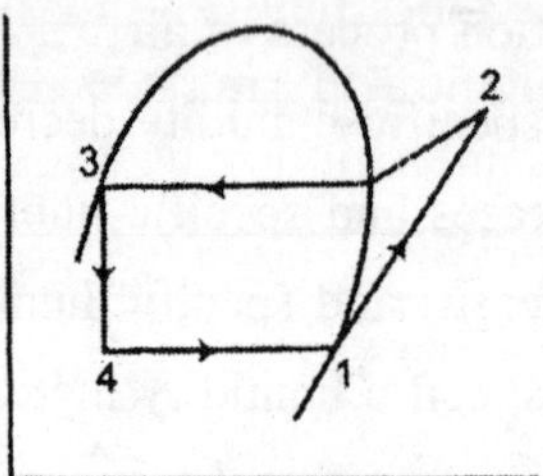

Fig. 2

(a) p-h (b) T-s (c) p-s (d) T-h

23. A vapour absorption refrigeration system is a heat pump with thermal reservoirs as shown in the Fig. 3. A refrigeration effect of 100 W is required at 250 K when the heat source available is at 400 K. Heat rejection occurs at 300 K. The minimum value of heat required (in W) is

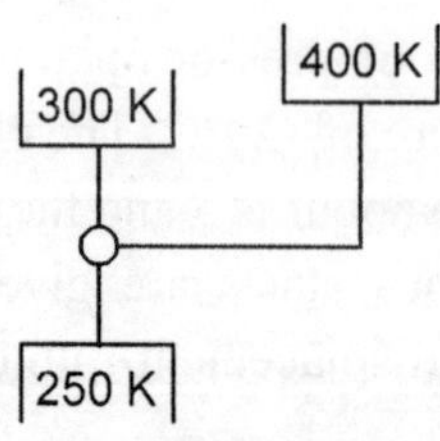

Fig. 3

(a) 167 (b) 100 (c) 80 (d) 20

24. Various psychrometric processes are shown in the Fig. 4.

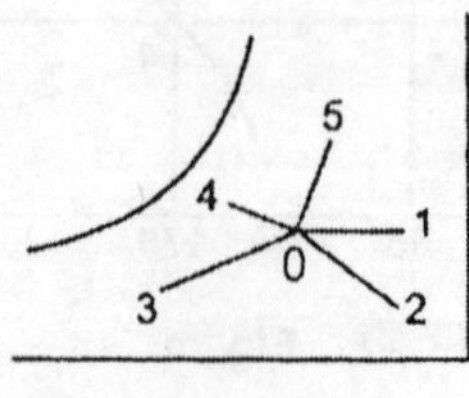

Fig. 4

Process in Figure		Process in Figure
P 0-1	(i)	Chemical dehumidification
Q 0-2	(ii)	Sensible heating
R 0-3	(iii)	Cooling and dehumidification
S 0-4	(iv)	Humidification with steam injection
T 0-5	(v)	Humidification with water injection

The matching pairs are

(a) P-(i), Q-(ii), R-(iii), S-(iv), T-(v) (b) P-(ii), Q-(i), R-(iii), S-(v), T-(iv)

(c) P-(ii), Q-(i), R-(iii), S-(iv), T-(v) (d) P-(iii), Q-(iv), R-(v), S-(i), T-(ii)

Statement for Linked Answer Questions 25 and 26 : The following table of properties was printed out for saturated liquid and saturated vapour of ammonia. The titles for only the first two columns are available. All that we know is that the other columns (columns 3 to 8) contain data on specific properties, namely, internal energy (kJ/kg), enthalpy (kJ/kg) and entropy (kJ/kg. K.).

T(°C)	*p(kPa)*						
–20	190.2	88.76	0.3657	89.05	5.6155	1299.5	1418.0
0	429.6	179.69	0.7144	180.36	5.3309	1318.0	1442.0
20	857.5	272.89	1.0408	274.30	5.0860	1333.2	1460.2
40	1554.9	368.74	1.3574	371.43	4.8662	1341.0	1470.2

25. The specific enthalpy data are in columns

(a) 3 and 7 (b) 3 and 8 (c) 5 and 7 (d) 5 and 8

26. When saturated liquid at 40°C is throttled at 20°C, the quality at exit will be

(a) 0.189 (b) 0.212 (c) 0.231 (d) 0.788

27. Dew point temperature is the temperature at which condensation begins, when the air is cooled at constant

(a) volume (b) entropy (c) pressure (d) enthalpy

28. The statements concern psychrometric chart.

1. Constant relative humidity lines are uphill straight lines to the right.
2. Constant wet bulb temperature lines are downhill straight lines to the right.
3. Constant specific volume lines are downhill straight lines to the right.
4. Constant enthalpy lines are coincident with constant web bulb temperature lines.

Which of the statements are correct?

(a) 2 and 3 (b) 1 and 2 (c) 1 and 3 (d) 2 and 4

29. A building has to be maintained at 21°C (dry bulb) and 14.5°C (wet bulb). The dew point temperature under these conditions is 10.17°C. The outside temperature is –23°C (dry bulb) and the internal and external surface heat transfer co-efficients are 8 $W/m^2 \cdot K$ and 23 $W/m^2 \cdot K$ respectively. If the building wall has a thermal conductivity of 1.2 $W/m \cdot K$, the minimum thickness (in m) of the wall required to prevent condensation is

(a) 0.471 (b) 0.407 (c) 0.321 (d) 0.125

30. Atmospheric air at a flow rate of 3 kg/s (on dry basis) enters a cooling and dehumidifying coil with an enthalpy of 85 kJ/kg of dry air and a humidity ratio of 19 grams/kg of dry air. The air leaves the coil with an enthalpy of 43 kJ/kg of dry air and a humidity ratio of 8 grams/kg of dry air. If the condensate water leaves the coil with an enthalpy of 67 kJ/kg, the required cooling capacity of the coil in kW is

(a) 75.0 (b) 123.8 (c) 128.2 (d) 159.0

31. Moist air at a pressure of 100 kPa is compressed to 500 kPa and then cooled to 35°C in an after cooler. The air at the entry to the after cooler is unsaturated and becomes just saturated at the exit of the after cooler. The saturation pressure of water at 35°C is 5.628 kPa. The partial pressure of water vapour (in kPa) in the moist air entering the compressor is closest to

(a) 0.57 (b) 1.13 (c) 2.26 (d) 4.52

32. Air (at atmospheric pressure) at a dry bulb temperature of 40°C and wet bulb temperature of 20°C is humidified in an air washer operating with continuous water recirculation. The wet bulb depression (i.e., the difference between the dry and wet bulb temperatures) at the exit is 25% of that at the inlet. The dry bulb temperature at the exit of the air washer is closest to

(a) 10°C (b) 20°C (c) 25°C (d) 30°C

33. In an ideal vapour compression refrigeration cycle, the specific enthalpy of refrigerant (in kJ/kg) at the following states is given as:

Inlet of condenser : 283

Exit of condenser : 116

Exit of evaporator : 232

The COP of this cycle is

(a) 2.27 (b) 2.75 (c) 3.27 (d) 3.75

34. A moist air sample has dry bulb temperature of 30°C and specific humidity of 11.5 g water vapour per kg dry air. Assume molecular weight of air as 28.93. If the saturation vapour pressure of water at 30°C is 4.24 kPa and the total pressure is 90 kPa, then the relative humidity (in%) of air sample is

(a) 50.5 (b) 38.5 (c) 56.5 (d) 68.5

35. If a mass of moist air in an airtight vessel is heated to a higher temperature, then

(a) specific humidity of the air increases (b) specific humidity of the air decreases

(c) relative humidity of the air increases (d) relative humidity of the air decreases

36. A thin layer of water in a field is formed after a farmer has watered it. The ambient air conditions are temperature 20°C and relative humidity 5%.

An extract of steam tables is given below,

Temperature (°C)	–15	–10	–5	0.01	5	10	15	20
Saturation Pressure (kPa)	0.10	0.26	0.40	0.61	0.87	1.23	1.71	2.34

Neglecting the heat transfer between the water and the ground, the water temperature in the field after phase equilibrium is reached equals

(a) 10.3°C (b) –10.3°C (c) –14.5°C (d) 14.5°C

37. If a mass of moist air in an airtight vessel is heated to a higher temperature, then

(a) specific humidity of the air increases (b) specific humidity of the air decreases

(c) relative humidity of the air increases (d) relative humidity of the air decreases

38. A room contains 35 kg of dry air and 0.5 kg of water vapour. The total pressure and temperature of air in the room are 100 kPa and 25°C respectively. Given that the saturation pressure of water at 25°C is 3.17 kPa, the relative humidity of air in the room is

(a) 67% (b) 55% (c) 83% (d) 71%

39. The pressure, dry bulb temperature and relative humidity of air in a room are 1 bar, 30°C and 70% respectively. If the saturated steam pressure at 30°C is 4.25 kPa, the specific humidity of the room air is kg water vapour/kg dry air is

(a) 0.0083 (b) 0.0101 (c) 0.0191 (d) 0.0232

Answers

1. (d) **2.** (a) **3.** (d) **4.** (a-2, b-4, c-6, d-1) **5.** (b) **6.** (d)
7. (1-c, 2-d) **8.** (c) **9.** (b) **10.** (a, c) **11.** (a) **12.** (c) **13.** (a)
14. (b) **15.** (a) **16.** (d) **17.** (b) **18.** (b) **19.** (a) **20.** (b) **21.** (c)
22. (d) **23.** (c) **24.** (b) **25.** (d) **26.** (b) **27.** (c) **28.** (d) **29.** (b)
30. (c) **31.** (b) **32.** (c) **33.** (a) **34.** (b) **35.** (d) **36.** (d) **37.** (d)
38. (d) **39.** (c)

EXPLANATORY NOTES

1. (d): $Q_r = 120$ kW, $Q_s = 30$ kW

$$\text{COP} = \frac{Q_r - Q_s}{Q_s} = \frac{120-30}{30} = \frac{90}{30} = 3$$

2. (a): ITR = 211 kJ/min

$$\text{Refrigeration capacity} = \frac{300 \times 10^3}{24 \times 60 \times 211} = 0.9874 = 1 \text{ ton}$$

3. (d) : R-22 is not recommended below – 30°C due to its high compressor discharge temperature.

4. A-2, B-4, C-6, D-1

5. **(b) :** $(COP)_{ref} = \dfrac{T_2}{T_1 - T_2} = \dfrac{250}{300 - 250} = \dfrac{250}{50} = 5$

6. **(d) :** DBT = 45°C, WBT = 27°C

For an evaporatively cooled condenser the lowest possible condensing temperature shall be 4°C.

7. I-C, 2-D.

8. **(c) :** In vapour compression refrigeration system, liquid to suction heat exchanger is used to subcool the liquid refrigerant leaving the condenser.

9. **(b) :** Global warming is caused by carbon dioxide.

10. **(a, c) :** When atmospheric air is heated at constant pressure then its humidity ratio and DPT do not change.

11. **(a) :** As relative humidity is increased isothermally, the WBT and specific enthalpy increase.

12. **(c) :** For air with RH = 80%, DPT = WBT

13. **(a) :**

$$h_3 = h_{f3} = h_4 = 80 \text{ kJ/kg}$$

$$h_2 = h_{g2} = 200 \text{ kJ/kg}$$

$$h_4' = 20 \text{ kJ/kg}, \; h_1' = 180 \text{ kJ/kg}$$

$$s_2 = s_{g2} = 0.67 \text{ kJ/kg}\cdot\text{K} = s_1$$

$$s_1' = s_{g1}' = 0.7366 \text{ kJ/kg}\cdot\text{K}$$

$$s_4' = s_{f4}' = 0.07 \text{ kJ/kg}\cdot\text{K}$$

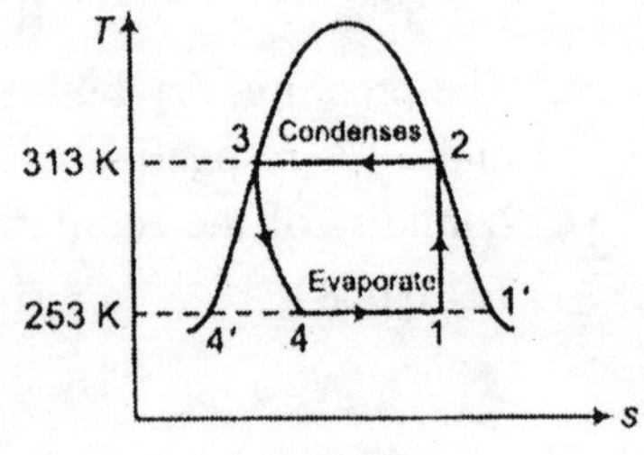

Fig. 5

$$s_2 = s_1 = s_4' + x_1\,(s_1' - s_4')$$

$$0.67 = 0.07 + x_1(0.7366 - 0.07)$$

$$x_1 = \frac{0.60}{0.6666} = 0.9$$

$$h_1' = h_4' + x_1(h_1' - h_4')$$

$$= 20 + 0.9\,(180 - 20) = 20 + 144 = 164 \text{ kJ/kg}$$

$$\text{Refrigeration effect} = \dot{m}(h_1 - h_4) = \dot{m}(h_1 - h_3)$$

$$= 0.025\,(164 - 80) = 2.1 \text{ kW}$$

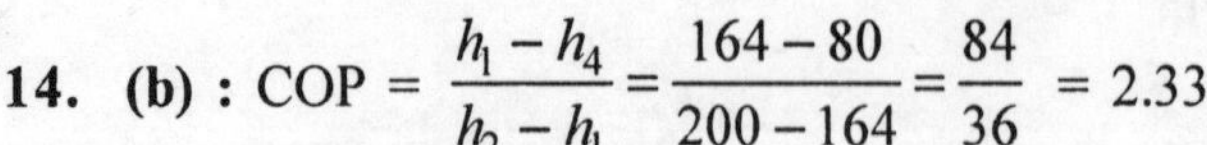

14. **(b) :** $\text{COP} = \dfrac{h_1 - h_4}{h_2 - h_1} = \dfrac{164 - 80}{200 - 164} = \dfrac{84}{36} = 2.33$

15. **(a) :** Capillary tube is used in window air-conditioner.

16. **(d) :** R134*a* :

1 : Carbon atoms, 1 + 1 = 2 *i.e.*, C_2

3 : Hydrogen atoms – 1, 3 – 1 = 2 *i.e.*, H_2

4 : Fluorine atoms, 4, *i.e.*, F_4

Formula R134*a* → $C_2H_2F_4$

17. (b) : During chemical dehumidification process DBT increases and specific humidity decreases.

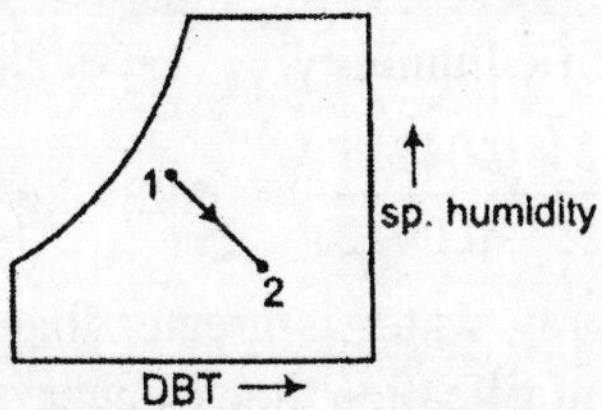

Fig. 6 Chemical dehumidification process

18. (b): Specific humidity,

$$w = \frac{0.622\, p_v}{p_a - p_v} - \frac{0.622 \times 0.02062}{0.013 - 0.02062} = 0.01292 \text{ kJ/kg d.a.}$$

Specific enthalpy of moist air.

$$h = c_{pa} \times DBT + wL + wc_{pw}\ (DBT - DPT)$$
$$= 1.005 \times 30 \times 30 + 0.01292 \times 2500 + 0.01292 \times 1.88\ (30 - 18)$$
$$= 63.15 \text{ kJ/kg d.a.}$$

19. (a): Clearance ratio, $c = 0.03$

Specific volume at suction, $v_1 = 0.1089 \text{ m}^3/\text{kg}$

Net refrigeration effect $= 2$ ton

$= 2 \times 3.516 = 7.032$ kJ/s

Specific heat ratio, $n = 1.15$

Net refrigerating effect $= \dot{m}(h_1 - h_2)$

$$7.032 = \dot{m}(176 - 65)$$
$$\dot{m} = 0.06335 \text{ kg/s}$$

Volume rat, $\dot{V} = \dot{m}v_1 = 0.6335 \times 0.1089$

$= 6.899 \times 10^{-3} \text{ m}^3/\text{s}$

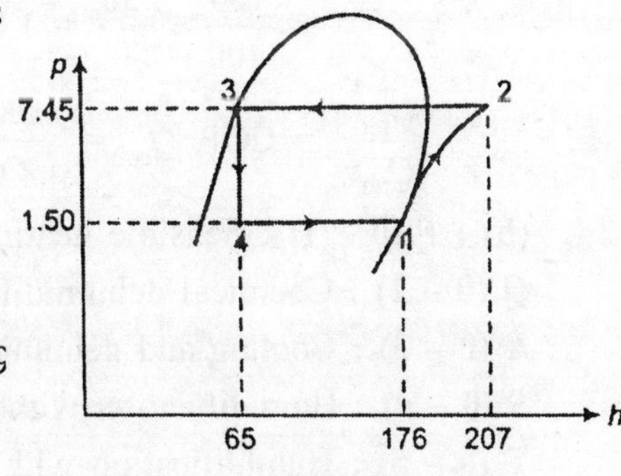

Fig. 7

Volumetric efficiency

$$\eta_{vol} = 1 + c - c\left(\frac{p_2}{p_1}\right)^{1/x}$$
$$= 1 + 0.03 - 0.03\left(\frac{7.45}{1.5}\right) = 0.9091$$

Volume displacement $= \dot{V} \times \eta_{vol}$

$= 6.899 \times 10^{-3} \times 0.9091$

$= 6.27 \times 10^{-3} \simeq 6.35 \times 10^{-3} \text{ m}^3/\text{s}$

20. **(b) :** $t_a = 35°C$, $p_a = 1.01325$ bar, $\phi = 0.75$. At 35°C,

$$p_s = 0.05628 \text{ bar}; \ \phi = \frac{p_v}{p_s} = 0.75$$

$$p_v = 0.75 \times 0.05628 = 0.0422 \text{ bar}$$

Amount of moisture *i.e.*, specific humidity,

$$w = \frac{0.622\, p_v}{p_a - p_v} = \frac{0.622 \times 0.04221}{1.01325 - 0.04221} = 0.027 \text{ kg/kg d.a.}$$

21. **(c):** As the temperature of spray water is greater than WBT and DBT of air, so cooling does not take place but humidification does occur. Thus, it is a case of heating and humidification.

22. **(d) : Process 3-4 :** It is a throttling process in which h = constant. Line 3-4 is a vertical line.

Process 2-3 : p = constant

Process 4-1 : p = constant.

Hence, it is a T-h line.

23. **(c) :**

$$\text{COP} = \left[\frac{T_3 - T_2}{T_3}\right] \times \left[\frac{T_1}{T_2 - T_1}\right]$$

$$= \left[\frac{400 - 300}{400}\right] \times \left[\frac{250}{300 - 250}\right]$$

$$= \frac{100}{400} \times \frac{250}{50} = 1.25$$

$$\frac{Q_{\text{ref}}}{Q_{\text{ref}}} = \text{COP}, \ Q_{\text{ref}} = \frac{100}{\text{COP}} = \frac{100}{1.25} = 80 \text{ W}$$

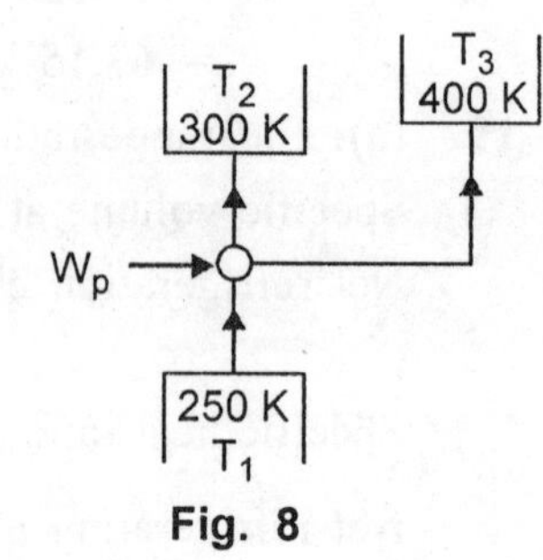

Fig. 8

24. **(b) :** ***P* (0 – 1) :** Sensible heating as DBT increases at specific humidity, ω = constant.

***Q* (0 – 2) :** Chemical dehumidification at WBT = constant.

***R* (0 – 3) :** Cooling and dehumidification at WBT = constant.

***S* (0 – 4) :** Humidification with water injection, DBt and w both decrease.

T (0 – 5) : Humidification with steam injection. Both DBT and w increase.

P-(iii), *Q*-(ii), *R*-(iii), *S*-(v), *T*-(iv)

25. **(d) :** $h = u + pv$, *i.e.*, $h > u$. Also s has very less values as compared to the values of h and u. Therefore specific enthalpy values are in columns 5 and 8.

26. **(b) :** $h_1 = h_{f1} = h_2$

$$h_{f1} = h_{f2} + x_2\,(h_{g2} - h_{f2})$$

$$371.43 = 89.05 + x_2\,(1418.0 - 89.05)$$

$$x_2 = \frac{371.43 - 89.05}{1418.0 - 89.05} = \frac{282.38}{1328.95} = 0.212$$

27. (c) : DPT is the temperature at which condensation begins when the air is cooled at constant pressure.

28. (d) : Statements 2 and 4 are correct.

29. (b) : Heat transfer rate due to convection at inner surface of building.

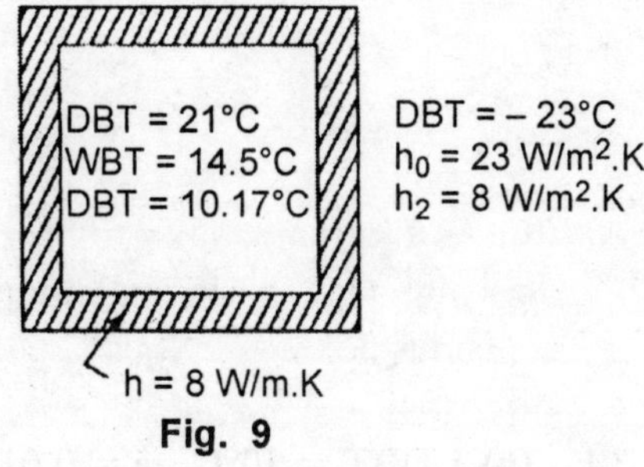

Fig. 9

$$\dot{q} = \frac{\dot{Q}}{A} = h_i(DBT - DPT)_i$$

$$= 8\ (21 - 10.17) = 86.64\ \text{W/m}^2$$

Heat transfer rate due to conduction through the wall,

$$\dot{q} = \frac{(DPT)_i - (DBT)_0}{\dfrac{x}{k} + \dfrac{1}{h_0}}$$

$$86.64 = \frac{10.17 - (-23)}{\dfrac{x}{1.2} + \dfrac{1}{23}}$$

$$\frac{x}{1.2} + \frac{1}{23} = \frac{33.17}{86.64} = 0.3823$$

$$\frac{x}{1.2} = \frac{33.17}{86.64} + 0.3823$$

$$\frac{x}{1.2} = 0.3393$$

$$x = 0.4072\ \text{m}$$

30. (c) : $\dot{m}_a = 3$ kg/s, $h_1 = 85$ kJ/kg, $w_1 = 0.019$ kg/kg d.a.,

$h_2 = 43$ kJ/kg $w_2 = 0.08$ kg/kg d.a., $h_{fg} = 67$ kJ/kg

Cooling capacity of coil.

$$Q = \text{S.H.} + \text{L.H} = m_a\ (h_1 - h_2) + \dot{m}_a (w_1 - w_2)\ h_{fg}$$

$$= 3 \times (85 - 43) + 3 \times (0.019 - 0.008) \times 67$$

$$= 126 + 2.21 = 128.21\ \text{kW}$$

Fig. 10

$$p_1 = 100\ \text{kPa},\ p_2 = 500\ \text{kPa}$$

31. (b) : $p_2 = p_3$, $p_2 = 5p_1$

$$p_1 = \frac{5.2628}{5} = 1.126\ \text{kPa} = 1.13\ \text{kPa}$$

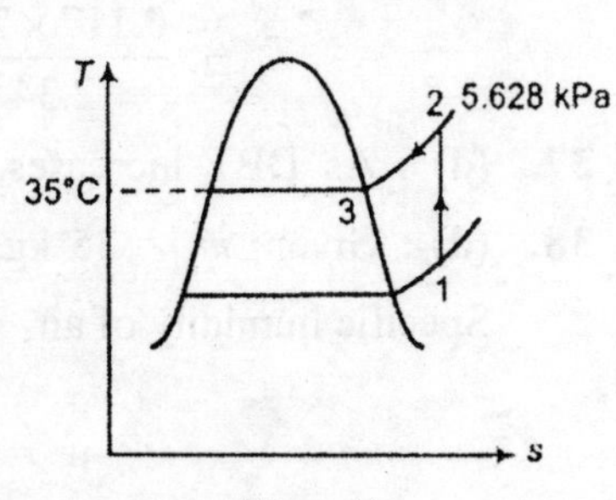

Fig. 11

32. (c) : $p_1 = 1.01325$ bar, $(DBT)_1 = 40°C$, $(WBT)_1 = 20°C$

$$(DBT - WBT)_2 = 0.25\ (DBT - WBT)_1$$

$$= 0.25\ (40 - 20) = 5°C$$

$$(DBT)_2 = 5 + (WBT)_2$$

$$(WBT)_2 = (WBT)_1$$

$$(DBT)_2 = 5 + 20 = 25°C$$

33. (a) : $h_2 = 283$ kJ/kg, $h_3 = 116$ kJ/kg $= h_4$,

$$h_1 = 232 \text{ kJ/kg}$$

$$\text{COP} = \frac{h_1 - h_4}{h_2 - h_1} = \frac{232 - 116}{283 - 232} = \frac{116}{51} = 2.27$$

Fig. 12

34. (b) : DBT = 30°C, $w = 0.0115$ kg/kg d.a., $M_a = 28.93$

$$p_s = 4.24 \text{ kPa}, \ p_t = 90 \text{ kPa}$$

Relative humiditiy,

$$\phi = \frac{w}{0.622} \times \frac{p_t}{p_s} = \frac{0.0115}{0.622} \times \frac{90}{4.24} = 0.385 \text{ or } 38.5\%$$

35. (d) : On heating moist air in an air tight ressel its relative humidity decreases.

36. (d) : $\phi = 5\%, = 0.05$, $p_s = 2.34$ kPa at 20°C

$$\phi = \frac{p_v}{p_s}, \ p_v = 0.05 \times 2.34 = 0.117 \text{ kPa}$$

Now, $$\frac{p_1 v_1}{T_1} = \frac{p_2 v_2}{T_2}$$

$$v_1 = v_2$$

$$\frac{p_s}{T_s} = \frac{p_v}{T_v}$$

$$\frac{2.34}{293} = \frac{0.117}{T_v}$$

$$T_v = \frac{0.117 \times 293}{2.34} = 14.65$$

37. (d) : As DBT increases, then RH decreases.

38. (d) : Given: $m_a = 35$ kg, $m_v = 0.5$ kg, $p_t = 100$ kPa, $t_a = 25$°C, $p_s = 3.17$ kPa

Specific humidity of air,

$$w = \frac{m_w}{m_a} = \frac{0.5}{35} = 0.01428 \text{ kg/kg d.a.}$$

Now $$w = \frac{0.622\, p_v}{p_t - p_v}$$

$$0.01428 = \frac{0.622\, p_v}{100 - p_v}$$

$$1.428 - 0.01428\, p_v = 0.622\, p_v$$

$$p_v = 2.244 \text{ kPa}$$

$$\text{RH}_1(\phi) = \frac{p_v}{p_s} = \frac{2.244}{3.17} = 0.708 \simeq 71\%$$

39. **(c)** : Given : $p_t = 1$ bar, $t_d = 30°C$, $\phi = 70\%$, $p_s = 4.25$ kPa

$$\phi = \frac{p_v}{p_s}$$

$$0.7 = \frac{p_v}{4.25},\ p_v = 2.975 \text{ kPa}$$

$$w = \frac{0.622\, p_v}{p_t - p_v} = \frac{0.622 \times 2.975}{(100 - 2.975)} = 0.0191 \text{ kg/kg d.a.}$$

APPENDIX–II

MCQ'S FROM CIVIL SERVICES EXAMINATIONS

1. Match List I with List II and select the correct answer using the codes given below lists.

List I | **List II**

A. 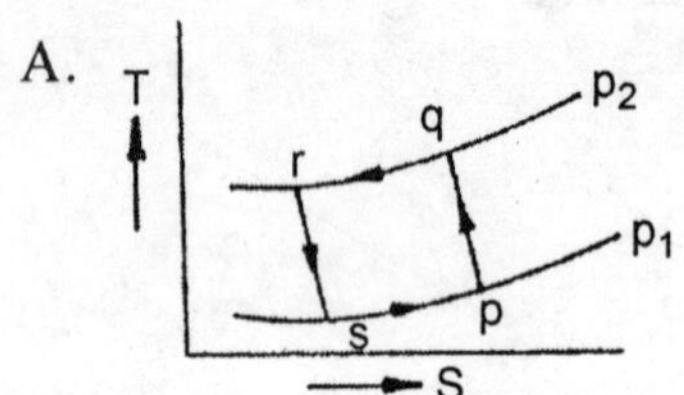

1. Vapour compression cycle using expansion valve.

B. 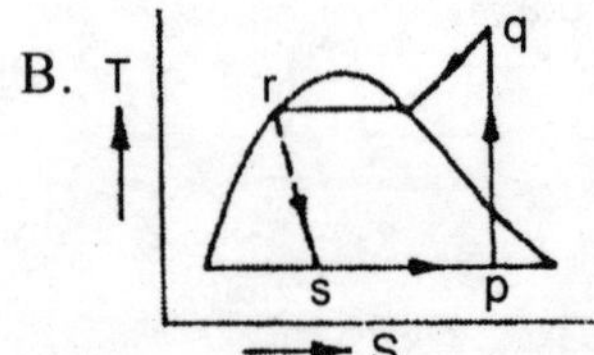

2. Bell-Coleman cycle (gas compression cycle)

C. 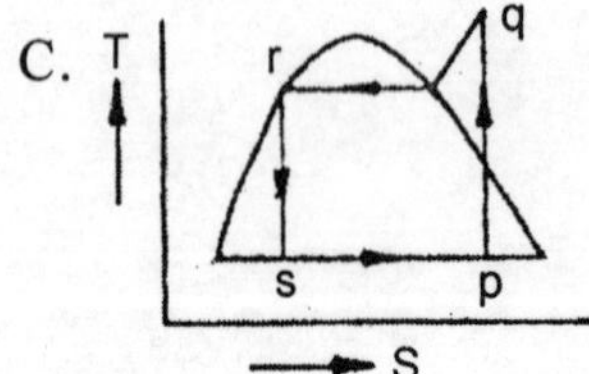

3. Vapour compression cycle using expansion engine

Codes:	A	B	C
(a)	1	2	3
(b)	2	3	1
(c)	1	3	2
(d)	2	1	3

2. In the absorption refrigeration cycle, the compressor of the vapour compression refrigeration cycle is replaced by

(a) liquid pump (b) generator

(c) absorber and generator (d) absorber, liquid pump and generator.

3. The C.O.P. of a Carnot refrigeration cycle decreases on

(a) Decreasing the difference in operating temperatures

(b) Keeping the upper temperature constant and increasing the lower temperature

(c) Increasing the upper temperature and keeping the lower temperature constant

(d) Increasing the upper temperature and decreasing the lower temperature

4. Desert coolers are suitable for hot very dry outside conditions because

(a) Water is recirculated in the spray

(b) Heat is neither added nor removed from the water

(c) Wet bulb depression ($t - t'$) is very large

(d) Large quantity of air can be conditioned

5. In an auditorium, the heat generated due to the occupants and the electric lights and other equipments is 100 kW. The rate of generation of excess moisture is 60 kg/hr. if an air-conditioner is supplying conditioned air to the auditorium at the rate of 500 m^3/min, then the sensible heat factor (SHF) for the auditorium is

(a) 0.27 (b) 0.40

(c) 0.73 (d) 0.95

6. A room air is at a DBT of T_r and relative humidity ϕ_r. The effective temperature of the room is

(a) The temperature at which the room air is saturated but gives the same feeling of comfort as the actual state of the room air

(b) The temperature at which the room air is at 50% relative humidity but gives the same feeling of comfort as the actual state of the room air

(c) The temperature at which the room air is completely dry but gives the same feeling of comfort as the actual state of the room air.

(d) None of the above.

7. Consider the following statements

I. Low value of the bypass factor for an air-conditioning equipment signifies higher performance of the equipment

II. Bypass factor for an air-conditioning equipment signifies the fraction of ambient air mixed with the air to be conditioned.

III. Bypass factor for an air-conditioning equipment signifies the fraction of the air to be conditioned coming in contact with the conditioning surface.

Of these statements:

(a) I and II are correct (b) I and II are correct

(c) III alone is correct (d) II alone is correct

8. It is desired to condition the outside air from 70% RH and 45°C dry bulb to 50% RH and 25°C dry bulb room condition. The practical arrangement would be

(a) cooling and dehumidification (b) dehumidification and pure sensible cooling

(c) cooling and humidification (d) dehumidification

9. To cool and dehumidify a stream of moist air, it must be passed over the coil at a temperature

(a) Which lies between the dry bulb and wet bulb temperature of the incoming stream

(b) Which lies between the wet bulb and dew point temperature of the incoming stream

(c) Which is lower than the dew point temperature of the incoming stream

(d) Of adiabatic saturation of incoming steam

10. Consider the following statements:

1. Dew point is reached by cooling air at constant moisture content.
2. Wet bulb temperature changes by addition of moisture at constant enthalpy.
3. For saturated air, the dry bulb temperature, wet bulb temperature and dew point are the same.
4. Dehumidification of air is achieved by heating.

Of these statements:

(a) 1 and 3 are correct (b) 1 and 2 are correct

(c) 3 and 4 are correct (d) 1 alone is correct

11. Which one of the following statements is correct?

(a) The sensible heat gain is due to the difference in humidity

(b) The latent heat gain is due to the temperature difference between the fresh air through unconditioned space in the building adds to the sensible heat gain

(c) The heat gain through the walls of ducts carrying conditioned air through unconditioned space in the building adds to the sensible heat gain

(d) Maximum heat gain toa building occurs through walls

12. For air-conditioning the operation theatre in a hospital, the percentage of outside air in the air supplied is

(a) zero (b) 20 (c) 50 (d) 100

13. The schematic diagram of a vapour compression refrigeration system can be represented as

(a)

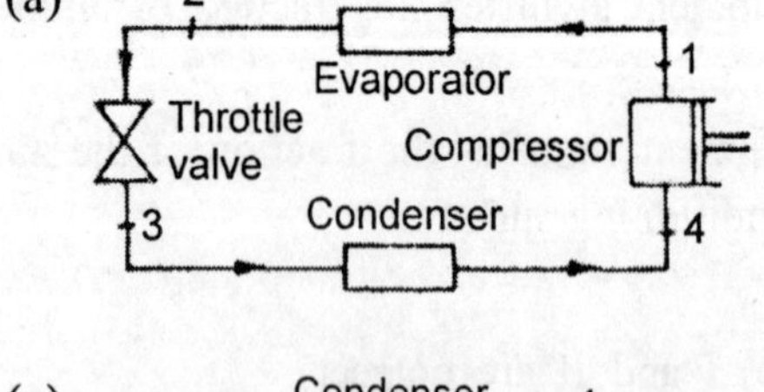

(b)

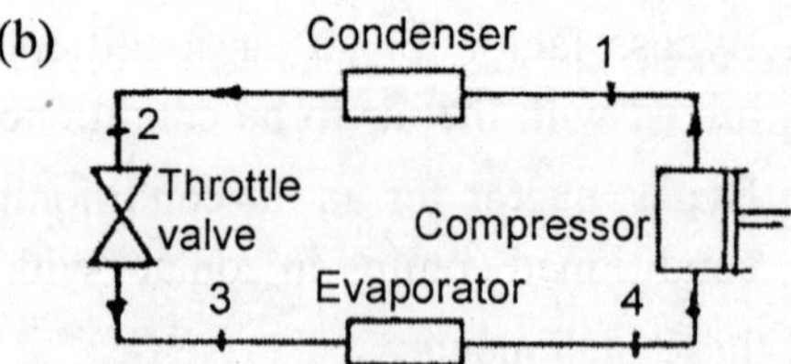

(c)

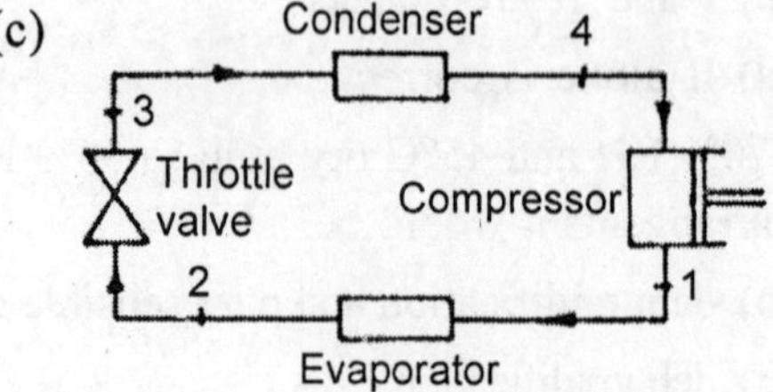

(d)

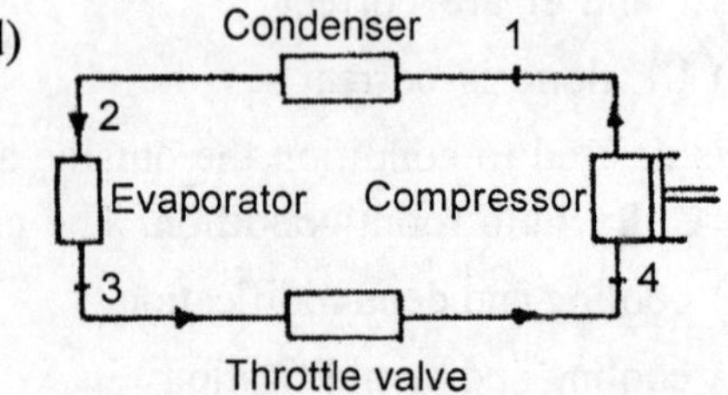

14. The colour of the flame of halide torch, in a case of leakage of Freon refrigerant, will change to

(a) bright green (b) yellow (c) red (d) orange

15. Which one of the following refrigerants has the highest critical temperature?

(a) Water (b) Carbon dioxide (c) Freon 12 (d) Ammonia

16. The significant advantages of using ammonia as a refrigerant is its

(a) characteristic odour (b) high latent heat

(c) solubility (d) inflammability

17. The condition of air for a cooling and dehumidification system is given by the point A at intake, B at discharge as marked on a psychrometric chart. If C is the apparatus dew point, the bypass factor is given by (see Fig. 13)

(a) $\dfrac{CA}{AB}$ (b) $\dfrac{CA}{BC}$

(c) $\dfrac{BC}{AB}$ (e) $\dfrac{BC}{CA}$

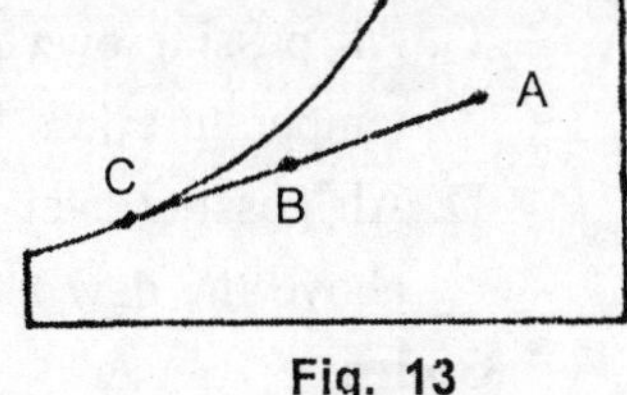

Fig. 13

18. With respect to the following figure 14, which shows four processes on the psychrometric chart, match List I with List II and select the correct answer using the codes given below the lists:

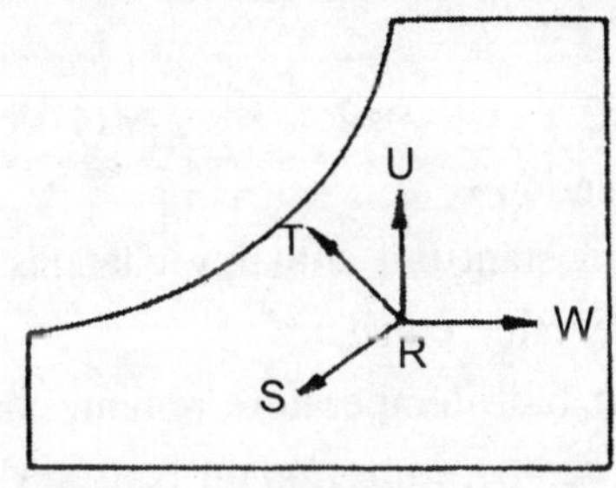

Fig. 14

List I	**List II**
A. Process RS	1. Cooling and humidifying
B. Process RT	2. Sensible heating
C. Process RU	3. Cooling and dehumidifying
D. Process RW	4. Humidifying

Codes:	A	B	C	D
(a)	2	3	1	4
(b)	1	4	2	3
(c)	3	1	4	2
(d)	3	2	1	4

19. Consider the following statements:

Effective temperature

1. Is a measure of the sensation of warmth or coldness.

2. Is the uniform temperature of an imaginary enclosure with which man will exchange the same dry heat by radiation and convection as in the actual environment.

3. Combines the effects of dry bulb temperature, wet bulb temperature and air movement.

On these statements:

(a) 1 and 2 are correct (b) 1 and 2 are correct

(c) 2 and 3 are correct (d) 1 and 3 are correct

20. Match List I with List II and select the correct answer using the codes given below the lists:

List I	List II
A. Steam spray into air	1. Sensible cooling
B. Air passing over a coil carrying steam	2. Cooling and dehumidification
C. Air passing over a coil having temperature less than dew point.	3. Heating an humidification
D. Air passing over a coil having temperature above the dew point but below the wbt	4. Sensible heating

Codes:	A	B	C	D
(a)	2	1	3	4
(b)	3	1	2	4
(c)	3	4	2	1
(d)	4	3	2	1

21. During chemical dehumidification

(a) Wet bulb temperature constant but enthalpy changes

(b) Dry bulb temperature remains constant

(c) Both dew point and wet bulb temperature remain constant

(d) Enthalpy and web bulb temperature remain constant

22. Theoretical vapour compression refrigeration cycle is represented on a T-8 diagram as

(a)

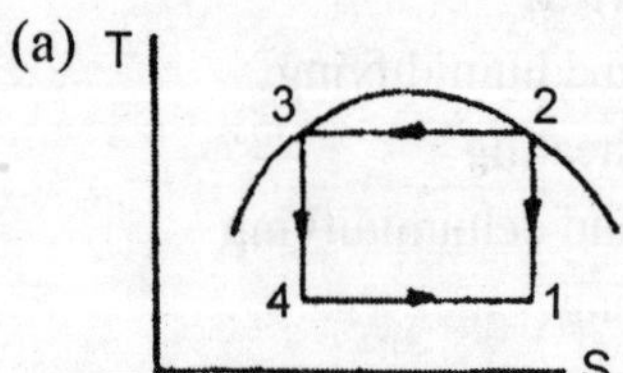

(b)

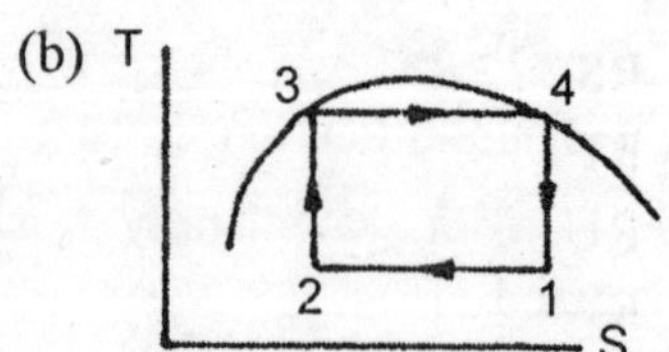

(c)

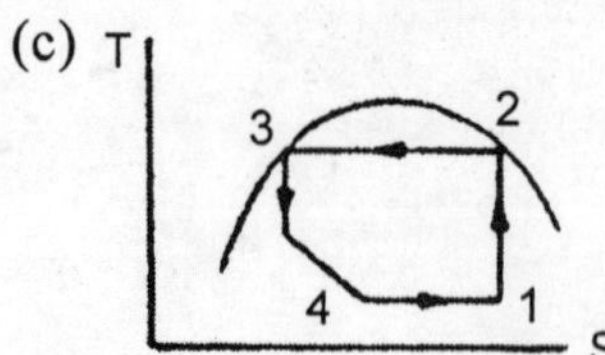

(d)

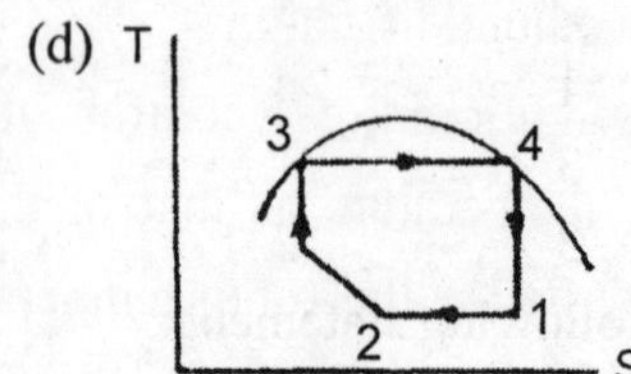

23. Match List-I (Effect) with List -II (Process) in the case of an ideal refrigeration cycle and select the correct using the codes given below the Lists:

List-I	List-II
A. Work input	1. Constant pressure at higher temperature
B. Heat rejection	2. Isentropic compression
C. Expansion	3. Constant temperature at lower pressure
D. Heat absorption	4. Adiabatic

Codes:	A	B	C	D
(a)	4	1	2	3
(b)	2	3	4	1
(c)	2	1	4	3
(d)	4	2	3	1

24. A rotary compressor is used when a refrigerating system has to handle a refrigerant with

(a) Low specific volume and high pressure difference

(b) Low specific volume and low pressure difference

(c) Large specific volume and high pressure difference

(d) Large specific volume and low pressure difference

25. For the same condenser and evaporator temperature, the COP of absorption refrigeration system is less than that of mechanical vapour compression refrigeration system, since in the absorption refrigeration system,

(a) A liquid pump is used for compression (b) A refrigerant as well as a solvent is used

(c) Absorber requires heat rejection (d) Low grade energy is used to run the system

26. If air at dry-bulb temperature of 35°C and dew point temperature of 20°C passes through a cooling coil which is maintained at 25°C, then the process would be

(a) sensible cooling (b) cooling and dehumidification

(c) cooling and humidification (d) cooling at constant wet bulb temperature

27. The process in a lot water spray washer maintained at a temperature of 40°C, through which unsaturated air at 10°C dry bulb temperature and 50% relative humidity passes, is

(a) sensible heating (b) humidification

(c) heating and humidification (d) heating and dehumidification

28. In room air-conditioning for comfort, the supply air in summer should be at

(a) the same temperature as that of the room

(b) 5 to 10°C below the room temperature

(c) 2 to 3°C above the room temperature

(d) at 0°C

29. In the case of a cooling coil with non-zero by pass factor, the apparatus, dew point temperature lies at the intersection point of

(a) Room DB line with the saturation curve (b) RSHF and GSHF lines
(c) RSHF and ESHF lines (d) GSHF line with the saturation curve

30. In an air-conditioning process, 5 kJ/min heat is extracted from a room. If the sensible heat factor is 0.8, then the latent heat extracted will be

(a) 4 kJ/min (b) 2 kJ/min
(c) 1 kJ/min (d) 0.25 kJ/min

31. A sample of moist air is at a temperature T and relative humidity 50%. A part of the moisture is removed adiabatically by using an adsorbent. If the heat of adsorption is negligible, the resulting air will have the same.

(a) Dry bulb temperature but a lower wet bulb temperature
(b) Wet bulb temperature but a higher dry bulb temperature
(c) Dry bulb temperature but a higher wet bulb temperature
(d) Wet bulb temperature but a lower dry bulb temperature

32. In summer air-conditioning the conditioned air passing through the space undergoes a process of

(a) sensible cooling (b) sensible heating
(c) cooling and dehumidification (d) heating and humidification

33. Suppose, the mean body surface temperature of a certain clothed person is 30°C, the mean temperature of surrounding surfaces is 21°C and the air within the room is at 32°C, 90 per cent RH, then the person will lose maximum body heat by

(a) convection (b) radiation
(c) mass transfer (d) convection, radiation and mass transfer

34. The difference between the comfort air-conditioning and industrial air-conditioning lies in the

(a) equipment used (b) process adopted
(c) indoor requirements (d) ambient conditions

35. The state of air supplied by a cooling coil with a by-pass factor 'X' lies on the psychrometric chart at the

(a) Intersection of RSHF line with saturation curve
(b) Intersection of GSHF line with saturation curve
(c) Point which divides RSHF line in proportion to 'X' and (1—X)
(d) Point which divides ESHF line in proportion to 'X' and (1—X)

36. Consider the following statements:

High condenser pressure in a refrigeration system can occur because

1. The water flow rate is lower than the desired value.
2. Non-condensible gases are present in th system.
3. Of accumulation of lubricating oil in condenser
4. Of low charge of refrigerant in the system

Of these statements

(a) 1, 3 and 4 are correct (b) 1, 2 and 3 are correct

(c) 1, 2 and 4 are correct (d) 2, 3 and 4 are correct

37. Consider the following statements:

Dry compression in reciprocating compressor is preferred because it

1. Prevents value damage.
2. Enables use of thermostatic expansion value.
3. Minimizes irreversibilities in the compressor.
4. Prevents washing out of the lubricating oil from cylinder walls.

Of these statements

(a) 1 and 2 are correct (b) 2 and 3 are correct

(c) 1 and 4 are correct (d) 3 and 4 are correct

38. Match List-I with List-II and select the correct answer using the codes given below the lists:

List-I (Refrigerant)	List-II (Designation)
A. Dichloro-difluoromethane	1. R 718
B. Water	2. R 22
C. Methyl chloride	3. R 40
D. Monochlorodifluoromethane	4. R 12

Codes:	A	B	C	D
(a)	4	1	2	3
(b)	1	4	3	2
(c)	1	4	2	3
(d)	4	1	3	2

39. Consider the following statements:

1. In Freon 22 system, moisture choking generally does not occur.
2. Freon 11 is mainly used in large capacity air-conditioning plants with centrifugal compressors.
3. Presence of lubricating oil in evaporator will increase the heat transfer coefficient.
4. Refrigerants that are completely miscible with oil, do not cause oil choking.

Of these statements

(a) 1, 2 and 3 are correct (b) 1, 2 and 4 are correct

(c) 2, 3 and 4 are correct (d) 1, 3 and 4 are correct

40. If a sample of moist air of 50% relative humidity at atmospheric pressure is isothermally compressed to a pressure of two atmospheres then

(a) Its relative humidity will reduce to 25%

(b) Its relative humidity will remain unchanged

(c) The sample of air will become saturated

(d) Saturation pressure will increase to twice the value

41. Evaporative regulation of body temperature fails when the body temperature is
(a) More than wet bulb temperature but less than dry bulb temperature
(b) More than dew point but less than wet bulb temperature
(c) More than dew point but less than dry bulb temperature
(d) Less than dew point

42. Consider the following statements:
Effective temperature is NOT a true comfort index because.
1. Discomfort may be experienced at extremely high or low humidities.
2. The radiation effect of surrounding surfaces has not been taken into account.
3. It presumes the absence of drafts.
Of these statements:
(a) 1, 2 and 3 are correct (b) 1 and 2 are correct
(c) 1 and 3 are correct (d) 2 and 3 are correct

43. If wet bulb depression is equal to the sum of range and approach of a cooling tower, then the water
(a) Inlet temperature is equal to the wet bulb temperature of ambient air
(b) Outlet temperature is equal to the wet bulb temperature of ambient air
(c) Inlet temperature is equal to dry bulb temperature of ambient air.
(d) Outlet temperature is equal to dry bulb temperature of ambient air.

44. A Carnot refrigerator works between the temperatures of 200 K and 300 K. If the refrigerator receives 1 kW of heat, the work requirement will be
(a) 0.5 kW (b) 0.67 kW
(c) 1.5 kW (d) 3 kW

45. The refrigerant-12 (R-12) used in vapour compression refrigeration system is
(a) $CHClF_2$ (b) CCl_2F_2 (c) $CHCl_2F$ (d) $CClF_3$

46. An expression valve is NOT used for reducing pressure in the
(a) Vapour compression refrigeration cycle (b) Vapour absorption refrigeration cycle
(c) Steam-jet refrigeration cycle (d) Gas refrigeration cycle

47. The function of a compressor in a vapour absorption system is performed by the
(a) absorber (b) absorber and pump
(c) pump and generator (d) absorber, pump and generator

48. Consider the following statements:
In a reciprocating compressor, clearance volume is provided
1. So that piston does not hit and damage the values
2. To account for differential thermal expansion of piston and cylinder
3. To account for machining tolerances
4. To achieve isentropic compression

Which of these statements are correct?

(a) 1, 2 and 3 (b) 1, 2 and 4

(c) 1, 3 and 4 (d) 2, 3 and 4

49. Replacing a water-cooled condenser with an air-cooled one in a vapour compression refrigeration system with constant evaporator pressure result in

(a) Increase in condensation pressure (b) Decrease in pressure ration

(c) Increase in pressure ratio (d) Increase in condensation temperature

50. If the measured wet-bulb temperature and the thermodynamic wet-bulb temperature are equal, then the non-dimensional number with a value of unity is the

(a) Lewis number (b) Prandtl number

(c) Schmidt number (d) Sherwood number

51. The ration of weight of water vapour associated with unit weight of dry air to the weight of water vapour associated with unit weight of dry air saturated at the same dry-bulb temperature and pressure is known as

(a) specific humidity (b) relative humidity

(c) absolute humidity (d) degree of saturation

52. Consider the following statements:

1. If air is heated in a closed chamber, its dew point will increase
2. As relative humidity decreases. The difference between the wet-bulb temperature and dew point will increase
3. In spray humidification process. The enthalpy of air will decrease
4. The dew-point temperature is always an indication of moisture content of the air

Which of these statements are correct?

(a) 1 and 2 (b) 2 and 4 (c) 1 and 3 (d) 3 and 4

53. When moist air comes into contact with a wetted surface whose temperature is less than the dry-bulb temperature but more than the wet-bulb temperature.

(a) Sensible, latent and net heat transfers are from air to surface

(b) Both sensible and net heat transfers are from air to surface but latent heat transfer is from surface to air

(c) Sensible heat transfer is from air to surface but both latent and net heat transfers are from surface to air

(d) Sensible heat transfer is from surface to air but both latent and net heat transfers are from air to surface

54. In a certain ideal refrigeration cycle, the COP of heat pump is 5. The cycle under identical conditions running as heat engine will have efficiency as

(a) zero (b) 0.20 (c) 1.00 (d) 6.00

55. The power (kW) required per ton of refrigeration is $\frac{N}{COP}$, where COP, is the coefficient of performance, then N is equal to

(a) 2.75 (b) 3.50 (c) 4.75 (d) 5.25

56. Match List-I with List-II and select the correct answer using the codes given below the lists:

List-I	List-II
A. Refrigerant 11	1. CCl_2F_2
B. Refrigerant 12	2. $C_2Cl_2F_4$
C. Refrigerant 22	3. CCl_3F
D. Refrigerant 114	4. $CHClF_2$
	5. CH_2ClF

Codes:	A	B	C	D
(a)	2	1	5	3
(b)	3	4	5	2
(c)	3	1	4	2
(d)	5	1	4	3

57. The expression = $\frac{0.622\, p_v}{p_b - p_v}$, where p_v = partial pressure of water vapour; p_h = atmospheric barometric pressure is used for calculating.

(a) relative humidity (b) degree of saturation
(c) humidity ratio (d) pressure of air

58. A cooling oil with a bypass factor of 0.1 and apparatus dew point (*adp*) of 12°C comes in contact with air having a dry-bulb temperature of 38°C and dew point of 9°C. Over the cooling coil the air would undergo

(a) Sensible cooling (b) Cooling and humidification
(c) Cooling and dehumidification (d) Adiabatic saturation

59. The main process which takes place in a desert-cooler is

(a) Sensible cooling (b) Dehumidification
(c) Adiabatic saturation (d) Cooling and dehumidification

60. Consider the following statements pertaining to a vapour compression type refrigerator:

1. The condenser rejects heat to the surroundings from the refrigerant
2. The evaporator absorbs heat from the surroundings to be cooled
3. Both the condenser and evaporator are heat exchangers with refrigerant as a common medium.
4. The amount of heat exchanged in condenser and evaporator are equal under steady conditions.

Which of the above statements are correct?

(a) 1 and 2 (b) 1, 2 and 3 (c) 1, 2 and 4 (d) 2, 3 and 4

61. The throttling device used in the domestic refrigerator is

(a) Internally equalized thermostatic expansion valve
(b) Externally equalized thermostatic expansion valve
(c) Automatic expansion valve
(d) Capillary tube

62. In system A vapours are superheated by 10°C in the evaporator while in system *B* vapours are superheated by 10°C in a liquid vapour regenerative heat exchanger, other conditions being the same. The

(a) C.O.P. of A = C.O.P. of B

(b) C.O.P. of both *A* and *B* > C.O.P. of Reversed Carnot Cycle

(c) C.O.P. of *A* > C.O.P. of *B*

(d) C.O.P. of *A* < C.O.P. of *B*

63. Match List-I (Chemical formula of refrigerant) with List -II (Numerical Designation) and select the correct answer using the codes given below the lists:

List-I *(Chemical formula of refrigerant)*	**List-II** *(Numerical designation)*
A. NH_3	1. 12
B. CCl_2F_2	2. 22
C. $CHClF_2$	3. 40
D. CCl_2FCClF_2	4. 113
	5. 717

Codes:	A	B	C	D
(a)	4	1	5	2
(b)	5	3	2	4
(c)	4	3	5	2
(d)	5	1	2	4

64. Which one of the following statements is correct?

(a) Dehumidifier coil surface temperature is above the dew point temperature but below the freezing point temperature

(b) Dehumidifier coil surface temperature is below the dew point temperature but above the freezing point temperature

(c) Dehumidifier coil surface temperature is below both the dew point temperature and the freezing point temperature

(d) Dehumidifier coil surface temperature is above both the dew point temperature and the freezing point temperature

65. Total heat transfer from a wetted surface depends upon

(a) Difference in temperature between surface and air

(b) Difference in humidity ratio of air and air saturated at wet surface temperature

(c) Difference in enthalpy between saturated air at surface temperature and that of air

(d) Difference in entropy between saturated air at surface temperature and that of air

66. Moist air exists at a pressure of 1.01 bar. The partial pressure and saturation pressure of water vapour are 0.01 bar and 0.02 bar respectively. What are the relative humidity and humidity ratio of the moist air, respectively?

(a) 50% and 0.00622 (b) 100% and 0.0126

(c) 50% and 0.0126 (d) 100% and 0.00622

67. In order to cool and dehumidify a stream of moist air, it must be passed over a coil at a temperature

(a) Which lies between the dry bulb and wet bulb temperatures of the incoming stream

(b) Which lies between the web bulb and dew point temperatures of the incoming stream

(c) Which is lower than the dew point temperature of the incoming stream

(d) Of adiabatic saturation of incoming stream

68. Which of the features of expansion valves in the following lists are correctly matched?

Expansion Device	*Feature*
1. Capillary tube	Choking
2. Thermostatic expansion valve	Constant temperature
3. Automatic Expansion valve	Constant degree of superheat
4. Float valve	Mass flow rate of refrigerant is proportional to load

Select the correct answer using the codes given below:

(a) 1 and 2 (b) 1 and 3 (c) 1 and 4 (d) 3 and 4

69. If moist air is sensibly cooled its dew point, which of the following statements are correct?

1. Relative humidity decreases
2. Wet bulb temperature decreases
3. Wet bulb temperature increases
4. Humidity ratio remains constant

Select the correct answer using the codes given below:

Codes:

(a) 1 and 2 (b) 1 and 3 (c) 3 and 4 (d) 2 and 4

70. Water in an insulated evaporative cooler evaporates at the rate of 0.003 kg/s. Air flow rate is 1 kg/s. What is the air temperature decrease if the specific heat of humid air is 1 kJ/kg. K and latent heat of water is 2500 kJ/kg?

(a) 2.5°C (b) 3.0°C (c) 7.5°C (d) 10°C

71. Match List-I with List-II and select the correct answer using the codes given below the lists:

List-I	**List-II**
A. Sulphur candle test	1. Propane
B. Halide torch test	2. Ammonia
C. Soap and water test	3. Halocarbon refrigerants
D. Ammonia swab test	4. Sulphur dioxide

Codes:	A	B	C	D
(a)	2	3	1	4
(b)	4	1	3	2
(c)	2	1	3	4
(d)	4	3	1	2

72. An Electrolux system of refrigeration has

(a) No liquid pump (b) One liquid pump

(c) Two liquid pumps (d) Three liquid pumps

73. A Bell-Coleman air refrigeration system operates on which one of the following cycles?

(a) Reverse Diesel cycle (b) Reverse Carnot cycle

(c) Reverse Brayton cycle (d) Reverse Otto cycle

74. When the lower temperature is fixed, how can the c.o.p. of a refrigerating machine be improved?

(a) By raising the higher temperature

(b) By lowering the higher temperature

(c) By operating the machine at higher speeds

(d) By operating the machine at lower speeds

75. Evaporator is used sometimes to dehumidify the air for conditioning purpose. Which onc of the following is the correct statement?

In this situation the evaporator temperature should be kept

(a) Lower than the cooling fluid temperature

(b) More than the apparatus dew point temperature

(c) Lower than the apparatus dew point temperature

(d) Lower than the boiling temperature of fluid

76. To maintain comfort in an air conditioned space, how can a small increase in temperature be compensated?

(a) By humidification (b) By ventilation

(c) By increasing air velocity (d) By decreasing air velocity

77. Where is an oil separator in vapour compression refrigeration system installed?

(a) Between compressor and condenser (b) Between expansion valve and receiver tank

(c) Inside compressor machinery (d) Outside compressor machinery

78. Which one eco-friendly refrigerant out of tne following is used in many domestic and car air-conditioners nowadays?

(a) R-22 (b) R-113 (c) R-123 (d) R-134a

79. Which one of the following is the correct statement? Adiabatic saturation process for a moist air involves

(a) Cooling and humidification. (b) Cooling and dehumidification.

(c) Change in dew point temperature (d) Constant relative humidity.

80. Which one of the following is the correct statement? In domestic air conditions air undergoes the process of

(a) Sensible cooling
(b) Cooling and humidification
(c) Cooling and dehumidification
(d) Dehumidification

81. Which one of the following is the correct statement? Sensible heating of moist air will not affect the

(a) relative humidity
(b) absolute humidity
(c) specific humidity
(d) wet bulb temperature

82. Which one of the following is the correct statement? When moist air passes through a cooling tower, the process is

(a) Evaporative cooling
(b) Adiabatic saturation
(c) Cooling and dehumidification
(d) Heating and humidification

83. Which one of the following is the correct statement? During sensible cooling process for a moist air,

(a) Specific humidity will decrease
(b) Relative humidity will increase
(c) Relative humidity will decrease
(d) Specific humidity will increase

84. During a comfort air conditioning process, the moisture content of the air may start decreasing. Which one of the following is the correct statement?

In order to maintain the same level of comfort

(a) WBT should be decreased
(b) DBT should be increased
(c) Air velocity should be decreased
(d) Moisture content should be increased

85. For which one of the following DBT, WBT and DPT have the same value?

(a) 0 per cent relative humidity line
(b) 100 per cent relative humidity line
(c) 50 per cent relative humidity line
(d) None of the above.

86. Which of the following items related to infiltration of outdoor air in an air-conditioning system, are correctly matched?

1. Stack effect	:	Height of building
2. Crack length method	:	Wind velocity
3. Air change method	:	Floor area
4. Door opening	:	Occupancy in kitchen

Select the correct answer using the code given below:

(a) 1 and 2
(b) 1 and 3
(c) 1 and 4
(d) 2 and 4

87. Consider the following statements related to all air - conditioning system:

1. All air system uses air as heating or cooling fluid.
2. When hot air is circulated through rooms, dehumidification is necessary to control relative humidity.
3. Return air ducts are required for recirculation.

Which of the statements given above are correct?

(a) 1, 2 and 3 (b) 1 and 2 only (c) 1 and 3 only (d) 2 and 3 only

88. Ideal refrigerant mixture is one which

(a) Obeys Raoult's law in liquid phase and does not obey Dalton's law in vapour phase.

(b) Does not obey Raoult's law in liquid phase and does not obey Dalton's law in vapour phase

(c) Obeys Raoult's law in liquid phase and obeys Dalton's law in vapour phase.

(d) Does not obey Raoult's law in liquid phase and obeys Dalton's law in vapour phase

89. Consider the following statements:

1. The work of compressor in vapour compression refrigeration system increases with superheat of the suction vapour.
2. The work of compressor depends on the pressure difference rather than the temperature difference of evaporator and condenser
3. The coefficient of performance is within the range of 3 to 6 except at very low temperature when it may be less than 1.

Which of the statements given above are correct?

(a) 1, 2 and 3 (b) 1 and 2 only (c) 1 and 3 only (d) 2 and 3 only

90. Match List-I with List-II and select the correct answer using the code given below the lists:

List-I *(Commercial name of refrigerant)*	**List-II** *(Chemical formula of refrigerant)*
A. R 11	1. $CHClF_2$
B. R 12	2. CCl_2F_2
C. R 13	3. CCl_3F
D. R22	4. $C_2Cl_3F_3$

Codes:	A	B	C	D
(a)	3	1	4	2
(b)	4	2	3	1
(c)	3	2	4	1
(d)	4	1	3	2

91. In a centralized air conditioning system, what is the power of a fan directly proportional to?

(a) Fan Speed (b) $(\text{Fan speed})^2$ (c) $(\text{Fan speed})^3$ (d) $(\text{Fan speed})^4$

92. The inlet and outlet temperatures of air flowing over a heating coil having temperature of 41°C are 15°C and 28°C, respectively. What is the coil efficiency?

(a) 25% (b) 50% (c) 75% (d) 90%

93. The cooling loads for a building are:

RSH = 100 kW (Room sensible heat)

RLH = 25 kW (Room latent heat)

OASH = 10 kW (Outside air sensible heat)

OALH = 10 kW (Outside air latent heat)

The bypass factor of the cooling coil is 0.15. what are the values of the room sensible heat factor and external sensible heat factor respectively.

(a) 0.72 and 0.8 (b) 0.8 and 0.79
(c) 0.79 and 0.8 (d) 0.79 and 0.72

94. Which one of the following statement is correct? Dew point temperature of a moist air simple remains constant during

(a) a sensible heating or cooling process (b) an adiabatic saturation process
(c) a dehumidification process (d) a humidification process

95. A vapour compression refrigerator has a COP of 3, if the work absorbed by the compressor is 25 kW. What would be the amount of heat rejection in the condenser?

(a) 25 kW (b) 50 kW (c) 75 kW (d) 100 kW

96.

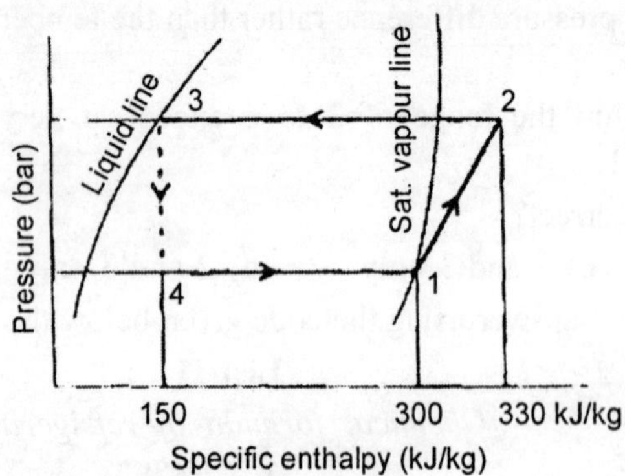

Fig. 15

A vapour compression refrigeration cycle as shown in the pressure ehtalpy diagram above (Fig. 15) what will be its COP?

(a) 3 (b) 4 (c) 5 (d) 6

97. High condenser pressure in a refrigeration system can occur due to which of the following reasons?

1. Cooling tower is inefficient or the water flow rate is less than the designed value.
2. Non-condensable gases are present in the system.
3. Accumulation of lubrication oil in the condenser
4. Low charge of refrigerant in the system.

Select the correct answer from the codes given below:

(a) 2 only (b) 1 or 2 or 3 (c) 1 or 2 or 4 (d) 2 or 3 or 4

98. Match List I with List II and select the correct answer using the code given below the lists:

List-I *(Refrigerant)*	**List-II** *(Colour)*
A. Ammonia	1. Purple
B. Freon 12	2. Green
C. Freon 22	3. Light blue
D. Methyl Chloride	4. White
	5. Orange

Codes:	A	B	C	D
(a)	1	2	4	3
(b)	3	2	4	5
(c)	1	4	2	3
(d)	3	4	2	5

99. Which one of the following refrigeration systems uses oil separator operating at – 10°C?

(a) System using NH_3
(b) System using R-12
(c) System using R-22
(d) System using R-502

100. Match List - I with List II and select the correct answer using the code given below the lists:

List-I *(Designation of refrigerant)*	**List-II** *(Chemical formula & name)*
A. R-11	1. SO_2 – Sulphur Dioxide
B. R-12	2. CHF_2C_1 – Difluorchloro Monochloro Methane
C. R-22	3. CF_2Cl_2 – Difluoro Dichloro Methane
D. R-764	4. $CFCl_3$ – Monofluoro Trichloro Methane

Codes:	A	B	C	D
(a)	1	3	2	4
(b)	4	3	2	1
(c)	1	2	3	4
(d)	4	2	3	1

101. In winter comfort air conditioning, the following changes may take place:

1. Temperature and humidity ratio both rise
2. The final relative humidity can be lower or higher than the initial value.
3. Both dry bulb and wet bulb temperature will increase.

Which of these statements is/are correct?

(a) 1, 2 and 3 (b) 1 and 2 only (c) 2 and 3 only (d) 3 only

102. The barometer pressure is 1.03 bar, the partial pressure of air is 1 bar, and the saturation partial pressure of water vapour at the same dry bulb temperature is 0.05 bar. What is the relative humidity?

(a) 5% (b) 50% (c) 60% (d) 97%

103. Moist air is a mixture of air and water vapour. Hence three independent thermodynamic properties are required to fix its thermodynamic state. Psychrometric chart however uses only two thermodynamic properties to fix the state. What is the reason?

(a) Psychrometric chart is an approximation to actual properties.
(b) Psychrometric chart assumes that water vapour and air behaves like perfect gases.
(c) Psychrometric chart is drawn for fixed atmospheric pressure.
(d) Psychrometric chart does not consider thermodynamic equilibrium states.

104. Consider the following statements. In measurement with a wet bulb thermometer

1. The wet bulb depression is an index of relative humidity.
2. The thermometer bulb should be wrapped by moist cloth and it is necessary to dip the end of the cloth in water.
3. Air is to flow at sufficiently low velocity past the thermometer bulb.
4. Dry bulb temperature will be equal to the wet bulb temperature when the air is completely dry.

Which of these statements is/are correct?

(a) 1 and 2 (b) 2 and 3 (c) 3 and 4 (c) 1 only

105.

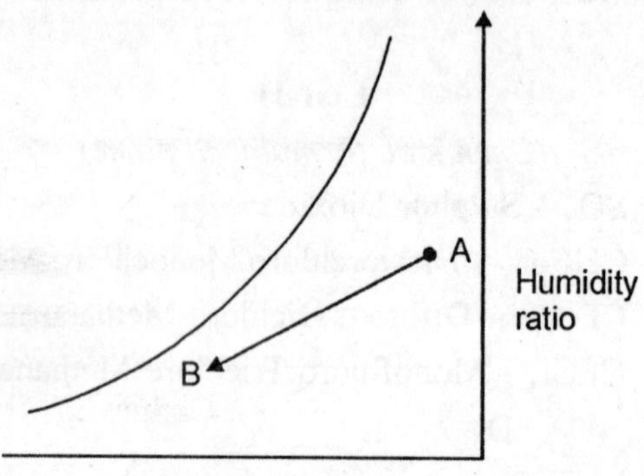

Fig. 16

What does the line AB as shown on the psychrometric chart (Fig. 16) represent?

(a) Sensible cooling with moisture addition (b) Sensible heating with dehumidification
(c) Sensible cooling with moisture removal (d) Sensible heating with moisture addition

106. Consider the following statements:

1. Effective temperature is an empirical index of warmth that takes into account the effect of temperature, humidity and air movement.
2. Effective temperature is not a true comfort index as it does not consider the radiation effect of surrounding surfaces.
3. It is the best comfort index for still air condition with 30% to 70% relative humidity.

Which of these statements is/are correct?

(a) 1, 2 and 3 (b) 1 and 2 only (c) 2 and 3 only (d) 3 only

107. The supply air temperature is 15 and apparatus dew point is 12 for the cooling coil with bypass factor of 0.15. what is the temperature at inlet of cooling coil?

(a) 32°C (b) 15.53°C (c) 14.45°C (d) 12.45°C

108. Consider the following statements:

Indoor design conditions for air-conditioning system for energy conservation are

1. The lowest condition of temperature, relative humidity and maximum air velocity are 20°C, 25% and 10 m/min respectively.
2. The maximum condition in summer is 25°C, 55% and 15 m/min.
3. When the building is not occupied during night, the temperature may be lowered in winter and increased in summer.

Which of these statements is/are correct?

(a) 1, 2 and 3 (b) 1 and 2 only (c) 2 and 3 only (d) 3 only

109. Match List I with List II and select the correct answer using the code given below the lists:

List-I	List-II
A. R – 1349	1. Propane
B. R – 152a	2. Isobutane
C. R – 290	3. Difluoroethane
D. R – 600a	4. Tetrafluoroethane

Codes:	A	B	C	D
(a)	4	3	1	2
(b)	2	3	1	4
(c)	4	1	3	2
(d)	2	1	3	4

110. What is the commonly used refrigerant for domestic refrigerator?

(a) R 11 (b) R 22 (c) Ammonia (d) Carbon dioxide

111. In a vapour compression refrigerating system it is desirable to have the refrigerant as

(a) saturated liquid at the suction to the compression

(b) wet vapour at the suction to the compression

(c) superheated vapour at the suction to the compression

(d) superheated liquid at the suction to the compression

112. Consider the following statements:

1. In a split type air-conditioning unit, the compressor alone is located outside the room.
2. In window type air-conditioning units water comes out of the unit as a result of cooling the air below the freezing point of water
3. Air may be cooled by blowing it over a wetted surface.
4. Air temperature increases when water vapour is separated by condensation.

Which of the above statements is/are correct?

(a) 1 only (b) 1, 2 and 3 (c) 2, 3 and 4 (d) 3 and 4 only

113. The psychrometric process the air undergoes while passing through a desert air cooler is

(a) Sensible cooling (b) Evaporative cooling

(c) Humidification (d) Dehumidification

114. The mass of water vapour in one m^3 of dry air is defined as

(a) Specific humidity (b) Humidity ratio

(c) Absolute humidity (d) Relative humidity

115. Which one of the following statements is correct fro sum of sensible and latent heat transfer from a wetted surface to another surface with air temperature t, humidity ratio W, enthalpy h, saturated air temperature t_s, humidity ratio W_s and enthalpy h_s?

(a) Proportional to $(t - t_s)$ (b) Proportional to $(W - W_s)$

(c) Proportional to $(h - h_s)$ (d) proportional to $(t - t_s) + 2500\ (W - W_s)$

116. Consider the following statements pertaining to comfort air-conditioning:

1. The temperature selected in comfort air-conditioning need not be that for optimum comfort
2. For each 5°C drop in outside temperature, 1°C should be reduced in effective temperature in design.
3. Heat insulated walls are not necessarily vapour barriers.

Which of the above statements are correcet?

(a) 1, 2 and 3 (b) 1 and 2 only (c) 1 and 3 only (d) 2 and 3 only

117. The expansion valve in a vapour compression refrigeration system expands

(a) Refrigerant vapour adiabatically

(b) high pressure liquid isentropically

(c) high pressure liquid by throttling process

(d) high pressure liquid by free expansion process.

Answers: Civil Services Examination

1. (d)	**2.** (d)	**3.** (d)	**4.** (c)	**5.** (c)	**6.** (a)	**7.** (b)	**8.** (a)
9. (c)	**10.** (a)	**11.** (c)	**12.** (d)	**13.** (b)	**14.** (a)	**15.** (a)	**16.** (b)
17. (d)	**18.** (c)	**19.** (a)	**20.** (c)	**21.** (d)	**22.** (c)	**23.** (c)	**24.** (d)
25. (d)	**26.** (a)	**27.** (c)	**28.** (b)	**29.** (d)	**30.** (c)	**31.** (b)	**32.** (d)
33. (a)	**34.** (c)	**35.** (d)	**36.** (b)	**37.** (c)	**38.** (d)	**39.** (b)	**40.** (b)
41. (d)	**42.** (b)	**43.** (c)	**44.** (a)	**45.** (b)	**46.** (c)	**47.** (d)	**48.** (b)
49. (d)	**50.** (a)	**51.** (d)	**52.** (d)	**53.** (c)	**54.** (b)	**55.** (b)	**56.** (c)
57. (c)	**58.** (c)	**59.** (c)	**60.** (b)	**61.** (d)	**62.** (d)	**63.** (d)	**64.** (c)
65. (d)	**66.** (a)	**67.** (c)	**68.** (c)	**69.** (d)	**70.** (c)	**71.** (a)	**72.** (a)
73. (c)	**74.** (b)	**75.** (c)	**76.** (c)	**77.** (a)	**78.** (d)	**79.** (a)	**80.** (c)
81. (c)	**82.** (b)	**83.** (b)	**84.** (c)	**85.** (b)	**86.** (b)	**87.** (d)	**88.** (c)
89. (d)	**90.** (c)	**91.** (c)	**92.** (b)	**93.** (b)	**94.** (a)	**95.** (c)	**96.** (c)
97. (b)	**98.** (b)	**99.** (a)	**100.** (b)	**101.** (c)	**102.** (c)	**103.** (c)	**104.** (c)
105. (c)	**106.** (c)	**107.** (a)	**108.** (c)	**109.** (a)	**110.** (b)	**111.** (c)	**112.** (c)
113. (b)	**114.** (c)	**115.** (d)	**116.** (c)	**117.** (c)			

EXPLANATORY NOTES

1. **(d) :** A – 2, B – 1, C – 3.
2. **(d) :** In absorption refrigeration cycle, the compressor of vapour compression refrigeration cycle is replaced by absorber, liquid pump and generator.
3. **(d) :** COP of Carnot refrigeration cycle $= \dfrac{T_2}{T_1 - T_2}$

   COP decreases as T_1 increases and T_2 decreases.
4. **(c) :** Desert coolers are suitable for hot very dry outside conditions because wet bulb depression is very large.
5. **(c) :** Sensible heat, $SL = 100$ kW or 100 kJ/s

$$\dot{m}_w = 60 \text{ kg/hr}, \; \dot{V}_a = 500 \text{ m}^3\text{/min}$$

$$c_{pa} = 1.022 \text{ kJ/kg·K}, \; p_a = 1.2 \text{ kg/m}^3$$

$$\dot{m}_a = \frac{p_a \dot{V}_a}{60} = \frac{1.2 \times 500}{60} = 10 \text{ kg/s}$$

$$(h_{fg}) = 2500 \text{ kJ/kg}$$

Sensible heat of air,

$$(SH)_a = c_{pa} \times \dot{m}_a = 1.022 \times 10 = 10.22 \text{ kJ/s}$$

Latent heat of moisture,

$$LH = \dot{m}_w \times (h_{fg})_w = \frac{60}{3600} \times 2500 = 41.67 \text{ kJ/s}$$

$$SHF = \frac{SH + (SH)_a}{SH + (SH)_a + LH} = \frac{100 + 10.22}{100 + 10.22 + 41.67} = 0.726 \simeq 0.73$$

6. **(a) :** Effective temperature of the room is the temperature at which the room air is saturated but gives the same feeling of comfort as the actual state of the room air.
7. **(b) :** Statements 1 and 2 are correct
8. **(a) :** Cooling and dehumidification shall condition at RH = 70% and DBT = 45°C to RH = 50% and DBT = 25°C.
9. **(c) :** The temperature of cooling coil must be lower than the dew point temperature of the incoming air stream.
10. **(a) :** Only statements 1 and 3 are correct
11. **(c) :** Statements (c) is correct
12. **(d) :** Infected air of operation theatre is replaced 100% by outside air.
13. **(b) :** Figure shown in choice (b) represents the vapour compression refrigeration system.
14. **(a) :** Flame colour of halide torch, in case of leakage of Freon refrigerant will change to bright green.

15. (a) : Water has the highest critical temperature.
16. (a) : The significant advantage of using ammonia as refrigerant is its high latent heat.
17. (d) : $BPF = \frac{BC}{CA}$
18. (c) : A – 3, B – 1, C – 4, D – 2
19. (a) : Statements 1 and 2 are correct
20. (c) : A – 3, B – 4, C – 2, D – 1
21. (d) : During chemical dehumidification, enthalpy and WBT remain constant.
22. (c) : Theoretical vapor compression refrigeration cycle is represented on T-s diagram as shown in choice (c).
23. (c) : A – 2, B – 1, C – 4, D – 3.
24. (d) : A rotary compressor is used in a refrigerating system to handle a refrigerant with large specific volume and low pressure difference
25. (d) : Statement (d) Is true
26. (a) : (DBT = 35°C and ADP = 20°C) $\xrightarrow[\text{Cooling coil}]{25°C}$ sensible cooling.
27. (c) : (unsaturated air at 10°C and 50% RH) $\xrightarrow[\text{Hot water spray}]{40°C}$ heating and humidification.
28. (b) : Supply air in summer for comfort room air-conditioning should be at 5°C to 10°C below the room temperature.
29. (d) : ADP lies at the intersection of GSHF line with the saturation curve.

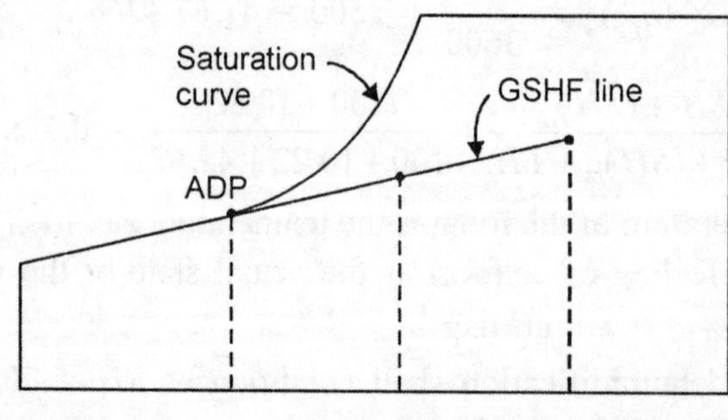

Fig. 17

30. (c) : $RSHF = \frac{RSH}{RSH + RLH}$

$$\frac{RSH}{S} = 0.8, \; RSH = 4 \text{ kJ/min}$$

$$RLH = 5 - RSH = 5 - 4 = 1 \text{ kJ/min}$$

31. (b) : During adiabatic dehumidification WBT will be same but DBT will be higher.
32. (d) : In summer air-conditioning, the conditioner air passing through the space undergoes heating and humidification.

33. (a) : Body temperature is more than the temperature of surrounding surfaces, therefore, the person will lose maximum body heat by convection.

34. (c) : The difference between comfort air-condition and industrial air-conditioning lies the indoor requirements.

35. (d) : The state of air supplied by a cooling will with a BPF = X lies on the psychrometric chart at the points which divides ESHF line in proportion to X and $(1 - x)$.

36. (b) : Statements 1, 2 and 3 are correct.

37. (c) : Statements 1 and 4 are correct.

38. (d) : A – 4, B – 1, C – 3, D – 2.

39. (b) : Statements 1, 2 and 4 are correct.

40. (b) : RH does not depend on pressure.

41. (d) : Evaporative regulation of body temperature fails when the body temperature is less then dew point temperature.

42. (b) : Statements 1 and 2 are correct.

43. (c) : If WBD = DBT – WBT

= sum of range and approved of a cooling tower

Then t_{iw} = DBT of ambient air

44. (a) :

$$\text{COP} = \frac{T_2}{T_1 - T_2} = \frac{200}{300 - 200} = \frac{200}{100} = 2$$

$$= \frac{Q_2}{W}$$

or

$$W = \frac{Q_2}{2} = \frac{1}{2} = 0.5 \text{ kW}$$

45. (b) : R12 : CCl_2F_2

46. (c) : An expansion valve is not used for reducing pressure in the steam-jet refrigeration cycle.

47. (d) : The function of the compressor in a vapour absorption system is performed by the absorber, pump and generator.

48. (b) : Statements 1, 2 and 4 are correct.

49. (d) : Replacing a water-cooled condenser with an air-cooled one in vapour compression refrigeration system with constant evaporator pressure result in increase in condensation temperature.

50. (a) : Lewis number, $\dfrac{h}{pc_p D}$.

51. (d) : Degree of saturation = $\dfrac{\text{Weight of water vapour with unit weight of dry air}}{\text{(Weight of water vapour with unit weight of dry air saturated at the same DBT and pressure)}}$.

52. (d) : Statement 3 and 4 are correct.

53. (c) : The correct choice is (c).

54. (b) : $(\text{COP})_{HP} = \dfrac{T_1}{T_1 - T_2} = 5$

$$T_1 = 5T_1 - 5T_2$$

$$\frac{T_2}{T_1} = 0.8$$

$$\eta_{HE} = 1 - \frac{T_2}{T_1} = 1 - 0.8 = 0.2$$

55. (b) : Power required per ton of refrigeration $= \dfrac{N}{COP}$

$$1\ TR = 3.5\ \text{kW}$$

$$COP = \frac{Q}{W} = \frac{1TR}{W}$$

$$W = \frac{1TR}{COP}$$

$$\therefore \quad N = 1\ TR = 3.5\ \text{kW}$$

56. (c) : A – 3, B – 1, C – 4, D – 2

57. (c) : Humidity ratio $= \dfrac{0.622\, p_v}{p_b - p_v}$

58. (c) : $BPF = 0.1, ADP, t_{d3} = 12°\text{C}, DBT, t_{d1} = 38°\text{C}, DPT = 9°\text{C}$

BPF for cooling coil $= \dfrac{t_{d2} - t_{d3}}{t_{d1} - t_{d3}}$

$$0.1 = \frac{t_{d2} - 12}{38 - 12} = \frac{t_{d2} - 12}{26}$$

$$t_{d2} = 2.6 + 12 = 14.6°\text{C} > \text{DPT}$$

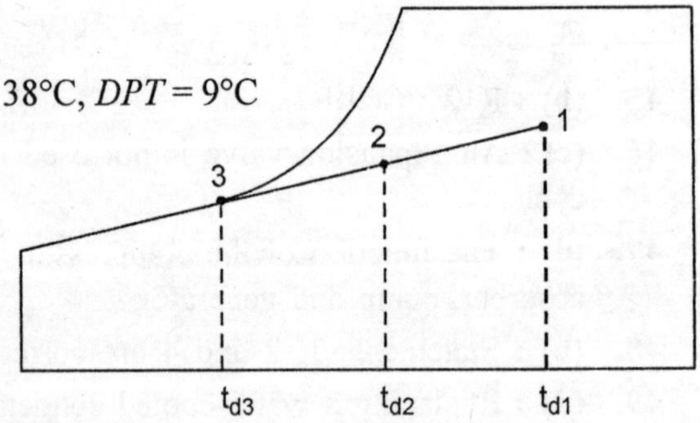

Fig. 18

Hence cooling and dehumidification takes place.

59. (c) : Adiabatic saturation takes place in desert cooler.

60. (b) : Both the condenser and evaporator are heat exchangers with refrigerant as a common medium in vapour compression type refrigerator. The condenser rejects heat and evaporator absorbs heat.

61. (d) : Capillary tube is used as the throttling device in domestic refrigerator.

62. (d) : COP od superheated vapour in evaporator is less than that of B superheated by the same amount of LVRHE.

63. (d) : A – 5, B – 1, C – 2, D – 4.

64. (c) : Dehumidification coil surface temperature is below both the DPT and freezing point temperature.

65. (d) : Total heat transfer from a wetted surface depends upon the difference in entropy between saturated air at surface temperature and that of air.

66. (a) : p_a = 1.01 bar, p_v = 0.01 bar, p_s = 0.02 bar

$$RH = \frac{p_v}{p_s} = \frac{0.01}{0.02} = 0.5 \text{ or } 50\%$$

$$w = \frac{0.622\, p_v}{p_a - p_v} = \frac{0.622 \times 0.01}{1.01 - 0.01} = 0.00622 \text{ kJ/kg d.a.}$$

67. (c) : In order to cool and dehumidification a stream of moist air, it must be passed over a coil heat at a temperature which is lower than the dew point temperature of the incoming stream.

68. (c) : 1 and 4 are correctly matched.

69. (d) : If moist air is sensible cooled above its dew point then WBT decreases and HR remains constant.

70. (c) : $\dot{m}_w$ = 0.003 kg/s, $\dot{m}_a$ = 1 kg/s, c_{pa} = 1 kJ/kg · K, h_{fg} = 2500 kJ/kg

$$\dot{m}_a c_{pa} \cdot \Delta t = \dot{m}_w \times h_{fg}$$

$$1 \times 1 \times \Delta t = 0.003 \times 2500$$

$$\Delta t = 7.5°\text{C}$$

71. (a) : A – 2, B – 3, C – 1, D – 4.

72. (a) : An electrode system of refrigeration has no liquid pump.

73. (c) : A bell-coleman air refrigeration system operates on reversed Brayton cycle.

74. (b) : COP of a refrigerating machine can be improved by lowering the higher temperature when the lower temperature is fixed.

75. (c) : When the evaporator is to be used as a dehumidification then its temperature should be kept lower than the apparatus dew point.

76. (c) : Small increase in temperature in air-conditioned space can be compressed by increasing air velocity.

77. (a) : Oil separator is installed between compressor and condenser in vapour compression refrigeration system.

78. (d) : R-134a is eco-friendly refrigerant.

79. (a) : Adiabatic saturation process involves cooling and humidification.

80. (c) : In domestic air-conditioners, air undergoes the process of cooling and dehumidification.

81. **(c) :** Sensible heating of moist air coil not affect the specific humidity.

82. **(b) :** When moist air passes through a cooling tower, the process is adiabatic saturation.

83. **(b) :** During sensible cooling relative humidity will increase.

84. **(c) :** If the moisture content of air start decreasing then air velocity should be decreased.

85. **(b) :** DBT = WBT = DPT when RH = 100%.

86. **(b) :** 1 and 3 are correctly matched.

87. **(d) :** Statements 2 and 3 are correct.

88. **(c) :** An ideal refrigerant misture obeys Raoult's law in the liquid phase and Dalton's law in the vapour phase.

89. **(d) :** Statements 2 and 3 are correct.

90. **(c) :** A – 3, B – 2, C – 4, D – 1.

91. **(c) :** Fan power (Fan speed)^3.

92. **(b) :** Efficiency of heating coil.

$$\eta_h = \frac{t_{d2} - t_{d1}}{t_{d3} - t_{d1}} = \frac{28-15}{41-15} = \frac{13}{26} = 0.5 \text{ or } 50\%$$

93. **(b) :** $\text{RSHF} = \frac{RSH}{RSH + RLH} = \frac{100}{100+25} = \frac{100}{125} = 0.8$

$$\text{ERSHF} = \frac{RSH + OASH \times BPF}{(RSH + OASH \times BPF) + (RLH + OALH \times BPF)}$$

$$= \frac{100 + 10 \times 0.15}{(100 + 10 \times 0.15) + (25 + 10 \times 0.15)}$$

$$= \frac{101.5}{128} = 0.79$$

94. **(a) :** DPT of moist air sample remains constant during sensible heating or cooling processes.

95. **(c) :** $(\text{COP})_{\text{ref}} = \frac{Q}{W}$

$$3 = \frac{Q}{25}, \; Q = 75 \text{ kW}$$

96. **(c) :** $\text{COP} = \frac{h_1 - h_{f3}}{h_2 - h_1} = \frac{300-150}{330-300} = \frac{150}{30} = 5$

97. **(b) :** 1, 2 or 3 are the reasons for high condenser pressure in a refrigeration system.

98. **(b) :** A – 3, B – 2, C – 4, D – 5.

99. **(a) :** System using NH_3 uses oil separator operating at –10°C.

100. **(b) :** A – 4, B – 3, C – 2, D – 4.

101. **(c) :** Statements 2 and 3 are correct.

102. (c) : p_b= 1.03 bar, p_a = 1 bar, p_s = 0.05 bar

$$p_v = p_b - p_a = 1.03 - 1 = 0.03 \text{ bar}$$

$$RH = \frac{p_v}{p_s} = \frac{0.03}{0.05} = 0.06 \text{ or } 60\%$$

103. (c) : Statement (c) is correct.

104. (c) : Statements 3 and 4 are correct.

105. (c) : Line *AB* represents sensible cooling with moisture removal.

106. (c) : Statements 2 and 3 are correct.

107. (a) : $BPF = \frac{15-12}{t_{d1}-12} = 0.15$

$$t_{d1} = 32°C$$

108. (c) :

	Lowest conditions	**Summer maximum conditions**
Temperature	20°C	25°C
RH	25%	55%
Velocity (max)	10 mpm	15 mpm

109. (a) : A – 4, B – 3, C – 1, D – 2.

110. (b) : R-22 is commonly used refrigerant in domestic refrigerators.

111. (c) : In a vapour compression refrigerating system, it is desirable to have the refrigerant as superheated vapour at the suction to the compressor.

112. (c) : Statements 2, 3 and 4 are correct.

113. (b) : Evaporative cooling takes place in a desert cooler.

114. (c) : The mass of water vapour in one m3 of dry air is defined as absolute humidity.

115. (d) : (SH + LH) → $(t - t_s) + 2500\ (w - w_s)$.

116. (c) : Statements 1 and 3 are correct.

117. (c) : Expansion valve in vapour compression refrigeration system expands high pressure liquid by throttling process.

APPENDIX–III

Table A-1	:	Properties of saturated steam (Temperature basis)
Table A-2	:	Properties of saturated steam (Pressure basis)
Table A-3	:	Moist air: Thermodynamic properties of saturated air at atmospheric pressure of 1.01325 bar
Table A-4	:	Properties of saturated liquid and saturated vapour of R-12
Table A-5	:	Properties of saturated liquid and saturated vapour of R-22
Table A-6	:	Properties of saturated liquid and saturated vapour of R-134a
Table A-7	:	Properties of saturated liquid and saturated vapour of R-717 (Ammoia)
Table A-8	:	General data of refrigerants?

Table A.1 : *Properties of saturated steam (Temperature basis)*

Temp. °C	Pressure bar	Specific Volume m^3/kg		Specific Enthalpy kJ/kg			Specific Entropy $kJ/kg.K$		
t	p	Sat. Liq. $v_f = v_s$	Sat. Vap. v_g	Sat. Liq. h_f	Evap. h_{fg}	Sat. Vap. h_g	Sat. Liq. s_f	Evap. s_{fg}	Sat. Vap. s_g
0	.006109	.00100	206.278	0.02	2501.4	2501.3	.0001	9.1566	9.1565
.01	.006113	.00100	206.136	0.01	2501.3	2501.4	.0000	9.1562	9.1562
1	.006567	.00100	192.577	4.16	2499.0	2503.2	.0152	9.1147	9.1299
2	.007056	.00100	179.889	8.37	2496.7	2505.0	.0305	9.0730	9.1035
3	.007577	.00100	168.132	12.57	2494.3	2506.9	.0457	9.0316	9.0773
4	.008131	.00100	157.232	16.78	2491.9	2508.7	.0610	8.9904	9.0514
5	.008721	.00100	147.120	20.98	2489.6	2510.6	.0761	8.9496	9.0257
6	.009349	.00100	137.734	25.20	2487.2	2512.4	.0912	8.9090	9.0003
7	.010016	.00100	129.017	29.39	2484.8	2514.2	.1062	8.8688	8.9751
8	.010724	.00100	120.917	33.60	2482.5	2516.1	.1212	8.8289	8.9501
9	.011477	.00100	113.386	37.80	2480.1	2517.9	.1362	8.7892	8.9253
10	.012276	.00100	106.379	42.01	2477.7	2519.8	.1510	8.7498	8.9003
11	.013123	.00100	99.857	46.20	2475.4	2521.6	.1658	8.7107	8.8765
12	.014022	.00100	93.784	50.41	2473.0	2523.4	.1806	8.6718	8.8524
13	.014974	.00100	88.124	54.60	2470.7	2525.3	.1953	8.6332	8.8048
14	.015983	.00100	82.848	58.80	2468.3	2527.1	.2099	8.5949	8.8048
15	.017051	.00100	71.926	62.99	2465.9	2528.9	.2245	8.5569	8.7814
16	.018181	.00100	73.333	67.19	2463.6	2530.8	.2390	8.5191	8.7582
17	.019376	.00100	69.044	71.38	2461.2	2532.6	.2535	8.4816	8.7351
18	.020610	.00100	65.038	75.58	2458.8	2534.4	.2679	8.4443	8.7123
19	.021975	.00100	61.293	79.77	2456.5	2536.2	.2823	8.4073	8.6897
20	.02339	.00100	57.791	83.96	2454.1	2538.1	.2966	8.3706	8.6672
21	.02487	.00100	54.514	88.14	2451.8	2539.9	.3109	8.3341	8.6450
22	.02645	.00100	51.447	92.33	2449.4	2541.7	3251	8.2979	8.6229
23	0.2810	.00100	48.574	96.52	2447.0	2543.5	.3393	8.2618	8.6011
24	.02985	.00100	45.883	100.70	2444.7	2545.4	.3534	8.2261	8.5794
25	.03169	.00100	43.360	104.89	2442.3	2547.2	.3674	8.1905	8.5580
26	.03363	.00100	40.994	109.07	2439.9	2549.0	.3814	8.1552	8.5367
27	.03567	.00100	34.774	113.25	2437.6	2550.8	.3954	8.1202	8.5156
28	.03782	.00100	36.690	117.43	2435.2	2552.6	.4093	8.0854	8.4946
29	0.4008	.00100	34.733	121.61	2432.8	2554.5	.4231	8.0508	8.4739

Table A.1 : *(continued) Properties of saturated steam (Temperature basis)*

Temp. °C	Pressure bar	Specific Volume m^3/kg		Specific Enthalpy kJ/kg			Specific Entropy $kJ/kg.K$		
t	p	Sat. Liq. $v_f = v_s$	Sat. Vap. v_g	Sat. Liq. h_f	Evap. h_{fg}	Sat. Vap. h_g	Sat. Liq. s_f	Evap. s_{fg}	Sat. Vap. s_g
30	.04246	.00100	32.894	125.79	2430.5	2556.3	.4369	8.0164	8.4533
31	.04496	.00100	31.165	129.97	2428.1	2558.1	.4507	7.9822	8.4329
32	.04759	.00101	29.540	134.15	2425.7	2559.9	.4644	7.9483	8.4127
33	.05034	.00101	28.011	138.33	2423.4	2561.7	.4781	7.9146	8.3927
34	.05324	.00101	6.571	142.50	2421.0	2563.5	.4917	7.8811	8.3729
35	.05628	.00101	25.216	146.68	2418.6	2565.3	.5053	7.8478	8.3531
36	.05947	.00101	23.940	150.86	2416.2	2567.1	.5188	7.8147	8.3336
37	.06281	.00101	22.737	155.03	2413.9	2568.9	.5323	7.7819	8.3142
38	.06632	.00101	21.602	159.21	2411.5	2570.7	.5458	7.7492	8.2950
39	.06999	.00101	20.533	163.39	2409.1	2572.5	.5592	7.7167	8.2759
40	.07384	.00101	19.523	167.57	2406.7	2574.3	.5725	7.6845	8.2570
41	.07786	.00101	18.570	171.74	2404.3	2576.1	.5858	7.6524	8.2383
42	.08208	.00101	17.671	175.91	2401.9	2577.9	.5991	7.6206	8.2197
43	.08619	.00101	16.821	180.10	2399.5	2579.6	.6123	7.5889	8.2012
44	.09111	.00101	16.018	184.27	2397.2	2581.4	.6255	7.5574	8.1829
45	.09593	.001010	15.260	188.45	2394.8	2583.2	.6390	7.5261	8.1650
46	.10101	.001010	14.540	192.61	2392.4	2585.0	.6523	7.4954	8.1470
47	.10622	.001011	13.861	196.82	2390.0	2586.8	.6650	7.4642	8.1292
48	.11181	.001011	13.223	201.00	2387.6	2588.6	.6782	7.4334	8.1110
49	.11753	.001012	12.612	205.12	2385.2	2590.3	.6914	7.4033	8.0942
50	.12354	.001012	12.034	209.31	2382.7	2592.1	.7040	7.3720	8.0761
52	.13632	.001013	10.972	217.70	2377.9	2595.6	.7301	7.3124	8.0423
54	.15021	.001014	10.013	226.00	2373.1	2599.1	.7552	7.2535	8.0082
56	.16530	.001015	9.149	234.41	2368.2	2602.6	.7810	7.1942	7.9753
58	.18172	.001016	8.372	242.82	2363.4	2606.2	.8063	7.1361	7.9420
60	.19941	.001017	7.671	251.13	2358.5	2609.6	.8312	7.0782	7.9092
62	.21863	.001018	7.037	259.54	2353.6	2613.1	.8562	7.0223	7.8783
64	.23934	.001019	6.463	267.90	2348.7	2616.5	.8810	6.9654	7.8462
66	.26172	.001020	5.943	276.24	2343.7	2619.9	.9063	6.9100	7.8164
68	.28590	.001022	5.471	284.62	2338.8	2623.4	.9302	6.8551	7.7850
70	.31192	.001023	5.042	293.00	2333.8	2626.8	.9554	6.8000	7.7552

Table A.1 : *(continued) Properties of saturated steam (Temperature basis)*

Temp. °C	Pressure bar	Specific Volume m^3/kg		Specific Enthalpy kJ/kg			Specific Entropy $kJ/kg.K$		
t	p	Sat. Liq. $v_f = v_s$	Sat. Vap. v_g	Sat. Liq. h_f	Evap. h_{fg}	Sat. Vap. h_g	Sat. Liq. s_f	Evap. s_{fg}	Sat. Vap. s_g
75	.38582	.001026	4,131	313.90	2321.4	2635.3	1.0150	6.6672	7.6823
80	.47391	.001029	3.407	334.91	2308.8	2643.7	1.0252	6.5373	7.6124
85	.57832	.001033	2.828	355.92	2296.0	2651.9	1.1343	6.4100	7.5445
90	.70144	.001036	2.361	376.90	2283.2	2660.1	1.1924	6.2871	7.4791
95	.84554	.001010	1.982	397.92	2270.2	2668.1	1.2500	6.1664	7.4162
100	1.01350	.001044	1.673	419.00	2257.0	2676.0	1.3072	6.0485	7.3553
110	1.4327	.00105	1.2102	461.30	2230.2	2691.5	1.4185	5.8202	7.2387
120	1.9853	.00106	0.8919	503.71	2202.6	2706.3	1.5276	5.6020	7.1296
130	2.7011	.00107	0.6685	546.31	2174.2	2720.5	1.6344	5.3925	7.0269
140	3.6132	.00108	0.5089	589.13	2144.7	2733.9	1.7391	5.1908	6.9299
150	4.7580	.00109	03928	632.20	2114.3	2746.5	1.8418	4.9960	6.8379
160	6.1781	.00110	0.3071	675.55	2082.6	2758.1	1.9427	4.8075	6.7502
170	7.9172	.00111	0.2428	719.21	2049.5	2768.7	2.0419	4.6244	6.6663
180	10.021	.00113	0.19405	763.22	2015.0	2778.2	2.1396	4.4461	6.5857
190	12.544	.00114	0.15654	807.62	1978.8	2786.4	2.2359	4.2720	6.5079
200	15.538	.00116	0.12736	852.45	1940.7	2793.2	2.3309	4.1014	6.4323
210	19.062	.00117	0.10441	897.76	1900.7	2798.5	2.4248	3.9337	6.3585
220	23.182	.00119	0.08619	943.62	1858.5	2803.8	2.5178	3.7683	6.2860
230	27.951	.00121	0.07158	990.12	1813.8	2804.0	2.6099	3.6047	6.2140
240	33.444	.00123	0.05976	1037.32	1766.5	2803.8	2.7015	3.4422	6.1437
250	39.734	.00125	0.05013	1085.36	1716.2	2801.5	Z7927	3.2802	6.0730
260	46.882	.00128	0.04221	1134.37	1662.5	2796.9	2.8838	3.1181	6.0019
270	54.993	.00130	0.03564	1184.51	1605.2	2789.9	2.9751	2.9551	5.9301
280	61.120	.00133	0.03017	1235.99	1543.6	2779.6	3.0668	7.7903	5.8571
290	74.361	.00137	0.02557	1289.07	1477.1	2766.2	3.1594	2.6227	5.7821
300	85.812	.00140	0.02167	1344.0	1404.9	2749.0	3.2534	2.4511	5.7045
310	98.363	.00145	0.01835	1401.3	1326.0	2727.3	3,3493	2.2737	5.6230
320	112.74	.00150	0.01548	1461.5	1238.6	2700.1	3.4480	Z0882	5.5362
330	128.15	.00156	0.01299	1525.3	1140.6	2665.9	3.5507	1.8909	5.4417
340	145.86	.00161	0.01079	1594.2	1027.9	2622.0	3.6594	1.6763	5.3357
350	165.13	.00174	0.00881	1670.6	893.4	2563.9	3.7777	1.4335	5.2112
360	186.51	.00189	0.00694	1760.5	720.5	2481.0	3.9147	1.1379	5.0526
370	210.32	.00221	0.00492	1890.5	441.6	2332.1	4.1106	0.6865	4.7971

Table A.2 : *Properties of saturated steam (Pressure basis)*

Pressure bar	*Temp °C*	*Specific Volume m^3/kg*		*Specific Enthalpy kJ/kg*			*Specific Entropy kJ/kg.K*		
p	t	*Sat. Liq.* $v_f = v_s$	*Sat Vap.* v_g	*Sat. Liq.* h_f	*Evap.* h_{fg}	*Sat Vap.* h_g	*Sat Liq.* s_f	*Evap.* s_{fg}	*Sat Vap.* s_g
.006113	0.01	.00100	206.136	0.01	2501.3	2501.4	.0000	9.1562	9.1562
.007	1.89	.00100	181.255	7.91	2496.9	2504.8	.0288	9.0775	9.1064
.008	3.77	.00100	159.675	15.81	2492.5	2508.3	0.575	8.9999	9.0573
.009	5.45	.00100	142.789	22.89	2488.5	2511.4	0.829	8.9312	9.0142
.010	6.98	.00100	129.208	29.30	2484.9	2514.2	.1059	8.8697	8.9756
.011	8.37	.00100	118.042	35.17	2481.6	2516.8	.1268	8.8140	8.9408
.012	9.66	.00100	108.696	40.58	2478.6	2519.1	.1460	8.7631	8.9091
.013	10.86	.00100	100.755	45.60	2475.7	2521.3	.1637	8.7162	8.8799
.014	11.98	.00100	93.922	50.31	2473.1	2523.4	.1802	8.6727	8.8529
.015	13.03	.00100	87.980	54.71	2470.6	2525.3	.1957	8.6322	8.8279
.016	14.02	.00100	82.763	58.87	2468.3	2527.1	.2102	8.5943	8.8044
.017	14.95	.00100	78.146	62.80	2466.0	2528.8	.2238	8.5586	8.7825
.018	15.84	.00100	74.030	66.54	2463.9	2530.5	.2368	8.5250	8.7618
.019	16.69	.00100	70.337	70.10	2461.9	2532.0	.2491	8.4931	8.7422
.020	17.51	.001001	67.004	73.48	2460.0	2535.5	.2607	8.4629	8.7237
.025	21.09	.001002	54.340	88.45	2451.6	2540.1	.3119	8.3330	8.6449
.030	24.10	.001003	45.670	101.00	2444.5	2545.5	.3545	8.2241	8.5778
.035	26.69	.001003	39.510	111.85	2438.4	2550.3	.3912	8.1329	8.5232
.040	28.98	.001004	34.803	121.45	2432.9	2554.4	.4234	8.0530	8.4759
.045	31.03	.001005	31.155	130.00	2428.2	2558.2	.4510	7.9829	8.4340
.050	32.90	.001005	28.195	137.80	2423.7	2561.2	.4759	7.9200	8.3958
.055	34.60	.001006	25.778	144.90	2419.6	2565.5	.5000	7.8619	8.3617
.060	36.18	.001006	23.741	151.50	2415.9	2567.4	.5208	7.8100	8.3310
.065	37.65	.001007	22.020	157.65	2412.4	2570.1	.5413	7.7621	8.3031
.070	39.03	.001007	20.530	163.40	2409.1	2572.5	.5589	7.7179	8.2768
.075	40.32	.001008	19.242	168.80	2406.0	2574.8	.5759	7.6760	8.2518
.080	41.54	.001008	18.104	173.90	2403.1	2577.0	.5932	7.6369	8.2300
.085	42.69	.001009	17.101	178.70	2400.3	2579.0	.6081	7.6000	8.2079
.090	43.79	.001009	16.203	183.30	2397.7	2581.0	.6220	7.5662	8.1877
.095	44.84	.001010	15.401	187.70	2395.2	2582.9	.6362	7.5330	8.1692

Table A.2 : *(continued) Properties of saturated steam (Pressure basis)*

Pressure bar	Temp °C	Specific Volume m^3/kg		Specific Enthalpy kJ/kg			Specific Entropy $kJ/kg.K$		
p	t	Sat. Liq. $v_f = v_s$	SatVap. v_g	Sat. Liq. h_f	Evap. h_{fg}	SatVap. h_g	Sat Liq. s_f	Evap. s_{fg}	Sat Vap. s_g
.10	45.81	.001010	14.674	191.83	2392.8	2584.7	.6493	7.5009	8.1502
.11	47.69	.001010	13.415	199.67	2388.3	2588.0	.6738	7.4430	8.1168
.12	49.42	.001010	12.361	206.92	2384.1	2591.1	.6963	7.3900	8.0863
.13	51.04	.001010	11.465	213.67	2380.2	2593.9	.7172	7.3412	8.0584
.14	52.55	.001010	10.693	219.99	2376.6	2596.6	.7366	7.2959	8.0325
.15	53.97	.001010	10.022	225.94	2373.1	2599.1	.7549	7.2536	8.0085
0.16	55.3	0.001015	9.43	231.6	2369.9	2601.5	0.772	7.214	7.986
0.17	56.6	0.001015	8.91	236.9	2366.8	2603.7	0.788	7.177	7.965
0.18	57.8	0.001016	8.45	242.0	2363.8	2505.8	0.804	7.141	7.945
0.19	59.0	0.001017	8.03	246.8	2361.0	2607.8	0.818	7.108	7.926
0.20	60.1	0.001017	7.65	251.4	2358.3	2609.7	0.832	7.077	7.909
0.22	62.1	0.001018	7.00	260.1	2353.2	2613.3	0.858	7.018	7.876
0.24	64.1	0.001019	6.45	268.1	2348.5	2616.6	0.882	6.964	7.846
0.26	65.9	0.001020	5.98	275.6	2344.1	2619.7	0.904	6.914	7.818
0.28	67.5	0.001021	5.58	282.6	2340.0	2622.6	0.925	6.868	7.793
0.30	69.1	0.001022	5.23	289.2	2336.1	26253	0.944	6.825	7.768
0.32	70.6	0.001023	4.92	295.5	2332.4	2627.9	0.962	6.784	7.746
0.34	72.0	0.001024	4.65	301.4	2328.8	2630.2	0.979	6.746	7.725
0.36	73.4	0.001025	4.41	307.0	2325.5	2632.5	0.996	6.710	7.706
0.38	74.6	0.001026	4.19	312.4	2322.3	2634.7	1.011	6.676	7.687
0.40	75.9	0.001027	3.99	317.6	2319.2	2636.8	1.026	6.644	7.670
0.45	78.7	0.001028	3.58	329.6	1312.0	2641.6	1.060	6.571	7.631
0.50	81.3	0.001030	3.24	340.6	2305.4	2646.0	1.091	6.503	7.594
0.55	83.7	0.001032	2.96	350.5	2299.3	2619.8	1.119	6.442	7.561
0.60	85.9	0.001033	2.73	359.9	2293.6	26535	1.145	6.387	7.532
0.65	88.0	0.001035	2.53	368.5	2288.3	2656.8	1.169	6.335	7.504
0.70	90.0	0.001036	2.37	376.7	2283.3	2660.0	1.192	6.288	7.480
0.75	91.8	0.001037	2.22	384.4	2278.6	2663.0	1.213	6.243	7.456
0.80	93.5	0.001039	2.087	391.7	2274.1	2665.8	1.233	6.202	7.435
0.85	95.1	0.001040	1.972	398.6	2269.8	2668.4	1.252	6.163	7.415
0.90	96.7	0.001041	1.869	405.2	2265.7	2670.9	1.270	6.125	7.395

Table A.2 : *(continued) Properties of saturated steam (Pressure basis)*

Pressure bar	*Temp °C*	*Specific Volume m^3/kg*		*Specific Enthalpy kJ/kg*			*Specific Entropy kJ/kg.K*		
		Sat. Liq.	*SatVap.*	*Sat. Liq.*	*Evap.*	*SatVap.*	*Sat Liq.*	*Evap.*	*Sat Vap.*
p	t	$v_f = v_s$	v_g	h_f	h_{fg}	h_g	s_f	s_{fg}	s_g
0.95	98.2	0.001042	1.777	411.4	2261.8	2673.2	1.287	6.090	7.377
1.00	99.6	0.001043	1.694	417.5	2258.0	26755	1.303	6.057	7.360
1.0132	100.0	0.001044	1.673	419.1	2256.9	2676.0	1.307	6.048	7355
1.05	101.0	0.001044	1.618	423.2	2254.4	2677.6	1.318	6.025	7.343
1.1	102.3	0.001045	1.549	424.8	2250.9	2679.7	1.333	5.994	7.327
1.2	104.8	0.001047	1.428	439.3	2244.2	2683.5	1.361	5.937	7.298
1.3	107.1	0.001049	1.325	449.1	2238.0	2687.1	1.387	5.884	7.271
1.4	109.3	0.001051	1.237	458.4	2232.0	2690.4	1.411	5.835	7.246
1.5	111.4	0.001053	1.159	467.1	2226.5	2693.6	1.434	5.789	7.223
1.6	113.3	0.001054	1.091	475.4	2221.1	2696.5	1.455	5.747	7.202
1.7	115.2	0.001056	1.031	483.2	2216.0	2699.2	1.475	5.706	7.181
1.8	116.9	0.001058	0.977	490.7	2211.2	2701.8	1.494	5.668	7.162
1.9	118.6	0.001059	0.929	497.3	2206.5	2704.3	1.513	5.631	7.144
2.0	120.2	0.001061	0.886	504.7	2201.9	2706.7	1.530	5.597	7.127
2.1	121.8	0.001062	0.846	511.3	2197.6	2708.9	1.547	5.564	7.111
2.2	123.8	0.001063	0.810	517.6	2193.4	2711.0	1.563	5.532	7.095
2.3	124.7	0.001065	0.777	523.7	2189.3	2713.1	1.578	5.502	7.080
2.4	126.1	0.001066	0.747	529.6	2185.4	2715.0	1.593	5.473	7.066
2.5	127.4	0.001067	0.719	535.4	2381.5	2716.9	1.607	5.446	7.053
2.6	128.7	0.001069	0.693	540.9	2177.8	2718.7	1.621	5.419	7.040
2.7	130.0	0.001070	0.669	546.3	2174.2	2720.5	1.634	5.393	7.027
2.8	131.2	0.001071	0.646	551.4	2170.7	2722.1	1.647	5.368	7.015
2.9	132.4	0.001072	0.625	556.5	2167.3	2723.8	1.660	5.343	7.003
3.0	133.5	0.001073	0.606	561.5	2163.8	2725.3	1.672	5.320	6.992
3.1	134.7	0.001074	0.588	566.3	2160.6	2726.8	1.684	5.297	6.981
3.2	135.8	0.001075	0.570	571.0	2157.3	2728.3	1.695	5.275	6.970
3.3	136.8	0,001076	0.554	575.5	2154.2	2729.7	1.706	5.254	6.960
3.4	137.9	0.001078	0.539	580.0	2151.1	2731.1	1.717	5.233	6.950
3.5	138.9	0.001079	0.524	564.3	2148.1	2132.4	1.728	5.213	6.941

Table A.2 : *(continued) Properties of saturated steam (Pressure basis)*

Pressure bar	*Temp* °C	*Specific Volume* m^3/kg		*Specific Enthalpy* kJ/kg			*Specific Entropy* $kJ/kg.K$		
p	t	*Sat. Liq.* $v_f = v_s$	*SatVap.* v_g	*Sat. Liq.* h_f	*Evap.* h_{fg}	*SatVap.* h_g	*Sat Liq.* s_f	*Evap.* s_{fg}	*Sat Vap.* s_g
3.6	140.0	0.001080	0.511	588.6	2145.1	2733.7	1.738	5.193	6.931
3.7	140.8	0.001081	0.498	592.8	2142.2	2735.0	1.748	5.174	6.922
3.8	141.8	0.001082	0.485	596.8	2139.4	2736.2	1.758	5.155	6.913
3.9	142.7	0.001083	0.474	600.8	2136.6	2737.4	1.767	5.137	6.904
4.0	143.6	0.001084	0.463	604.7	2133.8	2738.5	1.777	5.119	6.896
4.1	144.5	0.001085	0.452	608.6	2131.1	2739.7	1.786	5.102	6.888
4.2	145.4	0.001286	0.442	612.4	2128.4	2740.8	1.795	5.085	6.880
4.3	146.3	0.001086	0.432	616.1	2125.8	2741.9	1.804	5.068	6.872
4.4	147.1	0.001087	0.423	619.7	2123.2	2742.9	1.812	5.052	6.864
4.5	147.9	0.001088	0.414	623.3	2120.6	2743.9	1.821	5.036	6.857
4.6	148.7	0.001089	0.406	626.8	2118.9	2744.9	1.829	5.020	6.849
4.7	149.5	0.001090	0.397	630.2	2115.7	2745.9	1.837	5.005	6.842
4.8	150.3	0.001091	0.390	633.6	2113.2	2746.8	1.845	4.990	6.835
4.9	151.1	0.001092	0,382	639.9	2110.8	2747.8	1.853	4.975	6.828
5.0	151.9	0.001093	0.375	640.2	2108.5	2748.7	1.861	4.961	6.821
5.2	153.3	0.001094	0.361	646.7	2103.8	2750.5	1.876	4.932	6.808
5.4	154.8	0.001096	0.349	652.9	2099.3	2752.1	1.890	4.905	6.795
5.6	156.2	0.001097	0.337	658.9	2094.8	2753.8	1.904	4.879	6.783
5.8	157.5	0.001099	0.326	664.8	2090.5	2755.3	1.918	4.853	6.771
6.0	158.9	0.001101	0.316	670.6	2086.3	2756.8	1.931	4.829	6.760
6.2	160.1	0.001102	0.306	676.2	2082.1	2758.3	1.944	4.805	6.749
6.4	161.2	0.001104	0.297	681.6	2078.0	2759.6	1.956	4.782	6.738
6.6	162.6	0.001105	0.288	686.9	2074.0	2761.0	1.969	4.759	6.728
6.8	163.8	0.001107	0.281	692.1	2070.1	2762.2	1.981	4.737	6.718
7.0	165.0	0.001108	0.273	697.2	2066.3	2763.5	1.992	4.716	6.708
7.2	166.1	0.001119	0.266	702.2	2062.5	2764.7	2.004	4.695	6.699
7.4	167.2	0.001111	0.259	707.1	2058.8	2765.9	2.014	4.675	6.689
7.6	168.3	0.001112	0.252	711.8	2055.2	2767.0	2.025	4.655	6.680
7.8	169.4	0.001113	0.246	716.5	2051.6	2768.1	2.036	4.635	6.671
8.0	170.1	0.001115	0.240	721.1	2048.0	2769.1	2.046	4.617	6.663

Table A.2 : *(continued) Properties of saturated steam (Pressure basis)*

Pressure bar	Temp °C	Specific Volume m^3/kg		Specific Enthalpy kJ/kg			Specific Entropy kJ/kg.K		
p	t	Sat. Liq.	Sat Vap.	Sat. Liq.	Evap.	Sat Vap.	Sat Liq.	Evap.	Sat Vap.
		$v_f = v_s$	v_g	h_f	h_{fg}	h_g	s_f	s_{fg}	s_g
8.2	171.5	0.001116	0.235	725.6	2044.6	2770.2	2.056	4.598	6.654
8.4	172.5	0.001117	0.230	730.1	2041.1	2771.2	2.066	4.580	6.646
8.6	173.5	0.001119	0.225	734.4	2037.7	2772.1	2.076	4.562	6.638
8.8	174.4	0.001120	0.220	738.6	2034.4	2773.0	2.085	4.545	6.630
9.0	175.4	0.001121	0.215	742.8	2031.1	2773.9	2.095	4.528	6.623
9.2	176.3	0.001122	0.211	747.0	2027.8	2774.8	2.104	4.511	6.615
9.4	177.2	0.001124	0.206	751.0	2024.7	2775.7	2.113	4.495	6.608
9.6	178.1	0.001125	0.202	755.0	2021.5	2776.5	2.122	4.479	6.601
9.8	179.0	0.001126	0.198	758.9	2018.4	2777.3	2.130	4.463	6.593
10.0	179.9	0.001127	0.194	762.8	2015.3	2778.1	2.139	4.448	6.587
10.5	182.0	0.001130	0.186	772.2	2007.7	2779.9	2.159	4.411	6.570
11.0	184.1	0.001133	0.178	781.3	2000.4	2781.7	2.179	4.374	6.553
11.5	186.1	0.001136	0.170	790.1	1993.2	2783.3	1.198	4.340	6.538
12.0	188.0	0.001139	0.163	798.6	1986.2	2784.8	2.217	4.306	6.553
12.5	189.8	0.001141	0.157	806.8	1979.4	2786.2	2.234	4.275	6.509
13.0	191.6	0.001144	0.151	814.9	1972.7	2687.6	2.251	4.244	6.495
13.5	193.1	0.001146	0.146	822.6	1966.2	2688.8	2.268	4.214	6.482
14.0	195.0	0.001149	0.141	830.3	1959.7	2690.0	2.284	4.185	6.469
14.5	196.7	0.001151	0.136	837.6	1953.5	2691.1	2.300	4.157	6.457
15.0	198.3	0.001154	0.132	844.9	1947.3	2692.2	2.315	4.130	6.445

Table A.3 Moist air : *Thermodynamic properties of saturated air at atmospheric pressure of 1.01325 bar*

t °C	*Vapour pressure* bar	*Humidity ratio* kg/kg dry air	*Specific volume* m^3/kg dry air	*Enthalpy* kJ/kg dry air
– 40	.0001283	0.000079	0.6597	– 40.041
– 35	.0002233	0.000138	0.6740	– 34.868
– 30	.0003798	0.000234	0.6884	– 29.600
– 25	.0006324	0.000390	0.7028	– 24.187
– 20	.0010318	0.000637	0.7173	– 18.346
– 18	.0012482	0.000771	0.7231	– 16.203
– 16	.0015056	0.000930	0.7290	– 13.795
– 14	.0018107	0,001119	0.7349	– 11.314
– 12	.0021716	0.001342	0.7409	– 8.745
– 10	.0025971	0.001606	0.7469	– 6.073
– 8	.0030975	0.001916	0.7529	– 3.285
– 6	.0036846	0.002280	0.7591	– 0360
– 4	.0043716	0.002707	0.7653	2.724
– 2	.0051735	0.003206	0.7716	5.991
0	.0061072	0.003788	0.7781	9.470
1	.006566	0.00407	0.7813	11.200
2	.007055	0.00438	0.7845	12.978
3	.007575	0.00471	0.7878	14.807
4	.008130	0.00505	0.7911	16.692
5	.008719	0.00542	0.7944	18.634
6	.009347	0.00582	0.7978	20.639
7	.010013	0.00624	0.8012	22.708
8	.010722	0.00668	0.8046	24.848
9	.011474	0.00716	0.8081	27.059
10	.012272	0.01766	0.8116	29.348
11	.013119	0.00820	0.8152	31.716
12	.014017	0.00876	0.8188	34.172
13	.014969	0.00937	0.8225	36.719
14	.015977	0.01001	0.3262	39.362
15	.017044	0.01069	0.8300	42.105
16	.018173	0.01141	0.8338	44.955
17	.019367	0.01218	0.8377	47.918
18	.020630	0.01299	0.8417	50.998
19	.021964	0.01384	0.8457	54.205

Table A.3 Moist air : *(continued) Thermodynamic properties of saturated air at atmospheric pressure of 1.01325 bar*

t °C	*Vapour pressure* bar	*Humidity ratio* *kg/kg dry air*	*Specific volume* m^3*/kg dry air*	*Enthalpy* *kJ/kg dry air*
20	.023373	0.01475	0.8498	57.544
21	.024861	0.01572	0.8540	61.021
22	.026431	0.01674	0.8583	64.646
23	.026086	0.01781	0.8626	68.425
24	.029832	0.01896	0.8671	72.366
25	.031671	0.02016	0.8716	76.481
26	.033609	0.02144	0.8763	80.777
27	.035649	0.02279	0.8811	85.263
28	.037797	0.02422	0.8860	89.952
29	.040055	0.02572	0.8910	94.851
30	.042431	0.02732	0.8961	99.977
31	.044928	0.02900	0.9014	105.337
32	.047552	0.03078	0.9068	110.946
33	.050308	0.03266	0.9124	116.819
34	.053201	0.03464	0.9182	122.968
35	.056237	0.03674	0.9241	129.411
36	.059423	0.03895	0.9302	136.161
37	.062764	0.04129	0.9365	143.239
38	.066265	0.04376	0.9430	150.660
39	.069935	0.04636	0.9497	158.445
40	.073778	0.04911	0.9567	166.615
41	.077803	0.05202	0.9639	175.192
42	.082016	0.05509	0.9713	184.200
43	.086424	0.05833	0.9790	193.662
44	.091036	0.06176	0.9871	203.610
45	.095856	0.06537	0.9954	214.067
46	.100896	0.06920	1.0040	225.068
47	.106161	0.07324	1.0130	236.643
48	.111659	0.07751	1.0224	248.828
49	.117402	0.08202	1.0322	241.667
50	.123397	0.08680	1.0424	275.198
52	.136176	0.09720	1.0641	304.512
54	.150072	0.10887	1.0879	337.182
56	.165163	0.12198	1.1141	373.679
58	.181531	0.13674	1.1429	414.572

Table A.4 : *Properties of Saturated Liquid and Saturated Vapour* Refrigerant 12 (dichlorodifluoromethane)

Temp. °C	*Pressure, bar*	*Density kg/m³*	*Volume m³/kg*	*Enthalpy kJ/kg*		*Entropy kJ/(kg.K)*		*Specific Heat, c_p kJ/(kg.K)*	
		Liquid	*Vapour*	*Liquid*	*Vapor*	*Liquid*	*Vapor*	*Liquid*	*Vapor*
(1)	(2)	(3)	(4)	(5)	(6)	(7)	(8)	(9)	(10)
– 90.00	0.0288	1652.4	4.3671	121.33	310.39	0.6526	1.6849	0.853	0.470
– 80.00	0.0623	1626.0	2.1227	129.72	314.99	0.6972	1.6564	0.843	0.487
– 70.00	0.1235	1599.2	1.1226	138.19	319.69	0.7400	1.6334	0.850	0.504
– 60.00	0.2272	1572.0	0.63710	146.72	324.44	0.7809	1.6147	0.856	0.521
– 50.00	0.3925	1544.3	0.98362	155.32	329.23	0.8203	1.5996	0.863	0.537
– 45.00	0.505S	1530.3	0.30346	159.66	331.63	0.8395	1.5933	0.868	0.546
– 40.00	0.6426	1516.1	0.24281	164.01	334.03	0.8583	1.5876	0.873	0.554
– 35.00	0.6077	1501.7	0.10033	168.40	336.42	0.8769	1.6824	0.878	0.563
– 30.00	1.0044	1487.2	0.16020	172.81	338.81	0.8951	1.5779	0.884	0.572
– 29.80b	1.0132	1486.6	0.15899	172.99	388.90	0.8959	1.5777	0.884	0.572
– 28.00	1.0920	1481.4	0.14817	174.58	339.76	0.9024	1.5762	0.886	0.575
– 26.00	1.1872	1475.6	0.13716	176.36	340.70	0.9096	1.5745	0.888	0.579
– 24.00	1.2878	1469.6	0.12714	178.14	341.65	0.9167	1.5730	0.891	0.583
– 22.00	1.3949	1463.6	0.11800	179.93	342.59	0.9239	1.5715	0.894	0.586
– 20.00	1.5088	1457.6	0.10965	181.72	343.63	0.9309	1.5701	0.896	0.590
– 18.00	1.6296	1451.6	0.10202	183.52	344.46	0.9380	1.5688	0.899	0.594
– 16.00	1.7578	1445.5	0,09503	185.32	345.39	0.9450	1.5675	0.902	0.598
– 14.00	1.8937	1439.4	0,08862	187.14	346.32	0.9520	1.5662	0.905	0.602
– 12.00	2.0374	1433.3	0.08273	188.95	347.25	0.9589	1.5651	0.908	0.606
– 10.00	2.1893	1427.1	0.07731	190.78	348.17	0.9656	1.5639	0.911	0.611
– 8.00	2.3498	1420.9	0.07233	102.61	340.08	0.9727	1.5629	0.915	0.615
– 6.00	2.6190	1414.7	0.06773	194.45	349.99	0.9796	1.5618	0.918	0.619
– 4.00	2.6974	1408.3	0.06348	196.29	350.89	0.9864	1.5608	0.921	0.624
– 2.00	2.8851	1402.0	0.05956	198.14	351.79	0.9932	1.5599	0.925	0.628
0.00	3.0827	1395.6	0.05593	200.00	352.68	1.0000	1.5590	0.928	0.633
2.00	3.2902	1389.2	0.05256	201.87	353.57	1.0068	1.5581	0.932	0.638
4.00	3.5082	1382.7	0.04944	203.74	354.46	1.0136	1.5573	0.936	0.643
6.00	3.7368	1376.1	0.04654	205.62	356.32	1.0202	1.5565	0.940	0.648
8.00	3.9765	1369.5	0.04384	207.51	356.19	1.0269	1.5557	0.944	0.653
10.00	4.2276	1362.8	0.04134	209.41	357.06	1.0335	1.5550	0.948	0.658
12.00	4.4903	1356.1	003900	211.31	357.90	1.0402	1.5542	0.953	0.664
14.00	4.7651	1349.3	0.03682	213.23	358.75	1.0468	1.5535	0.957	0.669
16.00	5.0523	1342.5	0.03479	215.15	359.58	1.0534	1.5529	0.926	0.875
18.00	5.3521	1335.5	0.03289	217.09	360.41	1.0600	1.5522	0.966	0.881

Table A.4 : *(continued) Properties of Saturated Liquid and Saturated Vapour*
Refrigerant 12 (dichlorodifluoromethane)

Temp. °C	*Pressure, bar*	*Density kg/m³*	*Volume m³/kg*	*Enthalpy kJ/kg*		*Entropy kJ/(kg.K)*		*Specific Heat, c_p kJ/(kg.K)*	
		Liquid	*Vapour*	*Liquid*	*Vapor*	*Liquid*	*Vapor*	*Liquid*	*Vapor*
(1)	(2)	(3)	(4)	(5)	(6)	(7)	(8)	(9)	(10)
20.00	5.6651	1328.6	0.03111	219.03	361.23	1.0666	1.5516	0.971	0.687
22.00	5.9914	1321.5	0.02945	220.98	362.04	1.0731	1.5510	0.976	0.693
24.00	6.3315	1314.4	0.02789	222.34	362.83	1.0796	1.5504	0.981	0.700
26.00	6.6857	1307.2	0.02643	224.92	363.62	1.0862	1.5498	0.987	0.707
28.00	7.0544	1299.9	0.02505	226.00	364.40	1.0927	1.5493	0.992	0.714
30.00	7.4379	1292.5	0.02376	228.89	365.16	1.0992	1.5487	0.998	0.721
32.00	7.8366	1285.0	0.02255	230.90	365.92	1.1057	1.5481	1.004	0.728
34.00	8.2509	1277.4	0.02141	232.91	366.66	1.1121	1.5476	1.010	0.736
36.00	8.6811	1269.8	0.02034	234.94	367.39	1.1138	1.5470	1.017	0.744
38.00	9.1277	1262.0	0.01932	236.98	368.10	1.1251	1.5465	1.023	0.753
40.00	9.5909	1245.2	0.01837	239.03	368.81	1.1315	1.5459	1.030	0.762
42.00	10.071	1246.2	0.01747	241.10	369.49	1.1380	1.5454	1.038	0.771
44.00	10.569	1238.1	0.01661	243.18	370.16	1.1444	1.5448	1.045	0.780
46.00	11.085	1229.8	0.01581	245.27	370.82	1.1509	1.5443	1.053	0.791
48.00	11.618	1221.5	0.01505	247.38	371.45	1.1573	1.5437	1.062	0.801
50.00	12.171	1213.0	0.01432	249.51	372.07	1.1638	1.5431	1.071	0.812
52.00	12.742	1204.3	0.01364	251.65	372.67	1.1703	1.5425	1.080	0.824
54.00	13.333	1195.6	0.01299	253.80	373.25	1.1767	1.5418	1.090	0.837
56.00	13.944	1186.6	0.01237	255.97	373.31	1.1832	1.5412	1.100	0.850
58.00	14.575	1177.5	0.01179	258.16	374.35	1.1896	1.5405	1.111	0.865
60.00	15.227	1168.2	0.01123	260.37	374.86	1.1961	1.5398	1.123	0.880
62.00	15.901	1158.7	0.01070	262.60	375.35	1.2026	1.5390	1.135	0.896
64.00	16.595	1149.0	0.01020	264.85	375.81	1.2091	1.5382	1.148	0.914
66.00	17.312	1139.0	0.00972	267.12	376.34	1.2157	1.5374	1.163	0.933
68.00	18.052	1128.9	0.00926	269.42	376.64	1.2222	1.5365	1.178	0.954
70.00	18.814	1118.5	0.00882	271.73	377.01	1.2288	1.5356	1.194	0.977

Table A.5 : *Properties of Saturated Liquid and Saturated Vapour*
Refrigerant 22 (dichlorodifluoromethane)

Temp. °C	Pressure, bar	Density kg/m^3	Volume m^3/kg	Enthalpy kJ/kg		Entropy kJ/(kg.K)		Specific Heat, c_p kJ/(kg.K)	
		Liquid	Vapour	Liquid	Vapor	Liquid	Vapor	Liquid	Vapor
(1)	(2)	(3)	(4)	(5)	(6)	(7)	(8)	(9)	(10)
– 150.00	—	1701.5	—	26.01	335.85	0.0566	2.5752	—	0.434
– 140.00	—	1675.3	—	43.84	340.24	0.1961	2.4222	—	0.445
– 130.00	0.0006	1649.7	229.29	57.00	344.75	0.2916	2.3017	—	0.458
– 120.00	0.0023	1624.0	63.648	68.51	349.38	0.3694	2.2033	—	0.470
– 110.00	0.0074	1598.0	21.311	79.47	354.11	0.4386	2.1220	—	0.483
– 100.00	0.0200	1571.7	8.2080	90.24	358.93	0.5027	2.0545	—	0.497
– 90.00	0.0480	1545.1	3.6548	100.95	263.82	0.5629	1.9982	1.070	0.511
– 80.00	0.1035	1518.3	1.7816	111.66	368.75	0.6197	1.9508	1.070	0.527
– 70.00	0.2044	1491.1	0.94476	122.36	373.68	0.6738	1.9109	1.072	0.544
– 60.00	0.3747	1463.6	0.53734	133.11	738.58	0.7253	1.8770	1.076	0.563
– 50.00	0.6449	1435.5	0.32405	143.91	383.39	0.7748	1.8480	1.083	0.584
– 48.00	0.7140	1429.8	0.29469	146.08	384.35	0.7844	1.8427	1.085	0.589
– 46.00	0.7890	1424.1	0.26849	148.25	385.29	0.7940	1.8376	1.087	0.594
– 44.00	0.8700	1418.4	0.24507	150.43	886.23	0.8035	1.8326	1.089	0.598
– 42.00	0.9575	1412.6	0.22410	152.61	387.17	0.8130	1.8277	1.091	0.603
– 40.80b	1.0132	1409.1	0.21256	153.93	387.72	0.8186	1.8249	1.092	0.606
– 40.00	1.0518	1406.8	0.20536	154.80	388.09	0.8224	1.8230	1.093	0.608
– 38.00	1.1533	1401.0	0.18832	156.90	389.01	0.8317	1.8184	1.096	0.614
– 36.00	1.2623	1395.1	0.17306	159.19	389.93	0.8410	1.8140	1.098	0.619
– 34.00	1.3793	1389.2	0.15927	161.40	390.84	0.8502	1.8096	1.101	0.624
– 32.00	1.5045	1383.3	0.14680	163.61	391.74	0.8594	1.8054	1.104	0.630
– 30.00	16384	1377.3	013551	166.82	392.63	0.8694	1.8013	1.107	0.636
– 28.00	1.7816	1371.3	0.12525	168.04	393.52	0.8776	1.7973	1.110	0.642
– 26.00	1.9340	1365.2	0.11593	170.27	394.36	0.886	1.7934	1.114	0.648
– 24.00	1.0965	1359.1	0.10744	172.51	395.26	0.8955	1.7896	1.117	0.654
– 22.00	2.2683	1352.9	0.09970	174.75	396.12	0.9044	1.7869	1.121	0.660
– 20.00	2.4529	1346.8	0.09262	177.00	396.97	0.9133	1.7822	1.125	0.667
– 18.00	2.6477	1340.5	0.08615	179.26	397.81	0.9222	1.7787	1.129	0.674
– 16.00	2.8542	1334.2	0.08023	181.53	398.64	0.9309	1.7752	1.133	0.881
– 14.00	3.0729	1327.9	0.07479	183.81	399.46	0.9397	1.7719	1.137	0.688
– 12.00	3.3040	1321.6	0.06979	186.09	400.27	0.9484	1.7686	1.141	0.695
– 10.00	3.3482	1315.0	0.06520	188.38	401.07	0.9571	1.7653	1.146	0.703
– 8.00	3.8059	1308.5	0.06096	190.69	401.85	0.9657	1.7621	1.151	0.710
– 6.00	4.0785	1301.9	0.05706	193.00	402.63	0.9743	1.7590	1.156	0.718
– 4.00	4.3636	1295.3	0.05345	195.32	403.39	0.9829	1.7560	1.161	0.727
– 2.00	4.6646	1288.6	0.05012	197.66	404.14	0.9915	1.7530	1.166	0.735

Table A.5 : *(continued) Properties of Saturated Liquid and Saturated Vapour*
Refrigerant 22 (dichlorodifluoromethane)

Temp. °C	Pressure, bar	Density kg/m^3	Volume m^3/kg	Enthalpy kJ/kg		Entropy kJ/(kg.K)		Specific Heat, c_p kJ/(kg.K)	
		Liquid	Vapour	Liquid	Vapor	Liquid	Vapor	Liquid	Vapor
(1)	(2)	(3)	(4)	(5)	(6)	(7)	(8)	(9)	(10)
0.00	4.9811	1281.8	0.04703	200.00	404.87	1.0000	1.7500	1.171	0.744
2.00	5.3134	1275.0	0,04417	202.35	405.59	1.0085	1.7471	1.177	0.753
4.00	5.6622	1268.1	0.04152	204.72	406.30	1.0170	1.7443	1.183	0.762
0.00	6.0279	1261.1	0.03906	207.10	406.99	1.025.4	1.7415	1.189	0.772
8.00	6.4109	1254.0	0.03676	209.49	407.67	1.0338	1.7387	1.195	0.782
10.00	6.8119	1246.9	0.03463	211.89	408.33	1.0422	1.7360	1.202	0.792
12.00	7.2314	1239.7	0.03265	214.31	408.97	1.0506	1.7333	1.208	0.802
14.00	7.6698	1232.4	0.03079	216.74	409.60	1.0590	1.7306	1.215	0.813
16.00	8.1277	1225.0	0.02906	219.18	410.21	1.0673	1.7280	1.223	0.825
18.00	8.6056	1217.6	0.02744	221.63	410.80	1.0756	1.7254	1.230	0.837
20.00	9.1041	1210.0	0.02593	224.10	411.38	1.0840	1.7228	1.238	0.849
22.00	9.6236	1202.4	0.02451	226.59	411.93	1.0923	1.7202	1.246	0.862
24.00	10.165	1194.6	0.02319	229.09	412.46	1.1006	1.7177	1.254	0.875
26.00	10.728	1186.8	0.02194	231.60	412.98	1.1088	1.1751	1.263	0.889
28.00	11.314	1178.8	0.02077	234.14	413.46	1.1171	1.7126	1.272	0.904
30.00	11.924	1170.7	0,01968	236.69	413.93	1.1254	1.7101	1.282	0919
32.00	12.557	1162.5	0.01864	239.25	414.37	1.336	1.7075	1.292	0.935
34.00	13.215	1154.2	0.01767	241.84	414.79	1.1419	1.7050	1.302	0.952
36.00	13.898	1145.7	0.01657	244.44	415.18	1.1501	1.7024	1.313	0.970
38.00	14.606	1137.1	0.01589	247.06	415.54	1.1584	1.6999	1.325	0.989
40.00	15.341	1128.4	0.01507	249.71	416.87	1.18S7	1.6973	1.338	1.009
42.00	16.103	1119.5	0.01430	252.37	416.17	1.1749	1.6947	1.351	1.030
44.00	16.892	1110.4	0.01357	255.06	416.44	1.1832	1.6921	1.365	1.052
46.00	17.709	1101.2	0.01288	267.77	416.68	1.1915	1.6894	1.380	1.076
48.00	18.555	1091.8	0.01223	260.51	416.87	1.1998	1.6867	1.396	1.102
50.00	19.431	1082.1	0.01161	263.77	417.03	1.2081	1.6840	1.414	1.129
55.00	21.753	1057.1	0.01020	270.31	417.24	1.2291	1.6768	1.464	1.209
60.00	24.274	1030.5	0.00895	227.56	417.14	1.2503	1.6692	1.528	1.307
65.00	27.008	1001.8	0.00784	285.08	416.65	1.2718	1.6610	1.611	1.435
70.00	29.967	970.4	0.00684	292.90	415.69	1.2940	1.6518	1.726	1.609

Table A.6 : *Properties of Saturated Liquid and Saturated Vapour*
Refrigerant 134a (dichlorodifluoromethane)

Temp. °C	*Pressure, bar*	*Density kg/m³*	*Volume m³/kg*	*Enthalpy kJ/kg*		*Entropy kJ/(kg.K)*		*Specific Heat, c_p kJ/(kg.K)*	
		Liquid	*Vapour*	*Liquid*	*Vapor*	*Liquid*	*Vapor*	*Liquid*	*Vapor*
(1)	(2)	(3)	(4)	(5)	(6)	(7)	(8)	(9)	(10)
–103.30a	0.0039	1591.2	35.263	71.89	335.07	0.4143	1.9638	1.147	0.585
–100.00	0.0056	1581.9	25.039	75.71	337.00	0.4366	1.9456	1.168	0.592
90.00	0.0153	1553.9	9.7191	87.59	342.94	0.5032	1.8975	1.201	0.614
– 80.00	0.0369	1528.2	4.2504	99.65	849.03	0.5674	1.8585	1.211	0.637
– 70.00	00801	1498.6	2.0528	111.78	355.23	0.6286	1.8269	1.215	0.660
– 60.00	0.1594	1471.0	1.0770	123.96	361.51	0.6871	1.0816	1.220	0.685
– 50.00	0.2948	1443.1	0.60560	136.21	367.83	0.7432	1.7812	1.229	0.712
– 40.00	0.5122	1414.8	0.36095	148.57	374.16	0.7973	1.7649	1.243	0.740
– 30.00	0.8436	1385.9	0.22596	161.10	380.45	0.8498	1.7519	1.260	0.771
– 23.00	0.9268	1380.0	0.20682	163.62	381.70	0.8601	1.7497	1.264	0.778
– 26.07b	1.0132	1874.3	0.19016	166.07	382.90	0.8701	1.7478	1.268	0.784
– 26.00	1.0164	1374.1	0.18961	166.16	382.94	0.8704	1.7476	1.268	0.785
– 24.00	1.1127	1368.2	0.17410	168.70	384.19	0.8806	1.7455	1.273	0.791
– 22.00	1.2160	1262.2	0.16010	171.26	385.43	0.8908	1.7436	1.277	0.798
– 20.00	1.3268	1356.2	0.14744	173.82	386.66	0.9009	1.7417	1.282	0.805
– 18.00	1.4454	1350.2	0.13597	176.39	387.89	0.9110	1.7399	1.286	0.812
– 16.00	1.721	1344.1	0.12556	178.97	389.11	0.9211	1.7383	1.291	0.820
¬ 14.00	1.7074	1338.0	0.11610	181.56	390.33	0.9311	1.7367	1.296	0.827
– 12.00	1.8516	1331.8	0.10749	184.16	391.55	0.9410	1.7351	1.301	0.835
– 1000	2.0052	1325.6	0.09963	186.78	392.75	0.9509	1.7337	1.306	0.842
– 8.00	2.1684	1319.3	0.09246	189.40	393.95	0.9608	1.7323	1.312	0.850
– 6.00	2.3418	1313.0	0.08591	192.03	395.15	0.9707	1.7310	1.317	0.858
– 4.00	2.5257	1306.6	0.07991	194.68	396.33	0.9805	1.7297	1.323	0.866
– 2.00	2.7206	1300.2	0.07440	197.33	397.51	0.9903	1.7285	1.329	0.875
0.00	2.9269	1293.7	0.06935	200.00	398.68	1.000	1.7274	1.335	0.883
2.00	3.1450	1287.1	0.06470	202.68	399.84	1.0097	1.7263	1.341	0.892
4.00	3.3755	1280.5	0.06042	205.37	401.00	1.0194	1.7252	1.347	0.901
6.00	3.6186	1273.8	0.05648	208.08	402.14	1.0291	1.7342	1.353	0.910
8.00	3.8749	1267.0	0.05284	210.80	403.27	1.0387	1.7233	1.360	0.920
10.00	4.1449	1260.2	0.04948	213.53	404.40	1.0483	1.7224	1.367	0.930
12.00	4.4289	1253.3	0.04636	216.27	495.51	1.0579	1.7215	1.374	0.939
14.00	4.7276	1249.3	0.04348	219.03	406.61	1.0674	1.7207	1.381	0.950
16.00	5.0413	1239.3	0.04081	221.80	407.70	1.0770	1.7199	1.388	0.960
18.00	5.3706	1232.1	0.03833	224.59	408.78	1.0865	1.7191	1.396	0.971

Table A.6 : *(continued) Properties of Saturated Liquid and Saturated Vapour*
Refrigerant 134a (dichlorodifluoromethane)

Temp. °C	*Pressure,* bar	*Density* kg/m^3	*Volume* m^3/kg	*Enthalpy* kJ/kg		*Entropy* $kJ/(kg.K)$		*Specific Heat,* c_p $kJ/(kg.K)$	
		Liquid	*Vapour*	*Liquid*	*Vapor*	*Liquid*	*Vapor*	*Liquid*	*Vapor*
(1)	(2)	(3)	(4)	(5)	(6)	(7)	(8)	(9)	(10)
20.00	5.7159	1224.9	0.03603	227.40	409.84	1.0960	1.7183	1.404	0.982
22.00	6.0777	1217.5	0.03388	230.21	410.89	1.1055	1.7176	1.412	0.994
24.00	6.4566	1210.1	0.03189	233.05	411.93	1.1149	1.7169	1.420	1.006
26.00	6.8531	1202.6	0.03003	235.90	412.95	1.1244	1.7162	1.429	1.018
28.00	7.2676	1194.9	0.02829	238.77	413.96	1.1338	1.7155	1.438	1.031
30.00	7.7008	1187.2	0.02667	241.65	414.94	1.1432	1.7149	1.447	1.044
32.00	8.1530	1179.3	0.02516	244.55	415.90	1.1527	1.7142	1.457	1.058
34.00	8.6250	1171.3	0.02374	247.47	416.85	1.1621	1.7135	1.467	1.073
36.00	9.1172	1163.2	0.02241	250.41	417.78	1.1715	1.7129	1.478	1.088
38.00	9.6301	1154.9	0.02116	253.37	418.69	1.1809	1.7122	1.478	1.104
40.00	10.165	11465	0.01999	256.35	419.58	1.1903	1.7115	1.500	1.120
42.00	10.721	1137.9	0.01890	259.35	420.44	1.1997	1.7108	1.513	1.138
44.00	11.300	1129.2	0.01786	262.38	421.28	1.2091	1.7101	1.525	1.156
46.00	11.901	1120.3	0.01689	265.42	422.09	1.2185	1.7094	1.539	1.175
48.00	12.527	1111.3	0.01598	268.49	422.88	1.2279	1.7068	1.553	1.196
50.00	13.177	1102.0	0.01511	271.59	423.63	1.2373	1.7078	1.569	1.218
52.00	13.852	1092.6	0.01430	274.71	424.35	1.2468	1.7070	1.585	1.241
54.00	14.553	1082.9	0.01353	277.86	425.03	1.2562	1.7061	1.602	1.266
56.00	15.289	1073.0	0.01280	281.04	425.68	1.2657	1.7051	1.621	1.293
58.00	16.033	1062.8	0.01212	284.25	426.29	1.2752	1.7041	1.641	1.322
60.00	16.815	1052.4	0.01146	287.49	426.86	1.2847	1.7031	1.663	1.354
62.00	17.652	1041.7	0.01085	290.77	427.37	1.2943	1.7019	1.686	1.388
64.00	18.464	1030.7	0.01026	294.08	427.84	1.3039	1.7007	1.712	1.426
66.00	19.334	1019.4	0.00970	297.44	248.25	1.3136	1.6993	1.740	1.568
68.00	20.234	1007.7	0.00917	300.84	428.61	1.3234	1.6979	1.772	1.515
70.00	21.165	995.6	0.00867	304.29	428.89	1.3332	1.6963	1.806	1.576
72.00	22.130	983.1	0.00818	307.79	429.10	1.3430	1.6046	1.846	1.626
74.00	23.127	970.0	0.00772	311.34	429.23	1.3530	1.6926	1.890	1.693
76.00	24.159	956.5	0.00728	314.96	429.27	1.3631	1.6905	1.941	1.770
78.00	25.227	942.3	0.00686	318.65	429.20	1.3733	1.6881	2.000	1.861

Table A.7 : *Properties of Saturated Liquid and Saturated Vapour* **Refrigerant 717 (Ammonia)**

Temp., °C	Pressure, bar	Volume Vapor, m³/kg	Density, Liquid, kg/m³	Enthalpy Liquid kJ/kg	Enthalpy Vapor kJ/kg	Entropy Liquid kJ/kg K	Entropy Vapour kJ/kg K
–77.67	0.06060	15.689	733.86	–1110.11	380.09	4.2032	11.8278
–75	0.07481	12.874	730.90	–1097.37	385.08	4.2679	11.7507
–70	0.10911	9.0362	725.33	–1074.44	394.31	4.3822	11.6134
–65	0.15588	6.4694	719.66	–1052.24	403.38	4.4902	11.4845
–60	0.21848	4.7169	713.89	–1030.38	412.27	4.5939	11.3633
–55	0.30086	3.4973	708.00	–1008.63	420.97	4.6947	11.2491
–50	0.40756	2.6334	702.02	–986.88	429.45	4.7932	11.1413
–45	0.54379	2.0114	695.97	–965.06	437.70	4.8898	11.0394
–40	0.71540	1.5567	689.86	–943.12	445.70	4.9849	10.9428
–38	0.79538	1.4099	687.40	–934.31	448.83	5.0224	10.9056
–36	0.88251	1.2794	684.94	–925.48	451.91	5.0598	10.8692
–34	0.97725	1.1630	682.46	–916.63	454.96	5.0969	10.8334
–33.33	1.01325	1.1244	681.57	–913.44	456.04	5.1102	10.8208
–32	1.0801	1.0592	679.98	–907.75	457.95	5.1338	10.7984
–30	1.1915	0.96625	677.49	–898.86	460.91	5.1704	10.7641
–28	1.3120	0.88294	674.99	–889.94	463.81	5.2069	10.7304
–26	1.4421	0.80812	672.47	–881.01	466.67	5.2431	10.6974
–24	1.5825	0.74081	669.95	–872.05	469.49	5.2791	10.6650
–22	1.7335	0.68013	667.42	–863.07	472.25	5.3149	10.6332
–20	1.8959	0.62535	664.88	–854.06	474.96	5.3505	10.6019
–18	2.0702	0.57580	662.32	–845.04	477.63	5.3859	10.5713
–16	2.2571	0.53092	659.75	–835.99	480.24	5.4212	10.5411
–14	2.4571	0.49019	657.16	–826.92	482.80	5.4562	10.5116
–12	2.6710	0.45318	654.57	–817.83	485.31	5.4910	10.4825
–10	2.8993	041949	651.95	–808.71	487.76	5.5257	10.4539
–8	3.1427	0.38879	649.32	–799.37	490.16	5.5601	10.4257
–6	3.4020	0.36076	646.68	–790.40	492.50	5.5944	10.3981
–4	3.6777	0.33514	644.01	–781.21	494.79	5.6285	10.3708
–2	3.9708	0.31168	641.33	–771.99	497.01	5.6625	10.3440
0	4.2818	0.29019	638.63	–762.75	499.18	5.6963	10.3176
2	4.6115	0.27046	635.91	–753.48	501.28	5.7299	10.2916
4	4.9607	0.25233	633.17	–744.19	503.33	5.7633	10.2660

Table A.7 : *(continued) Properties of Saturated Liquid and Saturated Vapour* Refrigerant 717 (Ammonia)

Temp., °C	*Pressure, bar*	*Volume Vapor, m^3/kg*	*Density, Liquid, kg/m^3*	*Enthalpy*		*Entropy*	
				Liquid kJ/kg	*Vapor kJ/kg*	*Liquid kJ/kg K*	*Vapour kJ/kg K*
6	5.3302	0.23564	630.40	–734.87	505.30	5.7966	10.2407
8	5.7206	0.22027	627.62	–725.52	597.22	5.8298	10.2158
10	6.1329	0.20610	624.81	–716.14	509.06	5.8628	10.1912
12	6.5678	0.19301	621.98	–706.74	510.84	5.8956	10.1669
14	7.0261	0.18091	619.13	–697.30	512.54	5.9283	10.1429
16	7.5087	0.16971	616.25	–687.84	514.18	5.9609	10.1193
18	8.0164	0.15933	613.35	–678.34	515.74	5.9934	10.0959
20	8.5501	0.14971	610.42	–668.82	517.22	6.0257	10.0727
22	9.1106	0.14077	607.46	–659.26	518.62	6.0578	10.0499
24	9.6987	0.13247	604.48	–649.67	519.95	6.0899	10.0272
26	1.0315	0.12474	601.47	–640.05	521.19	6.1218	10.0048
28	10.962	0.11755	598.43	–630.39	522.35	6.1536	9.9826
30	11.638	0.11084	595.36	–620.70	523.42	6.1853	9.9606
32	12.346	0.10458	592.27	–610.97	524.40	6.2169	9.9387
34	13.086	0.09874	589.14	–601.21	525.28	6.2484	9.9171
36	13.859	0.09328	585.98	–591.41	526.08	6.2798	9.8955
38	14.667	0.08817	582.78	–581.57	526.77	6.3111	9.8742
45	17.775	0.07270	571.34	–546.81	528.39	6.4198	9.8003
50	20.281	0.06357	562.88	–521.64	528.72	6.4970	9.7483
55	23.043	0.05573	554.17	–496.16	528.29	6.5737	9.6965
60	26.078	0.04896	545.18	–470.33	527.04	6.6501	9.6448
65	29.403	0.04311	535.87	–444.11	524.91	6.7264	9.5929
70	33.034	0.03801	526.21	–417.43	521.80	6.8027	9.5407
75	36.991	0.03355	516.15	–390.23	517.63	6.8792	9.4878
80	41.292	0.02963	505.62	–362.42	512.27	6.9561	9.4338
85	45.956	0.02681	494.55	–333.89	505.57	7.0337	9.3785
90	51.005	0.02312	482.83	–304.50	497.35	7.1123	9.3212
95	56.462	0.02039	470.33	–274.09	487.37	7.1924	9.2615
100	62.351	0.01795	456.87	–242.41	475.30	72744	9.1986
105	68.698	0.01574	442.19	–209.15	460.66	7.3592	9.1312
110	75.533	0.01373	425.94	–173.82	442.77	7.4479	9.0578
115	82.922	0.01188	407.49	–135.67	420.53	7.5422	8.9757
120	90.816	0.01014	385.79	–93.341	392.02	7.6454	8.8005
125	99.358	0.008456	358.47	–43.811	353.42	7.7648	8.7629
130	108.60	0.006701	317.39	23.636	294.86	7.9263	8.5993
*132.3	113.0	0.00426	235	142.7	141.7	8.217	8.217
**Triple Point							
*Critical Point							

Table A.8 : *General data of refrigerants*

Refrigerant	*Chemical Formula*	*Designation*	*M (Molecular Weight)*	t_s*(N.B.P.)* °C	t_c*(Critical Temperature)*°C	p_c*(Critial) Pressure)* bar	v_c*(Critical Volume)* L/kg	t_f*(Freezing Point)* °C
Inorganic Refrigerants								
Water	H_2O	R 718	18.016	100.0	374.15	221.3	3.26	0.0
Ammonia	NH_3	R 717	17.031	–33.313	133.0	112.97	4.13	–77.7
Carbon Dioxide	CO_2	R 744	44.01	–78.4	31.1	73.72	2.135	–56.6
Organic Refrigerants Monofluoro-Trichloro Methane	$CFCl_3$	R 11	137.39	23.7	197.78	43.7	1.805	–111
Difluoro-Dichloro Methane	CF_2Cl_2	R 12	120.92	–29.75	112.04	41.15	1.793	–136
Trifluoro-Monochloro Methane	CF_3Cl	R 13	104.47	–81.5	28.78	38.8	1.721	–180
Trifluoro-Monobromo Methane	CF_3Br	R 13 B 1	148.9	–58.7	67.8	40.5	—	–142
Tterafluoro Methane	CF_4	R 14	88.01	–128.0	–45.5	37.5	1.50	–194
Monofluoro-Dichloro Methane	$CHFCl_2$	R 21	102.92	8.9	173.5	51.6	1.915	–135
Difluoro-Monochloro Methane	CHF_2Cl	R 22	86.48	–40.81	96.02	49.88	1.949	–160
Trifluoro Methane	CHF_3	R 23	70.01	–82.0	25.83	48.2	1.748	–155
Difluoro Methane	CH_2F_2	R 32	52.024	–51.65	78.41	58.3	2.326	--
Pentafluoro Ethane	CF_3CHF_2	R 125	120.02	– 48.1	66.25	36.31	1.748	–
Tetrafluoro Ethane	CF_3CH_2F	R 134a	102.03	–26.07	101.06	40.56	1.942	–96.6
Monochloro-Tetra-fluoro Ethane	CF_3CHClF	R 124	136.475	–12.05	122.5	36.34	1.786	–
Dichloro-Trifluoro Ethane	CF_3CHCl_2	R 123	152.93	27.82	183.79	30.74	1.818	..
-Do-	$CHClFCClF_2$	R 123a	152.93	29.9	189.62	38.89	–	–
Tetrafluoro Ethane	CHF_2CHF_2	R 134	102.03	–19.8	118.9			
Monofluoro-Dichloro Ethane	CH_3CCl_2F	R 141b	116.9	32.1	208.0	43.39	--	–
Trifluoro-Trichloro Ethane	$CF_2Cl–CF_2Cl$	R 113	187.89	47.68	214.1	34.15	1.735	–36.6
Tetrafluoro-Dichloro Ethane	$CF_2Cl–CF_2Cl$	R 114	170.91	3.6	145.8	32.75	1.715	–94
Pentafluoro-Monochloro Ethane	CF_2ClCF_3	R 115	154.48	–38.0	80.0	32.36	1.68	–106
Difluoro-Monochloro Ethane	$CH_3–CF_2Cl$	R 142b	100.495	–9.25	137.1	42.46	2.3	–131.0
Trifluoro Ethane	$CH_3–CF_3$	R 143a	84.04	–47.24	73.1	38.11	2.305	–111.3
Difluoro Ethane	$CH_3–CHF_2$	R 152a	66.05	–24.02	113.3	45.2	2.717	–117.6
Ethyl Chloride	C_2H_5Cl	R 160	64.52	12.0	187.2	52.47	3.03	–138.7
n-perfluoro Butane	C_4F_{10}	—	238.04	–2.0	113.2	23.3	1.588	—
Difluoro Ethane	$CH_2F–CH_2F$	R 152	66.05	66.05	30.7			

Table A.8 : *(continued) General data of refrigerants*

Refrigerant	*Chemical Formula*	*Designation*	*M (Molecular Weight)*	t_s*(N.B.P.)* °C	t_c*(Critical Temperature)*°C	p_c*(Critial) Pressure)* bar	v_c*(Critical Volume)* L/kg	t_f*(Freezin Point)* °C
Hydrocarbons								
Ethane	C_2H_4	R 170	30.06	–88.6	32.1	49.3	4.7	–183.2
Propane	C_3H_8	R 290	44.1	–42.1	96.8	42.56	4.545	–187.1
n-butane	C_4H_{10}	R 600	58.1	–0.56	153.0	35.3	4.29	–135
Isobutane	$(CH_3)_3CH$	R 600a	58.13	–11.67	135.0	36.45	4.526	–159.6
Unsaturated Hydrocarbons								
Propylene	C_2H_4	—	42.08	–47.7	94.4	46.0	4.2	–185
Dihloroethylene	$C_2H_2Cl_2$	—	96.9	90.9	243	54.9	–	–56.6
Aliphatic Amines								
Methylamine	CH_3NH_3	—	31.06	–6.7	156.9	74.5	–	–92.5
Ethylamine	$C_2H_4NH_3$	—	45.08	7.0	164.6	54.7	–	–93

APPENDIX–IV

Figure A-1 : p – h chart for R – 11
Figure A-2 : p – h chart for R – 12
Figure A-3 : p – h chart for R – 13
Figure A-4 : p – h chart for R – 22
Figure A-5 : p – h chart for R – 134a
Figure A-6 : p – h chart for R – 717
Figure A-7 : Psychrometric chart for moist air

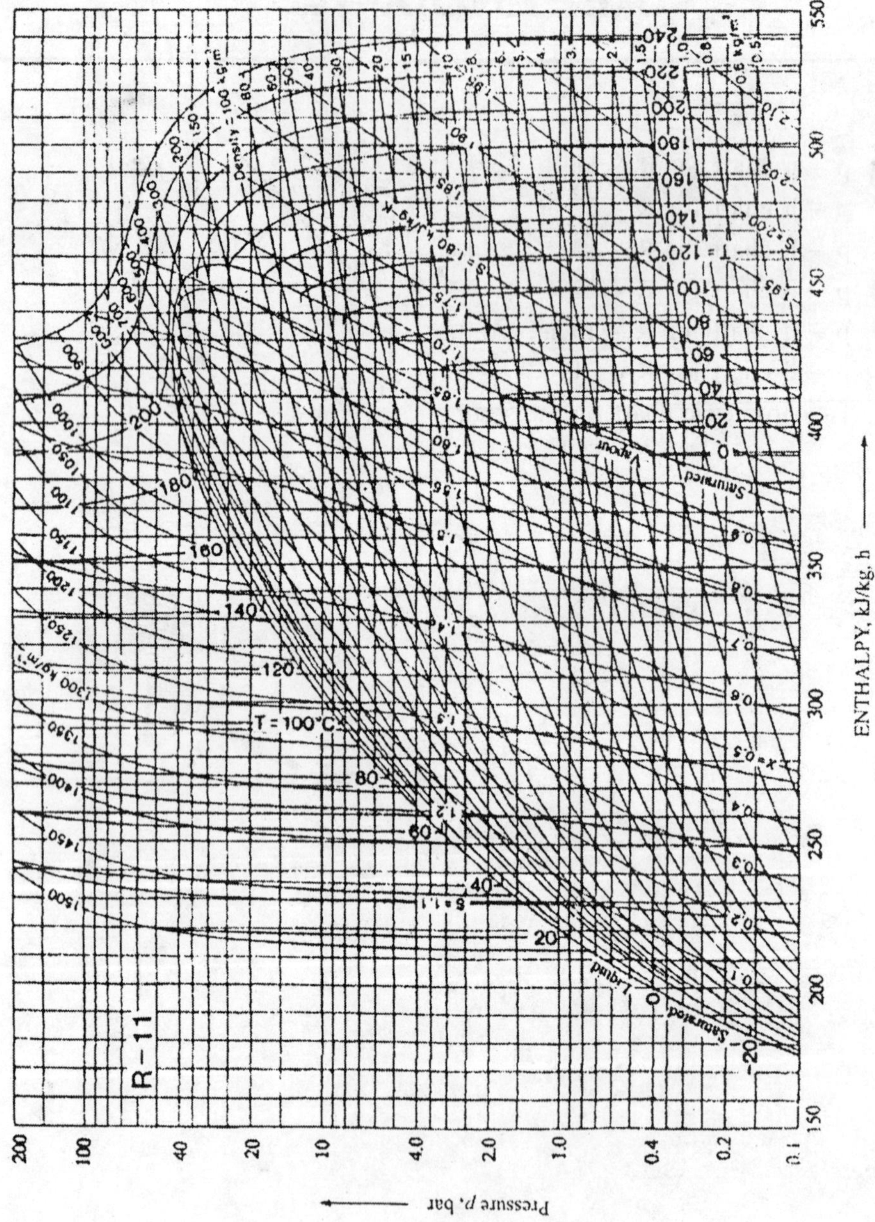

Fig. A-1. p-h chart for R-11 refrigerant

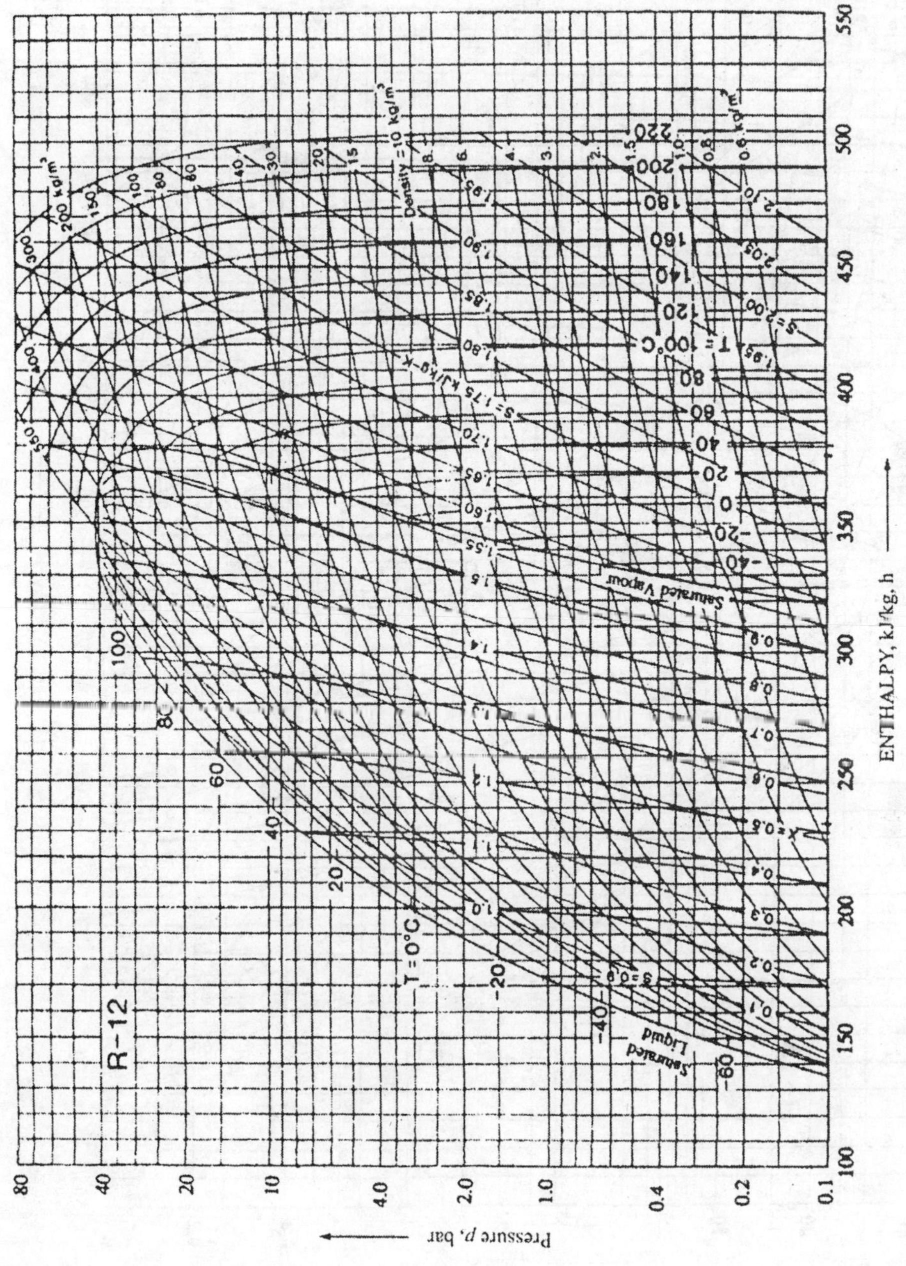

Fig. A-2. p-h chart for R-12 refrigerant

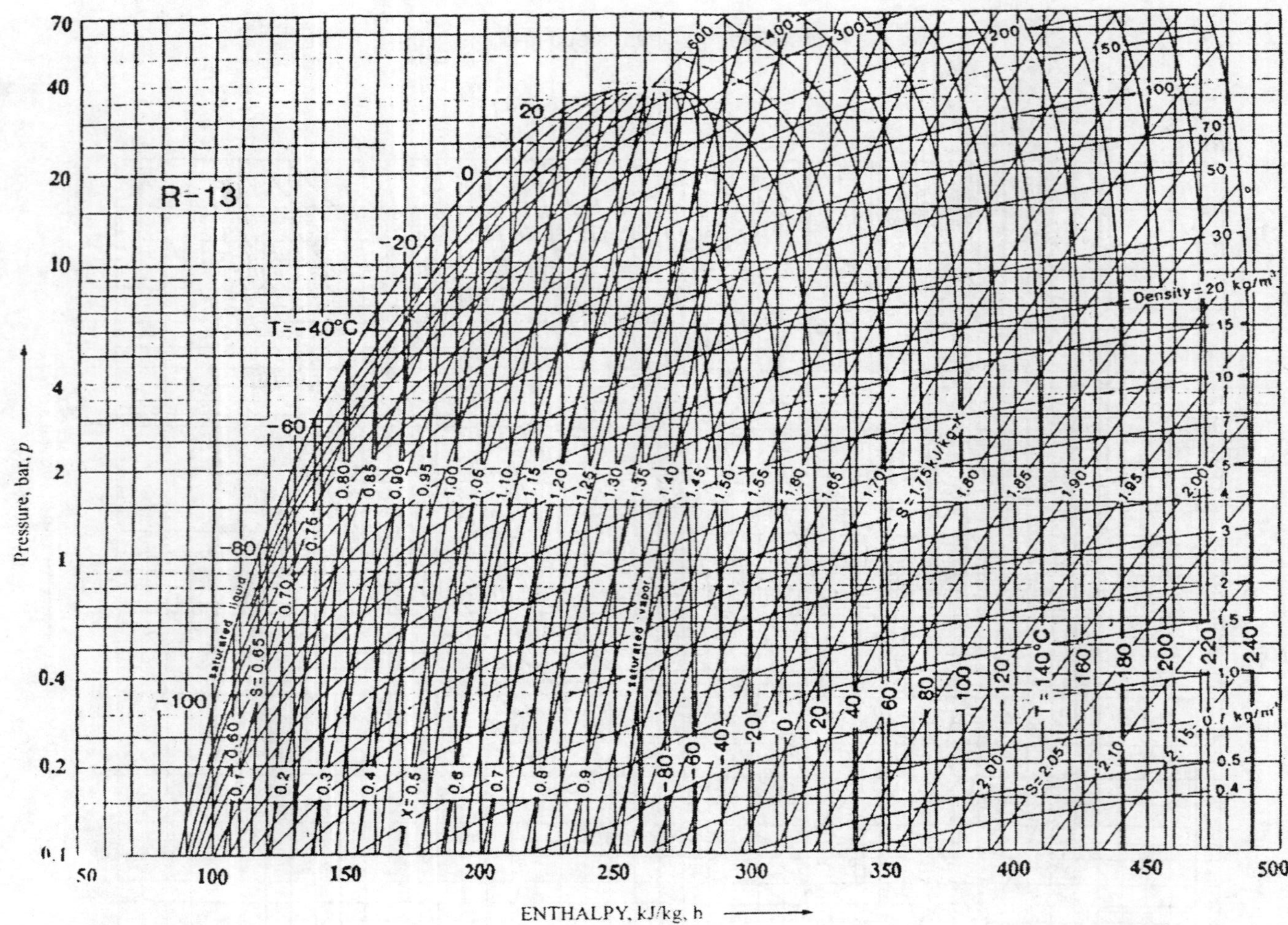

Fig. A-3. p-h chart for R-13 refrigerant

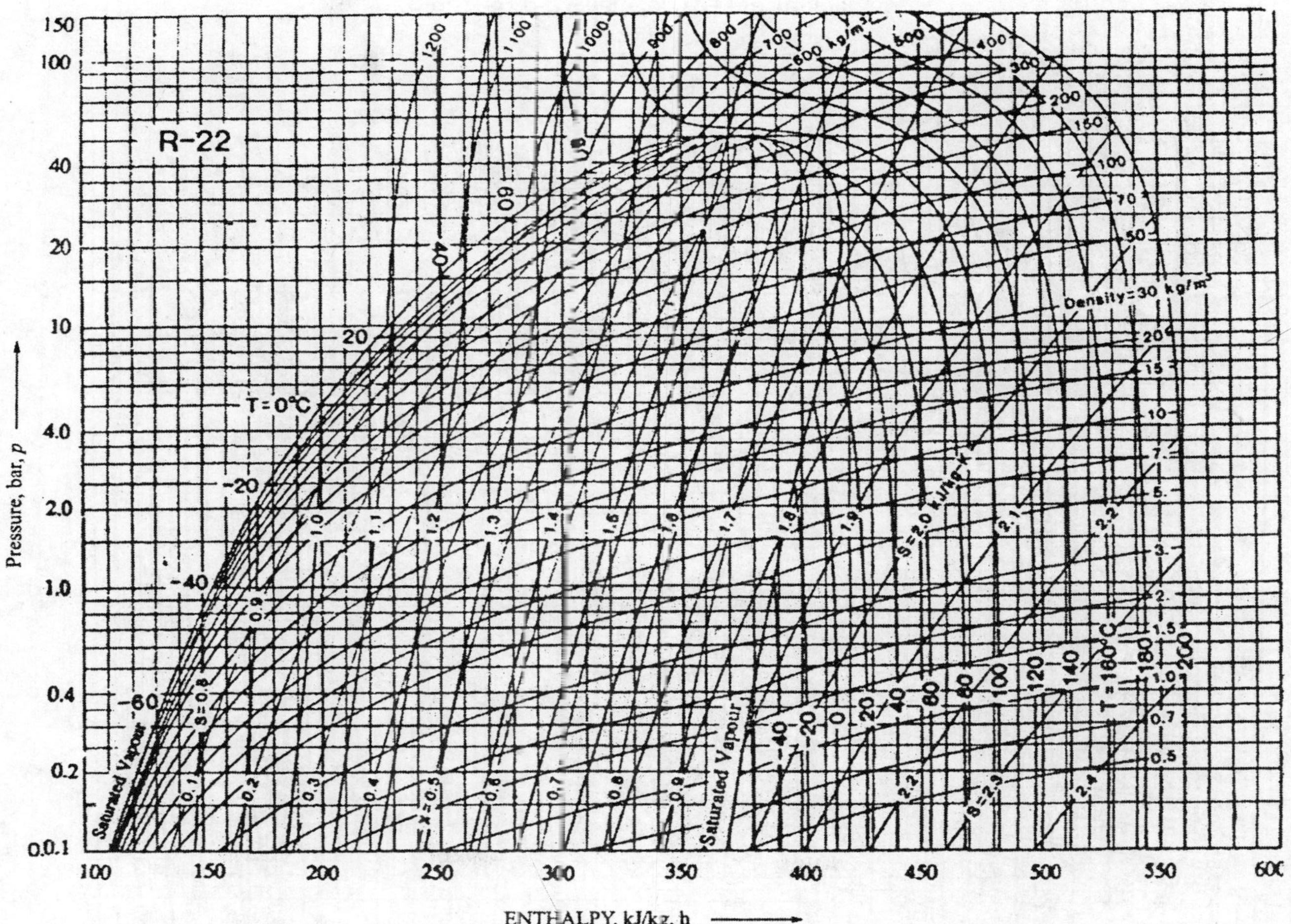

Fig. A-4. p-h chart for R-22 refrigerant

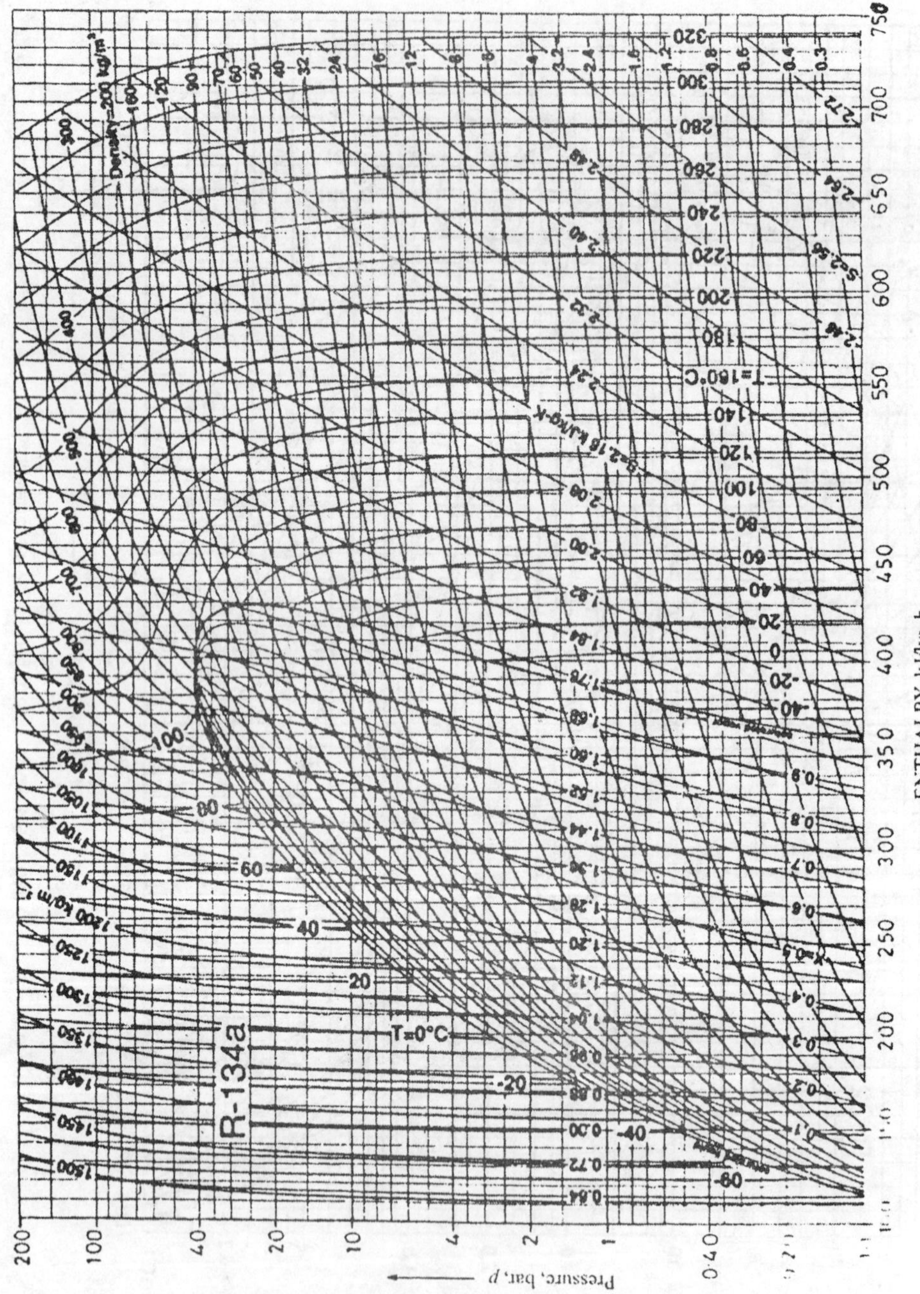

Fig. A-5. p-h chart for R-134a refrigerant

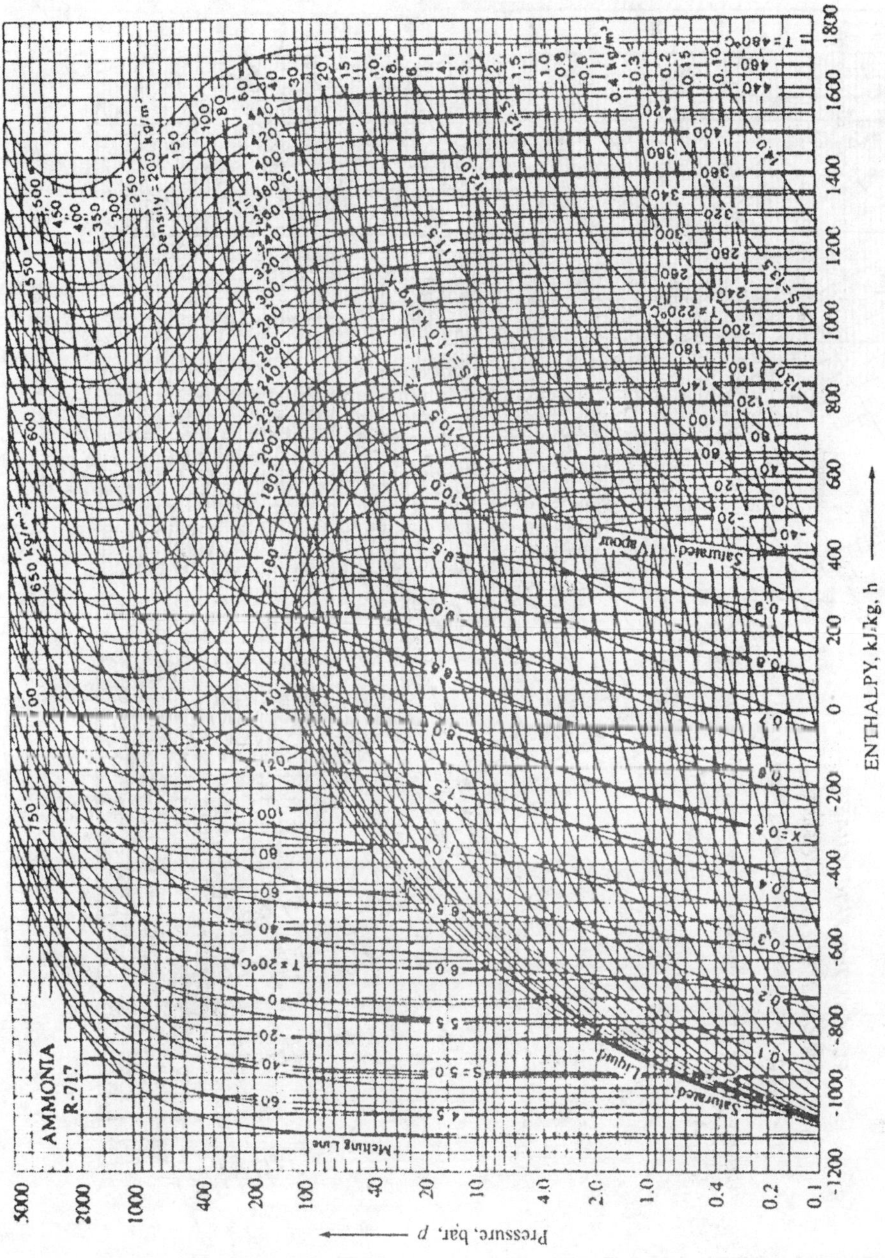

Fig. A-6. p-h chart for R-717 refrigerant

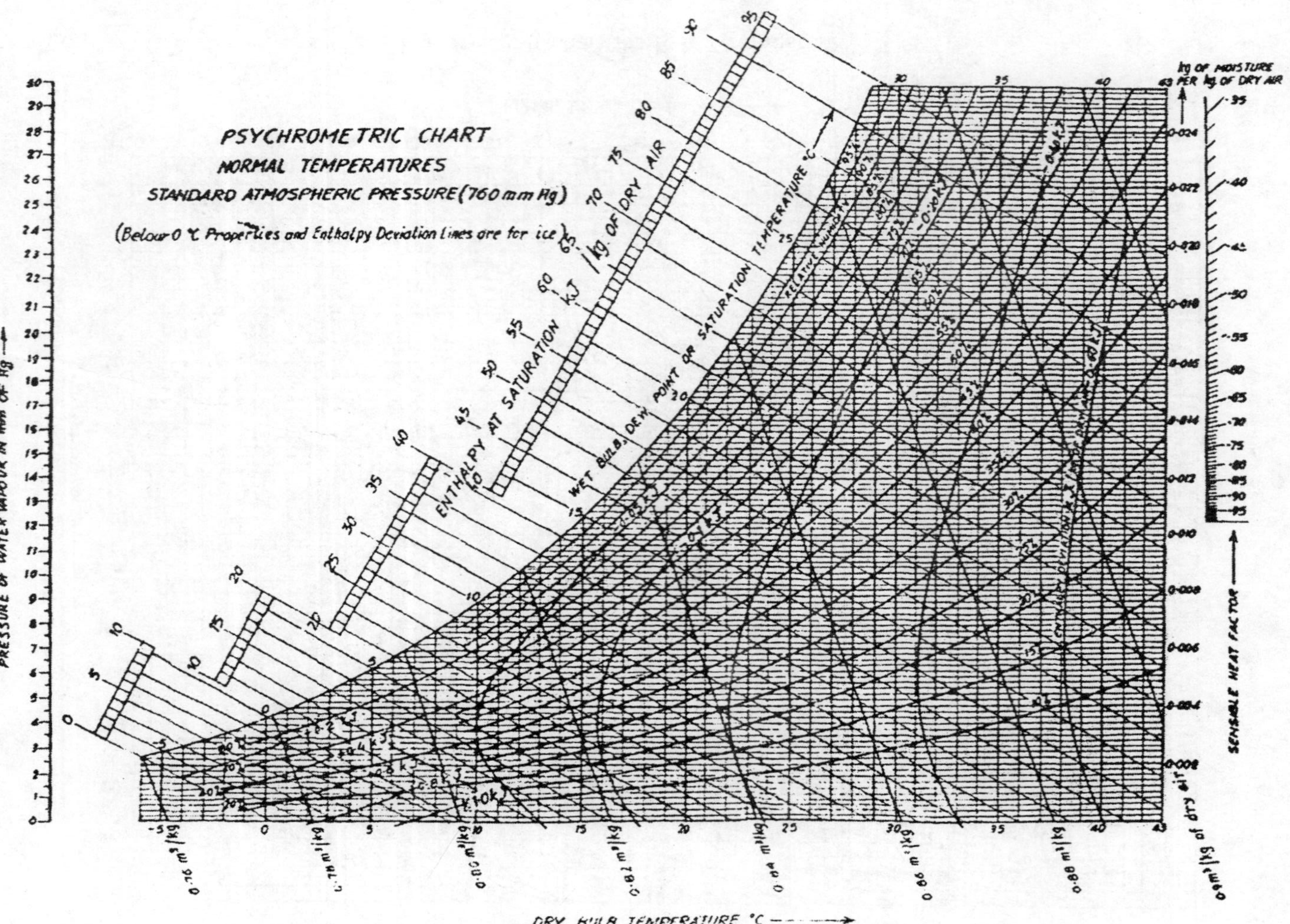

Fig. A-7. Isocherometric chart for moist air

INDEX

A

Absolute humidity, 499
Actual vapour compression refrigeration cycle, 220
Adiabatic mixing of two air streams, 521
Adiabatic saturation temperature (AST), 505
Air – duct design, 677
Air cycles for aircraft, 133
Air refrigeration cycles, 90
Air washer, 515
Aircraft refrigeration systems, 133, 134
Ammonia – water system, 364

B

Bell Coleman (or Brayton or Joule) cycle, 96
By – Pass Factor of heating and cooling coils, 646

C

Carnot cycle, 33
Carnot refrigeration cycle, 77
Cascade refrigeration systems, 338
Characteristics gas equation, 11
Claude system for air liquefaction, 470
Closed (or dense) air cycle, 90
Closed system, 2
Coefficient of performance, 76, 362
Cold storage, 720
Comfort air – conditioning, 580
Comfort chart, 582
Comparison of ammonia – water and lithium bromide – water system, 369
Compressors, 419
Condenser, 434
Conditions for reversibility, 7
Configurations of multi – stage systems, 317
Continuum, 3
Control devices, 453
Control volume, 2
Cooling and dehumidification, 514
Cooling and humidification by evaporative cooling, 518
Cooling load estimation, 615
Cooling towers, 440
Cooling with adiabatic humidification, 517
Cryogenics, 458
Cycle, 6
Cyclic process, 6

D

Dalton's law of partial pressures, 495
DBT, 496
Decrement factor and time lag method, 619
Degree of saturation, 496
Desert cooler, 723
Designation of refrigerants, 402
Desirable properties of refrigerants, 403
Dew point Depression (DPD = DBT – DPT), 497
Dew Point Temperature, (DPT), 497
Difference between heat engine, refrigerator and heat pump, 78

Distribution patterns of outlets, 670
Domestic refrigerator, 714
Duct design procedures 679
Ducts design, 666

E

Eco – friendly refrigerants, 408
Effect of discharge pressure, 214
Effect of subcooling of refrigerant vapour, 215
Effect of suction pressure, 213
Effect of superheating of refrigerant vapour, 214
Effective room sensible heat factor (ERSHF), 623
Effective temperature, 582
Efficiency of heating and cooling coils, 646
Ejector – compression system, 75
Electro – Lux refrigeration, 75
Electrolux refrigerator – domestic unit, 385
Elementary properties of binary mixtures, 371
Enthalpy – concentration diagram, 373
Enthalpy of moist air, 500
Enthalpy, 15
Entropy, 36
Equal friction method, 679
Equivalent temperature differential method, 620
Evaporators, 443
Expansion devices, 448
Extensive properties, 4

F

Factors affecting human comfort, 581
Fans in series and parallel, 703
Fans, 701
First law of thermodynamics, 13
Flash gas removal, 313
Flash intercooling, 314
Food preservation, 708
Free expansion process, 25
Friction chart for circular ducts, 692
Friction factor for ducts, 688

G

Governing equations for ducts, 680
Grand sensible heat factor (GSHF), 621

H

Heat engine, 33
Heat gain due to infiltration, 617
Heat gain from lights, 618
Heat gain from power equipments, 618
Heat gain from products, 618
Heat gain from solar radiation, 617
Heat gain through ducts, 619
Heat pump, 33
Heat transfer through walls and roofs, 615
Heat, 10
Heating and adiabatic chemical dehumidification, 520
Heating and humidification by steam injection, 519
Heating and humidification, 512
Heating load estimation, 620
High grade energy, 9
Humid specific heat, 500
Humidification or dehumidification, 512

I

Ice plant, 719
Intensive properties, 3
Internal energy, 8
Internal heat gains, 620
Irrerversible process, 8
Isolated system, 2

J

Joule's experiment, 14

L

Latent heat gain, 615
Leak detection methods, 413
Limitations of Carnot cycle, 96
Limitations of first law of thermodynamics, 29
Linde system for air liquefaction, 467
Liquefaction of gases, 467
Liquefaction of helium, 473
Liquefaction of hydrogen, 472
Lithium bromide – water system, 366
Locating return air openings, 673
Low grade energy, 9
Low temperature applications, 475

M

Methods of improving simple saturation cycle, 293
Methods of refrigeration, 73
Modes of heat transfer, 46
Modes of mass transfer, 57
Multi – stage vapour compression systems, 312

N

Newton – Rikhman law, 50
Non – flow processes, 19

O

Open air cycle, 90
Open system, 2
Ozone depletion potential, 408
Ozone layer depletion, 407

P

Path function, 11
Percentage humidity, 498
Perfect gas laws, 17
Performance curves, 702
Performance, 483
Phase, 5
Point function, 11
Process, 6
Production of dry ice, 460
Psychrometric chart, 508
Psychrometric processes, 510
Psychrometry. 495
Pure substance, 41

Q

Quasi – Static process, 7

R

Radiation, 55
Refrigerants, 401
Refrigerated freezers, 711
Refrigeration effect, 76
Refrigeration equipment, 419
Refrigeration systems, 72
Relative humidity, 496, 498
Reversed Carnot cycle, 91
Reversible process, 7
Room air conditioner, 724
Room air distribution, 667
Room sensible heat factor (RSHF), 621

S

Second law efficiency, 44
Second law of thermodynamics, 31
Selection and location of outlets, 669
Selection factors, 409
Selection of inside design conditions, 586
Selection of outside design conditions, 587
Sensible heat gain, 615
Sensible heating or cooling, 511

SHF, 513
Similarity laws, 703
Simple vapour compression systems, 189
Sol air temperature, 617
Solar heat gain through glass areas, 617
Solar radiation, 620
Solar refrigeration, 75
Solid CO_2 or dry ice, 459
Specific heats, 15
Specific humidity, 496
Specific speed, 702
State, 6
Static regain method, 680
Steady flow energy equation, 18
Steam jet refrigeration, 478
Superheat horn, 215
System components, 478

T

Temperature – concentration diagram, 373
Thermal analysis of human body, 580
Thermo – electric refrigeration, 76
Thermodynamic equilibrium, 10
Thermodynamic system, 1
Third law of thermodynamics, 40
Throttling process, 26
Transmission and distribution of air, 666
Transmission heat loss, 620
Transport refrigeration, 715
Types of supply air outlets, 668

U

Unit of refrigeration, 76

V

Vapour absorption refrigeration systems, 359
Vapour absorption system, 74
Vapour compression system. 73
Velocity reduction method, 679
Vortex tube refrigeration, 76

W

Water coolers, 720
WBT, 496
Work, 9

Z

Zeroth law of thermodynamics, 12

SOME BOOKS OF INTEREST

STRENGTH OF MATERIALS

Table of Contents

Chapter 1: Compound Stresses and Strains, **Chapter 2:** Theories of Failure, **Chapter 3:** Three Dimensional Stresses and Strains, **Chapter 4:** Stresses in Beams, **Chapter 5:** Deflection of Beams, **Chapter 6:** Fixed and Continues Beams, **Chapter 7:** Torsion of Shafts, **Chapter 8:** Columns and Struts, **Chapter 9:** Helical and Leaf Springs, **Chapter 10:** Thin Cylinders and Spheres, **Chapter 11:** Thick Cylinders and Spheres, **Chapter 12:** Rotating Dics and Cylinders, **Chapter 13:** Curved Beams, **Chapter 14:** Unsymmeterical Bending

Edition: 1th **ISBN:** 978-93-81068-80-9 **Pages (Approx.):** 600

FLUID MACHINERY (HYDRAULIC MACHINES)

Table of Contents

Chapter 1: Basic Concepts of Fluid Machinery, **Chapter 2:** Whirling of Fluids, **Chapter 3:** Hydro Power Generation Fundamentals, **Chapter 4:** Unit and Specific Quantities, **Chapter 5:** Pelton Turbine, **Chapter 6:** Francis Turbine, **Chapter 7:** Kaplan Turbine, **Chapter 8:** Draft Tube and Cavitation, **Chapter 9:** Governing, Selection and Performance of Hydraulic Turbines, **Chapter 10:** Reciprocating Pumps, **Chapter 11:** Centrifugal Pump, **Chapter 12:** Fluid Devices and Systems, **Chapter 13:** Fluid Laboratory Practice, **Chapter 14:** Conventional Questions From IES, **Chapter 15:** Multi-Choice Questions From IES

Edition: 1st **ISBN:** 978-93-82609-28-5 **Pages (Approx.):** 478

FLUID MECHANICS

Table of Contents

Chapter 1: Fluid Properties, **Chapter 2:** Fluid Statics, **Chapter 3:** Fluid Kinematics, **Chapter 4:** Fluid Dynamics, **Chapter 5:** Compressible Fluid Flow, **Chapter 6:** Viscous Flow, **Chapter 7:** Flow Through Pipes, **Chapter 8:** Boundry Layer flow, **Chapter 9:** Turbulent Flow, **Chapter 10:** Fluid Measurement, **Chapter 11:** Dimensional and Model Analysis, **Chapter 12:** Notches and Weirs, **Chapter 13:** Flow through Open Channels, **Chapter 14:** Potentional Flow

Edition: 1st **ISBN:** 978-93-81068-65-6 **Pages (Approx.):** 680

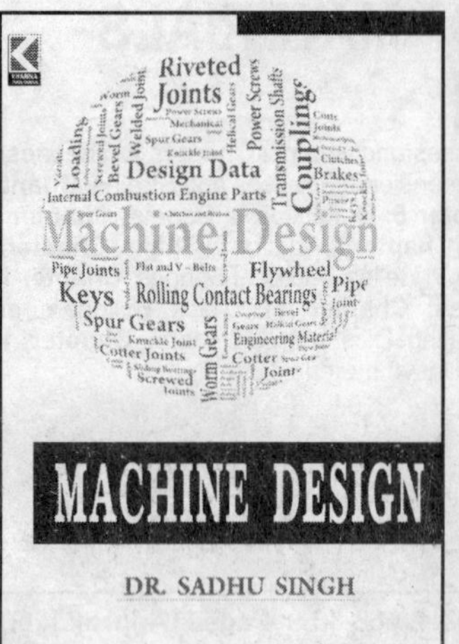

MACHINE DESIGN

Table of Contents

Chapter 1: Engineering Design, **Chapter 2:** Selection of Engineering Materials, **Chapter 3:** Design for Static Load, **Chapter 4:** Design for Variable Load, **Chapter 5:** Riveted Joints, **Chapter 6:** Welded Joints, **Chapter 7:** Screwed Joints, **Chapter 8:** Power Screws, **Chapter 9:** Transmission Shafts, **Chapter 10:** Keys and Couplings, **Chapter 11:** Levers, **Chapter 12:** Cotter and Knuckle-Joints, **Chapter 13:** Pine Joints, **Chapter 14:** Mechanical Springs, **Chapter 15:** Flat & V-Belts, **Chapter 16:** Chains and Ropes, **Chapter 17:** Spur Gears, **Chapter 18:** Helical Gears, **Chapter 19:** Bevel Gears, **Chapter 20:** Worm Gears, **Chapter 21:** Siding Bearings, **Chapter 22:** Rolling Contact Bearings, **Chapter 23:** Flywheel, **Chapter 24:** Clutches and Brakes, **Chapter 25:** Internal Combustion Engine Parts, **Chapter 26:** Lubrication in Transmission System, Multi-Choice Questions

Edition: 1th **ISBN:** 978-93-82609-57-5 **Pages (Approx.):** 800

MACHINE DESIGN DATA BOOK

Table of Contents

Chapter 1: Basic Design Data, **Chapter 2:** Engineering Materials, **Chapter 3:** Design for Static Loading, **Chapter 4:** Design for Variable Loading, **Chapter 5:** Riveted Joints, **Chapter 6:** Welded Joints, **Chapter 7:** Screwed Joints, **Chapter 8:** Power Screws, **Chapter 9:** Transmission Shafts, **Chapter 10:** Keys and Couplings, **Chapter 11:** Cotter Joints, **Chapter 12:** Knuckle Joint, **Chapter 13:** Pipe Joints, **Chapter 14:** Mechanical Springs, **Chapter 15:** Flat and V – Belts, **Chapter 16:** Chains and Ropes, **Chapter 17:** Spur Gears, **Chapter 18:** Helical Gears, **Chapter 19:** Bevel Gears, **Chapter 20:** Worm Gears, **Chapter 21:** Sliding Bearings, **Chapter 22:** Rolling Contact Bearings, **Chapter 23:** Flywheel, **Chapter 24:** Clutches and Brakes, **Chapter 25:** Internal Combustion Engine Parts

Edition: 1st **ISBN:** 978-93-82609-51-3 **Pages (Approx.):** 300

STRENGTH OF MATERIALS-I

Table of Contents

Chapter 1: Simple Stresses and Strains, **Chapter 2:** Compound stresses and strains, **Chapter 3:** Shear Force and Bending Moments, **Chapter 4:** Torsion of Circular Memders, **Chapter 5:** Bending and Shear Stresses in Beams, **Chapter 6:** Columns and Struts, **Chapter 7:** Slope and Deflection, **Chapter 8:** Fixed beams

Edition: 1st **ISBN:** 978-93-81068-61-8 **Pages (Approx.):** 536